Interdisciplinary Perspectives No. 4 (2026)

Series Editors:
Viktor J. Bruckman, Ernst Bruckmüller, Martin Gerzabek,
Gerhard Glatzel, Marianne Popp and Verena Winiwarter

Žiga Zwitter and Leonid Rasran

Historical Biodiversity in the Alps: Grassland Agroecosystems in the Last Millennium

To farmers in the Alps managing species-rich grasslands

AUSTRIAN ACADEMY OF SCIENCES PRESS

This publication was supported by the
University of Ljubljana, Faculty of Arts.

FF UNIVERSITY OF LJUBLJANA
Faculty of Arts

Images on the front cover (from left to right, from top to bottom):
Tomaž Zwitter, Ljubljana; Mojca Osvald, Borovnica; Žiga Zwitter, Ljubljana;
Mojca Odar, Archives of Tourism Bohinj, Stara Fužina; Žiga Zwitter, Ljubljana;
Boris Kuhar, Slovenian Ethnographic Museum, Ljubljana, Documentation,
Fieldwork campaign 20, photo 20/140; Tomaž Zwitter, Ljubljana; Pavla Štrukelj,
Slovenian Ethnographic Museum, Ljubljana, Documentation, Fieldwork campaign 20,
photo 20/918; Boris Kuhar, Slovenian Ethnographic Museum, Ljubljana,
Documentation, Fieldwork campaign 21, photo 21/413; Kärntner Landesarchiv,
Klagenfurt, AT KLA 129, Ansichtskartensammlung, E-Kahlkogel, 6.

Images on the back cover (from left to right, from top to bottom):
Žiga Zwitter, Ljubljana; Žiga Zwitter, Ljubljana; Rok Dovjak, Ljubljana;
Christian Berg, Graz; Žiga Zwitter, Ljubljana; Božidar Jožef Flajšman, Ljubljana;
Odilo Schmid, Außerberg in Wallis (Valais, Valês); Boris Orel, Slovenian Ethnographic
Museum, Ljubljana, Documentation, Fieldwork campaign 11, photo 11/53;
Matej Zwitter, Ljubljana; Charles Paris, Mediathek Wallis – Martigny, Martigny.

ISBN 978-3-7001-8718-9
Copyediting: Donald Francis Reindl, Ljubljana
Layout: Crossdesign, Graz
Printed: Prime Rate, Budapest
https://epub.oeaw.ac.at/8718-9
https://verlag.oeaw.ac.at
Made in Europe.

Accepted by the publication committee of the Division of Mathematics and
Natural Sciences of the Austrian Academy of Sciences by:

Andrea Bachmaier, Christa Cuchiero, Karlheinz Erb, Gerhard Herndl,
Sofia Kantorovich, Josef Smolen, Christoph Spötl, Miriam Unterlass

This publication was subject to international and anonymous peer review.
Peer review is an essential part of the Austrian Academy of Sciences Press evaluation process. Before any book can be accepted for publication, it is assessed by international specialists and ultimately must be approved by the Austrian Academy of Sciences Publication Committee.

The paper used in this publication is DIN EN ISO 9706 certified
and meets the requirements for permanent archiving of written cultural property.

Contact address for product safety:
Austrian Academy of Sciences Press
Dr. Ignaz Seipel-Platz 2, A-1010 Wien
Tel.: +43-1-51581-3420
E-Mail: verlag@oeaw.ac.at

Safety notice in accordance with Art. 9 Abs. 7 S. 2 of the GPSR is not required.

Table of Contents

1 Introduction

Grasslands have played a major role in the history of the Alps. This publication differentiates between meadows and pastures such that the word *meadow* is used only for regularly mowed grasslands; in addition to mowing, most meadows in the Alps were also grazed (Section 6). Grasslands that were not regularly mowed but grazed are designated as *pastures*, not as meadows, in this volume.[1]

Until the mid-twentieth century, the great majority of grasslands were seminatural. Various authors define a seminatural grassland in various ways.[2] Seminatural ecosystems are "modified by human influence but retai[n] many natural features", "a wide range of 'natural' processes continue to operate" there.[3] In this volume, a meadow or pasture is considered seminatural if it is either non-fertilized or moderately fertilized, and it is mowed and/or grazed by livestock. Seminatural grasslands have been species-rich and have harbored dynamic "combinations of grasses, sedges, rushes, broad-leaved herbs," a few bryophytes, lichens, and fungi. Grassland plants in seminatural meadows (and pastures) have rarely grown much more than one meter tall (Figure 1).[4] In addition, trees and shrubs have occurred in some seminatural meadows and pastures.[5] Seminatural grasslands have also supported a high diversity of animals, such as insects (e.g., butterflies, cicadas, bees and bumblebees, ground beetles, grasshoppers, and ants), spiders, and birds.[6] Highly species-rich seminatural grasslands are by far the most species-rich habitat types in Central Europe.[7]

---

[1] This is a usual manner of differentiating between meadows and pastures (e.g., Luick, Jones, & Oppermann, 2012, especially p. 52); nonetheless, this explanation is needed. The definition of a meadow in the Merriam-Webster dictionary differs from the one used in this volume; namely, it defines a meadow as "land that is covered or mostly covered with grass" (*Dictionary by Merriam-Webster*), and so Orland, 2016, for instance, also uses the word *meadows* for grasslands designated as *pastures* in this text.

[2] Peterken, 2013, 244, 358, 394.

[3] Luick, Jones, & Oppermann, 2012, 32.

[4] Wopfner, 1997; Keller, 2004; Küster & Keenleyside, 2009, 12–14; Peterken, 2013, 21 (quote), but see also the rest of the volume, pp. 29–31, 62 in particular; Babai, Molnár, & Molnár, 2014.

[5] Emanuelsson, 2009, 99, 155–158, 177, 179, 181–183, 187, 204–210, 293, 295, 330–332; Peterken, 2013, 24, 101–104.

[6] Pils, 1994; Keller, 2004, 27–28, 33, 40–42; Küster & Keenleyside, 2009, 14; Kaligarič et al., 2019, 127.

[7] Pils, 1994, 53.

Figure 1: A seminatural montane meadow at Zgornje Podjelje in Bohinj, eastern Julian Alps (Sln. *Julijske Alpe*), 1,020 m a.s.l., June 18th, 2017. Photo: Žiga Zwitter.

Moreover, regarding the spatial scale of several square meters, they belong to the most plant-species-rich biotopes in the world.[8]

The previous paragraph mentioned the diversity of seminatural grassland agroecosystems and the diversity of species in these habitats. The term *biodiversity* designates "the diversity of ecosystems, of species in these ecosystems[,] and of the genome within these species."[9] Seminatural grasslands have been ecologically and economically functionally tightly connected with other (adjacent) habitat types in the cultural landscape, such as hedges, stone walls, forests, arable land, and watercourses; thus, they have been (or were in landscapes heavily dominated by intensive agriculture or no longer farmed) integral parts of complex agricultural landscapes where connections and interrelations between ecosystems have significantly contributed to high levels of biodiversity.[10] Eco-

[8] Compare Heinz, Mayer, & Kuhn, 2012, 2.

[9] Leitgeb et al., 2023, 175.

[10] See, e.g., Achilles, 1989; Krausmann, 2008; Emanuelsson, 2009; Ellenberg & Leuschner, 2010; Oppermann & Hoffmann, 2012, 94–96; Peterken, 2013, 306–323, 326–329, 381–382; Poschlod, 2015, especially pp. 123–125; Machatschek, 2015, 42–67, 101–102; Leibundgut & Vonderstrass, 2016a; Leibundgut & Vonderstrass, 2016b; Tscharntke et al., 2021. Gratitude is expressed to the archaeologist Jana Horvat for recommending the cited Machatschek's volume.

logical connectivity in functional terms "entails the extent to which a species or population can move among landscape elements in a mosaic of habitat types."[11]

Seminatural meadows and pastures have provided "cultural ecosystem services along with a healthy environment."[12] From the anthropocentric perspective, seminatural grasslands have mainly yielded livestock fodder, pollen and nectar for bees, human food ingredients, healing plants and other culturally valued flowers, animal bedding, and some other raw materials (Sections 2 to 11). For instance, seminatural meadows not mowed before the pollination of flowers have been crucial habitats for pollinating insects; pollination of about 80% of all wild and cultivated plant species depends on these animals.[13] Species-rich grasslands have also harbored organisms that naturally control many organisms perceived as pests in farming (biological control agents).[14]

The concept of *social metabolism* appears in this volume. Social metabolism is "a functional equivalent of biological metabolism."[15] Social metabolism designates human-related material and energy flows: the transformation, distribution, use, and rejection of matter produced or taken from nature, as well as human-related energy conversion.[16]

The period dealt with in this volume encompasses the agrarian and the industrial sociometabolic regimes. The agrarian sociometabolic regime lasted from the Neolithic revolution until the industrial revolution. It was characterized by a high percentage of agricultural population (upward of 80%), and biomass was the principal type of material used and people's primary energy source. The agrarian sociometabolic regime relied on a "controlled" (i.e., modified) solar energy system, with farming as the principal factor modifying the natural production of biomass. The transition toward the industrial sociometabolic regime in Central Europe in the nineteenth century and further developments within the industrial regime gradually raised average human wealth, and resulted in increased population growth and general deagrarianization. The use of non-renewable, exhaustible minerals and biomass, fossil fuel–based energy conversion, and general environmental loading all increased by orders of magnitude and posed unprecedented challenges for the future of humans and other living beings. Two phases of the industrial sociometabolic regime are distinguished; the focus here is on farming. During the first phase of the industrial sociometabolic regime, when major changes in social metabolism were largely related to coal-based industrialization and increased trade flows, farming was

[11] Scheurer, 2016, 86.

[12] Kaligarič et al., 2019, 127 (quote); Peterken, 2013, 381–382.

[13] Crane, 1999, 472–477; Küster & Keenleyside, 2009, 13; Mlaker-Šumenjak, 2011, 39–41, 69–71; Peterken, 2013, 380; Bevk, 2018; Johansen et al., 2019; see also Martin et al., 2019.

[14] Peterken, 2013, 380; Kaligarič et al., 2019, 127; Tscharntke et al., 2021; for details, see Martin et al., 2019.

[15] *University of Natural Resources and Life Sciences, Vienna: Social Metabolism.*

[16] Padovan, Arrobbio, & Sciullo, 2022, 295.

still largely preindustrial.[17] Sections 3 to 10.1.1 and Section 11 deal with grasslands in the Alps during the agrarian sociometabolic regime and the first phase of the industrial regime. In the second phase of the industrial sociometabolic regime, starting in the mid-twentieth century, the nature of material and energy flows in farming thoroughly changed. In terms of energy conversion, the second phase of the industrial sociometabolic regime has been dominated by petroleum, natural gas, and electric power. Mass production, mass consumption, and global trade multiplied. Since then, farming has been increasingly intensified in an industrialized manner, reducing the labor input per unit of production and increasing yields, but also increasing farming's reliance on external inputs of energy and materials, and, in general, increasing the environmental load through farming[18] (see Sections 1.2, 9, 10.1.2, 10.2, and 11).

1.1 State of the art

Long-term environmental history and the historical ecology[19] of grasslands have received research attention in some parts of the world for many decades, most notably in environmental historiography written in the United States or by authors dealing with the steppes of the United States.[20] An example of recent scholarship from Europe and about Eurasia is David Moon's research on the Eurasian Steppe.[21] Indeed, among the various grassland ecosystems, steppes have received the most research attention from environmental historians that studied the topic from a long-term perspective.[22] Andrew C. Isenberg's recent global review of environmental history research on grasslands, ambitiously entitled *Seas of Grass: Grasslands in World Environmental History*,[23] is limited to steppe areas and completely omits grasslands in forest clearings such as those covered in our volume.

---

[17] Krausmann, 2016; Krausmann et al., 2016; Krausmann, Weisz, & Eisenmenger, 2016.

[18] Krausmann, 2016; Krausmann et al., 2016, 202–210; Krausmann, Weisz, & Eisenmenger, 2016, especially pp. 80–88.

[19] By using the terms *environmental history* and *historical ecology*, we wish to emphasize that past events, structures, and processes of human–nature interactions as well as the history of natural processes, structures, and events themselves have been dealt with (see, e.g., Winiwarter & Knoll, 2007; McNeill, 2010; Szabó, 2010). We by no means claim that this material does not belong to other disciplines as well (e.g., to anthropology, botany, ethnology, or geography). We fully support the notion that this field of research is interdisciplinary and that drawing any sharp boundaries between the domains of disciplines is a disadvantage, leaving the boundary zones insufficiently researched; for example, due to the lack of synergies between disciplines (compare Szabó, 2010).

[20] McNeill, 2003, especially pp. 14–16, 21, 34; Isenberg, 2017; see also Moon, 2020, and further literature listed there.

[21] Moon, 2013.

[22] McNeill, 2003, especially pp. 14–16, 21, 34; McNeill, 2010, 356, 371; Isenberg, 2017; Moon, 2020.

[23] Isenberg, 2017.

Focusing on Europe, in some regions where specific grassland or wetland environments played a highly significant role in economic and landscape history, long-term grassland or wetland environmental history research addressing a complex set of diverse research topics in the second millennium also has a rich tradition, such as the history of wetlands in the Netherlands[24] and the Po Valley.[25] Even as far as the Mediterranean is concerned, standard reference works providing a broader synthesis deal only with a narrow selection of long-term grassland environmental history topics.[26] In Central Europe, long-term environmental history and the historical ecology of grasslands, meadows in particular, have received much less attention than the environmental history and historical ecology of forests.[27]

There are only a few long-term studies on the history of grasslands in the Alps dealing with the last millennium and reaching back more than two centuries, also due to a lack of interdisciplinary cooperation. Nonetheless, solid starting points for this work were provided by Dirk Kauter's long-term interdisciplinary research into grassland history on the northwesternmost edge of the Alps (a region characterized by a special course of agrarian history; see Section 3.1),[28] Nikolaus Grass's long-term legal and agricultural history of grassland management practices with an emphasis on Tyrol,[29] the interdisciplinary synthesis of the long-term history of grassland irrigation in many parts of the Alps by Christian Leibundgut and Ingeborg Vonderstrass,[30] Nada Praprotnik's long-term investigation of herbarium collections kept by the Slovenian Museum of Natural History,[31] Georg Gentner's long-term analysis of Alpine valerian (*Valeriana celtica*) gathering and trade,[32] Alois Kapfer's synthesis of the long-term history of grassland management in Central Europe,[33] Romana Erhatič Širnik's

[24] See an overview in Winiwarter & Bork, 2019, 44–45, 184, and the publications cited there; see also Bakker 1989.

[25] See, for instance, the overview in McNeill, 2003, 20.

[26] An in-depth ecological history of grasslands gets short shrift in McNeill, 2002 (see also the review of the first edition: Butzer, 1996). Grove & Rackham (2003) present some more grassland history, but trees receive incomparably more research attention in their volume (see also the review: Butzer, 2003).

[27] See, e.g., McNeill, 2003; Winiwarter & Knoll, 2007; McNeill, 2010, 353, 363, 373–374. Péter Szabó's interdisciplinary team is currently working on the long-term environmental history of grasslands in the Jeseník Mountains (Cz. *Pohoří Jeseníky*, Sudetes; Szabó, 2023). For recent research on the history of plant knowledge in Central Europe, which does not usually focus on species-rich grasslands, see Boscani Leoni & Stuber (eds.), 2017. This edited volume also includes two overviews of earlier contributions to plant knowledge by historians (Boscani Leoni & Stuber, 2017; Popplow, 2017). See also Zwitter, 2020.

[28] Kauter, 2002.

[29] Grass, 1948.

[30] Leibundgut & Vonderstrass, 2016a; Leibundgut & Vonderstrass, 2016b.

[31] Praprotnik, 2015.

[32] Gentner, 1932.

[33] Kapfer, 2010.

research on bird protection in Slovenian territory in the nineteenth and twentieth centuries,[34] Gerhard Pils' volume on Upper Austrian meadows briefly including some historical aspects,[35] Christian Pfister's standard reference work on the impacts of climate change on Swiss agriculture from the sixteenth to the nineteenth century,[36] Jon Mathieu's studies of sixteenth- to nineteenth-century agriculture in the Alps[37] and his overview of the history of the Alps,[38] Barbara Orland's selected environmental history aspects of dairy farming in the Alps in the second half of the second millennium,[39] Herman Wopfner's history of the Tyrolean rural economy,[40] Pier Paolo Viazzo's social history of the Alps,[41] Werner Bätzing's (historical) geography of the Alps,[42] Brien Adrien Meilleur's ethnobotanical investigations of grassland management and gathering in the Allues (Francoprovençal *Alués*) Valley (French *Les Allues*; in the northern French Alps), with a strong emphasis on the first half of the twentieth century,[43] ethnobotanical investigation of Wallis (Francoprovençal *Valês*; French *Valais*) in late eighteenth to twentieth century by Christian Abbet, Romain Mayor, Didier Roguet, Rodolphe Spichiger, Matthias Hamburger, and Olivier Potterat,[44] Richard Palmer's research on northern Italian Renaissance medical botany,[45] ethnobotanical investigations in East Tyrol by Brigitte Vogl-Lukasser, Christian R. Vogl, Martina Bizaj, Susanne Grasser, and Christian Bertsch,[46] Michael Machatschek's ethnobotanical studies,[47] Erika Keller's study of hay meadows in montane, subalpine, and alpine vegetation belts in Carinthia linking historical ecology, ethnology, agrarian geography and biodiversity protection,[48] Inja Smerdel's ethnological publication on whetstones and haymaking in Slovenian territory,[49] Fanči Šarf's investigation of hay meadows and haymaking in the eastern Julian Alps from the early nineteenth century to the 1970s,[50] and studies of daily life in the Alps from the nineteenth to the mid-twentieth century by

[34] Erhatič Širnik, 2011.
[35] Pils, 1994.
[36] Pfister, 1985a.
[37] Especially Mathieu, 1996.
[38] Mathieu, 2019.
[39] Orland, 2016.
[40] Wopfner, 1997; on this author, see Meixner & Siegl, 2004.
[41] Viazzo, 1989.
[42] Bätzing, 2015; see also Tappeiner, 2003.
[43] Meilleur, 1982; Meilleur, 1986.
[44] Abbet et al., 2014.
[45] Palmer, 1985.
[46] Vogl-Lukasser et al., 2006.
[47] Especially Machatschek 1997a; Machatschek 1997b; Machatschek, 2002; Machatschek, 2006; Machatschek, 2007.
[48] Keller, 2004.
[49] Smerdel, 1994. Gratitude is expressed to the ethnologist Barbara Sosič for recommending this publication.
[50] Šarf, 1974.

Arnold Niederer,[51] Erika Hubatschek,[52] and Anka Novak.[53] The short synthesis of the history of species-rich meadows in the territory of Slovenia since the late nineteenth century by the botanists Jože Bavcon, Blanka Ravnjak, and Nada Praprotnik,[54] published when our volume was already in a mature stage, was also fully taken into account. Relevant context for the content of our volume was provided by the long-term study of changes in plant species occurrence and their most likely drivers at Piz Linard (a peak in southeastern Switzerland), now in the subnival belt above alpine grasslands, with the earliest (in 1835 CE) start of records on vascular plant occurrence among all high summits in the Alps, by Harald Pauli, Michael Gottfried, and Georg Grabherr.[55]

Parallels were drawn with studies from the Carpathians,[56] Great Britain,[57] Scandinavia, and some other parts of Europe[58] to improve the understanding of the grassland history presented. The interdisciplinary, agricultural-ecological volume on sustainable grassland management by Walter Dietl and Josef Lehmann,[59] and the volume *High Nature Value Farming in Europe*[60] improved our understanding of challenges that grasslands in the Alps have been facing recently and contributed to the integrated perspective on grasslands in recent centuries linking the views of animal husbandry and ecology.

Numerous publications dealing with more specific grassland history issues and those providing relevant contexts for grassland history are referred to in specific sections of this volume.

1.2 Present-day relevance of the topic

"[T]oday's landscapes are the result of many layers of past natural processes and human interventions, [and thus] a historical perspective is needed."[61] The topic of this volume is particularly relevant in the context of the current economic, environmental, and social conditions in the Alps and beyond. "In

[51] Niederer, 1996a; Niederer, 1996b; Niederer, 1996c; Niederer, 1996d.
[52] Hubatschek, 1992; see also Batzing, 2007; Jenin, 2007; Waibl, 2007; Andreae, 2022.
[53] Novak, 2024.
[54] Bavcon, Ravnjak, & Praprotnik, 2019.
[55] Pauli, Gottfried, & Grabherr, 2003.
[56] Břicháček & Beranová, 1993; Dahlström, Iuga, & Lennartsson, 2013; Babai & Molnár, 2014; Babai, Molnár, & Molnár, 2014; Babai & Molnár, 2016; Hartel, Craioveanu, & Réti, 2016; Łuczaj et al., 2015; Molnár et al., 2015; Iuga, 2016; Iuga, Iancu, & Stroe, 2016; Ivaşcu, Öllerer, & Rákosy, 2016; Sõukand & Pieroni, 2016; Pawera et al., 2017; Johansen et al., 2019; Janišová et al., 2023; Lenartsson et al., 2023.
[57] Thomas, 1987; Peterken, 2013.
[58] Mainly based on Henning, 1987, 1991, 1997, 2014; Austad & Hauge, 1989; Poschlod et al., 2008; Emanuelsson, 2009; Łuczaj et al., 2012; Hejcman et al., 2013; Peterken, 2013; Łuczaj et al., 2015; Poschlod, 2015; Fiol et al., 2016; Sucholas et al., 2022.
[59] Dietl & Lehmann, 2004.
[60] Oppermann, Beaufoy, & Jones (eds.), 2012.
[61] Bürgi, Hersperger, & Schneeberger, 2004, 857.

many countries there are conflicting attitudes to H[igh]N[ature]V[alue] farming within the agricultural sector, while the public may not understand that the familiar wildflower meadows and the birds and butterflies of farmland depend on farming systems and structures that are threatened by economic and societal changes."[62] Namely, "[o]ver the long history of integration of ... animal husbandry with the natural environment, many 'wild' plants and animals have come to depend upon the conditions in these semi-natural ecosystems."[63]

At the European level, seminatural grasslands, linking natural and cultural heritage, are still well represented only in small parts of the continent and they are still in serious decline. "Europe's endangered farmland biota now includes many species, and even whole species groups, that were formerly widespread" and "extensive pastures and meadows have a worse conservation status than other habitats of European conservation concern."[64] In Western and Northern Europe, "on all but the poorest land grass is normally managed as an intensive crop, regularly ploughed up and replanted with just a few specially bred species (such as *Lolium perenne* and clovers)."[65] "In Sweden, 99.7% of the semi-natural hay meadows were lost between 1870 and 2010" according to some estimates.[66] Swedish and Norwegian researchers nowadays need to perform investigations abroad, in the Eastern Carpathians for instance, where they fruitfully cooperate with Romanian researchers, and combine these results with Scandinavian historical records to improve their understanding of how a healthy cultural landscape with a significant proportion of species-rich meadows used to function in Scandinavia.[67] In England, 95% of species-rich meadows were lost only from 1950 to 1990,[68] or, taking into account not only meadows, but also pastures, and focusing on lowlands: "By the mid-1980s, the English and Welsh

---

62 Keenleyside & Oppermann, 2009, 296.

63 Luick, Jones, & Oppermann, 2012, 32.

64 Pils, 1994; Fischer et al., 2008; Emanuelsson, 2009, especially pp. 99, 155–158, 172–183, 187, 199, 204–211, 293–301, 303–315, 317–321, 323–333, 335–337, 340–343, 346–355; Oppermann, Beaufoy, & Jones (eds.), 2012 (second quote is from Beaufoy, Keenleyside, & Oppermann, 2012, 529); Peterken, 2013, 10–289, 302–405; Poschlod, 2015, 135–149, 154–155, 170–172, 176–177; Grass, Batáry, and Tscharntke, 2021 (first quote from p. 263); Tscharntke et al., 2021, 919; Leitgeb et al., 2023, 175.

65 Küster & Keenleyside, 2009, 12.

66 Dahlström, Iuga, & Lennartsson, 2013, 194; see also Emanuelsson, 2009, 159; but note that the Swedish definition of a seminatural grassland is stricter than in some other European countries, and that, according to the estimates published by Emanuelsson (2009, 173) and most likely mainly referring to the period from 1995 to 2000 (main author's correspondence with Urban Emanuelsson in December 2021 and January 2022), almost 20% of all agricultural land in Sweden was seminatural grassland, meaning not only meadows (mentioned in the text above) but also pastures.

67 Examples: Dahlström, Iuga, & Lennartsson, 2013; Johansen et al., 2019.

68 Kauter, 2002, 1; Peterken, 2013, 11–12, 24, 189–190, 195, 203–204, 209–210, 215, 357–377. The United Kingdom uses one of the broadest definitions of a seminatural grassland among all European countries (author's correspondence with Urban Emanuelsson in December 2021 and January 2022).

lowlands had lost 97% of the semi-natural grasslands they had possessed in the 1930s, and they have lost more ever since."[69] A list of European countries where seminatural grasslands are considered to have constituted well below 10% of all agricultural land in what is most likely the period from 1995 to 2000 includes, for instance, Belgium, Bulgaria, Denmark, Estonia, Finland, France, Germany, Latvia, Lithuania, Luxembourg, and the Netherlands, and the situation in many others has been worsening quickly.[70] In Germany, seminatural, species-rich grasslands cover around 5% of the used farmland or 14% of all farmed grasslands.[71] Thus, in the German federal state of Thuringia, 75% (359) of all endangered plant species (including flowering plants and ferns) are bound to extensively managed landscapes, and 69% of these (thus nearly half of all endangered species there) depend on extensively managed meadows and pastures, as revealed by a detailed study published in 2004.[72] Even in Baden-Württemberg, the German federal state with the highest percentage of seminatural farmland, 500 out of 870 plant species "listed on the red data files" in the early twenty-first century characteristically occur in seminatural grasslands.[73] Species from non-fertilized grasslands are most threatened,[74] but species from moderately fertilized comparatively extensively used meadows and pastures have also been disappearing from European landscapes.[75]

The following examples of well-known and easily recognizable meadow flowers, abounding in many moderately fertilized grasslands and representative of a large number of plant and animal species occurring in such grasslands, should increase readers' understanding of the statistics presented above. The examples come from the United Kingdom and northern Germany—areas where measures to preserve a significant percentage of seminatural meadows in landscapes have not been taken in time. In the United Kingdom, meadow clary (*Salvia pratensis*, still relatively frequent in some parts of the Alps:[76] medium-sized dark purple flowers in Figure 2 and Figure 166 in Section 10.2.2) became "near threatened." It is specially protected there,[77] although appro-

69 Peterken, 2013, 24, 357.
70 Emanuelsson, 2009, 173, 176–177; comparison there combines statistical data based on different criteria of what a "seminatural" grassland is, and the data include estimates and assumptions (author's correspondence with the plant ecologist and biodiversity specialist Urban Emanuelsson in December 2021 and January 2022).
71 Oppermann, 2012, 230; Grass, Batáry, and Tscharntke, 2021, 261.
72 Ellenberg & Leuschner, 2010, 77; on the spatial distribution of seminatural farmland in Thuringia, see Oppermann, 2012, 231.
73 Luick, Jones, & Oppermann, 2012, 55; on the spatial distribution of seminatural farmland in Baden-Württemberg, see Oppermann, 2012, 231.
74 Ellenberg & Leuschner, 2010, 77.
75 Pils, 1994; Jaksetič (interview with Kaligarič), 2019; *Results-based nature conservation plan*, [2019].
76 Aeschimann et al., 2004, vol. 2, 164 and authors' observations.
77 *Plantlife: Meadow clary*; the quote also is taken from this website.

priate site conditions (relatively dry, moderately fertilized to non-fertilized, base-rich to neutral soils on carbonate or silicate bedrock)[78] widely existed in several parts of England, for instance,[79] before the large-scale intensification of grassland management presented above. The decrease in the occurrence and abundance of meadow clary there started before 1950, and it is still going on. Nowadays, meadow clary is threatened in the United Kingdom principally by "habitat change, often caused by the ploughing, fertilising and re-seeding of slopes for 'improved' grassland, as well as the continual loss of grazing." This means that "[i]t is an offence to pick, uproot or damage" meadow clary in the United Kingdom, and that one can adopt a meadow clary specimen to support seminatural grasslands.[80] Another similar example from northern Germany is ox-eye daisy (*Leucanthemum vulgare*, Figure 166 in Section 10.2.2), which also tolerates nutrient-richer sites than meadow clary and is still widely present in several regions of the Alps. It has become almost entirely absent from significant areas of most intensive pastures in northern Germany.[81]

Fragmentation and isolation of species-rich grassland habitats make pollination, seed dispersal, and cross-fertilization difficult there, leading to continuous degeneration and extinction of insect and plant species; to make things worse, some insects (e.g., specialized butterflies) have limited mobility. Area losses and fragmentation are the results of two main types of processes: of changed land-use category (e.g., abandonment of grasslands and consequent forest regrowth, or conversion of grasslands into arable fields or built-up areas) and of widespread intensification of grassland management including, for instance, intensive and often rather one-sided fertilization, mowing up to five times annually in basins at the foot of the Alps to produce silage (first mowing in May, before seeds ripen), large-scale drainage, occasional plowing, sowing of grass monocultures, and the use of biocides in grasslands (Figure 2). This led to much higher yields than in seminatural grasslands (nonetheless, seminatural meadows have many merits in terms of hay quality, relevant for livestock and people; see Section 10.1.2.2)—but at major environmental costs. Namely, these processes have been causing large-scale homogenization of grassland species compositions in contrast to the previous small-scale diversity: grassland species compositions are becoming similar within the landscape and in different landscapes because site conditions that caused historical small-scale habitat heterogeneity have become outweighed by the impacts of intensive management (Section 2). Consequently, these grasslands do not have high natural value anymore; plant species diversity has decreased by up to 90% (data refer to the period before first-crop hay is mowed; i.e., the time when most mead-

---

[78] Aeschimann et al., 2004, vol. 2, 164.
[79] Wehling, 2007, 22–23.
[80] *Plantlife: Meadow clary*; quotes are also taken from this website.
[81] Pils, 1994, 213–214; Aeschimann et al, 2004, vol. 2, 164–165, 504–505.

ow plant species are visible). Whereas botanists counted up to ninety-eight plant species in 10 m^2 of an extremely species-rich traditionally managed dry meadow in Transylvania, belonging to the most species-rich habitats in the world, fewer than ten plant species live in 10 m^2 of highly intensively managed grasslands. Today's usual transition from meadows moderately fertilized with farmyard manure to intensive meadows heavily fertilized with slurry, liquid manure, and/or mineral fertilizers causes a decrease in plant species richness by more than half, from about thirty-five to fifty flowering plant species in 50 m^2 in moderately manured meadows to fifteen to twenty-five or even fewer than ten; at the same time, all rare species disappear from these meadows. As a consequence of intensive management, the number of animal species living in intensively managed meadows is also low. Instead of hundreds or even more than a thousand animal species in some historical meadow complexes (see Sections 6.3.1 and 10.1.2.2.4), only a few insects live in intensive meadows. Many insect species feeding on the flowers of seminatural grasslands, such as butterflies and bees, are deprived of food. Several insects lose shelter with more frequent mowing and earlier first cutting. Insects' evolutionary adaptations to light conditions in seminatural grasslands do not allow them to tolerate changes in grassland microclimates because the grassland becomes significantly denser with management intensification. Moreover, insects are poisoned by such heavy and frequent fertilizer application. The food chain causes further impacts on higher trophic levels, furthermore, some birds of prey primarily hunt in the sparse vegetation of non-fertilized grasslands. Thus, only a few birds that live in open habitats occur and abound in landscapes with intensive meadows. Mammals also experience a decline. Thus, many grassland species and entire habitat types have been vanishing (Sections 10.1.2 and 10.2.1). The European Farmland Bird Indicator shows that "populations of farmland bird species in Europe have undergone a widespread and rapid decline by as much as 57% since 1980 .... Recent evidence attributes much of this biodiversity loss to agricultural intensification" and farming abandonment (afforestation).[82] Even more drastically, populations of farmland butterflies decreased by 70% in two decades after 1990, as demonstrated by research on the environmental impacts of the European Common Agricultural Policy, and "farmland biodiversity continues to decline across the European Union."[83]

[82] Grabherr, Kusstatscher, & Mair, 1985, 271–274, 276–285, 288–289; Pils, 1994; Dietl & Lehmann, 2004, 9–10, 13, 24–25, 37–38, 42–43, 53–56, 67, 69, 94, 99–100, 102–103, 105–113, 118, 121–125; Keller, 2004; Keenleyside & Oppermann, 2009, 303; Küster & Keenleyside, 2009, 12–14; Kapfer, 2010; Peterken, 2013, 170, 189–191, 195, 203–204, 209–210, 222, 234–239, 241, 243–247, 261, 269–270, 273, 285, 287, 305, 308–309, 311–326, 328–329, 347, 355, 357–365, 380–381, 392, 394, 399–400; Kremen, 2015, 60; Geršič, Gabrovec, & Zwitter, 2018, 76; Bavcon, Ravnjak, & Praprotnik, 2019, 25, 43; Jaksetič (interview with Kaligarič), 2019, 20–21; Johansen et al., 2019; Grass, Batáry, and Tscharntke, 2021; Šumrada, Kmecl, & Erjavec, 2021 (quote from p. 1).

[83] Favilli, 2016, 115.

Figure 2: The sharp boundary in species compositions in one of the still numerous preserved species-rich moderately fertilized meadows in Bohinj[84] (lower part of the photo) and an intensively managed excessively fertilized silage meadow (above) corresponds to the boundary of cadastral parcels.[85] The area of the intensively fertilized and early-cut silage meadows in the valley bottom has been increasing recently, whereas the area of moderately manured, species-rich meadows has been shrinking (intensification; afforestation on slopes and at higher elevations). Whereas the species-rich grasslands of some farmers are under threat of excessive intensification, mainly motivated by higher quantities of livestock products, other farmers are aware of the importance of species-rich grasslands for healthy agroecosystems, for the best flavor and scent of milk, dairy products, and meat, for happy and healthy livestock, and for the environment, which has so far not been supported by adequately targeted and sustainable agri-environment payments, and sufficient interdisciplinary education on species-rich grasslands in this part of the Alps (see Section 10.1.2). Not only plant but also animal species diversity in the upper meadow is low.[86] In the case of similar recent management, the species composition in both parcels would not differ. May 30th, 2020. Photo: Žiga Zwitter.

Consequently, major efforts have recently been made in Western and Northern Europe to restore some species-rich grasslands, which in extreme cases of intensively farmed and urbanized landscapes costs up to several mil-

[84] Compare Dakskobler et al., 2019, 9, 11–12, 15–16, 19, 21–23, 25–26, 28, 69, 72, 77, 80, 82.

[85] *Environmental Atlas of Slovenia*, Spatial units, Land Cadastre.

[86] See, e.g., Pils, 1994; Machatschek, 2015, 25, 28–30, 32–34, 36–41; Machatschek, 2017, 20, 23, 25, 71–73, 77, 82 (with a romanticized presentation of the economic history of Damüls; compare Gross & Winiwarter, 2015, 492–498), 84, 86–95, 193–194. Gratitude is expressed to the archaeologist Jana Horvat for recommending the cited Machatschek's publications.

lion euros.[87] Moreover, as studies of wet meadows have demonstrated, there exists the threat of increased occurrence and dominance of some of ruderal species (i.e., species that quickly colonize disturbed sites) that cause a significant decrease in biodiversity and suppress target species of environmental protection in restored grasslands. Some of these "problematic" species are not as abundant in preserved species-rich grasslands.[88]

Successful preservation of a dense network of seminatural grasslands in landscapes where a significant percentage of species-rich grasslands has been retained to date, such as in several parts of the Alps, in a manner that is equitable to farmers, is thus a high priority.

The Alps are inhabited by "more than 30,000 animal and 13,000 plant species" altogether.[89] Cultural landscapes in the Alps belong to the most species-rich ones in Europe.[90] The general state of grasslands in the Alps, meadows in particular, is much better than on average in Western, Northern, and Central Europe, but it has been worsening quickly (Figure 2).[91] Habitat fragmentation and isolation, and species extinctions are also a challenge in the Alps.[92] Namely, biodiversity in the Alps is highest "in middle and even low altitudes,"[93] where species-rich grasslands are biodiversity hotspots in this mountain range.[94] Biodiversity in the Alps is not highest at high elevations, where nature protection areas are concentrated. Moreover, it is exactly at middle and low elevations that "all the conflicts of land use are at their highest potential" and biodiversity has been decreasing swiftly.[95] Thus, species-rich grasslands are one of the major topics of sustainable development in the Alps (see also Sections 10 and 11). For instance, during the twentieth century, eleven wild orchid species (i.e., almost 50% of all wild orchids) that had occurred in non-fertilized meadows in

---

[87] Kapfer, 2010, 133; Peterken, 2013, especially pp. 393–403; Jaksetič (interview with Kaligarič), 2019, 21.

[88] Rasran & Jeromin, 2010, 120–121, 126–127; Axmanová et al., 2021, especially pp. 2, 10–12, 14; but see Jogan, Bačič, & Strgulc Krajšek, 2012, 168, 173–174, 176, 178; and Zelnik, 2012a, 26–36.

[89] *Alpine Convention: Biodiversity & Nature Conservation.*

[90] Favilli, 2016, 115.

[91] Emanuelsson, 2009, 173, 181, 321, 347, and author's correspondence with the plant ecologist and biodiversity specialist Urban Emanuelsson in December 2021 and January 2022; Kauter, 2002, 1; Beaufoy, Oppermann, & Paracchini, 2012, 29–31; Jones & Poux, 2012, 452–453; Luick, Jones, & Oppermann, 2012, 51, 56; Trisorio, Borlizzi, & Povellato, 2012; Dahlström, Iuga, & Lennartsson, 2013, 194; Peterken, 2013, 11–12, 24, 189–190, 195, 203–204, 209–210, 215, 357–377; Bätzing, 2015, 279–280, 289, 358; Bavcon, Ravnjak, & Praprotnik, 2019, 25, 45, 85; Šumrada, Kmecl, & Erjavec, 2021, 2–3, 6, 10–12.

[92] Plassmann, 2016a, especially pp. 21–22.

[93] Plassmann, 2016b, 33.

[94] Favilli, 2016, 115.

[95] Plassmann, 2016a, 17–22; Plassmann, 2016b, 33 (quote); Poux & Ramain, 2012; Riedel, Walter, & Herzog, 2012.

Mühlviertel in the late nineteenth century became extinct in that region, which is located in Austria close to the Alps. Similarly, at least nearly 60% of butterfly species and about 70% of grasshopper species inhabiting farmed non-fertilized meadows below the subalpine vegetation belt (not including wet meadows) became extinct, threatened, or potentially threatened in Upper Austria by the late twentieth century – highly intensively managed grasslands were exceptionally rare in Upper Austria in the mid-1920s, but they are by far the most frequent farmed grassland plant communities there today.[96]

Seminatural meadows and pastures are relatively well represented in the upper montane, subalpine, and to a certain degree alpine vegetation belts Alps wide as long as farming was not abandoned there. At high elevations below the potential natural forest line (see Section 2) and at lower sites with significant natural constraints on farm mechanization, seminatural grasslands are increasingly abandoned and overgrown by woody vegetation, as a continuation of a long-term process that started in the first half of the twentieth century or in the mid-twentieth century (Sections 2, 3.1, and 9). In flat land and on gentle slopes in the colline and lower montane belts, where conditions for industrialized farming are more favorable whereas biodiversity in seminatural grasslands is very high as far as high nature value habitats are preserved at such sites, the percentage of seminatural grasslands (hay meadows at least) in the total agricultural area used is currently higher in several mesoregions of the Eastern Alps than in most mesoregions of the Western Alps, but the percentage of species-rich grasslands in cultural landscapes in such areas of the Alps has been decreasing quickly due to the intensification of farming.[97]

Consequently, the main emphasis of this volume is on the southeastern Alps, an area in agricultural transition, "where traditional farming has recently been collapsing [and also transforming, which frequently means intensifying] on an areawide basis."[98] According to late-twentieth- to early-twenty-first-century estimates by Urban Emanuelsson and probably mainly referring to the period from 1995 to 2000, about 20% of all agricultural land in Austria and about 50% (but no longer!) of all agricultural land in Slovenia was seminatural grassland.[99] "Since 2000, the intensification and restructur-

96 Pils, 1994, 27, 42, 69–70, 185. Ground-hoppers from the genus *Tetrix*, crickets, and mole crickets are not included in the grasshopper statistics above.

97 Cunder, 2012 (but see also Kaligarič et al., 2019, and Šumrada et al., 2024, 3, separate pagination); Oppermann, 2012; Poux & Ramain, 2012; Riedel, Walter, & Herzog, 2012; Suske, Gantner, & Gattermaier, 2012 (see also Leitgeb et al., 2023); Trisorio, Borlizzi, & Povellato, 2012; Hartmann et al., 2023; Machatschek, 2015, 25, 28–29, 32–33, 36–40; *Traditional Hay Milk Farming . . .*, [2023]. See also Emanuelsson, 2009, 347, and Bätzing, 2015, especially pp. 159 (but compare Leibundgut & Vonderstrass, 2016b, 148–163), 162, 277–280, 289, 322, 338–340, 358.

98 Bätzing, 2015, 280 (quote); see also Schmitzberger et al., 2005.

99 Emanuelsson, 2009, 173; and author's correspondence with Urban Emanuelsson in December 2021 and January 2022.

ing of livestock production in Slovenia have sped up. A significant share of farms has abandoned livestock breeding or has been abandoned altogether, whereas some farms have further increased in size and intensity of production."[100] This is part of the broader Alpine and European trend, where the number of small and economically often not viable "traditional farms has been decreasing, as they are replaced by larger, more modern [but see Section 11] and specialized farms."[101] Even though large-scale biodiversity loss is not the only option in such changes (Sections 10.1, 10.2, and 11), this is what usually happens. In grasslands, these processes in many places included management intensification, tilling permanent grassland parcels to grow fodder crops, and continuation of grassland abandonment in areas with environmental constraints.[102] The quickly worsening situation in Slovenian

Figure 3: A moderately fertilized montane meadow abounding in ragged-robin (*Silene flos-cuculi*, tall pink flowers). The Ši meadow complex above Srednja Vas in Bohinj, 1,100 m. The high abundance of ragged-robin is an indicator of humid conditions at this moderately fertilized site.[103] Photo: Žiga Zwitter, mid-June 2020.

[100] Šumrada, Kmecl, & Erjavec, 2021, 11.
[101] Favilli, 2016, 115 (quote); see also Trisorio, Borlizzi, & Povellato, 2012, 263, 271; *Traditional Hay Milk Farming . . .*, [2023], 128.
[102] Šumrada, Kmecl, & Erjavec, 2021.
[103] Pils 1994, Tafel 35, Photo 96; Aeschimann et al., 2004, vol. 1, 324–325.

grasslands is alarming (Figure 2).[104] From 2008 to 2018, populations of bird species that are grassland specialists declined in Slovenia by 40.8% according to the Slovenian farmland bird index, largely due to agricultural intensification on the one hand, and farming abandonment leading to afforestation (Section 2) on the other.[105] The area of moderately manured false oat-grass meadows, which were commonplace at lower elevations (e.g., in the bottoms of Alpine valleys and basins two decades ago), has recently been shrinking. Consequently, plants illusorily still perceived as common based on personal historical experience in the youth of the middle generation but occurring and abounding in moderately manured meadows, such as ragged-robin (*Silene flos-cuculi*; Figure 3), meadow clary (*Salvia pratensis* subsp. *pratensis*; Figure 166 in Section 10.2.2), ox-eye daisy (*Leucanthemum vulgare*; Figure 166), field scabious (*Knautia arvensis*; Figure 91 in Section 6.3.2.2), and goat's-beard (*Tragopogon pratensis*; Figure 91), have been disappearing from a significant part of the landscape, as part of the "dramatic recent declines of both rare and common species" from seminatural grasslands. Special efforts are necessary to make the grassland management more sustainable.[106] This means that contributions of environmental history and historical ecology to reflections about what has been occurring—and how to make the current transition more sustainable than experienced in most of Europe, by taking into account local specifics and including history-inspired innovations—are particularly welcome (Sections 10 and 11).

Building on years of interdisciplinary cooperation with natural scientists, the historian Adam Izdebski from the Max Planck Institute for the Science of Human History suggested in 2022 that, in the recent environmental crisis, historians should engage in social criticism, address "the fears of our times," and "help our contemporaries by sharing stories about the past that contribute to understanding what is going on around us in this decade and what will only become more powerful as time goes on."[107] This is one of the objectives of this volume as regards its current relevance (for further goals, see Section 1.4). However, focusing on this volume's twenty-first-century relevance, it attempts to do more than this. Namely, Jonatan Palmblad from the Rachel Carson Center for Environment and Society argued in 2024 that environmental history can also contribute to "integrative approaches" to "the current socio-ecological dilemmas" "that specialized research has rendered obsolete from its

[104] Kaligarič et al., 2019, 133.
[105] Šumrada, Kmecl, & Erjavec, 2021, 1–12.
[106] Bavcon, Ravnjak, & Praprotnik, 2019, 23, 25, 43; Jaksetič (interview with Kaligarič), 2019, 20–22; Kaligarič et al., 2019; Šumrada et al., 2021; Šumrada, Kmecl, & Erjavec, 2021; Grass, Batáry, and Tscharntke, 2021, especially p. 262 (quote); Šumrada et al., 2022.
[107] Izdebski, 2022, especially pp. 9, 11 (quotes), 16–17; see also McNeill, 2003, 34; McNeill, 2010, 362.

own fragmented vantage point."[108] Thus, our volume also identifies a selection of long-term-evidence-based good practices of grassland management and grassland plant uses aiming to bridge the divide between agronomy and nature conservation (Sections 10.1.2.2, see also Section 10.2). Good practices based on long-term-evidence is one of the core themes in applied historical ecology.[109] Namely, nowadays, efforts are being made to develop adequate and equitable, usually optimally (at least largely) result-based agri-environment payment schemes (Section 10.1.2.1) to preserve a significant percentage of species-rich grasslands in the Alps (Figure 2).[110] However, it has "not yet been sufficiently researched what kind of management is necessary for keeping the individual grassland types in favorable conservation status . . . . This knowledge is valuable because the present state of grasslands is usually a product of several centuries of traditional management and might exhibit long lag times in responding to some management changes," as recently emphasized by cooperating agronomists and biologists,[111] and by ethnoecologists.[112] This calls for contributions of environmental history and historical ecology to reflect on not only what has been going on, but also how to make the current transition more sustainable—thus socially equitable, ecologically sound, and economically viable—than experienced in most of Europe. This is a complex inter- and transdisciplinary endeavor. Environmental history and historical ecology are only two of the many contributing disciplines. They emphasize the global setting, and the regional and local specifics in a glocalized world that faces "simultaneous occurrence of both universalizing and particularizing tendencies"[113] and promote history-inspired innovations (Sections 10.1, 10.2, and 11).

1.3 Study areas

The Alps as the macroregion studied make this volume different from the many works of long-term grassland history scholarship worldwide that focus on steppes[114] (see also Section 1.1). To illustrate the relevance of this difference, a few fundamental dissimilarities between grasslands in the Alps and steppes are outlined first. Most grasslands dealt with in this volume receive significantly more precipitation than steppes (with the exception of some dry inner-Alpine valleys; on these, see Section 6.4).[115] To add the component of

[108] Palmblad, 2024, quotes from pp. 674–675.
[109] McClenachan et al., 2024; Santana-Cordero et al., 2024, 665–666, 668, 671, 673–675.
[110] Herzon et al., 2018; *Results-Based Nature Conservation Plan*, [2019]; compare Šumrada et al., 2021.
[111] Šumrada et al., 2021, 9 (quote); Šumrada et al., 2022, 9–10.
[112] Babai & Molnár, 2014.
[113] *Britannica Money: Glocalization.*
[114] Compare Isenberg, 2017.
[115] Compare Bätzing, 2015, 37, and Isenberg, 2017, 134–135, 138–139, 146–147.

climate variability (see Section 4), steppes "seem like better-watered deserts at one time and like slightly subhumid versions of their humid climate neighbors at another."[116] Although drought can also be a challenge for grassland farming in the Alps, in most cases it has never been comparably challenging as in steppes. In contrast to steppes, excessive moisture has been another challenge for grassland farming in the Alps in the long history of ever-changing climate, as presented in Section 4. Grasslands in this volume grow on various soil types, mostly different from steppes, and within the Alps they host highly diverse plant communities (see Section 2), which in most cases also differ from steppe vegetation. The abundance of forbs, for instance, especially broad-leaved forbs, in grasslands of the Alps is one of the significant differences between steppe vegetation and the study areas of this volume: whereas graminoids usually dominate in grasslands in steppes and most of the Alps, the abundance of forbs in the Alps can be high in many cases, in contrast to steppes,[117] and it used to be even higher in an average grassland in the Alps before the Green Revolution starting in the mid-twentieth century (Sections 6.3.1, 6.3.2, and 10.1). Moreover, whereas "[t]rees are rare outside of river valleys" in steppes,[118] they would colonize most grasslands in the Alps (Section 2) unless regularly suppressed by grassland management (especially Sections 6.1, 6.2, and 6.6): the final stage of ecological succession in the case of abandonment would be (with only a few exceptions) a woodland. Some fodder trees were a regular landscape element in many grasslands in the Alps until the twentieth century (Section 8.2.2.1). From the perspective of people and their economy, the site conditions and species compositions of grassland agroecosystems in the Alps that differ from those in steppes, together with dissimilar management practices and arrangements, resulted in different fodder quality and quantity (Sections 2, 3.2.1, 4, 6, 9, and 10), and dissimilar gathering opportunities (Section 8). Together with differences in human factors, this contributed to significantly dissimilar grassland histories in the Alps and in steppes.[119] Focusing on the second millennium AD, one of the fundamental differences is the very much later introduction of widespread grassland farming in the interior of steppes of the New World in comparison to the Alps.[120] Another key difference is the strong emphasis in the Alps not only on grazing but also on hay farming in seminatural grasslands throughout the last millennium, until the mid-twentieth century, as our study period.[121] Hay meadows are the main focus of this volume. In this regard, this work is not pioneering (see Sections 1.1 and 11), but it provides novel insights and complements existing mainstream grassland history, in many cases focused on steppes.

[116] Compare Isenberg, 2017, 138.
[117] Compare Isenberg, 2017, 138.
[118] Isenberg, 2017, 138.
[119] Compare this volume with Isenberg, 2017.
[120] Isenberg, 2017, 133, 139–145, 148.
[121] Compare Isenberg, 2017, 146.

The broader context of the Alps (Section 1.2) reveals that the southeast of this mountain range currently experiencing a strong grassland history transition is a particularly relevant study area. Moreover, the Eastern Alps have so far received relatively little historical ecology attention.[122] Favorable availability of archival documents (Section 1.5),[123] spatial allocation of research funds[124] (Section 1.6), and established contacts with local population led to the fact that Bohinj, located in the Julian Alps in Slovenia, was selected for the main study area. Data from many other parts of the Eastern Alps—partly as a result of primary investigations and partly summarized from the literature—were also widely used in this volume. Figure 4 makes it clear that the data are embedded in the context of the Western Alps as well.

An online map can be viewed by scanning this QR code:
(https://doi.org/10.1553/978OEAW87189-fig4)

Figure 4: Most of the settlements, mountains, and rivers or their valleys mentioned in this volume. In the text, toponyms designating small settlements, mountains, and microregions and some mesoregions are accompanied by a brief note designating the broader region the place belongs to or a larger settlement nearby to facilitate orientation for readers familiar with the region. Such additional geographical names that are only support for readers are not mapped. The locations mentioned in Sections 6.4.5 and 7.2, and in the introduction to Section 6.1.2.5, are shown on separate maps (Figures 59, 103, and 116), and so these locations are not mapped again here. Place names appearing only in archival signatures or in cited titles are not mapped. Figure 4 shows the locations mentioned in our text.
The cultural richness of the Alps includes their linguistic diversity. During the long period dealt with in our volume, the language or languages usually spoken by local inhabitants have not been constant in all regions of the Alps. Uniformly, though sometimes ahistorically, the use of toponyms in the languages of the Alps in this volume follows the Alpine Convention's early twenty-first century linguistic map of the Alps,[125] wherever a place-name appears in a book (sub)section for the first time. A partial exception is the Slovenian territory that appears in the volume most frequently; Figure 4 follows the same linguistic principles in the territory of today's Slovenia as in other regions, but in the text historical circumstances were taken into account in the sense that German toponyms were added next to Slovenian ones in the case of urban settlements and centers of seigneuries which produced sources studied in this volume, i.e. places where German was frequently spoken next to Slovenian by some highly relevant actors for a long time. This way, a certain degree of linguistic diversity and variability over time is vaguely alluded to without any degree of comprehensiveness; historically better-justified toponym language uses would take us too far from the topic of the volume.[126] In case of further mentions of a toponym within the same book (sub)section, only the place-name in (one of) the main language(s) of the country that covers that location nowadays is used, to save the space.

---

[122] Santana-Cordero et al., 2024, 665–668, 670, 672, 674.

[123] See also Zwitter, 2020, 52–53.

[124] See the explanation in Zwitter, 2020, 51–52.

[125] Ravazzoli & Laner, 2018, 39; thanks to the ethnologist Špela Ledinek Lozej for recommending this map.

[126] See, for instance, *Slovenski zgodovinski atlas*, 2011, 110–111, 134, 136, 143–144, 149.

1.4 Goals of this volume

In the volume *Grasslands in Europe of high nature value*, published in 2009, Hansjörg Küster and Clunie Keenleyside claimed: "The description of the history of grasslands in agricultural systems in Europe is complicated by the lack of clear evidence about this subject. Written, paleo-ecological or archaeological sources do not provide much information."[127] We decided to challenge this opinion and to provide a combination of analyses and syntheses of the long-term environmental history and historical ecology of seminatural grasslands in the Alps, including understanding; thus, not limited to descriptions. By focusing on archival records since the Late Middle Ages, historical publications, pictures, maps, oral history interviews, historical artifacts, and fieldwork, presented in an interdisciplinary and transdisciplinary context, this volume demonstrates that studies of sufficient quantities of diverse types of sources make it possible to draw interesting conclusions and novel interpretations of grassland history. Examples of long-term legacies of historical processes, structures, and events in present landscapes are given as well. The volume mainly deals with five centuries, from about 1500 to about 1960, but some older and more recent data are also included. This means that for most meadows in the Alps the period is covered (1) from almost the beginning of the existence of a species-rich grassland to its end (in the case of remote meadows in medieval forest clearings, abandoned in the mid-twentieth century) or (2) until the widespread gradual intensification of grassland management and progressive abandonment of remote meadows in the second half of the twentieth century (see also Sections 3, 9, 10.1, 10.2, and 11).[128]

In line with the topic and the standards of modern environmental history,[129] the main actors of this volume are people, largely farmers, livestock, and wild organisms, mainly grassland plants. Namely, the focus was primarily placed on floristic and vegetation topics, and less so on wild fauna.

This volume is a contribution to long-term socio-ecological research linking long-term process-oriented ecosystem research, biodiversity and conservation research, and socio-ecological research.[130] It aims to improve the understanding of historical grasslands in the Alps by creating fresh interdisciplinary historical-ecological links. This had an influence on the content of the volume because sections were omitted that would only be compilations of previous literature. Because the book focuses on human–nature interactions in the

---

127 Küster & Keenleyside, 2009, 9.

128 Regarding the 1950s and 1960s as a rupture in the agricultural history of the Alps, see Viazzo, 1989, especially pp. 35–36, 100–106, 112, 117–120; Jones & Poux, 2012, 451; Bätzing, 2015, especially pp. 314, 322; Orland, 2016, 46; Krausmann, 2016; Krausmann et al., 2016; Krausmann, Weisz, & Eisenmenger, 2016, especially p. 83.

129 Izdebski, 2022, 11.

130 Mirtl et al., 2015.

long-term perspective, it aims to improve the understanding of the history of grasslands in the Alps by educated farmers, agronomists, anthropologists and ethnologists, botanists, ecologists and other biologists, geographers, historians (not only agrarian and environmental, but also cultural, intellectual, military, and political), and landscape architects, for instance. Long-term insights, emphasizing examples from the pre-industrial period, aim to illustrate the fact that grassland history includes both dynamic and (relatively) constant components. This is highly relevant because expressions such as *traditional*, *earlier*, or *historical* grassland management are often uncritically used in a generalized way in ecological literature to refer to practices characteristic for the time from approximately 1900 to 1960, thereby ignoring important previous historical developments of grassland management.[131]

The volume also discusses the current relevance of the history of seminatural grasslands in terms of learning from the past for the benefit of present-day society and the environment (Sections 1.2, 10, and 11). In this way, it aims to contribute to "better informed management and policy decisions."[132]

1.5 Methods

From the methodological perspective, the volume combines historical results with ecological interpretations. Long-term grassland history analyzes and synthesizes data from different types of sources, and it must be interdisciplinary. Due to the long-term perspective, more time was invested in scarcer, older data collection to attain a more balanced representation of older and more recent data in the volume. Information from many sources referring to the nineteenth and twentieth centuries was also analyzed and integrated in the synthesis. The main types of sources useful for long-term grassland history investigations and studied in this volume include the following.

1) Archival records and historical publications:

– A variety of administrative records, produced mainly for economic and legal purposes. If sufficient quantities of these types of sources have been preserved and are studied, the analysis is likely to find data on a wide variety of grassland history topics, especially since the sixteenth or seventeenth centuries. The following were used:

- A selection of medieval and early modern rent-rolls[133] (Germ. *Urbare*) dealing with parts of the southeastern Alps. These rent-rolls are lists of

[131] Kapfer, 2010, 133, 139.

[132] Compare Szabó, 2010, 380.

[133] AT StLA, Archiv der Innerösterreichischen Hofkammer, Sachabteilung, box 69, folder 8: *Des ambts Capl bereüttung vnd beschreibung* 1567; SI NŠAL 100, Kapiteljski arhiv Ljubljana (Archives of the Ljubljana Cathedral Chapter), fasc. 98, unit 3: Rent-roll (*Urbar*) of the seigneury of Bleiburg (Sln. *Pliberk*) 1570; SI AS 173, Imenjska knjiga

holdings belonging to a certain seigneury, including, for instance, family farms, parcels of demesne land, lists of dues as sources of landlord's income resulting from the seigneury, data on corvée, and records documenting certain rights to use natural resources.[134] The 1567 rent-roll from the seigneury (specifically, *Amt* in this case) of (Eisen)Kappel (Sln. *(Železna) Kapla)*[135] is an exceptional early source providing data on first- and second-crop hay quantities in the Central Karawanks (Germ. *Karawanken*, Sln. *Karavanke*) at a holding (farm) level (Section 6.1.2.3), highly relevant for approximate calculations of the ratio of non-fertilized to moderately fertilized meadows at that time (Section 6.3).

Another type of analyzed rent-rolls are registers of beehive tithes,[136] in our case collected directly from farmers in the Kamnik–Savinja Alps (Sln. *Kamniško-Savinjske Alpe*) by the seigneury of Gornji Grad (Germ. *Oberburg*) as the tithe lord;[137] a selection of summary beehive tithe data from account lists of the same seigneury[138] was also used (Section 4.1.1.1.1). This study area was relevant because the current state of historical climatology research in the area provided the needed context, and thus it was possible to identify some impacts of historical weather variability on pollen and nectar resources. In this case, the tithe was collected directly by the tithe lord; the annual tithe quantity was not fixed because it depended on actual yields in apiculture.[139] The tithe was either collected in kind or, when a

za Kranjsko (*Gültbuch* for Carniola): Rent-roll of the seigneury of Radovljica (Germ. *Radmannsdorf*) 1579. Mlinar (ed.), 2018a and Mlinar (ed.), 2018b published the rent-rolls of the seigneury of Fusine (Friul. *Fusinis*, Germ. *Weißenfels*, Sln. *Bela Peč*) from 1498 and about 1636, and these editions were studied. In addition, Mlinar noted that an unpublished section of this 1636 rent-roll is a source for the history of mountain grasslands on the Golica (Germ. *Kahlkogel*) Ridge (Mlinar, 2018a, 27; Mlinar (ed.), 2018b, 163), and he kindly made available his copies of archival records (see Section 3.1). Another unpublished section of the same rent-roll (AT KLA 112, Allgemeine Handschriftenreihe, Ms. 2732) was relevant for the history of spring grazing in meadows (Section 6.2.1.2).

134 On rent-rolls as sources of grassland history, see Zwitter, 2013a and Zwitter, 2020; see also general considerations in Erni, 1999, 16; Schöggl-Ernst, 2004, 516–522; Umek, 1982; and Zwitter, 2016.

135 AT StLA, Archiv der Innerösterreichischen Hofkammer-Sachabteilung, box 69, folder 8: *Des ambts Capl bereüttung vnd beschreibung* 1567.

136 For some general mentions of tithes in beehives or honey in European history, see Crane, 1999, 228, 243, 245, 490.

137 SI NŠAL 20, Gornji Grad Avstrija (Gornji Grad Austria), box 26: Beehive tithe …, 1643, 1646, 1648: Zechend peinstockh in Tirossegg pergkh, 1646; SI NŠAL 20, box 40, seigneury's account by Georg Lorger, April 29th, 1686 – April 28th, 1687: Specification der gefhallenen[!] zehendt peünstökh de anno 1686, die herren schaffer Geörgen Lorger bezalth[!] vnd guet gemaht[!] seindt worden.

138 See the late-seventeenth-century accounts of the seigneury of Gornji Grad referred to in the next paragraph.

139 Based on comparison of the preserved detailed 1686 tithe register with summary beehive tithe data in the accounts of the seigneury of Gornji Grad from 1681 and 1682 (SI

pecuniary due was paid instead of beehives in kind, it was clearly recorded how many beehives would have been provided if the tithe had been collected in kind. This means that these tithe registers are relevant environmental history sources.[140] It is highly likely that farmers hid some beehives in order not to be expected to deliver the tithe,[141] but the percentage of hidden hives was probably similar from year to year, and so this does not have an important influence on relative comparisons of tithe quantities, except for rare situations.[142] It is important that contemporary guidelines for collection of tithes in beehives were studied, and so we can understand rather well what the tithe data—which are often unclear—tell us in this case.[143]

- Account lists and account books documenting revenues and expenditures;[144] in this case, we use the accounts of the seigneury of Eberndorf (Sln. *Dobrla Vas*) in the Jaun Valley (Germ. *Jauntal*, Sln. *Podjuna*), covering significant areas in the southeastern inner-Alpine Klagenfurt Basin (Germ. *Klagenfurter Becken*; Sln. *Celovška kotlina*) and some of its hilly and mountainous hinterland in the Karawanks, from 1684 AD,[145] and of the seigneury of Gornji Grad, including large territories in the Kamnik–Savinja Alps and a minor area in the Eastern Karawanks, from 1553, 1681/82, 1682/83, and 1686/87.[146] Namely, large sections of archives of these ecclesiastical

NŠAL 20, Gornji Grad Avstrija (Gornji Grad Austria), box 39, seigneury's accounts by Michael Gervason, with supplements, 1681–1686: account April 1st, 1681 – March 31st, 1682, tables: Peinstökh & account April 1st, 1682 – April 25th, 1683, tables: Peinstökh; SI NŠAL 20, box 40, seigneury's account by Georg Lorger, April 29th, 1686 – April 28th, 1687: Specification der gefhallenen[!] zehendt peünstökh de anno 1686, die herren schaffer Geörgen Lorger bezalth[!] vnd guet gemaht[!] seindt worden).

140 See the general methodological remarks on tithe registers in Landsteiner, 2004, especially pp. 543, 545, 551, 557; Zwitter, 2016, 130–135; beehive tithes are not dealt with in these works.

141 SI NŠAL 20, Gornji Grad Avstrija (Gornji Grad Austria), box 68, Bee tithe, 1690–1778: Löb[liches] kaÿ[serlich] könig[liches] creÿß-amt, autumn 1778 (archival register: NŠAL 20: Gornji Grad Avstrija [archival register]); Mihelič, 1970, 407.

142 Landsteiner, 2004, 549–551, 557.

143 Compare the general remarks on historical tithe data in Landsteiner, 2004, 545, 547.

144 For a general introduction, see Bizjak, 2014; see also Hoffmann & Winiwarter, 2010, 353, 371; on the use of account books in grassland history, see Zwitter, 2020.

145 SI NŠAL 100, Kapiteljski arhiv Ljubljana (Archives of the Ljubljana Cathedral Chapter), fasc. 103, folder 2: account of the seigneury of Eberndorf, 1684 (archival register: NŠAL 100: Kapiteljski arhiv Ljubljana [archival register]; the provenance of this source was corrected in the archival register based on previous investigations by this volume's main author: Zwitter, 2014a, 225); on the territorial range of the seigneury of Eberndorf see, for instance, Zwitter, 2014a and Zwitter, 2014e.

146 SI NŠAL 20, Gornji Grad Avstrija (Gornji Grad Austria), box 24, seigneury's account by Marx Textor, 1553; SI NŠAL 20, box 39, seigneury's accounts by Michael Gervason, with supplements, 1681–1686: account April 1st, 1681 – March 31st, 1682, tables and account April 1st, 1682 – April 25th, 1683, tables; SI NŠAL 20, box 40, seigneury's account by Georg Lorger, April 29th, 1686 – April 28th, 1687 (archival register: NŠAL 20: Gornji Grad Avstrija [archival register]).

seigneuries are well preserved, and several other sources of these seigneuries were also studied in this and other authors' research,[147] strengthening the likelihood that combining data would lead to a more complex understanding of the historical situation.

- Descriptions of agricultural land parcels belonging to farms produced before the state-initiated (partial) cadasters. Such hand-written descriptions were not produced as generally as rent-rolls, but officials of some seigneuries created them. Such a description was either a separate source, or it was integrated into a rent-roll.[148] In this volume, we refer to the 1664/65 description of holdings belonging to the seigneury of Eberndorf,[149] which had already been identified as a promising environmental history source by the author's previous research.[150]
- Records on natural hazards and disasters affecting agricultural land, produced as part of petitions for tax relief and/or reduced dues to the seigneury. The record on the devastation by locusts in Carniola in the early 1540s from the archive of the estates of that province,[151] and the record on the 1660 mudflow devastation from the archive of the estates of the province of Carinthia were studied. In the latter case, a combination of the handwriting[152] and a preserved sketch of this same mudflow's devastation[153] make possible quite a good understanding of the event. In this way, it was possible to solve an important terminological challenge in the archival record and improve our understanding of the environmental change this event brought about (Section 4.2). Several other data from records on natural hazards and disasters, produced in the context of petitions for reduced economic burdens were synthesized from the literature.[154]
- Some court minutes include valuable pieces of information on grassland history. In general, investigations of court proceedings are time-consuming, but they have the potential to cover a wide spectrum of grassland history, including the related traditional ecological knowledge of illiterate peasants in earlier centuries.[155] In this volume, a selection of court minutes

---

147 Zwitter, 2014a; Zwitter, 2014e; Zwitter, 2015d; Zwitter, 2016.

148 E.g., Zwitter, 2020, 53, 77–79.

149 AT StiAStP, Grundherrschaft Eberndorf (Handschriften), vol. 4, Grundbeschreibung 1664/65.

150 Zwitter, 2014e; Zwitter, 2015d.

151 SI AS 2, Deželni stanovi za Kranjsko (The Estates of the Province of Carniola), First Registry, box 322, fasc. 216, year 1542: September 4th, 1542.

152 AT KLA 207, Ständisches Archiv I, box 197, folder 3; see also Zwitter, 2018.

153 Fresacher, 1965.

154 Especially Rohr, 2007a; Rohr, 2007b; Zwitter, 2015b.

155 See the presentation of the use of this type of source in the environmental history of grasslands in Zwitter, 2020, 50–55, 58–73, 81.

from around 1500[156] and from the sixteenth, seventeenth, eighteenth, and nineteenth centuries were studied.

Most of them were recorded at the courts at Bled (Germ. *Veldes*),[157] covering a large part of Upper Carniola (Sln. *Gorenjska*; Germ. *Oberkrain*) in what is now northwestern Slovenia, and at the patrimonial court of the seigneury of Eberndorf in the Jaun Valley.[158] Villages consisting of farms with spatially dispersed patterns of parcel division (in contrast to compact holdings of isolated farms) strongly prevail in the study areas approached through court records. The higher the number of holdings to which neighboring land parcels belonged, the higher was the general potential for litigation over land and over the use of natural resources. Consequently, there is usually more relevant information about grassland history in court records in areas of spatially dispersed patterns of parcel division than in regions of isolated farms, and so the seigneuries of Bled and Eberndorf are highly suitable study areas.[159]

The author's previous research using patrimonial court records of the seigneury of Eberndorf from the first two-thirds of the seventeenth century revealed complex water uses in that seigneury. It demonstrated that the court records of this seigneury include valuable information on water meadows (irrigated meadows) and mowed wetlands.[160] Study of literature during the course of work on this grassland history synthesis identified a research deficit in environmental history research in wetland agroecosystems in the Alps.[161] Consequently, the author's research using patrimonial court records of the seigneury of Eberndorf, previously covering the time until 1665, was expanded to also cover the rest of the seventeenth century

---

156 Kaspret (ed.), 1889a; Kaspret (ed.), 1889b.

157 No detailed archival inventory of handwritten volumes of patrimonial court acts from Bled exists, and so the entire first three volumes and parts of the fourth and eleventh volumes of patrimonial court records from the late sixteenth to early seventeenth century and from the mid-seventeenth century were studied. Relevant new results were found in the second, third, fourth, and eleventh volumes (SI AS 721, Gospostvo Bled [The Seigneury of Bled, Germ. *Veldes*], Iustitialia, volumes 11, 12, 13, 14, and 21). Moreover, results from selected unbound territorial court acts are included (SI AS 721, Iustitialia, fascicles 25, 32, 33, and 34). The archival inventory (Verbič, 1960) provides only a brief general overview of topics dealt with in studied unbound court records. On the question of which court minutes within the fonds of the Seigneury of Bled were produced by which court, see Kos, 2016, 132–134.

158 AT StiAStP, Grundherrschaft Eberndorf (Handschriften): volumes 179, 181, 183, 186, 187, 188, 189, 190, 191, 192, and 193. Relevant cases in these volumes of court minutes were selected based on an early modern archival register (AT StiAStP, Grundherrschaft Eberndorf (Handschriften): volume 196a).

159 Zwitter, 2014e; Zwitter, 2020, 53.

160 Zwitter, 2014e, 342–348.

161 For a highly relevant ethnobotanical study dealing with traditional ecological knowledge about lowland wetlands in Eastern Europe in the twentieth century, see Sucholas et al., 2022.

and the eighteenth century, providing new results and increasing the understanding of the environmental history of wet meadows.[162]

Moreover, a relevant eighteenth-century court record from the seigneury of Idrija (Germ. *Idria*) and an important sixteenth-century court record from the seigneury of Gornji Grad dealing with the Kamnik–Savinja Alps in Lower Styria (Sln. *Spodnja Štajerska*, Germ. *Untersteiermark*) were taken into account;[163] the latter deals with a territory of isolated farms with relatively poor general potential for grassland environmental history analyses based on court records, as determined by the author's preliminary research.[164]

In addition, court records from the archive of the (agrarian) community of Koroška Bela in the Western Karawanks including long litigations over grasslands were studied.[165]

Information from the few nineteenth-century court records dealing with grasslands and kept in the archives of the district administrative authority (Germ. *Bezirkshauptmannschaft*) of Radovljica (Germ. *Radmannsdorf*)[166] was also analyzed and integrated into this synthesis. The district administrative authority of Radovljica covered the main study area and its broader surroundings in the Julian Alps, in the Western Karawanks, and at the foot of the Alps.[167]

Village bylaws, another valuable type of grassland history legal source, have received much attention in the literature, in Tyrol in particular,[168] and so results of previous research are taken into account in this synthesis. In contrast to many regions of the Habsburg Alps north and west of today's Slo-

---

[162] Sources selected based on the early modern archival register (AT StiAStP, Grundherrschaft Eberndorf (Handschriften): vol. 196a) focused mainly on entries related to Lake Gösselsdorf (Germ. *Gösselsdorfer See*, Sln. *Goslinsko jezero*), Lake Klopein (Germ. *Klopeiner See*, Sln. *Klopinjsko jezero*), and nearby wetlands and meadows, on large wetlands on the Rückersdorf Plateau (Germ. *Rückersdorfer Hochfläche*, Sln. *Na Vrhih*), and on the keyword *Sacher*.

[163] Idrija: SI AS 736, The seigneury of Idrija, box 1, unit 1: Gerichts prothocoll buch de anno 1754 biß 1759, 444–448; Gornji Grad: SI NŠAL 20, Gornji Grad Avstrija (Gornji Grad Austria), box 97, Various cases concerning the subjects of the seigneury of Gornji Grad (Germ. *Oberburg*), sixteenth–nineteenth century: 1588, Am S: Andreas tag . . .

[164] Zwitter, 2014a; Zwitter, 2014e; Zwitter, 2015d.

[165] Results relevant for the content of this volume were identified in SI AS 170, Soseska Koroška Bela (The Agrarian Community of Koroška Bela), box 2.

[166] SI AS 137, Okrajno glavarstvo Radovljica (District Administrative Authority (Germ. *Bezirkshauptmannschaft*) of Radovljica, Germ. *Radmannsdorf*), boxes 39 and 121; selection based on the archival inventory: Pivk, 2006.

[167] *Slovenski zgodovinski atlas*, 2011, 148.

[168] Mainly Grass, 1948; on village bylaws, see also Jaritz & Winiwarter, 1997; Birr, 2004; and Teuscher, 2014; thanks to the environmental historian Verena Winiwarter for recommending Teuscher's article. Elsewhere in Europe, grassland history has been studied based on village bylaws by Molnar et al., 2015, and De Keyzer, 2016, for instance.

venia, written documentation of village bylaws was very rare in Slovenian territory,[169] where the main study sites of this volume are located.

- Land cadaster record volumes: we studied cadaster records of the historical land cadaster that includes the most grassland history data in the main study area; namely, the Franciscan land cadaster. Its volumes created around 1830 for taxation purposes include extensive data on local farming and on some basic grassland plant characteristics, and so they are valuable grassland history sources.[170] These volumes were mainly studied for all cadastral municipalities in Bohinj as the main study area, for the surroundings of Bled, largely motivated by results from older court records from Bled, and for cadastral municipalities of the Western Karawanks above Jesenice due to the research interest in the history of montane meadows abounding in daffodils characteristic for the area (Section 6.2.2.3). Comparison of Franciscan cadastral record volumes dealing with cadastral municipalities in the upper Vellach (Sln. *Bela*) and Kokra catchments with the abovementioned 1567 rent-roll from the seigneury (*Amt*) of (Eisen)Kappel that dealt with the same territory two and a half centuries earlier contributed significantly to the calculations of the ratio of non-fertilized to moderately fertilized meadows (introduction to Section 6.3. and Section 6.3.0.1). In the case of

169 Vilfan, 1996, 89–90, 217–220.

170 See the general presentation of this cadaster in Bauer, 2017, and further literature given there; Drobesch, 2009; Granda, 1994; Krausmann, 2008; Kačičnik Gabrič, 2014; Geršič, Gabrovec, & Zwitter, 2018.
Final versions of cadaster record volumes, and preliminary questionnaires on agriculture were studied for the following cadastral municipalities in their entirety because they refer to the main study area: SI AS 176, Franciscejski kataster za Kranjsko (Franciscan Land Cadaster for Carniola), L30 (cadastral municipality of Nemški Rovt), L56 (cadastral municipality of Bohinjska Bistrica), L72 (cadastral municipality of Gorjuše), L107 (cadastral municipality of Češnjica v Bohinju [i.e., Bohinjska Češnjica]), L157 (cadastral municipality of Srednja Vas v Bohinju), L172 (cadastral municipality of Nomenj), L242 (cadastral municipality of Savica), and L284 (cadastral municipality of Studor), and largely or partly also L3 (cadastral municipality of Planina pod Golico), L10 (cadastral municipality of Jesenice), L16 (cadastral municipality of Hrušica), L22 (cadastral municipality of Podhom), L35 (cadastral municipality of Blejska Dobrava), L80 (cadastral municipality of Hraše pri Radovljici), L90 (cadastral municipality of Javornik), L102 (cadastral municipality of Koroška Bela), L139 (cadastral municipality of Dovje), L169 (cadastral municipality of Nova Vas), L179 (cadastral municipality of Zgornje Gorje), L210 (cadastral municipality of Poljšica), L227 (cadastral municipality of Rečica), L287 (cadastral municipality of Županje Njive), L310 (cadastral municipality of Spodnje Gorje), L322 (cadastral municipality of Višelnica), L335 (cadastral municipality of Bohinjska Bela), L343 (cadastral municipality of Selo pri Bledu); SI AS 178, Franciscejski kataster za Koroško (Franciscan Land Cadaster for Carinthia), K312 (cadastral municipality of Zgornje Jezersko), K470 (cadastral municipality of Spodnje Jezersko); AT KLA 75, Franziszeischer Kataster, cadastral municipalities nos. 76203 (Bad Vellach, Sln. *Bela*), 76205 (Ebriach, Sln. *Obirsko*), 76211 (Koprein Petzen, Sln. *Koprivna*), 76213 (Leppen, Sln. *Lepena*), 76214 (Lobnig, Sln. *Lobnik*), and 76218 (Remschenig, Sln. *Remšenik*).

missing older data, nineteenth-century information can in some cases be cautiously applied deeper into history based on rich comparative justification (but not in every case).[171] This cadaster was also a relevant source to identify some long-term environmental impacts of previous historical grassland management on early nineteenth-century grassland ecosystems (Section 6.3.1).
In systematic and extensive cadasters, such as the early nineteenth-century Franciscan cadaster, data on grasslands of the time are on the one hand presented quite systematically and can be collected much more quickly than environmental grassland history information from court proceedings or rent-rolls. However, grassland history information in cadasters is less diverse (e.g., the Franciscan cadaster hardly includes any direct information on traditional ecological knowledge about grasslands and it documents only a selection of grassland plant uses), and consequently it covers a narrower scope of grassland history than court minutes potentially do.

- Archives of commissions for abolition, redemption, and regulation of further existing *profits-à-prendre* offer detailed insights into some grassland management practices,[172] such as grazing and irrigation, and exceptionally some others, such as meadow spring cleaning. We studied spring grazing in meadows, meadow irrigation, and spring-cleaning in the mid-nineteenth century based on two fascicles of materials from the archives of the Commission for Redemption and Regulation of Profits-à-Prendre in the Province of Carniola (Sln. *Kranjska*, Germ. *Krain*), significantly improving our case studies.[173] Concerning the main study area, many data resulting from abolition, redemption, and regulation of *profits-à-prendre* had already been collected in previous literature cited in this synthesis.[174] Consequently, our selection of study areas for analysis of this type of source followed two main objectives. The first goal was to answer research questions concerning historical management of daffodil meadows that had not been answered by the volumes of the Franciscan land cadaster, and so our investigation focused on a detailed investigation of extensive quantities of preserved materials dealing with the study area above Jesenice. Selection of the second study area was mainly motivated by the existence of clearly localized sixteenth-century data on meadow irrigation, and so the main aim was to check whether the same meadows were still irrigated in the nineteenth century and which additional data on irrigation are available in those sources.

[171] For various aspects of such comparative rich additional justification, see, e.g., Blaznik, 1975; Hohensinner et al., 2013; Zwitter, 2020.
[172] See, e.g., examples in Grass, 1948.
[173] SI AS 48, Deželna komisija za odkup in uravnavo zemljiških bremen za Kranjsko (Provincial Commission for Redemption and Regulation of *Profits à Prendre* in Carniola), fasc. 56 (= original fasc. 9–1–5 and 9–2 were studied) and fasc. 163 (original fasc. 17–19).
[174] Novak, 1955.

- A volume documenting the irrigation sequence in the complex irrigation system at Lasa (Germ. *Laas*) in the Venosta Valley (Germ. *Vinschgau*) in 1949 and in the early 1950s,[175] rooted in a long historical tradition.
- Since the eighteenth and especially nineteenth century, fodder recommendations were sometimes spread by political authorities among the general public in the Alps of the Habsburgs Monarchy.[176] In our case, sources produced in times of scarcity during the First World War, kept in archives of the district administrative authority of Radovljica (regarding its territory, see the paragraph on court records) are taken into account.[177] A district administrative authority used to link the level of the province with the level of municipalities. Concerning the spread of expert knowledge, see also the paragraphs dealing with newspapers and with works by historical experts.
- Records from the archives of military commands are relevant, for instance, to understand the context of some alien grassland plant species introductions. In this volume, selected archives of the First World War military commands[178] and their correspondence with civil authorities[179] were studied because the numbers of livestock strongly relying on a grassland diet involved in military use reached a marked peak during that war (Section 7.2).
- Sources produced by institutions in charge of trade in livestock, other products of animal husbandry, and medicinal herbs, or in charge of trade taxes; we used a selection of mid-twentieth-century records dealing with trade in gathered medicinal herbs,[180] and a contract for long-distance trade in

---

175 Lechthaler family archive, Lasa (Germ. *Laas*), notebook Aufschrebung [!] für den Allitzbach im Jahre 1949.

176 Novak, 1970, 363–364, 383, 393; Valenčič, 1970b.

177 SI AS 137, Okrajno glavarstvo Radovljica (District Administrative Authority [Germ. *Bezirkshauptmannschaft*] of Radovljica), boxes 591 (November 10th, 1915, no. 30866, November 17th, 1915, no. 31141, and December 21st, 1915, no. 35377) and 592 (August 28th, 1915, no. 22751). Source selection is based on the archival inventory Volčjak (ed.), 2017, 60–62, providing a brief general overview of First World War topics dealt with in these archival boxes.

178 AT OeStA/KA FA NFA HHK AK/AGKdo, 5. Armee (HGK Boroević), box 1108; AT OeStA/KA FA NFA HHK HFK/HGK Kdo Südwestfront II, boxes 586, 587, and 621.

179 SI AS 137, Okrajno glavarstvo Radovljica (District Administrative Authority [Germ. *Bezirkshauptmannschaft*] of Radovljica), box 591, September 17th, 1915, no. 567; September 19th, 1915, no. 25220; September 21st, 1915, Feldpostamt no. 325; September 27th, 1915, no. 2569; and September 28th, 1915, no. 1071. Sources selection is based on the archival inventory Volčjak (ed.), 2017, 60–62.

180 SI AS 241, Ministrstvo za izvoz in uvod Ljudske republike Slovenije, 1945–1951 (Ministry of Export and Import of the People's Republic of Slovenia, 1945–1951), box 1, unit 16; SI AS 1126, Državni sekretariat za blagovni promet Ljudske republike Slovenije, 1958–1965 (State Secretariat for Trade of the People's Republic of Slovenia, 1958–1965), box 32, unit 2294; SI AS 1131, Republiški sekretariat za trgovino Socialistične republike Slovenije, 1962–1965 (Secretariat for Trade of the Socialist Republic of Slovenia,

spikenard valerian from 1547 was also studied.[181] The literature cited deals with trade in grassland-related items since the Late Middle Ages.[182]

- Livestock censuses; we used those from the second half of the twentieth century to improve the understanding of reforestation in the main study area (Section 9).
- Several other types of administrative records were also studied and yielded relevant information for the content of this volume, even though these are not main grassland history sources.
  - One example is manorial minutes, a source type with heterogeneous manor-specific and time-specific content. They were one of the coexisting types of sources produced by the administration of a manor, also including its court. Their structure is usually chronological.[183] In our case, a volume of seventeenth-century manorial minutes from the seigneury of Gornji Grad was used.[184]
  - Another example is the entire volume of the tax cadaster (Germ. *Gültbuch*, not to be confused with a land cadaster),[185] listing holdings belonging to freeholders (Germ. *Freisasse*) in mid-seventeenth- to mid-eighteenth-century Carniola.[186] *Freisasse* were usually "peasants with manorial rights to a certain source of agrarian income; that is, those exempt from obligations to the seigneury because they had bought their personal freedom as well as the freedom of their holdings" or of part of their holdings.[187] Some meadow toponyms and meadow sites men-

---

1962–1965), box 15, unit 623; SI AS 2164, Zadružništvo na Slovenskem, 1922–1962 (Cooperatives in Slovenian Territory, 1922–1962), box 2, unit 287, and box 3, units 445 and 451. Selection of sources was based on the archival inventories: Malbašić, 1998; Malbašič, 2005a; Malbašič, 2005b; Štrumbl, 2017.

181 AT StLA, Stadt Murau, box 85, folder 2201, Speikhandel 1547. This manuscript was found by the doctoral student Tadej Pavković during his livestock history research, and he kindly provided the main author with a scan.

182 These data are summarized in Sections 3.1, 5, 7.1, and 8.1. Ferdo Gestrin, for instance, studied late-medieval and sixteenth-century commercial flows between towns on the northeastern Adriatic coast and their broader hinterlands, including the Eastern Alps, partly based on this group of sources (Gestrin, 1965); Helfried Valentinitsch dealt with early modern trade in spikenard valerian (*Valeriana celtica* subsp. *norica*) based on this source type (Valentinitsch, 1989).

183 See the general remarks on volumes of manorial minutes in Just, 2004.

184 SI NŠAL 20, Gornji Grad Avstrija (Gornji Grad Austria), box 17, Manorial minutes (journal) 1613[–1630]. See also Zwitter, 2013b, for a historical climatology analysis of this particular volume of manorial minutes.

185 On *Gültbücher*, see Polec, 1936, 16–33; Fresacher, 1974, especially pp. 85, 88–93; Golec, 2015, 21; Zwitter, 2019.

186 SI AS 173, Imenjska knjiga za Kranjsko (*Gültbuch* for Carniola): *Gültbuch* No. 7 (1662–1757).

187 Quote from Zwitter, 2018, 166, based on Vilfan, 1980. See also Zwitter, 2019.

tioned in this source include relevant environmental history information.

– A selection of works by historical experts, including historical research publications (especially agronomic, botanical, ethnological and medical-pharmaceutical ones), publications produced or edited by agricultural societies, and notes of ethnological fieldwork in rural areas. Some works by experts of the time also deal with usual local grassland management practices and grassland uses, with ordinary farmers' knowledge, and with historical interactions between traditional ecological knowledge and experts' knowledge; nonetheless, vernacular knowledge was often poorly explicitly represented in works by agricultural experts.[188] A nineteenth-century ethnographic volume by Ludwig von Hörmann contributed insights into gathering plants and ants, and in the attitude to wild birds.[189] A mid-twentieth-century volume of Slovenian ethnologists' field notes provided valuable insights into the daily life in rural areas and its changes over time[190] (Sections 4.1.1.4 and 10.1.1). Richard Palmer's analysis of botanical works to present the genesis of botanical knowledge in sixteenth-century northern Italy[191] and Dirk Kauter's grassland history based on this type of source deserve special attention.[192] Our study improved the knowledge of a short section of Olivier de Serres' agricultural handbook published in 1600 (we used the 1603 edition),[193] one of the most extensive European agricultural handbooks of the time,[194] which had already been partly evaluated from the grassland perspective by Kauter.[195] An eighteenth-century treatise by the botanist, veterinarian, and physician Balthasar Hacquet,[196] published by the agricultural society of the province of Carniola,[197] a mid-nineteenth-century botanical publication by the botanist Valentin Plemel[198], data on Golica in the western Karawanks in the botanical monograph by Erwin Aichinger from 1930s[199] were also studied. The results of

[188] See, e.g., Kauter, 2002; Stuber & Lienhard, 2007; Hoffmann & Winiwarter, 2010; Abbet et al., 2014; Praprotnik, 2015; Dornheim, 2016; Kirchinger, 2016; Baumgartner, 2017; Knittel, 2017; Rippmann Tauber, 2017; on grassland history in a precise seventeenth-century general topography, see, e.g., Zwitter, 2014d, 613–614, 623–626.

[189] von Hörmann, 1877; gratitude is expressed to the Innsbruck historian Roland Köchl for recommending this work.

[190] Slovenian Ethnographic Museum, Documentation, Fieldwork campaign 21, Škocjan (1964), notebook no. 3. Many more such volumes kept by the Slovenian Ethnographic Museum include promising information for future environmental history research.

[191] Palmer, 1985.

[192] Kauter, 2002.

[193] de Serres, 1603, 242; history of sixteenth- and seventeenth-century editions of this work from Beutler, 1973, 1299–1300.

[194] Beutler, 1973, 1282–1283, 1293.

[195] Kauter, 2002, 127.

[196] Hacquet, 1779.

[197] Valenčič, 1970b, 550.

[198] Plemel, 1862.

[199] Aichinger, 1933.

analyses of historical herbarium collections and historical botanical studies were summarized especially from publications by the botanists Helmut Melzer,[200] Nada Praprotnik,[201] and Tone Wraber.[202] Janez Zalokar's mid-nineteenth-century agricultural publication contributed to understanding the history of farmers' attitudes to ants (Section 10.1.1).[203] Moreover, a mid-twentieth-century text on grassland management for agronomy students by the university professor Vinko Sadar was helpful.[204] Namely, Sadar commented on several grassland management practices in the mid-twentieth-century countryside. Agronomy was already quite well developed, and so Sadar understood the impacts of many such practices. Some of the practices he was referring to have had a long historical tradition.

– Historical topographies and narrative topographic data collections. This volume includes data from the published parts of responses to topographic questionnaires about Lower Styria collected in the first half of the nineteenth century at the initiative of Archduke John of Austria and his secretary Georg Göth.[205]

– Personal notes (e.g., of landlords, also including quotes from learned literature,[206] traces of readers,[207] and farmers' and merchants' notes). For this volume, the twentieth-century personal notes of the herder Anica Kavčič from her cottage in the Laz complex of mountain pastures in Bohinj was studied.[208] Quoted literature includes several other types of personal notes.[209]

– Newspapers were used (1) to better understand some historical practices of grassland and hay management, and (2) to prove the significant age of some specific hay management practices and sayings relevant for the environmental history of grasslands;[210] (3) to demonstrate the high likelihood of familiarity of literate farmers with advice from political authorities during the First World War;[211] (4) to improve the understanding of the history of gathering wildflow-

200 Melzer, 1965.
201 Praprotnik, 1993a; Praprotnik, 1993b; Praprotnik, 2012; Praprotnik, 2015.
202 Wraber, 1960; Wraber, 1964; Wraber, 1969; Wraber, 1974; Wraber, 1977–1978; Wraber et al., 1989–1990; Wraber, 1993; Wraber, 2005.
203 Zalokar, 1854.
204 Sadar, 1956.
205 Incomplete but quite extensive sources editions by Kuret (ed.), 1985, Kuret (ed.), 1987, and Kuret (ed.), 1989, were used; see the general remarks on the content in Kuret, 1985a, and Kuret, 1985b.
206 See Zwitter, 2020, and the literature cited there.
207 Winiwarter, 2009; Hoffmann & Winiwarter, 2010, 357–358.
208 *Bundrov stan* Summer Cottage in the Laz Alp in the Julian Alps: notebook starting with "Planšarska staja v Lazu."
209 E.g., Tremel, 1947; Zwitter, 2020, 55–57, 73–76, 82.
210 See Sections 4.1.1.3, 6.1.2.1, 6.1.2.2, 6.1.2.3, and 6.1.2.5.1.2.
211 See Section 8.2.2.

ers for decoration;[212] and (5) to cite a quantitative assessment of gathering wild herbs and fungi in the 1960s.[213]

– Belles-lettres are important sources for cultural grassland history and for increasing the level of understanding of mutual interrelations between cultural and material grassland history.[214] This publication provides a more complex environmental-history contextualization of a typical nineteenth-century poem that was widely accepted and helped impact changes in the average human attitude toward songbirds (Section 10.1.1).

The complexity of grassland history allows much creativity in identifying promising archival records and historical publications. Thus, the works cited in this volume also include analyses of several other types of archival records and historical publications, and future research will explore further sources and source types relevant in grassland history.

2) Oral history/ethnology interviews.[215] At least ten oral history interviews are considered recommendable when studying an environmental-history topic like grassland history to detect a local pattern.[216] Twenty-four long semi-structured oral history interviews (lasting from about one hour to more than two hours; conversations during visits to the interviewees after the long interview are not counted separately, unless each visit lasted at least one hour) and twenty-one short semi-structured oral history interviews were carried out. Twenty-two long and sixteen short oral history interviews took place in the main study area, in the northern Bohinj in 2015, 2017, 2018, 2019, 2020, 2021, 2022, and 2023, in the villages of Stara Fužina, Studor, Srednja Vas, Bohinjska Češnjica, Jereka, and Podjelje, and in the alps of Laz and Zajamniki in the eastern Julian Alps. Another long interview took place in the Jaun Valley, near the wetlands of Lake Gösselsdorf, in 2021, and the twenty-fourth long oral history interview took place in the Kamnik–Savinja Alps in 2012. Five more short oral history interviews in the Pohorje Hills, the Kamnik–Savinja Alps, the Upper Soča catchment, and the Western Karawanks helped in better understanding the grassland history of those locations.[217]

---

212 See Section 8.1.3.1.

213 See Section 8.1.2.

214 Especially Scheidegger, 2008; Erhatič Širnik, 2011.

215 Highly relevant transcisciplinary interviews with farmers relating to today's grassland farming are not counted here. Just like interviews with experts, they are cited in footnotes but absent from the final list.

216 Bürgi et al., 2017, 2100.

217 Long oral history interviews in Bohinj with A. Boltar, 2018; T. Ceklin, 2020; G. Gartner, 2017; J. Gašperin, 2017; A. Grm, 2017; A. Kavčič, 2017; K. Kranjc, 2018 and 2020; M. Kunavec, 2017, 2018, 2019, 2021 and 2023 (three long interviews); I. Mencinger, 2017, 2018, and 2020; C. Mlakar, 2015, 2017, 2018, 2019, and 2020; F. Mlakar, 2017 and 2018; L. and M. Odar, 2017; M. Šetina, 2018; J. Škantar, 2017; J. Smukavec, 2017; Milena Sodja, 2017; Minka Sodja, 2017; A. Soklič, 2017 and 2018; A. and M. Taler, 2017; Urh, 2017 and 2022. Long interview in the Jaun Valley with Plesnik, 2021, long interview

It is important for the quality of the results that all the interviewees were highly interested in grassland history and that the topics of these historical interviews were not traumatic.[218] The great majority of the interviewees were born between 1915 and the mid-1930s. They grew up as family members of farmers or smallholders, and they spent at least part of their life farming. They experienced times of economic crisis on the countryside, which lasted from 1930 to the early 1950s in the main study area, and/or they practiced traditional grassland management later that was gradually modernizing. The mentioned economic crisis contributed to the longer existence of some historical labor-intensive grassland management practices and uses of grassland resources. Due to the complex contemporaneity of grassland uses originating from various periods, a cautious analysis of some results of oral history interviews can even help in understanding previous centuries.[219]

3) Historical drawings, photographs, and maps. A ninth-century calendar drawing improved the knowledge of scythe evolution (Section 6.1.1.1). Fourteenth- to sixteenth-century Sunday Christ church murals contributed relevant data to understand historical haymaking dates, historical vulnerability of grassland farming to rainy weather during hay harvesting (Sections 4.1.1.4 and 6.1.2.1.2), and the knowledge about historical pruning implements for harvesting leaf fodder from trees (Section 8.2.2.1). An early to mid-twentieth-century painted decorative plate from Tyrol and an early twenty-first century wall painting of high-elevation flowers by the folk artist Angela Hodnik from the main study area were relevant contributions to cultural grassland history (Section 8.1.3.1).
Photos of grasslands and of practices and arrangements related to meadows and pastures are also highly useful. In our case, an early twentieth-century photograph of a blooming montane grassland from the study area of daffodil meadows was used;[220] its comparison with historical botanical data answered

in the Kamnik–Savinja Alps with A. and M. Prepotnik, 2012. Short oral history interviews in Bohinj with J. Cvetek, 2021; L. Frelih, 2021; L. Gartner, 2018 and 2022; A. Hodnik, 2019; L. Hodnik, 2018; M. Hodnik, 2018 and 2021; T. Hodnik, 2017; A. Hribar, 2017; M. Hribar, 2022; M. Markeš, 2020; I. Mlakar, 2020 and 2021; M. Rozman, 2020; M. Smukavec, 2021; V. Smukavec, 2017; R. Sodja, 2022; and M. Stergar, 2018. Short oral history interviews elsewhere with an anonymous farmer in Činžat, 2015; with an anonymous farmer in Planina north of Ljubno ob Savinji, 2011; with E. Lamot in Rakovec (Pohorje), 2019; with L. Janež, 2017, in Čadrg; and with J. Kavčič, 2021, in Dolina above Tržič. We thank the locals Martina Markeš, Marjana and Joža Hribar, and Marjeta Hodnik for helping identify the most promising interviewees in Bohinj: ordinary people that grew up on farms in the region and that were old but not forgetful. The snowball principle was followed after this.

218 See more in Zwitter, 2020.

219 Zwitter, 2020; see, for instance, the examples in this volume's Sections 6.1, 6.3, 6.6, and 8.

220 AT KLA 129, Ansichtskartensammlung, E-Kahlkogel, 6.

the question of the representativeness of the photo content (Section 6.2.2.3).[221] Moreover, all photos taken and technical drawings plotted during the fieldwork campaigns organized by the Slovenian Ethnographic Museum in the southeastern Alps in the mid-twentieth century and now available in the museum's online database, the photos taken in the early 1950s by the ethnologist Jernej Šušteršič and held by the Slovenian Ethnographic Museum, and a few other photos kept by the same museum[222] were studied, and a selection was integrated into this volume or cited (Sections 3.1, 4.1.1.3, 4.1.1.4, 6.1.1.1, 6.1.2.5.1.2, 6.3.1.2, 6.6.2, and 8.2.2.1). In addition, we used aerial photographs from the mid-twentieth century onward to illustrate reforestation in the main study area (Section 9).
Franciscan land cadaster maps—that is, the first precise land-use maps at the parcel level, commissioned by the Habsburg Monarchy in the early or mid-nineteenth century, depending on the location in the Eastern Alps[223]—solved many open questions of localization and helped understand the environmental conditions in several grasslands in the Eastern Alps.[224]

4) Historical artefacts; for example, archaeological or ethnological finds in terms of relevant (parts of) grassland management implements and arrangements. The most important artefacts studied are early medieval scythes from the southeastern Alps held by the National Museum of Slovenia, improving our understanding of ecological impacts of historical mowing (Section 6.1.1.1).[225] Another highly relevant artefact is a hayfork identified during fieldwork in a mountain barn in the Julian Alps; the hayfork type is the same as in an early fifteenth-century Sunday Christ church mural, and relevant in the context of historical vulnerability of grassland farming to rainy weather during hay harvesting (Section 4.1.1.4).

5) Results from "natural archives," in this case of pollen and tree-ring analyses, taken from the literature, are compared with data from historical records to provide added value by means of interdisciplinarity (Sections 3.2.2, 4.1.1.2,

[221] See some critical remarks on the use of old photographs in environmental history in Rohr, 2012, 552–556, 559, 565–569.

[222] E.g., using the filter "košnja" (haymaking) in the Slovenian Ethnographic Museum's online photo database (*Slovenski etnografski muzej*, Zbirke, Digitalne zbirke, Zbirke fotografij: Košnja).

[223] See Drobesch, 2009, 169, 172–174.

[224] See numerous references to *Arcanum Maps*, Cadastral Maps, Habsburg Empire – Cadastral Maps (XIX. [sic] Century); *KAGIS Maps*, Basiskarten, Franziszeischer Kataster; *Kärnten Atlas*, Darstellung, Kartenhintergrund, Franziszeischer Kataster. Especially in the case of cadastral municipalities not yet scanned, georeferenced, and uploaded to *Arcanum Maps*, scanned archival sources from the cadastral municipalities in footnote 170 are cited; in addition, *Archivio di Stato di Trieste*, Catasto franceschino, Mappe del Catasto franceschino, Distretto di Tolmino, Comune di Goriano di Circhina [Sln. *Gorje*] . . . (1822 – sec. XIX primo quarto).

[225] NMS: Inv. nos. S 2591 and S 2592.

4.1.3.1, and 8.1.2.1). This approach raised knowledge about the historical impacts of weather and climate variations on grassland species composition, about the changing size of grassland area, and about the history of cockchafer (*Melolontha hippocastani, Melolontha melolontha*) abundance.

6) Modern fieldwork, recently collected environmental data,[226] and a few discussions with farmers on grassland management in the Julian Alps and Western Karawanks at present.[227] Transdisciplinary work with farmers aimed at mutual learning is crucial to improve researchers' and farmers' understanding of grasslands in the Alps in early twenty-first century, and of related sustainability issues linking economy and ecology. In addition, grassland observations and recent environmental data allowed a better understanding of historical data and identification of historical legacies in today's landscapes. When environmental history fieldwork aims to understand a historical landscape better, researchers must pay attention to possible significant human- and/or nature-induced environmental changes between the time referred to in the historical data and the present. In some cases, interventions in environmental conditions in recent decades as far-reaching as major underground grassland drainage systems, leveling of hummocky meadows, removal of landscape legacies of historical grassland irrigation, or large-scale removal of rocks or trees from grasslands[228] are not evident in the landscape and must be identified based on more than only grassland observations.

7) To better understand today's marketing situation for history-inspired high-nature-value farming, (1) the recent document on certification of Traditional Speciality Guaranteed hay milk and hay meat in Slovenia[229] and (2) a recent handbook on quality hay production for Slovenian hay farmers[230] were analyzed from the perspective of history-inspired grassland biodiversity and its functional role.

---

[226] In this case, e.g., ARSO, Archives of Phenological Data: Koprivnik in Bohinj; a precise soil map and lidar-based elevation model in *Environmental Atlas of Slovenia*, Land and Soil, Soil Map (1:25,000) and *Environmental Atlas of Slovenia*, Backgrounds, Lidar hillshade.

[227] Interviews with T. Soklič 2021, in Spodnje Podjelje, on spring sheep grazing in meadows; with anonymized interviewees, farmers and tourism workers with excellent insights into farming in Bohinj, 2022; and an anonymous farmer from Javorniški Rovt, 2018.

[228] Odar, 1993; Keller, 2004, 55–56, 58, 90, 99; Bätzing, 2015, 159, 161, 261; Leibundgut & Vonderstrass, 2016a, 92; Leibundgut & Vonderstrass, 2016b, 158; Geršič, Gabrovec, & Zwitter, 2018; fieldwork at Činžat in the Pohorje Hills in the Eastern Central Alps, May 30th, 2015, and the author's interview conducted then with a farmer whose identity will be kept confidential.

[229] Kolar, [2019 or 2020]; see Section 10.2.1.

[230] Verbič & Verbič, 2014–2020; see Section 10.2.1.

Figure 5: *Cephalaria leucantha*, photo Leonid Rasran.

Every type of historical source has its pros and cons.[231] Critically studied court records, for instance, provide particularly relevant insights into the traditional ecological knowledge of illiterate people. However, the evidence of trials is often fairly reliable only if the plaintiff and the defendant agreed upon something (and torture was not applied), or that noted information was recorded in passing, although it was not a matter of the dispute.[232] Another example deserving special critical attention are comments on popular ecological knowledge in many eighteenth-, nineteenth-, and some twentieth-century works by experts in agriculture: they often (but not always) looked down on the traditional ecological knowledge of farmers and presented it in an overly pessimistic way.[233] Oral history interviews are also not neutral, they are "cultural artefacts influenced by the experiences, history, and culture of the interviewee, by his or her selective memory, under the influence of the

[231] See the general remarks in Winiwarter, 2016, and the critical presentation of sources for the history of sixteenth- and seventeenth-century non-expert knowledge about grasslands in Zwitter, 2020.

[232] Zwitter, 2020; see also the general remarks on court proceedings in Fuchs & Schulze, 2002, 9–10, 13–37, 40; Scheutz, 2004, 561–563, 567–569; Škrubej, 2012, 204–208, 212–224.

[233] Kauter, 2002, 12; Hoffmann & Winiwarter, 2010, 353–355; Dornheim, 2016; Kirchinger, 2016.

present reality and of the questions,"[234] which did not suggest the answers during the data collection for this volume. Nonetheless, in grassland history investigations dealing with non-traumatic topics, it can usually be assumed "that people, when they tell their stories, do attempt to get the facts of their lives and communities as correct as they can."[235] Even photographs and statistics have to be used cautiously to obtain valid results.[236] Every investigation of archival records and published texts, oral history interviews, drawings, maps, photographs, and artefacts thus has to follow the principles of the historical criticism; that is, it must fully take into account the contexts of production and historical use of analyzed data in order to understand as well as possible how reliable they are and how partial or complete the evidence is.[237] The data analysis in this volume respected the historical criticism. If there exist important uncertainties, they are noted. Following the principles of the historical criticism, botanists have, for instance, long been able to correct mistakes in works of earlier botanists, such as the identification of the famous herb *Scabiosa trenta*, which was observed, collected, and drawn by the natural scientist Balthasar Hacquet at a small number of sites in the Julian Alps, at least on the slopes above the Trenta Valley (which appears in Hacquet's Latin species name; it is part of the the Upper Soča Valley) in the late 1770s. His description and drawing of the plant were published in Vienna in 1782. Later, many botanists unsuccessfully tried to find this species in the Julian Alps. Fortunately, Hacquet's herbarium specimen has been preserved. In the late nineteenth century, this allowed the botanist Anton Kerner to determine that *Scabiosa trenta* was actually not a new species, but a specimen of the sub-Mediterranean *Cephalaria leucantha* (Figure 5) found in the eighteenth century deep in the Julian Alps, where it is not known to occur anymore.[238]

1.6 Acknowledgements

This publication is a result of the half-year postdoctoral project Traditional Ecological Knowledge of Grassland Management to Promote Alpine Sustainability (in the call Joint Excellence in Science and Humanities by the Austrian

234 Zwitter, 2020.
235 Grele, 2007, 66.
236 Rohr, 2012, 552–556, 559, 565–569; Winiwarter, 2016.
237 See, for instance, the short but thorough introduction to the historical criticism in Winiwarter, 2016; application of the historical criticism in the study of historical grassland experts' works by Kauter, 2002; the introduction to old paintings as sources for the history of material culture in Horvat, 1995; and critical remarks on early modern topographical portraits as sources for environmental history analyses in Zwitter, 2014d, 675–686.
238 Praprotnik, 2015, 60–62, 71–73; Wraber, 1993, 37, 41–42. These studies also explain that the northernmost known habitats of *Cephalaria leucantha* in the region are located far to the south, much closer to the Adriatic, on the Karst (Sln. *Kras*, Ital. *Carso*) Plateau and on the southern slopes of the Trnovski Gozd Plateau.

Academy of Sciences, at the Institute of Social Ecology, Vienna, supervised by Verena Winiwarter), of the research program Slovene History (P6-0235, Slovenian Research and Innovation Agency, at the Department of History, Faculty of Arts, University of Ljubljana), of the research project A Holistic Approach to the Earth Surface Processes Driven by Extreme Weather Events—ExtremEarth (J7-60124, Slovenian Research and Innovation Agency, at the Department of History, Faculty of Arts, University of Ljubljana), of the research project Post-Imperial Transitions and Transformations from a Local Perspective: Slovene Borderlands Between the Dual Monarchy and Nation States (1918–1923) (J6-1801, Slovenian Research Agency, at the Department of History, Faculty of Arts, University of Ljubljana), of the research project Motherhood and Reproductive Politics in 19th and 20th Century (J6-4602, Slovenian Research Agency, at the Department of History, Faculty of Arts, University of Ljubljana), of the research project The History of Women's Work in the 19th and 20th Centuries in Slovenia (J6-50190, Slovenian Research and Innovation Agency, at the Department of History, Faculty of Arts, University of Ljubljana), of the bilateral project Socio-Economic History, the Present Day and the Future of Alpine Pastures and Temporary Settlements in Montenegro and Slovenia (BI-ME/16-17-003), of a paid position at the Department of History and European Ethnology, Faculty of Philosophy and History, University of Innsbruck, combining university teaching and work on this volume from late September to the end of November 2024, and of activities related to expertise in plant ecology and conservation management at the Institute of Botany, University of Natural Resources and Life Sciences, Vienna. The authors are grateful for the project funding they received. Part of the finances for copy-editing was kindly provided by the young scholars research development fund at the University of Ljubljana's Faculty of Arts and by the institutional financing pillar at the same faculty. The authors also acknowledge support from the research project Social and Identity Mobility in the Slovenian Territory between the Late Middle Ages and the Disintegration of the Habsburg Monarchy (J6-7069, Slovenian Research Agency, at the Department of History, Faculty of Arts, University of Ljubljana).

The authors thank the Commission for Interdisciplinary Ecological Studies at the Austrian Academy of Sciences for accepting this publication for its series *Interdisciplinary Perspectives* and for repeatedly extended deadlines as the book evolved, Verena Winiwarter and Gertrud Haidvogl, who made our cooperation possible and kindly supported the project, members of the Institute of Social Ecology, Vienna, for their comments during the project workshops, the cartographers Lena Kropivšek, Ana Seifert Barba, and Tina Vrabič (Department of Geography, Faculty of Arts, University of Ljubljana) for maps, Viktor Bruckman and Karin Windsteig (Commission for Interdisciplinary Ecological Studies, Austrian Academy of Sciences) for editorial work and their flexibility, Donald Reindl, who improved our English, and Blaž Božič, who diligently translated the volume *"Ahogy gondozza, úgy veszi hasznát": Hagyományos ökológiai tudás és gazdálkodás Gyimesben* from Hungarian and thus made its full text available to the authors of this volume.

The authors thank the ethnologist Marija Makarovič that, at her own initiative, as an expression of gratitude for interpretative help with her archival records, performed three grassland history interviews for this volume in her study areas of the northern Pohorje Hills and Jazbina, and the botanist Martina Bačič, who determined the thistle specimen gathered by Makarovič before the interdisciplinary cooperation of the authors of this volume started, the historian Janez Mlinar, who kindly made available his copies of archival records referring to the mountain grasslands on the Golica Ridge in the sixteenth and seventeenth centuries, the historian Tadej Pavković, who identified and scanned a 1547 contract for long-distance trade in spikenard valerian that is analyzed in this volume, the Germanist and historian Marija Mojca Peternel for recommending a nineteenth-century mountaineering-club journal article dealing with edelweiss. Thanks to the geographer Jure Tičar for helping localize the meadows in Figure 11, the art pedagogue and design specialist Božidar Flajšman, who drew Figure 43, the astronomer Tomaž Zwitter, who drew Figure 70, and Matej Zwitter, who took the photo in Figure 52. We thank Roland Köchl from the Department of History and European Ethnology at the University of Innsbruck for the recommendation of Ludwig von Hörmann's nineteenth-century ethnographical volume on Tyrol,[239] and Thomas Wieser, of the Municipality of Umhausen in the Ötz Valley (Germ. *Ötztal*), for his kind information on mid-twentieth-century grassland management at Niederthai. We thank the archaeozoologist Borut Toškan for his review of Section 5, the archaeologist Dragan Božič for his review of scythe evolution in Section 6.1.1.1, the church historian Lilijana Žnidaršič Golec for her review of Section 4.1.1.4, the archaeologist Timotej Knific, who reviewed the chronological information concerning the scythes from the National Museum of Slovenia that are studied in Section 6.1.1.1, and anonymous book reviewers. We thank Barbara Lampič and Irma Potočnik Slavič for their geographical insights into Slovenian rural areas, Fridolin Krausmann and Stefan Hörtenhuber for explanations and recommendations concerning sustainability issues linking animal husbandry and greenhouse gas emissions, Johanna Huber from Suske Consulting for her responses to our questions concerning agri-environment payments for grasslands in Austria and the market for such products there, Heidi Trettler for her explanations concerning *ARGE Heumilch Österreich* hay milk farms, Dániel Babai and Anamaria Iuga for unpublished information about traditional grassland farming in Romania, and Helmut Pöll from the Municipality of Galtür for a guided tour through the museum exhibition at the Alpinarium Galtür outside opening hours, and his further help with photographs.

[239] von Hörmann, 1877.

Special thanks go to people from the Alps. The interviewees Angela Boltar, Tine Ceklin, Jožef Cvetek, Lojze Frelih, Gregor Gartner, Lucija Gartner, Janez Gašperin, Antonija Grm, Angela Hodnik, Lovro Hodnik, Marjeta Hodnik, Terezija Hodnik, Anton Hribar, Marjana Hribar, Ludvik Janež, Anica Kavčič, Janez Kavčič, Katarina Kranjc, Marjeta Kunavec, Edvard Lamot, Ivanka Mencinger, Cilka Mlakar, Franc Mlakar, Ivanka Mlakar, Minka and Lovrenc Odar, Sophie Plesnik, Ančka and Marija Prepotnik, Marija Rozman, Marija Šetina, Janez Škantar, Jože Smukavec, Matej Smukavec, Vida Smukavec, Milena Sodja, Minka Sodja, Roman Sodja, Anton Soklič, Metka Stergar, Anton and Matilda Taler, and Julka Urh generously shared their traditional ecological and agricultural knowledge with us. We thank the agricultural community of Bohinjska Češnjica—Cilka Mlakar in particular—and Lucija and Matevž Gartner from the Tourist Farm Gartner in Studor, Bohinj; both women shared from four days to several weeks a year in the Krstenica and Laz alps (summer settlements and complexes of mountain pastures) in the Julian Alps with the main author over the course of a decade (starting in primary school and continuing well into his university studies in history and geography, and again in 2021). We thank the Hodnik family from Stara Fužina, who made possible the main author's participant observation of manual turning and raking hay in a montane meadow on Vogar Hill, and Jožef Cvetek, who was joined by the volume's main author at traditional montane meadow spring-cleaning at Zebeja close to Stara Fužina (just below Zavaze in Figure 4). In these manners, the crucial experience of participant observation in grassland farming was provided. We thank Matevž Gartner, Irena Orešnik, Tomaž Soklič—who contributed his current experience with spring sheep grazing in meadows—and cited anonymized farmers from the Slovenian Alps, for their transdisciplinary dialogues on today's grassland farming.

We thank the personnel in all the archives and museums where work was carried out, in particular those at the Archives of the Republic of Slovenia, the Archdiocesan Archives in Ljubljana, the archives of the Benedictine abbey of Sankt Paul im Lavanttal, Carinthia, the Archives of the Province of Carinthia, the National Museum of Slovenia, and the Surveying and Mapping Authority of the Republic of Slovenia, who generously fulfilled our requests, such as the use of many sources per day, quick availability of working space despite the COVID-19 limits, and quick processing of reproductions. We thank the archivist Lilijana Žnidaršič Golec for drawing the author's attention to the archival records of the Commission for Redemption and Regulation of Profits-à-Prendre in the Province of Carniola, and Lea Lechthaler who brought the notebook from her family archive documenting the irrigation water rights at Lasa (Germ. *Laas*) in the Venosta Valley (Germ. *Vinschgau*) to Innsbruck so that it could be analyzed during the main author's stay at the University of Innsbruck.

We thank the Alpinarium Galtür and the Municipality of Galtür, the Archives of the Republic of Slovenia, the journal *Economic- and Ecohistory,* the

Slovenian Ethnographic Museum, Tourism Bohinj, Christian Berg, Rok Dovjak, Ewald Gabardi, Irmtraud Hubatschek, Rok Humerca, Borut Juvanec, Barbara Lampič, the National Museum of Slovenia, Christian Norz, Mojca Osvald, Odilo Schmid, Mateja Šmid Hribar, Eva Uršič Petkovšek, Giovanni Vauterin, and Ingeborg Vonderstrass for their copyright permissions.

Last but not least, we thank our families and friends.

2 Brief phytosociological overview

Seminatural meadows and pastures (see Section 1), the topic of this volume, are one of the three main categories of grasslands in Central Europe, all of which are also present in the Alps. The second category is natural (primary, non-farmed) grasslands, which develop where severe abiotic conditions limit the growth of woody species (e.g., in the alpine zone above the potential natural forest line, and elsewhere on some very dry, wet, sandy, or saline soils). The third category of Central European grasslands is meadows and pastures used intensively according to the recent standards of high land-use intensity.[240]

The vast majority of seminatural grasslands in the Alps have occurred at localities that are suitable for woods: potentially deciduous and alluvial in lowlands, deciduous in the colline vegetation belt, deciduous or mixed in the montane zone, and mainly coniferous or mixed forests (also including scrub, such as dwarf mountain pine, *Pinus mugo*) in the subalpine vegetation belt.[241] Thus, questions of deforestation, grassland "cleaning," and afforestation in various phases of human activity remain important in the background of the study (see Sections 3, 6.6, and 9). The remainder of seminatural grasslands were located in former natural grassland sites (originally the second category in the previous paragraph), where human activities, largely farming, converted natural grasslands into seminatural ones (thus into the first category). Grasslands in former forest clearings, ranging from valley bottoms to the subalpine vegetation belt, strongly prevailed within the total area of seminatural grasslands in the Alps, but farming and gathering at former natural grassland sites were also relevant (see, for instance, Sections 3, 4, 5, 6, 7.1, 7.2, 8, and 9). In the last millennium, the total area of farmed grasslands in the Alps that evolved from grassland categories one or two strongly varied over time, mainly due to anthropogenic impacts (Sections 3 and 9). Areas of farmed alpine grasslands above the potential natural forest line in the Central Alps, on igneous and metamorphic bedrock, must have generally been more extensive than in the mountain karst of the Southern or Northern Limestone Alps. However, human factors were also highly relevant in deciding whether to use a grassland as a pasture or a hay meadow.

[240] See Hejcman et al., 2013.
[241] Zukrigl, 1973.

Grassland communities of seminatural meadows and pastures are secondary formations. The current typical species compositions of plants, animals, fungi, lichens, and microorganisms of seminatural grasslands developed "over hundreds and sometimes thousands of years in a co-evolution of agriculture and nature" (Sections 3, 4, 5, 6, 7, 8, and 10.1).[242] The plant species in farmed grasslands originate mostly:[243]

- From forests (specialized to natural open-canopy forests), natural grazing areas of wild animals, and corridors of migrating large grazing mammals;
- From patches within regularly forested landscapes that are naturally without tree cover, such as banks of rivers and other water bodies, bogs, marshes, dunes, rocks, scree sites, and avalanche corridors; and
- From adjacent areas not covered by trees and bushes due to climatic conditions (alpine grasslands above the potential natural dwarf mountain pine (*Pinus mugo*) habitats, tundra, steppes, and Mediterranean grasslands).

The primary habitats of a low percentage of the plant species (about 10%, including some dominant species) occurring in seminatural grasslands, are not known for certain,[244] and it can be assumed that some of these species arrived in Central Europe as a consequence of historical human land use: they are archaeophytes or neophytes. In some cases, especially for species that arrived relatively recently, such plant movements and primary habitats are clearly documented, and so Section 7 deals with selected examples. Even false oat grass (*Arrhenatherum elatius*), eponymous for the entire vegetation class, is considered an archaeophyte in northwestern Europe (the British Isles and Scandinavia),[245] but presumably native in the Alps (see Section 6.3.2).[246]

The landscapes of the Alps are "characterized by strong natural gradients and large spatial . . . heterogeneity," such as the diversity of climate, lithology, land forms, and soils. Together with the diversity of human management practices (see a selection in Sections 6 and 8), this gave rise "to a great diversity of habitats and species."[247] Although seminatural grasslands differ significantly in their species composition and belong to several separate plant-sociological units (as presented below in this section), some important similarities can be mentioned concerning the structure of these communities. In nearly all cases, one or two dominant poaceous species (i.e., grasses) with the highest

[242] Ellenberg & Leuschner, 2010; Oppermann & Paracchini, 2012, 17–18 (quote from p. 18).
[243] According to Ellenberg, 1952; Mucina, Grabherr, & Ellmauer, 1993, vol. 2; Ellenberg & Leuschner, 2010.
[244] Ellenberg, 1952.
[245] Pfitzenmeyer, 1962; Roehrs, Klooss, & Kirleis, 2013.
[246] Akeret & Geith-Chauviere, 2011.
[247] Schmitzberger et al., 2005, 274–275; Fischer et al., 2008, quotes from p. 148.

cover and/or abundance build(s) the matrix of the community and provide(s) the most biomass (yield). The following paragraphs reveal that such dominant grasses are often false oat grass (*Arrhenatherum elatius*, Figure 90 in Section 6.3.2.2), yellow oat grass (*Trisetum flavescens*), purple moor-grass (*Molinia caerulea*), erect brome (*Bromus erectus*), tor-grass (*Brachypodium pinnatum*, Figure 10), or matgrass (*Nardus stricta*, Figure 30 in Section 4.1.2 and Figure 42 in Section 6.1.1.2). However, there is still enough room for a variety of subordinated species, both mono- and dicotyledonous, which utilize gaps in the community structure for their recruitment and are able to coexist with dominant species due to the moderate disturbance level provided by historical or recent extensive or rather extensive agricultural land uses.

According to the commonly used and generally accepted phytosociological classification, the majority of meadows and pastures dealt with in this volume belong to the class of Molinio-Arrhenatheretea R.Tx. 1937.[248] This vegetation formation is characterized by dominance of grasses (Poaceae) but assembles a significant amount of subordinated herbaceous plants. It can only exist continuously as long as it is regularly farmed (grazed and/or mowed). The moderate levels of disturbance and biomass removal involved in such management are prerequisites for persistence of these plant communities over a significant time span. In the case of abandonment, ongoing secondary succession (Section 3.2) results in different types of reed, tall herb, grass, or sedge communities significantly differing from the seminatural grasslands still in use (Figure 119 in Section 7.2). In nearly all cases, the final stage of succession is forests (Section 9).

The following paragraphs mention the most important subdivisions of Molinio-Arrhenatheretea, differing from each other mainly due to the abiotic conditions of the sites:[249]

The order Arrhenateretalia R.Tx. 1931 (Figures 1, 2, 6, 7, 8, Figure 54 in Section 6.1.2.2, and Figure 91 in Section 6.3.2.2) includes comparatively nutrient-rich meadows and pastures, including meadows used twice or three times per year and fertilized with organic manure (see Section 6.3.2), but also more heavily used agricultural grasslands. More intensive management within the order Arrhenateretalia leads to different species occurrence and abundance than in the case of moderate management as mirrored in the loss of yellow blossoms (goat's-beard, *Tragopogon pratensis*, Figure 91 in Section 6.3.2.2) and/or purple blossoms (field scabious, *Knautia arvensis*, Figure 91; on drier sites meadow clary, *Salvia pratensis*, Figure 166 in Section 10.2.2), which are frequent in moderately fertilized meadows in May and June. Most economically relevant for the present study are the meadows and pastures with moderate

[248] See, e.g., Schubert, Hilbig, & Klotz, 2001; Mucina, Grabherr, & Ellmauer, 1993, vol. 1.
[249] Schubert, Hilbig, & Klotz, 2001; Mucina, Grabherr, & Ellmauer, 1993, vol. 1.

Figure 6: Mountain false oat grass meadow in the eastern Central Alps (Nock Mountains, Germ. *Nockberge*), mowed once a year in late summer. This site is particularly rich in thistles (melancholy thistle, *Cirsium heterophyllum:* purple blossoms; cabbage thistle, *Cirsium oleraceum:* green tall herb with pale yellow blossoms at lower left), which can be seen as a sign of low land-use intensity in recent years. Photo: Leonid Rasran.

nutrient and moisture levels on soils with a nearly neutral pH, mainly located in colline and montane zones (Figures 6 and 91). With increasing elevation, the dominant grass—false oat grass (*Arrhenatherum elatius*, Figure 90, Section 6.3.2.2)—is gradually replaced by yellow oat grass (*Trisetum flavescens*). This is a similar species, closely related to the former, but less productive (nonetheless, providing higher hay quality; Sections 6.3.2 and 10.1.2.2.3) and better adapted to the environmental conditions at higher elevations within the montane zone and above it, and at lower elevations within the subalpine zone. Plant communities dominated by yellow oat grass are commonly grouped into the order of *Poo alpinae-Trisetetalia* Ellmauer et Mucina 1993: mountain yellow oatgrass meadows (Figure 7). In the case of abandonment, false oatgrass and yellow oatgrass meadows become overgrown by forest, usually depending on the seed rain from the surroundings (also including possible trees and/or scrub that occurred within the grassland before) and on vegetative propagation of woody species occurring in the immediate vicinity (Section 3.2).

The order Molinietalia Koch 1926 represents grassland communities of wet and moist sites (both acidic and base-rich). It includes the comparatively nutrient-poor societies of litter meadows (Molinion, Figure 9), normally mowed once a year in late summer or autumn. The hay of these sites has low fodder quality and was most suitably used as animal bedding in barns, even though it

Figure 7: Mountain yellow oatgrass meadow in the Eastern Alps (Hochschwab). Note the European yellow-rattle (*Rhinanthus alectorolophus*, light green and yellow herbs in the foreground), a hemiparasitic species reducing the productivity of the meadow from the farmers' perspective (Section 4.1.1.1). Photo: Leonid Rasran.

Figure 8: Alpine pasture (*Poion alpinae* Oberd.) in the Eastern Alps (Hochschwab). This is a subdivision of Arrhenateretalia occurring above the tree line and dominated by alpine meadow-grass (Poa alpina) and several rosulate herbs (e.g., golden hawk's beard, *Crepis aurea*, with orange blossoms in this photo). Photo: Leonid Rasran.

was also widely used as fodder in the past (see Section 6.1.2.5 and the text on wild Illyrian gladiolus, *Gladiolus illyricus*, in Section 7.1). Dominant tussock grasses are purple moor-grass (*Molinia caerulea*) and tufted hairgrass (*Deschampsia cespitosa*), which are unattractive for grazing livestock where better fodder is available. Plant communities of the nutrient-rich flank of the order Molinietalia belong to the group of marsh marigold meadows (Calthion). These meadows are more attractive from the farmers' perspective regarding hay quality and productivity, which permitted one more annual use than Molinion (Section 6.1.2.5.1). The amount of dicotyledonous plants in their cover is significant, and it can additionally increase in the case of declining land use. The most characteristic plants of the nutrient-rich flank of Molinion include, among others, marsh marigold (*Caltha palustris*), marsh hawk's-beard (*Crepis paludosa*), and some thistle species such as cabbage thistle (*Cirsium oleraceum*; Figure 119 in Section 7.2). The last one used to be widely gathered (Section 8.2.2). An early abandonment stage of marsh marigold meadows is tall herb communities dominated by meadowsweet (*Filipendula ulmaria*, Figure 119) and also herbs sensitive to grazing and cutting. In general, Molinietalia turn into reed communities, tall sedge communities, and finally alluvial forests like alder (*Alnus* spp.) carrs when farming is abandoned (Section 3.2). The majority of characteristic species of wet meadows have their primary habitats within alluvial forests, carrs, and fens.

The order Potentilio-Polygonetalia R.Tx.1947, sometimes listed as a separate class within Agrostietea stoloniferae Oberd., corresponds to comparatively species-poor swards on moist, waterlogged, or periodically flooded sites with compacted soils. They are dominated by stress-tolerant specialists able to deal with low oxygen content and sometimes high salt concentration in the soil. In spite of comparatively high nitrogen availability, the biomass output and the fodder quality in such sites are low. This grassland form is often a consequence of overgrazing, trampling, and other careless land-use practices. Abandonment (Section 3.2) commonly leads to the development of reed communities dominated by canary grass (*Phalaris arundinacea*). Similarly, vegetation of the next order, Plantagini-Prunelletalia Ellmauer et Mucina 1993, is species-poor and contains specialists for moist, compacted soils. It mainly occurs in the montane zone and in comparatively small patches along the shaded forest margins and pathways, also within settlements. It was only of marginal importance for historical land-use practices.

Highly relevant was the opposite type: the driest and warmest (thermophilic) part of seminatural non-fertilized grasslands on calcareous soils belonging to the class of Eurasian steppe vegetation (Festuco-Brometea Br.-Bl. et R. Tx. 1949; Figure 10; see Section 6.3.1). Dominant grasses are erect brome (*Bromus erectus*) and tor-grass (*Brachypodium pinnatum*, Figure 10), but the species diversity can be very high (Sections 1.2, 6.3.1, 10.1.2.2.2, and 10.1.2.2.4) and various subordinate species (both dicot and monocot) are present. Significant

Figure 9: A litter meadow (Section 6.1.2.5) in Alpine foothills (the Vienna Woods, Germ. *Wienerwald*), with flowering Siberian iris (*Iris sibirica*). This comparatively nutrient-poor site is mowed once a year in late summer or autumn, now only for environmental conservation purposes. The litter is no longer used and is discarded. Photo: Leonid Rasran.

cover and abundance are commonly attained by sedges (spring-sedge, *Carex caryophyllea*, see Figure 14 in Section 3.2; glaucous sedge, *Carex flacca*) and herbs (e.g., greater knapweed, *Centaurea scabiosa*; cowslip, *Primula veris*; and salad burnet, *Sanguisorba minor*). Fabaceae such as mountain clover (*Trifolium montanum*) and kidney vetch (*Anthyllis vulneraria*, Figure 10) are also present in high quantities and play an important role in raising hay quality. The most species came from treeless continental and Mediterranean vegetation, but their distribution was largely extended by farming (see, e.g., Section 7.1). Nowadays these semi-dry grassland sites are only kept open by grazing and haymaking, and they would turn into thermophilic deciduous forest and shrubland if they were abandoned (Section 3.2).[250]

The former class Nardo-Callunetea (Priesing 1949) is named after matgrass (*Nardus stricta*) and heather (*Calluna vulgaris*). Today, this community class is mainly classified as heathlands (the class Calluno-Ulicetea Br.-Bl. et R. Tx. ex Klika et Hadač 1944) rather than agricultural grasslands, but they have a long history of agricultural use; see Sections 6.3.1 and 6.6.2.1. A specific or-

[250] Schubert, Hilbig, & Klotz, 2001; Mucina, Grabherr, & Ellmauer, 1993, vol. 1.

Figure 10: Three sections of the same non-fertilized montane meadow at the Ši meadow complex above Srednja Vas, Bohinj. Note kidney vetch (*Anthyllis vulneraria*; i.e., low herbs with tricolor yellow, brown, and white blossoms in the center), orchids (fragrant orchid, *Gymnadenia conopsea*, with purple blossoms in the center and lesser butterfly orchid, *Platanthera bifolia*, the white flower in the right lower figure section), columbine (*Aquilegia* sp., the tall herb with a large blue flower on the left), tor-grass (*Brachypodium pinnatum*; i.e., the grass in front of the bush in the upper right), and an ash tree (*Fraxinus* sp.) pruned a year before in late summer (Section 8.2.2.1) in the top right. Photo: Žiga Zwitter, mid-June 2020.

der (subdivision) of this class found in nutrient-poor and acidic, moderately moist meadows and pastures in the montane and subalpine zone are matgrass swards in the narrower sense: Nardetalia Oberd. ex Preising 1949 (see Figure 42 in Section 6.1.1.2). Similar to most grassland types dealt with in this book, they are products of clearing and regular haymaking at (primarily forested) sites on silicate bedrocks and moraine deposits, also in the Limestone Alps. Although they provide low-quality fodder, in many places they were attractive to people due to the occurrence of characteristic dicotyledonous plants of acidic sites—for example, arnica (*Arnica montana*), spikenard valerian (*Valeriana celtica* subspec. *norica*), cowberry (*Vaccinium vitis-idaea*), and bilberry (*Vaccinium myrtillus*)—which were gathered on a large scale, although with significant changes in the quantities gathered over the course of history (see Sections 6.6.2, 8.1.1, 8.1.2, 8.1.3.2, 8.2.1.1, 8.2.1.2, and 8.2.1.3). Bilberry belongs to characteristic species of matgrass swards, like some other semi-shrubs and dwarf shrubs, such as heather (*Calluna vulgaris*) and pennate greenweed (*Genista sagittalis*). Reduction in land-use intensity leads first to an increase in the cover of dwarf shrubs and later to afforestation, mainly with conifers, as far as the nutrient status of the area allows it.[251]

Another order from the class Calluno-Ulicetea—namely, the mountain heathlands of Vaccinio-Genistetalia Schub. 1960—was used as subalpine or alpine pasture that required specific management in some regions of the Alps and historical periods (see Section 6.6.2.1). In this order, the dominant grasses are wavy hair-grass (*Avenella flexuosa*), bents (*Agrostis* spp.), and fescues (*Festuca* spp.), in contrast to Nardetalia Oberd. ex Preising 1949, which is dominated by matgrass (*Nardus stricta*).

A further class including dwarf shrub heathlands is Loiseleurio-Vaccinietea Eggl. 1952 ex Schub. 1960. Within this class, the order Rhododendro-Vaccinietalia Br.-Bl. et Jenny 1926 is relevant for this volume. Alpenrose (*Rhododendron ferrugineum*), juniper (*Juniperus communis* subsp. *alpina*), heather (*Calluna vulgaris*), bilberry (*Vaccinium myrtillus*), and cowberry (*Vaccinium vitis-ideae*) belong to characteristic species of this order. In these plant communities of the open subalpine and alpine zone, the amount of plant material consumable by grazing livestock is very low, and woody dwarf shrubs and lichens provide the main cover. These sites were only rarely used as pastures, unless burned (Section 6.6.2.1), but they could play a role as sites for gathering bilberries or cowberries (see Sections 6.6.2.1 and 8.2.1.1).[252]

The orders of plant communities presented above are mainly based on recent grassland studies. The oldest sufficiently complete grassland descriptions were usually only created in the nineteenth century, when a significant grassland history transition was in progress (see especially Sections 3.1, 6.1.2.1, 6.1.2.2, 6.1.2.3, 6.1.2.5, 6.2, and 6.3). Available evidence for eventual changes in widely present plant communities or even orders of plant communities before the nineteenth century is poor. There are some indirect pieces of information about earlier centuries, such as localized historical botanical, archaeobotanical, and less clearly localized palynological data on occurrence of a limited number of species at a selection of locations. In addition, there are some data on site conditions where these species occurred at that time, and fewer data on selected species' abundance in the same grasslands before the nineteenth century. Moreover, knowledge about earlier historical grassland management practices as well as plant information in agricultural works is relevant in this context; nonetheless, historical agricultural publications focus on agriculturally specifically relevant plants and not on entire species compositions. Combinations of different types of these data with in-depth present grassland-ecology knowledge clearly point to historical plant communities before the nine-

[251] Schubert, Hilbig, & Klotz, 2001; Mucina, Grabherr, & Ellmauer, 1993, vol. 1.

[252] Both paragraphs are based on Schubert, Hilbig, & Klotz, 2001; Mucina, Grabherr, & Ellmauer, 1993, vol. 1; on alpenrose as fodder, see also Machatschek, 2002, 112–113, 331–332, 394.

teenth century (indeed, such indirect conclusions about earlier historical plant communities are close to historical reality in many cases), but they cannot prove eventual changes in plant community structure (exceeding the issues of long-distance migrations of species) in history. In some cases, we can expand the categories of modern plant sociology units into earlier centuries of the last millennium under the assumption that the characteristics of species and their responses to biotic and abiotic conditions influencing them have remained similar. However, there may have also been some changes in the plant community structure within this observed period.[253]

This volume deals with agroecosystems, and so grasslands in the alpine vegetation belt are only dealt with if they were farmed (i.e., seminatural) and data on them fulfill the criteria given in Section 1.4, including novelty of interpretation. Primary dry grasslands and salt marshes restricted to very special locations and usually occurring in very small patches within the area dealt with in this book are hardly mentioned in the text, but they are important primary habitats and sources of species, contributing to species diversity in secondary grasslands.[254]

Intensive agricultural grasslands in the narrower sense (the third grassland category in the first paragraph of this section) appeared first within the last century and spread extensively in the landscape only due to broad access to mineral fertilizers, unprecedented availability of (liquid) organic manure, and commercially produced seed mixtures. Their large-scale spread in landscapes of the Alps exceeds the chronological scope of this volume. With increasing intensity of land use, human activities became a predominant factor determining species composition and grassland community structure. Thus, agricultural grassland sites of high land-use intensity turned into uniform, species-poor plant communities dominated by a few high-yield grasses and legumes (Figure 2; Sections 1.2, 10.1.2, and 11). In contrast, moderate and low levels of land-use intensity, strongly prevalent in the times dealt with in this book, provide comparable multilateral interactions with various abiotic factors such as moisture, temperature regime (elevation), and soil pH. The high variety of ecological niches resulting from these interactions is an important prerequisite for high biodiversity on all spatial scales, and it also brings grasslands managed in a manner inspired by history into the focus of sustainable grassland management (Section 10.1) and environmental conservation.[255]

---

253 Compare Kauter, 2002, especially pp. 161–171, 174–176; Kapfer, 2010; Peterken, 2013.
254 See, e.g., Wagner, 1989; Ellenberg & Leuschner, 2010.
255 E.g., Eriksson, Cousins, & Bruun, 2002; Sitzia & Trentanovi, 2011.

3 Fluctuations in grassland area in the Alps over the last millennium

3.1 Long-term prevalent increase in grassland area until the nineteenth century

Grasslands in the Alps were historically frequently located where site characteristics would not allow the creation of quality (arable) fields—for example, because the sites were at least occasionally too wet or too dry (due to a combination of climatic, geological, and soil conditions), or where the soil was too shallow or too stony, the slopes were too steep, the site was too shady, the temperatures too low and the growing season too short, or the flood hazard too serious. However, the distribution of grasslands in the landscape was influenced not only by the environment but also by social factors and processes, such as the area and distribution of agricultural land at the farm level, the orientation of agricultural production, field rotation systems, farming methods, and broader economic and demographic history.[256] Focusing on meadows, in past centuries most of the hay in the Alps was used by the farmers that harvested it, and a lot of hay produced on demesne land was also used by the seigneury whose demesne it was. However, "some hay was also bought and sold, both to alleviate temporary shortages and as a planned source of income."[257] As this section demonstrates, hay purchases influenced the area of grasslands. The variety of complexly interrelated environmental and human factors affecting the spatial distribution of grasslands in the Alps importantly contributed to the great diversity of grassland site conditions and species compositions (Sections 2, 3, 4, 5, 6, 7, 8, 9, and 10.1).

By the end of the first millennium as the beginning of the study period in this book, the landscapes of the Alps were patchworks of grasslands, fields, and extensive areas of sparser and denser forests, with a much higher percentage of grasslands than without human influence.[258] The bottoms of larger valleys and

[256] Adapted from Sadar, 1956, 3, 5–10, 12–13, Orland, 2016, 26–27, 43, Kapfer, 2010, 136, Peterken, 2013, 151–152, 153–155, 172, 249, and Winiwarter et al., 2013, 114.

[257] Peterken, 2013, 144 (quote), 145.

[258] Mathieu, 2019, 37, 49; van der Knaap et al., 2000, 107–108, 118–119; Hejcman et al., 2013, 351–352, 357; these publications present archaeobotanical and palynological results. Information from various types of early archival records is presented in Kleindinst, 1999, 22, 24, 26 and Štih, 2014b; see also Emanuelsson, 2009, 45–51, 59, 89–91, 98, 163–164; Bätzing, 2015, 58–59; Poschlod, 2015, 31–32, 64–67. Studies of the

basins were populated, and the lower parts of some side valleys were also settled, but many other side valleys had not been reached by settlements yet, in the Eastern Alps in particular. "The centers of settlement and land use were located in the midst of forested and little-used areas." Many complexes of distant summer pastures already existed,[259] especially above the forest line and in the sparse forest just below it.[260] Later in the Middle Ages, the population in the Alps grew, especially in the twelfth and thirteenth centuries. This corresponded to broader European demographic developments. Nonetheless, in the context of mountains, the rate of population growth in the Alps was atypically high at that time, and so from the Late Middle Ages onward "the Alps were heavily settled mountains … [i]n comparison to many European and non-European mountain regions." The size of villages increased, and some new settlements and patches of agricultural land were established in the valleys and basins; many settlements were established at higher elevations. This partly corresponded to the predominately relatively mild but variable climate of the High Medieval Climate Anomaly, often called the Medieval Warm Period (c. 800/1000–1250/1300). Nonetheless, the creation of new permanent settlements in remote areas sometimes lasted longer, in many places until the fourteenth century, although the beginning of a less favorable climate phase around 1300 is attested. Namely, that was the onset of one of the three major phases of advance of several glaciers in the Alps ever since. Around 1300, the Little Ice Age (or the period of transition toward the Little Ice Age, until the mid-sixteenth century) started, lasting until the mid-nineteenth century. Nonetheless, the Little Ice Age also included subperiods when the climate was favorable for agriculture, but the creation of new permanent settlements was not limited to them. In some areas, new remote permanent settlements largely based on agriculture were established until the mid-seventeenth century, for instance.[261]

---

shells of grassland mollusks also have the potential to reveal at a small-scale resolution whether forest-free patches (grasslands) ever existed in an area where there has been forest since the earliest preserved archival records. Such shells are well preserved especially in alkaline soils (Hejcman et al., 2013, 347–349). Similarly, dated representative samples of the skeletal remains of some mammals, especially some small rodents—field mice (*Apodemus*) and voles (*Microtus arvalis* and *M. agrestis*)—that require grasslands can reveal whether there were already grasslands in a certain landscape in the Alps at a certain time within the last millennium. The mobility of such small rodents is limited, as is their adaptability to habitat conditions (Toškan, 2016b, 221–222, 224–225).

[259] Viazzo, 1989, 124; Rösener, 1991, 1644; Štih, 1998, 135–137; Kleindinst, 1999, 22, 24, 27; Mathieu, 2019, 43–44, 49 (quote).

[260] For example, Jordan, 1945, 63–65; Melik, 1950, 50–53; Zwitter, 2013a, 91.

[261] For example, Grass, 1948, 164–165, 183–187, 191–192, 196; Blaznik, 1970c, 88–89; Schneiter, 1970, 61; Pickl, 1982, 27–28; Pfister, 1985a, 86, 127–128; Viazzo, 1989, 124–125, 127–128; Rösener, 1991, 1644–1645; Wopfner, 1997, 273, 402, 424; Štih, 1998, 136–137, 141; Wunderli, 2010, 25; Nicolussi, 2013, 71, 87, 90–91; Zwitter, 2013a, 85–88, 96; White, 2014; Moinuddin et al., 2013 and Corrigendum, 2015; Bätzing, 2015, 59–62, 68–70; Scharr, 2001, 147–154; Poschlod, 2015, 67–70; Rohr, 2017, 76; Mlinar, 2018d, 20; Pfister et al., 2018, 268–269; Rohr, Camenisch, & Pribyl, 2018, 248, 254–255;

Starting in the fifteenth century, fewer new settlements were established, but many recently founded and older ones grew; an important mechanism of older settlements' growth was the increased numbers of smallholders that gradually lived there.[262] Namely, the human population of the Alps, as nowadays delimited by the Alpine Convention, increased from an estimated 3.1 million people in 1500 to about 5.8 million in 1800 (this increase was largely a consequence of population growth in the sixteenth century and the second half of the eighteenth century) and 8.4 million in 1900. In 1870, more than two-thirds of the economically active population was farmers; thus, despite deagrarianization, the number of farmers in the Alps has probably never been as high as around 1870. In general, there existed major differences in the demographic history of the nineteenth century as the century of strongest general population growth. Human population increase was, for instance, much more intensive in the Swiss Alps than in the Austrian Alps, with pronounced diversity within both regions.[263] More farms often led to a locally and regionally increased livestock feed demand, and the weather and climate have also been changing, with years and clusters of years more and less favorable for livestock farming. Focusing on the spatial distribution of grasslands, these processes had the following main consequences:

First, there was the creation of new meadows and pastures on former forest sites. This large-scale high medieval to late medieval expansion of grasslands is proven by archival records and pollen diagrams. It included the creation of entire complexes of distant meadows in forest clearings between the settlements and the upper forest line, and several grasslands at other locations were also created. Manuscripts make it clear that in some areas, such as in the Captaincy of Tolmin (Germ. Tolmein; in the southern Julian Alps [Sln. *Julijske Alpe*, Ital. *Alpi Giulie*, Friul. *Alps Juliis*] and the pre-Alps), not only smallholders but even some large farms, and not only new farms but also old ones, created new permanent grasslands in forest clearings in later centuries as well. In many parts of the Alps, written records only provide the first mentions of distant grasslands in forest clearings, and in some others the chronology of grassland extension is better documented. For instance, it was recorded that some meadows at Ukanc and Zagradec, in the Voje Valley, and at Uskovnica (all in the Julian Alps close to Lake Bohinj, Sln. *Bohinjsko jezero*) were just being created anew in the 1490s, or that they were made (or recreated; see Section 3.2.2) only slightly earlier than when the record was produced in 1498.

Mathieu, 2019, 49 (first quote), 152, 154–155 (second quote from p. 154). On the concept of the Little Ice Age, see Collet et al., 2025.

262 Grass, 1948, 184; Gestrin, 1962, 684–685; Blaznik, 1970a, 167–169; Blaznik, 1970b, 154–156; Viazzo, 1989, 127–135, 264; Wopfner, 1997, 314–315, 395; Zwitter, 2013a, 87–88.

263 Viazzo, 1989, 129, 132–135, 185–189, 191–196, 262–263, 270–275, 291; Bätzing, 2015, 71, and map of the area dealt with by the Alpine Convention on the frontispiece; Mathieu, 2019, 49, 124–125, 155; map of the Alps as delimited by the Alpine Convention: Bätzing, 2015, frontispiece.

Area of pastures also increased, mainly through the descent of the upper forest line or through new forest clearings at lower elevations. Within the last millennium, the upper forest line varied considerably in many places, and consequently grasslands located above the actual forest line expanded by up to a couple of hundred meters (e.g., 200 to 400 m) down the slopes. Exceptionally, the highest permanent settlements (such as Damüls in Vorarlberg and settlements in the uppermost Tyrolian Inn Valley, Germ. *Inntal*) were even located above the upper forest line. In general, the descent of the upper forest line was the result of three often interrelated main driving forces: grazing in sparse forests, forest clearing, and climate change. Not only temperature, but also precipitation and wind significantly affected the upper forest line in the Alps in terms of climate change. There have been time lags between climate change and (rather small-scale) variations of the potential climate-determined upper forest line. However, in many places the real upper forest line had already previously been so much lower than a potentially climate-determined one that climate change did not have significant impacts on the upper forest line during the Medieval Warm Period and the following Little Ice Age there.

The second consequence was that some former pastures were transformed into meadows.

A third consequence was that some former grasslands were converted into arable fields; for instance, where villages, hamlets, or isolated farms at high elevations developed from former temporary settlements.[264]

The increase in economic and demographic pressures also led to larger areas where wild hay was mowed in the montane, subalpine, and, at some places, alpine belt. This means that wild hay mowing in the Alps took place below and above the actual forest line, and in some places above the potential natural forest line, including sites characterized by very low productivity (but see Section 6.3.1.2 for very diverse hay qualities, also including high-quality fodder from such meadows). Within the agricultural system, hay was important not only for animal husbandry; a higher number of wintered livestock, partly made

[264] Grass, 1948, 164–165, 183–187, 196; Blaznik, 1970b, 153, 155; Blaznik, 1970a, 167–169, 172; Blaznik, 1975, 84–85; Viazzo, 1989, 124–129; Lovrenčak, 1984; Pfister, 1985a, 86, 98–99, 127–128; Meilleur, 1986, 2, 22, 25, 27, 31, 41, 369, 374; Hubatschek, 1992, 28, 43; Lang, 1993; Wopfner, 1997, 230–231, 395–399, 402–404, 428–429 and photograph 54 between pp. 388 and 389; Štih, 1998, 137; van der Knaap et al., 2000, 86, 107–108, 118–119, 134, 137; Scharr, 2001, especially pp. 120–121, 123; Körner, 2003, 88; Tinner & Theurillat, 2003; Nicolussi et al., 2005; Jäger, 2006, 29–31, 36–37, 41–44, 55, 66, 78; Vogl-Lukasser et al., 2006, 31; Jäger, 2007b, especially pp. 104–106, 108; Rohr, 2007a, 417; Schreg, 2008, 135–136; Andrič et al., 2010, 1027–1029, 1033–1034; Löffler et al., 2011, 195–196; Zwitter, 2013a, 85–86, 91–94; Bätzing, 2015, 70, 77, 89–90; Beguš, 2015, 112–113; Panjek, 2015a, 97, 100, 113; Oitzl, 2021, 300–310; see the presentation of the forest clearance technique in Babai, Molnár, & Molnár, 2014, 24, 27–30. This footnote also refers to previous paragraphs without separate footnotes.

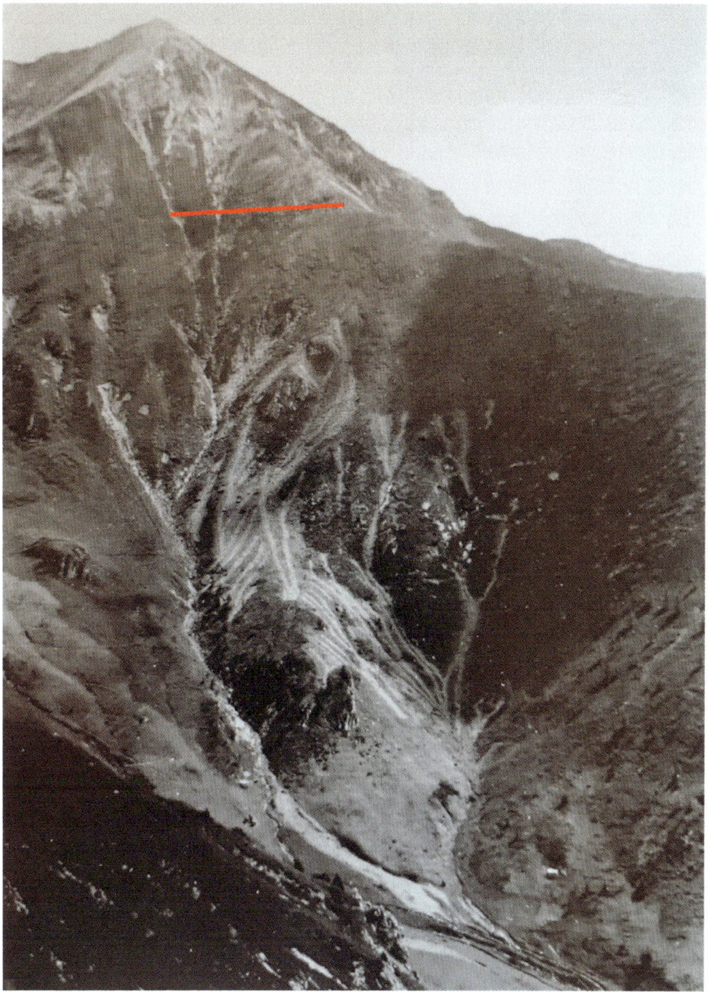

Figure 11: Mowed remote steep and high meadows on the western slopes of Vorderes Weißeneck,[265] southeast of the province of Salzburg, in 1939. Photo: Erika Hubatschek. Hubatschek, 1992, 172, © Irmtraud Hubatschek. A historical perspective reveals that in the century or centuries before this photo was taken, mowing also took place significantly higher in this area than in 1939, as documented on a precise cadastral map from 1830. The red line in Figure 11 designates the approximate previous historical upper boundary of a meadow complex at the center of this figure, as mapped in 1830: the highest meadow there reached an elevation of 2,325 m.[266] A century later, in 1939, only the highest mowed mead-

[265] Thanks to Jure Tičar for convincingly confirming our preliminary localization of the mountain in this photo. His comparison of the photo with *Google Earth* demonstrated that Vorderes Weißeneck's western slopes appear in this figure.

[266] *Arcanum Maps,* Cadastral Maps, Habsburg Empire – Cadastral Maps (XIX. [sic] Century), cadastral municipality of Lamm, 1830 and present-day cadaster showing the same cadastral parcels and topography in *SAGISonline*, Basiskarten, Alle Themen, Digitale Katastralmappe, DKM Grundstücke & Höheninformation, Höhenlinien 1 m; *Google Earth* was used to localize Figure 11, and combined with *SAGISonline*, Basiskarten,

ow section in the center of the photo exceptionally reached nearly 2,200 m.[267] In addition, the 1830 cadaster makes it clear that meadows had covered the entire west-facing slopes of the lower ridge that reaches the right end of the red line in Figure 11 and is the main feature at center right in this photo; in 1939, only relatively few meadows were seen on that slope, and they ended roughly half way to the top of the ridge.[268] Mowing might not have been fully completed by the time this photo was taken; nonetheless, Figure 11 proves that ecological succession in terms of woody overgrowth had already progressed significantly in some grassland sections by 1939. In addition, vegetation demonstrates that pastures had spread on some former meadows before 1939.[269] These facts prove that mowing had been abandoned in significant meadow sections at high elevations. Abandonment of meadows at high elevations or conversion of meadows into pastures was characteristic of the broader region at that time: Vorderes Weißeneck is located in the western part of the Lower Tauern (Germ. *Niedere Tauern*), and research in the nearby High Tauern and some other places in the Eastern Alps such as the Ötz Valley (Germ. *Ötztal*) identified widespread abandonment of meadows at extreme sites and those most difficult to access, at elevations of more than 1,900 m for instance, as early as between the two world wars[270]—thus, when this photo was taken. Abandonment of many mountain meadows at less extreme sites or their conversion into pastures followed from the 1950s onward (Section 9).[271]

possible by hay from remote places, also provided more manure for arable fields. In western Tyrol, for instance, increased demographic pressures led to a significant extension of wild hay meadows starting in the sixteenth century. In several places, the highest mowed meadows in the Alps reached an elevation of about 2,000 m or more (see Figure 11; Section 6.3.1; and Figure 93 in Section 6.4), and exceptionally even 2,600 or 2,700 m (such as at Vent in the Ötz Valley, Germ. *Ötztal*). In the Wipp Valley (Germ. *Wipptal*, Ital. *Val Isarco*), for instance, meadows at elevations of 1,800 to 2,100 m and in the Villgraten Valley (Germ. *Villgratental*) mowing at elevations of 1,900 to 2,200 m used to be frequent. The highest haymeadows above the Zadnjica Valley, in the uppermost Soča River catchment in the Julian Alps, exceeded 2,000 m.[272] At exceptional

---

Alle Themen, Orthofotos, Orthofotos 2021. Note that meadows reaching similarly high elevations were also located to the left, i.e. north, of the red line in Figure 11 as mapped in 1830. However, contrary to the meadow complex reaching the red line, meadows located further north did not range from the valley bottom in the center of the photo to such high elevations, there was only a narrow meadow belt north of the red line, so its location is not presented in Figure 11.

[267] Figure 11; *Google Earth*; *SAGISonline*, Basiskarten, Alle Themen, Höheninformation, Höhenlinien 1 m & Orthofotos, Orthofotos 2021.

[268] *Arcanum Maps*, Cadastral Maps, Habsburg Empire – Cadastral Maps (XIX. [sic] Century), cadastral municipality of Lamm, 1830; *SAGISonline*, Basiskarten, Alle Themen, Digitale Katastralmappe, DKM Grundstücke & Orthofotos, Orthofotos 2021; *Google Earth*.

[269] Figure 11 and *Arcanum Maps*, Cadastral Maps, Habsburg Empire – Cadastral Maps (XIX. [sic] Century), cadastral municipality of Lamm, 1830.

[270] General trends in Keller, 2004, 22, and Hessenberger, 2022, 170.

[271] General trends in Keller, 2004, 22.

[272] Grass, 1948, 146–147, 163, 182–183, 206–209; Hubatschek, 1992, 5, 17, 34, 48, 51, 120–123, 172–179; Pils, 1994, 193–194; Patzelt, 1996, 65–66; Wopfner, 1997, 310, 312,

places, like at Vals in Graubünden (Ital. *Grigioni*; Romansh *Grishun*), Switzerland, mowing in the Alps still takes place at elevations up to around 2,100 m nowadays, 850 meters above the village.[273] In Carinthia, meadows at elevations of 1,800 to 2,300 m are still mowed in the Lesach Valley (Germ. *Lesachtal*), at Heiligenblut (in the High Tauern, Germ. *Hohe Tauern*), and above Rennweg (at Katschberg), for instance; this is not the historical maximal elevation – the highest meadows in Carinthia were still located at 2,400 m after the Second World War.[274] These data make it clear that the highest hay meadows in many parts of the Alps were located incomparably higher than in the Carpathians of what is now Romania, for instance.[275]

It is important to be aware of the fact that some grasslands that are very difficult to access or mow also used to be mowed, usually on sunny sites (see Section 6.3.1).[276] Concerning the issue of access, mowing several meadows inaccessible to grazing livestock made agricultural use of these grasslands possible.[277] Hay had to be transported on one shoulder only (compare Figure 87 in Section 6.3.1.2, but a hay burden had to be smaller if borne on one shoulder only) to pass through some narrow path sections in extreme cases; in such conditions some smallholders that lacked hay mowed places between the temporary mountain settlement of Vodični Vrh (1,475 m) and Mount Pršivec (1,760 m) in the eastern Julian Alps according to information from the first half of the twentieth century.[278] Where small grasslands on ledges on extremely steep slopes were mowed, people descended from or to several small hardly accessible meadows fastened to a rope[279] or they climbed there using ladders and threw hay, sometimes bound to tree branches, to lower, more accessible places. This happened not only in the mountains but also in several hilly areas, such as above the village of Čadrg in the Upper Soča catchment, Julian Alps.[280] Similarly, hay bundles from Macesnovec (just north of Dolina in Figure 4, in the Karawanks [Sln. *Karavanke*, Germ. *Karawanken*]), at about 1,250 m, were thrown over a short steep slope section;[281] the lowest steep section of these meadows has by now turned into scree slopes.[282] The aim of these micro-histo-

314–315; Hinterwaldner, 2002, 103–104; Keller, 2004, 13, 27–28, 67; Bätzing, 2015, 156; Kravanja & Komac, 2023, 18.

[273] von Glasenapp & Thornton, 2011, 773.

[274] Keller, 2004, 13, 27, 32.

[275] Thanks to the Romanian ethnologist Anamaria Iuga for this explanation.

[276] Novak, 1970, 373; Wopfner, 1997, 310, 317, 330–333, 368–369; Keller, 2004, 15–16, 20, 65–67, 137; Bätzing, 2015, 99, 244.

[277] Wopfner, 1997, 317, 332.

[278] Author's interview with Kranjc, 2018, in Stara Fužina.

[279] Wopfner, 1997, 331; Bätzing, 2015, 99.

[280] Author's interview with Janež, 2017, in Čadrg.

[281] Author's interview with J. Kavčič, 2021, in Dolina.

[282] *Arcanum Maps*, Cadastral Maps, Habsburg Empire – Cadastral Maps (XIX. [sic] Century), cadastral municipality of Sv. Katarina (Germ. *St. Katarina*), 1826 and digital orthophoto (*Environmental Atlas of Slovenia*, Backgrounds, Dof 2019). In order to deter-

ry examples representative of many similar cases in the Alps is to demonstrate their reliability; these are no abstract data, they are reliable and linked to precise locations. Storing small amounts of hay this way was no exception in the broader context of the Alps.[283] Our aim is not to over-emphasize the role of hay harvested at such remote places, but to state that even many such places have a (long) history of human use (see also the fourth paragraph of Section 6.2).[284]

Concerning meadow sites difficult to mow due to landforms, slope steepness is discussed here (but see also Sections 3.2 and 4.1.2 for another kind of challenging landform). In quantitative terms, the hay harvest on steep slopes was usually lower than it would be on gentler slopes.[285] It was usual that some grasslands that were too steep for grazing were mowed in order to allow agricultural use of them.[286] At Stara Fužina and Srednja Vas v Bohinju, for instance, haymaking with the aim of winter fodder production was allowed in those sections of summer mountain grassland commons that were hazardous and unsuitable for grazing, as reported in the period of high agricultural density in the first half of the twentieth century and the middle of the century. Consequently, the inhabitants of Srednja Vas v Bohinju, especially those lacking other meadows, mowed remote and possibly steep grasslands at Avšje southeast of Mount Viševnik (mapped in Figure 4) in the Eastern Julian Alps, reportedly at an elevation of 1,600 m,[287] where the slope inclination is often more than 30°.[288] The hay was transported to the valley, to farms located at about 600 m.[289] Another example of mowing very steep meadows high in the mountains, which was usual in many places in the Alps, was mowing the steepest meadow sections on the ridge linking Mt. Vorderes Weißeneck and Mt. Wurmkogel in the province of Salzburg, where the steepest meadow sections reached an inclination of (close to) 50°.[290] Similarly to hardly accessible meadows, extremely

---

mine the locations, this early-nineteenth-century cadastral map was compared with the georeferenced present cadaster (*Environmental Atlas of Slovenia*, Backgrounds, Topographic maps & Spatial units, Land cadaster).

283 E.g., a short selection of examples from Tyrol in Wopfner, 1997, 331–332, and from Carinthia in Keller, 2004, 16, 66–67.

284 Compare Patzelt, 1996, 65–66.

285 Brugger & Wohlfarter, 1983, 180.

286 Wopfner, 1997, 310, 330–333; Keller, 2004, 20.

287 Novak, 2024, 96.

288 This calculation is based on a detailed topographic map with a 10-meter contour interval in *Environmental Atlas of Slovenia*, Backgrounds, Overview Maps.
Novak's figure of 1,600 metres above sea level (Novak, 2024, 96) serves as a starting point for this calculation. If instead we consider the area where the toponym Avšje is recorded on the detailed topographic map (*Osnovna državna karta*, 1972), it is remote grassland on moderate slopes (about 15° to 25°) at about 1,750 to 1,800 m.

289 Novak, 2024, 96.

290 Significantly long (e.g., 40 or 50 m) and relatively wide areas of the steepest meadow sections there reached an inclination of (approximately) 50°, as documented in 1830. Note that an area of a relatively similar size to the one on an extremely steep meadow slope forms a separate parcel mapped as infertile land on the slope of Vorderes

steep meadow sites were also characteristic not only for mountains (Figure 11) but were located on some hills as well. An example is the complex of very steep meadows at the uppermost part of the slope of the Hudi Konec Ridge (highest peak c. 1,150 m) above the village of Županje Njive north of Kamnik (Germ. *Stein*). These meadows ranged from about 1,000 or 1,100 m to the top. Our measurements of slope inclinations in the western part of Hudi Konec showed that the slopes of many of the steepest sections in the meadow plots reach inclinations of 35 to 40°, and locally even more.[291] One of the earliest proofs of mowing at Hudi Konec is a record from 1686.[292]

Historical haymaking in extremely steep grasslands was hazardous and only people that had become accustomed to working on such steep meadows in their youth usually dared do it. Sometimes, haymaking at extreme sites led to fatal accidents. As far as possible, many haymakers mowing very steep meadows in the Alps were fastened to trees, rocks, or the rope linked several haymakers, as a general strategy of increasing their security. Another even more usual way of coping with slope steepness was wearing climbing irons or crampons to decrease the likelihood of dangerous falls on steep slopes (Figure 12). Crampons worn by haymakers are attested since the first half of the sixteenth century at least and they must have also been worn earlier. They were particularly crucial when grassland plants were wet (a usual phenomenon early in the morning, the optimal mowing time, see Sections 6.1.1.2, and 6.3.1.2) or the soil was very dry because the slipping hazard was highest in these cases. Mowing such steep meadows also had cultural history consequences in terms of female clothing; long puffy skirts (note the clothing of the old woman behind in the Figure 87, Section 6.3.1.2) were too hazardous as one could step on a skirt, and a fall on the steep slope could be lethal. So, early on it became usual for women, in

---

Weißeneck in the same cadastral municipality, and so (large sections) of the approximately 50° slope belonging to a larger meadow parcel were in fact mowed, at least in some years; it was not just an unmowed area surrounded by mowed meadow and not excluded from the meadow parcel. However, it is possible that some minor sections within this approximately 50° slope were not mowed if the slope was locally too steep and/or if there existed smaller sections where grassland vegetation was not contiguous. *Arcanum Maps*, Cadastral Maps, Habsburg Empire – Cadastral Maps (XIX. [sic] Century), cadastral municipality of Lamm, 1830, and present-day cadaster showing the same cadastral parcels and topography in *SAGISonline*, Basiskarten, Alle Themen, Digitale Katastralmappe, DKM Grundstücke & Höheninformation, Höhenlinien 1 m & Hangneigung, Hangneigung 1 m & Werkzeuge: Höhenprofil.

291 Measurements were carried out in steep plots designated as meadows in the west of Hudi Konec on the cadastral map from 1826 (SI AS 176, Franciscejski kataster za Kranjsko (Franciscan Land Cadaster for Carniola), L287, cadastral municipality of Županje Njive, map sheet A08). In order to determine the locations, this map was compared with the georeferenced present cadaster (*Environmental Atlas of Slovenia*, Backgrounds, Topographic maps & Spatial units, Land cadaster) and with the digital orthophoto (*Environmental Atlas of Slovenia*, Backgrounds, Dof 2019).

292 SI AS 173, Imenjska knjiga za Kranjsko (*Gültbuch* for Carniola): *Gültbuch* No. 7 (1662–1757), sheets 306–307.

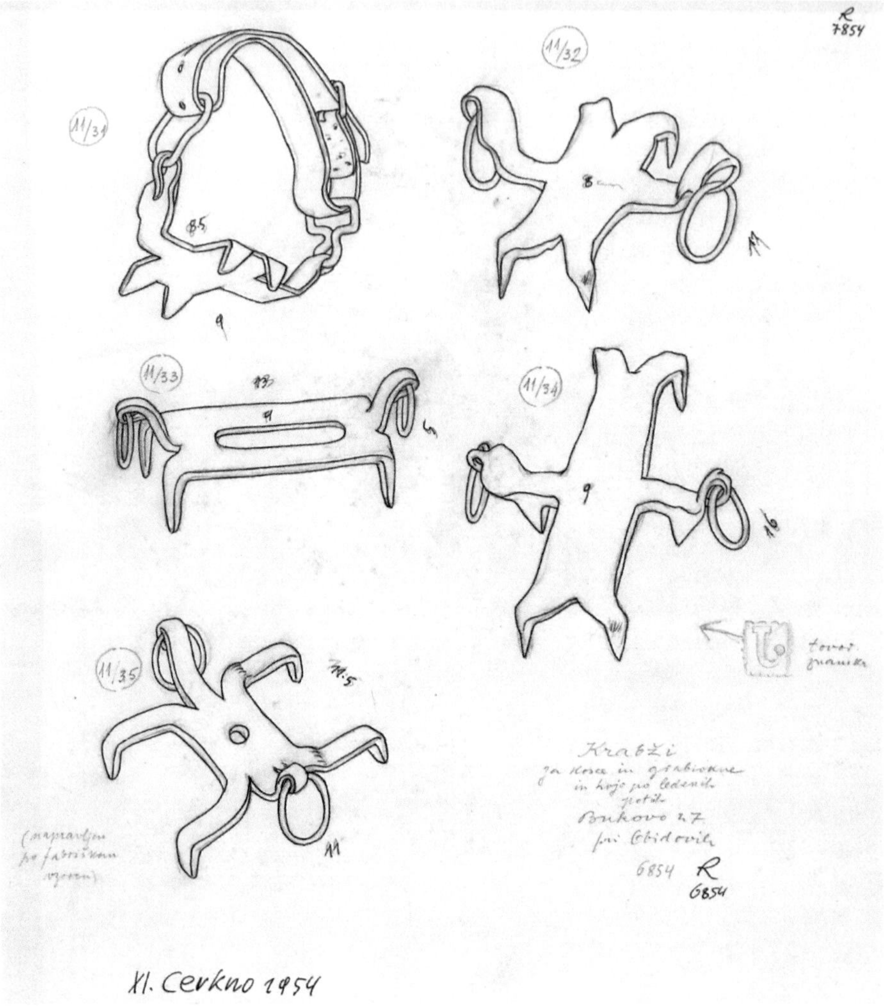

Figure 12: Crampons worn for mowing and raking the hay on steep slopes in the haymaking season (Sections 6.1.2.1 and 6.1.2.2) and also on icy roads in winter (the design of the crampon at lower left is inspired by industrial products; it is younger than the shape of the others), Bukovo near Cerkno, pre-Alpine hills south of the Julian Alps. Very similar crampons were worn elsewhere in the (Eastern) Alps.[293] Drawn by Ivan Romih. Slovenian Ethnographic Museum, Documentation, Fieldwork campaign 11, Cerkno Region (1954): drawing 11/31 & *Slovenski etnografski muzej* . . . Cerkljansko / 100 R0000011/31. © Slovenian Ethnographic Museum.

several places at least, to wear knee-length skirts for raking or mowing hay on extremely steep meadows. The example of Fundres in the present-day region of Trentino–Alto Adige (Germ. *Pfunders* in South Tyrol) demonstrates that a

[293] See, e.g., Wopfner, 1997, 331; the shape of the crampon from Tyrol is the same as the one in the upper right part of Figure 12.

reckless prohibition of such shorter skirts by eighteenth-century authorities took its toll in deaths, and knee-long skirts were soon allowed again. At least since the nineteenth century some women wore trousers for haymaking on such steep meadows—for instance, in inner Switzerland.[294] In cultural history terms, haymaking in extremely steep meadows was also a source of inspiration for exaggerated stories at some places in the Alps, such as one about men from a farm in the Padaun Valley (Germ. Padauntal) north of the Brenner (Ital. Brennero) Pass who allegedly only found mowing intriguing if the meadow was so steep that they could not turn around without touching the slope with the whetstone holder fastened to the belt on the haymaker's back.[295]

Animal husbandry in the Alps has served more than local and regional demands, especially since the Late Middle Ages and the sixteenth century. Livestock and other products of animal husbandry—such as cheese, clarified butter and salted butter, wool, and hides—were sold in significant quantities to the urban population all around the Alps, in cities in northern Italy in particular. Not only human factors, such as the aforementioned urban demand, promoted marked-oriented farming; it was often an interplay of human and environmental factors. Thus, villages, hamlets, and isolated farms at high elevations, for instance, which in many places were permanently settled only during the Late Medieval increase in trade flows, were often similarly or more involved in medium- or long-distance trade than lower settlements in the Alps. Namely, opportunities for self-sufficient diversified farming involving not only animal husbandry, but also more than moderate cereal cultivation and some other branches of farming, were more limited at higher elevations. Moreover, in some places, and not only exceptional ones, pastures made it possible to keep more livestock in the warm half of the year than farmers could overwinter with their hay, and so a significant percentage of the livestock were sold in the fall. Particular qualities of some animal husbandry products in the mountains also favored medium- and long-distance trade and thus co-influenced the orientation of the farm economy in the Alps. As early as the sixteenth century, the humanist Hippolytus Guarinonius, born in the town of Trento in the Alps of the present-day region of Trentino–Alto Adige (Germ. South Tyrol) and active as a physician and botanist, wrote that "[t]he products of the lowlands could not match the quality of Alpine products," and "until the nineteenth century it was a widely accepted belief that the best dairy products derived from the mountains."[296] Indeed, investigations from a testing station at 2,000 m in the Upper

---

294 Mathieu, 1996, 145; Wopfner, 1997, 317, 330–333; Keller, 2004, 15, 65–66, 137; Torkar Tahir, 2012, 133; Bavcon, Ravnjak, & Praprotnik, 2019, 151; Hessenberger, 2022, 168.

295 Wopfner, 1997, 332, and *Deutsches Wörterbuch von Jacob Grimm und Wilhelm Grimm*: Kumpf.

296 Novak, 1955, 311; Gestrin, 1965, especially pp. 23, 36–37, 42–43, 86–87, 107, 117, 148, 171–176, 193, 197, 211–212, 233, 236; Dinklage, 1966b; Dinklage, 1966c; Pfister, 1985a, 29, 57, 63, 96–97, 123–125; Riedmann, 1985, 496; Blanchard, 1986; Meilleur, 1986, 45–46;

Inn Valley proved that milk from the summer pastures has a richer aroma and more active substances, and is significantly richer in dry matter (see Sections 6.3.1.2, 10.1.2.2.2, 10.1.2.2.4, and 10.1.2.2.7); the color of cheese and butter is more intense than in the valley, but the quantity of milk is lower.[297]

Significant areas where environmental conditions made animal husbandry possible but not arable farming, high dairy-product quality in many places, and additional human factors (e.g., demand for animal husbandry products promoting a stronger orientation toward or even specialization in livestock farming in several regions in the Alps where environmental conditions would allow more diversified farming) led to the fact that the grassland area in the Alps and intensity of its use were influenced by broader external and internal economic factors. The nexus between local and external economic conditions had an influence on the history of all types of farmed grasslands in the Alps, on the orientation of the farm economy, and on its changes through history. In the northernmost Alps, from Savoy (French *Savoie*, Francoprovençal *Savouè*) across Switzerland and Vorarlberg to Allgäu (now in southern Germany), the transition from subsistence farming to specialized livestock farming was particularly early and particularly strong. In the Late Middle Ages, a combination of cereal cultivation with livestock rearing was still typical for those regions. However, starting in the Late Middle Ages, many arable fields were converted to grasslands due to increased sheep herds. Gradually, sheep were replaced with cattle. Extension of grasslands in the fifteenth to sixteenth centuries and another strong grassland expansion in the nineteenth century is also reflected in the pollen diagrams for Schöpfenwaldmoor and Trogenmoos in the northern Swiss Pre-Alps (north and northeast of Interlaken in Figure 4). In some places, this transition was completed earlier, such as in Appenzell Innerrhoden, where an increase in animal husbandry replaced the former cereal fields by 1700, so that the arable area was very limited. Another example is the surroundings of Bern: in the mid-eighteenth century, it was observed there that many previous fields had already been converted into grasslands. By 1900 almost all historical arable fields in the extensive area of the northwesternmost Alps had become grasslands.[298]

Žontar, 1987, 63; Viazzo, 1989, especially pp. 36–38, 142–143, 264; Mathieu, 1996, 156; Wopfner, 1997, 9–12, 17–19, 189–190, 217–222, 227–236, 473–474, 478–479; Flückiger & Radeff, 2000, 5, 8–11, 13–14, 33–35; Kauter, 2002, 117–120; Orland, 2016, 24, 27–28, 31, 34–39, 44, 48–49, 51 (quotes from pp. 28 & 34); Bätzing, 2015, 60, 69, 71–76, 80; Beguš, 2015, 106, 158; Bizjak, 2018, 127, 133–134, 137–138, 141–148; Mathieu, 2019, 55–56, 63, 66–67; Pavković, 2021, 114; Novak, 2024, especially pp. 180–190, 196–211; for Hippolytus Guarinonius's biography, see Grass, 1966. See also Sections 6.6.2.2 and 8.2.2 in this volume.

[297] Pils, 1994, 221; compare Orland, 2016, 27.

[298] Viazzo, 1989, 272–273, 282; Mathieu, 1996, 154–157; Wopfner, 1997, 301; van der Knaap et al., 2000, 86, 107–108, 118–120, 134, 137; Kauter, 2002, 118–119; Orland, 2016, 38–39; Mathieu, 2019, 54; Hanhart & Sonderegger, 1998, 41–42, 46–47, 58–59.

Elsewhere in the Alps, examples of early conversion of arable fields to grasslands due to social and/or harsher environmental conditions were more of an exception than a rule; only in some regions was this a relatively significant and more than temporary local phenomenon. In the court district of Villandro (Germ. *Villanders*) west of the Isarco River in the present-day region of Trentino–Alto Adige (Germ. *Eisack* in South Tyrol), for instance, about twenty former farms were allegedly transformed into temporary settlements, as recorded in 1523. Many of their former fields were converted into meadows.[299] This process had some long-term impacts, although the question of (dis)continuity remains open. Namely, it was still characteristic for the community of Villandro in the nineteenth century that several farmers owned not only a main farm at lower elevations, but also an additional grassland farm at elevations of 1,300 to 1,500 m.[300] However, according to data from about 1900, not only cattle husbandry, but also arable farming played major roles in this region. The role of grasslands in the farm economy in the territory of Villandro was significantly less than in the northernmost Alps, from Savoy to Allgäu.[301] Looking at the Alps as a whole, the large-scale transformation of numerous parcels that had for centuries been used as fields under marginal environmental conditions or at distant locations into permanent meadows took place between the second half of the nineteenth century and the mid-twentieth-century or in the following decades (Figure 13). Such grasslands also became species-rich, if (rather) extensively managed (no more than moderately fertilized; Section 6.3.2), which was often the case. At least in some cases they also "contain rare and/or endangered species," although these grasslands only have a relatively short history.[302] A study of ancient grasslands (at least 180 years old as suggested by land-use maps from about 1830, the early 1950s, and 2002) and relatively recent (field crops were grown in those parcels around 1830 and/or in the early 1950s, but not in 2002 anymore) calcareous grasslands in the Swabian Jura (Germ. *Schwäbische Alb*) determined that "[t]here was no difference in the species richness of ancient and [relatively] recent [seminatural] grasslands, however, their species compositions differed. This difference could be assigned to both habitat properties and history"[303] (see Sections 2, 3.2, and 6).

[299] Hinterwaldner, 2002, 21, 30–31, 48; Blaznik, 1970a, 167; Wopfner, 1997, 10–11, 402–404, 446; Scharr, 2001, 73; Jäger, 2006, 9–10, 55–58, 60, 62, 69, and Figures 17 and 18 between pp. 64 and 65; Zwitter, 2015d, especially pp. 144–165, 167–177 and the map on p. 142; see also Camenisch & Rohr, 2018. Jäger, 2006, 12 also deals with sixteenth-century conversion of vineyards at Merano (Germ. *Meran*) into meadows.

[300] Wopfner, 1997, 230–231; the chronological definition of the time of Wopfner's writing follows Meixner & Siegl, 2004, 232–233, 235.

[301] Conclusions based on Emmanuel de Martonne's map of farming types in the Alps, reproduced in Mathieu, 1996, 142–143.

[302] Pils, 1994, 165; Pärtel, Mändla, & Zobel, 1999, 194; Dietl & Lehmann, 2004, 20, 25, 124–125; Poschlod et al., 2008, 133–136, 138–140 (including the quote); Karlík & Poschlod, 2009, 322–323.

[303] Karlík & Poschlod, 2009 (quote from p. 333).

Roughly at the same time as conversion of many fields under marginal environmental conditions or at distant locations into permanent meadows, between the second half of the nineteenth century and the mid-twentieth century, although with huge differences in chronology within the Alps, the process of land-use changes from fields to meadows also became characteristic for valley bottoms. Nonetheless, meadows created in valleys or close to farms at that time were in many cases more intensive than at more remote locations at higher elevations. The establishment of medium-intensive (by present standards) meadows became a significant phenomenon in the Alps starting in the nineteenth century, in the main areas of dairy production in particular. Out of the core areas of the emerging dairy industry in the Alps, conversion of former fields into meadows often created moderately fertilized meadows by the mid-twentieth century (Section 6.3.2). The process of agricultural specialization into animal husbandry in the valleys was stimulated by broader European economic developments. It was rooted in agricultural reforms of the eighteenth and nineteenth century (e.g., the introduction of fodder crops, significantly prolonged keeping of livestock in stalls in several valleys, more intensive grassland fertilization related to this, and later also the introduction of mineral fertilizers), which gradually gained importance, especially at lower elevations and at more easily accessible sites. The process discussed was significantly influenced by further economic and demographic transformations. For instance, the establishment of the railway network, making possible the import of cheap cereals, and export of fresh milk and butter over longer distances in the second half of the nineteenth century belonged to important early factors contributing to the large-scale conversion of fields into meadows, especially at more accessible places. Moreover, starting in the 1870s, the introduction of intensive dairy farming and the emerging dairy industry were also characteristic for many regions in Europe where dairy farming had previously been much less important than in the Alps, including the plains. In the Alps, dairy farming based on more intensive grassland management in the valleys could succeed more easily than more extensive highland farming in quantitative terms when faced with the new type of market competition from elsewhere in Europe in times of increased international commerce. At the same time, the demand for fresh milk rose significantly due to the changed human diet. Nonetheless, more intensive grassland management at more accessible sites, which was gradually leaving the sphere of moderately fertilized meadows in some parts of the Alps, also led to a decline in the quality of dairy products.[304]

[304] Pfister, 1985a, 123–125; Niederer, 1996c, 50; Niederer, 1996d, 158, 175; Wopfner, 1997, 184, 190, 222–223, 259, 315, 320; Brückmüller & Stekl, 1998, 67; Dietl & Lehmann, 2004, 20, 25, 124–125; Orland, 2016, 25–27, 39–46, 50, 52; Vogl-Lukasser et al., 2006, 233; Kapfer, 2010, 138–139; Vojvoda, 1965, 22–25, 59, 116, 133; Karlík & Poschlod, 2009, 322–323; Bätzing, 2015, 158–159; Ledinek Lozej, 2021, 158, 162.

Figure 13: A species-rich grassland in a former montane arable field at 1,040 m, between the Javornica and Četeže complexes of montane meadows, 450 m above the farm in the valley that used to cultivate it.[305] The field was converted into a permanent grassland in the first half or middle of the twentieth century. The grassland is no longer used. Note the yellow blooms of the tall herb perforate St. John's wort (*Hypericum perforatum*) in the center of the photo, indicating the transition from a grassland to an abandoned tall forb community.[306] Perforate St. John's wort is a perennial ruderal rather than a meadow species, except for very late-mowed meadows. It is especially sensitive to regular mowing (regeneration of stems after cutting is limited) and occurs more often on forest margins or in abandoned pastures together with other tall herbs (the grassland class *Epilobietea angustifolii*).[307] It is less sensitive to grazing pressure than to mowing due to a high concentration of ethereal oils and hypericin; the fodder value of St. John's wort (*Hypericum* sp.) is very low, and most grazers avoid eating it. Nevertheless, it appears to be a weak light competitor due to its moderate height growth and limited capacity to form dense stands.[308] Photo: Žiga Zwitter, 2018.

Another process closely related to historical transformations discussed in the previous paragraph and leading to increases in meadow area were divisions of some common pastures in the Alps, mainly in the eighteenth and nineteenth centuries. Following these divisions, some sections of former pas-

[305] SI AS 176, Franciscejski kataster za Kranjsko (Franciscan Land Cadaster for Carniola), L157, cadastral municipality of Srednja Vas v Bohinju, map sheet A15; SI AS 176, L157, cadastral municipality of Srednja Vas v Bohinju, Protocoll der Grund-Parcellen, parcel nos. 943, 952.

[306] Pils, 1994, 182; Peterken, 2013, 64–72.

[307] Schubert, Hilbig, & Klotz, 2001.

[308] Grime, Hodgson, & Hunt, 1988.

tures were converted into meadows; nonetheless, the process of division of commons was a spatially much more limited phenomenon in the Alps than in their surrounding lowlands. In Appenzell, belonging to the northwestern edge of the Alps with a specific course of agricultural history as presented above, the conversion of pastures into meadows took place in the eighteenth century; when farming there specialized into increasingly intensive animal husbandry, meadow management also became more intensive, and fertilized meadows became common there. The example of the Slovenian Alps, representative of many parts of the Alps not as strongly specialized in animal farming as the north of the mountain chain from Savoy to Allgäu, demonstrates that division of commons first took place in some more accessible pre-Alpine valleys, where this process was also more intensive than in settlements at higher elevations.[309]

Further local economic factors importantly contributed to the creation of new meadows on former common pastures. In the Bohinj Basin in the Julian Alps, for instance, the division of several commons in the 1880s corresponded to the general crisis of the rural economy in the broader region, and also with the crisis and end of the operation of ironworks, which had been a significant economic factor there for several centuries. Emigration only partly mitigated the challenges of increasing agricultural density, and the railway connection, which made it possible to commute to industrial centers, was only built in the valley in the early twentieth century. Consequently, agricultural density in late-nineteenth-century Bohinj was high when many to a significant degree overgrown common pastures had just been divided. In this situation, new meadows were locally created from former pastures, but the total area affected through this process was spatially very limited.[310] This is a typical example in the sense that it was usual for many settlements in the Alps heavily dependent on mining and/or metallurgy to increase agricultural production during economic depressions in non-agricultural activities. The population of the settlement of Riva D'Agordo (now Rivamonte Agordino) in the Dolomites, for instance, increased from fewer than 150 inhabitants in 1584 to 1,130 in 1766, largely due to mining. "[I]n 1766 no less than 315 out of 347 adult men were listed as mine workers, and just eleven as agricultural workers." However, a major crisis of mining followed there, and so in the 1770s "[i]t would seem that … the whole community of Riva 'is moving into the agropastoral sector to exploit fields, meadows and animals as fully as possible, this sector functioning as a sort of *lung* in times of economic crisis.'" Increased agricultural activities were joined by emigration, in this case temporary, in times of hardship.[311] These examples

309 Blaznik, 1970a, 173–174, 176; Blaznik, 1970b, 156–158; Grafenauer, 1970c, 243; Meilleur, 1986, 313, 315; Pfister, 1985a, 85, 115; Wopfner, 1997, 234, 260, 280–288, 300, 395; Kauter, 2002, 119; Bätzing, 2015, 135; Leibundgut & Vonderstrass, 2016a, 76–77.
310 Melik, 1950, 78, 85; Vojvoda, 1965, 14, 26, 28, 48; Blaznik, 1970b, 158; Grafenauer, 1970a, 631–639; Petek, 2005, 51, 77, 81, 90, 167, 178; Zwitter, 2008, 3–7, 10–13, 26–27.
311 Viazzo, 1989, 158–159, 171, 173 (quotes), 177.

demonstrate that the crisis in the non-agrarian economy, which increased the importance of the agrarian sector, also led to increased pressures on grassland ecosystems and the formation of new meadows in certain cases.[312]

In addition, the nexus between the non-agrarian economy and agriculture has impacted the spatial distribution of grasslands in the Alps in several other ways. These include influences of broader external and internal economic factors related to mining and metallurgy, much exceeding the phases of decline in these activities discussed in the previous paragraph. Long-term impacts of historical metal extraction and forging in the Eastern Alps, demanding huge quantities of wood, illustrate this. There were major centers of mining and metallurgy in Tyrol, Salzburg, Styria (Germ. *Steiermark*), and Carinthia (Germ. *Kärnten*, Sln. *Koroška*), and in the Alps of the Republic of Venice, mainly in the valleys above Brescia and Bergamo. At least a quarter of the inhabitants of Tyrol lived on mining in the early sixteenth century according to some estimates, at the time of a mining boom when Tyrolian silver and copper mines were among the most important in Europe (1470–1535). Silver mining in the Alps declined quickly after the mid-sixteenth century.[313] Focusing on iron, several mining and ore-processing sites made the Eastern Alps one of the key regions of iron extraction in Europe in the Late Middle Ages and a significant part of the modern period. The area of Erzberg (including Vordernberg and Innerberg—now Eisenerz) in Styria (see Eisenerz and Vordernberg in Figure 4) was the largest iron-smelting and processing territory in the Eastern Alps and one of the largest in Europe at that time. In general, iron extraction and forging centers in Carniola (Sln. *Kranjska*, Germ. *Krain*; it covered large parts of today's Slovenia and a small but relevant area of today's northeasternmost Italy) were far behind. Nonetheless, iron-forging facilities located in the seigneury of Fusine (Friul. *Fusinis*, Germ. *Weißenfels*, Sln. *Bela Peč*) in the very northwest of Carniola probably made that territory one of the second-order regions of iron extraction and forging in the Eastern Alps around 1500. Within the seigneury of Fusine, seven iron smelteries on the western slopes of the Karawanks above Jesenice were the main center of iron ore smelting in the territory of what is now Slovenia, as suggested by late-fifteenth-century data, with a roughly estimated annual iron extraction of about 350 metric tons.[314] A marked sixteenth-century peak of iron extraction in Carniola followed. It corresponded to the important sixteenth-century peak of metal ore smelting in the Eastern Alps in general. In 1569, about 2,000 metric tons of local iron is said to have been extracted in Carniola, and an additional approximately 2,000 metric tons of Carinthian iron was reportedly forged in Carniola, mainly in its Alps. Such quantities seem high and may have been achieved in some very good years, or

[312] Compare Panjek, 2015a, 117.
[313] Viazzo, 1989, especially pp. 157–159; Mathieu, 2019, 61–62; Oitzl, 2021, 26, 46, 175, 304–305.
[314] Mathieu, 2019, 62; Oitzl, 2021, 101, 169, 292, 304–305, 401–402, 404–408, 424, 426.

perhaps the figures are somewhat exaggerated. In any case, a marked decline in iron extraction and forging followed in the second half of the sixteenth century in both the Alps and Carniola. Consequently, in 1581, up to 1,700 metric tons of iron and iron products are estimated to have been extracted and/or fabricated in the provinces of Carniola and Gorizia (Sln. *Goriška*, Germ. *Görz*) from local and imported iron, with merchandise from Carniola strongly prevailing over that from Gorizia within this figure. Most iron and iron products from Carniola were exported to Italian markets. Intermediate iron products from Jesenice and its surroundings, for instance, were largely sold to the region of Marche on the western Adriatic coast in the Papal States and also elsewhere on the Apennine Peninsula.[315] During the High to Late Middle Ages and the sixteenth century there was an increase in trade flows, increased demand for metals, and progress in smelting and metal-processing techniques; as this gradually spread across the Eastern Alps, mining, ore smelting, and fabrication of metal objects resulted in the establishment of some settlements and the growth of several older ones. The following lines illustrate this based on data from Carniola. Like many other places in the Alps, the fourteenth century saw the beginning and further development of several settlements in this province due to activities related to mining and metallurgy, and these processes continued in the fifteenth and sixteenth centuries.[316] Apart from the increase in population density, corresponding to higher food demand that was partly covered locally, the number of horses in the area involved in transport also increased during times of rising production in mining and/or metallurgy. Thus, in general, increased food and feed demand went hand in hand with wood-felling for metalworks, and so grasslands were created and retained on some new forest clearings.[317] This is the main type of long-term impact of metalworks, which sold their intermediate or final products far beyond the region, on the local cultural landscape in terms of grasslands' spatial distribution. In the fourteenth century, iron smelting in Carniola was already known to have led to the creation of several new grasslands.[318] In 1636 (the rent-roll reflects conditions in a certain year between 1617 and 1636), it was documented that some former iron smelteries on the slopes of the Karawanks above Jesenice (mentioned

---

[315] Gestrin, 1962, 671–672, 689–690 (and the historian Gašper Oitzl, personal communication); Gestrin, 1965, 179–188; Viazzo, 1989, 159; Rohr, 2007a, 368; Mathieu, 2019, 61–62; Oitzl, 2021, 399–408, 413–414, 424–426, 428, 430–432; see also Schneiter, 1970, 61.

[316] Oitzl, 2021, 34–36, 38, 40–51, 304–310, 405–408.

[317] Schneiter, 1970, 61–64; Viazzo, 1989, 156, 168–169; Wopfner, 1997, 9–11, 190; Oitzl, 2021, 300–310, 348–353; the question of grazing rights for horses involved in transport for ironworks in the Sava Dolinka Valley (Sln. *Zgornjesavska dolina*) in the seventeenth century is dealt with in AT KLA 112, Allgemeine Handschriftenreihe, Ms. 2435, Rent-roll (*Urbar*) of the seigneury of Fusine 1636, 148, and AT KLA 112, Ms. 2732, Rent-roll (*Urbar*) of the seigneury of Fusine 1636, 471, for instance. For relations between these rent-rolls and the questions of datation, see Mlinar, 2018a, 27–28 and Mlinar, 2018c, 30–32.

[318] Štih, 1998, 137–138.

above as a particularly relevant iron-smelting site in the late fifteenth century) had been abandoned, but about fifteen (or seven, at least)[319] farms (many or all of them smallholdings) were reportedly using meadows and other types of agricultural land that had been created close to these smelteries when the forest had been cleared to produce charcoal for iron ore smelting. Moreover, some additional grasslands were most likely created there when wood was felled in that microregion to supply ironworks with charcoal, but they were not mentioned in this notice, which refers only to abandoned smelteries, and not to five smelteries existing in the microregion at that time.[320] In the late fifteenth century, a smeltery or a couple of them in the Karawanks above Jesenice were probably located very close to the central section of the settlement of Planina pod Golico, with mining and iron smelting possibly belonging to the very motives for earliest Late Medieval inhabitants to settle there,[321] and so in many cases the introduction of agricultural land use in the microregion was related to (mining and) ore-processing. This case study of new grassland creation is typical for many places in the Eastern Alps[322] in times of increased fodder and

[319] Information comes from an entry following the list of subjects of the seigneury of Fusine on the slopes of the Karawanks above Jesenice, and these data refer only to "Dise[!] obverstandnen undtersassen" (Mlinar (ed.), 2018b, 162)—i.e., *Untersasse* listed above—but it is not clear where on the list of holdings *Untersasse* start in this case. Namely, there is some inconsistency in the use of the term *Untersasse* in this rent-roll. Usually, this source makes a distinction between *Hubleüth* and *Undtersasse* (Mlinar, 2018b, 44–45; the spelling follows Mlinar (ed.), 2018b, 113, 118). Nonetheless, it also uses the term *Untersasse* in a broader sense, for subjects in general (see *Deutsches Wörterbuch von Jacob Grimm und Wilhelm Grimm*: Untersasse), and so *huebsassig[e] undersass[e]* also appear in this rent-roll (Mlinar (ed.), 2018b, 113). Counting only the holdings delivering very low dues, typical for lists of *Untersasse* in this rent-roll (Mlinar (ed.), 2018b, 93–95, 109–111, 118–119, 125–128, 138–140, 148–156, 170–171), and excluding the holdings reportedly consisting only of a cottage, without any agricultural land, the notice in question refers to seven holdings (Mlinar (ed.), 2018b, 161–162). Nonetheless, some holdings delivering higher dues also stand on lists of *Untersasse* in some other settlements in the same rent-roll (see pages cited above). Adding such holdings on the slopes of Karawanks above Jesenice (none of them is titled a *hueben*) makes the number fifteen. The notice cited above clearly refers to land parcels cleared of forest to produce charcoal for iron-ore smelteries that had already been abandoned by 1636. For this reason, three units documented in this rent-roll and delivering very high dues were excluded from these calculations because these entries still include a link to smelteries that existed at the time. The chronological orientation of data from this rent-roll is based on Mlinar, 2018a, 27–28.

[320] Mlinar (ed.), 2018b, 162; for terminological issues regarding iron ore smelteries, see Oitzl, 2021, 51–54. Abandoned smelteries were in this case largely a result of the introduction of more efficient smelting technology and the related partial displacement of smelting to the bottom of the main valley during the sixteenth century (Oitzl, 2021, 46, 49, 102–103, 170–171, 405); the number of smelteries there in the early seventeenth century: Oitzl, 2021, 102.

[321] Oitzl, 2021, 101, 305, 307.

[322] See, for instance, a parallel from Central Europe north of the Alps in Poschlod et al., 2008, 133.

food demand in mining and metal-processing centers, where and when wood supplies and other interest in forests allowed for new forest clearings. In the modern period, large-scale iron mining and smelting was more constant and the extracted quantities were higher than extraction of most other metals at several locations in the Alps, and so its environmental impacts (in addition to the environmental legacies of past processes and events) were significant not only in the sixteenth century. After the crisis of the ironworks in the second half of the sixteenth century, the pre-crisis production level was again reached in the mid-eighteenth century, when "Styria alone seems to have outstripped even England."[323] Dependence of ironworks on charcoal, also partly relevant for the history of grasslands, decreased during the following century of iron production growth, mainly in the Alps of Styria and Carinthia, when coal was gradually introduced as an energy source.[324]

Another important non-agrarian economic factor was traffic, with a major influence on mowed grassland areas close to important trade routes, where particularly large quantities of wild hay were sold. The example of the route crossing the Gotthard Pass (Germ. *Gotthardpass*; Ital. *Passo del San Gottardo*) as one of the main trans-Alpine traffic connections linking northern Italian cities and Central Europe demonstrates this. The Gotthard route has been important since the Late Middle Ages, and there was a general trend of growth in traffic there from the sixteenth to the late eighteenth centuries, with fluctuations due to broader economic and political reasons. The traffic led to significant expansion of meadows in the surroundings. Wild hay was bought there mainly to feed packhorses crossing the Alps in both directions, and to feed extensive herds of livestock crossing the Alps from the Late Middle Ages onward on their way to Italian markets. Among other trade items, packhorses carried great quantities of textiles, wine, and cereals to the north (cereal transport was especially active after harvest failures decreased the supply on the markets in Lucerne (Germ. *Luzern*) and Zürich), and they transported dairy products, such as cheese, and textiles to the south. Moreover, as mentioned, herds of livestock were also driven there. At the peak of the transport volume around 1790, approximately 31,000 packhorses crossed the Gotthard Pass yearly. Apart from packhorses, official statistics record that on average 3,221 head of cattle, 631 horses, and 1,290 sheep and goats crossed the Gotthard Pass yearly between 1790 and 1797 to be sold on Italian markets. These were not the highest early modern numbers of livestock driven across the Gotthard Pass to Italian markets, and the herds driven across this pass even became much larger in the first half of the nineteenth century. In addition, based on a 1796 estimate, about 15,000 to 16,000 people crossed the pass yearly—some of them on horseback, representing another kind of fodder demand. Traffic of-

[323] Mathieu, 2019, 61–62, 115 (quote from p. 62).
[324] Schneiter, 1970, 61, 63–64; Mathieu, 2019, 62, 115.

fered an important source of income in the integrated peasant economy of the local farmers, especially in settlements with inns. These included engagement in transport, and also sales of wild hay as well as standing grassland fodder in pastures.[325] The enormous historical demand for wild hay can be considered one of the factors due to which the high Ursern Valley just below the Gotthard Pass was exceptionally free of forest in the broader context of the Alps, as observed, for example, in 1811. Consequently, even dwarf shrubs from mountain pastures, such as alpenrose (*Rhododendron ferrugineum*), heather (*Calluna vulgaris*), and bilberry (*Vaccinium myrtillus*), were used as firewood for cheese production in the mountain huts there. Such use of dwarf shrubs again had ecological impacts because dwarf shrubs were regularly removed from grasslands (see also Section 6.6.2).[326] Human impacts in remote grasslands close to important traffic connections in the Alps were thus particularly strongly related to broader economic conditions. The growth of urban settlements in both northern Italy and Switzerland, general population growth in Uri, and higher numbers of livestock in the countryside all contributed to greater demand for livestock fodder along the Gotthard route, which led to an increase in the area where wild hay was mowed from the mid-seventeenth century onward. The demand for wild hay there generally remained high until the nineteenth century, followed by a significant decrease in mowed remote grasslands.

The main reasons for this long-term change in grassland management along the Gotthard route were the following, and, with some differences in timing, they can be considered representative of several important traffic routes crossing the Alps.

First, packhorses started being replaced by draft horses in the 1830s because (following the examples of passes such as Simplon, Splügen (Germ. *Splügenpass*, Ital. *Passo dello Spluga*), and San Bernardino (Germ. *Bernhardinpass*, Ital. *Passo del San Bernardino*) a road was built over the Gotthard Pass between 1820 and 1836. This replaced the former bridleway and made transport with wagons possible. Consequently, the number of horses per unit of cargo was significantly reduced, and so, despite the initial growth in traffic

---

[325] Stadler-Planzer, Stadler, & Aschwanden, 2015, 52–53, 106–107, 109, 123, 271; Stadler-Planzer, Stadler, & Degler-Spengler, 2015, 20, 22, 81, 97, 99–103, 122–126, 131–135; Wopfner, 1997, 9–11, 19. On the concept of the integrated peasant economy, see Panjek, 2017.

[326] Compare Wunderli, 2010, especially pp. 20–23, 25. Use of alpenrose as firewood for cheesemaking in high summer pastures has parallels elsewhere in the Alps, not limited to exceptional sites like the ones along the Gotthard route. Where alpenrose abounded in pastures, removal of this species (which is unattractive to cattle) from pastures could provide firewood. This was most likely the case in the Allues (Francoprovençal *Alués*) Valley (French *Les Allues*) in what is now the northern French Alps, for instance (Meilleur, 1986, 354). In some parts of the Alps, dry livestock feces were also used as fuel (Niederer 1996d, 165, 200).

flows, there were fewer horses on the Gotthard road. This means that less hay was needed than in the era of pack horses, although the demand for local wild hay remained relatively high until the railway connection started operating. Second, with the opening of the railway connection in 1882, the great majority of cargo was no longer transported by horses. Livestock also no longer walked over the Gotthard Pass on the way to Italian markets, and trains replaced the former role of carriages in passenger transport (on average, 80,000 passengers a year crossed the Gotthard, Furka, and Oberalp Passes (Germ. *Oberalppass*, Romansh *Alpsu*) in postal carriages in the 1870s and 1880s and, in addition, there were several thousand in private carriages). Third, there was a strong decrease in numbers of goats, important wild-hay consumers, from around 1870 onward, and so less wild hay was needed (see Section 6.3.1). Fourth, intensification of agriculture on private land led to a decrease in the demand for hay from remote commons. Fifth, deagrarianization took place because especially the railway connection helped industry develop, which offered many more non-agricultural jobs and led to a decrease in the workforce available for farming. Finally, intensified emigration took place. In all of Switzerland, this was the strongest precisely in rural areas of Uri in the last quarter of the nineteenth century, especially after the railway connections in the area were established.[327]

The Gotthard route had an impact on the hay trade that affected meadow area. In general, hay as a relatively cheap commodity needed in significant quantities was usually economically feasibly transported overland only over a few kilometers before railway connections (but see also the next paragraph); some inland waterways made trade in hay possible over longer distances.[328] Hay, sold locally, and in the case of specific demand over longer distances (next paragraph), has long been a source of income for some farmers and institutions,[329] and this also had an influence on areas of agriculturally used grasslands in the Alps. Core areas of dairy farming in Switzerland stand out for the early specialization of several farms in terms of "the production of fodder that was sold to cattle breeders, cattle traders and dairymen"; that is, those farmers' sons "who did not inherit enough land to live on," for instance. These buyers "tried to maintain their herds through the winter," and the hay they bought played a prominent role in fodder for their livestock. This type of farming specialization was identified

327 Stadler-Planzer, Stadler, & Aschwanden, 2015, 76–77, 92–93, 100, 102, 106, 109, 121, 123–128, 130–131, 135–136, 265, 267–268, 270–271; Stadler-Planzer, Stadler, & Degler-Spengler, 2015, 15, 22–23, 99–103.

328 Krausmann, Weisz, & Eisenmenger, 2016, 73, 77; Lane, 1973, 305.

329 Orland, 2016, 29, 37; Peterken, 2013, 144–145, 158–159; for instance, second-crop hay from demesne meadows in the Dreta Valley (at least close to Bočna) was a minor source of income for the seigneury of Gornji Grad (Germ. *Oberburg*) in 1686 (SI NŠAL 20, Gornji Grad Avstrija (Gornji Grad Austria), box 40, seigneury's account by Georg Lorger, April 29th, 1686 – April 28th, 1687: Gelt[!] empfang an unterschidlichen[!] geföllen[!]).

there "[a]s early as the late Middle Ages," and it became more frequent between the sixteenth and eighteenth centuries. These dairymen bought not only hay but also stall space from farmers in the valleys.[330] In terms of grassland history, this means that the system had some common characteristics with some other places in the Alps where meadow areas in the mountains were insufficient for regular late-autumn to early-spring hay supplies for local livestock, and the solution was having livestock wintered elsewhere. For instance, "[i]n the highest communes of the Tarentaise [Francoprovençal *Tarentèsa*] and the neighboring Upper Maurienne [Fr. Haute Maurienne, Francoprovençal *Hiôta Môrièna*], large numbers of animals were commonly wintered outside the communes of their owners, in lower elevation communes with more plentiful hay supplies. In Termignon [Francoprovençal *Tèrmegnon*] (Haute Maurienne), over 90% of the sheep and milk-cows were over-wintered outside the commune, sometimes at considerable distances," as determined by Meilleur's investigation focused on the late nineteenth and early twentieth centuries. In the nearby Allues Valley, wintering the livestock at lower elevations was probably only practiced in the case of (consecutive) hay harvest failures.[331]

These examples from Switzerland and Savoy demonstrate that there existed areas, predominately lowland areas in the Alps or at the foot of the Alps, "able to share in the profitability of Alpine farming through leasing, renting out land or selling hay" until the agricultural reforms and changed market conditions of the late eighteenth or the nineteenth century.[332] These data suggest that the existence of microregions in the Alps dependent on regular purchases of a significant percentage of hay elsewhere, namely in fairly near lowlands within or at the foot of the Alps, must have been more widely present, at least at a smaller scale. This thesis finds confirmation in sources dealing with the Slovenian Alps. In the settlement of Gmajna west of Slovenj Gradec (Germ. *Windischgraz*), a lack of hay in isolated low mountain farms was documented as a general phenomenon in the 1840s, and so farmers purchased hay in the plain in the nearby valley bottom.[333] As early as 1570 it was reported that the isolated Pop farm above Leše in the Meža (Germ. *Mieß*) Valley, relatively close to Gmajna, had to buy all its hay.[334]

Not only hay from valleys and basins, but also hay from high elevations was sold in some microregions and time intervals, as already demonstrated

330 Orland, 2016, 29, 37–38 (quotes from pp. 29 and 37).

331 Meilleur, 1986, 277–282, 307 (quote). See also wintering cows elsewhere and transhumant sheep pastures in Section 7.1, which all had an influence on grassland uses in the mountains and lowlands.

332 Orland, 2016, 37–46 (quote from p. 39).

333 Kuret (ed.), 1989, 250.

334 SI NŠAL 100, Kapiteljski arhiv Ljubljana (Archives of the Ljubljana Cathedral Chapter), fasc. 98, unit 3: Rent-roll (*Urbar*) of the seigneury of Bleiburg (Sln. *Pliberk*) 1570, fol. 176v; Zwitter, 2016, 125; another example: Mlinar (ed.), 2018b, 132. The distance between the Pop farm and Gmajna is about 11 kilometers in a straight line.

by meadows along the Gotthard Pass route. The aforementioned hay business in Switzerland with a long historical tradition also involved hay from meadows at higher elevations; selling hay from high elevations there became more frequent with the decrease in dairy farming at high elevations, a process that affected the Bernese Highlands (Germ. *Berner Oberland*) in the 1860s.[335] Again, more limited parallels of hay from remote meadows as a trade item can be identified in other regions of the Alps. In the Western Karawanks, wild hay from the grassland complex of Jekelj (Germ. *Eckele*) at approximately 1,500 m and higher, west of the peak of Mount Velika Golica (Germ. *Kahlkogel*), was partly mowed to become a trade item. Namely, according to information from 1616, it was already a practice to sell any wild hay from Jekelj that was not needed by people from Hrušica (680 m) at the southern foot of the mountain chain, who had mowing rights there, or which people from Hrušica considered too complicated to transport to their farms (5 km in a straight line, with an elevation difference of 800 m or more). At least some of this hay was sold to Carinthians; that is, to buyers from the opposite, northern side of the Karawanks;[336] namely, the quality of the hay from this area was very high (see Section 6.3.1.2). Data from Tyrol also demonstrate that so much wild hay was stored in some high valleys, such as the Valles (Germ. *Vals*) Valley, a side valley of the western Puster Valley (Ital. *Val Pusteria*, Germ. *Pustertal*), that surplus hay was sold to the main valley, according to nineteenth or early twentieth-century information.[337]

Leaving the sphere of local and short-distance hay purchases, medium-distance overland hay transport was carried out in the case of specific demand and adequate road connections in times predating railway connections, and this also had an impact on meadow sites and area. This is demonstrated by the

---

[335] Orland, 2016, 37, 41, 52.

[336] AT KLA 112, Allgemeine Handschriftenreihe, Ms. 2435, Rent-roll (*Urbar*) of the seigneury of Fusine 1636, 152; AT KLA 112, Ms. 2732, Rent-roll (*Urbar*) of the seigneury of Fusine 1636, 476–477; for relations between these rent-rolls and questions of datation, see Mlinar, 2018a, 27–28, and Mlinar, 2018c, 30–32. For localization of *Eggele*, see Wutte, 1924, 99, whose localization is in line with the name *Jekel* on the 1826 map (SI AS 176, Franciscejski kataster za Kranjsko (Franciscan Land Cadaster for Carniola), L16, cadastral municipality of Hrušica, map sheet A01); the German name for the col at the westernmost edge of this grassland complex is still *Eckelesattel* (Sln. *Jekljevo sedlo*; *Atlas Slovenije*, 2005, map 50). Even though the early nineteenth-century cadaster mapped the grasslands at Jekel as a pasture and no hay meadows, mowing there is recorded not only in the cited seventeenth-century record but also later, when it was carried out either sporadically or more frequently. The right of mowing and grazing at Jekel was documented in 1868 (SI AS 48, Deželna komisija za odkup in uravnavo zemljiških bremen za Kranjsko (Provincial Commission for Redemption and Regulation of *Profits à Prendre* in Carniola), fasc. 56, 9–2: no. 691/II de 1876, Beilage A: Protokoll, December 31st, 1868), and it is clear that part of the grassland at Jekel was mowed in 1942 (Smolej, 2008, 15).

[337] Wopfner, 1997, 313.

following parallels from the Dinaric Karst south of the Alps and from the Carpathians. In the last quarter of the sixteenth century, a large princely stud farm was founded at Lipica (Ital. *Lipizza*) on the Karst (Sln. *Kras*, Ital. *Carso*) Plateau in the northwestern Dinaric Karst. It became an enterprise with significant local and regional economic and environmental impacts.[338] In 1728, this stud farm bought the estate of Prestranek (Germ. *Pröstranegg*) in the Pivka Valley, about 40 km road distance from Lipica. Prestranek, together with its branch estates, where environmental conditions for (most) meadows were more suitable than usual on the Karst Plateau, provided significant quantities of quality hay to Lipica at the onset of rapid eighteenth-century development of the Habsburg port of Trieste, which sharply increased the regional hay demand in the broader surroundings of Lipica.[339] Both estate centers involved in this example of hay transport, Prestranek and Lipica, were located in the immediate vicinity of the main commercial roads linking the Habsburg ports of Trieste (Germ. *Triest*, Sln. *Trst*) and Rijeka (Ital. *Fiume*, Germ. *St. Veit am Pflaum*) with Ljubljana (Germ. *Laibach*), the provincial capital, and Vienna (Germ. *Wien*), the capital of the Habsburg Monarchy, and so the road infrastructure linking Prestranek and Lipica was exceptionally good,[340] which was favorable for road transport of hay over such long distances. One can assume similar medium-distance land-based hay transport in the case of large towns in the Alps and their near surroundings in periods of flourishing trade and a solid state of road connections.[341] A twentieth-century parallel from the Carpathians demonstrates that in exceptional situations, to alleviate shortages, ordinary people also transported hay on roads over similar, relatively long distances before railway construction or in areas distant from railway connections. Traditional farmers in Ghimeş (Hung. *Gyimes*) in the Eastern Carpathians traveled by horse-drawn wagon to buy hay in the Ciuc basin, an area 45 to 50 km from their farms, when the hay harvest in Ghimeş failed and the hay harvest in the Ciuc basin produced hay surpluses.[342]

Moving further in time, since the second half of the nineteenth and early twentieth century, railway connections have made hay transport possible from areas relatively close to railway stations over unprecedented distances, which

---

[338] Zwitter, 2015a.

[339] *Das k.k. Hofgestüt zu Lippiza 1580–1880*, 1880, 30–31, 33–34, 90–96; Klemenčič, 1959, 136–137; on environmental conditions at Prestranek and its branch estates, see also Klemenčič, 1959, 40 & 45, 116–117, and *Arcanum Maps*, Cadastral Maps, Habsburg Empire – Cadastral Maps (XIX. [sic] Century), cadastral municipalities of Matenja Vas (Germ. *Mautersdorf*), 1823, Trnje (Germ. *Dorn*), 1823, and Velika Brda (Germ. *Grossberdu*), 1822; *Environmental Atlas of Slovenia*, Land and Soil, Soil Map (1:25,000).

[340] Zwitter, 2014c, 146–149, 205–206; *Arcanum Maps*, Maps of Europe, Europe in the XVIII. [sic!] Century, Inner Austria 1784–1785 – First Military Survey.

[341] Compare Peterken, 2013, 145.

[342] Carpathians: the ethnoecologist Daniel Babai, personal communication; general considerations from Peterken, 2013, 144–145.

locally again affected mowed areas in the Alps.[343] Some hay from montane meadows on the southern slopes of pre-Alpine Mount Porezen, south of the Julian Alps, was sold at least since Italian rule in the interwar period in the first half of the twentieth century. After the Second World War, when the Porezen area as a whole belonged to Yugoslavia and railway traffic connections already had a tradition of easy long-distance inland cargo transport, highly fragrant hay from Mount Porezen or its surroundings is recalled to have been transported by train as far as Macedonia (over 750 km in a straight line) to feed military horses due to a severe drought there. The organizational environment also mattered: this long-distance hay trade was supported by a broader system in the state of Yugoslavia that ranged from Slovenia to Macedonia, involving the local agricultural cooperative.[344]

Improved traffic connections, increased traffic flows, developments in agriculture, and demographic history all had an influence on the hay trade in the Alps. "Between 1870 and the beginning of the First World War, cow's milk became a mass consumer good which was increasingly being sold as fresh produce .... The farmers based in the immediate vicinity of large consumer centers were understandably the first to react to the observation that the price of fresh milk was rising faster than that of other agricultural products. Many of them sold their farmland as construction sites, often involuntarily, and only practiced pure dairy farming on their remaining property. In extreme cases, these farms had no independent agriculture anymore but instead purchased nearly all the necessary fodder on the fodder markets, which were growing fast due to imports."[345]

Exceptional processes and events involving a large number of livestock, such as some major wars, created extremely high but short-term hay demand in many areas that was covered by hay from valleys and mountains in the Alps and, at least in the case of favorable traffic connections, also by hay from beyond this mountain chain. In such cases, huge amounts of hay from elsewhere left long-term environmental legacies in grasslands at several places in the Alps (see Sections 7.2 and 10.3).[346]

---

343 Krausmann, Weisz, & Eisenmenger, 2016, 82, 86. A map showing the history of railroad connection development in Austria-Hungary until 1914 is available in Bruckmüller & Stekl, 1998, 67, but the location of the railroad route linking the Julian Alps and the Adriatic (which is relevant for the next sentence of this text) is presented erroneously; thus, see also the map in Černe, 2001.

344 Štucin, 2011, 70; Bavcon, Ravnjak, & Praprotnik, 2019, 171; see also *Geografski atlas i statističko-geografski pregled svijeta*, 1953, 42–43 (Yugoslav traffic statistics in 1951), and map sheets 9, 10, 11, & 12, showing the railroad network in Yugoslavia.

345 Orland, 2016, 43–45.

346 Bavcon, Ravnjak, & Praprotnik, 2019, 171.

3.2 Grasslands "have come and gone"[347]

Many grasslands in the Alps have not existed continuously due to the characteristics of historical crop rotation systems and other socioeconomic and environmental factors. This section focuses on discontinuously existing grasslands, which is briefly addressed in Section 3.1 based on historical fluctuations in mowed grassland areas along the Gotthard Pass route. The discontinuous existence of grasslands dealt with in Section 3.2 was triggered by farming or its absence. Discontinuities in grassland history caused or stimulated by environmental processes are dealt with in Section 4.

The British ecologist George Peterken argued that "[i]t would be valuable to be able to establish how many years an existing meadow has been meadow, or at least how long the grassland has been permanent. Further, it would be helpful to identify any ancient meadows, which … would be late-Medieval meadows that have remained meadow ever since." It "helps us to understand the ecology and biology of those that remain."[348] In general, the question of grassland (dis)-continuity is best approached interdisciplinarily. Although historical land-use records can provide more reliable long-term data than biotic indicators, long uninterrupted series of historical data at the level of land parcels usually do not exist. Some ecological analyses showed that species compositions of sixty-year-old grasslands, for instance, can significantly differ from those at least two hundred years old, or that there can be significant differences in species composition between grasslands 1) thirty to forty years old, 2) eighty years old, and 3) one hundred to 130 years old. Spring-sedge (*Carex caryophyllea*; Figure 14) is an example of a species identified as an indicator of grasslands more than a century old by at least three studies, although it is not an exclusive indicator of ancient grasslands, and viper's grass (*Scorzonera humilis*) is "rare and attached to old meadows in much of Europe."[349] Statistical analysis of dry calcareous grasslands near Kallmünz, Bavaria, comparing grasslands at least 175 years old, defined as ancient, and about sixty years old, defined as recent, came to the conclusion "that history (occurrence on ancient or recent grassland) was obviously the most important factor affecting the actual vegetation pattern." Nonetheless, based on adequate management, no more than limited plant nutrient availability at the site as a legacy of historical management of the land parcel, and the availability of seeds from old grasslands made available by means of natural dispersal processes and/or farming (see Section 7.1), some relatively recent grasslands can also "contain rare and/or endangered species."[350]

[347] Peterken, 2013, 166.
[348] Peterken, 2013, 166, see also pp. 162–165.
[349] Poschlod et al. 2008, 139–140; Karlík & Poschlod, 2009; Peterken, 2013, 239–241, 285, 287 (quote), 399.
[350] Poschlod et al. 2008, 136–140 (quotes from p. 140); Karlík & Poschlod, 2009 (the second quoted passage also appears here, on p. 335); see also Peterken, 2013, 239–241, 396.

Figure 14: Blooming spring-sedge (*Carex caryophyllea*)[351] in an old, extensively managed meadow (on rendzina on moraine or on eutric cambisol on moraine) on a northwest-facing slope at Zavaze, south of Stara Fužina in Bohinj.[352] A meadow was already mapped here in the early nineteenth century,[353] and it is highly likely several centuries old, possibly of medieval origin. Photo: Žiga Zwitter, April 2020.

Determining whether a certain meadow has ever been used as an arable field or not is an easier task in the Northern or Southern Limestone Alps than in the United Kingdom, the main area of George Peterken's research emphasizing the relevance of knowledge about (dis)continuities in grassland history. Apart from precise early-nineteenth-century land cadaster, created in times of high agricultural density, thus often showing (close to) the maximal historical extent of arable land (see also Section 3.1), landforms are particularly helpful in the Southern and Northern Limestone Alps. Namely, some hummocky meadows (Figure 15, see also Section 4.1.2) have been preserved there.[354]

351 In this case, species identification in the photo is almost certain.

352 *Environmental Atlas of Slovenia*, Land and Soil, Soil Map (1:25,000).

353 *Arcanum Maps*, Cadastral Maps, Habsburg Empire – Cadastral Maps (XIX. [sic] Century), cadastral municipality of Studor (Germ. *Studorf*), 1826, parcel no. 1563.

354 Cvetek, 1971, 67–68; Odar, 1993, 3, 48; Embleton-Hamann, 2004, 176–177; Koblar, 2013, 1–2. The conversion of hummocky meadows into arable fields was practiced to a certain degree in history because an unconsolidated moraine was not difficult to level, but this left clear geomorphic legacies because there are no longer any hummocks or pits, or they are at least lower than in the surrounding parcels if uneven relief formed

Figure 15: Hummocky meadow on the Uskovnica Plateau above the Bohinj Basin. The creation of meadows there probably started in the fifteenth century (Zwitter, 2013a, 92). Note the slightly curved traces of the scythe running parallel and ascending from the bottom left of the photo. Preservation of agroecosystems requiring such labor-intensive management is often a challenge in Central Europe.[355] Inadequate agri-environmental payment is one of the reasons for the virtual disappearance of hummocky meadows as Slovenian biodiversity hotspots of global relevance, hosting many specialist plant and animal species. Further reasons for their abandonment or leveling include widespread deagrarianization, farm mechanization, the changed socioeconomic structure of farm households, and a changed farming style at existing farms (see also Section 10.1.2.1). The area where farmers applied for agri-environmental payments aimed at highly labor-intensive haymaking in hummocky

---

again. However, in the mid-twentieth century hummocky morphology was still characteristic for distant montane meadows, revealing that those meadows have never been plowed. Mowing in hummocky meadows has always been much more time-consuming than mowing in flat meadows, and drying hay has also been more labor-intensive because mowed plants were (or have been) raked to the tops of hummocks to promote drying in the humid climate of the region. Severe limits posed by these landforms on mechanical mowing have exacerbated farming challenges in hummocky meadows. Consequently, many hummocky meadows have been levelled or abandoned more recently (Cvetek, 1971, 69–70, 76; Odar, 1993; Koblar, 2013, 16–21, 27–28, 30). Maps of the distribution of hummocky meadows in contemporary history (e.g., Koblar, 2013, 17; Odar, 1993, maps between pp. 12 and 13, between pp. 52 and 53, and in the appendix) reveal the young age of grasslands in several areas of previously hummocky meadows, which have been leveled in recent decades. In some cases, the aim of leveling was to create intensive meadows; in some others, the goal was to solve urgent issues of the social and economic pillar of sustainability in extensive meadow management.

355 Compare, for instance, Suske, Gantner, & Gattermaier, 2012, 125.

meadows in Triglav National Park, in the eastern Julian Alps, declined from 22 ha to 9 ha between 2011 and 2017, which reflects real changes in the landscape[356]—a continuation of a large-scale process resulting in abandonment or leveling of former hummocky meadows that started by the mid-twentieth century[357] (see also Section 9). Photo: Žiga Zwitter, July 2017.

Hummocky meadows exist in flatland and on slopes with gradients that are not too steep (e.g., up to 20°). The density of hummocks is often between 500 and 2,000 per hectare. Lower hummocks are 0.3 m high in many places; the highest are up to 1.5 meters high and usually one to four meters wide. The dimensions of nearby hummocks and pits differ significantly, and the size of average and median hummocks and pits in nearby meadows often varies.[358] These landforms prove by their morphology that the grassland parcel has never been used as an arable field in recent centuries, and so hummocky meadows could become suitable study sites for identification of some further species characteristic of old meadows.

3.2.1 Grasslands in field rotation systems

Historical field rotation systems included short-term pastures or meadows in fallow fields. A single fallow interval often did not last longer than one annual vegetation period within one rotation cycle; the entire rotation cycle very frequently covered three years. Nonetheless, in a significant percentage of all fields in the Alps, grassland fallows lasted longer. This was especially frequent at isolated farms in the eastern part of the Eastern Alps and in the northern part of the Alps, where fallows lasted, for example, three to fifteen years after a year or two, or more, of cereal cultivation; in some cases, both types of land use lasted more or less equally long, alternating every three to eight years, for instance. Factors leading to long grassland fallows differed, ranging from environmental (soils less fertile for field crops required longer fallows) to socioeconomic. These included, for instance, agricultural density that varied significantly over

356 Bertoncelj & Travnikar, 2024, 78–80, 82.

357 See, e.g., the maps of hummocky meadows in Bohinj in the early 1990s in Odar, 1993.

358 Cvetek, 1971, 67, 69; Odar, 1993; Embleton-Hamann, 2004; Koblar, 2013, 6–7, 20–21, 25–27, 30; Prezelj, 2012, 30, supplements C4 and C5. Various types of hummocky meadows are the result of an interplay of various factors that are not yet totally understood. There are two important processes leading to the most spatially extensive areas of persistent hummocks and depressions. The first is corrosion affecting unconsolidated carbonate moraines (where there was a larger amount of fine particles, such as fine sand, a pit formed), and in some places even unconsolidated old fluvial deposits that have been out of reach of floods at least in recent centuries. The second process is windthrows on unconsolidated carbonate sediments, so that unconsolidated material was removed with roots, forming depressions. This material was later deposited, forming hummocks, and later corrosion acted there as well (Cvetek, 1971, 67–69, 71–76; Odar, 1993; Embleton-Hamann, 2004; Koblar, 2013, 1, 3, 7–8, 23–24, 27–28, 30).

time. In general, long fallows were more feasible for many mid-sized and large farms in the Eastern Alps in some periods than for the typically small farms in the Central and Western Alps. Another human factor promoting long fallows was the interest of animal husbandry in fallow grasslands.[359] Namely, high nutrient availability in fallow grasslands as a legacy from the stage of manured field-crop cultivation on the same land parcel and livestock excrement from periods of grazing in the fallow resulted in high grazing resource availability and/or high hay yields for the standards of the time (Section 6.3.2). The experiences of traditional Csángó farmers in the Eastern Carpathians demonstrate that even about a decade after the last manure application to the same (arable) parcel, yields in fallow grasslands remain (relatively) high[360] (under appropriate site conditions). Hay from fallow fields or the grazing resources there were usually rich in proteins, in the first years of the grassland phase in particular.[361] From the perspective of arable farming, grazing livestock left their excrement there, which was rich in nutrients in forms available to plants. Moreover, grassland fallows produced much more humus than cereal fields on the same parcel. The grassland stage positively impacted the soil structure because grass roots (in certain soil types) tended to break up big compact soil aggregates, but they helped bind smaller soil aggregates. Positive grassland impacts on soil structure were maintained during the next couple of years of arable farming. A longer grassland fallow successfully suppressed several soilborne diseases in arable land as well as several species considered weeds and pests in farmers' eyes that require a certain cultivar or that were unable to compete in a dense grassland or tolerate grazing.[362] Seeds of a couple of frequent cereal "weeds" were toxic to people, and their weight was similar to that of cereal grains, and so until roughly a century ago their seeds were often not successfully separated from grain seeds; nonetheless, the abundance of these species was reduced by means of cereal-field weeding in May or June, and/or with a grassland fallow.[363] Nonetheless, grazing could also foster some field weeds avoided by livestock, such as thistles (Section 6.6.2).[364]

---

359 Meze, 1965, 215; Grafenauer, 1970c; Pickl, 1982, 38; Pfister, 1985a, 50, 99, 106–107; Meilleur, 1986, 100–102, 105, 107–109, 145, 267; Mathieu, 1996, 147–148, 156, 159; Wopfner, 1997, 293–294; Keller, 2004, 13, 27; Emanuelsson, 2009, 124, 128, 130; Poschlod, 2015, 71; Kauter, 2002, 131–132, 143–145; Kačičnik Gabrič, 2008, 298–299, 301; Peterken, 2013, 165, 170–172; Bätzing, 2015, 58, 62, 66–67, 77; Mathieu, 2019, 51, 66–67. Note also the relatively intensive agriculture documented in a 1584 rent-roll at some farms in eastern Carinthia, which, nonetheless, does not reveal what a usual fallow duration was (Dinklage, 1966b, 106–107 and *Arcanum Maps*, Cadastral Maps, Habsburg Empire – Cadastral Maps (XIX. [sic] Century), cadastral municipality of Großenegg, 1827).

360 Babai, Molnár, & Molnár, 2014, 74.

361 Ellenberg & Leuschner, 2010, 46–47.

362 Dietl & Lehmann, 2004, 77–79; Grafenauer, 1970c, 243, 245.

363 Maček, 1979, 304–305; Makarovič, 1982, 137, 143; Hubatschek, 1992, 34, 150–151.

364 Poschlod, 2015, 71.

Throughout history, repeated creation of species-rich grasslands in fallow fields was easy. At the beginning of the grassland stage, the fallow nutrient availability inherited from the arable stage was considerable. Some perennial plants survived there from the arable stage, and there was a seed pool in the soil, which had arrived there by means of wind transport, stubble grazing (Section 7.1), and application of barn manure to that parcel during the phase of arable use, for instance. Moreover, local barn-sweepings were sown there to foster a transition from arable land to grassland, if available in sufficient quantities (see Section 6.5). During the grassland phase, "seed could enter easily from nearby [grasslands], aided by the regular movements of [livestock]"[365] (see Section 7.1) and by other processes.

Data from the humid areas of northeastern Tyrol and from many Tyrolian valleys at high elevations, where enduring grassland fallows were a characteristic component of the rotation cycle in the long run, demonstrate that many such meadows were distinctly blue and yellow when wild pansy (*Viola tricolor*) blossomed in the first year of a grassland stage. *Viola tricolor* used to be a frequent "weed" in arable land, and so it also abounded in the early grassland stage.[366] Wild pansy occurs in all regions of the Alps at moderately fertilized mesic sites, with soil pH close to neutral, and it abounds at some places in the montane vegetation belt[367] that used to be characterized by significant areas of arable land,[368] and so the relevance of this data is not limited to Tyrol. Thistles (*Cirsium* spp.) thriving at relatively nutrient-rich sites also abounded in some fallow grasslands.[369] Similarly to wild pansy, some thistles, especially creeping thistle (*Cirsum arvense*), a perennial species with strong vegetative propagation by means of sprouting shoots from belowground fragments,[370] also survived in a grassland fallow from the arable stage; ploughing was shallow, and so this is one of the species that lived in the parcel continuously, in the arable and grassland stages.[371] Creeping thistle, widely present in the Alps and abounding in nutrient-rich, mesic soils with a pH close to alkaline in colline and montane vegetation belts,[372] was a frequent species in arable land, often removed from cereal fields by weeding but nonetheless occurring or even abounding there. Weeding could not remove its roots, which can easily reach a depth and breadth of a couple of meters or more, and hundreds or thousands

---

365 Wopfner, 1997, 294; Karlík & Poschlod, 2009, 334; Peterken, 2013, 140, 165, 171, 395 (quote); sowing barn sweepings on fallows in Tyrol: Wopfner, 1997, 294, in the Allues Valley: Meilluer, 1986, 282; similarly in Ghimeş in the Eastern Carpathians: Babai, Molnár, & Molnár, 2014, 92–94.

366 Wopfner, 1997, 293.

367 Aeschimann et al., 2004, vol. 1, 450–451

368 E.g., Pfister, 1985a; Mathieu, 1996; Zwitter, 2015d.

369 Babai, Molnár, & Molnár, 2014, 74; Aeschimann et al., 2004, vol. 2, 582–591.

370 Aeschimann et al., 2004, vol. 2, 590–591; *Weed of the Month: Canada Thistle*, 2019.

371 Compare Lenartsson et al., 2023.

372 Aeschimann et al., 2004, vol. 2, 590–591.

of stems can emerge from such a root system. At one of the isolated farms at Jazbina on the slopes of *Uršlja Gora* in the Eastern Karawanks, one of the arable fields was called *Oscava njiva* (literally, 'creeping thistle field') due to the high abundance of creeping thistle (*osc*) there,[373] surely also occurring in the historical fallow grassland there and at many other places. Further historical "weeds" from arable land occurring in fallow grasslands included field bindweed (*Convolvulus arvensis*), field mouse-ear (*Cerastium arvense*), and two hemiparasitic species, European yellow-rattle (*Rhinanthus alectorolophus*) and field cow-wheat (*Melampyrum arvense*), for instance.[374]

Dandelion (*Taraxacum officinale*) was one of the first grassland plants that appeared visibly after a field was left fallow[375] and nutrient availability was still relatively high. Its importance in the human diet in early spring was significant (see Section 8.2.1.1), and slightly later, when its juicy stems grew, it was an excellent fodder plant in fallow pastures.[376] Coltsfoot (*Tussilago farfara*) also belongs to early plants appearing in several grassland fallows. Still in the early grassland stage, nitrophilous docks (*Rumex* spp.) colonized some fallows.[377] Common couch (*Elymus repens*) and other grasses of moderately fertilized sites, field scabious (*Knautia arvensis*, Figure 91 in Section 6.3.2.2), goat's beard (*Tragopogon pratensis*, Figure 91), creeping bellflower (*Campanula rapunculoides*), oxeye daisy (*Leucanthemum vulgare*, Figure 166 in Section 10.2.2), corn marigold (*Glebionis segetum*), ground-elder (*Aegopodium podagraria*),[378] and hogweed (*Heracleum sphondylium*, see Sections 6.2.2.2, 6.6.1.1, 8.2.1.1, 8.2.2, including Figure 142, and 8.2.3), for instance, were likely to occur in several grasslands in fallows if enough nutrients were still available from the last manuring (see also Section 6.3.2 on moderately fertilized meadows).[379] Many of the plants abounding in several fallow grasslands either have a high dispersal potential (e.g., coltsfoot, dandelion, and goat's beard) or are highly likely to persist in the soil either as a germinable seed bank or as vegetative propagules (docks, ground-elder, and common couch). Occasional ploughing contributed to suppression of mosses (unwanted from the perspective of livestock farming) and some other plant species in fallow grasslands.[380] Biodiversity in grassland

---

373 Interview by Marija Makarovič, with M. Plesnik, in Jazbina (Makarovič, 2016); the species was collected by M. Makarovič and so the folk taxon *osc* was deciphered by the botanist Martina Bačič. See also *Weed of the Month: Canada Thistle*, 2019.

374 Karlík & Poschlod, 2009, 334.

375 Kauter, 2002, 143–144; Babai & Molnár, 2013, 10; Babai, Molnár, & Molnár, 2014, 74.

376 Pils, 1994, 173; Dietl & Lehmann, 2004, 43.

377 Kauter, 2002, 143–144; Babai, Molnár, & Molnár, 2014, 74, 77; Aeschimann et al., 2004, vol. 1, 380–391.

378 Machatschek, 2015, 34–35; identification of *Acker-Quecke*, *Giersch*, and *Saatwucherblume* is based on *Langenscheidt Dictionary*: Acker-Quecke, Giersch & Saatwucherblume.

379 Meilleur, 1986, 297; Kauter, 2002, 143–144.

380 Emanuelsson, 2009, 206–207; Peterken, 2013, 140, 165, 171.

fallows was lower than in non-ploughed grasslands at similar site conditions. There were several plant species belonging to habitat generalists and few if any specialists in fallow grasslands.[381]

Mainly in the eighteenth and early nineteenth century, sowing clover and/or grass seed mixtures in fallows at the beginning of their grassland stage gradually spread in the Alps. Fodder legumes, or "clovers" (actually not limited to the clover genus, *Trifolium*), are rich in proteins. Moreover, symbiotic bacteria in legumes' root nodules enrich the soil with airborne nitrogen.[382] Growing "clovers" (such as red clover, *Trifolium pratense* subsp. *pratense*, a native species in Central Europe, and introduced alfalfa, *Medicago sativa*, or bastard alfalfa, *Medicago* × *varia*) as quality fodder plants was not an unprecedented novelty in medieval and early modern Europe as a whole. Nonetheless, the number of legume species and varieties involved increased, and the array of ecological niches in the colline and montane vegetation belts they were able to inhabit diversified. From the Middle Ages to the mid-twentieth century, "clover" sowing spread significantly around Central Europe, to areas where it had not been practiced before, except for wild legume seeds in barn sweepings (Section 6.5). At the local level, first "clover" fallows usually appeared in the Eastern Alps in early modern times. (Bastard) alfalfa, for instance, has been continuously grown in the Eastern Alps since early modern times.[383] The example of the Stubai Valley (Germ. *Stubaital*) in Tyrol demonstrates that seeding clover in fallows was introduced by the early nineteenth century. Nonetheless, a century later, in the late 1920s, 240 km^2 (or the equivalent of three-fourths of the total fallow area in Tyrol) was not sown with anything but barn sweepings.[384] Even if sown with clover and/or grass mixtures in the first year of their grassland stage, extensively managed long-lasting fallows gradually become relatively species-rich due to the seed pool in the soil and seeds entering the parcel from the surroundings,[385] unless a thorough weeding (recorded in (medium-)intensive meadows close to farmsteads in the Villgraten Valley in the first half of the twentieth century and probably introduced only with more intensive meadow management)[386] significantly decreased the species diversity. The parallel of traditional twentieth-century Csángó farmers in Ghimeş in the Eastern Carpathians demonstrates that non-weeded fallow fields sown

---

381 Peterken, 2013, 285, 395–396, 398–399.

382 Britovšek, 1958, 141; Hayot Carbonero et al., 2011, 73–74; Poschlod, 2015, 120.

383 Britovšek, 1958, 141–143, 145–146, 149; Kauter, 2002, 93–95; Hayot Carbonero et al., 2011, 72–73; Poschlod, 2015, 85–86, 119–123; Aeschimann et al., 2004, vol. 1, especially pp. 906–907, 926–929, 948–949, 954–955.

384 Dinklage, 1966b, 131–132, 134, 139–140; Valenčič, 1970a, 260, 262; Viazzo, 1989, 191; Pfister, 1985a, 109–111; Mathieu, 1996, 152; Wopfner, 1997, 293–294.

385 Bavcon, Ravnjak, & Praprotnik, 2019, 41, 205.

386 Wopfner, 1997, 294; the ethnologist Maria Lang-Reitstätter, cited by Wopfner, performed fieldwork in the Villgraten Valley in the 1930s (Mühlburger, 2020).

with clover turn into a significantly species-richer grassland in about three years.[387]

In general, the duration of grassland fallows was reduced and fallows were gradually excluded from the rotation cycle during the modern period, in particular at lower elevations. In most places in the Alps, these processes became significant by the late nineteenth century, but there exist earlier and later examples. An early example is numerous fields in Engadin (Romansh *Engiadina*) with arable use every year at least since the sixteenth century, despite the high elevations. A late example is the French Southern Alps, characterized by low population density, where fallows persisted beyond the end of the nineteenth century.[388] Similarly, grassland fallows existed in several areas of isolated farms in the Eastern Alps until the mid-twentieth century.[389]

Occasional examples of field crops grown in some parcels, which were usually used as meadows, were not necessarily phases of more or less regular rotation cycles. Agricultural handbooks from Classical Antiquity, the sixteenth century, and later recommended exceptional temporary conversion of low-quality meadows into (fertilized) fields for a couple of years to improve soil fertility, but it is not clear how common this practice was.[390] It certainly was not practiced in many meadows in the Alps where environmental conditions were unsuitable for fields and human factors also importantly influenced the decision on whether to plough up (or hoe) a certain grassland or not. This practice increased soil nutrient availability through the short-term introduction of more intensive management; some extensive land-use types were also an option to attain a similar goal (see Sections 3.2.2 and 6.1.2.6).

Another type of regularly recreated but discontinuous grassland was pastures as parts of the swidden cultivation cycle. After forest was cleared, bushes, tree branches, and thinner trunks were burned, a couple of years (or a year at least) of hoe-farming field-crop cultivation followed in a typical swidden, and then the parcel served as pasture. This pastoral stage usually lasted for several years (three to twelve, fifteen years or even longer). The pasture was increasingly overgrown by woody species. Later, the area was left ungrazed for some years (e.g., from two to roughly ten or more, up to a few decades) to allow further growth of trees and bushes, until the cycle was repeated. In many cases, pasture was included in such a typical cycle in the long run, as proven for instance by a description from Carniola in 1577, early-nineteenth-century descriptions from the Karawanks in Carinthia, and later observations. Focus-

387 Babai, Molnár, & Molnár, 2014, 74.
388 Grafenauer, 1970c; Pfister, 1985a, 106–116; Meilleur, 1986, 100–102, 105, 107–109, 145, 267; Mathieu, 1996, 147–148, 156; Wopfner, 1997, 293–294; Bätzing, 2015, 77.
389 E.g., Meze, 1963, 254–255; Meze, 1965, 215–216 (see early-nineteenth-century data in a comparable territory in Kačičnik Gabrič, 2008, 298–299, 301); Keller, 2004, 13.
390 Kauter, 2002, 105–106, 131; Emanuelsson, 2009, 207.

ing on the modern period, areas of temporary swidden pastures expanded in the sixteenth and seventeenth centuries, and they were still important in some parts of the Styrian Alps until the nineteenth or even mid-twentieth century. In Styria (Germ. *Steiermark*, Sln. *Štajerska*) and neighboring parts of the Eastern Alps in particular, an enormous total area of such temporary pastures was characteristic over the long run. There, in many (but not all) cases of isolated farms, pastures in a swidden cycle were the only summer pastures. An example is the Karawanks in the Vellach (Sln. *Bela*) catchment in Carinthia, where most pastures were part of the swidden cycle, as reported in the first half of the nineteenth century. Temporary swidden pastures were also relevant for a long time in the western Vercors Massif, for example; in many other parts of the Alps, they disappeared sooner. In the upper Emme Valley (Germ. *Emmental*), for instance, they were still relevant in the nineteenth century and they had a longer historical tradition.[391]

Site conditions and species compositions in such temporary pastures varied due to differences in local environments, local practices of swidden cultivation, and their environmental impacts. Consequently, the quality of subsequent pastures differed significantly. Couch grass (*Elymus* spp., attractive for livestock when young and providing good fodder) was usually one of the genera abounding in swidden pastures in Styria, and common sorrel (*Rumex acetosa*) was also frequent. Some highly esteemed grass species occurred there as well.[392] The fodder value of common sorrel in pastures is (relatively) low,[393] and so, if its abundance was high, such pastures were not attractive for livestock and livestock's performance was limited. Thus, data from Upper Styria in the first half of the twentieth century reveal that most farmers sowed cereals and barn sweepings (see Section 6.5) on the same parcel (grain grassland plants mix) in the last year of a cereal field within a swidden cycle. The goal of this practice was to make the grassland species composition in the swidden

[391] AT KLA 75, Franziszeischer Kataster, cadastral municipality no. 76203, Bad Vellach (Sln. *Bela*), cadastral appraisal protocols: Commissions Protocoll, August 21st, 1837; AT KLA 75, cadastral municipality no. 76205, Ebriach (Sln. *Obirsko*), cadastral appraisal protocols: s.n., 47/4, 84–85 and Protocoll, July 12th, 1836; AT KLA 75, cadastral municipality no. 76213, Leppen (Sln. *Lepena*), cadastral appraisal protocols: Protocoll, June 9th, 1836; AT KLA 75, cadastral municipality no. 76214, Lobnig (Sln. *Lobnik*), cadastral appraisal protocols: s.n., 246/4, 84–85 and Protocoll, July 4th, 1836; AT KLA 75, cadastral municipality no. 76218, Remschenig (Sln. *Remšenik*), cadastral appraisal protocols: s.n., 371/4, 84–85 and Protocoll, July 11th, 1836; Predan, 1956b; Vršnik, 1962 (for a biography of this author, see Kumer, 2005); Schneiter, 1970; Pickl, 1982, 27–36, 38–40, 43, 45–55; Mathieu, 1996, 146; Wopfner, 1997, 263; Zwitter, 2013a, 95; Zwitter, 2014a, 221–223; Panjek, 2015a, 99; Zwitter, 2015d, 142, 176.

[392] Vršnik, 1962; Schneiter, 1970, 32–39, 80–81, 86–87, 90; Makarovič, 1982, 55 (information from Strojna in northern Slovenia); Kerschbaumer, 2007; Poschlod, 2015, 75–76; Lauk & Erb, 2016; see also Krausmann, Weisz, & Eisenmenger, 2016, 74–75.

[393] Brugger & Wohlfahrter, 1983, 23, 26; Dietl & Lehmann, 2004, 42–43; on docks see also Section 8.2.2.

pasture that followed cereal harvest more favorable for livestock. At certain sites, high-quality pastures lasting for a couple of years are recorded; they appeared, for instance, on dystric cambisols, especially if cereal cultivation in swiddens only lasted one year, and thus not as many plant nutrients were taken from the soil by field crops and leached as in the case of two years of cereal cultivation. Under certain environmental conditions and swidden practices, two years of cereal cultivation were still followed by a phase of high-quality pasture.[394] In Bohinj in Carniola, in 1694 the opinion was recorded that higher-quality pastures would develop above the southeastern shore of Lake Bohinj due to temporary clearings (probably swidden fields)[395] in the common land there.[396] Similarly, a century and a quarter later (in the early nineteenth century), pasture was recorded as the economically most important phase of the swidden cycle in the nearby cadastral municipality of Bohinjska Češnjica in the northern Bohinj Basin.[397] Later data, referring to the first half of the twentieth century, reveal that this was no exception where swidden cultivation was still practiced: in Upper Styria, for instance, swidden pasture (a land-use type that lasted longer within a swidden cycle than a hoe-farmed field) was in general economically more relevant for the farmer than swidden fields. This was particularly the case where (unlike Bohinj) no other non-forested summer grazing grounds at higher elevations were available.[398] Nonetheless, the added value of seventeenth- and early-nineteenth-century data presented from Bohinj lies in their potential to explain that swidden grazing had already been the main motivation for swidden cultivation at some places in the Alps before railway connections made long-distance cheap cereal imports possible.

---

[394] Vršnik, 1962; Schneiter, 1970, 32–39, 80–81, 86, 90; Makarovič, 1982, 55 (information from Strojna in northern Slovenia); Makarovič, 2005, 105–106; soils in Strojna: *Environmental Atlas of Slovenia*, Land and Soil, Soil Map (1:25,000).

[395] The description of farming in the cadastral municipality this area belonged to, from the 1830s, still documented swidden cultivation there. At that time, swidden cultivation was practiced at some locations in seven out of eight cadastral municipalities in the Bohinj Basin (with the exception of Bohinjska Bistrica) (SI AS 176, Franciscejski kataster za Kranjsko (Franciscan Land Cadaster for Carniola), L30, cadastral municipality of Nemški Rovt, cadastral appraisal protocols: s.n., 30/4, 85; SI AS 176, L72, cadastral municipality of Gorjuše, cadastral appraisal protocols: s.n., 72/4, 85; SI AS 176, L107, cadastral municipality of Češnjica v Bohinju, cadastral appraisal protocols: s.n., 107/4, 85; SI AS 176, L157, cadastral municipality of Srednja Vas v Bohinju, cadastral appraisal protocols: s.n., 157/4, 85; SI AS 176, L172, cadastral municipality of Nomenj, cadastral appraisal protocols: s.n., 172/4, 85; SI AS 176, L242, cadastral municipality of Savica, cadastral appraisal protocols: s.n., 242/4, 85; SI AS 176, L284, cadastral municipality of Studor, cadastral appraisal protocols: s.n., 284/4, 85).

[396] SI AS 721, Gospostvo Bled (The seigneury of Bled, Germ. *Veldes*), Iustitialia, fasc. 33, September 6th, 1694.

[397] SI AS 176, Franciscejski kataster za Kranjsko (Franciscan Land Cadaster for Carniola), L107, cadastral municipality of Češnjica v Bohinju, cadastral appraisal protocols: s.n., 107/4, 85.

[398] Schneiter, 1970, 41, 86.

In general, the quality of swidden pastures decreased so markedly over time that the number of livestock grazing in the same land parcel within the same swidden cycle was usually higher in the first years than in the last years of the pastoral phase. Increased swidden-pasture overgrowth with ferns, bramble (*Rubus* spp., avoided by most livestock), and, most importantly, woody species played a prominent role in this context. Swidden cultivation included hoeing, usually no plowing, and so tree roots and stumps were not removed during the phase of field crop cultivation. Consequently, birch (*Betula pendula*) and alder (*Alnus* spp.) roots, for instance, usually survived swidden burning, and so these woody species, abundant in many swiddens, resprouted. Alder and birch were exactly the woody species dominating in many overgrown swiddens in the Vellach catchment in the Karawanks, as reported in the first half of the nineteenth century. Alders were particularly relevant due to the nitrogen-fixing bacteria in their root nodules. In limited quantities, alders also provided relevant leaf fodder (see Sections 6.1.2.6 and 8.2.2.1). Although the average quality of swidden pastures was relatively high in several cases (according to the standards of the time) and swidden cultivation there was sustainable in the long-run, at some other places, plant-nutrient availability in swidden after the phase of cereal cultivation was so low that only sparse pastures could develop there. In such grasslands, hard matgrass (*Nardus stricta*) and bilberry (*Vaccinium myrtillus*) became dominant. In certain cases, this was a result of unsustainable swidden management causing degradation of a swidden pasture as a consequence of too short a phase of forest regrowth or of improper swidden techniques. A few years sufficed for such a swidden pasture to become a nutrient-poor *Vaccinieto-Callunetum* community (named after the genus *Vaccinium*, including bilberry (*Vaccinium myrtillus*) and cowberry (*Vaccinium vitis idaea*), and after *Calluna vulgaris*, i.e., heather), providing very low-quality fodder (Section 6.3.1.2).[399]

3.2.2 Temporarily abandoned grasslands

Another reason for fluctuations in grassland area was temporary or permanent abandonment of permanent grasslands, which can be traced in the history of the Alps before the large-scale abandonment of grasslands that started in the nineteenth century (Section 9). In former times, grassland abandonment was often a consequence of a nexus of environmental and human factors. For instance, some parts of the Alps were severely affected by the mid-fourteenth-century Black Death, but there existed significant regional diversity in the Black Death's

[399] Predan, 1956b (and *Slovar slovenskega knjižnega jezika*: ostrogovina); Schneiter, 1970, 32–43, 80–81, 86; Kerschbaumer, 2007; Poschlod, 2015, 75–76; Lauk & Erb, 2016; nineteenth-century data from the Vellach catchment: AT KLA 75, Franziszeischer Kataster, cadastral municipality no. 76203, Bad Vellach (Sln. *Bela*), cadastral appraisal protocols: Commissions Protocoll, August 21st, 1837.

impacts within the mountain range. In severely affected areas, several meadows and fields were converted into pastures, and in many other grasslands ecological succession took place, and many of them became (partly) reforested. Apart from several waves of plague and other human epidemics,[400] animal diseases, especially epizootics,[401] and wars were also likely to lead to temporarily abandoned grasslands at a micro- to mesoregional level,[402] as were, locally, exceptional personal or family situations.[403] At the isolated Retko farm in the settlement of Planina in the Kamnik–Savinja Alps (Sln. *Kamniško-Savinjske Alpe*) in Lower Styria, on a south-facing slope at 1,000 m, the seigneury determined in 1613 that the farmer that had become the head of that farm fifteen years earlier had not even gathered in the hay.[404] In some cases, seigneuries forced their subjects to offer an overgrown grassland for free to another farmer so that the meadow would again be used and the seigneury would not lose its income (the dues) related to that parcel. This was the case of an overgrown distant montane meadow at Javorski Vrh, west of the village of Podjelje in Bohinj, in 1621.[405]

Relatively short-term abandonment of grasslands could have various ecological impacts. At an early abandonment stage, up to four years since the last mowing, plant and animal diversity increased. Such short-term abandonment resulted in higher yields in times of restored use. Namely, nutrients released from decaying grassland plants increased plant nutrient availability in soils in comparison with regularly mowed meadows, and mowing did not disturb the development cycles of some grassland animals (Section 6.1.2.6).[406] However,

[400] Fornasin, 2017, 64–67; Izdebski et al., 2022 (focusing on mesoregional diversity of the Black Death's demographic impacts in Europe); Dinklage, 1966a, 80–81; Dinklage, 1966b, 125–126; Koropec, 1972, 89, 93–95, 97, 99, 101–102 (Koropec's results suggest significant microregional diversity of the Black Death's demographic impacts in the southeastern Central Alps of the Pohorje Hills and Kozjak); Viazzo, 1989, 129, 210–211; Scharr, 2001, 88–89, 93; Emanuelsson, 2009, 40, 225, 227–229; Hejcman et al., 2013, 355; Peterken, 2013, 152; Bätzing, 2015, 69–70; Poschlod, 2015, 84, 87, 97; Mathieu, 2019, 57. These works on the one hand make it clear that the fourteenth-century Black Death affected at least some parts of the Alps as strongly as lowlands (but not all parts of the Alps were strongly affected); on the other hand, later demographic crises were less frequent in the Alps than in lowlands. Plague again severely struck some parts of the Alps about 1630.

[401] Emanuelsson, 2009, 146, 227, 229; Hinterwaldner, 2002, 158, 200.

[402] Wopfner, 1997, 222; Zwitter, 2013a, 87–88; Emanuelsson, 2009, 40, 42, 227, 229, 231–233; Hejcman et al., 2013, 355; Poschlod, 2015, 98–99, 101–102. Impacts of wars on livestock numbers were complex and diverse; wars often decreased livestock numbers in many territories, but the First World War led to an increase in sheep keeping in North and East Tyrol (thus in the hinterland), for instance, because of limited wool import (Wopfner, 1997, 235–236; Leisering, 1998, 74).

[403] Emanuelsson, 2009, 227, 229; Zwitter, 2014e, 319–320.

[404] SI NŠAL 20, Gornji Grad Avstrija (Gornji Grad Austria), box 17, Manorial minutes (journal) 1613[–1630], 11 (February 22nd, 1613).

[405] SI AS 721, Gospostvo Bled (The Seigneury of Bled, Germ. *Veldes*), Iustitialia, vol. 12, July 26th, 1621.

[406] Keller, 2004, 19, 33, 40–42, 86, 110.

mowing (Section 6.1) and spring cleaning (Section 6.6) no longer suppressed woody plants and some others that were unwanted from the livestock's perspective, and so species composition and abundance changed quickly. If the grassland was abandoned for a longer period, biodiversity decreased; some species, including several esteemed fodder plants, disappeared.[407] For instance, in many regions of the Alps, relatively productive non-fertilized species-rich meadows abounding in upright brome (*Bromus erectus*, see Sections 2 and 6.3.1) and growing at sites characterized by a high percentage of fine-grained soils were likely to turn into meadows abounding in tor-grass (*Brachypodium pinnatum*) as the first stage of ecological succession after the abandonment of mowing. Tor-grass had already occurred in these meadows earlier, but mowing had decreased its abundance. Tor-grass is a hard-leaved plant that livestock does not like unless it is young (but see Section 9). Similarly unfavorable from the farmers' perspective was the further stage of ecological succession in these relatively dry meadows characterized by a (relatively) high abundance of tall moor-grass (*Molinia arundiacea*) or purple moor-grass (*Molinia caerulea*), a species more suitable for animal bedding than for fodder (Sections 6.1.2.5.1.2 and 6.1.2.5.2). Another example of changing the species composition in abandoned meadows for worse from the perspective of reintroduced farming was extremely nutrient-poor matgrass swards providing low-quality livestock fodder (Section 6.3.1.2). There, the abundance of dwarf shrubs (such as heather, *Calluna vulgaris*; bilberry, *Vaccinium myrtillus*; and cowberry, *Vaccinium vitis-idaea*), making the fodder quality extremely low, increased significantly very soon after mowing was abandoned because these shrubs had already occurred there earlier but they had been kept short by the scythe.[408]

Another human challenge for restoration of several mowed grasslands after a moderate phase of abandoned mowing was the anthills of some grassland ant species (such as the yellow meadow ant, *Lasius flavus*, and to a lesser degree the pavement ant, *Tetramorium caespitum*). Unless removed during spring cleaning (Section 6.6.1), these anthills became tough structures, reinforced by grassland vegetation roots, and in abandoned grasslands the ants built hills that were larger than those in meadows mowed every year.[409]

Roughly similarly to meadows, the species compositions in abandoned or less frequently used pastures changed as well. In addition, an abundance of grazing resources in times of reduced grazing pressures allowed more selective livestock grazing behavior (consumption more limited to more palatable plant species); several species less attractive for livestock profited from such a

[407] Pils, 1994; Orland, 2016, 32.
[408] Pils, 1994, 223–224, 227, 230–232, 234–235, 240–241; all these species occur throughout the Alps or in the great majority of the Alps: Aeschimann et al., 2004, vol. 1, 618, 624, 626; Aeschimann et al., 2004, vol. 2, 916, 926, 994, 996; *Global Biodiversity Information Facility: Brachypodium pinnatum* (L.) P.Beauv.
[409] Compare Pils, 1994, 223, 229–230, 253, Tafel 74: Photo 179.

grazing regime. At the same time, an abundance of grassland resources must in some cases have coincided with less motivation among farmers and landowners to remove species that were unwanted from the viewpoint of animal husbandry (Section 6.6). Consequently, species composition changed quickly and biodiversity decreased.[410]

When grassland use was restored after ecological succession had significantly changed the species composition, even before afforestation, the farmers' aim was to reestablish meadows or pastures providing higher quality livestock fodder than the existing grassland, reed, tall-herb, or dwarf-shrub community. Usually, this also meant that changed grassland habitats again became richer in plant and animal species than in times when that parcel was not farmed. From the farmers' perspective, such recreation of a productive grassland was already a labor-intensive process before the succession reached the stage of bushes or young forest. Knowledge about methods historically applied to recreate a more productive farmed grassland is poor. Practices such as mowing, burning, and other manners of (spring-)cleaning much more labor-intensive than usual (Section 6.6) must have been applied. Grazing must have also played a role in this process. The time needed for approximate restoration of previous historical grassland fodder quality differed significantly among the various communities and stages of plant succession. The following example of purple moor-grass (*Molinia caerulea*) illustrates that, after a short phase of abandonment, introduction of the same management type could have diverse ecological impacts in different habitats; this might have been a serious challenge for management based on traditional ecological knowledge. Reintroduction of summer mowing, for instance, significantly decreased the abundance of purple moor-grass in nutrient-poor moist meadows. However, the abundance of purple moor-grass in abandoned moist meadows with higher nutrient availability increased in several sites in the case of summer mowing; namely, purple-moor grass is better adapted to summer mowing at such sites than tall herbs. Several tall herbs that spread in abandoned grasslands were suppressed relatively quickly by restored grassland management, such as summer mowing.[411] Focusing on a different example, false-hellebore (*Veratrum* spp., Figure 16) was more persistent. *Veratrum*, occurring Alps-wide in various colline, montane, subalpine, and alpine (calcareous) meadows and pastures that were neither dry nor extremely nutrient-poor, spread rapidly in calcareous grasslands if seeds were allowed to ripen in (late) summer.[412] False-hellebores are poisonous to

---

[410] Ellenberg & Leuschner, 2010, 33–34.

[411] Compare Austad & Hauge, 1989 (thanks to the environmental historian Verena Winiwarter for recommending this article); Pils, 1994.

[412] *Results-based nature conservation plan*, [2019], 12; author's interview with Kunavec, 2018, in Stara Fužina, and personal observations in the montane meadows of Zajamniki and Ši in Bohinj. The same is proven in the Eastern Carpathians: Babai, Molnár, & Molnár, 2014, 97–98. On site conditions and occurrence in the Alps, see Aeschimann

Figure 16: False-hellebore (*Veratrum* sp.) has spread massively in a continuously used pasture in the eastern Julian Alps (1,570 m) since it has no longer been removed by maintenance mowing on an annual basis due to the limited available workforce. Note the selective grazing behavior of cattle (almost only false-hellebore remained ungrazed). The pasture has been becoming increasingly unproductive. Photo: Žiga Zwitter, July 29th, 2022.

livestock and their stalks became very hard and difficult to mow by historical mowing times (Sections 6.1.2.1 and 6.1.2.2). Thus, from the farmers' perspective, abandoned meadows and pastures became increasingly unproductive with the spread of *Veratrum* (Section 6.6). False hellebores are perennial tall herbs[413] storing energy in rhizomes. Thus, after a relatively short phase of absence of grassland farming (or at least absence of maintenance mowing in grazed pastures and late-mowed meadows), spring cleaning in terms of maintenance mowing (see Section 6.6) had to be significantly more labor intensive

et al., 2004, vol. 2, 1024–1027; *Global Biodiversity Information Facility*: *Veratrum album* L., *Veratrum lobelianum* Bernh., and *Veratrum nigrum* L.

413 Aeschimann et al., 2004, vol. 2, 1024–1027. According to information from the second half of the eighteenth century, white false-hellebore (*Veratrum album*) blossomed in July in the pre-Alps and at moderate elevations of the mountains of Carniola (relevant for this volume are the Alps and Pre-Alps in what is now western, northern, and central Slovenia), and its seeds ripened there in September (Hacquet, 1779, 40; species identification by Praprotnik, 2015, 80). This, however, does not guarantee that cutting earlier than September prevented ripe seed dispersal because seeds can usually mature during hay drying, unless the meadow was cut too early.

if the abundance of *Veratrum* had increased significantly. In this case, higher intensity of maintenance mowing once annually had to persist for years.

Afforestation of abandoned grasslands was thus not the only type of ecological succession challenge that farmers faced when agricultural grassland use was restored. In some grassland plant communities, afforestation progressed extremely slowly (but see a different case in Section 9). An example is very dry and nutrient-poor sites, not located at forest edges and without trees or bushes in the grassland when it had been used. Significantly more than half a century since the end of agricultural use was needed for shrubs and trees to colonize such a parcel, but the grassland species compositions thoroughly changed much more quickly after the abandonment of farming than afforestation progressed there.[414]

Scrub and forest overgrew the great majority of abandoned meadows and pastures below the potential natural forest and tree line (see also Sections 2 and 9) unless grassland farming was reintroduced on time to prevent this process.[415] Numerous examples of ecological succession temporarily reaching this stage, especially in a limited grassland area, can be found in archival sources,[416] and natural archives provide highly relevant data: big-data paleoecology has recently made a great contribution to the knowledge of Black Death-related succession in Europe.[417] The following examples approach this question interdisciplinarily, linking written records, palynological analyses, and ecological interpretations at the micro-history level. Pollen analysis of Lake Bohinj by Maja Andrič et al. proved two late-medieval demographic crises leading to significant afforestation of grasslands (a decrease in the pollen of grasses, *Poaceae*, and sedges, *Cyperaceae*) and spread of forest, especially beech (*Fagus sylvatica*) and partly spruce (*Picea abies*) forest. Beech is not a pioneer plant and beech succession is a slow process, so that the beech pollen increase could only be a consequence of a longer abandonment of a significant percentage of grasslands. These stages of ecological succession, the second one dating to the fifteenth century, were followed by repeated phases of significant forest clearing. Palynological data suggest that the population of the nearby villages, such as Stara Fužina and Studor, recuperated after each of these serious phases

[414] Pils, 1994, 223–245; Bätzing, 2015, 265–268.
[415] Pils, 1994; Pickl, 1982, 28; Zwitter, 2013a, 87–88.
[416] Some examples of recorded overgrowth with bushes in mountain pastures include lower sections of the complex of pastures in the territory of Renon in the present-day region of Trentino Alto Adige (Germ. *Ritten* in South Tyrol) in the mid-seventeenth century. In sections of the Renon Alp (Ital. *Altopiano del Renon*, Germ. *Rittener Alm*), overgrowth with dwarf mountain pine (*Pinus mugo*) again strongly advanced before 1793. Reforestation of sections of the nearby Villandro Alp (Ital. *Alpe di Villandro*, Germ. *Villanderer Alm*) was also recorded in the mid-seventeenth century (Hinterwaldner, 2002, 49, 210–211).
[417] Izdebski et al., 2022.

of depopulation.[418] An interdisciplinary approach reveals complementary processes at higher elevations. Palynological analysis of Lake Ledvica (Sln. *Jezero v Ledvicah*) in the Triglav Lakes Valley (Sln. *Dolina Triglavskih jezer*) by Nina Caf et al. revealed a tremendous increase in pollen from grassland indicator species roughly at the end of the first millennium. This suggests large-scale forest and/or scrub clearing to create pastures. Archaeological finds from the Triglav Lakes Valley include charcoal from roughly the same time. For a short period after this clearing, the extent of grasslands in the area was most probably the largest in the last six millennia. Palynological analysis of Lake Ledvica also demonstrated that the percentage of non-arboreal pollen declined again after this large-scale grassland expansion. Nonetheless, pastures in the Triglav Lakes Valley in general remained significantly larger after this clearing than prior to it. This was the case until the ecological succession that followed the twentieth-century grazing abandonment in the Triglav Lakes Valley.[419] Despite these general trends, archival records make it clear that grazing pressures in summer pastures at higher elevations decreased, in a part of the Triglav Lakes Valley and east of it at least, during the late-medieval afforestation at lower elevations discussed above in this paragraph based on palynological results from Lake Bohinj. As revealed by written evidence, parts of this extensive grazing complex became overgrown with trees and/or with dwarf mountain pine (*Pinus mugo*).[420] The short-term late-medieval significant increase in total arboreal pollen in Lake Ledvica (e.g., pine, *Pinus* sp., and spruce, *Picea abies*)[421] probably corresponds to this phase of pasture abandonment.[422] This overgrowth

[418] Andrič et al., 2020, especially pp. 2, 8, 12–14; see also Izdebski et al., 2022. In this way, the paleoecological approach improves the understanding of the history of the villages of Stara Fužina and Studor. Information in preserved written records only allows the conclusions that Studor was first mentioned in preserved archival records in the late fifteenth century even though this village was probably founded significantly earlier, and that the likely foundation or significant growth of Stara Fužina dates back to the fifteenth or early sixteenth century. Gašper Oitzl has recently emphasized a lack of known records useful for these settlements' older history (Oitzl, 2021, 309–310), and so palynological results have significantly improved the state of knowledge. Plague is one of the highly likely reasons for the aforementioned phases of late medieval depopulation (Andrič et al., 2020, especially pp. 8, 12; Izdebski et al., 2022; Koropec, 1972, 89, 93–95, 97, 99, 101–102). If a large undated rockfall in fact buried much of the village of Studor (hypothetically, for instance, during the 1348 earthquake: see Rohr, 2007a, 131–158, and Merchel et al., 2014), as folklore has it (Komac & Zorn, 2007, 71–72), this natural disaster is yet another likely reason for one of the late-medieval grassland afforestation phases.

[419] Caf et al., 2025.

[420] Kaspret (ed.), 1889b, 116; Zwitter, 2013a, 91–92.

[421] Caf et al., 2025.

[422] Differences in chronological trends of beech pollen percentages in Lake Bohinj and Lake Ledvica (the beech pollen percentage was decreasing in the latter during this late medieval phase of forest and shrub regrowth) are most likely a consequence of several factors. One of them might have been the local occurrence of beech on the slopes above

was later cleared again: in (much of) the western part of the area in the second half of the fifteenth century, and in the east by the mid-sixteenth century.[423]

At that time, extensive clearing at high elevations of the Triglav Lakes Valley mainly resulted in a larger grazing area,[424] but to a certain degree also in a long-term increase in bare rocks, which are characteristic for parts of the sensitive mountain karst environment in the lower and upper part of the Triglav Lakes Valley, although this was neither the first (mountain pastures had existed there before) nor the last clearing of forest there. Almost complete clear-cutting followed in 1883.[425] The botanist Tone Wraber hypothesized that more dwarf mountain pine had probably grown in the upper part of the Triglav Lakes Valley, in the vicinity of Lake Ledvica, and that it seems that it was cleared there for grazing. Consequently, one finds rocky karst features instead of soil there. Wraber also stated that this environment was evidently severely devastated by people.[426] Namely, the depth of soil on limestone changes over very short distances in many cases, and so historical clearings on patchworks of shallow and deep soil could easily lead to long-term ecological side-effects in terms of creation of bare rock where dwarf mountain pine had grown. This species grows on both shallow and deep soils on limestone. Whereas areas of dwarf mountain pine on deep soil sections can be converted into pastures, the ecosystem functions of shrubs on shallow, stony soils, such as soil fixation through roots, are crucial. The bushes' shade also plays an important role in lessening heat extremes and mitigating shallow-soil drying. Today it can still

---

Lake Bohinj and the probable absence of beech forests in (higher parts of) the Triglav Lakes Valley; consequently, in times of forest regrowth in the Triglav Lakes Valley, the pollen percentage of local woody genera and species, such as pine and Norway spruce (*Picea abies*), increased. A likely stronger influx of beech pollen from the Soča River catchment (where beech forest reaches the upper forest line) into the Triglav Lakes Valley than into Lake Bohinj might also have been relevant. Comparison of two pollen diagrams both based on samples from the Triglav Lakes Valley—a recent one by Caf et al. (2025: Lake Ledvica) and Alojz Šercelj's pollen diagram (1961: a moor close to the alpine lodge in the Triglav Lakes Valley)—seems to suggest significant differences in beech pollen sediments even within the Triglav Lakes Valley. Šercelj's results on his early pollen diagram from 1960s are not dated, his speculative dating attempt of those palynological results has never been confirmed; the recent summary of Šercelj's findings by Zorn et al., 2015, 77, is equally uncertain. Similar trends in pine and Norway spruce pollen in the diagram by Šercelj (1961) and Caf et al. (2025) seem to challenge Šercelj's dating. They suggest that Šercelj's (1961) diagram starts between the twentieth and thirtieth centimeter of the diagram by Caf et al. (2025). If this tentative assumption is correct, the beech pollen on Šercelj's diagram remained constant during the late medieval phase of afforestation discussed above, and so changes in beech pollen percentage trends in different sections of the Triglav Lakes Valley (and not only between Lake Bohinj and the Triglav Lakes Valley) differed at that time.

423 Kaspret (ed.), 1889a, 128–129; Kaspret (ed.), 1889b, 116; Zwitter, 2013a, 92.

424 Caf et al., 2025; Zorn et al., 2015, 77–78, 84, 89–92.

425 Zorn et al., 2015, 78, 89, 93–94.

426 Wraber, 1974, 407.

be a challenge to decide which areas can be cleared of *Pinus mugo* to restore pastures. Information on successful historical land use currently provides an important background for making the right decisions about where it makes sense to clear dwarf mountain pine to extend productive pastures.[427] Consequently, overly extensive clearing of dwarf mountain pine at several locations in the history of the Alps can easily be understood, especially due to the fact that it took time before ecological succession at high elevations had proceeded far enough for the mountain pasture to be overgrown by dwarf mountain pine, and so it was difficult for people to recall exactly where pastures had previously been located. Moreover, it is likely that the clearing in the Triglav Lakes Valley in the fifteenth century was carried out by people that had no experience in former land use there because they came from the Upper Soča catchment (the territory of Tolmin in the Republic of Venice), and not from the nearby Bohinj Basin.[428] Consequently, that clearing of dwarf mountain pine in a sensitive karst environment was likely excessive.

Another historical example of the relatively widespread process of dwarf mountain pine clearance at inappropriate sites on carbonate rocks, leading to soil erosion, deals with the Belščica grazing complex in the Karawanks. Soil erosion and talus formation triggered by nineteenth-century dwarf mountain pine clearance (burning, at least) was clearly recorded on shallow soils on steep slopes of limestone and dolomite in some parts of Belščica, in times of extremely high agricultural density in the second half of the nineteenth century (Section 3.1).[429] As soon as shortly after the First World War, similar pasture expansion would already be unimaginable in this case; the number of sheep there decreased from over five hundred before that war to around two hundred soon after the war, and by the 1940s grazing in most of this large mountain pasture complex was fully abandoned (see Section 6.3.1.1 on reasons to abandon sheep rearing).[430]

---

427 Wopfner, 1997, 401; Machatschek & Kurz, 2006, 163–164, 176–181.

428 Kaspret (ed.), 1889a, 128–129; Kaspret (ed.), 1889b, 116.

429 SI AS 48, Deželna komisija za odkup in uravnavo zemljiških bremen za Kranjsko (Provincial Commission for Redemption and Regulation of *Profits à Prendre* in Carniola), fasc. 56, 9–5: Tagebuch über den Verhandlungsact in Sachen der von der Gemeinde Bach angemeldeten Ablösung oder Reguli[!]rung des Weiderechtes an der Alpe Beušca der Berechtigten von Hrašach, Studenčič, Rann, Velben et Hlebic: Erhebungs- respective Vermarkungs-Protokoll, June 18th, 1881, 3 & Schlussverhandlungsprotokoll, July 5th, 1886 & An die löbliche k.k. Bezirkshauptmannschaft als Grundlasten-Ablösungs- und Regulierungs Local-Commission in Radmannsdorf, April 27th, 1887 & Protokoll, September 2nd, 1887; and the map kept in this archival bundle; *Environmental Atlas of Slovenia*: Spatial Units, Land Cadaster and Land and Soil, Soil Map (1:25,000).

430 Twentieth-century data: Jordan, 1945, 52, 78, 92, 96, 102. On twenty-first-century sheep grazing in at least part of Belščica, involving sheep from the village of Javorniški Rovt, located much closer to Belščica than the villages with grazing rights there before the mid-twentieth century, see Klinar, 2012, 109, 111.

When agricultural density was moderate, temporary pasture abandonment was an optional strategy to increase the soil nutrient availability, where green alder (*Alnus alnobetula*) overgrowth was highly likely. Namely, a longer phase of green alder overgrowth followed by grassland pasture restoration had particularly favorable impacts on overused, degraded grassland sites where only low-quality plant species, from the viewpoint of animal husbandry, could abound, such as matgrass (*Nardus stricta*), tufted hair-grass (*Deschampsia cespitosa*), and/or dwarf shrubs. In the case of green alder succession, not only plant nutrients from decomposed biomass accumulated in the soil, as in usual ecological succession patterns, but symbiotic bacteria in alder root nodules also bound air-borne nitrogen, one of the primary plant macronutrients, and made it plant-available in the soil. The practice of improving degraded pastures this way is proven in the Eastern Alps in the twentieth century, at least, but it is likely older. Namely, the soil-fertility-raising impact of alders might have been known to local inhabitants in earlier centuries because alders were favored in some agroecosystems in various parts of preindustrial Europe.[431] Green alder (*Alnus alnobetula*) overgrowth was thus not necessarily an indicator of decreased interest in use of a permanent grassland. This was not the only way of increasing soil nutrient availability through an extensive grassland land-use type. A widespread practice was less frequent mowing intervals in meadows (see Section 6.1.2.6).

[431] Machatschek, 2002, 436–437; Machatschek & Kurz, 2006, 166–175, 184–190; Schneiter, 1970, 33–39; Emanuelsson, 2009, 119, 189, 192, 206.

4 Natural and nature-induced environmental changes in grasslands

Species compositions in seminatural grasslands have not been constant over time—natural developments, changes in management, and their multiple interactions and coincidence have all contributed to these dynamics.[432] "[T]he multiplicity of ever-changing factors that influence outcomes, producing systems that are unpredictably complex," which makes changes impossible to anticipate in detail, is called "chaos" in ecological theory. However, in ecology, "chaos is not to be confused with randomness"; there are "certain rules and order" in the dynamics of ecosystems, "context influences the direction of change."[433] For instance, there exists a dormant seed bank in grassland soils; it depends largely on coincidence, which of the species from the seedbank will be "the first to sprout and get established at that site, for the following vegetation period or until the next disturbance at least."[434] Still, there are some rules and order—this will be a species (rather) adapted to the site, favored by the weather of that spring, and able to interact (compete with, and possibly also synergistically benefit from) the others present. The quantity and longevity of seeds in the seedbank also differ significantly between species and (micro-)sites. Thus, species-rich grassland ecosystems have always been highly dynamic over a short (e.g., annual) timescale. Moreover, they have seen certain dynamics over longer periods.[435] In some cases, changes in grassland environments have significantly altered possible grassland uses (see also Sections 1.2, 2, 3.2, 6, 7, 8, 9, and 10).

This section focuses on natural and nature-induced changes in grassland ecosystems. A selection of manmade and human-induced changes in grassland environments in terms of grassland management is given in Sections 1.2, 2, 6, 7, 8, and 10.

Section 4 deals with the historical impacts of weather and climate fluctuations, changing landforms in grassland environments, and a selection of economically relevant ecological disturbances in grassland agroecosystems posed by insects, rodents, and fungi. Some further factors, such as interactions be-

[432] Kapfer, 2010; Peterken, 2013.
[433] Isenberg, 2017, quotes from pp. 136 and 137.
[434] Pils, 1994, 56.
[435] Adapted from Peterken, 2013, 221–247, and Stampfli, 1995.

tween organisms in the local environment, and feedback loops linking various factors have also co-influenced the temporal dynamics of species abundance and occurrence in grasslands, apart from coincidence. From a long-term perspective, environmental dynamics have raised grassland species diversity in general, especially if processes in the landscape allowed easy transport of germinable diaspores of species that can become naturalized there (e.g., Sections 6.5 and 7). Namely, a range of environmental conditions in the same grassland has favored different species in different years, which has decreased the likelihood of full competitive exclusion of a species in the long run.[436]

4.1 Continuously existing grasslands in naturally dynamic environmental conditions

4.1.1 Impacts of weather and climate variations in grasslands

4.1.1.1 Impacts of short-term weather variability

Weather dynamics have had important impacts on plant development, influencing species competition, and consequently they have co-shaped the species composition in terms of occurrence and abundance, and yield quantities. Species richness partly mitigates the negative impact of temperatures and soil moisture unfavorable for some economically highly relevant meadow plants on hay quantity because such conditions are not equally unfavorable for all of the species covering a significant percentage of the meadow. This decreases the variations in yield quantities in comparison with recent intensively managed meadows. Nonetheless, some basic weather situations positively or negatively affect the great majority of grassland plants in species-rich meadows in certain parts of the year, such as dry weather leading to an agricultural drought in May as the month when grassland plant growth is usually quickest.[437] Agricultural drought in a sensitive phenophase (in May, for instance) significantly reduces hay quantities as the growing plants become thinner and shorter. Partly similar are the predominant impacts of an agricultural drought in summer, during second-crop hay growth (Section 6.1.2.3).[438] A dry summer results in significantly less second-crop hay (less than one-third of the amount of first-crop hay from the same meadow), even in the wettest parts of the Alps,[439] and no second-crop hay at drier sites.[440] The impacts of an agricultural drought are rel-

[436] Peterken, 2013, 221–247; Stampfli, 1995.
[437] Pfister, 1985a, 39–40, 47, 51, 62; Stampfli, 1995; Ellenberg & Leuscher, 2010, 916–927, 1011–1012, 1014, 1016–1017; Peterken, 2013, 223–247; Verbič & Verbič, 2014–2020, 19.
[438] Stampfli, 1995; Ellenberg & Leuscher, 2010, 923–925; Babai, Molnár, & Molnár, 2014, 109; Zwitter, 2015a, 470.
[439] Author's interview with Kunavec, 2023, in Stara Fužina.
[440] Compare Iuga, 2016, 72.

evant in all vegetation belts—even though morning dew is more abundant at higher elevations, in some cases summer drought also significantly decreases grazing resources in summer mountain pastures and hay yields in high meadows.[441] In such cases, farmers would make more tree-leaf fodder than usual to partly compensate for the little grassland hay available; in addition to leaves from usual fodder trees, lower-quality tree-leaf fodder was also used (Section 8.2.2.1).[442] Apart from such basic general facts, the impacts of weather depend on site conditions, on individual species, on interactions between organisms, and also on plants' development stage.[443]

Agricultural drought in spring significantly suppresses the growth of several productive grasses, which are good fodder plants and prefer mesic sites (Section 6.3.2). This is not a general rule, however; false oat-grass (*Arrhenatherum elatius*), a highly relevant fodder grass in moderately manured meadows, is remarkably drought-resistant: due to its long roots, which can reach up to 2.8 m deep, it suffers no visible damage even when the foliage of several small herbs dries out in some meadow sections. In contrast to the majority of productive grasses, the abundance of xerophytes, plants able to recover soon after they have been harmed by drought, and grass and forb species taking advantage of gaps in meadow vegetation created by a drought or dry freeze (Sections 6.3.1.1 and 6.5) increases as a consequence of drought events. Several relatively drought-tolerant grassland plants are in fact affected by agricultural droughts during the vegetation period such that parts of their fine roots die; however, their root systems regenerate soon after the drought ends, and so these species regenerate quickly and their abundance increases.[444]

Two conspicuous examples of species occurring Alps-wide and increasing their abundance in grassland ecosystems during drought events are meadow clary (*Salvia pratensis*, Sections 2, 6.3.2, and 10.1.2.2.3) and yarrow (*Achillea millefolium*). Spring drought followed by a usually damp late spring (or summer) leads to an increased percentage of meadow clary in hay from moderately manured meadows at sites that are relatively dry under normal conditions, on neutral to alkaline soils in colline and montane vegetation belts. Namely, relatively dry conditions are optimal for this herb, and it also tolerates dry sites. Drought does significantly negatively affect meadow clary, but it recuperates soon when soil moisture increases. Decreased yields of productive grasses are nonetheless not largely compensated by the increased abundance of meadow clary, and so hay yields remain lower as a consequence of drought. In addition, the nutritive

441 Fajgelj, 1953, 130.
442 Machatschek, 2002, 18, 34, 48–49, 59, 73–75, 112, 252, 370, 411; Hartel, Craioveanu, & Réti, 2016, 58–59, 61–62.
443 Pfister, 1985a, 34; Ellenberg & Leuscher, 2010, 916–927, 1011–1012, 1014, 1016–1017.
444 Pils, 1994, 167, 179; Stampfli, 1995, 122, 124–129; Ellenberg & Leuschner, 2010, 927; Peterken, 2013, 223; Babai, Molnár, & Molnár, 2014, 38, 40; Aeschimann et al., 2004, vol. 2, 950–951, 956–957.

value of meadow clary is low, but this herb is relevant for the health of livestock (Section 10.1.2.2.3). When an agricultural drought in spring continues too long (well into the summer, for instance), a significantly increased abundance of meadow clary the next year is likely. Drought in moderately manured grasslands on relatively dry sites at all meadow elevations has often also led to an increased abundance of yarrow, especially in meadows manured more heavily than suitable for meadow clary. Namely, yarrow is able to quickly spread vegetatively under dry conditions, with its robust shoots.[445] Similarly to meadow clary, yarrow is sensitive to agricultural drought in spring, and it recuperates soon after the drought. This also means that a long agricultural drought in spring and summer leads to a delayed increased abundance of yarrow the next year.[446] Yarrow can be excellently stored as hay, its nutritive value is relatively good (only slightly worse than yellow oat-grass, *Trisetum flavescens*, for instance), and yarrow yield quantities are also similar to productive grasses of moderately manured meadows,[447] and so this type of drought-posed change is not a serious challenge to the farm economy.

Winter cold without snow cover can also kill some grassland plants; many others can resist it well, and the abundance of some species increases significantly after such events. The abundance of tufted hair-grass (*Deschampsia cespitosa*) in moist meadows at relatively nutrient-rich sites, for instance, is likely to decrease; the abundance of foxtail (*Alopecurus*), creeping buttercup (*Ranunculus repens*), and some other species is likely to increase as a consequence of cold winters without snow cover.[448]

Yellow rattle (*Rhinanthus* sp., Figure 17), a plant genus that can abound in moderately fertilized or non-fertilized meadows on base-rich to acidic, mesic sites Alps-wide,[449] is one of the species profiting from gaps that come into existence in exceptionally cold springs.[450] It used to be a common meadow species (Section 6.1.1.3). Namely, it is an annual plant that lacks dormant seeds, but its recent seeds can become established in gaps in a meadow. Yellow rattle is a hemiparasite that "parasitises grasses and some broadleaved herbs strongly enough to suppress their growth by about half," thus strongly reducing their competitive abilities. Nonetheless, a model covering five hundred years showed

---

445 Stampfli, 1995, 122–126; Dietl & Lehmann, 2004, 38, 42–43; Ellenberg & Leuschner, 2010, 920; Heinz, Mayer, & Kuhn, 2012, 19; Peterken, 2013, 225; Babai, Molnár, & Molnár, 2014, 38, 40, 42; Aeschimann et al., 2004, vol. 2, 164–165, 490–491.

446 Stampfli, 1995, 122, 124–126.

447 Dietl & Lehmann, 2004, 39, 42–43.

448 Ellenberg & Leuschner, 2010, 1017; Aeschimann et al., 2004, vol. 1, 160–161; Aeschimann et al., 2004, vol. 2, 922–923, 958–959, 976–983. See also the following text on yellow rattle.

449 Aeschimann et al., 2004, vol. 2, 272–279; Pils 1994, 46; see, e.g., the photo of a grassland rich in yellow rattle in Bavcon, Ravnjak, & Praprotnik, 2019, 102–103.

450 Meilleur, 1986, 304.

that "meadows with yellow-rattle would oscillate between grass-rich and forb-rich phases. Within a single meadow, different patches would be at different points in the oscillation . . . . Actual experience with greater yellow-rattle [*Rhinanthus alectorolophus*] over three decades showed wild fluctuations, owing mainly to the plant's vulnerability to spring drought." Under favorable conditions, such as following exceptional cold in springs that created gaps in meadows, "yellow-rattle can expand very quickly after just a few seeds are scattered in a meadow; but populations crash just as quickly" if seeds cannot form or ripen in a single year.[451] A successful historical strategy of decreasing its abundance was grazing in meadows in May (Section 6.2.2.2). Farmers noted the connection between the abundance of yellow rattle and weather: according to traditional ecological knowledge proven in the Allues Valley (Francoprovençal *Alués*, French *Les Allues*) in the northern French Alps in the first half of the twentieth century, the abundance of greater yellow rattle (*Rhinanthus alectorolophus*) was the highest in years beginning with an exceptionally cold spring,[452] and traditional ecological knowledge about yellow rattle (*Rhinanthus* sp.) in Bohinj includes the same well-founded observation.[453] The huge distance and cultural diversity of Bohinj and the Allues Valley, where such knowledge must have evolved independently, makes it clear that this type of knowledge was widely present in the Alps, and farmers most likely learned this connection long ago. A parallel from the Eastern Carpathians suggests that gaps created through a serious drought at the right time (not in spring; see above) could equally lead to an increased abundance of yellow rattle (*Rhinanthus minor*), as traditional ecological knowledge by traditional Csángó farmers has it.[454] A sudden significant increase in the abundance of yellow rattle (*Rhinanthus* sp.) in meadows triggered by the impacts of weather dynamics was a great misfortune for farmers because the quantity and quality of hay including much yellow rattle are low.[455] Due to its hemiparasitic life strategy, yellow rattle significantly lowers the yields of grasses and some other herbs in the meadows it dominates.[456] Yellow rattle itself produces little biomass,[457] and its small leaves fall off and decompose very quickly, so that much of the plant is already lost or falls away when hay is mowed, shaken out, turned, and stored.[458] In most cases, the empty seed capsules are the only remnants of yellow rattle in the hay. These pieces of information are reflected in an old German proverb from the Eastern Alps: "Either the yellow rattle or the farmer goes away"; that is, either the abundance of yellow rattle will be reduced, or a farm family will lose its

[451] Peterken, 2013, 226–227 (quote); *Enoletnice – Index seminum 2018*.
[452] Meilleur, 1986, 304.
[453] Author's interview with Kunavec, 2021, in Stara Fužina.
[454] Babai, Molnár, & Molnár, 2014, 46.
[455] Meilleur, 1986, 304; Babai, Molnár, & Molnár, 2014, 46; Machatschek, 2017, 19.
[456] Peterken, 2013, 226; *Enoletnice – Index seminum 2018*.
[457] Aeschimann et al., 2004, vol. 2, 272–279.
[458] Author's interview with Kunavec, 2021, in Stara Fužina.

Figure 17: Yellow rattle (*Rhinanthus* sp., the low plant with two-color inflorescence; discernible are yellow blossoms and pale green leaves within the inflorescence) abounding in a moderately fertilized meadow in Bohinj, at the western foot of Rudnica Hill. Photo: Žiga Zwitter, May 30th, 2022.

subsistence possibilities[459] (but see a viable suppression strategy in Section 10.1.2.2.3). According to traditional ecological knowledge from East Tyrol, a high abundance of yellow rattle in hay is believed to cause health problems for cattle;[460] this is correct, but it is not an issue in the hay[461] because hardly any yellow rattle is left in the hay. Farmers could exert a certain degree of influence on the species colonizing gaps created by dry freezes, for example, if they sowed barn sweepings there (see also Section 6.5). This way, species abounding in the meadows of the farm were likely to also become established in meadow gaps if site conditions were appropriate.[462] Nonetheless, quantities of barn sweepings only allowed for their application to a limited meadow area (see also Section 6.5), and the hemiparasitism of yellow rattle also significantly influenced the species composition in many species-rich meadows at appropriate site conditions.

459 Machatschek, 2017, 19, 370.
460 Vogl-Lukasser et al., 2006, 70–72, 91.
461 *Notranjski regijski park:* Mali škrobotec.
462 Peterken, 2013, 232; Babai, Molnár, & Molnár, 2014, 92–94.

Not only dry freezes and drought, but heavily increased precipitation during the vegetation period also impact plant development in a manner that is relevant for livestock farming. In the relatively wet climate of the Solčava region in what is now northern Slovenia (about 1,600 to 2,000 mm mean annual precipitation from 1971 to 2000),[463] there is the folk saying: "In rainy years, when forage is abundant, livestock farming is not something to look forward to. Long grass . . . is leached, and so it is consumed in large quantities because it is not as calorific as short fodder in dry years."[464] In grassland farming in general, wet weather causes the most damage at the time hay is harvested (Section 4.1.1.3). Moreover, a rainy and cool May (and early June) leads to lower hay protein content,[465] which is likely also reflected in the folk saying above. However, although the saying correctly mentions generally high hay quantities in wet springs, this was not the case in all meadow agroecosystems. In sites prone to becoming waterlogged, hay quantity declines in such weather. Namely, focusing on grasses, many of them tolerate a high groundwater level without noticeable damage. Some agriculturally relevant grasses, however, are highly sensitive to waterlogged conditions in the soil. The lack of oxygen causes damage to the fine roots. Yields of grasses such as red fescue (*Festuca rubra*) in non-fertilized or very moderately fertilized meadows, false oat-grass (*Arrhenatherum elatius*) in moderately fertilized meadows, and cock's-foot (*Dactylis glomerata*) in medium-intensive or moderately fertilized meadows decline significantly when the groundwater level lies less than half a meter below the surface.[466] Another type of impact of weather on grassland agroecosystems is the connection between wet weather in the cold part of the year and pathogenic fungi; see Section 4.1.3.3.

At avalanche-prone sites, avalanches have importantly contributed to the pronounced dynamics of many grasslands in the Alps. The frequency of avalanches has fluctuated considerably across space and time,[467] and so their impacts on grasslands over time have been changing accordingly. In many cases, the affected areas, the dimensions of (avalanche) events, and changes in grassland ecosystems triggered by avalanches have been strongly influenced by past and contemporaneous human economic activities in the landscape; for example, by the clearing of dense forest on slopes.[468] In general, the avalanche hazard used to be lower in grazed or mowed grasslands than it is in many abandoned grasslands on slopes, which are now susceptible to avalanches.

---

[463] *Environmental Atlas of Slovenia*, Climate, Precipitation, Mean annual corrected precipitation 1971–2000.
[464] Vršnik, 2005, 170–171.
[465] Pfister, 1985a, 62.
[466] Ellenberg & Leuschner, 2010, 1014; Aeschimann et al., 2004, vol. 2, 854–855, 892–893, 950–951; Dietl & Lehmann, 2004, 20, 25, 124–125.
[467] Pavšek, 2002; Rohr, 2007a, 417; Komposch, Frieß, & Kreiner, 2013; Zwitter, 2024.
[468] Pavšek, 2002.

Namely, the decomposing leaf remains of grassland monocots lie flat in the autumn and their surface is often frozen, forming a slippery and fragile substrate for a snow blanket on slopes. In contrast, mowed or grazed grassland plants do not lie flat; they fix the lowest part of the snow blanket,[469] which requires a more favorable complex set of other factors with an influence on avalanche hazard for an avalanche to happen. To give just one of the many examples of former highly avalanche-prone meadows, in 1605 the community of Haggen in the Sellrain Valley (Germ. *Sellraintal*), Tyrol, east of the Ötz Valley (Germ. *Ötztal*), received the right to mow (sections of) avalanche tracks in the forest at the Zirmbach Alp (*Zirmbachalm*). The territory of this complex of summer pastures and meadows crosses the upper forest line, and so several large avalanches must have started at higher elevations. The document was valid at least until the first half of the eighteenth century.[470] The role of avalanches in the history of grasslands in the Alps has been diverse (see also Section 4.2). In certain places, the fact that avalanches keep land open has locally retained narrow patches of open grasslands hundreds of meters below the climatic forest line.[471] Avalanches have influenced soil properties, have transported propagules from higher elevations down the slope, and have provided regular or occasional habitat disturbance.[472] In lower parts of avalanche tracks, nutrients from higher parts of the track are deposited,[473] which has been favorable for fodder quantities in lower avalanche track sections, especially in non-fertilized grasslands (Sections 6.3.1 and 6.3.2). In addition to nutrient availability, late melt of avalanche-deposited snow makes such nutrient-rich sites damp, and so in early and high summer they often abound in high-quality fodder.[474] Farmers have noted significantly higher hay quantities (and at some sites also higher fodder quality) in the lower parts of avalanche tracks.[475] However, the same site conditions (high moisture and nutrient supply) at the end of avalanche tracks are also suitable for various poisonous plants, which reduce the fodder quantity or require agricultural grassland cleaning (Section 6.6). Moreover, not only soil but also coarser material is deposited there by avalanches, which called for labor-intensive spring cleaning in such grasslands (Section 6.6). Unique vegetation, characterized by a high diversity of plant species that otherwise belong to different elevation zones and vegetation types, develops in lower sec-

[469] Pils, 1994, 246; Pavšek, 2002, especially 76, but also 75–76, 78, 81, 110, 144, 171–172, 178, 185, 188–189; Keller, 2004, 47; Groß, 2015, 192.

[470] Grass, 1948, 87; *Arcanum Maps*, Cadastral Maps, Habsburg Empire – Cadastral Maps (XIX. [sic] Century), cadastral municipality of Stams, 1856, and *Google Maps*, Layers, Satellite.

[471] Pavšek, 2002, 73, 77, 81, 89, 91, 117, 186; Bohner, 2008, 83, 86; Bohner et al., 2010, 53.

[472] Bohner, 2008, 83, 86; Bohner et al., 2010, 55–56; Komposch, Frieß, & Kreiner, 2013, 389, 391, 395–395.

[473] For example, Hejcman et al., 2013, 351.

[474] Wopfner, 1997, 263, 387–388.

[475] Keller, 2004, 66.

tions of long avalanche tracks.[476] Namely, some plant species from grasslands above the forest line can grow at lower elevations, but they are almost unable to reproduce there, and so a frequent influx of germinable diaspores through avalanches is needed to maintain or recreate such unique grasslands. A few other species from the alpine and subalpine vegetation belt can reproduce well at lower sites but cannot compete with other species in those ecosystems without the disturbance caused by avalanches. In contrast, several species from the alpine belt cannot grow below the forest line, not even in avalanche tracks.[477]

4.1.1.1.1 Tithe in beehives as an indicator of impacts of annual weather variations on insect-pollinated herbs and bees

The annual impacts of weather fluctuations on several insect-pollinated herbs and grassland-pollinating insects are partly reflected in registers of tithes in beehives, which partly reflects the dynamic history of insect-pollinated grassland flowers. Namely, "[g]rasses and sedges are mainly wind pollinated and of little importance for pollinating insects,"[478] and so they are not represented in the data on beehives.

Beekeeping used to be a relevant economic activity for several farms in many parts of the Alps at least since the Middle Ages, and most probably since much earlier.[479] In the first half of the second millennium, the area of grassland offering bee forage in the Alps expanded significantly (Section 3.1), which must have positively influenced apiculture; moreover, it might have increased the role of grassland pollen and nectar in the diet of farmed bee populations in this mountain chain. Later, numbers of beehives increased significantly during the modern period, as demonstrated by data from the Slovenian Alps,[480] but there were ups and downs in apiculture over time.[481]

Before presenting contextualized fluctuations in tithe in beehives, it is important to emphasize that grasslands used to be a more continuous source of pollen and nectar than they are now, when first mowing in the majority of meadows takes place sooner, it is completed more quickly, and many meadows are mowed more frequently than they used to be. Namely, mowing that was late but characterized by considerably different mowing dates in various meadows or in various meadow complexes belonging to the same settlement, between June and August/September (see Section 6.1.2), contributed to the

[476] Bohner, 2008, 83–84, 86; Bohner et al., 2010, 55–56; Komposch, Frieß, & Kreiner, 2013, 389–395.
[477] Ellenberg & Leuschner, 2010, 672–673.
[478] Johansen et al., 2019, 4.
[479] For an overview of history of beekeeping, see Crane, 1999; Mihelič, 1970, 400, 403, 407; Niederer, 1996d, 188.
[480] Mihelič, 1970, 400–401, 403, 407.
[481] Crane, 1999.

structural heterogeneity of habitats. Such mowing timing resulted in different blooming dates as a consequence of differences in species compositions at different meadow sites and of plant adaptations to the usual mowing time (Section 6.1.2). Investigations of grasslands in the village of Botiza in the Romanian Carpathians with mowing dates of first-crop hay similarly late (ranging from late June to later than mid-August) as historical mowing in the Alps (Section 6.1.2) and preceded by early spring sheep grazing (which used to be characteristic for the Alps as well; see Section 6.2) help us understand this aspect of grassland history in the Alps. The results from Botiza explain that mowing later than mid-August "is important for providing flower resources in June and July," "whereas meadows mowed [in late June] have considerably more flower resources [in second-crop hay plants] from around 1 August onwards"; see Section 6.1.2.3. Namely, "[b]y early August, meadows that were mowed around 25 June had three times as many flowers per area unit as the meadows that were not yet mowed." "By 1st of August about 87% of the reproductive units [of meadow flowers; grasses and sedges were not analyzed] were in flower or bud stage in . . . meadows [mowed in late June], compared to ca. 18% in late-mowed meadows [characterized by "the highest proportion of [flower] species in the fruit stage" in early August], indicating a vigorous re-flowering after . . . mowing [that took place in late June]." Consequently, late but significantly diverse mowing dates "enhance[d] flowering throughout the summer, thus providing more continuous flower resources for pollinators," such as bees and butterflies.[482]

Bees are extremely sensitive to weather fluctuations with an influence on (grassland flowers' and other) pollen and nectar resources. Consequently, weather is one of the main reasons for strong annual variations in numbers of hives, as well as in honey and wax yields.[483] Impacts of weather in spring and early summer on quantities of honey produced are indirect: at that time "rather little nectar may be converted into honey, because most of it is consumed by the many young non-foraging bees in the colony, and by the whole colony during spells of inclement weather when nectar cannot be collected." The impacts of weather on honey production become direct from early summer or mid-summer onward, when "the rate of brood rearing and the number of young bees decrease, and nectar collected that is surplus to the colony's immediate requirements is made into honey and stored . . . . The supply of nectar is not constant even during the summer. It is available in large quantities only during a 'honey flow,' when a great number of individual nectar-yielding flowers are open at the same time. In the course of a summer any one area may have one, two or three major honey flows; each lasts for a few days or up to a few weeks,

[482] Johansen et al., 2019, 2–9 (quotes from pp. 7 and 8).
[483] Mihelič, 1970, 396, 403; Crane, 1999, 22–23, 69, 252–253; Mlaker-Šumenjak, 2011, 27, 36, 38–41, 45–46, 49–50, 68.

during which most of a colony's honey is made and stored."[484] As explained in the previous paragraph, there were most likely more honey flows relevant for beekeeping as long as landscapes were full of species-rich meadows mowed relatively late and very late, with significant differences in mowing timing. To make these differently timed honey flows accessible to bees, farmers had to transport them to optimal foraging grounds;[485] namely, bees only fly about two or three kilometers or somewhat further.[486] According to seventeenth-century information from the Southeastern Alps, farmers usually kept beehives indoors (in the attic) during the winter. This practice strongly reduced the negative impact of cold winters and fluctuating winter temperatures on farmed bee populations,[487] thus strengthening the connection between pollen and nectar resources and numbers of beehives.

The impacts of the dynamic history of weather and climate on apiculture have been studied in southern Sweden. In the late sixteenth century, a partly similar crisis in apiculture in several parts of the Alps can be assumed as found there: "In the late 1500s—and especially after 1570—the climate deteriorated, and there were fewer swarms in summer [meaning fewer new beehives] and greater winter losses of colonies in both hives and natural nests."[488] Here, a few data are presented on the impacts of weather and climate on apiculture in the Alps just before the next major peak of the Little Ice Age in the late seventeenth century.[489] They refer to the Alps of the Upper Savinja Valley in Lower Styria (Germ. *Untersteiermark*; Sln. *Spodnja Štajerska*)[490]—more precisely, to the administrative units (Germ. *Amt*) of Ljubenska Gora (most of the slopes north and south of Ljubno ob Savinji between Smrekovec and Arničev Vrh in Figure 4), Luče (the slopes northeast and southwest of the village of Luče between the Volovljek Pass and Mt. Raduha in Figure 4), and Tirosek (most of the pre-Alpine Dreta River catchment in Figure 4),[491] all in the episcopal

484 Crane, 1999, 22 (both quotes), 23, 239.

485 History of movements of beehives: Mihelič, 1970, 401, 408, and page 50 of illustrations: images 2 and 3; Niederer, 1996d, 188; Wopfner, 1997, 375; Crane, 1999, 244, 347–352; Mlaker-Šumenjak, 2011, 43.

486 Crane, 1999, 313; Mlaker-Šumenjak, 2011, 39.

487 Zupan, 1934, 96; Mihelič, 1970, 401; on general statements on negative impacts of severe winters on bees in European history and an apicultural strategy for coping with severe winters in Northern Europe, see Crane, 1999, 24, 227, 235–236, and Zupan, 1934, 96.

488 Crane, 1999, 69–70, 235 (quote), 239.

489 See the climate history chronology in Pfister et al., 2018, 273–274, 289–290.

490 For general climate history trends in the last five hundred years, see Dobrovolný et al., 2010; local data on seventeenth-century weather and climate in the Alpine area of the Upper Savinja Valley are presented in Zwitter, 2015d.

491 In the administrative units of Luče and Ljubenska Gora, farms reach higher elevations than in Tirosek. Keeping of many beehives also reached higher elevations in the administrative units of Luče and Ljubenska Gora—in both of them, the average elevation of farms keeping more than ten beehives in 1686 was about 700 m, and the highest farms

seigneury of Gornji Grad (Germ. *Oberburg*) in the Kamnik–Savinja Alps. According to historical information from 1677 sent to the district office (*Kreisamt*) in Celje a century later, in 1778, a beehive tithe had to be delivered in the seigneury of Gornji Grad only by those farmers that kept more than ten hives.[492]

Figure 18 shows that in 1681, 1682, and 1686 those farmers in the administrative units of Luče, Ljubenska Gora, and Tirosek that kept more than ten beehives had a total of more than three hundred (in 1681), more than 180 (in 1682), and more than 240 (in 1686) beehives altogether. These are the numbers of all hives delivered as a tithe multiplied by ten (see figure caption). The graph suggests that the weather in 1681 was (relatively) favorable for bees (and many grassland flowers) in the entire Upper Savinja Valley. Indeed, the weather in August 1681 in Central Europe, including the Alps, was known for its extreme warmth and in many places also dryness, under the influence of the Azores anticyclone; September 1681 also was warm and dry in many places.[493]

---

in both these administrative units providing a beehive tithe in 1686 were located at an elevation of about 950 m (SI NŠAL 20, Gornji Grad Avstrija (Gornji Grad Austria), box 40, seigneury's account by Georg Lorger, April 29th, 1686 – April 28th, 1687: Specification der gefhallenen[!] zehendt peünstökh de anno 1686, die herren schaffer Geörgen Lorger bezalth[!] vnd guet gemaht[!] seindt worden. Data on elevations of farms were taken from Meze, 1965, 186, and Meze, 1969, 11–13; elevations of farms missing there: *Arcanum Maps,* Cadastral Maps, Habsburg Empire – Cadastral Maps (XIX. [sic] Century) in combination with *Environmental Atlas of Slovenia*, Spatial Units, Land Cadastre & Backgrounds, DEM). It is highly likely that beehives were moved to pastures located much higher than the farms in late spring or summer (Mihelič, 1970, 401, 408, and page 50 of illustrations: images 2 and 3; Niederer, 1996d, 188; Wopfner, 1997, 375; Crane, 1999, 244, 347–352). In the administrative unit of Tirosek, the average elevation of farms keeping more than ten beehives in 1686 was about 600 m (SI NŠAL 20, box 40, seigneury's account by Georg Lorger, April 29th, 1686 – April 28th, 1687: Specification der gefhallenen[!] zehendt peünstökh de anno 1686, die herren schaffer Geörgen Lorger bezalth[!] vnd guet gemaht[!] seindt worden); nonetheless, another tithe register from 1646 referring only to Tirosek reveals that the average elevation of farms delivering a beehive tithe in that year was about 680 m (SI NŠAL 20, box 26, Beehive tithe . . . , 1643, 1646, 1648: Zechend peinstockh in Tirossegg pergkh, 1646).
See the map of these administrative units of the seigneury of Gornji Grad in Zwitter, 2010, 10.

492 SI NŠAL 20, Gornji Grad Avstrija (Gornji Grad Austria), box 68, Bee tithe, 1690–1778: Copia einer untern[!] 6ten November 1778 an die stift oberburg[er] unterthanen als bienen zehends holden erlassenen verordung. Numbers of beehives in the Upper Savinja Valley fluctuated significantly, but detailed data at a farm level from 1686 make clear that only a very limited percentage of all farms were known to keep more than ten beehives, and so they are presented in Figure 18 (SI NŠAL 20, box 40, seigneury's account by Georg Lorger, April 29th, 1686 – April 28th, 1687: Specification der gefhallenen[!] zehendt peünstökh de anno 1686, die herren schaffer Geörgen Lorger bezalth[!] vnd guet gemaht[!] seindt worden. Surnames suggest that data refer to isolated farms, but it is possible that some of these farmers were in fact cottagers in the village of Luče).

493 Pfister, 1999, 139–140, 166.

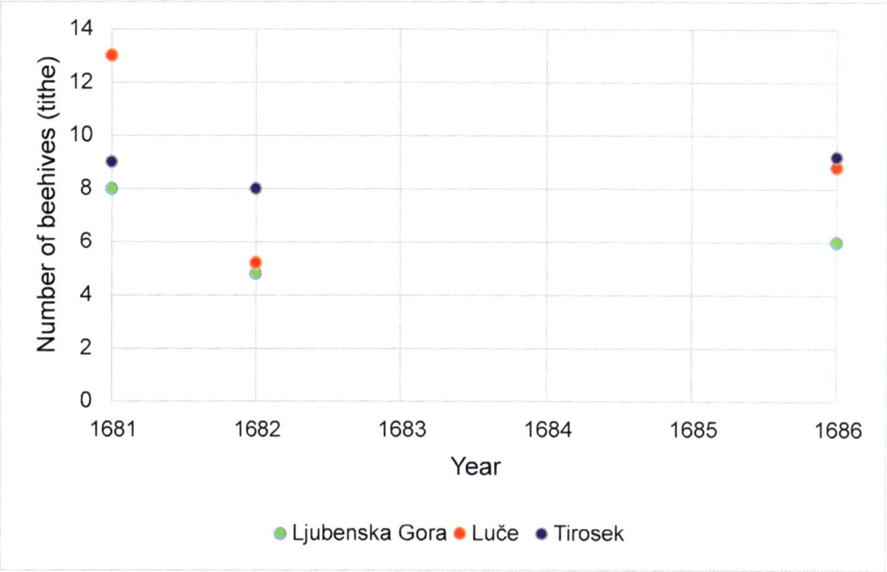

Figure 18: Tithe in beehives in the seigneury of Gornji Grad's Alpine administrative units of Ljubenska Gora, Luče, and Tirosek in 1681, 1682, and 1686 in the Southeastern Alps. Note that the number of beehives is always an integer; pairs of overlapping values in 1682 (Ljubenska Gora and Luče: five hives) and 1686 (Luče and Tirosek: nine hives) are slightly shifted on the graph so that all values are clearly presented. Because farmers delivered 10% of beehives as a tithe in this area and only those keeping more than ten hives were expected to provide a tithe, the total numbers of hives were more than ten times the values on this graph. In 1681 and 1682, the seigneury as the tithe lord collected the beehive tithe in kind. In 1686, pecuniary dues were handed over to the seigneury instead of the tithe in beehives in kind; nonetheless, it was clearly recorded how many beehives would otherwise have to be collected in kind. Data sources: SI NŠAL 20, Gornji Grad Avstrija (Gornji Grad Austria), box 39, seigneury's accounts by Michael Gervason, with supplements, 1681–1686: account April 1st, 1681 – March 31st, 1682, tables: Peinstökh & account April 1st, 1682 – April 25th, 1683, tables: Peinstökh; SI NŠAL 20, box 40, seigneury's account by Georg Lorger, April 29th, 1686 – April 28th, 1687: Gelt empfang von denen zehent- vndt heimischen peinstökhen & Specification der gefhallenen[!] zehendt peünstökh de anno 1686, die herren schaffer Geörgen Lorger bezalth[!] vnd guet gemaht[!] seindt worden; SI NŠAL 20, box 40, seigneury's account by Georg Lorger, April 29th, 1686 – April 28th, 1687, tables: Peinstökh. Author of the graph: Žiga Zwitter.

Readers are familiar with the high importance of mild weather in late summer and early autumn for honey production (and blossoming of second-crop hay flowers), and so it is easy to understand the high number of hives in the graph in 1681. In contrast, the next year (1682), when unfavorable weather is locally indicated by a likely cereal harvest failure, at least of oats,[494] was worse for bees in the entire Upper Savinja Valley than any other study year. The relatively low number of beehives that had to be paid for as a tithe in 1686 in the administra-

[494] Zwitter, 2015d, 151, 155, 168, 171–172.

tive units of Luče and Ljubenska Gora was probably at least partly influenced by the late frost on June 22nd, 1685, by low temperatures that preceding summer, and by early frosts from September 9th to September 13th, 1685. They adversely influenced bees, and cold spells in the summer had similar impacts on (several grassland) flowers. Spring frosts in early May 1686[495] were likely too early to cause considerable damage to bees and grassland flowers at these Alpine farms; nonetheless, their adverse impact on bees cannot be excluded. Namely, it is known that seventeenth-century farmers in the broader region (in milder climate at the foot of the Alps at least) usually moved their hives that overwintered in the attics outdoors on Palm Sunday[496] (in 1686: April 7th).[497]

As is characteristic of proxy data in historical climatology (i.e., indirect information about past weather and climates), "other climatic and non-climatic factors" influenced the results of an individual proxy to a certain degree.[498] This also applies to the tithe in beehives. Although the weather in the years in question and the availability of grassland pollen and nectar during the honey flows certainly played an important role, it is clear that further factors also had an impact on the results in Figure 18. For instance, according to the late seventeenth-century stage of beekeeping, when apiculturists did not use sugar, young hives were kept alive and only any surplus honey that they produced was taken. In contrast, bees from two- to three-year-old hives and most probably also from hives that produced too little honey for the approaching winter (e.g., due to unfavorable weather, and so such hives would likely perish) were killed in the autumn and all their honey was taken.[499] This, for instance, means that the weather in preceding years also had an influence on numbers of beehives.[500] In addition, further factors, such as non-grassland bee forage resources, bee diseases, pests, predators, and bees' behavior impacted the numbers of beehives in history.[501]

---

495 Zwitter, 2014d, 665–666.

496 End date of keeping bees in attics: Zupan, 1934, 96.

497 Grotefend & Ulrich, 1971, 190–191.

498 Brönnimann, Pfister, & White, 2018, quote from page 34. Many proxy data used in historical climate reconstructions "have correlation coefficients of around 0.5–0.6 with the climate variable they are trying to reconstruct" (Brönnimann, Pfister, & White, 2018, 34). The impacts of unwanted factors on results can be strongly reduced through comparisons of several independent data series from different kinds of proxies and/or pre-instrumental or early instrumental direct weather observations (Brönnimann, Pfister, & White, 2018; Camuffo, 2018; Pfister, 2018a; Pfister, 2018b; Pfister, Camenisch, & Dobrovolný, 2018; Pfister & White, 2018).

499 Valvasor, 1689, vol. 1, book 3, 455; Zupan, 1934, 96 (erroneously stating that according to Valvasor bees from three- to four-year-old hives were killed); Mihelič, 1970, 396–397, 401; Crane, 1999, 21, 66–67, 227, 235, 239–240, 243, 253, 493–494.

500 Crane, 1999, 236.

501 Crane, 1999, 20, 22–28, 30–31, 132–133, 232, 235, 239, 244, 253, 573–575; Mlaker-Šumenjak, 2011, 38–42, 55–61.

4.1.1.2 Impacts of climate variability

The changed time perspective from short-term variations (including weather) to long-term (including climate change) demonstrates that, on the one hand, "[s]easonal and year-to-year changes may simply be fluctuations round a long-term, stable mean, i.e. they may have no long-term impact. On the other hand, and for a variety of reasons, [grasslands] often change more fundamentally over the years."[502] In several grasslands in the Alps, historical climate dynamics, together with grassland management and other factors, co-influenced the changes in species composition. The impacts of climate change exceeded the migration of species up and down the slope, largely because there are winners and losers among species, and climate is only one of the factors determining the suitability of a site for plant species.[503] Climate change mattered (most) when and where the climate parameter(s) that changed had already previously been marginally suitable for a species or some species occurring or even abounding there. For instance, where most species "are far below their altitudinal limits," such as alpine plants at several places relatively close to the potential natural upper forest line, "local vegetation appears well buffered against climate coolings."[504]

The impacts of historical climate change on grassland plant species composition were studied; for example, in the Lake Hagelseeli in the northern Swiss Pre-Alps (palynological sampling of lake sediments at 2,339 m). A pollen diagram covering the period from 1410 to 1990 suggests that snow-bed indicators—milfoil (*Achillea* spp.), lady's-mantle (*Alchemilla* spp.), moonwort (*Botrychium* spp.), the *Cichorioideae* subfamily, cudweed (*Gnaphalium* spp.), and alpine spicknel (*Ligusticum mutellina*)—were most abundant from the last quarter of the sixteenth century to about 1730. Palynologists consider this pollen to have mainly been of local origin. The higher abundance of snow-bed indicators suggests that the annual duration of snow cover in the last six hundred years was most likely the longest between the late sixteenth century and about 1730,[505] which also includes two peaks of the Little Ice Age in Central Europe and the Alps: first, the one in the last third of the sixteenth century and early seventeenth century, and, second, during the Maunder Solar Minimum, which was particularly strong in the late seventeenth and early eighteenth century.[506] Christian Pfister's analysis of reliable observations of snowmelt on Mount Pilatus close to Lucerne (Germ. *Luzern*), only about 30 km from Hagelseeli, nicely matches with the palynological data presented above. Namely, Pfister found

---

[502] Peterken, 2013, 224.

[503] Pauli, Gottfried, & Grabherr, 2003; Nagy et al., 2003, 459–460; Jäger, 2006, 65; Nicklas, 2018.

[504] Tinner & Theurillat, 2003, 166.

[505] van der Knaap et al., 2000, 86–87, 108–109, 121, 138.

[506] Mathieu, 1996, 144; Rohr, 2007a, 409, 450; Jäger, 2006; Jäger, 2007b, 91–98; Dobrovolný et al., 2010; Pfister et al., 2018, 276–281; for an overview of historical climatology in the Alps in early modern times, see Rohr, 2017, and also Camenisch & Rohr, 2018.

that "around 1600, the snow cover above 1,800–2,000 m [usually] melted [in July; i.e.,] more than 35 days later than it has done during the climatic optimum of the twentieth century" (this research was published in 1985, and so the "climatic optimum" refers to the period from 1930 to 1941, when observations for Mt. Pilatus are available).[507] Similarly, in the late seventeenth century, spring and autumn temperatures dropped simultaneously, leading to the cold autumns of 1688–1700 (on average about 1.1 °C colder in Central Europe than in 1961–1990) and consistently cold springs from 1687 to 1701 (in 1687–1700 on average about 1.4 °C colder in Central Europe than in 1961–1990), making the long duration of the snow cover understandable.[508] Although the long duration of snow cover caused serious problems to the farm economy (most relevant aspects for grassland farming are presented lower in this section, and in Sections 4.1.1.3 and 6.2), in some cases long-lasting snow positively affected the fodder quality in high pastures (while at the same time probably reducing the quantity of grazing resources): lady's-mantle (*Alchemilla* spp.) and alpine spicknel (*Ligusticum mutellina*) growing in the pollen catchment area of Hagelseeli between the late sixteenth and early eighteenth century more abundantly than earlier or later belong to most esteemed fodder plants of high summer pastures according to traditional ecological knowledge in East Tyrol.[509] Similarly, folk tales from the Ötz Valley and the Swiss canton of St. Gallen relate the "milk-richness" of some complexes of mountain pastures with lady's mantle, alpine spicknel, and a couple of other organisms.[510] Lady's-mantle is also considered a particularly good fodder plant by traditional Csángó farmers in Ghimeș (Hung. *Gyimes*) in the Eastern Carpathians; Csángós know that cows and sheep produce more milk if fodder contains much lady's mantle.[511] Historical herders spending whole days in grasslands with livestock had time to observe the eating habits of the animals they were keeping, and these pieces of traditional ecological knowledge from distant regions correspond with each other in this case. Indeed, these observations are well founded: according to the standards of current agronomy, an integrated quality indicator (linking nutritive value, palatability, and livestock health impacts) of common lady's mantle (*Alchemilla vulgaris*) is high, comparable to false oat-grass (*Arrhenatherum elatius*) and yellow oat-grass (*Trisetum flavescens*) from moderately manured meadows (see Section 6.3.2);[512] common lady's mantle is one of the few species

[507] Pfister, 1985b, 154, 157–158, 161, 164–165, 167–172 (quote from p. 164). Concerning the designation of the "climatic optimum," Mt. Säntis in eastern Switzerland (2,500 m) became free of snow about a month earlier in 1920–1953 than in 1886–1919 and 1954–1980 (Pfister, 1985b, 161).

[508] Pfister et al., 2018, 276–281.

[509] Vogl-Lukasser et al., 2006, 91, 183.

[510] Machatschek, 1997b, 112.

[511] Babai, Molnár, & Molnár, 2014, 26, 36, 43, 49, 110.

[512] Dietl & Lehmann, 2004, 39, 42–43 (identification of the species "gewöhnlicher Frauenmantel" is based on Poppendieck et al., 2010, and *Montana Field Guides: Common Lady's Mantle*); Brugger & Wohlfarter, 1983, 26, 31; Machatschek, 2006, 134.

that remains highly attractive to livestock even if it abounds in fodder from a species-rich grassland.[513] Similarly, agronomy confirms that the fodder value of alpine spicknel (*Ligusticum mutellina*) is particularly high, comparable to the best clovers and even higher than that of common lady's mantle.[514]

During the Little Ice Age, at its peaks in particular, low temperatures in the warm part of the year were more frequent than earlier or later. Unusually, low temperatures, in some cases related to advancing glaciers (see also Section 4.2), slowed down the growth of grassland plants. Cold summer weather not only posed the problem of a lack of fodder and made the use of snowy paths hazardous; it also had further negative impacts on livestock health. Cattle were likely to fall ill in such situations, and all this negatively affected livestock's performance. As late as in 1864, one of the last cool Little Ice Age summers, deep snow in the mountain pastures in early July, where farmers from the Sava Dolinka Valley had grazing rights, caused hundreds of sheep and cattle, and many horses, to perish.[515] In addition, temperature stress was one of the factors leading to lower milk quantities; lactation stopped in a cow suffering from temperatures that were too low, and it did not resume when the weather warmed up again.[516]

One type of adaptation to scarcer grazing resources and other impacts of significantly lower temperatures was shorter grazing periods, such as in the inner Pitze Valley (Germ. *Pitztal*, west of the Ötz Valley) below the Mittelberg Glacier (Germ. *Mittelbergferner*) in the uppermost Pitze Valley section, where the 1689 grazing season (during the harsh climate of the Late Maunder Minimum) in mountain pastures started with a delay of three weeks, on July 2nd, and after the next two weeks there was hardly any fodder available for grazing livestock. Another type of human response to a harsher climate was reductions of grazing rights as an adaptation to the lowered carrying capacity of pastures. An example is the Outer Stein Alp (Germ. *Äußere Steiner Alm*) near the Gradötz Glacier (Germ. *Gradötzkees* or *Gradetzkees*) in east Tyrol, where grazing rights were reduced in 1609 during the late phase of a serious Little Ice Age-peak lasting from the late 1560s. More precisely, this happened after the sixteen cold summers of 1585–1589 and 1591–1601, most of them snowy in high summer pastures in Central Europe, were followed by another cold summer in 1608.[517] A third type of human response to harsher climate was the conversion of some former dairy cattle grazing complexes into pastures for young and other non-dairy cattle; several other former cattle pasture complexes were used only as sheep grazing grounds as a similar adaptation. The

513 Pils, 1994, 173–174.
514 Brugger & Wohlfarter, 1983, 25–26, 188–189.
515 Studen, 2024; Zwitter, 2015d, 151; compare Fajgelj, 1953, 129.
516 Pfister, 1985a, 38–39, 57, 62–63, 92.
517 Pfister, 1985a, 62–63, 92; Jäger, 2006, 16, 20, 24–27, 30, 36–37, 45–50, 68–71, 76, 78–79, and Figure 7 between pp. 16 and 17; Jäger, 2007b, 92, 95–96, 99, 106–107, 110; Pfister, Camenisch, & Dobrovolný, 2018, 126; Pfister et al., 2018, 277–280.

upper Agls Alp (Germ. *Aglsalm*, about 2,110 m) by the Malevalle Glacier (Ital. *Ghiacciaio di Malevalle*, Germ. *Übeltalferner*) in the Ridanna Valley (Ital. *Val Ridanna*, Germ. *Ridnauntal*), now in Trentino–Alto Adige (South Tyrol), for instance, was a mountain pasture complex for cattle and sheep in the second half of the eighteenth century. Starting in 1817, following several wet and/or cold summers, including 1816 (i.e., the "year without a summer"), it became a pasture complex for a few sheep only.[518] Due to the shift from cattle to sheep, plants were grazed much closer to the sod,[519] adding to climate-related stress. In comparison to cattle, sheep are generally much more selective in their feeding, and they also prefer grazing at raised structures in the landscape—the tops of hills and hummocks, for instance—and thus additionally increase the grassland heterogeneity.[520] Thus, the climate-change-induced transition from cattle to sheep grazing caused changes in grassland agroecosystems, which were the result of interrelated impacts of natural dynamics and altered grassland management. A fourth type of farmers' response to significantly harsher environmental conditions was the partial abandonment of a grazing complex due to the hazards posed by the climate change, even though the grazing rights were not formally reduced, as presented in the next paragraph.

The impacts of climate cooling on the grazing regime and species composition at higher elevations also affected regions of the Alps far from glaciers. From 1568 to 1579, for instance, very snowy winters, cold spells in the spring, and low temperatures and abundant precipitation in the summer were typical.[521] A notice from 1579 reflects these conditions, documenting that snow often covered the Belščica mountain pasture complex (Figure 19) in the Karawanks (Sln. *Karavanke*, Germ. *Karawanken*) west of Mount Stol (Germ. *Hochstuhl*) at an elevation of approximately 1,900 to 2,100 m throughout the summer. When covered with snow, the area also became dangerous due to the steep slopes below the pastures. Consequently, many animals perished and only a few farmers with grazing rights still kept their livestock there.[522] The coldest sixteenth-century summers were yet to come: the average Central European summer temperature in 1585–1598 was 1.2 °C below the 1961–1990 average; with the exception of 1590, all these summers were cold or very cold, and summer snow covered Alpine pastures in most of them.[523] Another example of a grazing complex located far from glaciers—in this case characterized by a (generally speaking) relatively

[518] Jäger, 2006, 9–10, 36, 44–50, 52, 61, 68, 70–71, 78–79; Jäger, 2007b, 91, 99–100, 103, 105–111; Nicolussi, 2013, 78–79, 90–91.

[519] Ellenberg & Leuschner, 2010, 1007.

[520] Agten, 2010: for instance, sheep avoid some highly productive grasses such as tor-grass (*Brachypodium pinnatum*) and upright brome (*Bromus erectus*), which can also abound in the subalpine vegetation belt (Aeschimann et al., 2004, vol. 2, 916–917, 926–927).

[521] Pfister, Holzhauser, & Zumbühl, 1994, 65; Jäger, 2006, 13, and Figure 5 between pp. 16 and 17.

[522] Zwitter, 2020, 79.

[523] Pfister et al., 2018, 277–279.

Figure 19: The Belščica mountain pasture complex (at the top of the long ridge, about 1,900 to 2,100 m) covered with snow during a cold spell in late summer. Photo: Žiga Zwitter, August 26th, 2018.

mild climate, but nonetheless seriously affected by the peaks of the Little Ice Age—is the pasture at the top of Mount Blegoš (1,562 m), a pre-Alpine hill. In the summer of 1628 (or 1630), livestock could graze there for four weeks or even less in the entire year due to severe frost and snow.[524] According to information from Switzerland, 1628 was one of the three years with the coldest warm half year during the last five hundred years.[525] An extremely cold summer was also reported in Tyrol,[526] and in part of Slovenian territory the grapes were not harvested until the beginning of November, even though they were still unripe—similar to reports from Switzerland.[527] Snow in mountain pastures in late July forcing the retreat of livestock in Tyrol was reported again in 1630.[528] This means that low carrying capacity of mountain summer pastures was an Alps-wide phenomenon in many cold summers during the peaks of the Little Ice Age, and it was not limited to the highest sections of the mountain chain.

In many places in most of the Alps, the challenges posed by untimely summer snow in grazing complexes were partly mitigated by the right for herders to

[524] Zwitter, 2020, 79–80; compare Jäger, 2006, 16.
[525] Pfister, 1999, 158, 196.
[526] Jäger, 2007b, 92.
[527] Zwitter, 2013b, 342, 345, 362, 384–385.
[528] Jäger, 2006, 16.

escape with their livestock to an area with milder climatic conditions. In many cases they would otherwise not have grazing rights there, at least during that part of the year. Such escape rights were a frequent, although not ubiquitous, component of Alpine "cultures of risk management." Although such rights were highly relevant in the context of short-term weather variability, they mattered even more in longer phases of a cool or cold climate in the warm part of the year. An escape ground from a summer mountain pasture was usually a lower open pasture, a lower forest grazing ground, or a montane meadow complex, provided that the hay had already been stored (see Section 6.1.2.1). Hay from montane meadows that played the role of escape grazing grounds in case of summer snow often belonged to other farmers than those whose livestock had the right to escape there from high summer pastures. Rights to retreat in case of untimely snow were not limited to high or relatively high summer grazing complexes: for instance, according to a source compiled or copied in the eighteenth century, herders from Rečica Alp (Sln. *Rečiška planina*, 1,270 m) or Grad Alp (Sln. *Grajska planina*, 1,200 m) in the eastern Julian Alps (Sln. *Julijske Alpe*) had the right to escape with livestock to the Rčitno meadow complex (900 m) if forced by weather after haymaking at Rčitno had been completed. Even some permanent settlements had the right to keep their livestock in lower pastures in the colder part of the year if snow rendered grazing on their usual grazing grounds impossible (see Sections 6.1.2.5.1.1 and 6.2). Grazing communities entitled to escape to a lower grazing ground often did not have any grazing rights there under normal weather conditions. Usually, the right to escape from snowy pastures was not free of charge or without mutual favor agreements. Temporary escape rights in case of summer snow are clearly proven to have existed for centuries, in many cases since the Middle Ages or since the early modern period at least. In some cases, only some livestock types, such as cattle, were allowed to leave the snowy pastures. In many cases, the permitted duration of such a retreat was shorter than an optimal adaptation to severe weather would have required (e.g., livestock were only allowed to be kept at such a lower retreat for half a day, one day, a day and a half, or three days, or in some cases longer; at many places, livestock were only allowed to reach the escape ground, for example, eighteen, twenty-four, or forty-eight hours after the snowfall). In some other cases, livestock were allowed to stay at the temporary retreat ground until grazing at the higher grazing complex could resume. Not only the frequently limited duration of allowed grazing in an escape ground was sometimes challenging—the descent of hungry cattle on a snowy path was tricky, and sheep were inclined to become stuck in rocks in snowy weather unless gathered in time, which was impossible in the case of snowfall during the night. A summer snow blanket that lasted too long or a lack of retreat rights forced an earlier return from summer pastures to more distant places; for example, to the valley.[529] In July 1948,

[529] Data from several places in the Alps: Grass, 1948, 100–124, 136, 189; Novak, 1955, 297, 304–305; Pfister, 1985a, 39; Wopfner, 1997, 437–439; Jäger, 2006, 13–14, 27; Kačičnik

for instance, cattle from the high summer pasture complex of Velo Polje (1,680 m) and the lower, early summer grazing complex of Vodični Vrh (1,486 m) returned to the valley when a thick snow blanket formed (about 0.5 m at Velo Polje and about 0.25 m at Vodični Vrh, as far as recalled). Cattle from the nearby Krstenica Alp (1,655 m) stayed in a montane forest, and some fodder was brought to them from the valley (see below); a farmer's son that brought the fodder to their cow recalled that the cows were bellowing in distress.[530] The risk of challenges posed by untimely snow was decreased where enough hay was mowed at nutrient-rich sites around alpine cottages and/or enough hay surrogates were gathered (Section 6.6.2.2) to be used as emergency fodder. In general, limited mowed areas close to alpine cottages expanded somewhat in the nineteenth and early twentieth century, even though such mowing sites were not an entirely new phenomenon at that time (Section 6.1.2.4). Making enough hay and, optionally, other emergency fodder made it possible to stay in the alp with livestock in the case of snow, provided that there were shelters for the livestock or that domestic animals could be driven to a place in a forest without snow cover. This was the case only in some high summer grazing complexes.[531]

Apart from the rights of retreat and locally made hay to increase the resilience of the mountain pasture complex in the case of summer snow, some fodder was also brought from villages in the valley to high summer mountain pastures in case of summer snow. This is recalled about July 1948 for Krstenica (1,655 m) and for the remote summer grazing complex of Govnjač (1,474 m); the livestock were from the village of Bohinjska Češnjica (elevation 600 m), Govnjač is 14 km away as the crow flies.[532]

Important from the perspective of environmental dynamics in terms of weather and climate were also watering rights for cattle and horses in case of drought in neighboring grazing complexes with (more) abundant water resources. For instance, herders from the karst Ovčarija summer pasture com-

Gabrič, 2014, 114, 221; Zwitter, 2015d, 148; Zwitter, 2020, 71; Zwitter, 2024. A "culture of risk management" (Rohr, 2013, 135) "designates a bundle of well-rehearsed practices of risk minimization and ways of coping with extraordinary situations that people have integrated into their daily lives so as not to be taken by surprise" (Pfister, 2011, 4–5).

530 Author's interviews with F. Mlakar, 2017, in Bohinjska Češnjica; with L. and M. Odar, 2017, and with A. and M. Taler, 2017, both in Studor; see also data on usual grazing periods in these pasture complexes in Novak, 1955, 290–292.

531 Grass, 1948, 100, 102, 115, 120, 123, 208–213; Novak, 1955, 304; Wopfner, 1997, 439–441; Novak, 2024, 249; see also Jäger, 2006, 13. At some other summer mountain pasture complexes, only small quantities of hay were made for ill livestock, for calves, and as human bedding (see also Section 8.2.3), and so this did not suffice if there was untimely snow (author's interviews with C. Mlakar, 2015, in the Krstenica Alp in the Julian Alps; with Kunavec, 2017, with Mencinger, 2017, and with Gašperin, 2017, all in Stara Fužina).

532 Author's interview with F. Mlakar, 2017, in Bohinjska Češnjica. The interviewee recalled how angry his fellow villager was when he had to bring some fodder to his ox at the distant Govnjač Alp in 1948.

plex in Bohinj in the Southern Limestone Alps, where cattle strongly depended on waterholes, had the right to water their livestock in a lake of the neighboring mountain pasture complex at the Triglav Lakes if required by drought. This was recorded in the 1870s and was regarded as a relevant custom at that time, rooted in older tradition.[533] Old farmers with grazing rights at Ovčarija still recall watering their livestock at the Triglav Lakes (about 1.5 km away in straight line) in cases of drought.[534] In serious drought, livestock from the Na Kraju Alp on the Komna Plateau in western Bohinj were watered at Črno Lake (Sln. *Črno jezero*, about 3 km away in a straight line), meaning a walk of an hour and a half with livestock.[535] In some grazing complexes in the limestone Alps, herders not only maintained waterholes but also stored and kept snow covered with litter to reduce its melting in karst depressions, which were accessible with a ladder; they used it in the summer as a water source.[536] Alternatively, they used the snow that accumulated naturally in karst depressions, often a quarter of an hour's walk from the temporary settlement, and up to an hour's walk in some places after the accessible snow in closer depressions had been used. They cut the snow in depressions with an axe, carried it to the temporary settlement, and let it melt in a bucket in a sunny place. In this manner, they obtained drinking water and water for cooking, and in some cases also drinking water for livestock. This was a frequent practice at several summer alps in the Southern Limestone Alps without local water springs before reservoirs or water supply systems were built in the twentieth century.[537]

Where not enough water (and accessible snow in depressions) was available within a fairly long distance, severe droughts forced an earlier return from summer pastures, reducing the grazing period there. In the first half of the twentieth century, for instance, cattle sometimes had to leave the large grazing complex of the Velika Alp (Sln. *Velika planina*, more than 5.5 km^2) in the Kamnik–Savinja Alps due to drought before the usual end of the grazing season. Nearby, horses, the only livestock type kept at Jezerca Alp close to Podvežak on the Dleskovec Plateau (Sln. *Dleskovška planota*), had to leave this grazing complex earlier due to the 1952 drought. As early as mid-summer, drought led to migration of horses several times from Bela Peč below Mt. Raduha because priority was given to non-milk cattle also kept in the same mountain pasture and they could stay there longer.[538] All these examples are at least largely located in karst areas.[539]

533 Novak, 1955, 297–298, 303, see also Grass, 1948, 134–136.
534 Author's interview with Kunavec, 2021, in Stara Fužina.
535 Novak, 2024, 99.
536 Fajgelj, 1953, 125, 128, 154, 156; Novak, 1970, 358; Novak, 2024, 100; compare Cvijić, 1925, 3–4, and photo 14.
537 Novak, 2024, 98–103, 244.
538 Jordan, 1945, 55; Čerček, 1948–1949, 50, 52; Fajgelj, 1953, 126, 128, 154, and the map between pp. 128 and 129.
539 *Environmental Atlas of Slovenia*, Land and Soil, Soil Map (1:25,000).

4.1.1.3 Evolution of material strategies for coping with wet weather after mowing

Rainy summers, especially cold or cool ones, were a significant challenge to drying mowed hay.[540] Precipitation after mowing significantly decreased hay's nutritive value, worsened its aroma and taste, allowed the spread of pathogenic mold, and caused much additional work. Namely, the labor-intensive actions to protect the hay from getting wet dealt with in this section had to be taken, and, if hay nonetheless repeatedly became wet, additional turning was needed several times. More turning and spreading also increased the loss of dry matter in hay, meaning that less of the protein-rich herbs such as clovers were gathered in because they are more fragile than grasses (Section 6.3.1.1).[541] Significantly diverse mowing dates in meadows in different grassland complexes located in different microenvironments but belonging to the same farm decreased the likelihood of a harvest failure of a high percentage of the total hay at the farm level; for example, due to a rainy period after mowing. Thus, diversity of mowing dates and a limited meadow area mowed on the same day[542] strengthened the resilience of the farm economy (see Sections 6.1.2.1 and 6.1.2.2). Still, according to traditional ecological knowledge from Bohinj—which belongs to the wettest parts of the Alps, with about 2,000 to 2,600 mm of annual precipitation in the valleys and many montane meadows, and slightly more than 2,600 mm in the wettest montane meadows (in the period of 1971–2000)[543]—on average wet weather resulted in twice the amount of economic damage in meadows caused by drought.[544]

Nutrient leaching from hay by rain is still moderate in times of mowing carried out when seeds just ripen.[545] Consequently, depending on the agroecosystem, it sometimes happens that "[i]f hay is [slightly] rained on [immediately after mowing] and then you dry it well . . . the cows would eat it better," as suggested by a parallel from the Carpathians.[546] This surprising observation from the times of mowing by scythe[547] might be related to the specific aroma of the hay from that very site because cows pay much attention to the scent;

540 Pfister, 1985a, 119–120; Makarovič, 2007, 216, 219, 226–228; Slavin, 2018; and synthesis of European historical climatology in Rohr, Camenisch, & Pribyl, 2018; Pfister et al., 2018; and Brönnimann, White, & Slonosky, 2018.

541 Makarovič, 1982, 151; Pfister, 1985a, 41–43, 47, 57, 62–63, 86, 91, 103, 120; Keller, 2004, 15, 60, 93; Peterken, 2013, 125, 141; Bavcon, Ravnjak, & Praprotnik, 2019, 137; Verbič & Verbič, 2014–2020, 20–22, 24.

542 Compare Iuga, 2016, 74.

543 *Environmental Atlas of Slovenia,* Climate, Precipitation, Mean annual corrected precipitation 1971–2000; for a basic comparison, see the map of mean annual precipitation in the Alps in 1971–1990 in Bätzing, 2015, 37.

544 Author's interview with L. and M. Odar, 2017, in Studor.

545 Pfister, 1985a, 41–42; Bavcon, Ravnjak, & Praprotnik, 2019, 149.

546 Iuga, 2016, 74.

547 Thanks to the Romanian ethnologist Anamaria Iuga for this explanation.

Figure 20: Manually turning hay was not just placing it on the ground upside down; it was done in such a manner that the hay was deposited in airy, slightly elevated structures when it fell from the rake to promote drying, as seen in this photo. As the text explains, this was usually done in a direction so that the sun dried such slightly elevated forms more quickly.[548] This photo was taken near Spodnji Dolič in the foothills of Pohorje in the first half of July 1963. Slovenian Ethnographic Museum, Documentation, Fieldwork campaign 20, Vitanje (1963): photo 20/918 (by Pavla Štrukelj). *Slovenski etnografski muzej* . . . Vitanje / 18 F0000020/918. © Slovenian Ethnographic Museum.

where cattle like the aroma of the hay and the hay is clean (as it usually was in times of mowing by scythe—in contrast to motorized mowers, resulting in much more soil dust in the hay),[549] no rain at that time is optimal, as a general rule.[550]

To promote quicker drying, (1) hay from shady sites was regularly carried to sunny sections within the same meadow immediately after mowing;[551] in wooded meadows, such as larch meadows (Section 6.6.1), this meant a lot of work;[552] (2) it was turned toward the sun, as reported in Switzerland and Slovenia, and so this practice was widespread[553] (see Figure 20); and (3) in the

[548] Participant observation of manual hay turning with the Hodnik family in a montane meadow at Vogar, July 2018.
[549] Babai, Molnár, & Molnár, 2014, 105–106; Iuga, 2016, 73–74.
[550] Pfister, 1985a, 41–43, 47.
[551] Šarf, 1974, 139; Makarovič, 2005, 127.
[552] Norz, 2014, 22–23.
[553] Switzerland: von Glasenapp & Thornton, 2011, 775; Bohinj: author's interview with M. Hodnik, 2018, in the Vogar complex of montane meadows, recalling how they made hay with her great-aunt, born in the early twentieth century.

case of hummocky meadows (see Sections 3.2 and 4.1.2), hay was raked to the tops of the hummocks and ridges immediately after mowing; it was not left in the moister and shadier depressions.[554] All these practices are documented through ethnological data, which certainly reach back to the late nineteenth or early twentieth century, but it is clear that many of them are much older.

In sunny summer weather without rain, hay is dry on the ground the next day in non-fertilized sunny meadows.[555] In moderately fertilized meadows, grassland plants grow taller and denser (Section 6.3.2), and so such hay dries on the ground at least one day longer than hay from non-fertilized grasslands.[556] Hay on shady, northern slopes dries significantly longer than on sunny slopes.[557] The drier the hay is, the more it is prone to nutrient leaching in case of rain, and so a decrease in quality was particularly strong if the hay was already half dry or dry when it was raining.[558] A short heavy rain event was less problematic than several rainy days, even if it did not rain comparably heavily. Wetted half dry or dry hay became significantly less nutritive and less attractive. Too many days with precipitation during hay drying on the ground (drying lasting for a week) caused rotting of exposed grassland plants.[559] An example from the broader cool climate anomaly of 1880s and 1890s[560] demonstrates this: in the extremely rainy summer of 1882, the chronicle of the parish of Ljubno ob Savinji in the Alps of Lower Styria, where drying hay in a roofed hayrack (see below) has never become dominant, recorded that a lot of hay rotted at high isolated farms in that area.[561] "In many cases, putrefying hay encouraged the activity of parasitic fungi, bacteria, worms and gastropods"—snails in wet habitats were relevant not only as consumers of leaves; more relevant for livestock farming is that especially the snail *Galba truncatula* can host the liver fluke (*Fasciola hepatica*), a serious parasite for livestock (Sections 6.1.2.5.1.1, 6.1.2.5.1.2, and 6.4.1).[562]

During the High Medieval Climate Anomaly (or Medieval Warm Period), lasting roughly from the beginning of the second millennium to the thirteenth

[554] Odar, 1993, 6.
[555] Šarf, 1974, 139–140, 144.
[556] Compare Verbič & Verbič, 2014–2020, 26.
[557] Wopfner, 1997, 334.
[558] Pfister, 1985a, 41–43, 47.
[559] Makarovič, 1982, 151; Pfister, 1985a, 42, 98, 120; Slavin, 2018, 499; author's interview with A. & M. Prepotnik, 2012, at Haudej farm in the Solčava region; these data from Haudej farm and from Strojna in Carinthia come from areas where a roofed hayrack has never become usual (Melik, 1931, 33–34, 36–37, 48–49, and the map Répartition du *kozolec* en Slovénie).
[560] On this anomaly: Brönnimann, White, & Slonosky, 2018, 312; Ogrin, 2005, 63.
[561] Zwitter, 2015c, 179, 181; compare Melik, 1931, 33–34, 36–37, 48–49, and the map Répartition du *kozolec* en Slovénie; Baš, 1984, 335–336; Zwitter, 2010, 53.
[562] Slavin, 2018, 499–500, 504–505; on liver flukes and snails, see also Novobilský et al., 2014 (gratitude is expressed to the agronomists Martin Škrlep and Jože Verbič for recommending this article) and Ballweber, 2024.

century, there were occasionally cold wet summers, but the frequency of such summers was low at that time.[563] Later, peaks of the Little Ice Age brought about clusters of wet and cold or cool summers, especially in the 1310s, in the late fifteenth century, in the last third of the sixteenth century (a series of such summers: 1569–1571, 1585–1589, 1591–1601), in the late seventeenth century, and in 1816, belonging to the broader interval of cold summers from 1812 to 1817. For instance, there were seventy-seven rainy days in June, July, and August of 1588 at Lucerne and seventy-five in 1596. In June and July of 1816, belonging to one of the coldest summers of the last millennium, there were "eight successive weeks of rain" in Switzerland; it snowed there "down to 800 metres elevation in every summer month" and the average summer temperature there was 3 °C lower than the average of preceding twenty years.[564]

In general, a significant decrease in hay quality and useable hay quantity led to much lower milk quantities, even up to entire cessation of cows' lactation in the winter. In the past, cows usually delivered much less milk than today (Section 5), and this led to a more significant decline in milk production in the case of low-quality fodder than for more recent cows normally producing more milk.[565] By the end of the seventeenth century—and, less frequently, by the early nineteenth century—summers when a major part of hay rotted and/or clusters of years when the summer allowed only production of small quantities of low-quality hay were followed by a very long winter (such as 1569–1574, 1689–1694, and some others), occasionally corresponded to general subsistence crises. The connection between hay harvest failures and subsistence crises was particularly relevant among the poorer population, in less accessible areas strongly dependent on agriculture (i.e., by no means in all less accessible areas in the Alps), unless human factors prevented the crisis. In such cases of general scarcity, many farmers also used partly rotted hay. Consumption of hay that had partly rotted, was mold-infested, or had simply grown in much wetter conditions than normal, which allowed expansion of the liver fluke habitat, easily led to livestock infections with diseases that were either lethal or at least weakened the livestock to a degree to cause severe suffering among the animals and serious economic impacts. This was particularly relevant if the immunity of the livestock that consumed such fodder was weakened due to starvation and unusually cold and wet weather, which was possibly the case during the great subsistence crisis of the 1310s.[566]

---

563 Rohr, Camenisch, & Pribyl, 2018, 254; White, Brooke, & Pfister, 2018, 333; on 1257–1258, see Duh, 2023, 62–79, 98–99.

564 Brönnimann, White, & Slonosky, 2018, 312–315; Pfister et al., 2018, 268–269, 274, 276–281; Pfister & White, 2018, 552 (quotes), 553; Rohr, Camenisch, & Pribyl, 2018, 255; Slavin, 2018; White, Brooke, & Pfister, 2018, 333, 338–340, 342; Pfister, 1985a, 63, 121.

565 Pfister, 1985a, 38, 42–45, 47, 62–63, 84, 90–92, 103, 118–119; Jäger, 2006, 12–13.

566 Pfister, 1985a, 47, 53–54, 60–64, 83–84, 86, 90–96, 102–104, 107–116, 124–129; Viazzo, 1989; Zwitter, 2015d; Slavin, 2018, 495, 499, 500, 503–508; White, Brooke, & Pfister,

There existed several varyingly efficient historical strategies for materially coping with the hazard of rain during mowing time. In the case of imminent rain or in the evening, before the dew, hay was usually arranged in heaps called haycocks (Figure 21). In many places, their surface was neatly combed down with a rake to promote water runoff, thus increasing the likelihood that partly dry hay inside the cock would not become wet. Attention was also paid to haycocks' form: the lowest part of a haycock was slightly narrower than its section slightly above the ground in order to decrease as much as possible the quantity of hay in contact with rain and with potentially wet ground. However, cautious haycock building was not a general rule: a comment from 1778 explains that haycocks in the parish of Laupen (west of Bern) were built carelessly. Consequently, longer rainy episodes caused more damage to hay there than in Highlands (Germ. *Oberland*), where farmers paid more attention to haycock forms. When the weather turned better again or the next morning, the haycocks were spread out, as a general rule. The practice was labor intensive but effective if the rain did not last long.[567] Haycocks have a long history. Haycocks with a narrower bottom, thus perfected in shape, are presented in the vicinity of Paris in an early fifteenth-century calendar by the Limborg brothers. A June landscape with haycocks was drawn around 1500 in the workshop of the French Renaissance illustrator Jean Poyer, who lived in Tours. Pieter Bruegel the Elder drew haycocks in an Italian landscape in his painting *Haymaking* (1565). Haycocks stand in a drawing showing the surroundings of Laxton, northeastern England, in 1635.[568] Archival records prove that they were made in Bohinj, as recorded in the early eighteenth century,[569] to name just a few early preserved proofs making it clear that haycocks had already been usual in different and distant parts of Europe well before the developments in agriculture in the second half of the eighteenth century and in the nineteenth century. The wide territorial distribution[570] and conventionality of haycocks suggest that the practice of stacking hay in cocks is in fact much older.[571]

2018, 331–334, 339–342, 346; on common liver fluke, see also Novobilský et al., 2014; Ballweber, 2024; and *Veterina Novo mesto:* Fasciola hepatica – *veliki metljaj.*

567 Melik, 1931, 50; Šarf, 1974, 139–140; Pfister, 1985a, 42, 120; Peterken, 2013, 125–127, 130–131, 134–135, 336, 338–339; Bavcon, Ravnjak, & Praprotnik, 2019, 134, 137, 146–147; compare Wopfner, 1997, 329.

568 Peterken, 2013, 118–119, 123, 142, 333–336; on Poyer's biography: *The Morgan Library and Museum: Jean Poyer: Artist to the Court of Renaissance France Opens January 25, 2000*; *The British Museum: Jean Poyer.*

569 SI AS 721, Gospostvo Bled (The seigneury of Bled, Germ. *Veldes*), Iustitialia, fasc. 33, Lawsuit between the communities of Studor and Savica concerning grazing rights in Ukanc, 1697–1717: *Die nahberschaft zu Sauiz allegiert . . .*, no date.

570 Apart from examples given above, there is more recent evidence from the Carpathians; e.g., in Babai, Molnár, & Molnár, 2014, 106, 108, and in Zwitter, 2015a, 470.

571 Compare Makarovič, 2007, 214.

Figure 21: Due to the imminent rain, hay was cocked. Stara Fužina, Bohinj. Nowadays, the great majority of farmers do not make haycocks anymore and farmers' time is limited, and so the form of the lower parts of haycocks is not as perfect anymore as the ones that the farmers belonging to the middle generation that made the cocks in the photo recall from their youth. Photo: Žiga Zwitter, June 10th, 2017.

Partly dry second-crop hay started rotting much sooner than partly dry first-crop hay (see also Section 6.1.2.3). Consequently, cocks made of second-crop hay could not remain untouched on the next day even if there was continuous rainy weather: decomposition was so intense that cocks of partly dry second-crop hay warmed up so much that they started smoking, causing a strong decrease in fodder quality; the color of wet second-crop hay in a cock also changed, it became red, and the interior of the cock turned white. Usually, farmers would do their best to move second-crop hay to any kind of roofed place in the case of imminent rain in the following days, such as small roofed spaces in front of a remote livestock barn; they would put it in the remote livestock barn if the livestock were kept elsewhere at that time. Some would even spread the second-crop hay about 0.3 meter deep inside their cottage (a human habitation next to a barn in a remote meadow complex) to save this valuable fodder, or they transported it home as early as drying time and put it on a hayrack.[572]

[572] Author's interview with Kranjc, 2018, and with Kunavec, 2023 (with personal experience of saving second-crop hay by drying it in the cottage in the Voje Valley), both in Stara Fužina.

Another way of protecting partially mowed hay against rain, most likely relevant in several regions of the Alps during much of last millennium, was stacks for drying hay that were built around a pole. The simpler and probably older form is the top of a coniferous tree, often two to three meters tall, at most six meters, with the bark peeled off and the branches cut off, or with short stumps of branches (Figure 22). The form of the frame in stacks for drying hay underwent some evolution over time. The plain type presented here still prevailed in some regions of the Alps in the mid-twentieth century, such as parts of Carinthia (Germ. *Kärnten*, Sln. *Koroška*) and Styria (Germ. *Steiermark*, Sln. *Štajerska*). Stacks for drying hay with a pole with holes drilled into it, into which sticks were inserted replacing the natural stumps of branches, were frequent in some other regions at that time, such as Vorarlberg and Tyrol. Such haystacks were broader and lower (about 1.5 to 2.5 m) than the type in Figure 22. In any case, the hay had to be partly dried on the ground before stacks were made from it. Where usual, stacks for drying hay were built in moderately fertilized meadows (Section 6.3.2), where the denser grassland and longer plants than in non-fertilized meadows prolonged the process of drying hay on the ground, especially where the soil was often moist due to rain. In addition, the long-fibered hay from fertilized meadows facilitated the construction of stacks. Limiting the use of the stacks for drying hay to moderately fertilized sites was a general rule, as confirmed by data from the Slovenian and Swiss Alps, for instance.[573]

In comparison with haycocks, stacks greatly reduced the percentage of hay in contact with potentially moist ground during the drying process. In addition, stacks made better use of wind than drying on the ground. Hay dried in the stacks, whereas the hay in cocks dried spread on the ground.[574] Hay drying in a stack was significantly better protected against moderate precipitation than spread on the ground because the shape of the stack fostered quick runoff along the outer layer of hay;[575] for the same reason, drying hay in a stack also protected better against moderate rain than drying hay in cocks because the area covered by a stack per given volume of hay is much smaller than in a cock. Stacks required longer drying than in the case of hay spread on dry ground, but no additional turning and cocking was needed, except for very rainy weather. Consequently, drying hay in stacks reduced the losses of fragile but nutritious, protein-rich herbs. If a two-meter-tall stack built around a pole with natural branch stumps was well made and the weather not challenging, the hay was dry in a week—these data refer to the moist climate at Jelje and Konjska Dolina (about 1,300 m) in the Eastern Julian Alps.[576] A stack for drying hay is not to be

[573] Melik, 1931, 36–37, 52, 92–93, 95, and Figure 44; Escher, 1958, 402, 405; Kretschmer & Nestroy, 1974, 9, 11–12; Cevc & Čop, 1993, 28, 59; Makarovič, 2007, 214–216; Iuga, 2016, 71–72; Klinar, 2020, 56; Kravanja & Komac, 2023, 19.

[574] Melik, 1931, 36–37; Pfister, 1985a, 119; Cevc & Čop, 1993, 28.

[575] Pfister, 1985a, 119.

[576] Author's interview with A. Soklič, 2017, in Zgornje Podjelje.

Figure 22: Hay drying in stacks at Ljubnica, at the foot of the Pohorje Hills, first half of July 1963. There is a haypole (Sln. *ostrv*, Germ. *Hiefel*) in every stack. Slovenian Ethnographic Museum, Documentation, Fieldwork campaign 20, Vitanje (1963): photo 20/140 (by Boris Kuhar). *Slovenski etnografski muzej* . . . Vitanje / 63 F0000020/140. © Slovenian Ethnographic Museum.

confused with a different, densely pressed, larger haystack of a different shape intended to store hay in a meadow (Figure 65 in Section 6.1.2.5.1.2).[577]

It is not clear when stacks for drying hay became usual in the first regions in the Alps. The sophisticated hayrack, a much more complex construction, is clearly proven in the Southeastern Alps since the middle of the second millennium (see below), and so simple hay stacks such as the ones in Figure 22 were likely not invented later. It is clear, however, that they were only introduced in the eighteenth to twentieth centuries to some regions of the Alps,[578] which roughly corresponded with a significant increase in moderately ferti-

[577] Such confusion occurs, e.g., in Bavcon, Ravnjak, & Praprotnik, 2019, 159, presenting only the haystack for storing hay in the text, but the photo on the right side of page 159 shows a different stack for drying hay. Compare the text above and Melik, 1931, 92–93, 95; Iuga, 2016, 71–72, 75–76, 85; Janišová et al., 2023, 11. Thanks to the agronomist Janko Verbič for explaining the impact of hay stacks on losses of protein-rich herbs.

[578] Compare Melik, 1931, 95, and a late medieval proof of stacks for drying cereals in Makarovič, 2007, 212–216, 220, 227 (see also Klinar, 2020, 43–45). On stacks for drying hay since the eighteenth century, see Escher, 1958, 402–405.

lized meadows (see the introduction to Section 6.3 and Section 6.3.2), where stacks for hay drying were used. For instance, from Montafon, they reached St. Antönien in Prättigau (in eastern Switzerland) around 1740 and were used there until the twentieth century. Prättigau belonged to the core area of stacks for hay drying in mid-twentieth-century Switzerland: at that time, a farm at St. Antönien would keep three hundred to seven hundred poles for such stacks; similarly, a farm at Davos had several hundred. Stacks for drying hay were not constantly used for a long time in every region—in some cases, the use of stacks for drying hay was abandoned after some time.[579] In Montafon and Tyrol, they seem to have a longer continuous tradition than in Prättigau.[580]

Until the twentieth century, stacks remained important for drying hay in some regions of the Alps where the hayrack never became usual, and also locally in regions with hayracks—in (moderately fertilized) meadows without a free hayrack in the vicinity that could be used by the same farmer.[581] Major parts of the Slovenian Alps belong to the area of hayrack dominance; nonetheless, mid-twentieth-century farmers still made stacks for hay drying in the Upper Soča catchment area, the uppermost Savinja Valley, the Meža Valley, the Pohorje Hills (Figure 22), and some complexes of remote montane meadows in regions where hayracks dominated in the valley, such as the hay stacks used in montane meadows at Jelje and Konjska Dolina in Bohinj.[582]

In very rainy, cold, or cool summers, which sometimes occurred in successive years, neither cocks nor stacks for hay drying efficiently saved the hay, and, because drying cereals also failed, some core pillars of the farm economy were challenged. Such conditions, but also much shorter rainy periods during mowing time, promoted the spread of hayracks in many parts of the Alps.[583] A sophisticated hayrack "is a freestanding, roofed, permanent building with a load-bearing column-construction (or two such [parallel] load-bearing constructions), into which horizontal poles are set at intervals equal to the thickness of a grain sheave [or two sheaves], and whose purpose is to dry . . . various crops and fodder"[584] (Figure 36 in Section 5). First-crop hay and second-crop hay (which is shorter than first-crop hay; nonetheless, it can be loaded on the

[579] Escher, 1958, 402–405.
[580] Escher, 1958, 402.
[581] Melik, 1931, 36–37; Cevc & Čop, 1993, 59.
[582] Melik, 1931, 37, 45; Cevc & Čop, 1993, 23, 28; Bavcon, Ravnjak, & Praprotnik, 2019, 159; Kravanja & Komac, 2023, 19; author's interview with A. Soklič, 2017, in Zgornje Podjelje.
[583] Adapted from Pfister, 1985a, 119–120, and Makarović, 2007, 216, 219, 226–228 (neglecting the year-to-year weather variability).
[584] Makarovič, 2007, especially page 211 (the definition of a sophisticated hayrack is taken from this section of the Slovenian text, and the quote partly follows the same definition in the English abstract on page 247; the section in square brackets was added from Melik, 1931, 7).

horizontal poles of a hayrack) from meadows still dry in some hayracks in the Southeastern Alps[585] (Figures 23 and 24), and drying both hay harvests in a hayrack has a long history. For instance, drying both hay harvests in hayracks was recorded at Spodnje Gorje, in the eastern Julian Alps, around 1830.[586] A single hayrack (Figure 23) is a single construction, and a double hayrack consists of two such joined parallel constructions, usually covered by a shared roof together with the space in between providing additional storage capacity (Figures 24 and 25).[587] Roofed hayracks (with sufficiently long roofs to protect the hay as usual in the Southeastern Alps) solved the challenge of precipitation (not only rain, but also morning dew) and soil moisture during the hay-drying period; neither haycocks nor hay-drying stacks protected the hay in the case of longer rainy periods.[588] Even if rainy weather lasted for several weeks, when hay was drying in a hayrack, fodder retained its nutritive value and flavor; only the quality of its thin outer layer declined.[589] When a hayrack is filled, a farmer rakes the outer part of the hay in the hayrack from top to bottom to make sure that any rain reaching the hay despite the roof will drain optimally along the outermost hay layer and also to check that no hay will fall from the hayrack in case of windy weather.[590] Gradual drying at air temperatures in a (roofed) hayrack ensures preservation of high hay quality. Drying in a roofed hayrack is safe and it is based on sun and wind, without non-renewable energy input, but the practice is labor-intensive (however, less labor intensive than repeatedly cocking and spreading the hay or building haystacks for drying), and hayrack maintenance requires investment.[591] Hay is usually already partly dry when placed in a hayrack. If hay, second-crop hay in particular, is too moist to dry well when placed into a hayrack (very fresh or rain-wetted), it is taken out after a while and filled into the hayrack again in a different order to make sure that the entire hay harvest dries well and does not become moldy.[592] Some farmers always shake the stored hay that is only partly dry before they place it in a hayrack and let it dry completely there as a preventive measure against a potential

[585] E.g., Klinar, 2012, 107.

[586] SI AS 176, Franciscejski kataster za Kranjsko (Franciscan Land Cadaster for Carniola), L310, cadastral municipality of Spodnje Gorje, cadastral appraisal protocols: s.n., 310/4, 53, 97. Erroneously, Franjo Baš stated that second-crop hay has never been dried in hayracks (Baš, 1984, 336).

[587] Makarovič, 2007, 225–226, 228; Cevc & Čop, 1993.

[588] Melik, 1931, 53; Pfister, 1985a, 119; Juvanec, 2020, 15; author's interview with A. and M. Prepotnik, 2012, at Haudej farm in the Solčava region.

[589] Pfister, 1985a, 120.

[590] Author's observations in Bohinj in 2022, and explanation by the farmer M. Hribar, 2022, in Bohinjska Češnjica.

[591] Melik, 1931, 58–59; Pfister, 1985a, 42, 119; Juvanec, 2007, 36, 60, 91–93; Peterken, 2013, 251; Juvanec, 2020, 14–16.

[592] Pfister, 1985a, 120; Bavcon, Ravnjak, & Praprotnik, 2019, 146, 149, 168–169; Klinar, 2020, 56; author's interviews with Urh, 2017, in Stara Fužina, and with I. Mlakar, 2021, in the Zajamniki Alp.

Figure 23: Hayracks have existed in some remote montane meadows—for instance, in the Julian Alps (here a roofed single hayrack in the Javornica montane meadow complex in Bohinj, 1,160 m) and in Tyrol—but they have never been characteristic of remote montane meadows in any part of the Alps. Consequently, drying hay in remote montane meadows has often remained more sensitive to rainy weather than drying hay close to permanent settlements.[593] Photo: Žiga Zwitter, July 2021.

mold infestation in a hayrack.[594] Hayracks are not Alps-specific: they have also been (or were) used in some other parts of Central, Eastern, and Northern Europe, especially if unroofed constructions are also taken into account.[595]

In the (second half of the) last millennium, hayracks gradually spread to many places in the Eastern Alps and to much more limited areas in the Western Alps (mainly to southeastern and southwestern Switzerland). Within the Alps, the use of a hayrack has become most usual in the southeast.[596] Hayrack forms also developed over time.[597] Highly likely early evidence of hayracks in northwestern Tyrol dates back to the fifteenth century.[598] In the Alps of Carinthia and Upper Carniola (Germ. *Oberkrain*, Sln. *Gorenjska*; now northwest Slovenia), roofed hayracks were introduced roughly in the sixteenth century, when archival evidence makes it clear that they already existed (thus, they must have played a crucial role during the cold and wet late sixteenth-century summers), and they became usual there by the mid-seventeenth century at the

593 Melik, 1931, 9, 24–27; Cevc & Čop, 1993, 40, 48–49, 90, 136–137, 139–140; Juvanec, 2007, 87; Novak, 2024, 39, 105–106, 126, 131, 133.
594 Author's interview with Kunavec, 2021, in Stara Fužina.
595 Melik, 1931; Dahlström, Iuga, & Lennartsson, 2013, 196, 198; Peterken, 2013, 251.
596 Melik, 1931, 79, 89, 99; Cevc & Čop, 1993, especially pp. 7, 13–14, 22.
597 Makarovič, 2007, especially pp. 226, 228–230, 232; Melik, 1931; Cevc & Čop, 1993.
598 Melik, 1931, 60; Rhamm, 1908, 355; compare Makarovič, 2007, 217.

latest. Roofed hayracks gradually spread to further territories, and their density increased and reached its peak in the late nineteenth century.[599] In 1879, in the golden age of hayracks, there were seventy-six double and sixty-four single hayracks in the villages of Savica, Brod, Polje, Kamnje, Ribčev Laz, Laški Rovt, and Žlan in Bohinj (there were 106 houses in these villages in 1880) or in the immediate vicinity of these villages. At the same time, the number of hayracks per farm in the nearby villages of Češnjica, Jereka, Podjelje, and Koprivnik (there were 193 houses in these villages in 1880) was considerably higher: 217 double and 198 single hayracks belonged to farmers from these villages (on average more than one double plus one single hayrack per farm), including sixty-six double and thirty-one single hayracks located further away from villages.[600] The late nineteenth-century density of freestanding hayracks in parts of the Slovenian Alps was much higher than even the density of temporary hayrack-like constructions in late nineteenth-century Tyrol, whereas similar smaller structures attached to outbuildings were frequent in Tyrol.[601] In the Alps of southeastern Switzerland, the introduction of the hayrack was late in comparison with Tyrol, Carinthia, or the Alps of what is now northwestern Slovenia; it did not happen before the eighteenth or nineteenth century, and the hayrack was certainly only introduced to some Swiss microregions in the nineteenth (and possibly early twentieth) century.[602] The roofed hayrack

599 Makarovič, 2007, 212–223, 226–242; Baš, 1984, 336–338; Juvanec, 2007, 86–88; Zwitter, 2014d, 685; for further evidence of permanent hayrack constructions from the sixteenth and seventeenth centuries, see SI AS 721, Gospostvo Bled (The Seigneury of Bled, Germ. *Veldes*), Iustitialia, vol. 11, May 27th, 1597, November 17th, 1598, and January 21, 1603; SI AS 721, Iustitialia, vol. 12, May 1st, 1606, and May 25th, 1621; SI AS 721, Iustitialia, vol. 13, June 12th, 1606; Mlinar (ed.), 2018b, 92; and recent findings in Pečnik, 2024, 39, 72.

600 Novak, 1955, 300, and numbers of houses from the 1880 census in *Special-Orts-Repertorium von Krain*, 1884, 102, 104.

601 Makarovič, 2007, 228.

602 See the map in Cevc & Čop, 1993, 14. Concerning the chronology of hayrack introduction to southeastern Switzerland, compare Escher, 1958, 402–405, and Pfister, 1985a, 119–120, and note that the word *Heinze* can designate any type of wooden object outside where hay, cereals, or some other crops are loaded to dry there. A *Heinze* can be a haystack, a hayrack, or some other, less frequently used object types. The tentative general chronology of introduction and spread of *Heinzen* in southeastern Switzerland in the 1810s and 1820s as presented by Pfister 1985a, 119–120, is partly erroneous because it is too uniform; *Heinzen* were introduced earlier than that to some places of southeastern Switzerland and later than that to some others, as found by Escher in the mid-twentieth century (see the paragraph above on the chronology of introduction of stacks for drying hay). Pfister (1985a) referred only to Angst's research published back in 1930 in this respect. Consequently, Pfister's link between the spread of *Heinzen* and the 1816/1817 weather-related crisis is somewhat exaggerated; the connection was not so strong in general. However, it is likely that the phase of cool climate from the 1810s to the 1830s that culminated in the cold summers of 1812–1817, most prominently in 1816, the "year without a summer" (on climate and weather at that time, see also Brönnimann, White, & Slonosky, 2018, 312–315; Pfister & White, 2018; see also the introduction to Section 4.1.1.3) in fact fostered the spread of *Heinzen* to some places

Figure 24: A typical representative of the many double hayracks in Gorenjska, northwestern Slovenia. The primary function of the parallel pair of vertical column-constructions with horizontal poles on the longer sides of the building is drying hay, cereals, beans, clover, and some other crops.[603] The attics were usually used for hay storage; the same applies to the central part of the upper floor in the case of appropriate construction (the upper floor longitudinally partitioned off twice, so that rain did not reach the central part).[604] The closed central part of the upper-floor front of the large double hayrack in the rear part of this picture reveals the existence of such a (hay) storage facility there. In the front of the picture: drying second-crop hay, Studor in Bohinj, August 19th, 2021. Photo: Žiga Zwitter.

was not limited to Central Europe; it also spread in Sweden in the early modern period. In the nineteenth century, roofed hayracks were more widespread than in the mid-twentieth century and later; in the Alps, they (freestanding or attached to outbuildings) were generally used in most of the Alps and Pre-Alps of what is now northwestern, central, and partly eastern Slovenia (Figures 23–25), rather generally in many parts of Carinthia, at several places in East Tyrol, in parts of Carnia (Friul. *Cjargne*), and further west in Cadore (Ladin: *Ciadòre*), Alto Adige (South Tyrol), and some places in southern Switzerland, for instance.[605] These areas include, but are not limited to several (not all) of the wettest sections of the Alps.[606]

---

or stimulated the introduction of a more sophisticated *Heinzen* type, like a hayrack, to some parts of southeastern Switzerland.

[603] Melik, 1931; Makarovič, 2007.

[604] Melik, 1931, 13–14, and Figures 15 and 16; Juvanec, 2020, 16.

[605] Melik, 1931; Kretschmer & Nestroy, 1974, 14; Cevc & Čop, 1993; Juvanec, 2007.

[606] Compare the map of precipitation in the Alps in 1971–1990 in Bätzing, 2015, 37.

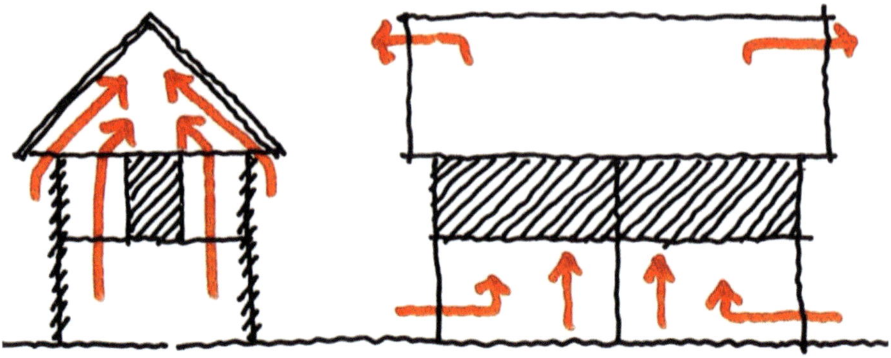

Figure 25: Scheme of air circulation in a double hayrack full of drying hay between the horizontal poles on both parallel sides on a sunny day.[607] Drawing by Borut Juvanec. © Borut Juvanec.

Double hayracks are a more recent invention than single ones. Air circulation in a double hayrack (Figure 25) additionally stimulates hay drying, especially if a hayrack is built in a west–east or northwest–southeast direction, the latter making best use of sun in the afternoon as the warmest part of the day; not all hayracks are orientated this way due to human and environmental factors. The Sun's rays warm the double hayrack's steep roof, and consequently warm air exits the highest part of the hayrack below the roof, and these air movements cause air to rise from lower parts of the hayrack, resulting in air circulation around drying hay even during calm weather.[608] In contrast to single hayracks, double hayracks (Figures 24 and 25) were most likely still a rarity in Slovenian Alps in the seventeenth century. In the second half of the eighteenth century and in the nineteenth century in particular, thousands of double hayracks were built in the Southeastern Alps (but they remained absent from the rest of the Alps), hence double hayracks became characteristic of some places in the southeasternmost Alps as the territory where hayrack construction was most developed, such as Bohinj and the Gail (Sln. *Zilja*) Valley in southern Carinthia.[609] According to a general comment, the double hayrack was recorded to have already been in use in the provinces of Carniola (Germ. *Krain*, Sln. *Kranjska*; it included the major part of the Alps of present northwestern Slovenia) and Carinthia around 1820.[610] This can be confirmed by the

[607] Juvanec, 2007, 56.

[608] Juvanec, 2007, 56, 92; Juvanec, 2020, 15; Klinar, 2020, 57.

[609] Makarovič, 2007, 219, 221–222, 228–230, 232–242; Melik, 1931, 14, 27–28, 46, 57, 60, Figures 11–12 and the map Répartition du *kozolec* en Slovénie; Juvanec, 2007, 30; Cevc & Čop, 1993, 7 (the English translation is inaccurate: the original Slovenian text says that examples of double hayracks can be found as far to the west as the source of the Drau/Drava River and also in Styria), 14, 79; Klinar & Roškar, 2020, 103.

[610] Juvanec, 2007, 26, 28.

example of the village of Bohinjska Bela, nowadays belonging to the territory of double hayracks;[611] they must have already existed there by around 1830 because hay storage (not only drying, thus requiring a double hayrack) in hayracks was recorded there at that time.[612] Similarly, the oldest still existing and dated double hayrack in the village of Studor in Bohinj, well known for a group of double hayracks below the settlement, was erected in 1825.[613]

Although drying hay in hayracks is environmentally friendly, it is highly labor intensive,[614] and so it has experienced a marked decline that started early in the twentieth century in areas where hayracks were less sophisticated, such as southern Switzerland, and this reached the territories of the most advanced hayracks in the southeastern Alps by the last third of the twentieth century, often related to a marked decline in milk flavor and meadow species richness (the transition from hay to silage) or additional energy input to the hay-drying process.[615] Some small and rare middling farms in the region still use them.

In addition to the constructions presented for drying hay and the hayrack without a roof, there existed a few other unroofed construction types for drying hay outdoors, most of them without a long historical tradition of being widely used in the Alps.[616]

Material strategies of coping with damp weather after mowing also include salting hay that was almost dry, but not entirely dry, when it was put in hay sheds in remote montane meadows (where hayracks were usually lacking). This practice decreased the likelihood of mold developing. In addition, salt increased the attractiveness of such fodder for livestock; namely, such hay had often previously been exposed to rain, and so nutrients were partly leached and its aroma and taste became less pleasant.[617] Through salted hay, livestock also received the salt ration that would otherwise have been provided separately.[618] In the haymaking season of the rainy and cold Little Ice Age summer of 1864,[619] an article on haymaking, with advice on how to reduce damage in case of rainy weather during haymaking season, was published in the news-

611 Map of the present distribution of the main hayrack types in Gorenjska: Klinar & Roškar, 2020, 103.
612 SI AS 176, Franciscejski kataster za Kranjsko (Franciscan Land Cadaster for Carniola), L335, cadastral municipality of Bohinjska Bela, cadastral appraisal protocols: s.n., 335/4, 59.
613 Cevc & Čop, 1993, 91; Juvanec, 2007, 32.
614 Pfister, 1985a, 43.
615 Juvanec, 2007; Makarovič, 2007, 228; Juvanec, 2020.
616 Escher, 1958, 404–405; Kretschmer & Nestroy, 1974, 12–14; Cevc & Čop, 1993, 24, 36, 38.
617 Bavcon, Ravnjak, & Praprotnik, 2019, 137; author's interviews with F. Mlakar, 2017, in Bohinjska Češnjica, and with Gašperin, 2017, in Stara Fužina; the latter mentioned application of c. 10 kg of salt to the hay when filling a small hay barn.
618 Pavkovič, 2021, 101–104; see also Section 8.2.2.1.
619 Zwitter, 2015d, 151; Studen, 2024.

paper *Novice*.[620] *Novice* was the leading Slovenian newspaper of the time; it was founded by the Carniolan Agricultural Society and thus also aimed at farmers, although many of them were still illiterate (see Section 8.2.2) and farmers were few among the subscribers to this newspaper (but many priests spread the content of the newspaper among farmers).[621] That article recommended salting not entirely dry hay before storing it; more precisely, the application of 0.12 to 0.14 kilograms of salt to a quintal (56 kg) of hay was recommended, with a reference to *Wochenblatt für Land-, Forst- und Hauswirthschaft*.[622] It is an open question whether the practice of salting not fully dry hay was indeed new for Slovenian farmers at that time.[623] Data from Bohinj also reveal another practice that might have been older or was at least cheaper but also less efficient: farmers placed dry limestone rocks in hay sheds in montane meadows, so that the limestone would absorb some moisture from hay that was almost completely dry, although not entirely dry.[624] All in all, such practices mitigated the hazard of mold infestation, but they were not always sufficient,[625] depending on the weather at hay drying times, microclimate of the meadow, plant species composition (see Sections 6.3.1 and 9), and hay storage facilities. A bit of hay stored almost, but not completely, dry was more easily accommodated in a manner that prevented mold infestation, where an at least medium-size barn was available, which was the case in permanent or temporary settlements—such a barn was usually not entirely packed with hay, so that air flows below the roof allowed final drying of limited hay quantities on top of the stored hay.[626] The hazard of mold infestation in almost dry hay was higher in many montane, subalpine, or alpine meadows, where a lot of hay had to be crammed into a small shed in the case of a good hay harvest[627] (such sheds have been documented for centuries[628]) or haystacks intended to store hay were erected that had to be particularly firmly pressed to withstand periods of windy and wet weather. Nonetheless, there were reasons why farmers kept much hay in distant sheds, such as increasing resilience in case of a fire at home when

[620] Studen, 2024.

[621] On *Novice*, see Rozman, 1970; Amon, 1991; Štih, Simoniti, & Vodopivec, 2008, 275, 280, 282, 284, 285, 288, 290, 292–293, 298–299, 308; and Selišnik & Verginella, 2021, 156, 159–161, 163–164, 168.

[622] "Košnja," 1864, 199. Conversion of a *Lot* to kilograms follows Vilfan, 1954, 68.

[623] A similar practice was currently documented in the Romanian Carpathians: when haystacks intended to store hay are built, salt is added "to better preserve the hay" (Iuga, 2016, 75–76, quote from p. 76).

[624] Author's interviews with Gašperin, 2017, in Stara Fužina; with A. Soklič, 2017, in Zgornje Podjelje; and with Frelih, 2021, in the Zajamniki montane meadow complex, where the author helped him put the hay in a hay barn.

[625] Author's interview with Gašperin, 2017, in Stara Fužina.

[626] Author's interview with Mencinger, 2018, in Stara Fužina.

[627] Author's interview with F. Mlakar, 2017, in Bohinjska Češnjica.

[628] E.g., SI AS 721, Gospostvo Bled (The Seigneury of Bled, Germ. *Veldes*), Iustitialia, vol. 12, April 6th, 1609, December 8th, 1610, April 18th, 1611, and July 2nd, 1620; SI AS 721, Iustitialia, vol. 13, October 1st, 1609.

roofs were made from combustible materials.[629] Remote hay sheds were small because this sufficed in usual years, and maintenance of many structures in a good state was a burden to the farmer[630] that would have increased in case of larger remote barns.

4.1.1.4 The cultural-history calendar impact on haymaking resilience to weather events problematic for haymaking

Whereas Section 4.1.1.3 focused on material strategies increasing the resilience of haymaking to rainy weather, this section looks at the work-free days in the calendar from the perspective of the resilience of haymaking when the weather turned rainy. Namely, other weather events during haymaking rarely damaged the hay severely compared to the processes frequently triggered by rain. Only at some particularly jeopardized sites and/or in exceptional years was there a significant risk during haymaking of adverse weather events other than rain or of weather-induced events that caused serious damage to hay. Examples are strong wind that blew the mowed hay away from some exposed sites in the mountains—thus, in the worst case, the hay was lost;[631] floods that carried away mowed hay; and severe hail storms that tore off grassland plants not mowed yet and washed the biomass away.[632] Where such hazards jeopardized haymaking, many work-free days in the calendar also decreased the resilience of haymaking to all such events, although only the frequently relevant connection between haymaking, rain, and holy days is mentioned in the continuation of this section.

Christianity spread relatively quickly around the Roman Empire and became the predominant religion there, especially in the fourth and fifth centuries.[633] In the majority of the easternmost Alps, Christianity faced a discontinuity, with a break mainly in the sixth century,[634] followed by re-Christianization especially in the second half of the eighth century and in the ninth century. A shift in religion entailed changes in cultural values and daily life—a process that took generations.[635] For a millennium and a half, the Christian calendar (later Roman Catholic, as prevalent in the Alps) required a significant percentage of farm-work-free days, with the exception of the unavoidable daily routine. This differed from the Christian rules in antiquity and from the

629 Author's interviews with Gašperin, 2017, and with Kranjc, 2018, both in Stara Fužina.
630 E.g., Ledinek, 1999, especially the section Kuharjev Joža; Novak, 2024, 23.
631 This was, for instance, a relevant hazard in the subalpine meadows at the very top of the Golica mountain ridge (Smolej, 2008, 15); on these meadows, see also Sections 3.1, 6.2.2.3, and 6.3.1.2.
632 Compare Šifrer, 1962, 148, 150, 154, 170–171.
633 Trombley, 2006; Bratož, 2000, 16, 25.
634 Štih, 2010, 90–91, 93–96; Bätzing, 2015, 58.
635 Štih, 2014a, 53–54, 58–59; Štih, 2018; Makarovič, 1995, 27–30, 41–42, 46, 48, 50, 52–54, 101; see also Štih, Simoniti, & Vodopivec, 2008, 95–101, 150–153.

official Roman Catholic norms since the second half of the twentieth century, as explained in this section. In practice, the impacts of largely work-free holy days on the lives of the rural population and on the farm economy were diverse, ranging from positive in terms of physical and mental (through religion and socializing) relaxation and recovery, and information exchange among the gathered people, to negative, such as (1) widespread excessive alcohol consumption, frequent enough to risk addiction (see also Section 8.1.2), (2) general economic impacts of an unnecessarily high number of largely work-free days, and (3) adverse weather impacts on harvest when sensitive crops, such as mowed hay, were not stored in time.[636]

In the fourth century, weather-dependent farming tasks outside were still exempt from work prohibited by Christianity on Sundays.[637] After the opposing decisions by the Council of Orléans (AD 538) prohibiting much farmwork on Sundays, and the Council of Rouen (AD 650) returning to the older clause exempting farming,[638] this changed in the following centuries. In several parts of the Eastern Alps, the prohibition on farmwork on Sundays and holidays roughly corresponded with general re-Christianization. For instance, the Synod of Cividale (in Friuli, Patriarchate of Aquileia; Friul. *Cividât*) in AD 796 emphasized that Christians were not allowed to carry out farmwork on Sundays (except for the unavoidable daily routine); it is clear that the Church made sure at that time already that synods' conclusions reached many believers.[639] Roughly contemporaneously, disobedience of the prescribed Sunday behavior was listed in the catalog of sins in a confession prayer, originating in the mid-ninth century at the latest, and written down and partly reworked as the First Freising Manuscript around AD 1000, most likely in what is now (upper) Carinthia.[640]

In addition to Sundays, the number of largely work-free holy days increased significantly in the Middle Ages.[641] Carrying out most work on those days was considered a deadly sin, except for the unavoidable daily routine.[642] As early as the transition from the eighth to ninth century, there were thirty-six holy days

[636] See the continuation of this chapter and Pribyl, 1998; Makarovič, 1982, 254, 258, 412–415; Spehr, 2010; Ambrožič, 2018, 227; see also von Hörmann, 1877, 149–150, 152–153; Ušeničnik, 1945, 164 (thanks to the church historian Lilijana Žnidaršič Golec for recommending this work); Grass, 1950, 159, 169–170; Netting, 1974, 69; Makarovič, 1995, 20–21, 101, 225, 227, 238–240, 254, 256–257, and caption to Figure 20; Štih, Simoniti, & Vodopivec, 2008, 152; Selišnik & Verginella, 2021, 159; *Meja za malo tvegano pitje alkohola* and the physician Luka Zwitter, personal communication.
[637] Wildhaber, 1956, 27; Pribyl, 1998, 100–101; Peršič, 2022, 62–63.
[638] Pribyl, 1998, 102.
[639] Štih, 2014a, 54–56; see also Makarovič, 1995, 240, and Pribyl, 1998, 117.
[640] Štih, 2018, 347, 350; compare Pribyl, 1998, 103.
[641] Ušeničnik, 1945, 164; Dohrn-van Rossum, 2006, 869; Spehr, 2010, 209–210; Štih, 2014a, 55.
[642] Kranemann, 1995, 2050; see also Makarovič, 1995, 240, and Pribyl, 1998, 117.

annually in addition to Sundays, as recorded in the Archbishopric of Salzburg, which covered the great majority of the Eastern Alps north of the Drau (Sln. *Drava*) River.[643] The number of holidays kept rising in the following centuries. On average, this made approximately five workdays per week in the Late Middle Ages; however, holidays were not evenly distributed throughout the year.[644] In the sixteenth century, as many as around 120 days annually were festal:[645] in the parish of Radovljica (Germ. *Radmannsdorf*), for instance (now in northwest Slovenia), there were 122 holy days when most work was prohibited, meaning seventy in addition to Sundays (as recorded in 1468), and around 110 holy days (including Sundays) in the parish of Kranj (Germ. *Krainburg*) in the late fifteenth century.[646] In addition, several regions in the Eastern Alps knew work-free afternoons on Saturdays and days preceding other holy days (see the paragraph on Sabbatism below). Moreover, the farm workload on rainy days was often reduced significantly, allowing additional relaxation and recovery,[647] and the amount of farmwork in winter was in general much lower than in other seasons.[648] However, working days during the summer peak of agricultural work coinciding with long daylight hours were often long and lasted, for instance, from 4 am until 9 pm, with relatively short breaks in between.[649] In general, such long working days were more frequent with the increased farm workload beginning in the second half of the eighteenth century and the nineteenth century.[650] Nonetheless, several factors must have also stimulated long working days in the haymaking season in earlier centuries, among more ambitious farmers in particular. These included long daylight time; a desire to make use of nice weather with highly positive impacts on hay quality; imminent rain events; a possible high number of rainy days in the summer resulting in less time suitable for haymaking and cereal harvest and in a heavier workload in terms of additional cocking, spreading, and turning wetted hay (see Section 4.1.1.3); and a higher number of holidays in the same week.[651]

Although the activities of the Church itself belonged to the most important ways of enforcing the prohibition on farmwork on Sundays and other holy

643 Štih, 2014a, 55. For the spatial coverage of that archbishopric, see *Slovenski zgodovinski atlas*, 2011, 56; see also Štih, 2018, 339.
644 Dohrn-van Rossum, 2006, 869.
645 Hiller, 2020, 82; see also Makarovič, 1995, 46, 52.
646 Makarovič, 1995, 101, and the text in front of Figure 16 in the unpaginated final section of the book; Štih, Simoniti, & Vodopivec, 2008, 237–238.
647 Seventeenth century: SI AS 721, Gospostvo Bled (The Seigneury of Bled, Germ. *Veldes*), Iustitialia, vol. 13, October 1st, 1609; early twentieth century: Šarf, 1974, 143 and Slovenian Ethnographic Museum, Documentation, Fieldwork campaign 21, Škocjan (1964): notebook no. 3, 132.
648 Makarovič, 1995, 87, 240, 256, and the text in front of Figure 16 in the unpaginated final section of the book.
649 Makarovič, 1995, 88; examples: Šarf, 1974, 139–140; Hessenberger, 2022, 166–167.
650 Makarovič, 1995, 87–88, 125–126, 227–228, 240, 244.
651 Compare Makarovič, 1995, 88, 225–228.

days, except for the unavoidable daily routine, religion and ecclesiastical rules also impacted some legal norms prescribed by the state and local communities that were steeped with faith.[652] An example is the prohibition on Sunday work in *Lex Baiuvariorum*,[653] originating in the seventh or eighth century at the latest. Focusing on meadows, *Lex Baiuvariorum* prohibited the use of ox-drawn carts, mowing hay, and gathering hay on Sundays.[654] Similarly, Charlemagne's law from 789 prohibited mowing hay on Sundays.[655] Although the territory where these rules were in force within the Eastern Alps was initially more limited, Franco-Bavarian legal rules were gradually introduced in the great majority of the Eastern Alps from the second quarter of the ninth century onward. Nonetheless, it took time for the new legal order to actually be selectively respected.[656]

In the first centuries of the second millennium, there was diversity in observance of feast days across time and space within the Alps. (1) The behavior of part of the population differed from their belief and words because some Christian rules posed serious challenges to daily life,[657] such as rigorous limitations on work during the summer peak of the farm workload in times of haymaking.[658] Thus, in the bishopric of Bressanone (Germ. *Brixen*; the diocese roughly covered Tyrol),[659] the synod of 1438 took seriously the distress posed by occasional frowning upon the behavior of those that could not take time for many work-free holy days during the summer peak of (farm)work,[660] and so, at least occasionally, not everyone respected the prohibition on work. (2) In addition, there was a "lack of uniformity in the medieval church regarding festal day observance[.]"[661] In the Eastern Alps, the bishopric of Bressanone stood out for tolerance toward work on feast days in the summer, at least in part of the fifteenth century. Namely, in 1438, the Bressanone synod converted some summer holy days (not all of them and no Sundays) to "half holy days," and so the population was obliged to attend mass on those days but they were allowed to work after the service.[662] (3) Moreover, the list of work considered

[652] Štih, 2014a; Štih, 2018; see also Grass, 1950, 155–169; Makarovič, 1995, 240; and Pribyl, 1998, 117.
[653] Schmidt-Wiegand, 1991.
[654] Wildhaber, 1956, 27–28; Schmidt-Wiegand, 1991; Pribyl, 1998, 103–104.
[655] Wildhaber, 1956, 28.
[656] Štih, 2014a, 45–47; Spehr, 2010, 210; compare Elias, 1994, 181.
[657] Wildhaber, 1956, 30; Makarovič, 1982, 438–439, 516; Pribyl, 1998, 105–106; Štih, 2014a, 58; Kos, 2015, 286–287, 346; see also Section 8.2.1.3.1 in this volume.
[658] Compare Makarovič, 1995, 87, 225, 227, and Iuga, 2016, 79–82.
[659] *Die Grenzen der Bistümer des Alten Reiches um 1500*. Thanks to the historian Matjaž Bizjak for recommending this map.
[660] Grass, 1950, 167.
[661] Hiller, 2020, 80 (quote).
[662] Grass, 1950, 167. Thanks to the classical philologist Matej Petrič and the church historian Lilijana Žnidaršič Golec for their confirmation that Grass's interpretation of the formulation "unde sub avisamento opportunum nobis videretur, fore statuendum" is

forbidden on Sundays and other holy days may have varied to some extent from one area to another.[663]

In the thirteenth and fourteenth centuries, there were "important developments in popular religious practices," including "an increased emphasis on individual and private devotion among the laity" and "[t]he growth of affective piety" in times of "an upsurge in blood piety and devotion to Christ's wounds." The religious position of Christ became exceptionally central in the fourteenth century; the cult of Mary faded. Roughly by that time, the density of church buildings where service was held relatively frequently became dense enough to make possible (relatively) regular mass attendance in most microregions throughout the Eastern Alps. In such a spiritual atmosphere and ecclesiastical history context, Sunday Christ wall paintings were introduced in Europe in the fourteenth century and were painted until mid-sixteenth century.[664] They depict the central figure of a standing Christ surrounded by either farming implements and artisanal tools (Figure 26) or by paintings of work procedures and other activities considered immoral on Sundays and other holy days; the scenes are sometimes joined by devils.[665] The tools on Sunday Christ paintings were believed to cause Christ to suffer (i.e., bleed) anew if believers worked on Sundays and other holy days.[666] Thus "[a]lmost all of the agricultural implements depicted . . . involve cutting, chopping, slicing, . . . [and] have blades of some type,"[667] or pointed ends.[668] "In some instances[,] the tools penetrate Christ's near-naked body and blood spurts from the wounds"; the implements are sometimes "linked to his limbs and chest with arrow-like lines" (Figure 52 in Section 6.1.2.1.2).[669] On some Sunday Christ paintings, there are drops of Christ's blood on the tools, or the tools are presented on a red background of Christ's blood.[670] "Further examples simply depict the implements around his body[,] and their connection with Christ's suffering is more nuanced."[671] Sunday Christ was a monitory motif, a perpetual didactic and andragogical reminder "or pictorial sermon against participating in work-day activities on a

correct: this is one of the conclusions of that synod, not its recommendation for the future.

663 Moser, 1990, 334; Pribyl, 1998, 103–104.

664 Wildhaber, 1956, especially pp. 1, 32–34; Kos, 2015, 346; Makarovič, 1995, 29, 41, 101 (compare Makarovič, 2005, 242, 245–246, 252–253, 260–264; Štih, Simoniti, & Vodopivec, 2008, 253; and Ambrožič, 2018, 223, 225–227); Kretzenbacher, 1997, especially pp. 158, 178; Hiller, 2020, 80, 86, 97, 107 (quotes). See the inscription from St. Roch's Church in Tesero, Trentino (South Tyrol) in Moser, 1990, 339.

665 Wildhaber, 1956; Hiller, 2020.

666 Wildhaber, 1956, especially pp. 8, 19–21, 24–25; Hiller, 2020, 79, 82–85, 96–100, 103, 106, 111.

667 Hiller, 2020, 98.

668 Wildhaber, 1956, 24.

669 Wildhaber, 1956, especially p. 24; Kretzenbacher, 1997, 172; Hiller, 2020, 82 (quote).

670 Hiller, 2020, 98.

671 Hiller, 2020, 82.

Sunday" or another holy day, not to risk the afterlife recklessly.[672] Church wall paintings were highly relevant media to reach illiterate believers, as the great majority of farmers were at that time. There is every reason to assume that priests explicitly referred to Sunday Christ paintings in their sermons.[673]

Sunday Christ paintings did have an impact on community behavior.[674] Frequently located on a prominent part of an external church wall, at the top of the hill, and often using strong color contrasts, Sunday Christ paintings were discernible from some distance. To increase their visibility from a distance, they were often painted high on the church wall.[675] In addition, the painting of St. Christopher, the saint believed to prevent a sudden death if looked at in the morning—thus regularly and purposefully looked at by many—was often placed nearby on the external church wall. Thus, in many cases, Sunday Christ was glanced at simultaneously with St. Christopher.[676] More than seventy Sunday Christ wall paintings are nowadays known to be kept in Europe *in situ* or were documented but deteriorated later; "many of the paintings cluster around the [A]lpine regions, in particular the Trentino-Alto Adige" (South Tyrol). Many more Sunday Christ paintings existed, and some are still covered by later plaster.[677] The gravity zone of some churches with a Sunday Christ painting was rather broad, especially in the case of pilgrim churches, such as Maria Rojach in the Lavant Valley, Carinthia,[678] or Crngrob (Germ. *Ehrengruben*), now in west-central Slovenia,[679] and so the motif was observed not only by the locals. Sunday Christ, a non-dogmatic motif, became unpopular in the context of Catholic renewal, with the implementation of the relevant decree of the Council of Trent (Ital. *Trento*, Germ. *Trient*; the decree from early December 1563) in particular. Namely, there are many profane tools and/or quotidian scenes on Sunday Christ paintings, and this motif is not directly rooted in the Bible. Consequently, Sunday Christ could be perceived as "an insult to the holy figure of Christ," and so most paintings of this type were later plastered over or destroyed, often in the seventeenth century.[680]

---

672 Wildhaber, 1956; Hiller, 2020 (quote from p. 82).

673 Compare the general discussion in Makarovič, 1995, 46, 52, 94–95, 240, not focused specifically on the Sunday Christ motif.

674 Wildhaber, 1956, especially, p. 34; Hiller, 2020, 82, 85, 106, 111.

675 Hiller, 2020, 82, 85, 103–106, 111.

676 Wildhaber, 1956, 23, 33–34; Höfler, 2001, 155–156.

677 Wildhaber, 1956, especially pp. 1, 22–23; Moser, 1990, 336, 340–341, 343; Kretzenbacher, 1997, 159, 170, 181; Höfler, 2001, 155–156; Kos, 2015, 346; Hiller, 2020, 80–81, 86 (quote from p. 80).

678 Compare Kretzenbacher, 1997, 176.

679 Höfler, 1996, 86, 91–92.

680 Wildhaber, 1956, 1, 9, 16, 19, 34, and Table 6, Figure 19; Makarovič, 1995, 52; Kretzenbacher, 1997, 176, 181; Hiller, 2020, 80–81, 86, 90–91, 97 (quote from p. 81); Wohlmuth (ed.), 2002, 775 (thanks to the church historian Lilijana Žnidaršič Golec for letting me know which the relevant decree was).

The following analysis of Sunday Christ paintings from the viewpoint of vulnerability of grassland farming to rainy weather builds on photos and descriptions of Sunday Christ paintings in cited publications[681] and on a visit to the known Sunday Christ paintings in northwest Slovenia,[682] an area characterized by a relatively high density of representations of this motif.[683] The painting at Koreno nad Horjulom (Figure 26), usually only casually mentioned in Sunday-Christ publications,[684] proved particularly relevant. None of the publications analyzed these paintings from the perspective in this volume.

The sensitivity of mowed hay to nutrient leaching through precipitation increases significantly during the drying process (Section 4.1.1.3). This fact is easily recognized by traditional farmers because it is reflected in the color of the hay, which becomes undesirable much more quickly in the case of rain on relatively dry hay than in the case of rain on hay that had just been mowed.[685] Consequently, this section focuses on Sunday Christ representations of storing hay or implements clearly used for storing hay, when the nutritive value and palatability of hay would decrease most significantly if wetted. Focusing on storing the hay, most prominently, a farmer driving an ox-drawn hay wagon down a steep slope, aided by a devil, stands in a Sunday Christ painting at Sts. Philip and James Church at Campitello di Fassa (Ladin *Ciampedél* in Trentino-Alto Adige – South Tyrol, drawn about 1455–1460).[686] Likewise, summer hay raking and hay wagon transport with a pair of draft animals appears in the Sunday Christ on the church of Schlans in Graubünden (Ital. *Grigioni*; Romansh *Grishun*), possibly from late fourteenth century.[687]

Moreover, the analysis of tools in Sunday Christ paintings proved relevant for the analysis of the content of these works of art from the perspective of vulnerability of grassland farming to rainy weather. Local farmers were supposed to clearly recognize the purpose of all tools painted in a Sunday Christ motif, should the image have the potential to reach its goal—and at that time realism was generally well present in church wall paintings. Thus, local specifics in implement types known from other sources are sometimes evident in Sunday Christ paintings. True, it is also clear that the shape of tools is sometimes adjusted to fit better into the composition, or the artist did not look at the details.

---

681 Wildhaber, 1956; Moser, 1990; Kretzenbacher, 1997; Hiller, 2020.
682 See the list in Kos, 2015, 346.
683 Compare Kretzenbacher, 1997, 178–179, and Kos, 2015, 346 (note that the Sunday Christ paintings at Gosteče and Mošnje have fully deteriorated).
684 E.g., Kretzenbacher, 1997, 179.
685 Author's interview with J. Kavčič, in Dolina above Tržič, 2021, aware of the increased rate of nutrient leaching when the drying of hay progresses.
686 Description of the Sunday Christ content at Campitello di Fassa, without further conclusions concerning storing hay: Wildhaber, 1956, 16, and Table 6, Figure 20, and Hiller, 2020, 86, 91.
687 Wildhaber, 1956, 13, and Table 4, Figure 13.

Figure 26: Sunday Christ on the south external wall of Sts. Hermagoras and Fortunatus Church at Koreno nad Horjulom, pre-Alpine hills west of Ljubljana, probably painted around 1410.[688] The black arrow points to a hayfork with two long coplanar tines and a shorter protruding tine (see the text below and Figure 27). It is typical for Sunday Christ paintings that many sharp implements, such as the hayfork of interest here, are projected toward Christ's body to demonstrate that he is hurt if these implements are used on Sundays or other holy days. Characteristically, the lower edge of this wall painting is located on the uppermost section of the outer church wall (thus, note the shadow of the modern church roof on the upper part of the painting), at the top of the hill. The strong contrast between the implements and the dark background (which might have originally been blood red) contributed to the visibility of the content.[689] A painting of St. Christopher used to stand on the same south side of this church.[690] Photo: Tomaž Zwitter, 2022.

In addition, the farm implements in Sunday Christ paintings alone do not provide reliable insights into the forms of locally used implements because the artist might have been inspired by painted examples from elsewhere or by printed figures, or he might have observed farm implements at distant places where he might have spent a significant part of his life. Nonetheless, comparisons of the farm tools in Sunday Christ paintings with ethnologically documented implements demonstrate that several tools of such or similar shapes were indeed used in the southeastern Alps,[691] and this methodological approach also proved rewarding in this analysis.

688 Höfler, 2001, 104.
689 Compare Hiller, 2020, 104.
690 Peskar, 1991, 72.
691 Moser, 1952, 475, 485–486, 489–490; Wildhaber, 1956, 25; Moser, 1974, 14–15; Moser, 1990, 336–338, 341, 343–350, 362–363.

Among the farm implements in Sunday Christ paintings, this research focuses on a hayfork with two long coplanar tines and a third, shorter tine protruding upward between the coplanar ones. It appears in the Sunday Christ painting at Sts. Hermagoras and Fortunatus church at Koreno nad Horjulom (in the pre-Alpine hills west of Ljubljana, probably painted around 1410; Figure 26),[692] and it is relevant for our topic. The wooden hayfork type with two coplanar tines and a third tine protruding upward between the coplanar ones was territorially widespread, as proven by mid-twentieth century ethnological documentation.[693] This hayfork type was also used in the Eastern Alps, even though in certain regions at least it was relatively rare, such as in mid-

---

[692] Höfler, 2001, 104. Höfler does not deal with farm tools. Motifs of this kind might have also existed in some other known Sunday Christ paintings because several of them are only partially preserved.

[693] E.g., in Slovenia at rather randomly distributed places of thorough fieldwork campaigns: at Lopar near Marezige, at Budanje near Vipava, at Materija at the foot of the Brkini Hills, at Male Lipljene near Grosuplje, at Veliki Vrh pri Šmarju, at Šentvid pri Stični, at Gradenc near Žužemberk, and at Drašiči in Bela Krajina (Slovenian Ethnographic Museum, Documentation, Fieldwork campaign 1, Šentjurij, Škocjan, Turjak (1948): drawing 1/52b; Fieldwork campaign 2, Šmarje–Sap, Polica (1949): drawing 2/14; Fieldwork campaign 4, Marezige (1950): drawing 4/56; Fieldwork campaign 5, Šentvid pri Stični (1950): drawing 5/87 (in less detail, this drawing is published in Grafenauer, 1970c, 242); Fieldwork campaign 12, Brkini (1955): drawing 12/7 (this drawing is published in Šarf, Habinc, & Smerdel, 2011, 671–672); Fieldwork campaign 14, Žužemberk (1957): drawing 14/18; Fieldwork campaign 15, Vipava (1958): drawing 15/13; Fieldwork campaign 22, Drašiči (1965): photo 22/264). Most of these figures can be accessed in the museum's online database, at *Slovenski etnografski muzej* . . . Šentjurij – Škocjan – Turjak / 49 R0000001/052b; *Slovenski etnografski muzej* . . . Šmarje-Sap, Polica / 27 R0000002/14; *Slovenski etnografski muzej* . . . Marezige / 86 R0000004/56; *Slovenski etnografski muzej* . . . Brkini / 178 R0000012/7; *Slovenski etnografski muzej* . . . Žužemberk / 16 R0000014/18; *Slovenski etnografski muzej* . . . Vipava / 98 R0000015/13; *Slovenski etnografski muzej* . . . Drašiči / 57 F0000022/264.
In addition, such a hayfork is documented at Dobrepolje (Novak, 1944, 149), and the Nose ethnological collection at Bojanji Vrh near Muljava keeps such forks from the village or its surroundings (*Etnološka zbirka Nose*).
All these data come from the southern half of Slovenia. In the Slovenian Alps, historical use of this hayfork type, on some farms at least, is documented in Bohinj and at Češnjevek near Cerklje, at the foot of the Kamnik–Savinja Alps, for instance (one example from Bohinj is shown in this volume's Figure 27; another one from Bohinj that was used in a mountain meadow at Zajamniki in the first half of the twentieth century and the one from Češnjevek are kept by the Gorenjska Museum—see Dolžan Eržen, 2024, 268, 280, and Dolžan Eržen, personal communication).
I thank the ethnologist Barbara Sosič, curator for folk livelihood at the Slovenian Ethnographic Museum, who identified this fork type in the cited documentation from Šentvid (and possibly Šentrupert, see Figure 28), the ethnologist Tatjana Dolžan Eržen, curator at the Gorenjska Museum, who checked the provenance of the exhibited hayfork from Zajamniki and identified this hayfork type from Češnjevek in the museum's inventory, and Marjeta Hodnik, who identified this same fork type in the Nose collection.

twentieth-century Bohinj[694] (Figure 27). This hayfork type was an implement typically used for collecting and storing hay, and so it was used in the most rain-sensitive phase of haymaking, which is most relevant for this section. Strong people lifted much hay at the same time with such a fork when uploading or unloading a wagon (Figures 28 and 29) or throwing collected hay into

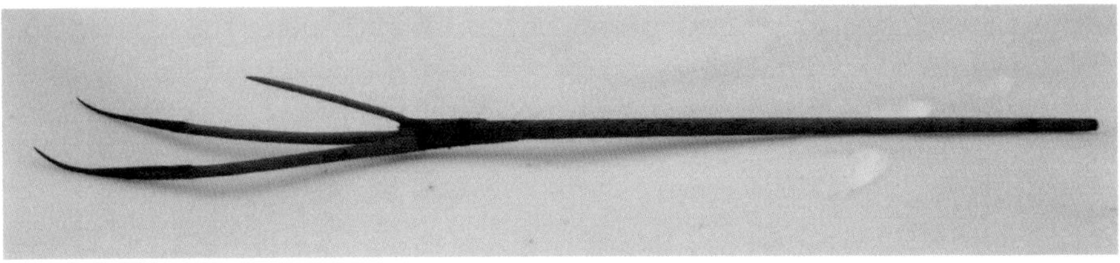

Figure 27: A wooden hayfork with two coplanar tines with iron tine points and a third entirely wooden tine protruding upward, kept in a haybarn in the remote montane meadow complex on Vogar Hill in Bohinj. This exemplar has not been used for decades. This fork type is shown on the early fifteenth-century Sunday Christ painting at Koreno nad Horjulom, Figure 26. Gratitude is expressed to Lovro and Marjeta Hodnik for showing this hayfork to me while making hay, and for the joint ascent of Vogar Hill in winter, which allowed me to take this photo. Photo: Žiga Zwitter, 2022.

Figure 28

694 Author's interview with Kunavec, 2023, in Stara Fužina: the knowledgeable interviewee was not familiar with this hayfork type; but it was used by a farm (or a few farms) from the same village in a meadow complex where the interviewee's farm had no meadow; see Figure 27.

Figures 28 and 29: Use of a hayfork, possibly the type with two coplanar tines and a third tine protruding upward between the coplanar ones. Characteristically, longer hay was placed on the outer side of the wagon and carefully distributed, and shorter hay was put to the middle to make sure that the load would not collapse during transport.[695] Photo 28 was taken at Šentrupert in the pre-Alpine hills of eastern Slovenia in August 1961, and photo 29 in the northern Dinaric Karst at Male Lipljene in summer 1964. Another photo of haymaking at Male Lipljene in 1964[696] and the huge hay quantity lifted in Figure 29 at the same time[697] suggest that the hayfork type of interest here was probably used in Figure 29. Slovenian Ethnographic Museum, Documentation, Fieldwork campaign 18, Šentrupert (1961): photo 18/136 (by Boris Orel); Fieldwork campaign 21, Škocjan (1964): photo 21/413 (by Boris Kuhar). *Slovenski etnografski muzej ...* Šentrupert / 3 F0000018/136; *Slovenski etnografski muzej ...* Škocjan / 138 F0000021/413. © Slovenian Ethnographic Museum.

695 Author's interview with J. Kavčič, in Dolina above Tržič, 2021.

696 Slovenian Ethnographic Museum, Documentation, Fieldwork campaign 21, Škocjan (1964): photo 21/253, documenting the use of the hayfork type of interest here at Male Lipljene at that time (*Slovenski etnografski muzej ...* Škocjan / 66 F0000021/253). The date June 1st, 1964, in the database on the website of the Slovenian Ethnographic Museum only means that the photo was taken during the fieldwork campaign that started on that date. The e-system requires an exact date, and so the museum enters the first day of the campaign when a precise date is not known (the ethnologist Miha Špiček, personal communication). Obviously, haymaking started later than June 1st in 1964 (see Section 6.1.2.2).

697 Thanks to Barbara Sosič for the general confirmation of this observation.

a nearby barn.[698] Ethnological data explain that the pair of coplanar tines was coated in iron or strengthened with horn tips in some cases,[699] as in Figure 27, to make the fork even more robust and able to carry heavy loads of hay. Alternatively, the tine tips of entirely wooden hayforks were thermally strengthened by being scorched in embers every year to reach a similar goal.[700] The protruding tine exposed to heavy pressure was made from very hard and resistant wood, such as cornelian cherry (*Cornus mas*).[701] The author's participant observation in haymaking on Vogar Hill in Bohinj, when rain was approaching, demonstrated that the forks in Figures 26 and 27 were also very handy tools for collecting hay that had been raked into windrows. Namely, when one pushes such a hayfork to the front, the protruding tine blocks the hay, causing it to roll, and so this tool also allowed quick hay collection (to build hay cocks or store the hay), followed by fine raking. For this use to be possible, the projecting tine had to be not too long, and it had to project upward sufficiently steeply—a condition often fulfilled (e.g., in Figures 26 and 27), but not in all forks of this type.[702]

Other farm implements in Sunday Christ paintings connected or potentially connected with haymaking, such as a scythe (see Section 6.1.2.1.2), a wooden hayfork with three (or four) tines, all of them coplanar in a fan-like form,[703] or a wooden rake,[704] appear more frequently on this church wall-painting

[698] Personal experience by the ethnologist Barbara Sosič and her analysis of use of this hayfork type in the photos held by the Slovenian Ethnographic Museum. This is perfectly in line with the presentation of the hayfork use in montane meadows in the eastern Julian Alps in Šarf, 1974, 139–140, and with the marked general predominance of the hay rake over hayfork at haymaking in Carinthia (Moser, 1952, especially pp. 463, 466, 471, and 485).

[699] Examples of hayforks of this type with coplanar tines coated in iron: at Dobrepolje (Ložar, 1944, 149), at Male Lipljene near Grosuplje, at Veliki Vrh pri Šmarju (Slovenian Ethnographic Museum, Documentation, Fieldwork campaign 1, Šentjurij, Škocjan, Turjak (1948): drawing 1/52b; Fieldwork campaign 2, Šmarje–Sap, Polica (1949): drawing 2/14), and at Bojanji Vrh near Muljava from the village or its surroundings (*Etnološka zbirka Nose*). A hayfork of this type with horn tips on coplanar tines from Češnjevek is held by the Gorenjska Museum (inventory number 510:KRN;E-0000424 – Dolžan Eržen, personal communication).

[700] Kuret, 1989, 439.

[701] Ložar, 1944, 149; on cornelian cherry wood, compare Brus, 2008, 145.

[702] An example of a fork not fulfilling this condition: Slovenian Ethnographic Museum, Documentation, Fieldwork campaign 21, Škocjan (1964): photo 21/253 (*Slovenski etnografski muzej . . .* Škocjan / 66 F0000021/253).

[703] E.g., on Sunday Christ paintings at Saak (Sln. *Čače*) in the Gail Valley (Germ. *Gailtal*, Sln. *Ziljska dolina*; Moser, 1990, Figure 6), at Mauthen in the same valley, at Tesero in Trentino (South Tyrol; both locations: Wildhaber, 1956, 16–17, and Table 7, Figures 21 and 22), and at St. Mary's Church at Biella in Piedmont (Hiller, 2020, 86, 92).

[704] On rakes in Sunday Christ paintings, see Moser, 1952, 490; Wildhaber, 1956, 9–10, 13, 15, 17, 20–21, and figures; Moser, 1990, 338, and figures; Kretzenbacher, 1997, 171–173; Hiller, 2020, 82, 99, and figures.

motif in the Alps than the hayfork with two long coplanar tines and a third, shorter tine protruding upward. Nonetheless, the scythe, rake, and hayfork with a coplanar fan-like form are not so clearly connected with the prohibition on storing dry hay on Sundays and other holy days in the most rain-sensitive phase of haymaking, and so they are not a matter of this analysis. A rake, for instance, was a multipurpose tool; in haymaking, it was typically used not only for storing hay but also to spread the freshly mowed hay, to turn the hay, and for other work.[705] Likewise, a fan-like hayfork was typically also used for turning the hay in some regions of the Alps.[706]

Even though the prohibition on gathering in dry hay on Sundays is clearly represented on only a few Sunday Christ images, ethnological documentation proves that even this type of work was usually considered banned, although rain would result in significant loss of the hay's nutrient value and taste. In 1977, the ethnologist Marija Makarovič studied the Alpine settlement of Strojna in the Meža Valley, Carinthia, consisting of isolated mountain farms. In 1901, Strojna's inhabitants were praised by the local priest for exemplary mass attendance that was considered the best in the parish. By 1977, average religious commitment there was far from comparable to the early twentieth century, ranging from deep piety rooted in tradition all the way to atheism. Still, only some farmers stored dry hay on Sundays and other traditional religious holy days, although many occasionally broke some other religious constraint on daily life. Some deeply pious farmers repeatedly did not gather in dry hay on Sundays and holidays in 1977, even though this caused repeated wetting of their dry hay. They claimed that, if God permitted the hay to get wet, he would also dry it later[707] (in fact, hay quality decreased significantly; see Section 4.1.1.3). Exactly the same is documented in Plavški Rovt in the Karawanks

[705] Spreading and turning the hay belong to the work typically performed with a rake (see Figure 20 in this volume; abundant evidence from Carinthia: Moser, 1952, especially pp. 463, 466, 471, and 485; various sites in Slovenia: Slovenian Ethnographic Museum, Documentation, Fieldwork campaign 2, Šmarje–Sap, Polica (year 1949): photo 2/242 (at Dolenja Vas pri Polici); Fieldwork campaign 10, Goriška Brda (year 1953): photo 10/194 (at Slavče); Fieldwork campaign 13, Kostanjevica (year 1956): photo 13/266 (at Orehovec); Fieldwork campaign 16, Črni Vrh, Vojsko (year 1959): photo 16/135 (at Vojsko); Fieldwork campaign 21, Škocjan (year 1964): photo 21/436 (at Male Lipljene); Individual fieldwork by curators of the Slovenian Ethnographic Museum: photos 1685/I and 1785/I (both taken in 1975 at Zelen Breg in the Alps of southeastern Carinthia). Most of these figures are accessible online: *Slovenski etnografski muzej* . . . 630:LJU;F-0001685/I; *Slovenski etnografski muzej* . . . 630:LJU;F-0001785/I; *Slovenski etnografski muzej* . . . Šmarje-Sap, Polica / 220 F0000002/242; *Slovenski etnografski muzej* . . . Kostanjevica / 46 F0000013/266; *Slovenski etnografski muzej* . . . Črni Vrh, Vojsko / 151 F0000016/135.

[706] Author's interview with Kunavec, 2023, in Stara Fužina: in the Upper Bohinj Valley (i.e., the villages in the Ribnica River catchment area and east of there) a rake would be used, and in the Lower Bohinj Valley (i.e., south of the Upper Bohinj Valley) a fan-like coplanar hay fork to turn the hay in the mid-twentieth century.

[707] Makarovič, 1982, 428–446.

above Jesenice.[708] Similarly, when the ethnologist Fanči Šarf studied haymaking in montane meadows above the village of Bohinjska Bela in the eastern Julian Alps in the first half of the 1970s and collected memories related to this work type from the late nineteenth century onward, she noted that in former times haymakers descended from distant meadows to the village on Saturday evening regardless of whether the work was finished or not. Where needed, they returned to the distant meadow on Monday morning. Continuation of work on Monday was usually needed only in the case of unfavorable weather (see Section 4.1.1.3) because farmers tried to secure enough people to finish haymaking by Saturday evening if the weather was favorable.[709] These examples demonstrate that the absence of any kind of haymaking tasks on Sundays and other holy days was widespread. They suggest that gathering in hay is hidden behind the scythe, mowing, or multipurpose tools such as rakes depicted in several Sunday Christ paintings.

However, although Sunday strongly prevailed in the Alps as a regular weekly holy day, largely free of farmwork in the long run, this was not the case in all regions. Heterodox Christian Sabbatism (celebration of Saturdays) is documented in medieval sources in rare areas from Central Europe to southern Italy. Thus, also in some parts of the Alps, Sabbatism was usual in the Middle Ages or even in the Early Modern Period. In Friuli (Friul. *Friûl*), it is attested as a usual practice in the countryside in the late eighth century. Evidently, this tradition was retained in the long run; in 1499, the patriarch of Aquileia declared that Saturday was not allowed to be considered a holiday anymore, the "erroneous habit" in Friuli of starting the weekly holiday every Saturday with the ringing of the noon bell was officially banned, and it was proclaimed that farmwork (exceeding unavoidable everyday tasks) was allowed until sunset on Saturday. However, the practice was not eradicated before the Tridentine reforms were fully implemented, or it was abolished even later. In 1602, a visitation in Carnia (Friul. *Cjargne*) ordered by the patriarch of Aquileia revealed that people stopped working with the Saturday (noon) bell and a holiday started, but they did not mind working on Sundays. Contemporaneously, in 1604, the start of the holiday on Saturday morning or at noon was documented in the Alpine region of Cadore (Ladin: *Ciadòre*) just west of Carnia. This heterodoxy was considered a grave infringement by the official Church. Similar practices were also found in several other regions of the Eastern Alps, at least in parts of the Slavic east of the Patriarchate of Aquileia,[710] in part of

[708] Ravnik, 2008, 37.

[709] Šarf, 1974, 143.

[710] Peršič, 2022, especially pp. 54, 56–58, 62–64, 66–67. E.g., there was a Holy Sabbath Church in the Poljane Valley (Sln. *Poljanska dolina*), in Bukov Vrh above Poljane in the patriarchate of Aquileia, in the pre-Alpine hills south of the Julian Alps, recorded in 1501 as *kh[irche] zum heiligen Sambstag*, but a church consecration feast was (already) celebrated on Sunday at that time (Blaznik (ed.), 1963, 318, 361). Characteristically, the patron saint was changed in the sixteenth or seventeenth century because Sabbatism

the diocese of Ljubljana,[711] and at several places in Tyrol, thus in the dioceses of Trento, Bressanone, and Freising.[712] At Caldaro (Germ. *Kaltern*) south of Bolzano (Germ. *Bozen*), work stopped at 1 pm on Saturdays and days before other holy days, at Umhausen in the Ötz Valley (Germ. *Ötztal*) at 3 pm, and at Telfs near Innsbruck at 4 pm in the warm half of the year (as documented in 1631). Farmers, farmhands, and maids gave up work and headed to the village. At Kundl and Liesfeld (in the Inn Valley close to Alpbach), this habit was kept until the eighteenth century, at least,[713] and probably as late as the mid-twentieth century in several places in Tyrol.[714] Similarly, at Plavški Rovt above Jesenice in the western Karawanks, work on Saturdays still stopped as early as 3 pm in the mid-twentieth century.[715] At an annual level, the end of work with the exception of the unavoidable routine as early as Saturday afternoon was a significant extension of the already long festive period. Nonetheless, at least at some of the places where farmwork on the days preceding all holy days had to stop before sunset, farmers were allowed to work longer on Saturday afternoon or the afternoon before a holy day at times of haymaking and cereal harvest if bad weather prevailed on other workdays, but it turned better on the day before a holy day—but such an extension of the work period was only permitted if the priest consented, as documented in Telfs (1631), and in Kundl and Liesfeld (1730).[716]

During the Reformation, Lutheranism, not a significant phenomenon in the Alps in the long run, reversed the theologically founded prohibition on Sunday work in limited parts of the Alps. Nonetheless, the tradition of largely work-free Sundays was usually by and large kept by the Lutherans because people were accustomed to it, to attending mass, and to resting. This was especially true because several public holidays were abolished at the same time by the Protestant authorities, and so the role of Sunday relaxation increased where the Protestant Reformation took root. Calvinism, which became prev-

was eradicated in the post-Tridentine Church. Thus, the same edifice was recorded as Our Lady of Sorrows Church in 1704 (compare Vale, 1943, 178), and the Virgin Mary is still its official patron saint (*Župnija Poljane nad Škofjo Loko: Cerkev Žalostne Matere božje na Bukovem Vrhu*). This was a usual historical development path: the Virgin Mary was one of the most frequent patron saints to replace Holy Sabbath (Peršič, 2022, 64). Nonetheless, the tradition of its former heterodox patron saint is seemingly so strong that it is recorded as Holy Sabbath Church (*Sv. Sobota*) in a recent atlas (*Atlas Slovenije*, 2005, 119).

711 Compare Ravnik, 2008, 37, and the maps of dioceses since the late Middle Ages in *Slovenski zgodovinski atlas*, 2011, 100, 116, 122, 124, 154, 178.

712 *Die Grenzen der Bistümer des Alten Reiches um 1500.*

713 Grass, 1950, 165–167.

714 E.g., in the second half of the nineteenth century, farmwork, with the exception of the unavoidable daily routine, stopped at noon at several places in Tyrol on the day before major holidays, such as the feast of Corpus Christi (von Hörmann, 1996, 107–108).

715 Ravnik, 2008, 37.

716 Grass, 1950, 166–167.

alent in limited sections of the Alps in the long term, required stricter obedience of the Sunday work interdiction than Lutheranism.[717]

Returning to the territories of prevalent mainstream Roman Catholicism, in the long run, sermons were another important medium to remind ordinary people to abstain from work on Sundays and holy days, unless the work truly could not wait, and storing the hay was sometimes explicitly mentioned as a forbidden task in a sermon.[718] In addition, supervisors monitoring whether people obeyed the ban on work on Sundays and other holy days, and people that carried out punishment are known from Tyrol.[719] Such monitoring might have been more relevant in the case of isolated farms. Namely, in most compact settlements, there was spontaneous community control, once carrying out certain work on Sundays and holy days became a generally accepted taboo—people sought "to avert any social risk that might [have led] to their becoming marginalized within the community, a risk resulting from the incorrect execution of . . . work," such as breaching the taboo of work on a holy day[720] (see the aforementioned distress posed by occasional frowning upon breaking the work interdiction because some people could not take time for many work-free holy days in summer, as reported by the 1438 Synod of Bressanone). In addition, local traditions knew of many cases of accidents claimed to be caused by work on Sundays or other holy days; these functioned as didactic and andragogical cautionary examples "considered to be true." Some of them explicitly referred to haymaking, and some of them were based on real experience of unusual events, which by alteration of reality (moralizing "in the sense of 'a retrospective rationality based on the sacred order attributed to the world'") and exaggeration became "part of the system of local beliefs."[721] An example is that nobody should gather hay in on Prophet Elijah's Day (July 20th), otherwise lightning will strike it.[722] In this case, the threat follows the prophet's specific attributes: his relation to lightning.[723] Similar tales had already been documented centuries earlier.[724]

Nonetheless, work prohibitions on Sundays and festal days were "not always observed" in early modern times.[725] In addition, a certain degree of di-

[717] Pribyl, 1998, 106–107, 117–118; Dohrn-van Rossum, 2006, 869–870; Spehr, 2010, 210.
[718] Wildhaber, 1956, 30–31; Makarovič, 1995, 52.
[719] Grass, 1950, 162–163.
[720] Iuga, 2016, 75 (quote), 78.
[721] Pribyl, 1998, 118; Iuga, 2016, 67–68, 71–72, 77–82 (first quote from p. 79, second and third quotes from p. 82).
[722] Kuret, 1989, 509.
[723] Compare Kuret, 1989, 507–509, and Iuga, 2016, 82.
[724] An example is the etiological tale concerning a large speleothem structure in Tkalca Cave (Sln. *Tkalca jama*, now in central southern Slovenia). It claims that a weaver was petrified there because he worked on a holy day, as reported by local farmers and written down in the seventeenth century (Carmichael, 1995, 100).
[725] Pribyl, 1998, 105; Hiller, 2020, 82 (quote).

versity in likely interpretations of what was prohibited to do on those days has already been mentioned; this diversity seems to have persisted until the standards set by the Council of Trent came into force,[726] and partly even longer (see Section 6.4.3.1). In the Urbeis Valley in Alsace, north of the Alps, it was recorded in 1536 that more than a few people were actively engaged in market activities on Sundays and other holy days, which was then prohibited, with the exception for those in distress.[727] Thus, exceptions were allowed, in several cases at least. According to a ruling from Zürich in 1530, haymaking on Sundays and other holy days was allowed if farmers were heavily pressed by weather or other business of the integrated farm economy.[728] Similarly, in the province of Tyrol, physical work on Sundays and holy days was allowed if abstaining from it would lead to serious distress, according to the ordinances from 1532 and 1573.[729] This is relevant in the context of the Little Ice Age peak in the last third of the sixteenth century (Sections 4.1.1.2 and 4.1.1.3). All this proves that experience of a significant percentage of hay decaying or its quality declining significantly due to abstinence from work on Sundays and other holy days led to partial distancing from the Sunday work taboo if the situation was critical.

In the mid-seventeenth century (1642), the number of holy days was significantly reduced; around ninety days annually with a prohibition on work were kept in the official Catholic calendar since then; at many places in the Eastern Alps, this amounted to thirty-nine days in addition to Sundays. In the bishopric of Ljubljana (Germ. *Laibach*), for instance, this meant only half the former number of holy days with a prohibition on work, apart from Sundays.[730] Further reductions followed in the second half of the eighteenth century: from 1754 onward, there were only sixty-five days annually when most farmwork was prohibited in the Alps of the Habsburg Monarchy, thus thirteen in addition to Sundays; in the bishopric of Ljubljana this was slightly more than half the figure from before 1642, including Sundays. From 1754 onward, attending mass was also still compulsory on twenty-six other former full holy days, but the prohibition on work during the rest of the day was abolished (these holy days were reduced to "half holy days"). In 1771, the number of full holy days was slightly increased in the Alps of the Habsburg Monarchy to reach sixty-eight holy days annually, including Sundays, but at the same time "half holy days" were converted into usual workdays without

[726] Moser, 1990, 334; compare Wildhaber, 1956, 30.
[727] Wildhaber, 1956, 29–30; see also Makarovič, 1995, 89.
[728] Wildhaber, 1956, 30; Pribyl, 1998, 115.
[729] Wildhaber, 1956, 30.
[730] Ušeničnik, 1945, 164–165; Dolinar, 1980, 36; Makarovič, 1995, 101, 225; Dohrn-van Rossum, 2006, 870; Ambrožič, 2018, 226; Lilijana Žnidaršič Golec, personal communication. Thanks to Lilijana Žnidaršič Golec for recommending the cited publications by Ambrožič, Dolinar, and Ušeničnik.

compulsory mass attendance. Thus, farmwork exceeding the unavoidable daily routine was prohibited on Sundays and sixteen other feast days, and this rule was in force for more than a century.[731] However, many farmers resisted the reductions in numbers of holy days,[732] and most farmers continued to respect the historical habit of some previous official work-free holy days not requested to be free of work by the ecclesiastical authorities anymore.[733] The gradual real abolition of the great majority of formally canceled historically largely work-free holy days went hand in hand with a significant general increase in workload on farms as a consequence of agricultural modernization in the second half of the eighteenth century and in the nineteenth century, such as the introduction of fallow-free crop-rotation systems in arable land, the introduction of new crops (including labor-intensive ones), and the gradually increased duration of barn-keeping in animal farming. This also demanded increased efficiency of the laborers and longer work days on farms on average, often extremely long in the warm part of the year, while keeping the work limited to the unavoidable routine on Sundays and other prescribed work-free holy days.[734] Thus, starting in the (seventeenth or) eighteenth century or later, when farmers started largely following the reduced number of holy days,[735] the ban on Sunday and holy day work in meadows did not increase the vulnerability of grassland farming to rainy weather at times of mowing as much as earlier in history, but it retained some impact. Interestingly, a similar phenomenon (reduction of holidays, then in terms of only a half holiday regularly free of work at the weekly level) was recorded in the case of Sabbatism in Carnia as early as the transition from the sixteenth to seventeenth century (see above).

Largely work-free Sundays and other holy days on farms were challenged by interactions with the (rural) non-farm economy and its complex impacts on the farm economy (see, e.g., the abovementioned 1530 ruling from Zürich, and the cases of Sunday grassland irrigation in Sections 6.4.3.1 and 6.4.3.2.1). In general, a significant increase in non-agrarian jobs also affected farmwork because numerous employees in the non-farm economy were also involved in farming as part-time farmers, as workforce helping their relatives on farms, or as day laborers working on farms during the peaks of farmwork, such as summer haymaking. Consequently, haymaking timing has adjusted to a certain degree to focus on the days when such workers have been free of their main

731 Ušeničnik, 1945, 166–167; Makarovič, 1995, 102; Štih, Simoniti, & Vodopivec, 2008, 238, 251; Dohrn-van Rossum, 2006, 871; Ambrožič, 2018, 217–219, 222–223, 227; Lilijana Žnidaršič Golec, personal communication.

732 Dolinar, 1980, 36.

733 Dohrn-van Rossum, 2006, 871; Ambrožič, 2018, 227; see also Makarovič, 1982, 428.

734 Makarovič, 1995, 87–88, 93, 125–126, 225–228, 240, 244, 254, 256, 329, and the text in front of Figure 16 in the unpaginated final section of the book; Makarovič, 2007, 228.

735 Compare Ambrožič, 2018, 227, and Makarovič, 1982, 428.

job.[736] Developments in the Church following the Second Vatican Council, which was held in the 1960s, put an end to farmwork officially forbidden on Sundays and other holy days in Roman Catholicism. Although attending mass was kept compulsory on those days, the prohibition on work was replaced by a recommendation that these days be largely work-free for the recovery of the mind and body.[737] The older tradition was partly maintained in the subsequent decades, primarily by older generations.[738] For instance, some farmers in Strojna in the Alps of northern Slovenia would still not gather in dry hay on Sts. Peter and Paul Day (June 29th) in the 1970s despite imminent rain[739] (until the changes following the Second Vatican Council, Sts. Peter and Paul Day was kept on the list of holy days with farmwork prohibited in Roman Catholicism, except for the unavoidable daily routine[740]). However, such strict observance of this feast was rare in Slovenian territory at that time,[741] and haymaking on Sundays also gradually gained importance.[742] Thus, in practice, such rigorous observance of feast days had already been abandoned in many places in the countryside before the Second Vatican Council[743] because non-obedience of the strict prohibition on work in many places must have been rooted in a longer development path as outlined above, and was thus not as simple as a quick adaptation to the reform spirit in the Church. With the gradual spread of work on Sundays and other holy days, on the one hand, the ban on work on at least one day per week has not increased the vulnerability of grassland farming to rainy weather anymore. On the other hand, a full workload seven days a week, not only during the peaks of farmwork but as a usual practice, has decreased the quality of life of some farmers.[744]

---

[736] Compare Grass, 1950, 167–168 (footnote 80); Šarf, 1974, 142, 144; Makarovič, 1995, 52, 93, 101, 225–226, 252, 329; Pribyl, 1998, 108–111, 117–119; see also Section 3.1 in this volume.

[737] Kretzenbacher, 1997, 160, 162.

[738] Compare Pribyl, 1998, 118; Makarovič, 2005, 246. As late as the transition from the twentieth to twenty-first century, when a calf in a summer mountain pasture in the Julian Alps became lost, one attempt at an explanation identified the following possible reason: the calves had been taken to that mountain pasture complex on a Sunday (main author's participant observation).

[739] Makarovič, 1982, 428–429.

[740] Compare Ušeničnik, 1945, 165–167; Kretzenbacher, 1997, 160, 162; and Pribyl, 1998, 117.

[741] Makarovič, 1982, 428.

[742] Šarf, 1974, 144, see also p. 142.

[743] E.g., Slovenian Ethnographic Museum, Documentation, Fieldwork campaign 21, Škocjan (1964): notebook no. 3, 132.

[744] Compare Makarovič, 1982, 428; Makarovič, 1995, 227–229, 244, 254, 315, 329–332, 337–338, 341–342; author's interview with an anonymized interviewee from Stara Fužina in 2018, and recent observations in the Slovenian Alps.

4.1.2 Changing landforms in continuously existing grasslands

Landforms in some continuously existing grasslands have been so dynamic to have caused significant changes in site conditions within the last millennium. The difference in elevation between the bottoms of pits and tops of hummocks in hummocky meadows (Section 3.2), for instance, has increased since the late medieval creation of meadows, but it remains unclear at what pace; such dynamics have also differed strongly from one area to another. Only preliminary calculations based on various kinds of measurements and observations exist, and these figures differ too much to reflect only different environmental conditions. Nonetheless, they demonstrate relevant changes at a millennial scale. The pits in a hummocky meadow in dolomite sediments on the Kräuterin Massif in Styria are estimated to have decreased due to corrosion by 77 mm per millennium, but hummocks decrease at a lower pace as well. The geomorphogenetic dynamics in the Radovna Valley in the eastern Julian Alps, on limestone fluvial and moraine sediments, is significantly more pronounced.[745]

These morphological changes have had important ecological consequences; they have caused increasing habitat diversity in hummocky meadows. Namely, patchworks of sites have developed in hummocky meadows due to differences in soil properties and soil depth as well as due to differences in insolation, and the patchwork character has increased when the elevation differences between the pits and tops of hummocks have been increasing. Shallower—and at least in some places sandier—soils on hummocks dry much sooner than deeper soils in pits, which at least in some places contain more silt and clay. The higher the hummocks, the drier their tops. Sunny sides are drier than shady sides. Observations on Vogar Hill in 2020 showed that if there is deciduous forest or extensive deciduous hedges in the relative vicinity, the wind deposits fallen leaves in pits of mowed hummocky meadows in late autumn and they decompose there, leading to more nutrient intake to soils in nutrient-richer pits than on nutrient-poorer hummocks. An inventory of 116 plant species in a meadow (*Mesobrometum*) with low hummocks on carbonate sediments in the Radovna Valley in the eastern Julian Alps, which is regularly mowed once a year, could prove differences in vegetation despite the relatively low heights of hummocks in a broader Alpine context.[746] Among the five most frequent species on the tops of six hummocks rising 33 to 55 cm above pits, there are two that do not belong to the most frequent species in the pits: sheep's-fescue (*Festuca ovina*) and upright brome (*Bromus erectus*), which both prefer drier soils. The percentage of mosses is also higher on the tops and slopes of hummocks than in pits, and it is lower on low hummocks than on higher ones. Mosses can tolerate drought in a state of anabiosis, and the moss species identified are relatively

[745] Embleton-Hamann, 2004, 181–182, 187; Koblar, 2013, 6, 9, 11–12, 18, 28–30.

[746] Prezelj, 2012, 7, 9, 12, 19–36; for a much less detailed discussion, also lacking changes over time, see Peterken, 2013, 282.

Figure 30: A patchwork of sites in a hummocky meadow in the Voje Valley. In this case, differences in plant communities are well developed, ranging from wild thyme (*Thymus* sp.) and matgrass (*Nardus stricta*) on the top and slopes of the hummock (lower and middle central section of the photo) to plants yielding much more hay in the pit (upper as well as central right and left sections of the picture). The quality of hay in the pit is better than matgrass. The abundance of thyme on some hummocks is a rare example of a plant that could help motivate some farmers to keep mowing hummocky meadows at present. Namely, even though the nutritive value of thyme is low,[747] thyme is highly relevant in extensive dairy cattle farming: thyme consumption leads to a particularly pleasant milk flavor,[748] and thyme has favorable impacts on livestock health.[749] Photo: Barbara Lampič, September 11th, 2019. © Barbara Lampič.

heliophilous, and so they cannot tolerate the shade of the higher plants in the pits. Similarly, among the five most frequent species in the pits, there are two that do not also grow most frequently on the tops of the hummocks: tormentil (*Potentilla erecta*) and alpine bistort (*Bistorta vivipara*), which both prefer medium soil water availability, and thus more humid soils than on the tops of hummocks. The percentage of grasses is significantly higher in pits than on slopes and on the tops of hummocks.[750] Early spring observations in montane hummocky meadows at the southern Zajamniki (eastern Julian Alps) showed that spring crocus (*Crocus vernus*) grows only in pits and on slopes, but not on

[747] Heinz, Mayer, & Kuhn, 2012, 18.
[748] Author's interview with C. Mlakar, about 2018, in the Zajamniki Alp in the Julian Alps, commenting on montane hay harvest at Zajamniki.
[749] Vogl-Lukasser et al., 2006, 71, 73, 84, 124, 178, 189; Heinz, Mayer, & Kuhn, 2012, 18.
[750] Prezelj, 2012, 24, 27–28, 32, 35–36, and appendix C5; Aeschimann et al., 2004, vol. 1, 372–373, 776–777, vol. 2, 860–867, 870–871, 916–919.

the tops of hummocks.[751] The patchwork of habitats in hummocky meadows is clearly seen in Figure 30. Gradual historical increases in the depth of pits and height of hummocks have had not only significant ecological impacts. From a long-term perspective, this type of geomorphic dynamics gradually rendered haymaking in hummocky meadows increasingly labor-intensive because, even in times of mowing by scythe and manual raking, mowing and hay drying was significantly more labor intensive in a hummocky meadow than in a flat one (Sections 3.2 and 4.1.1.3). In major parts of the Julian Alps and also in parts of the Northern Limestone Alps, studies have demonstrated that the soil is shallower on the top of hummocks and deeper in pits.[752] This synthesis is relevant for such hummocky meadows and not for those whose formation was triggered through a windthrow, where the distribution of various soil depths is more heterogeneous, with deeper soils on hummocks and thin rendzinas in pits.[753]

Geomorphically relevant impacts of floods and mass movements are presented more extensively below because they often caused discontinuity in the existence of the grasslands affected. However, in some cases they caused only environmental change in continuously existing grasslands. In the lower Stubach Valley in Pinzgau, for instance, a forest was cleared at Uttendorf before 1530 to obtain timber and firewood for a large-scale wood consumer: the salt works at Hallein, located about 90 km downstream on the Salzach River. In 1585, 2.5 km^2 of forest was again cleared there, and the wood was used elsewhere. Consequently, there were repeated rockfalls from the clearing, and so the pasture below was converted from a cattle pasture into a pasture for goats.[754] The transition from a cattle pasture to a goat pasture must have seen a limited decrease in grassland plant diversity. Namely, among all domestic ungulates, grazing cattle induce the highest grassland species richness; the impacts of goat grazing on biodiversity are not as favorable.[755]

4.1.3 Grassland dynamics co-influenced by insects, rodents, and fungi

Many wild animals and fungi had an impact on grassland dynamics. The focus of this section is on those with the most significant impacts on grassland yields.

---

751 Personal observations on April 12th, 2020, at 1,270 m.

752 Cvetek, 1971, 75–76; Embleton-Hamann, 2004, 176–177; Prezelj, 2012, 30, 36; Koblar, 2013, 5, 29.

753 Embleton-Hamann, 2004, 177, 183–184; Koblar, 2013, 3.

754 Rohr, 2007a, 318–319.

755 Ellenberg & Leuschner, 2010, 1007.

4.1.3.1 Insects

The large-scale local reproductive success of animals with a strong influence on grassland ecosystems, perceived as pests by farmers, or their large-scale arrival from elsewhere was sometimes recorded in preserved historical sources.

In many grassland ecosystems in valleys and basins in the Alps in the colline and lower montane elevation belts, the hazard of extensive economic damage in grasslands posed by cockchafer (*Melolontha melolontha* and *Melolontha hippocastani*) larvae every third year, and less frequently every fourth or fifth year, has been serious, provided that site conditions were suitable for this animal and that there were large populations of cockchafers in the microregion.[756] Cockchafer larvae do not cause significant damage to grassland yields in mountain grasslands at higher elevations, where the average air temperature from April to October is less than about 12.5 °C. The highest sites where they cause considerable damage in grasslands varies in the Alps, reaching 1,100 m on sunny slopes in Tyrol (these data are probably from the early twentieth century), slightly more than 950 m in Bohinj, and about 700 to 800 m at some other places in the southeasternmost Alps in the mid-twentieth century. Thus, many grasslands in the Alps were not threatened: those close to the highest (or high) permanent settlements in many regions, most of the distant mountain grassland complexes, and some areas at lower elevations, such as all wet meadows.[757] For example, in Bohinj, only the lower complexes of some distant meadows have sometimes been affected, and higher grasslands have been safe, according to mid-twentieth century information.[758] Where grassland irrigation was practiced, farmers could prevent the negative impact of cockchafers on the hay harvest by properly timed watering of meadows (Section 6.4.1). Elsewhere, gathering cockchafer beetles in early morning, as recorded in recent centuries, might have at least partly reduced the local impacts:[759] a farmer would, for instance, place linen sheets below a tree full of cockchafers in the crown and shake the tree early in the morning when the beetles were still sleeping; in such cases, the beetles fell down and did not fly away.[760] Cockchafer beetles feed on many tree species, but they show a clear preference for some tree genera and species, such as oak (*Quercus* spp.) and European larch (*Larix decidua*). Slightly less preferred but still consumed on a large scale are

---

756 Janežič, 1971; Billamboz, 2014, 115.

757 Janežič, 1971, 4, 18–19, 21–28, 31, and the supplement Karta majskega hrošča za Slovenijo (Cockchafer Map of Slovenia).

758 Author's interviews with Grm, 2017, and with Škantar, 2017, both in Srednja Vas; with Kunavec, 2017, and with Mencinger, 2017, both in Stara Fužina; with F. Mlakar, 2017, in Bohinjska Češnjica; and with L. and M. Odar, 2017, in Studor.

759 Janežič, 1971, 7, 11–12, 15, 30; see also Nowosadtko, 2007, 89.

760 Author's interviews with A. Soklič, 2017, in Zgornje Podjelje, and with Kranjc, 2018, in Stara Fužina.

hazel (*Corylus avellana*) and maple (*Acer*) leaves, and others.[761] Thus, in microregions where these species were not too frequent in ecosystems suitable for cockchafers (in the case of a forest, cockchafers prefer the trees at forest edges),[762] farmers knew where to look for them.[763] The gathered cockchafer beetles were killed with boiling water.[764] In the district of Radovljica (Germ. *Radmannsdorf*), covering several dozen settlements in Bohinj, part of the Sava Dolinka Valley (Sln. *Zgornjesavska dolina*) and upper Ljubljana Basin northwest of Ovsiše (located east of Kropa), eight metric tons of cockchafer beetles were destroyed in 1951, for instance.[765] Such campaigns are known from the eighteenth or nineteenth to the twentieth century,[766] and they were sometimes financially stimulated.[767] Another widespread historical means of cockchafer suppression was gathering the larvae in the years when they abounded; this was usually done in arable land at plowing time.[768] For instance, every farm at Stara Fužina had to deliver a certain amount (e.g., 0.5 *Metzen*; i.e., about thirty liters) of cockchafer larvae when cockchafer populations in the region were large in the mid-twentieth century; they were killed with quicklime.[769] This type of cockchafer suppression is also reported in Bavaria,[770] and so it was carried out in many parts of the Alps. Until the mid-twentieth century as the timeframe studied here, chemical suppression *in situ* was not carried out. In this respect, reports from Slovenia[771] match those from Bavaria.[772]

The risk of a cockchafer plague fluctuated considerably in history, and it has been locally to regionally specific. To a certain degree, this was the result of weather and climate. Some factors with a strong influence on the abundance of cockchafers are frost and snow in the second half of April or in May, or low temperatures and heavy rain at that time, which kill the beetles; low

[761] Janežič, 1971, 5–6, 28–29.
[762] Billamboz, 2014, 117.
[763] For instance, author's interviews with Kranjc, 2018, in Stara Fužina; the interviewee mentioned exactly oak and larch as their preferred tree species, which corresponds with the research cited above. Author's interview with Mencinger, 2017, in Stara Fužina, mentioning cockchafers' preference for hazel, was again in line with scientific knowledge. Author's interview with F. Mlakar, 2017, in Bohinjska Češnica, who mentioned oak, larch, and birch (*Betula pendula*)—of course, cockchafers also feed on birch (Janežič, 1971, 28).
[764] Poženel, 2007, 73.
[765] Janežič, 1971, 15; on the territory of this district, see *Virtualna arhivska čitalnica: SI_ZAL_RAD/0014 Okrajni ljudski odbor Jesenice-Radovljica, 1945–1955 (Fond/zbirka).*
[766] Nowosadtko, 2007, 89, 94; Janežič, 1971, 11–12.
[767] Janežič, 1971, 11–12.
[768] Nowosadtko, 2007, 89.
[769] Author's interview with Gašperin, 2017, in Stara Fužina. Conversion into liters: Vilfan, 1954, 49, 51.
[770] Nowosadtko, 2007, 89.
[771] Janežič, 1971, 30.
[772] Nowosadtko, 2007, 89.

temperatures in late spring kill the larvae close to the surface; excessive water forces the larvae to crawl to the surface, where they suffer from insolation and are easy prey for birds, for instance; serious drought kills eggs and young larvae in the soil; sudden unusually early and persistent cold weather in autumn before the larvae burrow deep into the soil also kills large numbers of them. Severe weather-related stress in the case of very large cockchafer populations increases the likelihood of infectious diseases spreading among the weakened organisms, such as the fungus *Beauveria brongniartii*, which kills the cockchafer larvae in the soil. Other factors include, for instance, decades-long "cyclic, high density variations . . . which show a mass development between two phases of latency."[773] Cockchafer populations move slowly, and so there have been patchwork-like areas of cockchafer tribes in the same stage of development.[774] It usually takes a decade or several decades before cockchafers again pose a serious threat to local grassland farming after a severe decrease in populations, caused, for example, by a sudden cold outbreak and snow in late spring in the year of chafer flight, such as in 1957.[775]

It is known that in the late thirteenth century cockchafers were observed in the Drau Valley in Carinthia close to Sachsenburg above Spittal every third or fourth year.[776] This diversity in the duration of the development of this insect recorded in the final stage of the High Medieval Climate Anomaly (Medieval Warm Period) just preceding the Little Ice Age is in line with the twentieth-century observation that there was a transition from a four-year cockchafer development cycle (favored by a slightly colder climate) to a three-year development cycle in the Upper Drau Valley, as observed between the early and mid-twentieth century.[777] A large-scale devastation of grasslands in several parts of the Swiss Confederacy and its surroundings might have taken place in the 1470s.[778] Cockchafers were observed in the Alps in the northwest of what is now Slovenia prior to 1579, when their regular presence every third year is reported; the seigneuries of Radovljica and Bled (Germ. *Veldes*) and the farmers counted triennial intervals based on cockchafer flight.[779] This corresponds with observations of a three-year cockchafer development cycle in the area, which started in the early twentieth century.[780] Although written sources only provide very partial insights into a narrow selection of cockchafer plagues, a

[773] Janežič, 1971, 4–5, 16, 28–29; Billamboz, 2014, 115–117, 120–121 (quote from p. 117); Poženel, 2007, 73, 75–77.
[774] Janežič, 1971 (see also Studen, 2024); Billamboz, 2014, 115–117.
[775] Janežič, 1971, 5, 16, 23–25, 29; Poženel, 2007, 73.
[776] Dinklage, 1966a, 65, 67.
[777] Janežič, 1971, 6–7 (see also p. 13), 26–28.
[778] Rohr, 2007b, 116–118.
[779] Zwitter, 2016, 140.
[780] Janežič, 1971, 13–14, 18, and the supplement Karta majskega hrošča za Slovenijo (Cockchafer Map of Slovenia).

dendroecological investigation in the area of Lake Constance (Germ. *Bodensee*) suggests that turf devastation due to cockchafer larvae was more frequent in the High Medieval Climate Anomaly and, in particular, since the late nineteenth century than during the Little Ice Age. Occasional severe outbreaks or short series of severe outbreaks during the Little Ice Age were followed by an increased frequency of grassland harvest failures due to cockchafers since the mid-nineteenth century.[781] Nonetheless, in line with the previous paragraph, the same region was not continuously affected; twentieth-century data demonstrate that sometimes entire decades or longer periods passed without any serious cockchafer plague at least at a micro- to mesoregional level, even though there are sites suitable for this animal.[782]

At heavily affected sites, cockchafer larvae fully devastated the most affected grassland complexes by feeding on the roots.[783] Natural regeneration of such meadows led to changes in species composition.[784] Namely, cockchafer larvae feed selectively, with a preference for fleshy roots of herbs and fine roots of grasses; they do not consume the belowground parts of every plant species. Cockchafers usually avoid herbs and grasses with tough belowground rhizomes. Cockchafer excrement and decaying plants increase the plant nutrient availability in the soil, there are many gaps in the grassland that are partially or fully devastated, and so the abundance of species such as smooth meadow-grass (*Poa pratensis*) with mainly aboveground runners is likely to increase after a cockchafer plague.[785] It can be assumed that farmers tried to improve the devastated grasslands or a selection of grassland sections by sowing the limited available quantities of barn-sweepings there (see Section 6.5).[786] From a long-term perspective, a cockchafer plague might have contributed to the occurrence of certain species in local environments due to an exchange in the seed pool over medium distances. Namely, when cockchafer larvae devastated the meadows, farmers were more likely than usual to have purchased hay from elsewhere (Section 3.1). Barn sweepings from such hay were later sown in meadows, unless they were all used as fodder (Section 6.5), which was scarce when cockchafer larvae devastated a significant percentage of local grasslands.

Another insect that was occasionally a challenge to grasslands in the Alps was migratory locusts. They were incomparably more mobile than cockchafers. Similarly to cockchafers, they mainly affected lower elevations, up to about the middle of the montane belt. In the last six hundred years, their swarms particularly frequently reached this mountain chain in parts of the fourteenth

[781] Billamboz, 2014, 115, 118–123, 127; Janežič, 1971.
[782] Poženel, 2007, 73; Billamboz, 2014, 122.
[783] E.g., Poženel, 2007, 73.
[784] Sadar, 1956, 129.
[785] Dietl & Lehmann, 2004, 32–33, 35; Aeschimann et al., 2004, vol. 2, 880–881.
[786] Compare Emanuelsson, 2009, 160.

century, as well as between 1477 and 1480 and in the 1540s. A large swarm could have up to two billion animals. Locusts usually reached the Alps in the second third of August or later; that is, when the first-crop hay from the great majority of meadows had already been mowed and stored (see Section 6.1.2.1). First-crop hay was thus threatened by locusts where it was stored outside in stacks; only in the case of an exceptionally early occurrence of migratory locusts (in the Eastern Alps from late May onward) could they cause significant damage to first-crop hay in meadows. From the ecological perspective, locusts mainly grazed on second-crop hay species at times not much differing from usual second-crop hay mowing. They did not graze heavily in meadows and pastures in every region they crossed, but only where a swarm of flying adult animals settled on the ground. Their grazing seems to have been no greater than that of heavy grazing pressures posed by livestock. Consequently, there seem to have been no major consequences for the grasslands in the future.[787] Where the locusts grazed heavily, their fertilizing impact was likely relevant: a report on the impacts of their large-scale grazing in Carniola (mainly central, northwestern, and most of southern Slovenia nowadays), including mountainous regions, in 1542 emphasized that locust grazing grounds were dirty,[788] thus full of their excrement. From the grassland economy perspective, locust swarms at places mainly consumed the valuable second-crop hay (Sections 6.1.2.3 and 6.3.2) and autumn grazing resources in pastures and meadows after mowing. This caused serious economic damage in many microregions, especially where they also consumed the buckwheat or millet sown after the harvest of autumn-sown cereals and in colder regions, where they consumed the first cereal harvest.[789]

4.1.3.2 Rodents

Impacts of rodents on grasslands are no more than exceptionally mentioned in older historical records, and so only voles are presented here. An exceptionally high abundance of voles (*Arvicolinae*) sometimes contributes significantly to grassland dynamics. Where their activities are strong, this results in a shift in meadow species composition, in non-fertilized meadows in particular (Section 6.3.1)—there, voles could increase the number of species by more than 15%. This increases the variegated character of the grassland. Voles burrow in meadows and pastures, and they consume underground plant organs, especially of species with fleshy roots, and so their feeding is selective. Similar to cockchafers, voles like fine grass roots, whereas plants with rhizomes, espe-

[787] Rohr, 2007a, 488–491; Rohr, 2017, 82, 93–100.

[788] SI AS 2, Deželni stanovi za Kranjsko (The Estates of the Province of Carniola), First Registry, box 322, fasc. 216, year 1542: September 4th, 1542.

[789] SI AS 2, Deželni stanovi za Kranjsko (The Estates of the Province of Carniola), First Registry, box 322, fasc. 216, year 1542: September 4th, 1542; see also Rohr, 2017, 98.

cially tough, lignified ones, are usually avoided. Leguminous species, which are good fodder plants (see Sections 6.3.1, 6.3.2, and 10.1.2.2.3), for instance, are suppressed by voles. Root consumption kills the plants, which can cause patchy or general but selective grassland devastation. This leads to low hay yields in the case of strong vole activities. Moreover, heaps of excavated soil rendered mowing by scythe more time consuming because haymakers avoided the buried soil to prevent impurities in the hay—in contrast to motorized mowers, mowing with a scythe made it possible to see mounds from buried soil through the standing grass in front of the scythe. Whereas voles' negative impacts have been mentioned so far, their selective herbivory and their creation of gaps in meadows by digging foster some species. On the one hand, those avoided by voles benefit from their activities, such as hairy violet (*Viola hirta*), cowslip (*Primua veris*), and perforate St. John's wort (*Hypericum perforatum*), and so the abundance of such species increases. On the other hand, species profiting from gaps in grasslands are promoted. For instance, sites dug out by voles are frequently colonized by field forget-me-not (*Myosotis arvensis*), a small-seeded annual dependent on recruitment from the soil seedbank and germinating when its seeds are brought to an open soil surface by mammals. In addition to annuals, some perennials also use open soil gaps in meadows created by voles for generative reproduction, such as wild mint (*Mentha arvensis*) and nettle (*Urtica dioica*), both likely to colonize loosened soils; once these perennials become established, they are resistant. Vole-affected sites are also suitable for meadow fescue (*Festuca pratensis*), cock's-foot (*Dactylis glomerata*), smooth meadow-grass (*Poa pratensis*), rough meadow-grass (*Poa trivialis*), upright hedge-bedstraw (*Galium album*), and hairy rock-cress (*Arabis hirsuta*), for instance.[790]

Voles could impact the hay quality in diverse ways. Some species promoted by them are good fodder plants, such as meadow fescue, cock's-foot, smooth meadow-grass, and rough meadow-grass. On nettles, see Sections 6.3.2.2, 6.5., 6.6.2.2, 8.2.1.1, 8.2.2, 10.1.2.2.7, and 10.2.1. At some other sites, negative impacts prevailed (e.g., wild mint was not useful as cattle fodder),[791] and suppression of leguminous species was mentioned above.

Old historical records rarely explicitly mention cases of increased abundance of voles, but the ecological processes presented might have been experienced by the residents of Stelvio (Germ. *Stilfs*) in South Tyrol in the early sixteenth century, with second-crop hay more seriously affected than first-crop hay, to name but one potential early example from historical records.[792]

---

790 Sadar, 1956, 129, 135–136; Pils, 1994, 58, 204; Dietl & Lehmann, 2004, 20, 32–35; Ellenberg & Leuschner, 2010, 1020; author's interview with Kunavec, 2023, in Stara Fužina. On seeing through the plants in the case of mowing with a scythe, see Iuga, 2016, 73.

791 Sadar, 1956, 129, 135–136; Pils, 1994, 204; Dietl & Lehmann, 2004, 20, 32–35.

792 Rohr, 2007b, 120–124.

4.1.3.3 Fungi

The impacts of fungi on grassland dynamics are complex and often significant. Their impacts on grassland plants through mycorrhiza are dealt with in Section 6.3.1. Section 4.1.3.1 mentions the impacts of a fungus on cockchafer populations. This section focuses on obvious negative effects of pathogenic fungi on grassland plants.

Pink snow mold (*Microdochium nivale*) and gray snow mold (*Typhula* sp.) have significantly impacted grassland species compositions in cases of a wet autumn, followed by persistent snow cover and a wet and cold spring, at shady sites in particular. Most vulnerable have been grasslands overused in late autumn (in history, mainly intensive late autumn grazing, so that plants were bitten off close to the ground at the end of the vegetation period) or grasslands not used in the autumn at all; the latter were rare in history. To focus only on species of historical grasslands moderately manured with barn manure, meadow fescue (*Festuca pratensis* subsp. *pratensis*), cock's foot (*Dactylis glomerata* subsp. *glomerata*), and bents (*Agrostis* sp.), all widely present in the Alps, are plants highly vulnerable to pathogenic fungi, which at present also (or even more) threaten species of intensive grasslands. Pink snow mold and gray snow mold killed vulnerable species at several microlocations within a grassland parcel or in the entire grassland. The abundance of some other species such as meadow buttercup (*Ranunculus acris*) in the meadow was likely to rise to a certain degree due to the increased cover of pathogenic fungi, in some cases at least.[793]

There is every reason to assume that significantly increased cover of pathogenic fungi in grasslands was an issue several times for grassland farming during the Late Maunder Minimum; in Central Europe, autumn temperatures between 1688 and 1700 were on average about 1.1 °C colder than between 1961 and 1990, and wet autumns were frequent; winter temperatures between 1691 and 1700 were on average 2.6 °C lower than between 1961 and 1990, and spring temperatures between 1687 and 1700 were on average 1.4 °C colder than in the same twentieth-century reference period. Nonetheless, data from Switzerland demonstrate that spring precipitation was below average at that time,[794] which might have mitigated grassland mold development there to a certain degree. The hazard of spread of pathogenic fungi was highest in exceptional years with extremely long snow cover, such as the long winter of 1613/1614, which lasted for almost five months, from November to April, and a considerable part of May was also cold and rainy. This is demonstrated by data from the Slovenian Alps, very much resembling the situation in Switzerland, and it was thus representative of most of the Alps. The snow blanket

793 Dietl & Lehmann, 2004, 24–25, 32–34; Aeschimann et al., 2004, vol. 1, 160–161; Aeschimann et al., 2004, vol. 2, 840–841, 892–893, 966–973.

794 Pfister et al., 2018, 277, 279–280.

probably lasted for up to 130 days or more even in many low valley bottoms in the Eastern Alps in that winter.[795] In such cases, pathogenic fungi added to the fodder challenge that was already serious due to shorter spring and autumn grazing periods because of low temperatures; they contributed to the dynamic changes in grassland species composition over time.

The processes presented in the next section partly also resulted in environmental change in continuously existing grasslands, but more frequently they caused discontinuities.

4.2 Discontinuities in the existence of grasslands due to natural and nature-induced environmental change

A good example of particularly dynamic grasslands is those in former beds of rivers or creeks. Such grasslands were frequent in sections of broader Alpine valleys and basins where the slope of the river or creek was gentle and its bed was incised into fluvial sediments. Consequently, beds of rivers or creeks frequently moved during high discharges until the large-scale channelizing of rivers (see the introduction to Section 6.1.2.5). Locations with a high groundwater level close to the riverbed (e.g., the former riverbed) were usually used either as highly productive (damp) grasslands, frequently as meadows (Figure 31), or as forests.[796] Such meadows were regularly flooded during high discharges because the topography of the former riverbed was lower than that of the surroundings. Such biotopes were particularly dynamic due to changes during flood events, such as erosion and sediment deposition, also including seed transport.[797] The majority of such meadows mainly disappeared between the eighteenth century and the first half of the twentieth century due to manmade changes in riverbeds.[798]

The magnitude and frequency of serious floods affecting grasslands has changed significantly over time.[799] Moreover, the chronology of major floods varies within the Alps due to the "heterogeneity of catchments and landscapes, whose [climatological,] hydrological and [other] environmental settings differ … and in which the sensitivity of the system to climatological and hydrolog-

[795] Pfister, 1985a, 61, 117; Zwitter, 2013b, 342–344, 348, 362–363, 371–375.

[796] Pfister, 1985a, 26, 83; Kauter, 2002, 116–117, 171; Emanuelsson, 2009, 159, 277–278, 281, 304; Hohensinner et al., 2013; Winiwarter et al., 2013, 110; compare Sadar, 1956, 5–13. Some examples from the Alps in the sixteenth and seventeenth centuries in Rohr, 2007a, 357 (the changed bed of the Salzach River); Zwitter, 2014e, 325–332, 346 (three examples from the Jaun Valley [Germ. *Jauntal*, Sln. *Podjuna*], Carinthia: the Drau [Sln. *Drava*] and the Vellach [Sln. *Bela*] Rivers, and Globasnitz Creek [Germ. *Globasnitzbach*, Sln. *Globasnica*]); and Zwitter, 2015b, 19 (example from Bohinj in the Julian Alps).

[797] Komposch, Frieß, & Kreiner, 2013, 389; Poschlod, 2015, 83.

[798] Emanuelsson, 2009, 304, 307.

[799] Wirth et al., 2013; Schulte et al., 2019; Blöschl et al., 2020.

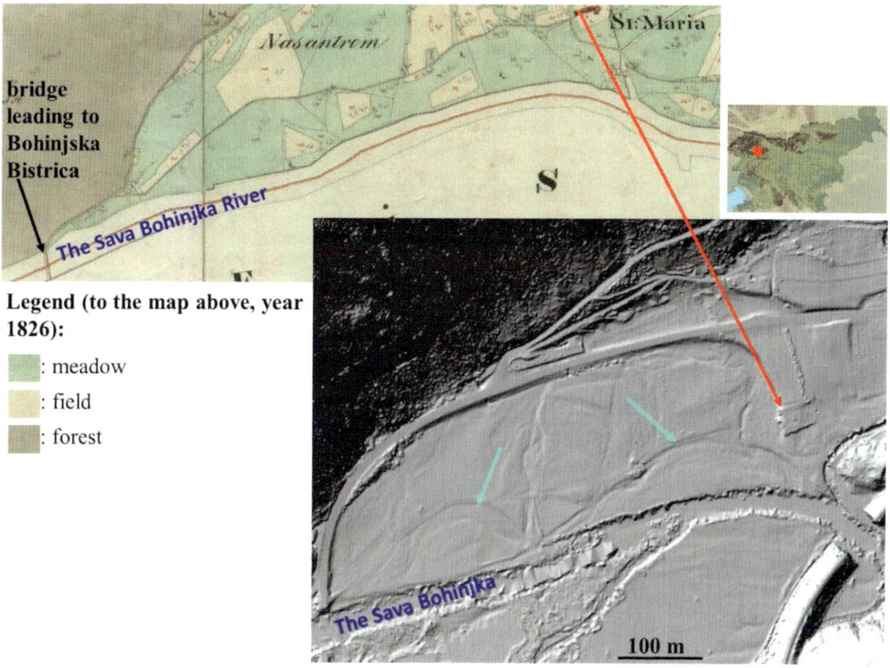

Figure 31: In 1606, meadows called *V mlaze*[800] (i.e., *V mlace*, literally, 'in the pool', a traditional Slovenian designation for wet meadows)[801] and *V logu*[802] (literally, 'in the (wetland) meadow close to a waterbody, usually a wooded meadow')[803] were located below the church at the village of Bitnje in Bohinj, between the church and the bridge leading to the village of Bohinjska Bistrica.[804] The location of the church (red) is constant and, although the position of the bridge might have changed by the early nineteenth century, it is clear that in 1606 there were wet meadows in the former riverbed there. The blue arrows point to the former riverbed, evidently from before the 1826 cadaster, as seen on the lidar-based digital elevation model. *Environmental Atlas of Slovenia*, Backgrounds, Lidar hillshade; SI AS 176, Franciscejski kataster za Kranjsko (*Franciscan Land Cadaster for Carniola*), L172, cadastral municipality of Nomenj, map sheet A09.

ical extremes can vary."[805] Nonetheless, some common main flood-rich periods in most catchments can be identified. Focusing on the second half of the second millennium before the late twentieth century, such a period in (many parts of) the Eastern Alps happened in the second half of the eighteenth century, when floods were frequent not only throughout the Alps but also in large

[800] SI AS 721, Gospostvo Bled (The Seigneury of Bled, Germ. *Veldes*), Iustitialia, vol. 12, June 2nd, 1606.
[801] Sadar, 1956, 1.
[802] SI AS 721, Gospostvo Bled (The Seigneury of Bled), Iustitialia, vol. 12, May 30th, 1606.
[803] *Slovar slovenskega knjižnega jezika*: *log*.
[804] SI AS 721, Gospostvo Bled (The Seigneury of Bled), Iustitialia, vol. 12, May 30th, 1606, June 2nd, 1606.
[805] Wirth et al., 2013, 114, 119, 126; Schulte et al., 2019, 67 (quote), 75–78, 83, 86.

sections of Central, Western, Southern, and Eastern Europe; the second main flood-rich period occurred around 1560–1580; that is, during one of the main peaks of the Little Ice Age when floods were frequent not only in the Eastern Alps but also throughout Central Europe and in the eastern part of Western Europe.[806] Another common characteristic in the Alps is a long flood-disaster gap in most catchments, lasting from the very beginning of the twentieth century until the late twentieth century, which is now followed by the recent flood-rich period.[807] Several other flood-rich periods and serious or disastrous single flood events also shaped the landscapes in the Eastern Alps in the second half of the second millennium, with relevant impacts in grasslands. Not to focus on local events, only two large-scale examples from the sixteenth century are given. In 1501, a high-magnitude flood affected major parts of the Eastern Alps, at least the catchments of the Inn, the Salzach, the Enns, and the Traun, although only scant preserved information refers to inner-Alpine territories.[808] This was part of the broader flood-rich period in Central Europe and the northeastern part of Western Europe between 1500 and 1520. The historical hydrology of the Traun explains that this was one of the main flood-rich periods in the catchment since AD 1500, although no more than the fifth flood-richest interval in that area ever since.[809] Another example occurred during the late sixteenth-century Little Ice Age peak: in August 1598, major parts of the Eastern Alps were affected by large floods. It is clear that such events had profound impacts on many grasslands, but even the general influences on grasslands are rarely clearly documented in such early records. Devastation of meadows during the major 1598 flood is reported, for example, in the Gastein Valley (Germ. *Gasteinertal*).[810]

Apart from floods triggered by similar processes as are usual nowadays, past centuries also experienced many ice jam floods and some glacial lake outburst floods. An ice jam flood was caused by a break in an ice jam or a couple of ice jams. Namely, when the ice cover on rivers was melting, ice sheets accumulated at obstacles, such as bridges. Ice cover on rivers used to be a frequent feature not only due to colder winters, but also due to the different morphology of much more natural riverbeds (freezing could start in shallows with a slow current) and the absence of thermal water pollution. Another type of historical flood in the Alps that could locally affect a high percentage of good agricultural land happened due to advancing glaciers, which dammed streams. In this way,

[806] Blöschl et al., 2020, 561–564; but compare also Schulte et al., 2019, 81, and Wirth et al., 2013, 120–121, 126 from the Western Alps.
[807] Blöschl et al., 2020, 561–565; see also Schulte et al., 2019, 77, 80–81.
[808] Rohr, 2007a, especially 235–236, 275.
[809] Blöschl et al., 2020, 561–564.
[810] Rohr, 2007a, 257–273; for further examples of floods in the Eastern Alps from the thirteenth to sixteenth centuries, see pp. 201–256, 276–278, 327–332, 336–342, 346, 349–358, 366–368 in Rohr's volume; see also Zwitter, 2015b.

a temporary lake formed. In many such cases, failure of the ice dam caused a severe glacial lake outburst flood downstream. The glacier-dammed lake from the Rofen Valley (Germ. *Rofental*) in the Ötz Valley Alps (Germ. *Ötztaler Alpen*), for instance, covered pastures, and it was sometimes more than a kilometer long; exceptionally, its length allegedly exceeded 2.3 km. Since 1600, breaks of ice dams there caused downstream devastation in the Ötz Valley (*Ötztal*) on July 20th, 1600, in mid-July 1678, to a certain degree on June 14th, 1680, in 1847, and on June 13th, 1848. In addition, the Ötz Valley was affected by glacial lake outburst floods that started above the village of Obergurgl (Figure 4).[811] Such floods affected many grasslands, caused severe erosion, and often deposited a thick layer of material of various sizes downstream in the valley.[812]

Human interventions into rivers and creeks in the Alps became significant long before the industrial period, but the scale of these interventions was usually much more limited than in industrial times. Apart from small-scale work, there also existed large-scale interventions in a selection of streambed sections with significant impacts on grasslands, already prior to the eighteenth century (see the introduction to Section 6.1.2.5). In several cases, human interventions into beds of rivers and creeks led to flood-inflicted damage in (grassland) parcels, which would have been safe without such interventions. These included, for instance, mill-dams raising the water level, and displacement of shorter or longer streambed sections.[813] The following example of complexity of such socio-natural sites is representative of many cases of artificial stream-bed section displacements in the Alps that locally had relevant impacts on grasslands. Sucha Creek (Germ. *Sucha*, Sln. *Suha*), a small stream in the Jaun Valley (Germ. *Jauntal*, Sln. *Podjuna*), Carinthia, used to make curves in the pasture below Gösselsdorf (Sln. *Goselna Vas*) until the early eighteenth century, and its surface flow often ended in a demesne meadow of the seigneury of Eberndorf (Sln. *Dobrla Vas*) close to Lake Gösselsdorf (Germ. *Gösselsdorfer See*, Sln. *Goslinsko jezero*). This meadow used to be irrigated (see Section 6.4) with the water of Sucha Creek. During flood events, when the current of Sucha Creek was stronger, it transported larger material, and it deposited its load of sand and pebbles at (the outer side of) the curves (in the pasture), and so material unproductive from the agricultural perspective did not reach Eberndorf's demesne meadow. Some years before 1713, the seigneury of Sonnegg (Sln. *Ženek*) straightened the bed of Sucha Creek below Gösselsdorf. Consequently, during high discharges, Sucha Creek did not deposit its sand and pebbles in the pasture; instead, its deposits covered large sections of the demesne meadow of the seigneury of Eberndorf (and of another meadow) that had not been prone to infertile deposits before. Consequently,

[811] Wopfner, 1997, 82; Rohr, 2007a; Nicolussi, 2013, 71–83, 87; Scharr, 2001, 21; Jäger, 2006, 9–10, 21–23, 58–59, 75, Figure 19 between pp. 64 and 65.

[812] Wopfner, 1997, 73–74, 82.

[813] Rohr, 2007a, especially 365–366; Winiwarter et al., 2013; Zwitter, 2014e, 332; Zwitter, 2018, 168–169, 173; Zwitter, 2024.

this seigneury built a small construction, saving its meadow from accumulation of most sand and pebbles, but still allowing irrigation. This structure directed Sucha Creek's load to the neighboring meadow in possession of a subject of the seigneury of Sonnegg, causing a dispute. In 1713 the seigneury of Eberndorf responded that this subject of Sonnegg was allowed to build a similar construction in the meadow that he was using to mitigate the flood hazard if he wanted to reduce the damage there.[814] All this probably would have been avoided had the bed of Sucha Creek not been straightened, whereas the damage upstream would have been more serious.

Watercourses deposited part of their load in flatter parts of Alpine valleys and basins. One consequence thereof was an increase in wetland areas in floodplains in some broader Alpine valleys and basins, and so meadow sites gradually became wetter there. A significant percentage of fine (thus highly watertight) material in deposits mattered greatly in this process. Fine sediments were deposited in standing or very slowly flowing water, thus by some floods in bottoms of flat valleys and basins, during some lake floods, and/or where coarser flood deposits hindered water outflow. Historical large-scale deforestation on slopes led to an increase in sediment transport in creeks and rivers, and so forest clearing contributed to the increase in wetland areas downstream. The process must have been significant at some places during late medieval large-scale deforestation, for example, when agricultural land on forested terrain at high elevations was greatly expanded (Section 3.1). This was also highly relevant later: the increase in the area of wetland in some Alpine valleys, such as the Alpine Rhine in the early nineteenth century, was linked to excessive wood harvesting on slopes higher in the river catchment. In addition, erosion and downstream deposition of water-transported material causing an increase in wetland area was importantly influenced by the dynamic history of precipitation. In the Emme Valley (Germ. *Emmental*), for instance, severe floods appeared in the interval from 1567 to the end of the sixteenth century more frequently, and their magnitude was higher than in the mid-sixteenth century (see the note above on the general flood-rich period of 1560–1580), even though intensive forest clearing started earlier, and so the influence of the important peak of the Little Ice Age was significant. To the east of Emme Valley, deposits of the major 1762 and 1764 floods of the Linth River led to an expansion of wetlands at Lake Walen (Germ. *Walensee*) in the northern Swiss Alps.[815]

In many places creeks, where deposits were easier to remove than from large riverbeds, experienced a long history of human prevention of a significant increase in the adjacent wetland area. The example of Seebach (literally,

[814] AT StiAStP, Grundherrschaft Eberndorf (Handwritings), vol. 189, Gerichtsprotokoll 1711–1717, 126–127 (November 10th, 1713).

[815] Wopfner, 1997, 399–400; Pfister, 1985a, 86–88, 98–99, 127–128; Haidvogl, Pont, & Zwitter, 2019.

'lake creek'; Sln. *Jezernica*), flowing out of Lake Gösselsdorf in the Jaun Valley to the north, is representative of many cases in terms of stream bed clearing, but it also has its local specifics in the context of carp breeding. On May 16th, 1743, farmers from the village of Gösselsdorf complained to the seigneury of Eberndorf that the Seebach's bed this seigneury was in charge of, near the outflow point from the lake, was too filled with deposits, and so the creek flooded the adjacent meadows that these farmers were using. They demanded the removal of some deposits from the creek bed so that they would be able to make hay for cows in these meadows. The seigneury of Eberndorf agreed to order clearing some deposits from the creek bed as soon as the carp spawning period passed.[816] It is no coincidence that farmers mentioned the type of livestock (cows) they wanted to continue being able to feed with hay grown in these meadows. Namely, if deposits in the creek bed led to a well-developed wetland ecosystem, it would either still be possible to harvest fodder there, but such hay would only be suitable to feed horses (see Section 6.1.2.5.1.2), or the hay would only be usable as animal bedding (Section 6.1.2.5.2). The seigneury recognized the farmers' claims, but it also acted in favor of its carp breeding by ensuring a limited postponement of work in the creek bed stopping the process of wetland formation (see some seventeenth- and eighteenth-century data on carp spawning there in Section 6.4.3.2.2). Namely, limiting the flooded meadow area in times before the water reached 18 °C and then remained at least 17 °C for some days would have negatively impacted the seigneury's carp breeding.[817] However, this also means that the quality of hay from those meadows was still poor that year.

Some mass movements also significantly influenced the history of grasslands, such as the massive rockfall on the southern slopes of Mount Dobratsch (Sln. *Dobrač*) in the Gail Valley (Germ. *Gailtal*, Sln. *Ziljska dolina*) in 1348, following a major earthquake. The largest among several major rockfalls that took place there on that occasion is considered the largest rockfall in the Eastern Alps ever since. It resulted in the deposition of tens of millions of cubic meters of material. The deposits from this largest rockfall reached and blocked the Gail (Sln. *Zilja*) River.[818] A predominantly shallow lake formed behind the barrier. This water body was probably about 3 km long and covered approximately 2 km^2; its maximum depth was about 15 m. Soon afterward, still in

[816] AT StiAStP, Grundherrschaft Eberndorf (Handwritings), vol. 193, Gerichtsprotokoll 1743–1765, 11 (May 16th, 1743).

[817] Carp spawning starts when the temperature reaches 18 °C and is interrupted if the water temperature falls below 17 °C. Carp spawns in shallows, on flooded meadows; they fix eggs to the stems and leaves of flooded grassland plants, and their larvae start swimming about six to ten days later (Povž & Sket, 1990, 47, 116–117; Balon, 1995; Govedič, 2004, 138–139, 148, 150; *Gradivo za strokovni izpit za ribogojca*, 2009, 45–46, 63–65; *Carp Reproduction*, 2007).

[818] Rohr, 2007a, 184–185; Zorn, 2002, 12–14.

Figure 32: *Seewiesen* (literally, 'lake meadows'), an example of a relatively young grassland complex on historical lake deposits. The map shows land-use categories in the early nineteenth century (dark green: meadow, light green: pasture, dark brown: forest, light blue: water). Contour lines are added in red, and the contour interval is 10 m. The elevation of the lake after the partial medieval break of the dam, mentioned above, was close to 540 m. The north–south direction of the contour lines east of the meadows marks the barrier blocking the Gail River in 1348; the material released through the rockfall moved from north to south.[819] *Kärnten Atlas*, Darstellung, Kartenhintergrund, Franziszeischer Kataster & Geländehöhe, Höhenschichten (10 m).

[819] Rohr, 2007a, 185–186, 188.

Figure 33: As a complex of almost entirely flat land distant from villages but located on deep lake deposits—in contrast to rockfall deposits with shallow, stony soils on the neighboring parcels abounding in rocks on the surface (Figure 34)—the environmental conditions at *Seewiesen* have been perfect for meadows. Their youngest parts are two to three centuries old. In September 2019, the grassland was grazed. Photo: Žiga Zwitter, September 2019.

1348, the barrier partly broke, so that a smaller shallow lake, about 10 to 12 m at its deepest part, remained there. The at least temporary flood and/or long-term change in soil properties—possibly turning formerly better-quality agricultural land into marshes—caused the permanent abandonment of two small villages above the confluence of the Gailitz (Sln. *Ziljica*) and the Gail. The last remnants of the fourteenth-century lake disappeared in the eighteenth century.[820] As the lake vanished, some grasslands formed, among them the complex of meadows known as *Seewiesen* (literally, 'lake meadows') and the adjacent pastures (Figures 32–34).

Although evidence is incomplete, it is clear that several other lakes also formed in the Alps in the last millennium as a consequence of rockfall deposits, such as Lake Kummer (Ital. *Lago Kummer*; Germ. *Kummersee*, literally, 'lake of sorrows'[821]) in the upper Passiria Valley (Ital. *Val Passiria*, Germ. *Passeiertal*) north of Moso (Germ. *Moos*). This lake might have formed due to a rockfall

[820] Rohr, 2007a, 185–186, 188, 190–191, 382–383.

[821] We thank the language copyeditor Donald Reindl for the addition of this translation.

Figure 34: Rockfall deposits from 1348 on the neighboring parcels. Photo: Žiga Zwitter, September 2019.

from Mount Gspell in 1401. The Lake Kummer's dam partly broke nine times until the late eighteenth century, always causing a flood down the valley, as far as Merano (Germ. *Meran*). In the eighteenth century, the lake was still 1.1 km long, 250 m wide, and 36 m deep.[822] The great majority of the bottom of the former Lake Kummer is now grassland.[823] The still-existing Hinter Lake (Germ. *Hintersee*) at the mountain pasture complex Meilinger Alp (Germ. *Meilingeralm*) south of Mittersill in the province of Salzburg possibly formed as a consequence of a rockfall in the 1490s blocking a stream.[824] At present, there is an extensive wet grassland south of the lake.[825] In the immediate vicinity, in the Felber Valley (Germ. *Felbertal*), the deposition of the 1569 rockfall covered meadows and moved the bed of Amer Creek (Germ. *Amerbach*), and consequently a lake formed at Öd, flooding former meadows and pastures.[826]

Debris flows, mudflows, alpine debris avalanches, and landslides affecting grasslands also significantly changed site conditions. Written records sever-

---

822 Rohr, 2007a, 194–195, 200, 336–340.
823 *Satellite World Map*.
824 Rohr, 2007a, 197–198.
825 *Satellite World Map*.
826 Rohr, 2007a, 198.

al centuries old provide relevant information on such events in some cases, although these events are poorly represented in old records.[827] Recently, investigation combining analysis of archival records, geological data, and fieldwork was able to identify an alpine debris avalanche from Mount Olševa in the Eastern Karavanks as the most likely process that covered an economically important meadow of the isolated Macesnik mountain farm in the Upper Savinja Valley (Sln. *Zgornja Savinjska dolina*) before 1665; further information from the eighteenth century makes it clear that this event caused the long-term abandonment of that meadow.[828] In a few cases, early documents already make it possible to precisely locate some affected grasslands; elsewhere, dated deposit analyses help in understanding these historical processes.[829] On June 7th, 1660, heavy rain triggered a debris flow in the Lavant Valley (Germ. *Lavanttal*) in Carinthia, from the Koralpe down Gemmersdorf Creek (Germ. *Gemmersdorfer Bach*). There were twenty-nine casualties, and parcels, including meadows, used by farmers from three villages were devasted through erosion and deposition.[830] The report on the 1660 event—which was written in 1662 or 1663 and referred only to parcels belonging to a single farm, not to all affected parcels—explicitly documented that it would not be possible to restore certain meadows due to the very heavy rocks deposited there. Comparing a preserved sketch of the 1660 devastation[831] with a precise early nineteenth-century land-use map (Figure 35), several parcels can be recognized where grassland was no more than about a century and a half old when the early nineteenth-century map was drawn. They were located north, northeast, and east of the church in Gemmersdorf, and they also included the quite extensive pasture along the stream in the eastern part of the settlement: the shape of this pasture also resembles the shape of an alluvial fan, (largely) the 1660 mudflow deposit. This pasture within the settlement probably replaced some houses, meadows, and fields, but the area was used as pasture from 1660 onward due to both erosion

[827] See, for example, Rohr, 2007a, 244, 247, 278, 332, 358, 409; Zenhäusern, 2008, 184–185; Zwitter, 2015d, 149; Zwitter & Mrak, 2017; Wopfner, 1997, 73–74, 86–90, 399–400. Identification of the actual process of mass movements behind the terminology used in historical records is sometimes a challenge that should be addressed interdisciplinarily.

[828] Zwitter & Mrak, 2017. The record from 1665 says only that "the mountain / [mountain] pasture buried" that meadow.

[829] To give but one example from the southeastern Alps dealing with the last millennium: Merchel et al., 2014.

[830] Fresacher, 1965; information from further archival records was added in Zwitter, 2018, 166–167. Mudflows and debris flows are sometimes hidden behind expressions that can also designate floods; it is entirely clear that the 1660 event in the Lavant Valley at Gemmersdorf was a debris flow, and a contemporaneous sketch also proves it (Fresacher, 1965). However, another record referring this same event designates the devastation simply as *Wasserschaden* 'water damage' (AT KLA 207, Ständisches Archiv I, box 197, folder 3). Thanks to the possibility of comparison with other contemporary sources, it is known that the expression in this case does not designate a usual flood, but a debris flow.

[831] Fresacher, 1965, 211–212.

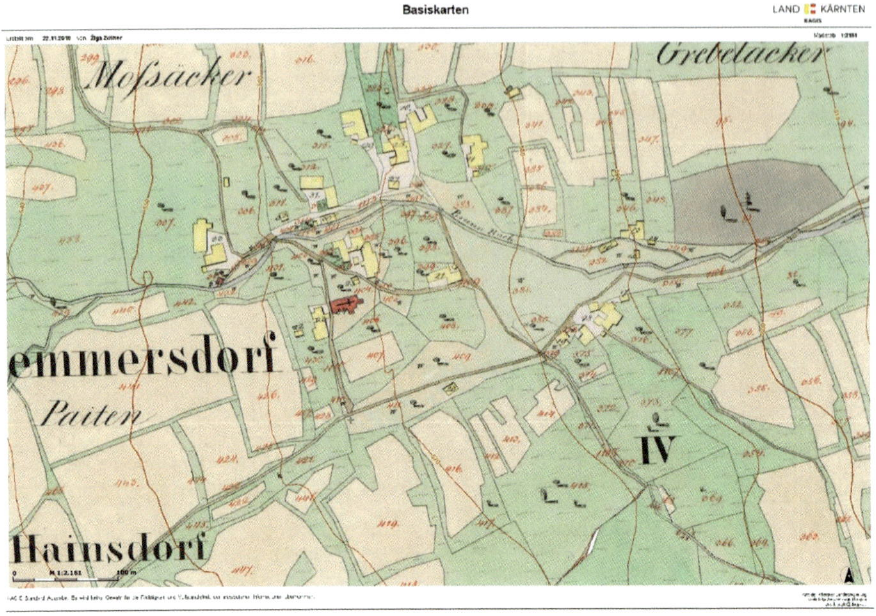

Figure 35: Land use in Gemmersdorf in the Lavant Valley of eastern Carinthia in the early nineteenth century was partly a consequence of long-term environmental changes due to the 1660 debris flow. Light grayish blue designates Gemmersdorf Creek; the 1660 debris flow arrived from the east down the bed of this stream. Light green parcels with the letter "W" in the eastern part of the village were pasture after the 1660 mudflow, and the red building with the cross is the church at the edge of the 1660 mudflow devastation: mud reached the church wall and the event fully destroyed the vestry, but the church, located at a slightly elevated microlocation, did not collapse.[832] Yellow represents wooden buildings, dark green is meadows, dark gray is forest, and beige is arable land in the early nineteenth century. Contour lines are added in red, and the contour interval is 10 m. *Kärnten Atlas*, Darstellung, Kartenhintergrund, Franziszeischer Kataster & Geländehöhe, Höhenschichten (10 m).

and deposition (the reported heavy rocks) on the one hand and awareness of the potential hazard of a repeated similar event on the other. At present, there are few remains of this pasture because the majority of it has been built up.

In some places, new grassland parcels were created after severe floods or mass movements;[833] in many other places these processes changed site conditions in existing grassland parcels significantly, and thus the grassland species composition changed accordingly.[834] After the problems that such processes posed to the farm economy in the first year(s) after the flood or mud-flow,[835]

[832] Fresacher, 1965, 210–212.
[833] Rohr, 2007a, 384.
[834] Compare Peterken, 2013, 194.
[835] E.g., Jäger, 2006, 46.

the long-term impacts of these environmental changes were either favorable or unfavorable from the perspective of agricultural grassland uses. This depended on the size of the material deposited and on the nutrient content in it available to plants, which was particularly relevant when non-fertilized grasslands were prevalent (Section 6.3).[836] In 1737, a layer of mudflow deposits up to 4 m thick on the bottom of the valley near Sluderno (Germ. *Schluderns*) in Trentino-Alto Adige (South Tyrol) improved the agricultural quality of grasslands close to the Adige (Germ. *Etsch*) River, where a wetland had existed. Gravel, allowing groundwater flow and thus drainage of soils, is well represented in the deposits that covered the former wetland, and the high content of sand and silt in the upper layer of the deposits is favorable for agricultural grasslands.[837]

Many (larger) avalanches also eroded much of the turf and soil in some sections of the avalanche track, causing discontinuities in the existence of some higher grasslands, and they transported this and other eroded material down the slope, including large objects, such as rocks (Figure 105 in Section 6.6.1) and tree trunks. The deposits changed the site conditions at the end of the avalanche track (see also Section 4.1.1.1).[838] To give but one early example from archival records, the devastation of a considerable meadow area at Vomp, east of Innsbruck, which was recorded in mid-April 1422, was likely the result of an avalanche.[839] In winters when the avalanche hazard was unusually high, and so they heavily affected more grassland sites than normal, avalanches also destroyed some hay waiting in the Alps (e.g., stored in haystacks in mountain meadows to be transported to the farms in the winter), as reported in the unusually snowy winter in early 1608 in the Slovenian Alps: some farmers kept hay in places that were avalanche-prone in the case of extreme conditions despite the widespread Alpine cultures of risk management that in general decreased risks of this kind (see also Sections 4.1.1.2 and 4.1.1.3). They did so either due to a lack of awareness of the site being potentially avalanche-prone or in the hope that avalanches would not affect the site in the following winter, where they were a rare phenomenon.[840] Exceptionally, avalanches also claimed the lives of farmers transporting hay home[841] or visiting their livestock kept in remote mountain barns (see Section 6.3.2.1).[842]

---

836 Inspired by the generalized statement on the relevance of flood-deposited nutrients in Winiwarter et al., 2013, 114.

837 Leibundgut & Vonderstrass, 2016b, 159.

838 Hubatschek, 1992, 39, 132; Pavšek, 2002, 77–80, 118; Rohr, 2007a, 399–400; Bohner, 2008, 85.

839 Rohr, 2007a, 416, the expression *len* used in the passage might also refer to mass movements: for instance, a landslide or a debris flow; see Wopfner (1997, 91) for a general explanation and Zwitter (2015b, 158) for a comparative result from archival records.

840 Zwitter, 2020, 72–73.

841 For example, Grass, 1948, 206, but see the related traditional ecological knowledge in Keller, 2004, especially page 67.

842 Hlebanja & Hlebanja, 2006, 54, 69.

In general, deposits of floods, mass movements, and larger avalanches in the Alps also included much coarse material, which would not be suitable for mowing, except for some sections of floodplains in the flatlands of broader valleys and basins. Pebbles, gravel, wood, and most of the deposited rocks could not be left there if the aim was to create or restore meadows (not just poorly productive pastures) on deposits. This was particularly relevant because several of these processes were recurring, and so avalanche-deposited rocks, for instance, would have covered an increasingly high percentage of the grassland had they not been removed regularly in spring (Sections 6.6.1 and 6.6.2). Consequently, in many cases people significantly changed the site conditions for restored grasslands after floods, mass movements, or avalanches caused deposition or even erosion. In 1611, the application for tax relief by subjects of St. Lambrecht's Abbey in Upper Styria documented that a flood in the spring of 1611 swept away a very considerable area of fields and meadows, and some grasslands were covered with debris. However, the possibility of making the devastated land suitable for crops (including hay) again through enormous labor investment is mentioned.[843] Although such an application for tax relief was likely exaggerated, it is clear that such work was in reality carried out—for example, in the Ötz Valley, where debris flows and floods affected much agricultural land on several occasions (see the examples of glacial lake outburst floods above). Three methods of making the devastated parcels suitable for meadows or fields were used there. The most labor-intensive one consisted of removing all deposits from a small patch, excavating the previous soil, filling the excavated hole with coarser deposits, and placing the previous soil on the surface. Due to this extremely labor-intensive human farmland restoration method, one does not find fossil soils there despite the fact that soils were covered with deposited material. However, this method was predominately or exclusively used to restore arable land. To restore grasslands, in the case of deposition of material of different sizes, deposited stones were removed and piled into mounds. Larger rocks could be removed using a sled pulled by horses, and young men might even have competed in their ability to remove rocks that could barely be removed. Before removal, rocks that were too large were smashed to pieces in some places. The third method, also important in restoration of grasslands, was practiced where former soil had been eroded, for example. In this case, a small channel of water was directed into a depression to deposit its fine load there. Such restoration of soils for a grassland was possible only in certain locations.[844]

---

843 Hubatschek, 1992, 39, 132; Wopfner, 1997, 298–299, 318; Keller, 2004, 20; Zwitter, 2015b, 17.

844 Wopfner, 1997, 87–90; "Bildteil," 1996, 439 (i.e., the photo *Zerschlagen von Steinblöcken* 'breaking apart boulders'); Niederer, 1996b, 342, 344; Niederer, 1996c, 41, 70–71, 79–80; Jäger, 2006, 58–59.

In other cases, deposited course materials were left in situ, and this often led to abandonment of former grasslands: former agricultural land was turned into part of the riverbed, into bare deposits of stones, or covered with a mixture of predominately coarse deposits, for instance. An example of such locally significant abandonment of grasslands is the bottom of the Ötz Valley between Huben and Aschbach, where the aforementioned glacial lake outburst floods in 1678, 1680, and the 1840s turned much former agricultural land into land unproductive for farming. Sooner or later, unproductive debris gradually became reforested due to ecological succession unless environmental dynamics prevented establishment of vegetation.[845]

Some meadows were created elsewhere to compensate for the land lost due to natural hazards. In this way, floods, for instance, not only affected flood-prone areas but also led to the creation of some grasslands at flood-safe sites. Nonetheless, the total area of meadows created in safe locations due to natural hazards was limited.[846]

During the Little Ice Age, advancing glaciers covered several grasslands at high elevations (for further impacts of colder weather, see also Section 4.1.).[847] Far-reaching general phases of Alpine glaciers advances were "the late 1200s–*c.* 1380, the 1580s–*c.* 1660, and 1810s–*c.* 1860."[848] Looking at the Alps as a whole, the area of glacier-covered pastures and meadows was spatially very limited,[849] but many glaciers covered considerable areas of exposed grasslands, which had significant impacts on the local economy. Between 1790 and 1820, for instance, the Schwarzstein Glacier (Germ. *Schwarzsteinkees*) in the Tyrolian Zillertal Alps covered more than five hectares of pastures, and the Solda Glacier (Ital. *Vedretta di Solda*, Germ. *Suldenferner*) in South Tyrol progressed by 1.2 km in 1817 and 1818 together, in a phase of its rapid advance.[850] In an opposite process of glacier retreat, after the Little Ice Age (which ended around 1860), about 60 km^2 of land that had been agriculturally unproductive turned largely to pastures in the Swiss canton of Vaud (Francoprovençal *Vôd*) alone, partly as an immediate consequence of glacier retreat, and largely also as a matter of more intensive pasture clearing (see Section 6.6.2). Namely, in the period

[845] Wopfner, 1997, 90 (and *Arcanum Maps*, Cadastral Maps, Habsburg Empire – Cadastral Maps (XIX. [sic] Century), cadastral municipality of Längenfeld, 1856), 399–400; Jäger, 2006, 59; see also Orland, 2016, 49; Rohr, 2007a, 358; Zwitter, 2015b, 17–21; Mlinar (ed.), 2018b, 138.

[846] Jäger, 2006, 46, 59; Zwitter, 2015b, 12, 18–19.

[847] Jäger, 2006, 9–11, 16–34, 36–48, 59–62, 67–71, 75–79, Figures 1–14 between pp. 16 and 17, and Figures 19–27 between pp. 64 and 65; Jäger, 2007b, 91, 96–100, 103, 105–111; Nicolussi, 2013.

[848] Pfister et al., 2018, 268; Nicolussi, 2013. See also promising methodological advances in glacier ice dating: Feng et al. 2019.

[849] Mathieu, 2019, 56–57.

[850] Jäger, 2006, 27–33, 42; Jäger, 2007b, 105, 109.

of glacier retreat that coincided with high agricultural density (Section 3.1), pasture clearing was also intensive in areas earlier perceived to be threatened by glaciers. The canton of Vaud extends from the Jura Mountains (which reach only moderate elevations) over the basin of the Swiss Plateau (Germ. *Schweizer Mittelland*, Fr. *Plateau suisse*, Ital. *Altopiano svizzero*) to a small territory in the northern Alps, thus including only a fairly limited area of high mountains and glaciers,[851] and so the Little Ice Age glacier expansion and later retreat and the related pasture expansion in many colder regions of the Alps were even much more significant.

Some other processes of environmental dynamics also caused discontinuities in the history of grasslands or they helped establish new grasslands. For instance, tree throws as results of wind significantly changed soil properties and geomorphology as material was lifted with the roots of fallen trees and it gradually formed hummocks on the ground, unless material was leveled. If kept open by farming, grasslands developed there. The windthrow that possibly took place in the twelfth or thirteenth century in the Kräuterin Massif in the Northern Limestone Alps, for instance, was followed by the development of a hummocky grassland.[852]

Section 4 deals with natural and nature-induced environmental changes in grasslands. Before focusing on environmental impacts of grassland management in Section 6, an outline of historical changes in livestock size and feed demand is needed—the topic of Section 5.

[851] Pfister, 1985a, 92; Nicolas-Obadia, 1974, 16.
[852] Embleton-Hamann, 2004, especially 177–178, 183–187.

5 Historical changes in livestock size and feed demand

Grassland management in the Alps used to be (and largely still is) motivated by animal husbandry. Not only quantities of livestock, but also their qualities (e.g., the average size of grazing animals) have changed significantly throughout history. Thus, before discussing grassland management practices in Section 6, the historical development of cattle and sheep size is briefly presented in Section 5. In terms of grassland history, livestock size is ecologically particularly relevant in terms of feed demand (important for Sections 6.1, 6.2, 6.3, 6.4, 6.5, and 6.6) and demand for bedding (Section 6.1.2.5.2), as well as from the viewpoint of trampling effects (Section 6.2) and quantities of manure produced (Section 6.3). It also influenced seed transport between grasslands (Sections 7.1 and 7.2).[853]

An investigation of 8,579 cattle remains from twenty-nine archaeological sites in what is now Slovenia (the southeastern Alps and their broader surroundings) from the middle of the first century BC to the sixth century AD by Borut Toškan demonstrated the huge economic importance of cattle from the first century BC to the fourth century AD. At that time, the average height at the withers of 101 cattle investigated was slightly less than 125 cm, a result that agrees with data from neighboring territories (compared to only up to 110 cm in the pre-Roman period). This proves the prevalence of a quality Italian cattle breed (or breeds), which spread from the Apennine Peninsula into the territory investigated.[854] Large cattle from Roman Antiquity were also found at some other archaeological sites in the Eastern Alps; for example, on Hemmaberg (Sln. *Junska gora*) in what is now Carinthia (Germ. *Kärnten*, Sln. *Koroška*) from about AD 400, at Invillino (Friul. *Divilin*) in the Tagliamento (Friul. *Tiliment*) Valley, and at Volano near Trento.[855]

Nonetheless, smaller breeds were also kept in the Alps throughout Roman Antiquity and later.[856] From the late fourth to sixth centuries AD, in Late Antiquity, when the settlement pattern changed significantly (many fortified hilltop

[853] Krausmann, 2008, 3–4, 10–11, 21, 38, Table 14; Véron, 2009, 165–167.
[854] Toškan, 2013; Toškan, 2016a, 239–241; Erich Pucher's sketch of a small pre-Roman cow on the front cover of the journal *Forschungsberichte der ANISA* 5, 2014.
[855] Glaser, 2013, 47, 50.
[856] Toškan, 2013, 59; Pucher, 2014, 93–94.

settlements were built; the east of the Eastern Alps experienced more discontinuities in terms of settlement, land-use, and culture at that time than many other parts of the Alps), the importance of raising cattle decreased in the local economy and cattle size declined as well.[857] The significant decrease in cattle size at that time was not regionally specific; it has been proven at several places in the Western Roman Empire in times when "[t]he rural population was thin and widely dispersed, and it was far less capable of supporting a non-agricultural population than its fourth-century predecessor had been."[858]

Small and large cattle of the time had their merits, as revealed by contemporary reports. Around AD 507 Theodoric the Great, the king of the Ostrogoths, recommended exchanging small cattle, especially small oxen, of the locals from the province of Mediterranean Noricum (Lat. *Noricum Mediterraneum*) in the Eastern Alps for the larger cattle of a group of Germanic Alemanni, who were under Theodoric's protection and who stopped in Mediterranean Noricum during their migration. Theodoric considered such an exchange of oxen mutually beneficial. Namely, the larger cattle of the Alemanni were better in terms of (commercial) meat production, and they were also more prestigious; nonetheless, they were exhausted due to the long journey. The small cattle prevailing in the province of Mediterranean Noricum were stronger and healthier, and thus better for draft work than the (exhausted) larger cattle of the Alemanni. Records from Classical Antiquity confirm not only the interpretation that small cattle in the Alps were excellent draft animals, but also the fact that the milk production of such small Alpine cattle was high, whereas the milk production of the large Italian breed was lower.[859] Similarly, in the late nineteenth and early twentieth century, small cattle breeds producing high quantities of milk for the standards of the time were recorded in some parts of the Alps.[860] It is likely that the milk quality of small cattle in Late Antiquity was also high.[861] Similarly, archaeozoological investigations, based on limited evidence, suggest that milk production and draft labor were among the main goals of keeping cattle in the southeastern Alps and their surroundings in Late Antiquity.[862] Small cattle, adapted to local environmental conditions, also had further advantages. Their metabolic needs were lower,[863] and so they might have made better use of the fodder from those non-fertilized grasslands that

857 Pucher, 2014, 92–93; Toškan, 2016a, 242–243; see also Glaser, 2013. Issues of (dis)-continuities between the Classical Antiquity and the Early Middle Ages: Staab, 1991; Cheyette, 2008; Štih, 2010; Bätzing, 2015, 58; and further literature quoted there.

858 Cheyette, 2008, 152–153 (quote from p. 153).

859 Glaser, 2013, 47–48, 50, 54; Henning, 2014, 337–339; Bratož, 2017, 220–225, 230–231, 234; Pucher, 2014, 81–82, 92–93; see the political map in Daim & Glaser, 1998, 19.

860 Šalehar et al., 2014, 100–101, 114, 117–118; Novak, 2024, 27.

861 Pucher, 2014, 81.

862 Toškan, 2013, 49, 54–55, 60; Toškan, 2016a, 242–243; see also Glaser, 2013, 48.

863 Cheyette, 2008, 153; Glaser, 2013, 48; Pucher, 2014, 73–82; Novak, 2024, 27.

produced low-quality fodder (see Section 6.3.1.2).[864] Due to their lower metabolic needs, in some cases small but well-fed cows provided more milk than larger but starving cows would have, or small but well-fed cows provided more milk than larger but inappropriately fed cows lacking the relevant nutrients would have. Potentially, smaller cattle might have better coped with fodder scarcity and feed surrogates in times of general scarcity and subsistence crises.[865] This was likely relevant, for instance, in times of the harsher climate of Late Antiquity and the Early Middle Ages, and later—during the peaks of the Little Ice Age (see Section 4.1.1.2).[866] Moreover, small cattle were able to graze in steeper pastures (not only were large Italian cattle heavier, but their hooves were also sharp, and thus acceleration of slope processes through grazing was a more serious issue than in the case of smaller cattle) and in a harsher climate, and to (safely) walk on significantly more difficult trails than the large breed(s) (Figure 37).[867] In the second half of the nineteenth century, for instance, the agronomist Fran Povše recommended that cattle in Bohinj retain their physical characteristics, including their small size, to remain suitable for grazing in mountain pastures.[868] At that time, grazing still took place at several very steep and barely accessible sites, which were later abandoned (Section 3.1), and so the Simmental breed grazes successfully in some mountain pastures today after more challenging grazing sites have been abandoned.[869] An additional factor promoting smaller cattle was farmers' mentality in some regions of the Alps and historical periods. As emphasized by Arnold Niederer, in the traditional mentality of part of the twentieth century, the quantity of livestock had a greater impact on a farmer's reputation than the livestock's performance in some parts of this mountain range. Thus, eight meagre cows would make the farmer more respected and increase the chance of economically favorable marriage of the daughters in comparison to four larger cows, even if four larger cows would possibly make more economic sense.[870] However, it is questionable whether the four larger cows really could have performed so well based on the feed (hay quality, among other factors) that sufficed for eight meager ones, and whether they would have been able to make good use of the same pastures suitable for the less productive breed.

[864] Compare Peterken, 2013, 191.
[865] Pfister, 1985a, 95–96.
[866] Cheyette, 2008; Toškan, 2013, 54–55; and a novel comparison of early modern data and legends of common lore in Pfister, 1985a, especially pp. 95–96, with historical climatology overviews in Brooke, 2018, 180–181, Horden, 2018, 184–186, and Rohr, Camenisch, & Pribyl, 2018, 254, 256.
[867] Pucher, 2014, 81, 93; Wopfner, 1997, 185, 395; Jäger, 2006, 22; Hodnik, 2016, 41.
[868] Šalehar et al., 2014, 106.
[869] Main author's observations, mainly participatory observation in the Krstenica Alp, Bohinj.
[870] Niederer, 1996a, 127.

Figure 36: Small cattle and a hayrack (see Section 4.1.1.3) being loaded with cereal sheaves in Carniola (Germ. *Krain*, Sln. *Kranjska*) in the seventeenth century. Graphic materials often do not reveal whether young or adult cattle were presented; in this case, it is clear that the draft animals that pulled the cart were not young livestock, and so small adult cattle were presented. Valvasor, 1689, book II, 105.

In the Middle Ages and the first centuries of the Modern Period, cattle were usually small. Archaeozoological analysis of the remains of seven cattle dated to the twelfth to fourteenth centuries from the settlement of Otok (Germ. *Gutenwerd*), located south of the southeasternmost pre-Alps and abandoned in the fifteenth century, calculated the average height at the withers as 106 cm (individual values: 96–117 cm); in this case, the average equaled the median.[871] The estimated height at the withers of nine cattle from the fourteenth to mid-nineteenth centuries in the Alpine town of Slovenj Gradec (Germ. *Windischgraz*) was about 105 to 120 cm (median c. 110 cm). The highest specimen from Slovenj Gradec belongs to the layer dated from the fourteenth to sixteenth century, corresponding to further achaeozoological data proving higher cattle in fourteenth–sixteenth-century Slovenj Gradec than in twelfth–thirteenth century Otok. Comparison of these data from Otok and Slovenj Gradec suggests diver-

[871] Bartosiewicz, 2006, 457–464, 470–473; on the history of the settlement of Otok, see Golec, 2014, 174 and *Slovenski zgodovinski atlas*, 2011, 76–77, 96.

Figure 37: Difficult passages on the way to some mountain pastures require limited cattle size (however, the cows in the picture are larger than in Figure 36; this passage is passable for Simmental-size cattle—even narrower passages could be passed through by smaller historical cattle breeds) and robust hooves. The Krstenica Alp in the Julian Alps (Sln. *Julijske Alpe*): a narrow passage just before the *Trdinovc* pasture, recently called also *Zgorna Nedelska*, on the eastern slope of Mt. Krsteniški Stog. Gratitude is expressed to the alp personnel, Cilka Mlakar, and young herders, who made it possible for me to take this photo. Photo: Žiga Zwitter, August 2015.

sity in cattle sizes. Moreover, together with other archaeozoological information from the broader region, this comparison suggests a moderate increase in cattle size in some cases in the centuries before the major late modern increase in cattle size; nonetheless, the paucity of data does not make it possible to draw firm conclusions.[872] Similarly, the height at the withers of two cows from Šentvid pri Stični in the contact zone between the southeastern pre-Alps and the Dinaric Karst, approximately dated to the early seventeenth century, was about 116 cm, and these cows were large in the context of other archaeozoological cattle finds from that site.[873] In contrast to these results, the median of estimated height at the withers of seven fifteenth–seventeenth-century cattle from Novo Mesto (Germ. *Rudolfswert*; located close to Otok) was still as low as 106 cm (individual values: 98–122 cm; it is not clear if these are fifteenth-, sixteenth-, and/or seventeenth-century cattle remains, and so an increase in cattle size over time

[872] Toškan 2016c, 194; Toškan, 2017, 84–89, 91; the archaeozoologist Borut Toškan, personal communication.

[873] Porenta et al., 2015, 335, 340, 358–359, 370–373, 381–383, 386.

cannot be excluded).[874] Similarly, the 1689 copperplate print with a hayrack in Carniola as the main motif includes cattle with as low a height at the withers as the large wheels of a farmer's cart (Figure 36),[875] which either presents local reality or was inspired by the usually small cattle in many parts of Europe.[876] Results of approximate calculations for eleven cows and three oxen kept by the seigneury of Gornji Grad (Germ. *Oberburg*) in the Kamnik–Savinja Alps (Sln. *Kamniško-Savinjske Alpe*) in 1680s, based on account books, show that the average live weight of cows was only slightly over 140 kg.[877] About 1804, small cattle were presented on an artistic presentation of the Alpein Alp (Germ. *Alpein Alm*) in the Stubai Valley (Germ. *Stubaital*), Tyrol.[878] This information corresponds well to the data from the province of Carniola (i.e., major parts of present-day Slovenia, also including parts of the Alps and pre-Alps) from about 1840, when the live weight of adult cattle was still between 140 and 222 kg.[879] In the first half of the nineteenth century, an average cow at Großarl in the Alps of the province of Salzburg weighed about 230 kg.[880] In 1864, the average *Fleckvieh* cattle live weight at Interlaken in the Swiss Alps was 175 kg, that of *Braunvieh* cattle in the canton of Uri 188 kg; these figures are very low in the broader Swiss context of the time, and so the mid-nineteenth century cattle were very small there.[881] They did not provide optimal milk quantities for the standards of the time; nonetheless, these small cattle were robust, and so they were esteemed more than locally, as reported for the cattle from Uri.[882] Relatively small cattle were still characteristic for several parts of the Eastern Alps in mid-twentieth century, nonetheless, they were not the same as a century earlier.[883]

In contrast to these findings, the average live weight of twelve cows sold from Switzerland's Simme Valley (Germ. *Simmental*) and Vaud (Francoprovençal *Vôd*) Highlands (French *Pays-d'Enhaut*) to the slaughterhouse in Geneva (French *Genève*) in 1728/29 was no less than about 420 kg (compared to about 250 kg as the average live weight of 1,476 cows from Burgundy (Burgund. *Bregogne*, French *Bourgogne*) delivered to the same slaughterhouse in the same years), with only minor differences in weights between these animals.[884] Even though these cows from Simme Valley and Vaud Highlands were probably well fattened and

[874] Toškan, 2017, 91, and Borut Toškan, personal communication.
[875] Valvasor, 1689, II, 105; see also Pavković, 2021, 85. We thank the historian Stanko Granda for his comment on the small cattle size in this copperplate print.
[876] See critical comments on the content of these copperplate engravings in Zwitter, 2014d, 675–686.
[877] Zwitter, 2015d, 144–145.
[878] Jäger, 2006, Figure 2 between pp. 16 and 17.
[879] Novak, 1970, 387.
[880] Krausmann, 2008, 27.
[881] Pfister, 1985a, 43.
[882] Stadler-Planzer, Stadler, & Aschwanden, 2015, 106–107.
[883] Hodnik, 2016, 25–26, 37–41, 87; Šalehar et al., 2014, 100–101, 114, 117–118.
[884] Pfister, 1985a, 43.

do not represent an average cow in the area, it is important to notice that their live weight was three times the weight of cows of the seigneury of Gornji Grad half a century earlier, and 2.25 times the average *Braunvieh* cattle weight in Uri in the mid-nineteenth century. Cows weighing about 400 kg were not only introduced to the Bernese Highlands (Germ. *Berner Oberland*) and Central Switzerland in the early eighteenth century; a limited percentage of such cattle have a longer history in that region[885] belonging to the northernmost Alps between Savoy (French *Savoie*, Francoprovençal *Savouè*) and Allgäu, which has long been exceptionally strongly oriented to animal husbandry in the broader Alpine context (Section 3.1). An extensive area of moderately fertilized meadows in that region was one of the factors making it possible to rear large cattle (Sections 6.1.2.3 and 6.3.2).

Oxen were usually heavier than cows. In the seigneury of Gornji Grad in 1680s, the live weight of two oxen was about 250 to 300 kg, and that of a lighter ox about 170 to 190 (or 220) kg.[886] An ox slaughtered in the Carinthian seigneury of Waldenstein in 1678 belonged to the same weight class as the two oxen in the seigneury of Gornji Grad, with the most likely live weight being about 270 to 310 kg.[887] The heavy oxen kept by seigneuries were not representative of the oxen at an average farm.[888] A contract for oxen export from Carinthia (Germ. *Kärnten*, Sln. *Koroška*), signed in 1675, for instance, regulated general export of oxen heavier than about 200 kg and also annual export of 600 oxen lighter than this weight.[889] Not only prominent economic actors such as several seigneuries, but also some ordinary farmers kept oxen that were considered large for the standards of the time.[890] A similar notice of large oxen kept west of the Carpathian basin comes from the sixteenth-century Habsburg Adriatic port of Rijeka (Ital. *Fiume*, Germ. *St. Veit am Pflaum*). At that time, one of the types of cattle hides available there as merchandise from the broader surroundings were *pelles magnae teotonicae*, large cattle (oxen) hides from the Holy Roman Empire of the German 'Nation', most likely from the Habsburg territories west of Hungary, west of Slavonia and Croatia.[891]

The data presented in this section make it clear that, despite the prevalence of small cattle (which nonetheless had some merits, as explained in the pre-

[885] Pfister, 1985a, 96.
[886] Zwitter, 2015d, 144–145; Pavković, 2021, 81–85. The assumption was made that oxen slaughtered at seigneuries were either well fattened or at least moderately fattened; only in the case of the lightest ox, the figure in brackets refers to the approximate live weight if the recorded slaughter weight refers to an unfattened animal; in this case, the slaughter weight represented a lower percentage of the live weight, and thus the calculated live weight is higher (Pavković, 2021, 81–82).
[887] Pavković, 2021, 81–82, 84–85.
[888] Zwitter, 2015d, 144–145.
[889] Dinklage, 1966b, 136.
[890] Pavković, 2021, 81.
[891] Gestrin, 1965, 174.

sentation of Late Antiquity) in the Eastern Alps, late medieval or early modern cattle sizes at the local and regional level were significantly diverse.[892] Diversity in cattle sizes could be related to several factors or their interrelations. Larger animals could, for instance, have been a result of the introduction of new breeds, improved winter feeding, longer duration of the suckling period, and/or cattle not being too young at first mating or when they started being used for heavy physical work.[893] In the Late Middle Ages and the Early Modern Period, foreign breeds were introduced to the Eastern Alps from various directions, as illustrated here mainly based on imports from the east and west. Concerning the introduction of foreign breeds to the Alps from the east, Hungarian cattle deserve special attention. Late medieval cattle types in Hungary were diverse, with a general significant increase in withers height between the Early and Late Middle Ages. The average cattle weight in the slaughterhouse at Sopron around 1570 was as much as 387 kg, even though a highly significant early modern increase in average Hungarian cattle size was yet to come.[894] As early as about 1320, Hungarian cows, larger than local ones, were imported to the Alps of the present-day region of Trentino–Alto Adige (South Tyrol) and kept in a princely holding at Terlano (Germ. *Terlan*).[895] A Hungarian ox weighing about 340 to 390 kg might have been bought by the seigneury of Stična (Germ. *Sittich*), with its center in the contact zone between the pre-Alpine hills south of the southeasternmost Alps and the Dinaric valley systems and corrosion plains, in 1683. Nonetheless, the nexus between environmental and economic conditions in the Eastern Alps probably did not allow any high-scale rearing of large Hungarian cattle there, as suggested by Tadej Pavković's investigation of samples of partially preserved archival data on cattle in the Alps of the provinces of Carniola, Styria (Germ. *Steiermark*, Sln. *Štajerska*), and Carinthia—today parts of Austria and Slovenia—in the sixteenth and seventeenth century.[896] In contrast to these findings, early modern imports of Swiss cattle to the Eastern Alps motivated by its economic advantages (high prices of slaughter animals clearly suggest a large body size: note the aforementioned weight of cows from the Simme Valley in 1720s; at least some Swiss cows also provided high milk quantities for the standards of the time) seem to have been relatively important.[897] The nexus between environmental and economic conditions in the Eastern Alps was more similar to the circumstances in Switzerland than in the Carpathian basin. Consequently, introductions of

[892] Pfister, 1985a, 43, 96; Hanhart & Sonderegger, 1998, 8–9, 38, 48, 50, 52; Doll, 2003, 19, 54–55, 156, 195–199; Pucher, 2014, 93–94; Makowiecki et al., 2019, 345, 350, 364–366, 369; Hinterwaldner, 2002, 165–166; Bartosiewicz, 2006, 461–462, 464, 470–473; Pavković, 2021, 64–85, 95–100.

[893] Pfister, 1985a, 43; Šalehar et al., 2014, 97; Pavković, 2021, 64–85, 95–100.

[894] Bartosiewicz et al., 2018, 132–133; Csippán & Ferenczi, 2020.

[895] Riedmann, 1985, 494.

[896] Pavković, 2021, 74–76, 83–84, 114–115.

[897] Pavković, 2021, 65–74.

Swiss cattle to the Eastern Alps were more likely to succeed than introductions of cattle from the Carpathian basin. For instance, in the second half of the sixteenth century, the convent in Hall, Tyrol, imported "Swiss" cattle, and in 1603 people from the Ulten Valley (Ital. *Val d'Ultimo*, Germ. *Ultental*) in the present-day region of Trentino–Alto Adige (South Tyrol) bought breeding cattle in Prättigau, about 100 km away in a straight line.[898] In the second half of the seventeenth century, rearing "Swiss" cattle and hybrids between local and "Swiss" cattle is documented on demesne land of some seigneuries in Styria; for example, at Murau in the Alps, and in a couple of Lower Styrian seigneuries east of the Eastern Alps; in the case of the demesne at Črnci (Germ. *Schirmdorf*), an inventory from 1705 makes clear that "Swiss" cattle were numerous (twenty-three "Swiss" cattle and thirty-two usual cattle were kept there). These data on "Swiss" cattle might be representative of those seigneuries that kept large cattle herds in several parts of the Eastern Alps. "Swiss" cattle reared in the Eastern Alps were diverse,[899] which is explained by the fact that cattle in Switzerland were also diverse.[900] Furthermore, cattle of other provenances already played a certain role in the medieval and early modern history of the Eastern Alps. For instance, the occurrence of "Bavarian" and "Italian" cattle is documented in Tyrol about 1300.[901]

As early as the mid-nineteenth century (1864), average *Braunvieh* weight in the Swiss cantons of Schwyz and Obwalden reached 475 kg;[902] note the proximity of Interlaken, where very light *Braunvieh* was mentioned above. The results from several diverse areas of the Alps, such as Appenzell (where the nineteenth-century development of cattle breeds was delayed in the Swiss context) and from Slovenian territory prove that largely since the mid-nineteenth century advanced cross-breeding and better feeding has resulted in much larger animals. This was a general phenomenon in the Alps and beyond. In Carniola, for instance, the live weight of cattle on average doubled between the mid and late nineteenth century.[903] Similarly, in Tyrol, average adult cattle liveweight increased from 300–400 to 500 kg at the end of the long nineteenth century. With the increase in body size, milk production rose significantly, despite the characteristic diversities among breeds.[904] Focusing on the contribution of meadows to the complex of factors allowing the increase in cattle size, this process roughly corresponded to a significant increase in meadow area

[898] Wopfner, 1997, 184, 187.
[899] Pavković, 2021, 65–74.
[900] Pfister, 1985a, 43; it can be assumed that only more productive or exceptionally robust cattle were exported over long distances.
[901] Riedmann, 1985, 629.
[902] Pfister, 1985a, 43.
[903] Novak, 1970, 387–389; Wopfner, 1997, 187; Hanhart & Sonderegger, 1998, 48, 51–54; Kačičnik-Gabrič, 2014, 77–78; Hodnik, 2016, 41; Šalehar et al., 2014, 97–119; Orland, 2016, 25.
[904] Wopfner, 1997, 223–224, 239.

at better-accessible sites within the Alps (Section 3.1). Moderately fertilized meadows were typically created there, providing higher yields than non-fertilized meadows, including higher quantities of relatively high-quality first-crop hay, and, in addition, valuable second-crop hay (Sections 6.1.2.3 and 6.3). The nutritive value of hay from average non-fertilized meadows, the winter fodder basis for historical small cattle, was lower than that from moderately fertilized meadows (Section 6.3). Whereas relatively poorly nutritive hay had sufficed for the good performance of some small cattle breeds for the standards of the time (but massive cereal straw feeding often decreased their performance, see also Sections 8.2.2 and 8.2.2.1), larger cattle required a higher percentage of more nutritive fodder to perform well. In the late nineteenth and first half of the twentieth century, such winter fodder largely came from moderately fertilized meadows (Sections 6.1.2.3 and 6.3.2) and fields of fodder legumes, "clovers" (red clover – *Trifolium pratense* subsp. *pratense*, alfalfa – *Medicago sativa* or bastard alfalfa – *Medicago x varia*, for instance).[905] In the second half of the twentieth century, cattle's dietary requirements and human expectations about milk and meat production gradually changed to the degree that green plants, first-crop and second-crop hay from seminatural grasslands sufficed for historical breeds, whereas only some current cattle breeds in the Alps can perform relatively well largely on such a diet at present (see also Section 10.1).[906] Several old sheep breeds are similarly better adapted to rougher fodder than modern ones.[907]

It was not only cattle sizes that changed significantly over time. Some archeozoological finds indicate the existence of large sheep (with a shoulder height exceeding 70 cm) in the Eastern Alps in Roman Antiquity, which was, similar to cattle, followed by smaller, older breeds after Classical Antiquity.[908] This was again not only a regional phenomenon of the Eastern Alps; the decrease in sheep size at that time has been proven at several places in the (former) Western Roman Empire.[909] Nonetheless, according to investigations in Slovenian territory, this decrease in the size of sheep and goats was not as significant as in the case of cattle.[910] Later, numerous predominately small sheep breeds existed in Europe by the late eighteenth century.[911] In Carinthia, for instance, the main

[905] Compare Britovšek, 1958, 141–143, 145–146, 149; Novak, 1970, 387–388; Pfister, 1985a, 43–47, 106–116; Kauter, 2002, 74–105; Hayot Carbonero et al., 2011; Šalehar et al., 2014, 97, 100–101, 106, 112–114, 117–118; Poschlod, 2015, 85–86, 119–123; Novak, 2024, 27.

[906] Pils, 1994, 206–207, 222; Peterken, 2013, 188, 191, 360; Vogl-Lukasser et al., 2006, especially pp. 98–99, 239–240; Orel (interview with S. Kejžar), 2020, 8; author's personal observations in Bohinj.

[907] Pils, 1994, 207.

[908] Pucher, 2014, 82.

[909] Cheyette, 2008, 152.

[910] Toškan, 2016c, 194.

[911] Schöller, 2010, 636.

study area in the following paragraphs, small sheep yielding rough wool were most widespread according to reliable data from 1815.[912]

Nonetheless, since the Late Middle Ages, information on intentional sheep crossbreeding exists in some parts of Europe,[913] and the impacts on sheep size were significant at a micro- to mesoregional scale.[914] This process was also relevant in the Eastern Alps.

The large sheep breed from the Bergamo region spread to several places in the Eastern Alps, such as the territory reaching from the catchments of the Kokra and Vellach (Sln. *Bela*) Rivers in the Karawanks (Sln. *Karavanke*, Germ. *Karawanken*) and Kamnik–Savinja Alps in the west to Uršlja Gora in southeastern Carinthia in the east, the Canale Valley (Ital. *Val Canale*, Friul. *Val Cjanâl*, Germ. *Kanaltal*, Sln. *Kanalska dolina*),[915] and the Senales Valley (Ital. *Val Senales* in the present-day region of *Alto Adige*; Germ. *Schnalstal*, South Tyrol).[916] Sheep specimens with obvious characteristics of the Bergamo breed were evidently not introduced directly to the regions mentioned in the Canale Valley, the Karawanks, and the Kamnik–Savinja Alps, but from Friuli (Friul. *Friûl*, Sln. *Furlanija*), and thus the large sheep breed was known in the Karawanks as a type of Friulian sheep, as documented in the early nineteenth century.[917] The introduction and wide distribution of large sheep in these areas was clearly a significant phenomenon before the agricultural improvements of the second half of the eighteenth century. This is demonstrated based on the example of the well-known Jezersko–Solčava sheep breed (Germ. *Seeländer Schaf*), which was developed in the Karawanks and the Kamnik–Savinja Alps. A report from 1767 mentioned particularly well-developed sheep breeding in the regions of above (Eisen)kappel (Sln. (*Železna*) *Kapla*) and Mt. Peca (Germ. *Petzen*), and in the hills of the Meža (Germ. *Mieß*) Valley around Črna (Germ. *Schwarzenbach*),[918] belonging to the territory between the Vellach River and Uršlja Gora. That area was located far from long-distance sheep transhumance routes (Section 7.1) where one would expect that crossbreeding with sheep breeds considered more advanced required less effort. Potential activities of agricultural improvers since the mid-eighteenth century would not have resulted in such widespread results by 1767. Earlier historical sheep breeding developments in the area involved crossbreeding, as suggested by medieval and, more clearly, early modern sources. Namely, in 1146, the new Benedictine monastery in Gornji

[912] Dinklage, 1966c, 179.
[913] Schöller, 2010, 637.
[914] A likely example: a slight increase in sheep size in Late Medieval Prussia: Makowiecki et al., 2019, 353, 364, 368.
[915] Dinklage, 1966c, 179–180; Šalehar et al., 2014, 166–172.
[916] Wopfner, 1997, 233; Šalehar et al., 2014, 171.
[917] Compare Kačičnik Gabrič, 2008, 295.
[918] Dinklage, 1966c, 180.

Grad, the center of a large seigneury in Lower Styria, was granted five farms at Kojca, located 70 km in a straight line to the west-southwest, in the Pre-Alps of the Idrijca River Basin, with the aim of provisioning the monastery in Gornji Grad with sheep.[919] This example strongly suggests that crossbreeding over long distances might have taken place in the seigneury of Gornji Grad in the High Middle Ages, but it is an open question whether the results of such crossbreeding were already significant at that time. Namely, as explained by more recent data, the introduction of a better appreciated breed did not necessarily lead to its local establishment and abundance. Breeding had to be well considered, and the breed had to adapt successfully to local environmental conditions to succeed. This is demonstrated in the following lines, focusing on wool quality. Mating involving a ram of a more prestigious breed with a local ewe belonging to a small breed with rough wool was likely to result in higher wool quality of the offspring. In the case of repeated mating between offspring from such mixed pairs and a (rather) pure high-quality breed, the wool quality was likely to increase; mating between mixed offspring and the old local breed would result in loss of the appreciated characteristics of the introduced breed.[920] The challenge of adaptability of an introduced breed to local environmental conditions is demonstrated by the failure of merinos, originating from Spain, a sheep breed highly esteemed for its fine wool. After the initial introduction of merinos to several places in the Eastern Alps in the late eighteenth and early nineteenth centuries, this breed turned out to be unsuitable for these mountain environments, and so its introduction was short-lived in many places, possibly leaving some traces in local breeds.[921] These general explanations were needed to understand the sheep history in the study area in the Karawanks and the Kamnik–Savinja Alps.

In the seventeenth century, flocks of "Italian" sheep were recorded far from Italy, at a couple of farms in the Upper Savinja Valley (e.g., above Luče), belonging to the seigneury of Gornji Grad, and in the nearby Karawanks: in Leppen (Sln. *Lepena*) above (Eisen)kappel, at least. These records do not provide insights into how widespread the Italian breed(s) or their descendants with obvious "Italian" characteristics were in the area at that time. However, it was at these places that the Jezersko–Solčava breed, partly originating in the Bergamo region, gradually evolved. The "Italian" sheep breed(s) recorded in the

919 Source edition: Bernhard (ed.), 2006, 320–321. General interpretation: Bizjak, 2003, 213 (and convincing rejection of localizations in Bernhard's publication: the historian Matjaž Bizjak, personal communication, taking into account a rent-roll from 1377: see Bizjak et al., 2022, 183); link to sheep breeds: Zwitter, 2014a, 216.

920 This more recent general explanation (Šalehar et al., 2014, 163, 165, 168, 172) has not been applied to the medieval Upper Savinja Valley (Sln. *Zgornja Savinjska dolina*) so far; see also Wopfner, 1997, 233, and Schöller, 2010, 636–637.

921 Dinklage, 1966c, 172, 179–181, 201; Novak, 1970, 390; Wopfner, 1997, 233; Šalehar et al., 2014, 161–169, 171.

area in the seventeenth century was (or were) evidently more esteemed than the local one(s). Nonetheless, mentions of "Italian" sheep make it clear that sheep with obvious Italian characteristics were not the only breed present in the microregion in the seventeenth century, and so the Italian sheep characteristics were worth being mentioned.[922] The introductory general comment on the large sheep breed from the Bergamo region explains that those "Italian" sheep in the Upper Savinja Valley and the nearby Karawanks were larger than the small breed with a longer tradition there. Crossbreeding with Italian sheep resulted in much larger animals, still able to perform well on fodder from species-rich grasslands.[923]

Nonetheless, by the second half of the seventeenth century, large sheep had not yet prevailed in the region. In the seventeenth century, many sheep in the area were small. The average live weight of twelve wethers (probably much heavier than average sheep) of the seigneury of Gornji Grad in 1694 was still only about 33 kg,[924] and so they were not closely related to Bergamo sheep. Similarly, according to 1665 data, the sheep-to-cow ratio for the grazing tax in the Solčava region was 1 to 10, despite the small size of the cows; note the weights of eleven seventeenth-century cows from this area mentioned above.[925] These data suggest that large sheep were not prevalent in the area at that time. In the late seventeenth century and/or the first half of the eighteenth century, the average sheep size in the mesoregion increased significantly. It was mentioned above that the 1767 report documented a particularly good state of sheep farming as a general characteristic of the broader region. An expert report on Carinthian agriculture published in 1815 claimed that the size of a large sheep breed kept in some places in southeastern Carinthia was at that time almost twice the size of sheep usual in other parts of the province. The large breed was reportedly reared in several parts of the Meža Valley (for instance, at the foot of Uršlja Gora and around Črna, and also in side valleys, such as Topla) and to the west of it, in the surroundings of Bleiburg (Sln. *Pliberk*) and (Eisen)kappel—thus in the study area dealt with here. The same sheep size was also characteristic for Jezersko, the most important microregion of sheep breeding in Carinthia.[926] In the mid-nineteenth century, well-fattened Jezersko rams (two years old) weighed as much as 70 kg, and a ewe up to 45 kg. At that time, the Jezersko sheep breed won a prize at the

922 Zwitter, 2014a, 216; Zwitter, 2015d, 150.

923 Kačičnik Gabrič, 2008, 295–296; Kačičnik Gabrič, 2009, 34–35; Šalehar et al., 2014, 170.

924 Zwitter, 2015d, 144, 146–147.

925 Zwitter, 2015d, 145.

926 Dinklage, 1966c, 179–180; Kačičnik Gabrič, 2008, 295; Kačičnik Gabrič, 2009, 34–35; SI AS 178, Franciscejski kataster za Koroško (Franciscan Land Cadaster for Carinthia), K312, cadastral municipality of Zgornje Jezersko, cadastral appraisal protocols: Catastral-Schätzungs-Elaborat, Einleitung, § 4.

1856 Paris exhibition.[927] This large sheep breed spread widely in the Eastern Alps. Its spread partly predated the mid-nineteenth century, like the introduction of this breed to the Lesach Valley (Germ. *Lesachtal*) in Carinthia. The spread of the Jezersko–Solčava sheep breed was intensive in the second half of the nineteenth century and the first half of the twentieth century.[928] The sheep size kept increasing, and so the late-twentieth-century live weight of the Jezersko–Solčava sheep breed was 65 to 75 kg or even more than 80 kg, and that of wethers more than 100 kg.[929]

Contrary to the early example of larger sheep largely resulting from crossbreeding with specimens or their predecessors originating in the Bergamo region, significant increases in sheep size were a late phenomenon in some regions of the Alps, comparable to small cattle kept in several places until mid-twentieth century. In what are now the Alps of southeastern France, for instance, a territory where sheep rearing remained a highly relevant part of the farm economy for an exceptionally long time (see Section 6.3.1.1), sheep weight increased from about 25 kg in the eighteenth century to only "30 kg during the first half of the twentieth century."[930] Some of the reasons for the persistence of such small sheep probably lie in the specific broader demographic and economic history of that area. These include the long tradition of low agricultural density—"[e]ven at its high point in the early nineteenth century, the population density in the French Southern Alps and their surrounding areas was strikingly low" and agricultural density remained relatively low throughout the nineteenth century—and a general lack of nineteenth-century agricultural modernization in the French Southern Alps,[931] apart from the adaptation of the local small sheep breed with a long historical tradition to local environmental conditions.[932] Later, sheep in that area "grew up to 35 and 40 kg during the 1960s and 1970s." At present, they weigh about 45 kg.[933]

Horses also were highly diverse in size, which depended to a high degree on the diversity of their uses and the social status of their owners.[934]

Figure 38 considers seventeenth-century data on historical livestock weight in the Upper Savinja Valley's region of Solčava. It shows a calculation of seventeenth-century fodder availability and fodder demand at the farm level in the barn-feeding period, based on archival records. "Especially situations where

927 Šalehar et al., 2014, 166–167, 172.
928 E.g., Dinklage, 1966c, 180; Šalehar et al., 2014, 167–168, 170–171; Kačičnik Gabrič, 2009, 34–35.
929 Kompan, 1999, 20.
930 Véron, 2009, 164–167 (quote from p. 166); see also Mathieu, 1996, 142–143.
931 The connection between Mathieu's general syntheses (Mathieu, 1996, 156; Mathieu, 2019, 55 (quote)) and sheep history is a novel contribution.
932 E.g., Schöller, 2010, 636.
933 Véron, 2009, 166.
934 Novak, 1970, 381–382; Thomas, 1987, 29, 59–60, 133.

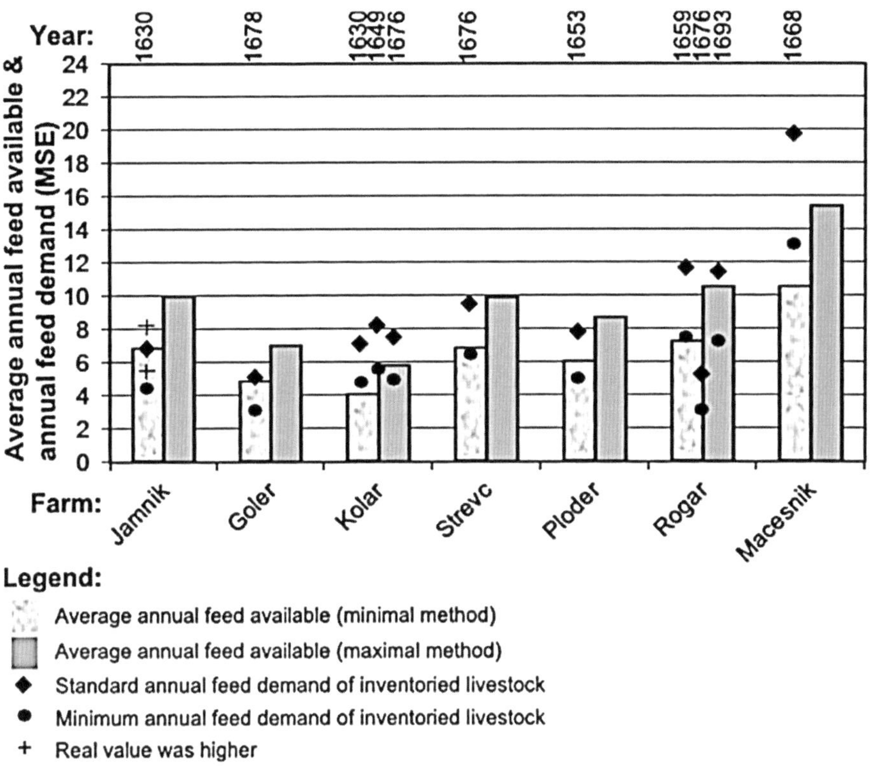

Figure 38: Approximations of fodder availability and feed demand (in mega starch equivalents, i.e. millions of starch equivalents, MSE[935]) by livestock at seven isolated mountain farms in the region of Solčava in the Upper Savinja Valley in the seventeenth century. Approximations of feed demand were calculated on the basis of probate inventories from the barn-feeding part of the year and local data on livestock weights. Feed demand is designated by circles (minimum annual feed demand) and diamonds (standard annual feed demand for the same livestock); the data refer to the years written above the graph. Where the information is available for the farm in question for more than one year in the seventeenth century, pairs of circles and diamonds are presented on the graph chronologically from left to right. Columns designate approximations of the nutritive value of fodder available on these same farms according to a description of the farms from 1665. It is not clear whether the description considers the minimum feed demand (left column) or the standard feed demand (right column) while recording fodder availability. In reality, fodder availability fluctuated every year, depending especially on weather conditions.[936] Zwitter, 2015d, 147.© Economic- and Ecohistory.

the spot [circle] is close to the [top of the] right column probably already resulted in loss of weight due to malnutrition …, but also in exceeded carrying capacity and important environmental consequences unless the herd was con-

[935] "One starch equivalent (SE) equals the digestive value of one gramme of starch and allows the different feed stuff to be aggregated according to its specific digestive value" (Krausmann, 2008, 11).

[936] Zwitter, 2015d, 146–148.

siderably reduced or additional fodder [was] bought. Among those rearing close to as much or more livestock than their farm could sustain there was no uniformity." On the one hand there is the low-quality Kolar farm, which repeatedly faced economic difficulties in the seventeenth century. "The situation of [the] Macesnik farm …, where the herd present in December 1668 could possibly also not have met [the] standard feed demand until the end of the barn-feeding period, was entirely different. This well-situated farmer … had additional four oxen not included [in Figure 38], reared at other farms.… He could afford to buy additional fodder or use a part of [his] extensive cereal stocks to feed selected animals."[937] The fact that the minimum feed demand of an additional four of Macesnik's oxen (1.8 mega starch equivalents more),[938] kept elsewhere, would move the circle in the case of Macesnik in 1668 to the immediate vicinity of the top of the rightmost column in Figure 38, demonstrates that these calculations are in fact good, despite the approximations they include; these Macesnik's oxen were kept elsewhere due to the lack of home-grown fodder at that farm.

937 Zwitter, 2015d, 148.
938 Zwitter, 2015d, 146.

6 Selected historical practices of species-rich grassland management and their ecological impacts

The majority of species from seminatural grasslands were already present in Europe before the Neolithic. Nonetheless, none of the existing seminatural meadow types "is likely to be an unchanged survival of communities present before the Neolithic."[939] Grassland management has had fundamental ecological implications for flora and fauna, including legacies of historical management in present grasslands (Sections 1.2, 2, 6, 7, 8, 9, 10, and 11). Millennia or centuries (see Section 3) of society-nature interactions in non-fertilized and moderately fertilized meadows changed site conditions, modified the life-cycle rhythms of some non-cultivated plant and animal species, and probably also led "to many nutrition physiology adaptations" of wild living beings.[940] Thus, not only farming adapted to environmental conditions in several aspects, but wildlife also gradually adapted to a certain degree to human management in the long term, as long as the management was moderate, which was typical before the second half of the twentieth century as the study period of this volume.

Particular management regimes significantly influenced various kinds of interactions between species, ranging from competition to symbiosis, and gave some species a chance to persist or colonize an area, whereas others were outcompeted and became locally extinct.[941] For instance, sheep, which graze very selectively, gradually led to the formation of specific plant communities in pastures rich in thorny, bitter, and aromatic plant species that sheep avoid. Thorny stemless thistle (*Carlina acaulis*), carline thistle (*Carlina vulgaris*), common juniper (*Juniperus communis*), large thyme (*Thymus pulegioides*), cypress spurge (*Euphorbia cyparissias*), rough-fruited spurge (*Euphorbia verrucosa*), spring gentian (*Gentiana verna*), Chiltern gentian (*Gentiana germanica*), and fringed gentian (*Gentianella ciliata*) are examples of species more frequent in extensively managed sheep pastures.[942] Some of these organisms have now turned

[939] Peterken, 2013, especially 108 (quote), 117; see also Poschlod, Baumann, & Karlík, 2009, 15–17.

[940] Pils, 1994, especially p. 169 (quote); Peterken, 2013, 88–90, 194, 202–203, 206, 230–231, 239–240, 243, 285; Kapfer, 2010; Jonstrup et al., 2016; Janišová et al., 2023. See this volume's Section 6.

[941] Peterken, 2013, 88–90, 194, 202–203, 206, 230–231, 239–240, 243, 285; Kapfer, 2010; Janišová et al., 2023; Praprotnik 1993a, 218–219; Praprotnik 1993b, 261; see also all subsections of Section 6 and Sections 1.2, 10.1, and 10.2.1.

[942] Poschlod, 2015, 179.

from agriculturally unwanted grassland "weeds" into rather rare plants and targets of nature conservation measures since the former management regime became uncommon or disappeared, and thus appropriate sites and habitats became rare or vanished (Sections 1.2 and 10.1, 10.2.1).

Long-term data on the management of grasslands are often lacking in ecological literature,[943] and so Section 6 presents selected historical practices of grassland management and their main economic and environmental impacts. It deals with the environmental histories of mowing, spring grazing in hay meadows, moderate fertilization or absence of fertilization, grassland irrigation, the history of ensuring the long-term renewal of meadows, and spring cleaning and other removal of undesired plants and plant remnants from grasslands.

It is not the aim of this section to provide an overview of all ecologically relevant historical practices of grassland management. The focus was on those where new research findings from the historical sources studied (on them, see Section 1.5) or novel historical ecological interpretations offer something new. In line with the focus of the volume, management of hay meadows and the resulting ecological impacts receive the most attention. Regarding grazing, we present spring grazing in hay meadows and its economic and ecological relevance (Section 6.2) because some spring-grazing regimes had a stronger influence on species composition than autumn grazing in the same meadows.[944] Permanent pastures, often characterized by slightly lower biodiversity than species-rich hay meadows at comparable sites,[945] are not the focus of this volume, and so they are only dealt with in Chapter 6 if the sources analyzed provided ecologically relevant information.

6.1 Mowing

6.1.1 Scythe evolution and its ecological implications

6.1.1.1 Evolution and maintenance of the scythe, with an emphasis on medieval and later implements

The scythe is an Iron Age invention, and during that period it also started being used in the Alps to harvest considerable quantities of hay.[946] Experimental archaeology has proven that a replica scythe combining a copy of an Iron Age

943 Kapfer, 2010, 139.
944 Sadar, 1956, 148; Peterken, 2013, 392; Janišová et al., 2023, 9, 11–13, 17–18.
945 Bakker, 1989.
946 Nothdurfter, 1979, 41, 45–47, Table 21; Břicháček & Beranová, 1993, 252–256; Kauter, 2002, 126; Hejcman et al., 2013, 354; Hanemann, 2014, 190–191, 193–194, 489; Poschlod, 2015, 54; Laharnar, 2018, 217; gratitude is expressed to the archaeologist Daša Pavlovič for the recommendation of works by Hanemann and Laharnar.

scythe blade found in the Carpathians (at Nižná Myšľa in Slovakia) with a copy of a preserved Iron Age scythe snath from Switzerland was a handy tool. Old grasses and some grasses and herbs that are not easy to mow were mowed even more easily with this replica than with a modern scythe. The replica also made it easier for the haymaker to avoid obstacles in meadows, such as stones. However, this mowing experiment on a slope made it clear that such a scythe with a coplanar blade and snath did not yet allow mowing as close to the ground as most scythes from Late Antiquity or the Early Middle Ages onward, which have an angle between the blade and the snath. Thus, mowing with the replica Iron Age scythe could only be carried out where grassland plants were tall enough.[947] Such mowing had fewer ecological impacts than modern mowing (Section 6.1.1.3). Examples of the continuity of use of short scythe blades following the Celtic tradition in peripheral areas of the Roman Empire in late Roman and post-Roman times were proven, even though more advanced scythes already existed in (some parts of) Europe by that time.[948] These types of scythes make cutting possible, but no swinging mowing; thus, the scythes similar to the Iron-Age implements did not fully disappear from the Eastern Alps until the end of the widespread use of the scythe in the twentieth century, but they played a fully subordinate role and served specific purposes because they were better suited for mowing in specific environmental conditions. In the last millennium they were never the only scythe available at a farm. They were used in particular to cut dwarf shrubs for livestock bedding, such as heathland.[949]

Focusing on the history of typical grassland scythes after prehistory, in the earlier part of Roman Antiquity, the use of implements that only allowed cutting grassland plants significantly higher than a later scythe is still documented. Pliny mentioned long scythes in Gaul, and starting about the beginning of the Common Era improved scythes with long blades were also used in the Eastern Alps. As in Gaul, great quantities of hay were also needed here, in contrast to regions with a Mediterranean climate, where grazing could take place all year round, also in winter. Nonetheless, these long scythes still did not allow mowing as close to the ground as later ones. The development of scythes gradually allowed real swinging mowing, which is a much easier action than the previous cutting of grassland plants using less-developed scythes. This fact can easily be understood by those of us that have tried to mow using a twentieth-century scythe but did it in the wrong way and were observed by traditional farmers, who showed how to mow much more efficiently, based on waist movements. To make such mowing possible, scythes with the following

947 Břicháček & Beranová, 1993, 256, 264, 358.

948 Hanemann, 2014, 191, 193–194, 196, 465, 468, 472–473, 489, 493; on the length of the blade, see Myrdal, 2005, 11–15.

949 Moser, 1974, 5, 12–13, 31; Mlakar, 2015, 68; and see the paragraphs below on scythe evolution.

basic characteristics had to be developed. First, the sharp edge of the blade had to become almost straight; namely, some earlier scythes had a bent blade, also including a strongly bent sharp edge close to the snath. Harvesting hay using such bent blades was cutting rather than real mowing, and such cutting could only be carried out further from the soil than later mowing, and so it produced less hay per area unit. Second, the blade and the snath were no longer coplanar; the great majority of scythes from Roman times still had a coplanar blade and snath. Finally, additional handles were added to the snath.[950]

In Late Antiquity, by the late fifth century, the first scythes allowing swinging mowing close to the ground were gradually developed,[951] although exceptional finds from Belgium already prove the existence of scythes permitting mowing close to the ground there in the second to third centuries AD.[952] Some examples of advanced long scythes from AD 350–650 have been found in the Alps, pre-Alps, and their surroundings, such as in Benken close to Zürich, in Tutzing-Monatshausen southwest of Munich, in Maiersdorf close to Wiener Neustadt, and at Tinje close to Loka pri Žusmu in eastern Slovenia.[953] Focusing on the southeasternmost part of the Alps, a long scythe blade with a diagonal length of 91 cm of the almost straight sharp edge, which used to form an acute angle with the snath, was found in the surroundings of the village of Naklo in the inner-Alpine part of the Ljubljana Basin, in west-central Slovenia. This scythe was made in Late Antiquity or in the Early Middle Ages. It was used for real swinging mowing, and it made mowing close to the ground possible.[954] These examples of advanced scythes prove that they were in use at least in parts of the Alps at the beginning of the Early Middle Ages.[955]

---

[950] Nothdurfter, 1979, 45–46, Tables 21–22; Henning, 1987, 43, 91–92, 175, Tables 42–44; Henning, 1991, 51–53, 56; Henning, 1997, 780; Břicháček & Beranová, 1993, 252; Kauter, 2002, 126–127; Hanemann, 2014, 191–196, 465, 467–468, 472, 489–490, 493. It is erroneous that additional handles on the snath can only be proven since the eleventh century AD, as Hanemann (2014, 196) summarizes from older literature. A scythe with a curved snath may have been relatively handy even without a handle close to its top (Moser, 1974, 6, 16–17, 19, 22–24, 26, 33, 40). At least the handle close to the middle of the snath—or two handles (close to the middle and near the top); the illustration is not clear—can already be observed in the version of the calendar drawn in Salzburg c. 818 kept in Vienna (a color reproduction is published in Štih, 2016, 170). Two handles on the scythe snath are clearly proven at least since the ninth century (Henning, 1991, 55).

[951] Henning, 1991, 49, 51, 53; Henning, 1997, 775, 780; Hanemann, 2014, 196, 475.

[952] Henning, 1991, 52–53, 56.

[953] Henning, 2014, 345–346, 351, 354. Gratitude is expressed to the archaeologist Dragan Božič for his explanation that the archaeological site of Tinje (see Figure 4), mentioned by Henning, is located close to Loka pri Žusmu; it is not the village of Tainach (Sln. *Tinje*) located at the foot of the Alps in Carinthia.

[954] Henning, 1987, 43, 87, 91–92, 140, 175, Table 44; Valič, 1968, especially 485, 487, 492, 494, Table 5.

[955] Henning, 2014, 346, 348.

Figure 39: Fragments of early medieval scythes, most probably from the late eighth or early ninth century, from Gradišče Hill above Bašelj in the Kamnik–Savinja Alps (Sln. *Kamniško-Savinjske Alpe*). It is clear that these finely forged thin blades were not coplanar with the snaths and these scythes made it possible to mow close to the ground. The acute angle between the ground and the snath is similar in both cases (around 20° to 25° in one case and around 30° in the other; this angle of the scythe on the left looks larger, but this appearance is erroneous, a consequence of that scythe being reforged), and they resemble the one usual in Austria in the mid-twentieth century (around 30°).[956] The blade was almost straight. The larger preserved part of the blade is 32 cm[957] long, and the original length must have been at least 40 to 50 cm. © National Museum of Slovenia, inv. nos. S 2591 and S 2592 (photo: Tomaž Lauko).

The use of advanced non-coplanar forms of scythes in the Eastern Alps in the Early Middle Ages after the Slavic settlement[958] is proven, for example, through scythe blades found at the archaeological site at the top of Gradišče Hill (873 m), rising on the southern slopes of Mount Storžič above the village of Bašelj (Figure 39), in north-central Slovenia. They were found there together with other metal artefacts, mainly from the late eighth century and first half of the ninth century, which show clear Frankish influence in terms of their forms and functions.[959] At that time, Gradišče Hill belonged to the principality

[956] Compare Moser, 1974, 20–21.

[957] The guidelines by Myrdal (2005, 12) were followed, measuring "the longest length of the [preserved part of the] blade, a straight line from the [end of the preserved edge] to the end of the back."

[958] See the general remarks on (dis)continuity in the Alps in the Early Middle Ages in Section 5.

[959] Knific, 1999, 55, 60–67; Knific, 2004, 193; in both publications only the find of (short) scythes there is recorded, without their characteristics, and so they were checked in the depository of the National Museum of Slovenia, where these artifacts are kept (NMS, inv. nos. S 2591 and S 2592), by the main author and archaeologist Daša Pavlovič. Gratitude is expressed to the archaeologist Timotej Knific, an expert on the Gradišče site, who checked the chronological information concerning these scythes. See also the note

of Carniola, which had been under Frankish overlordship since the 790s but was still relatively independent until the 820s, when a Frankish count was introduced.[960] These scythes allowed swinging mowing close to the ground (Figure 39). Thus the scythe that was as functional as in later periods has certainly been continuously used in the Eastern Alps since the Early Middle Ages at least, and not only since the High Middle Ages, as older agricultural history literature has it, which is erroneously still cited in some recent publications.[961] In this respect, this finding from the Eastern Alps corresponds with the archaeological findings in some other parts of Central and Western Europe.[962] Approximately at the time when the scythe blades were left on Gradišče Hill, a preserved calendar was drawn in Salzburg (about 818; see Figure 51), and it "may reflect the local rural situation," representing a long straight scythe blade in the same plane as the snath[963]—probably a consequence of awkward drawing rather than a reflection of reality.[964] If not, this would mean that there were still great regional differences in scythe types.

Production of metal scythe blades required specialized knowledge, and so since the Late Middle Ages, at least, this was concentrated in specialized production centers at a macroregional scale. Thus, since the Late Middle Ages local smiths have not significantly contributed to scythe blade diversity. The scythe blade production centers in the Eastern Alps (especially in Styria [Germ. *Steiermark*] and Upper Austria [Germ. *Oberösterreich*]) belonged to the main ones in the wider European context. There was a general increase in scythe blade quality from the sixteenth to nineteenth centuries, made possible by improvements in iron and steel processing. Because these improvements were not introduced in all relevant production centers contemporaneously, there was some diversity in scythe blade quality.[965]

There were relevant differences in scythe blade models, mainly in terms of the diversity of their lengths, influenced by the nexus between environmental and human factors. Examples are a relatively short scythe blade (40 to 50 cm)

---

on contemporaneous early ninth-century rings to fix the scythe blade to the snath, found at Sebenje near Bled, in northwestern Slovenia, in Smerdel, 1994, 84. The use of a ring binding the scythe blade and snath is proven in the Eastern Alps since the Iron Age (Moser, 1974, 22). Concerning the questions of (dis)continuity in farming implements in the Eastern Alps at the transition from Late Antiquity to the Early Middle Ages, see Grafenauer, 1970d, 211–213.

960 Štih, Simoniti, & Vodopivec, 2008, 34–37, 39–41.

961 For instance, the High Middle Ages is still given as the time of the introduction of a scythe allowing a swinging mowing in the syntheses by Kapfer, 2010, 135–136, and Bätzing, 2015, 53.

962 E.g., Henning, 1991, 50–53, 56; Hanemann, 2014, 196.

963 Henning, 1991, 51, 55, 57; Henning, 1997, 776–777; Henning, 2014, 333–335, 339, 346.

964 Henning, 1991, 51; Henning, 1997, 783.

965 Moser, 1974, 7–8, 11, 17, 19, 38; Pfister, 2007, 908–909; Bingener, 2010; Gorißen, 2010; Meiners, 2010; Reith, 2010a; Reith, 2010b.

to mow stony and/or hummocky meadows (for instance; see Section 6.1.1.2), or long (e.g., one meter) but light blades to mow flat wetlands with thin plant cover.[966]

Less clear than the evolution of the scythe blade from iron and steel is the historical development of the wooden scythe snath and its handles, which also influenced the handling, use, and efficiency of the implement. It is clear, however, that there were regionally diverse types of scythe snaths and handles within the Eastern Alps, which underwent different historical evolution. Moreover, local differences mattered because the snath and handles were usually products of a local craftsman or the people working at a farm. Some continuities since the Middle Ages and significant evolution of wooden snaths and handles that ended as late as the first half of the twentieth century have been proven. For instance, in inner (southern) Carinthia, a scythe snath type with two bent handles pointing in the mowing direction succeeded a scythe snath type with two handles pointing in the direction opposite to mowing (the scythe blade pointing left from the snath, and handles to the right, similar to Figure 44). This change made mowing significantly easier. The handle in the middle of the novel snath, where the right hand provides the driving force for mowing, was longer (partly inspired by the older scythe type). This improved snath type spread significantly in inner (southern) Carinthia in the first half of the twentieth century.[967]

Scythes required regular sharpening with a wet whetstone (Figure 40)—every twenty steps, for instance, depending on the hardness of the plants and on the weather—very dry weather required more frequent whetting than humid or wet conditions. A whetstone's quality, strongly dependent on natural stone properties, had a relevant impact on the sharpness of the scythe blade. Although whetstone production for local, microregional, or mesoregional use existed at several places (Figure 40 probably shows a local product), some of the dominant grass species in meadows in the Alps required optimal whetstones (see the trickiness of mowing matgrass in Section 6.3.1.2): "We have to buy Bergamo whetstones," some farmers in Slovenian territory used to claim, "because our grass is naughty," as remembered in the second half of the twentieth century.[968] Whetstones from Pradalunga near Bergamo were most esteemed; the tradition of whetstone production there reached back to the fourteenth century, and trade in this product increased later.[969] Because several merchants and other businessmen from the Bergamo region were active in the southern part of the Eastern Alps and their surroundings in the fifteenth and following centuries, and some of them sold their merchandise from the Apennine Pen-

[966] Myrdal, 2005; Zwitter, 2020, 77.
[967] Moser, 1974, 5–8, 11–19, 22–26, 32–43.
[968] Smerdel, 1994, 45–50, 82–83, 100–104 (quote from p. 103).
[969] Smerdel, 1994, 100, 102–103.

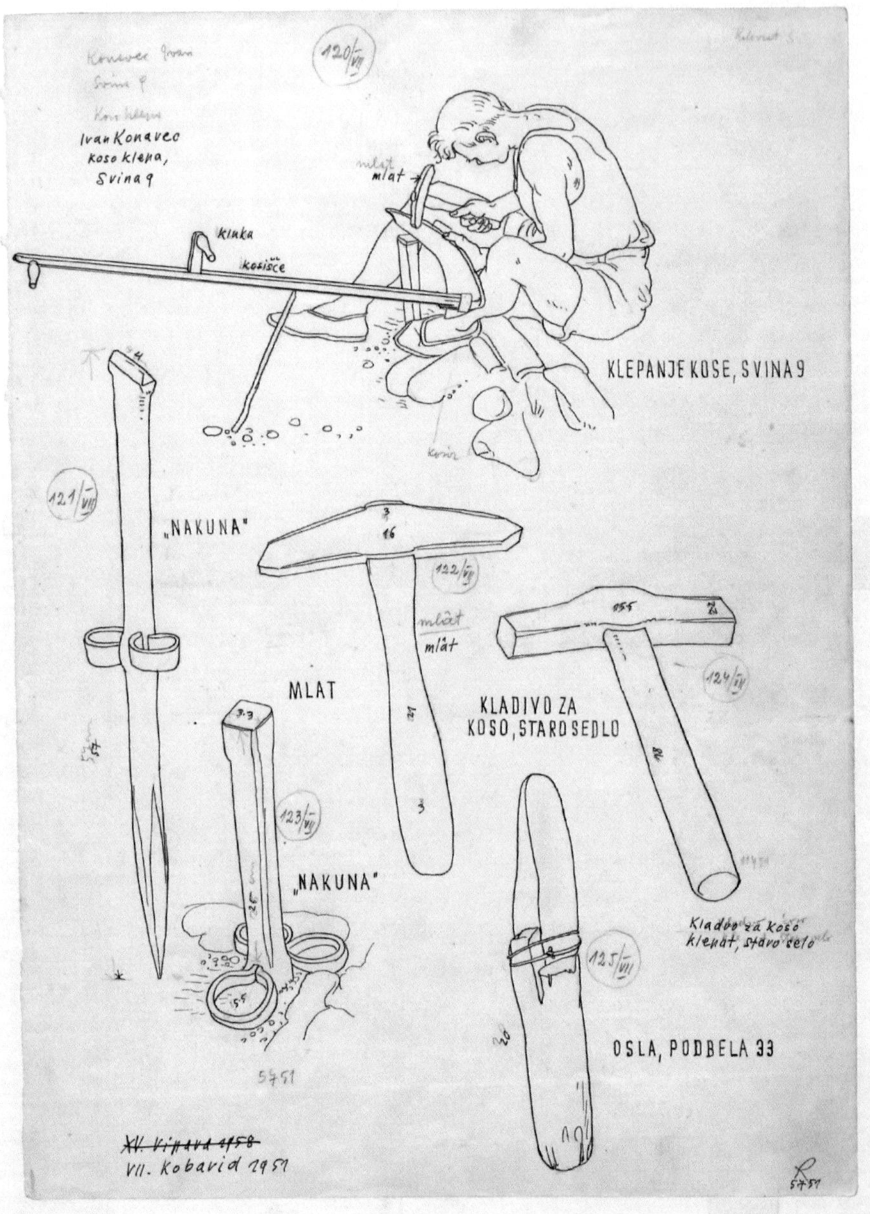

Figure 40: Implements regularly used for scythe maintenance: (1) a wet whetstone (lower right); (2) a peening hammer (two such hammers are drawn above the whetstone) and anvil for occasional peening (two anvils stand at the lower left, one seen as a whole, the other shown hammered into the meadow ground, thus ready to be used). Rings on the anvil prevented the penetration of the anvil deeper into the soil than wanted, allowing efficient use of this accessory.[970] The upper part of the drawing shows the practice of peening a scythe.

[970] Keller, 2004, 15, 48, 64–65, 69, 103, 107, 115; Bavcon, Ravnjak, & Praprotnik, 2019, 149.

Svino in the Kobarid region in the Julian Alps (Sln. *Julijske Alpe*; Ital. *Alpi Giulie*; Friul. Alps *Juliis*), 1951. All these types of implements for scythe blade maintenance have a long historical tradition;[971] however, the shapes of the same implement types and the methods of scythe sharpening and peening differed vastly within the Eastern Alps.[972] Drawn by Ivan Romih. Slovenian Ethnographic Museum, Documentation, Fieldwork campaign 7, Kobarid (1951): drawing 7/120. *Slovenski etnografski muzej* . . . Kobarid / 101 R0000007/120. © Slovenian Ethnographic Museum.

insula eastward,[973] a longer history of importing whetstones from that area to some regions in the Eastern Alps would not be surprising. Nonetheless, local products and those from some other places also clearly played a highly significant role in the Eastern Alps.[974]

Additional measures to increase whetstones' quality are proven from the late nineteenth century and the first half of the twentieth century, some with a longer historical tradition, demonstrating the attention that haymakers paid to whetstones. They range from effective in some cases (e.g., addition of vinegar

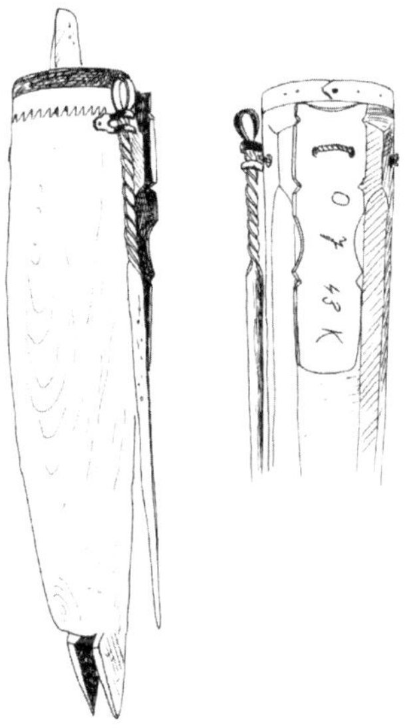

Figure 41: In addition to the implements in Figure 40, a steel straightener was used between peenings in stony meadows. This is the long straight object attached to the side of the whetstone holder (on the right in the left part of figure, and on the left in the right part of the figure). The figure shows the same whetstone holder and steel straightener from two perspectives, documented at Vojsko in the Idrija Hills in 1959. Drawn by Slavka Čufer. Smerdel, 1994, pp. 226, 297. Slovenian Ethnographic Museum, Documentation, Fieldwork campaign 16, Črni Vrh, Vojsko (1959): drawing 16/97. © Slovenian Ethnographic Museum.

971 For instance, they correspond with the data on whetstones from the early Iron Age to the twentieth century (Moser, 1974, 8–10, 18; Smerdel, 1994; for historical data from prehistory to the eighteenth century, see pp. 15, 17, 27, 81–85, 104) and with data on peening equipment from Roman Antiquity (today's Belgium, westernmost Germany, and France; Henning, 1991, 51–53) and from the fifteenth, sixteenth, and twentieth centuries (see Figure 52 in this volume's Section 6.1.2.1.2; Moser, 1974, 8–10, 18, 21; Smerdel, 1994, 59–60, 109, 290–291). Legacies of a kind of peening are already evident on prehistoric bronze sickles (Smerdel, 1994, 13, 81).

972 Moser, 1974, 21.

973 Gestrin, 1981.

974 Smerdel, 1994, 100–104.

to the water in the whetstone holder if a whetstone was too hard) to superstitious, such as the widespread belief in the southeastern Alps that the addition of a venomous snake's head (*Vipera* spp., living in many rocky meadows in the Alps) to the water in a whetstone holder would result in optimal scythe blade sharpness. This opinion had older historical roots and might have been based on the belief that the sharpness of the snake's teeth or tongue (!) would contribute to the sharpness of the scythe.[975]

In stony meadows, sharpening with a whetstone was preceded by straightening and tightening the scythe's blade with a steel straightener (Figure 41). "[I]f a scythe is jagged from hitting a stone and a whetstone is applied to it, the damaged, slightly bent spot will be ground off and the scythe will be notched. The [haymaker], therefore, has to take his straightener to straighten and tighten the scythe's blade with it before whetting."[976] Steel straighteners also have a long historical tradition.[977]

When the scythe was not sufficiently sharp after whetting, peening followed (Figure 40).[978] In some regions, the blade was usually removed from the scythe every time it was peened, and in the others the blade was peened fixed to the snath (see Figure 40). Where the blade was removed for each peening, the blade's closeness to the ground while mowing could change every time the scythe was peened, unless the blade was fixed skillfully after peening.[979] The latter was usually the case because peening was performed by an experienced worker, if available.[980] Namely, the details of how the scythe was attached to the handle every time after being removed affected how close to the ground it scythed; a blade working too close to the ground damaged the vital organs of more plants, and thus it created more gaps in the grassland ecosystem, and the start of growth of second-crop hay or grazing resources after mowing was delayed. Another extreme was a scythe working relatively high from the ground, and so it cut the plants significantly higher than usual.[981] Even though mowing close to the ground was widespread from the times when the scythe allowed it onward,[982] history also saw slightly higher mowing in some regions and periods, at least; hence the traditional saying that "Mowed at height is half

975 Smerdel, 1994, 85, 102, 108–109 (see also 59–60 because the English translation is not fully accurate).
976 Smerdel, 1994, 61, 104–105 (quote from p. 104; thanks to the ethnologist Barbara Sosič for recommending this work); Moser, 1974, 9–10; Šarf, 1974, 138–139.
977 For sixteenth- to twentieth-century straighteners, see Moser, 1974, 9–10, 22, and Smerdel, 1994, 52–53, 61, 104–105, 221–222, 224–226, 286–289, 291–292, 294, 297–298 (and further whetstone holders with a holder for a straightener on pp. 241–299).
978 Smerdel, 1994, 109 (the English translation is inexplicit; the Slovenian text is on pp. 59–60).
979 Moser, 1974, 21.
980 Smerdel, 1994, 109.
981 Moser, 1974, 21; Verbič & Verbič, 2014–2020, 13–14, 25.
982 E.g., see Figure 15 in Section 3.2; see also Babai, Molnár, & Molnár, 2014, 105–106.

manured" was documented in Austria in the mid-twentieth century as an old farmers' rule.[983]

The general use of the scythe as the most important implement for harvesting hay declined in the Eastern Alps from the first third of the twentieth century onward with the introduction of livestock-powered mowers on relatively flat grasslands in most technologically advanced regions or farms.[984] Mowers spread much later in many regions of the Eastern Alps and in steep meadows in particular. Since the mid-twentieth century, mowers have been motorized. A major decrease in the use of scythes in the Slovenian Alps, for instance, occurred in the 1960s, when they ceased to be the prevalent mowing implement, with the main exception of some very steep or hummocky meadows.[985] A detailed anthropological investigation of the settlement of Strojna in the Alps of the Meža catchment, in southeastern Carinthia, with a lot of steep farmland, documented that around half of all meadows there were still mowed by scythe around 1980. Even the farm with second-largest area of motorized mowing at Strojna mowed a quarter of all meadows by scythe at that time.[986] The scythe's role in grassland management has strongly declined ever since.[987]

6.1.1.2 Mowing in various grassland environments

Site conditions and species composition had a strong impact on the workload at mowing, and so the area mowed per haymaker in a unit of time differed significantly.[988] Many molehills or vole tunnels created after spring cleaning (Section 6.6.1) slowed down the scything; namely, they were avoided by haymakers to prevent the soil from spoiling the hay.[989] In mesic meadows on many northern slopes where the availability of plant nutrients in the soil was not very low, and in humid depressions elsewhere, grasses and herbs are significantly softer at many meadow sites, which rendered mowing easier than at very nutrient-poor or dry sites. In addition, morning dew lasted longer on northern slopes, and so the scythe required less frequent sharpening. In contrast, mowing non-fertilized meadows on southern slopes was usually significantly more labor intensive in the case of a significant abundance of dwarf shrubs (such as bilberry, *Vaccinium myrtillus*, and heather, *Calluna vulgaris*) and Iceland moss (*Cetraria islandica*) in particular (see Section 8.2.2).[990] Matgrass (*Nardus*

[983] Moser, 1974, 21.
[984] Moser, 1974, 1; Šarf, 1991, 333.
[985] Makarovič, 1982, 149; Šarf, 1991, 333; Smerdel, 1994, 61, 86, 101; Habinc, 2011a, 212; Habinc, 2011b, 238; Šarf, 2011, 240; see also Šarf, 1974, 144.
[986] Makarovič, 1982, 149–150, 152, 187.
[987] Smerdel, 1994, 86; Habinc, 2011a, 212; Keller, 2004.
[988] Wopfner, 1997, 333; Sucholas et al., 2022, 19–21, 23, 29.
[989] Author's interview with Kunavec, 2023, in Stara Fužina, with personal experience of such mowing in a remote meadow on Rudnica Hill. On seeing molehills through the plants in the case of mowing with a scythe, see Iuga, 2016, 73.
[990] Wopfner, 1997, 333.

Figure 42: Montane matgrass sward (*Nardetum strictae*) with Salisbury's daffodil (*Narcissus poeticus* subsp. *radiiflorus*) and heather (*Calluna vulgaris*) in early spring in the Western Karavanks, the Lepenje meadow above Javorniški Rovt. This plant community was favored by the long history of mowing in this non-fertilized meadow (see Section 6.3.1). Photo: Žiga Zwitter, May 1st, 2018.

stricta; Figure 42), dominating many very nutrient-poor acidic sites regardless of the water availability, was extremely difficult to mow (Section 6.3.1.2). Nevertheless, it was the species providing the most hay in these plant communities, and so it was necessary to harvest it as thoroughly as possible. Some wetland monocots also required an excellently peened and sharpened scythe blade and a skilled haymaker to be mowed successfully.[991]

In the Alps, a swinging mowing with the scythe was usually performed in a manner such that the implement moved in a direction describing an arc of a circle around the haymaker's stance (see Figure 15 in Section 3.2), as far as observed in recent centuries, unless obstacles had to be avoided. Such a mowing direction was usual in Central Europe, but there were differences across the continent.[992]

History saw small but important innovations in the mowing technique, adapted to plant characteristics.[993] Slopes were usually mowed in a downhill

[991] Sucholas et al., 2022, 10, 19–21, 23, 29.
[992] Moser, 1974, 6–7.
[993] See below in this paragraph.

Figure 43: The usual (left) and upward (right) mowing direction adapted to specific species composition in meadows abounding in upright brome (*Bromus erectus*) or matgrass (*Nardus stricta*) as a mid-twentieth-century innovation at a smallholding at Dolina. Author's interview with J. Kavčič, Dolina above Tržič. Drawing by Božidar Jožef Flajšman, following J. Kavčič's sketch. © Božidar Jožef Flajšman.

Figure 44: Demonstration of how the scythe was held while mowing up the hill if upright brome (*Bromus erectus*) or matgrass (*Nardus stricta*) were mowed at this farm at Dolina in the Karavanks since the mid-twentieth century. In this case, the scythe was not adjusted to the specific mowing technique; it was simply held differently, as seen in the picture. Photo: Žiga Zwitter, October 2021.

direction (Figure 43). In contrast, mowing by scythe up the hill (Figures 43 and 44) was introduced in some places, such as at a small holding at Dolina above Tržič in the Karavanks in the mid-twentieth century (around 1960), only on those non-fertilized steep meadow sections where matgrass (*Nardus stricta*; see also Section 6.3.1.2) or upright brome (*Bromus erectus*, a relatively esteemed fodder plant in non-fertilized meadows; see Sections 6.3.1.1 and 6.3.1.2) abounded. These species were difficult to mow down the hill. The leaves of upright brome tussocks are bent in a downhill direction (Figure 45), and so mowing up the hill was more efficient. In the mid-twentieth century, usual mowing of steep, non-fertilized meadows up the hill was also recorded at isolated farms at Florijan pri Gornjem Gradu in the Kamnik–Savinja Alps,[994] and therefore this kind of adaptation was not entirely exceptional in the wider context of the Alps, despite the paucity of evidence. Nonetheless, exactly the late introduction of this practice at Dolina warns against deterministic generalizations: such species compositions were not always mowed this manner; the mowing direction depended not only on the meadow environment, but also on the local mowing culture.

Another widespread adaptation was required in rocky meadows. A scythe that would otherwise allow swinging mowing had to be used in such meadows for less labor-efficient cutting of fewer grassland plants at the same time. Thus, shorter scythe movement than swinging mowing was carried out as an adaptation to obstacles. This was usual until the twentieth century in rocky meadows where the density of rock outcrops did not allow swinging mowing but the area of the meadow covered by stones was at the same time not so great as to prevent such scythe use[995] (see the rockiest meadow sections in Figure 46). The practice of short scythe movement in such meadows has a long history. According to the 1567 rent-roll of the seigneury of (Eisen-)Kappel (Sln. *(Železna) Kapla*) in the Central Karawanks: in stony, steep meadows "no real scythe stroke happens on many farms"; grass was also harvested there around the stones.[996] In such meadows, where one had to avoid many rock outcrops or numerous trees, a relatively short scythe blade was more convenient than a longer one. For example, at Pölling above Treffen (Sln. *Trebinje*) north of Villach, Carinthia, where many extremely steep meadows were mowed, a usu-

---

994 Predan, 1956a. In all but exceptional cases, meadows in the area were mowed once annually. This is the background for the above statement that mowing up the hill at Florijan pri Gornjem Gradu refers to non-fertilized grasslands (see Section 6.3). These farms are located in pre-Alpine hills at about 700 to 800 m.

995 Author's interview with Kunavec, 2021, in Stara Fužina, with personal experience in such mowing. The information refers, for instance, to the former rocky meadows on the west-facing slope above the bottom of the Voje Valley in Bohinj.

996 ". . . dan in gepurg vmb die stain, vnnd in den stickhln, nur gleich abkhlauben, vnnd khain rechten senssen straich auf vill huebnen beschicht" (AT StLA, Innerösterreichische Hofkammer, Sachabteilung, box 69, folder 8: Des ambts Capl bereuttung und beschreibung, 1567, 44).

Figure 45: Upright brome (*Bromus erectus*) in a meadow where it used to be mowed by scythe up the hill. The height of the mobile phone case in the picture is 7.5 cm. Photo: Žiga Zwitter, October 2021. The specimen was identified by Leonid Rasran.

al scythe blade for grassland mowing in 1948 was 40 to 50 cm long, and the longest scythe blades used in grasslands in the same settlement at that time were around 90 cm. Thus, a short scythe blade was more useful in particular meadow environments. Comparison of these data with older information from post-Roman history demonstrates that the shortness of scythe blades, which became more frequent in post-Roman times, was not in itself an indicator of an inferior scythe.[997]

However, despite the widespread use of scythes in meadows in the Alps, limited quantities of grass continued to be cut with a sickle until the twentieth century in some extremely stony mountain meadows; for example, in the karst of the Upper Soča catchment where average farmers faced a scarcity of good meadows, but also in many other places (see also Section 6.3.1.1).[998]

[997] Moser, 1974, 12; Myrdal, 2005, 11–15; Hanemann, 2014, 191, 193–194, 196, 465, 468, 472–473, 489, 493.

[998] Melik, 1950, 116–117; Novak, 1970, 362–363; Smerdel, 1994, 57; Niederer, 1996d, 174; Wopfner, 1997, 368–369; Keller, 2004, 20, 65; Mlakar, 2015, 50–51, 70; Laharnar, 2018, 217–218; Bavcon, Ravnjak, & Praprotnik, 2019, 193; author's interview with F. Mlakar, 2017, in Bohinjska Češnjica.

Figure 46: Only the most diligent and not excessively busy farmers have been encouraged through insufficiently targeted, thus low, agri-environment payments to preserve the cultural landscape by still mowing such stony meadows. Note the hay drying between rock outcrops in the upper right part of the figure (beige color). Rocky meadows used to be frequent in the Limestone Alps[999] and, depending on geology and slope gradient, in some other parts of the Alps as well.[1000] They were mostly abandoned in the twentieth century. It was perceived as improper and shameful for the haymaker not to mow the tall grass next to the stones.[1001] Rock outcrops contributed significantly to the site diversity by providing additional microhabitats and thus raising the plant and animal species diversity.[1002] Vogar Hill in the Julian Alps, 1,120 m, July 2018. Photo: Žiga Zwitter.

Cutting with sickles had to be performed significantly higher than mowing.[1003] In previous centuries—from the fifteenth to early nineteenth century, for instance—poor smallholders in the Alps collected significant hay quantities at remote sites in the mountains (e.g., on the slopes of Mount Tosc in the Julian Alps in times of high agricultural density). Often they had to use sickles at such

[999] Observations in the study area and Wopfner's (1997, 330, 333) parallel from South Tyrol.

[1000] Pils, 1994, 260–261 and Table 50, photo 130; "Bildteil," 1996, 409.

[1001] Wopfner, 1997, 334.

[1002] Pils, 1994, 261; Oppermann & Hoffmann, 2012, 85; Oppermann & Paracchini, 2012, 19.

[1003] Hanemann, 2014, 190.

places.[1004] These were also usually used in meadows to collect plants growing in the immediate vicinity of rocks and trees, where the use of a scythe would not be possible.[1005] Similarly, sickles were used to collect grass in sparse forest, and alongside paths and roads.[1006] Gathering (i.e., pulling) grassland plants by hand around rocks in meadows (compare Figure 46) and in the forest as well as in barely accessible sites below bushes was also still practiced by some farmers until the mid-twentieth century.[1007]

6.1.1.3 Ecological impacts of mowing

In contrast to livestock grazing, which usually has similar impacts as game grazing on the species composition and vegetation structure of grassland communities, the effects of haymaking are significantly different. Poisonous and woody species, which are normally avoided by livestock and thus become dominant in colline, montane, and subalpine pastures (e.g., dwarf mountain pine, *Pinus mugo*, at high elevations) unless removed (Section 6.6.2), have fewer opportunities to become established in regularly mowed areas. The ecosystem disturbance through mowing, which is less frequent but more thorough than grazing, leads to the development of a higher niche diversity and, consequently, to higher species diversity than at grazed sites.[1008] The main plant communities of hay meadows in the Alps that originate in and have been preserved through mowing are (dependent on soil nutrient and water availability; see Sections 2, 6.1.2.5, and 6.3) false oatgrass meadows (Figure 6 in Section 2, Figure 91 in Section 6.3.2.2), yellow oatgrass meadows (Figure 7 in Section 2; Figure 13 in Section 3.1), dry calcareous grasslands (Figure 10 in Section 2), purple moor-grass meadows (Figure 9 in Section 2 and Figure 66 in Section 6.1.2.5.1.2), marsh marigold meadows, and matgrass swards below the alpine belt (Figure 42).[1009] These vegetation types can also exist as pastures, but with slightly different species compositions.[1010]

Similar to grazing, mowing supports plants with their main assimilatory organs close to the soil surface because a scythe (and even more a motor mower) principally works above them, and so mowing does not (thoroughly) affect these organs.[1011] Such plants include rosulate herbs, and tussock grasses and sedges. Lower sections of the tussocks of several monocots, such as purple

---

1004 Wopfner, 1997, 333, 368–369; Hodnik, 2016, 79.
1005 Kauter, 2002, 133–134; Stadler-Planzer, Stadler, & Degler-Spengler, 2015, 100.
1006 Niederer, 1996d, 186, 190.
1007 Niederer, 1996d, 186, Hubatschek, 1992, 173; Stadler-Planzer, Stadler, & Degler-Spengler, 2015, 100; author's interviews in Stara Fužina with Boltar, 2018, and with Stergar, 2018, recalling how her grandfather collected hay in the mid-twentieth century.
1008 Begon, Harper, & Townsend, 1991.
1009 Ender, 1998; Schubert, Hilbig, & Klotz, 2001.
1010 See also Pils, 1994.
1011 Bakker, 1989; Begon, Harper, & Townsend, 1991.

moor-grass and some tall sedges (*Carex* spp.), are hard, and the scythe worked above them.[1012] Focusing on rosulate herbs, the rosette of stemless thistle (*Carlina acaulis*; see Figure 127 in Section 8), for instance, was not thoroughly mowed by the scythe, not even by experienced haymakers due to its flat shape,[1013] even though cattle consume it in hay.[1014] The mowing technique with the help of a broom-like bound bunch of branches—enabling the haymaker to mow rosulate herbs more thoroughly, but historically usual only in some microregions in the Eastern Alps—is presented in Section 6.3.1.1.

The swinging mowing close to the ground, first introduced in Late Antiquity, significantly increased the working efficiency of haymakers, but it also had a stronger impact on grassland ecosystems[1015] than earlier historical mowing more distant from the ground. In general, the differences in historical mowing's ecological impacts were more pronounced in the case of a change in mowing frequency or timing (Sections 1.2, 6.1.2.1, 6.1.2.2, 6.1.2.3, 6.1.2.4, 6.1.2.5, 6.1.2.6, 6.1.2.7, 6.3, 10.1.2.1, 10.1.2.2.3, 10.1.2.2.4, 10.1.2.2.5, 10.2.1, 10.2.2) than in the case of a moderate difference in removed plant biomass caused by more thorough mowing that was made possible by an advanced scythe.[1016] Nevertheless, some relevant effects of mowing closer to the ground since Late Antiquity or the Early Middle Ages than earlier can be mentioned. Mowing closer to the ground and the related meadow management (the third point below) led to a shift in the competition ability among grassland plants in terms of suppression of some species and promotion of some others.

First, swinging mowing close to the ground usually causes numerous cases of small-scale damage to the sward and soil surface (see the foreground of Figure 15 in Section 3.2). Such gaps are essential for biodiversity because they are important regeneration niches, especially for short-lived herbs, such as yellow-rattle (*Rhinanthus* spp., undesired from the perspective of livestock farming; see Section 4.1.1.1).[1017] Consequently, swinging mowing close to the sod increases the species diversity of the plant community in comparison to other mowing techniques. In some cases, mowing helped species compete that naturally grew in different site conditions.[1018] Until the mid-twentieth century or later, gaps caused by mowing were usually joined by the gaps caused by livestock grazing in hay meadows in autumn (and in spring, until the eigh-

[1012] Sucholas et al., 2022, 18, 21, 36.
[1013] Babai, Molnár, & Molnár, 2014, 46.
[1014] Author's participant observation at hay raking in the montane meadow complex of Javornica, west of Zgornje Podjelje in Bohinj, where stemless thistle occurs in relatively large numbers, and interview with the hay owner M. Smukavec on the same occasion, July 2021.
[1015] Henning, 1991, 53; Henning, 1997, 780.
[1016] Compare Liira et al., 2009; Tälle et al., 2014.
[1017] Coulson et al., 2001.
[1018] Emanuelsson, 2009, 160.

teenth or nineteenth century; see Section 6.2), by the gaps caused by digging mammals (see also Section 6.5), and in manured meadows by the gaps caused by rubbing the manure into the sward (Section 6.3.2.1.1).

Second, focusing on the impacts of removal of slightly more biomass, the higher quantity of nutrients removed by thorough mowing closer to the ground mattered in the long run. Lower mowing meant somewhat fewer grazing resources in the same meadow in the autumn, and so slightly less nutrients returned to the meadow unless it was fertilized. In addition, a comparison with a study of stubble weeds explains that grasslands regrew more slowly because they were mowed closer to the sod.[1019] In this respect, it mattered that meadows mowed close to the soil are also more drought-sensitive during the regeneration after mowing.[1020] All this means that, in the autumn, fewer nutrients usually returned to non-fertilized meadows in the form of excrement from the livestock grazing there.

Third, removal of stones from meadows was economically and ecologically relevant. Removal of stones from grasslands was already recommended by the Roman agricultural writer Columella in the first century AD to reduce the damage to mowing implements.[1021] Most likely, stone removal became more thorough with the development of the scythe – with mowing closer to the ground, the scythe was more likely to hit small stones, and so it would have to be repaired with a steel straightener (Figure 41) or peened (Figure 40). Thus, even small stones that appeared from molehills, for example, were removed from meadows every year to make sure that scythes would not hit them, as recalled for the mid-twentieth century, and this was a continuation of an older tradition.[1022] Removal of larger stones left more traces in historical records and was ecologically more relevant. In the eastern Julian Alps, stone removal from meadows was recorded in 1610, for instance, but the practice is much older. According to a decision of the court of the seigneury of Bled (Germ. *Veldes*) from that year, somebody had to collect stones in the meadow(s) of a farm from the village of Višelnica in order to settle a debt of eight guilders.[1023]

1019 Molnár et al., 2015, 5, explaining that if stubble is cut at a height of approximately one span weeds are able to "regenerate rapidly, resulting in high-quality pastures."

1020 Sadar, 1956, 146–147, 196.

1021 Kauter, 2002, 105–106.

1022 Author's interviews with Gašperin, 2017, with Kranjc, 2018, with Mencinger, 2017, and with Urh, 2017, all in Stara Fužina; with Grm, 2017, and with Škantar, 2017, in Srednja Vas; with L. and M. Odar, 2017, in Studor; and with Minka Sodja, 2017, in Jereka.

1023 SI AS 721, Gospostvo Bled (The Seigneury of Bled, Germ. *Veldes*), Iustitialia, vol. 12, February 14th, 1610. The type of work is specified as "die wisen . . . ablesen." The meaning of the verb *ablesen* is "colligere, … [e.g.] steine vom acker . . . , raupen vom baum ablesen[,] beeren von der staude" 'to gather, . . . for example, remove stones from a field, caterpillars from a tree, berries from a plant' (Grimm & Grimm, *Deutsches Wörterbuch*) – in meadows ("wisen", i.e. *Wiese(n)*), in this case, and so stone removal is meant.

Figure 47: On sunny days, stones warm up quicker than the surroundings and thus clearance cairns are among the first sites where spring flowers blossom. Primroses blossomed in the valley bottom in Bohinj at 555 m at such a west-facing site lit by the afternoon sun in the exceptionally warm January of 2024. This rock pile is located at the edge of a species-rich grassland; in this case, a former arable field.[1024] The specific microclimate in the warm part of the year is shown by the historical practice of drying some field crops, such as green pulses, on clearance cairns consisting of stones removed from arable land in the Allues (Francoprovençal *Alués*) Valley (French *Les Allues*).[1025] It is, however, not highly likely that clearance cairns played a prominent role in historical hay drying due to the well-known preference of venomous snakes for hot stones.[1026] Photo: Žiga Zwitter.

Because a hay shed in a big meadow was sold for two guilders in 1614 in the office of the seigneury of Bled[1027] (i.e., the same seigneury that formalized the duty of stone removal in 1610), this means that the work to remove the stones in 1610 (and possibly soon thereafter), worth eight guilders, was extensive. Two centuries later, in the first half of the nineteenth century, removal of rocks was still performed regularly in some remote meadows in the eastern Julian Alps. This is proven by the description of agriculture in the cadastral municipality of Spodnje Gorje, thus close to Višelnica (probably in montane meadows on

1024 *Arcanum Maps*, Cadastral Maps, Habsburg Empire – Cadastral Maps (XIX. [sic] Century), cadastral municipality of Studor (Germ. *Studorf*), 1826.

1025 Meilleur, 1986, 98–99, 120, 130–131.

1026 Compare Meilleur, 1986, 99.

1027 SI AS 721, Gospostvo Bled (The Seigneury of Bled), Iustitialia, vol. 12, May 10th, 1614. This meadow was sold for 105 guilders, and so it was evidently a big meadow, as supported by the following data: on January 10th, 1615, a small meadow ("wüsmatl") was sold for 12.5 guilders; on the same day, 90 guilders was the purchase price for a small arable field and the adjacent meadow ("akherlle vnd madt daselbst") and four small meadows ("wüsmatl", "wüsl") together (SI AS 721, Iustitialia, book 12, January 10th, 1615). Thus, 105 guilders was the price of a big meadow.

Figure 48: Stone walls surrounding many montane meadow complexes and some permanent pastures were usually built from stones cleared from those grasslands. The same was the case of many platforms built from stone below hay barns in meadows; elsewhere, removed stones were piled into clearance cairns.[1028] The Konjska Dolina mountain pasture and meadow complex with a wall made of removed stones, 2018. Photo: Žiga Zwitter.

the Mežakla Plateau),[1029] presentations of farming in some montane meadows on the Pokljuka Plateau managed by inhabitants of the cadastral municipality of Gorjuše,[1030] and in the cadastral municipality of Selo pri Bledu (montane meadows on the Jelovica Plateau), all from the 1830s.[1031] Ethnological investigation in the village of Bohinjska Bela by Fanči Šarf made it clear that people born in the late nineteenth and early twentieth century no longer participated in the removal of large stones from meadows; they only observed rock piles or

1028 For example, Niederer 1996c, 69; Niederer 1996d, 163; Hodnik, 2016, 61, 87; author's interview with Škantar, 2017, in Srednja Vas, mentioning their high-quality meadow Za Robom (among the Uskovnica meadows), which is free of stones, which were used to build a wall.

1029 SI AS 176, Franciscejski kataster za Kranjsko (Franciscan Land Cadaster for Carniola), L310, cadastral municipality of Spodnje Gorje, cadastral appraisal protocols: s.n., 310/4, 51.

1030 SI AS 176, Franciscejski kataster za Kranjsko (Franciscan Land Cadaster for Carniola), L72, cadastral municipality of Gorjuše, cadastral appraisal protocols: s.n., 72/4, 51.

1031 SI AS 176, Franciscejski kataster za Kranjsko (Franciscan Land Cadaster for Carniola), L343, cadastral municipality of Selo pri Bledu, cadastral appraisal protocols: s.n., 343/4, 51.

Figure 49

regularly built stone structures in the landscape made of stones that had been removed from montane meadows by their predecessors (see Figures 48, 49, and 50).[1032]

Where many rocks were removed, this had an impact on the quantities of hay yielded.[1033] Moreover, stone removal allowed easier swinging mowing without taking care to avoid the rocks. Stone removal reduced the time used for repairing scythes or even for sickle-cutting or pulling plants around rocks during haymaking (see Section 6.1.1.2). Thus, rocks in meadows have been unwanted from an economic perspective. In contrast, they are appreciated from an ecological perspective, contributing to habitat diversity.[1034] Especially the removal of big stones (Figures 48, 49, and 50) as well as accumulations of smaller ones were ecologically relevant for meadows because this affected soil properties and the microclimate (Figure 47). The historical practice of piling rocks in meadows or using them to build fences achieved the economic benefits of more hay and easier mowing by hand, while at the same time the connection between the meadow and rocky habitats was preserved.

1032 Šarf, 1974, 138.

1033 This was also the goal of a similar practice of stone removal from pastures to increase the productive area: Bätzing, 2015, 90; Machatschek, 2015, 279; Hodnik, 2016, 61, 87.

1034 Pils, 1994, 261; Keller, 2004, 41, 99, 109; Oppermann & Hoffmann, 2012, 85.

Figures 49 and 50: In order to save meadow area, in some meadows in stony sites removed stones were not simply accumulated in irregular clearance cairns but built into more regular structures occupying less space than a pile from the same amount of stones. Figure 49 shows them in a montane meadow in the southern part of the Zajamniki grassland complex, at 1,260 m. The same type of structures can be found in the nearby forest that overgrew historical meadows in the twentieth century (Figure 50), and so a nineteenth-century cadastral map shows meadows there.[1035] These stony structures in the forest are evidently a historical human legacy. However, one can only understand the purpose of creating these stone structures based on the understanding of historical land use and observation of the same structure types in nearby meadows (Figure 49). Photo: Žiga Zwitter, April 2020.

Stones removed from grasslands were used as building material or they were piled, and rock piles and stone walls themselves developed into specific habitats, contributing so significantly to the mosaic of landscape structures relevant for biodiversity that dry stone walls are nowadays one of the ecological focus areas eligible for agrienvironmental payments in some European Union countries. Because there is very little soil, clearance cairns are usually dry and on sunny days very warm if exposed to the rays of the Sun (Figure 47).[1036] In the nineteenth century, the botanist Valentin Plemel observed the

[1035] *Arcanum Maps*, Cadastral Maps, Habsburg Empire – Cadastral Maps (XIX. [sic] Century), cadastral municipality of Srednja Vas v Bohinju (Germ. *Mitterdorf*), 1826.

[1036] Meilleur, 1986, 98–99, 120; Oppermann & Hoffmann, 2012, 85–86, 93; Oppermann & Paracchini, 2012, 19; Pe'er et al., 2017, 518–519, 521–523; Šumrada et al., 2024, 13, 18–19.

frequent occurrence and high abundance of orpine (*Sedum telephium* subsp. *maximum*) in such sites around Bled,[1037] at the foot of the eastern Julian Alps. Hazel (*Corylus avellana*; Figures 47 and 50) and juniper (*Juniperus communis*) frequently grow at such sites, and barberry (*Berberis vulgaris*) also occurs there.[1038] In the Allues Valley (now in the northern French Alps), common elder (*Sambucus nigra*), ash (*Fraxinus excelsior*), wild cherry (*Prunus avium*), mezereon (*Daphne mezereum*), the edible great pignut (*Bunium bulbocastanum*), raspberry (*Rubus idaeus*), bramble (*Rubus fructicosus*), stone bramble (*Rubus saxatilis* – see Figure 139 in Section 8.2.1.1), dog rose (*Rosa canina*), and several lichens, for instance, were recorded on clearance cairns.[1039] All these data are representative of many parts of the Alps, and this is only a brief selection of plant species colonizing clearance cairns in meadows.

Historically usual connections between rock piles or stone walls and the adjacent species-rich grasslands are crucial, but they are becoming increasingly rare nowadays (removal of such stony landscape structures is an ongoing process,[1040] apart from the intensification or abandonment of species-rich grasslands; see Section 1.2). Namely, the habitat quality and habitat functions of rock piles and stone walls increase highly significantly for many animal species if such a stone structure is located adjacent to high-nature-value farmland, such as an extensively managed meadow.[1041]

6.1.2 Mowing dates and frequency

6.1.2.1 Dates of first-crop hay mowing in the last millennium until the eighteenth or nineteenth century

6.1.2.1.1 Mowing timing

Mowing dates in different meadows belonging to a typical farm were distributed through most of the summer: first-crop hay in the last meadows was mowed several weeks, often a couple of months, later than in the first meadows (see the reasons in Section 6.1.2.1.2). Thus, it was important that meadow sites belonging to a farm in the Alps from a compact village (in contrast to some isolated farms) were usually significantly diverse (e.g., in different elevation zones, fertilized or non-fertilized, at sunnier and shadier, drier and moister

[1037] Plemel, 1862, 157; the subspecies was identified based on Aeschimann et al., 2004, vol. 1, 676.
[1038] Machatschek, 2015, 227–236, 279; Machatschek, 2017, 274–275.
[1039] Meilleur, 1986, 98–99, 146, 343, 347–348, 350, 353, 359, 361.
[1040] For instance, Šumrada et al., 2024, 5, 13, 15, 19, 27. The uptake of the agrienvironmental payment for stone walls is low and it only exists in some European Union member states (Pe'er et al., 2017, 518–525), not in Slovenia (Šumrada et al., 2024, 19).
[1041] Oppermann & Hoffmann, 2012, 94–96.

places, and on more or less productive soils, but also influenced by social rules concerning mowing dates; see below), which resulted in different timing of grassland plant seeds ripening. Thus, the diversity of meadow sites reduced the percentage of those meadows that, due to the lack of time of farm personnel, were mowed significantly later than when most plant seeds matured; that is, of those meadows where by the time of mowing the quality of the hay had already declined beyond what was considered required to maintain productive species-rich meadows in the long run. Therefore, in general, site diversity within a typical farm contributed to higher hay quality. Moreover, significantly diverse mowing dates decreased the likelihood of hay harvest failure in many meadows belonging to the same farm in the same year; for example, due to a long rainy period after mowing (Section 4.1.1.3). Due to the long duration of haymaking, it was more likely that only hay from a fairly limited percentage of meadows was negatively affected, and so diversity of mowing dates in general strengthened the resilience of the farm economy.[1042] Diverse mowing timing in various meadow complexes, in interaction with diverse site conditions, raised the diversity of meadow species compositions in the landscape.[1043]

Haymaking in the same meadow complex was usually carried out by all farms involved on the same days. In Parcines (Germ. *Partschins*) in South Tyrol, for instance, the beginning of haymaking was publicly announced every year, as documented in 1431.[1044] This practice used to be widespread in the Alps, as proven by the following examples documented in the twentieth century and referring to the period from the late nineteenth century onward, but clearly rooted in older traditions. In Bohinj, people from the community consisting of farmers from the villages of Studor and Stara Fužina started mowing according to a decision by the Pri Andreju farm in Studor[1045] (which probably had a special function in the community of farmers from these two villages and/or linked the seigneury and its subjects before the abolition of serfdom in 1848), and others followed suit.[1046] Not only the general beginning of mowing there was ordered: when it was the time to mow montane meadows in the slightly higher Voje Valley (around 700 m), a designated farmer from this community attracted farmers' attention by producing noise hitting an iron pan with a wooden spoon and shouting: "Let's go to Loje [i.e., Voje]!"[1047] Thus, mowing dates were determined much less individually than today.[1048] Another example:

1042 Based on general consideration on distribution of land parcels in the Alps by Viazzo, 1989, 21, 227, here adapted to meadows. See also Section 6.1.2.1.2.
1043 Oppermann & Paracchini, 2012, 18–19
1044 Wopfner, 1997, 362.
1045 See the location of the farm in Klinar, 2011, 23, and *Environmental Atlas of Slovenia*, Spatial Units, Address.
1046 Author's interview with Mencinger, 2017, in Stara Fužina.
1047 Author's interview with Kranjc, 2018, in Stara Fužina: "Gremo na Loje [i.e., Voje]!"
1048 Author's interview with L. and M. Odar, 2017, in Studor.

in the village of Bohinjska Bela in the Eastern Julian Alps (Sln. *Julijske Alpe*) it was compulsory for every household to go to the distant montane meadows on the Sunday after Saint James's Day (July 25th) to start work on the following Monday, as recorded for the late nineteenth and early twentieth century.[1049] The preparations for going to the distant meadow complexes "were not only the concern of individuals, but of the whole village."[1050] Help was offered to neighbors if they were still busy with other urgent work at the farm, such as reaping autumn-sown cereals in the lowest section of the montane vegetation belt, when it was time to go to higher meadows.[1051] This made it easier to respect the uniform mowing time.

In some microregions, only a day or two was allotted to mowing in every meadow complex, such as in the Allues (Francoprovençal *Alués*) Valley (French *Les Allues*).[1052] Such intervals were longer in some other microregions.[1053] If a farm was not finished with haymaking during that time slot, people returned there later from a higher and/or shadier meadow complex, where their plot was so small that they saved some time. This was the practice in the Allues Valley by about the First World War, and it had a long tradition. In neighboring Bozel (Francoprovençal *Bosél*), for instance, such uniform mowing time can be proven since the mid-eighteenth century, but it is most likely that it developed earlier.[1054]

The issue of access and reduction of the likelihood of mowing over the parcel boundary (less likely when all farmers possessing meadows in a certain meadow complex were present there at the same time) belonged to the main reasons for much more uniform mowing dates, apart from ensuring that all the hay would be stored by the onset of autumn grazing, when the meadow complex functioned as a common. Concerning access, a path or cart track did not reach many meadow parcels, and so haymakers had to cross other meadows. This was characteristic of meadow complexes in the Alps and beyond. Consequently, mowing at similar dates prevented trampling a path in meadows not yet mowed when accessing other meadow parcels.[1055] In the Allues Valley, for instance, "[t]hose people whose hay parcels were toward the outside of the [meadow complex] would cut paths through their parcels creating the access necessary to those [hay] fields located toward the center" as part of preparation for large-scale mowing.[1056]

1049 Šarf, 1974, 138–139.
1050 Šarf, 1974, 138.
1051 Šarf, 1974, 138; see also Hodnik, 2016, 80.
1052 Meilleur, 1986, 271, 274.
1053 Compare Hodnik, 2016, 78–80.
1054 Meilleur, 1986, 271–272, 274.
1055 Grass, 1948, 201; Meilleur, 1986, 271–272; Wopfner, 1997, 362; Kapfer, 2010, 136, 138.
1056 Meilleur, 1986, 273.

6.1.2.1.1.1 Start of mowing in June

At some climatically more favorable and nutrient richer sites, mowing started in June. Some medieval paintings and sculptures representing mowing in June are preserved in France (see Section 4.1.1.3). Medieval representations of mowing in June correspond with (1) a poem written by Wandalbert of Prüm (now in western Germany) before 848 that mentions many farm tasks and was inspired by the authors of Classical Antiquity (especially Virgil), and (2) with Matfre Ermengaud's encyclopedic treatise on seasons and months written in Provence (Occitan *Provença*) around 1288.[1057] Rarely, the expression *Heumonat* (literally, 'hay month') would designate June in Central Europe in the Middle Ages and later because the word has usually meant 'July'.[1058]

At lower elevations in the Eastern Alps, mid-June was a relevant start of mowing. Data from the early nineteenth century, before the abolition of spring grazing in meadows (Section 6.2), document the start of mowing in mid-June in some of the most productive meadows mowed twice annually in several valleys in the lowest section of the montane vegetation belt in the Julian Alps and their surroundings, such as in the cadastral municipalities of Bohinjska Bistrica, Bohinjska Češnjica, Srednja Vas, and Studor in Bohinj (at least around 500–600 m)[1059] and in the Sava Dolinka Valley (Sln. *Zgornjesavska dolina*) at Dovje (around 700 m), today in northwestern Slovenia.[1060]

An early historical example of mowing in late June is found for common meadows at Bolzano (Germ. *Bozen*) in South Tyrol, which was allowed after June 24th according to a ruling from 1190,[1061] corresponding to June 30th in today's calendar.[1062]

Analysis of similar mowing dates in the traditional community of Botiza in the Romanian Carpathians (late June in the present calendar) showed that

---

[1057] Golob, 1985, 9, 12, 30, 36–38, 50, 57.

[1058] *Frühneuhochdeutsches Wörterbuch*: heumonat. See also Section 6.1.2.1.1.2.

[1059] SI AS 176, Franciscejski kataster za Kranjsko (Franciscan Land Cadaster for Carniola), L56, cadastral municipality of Bohinjska Bistrica, cadastral appraisal protocols: Catastral-Schätzungs-Elaborat, Einleitung, § 10; SI AS 176, L107, cadastral municipality of Češnjica v Bohinju [i.e., Bohinjska Češnjica], cadastral appraisal protocols: Catastral-Schätzungs-Elaborat, Einleitung, § 10; SI AS 176, L157, cadastral municipality of Srednja Vas v Bohinju, cadastral appraisal protocols: Catastral-Schätzungs-Elaborat, Einleitung, § 10; SI AS 176, L284, cadastral municipality of Studor, cadastral appraisal protocols: Catastral-Schätzungs-Elaborat, Einleitung, § 10.

[1060] SI AS 176, Franciscejski kataster za Kranjsko (Franciscan Land Cadaster for Carniola), L139, cadastral municipality of Dovje, cadastral appraisal protocols: Catastral-Schätzungs-Elaborat, Einleitung, §10 (note the mistake in Kačičnik Gabrič, 2016, 506, who, erroneously, wrote mid-July instead of mid-June, with a reference to the same source).

[1061] Grass, 1948, 168.

[1062] The conversion is based on *Julian Day and Civil Date Calculator* and the calculations by the astronomer Tomaž Zwitter.

numerous (but not all) species in meadows mowed in late June re-flowered after first-crop hay was mowed, and they mostly blossomed for the second time in August (see Section 4.1.1.1.1).[1063] Haymaking at times of seed ripening, such as in late June in the lowest section of the montane vegetation belt in the Alps, favorably impacts meadow plants, with their main organs carrying out photosynthesis at various heights, and not only those with many leaves high on stalks. Namely, the impacts of removal of biomass through mowing at that time are sufficient to promote rosette plants through much better light conditions close to the turf after mowing.[1064]

6.1.2.1.1.2 July: the month of haymaking

The word *Heumonat* has usually designated July. The use of the term *Heumonat* is proven since the Early Middle Ages. It was a frequently used expression for July in the Middle Ages and the Early Modern Period, but today it is antiquated. Even *Heuet*, literally 'haymaking', was a synonym for July in the Middle Ages and at least until the sixteenth century, later leaving the space to *Heumonat*.[1065] Similarly, mowing is a typical motif of July in late medieval calendars from environmentally diverse parts of Central Europe from what is now central Germany to the north.[1066] Similarly, mowing in July is represented in a twelfth-century chronicle from the Zwiefalten monastery at the foot of the Swabian Jura (Germ. *Schwäbische Alb*) in what is now southern Germany.[1067]

Evidently, most first-crop hay mowing usually also took place in July in the Alps, where the elevation and climatic impact of nearby high mountains led to similarly late and later haymaking dates compared to central Germany, as mentioned in the previous paragraph. This thesis finds confirmation in mowing as the most typical farm work representing July in a codex of astronomical, computational, and naturalist texts written in Salzburg about 818[1068] that corresponds with mowing in July mentioned in two poems written in the same town in the Alps around 859 but inspired by Virgil in their metaphors.[1069]

Despite the possible inspiration from older works during the selection of the month when mowing is mentioned or represented in some of the sources dealt with in the two previous paragraphs, it is clear that the month when mowing was mentioned or represented as one of the most typical activities did not as a general rule differ from contemporaneous reality in the region

1063 Johansen et al., 2019, 4–9.
1064 Pils, 1994, 163; Ellenberg & Leuschner, 2010, 1017–1018.
1065 Compare *Frühneuhochdeutsches Wörterbuch*: heumonat, heuet; Weinhold, 1869, 27, 43–44, 65–68; *Duden Wörterbuch*: Heumonat; Dietl & Lehmann, 2004, 124.
1066 Golob, 1988, 38; see also Golob, 1985, 12, 37, 51, 56.
1067 Golob, 1985, 27, 51.
1068 Golob, 1985, 32–33, 51; Štih, 2016, 170.
1069 Golob, 1985, 36–37, 51.

Figure 51: Mowing as typical farm work in July (on the left) in a codex of astronomical, computational, and naturalist texts written in Salzburg about 818. *Wikimedia Commons:* File:Monatsbilder Salzburger Handschrift 818.

observed by the authors of these pieces. A confirmation is the common use of the names *Heumonat* and *Heuet* to designate the month.

The beginning of July was very typical timing of first-crop haymaking in meadows in the colline and lower montane vegetation belts of Central Europe, which were grazed no longer than in early spring (see Section 6.2, especially Sections 6.2.1.2 and 6.2.2.1) and where a second mowing would follow in late August or September (Section 6.1.2.3).[1070] Despite the harsher climate in several meadows at higher elevations mowed twice annually compared to the average in colline and lower montane vegetation belts of Central Europe (see Section 6.1.2.3), the first half of July was the last acceptable timing of first-crop haymaking in the meadows in the Alps mowed twice annually (meaning moderately fertilized; Section 6.3.2). Such timing of first-crop hay mowing allowed adequate plant regrowth before second-crop haymaking, which could not take place later than in September because the risk posed by the climate to second-crop hay harvest later in the autumn would be unacceptable (Sections 4.1.1.3 and 6.1.2.3). Mowing first-crop hay in mid-July made a second-crop hay harvest in early September uncertain, as experienced at the mountain farms in the northern part of the Solčava region in the Eastern Karavanks, at around 1,150 to 1,300 m, in the mid-twentieth century.[1071] Indeed, many of the meadows in the Alps mowed twice annually were mowed for the first time exactly in early July in previous centuries, as proven by a case study in the following lines, considered representative for many microregions of the Alps. Figure 82 in Section 6.3 makes it clear that moderately fertilized meadows mowed twice annually usually covered roughly between 15% and 30% of the total meadow area per typical isolated mountain farm in the Central Karawanks (Germ. *Karawanken*, Sln. *Karavanke*) and the nearby Kamnik-Savinja Alps (Sln. *Kamniško-Savinjske Alpe*), located at around 550 to 1200 m, in the mid-1560s, and about 5% to 20% of the total hay

[1070] Wopfner, 1997, 295; Kapfer, 2010, 136, 138.
[1071] Meze, 1963, 226, 234.

per farm in that region was second-crop hay in the mid-1560s (Figure 57 in Section 6.1.2.3). In the early nineteenth century, mowing in the same region in the Central Karawanks and the nearby Kamnik-Savinja Alps started in early July.[1072] Due to the climatic limits on second-crop haymaking in the autumn (Section 6.1.2.3), the beginning of first-crop haymaking no later than early July can be projected back to the 1560s in the same region, knowing the relevance of second-crop hay there, as in many other regions of the Alps.

Mowing in non-fertilized meadows that were cut once a year continued later in July, especially where spring grazing ended early (around May 1st), but also in some places where spring grazing ended later. Although mowing would usually be completed in such meadows by mid-July in colline and lower montane vegetation belts in Central Europe,[1073] the harsher climate and/or later end of spring grazing often postponed the only mowing in non-fertilized montane meadows in the Alps to the second half of July or early August (see also Section 6.1.2.1.1.3). An example is mowing in late July (and early August) in the Talež complex of montane meadows (about 700 m) on the Jelovica Plateau. That meadow complex had to be mowed before the first Sunday after August 1st, in line with information from 1605, so that unfenced grazing could begin there afterward.[1074] By this date, hay was cut and stored, and so mowing had usually taken place somewhat earlier to ensure that it was carried out on time even if the weather was not optimal. The end of spring grazing in these meadows in May (or between April 30th and June 6th, with regular annual alterations between early, middling, and late timing of the end of spring grazing) is most likely (see Section 6.2.1.2).

Flowering times and/or the plant habitus of some herbs were exceptionally well adapted to late mowing, which was usual in the period dealt with in this section and later (Section 6.1.2.2). European cyclamen (*Cyclamen purpurascens*), a late flowering species of low habitus, occurring on basic and neutral,

---

[1072] AT KLA 75, Franziszeischer Kataster, cadastral municipality no. 76205, Bad Vellach (Sln. *Bela*), cadastral appraisal protocols: s.n., 476/4, 48–49, 54–57; AT KLA 75, cadastral municipality no. 76205, Ebriach (Sln. *Obirsko*), cadastral appraisal protocols: s.n., 47/4, 48–49, 56–57; AT KLA 75, cadastral municipality no. 76213, Leppen (Sln. *Lepena*), cadastral appraisal protocols: s.n., 248/4, 48–49, 54–57; AT KLA 75, cadastral municipality no. 76218, Remschenig (Sln. *Remšenik*), cadastral appraisal protocols: s.n., 371/4, 48–49, 54–57; SI AS 178, Franciscejski kataster za Koroško (Franciscan Land Cadaster for Carinthia), K312, cadastral municipality of Zgornje Jezersko, cadastral appraisal protocols: s.n., 312/4, 54–57; SI AS 178, K470, cadastral municipality of Spodnje Jezersko, cadastral appraisal protocols: s.n., 470/4, 48–49, 54–57.

[1073] Kapfer, 2010, 136.

[1074] SI AS 721, Gospostvo Bled (The Seigneury of Bled, Germ. *Veldes*), Iustitialia, book 13, [August 25th], 1605. According to these records, these meadows had to be mowed before the feast of the parish church in the town of Radovljica (Germ. *Radmannsdorf*). A seventeenth-century encyclopedic topographic description makes it clear that at that time the summer feast for this parish church, whose patron is Saint Peter, took place on the first Sunday after "Saint Peter in Chains" (Valvasor, 1689, book VIII, 794), thus on the first Sunday after August 1st (Grotefend & Ulrich, 1971, 88).

mesic soils, was able to quickly make use of the fact that it was exposed to more light in the course of late mowing. By the time its wintergreen leaves from the previous year decomposed in late May, they had synthesized energy for the late summer blossoming that took place after the mowing; this energy was stored in the tuber. Since the abandonment of many montane meadows, earlier mowing dates, and much more intensive fertilization, European cyclamen usually does not grow in hay meadows anymore; in the past it used to blossom in meadows only after late mowing, in some cases as soon as three days after hay was cut in July or August.[1075] Unlike European cyclamen, which blossoms only once annually, some other herbs are adapted to late mowing in the manner that their habitus in the dense grassland before mowing was much weaker than the more vigorous herb and blossoms that developed in much lighter conditions after first mowing. This is the case of some eyebright (*Euphrasia*), cow-wheat (*Melampyrum*), yellow-rattle (*Rhinanthus*), red bartsia (*Odontites*), and gentian (*Gentianella*) species.[1076]

6.1.2.1.1.3 August: the last typical month of the only hay harvest of the year

One single annual mowing taking place in August was typical of non-fertilized meadows in Central Europe that were regularly grazed until mid-May or later; thus, where spring grazing ended late (see Sections 6.2.1.2 and 6.2.2.2).[1077] For instance, in meadows at Wieting in Carinthia west of Saualpe, located at 700 m, spring grazing lasted until May 25th, as reported in 1471 (corresponding to June 2nd in today's calendar), and the single mowing was not allowed to take place earlier than August 10th (corresponding to August 18th in today's calendar).[1078] This means that around eleven weeks passed between the end of spring grazing and mowing, which is quite typical for Central Europe, where this interval lasted for nine weeks in eutrophic meadows and thirteen weeks in oligotrophic meadows in many cases.[1079]

In another example, a meadow on a north-facing slope of Mount Vogel above Holy Spirit Church at Lake Bohinj (Sln. *Bohinjsko jezero*), at an elevation of about 750 m, was mowed in July 1673, allegedly illegally,[1080] and so the

1075 Bavcon, Ravnjak, & Praprotnik, 2019, 77; Aeschimann et al., 2004, vol. 1, 660–661.

1076 Bavcon, Ravnjak, & Praprotnik, 2019, 77; Peterken, 2013, 227.

1077 Kapfer, 2010, 136, 138.

1078 Dinklage, 1966a, 82, 84; the conversion is based on *Julian Day and Civil Date Calculator* and the calculations by the astronomer Tomaž Zwitter.

1079 Janišová et al., 2023, 9, 11.

1080 SI AS 721, Gospostvo Bled (The Seigneury of Bled), fasc. 33, Iustitialia, Lawsuit between the communities with centers in Savica and in Studor over grazing rights in Ukanc 1673/74: *Edl Gestrenger sonders fr[eundlicher] lieber Herr*, January 16th, 1674. Due to the absence of toponyms in this area on nineteenth-century cadastral maps as well as on most detailed recent maps, localization of this former westernmost meadow in that meadow complex is based on interviews (1) with a local born in the 1930s at the closest farm located next to Holy Spirit Church, and (2) with an owner of a nearby legal mowing surely would not have taken place earlier. Spring grazing in that meadow complex ended on May 24th or 25th,[1081] and so usual mowing in August (the thirteen weeks usual in oligotrophic meadows would bring us to the second half of August) is most plausible.

Returning to the transition from July to August, mowing in the Jezerca montane meadow complex northeast of Javorniški Rovt in the Western Karavanks, at an elevation between about 930 and 1,260 m, is instructive because it illustrates the significantly later mowing dates than usual in Central Europe after an early end of spring grazing in non-fertilized meadows mowed once annually. The most likely reason is the harsher climate at these elevations, on slopes of high mountains. At Jezerca, the first grassland use in farming was grazing until May 1st (see Section 6.2.2.3). Mowing in (late July and) early August followed (as compared to the mowing in the first half of July in many meadows in Central Europe grazed until around May 1st[1082]). Namely, before the abolition of spring grazing in meadows, mowing at Jezerca began either on July 25th or in early August,[1083] meaning that mowing was definitely oc-

---

meadow born in the 1940s. The cited archival record gives the toponym *Nauisnam Werdo*, pronounced **Na Višnam brdo*, (i.e., *Na Višnjem brdu*) in the Gorenjska dialect (meaning 'on the higher / highest [grassland] lower hill' (see the general explanation of the archaic adjective *višnji* in Snoj, 2009, 459–460), in this case (one of) the former highest meadow(s) behind Holy Spirit Church, on the slope below the forest on a steeper slope (*Arcanum Maps*, Cadastral Maps, Habsburg Empire – Cadastral Maps (XIX. [sic] Century), cadastral municipality of Savica (Germ. *Saviz*), 1826; *Environmental Atlas of Slovenia*, Spatial Units, Land Cadastre & Backgrounds, Topografic [!] maps); another potential explanation, that the toponym evolved from Sln. *višnja* 'sour cherry' (*Prunus cerasus*; compare Šekli, 2024), is unlikely because sour cherry does not grow there wild and no farmer would plant a heliophilous fruit tree that is rarely grown in the Slovenian Alps (compare Brus, 2012, 208) in a remote meadow on a northern slope at an elevation of c. 750 m, under the influence of the yet harsher climate of over 1,600 m high Mt. Vogel and of mountains in the central Julian Alps in general). The interviewees recalled the toponym *Višnovc* (author's interviews with Šetina, 2018, in Srednja Vas in Bohinj, and with Kunavec, 2019, in Stara Fužina), which linguistically evolved from the same lexical root and is located in the same meadow complex, and so **Na Višnam brdo* and *Višnovc* designate the same meadow or two meadows located very close to each other. Gratitude is expressed to the Slavicists Rok Dovjak and Matej Šekli for their linguistic advice.

1081 Novak, 1955, 270.

1082 Kapfer, 2010, 136.

1083 Mowing started on July 25th at least in some mid-elevation montane meadows owned by farmers from Koroška Bela and Javorniški Rovt according to data from around 1870 (SI AS 48, Deželna komisija za odkup in uravnavo zemljiških bremen za Kranjsko [Provincial Commission for Redemption and Regulation of *Profits à Prendre* in Carniola], fasc. 56, 9–5: *Tagebuch über den Verhandlungsact in Sachen der von der Gemeinde Bach angemeldeten Ablösung oder Reguli[!]rung des Weiderechtes an der Alpe Beušca der Berechtigten von Hrašach, Studenčič, Rann, Velben et Hlebic*: *Protokoll*, December 16th, 1871, 4, and *Protokoll*, December 27th, 1871, 3). Data from 1870 refer to the times after the abolition of spring grazing at Jezerca that might have led to an earlier start of mowcurring in several meadows at Jezerca in early August or at the transition from July to August. A confirmation is provided by the dates when autumn grazing (making use of the biomass regrown after the mowing) started in these meadows: autumn grazing there was allowed to start in the third quarter of August according to data from about 1780, from the 1820s, and from 1861, before the abolition of spring grazing there (see Section 6.2.2.3).[1084] This means that mowing at Jezerca was evidently completed by mid-August or earlier. The start of mowing in the highest meadow complex in the region, located higher than Jezerca, on Golica, around August 10th (see below) suggests that hay at Jezerca had been stored by that time. An example from the Jezerca meadow complex recorded in 1609 due to its exceptionality proves that the beginning of autumn grazing in the third quarter of August (and therefore also the start of mowing in late July or early August[1085]) probably had a very long tradition there. Namely, one of the meadows at Jezerca was still not mowed on August 20th, 1609, but the records make it clear that this was an atypically late mowing caused by an exceptional situation: the possession of that meadow was disputed at that

ing (Section 6.1.2.2), and so mowing must have previously started either on July 25th or in early August; see Sections 6.1.2.2 and 6.2.2.3 (SI AS 48, fasc. 56, 9–4: no. 3309 de 1861).

1084 Autumn grazing in meadows at Jezerca was allowed to start on the day after the feast of the church in Koroška Bela according to data from about 1780 and from the 1820s (SI AS 170, Soseska Koroška Bela [The Agrarian Community of Koroška Bela], box 2: unit 4, Lawsuit between the community of Koroška Bela as plaintiff and Blaž Justin as defendant, 1781; SI AS 170, box 2: unit 7, Lawsuit[s] between the community of Koroška Bela as plaintiff and the community of Javorni[š]k[i Rovt] as defendant, 1720–1828: 1826–1827). Late seventeenth-century information reveals that this feast was celebrated on the Sunday before Saint Bartholomew's Day (Valvasor, 1689, book VIII, 720); i.e., on the Sunday before August 24th (Grotefend & Ulrich, 1971, 37). Similarly, the sources from 1826–1827 and 1861 claim that autumn grazing in meadows at Jezerca was allowed to start on the first Monday after August 15th (SI AS 170, box 2: unit 7, Lawsuit[s] between the community of Koroška Bela as plaintiff and the community of Javorni[š]k[i Rovt] as defendant, 1720–1828: 1826–1827; SI AS 48, Deželna komisija za odkup in uravnavo zemljiških bremen za Kranjsko [Provincial Commission for Redemption and Regulation of *Profits à Prendre* in Carniola], fasc. 56, 9–4: no. 760, *Protocoll*, June 28th, 1861, and the list of parcels belonging to Jezerca); in 1826 and 1827, when the contents of this record were produced, or in 1860 relevant for the record written in 1861, the day following Sunday before August 24th corresponded to the first Monday after August 15th (Grotefend & Ulrich, 1971, 153, 179, 193). Localization is based on comparison of the toponym in the records quoted above with the toponym on the early nineteenth-century cadastral maps (SI AS 176, Franciscejski kataster za Kranjsko [Franciscan Land Cadaster for Carniola], L90, cadastral municipality of Javornik, map sheets A02 and A03; SI AS 176, L102, cadastral municipality of Koroška Bela, map sheets A02, A03, A05, and A06), with georeferenced cadastral parcel shapes, and with an elevation map (*Environmental Atlas of Slovenia*, Spatial units, Cadastral areas, & Land Cadastre; *Environmental Atlas of Slovenia*, Backgrounds, DEM [Digital elevation model]; see also the map *Ledinska imena v občini Jesenice*, 2015).

1085 Two weeks after first mowing, the meadow can be considered suitable for grazing (Babai, Molnár, & Molnár, 2014, 112).

time. In this case, the comment by the court was relevant that, if mowing could not wait any longer (suggesting that it was an urgent case), the hay should be mowed by impartial people, and this was stated even though the parties were expected to be present at the court again as early as seven days later, on August 27th, when the lawsuit was scheduled to continue.[1086] Thus, typically, mowing there would have taken place earlier. This means that pieces of information from 1609, from about 1780, and from 1826–1827 and 1861 all suggest usual mowing in montane meadows in this microregion in (late July and) the first half of August. In this case, the interval between the end of spring grazing and mowing corresponded with or was only slightly longer than that typical of many oligotropic meadows mowed once annually in Central Europe, where this interval often lasted for thirteen weeks.[1087] The climate at Jezerca, located on the slopes of high mountains, is harsher than average in meadows in Central Europe, and so a possibly slightly longer interval between the end of grazing and mowing is understandable. The next paragraph explains that this was not the latest mowing timing in that region, and that later mowing also took place at several other places in the Alps.

Meadows at the top of the Golica Ridge (1,500–1,800 m[1088]), located above Jezerca in the Karavanks, were usually mowed around August 10th (with a most likely possibility of postponement if needed[1089]) in the mid-nineteenth century; that is, at times of historical spring grazing, which lasted there until May 25th, and so it ended late (see Section 6.2.1.2 and 6.2.2.2).[1090] The quality of such hay was high (see Sections 6.1.2.2, 6.2.2.2, and 6.3.1.2). Similarly, many meadows in the large Siusi Alp (Ital. *Alpe di Siusi*, Germ. *Seiser Alm*, roughly

[1086] SI AS 721, Gospostvo Bled (The Seigneury of Bled), Iustitialia, vol. 13, [August 20th] 1609, [August 27th] 1609.

[1087] Janišová et al., 2023, 11.

[1088] Localization of these meadows is explained in Section 6.2.2.3.

[1089] Although it was usual to mow there around August 10th (SI AS 48, Deželna komisija za odkup in uravnavo zemljiških bremen za Kranjsko [Provincial Commission for Redemption and Regulation of *Profits à Prendre* in Carniola], fasc. 56, 9–2: *Protokoll*, June 1st, 1859), the first mowing could most likely easily have taken place by September 8th, when autumn grazing in these meadows started. According to a regulation from 1861 (i.e., the times of abolition of spring grazing there), mowing after September 8th was still allowed, but in this case it was the task of the meadow owner to prevent the livestock from grazing in that meadow (SI AS 48, fasc. 56, 9–2: *Protocoll*, December 21st, 1861). September 8th, the feast of the Nativity of Mary, is one of the usual historical dates introducing autumn grazing in high montane meadows in the broader region, as clearly documented in records from previous centuries. The same role of this holy day was in use, for example, in the meadows close to Gorjuše (somewhat above 1,000 m) in line with information from 1635 (Novak, 1975, 5–6). Consequently, it is highly likely that the start of autumn grazing on Golica after September 8th was a custom several centuries old.

[1090] SI AS 48, Deželna komisija za odkup in uravnavo zemljiških bremen za Kranjsko (Provincial Commission for Redemption and Regulation of *Profits à Prendre* in Carniola), fasc. 56, 9–2: *Protokoll*, June 1st, 1859.

more than 1,800 m; i.e., slightly higher than the example of Golica) in South Tyrol were mowed from August 10th onward prior to the sixteenth-century calendar reform (locals moved this date to August 21st after the calendar reform), roughly two to two and a half months after the end of the spring grazing season in these meadows (two and two-thirds months with the changed spring grazing regime after the calendar reform; see Section 6.2.1.2). Thus, the duration of the period of grassland plant growth between the end of spring grazing and mowing was comparable at Golica and the Siusi Alp. Another example of late mowing from South Tyrol, at higher elevations than Golica: a ruling from 1588 allowed the beginning of mowing in the common meadows at Vallelunga (Germ. *Langtaufers*) from August 18th onward. Mowing in a complex of montane meadows above Trins in Tyrol was allowed to take place from August 11th to August 23rd according to information from 1765. At Silz in Tyrol, the meadows in the mountains had to be mowed by or in early September according to a ruling from 1616.[1091]

6.1.2.1.1.4 September: the only mowing in the very last meadows

A few cases of possible mowing in September were mentioned in the previous paragraph, and September was the regular month of the only annual haymaking in some cases.

In parts of the Villandro Alp (Ital. *Alpe di Villandro*, Germ. *Villanderer Alm*) in South Tyrol, grazing in meadows lasted until July 25th (corresponding to August 4th in today's calendar). Consequently, mowing in the Villandro Alp was only allowed to take place after a date most probably corresponding to September 8th according to a ruling from mid-sixteenth century— corresponding to September 17th in today's calendar.[1092] This was an extreme example, emphasizing that in the period dealt with throughout Section 6.1.2.1 the mowing date alone tells nothing about the quality of the hay in a usual year, unless the end of grazing (usually in spring) is known; plants grazed until July 25th were not overly ripe by September, but lower temperatures and more precipitation made successful drying less certain (compare Section 6.1.2.3).

So far, the cases presented of mowing in September referred to meadows at (relatively) high elevations. The last example deals with late mowing in the lower montane belt. This is no exception in the wider Central European context, where poorly productive meadows at nutrient-poor sites mowed only in September (spring grazing lasted there until deeply in May) were nothing unusual, when the weather in the summer was unfavorable for haymaking in particular. Such late haymaking, preceded by spring grazing lasting no later than May,

[1091] Grass, 1948, 167, 169, 172, 192, 196–198, 200, 208; Wopfner, 1997, 316–317.

[1092] Grass, 1948, 84, 162, 200–201, 207; Hinterwaldner, 2002, 104, 120–121, 172–174, 186. The conversions are based on *Julian Day and Civil Date Calculator* and the calculations by the astronomer Tomaž Zwitter.

produced low-quality fodder.[1093] In Ukanc in western Bohinj (roughly 550–600 m), plaintiffs claimed that large quantities of first-crop and second-crop hay—partly mowed, but partly still standing—were consumed by the defendants' cows in September 1697, and so part of the hay was highly likely mowed there only in September.[1094] Whether first- or second-crop hay was consumed was not a matter of the dispute, and so this information is likely reliable.[1095] Later data make it clear that very late mowing regularly took place in some meadows in the lower montane vegetation belt if the hay quality was poor, and so the meadows delivering better hay were given priority (see Section 6.1.2.2). In addition, focusing on the Ukanc case in 1697, such late mowing might have been influenced by cold and wet weather and climate because this belonged to one of the most prominent peaks of the Little Ice Age (see Sections 4.1.1.2 and 4.1.1.3).[1096] Consequently, unfavorable weather during regular mowing and reaping time might have postponed the mowing.

6.1.2.1.1.5 More flexible mowing dates

Although the data presented in Sections 6.1.2.1.1.1 to 6.1.2.1.1.4 allowed some limited flexibility in mowing timing, the rules presented were more than merely orientations. More flexible mowing dates, allowing the community to pay more attention to the grassland plant development, are also known, even though the peak of farmwork in high summer, the amount of time that had to be invested in mowing, and further factors with an impact on mowing timing (Section 6.1.2.1.2) in reality posed limits on flexibility.[1097]

Medieval and early modern examples of relatively flexible mowing dates are Parcines (Germ. *Partschins*) in South Tyrol, where the beginning of haymaking was publicly announced every year, as documented in 1431,[1098] and Pfunds in the Inn Valley, Tyrol, where the date of mowing in a common meadow was proclaimed every year according to a ruling from 1615.[1099] They have parallels elsewhere in the Alps: in the Allues Valley, "the date each [meadow complex] was opened to haying and the number of days allotted to mowing at each [meadow complex] were prescribed by common accord of the property owners involved" "after evaluating the actual state of the vegetation." This is how the system rooted in older history functioned before the First World War.[1100]

1093 Kapfer, 2010, 136.
1094 SI AS 721, Gospostvo Bled (The Seigneury of Bled), fasc. 33, Iustitialia, Lawsuit between the communities with their centers in Studor and in Savica over grazing rights in Ukanc 1697–1717: May 21st, 1699.
1095 On this explanatory approach, see Fuchs & Schulze, 2002, 9.
1096 Zwitter, 2015d, 169.
1097 Compare Meilleur, 1986, 270–272, 276.
1098 Wopfner, 1997, 362.
1099 Grass, 1948, 169.
1100 Meilleur, 1986, 271.

Indicators such as blossoming or seeding of some plant species were often observed by farmers in Europe to determine the start of mowing.[1101] Caraway (*Carum carvi*; see also Section 8.2.1.2) is an example of a species that used to be observed to determine mowing dates in moderately fertilized meadows in some places in Central Europe; when its seeds started ripening, it was the time for mowing, as reported in the mid-nineteenth century.[1102] In Slovenia, this indicator was recently documented among traditional farmers in the east of the pre-Alpine hills, in the Kum microregion, providing the explanation that ripe caraway seeds are even more obvious than ripe seeds of most meadow plants because ripe caraway seeds have a distinctive odor.[1103] Another species widely observed for this purpose was yellow rattle (*Rhinanthus* spp., an undesired plant in hay; see Section 4.1.1.1), which occurred in so many meadows that farmers in several regions used it as an orientation for the onset of mowing. They started mowing when yellow rattle seeds were ripe, as reported in the mid-eighteenth century by Johann Christian Daniel von Schreber,[1104] a botanist living in what is now northeastern Germany, with a broad network of international scientific correspondents in Central, Western, Northern, and Eastern Europe.[1105] The same indicator for the start of mowing was widely used in Europe. It was historically reported in Sweden[1106] and it is still followed by some traditional farmers in the Carpathians: when the yellow rattle seeds fully ripen and begin to sound when one walks through the grassland, the hay is ready to be mowed. By that time, "the seeds of the flowers or the grass have fallen . . . [onto] the ground and sown themselves."[1107] As long as farmers regularly waited to start mowing until yellow rattle seeds ripened, this indicator enabled them to maintain a fairly constant mowing timing, mainly dependent on the impact of the weather and climate on plant development. Namely, yellow rattle is able to adapt its flowering time to the usual local mowing timing in June or later to a certain degree;[1108] nonetheless, farmers' attention to the constant phenophase of yellow rattle seeds ripening means that the genus had adapted to the locally usual mowing timing long ago. Mowing at times of ripe yellow rattle seeds is nowadays perceived a bit too late by many rather traditional farmers.[1109] In some regions of the Alps, slightly earlier mowing is not a novelty concerning the few most productive meadows that were mowed first (see Section 6.1.2.1.1.1).[1110]

[1101] Šumrada et al., 2021, 8.
[1102] "Košnja," 1864, 199.
[1103] Debeljak et al., 2020.
[1104] Kauter, 2002, 127.
[1105] Beyer-Thoma, 2007.
[1106] Ivaşcu, Öllerer, & Rákosy, 2016, 48–49.
[1107] Iuga, 2016, 72 (quote); compare Ivaşcu, Öllerer, & Rákosy, 2016, 45, 49.
[1108] Compare Jonstrup, Hedrén, & Andersson, 2016.
[1109] Ivaşcu, Öllerer, & Rákosy, 2016, 45, 49.
[1110] See also the example of the Allues Valley in Meilleur, 1986, 270–271.

In some other cases, mowing dates could be determined individually, such as at sites where wild hay was mowed in remote commons above St. Anton am Arlberg and St. Jakob am Arlberg according to a ruling from 1802. Mowing rights in these commons were allocated to individuals every year in July, and after 1804 in mid-June. Farmers were then allowed to schedule their mowing according to their preferences. It is important to emphasize that these meadows were too dangerous to serve as pastures,[1111] and so mowing dates did not have to be adjusted to collective spring and autumn grazing. Meadow management was also (more) individual in the case of those isolated farms where all meadows were located in the area of a compact holding of a farm, so that there were no (or hardly any) adjacent meadow parcels belonging to different farms.

The examples of mowing timing presented involved cases of very late mowing. First mowing in August in the lower montane belt, or in September at any elevation, unless preceded by extremely long spring grazing, leads to the conclusion that in past centuries mowing dates were not solely defined by plant development after spring grazing and appropriate weather for mowing and drying, as explained in the next section. The factors impacting mowing dates presented in Section 6.1.2.1.2 are reflected in the rules and customs mentioned above, defining the earliest allowed mowing date and at least roughly the date by which haymaking had to be completed, so that common autumn grazing in meadows could start.

6.1.2.1.2 Main factors impacting mowing dates

Historical late mowing dates were the result of the nexus between meadow management, plant ecology, farm workload, adaptations to weather, and cultural history.

Concerning plant ecology, the most productive grasses of moderately fertilized and non-fertilized meadows (see Section 6.3) can abound in such meadows and their populations are fairly stable only if they are cut no earlier than their first seeds are ripe. Namely, the grasses of non-fertilized and moderately manured meadows that are most relevant for livestock farming strongly depend on generative propagation (see Sections 6.3 and 6.5). The removal of nutrients from meadows with hay was also slower in the case of later mowing, which mattered in the long run in times of prevalent non-fertilized meadows (Section 6.3.1). Thus, farmers cared about the long-term preservation of productive grasslands by choosing mowing dates that were late enough to allow seed dispersal.[1112] Many seeds regularly fell off when the hay was shaken off and raked, so that hay making itself was a significant part of the seed dispersal process for meadow species. A great deal of mowing took place roughly

[1111] Grass, 1948, 176.

[1112] Dietl & Lehmann, 2004, 20, 24, 66, 83, 106, 118, 124–125; Peterken, 2013, 122.

at the time when many seeds had just ripened or later. The quality of such ripe hay was not optimal, and the nutritive value and digestibility of the hay and the crude protein content in it were lower than when the plant biomass in the meadow had just reached its highest quantities,[1113] but the hay quality was acceptable because it was impossible to preserve the high level of meadow productivity in the long run in the case of early mowing.[1114] The decline in hay quality in species-rich permanent meadows during the phenophases of flowering, seed set, and seed ripening of productive grasses is not as severe as in intensively fertilized, sown grassy plains.[1115] Any further postponement of mowing dates later than those simply allowing seed dispersal caused further significant decline in hay quality.[1116]

Focusing on the farm workload, mowing with a scythe is highly time consuming, and so first-crop haymaking at a typical farm often lasted for a few weeks, even where smallholdings prevailed, and longer at larger farms, unless very many haymakers were hired.[1117] The details of the labor intensity depended on the meadow environment (Sections 6.1.1.2 and 6.3.1), the state of scythes (Section 6.1.1.1), the mowing technique (Section 6.1.1.2), and the skillfulness of the haymaker (Section 6.3.1.2). Additional time was needed when rain posed challenges, and so the hay had to be cocked, spread, and turned several times (Section 4.1.1.3). Northerly slopes also often took more time due to longer hay drying.[1118] On some farms, mowing in the largest meadow already took a whole week (i.e., five to six working days; see Section 4.1.1.4) or longer even in the case of favorable weather, four people mowing, and additional workforce with rakes and hayforks.[1119] This means that, at many farms, mowing in some meadows when seeds were ripe was prevented by mowing going on elsewhere at the same time, and so some meadows had to be mowed later.[1120] In addition, mowing timing also had to allow the rest of the main agricultural work in the summer, such as reaping autumn-sown cereals, subsequent ploughing, and sowing stubble crops,[1121] which all delayed mowing in some meadows.

1113 Sadar, 1956, 102, 145; Pfister, 1985a, 41; Peterken, 2013, 20; Verbič & Verbič, 2014–2020, 21–22.

1114 Dietl & Lehmann, 2004, 20, 24, 66, 83, 106, 118, 124–125.

1115 Verbič & Verbič, 2014–2020, 22.

1116 Pfister, 1985a, 41–42.

1117 Makarovič, 1978, 141; Hodnik, 2016, 76–80; Meilleur, 1986, 272; Oppermann & Paracchini, 2012, 19; Hessenberger, 2022, 154, 166; Novak, 2024, 23.

1118 Wopfner, 1997, 334.

1119 Šarf, 1974, 143; Hodnik, 2016, 79–80; Novak, 2024, 23.

1120 Author's interviews with Kunavec, 2017, in Stara Fužina, and with F. Mlakar, 2017, in Bohinjska Češnjica.

1121 Examples from different regions in the Alps: Grass, 1948, 168, 177; Šarf, 1974, 138; Meilleur, 1986, 122, 272; Hodnik, 2016, 79–81 (see also older data in SI AS 176, Franciscejski kataster za Kranjsko [Franciscan Land Cadaster for Carniola], L284, cadastral municipality of Studor, cadastral appraisal protocols: s.n., 284/4, 11, 35). Compare Peterken, 2013, 122–123.

Concerning the impacts of meadow management on mowing dates, fertilization (Section 6.3) and irrigation (Section 6.4) fostered earlier ripening and mowing, whereas spring grazing in meadows delayed it until its regionally diversely timed abolition between the late eighteenth and mid-nineteenth century (Section 6.2). As a general rule, the earlier spring grazing ended, the sooner the meadow plants regrew and fruited; the later spring grazing lasted, the more the mowing had to be postponed (see Sections 6.1.2.1, 6.1.2.2, 6.2.1.2, and 6.2.2).[1122] Further site conditions mattered as well. Humid, relatively nutrient-rich (manured) meadow parcels, where spring grazing was practiced only for a short time early in spring or not practiced at all, were usually mowed first because the growth of most grassland plants there started earlier and was relatively quick. Moreover, the quality of hay was high, and so it was worth storing the good hay in time, before its quality declined (see Section 6.1.2.2). In contrast, at nutrient-poor, non-fertilized sites, the seeds of grasses ripened significantly later than at moderately fertilized sites at similar elevations, and grassland plants ripened even later at non-fertilized meadows at significantly higher elevations.[1123] In addition, less productive sites were often grazed later in spring than more productive ones (Section 6.2.1.2 and 6.2.2.2)—another reason for very late mowing.

Mowing timing tried to avoid rainy weather (see, e.g., the weather forecasting methods presented in the introduction to Section 8 and in Section 8.2.3), and so longer intervals of precipitation during the usual mowing season (see Sections 4.1.1.2 and 4.1.1.3 for historical clusters of such rainy summers) led to delayed mowing, causing lower nutritive value of hay, and worse hay taste and aroma.[1124]

Moreover, cultural history mattered. The prohibition of most work on Sundays and other holy days had a relevant impact on mowing timing because the mowing season involved several days when this task was considered indecent. This was particularly relevant from the Late Middle Ages until the mid-seventeenth century, when in many places on average a bit more than two days per week were free of most seasonal agricultural work, such as mowing (see Section 4.1.1.4). A scythe or mowing is represented in many Sunday Christ paintings (Figure 52)—depictions of "tasks people should not carry out on Sundays [and other holy days] lest they be devoured by the abyss of hell at the end of their lives"[1125] (see the context in Section 4.1.1.4). In the Alps and their broader surroundings, a scythe or mowing appear in many Sunday Christ paintings from the fourteenth to mid-sixteenth century, such as the following

[1122] Kapfer, 2010, 136–140; Peterken, 2013, 122; Grass, 1948, 162–163, 167, 192; Babai, Molnár, & Molnár, 2014, 103, 109.
[1123] Dietl & Lehmann, 2004, 124–125.
[1124] Pfister, 1985a, 41, 47, 91; Kuret, 1989, 439.
[1125] Horvat, 1995, 150.

Figure 52: Mowing and scythe peening (in the upper right part of the picture) on a Sunday Christ painting at Ortisei, Alto Adige (Germ. Sankt Ulrich, South Tyrol, around 1452–1460). Blood spurts from Christ's body; the work represented, such as haymaking, causes him to bleed anew, and haymakers are joined by devils. Like many Sunday Christ paintings (see Section 4.1.1.4), this one was plastered over later. In the second half of the eighteenth century, part of the figure of Christ and many tools were covered by a station of the Cross, seen on the left.[1126] Photo: Matej Zwitter, July 2024.

[1126] Hiller, 2020, 82, 86, 90, 96–97 (plastering in this case: Wildhaber, 1956, 16, and Table 6, Figure 19); see also Section 4.1.1.4 in this volume.

ones presented in geographical order: at St. Mary's Church at Biella in Piedmont (painted 1460–1471);[1127] at the church at Reutigen south of Bern (early fifteenth century);[1128] at the church of Ormalingen near Basel, drawn in the second half of the fourteenth century;[1129] at St. John the Baptist pilgrimage church at Frauenfeld-Kurzdorf northeast of Zürich (possibly first half of the fifteenth century);[1130] at the popular late-medieval pilgrimage parish church at Eriskirch at Lake Constance (possibly depicted about 1400);[1131] at the church at Waltensburg-Vuorz in Graubünden (Ital. *Grigioni*; Romansh *Grishun*; probably mid-fifteenth century);[1132] at St. Andrew's Church at Tarces (Germ. *Tartsch*) in Trentino-Alto Adige (South Tyrol; painted in the early sixteenth century);[1133] at St. Rocco's Church at Tesero (Trentino-Alto Adige, 1557);[1134] at St. Philip and James's Church at Campitello di Fassa (Ladin *Ciampedél*; Trentino-Alto Adige, drawn about 1455–1460);[1135] at St. James's Church at Ortisei, Val Gardena in Trentino-Alto Adige (Germ. *Sankt Ulrich in Gröden*, drawn about 1452–1460—in the hamlet of San Giacomo, Germ. *St. Jakob*; Figure 52);[1136] at St. Ursula's Church at Valgiovo, Trentino-Alto Adige (Germ. *Jaufental*, painted around 1400);[1137] at St. Mark's Church at Mauthen in the Gail Valley (Germ. *Gailtal*, Sln. *Zilja*), Carinthia (possibly early sixteenth century);[1138] at Saak (Sln. *Čače*) in the same valley (1462);[1139] at the parish church of Mary's Assumption at Maria Rojach, Lavanttal, Carinthia (fifteenth century);[1140] and at St. Elizabeth's Church at Podreber near Polhov Gradec, in the pre-Alps southeast of the Julian Alps (drawn around 1520–1530).[1141] Thus, avoiding mowing on Sundays and other holy days was a widespread phenomenon (compare the presentation of the history of storing hay on Sundays in Section 4.1.1.4), which is also directly proven.[1142]

1127 Hiller, 2020, 86, 92.
1128 Wildhaber, 1956, 10–11, and Table 2, Figure 10.
1129 Wildhaber, 1956, 9–10, and Table 2, Figure 9.
1130 Wildhaber, 1956, 8–9, and Table 2, Figure 8.
1131 Wildhaber, 1956, 7–8, and Table 1, Figure 6.
1132 Wildhaber, 1956, 11–12, and Table 4, Figure 12.
1133 Wildhaber, 1956, 15, and Table 6, Figure 18.
1134 Wildhaber, 1956, 16–17, and Table 7, Figure 21.
1135 Wildhaber, 1956, 16, and Table 6, Figure 20; Hiller, 2020, 86, 91.
1136 Moser, 1974, 14–15; Hiller, 2020, 90.
1137 Wildhaber, 1956, 15; reproduction at *Realonline: 002351: Feiertagschristus/Christ with craftsman's tools.*
1138 Wildhaber, 1956, 17, and Table 7, Figure 22.
1139 Moser, 1990, 343, 346, and Figure 6.
1140 Wildhaber, 1956, 17–18, and Table 7, Figure 23.
1141 Horvat, 1995, 150; Höfler, 2001, 155–156; a publication of the haymaker in that Sunday Christ painting that has deteriorated in the meantime: Grafenauer, 1970c, Figure sheet XXVI between pp. 240 and 241: Figure 2.
1142 E.g., Makarovič, 1982, 428–429; Meilleur, 1986, 272; see also a study in the Romanian Carpathians: Iuga, 2016, 67–72, 77–83.

In addition, phases of the Moon were taken into account for setting mowing dates, if possible, in several parts of the Alps at least, according to information documented later but surely rooted deeper in history. According to testimonies from East Tyrol, some farmers in the mid-twentieth century considered the hay cut during a waning Moon significantly tastier to livestock than hay cut during a waxing Moon. However, at least some farmers interviewed in East Tyrol considered hay cut during a waxing Moon to be significantly tastier.[1143] Either this confusion is a reflection of diversity in historical thought, or not all interviewees remembered well which phase of the Moon was considered better. In general, a waxing Moon and the Full Moon in particular were historically considered better for work related to any kind of growth, whereas the waning Moon was considered suitable for tasks that have something to do with decrease and dieback.[1144] This might have allowed some diversity in interpretation; for example, that during the waxing Moon and Full Moon grassland plants would grow and juices would increase in them, and so it was optimal to wait and cut the hay soon thereafter, during the following waning Moon; or, that during a waxing Moon, plants would grow and the livestock consuming them would grow more quickly or provide more milk, and the like. Mowing during a waning Moon was probably more widespread in the Alps because it has parallels in other regions, such as in Carinthia and in Bohinj. A waning Moon was considered particularly important if hay in non-fertilized meadows at nutrient-poor, acidic, mesic sites with a significant percentage of bilberry (*Vaccinium myrtillus*) and cowberry (*Vaccinium vitis-idaea*) dwarf shrubs was cut; cattle liked to eat these shrubs if they were cut during a waning Moon, according to testimony from Carinthia.[1145] In species-rich grasslands in Bohinj, this was evidently not (or perhaps no longer) a practice in the first half of the twentieth century, but at that time farmers (still) paid attention to the Moon there when the date for mowing clover was determined: unless cut during a waning Moon, clover was considered to cause such severe drooling in horses that saliva would dribble from their muzzles.[1146] Testimonies from East Tyrol in the mid-twentieth century explain that, due to the long duration of mowing entirely by hand and dependence on suitable weather for hay drying, farmers tried to partly adjust mowing timing to the phases of the Moon, but they did not mow only during the phase of the Moon (mainly a waning Moon) that they considered more suitable.[1147] Another shorter phase of the Moon with an influence on mowing timing in some parts of the Alps was the New Moon. Data from the Trenta Valley (the Upper Soča catchment in the Julian Alps)

[1143] Vogl-Lukasser et al., 2006, 69, 79, 88, 227.
[1144] Makarovič, 1995, 280.
[1145] Keller, 2004, 15, 36, 65, 87, and the biologist Erika Pirker-Keller, personal communication; Aeschimann et al., 2004, vol. 1, 624–627.
[1146] Author's interview with Gašperin, 2017, in Stara Fužina.
[1147] Vogl-Lukasser et al., 2006, 69, 79, 88, 227.

reveal that the New Moon was considered inappropriate for mowing: farmers believed that cows did not like to eat such hay and that they would not gestate if they did eat it. In the opinion of farmers from the Trenta Valley, the time suitable for mowing started again three days after the New Moon.[1148] Because the duration of this phase of the Moon considered unsuitable for mowing was much shorter than in the aforementioned case of the waning Moon, avoiding the New Moon seems to have been stricter where this rule was considered relevant.[1149]

Moreover, extremely late dates of first mowing in some grasslands (e.g., in September of today's calendar) were a consequence of the fact that mowing was also performed in some pastures where grazing was the main land-use type (see the example of the Villandro Alp, in Section 6.1.2.1.1.4). Consequently, in some pastures hay was allowed to be mowed after the end of the summer grazing period. In this way, the agricultural use of the grassland complex was more thorough if the intervals of stocking rates were well below the pasture's carrying capacity, or if grassland sections that were inaccessible to livestock due to their steepness or an inappropriate path were used agriculturally.[1150] A source from 1577 documents mowing on the Golica Ridge in the Western Karawanks, which was only allowed to take place there after the end of summer grazing.[1151] According to a document from 1580, in Nassereith, Tyrol, wild hay was also only allowed to be mowed after the end of grazing.[1152] Similarly, at Tannheim in Tyrol, after September 16th, every community member had the right to mow any part of the common he or she found attractive, according to a ruling from 1607.[1153] The same principle is reflected in extremely late mowing dates in several pastures in the Alps.

With the increase in agricultural density in the Alps in the second millennium (Section 3.1), such rulings were modified in several places. For instance, in the early seventeenth century it was still allowed in Uri for two mowers per household to mow common mountain pastures for sheep and goats any time of the year, whereas in cow pastures mowing was only allowed to take place after September 1st and, in the common in the bottom of the valley, mowing was allowed to start on September 14th. Starting in the mid-seventeenth century, due to an increased demand for wild hay, the rules regarding wild hay harvesting in Uri became more restrictive, and so from 1662 onward it was not

1148 Kuret, 1989, 439.
1149 Compare Kuret, 1989, 439.
1150 Grass, 1948, 208.
1151 AT-KLA 112, Allgemeine Handschriftenreihe, Ms. 2435, 147–148; AT-KLA 112, Ms. 2732, 467–468, 471: the historian Janez Mlinar noted that these are important sources for the mountain pasture observed (Mlinar (ed.), 2018b, 163, footnote 139), and he kindly allowed the use of his copies of archival records.
1152 Grass, 1948, 162–163.
1153 Wopfner, 1997, 317.

allowed to mow wild hay in remote grasslands in commons in the mountains unsuitable for cattle grazing before the beginning of August.[1154]

Abandonment of spring grazing in meadows between late eighteenth and mid-nineteenth century (Section 6.2.3) allowed earlier mowing dates. These are presented in Section 6.1.2.2.

6.1.2.2 Usual dates of first mowing from the late eighteenth / mid-nineteenth to mid-twentieth centuries

Non-fertilized meadows mowed only once annually, late enough to allow the dispersal of significant seed quantities at haymaking, were still the most common meadow type in Central Europe in the early twentieth century. Nonetheless, the moderately fertilized meadow area increased quickly in the period dealt with here (Section 6.3), and so the decades between the late nineteenth and mid-twentieth century were the golden period of moderately fertilized meadows in Central Europe mowed twice annually. Typically, non-fertilized and moderately fertilized meadows were cut significantly earlier in the period presented here than in the one dealt with in Section 6.1.2.1. Namely, abandonment of spring grazing in hay meadows between the late eighteenth and mid-nineteenth centuries usually led to earlier mowing dates, often by a week, two weeks, or more—especially in the case of spring grazing that had lasted long in May (see Section 6.2).[1155]

The example of the Senožeta meadow complex (Figure 53) in Bohinj (at about 580–750 m) illustrates the shift in mowing dates in meadows that had previously been grazed until the second half of May. Before the reorganization of meadow management, spring grazing lasted there until May 25th, followed by late mowing, evidently in late July or August, because unfenced autumn grazing in these meadows was again allowed to start on August 24th.[1156] This roughly corresponds with the mowing timing following spring grazing until the same date, May 25th, at Wieting in Carinthia, located at an elevation comparable to Senožeta, as presented in Section 6.1.2.1.1.3. Meadow management at Senožeta was reorganized in 1860, when spring grazing was abandoned,[1157] and so old farmers from the area only recall earlier mowing there: after June

[1154] Stadler-Planzer, Stadler, & Degler-Spengler, 2015, 81, 98–99, 101–103.

[1155] Kapfer, 2010, 137–140; Babai, Molnár, & Molnár, 2014, 103, 109. In general, the same dates sometimes continued to designate the timing of some tasks, despite the difference in plant development, and so the farm calendar did not always adapt; see Makarovič, 1995, 102, 108 and Sections 6.2 and 6.6.1 in this volume. However, the timing of haymaking had to adapt to the abolition of spring grazing there, at least in those meadows where spring grazing had lasted long in May, otherwise the hay mowed in late summer would be so old that cattle would reject it (compare Babai, Molnár, & Molnár, 2014, 103, 109; Kuret, 1989, 537; Bavcon, Ravnjak, & Praprotnik, 2019, 199).

[1156] Novak, 1955, 275.

[1157] Novak, 1955, 275.

Figure 53: The Senožeta meadow complex (in the middle of the photo) in Bohinj. Photo: Žiga Zwitter, 2018.

29th, usually around the first half of July, in the first half of the twentieth century, thus about one month earlier than before the abolition of spring grazing. It was only when farmers did not manage to mow their meadows at Senožeta prior to the majority of mowing in higher montane meadows that they individually postponed mowing in Senožeta to around the first half of August in the first half of the twentieth century,[1158] roughly corresponding with centuries-old mowing dates in that meadow complex, but no longer preceded by spring grazing. As a consequence of such late mowing in the first half of the twentieth century, the quality of hay was much lower than when it had been mowed at similar dates before 1860.[1159]

With some modifications, the factors causing relatively late mowing dates before the late eighteenth or mid-nineteenth century, in the era of spring grazing, presented in Section 6.1.2.1.2, were also relevant between the end of spring grazing and the mid-twentieth century. Some former collective rules and customs regarding mowing dates still retained importance,[1160] but it was gradu-

1158 Hodnik, 2016, 76, 78–80; author's interviews with Grm, 2017, and with Milena Sodja, 2017, both in Srednja Vas.

1159 Compare Babai, Molnár, & Molnár, 2014, 103, 109.

1160 For example, Makarovič, 1982, 428–429; Meilleur, 1986, 272.

ally decreasing (see also Section 4.1.1.4). According to late twentieth-century data referring to the warm climate (but not yet as warm as today), mowing in moderately fertilized meadows could usually start no earlier than in mid-June (false oat-grass meadows in the lower montane zone), and in late June at higher elevations (yellow oat-grass meadows). By that time, most grasses' and herbs' seeds are ripe. Mowing in non-fertilized meadows can only start later than in moderately fertilized ones to keep them productive in the long run; in the late twentieth century, agronomically adequate mowing timing in non-fertilized meadows at moderate elevations was not earlier than in (early) July.[1161]

This is reflected in historical mowing dates, which were sometimes even later than in line with the previous paragraph due to the factors presented in Section 6.1.2.1.2. In many places in Central Europe and the Alps, the first mowing under better site conditions between the abolition of spring grazing in hay meadows and the mid-twentieth century was carried out from mid-June onward (see also Section 6.1.2.1.1.1), and in non-fertilized meadows at nutrient-poor sites in late July and August, at high elevations in particular.[1162] Earlier mowing at manured sites is a consequence of quicker growth and high hay quality but also of higher lodging hazard in the case of tall grasses at manured sites compared to shorter grasses at non-fertilized sites (see Section 6.3.2). Abolition of spring grazing in hay meadows increased the lodging hazard in moderately manured meadows because the grasses grow taller in the absence of spring grazing.[1163] The practice of annual alternations between middling and late mowing dates at moderately fertilized sites that, apart from the practice of regular mowing as late as when seeds ripen, also allows long-term preservation of productive, moderately fertilized, species-rich grasslands, has its origins in the period dealt with here at the latest (see Sections 6.5 and 10.1.2.2.3).

The following data, mainly from the Alps of Slovenia and the northern French Alps, demonstrate the wide historical spatial range of mowing dates presented here that were relevant in most of the Alps. The presentation of mowing dates in the Slovenian Alps focuses first on mowing at the elevations of farms, followed by mowing dates in meadows at higher elevations.

In many places in the Slovenian Alps, characterized by a humid climate, the main mowing season in the first half and the middle of the twentieth century did not start before Midsummer Day (June 24th; this date had already been relevant in the Middle Ages, but it usually designated more advanced plant development during the High Medieval Climate Anomaly (the Medieval Warm Period) than in the mid-twentieth century—medieval people followed the Julian calendar; see Section 6.1.2.1.1.1) or just after St. Peter's Day (June 29th;

[1161] Dietl & Lehmann, 2004, 20, 24, 66, 83, 106, 118, 124–125.
[1162] "Košnja," 1864, 199; Smerdel, 1994, 106; Vogl-Lukasser et al., 2006, 69; Kapfer, 2010, 137–139; Norz, 2014, 115; Hodnik, 2016, 76–79.
[1163] Babai, Molnár, & Molnár, 2014, 36–37, 109.

Figure 54: Fully ripe grasses in a meadow at an elevation of 560 m on June 24th, 2018. After wet and unstable weather in June and early July 2018, many meadows in the bottom of the Bohinj Basin were mowed as late as the last decades of the nineteenth century and the first half of the twentieth century. In the mid-twentieth century, grasses probably ripened slightly later than today due to climate change. Trends in Europe show that the blooming of autumn-sown cereals has been starting 0.3 to 0.5 days per year earlier in recent decades.[1164] Photo: Žiga Zwitter.

see Figure 54), but the very start of mowing in mid-June was usual in some lower valleys—at 550 m, for instance (Sections 6.1.2.1.1.1 and 6.5).[1165] For earlier mowing, see the end of this section and Section 6.1.2.4. Even later mowing among the cited mowing dates, in late June, was common in moderately fertilized meadows in many places in Central Europe.[1166] Mowing in mid-June became more frequent in Central Europe with intensified fertilization in the grasslands still considered moderately fertilized, especially in the first half of the twentieth century or in the middle of the century (Section 6.3.2.2);[1167] however, Sections 6.1.2.1.1.1 and 6.5 explain that starting mowing in mid-June has

[1164] Žust, 2016, 16–19.

[1165] Kuret, 1989, 438; Hodnik, 2016, 76; author's interviews with Mencinger, 2017, in Stara Fužina, and with L. and M. Odar, 2017, in Studor.

[1166] Kapfer, 2010, 138. At traditional farms at the foot of Carpathians in Romania, mowing first-crop hay in the meadows near the village around June 29th is still usual (Ivașcu, Öllerer, & Rákosy, 2016, 44–45, 48).

[1167] Kapfer, 2010, 138–139.

a longer tradition in moderately fertilized meadows at moderate elevations in several microregions of the Alps.

Where permanent settlements were located at high elevations and on the slopes of high mountains, mowing started significantly later. At about 1,160 m, mowing at the highest farms in the Luče region in the Upper Savinja Valley (Sln. *Zgornja Savinjska dolina*) of northern Slovenia started in the first days of July in the mid-twentieth century. At the mountain farms in the north of the neighboring region of Solčava, with a similar amount of precipitation (about 1,500 to 1,600 mm annually between 1961 and 1990), mid-July was an important haymaking time at about 1,200 to 1,300 m, whereas at Bukovnik, the highest farm in the Solčava region and the Upper Savinja Valley (at 1,325 m), mowing only started in early August. Other major agricultural work did not delay mowing at all the locations mentioned in the Upper Savinja Valley because cereals were spring-sown at those farms at high elevations, and thus only reaped in September and October.[1168]

Mowing in distant meadows usually started later than close to permanent settlements. It began in the second half of July or even in August.[1169] Exceptions were highly nutrient-rich places around cottages in summer pasture complexes (Section 6.1.2.4) that were mowed around June 29th in the mid-twentieth century, before the livestock arrived there; that is, contemporaneously with meadows mowed relatively early around 1,000 m lower in the valley, as demonstrated by data from Bohinj.[1170] Moreover, fertilized distant montane meadows at elevations between the valley bottom and the summer pastures—reaching, for instance, elevations of up to 1,250 m—were mowed twice (Section 6.1.2.3), and so the first mowing there in early July took place sooner than in many lower meadows to allow second-crop haymaking in early September (see Sections 6.1.2.3 and 6.3.2).[1171]

A comparison of mowing dates in distant meadows and in grasslands close to high permanent settlements at similar elevations but in other regions shows that some distant meadow complexes were mowed earlier, some contemporaneously, and some later than those close to farms at similar elevations. At an elevation of slightly more than 1,000 m, above the village of Bohinjska Bela in the Julian Alps (Sln. *Julijske Alpe*, Ital. *Alpi Giulie*, Friul. *Alps Juliis*; the elevation of the permanent settlement in the valley is 480 m), mowing started on the Monday following the first Sunday after St. James's Day (July 25th) and lasted until the beginning of August. Thus, it started significantly later than at the farms in the Upper Savinja Valley at comparable elevations. In contrast, at the beginning of August, wild hay was mowed high on the slopes of Mount Begunjščica in

[1168] Meze, 1963, 226, 234–235, 268; Meze, 1965, 186, 202; Zupančič, 2001.
[1169] Šarf, 1974, 138; Kuret, 1989, 448–450; Hodnik, 2016, 76–79.
[1170] Author's interview with Mencinger, 2017, in Stara Fužina.
[1171] Hodnik, 2016, 76, 79.

the Karawanks (Germ. *Karawanken*, Sln. *Karavanke*), at an elevation of up to at least 1,800 m (see also Section 6.1.2.7),[1172] meaning contemporaneously with the highest farm in the Upper Savinja Valley (1,325 m), located significantly lower than meadows on Begunjščica. In contrast to the quality of the hay made near the farms at around 1,300 m in August (unless still preceded by spring sheep grazing, which is not impossible at isolated mountain farms where spring grazing would be an individual decision and would affect the farmer's land only; see Section 6.2), hay in high mountain meadows was still very fragrant in August, in the lower section of the alpine belt, for instance; that is, in meadows above the potential natural dwarf mountain pine zone (see Section 2).[1173] Similarly, the quality of the hay from subalpine meadows on the highest slope section of the Golica Ridge in the Karawanks was very high (see Section 6.3.1.2) according to the standards of traditional farmers in the first half of the twentieth century, although it was only mowed in August, often in the second half of that month.[1174]

A comparison with the Allues (Francoprovençal *Alués*) Valley (French *Les Allues*) in the northern French Alps, characterized by a drier climate (about 900 mm of annual precipitation) than the Alps of Slovenia, demonstrates that in the first half of the twentieth century haymaking at comparable elevations commenced at similar times in that part of the Western Alps and in the Upper Savinja Valley. Meadows in the Allues Valley were located at elevations of about 900 to 1,700 m. The data refer to the first half of the twentieth century, when spring grazing was only carried out in a few wet meadows in the Allues Valley and it lasted for a short time; most hay meadows were not grazed there in spring. Depending on grassland vegetation development, mowing first-crop hay in the Allues Valley usually started at the very beginning of July in hamlets at a lower elevation and by July 15th in hamlets with meadows at elevations of more than 1,300 m, with year-to-year fluctuations of the date when mowing commenced due to weather impacts on grassland plant maturation. Mowing first-crop hay lasted there until the end of August, interrupted only on Sundays (and major holy days) and in the short interval of winter rye harvest in lower hamlets.[1175] Partly similarly to the Allues Valley, high meadows belonging to Brandberg in Tyrol (a village at about 1,100 m), some of them reaching elevations of more than 2,000 m, were only mowed in late August, after the second-crop haymaking close to the village.[1176]

It was, however, not a general rule that the highest meadows were mowed the latest, as suggested above by the relatively early mowing in fertilized dis-

---

1172 Kuret, 1989, 438, 448–450; Šarf, 1974, 135, 138–139, 143.
1173 Pils, 1994, 193.
1174 Smolej, 2008, 15; Torkar Tahir, 2012, 133.
1175 Meilleur, 1986, 15–21, 30–31, 122, 263, 270, 271 (quote), 272, 276, 284, 296–297.
1176 Keller, 2004, 14; elevations are based on *Arcanum Maps*, Cadastral Maps, Habsburg Empire – Cadastral Maps (XIX. [sic] Century), cadastral municipality of Brandberg, 1855, and *Elevationmap.net*: Brandberg.

tant meadows mowed twice annually. In the meadows mowed once annually, differences in hay quality could lead to more profound diversity of mowing timing than the differences in microclimate, which, for instance, caused slightly earlier grass seed ripening at some higher sunny sites than at sites a bit lower but somewhat shadier.[1177] Focusing on the impact of hay quality, despite July 25th as the usual beginning of mowing in remote meadows in forest clearings in Bohinj,[1178] it was not exceptional for meadows in the Zavaze[1179] complex southeast of Stara Fužina, at elevations of no more than about 900 m, to usually be mowed in mid-August in the mid-twentieth century—at the end of the first-crop haymaking season, before the spread of mineral fertilizers. The reason is that hay from most meadows in Zavaze was of poor quality (see Section 9); consequently, more productive meadows producing better hay were mowed first because the workload did not allow mowing timing in all meadows at the desired plant phenophase (see Section 6.1.2.1.2).[1180] Similar cases of non-fertilized meadows in the lower montane belt (600–700 m) mowed in the mid-twentieth century later than non-fertilized meadows providing quality hay at twice that elevation (around 1,300 m) are known elsewhere.[1181] A parallel with the traditional ecological knowledge of Csángó farmers in the Eastern Carpathians adds another likely aspect to this explanation. It suggests that the dynamics of nutrient leaching in the case of delayed mowing depended on grassland species composition (see also Section 9). Matgrass (*Nardus stricta*), for instance, a grass species of little value in animal husbandry (see Section 6.3.1.2), is considered to be little affected by leaching. Thus, unlike the majority of grass species, clovers, and many other herbs esteemed in animal farming, extremely late matgrass mowing is considered by traditional farmers not to have a comparably relevant economic impact.[1182]

Exceptionally, specific environmental conditions or species compositions promoted relatively late or early mowing dates. The impact of specific species compositions is presented first, followed by an example of site conditions motivating the timing of mowing.

Species composition stimulated the mowing timing where some grassland monocots or dicots influenced the hay quality in a highly specific manner in certain phases of plant development and the abundance of such a species was high. Thus, the quality of hay from these meadows was holistically much better exactly at the relatively late or early mowing time.

1177 On the impact of differences in microclimate on historical mowing, see Meilleur, 1986, 276.

1178 For example, the author's interviews with Mencinger, 2017, and with Kunavec, 2019, both in Stara Fužina, and with L. and M. Odar, 2017, in Studor.

1179 The form *Za laze* on usual maps does not correspond to the local pronunciation.

1180 Author's interview with Kranjc, 2020, in Stara Fužina.

1181 Hodnik, 2016, 76, 79–80; author's interview with Škantar, 2017, in Srednja Vas.

1182 Babai, Molnár, & Molnár, 2014, 38–39, 108–109.

An example of meadows where relatively late mowing (no later than suitable for long-term meadow regeneration) was promoted by the species composition to increase the general hay quality was meadows where daffodils abounded, such as at Javorniški Rovt in the Karawanks, rich in Salisbury's daffodil (*Narcissus poeticus* subsp. *radiiflorus*; see Section 6.2.2.3). In the mid-twentieth century, mowing there started after (the later dates of) Whitsuntide (Whit Sunday moves between May 10th and June 13th: evidently, later Whitsuntide days are meant because mowing before June is unimaginably early for species-rich meadows). Mowing dates later than Whitsuntide are also reported there. Meadows where Salisbury's daffodil abounds can only be mowed after the daffodil leaves have dried out and decomposed because the species is toxic to livestock. If they nonetheless consume the daffodils, the milk becomes unpalatably bitter (this fact is advantageous for Salisbury's daffodil because by the time the plant leaves decompose the seeds have already matured and been dispersed). This requirement was often fulfilled by (the later dates of) Whitsuntide.[1183] Mowing dates in meadows at Jezerca above Javorniški Rovt by the mid-nineteenth century (Section 6.1.2.1.1.3) demonstrate that, as long as spring grazing was carried out in those hay meadows, daffodils did not impact the mowing dates at Jezerca because mowing there was carried out later than the daffodils would require.

An example of early mowing motivated by specific plant characteristics, because the hay quality would have declined more quickly than usual during further plant development, is wet meadows rich in common reed (*Phragmites australis*), as demonstrated by the example of the former wetland below Srednja Vas in Bohinj. According to twentieth-century data, these meadows were mowed in the first half of June, which is early in the context of other meadows belonging to the same farms. This way, hay largely consisted of sedges and it was used to feed horses (and, at farms lacking more appropriate fodder, also to feed cattle, sheep, and goats). Farmers learned from experience that, if these meadows were mowed later, the palatability of common reed (*Phragmites australis*) would decline (see 6.1.2.5.1.1) due to a quick increase in crude fiber content. Moreover, the common reed content in hay would have risen substantially because common reed would have grown much taller there (Figure 55).[1184]

Specific site conditions stimulating particular mowing timing include a serious drought hazard at the foot of the Alps in areas with a relatively warm climate leading to relatively quick plant development. According to information from the first half of the twentieth century, mowing in the meadows on drought-prone coarser alluvial deposits on the eastern foot of the Pohorje Hills

1183 Praprotnik, 1993a, 217, 219; Rogelj, 2012, 59; Bavcon, Ravnjak, & Praprotnik, 2019, 139.
1184 Author's interviews with Mencinger, 2017, and with Ceklin, 2020, both in Stara Fužina.

Figure 55: Relatively early mowing time in wet meadows below Srednja Vas in the first half of June prevented the prevalence of tall common reed (*Phragmites australis*) plants in hay. Common reed is poorly palatable for livestock if more than around a meter tall (see Section 6.1.2.5.1.1). The exceptional meadow in the photo was not mowed until mid-August 2020. Note the layer of plants about 60 to 70 cm tall that used to be mowed for hay, in this case overgrown by common reed about 270 cm tall. Photo: Žiga Zwitter, August 15th, 2020.

was carried out as early in June as possible. Namely, at that time it was still likely that water for meadow irrigation (see Section 6.4) after mowing was available. When irrigation water was not available after first-crop-hay mowing, no second-crop haymaking in those meadows was possible,[1185] but second-crop hay is a highly valuable hay type (see Section 6.1.2.3).

In general, June was a typical month of mowing in meadows at lower elevations cut twice annually (see also Section 6.1.2.1.1.1). Historical timing of second-crop haymaking and its impacts on hay quality are presented in the following section.

6.1.2.3 Mowing second-crop hay

Second-crop hay was only mowed in the Alps in fertilized (see Section 6.3) or irrigated meadows (see Section 6.4),[1186] with the likely exception of some

[1185] Baš, 1937, 333.
[1186] Kapfer, 2010, 136.

highly productive wetlands that were mowed twice although no fertilizer was applied (see Sections 2 and 6.1.2.5.1.2).[1187] Even though environmental conditions would allow a second mowing in some non-fertilized and unirrigated grasslands, widespread autumn grazing in hay meadows, which started in August or September, made second mowing at these sites impossible.[1188] As a general rule, farmers usually only fertilized more productive meadow sites within their farms (Section 6.3.2). Consequently, the example of the Allues (Franco-provençal *Alués*) Valley (French *Les Allues*) with its relatively dry climate and absence of irrigation in usual years demonstrates that humid sites were preferred for second-crop haymaking because of higher yields in comparison to drier meadows. Nonetheless, weather during first-crop hay growth and drying, reflected in first-crop hay quantity and quality, often also had an impact on the total area where second-crop hay was mowed at the farm level in the same year: the less first-crop hay was stored, the more second-crop hay was harvested,[1189] as far as allowed by the grazing rules in the meadows and site conditions.

Quantities of second-crop hay have been much lower than those of first-crop hay (Figure 57), often between one-third and half the quantity of first-crop hay from the same meadow (see Section 6.3.0.1), but second-crop hay's nutritive value has been higher. Second-crop hay is significantly more protein-rich than first-crop hay mowed at the time usual in the study period of this volume.[1190] Second-crop hay from traditional moderately fertilized meadows is considered high-quality or at least relatively high-quality feed even according to standards in present agronomy. In a moderately fertilized false oat-grass (*Arrhenatherum elatius*) meadow mowed twice annually and grazed in autumn, the net energy value of hay mowed in late June for lactation is slightly less than 5 MJ/kg of dry feed substance, and the corresponding value of second-crop hay mowed in early September is slightly less than 5.5 MJ/kg. At higher elevations, where moderately fertilized meadows are usually yellow oat-grass (*Trisetum flavescens*) meadows, the quality difference between first-crop and second-crop hay is not as significant, but with about 5.75 MJ/kg dry substance the net energy value of first- and second-crop hay for lactation is high there, comparable with the third and fourth cuttings in present-day intensive and low-biodiversity perennial rye-grass (*Lolium perenne*)–smooth meadow-grass (*Poa pratensis*) meadows.[1191] Due to the significant difference in hay quality, farmers have traditionally stored second-crop hay separately.[1192] Second-crop hay has always been a welcome addition to any kind of first-crop hay to increase the milk quan-

[1187] Sucholas et al., 2022, 9–10, 16, 19, 35–36.
[1188] Grass, 1948, especially 172, 189, 191.
[1189] Meilleur, 1986, 16, 269, 277, 279–280.
[1190] Meilleur, 1986, 277; Luick, Jones, & Oppermann, 2012, 38; Machatschek, 2015, 30; *Traditional Hay Milk Farming …*, [2023], 63.
[1191] Dietl & Lehmann, 2004, 122, 125.
[1192] Ivaşcu, Öllerer, & Rákosy, 2016, 46; Luick, Jones, & Oppermann, 2012, 38.

Figure 56: A moderately fertilized meadow in Bohinj just before second-crop hay harvest. Photo: Žiga Zwitter, August 2022.

tity.[1193] Before the mid-twentieth century as the study period of this volume, the share of second-crop hay in the total hay was usually significantly lower than the share of second (and often further) mowings in the total hay quantity today (Sections 6.3.0.1 and 10.1.2). Thus, if second-crop hay was not available throughout the period of barn feeding, farmers mixed second-crop hay with low-quality first-crop hay to maintain similar milk production as in times when their livestock received better first-crop hay.[1194] Second-crop hay was also of utmost importance when cereal (wheat) straw, characterized by low fodder value, was part of the fodder (see Section 8.2.2.1).

The long history of the high fodder value of second-crop hay is largely a consequence of the fact that in usual years second-crop hay has always been cut when grassland plants are (partly) still green, thus the German expression *Grummet*, recorded in the thirteenth century as *Gruemat* (i.e., *grüne Mahd* 'green grassland plant harvest'; Figure 56). This was the difference compared to the first-crop hay from species-rich grasslands, especially from meadows at

[1193] Luick, Jones, & Oppermann, 2012, 38; Hodnik, 2016, 42–43; author's interview with Milena Sodja, 2017, in Srednja Vas.

[1194] Bavcon, Ravnjak, & Praprotnik, 2019, 81.

moderate elevations mowed only in August or September for the first time and containing a significant part of dead plant litter, making the mowed biomass less green (see Sections 6.1.2.1.1.3, 6.1.2.1.1.4, and 6.1.2.2).[1195] Whereas historical species-rich meadows cannot tolerate early first-crop haymaking (see Sections 6.1.2.1.2 and 6.5), moderately-manured meadows can tolerate green second-crop hay mowing in late summer.

Second-crop hay has always been considerably less species-rich (containing a lower diversity of herbs) than first-crop hay from the same meadow (Figure 56). Nonetheless, a few species regrow more vigorously in the lighter conditions of second-crop hay than in the shadier stand producing first-crop hay (see Section 6.1.2.1.1.2). A few others simply regrow and reblossom, such as daisy (*Bellis perennis*), which blossoms in spring and then again repeatedly every time the shade of the stand is removed by mowing.[1196]

Historical data from the second millennium prove second-crop hay mowing (not just grazing on the regrowth in meadows after the first mowing) in Central Europe since the eleventh century.[1197] The documentation of medieval second-crop haymaking in the Eastern Alps is scant, but it is clear that it usually took place in a limited area of fertilized meadows at the farm level at least since the Late Middle Ages and probably earlier. In the Kauns Valley in Tyrol (Germ. *Kaunertal*, in the Upper Inn catchment—at Feichten, for instance, at an elevation of about 1,300 m or more), a list of fertilized meadows mowed twice annually was recorded in 1440.[1198] In some regions, the first sources allowing recognition of the role of second-crop hay at a large number of farms were recorded in the sixteenth century. Such a relatively early example providing mesoregional insights is the rent-roll, with a description of holdings belonging to the seigneury of (Eisen)Kappel (Sln. *(Železna) Kapla*) in Carinthia (Germ. *Kärnten*, Sln. *Koroška*), based on interviews with farmers and field observations in 1567. Figures 57 and 82 (in Section 6.3; see also Section 6.3.0.1) demonstrate that by the mid-sixteenth century (and most likely earlier) it had already been typical for a farm in the Alps to make second-crop hay in a significant percentage of all meadows: second-crop hay amounted to about 6% to 20% (often 10% to 15%) of the total hay quantity stored at a typical isolated mountain farm in the Central Karawanks (Germ. *Karawanken*, Sln. *Karavanke*) and the nearby Kamnik–Savinja Alps (Sln. *Kamniško-Savinjske Alpe*) at that time.

[1195] Pfister, 1985a, 41; Kauter, 2002, 126; Kapfer, 2010, 136; Machatschek, 2015, 30; Grimm & Grimm: *Deutsches Worterbuch*: Grummet. The thirteenth-century form *Gruemat* proves that by that time the word was already old because the *n* at the end of the first word forming this compound had already been lost in this case.

[1196] Pils, 1994, 163.

[1197] Kauter, 2002, 126.

[1198] Wopfner, 1997, 273–274, 393–394.

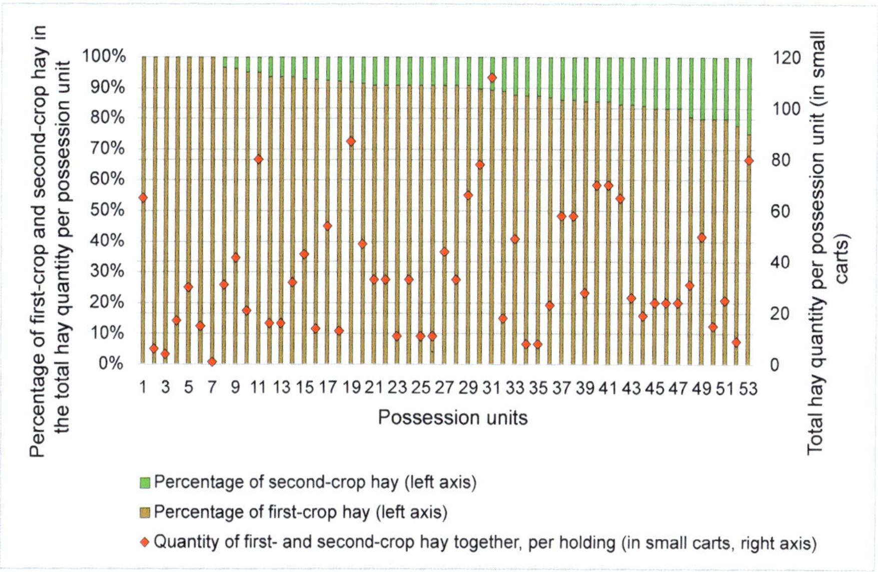

Figure 57: Percentage of first- and second-crop hay (left ordinate) in the total quantity of first- and second-crop hay together per holding (columns) for fifty-three holdings (on the abscissa) in the upper Vellach (Sln. *Bela*) and upper Kokra River catchments; that is, the Central Karawanks and the Kamnik–Savinja Alps nearby, as recorded in 1567. Diamonds (right ordinate) show the quantity of first- and second-crop hay together per possession unit to give an idea whether a large or a small total meadow area belonged to it. The great majority of holdings in the graph are isolated mountain farms from Spodnje Jezersko in the south to Lobnig (Sln. *Lobnik*), the surroundings of (Eisen)Kappel, and Ebriach (Sln. *Obirsko*) in the north. They are located at elevations of around 550 to 1,200 m;[1199] rarely, a few hay meadows were significantly higher than the farm, reaching an elevation of 1,300 m, for instance (see the example of the Anko farm in Section 6.3.0.1). The only holding entirely located higher is a demesne alp with hay meadows at about 1,300 m (i.e., column no. 5; see Section 6.3.0.1). A few of the isolated farms in Figures 57 and 82 had two to four heads instead of being managed by the members of one household, and so they were divided, but hay quantities were recorded together; consequently, they are presented as one unit on the graph. See further methodological remarks at Figure 57 in Sections 6.1.2.3.1 and 6.3.0.1. Data source: AT StLA, Innerösterreichische Hofkammer, Sachabteilung, box 69, folder 8: Des ambts Capl bereuttung und beschreibung, 1567. Author of the graph: Žiga Zwitter.

Later, the area of meadows where second-crop hay was cut greatly increased, especially since the second half of the nineteenth century, due to the increased availability of farmyard manure for meadows (see Section 6.3), and again since the mid-twentieth century, with further abandonment of arable farming in the Alps and increased availability of mineral fertilizers, so concentrated that they were not too heavy to be transported uphill even by livestock. From the

1199 Localization is based on a comparison of this rent-roll with the early nineteenth-century cadaster (*Arcanum Maps*, Cadastral Maps, Habsburg Empire – Cadastral Maps (XIX. [sic] Century)).

perspective of second-crop hay history, these processes strongly increased the area of fertilized meadows.[1200]

Nonetheless, despite these later general expansions of the meadow area mowed twice annually, mowing second-crop hay not only in meadows close to the farms but also in fertilized sections of remote grasslands near remote livestock barns in lower temporary mountain settlements (see Section 6.3.2.1) has a long tradition. For instance, second-crop hay was mowed in much[1201] of the montane meadow complex in the Voje Valley (at roughly 700 m above sea level), as recorded in the early seventeenth century,[1202] and in sections of some montane meadows in Ukanc (less than 600 m) west of Lake Bohinj (Sln. *Bohinjsko jezero*) as reported in the 1690s and in the early eighteenth century.[1203] Similarly, mowing second-crop hay was documented in western Bohinj in some sections of distant montane meadows at elevations higher than in Ukanc in the early eighteenth century.[1204] Indirect information indicates that second-crop hay was mowed at some places not only higher than Ukanc but

1200 Kauter, 2002, 126–127; Kapfer, 2010, 136–139; author's interviews with Kunavec, 2017 and 2023, in Stara Fužina.

1201 In this case, the plaintiff claimed that around ten farm wagons of second-crop hay were illegally grazed only in a certain meadow (or in meadows belonging to one holding; SI AS 721, Gospostvo Bled (The Seigneury of Bled, Germ. *Veldes*), Iustitialia, vol. 14, [November 26th, 1612]). Even though it is not clear how accurately the hay quantity was recorded, this makes it clear that the area mowed twice annually in the entire Voje Valley and not only in the meadow(s) belonging to the farm involved in this lawsuit was relatively large.

1202 SI AS 721, Gospostvo Bled (The Seigneury of Bled), Iustitialia, vol. 14, [November 26th, 1612]. An explanation of the localization: the toponym *Vlosech* is mentioned in the cited German record. *L* is currently pronounced as *w* in the local dialect in such cases, and the locative *v Vozeh* (pronounced *u̯ Wọzẹ́x*; *x* designates a voiceless velar fricative in phonetic transcription) was still used for Voje in the mid-twentieth century (see Melik, 1950, 56–57, 110; Snoj, 2009, 18–19, 463). The person acting as the plaintiff here (Anthony, i.e. sir Hans Corein's factor) on the same day acted as the plaintiff in another process dealing with an alleged theft of iron ore that had been sold to a person from Stara Fužina (SI AS 721, Iustitialia, vol. 14, [November 26th, 1612]; for the localization of the village "beim Alten Hamer," see Pančur, Blaj Hribar, & Györfi, 2017: Althammer). Thus, the localization of the aforementioned meadow complex is clear because Stara Fužina is the closest permanent settlement to the Voje Valley.

1203 SI AS 721, Gospostvo Bled (The Seigneury of Bled), Iustitialia, fasc. 33, Lawsuit between the communities with centers in Savica and Studor over grazing rights at Lake Bohinj and in Ukanc, 1655–1666 [!]: *Proiects puncta*, the document was sent to the territorial court in 1704; SI AS 721, Iustitialia, fasc. 33, Lawsuit between the communities with centers in Studor and Savica over grazing rights in Ukanc 1697–1717: *Herr landtshaubtman in Crain*, verso the later date May 21st, 1699; *Durchleichtiger Hörza[!]g …*, verso the later date June 1st, 1699; and *Hochgebohrner Raichß Graff, vnd Herr landtshaubtman in Crain etc.*, June 19th, 1717.

1204 SI AS 721, Gospostvo Bled (The Seigneury of Bled), Iustitialia, fasc. 33, Lawsuit between the communities with centers in Studor and Savica over grazing rights in Ukanc 1697–1717: *Ihr Gnaden*, verso September 23rd, 1704.

also significantly higher than the Voje Valley already at that time. Namely, the use of livestock barns in some relatively high meadow complexes is recorded in the Eastern Alps, and so local barn manure was applied the meadows there (see the example of Hudi Hlevi in the western Karawanks in 1777 in Section 6.3.2.1), which allowed second-crop haymaking at many sites. Information referring to the west of Bohinj in the first half of the twentieth century, before the large-scale introduction of mineral fertilizers, explains that the highest remote meadows where second-crop hay was made in limited meadow sections reached around 1,250 m above sea level in that region, and such distant montane meadows, partly mowed twice annually, were often located about 550 to 600 m higher than the farms (Section 6.3.2.1). However, the dynamics of plant regrowth and sensitivity of second-crop hay to precipitation, embedded in the wider socioeconomic historical context, posed limits for second-crop haymaking at high elevations, as explained lower in this subsection.

Second-crop hay harvest dates were relatively similar throughout the Alps. In general, second-crop hay was harvested between the Assumption (August 15th) and Nativity of Mary (September 8th) or (at lower elevations) until Michaelmas (September 29th) in Bohinj[1205] and in many places in Slovenia.[1206] It was made around September 1st on mountain farms above the Poljane Valley (Sln. *Poljanska dolina*) in the Pre-Alps south of the Julian Alps (Sln. *Julijske Alpe*; Ital. *Alpi Giulie*; Friul. *Alps Juliis*),[1207] in the second half of August and in early September at the bottom of the Ötz Valley (Germ. *Ötztal*) in Tyrol,[1208] and in September in the Allues Valley in the Western Alps.[1209] All these data refer to the early- and mid-twentieth century. Thus, second-crop haymaking in moderately fertilized meadows in the Alps usually started later than usual in much of the colline and lower montane vegetation belt elsewhere in Central Europe after the abolition of spring grazing in meadows (see Section 6.2); in many places in Central Europe outside the Alps, second-crop haymaking in moderately fertilized meadows started in early August.[1210]

Although the dates presented for second-crop haymaking in the Alps were similar at a macroregional scale, the details depended not only on plant development but also on the start of autumn grazing in the same meadows[1211] and further social factors, such as the saint's name day, locally serving as orientation to determine the time of second-crop haymaking. The data presented make it clear that two feasts of the Virgin Mary often served this purpose.

[1205] Author's interviews with Mencinger, 2017, in Stara Fužina; with Škantar, 2017, and with Milena Sodja, 2017, both in Srednja Vas.
[1206] Bavcon, Ravnjak, & Praprotnik, 2019, 199, 203.
[1207] Makarovič, 1995, 107, 414.
[1208] Hessenberger, 2022, 154.
[1209] Meilleur, 1986, 276–277.
[1210] Kapfer, 2010, 138–139.
[1211] Compare Janišová et al., 2023, 9, 11.

However, on mountain farms above the Poljane Valley, there was the impact of St. Giles's name day (September 1st), the patron saint of the local parish church at Javorje.[1212]

The start of second-crop haymaking in the second half of August and its finish around the Nativity of Mary or Michaelmas evidently had a long historical tradition, which is in line with the general picture of the timing of second-crop haymaking in many places in Central Europe in the colline and lower montane belts before the abolition of spring grazing in hay meadows. An early end of spring grazing in meadows (Sections 6.2.1.2 and 6.2.2.1) and first-crop haymaking in early July (Section 6.1.2.1) allowed for second-crop haymaking to start in mid-August or early September,[1213] thus the beginning of second-crop hay mowing in the second half of August was not a novelty of the period after the abolition of spring grazing in meadows (Section 6.2.3). For instance, the second mowing took place in late August (before September 3rd) 1728 in many, but not all, meadows mowed twice annually in the Jaun Valley (Germ. *Jauntal*, Sln. *Podjuna*), Carinthia.[1214] Similarly, illegal mowing of second-crop hay in the first days of September or the last days of August 1611 was documented not far from the village of Poljšica, which is located at the eastern foot of the Julian Alps, possibly at around 600 m, because the haymaker wanted to bring it in on September 3rd.[1215] In a large irrigated meadow at 500 m, at the bottom of the Ljubljana Basin southwest of the village of Hlebce (see Sections 6.2.1.2, 6.4.3.3, and 6.4.5; now northwestern Slovenia), spring grazing until April 24th was followed by two mowings and autumn grazing, regulated by specific grazing rights from the second mowing until September 29th, and general communal grazing rights for livestock owned by farmers from Hlebce from September 29th onward, as documented in the mid-nineteenth century.[1216] This means that second mowing must have been carried out there before late September. Second-crop hay had probably already been mowed in that meadow in September in the sixteenth century.[1217]

1212 Makarovič, 1995, 107.

1213 Wopfner, 1997, 295; Kauter, 2002, 127; Kapfer, 2010, 136–139.

1214 AT StiAStP, Grundherrschaft Eberndorf (Handwritings), vol. 191, Gerichtsprotokoll 1721–1728, 424 (September 3rd, 1728).

1215 SI AS 721, Gospostvo Bled (The Seigneury of Bled), Iustitialia, vol. 14, September 12th, 1611.

1216 SI AS 48, Deželna komisija za odkup in uravnavo zemljiških bremen za Kranjsko (Provincial Commission for Redemption and Regulation of *Profits à Prendre* in Carniola), fasc. 163, 17–19: no. 2277 de 1858, *Zur Sitzung am 14. Oct. 1858*.

1217 The 1579 rent-roll also documents spring grazing in this same meadow until April 24th, meadow irrigation, and grazing rights for the Hlebce farms from September 29th onward (SI AS 173, Imenjska knjiga za Kranjsko [*Gültbuch* for Carniola]: Rent-roll (*Urbar*) of the seigneury of Radovljica, 1579, 384–385). The end of spring grazing was not postponed in this case with the 1580s calendar reform (see Section 6.2.1.2), and so if the seigneury adjusted the mowing timing in this demesne meadow to the vegetation

A long historical tradition of the end of second-crop haymaking around September 8th or 29th is demonstrated by data from Bohinj. In 1704 or a little earlier, it was stated that second-crop hay in Ukanc (western Bohinj, less than 600 m) had to be stored by September 10th.[1218] Later in the same year, it was recorded that second-crop hay in Ukanc had to be stored by the Sunday after the Feast of the Cross (September 14th), but, if other farmers' tasks and the weather did not allow compliance with this rule, second-crop hay was mowed there until Michaelmas.[1219] This demonstrates the long tradition of finishing second-crop haymaking around September 8th in Bohinj; however, when the weather (in combination with other tasks) did not allow this, second-crop haymaking in meadows at low elevations was finished by September 29th. The cited records suggest that such late second-crop haymaking was relevant in some summers during the Late Maunder Minimum—that is, the peak of the Little Ice Age in the late seventeenth and early eighteenth century (see Sections 4.1.1.2, 4.1.1.3, and 6.1.2.1.2)—on the lowest flank of the montane vegetation belt at least.

The main reason for the dates around September 8th and 29th as final limits for second-crop haymaking is the hazard of weather-induced damage to second-crop hay. Namely, second-crop hay is too sensitive to moisture to be stored in October, and the weather later than early September would often significantly reduce its quality. To harvest quality second-crop hay, environmental processes required that farmers pay particular attention to the dryness of the stored biomass. Namely, whereas first-crop hay already has to be dry when stored and is very sensitive to precipitation during the drying period, second-crop hay is even more sensitive when it dries (see Section 4.1.1.3). If not stored totally dry, second crop hay's decomposition could release enough heat that in exceptional cases it caused spontaneous combustion, resulting in a fire in a haybarn.[1220]

Thus, it is understandable that old farmers from Bohinj and many other regions of the Slovenian Alps recall the saying that mowing second-crop hay after the Nativity of Mary (September 8th) would require drying it in a large

development, the first and second mowing might have been moved to a date roughly ten days earlier in the late sixteenth century, unless prevented by the peak of the Little Ice Age at that time (Sections 4.1.1.2 and 4.1.1.3) or social factors (Section 6.1.2.1.2).

1218 SI AS 721, Gospostvo Bled (The Seigneury of Bled), Iustitialia, fasc. 33, Lawsuit between the communities with centers in Savica and Studor over grazing rights at Lake Bohinj and in Ukanc, 1655–1666 [!]: *Proiects puncta*. This document was written later than July 27th, 1658, and it was sent to the territorial court on July 28th, 1704.

1219 SI AS 721, Gospostvo Bled (The Seigneury of Bled), Iustitialia, fasc. 33, Lawsuit between the communities with centers in Studor and Savica over grazing rights in Ukanc 1697–1717: *Ihr Gnaden*, verso September 23rd, 1704.

1220 Author's interview with Kranjc, 2018, in Stara Fužina; Verbič, 2014–2020, 27; *Oklahoma State University: Spontaneous combustion a possibility with wet hay.*

grain sieve (Sln. *reta*),[1221] or, alternatively, on top of the stove.[1222] However, it was not (always) possible to finish second-crop haymaking by that time; even at roughly 1,200 m (at Uskovnica in the Eastern Julian Alps) it happened that second-crop hay was made slightly later than September 8th, as recalled for the mid-twentieth century.[1223] Thus, in Bohinj and elsewhere, there also exist variations of this saying in which the Nativity of Mary is replaced by Michaelmas, a late September date (September 29th).[1224] Today these sayings are considered only didactic humor by several members of the oldest generation;[1225] nonetheless, they are likely reminiscences of actual attempts to save at least a bit of the valuable second-crop hay in rainy Septembers (Section 4.1.1.3). See a documented twentieth-century example of drying second-crop hay in a human habitation in a remote meadow complex in Section 4.1.1.3. A newspaper article from the very beginning of the haymaking season in the second half of June 1864, which belonged to the Little-Ice-Age–type summer of that year,[1226] mentioned the folk saying that cutting second-crop hay later than on sunny September days would require drying it on the stove.[1227] Thus, this saying is older than 1864 and it was already used during the Little Ice Age, when it was in fact relevant more frequently than later (see Sections 4.1.1.2 and 4.1.1.3).[1228]

The sensitivity of second-crop hay to precipitation leads to the point that there has always existed an upper elevation limit of second-crop haymaking in the Alps. It is located much lower than the upper limit of first-crop hay mowing (on which see Section 3.1). The upper limit of second-crop hay mowing was a consequence of several environmental and socioeconomic factors, such as (1) harsher climate (lower average temperatures matter directly, through their impact on plant growth, and indirectly through more intensive morning dew at higher elevations, where night temperatures are usually lower, making hay drying in early autumn riskier; see Section 4.1.1.3); (2) late dates of first-crop haymaking (Sections 6.1.2.1 and 6.1.2.2) reducing the height of plant regrowth; and (3) the demand for autumn grazing in hay meadows (see also Section 6.1.2.1.2) limiting the availability of sites for second mowing. At many places in the Alps, the upper limit of second-crop hay in moderately fertilized meadows has been at about 1,300 m, as revealed by twentieth-century data from the Julian Alps (Section 6.3.2.2), the Eastern Karawanks in the Solča-

---

1221 Author's interviews with Mencinger, 2017, in Stara Fužina, and with Milena Sodja, 2017, in Srednja Vas.

1222 Ložar, 1944, 148; Bavcon, Ravnjak, & Praprotnik, 2019, 199, 203.

1223 Hodnik, 2016, 37–38, 76, 79.

1224 Bavcon, Ravnjak, & Praprotnik, 2019, 199, 203; author's interview with Škantar, 2017, in Srednja Vas.

1225 Author's interviews with Mencinger, 2017, and with Urh, 2017, both in Stara Fužina.

1226 On the weather that summer and its impact on society in the southern Eastern Alps and the northern Dinaric Karst, see Zwitter, 2015d, 151 and Studen, 2024.

1227 "Košnja," 1864, 199.

1228 On the chronology of the Little Ice Age, see White, 2014.

va region in Lower Styria (Germ. *Steiermark*, Sln. *Štajerska*),[1229] the Alps of Carinthia,[1230] and the agricultural handbook by the Swiss authors Walter Dietl and Josef Lehmann, aimed at German-speaking farmers in various parts of the Alps.[1231] Nonetheless, similar to differences in the potential climatic upper forest line within the Alps, and to differences in the historical upper line of cereal cultivation within the Alps,[1232] there also existed differences in the upper line of second-crop haymaking within the mountain chain. As in the case of the upper line of cereal cultivation, the differences in the elevation of the upper limit of second-crop haymaking were a result of interrelated environmental[1233] and human factors. This regional diversity in the altitude of the highest meadows where second-crop hay was made is confirmed by a comparison of the aforementioned frequent upper second-crop haymaking limit at around 1,300 m with the data from the Ötz Valley in Tyrol, where the upper forest line[1234] and historical upper line of cereal cultivation (at more than 1,700 m)[1235] both reach(ed) high elevations in a wider Alpine context. Second-crop hay was mowed in meadows close to the village center of Niederthai (at a bit more than 1,550 m, in a side valley of the eastern Ötz Valley) in the mid-twentieth century.[1236] This explains why it was worth recording in 1629 (the previous year of 1628 is one of the three years with the coldest warm half year in the second half of the second millennium[1237]) that the highest isolated farm above Niederthai, located at 1,780 m, could not make any second-crop hay;[1238] farms located

[1229] Meze, 1963, 226, 231, 234, 237: in the mid-twentieth century in the Solčava region, second-crop hay was not mowed at Bukovnik—the highest farm (farmstead at 1,325 m; barley was harvested at this farm in early September, and other cereals were reaped in late September and early October; they were all spring-sown because the climate is too harsh for autumn sowing), and it was mowed in good years at farms located on sunny slopes at about 1,235 to 1,250 m near Holy Spirit Church, where cereals were mowed 10 to 15 days earlier than at Bukovnik. The second-crop hay harvest at the farms near Holy Spirit Church took place in early September.

[1230] Keller, 2004, 32.

[1231] Dietl & Lehmann, 2004, 121: two mowings annually at 1,200 m, one mowing at 1,300 m.

[1232] Bätzing, 2015, 40, 54; see also p. 70; Lovrenčak, 1984.

[1233] E.g., general considerations on environmental diversity in the Alps in Viazzo, 1989, 16–19.

[1234] Bätzing, 2015, 40.

[1235] Scharr, 2001, 22; *Arcanum Maps*, Cadastral Maps, Habsburg Empire – Cadastral Maps (XIX. [sic] Century), cadastral municipality of Sölden, 1856.

[1236] Griesser, 1999, 42; the author was born at Niederthai in 1914 and spent much of his life in the Ötz Valley (Riccabona, von Sauter, & Unterkircher, 2020); *Arcanum Maps*, Cadastral Maps, Habsburg Empire – Cadastral Maps (XIX. [sic] Century), cadastral municipality of Umhausen, 1856. Thomas Wieser, municipality of Umhausen, kindly confirmed that Griesser's sentence in fact relates to second-crop hay mowing in the mid-twentieth century, not only to second-growth grazing (Wieser, personal communication).

[1237] Pfister, 1999, 158, 196.

[1238] Wopfner, 1997, 293; including no context on the upper line of second-crop hay mowing in the Ötz Valley.

about 200 meters lower in the village of Niederthai[1239] did harvest a limited quantity of second-crop hay, in good years at least. In the mid-twentieth century, second-crop hay was made as high as between around 1,800 and 1,900 m in the east-facing, moderately manured meadows at Stabele west of Lehn in the Ötz Valley,[1240] certainly as a continuation of a practice rooted at least in times of high agricultural density since the nineteenth century. South of the upper Ötz Valley, in the climatically milder Senales Valley (Ital. *Val Senales*, Germ. *Schnalstal*) in Alto Adige (South Tyrol), where the highest ploughed or hoed fields were located at 1,800 to 1,900 m,[1241] the upper limit of second-crop hay mowing was even higher than in the Ötz Valley.[1242]

In several places, the upper second-crop hay line was somewhat lower when temperatures in the warm half of the year were low. For instance, in 1816, the year without a summer (Sections 4.1.1.2 and 4.1.1.3), no second-crop hay could be made in the Martello Valley (Ital. *Val Martello*, Germ. *Martelltal*) in South Tyrol. The main settlement agglomeration there lies at about 1,300 m, and second-crop hay was usually mowed in that valley.[1243]

Thus, the highest farms in some regions of the Alps did not have valuable second-crop hay because they only mowed their meadows once annually. Nonetheless, on the one hand, even at farms at lower elevations quantities of second-crop hay have always been much lower than those of first-crop hay (they were low until the mid-nineteenth century in most of the Eastern Alps), and, on the other hand, the quality of first-crop hay from moderately manured meadows at high farms (yellow oat-grass meadows) was likely higher than the quality of first-crop hay from moderately manured meadows (false oat-grass meadows) lower in the valleys (Section 6.3.2). In addition, the quality of some first-crop hay from non-fertilized meadows at higher elevations was high (Section 6.3.1.2). Thus, it depended on local microenvironmental conditions and species compositions whether the absence of valuable second-crop hay at the farms at extreme elevations was a serious disadvantage.

6.1.2.3.1 Methodological remarks on the comparison of first- and second-crop hay quantities in the mid-sixteenth century

Figure 57 compares the quantities of first- and second-crop hay and shows the total hay quantity per unit of possession, mostly a mountain family farm, in

[1239] See also the map of seigneuries that farms in Niederthai belonged to in 1627: Scharr, 2001, supplement between pages 48 and 49.

[1240] Hessenberger, 2022, 146, 154, 158; localization and elevation: *Arcanum Maps*, Cadastral Maps, Habsburg Empire – Cadastral Maps (XIX. [sic] Century), cadastral municipality of Längenfeld, 1856, and *Google Maps*, Terrain.

[1241] Scharr, 2001, 22.

[1242] The historian and geographer Kurt Scharr, personal communication, 2023.

[1243] Jäger, 2006, 29.

Figure 58: Driving second-crop hay home. Bohinj, second half of the twentieth century. In contrast to loading first-crop hay on a cart (see Figures 28 and 29 in Section 4.1.1.4), the short length of second-crop hay required the use of hay sheets to stabilize a large load. In this manner the quantity of second-crop hay loaded on a cart was only slightly smaller than first-crop hay, as explained by the driver of the ox-drawn cart in this photo.[1244] Cevc, 1992, 86, photo by Stane Klemenc. © Stane Klemenc.

the Central Karawanks and the nearby Kamnik–Savinja Alps, all in southern Carinthia, as documented in the rent-roll of the seigneury of (Eisen)Kappel in 1567.[1245] This subsection provides an additional methodological explanation for Figure 57.

The rent-roll analyzed measures first- and second-crop hay in carts. Based on twentieth-century experience, it can be assumed that the quantity of second-crop hay transported by a full cart was quite comparable to a full cartload of first-crop hay, despite the longer length of first-crop hay, making it easier to load more of it on a cart (see Figure 58 and Figure 29 in Section 4.1.1.4). Namely, the upper part of the second-crop hay load was put on a cart in linen

[1244] Author's interview with Kunavec, 2023, in Stara Fužina. Gratitude is expressed to M. Kunavec for recommending this photo in which she is the driver, for her explanation concerning the load size, and for explaining that second-crop hay is being transported in this photo, an important piece of information that Cevc failed to mention.

[1245] AT StLA, Innerösterreichische Hofkammer, Sachabteilung, box 69, folder 8: Des ambts Capl bereuttung und beschreibung, 1567.

sheets, which stabilized the load (Figure 58).[1246] The use of hay sheets in dealing with second-crop hay is clearly recorded in (the vicinity of) the Western Karavanks in 1605,[1247] less than four decades after the 1567 rent-roll of the seigneury of (Eisen)Kappel was written. Thus, it is safe to assume that hay sheets were also used in the sixteenth century (i.e., the time of the data shown in Figure 57) to upload second-crop hay to a cart in more than one layer, fully using the cart's and draft livestock's capacity. The size of small carts was considered comparable throughout the region by the officials that produced the rent-roll that Figure 57 is based on.[1248]

Most of the holdings in Figure 57 were the only land held by the family making a living there. Nonetheless, a minor percentage of these holdings were held by a family that also had another holding (e.g., another isolated farm). The families in charge of some holdings recorded in Figure 57 held another additional land complex not recorded in this rent roll. Figure 57 only presents the contents of the 1567 rent-roll of the seigneury of (Eisen)Kappel.

The explanation for atypically low and high second-crop hay percentages in Figure 57 is provided in Section 6.3.0.1. The order of possession units in Figures 57 and 82 in Section 6.3 is the same.

Note that the data from sheets 14 to 16 of the 1567 rent-roll were copied on sheet 23, so that they erroneously appear twice in the archival source. Thus, they were only taken into account once in Figure 57.

6.1.2.4 Early mowing dates

Even though the majority of meadows were mowed late until the mid-twentieth century (Sections 6.1.2.1 and 6.1.2.2), a minor share of the total meadow area in the Alps (has) had a long history of early mowing.

The number of times grasslands were mowed in the valley bottoms, at lower elevations in particular, already increased between the sixteenth and nineteenth centuries in several cases, especially in the nineteenth century (Section 6.3), and it rose again starting in the (second half of) the twentieth century. However, important differences in the number of mowings were already characteristic around 1500, at least in some places. According to information published in 1839, many meadows in the Inn Valley (Germ. *Inntal*), Wipp Valley

1246 Author's interview with Kunavec, 2023, in Stara Fužina.

1247 In 1605, the second-crop hay yield not far from Koroška Bela, a village at the foot of the Western Karavanks, was measured in hay sheets (SI AS 721, Gospostvo Bled (The Seigneury of Bled), Iustitialia, vol. 13, August 21st, 1605).

1248 Namely, there is the explanation in the rent-roll that three small cartloads in the area equaled one considerable flatland farmer's cart (AT StLA, Innerösterreichische Hofkammer, Sachabteilung, box 69, folder 8: Des ambts Capl bereuttung und beschreibung, 1567, 44r).

(Germ. *Wipptal*, Ital. *Val Isarco*), Puster Valley (Germ. *Pustertal*, Ital. *Val Pusteria*), and Venosta Valley (Germ. *Vinschgau*, Ital. *Val Venosta*) in Tyrol were mowed twice annually (see also Sections 6.1.2.3 and 6.3.2). In some places in these valleys, the very best meadows were mowed three times in the 1830s, and in the mild climate of places near Merano (Germ. *Meran*), Lana, and Bolzano (Germ. *Bozen*) up to four times a year.[1249]

In the meadows mowed more than twice annually, the first mowing took place significantly earlier than usual in the Alps, but later than in the wetlands dealt with in Section 6.1.2.5.1.1. In the spatially very limited area of intensively fertilized or irrigated meadows that were mowed three times a year, hay was mowed as early as in late May even before the widespread changes in meadow management took place between the late eighteenth and mid-nineteenth centuries. Spring grazing (Section 6.2) was not practiced in such meadows. At that time, the total area of these highly productive meadows in Central Europe was usually small.[1250] Parallels can be found in even more intensive meadows in the plains of Lombardy (Ital. *Lombardia*),[1251] but also relatively close to the northern foot of the Alps. Mowing in May in the Upper Rhine Valley in the sixteenth century and the start of mowing at Cottens (at the foot of the Jura Mountains) in late May in the mid-eighteenth century[1252] suggests that this happened before the seeds of most grassland plants were ripe. Yields and the nutritive value of hay in such meadows were usually higher and biodiversity lower than in other grasslands in the surroundings.[1253] However, due to the extremely low percentage of such intensive meadows in a typical landscape of the Alps, this was not a relevant challenge, neither for livestock health nor for biodiversity.

One source of inspiration for early mowing must have been the handbook *De re rustica* (On Agriculture) by Lucius Junius Columella from the first century AD—that is, the most systematic and comprehensive agricultural text from Roman Antiquity—and later agricultural handbooks (see Section 6.5). Columella already recommended that mowing should take place before the seeds of grassland plants ripened to keep the nutritive value of the hay high. Columella's work was printed several times in the fifteenth and sixteenth centuries, and passages on grassland management in agricultural handbooks written in the sixteenth and seventeenth centuries, less frequently even in the eighteenth century, were often copied or adapted from *De re rustica*.[1254] It is proven that some Alpine landlords and most probably some manorial officials and priests

1249 Mathieu, 1996, 144, 155; Wopfner, 1997, 292–293, 295.
1250 Kapfer, 2010, 137–138.
1251 Mathieu, 1996, 145.
1252 Kauter, 2002, 106–107, 127.
1253 Dietl & Lehmann, 2004.
1254 Ash, 1993, xiii–xiv, xvii, xxiii–xxiv; Brown, 1976, 173, 175–177, 180; Kauter, 2002, 11–12, 102, 105–106; Schindler, 2010, 263, 265–266.

in the late sixteenth century were familiar with Columella's agricultural handbook.[1255] Consequently, Columella's and later advice on early mowing dates possibly had some minor influence in the Alps, but late mowing strongly prevailed.[1256] Namely, focusing on the economic perspective, earlier mowing was most reasonable at nutrient-rich sites, where the biomass production is high and the productive grass species abounding there could tolerate early mowing in the long run (compare Section 10.1.2). Thus, the very limited quantities of manure available for grasslands in most of the Alps before the nineteenth and twentieth centuries (see Section 6.3)[1257] kept the area of meadows, mowed early, low in most places.

A typical type of fertilized and frequently mowed small grassland patch was a grassy garden, attested at least since the sixteenth century.[1258] A description of agriculture in the remote cadastral community of Gorjuše on the Pokljuka Plateau in the Julian Alps (Sln. *Julijske Alpe*; Ital. *Alpi Giulie*; Friul. *Alps Juliis*) from the 1830s, for instance, documents manuring of grassy gardens, which was intensive for the standards of the time.[1259] Information from the early twentieth century reveals that, unlike in other meadows, children were allowed to play in such grassy gardens and to walk on the grass any time because trampling helped preserve the desired plant communities; one can also imagine the positive impacts of children's urine—apart from poultry excrement.[1260] Trampling by children contributed significantly to the suppression of unwanted umbellifers (*Apiaceae*), which in the absence of trampling became abundant in nutrient-rich grassy gardens.[1261] In grassy gardens in valleys and close to cottages in high summer mountain pastures, trampling-resistant grassland communities consisting of small-growing plants tolerating frequent mowing have grown or grew. These include, for instance, supina bluegrass (*Poa supina*), annual meadow-grass (*Poa annua*), perennial rye-grass (*Lolium perenne*), wild marigold (*Matricaria discoidea*), daisy (*Bellis perennis*), knotgrass (*Polygonum aviculare*), greater plantain (*Plantago major*), and white clover (*Trifolium repens*).[1262] In some places, taller plants such as grass meadow foxtail (*Alopecurus pratensis*) were also frequent.[1263]

1255 Zwitter, 2020, 55–57, 73–76; see also Schlude, 2009; Winiwarter, 2009.

1256 Kauter, 2002, 127.

1257 See also Wopfner, 1997, 292–293.

1258 Kauter, 2002, 133–135.

1259 SI AS 176, Franciscejski kataster za Kranjsko (Franciscan Land Cadaster for Carniola), L72, cadastral municipality of Gorjuše, cadastral appraisal protocols: s.n., 72/4, 47.

1260 Griesser, 1999, 41.

1261 Author's interview with Mencinger, 2018, in Stara Fužina. On umbellifers in intensively fertilized grasslands, see Sadar, 1956, 57, 109–111, 129, 138–139, 147; Dietl & Lehmann, 2004, 28.

1262 Sadar, 1956, 1, 66–67; Pils, 1994, 208, 220, 259–260; Machatschek, 2006, 134; Babai, Molnár, & Molnár, 2014, 76.

1263 Kauter, 2002, 135.

One reason for the long history of mowing in grassy gardens many times a year is the need for a suitable pasture for poultry close to the farm. Poultry pasture requires dense turf and small-growing grassland plants, and one way to attain this goal is to mow frequently at nutrient-rich sites.[1264] Another reason for mowing grassy gardens many times a year is the dietary requirements of pigs. The first cutting of grassland plants for pigs had to be carried out early because, unlike cattle, pigs cannot digest crude fiber. The protein content in such early and frequently cut grassland fodder was high.[1265] Dairy cows also received the biomass mowed in grassy gardens as nutritious green fodder.[1266]

Apart from the immediate vicinity of permanently settled farms, nutrient-rich grassy gardens were located in high summer mountain pastures;[1267] for example, at Ovčarija (about 1,700 m), Viševnik (1,610 m), Dedno Polje (1,600 m), Laz (1,550 m), and Pri Jezeru (1,450 m) in north-western Bohinj, to name but a few. There, according to mid-twentieth-century information, nutrient-rich grassland close to cottages was mowed for the first time in late June, usually prior to the arrival of livestock (compare Section 6.1.2.2).[1268] A prohibition on the creation of grassy gardens in summer mountain pastures in the area of Rettenberg (close to Innsbruck, the location of the castle of Rettenberg is marked in Figure 4) in 1501, proves that grassy gardens in summer mountain pastures had already been a known phenomenon by that time. Sporadic data on the history of such places before the nineteenth century include the 1726 permission to create a grassy garden in the Kesselboden Alp above Landl, a side valley of the Lower Inn Valley,[1269] and a century and a quarter later a land-use map from 1855 shows a grassy garden there.[1270] However, in the Lower Inn Valley, grassy gardens were still an exception in summer mountain pastures in the early nineteenth century.[1271]

Mowing nutrient-rich grassy gardens, like those around mountain cottages where plant nutrients accumulated due to the frequent livestock presence there, contributed to the ability of the site to produce high-quality (or higher-quality) hay. Namely, nutrients were regularly removed from the soil with the hay, which rendered the nutrient-rich grassland sites around cottages less suitable for highly nitrophilous flora, such as Alpine dock (*Rumex alpinus*; but see Sec-

1264 Vidrih, 2005, 47–48.

1265 Machatschek, 2006, 133–134; Vogl-Lukasser et al., 2006, 66.

1266 Wopfner, 1997, 295; Vogl-Lukasser et al., 2006, 66.

1267 Grass, 1948, 192, 208–213; Wopfner, 1997, 295; Machatschek, 2006, 133–134.

1268 Novak, 1978, 58; Hodnik, 2016, 8–9, 34, 51; author's interviews with Mencinger, 2017, with Kunavec, 2017, and with Gašperin, 2017, all in Stara Fužina, and with A. and M. Taler, 2017, in Studor.

1269 Grass, 1948, 208–210.

1270 *Arcanum Maps*, Cadastral Maps, Habsburg Empire – Cadastral Maps (XIX. [sic] Century), cadastral municipality of Thiersee, 1855.

1271 Grass, 1948, 211.

tion 8.2.2). Mowing had to be combined with further management practices to cope with nitrophilous flora in limited sections where grass was mowed close to the cottages. Nitrophilous "weeds" were uprooted (Section 6.6.1) and dung was removed from such highly nutrient-rich sites not only to be applied elsewhere (Section 6.3.2);[1272] in some places the dung was discarded on rocks according to mid-twentieth century data (when Alpine dock no longer played the role of a fodder plant at Ovčarija, where these data refer to—see Section 8.2.2) because urine and small quantities of dung that remained in situ already made the site highly nutrient rich.[1273]

Hay from grassy gardens in high mountain pastures was mainly used as fodder for calves (because they were only allowed to suckle four to five weeks to save more milk for people), for ill livestock, for pigs, and as hay for the workforce to sleep on. When there were hailstorms or summer snow, such hay could also be used to feed livestock in general, but its quantities were fairly limited, where mowed grass gardens existed at all in high summer pasture complexes. Nonetheless, some farmers even transported some grass-garden hay from high summer mountain pastures to the valley. For example, in the mid-twentieth century, some hay from the Laz mountain pasture complex (about 1,550 m) and the lower pasture and meadow complex at Grintojca (around 1,200 m) was transported to the village of Stara Fužina, located 1,000 m (Laz) or 600 to 700 m (Grintojca) lower than the site the hay originated from.[1274] In many places, transport of hay from high mountain pastures was prohibited, at least as a matter of decisions made by agricultural offices since the nineteenth century.[1275]

The following section deals with mowing in wetlands, another biome where the biomass was cut early in some cases.

6.1.2.5 Various dates of wetland mowing

Mowing in wetlands used to play a much more important role in history than nowadays. It used to be widespread.[1276] From the point of view of ani-

---

1272 Pils, 1994, 208, 220.

1273 Author's interview with Kunavec, 2017, in Stara Fužina; data refer to the Ovčarija mountain pasture complex; compare Ellenberg & Leuschner, 2010, 37.

1274 von Hörmann, 1877, 147; Novak, 1978, 58; Hodnik, 2016, 8–9, 34, 51; author's interviews with Mencinger, 2017, with Kunavec, 2017, and with Gašperin, 2017, all in Stara Fužina, and with A. and M. Taler, 2017, in Studor.

1275 Grass, 1948, 212–213.

1276 Melik, 1946, 173, 184, 187–190, 192–195, 200, 206–209, 221; Meilleur, 1986, 264, 266–267, 280, 286–287, 295; Pils, 1994, 112–162; Emanuelsson, 2009, 157, 277–278, 280–281, 285–286; Peterken, 2013, 14, 17–18, 76–80, 153, 155; Orland, 2016, 26; Zwitter, 2020, 64–66, 68–69; Grass, 1948, 167–169; Mlinar (ed.), 2018b, 99, documenting a *möſſl* of economic use near Fusine in Valromana (Friul. *Fusinis*, Germ. *Weißenfels*, Sln. *Bela Peč*) in a rent roll from 1636.

mal husbandry, wetlands had their merits, mainly in early spring (see Section 6.1.2.5.1.1) or as sources of animal bedding (see Section 6.1.2.5.2). Moreover, wetlands were grazed or the biomass was used as hay made in the summer, especially if the farm was not entitled to mow enough other grasslands providing better fodder (see Section 6.1.2.5.1.2). Another aim of mowing in wetlands was obtaining common reed stalks as building material (see Sections 6.1.2.5.2.1 and 8.2.3). Mowed wetlands were located not only near permanent settlements, but also at many remote sites, despite the limited hay quality. The meadow name *Mlakauishah* (literally, 'puddle(s)') in the Talež montane meadow complex on the Jelovica Plateau in the eastern Julian Alps (Sln. *Julijske Alpe*), recorded in 1620,[1277] is one of the many toponyms indicating wet conditions in remote meadows in historical sources.

Different types of wetlands were used in livestock farming in the Alps, ranging from productive marshes all the way to some historically farmed raised bogs. Marshes are defined here as wetlands on soils, which are predominately mineral, even though the fraction of soil organic matter is higher than outside wetlands. Fens are another wetland type; they include a shallow peat layer but groundwater, or surface water during flood events, is still a relevant source of water and nutrients for fen vegetation. The productivity of fens and marshes varies, ranging from nutrient-poor marshes and fens to somewhat more productive fens and highly productive marshes. Fodder quality in marshes and fens was usually middling to low from the perspective of livestock farming if fully grown plants were used (Section 6.1.2.5.1.2), whereas it was often high in spring (Sections 6.1.2.5.1.1 and 6.1.2.5.1.2).[1278]

Another wetland type is raised bogs, characterized by an extensive peat layer. They are extremely nutrient-poor and acidic wetlands. A bog is dome-shaped and raised above the surroundings. Plants there do not have access to groundwater because their roots stand in the peat, and they usually depend on extremely nutrient-poor precipitation. Even at nutrient-poor fen sites, nutrient availability is higher than in raised bogs. Consequently, if farmed, raised bogs produced low quantities of extremely low-quality livestock fodder that was sometimes grazed if accessible for livestock, or, less usually, the biomass was mowed there.[1279]

1277 SI AS 721, Gospostvo Bled (The Seigneury of Bled, Germ. *Veldes*), Iustitialia, vol. 12, June 10th, 1620.

1278 Dierssen & Dierssen, 2001; Pils, 1994, 112–152, 162; Melik, 1946, 187–188, 207–209; Emanuelsson, 2009, 157, 277–278, 281, 285–286; Peterken, 2013, 18; Sucholas et al., 2022.

1279 Melik, 1946, 173, 184, 187–190, 192–195, 200, 206–209, 221; Emanuelsson, 2009, 277, 281; Peterken, 2013, 18, 155; Sucholas et al., 2022. Compare the maps of land-use categories in the Ljubljana Marsh in the early nineteenth century and the depth of peat layer in Melik, 1946, 187–190, 192–195, 200, 206–209, 221, and the enclosed maps *Kulturno-gospodarsko stanje Barja v letu 1825* (Land-Use Map of the Ljubljana Marsh

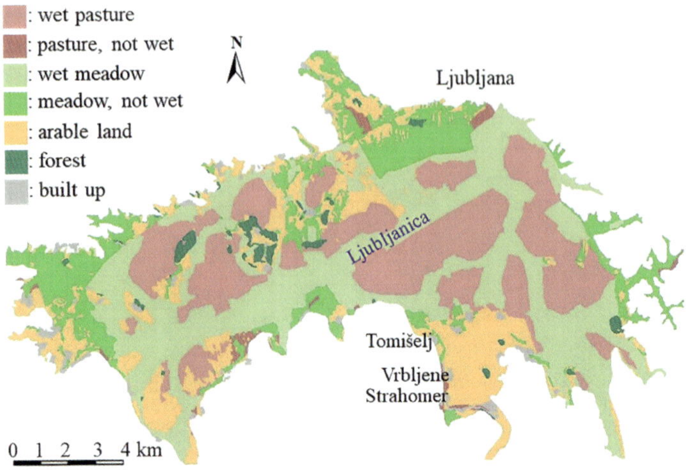

Figure 59: Land use in the Ljubljana Marsh and its near surroundings in 1825. Wet pastures were located in raised bogs, and wet meadows largely in fens.[1280] The toponym *Ljubljanica* is oversized to match the font size of other toponyms; it covers the river (some 40 to 50 m wide) and the adjacent wet meadows in fens. The 1415 dispute presented in the text most likely occurred in the southwest part of the location of the toponym of the Ljubljanica River on this map. Further east, the river flows through the fen to the north to reach the town of Ljubljana. Adapted from Šmid Hribar, 2016, 77. Map author: Mateja Šmid Hribar, © Anton Melik Geographical Institute.

Site conditions within the same wetland and between wetlands often differ greatly (see Figures 59 and 62), and so does the quality and quantity of biomass from the perspective of livestock farming.[1281] This is demonstrated by the following late-medieval example from the Ljubljana Marsh, the former southernmost wetland with large areas of raised bogs and fens in Europe (around 110 km^2 of peatlands),[1282] located just south of the southeasternmost Pre-Alps. To avoid misunderstandings, despite the usual English toponym Ljubljana Marsh, a landscape of bogs and fens used to cover the majority of this wetland, in line with the definition presented above. In 1415, a dispute about mowing rights was documented between the burghers from Ljubljana (Germ. *Laibach*), located north of the Ljubljana Marsh, and the inhabitants of three villages on the southern edge of this wetland, Tomišelj, Vrbljene, and Strahomer, which are located some 10 km straight-line distance from Ljubljana (Figure 59).[1283] As in

in 1825) and *Debelina šotne plasti na Ljubljanskem barju v letu 1881* (Map of the Depth of the Peat Layer in the Ljubljana Marsh in 1881); the peat depth in the Ljubljana Marsh in 1881, as shown on the oldest existing map, was no longer the maximum historical peat depth at several sites (Melik, 1946, 169–171; Zorn & Šmid Hribar, 2012, 52–54).

1280 Melik, 1946, 173, 184, 187–190, 192–195, 200, 206–209, 221.

1281 Pils, 1994, 112–162.

1282 Melik, 1946; Smrekar et al., 2014, especially pp. 18, 23–24; Zorn & Šmid Hribar, 2012, 46, 52.

1283 Kosi, 2014, 119.

so many wetlands, hay from the Ljubljana Marsh was usually fed to horses[1284] (see Section 6.1.2.5.1.2). Quite a large quantity of wetland hay from the Ljubljana Marsh was probably consumed by packhorses on the days when they were not active in transportation because there were extensive regional and long-distance traffic routes through Ljubljana (linking, beside other territories, the Republic of Venice with the Pannonian Basin and the easternmost Eastern Alps).[1285]

It would, however, be erroneous to conclude from the 1415 dispute that south of Ljubljana the wetland was mowed from the north to the south in the entire area between Ljubljana and Tomišelj, Vrbljene, and Strahomer because inhabitants of settlements from the north and the south of the wetland litigated over mowing rights. Namely, extensive poor pastures in raised bog, hazardous for livestock, documented there in the sixteenth century[1286] and until the nineteenth century,[1287] demonstrate that this was not the case. Hay was made mainly along the watercourses, in the fens in the Ljubljana Marsh,[1288] which was common in many wetlands consisting of both raised bogs and fens.[1289] Namely, there were significant differences in fodder quality and quantity within such wetlands. Along watercourses, where flowing water oxygenates the soil and deposits some nutrients during more frequent flood events, there were fens, providing significantly better hay and higher hay yields than the raised bog in the rest of the Ljubljana Marsh.[1290] Indeed, the boundary between the land farmed by the inhabitants of Ljubljana on the one hand and Vrbljene, Tomišelj, and Strahomer on the other was located close to the Ljubljanica River, near the contact zone between a raised bog and a fen at the later village of Lipe or further southwest (some 7 km straight-line distance from Ljubljana and 5 km from Vrbljene), at one of the more productive sections of the Ljubljana Marsh, as suggested by early nineteenth-century cadastral land-tenure evidence.[1291] In addition to more and at least slightly better hay than in a raised bog, the location close to the river allowed easy water transportation of hay

[1284] Melik, 1946, 200.

[1285] Kosi, 1998, 54, 56–58, 136–153, 206–224; see also Zwitter, 2014d, 614.

[1286] An example is grazing in a large raised bog north of Matena, where livestock were perishing (Zwitter, 2016, 154). There used to be a raised bog at that time: Smrekar et al., 2014, 24.

[1287] Melik, 1946, 187–188, 207–209, and the enclosed map *Kulturno-gospodarsko stanje Barja v letu 1825* (Land-Use Map of the Ljubljana Marsh in 1825).

[1288] Melik, 1946, 173, 188–190, 192–195, 200, 206, 209, 221, and the enclosed maps *Kulturno-gospodarsko stanje Barja v letu 1825* and *Nekdanja vodna mreža na Barju*; see also the map in Smrekar et al., 2014, 71 (and p. 72).

[1289] Emanuelsson, 2009, 281.

[1290] Melik, 1946, 173, 184, 188–190, 192–195, 200, 206, 209, 221; Emanuelsson, 2009, 281; Zorn & Šmid Hribar, 2012, 57.

[1291] Melik, 1946, 170, 177, and the enclosed maps *Posestno razmerje na Barju v letu 1825* (1825 Land Tenure Map of the Ljubljana Marsh) and *Debelina šotne plasti na Ljubljanskem barju v letu 1881* (Map of the Depth of the Peat Layer in Ljubljana Marsh in 1881;

across significant distance.[1292] Thus, settlements on the northern and southern edge litigated over mowing rights at relatively favorable sites near the Ljubljanica River, whereas large sections of the wetland south of Ljubljana were used as poorly productive and hazardous pastures (Figure 59).

Mowing was even carried out at some places where the water level required thorough work adaptation. For instance, eighteenth-century data explain that the wetland in shallower parts of Lake Gösselsdorf (Germ. *Gösselsdorfer See*; Sln. *Goslinsko jezero*; see Figure 61) in the Jaun Valley (Germ. *Jauntal*, Sln. *Podjuna*), Carinthia, was mowed up to the deep water ("bis an des[!] tieffen see"), so that hay was stored from water ("aus dem waßer bringen");[1293] fish and crawfish lived on these parcels.[1294] At least some haymakers in Lake Gösselsdorf wore boots, as already recorded in June 1639.[1295] Still in the Klagenfurt Basin, hay made in the wetland west of Lake Sablatnig/Turner (Germ. *Sablatnigsee,* later *Turnersee*, Sln. *Zablaško jezero*) in July 1679 was transported from the wetland by boats,[1296] not by carts. Meadows named *Pod vodo*[1297] (literally, 'under water, below water') and *Pod loko v mlakach*[1298] (literally, 'in pools or wet meadows below a (wet) meadow') were recorded in the pre-Alpine Posavsko Hills (Sln. *Posavsko hribovje*, now in eastern central Slovenia) at the foot of *Zasavska Sveta Gora* or in its vicinity in the seventeenth and eighteenth centuries. Mowing in more than knee-deep water was carried out in wetlands in mountain grasslands of the Villandro Alp (Ital. *Alpe di Villandro*, Germ. *Villanderer Alm*) in Trentino–Alto Adige (South Tyrol), as documented in the twentieth century.[1299] The data presented suggest that mowing from boats or standing in relatively deep water, where the hazard of getting stuck in the mud was not serious, was no exception, in flatter valleys and basins in the Alps in particular. This is not surprising because wetlands and aquatic environments were often much more accessible to a haymaker mowing by scythe than they are nowadays to farm mechanization,[1300] and the fodder quantity was often

note that the peat was covered by fluvial deposits along the main watercourses; Melik, 1946, 173, 194).

1292 Melik, 1946, 192–193, 205–207, 209, 212–213.

1293 AT StiAStP, Grundherrschaft Eberndorf (Handwritings), vol. 192, Gerichtsprotokoll 1728–1742, 79 (1730).

1294 AT StiAStP, Grundherrschaft Eberndorf (Handwritings), vol. 181, Gerichtsprotokoll (from the 1660s that begins in 1661), 40–41 (June 17th, 1645).

1295 Zwitter, 2014e, 342, 344.

1296 AT StiAStP, Grundherrschaft Eberndorf (Handwritings), vol. 183, Gerichtsprotokoll 1675–1683, 195–197 (August 1st, 1679).

1297 SI AS 173, Imenjska knjiga za Kranjsko (*Gültbuch* for Carniola), *Gültbuch* no. 7 (1662–1757), sheets 35, 38.

1298 SI AS 173, Imenjska knjiga za Kranjsko (*Gültbuch* for Carniola), *Gültbuch* no. 7 (1662–1757), sheets 38, 125; the translation of toponyms is based on vernacular terminology of grassland agroecosystems in Sadar, 1956, 1.

1299 See the photo in Hinterwaldner, 2002, 174, 176.

1300 Sucholas et al., 2022, 14.

insufficient or barely sufficient (Section 6.2.2.1). Similarly, mowing sedges in standing water and floating off the hay was recorded in Iceland, for instance.[1301] Mowing on frozen wetlands in winter to obtain animal bedding or building material (Section 6.1.2.5.2) made possible economic use of wetland sites that were otherwise inaccessible or barely reachable.

In the last millennium, wetland areas faced local increases and decreases, followed by major drainage work. Expansion of wetland areas was often a result of interactions between nature and humans. Climate and weather[1302] and the flood events and mass movements they helped trigger[1303] belonged to the principal natural and hybrid socio-natural driving forces relevant in this case. Among human factors contributing to more humid or wetter sites, building numerous dams is a type of intervention in the landscape that resulted in many local expansions of wetlands in the Alps.[1304] Deforestation to establish new agricultural land, causing a local rise in groundwater level, was another anthropogenic driving force relevant to a certain degree; it contributed to the expansion of wetlands. Deforestation-related increased erosion on slopes could result in more fluvial sediment transport and accumulation in the valley in some cases, which also led to the growth of wetlands adjacent to rivers.[1305] Expansion of wetlands as a result of interactions between several driving forces is demonstrated by an early example from the Upper Pinzgau region, on the Upper Salzach River. Sixteenth- and seventeenth-century records prove that the area of wetlands there had increased by that time. Consequently, some arable fields had to be converted into permanent grasslands, and the quality of hay in meadows declined.[1306]

The reverse process of wetland drainage is in principle an ancient practice, but before "the era of fossil-fuel powered earth-moving equipment," drainage and maintenance of the related infrastructure were highly labor intensive. Consequently, before fossil fuel–based technology, draining "gathered pace chiefly where and when plentiful labor could be focused by the coercive power of lords and states, and where land hunger pushed young men to desperation."[1307] Medieval, sixteenth-century, and seventeenth-century records on lo-

[1301] Peterken, 2013, 274.

[1302] Pfister & Brändli, 1999.

[1303] E.g., Wopfner, 1997, 73–74.

[1304] E.g., a very local example from Bohinjska Bistrica in western Bohinj: in 1607 a plaintiff claimed that a parcel along Bistrica Creek became wet and consequently, as the plaintiff also argued, it decreased in quality because the defendants had constructed or expanded a dam (SI AS 721, Gospostvo Bled [The Seigneury of Bled], Iustitialia, vol. 13, [July 2nd (?)] 1607). Even though the reliability of the argument is unclear in this case, the court minutes record a process that was occurring in many places.

[1305] Pfister & Brändli, 1999; Winiwarter & Sonnlechner, 2001, 14; Poschlod, 2015, 83; Haidvogl, Pont, & Zwitter, 2019, 41.

[1306] Rohr, 2007a, 320, 349–351, 365–367.

[1307] McNeill, 2010, 363 (the quotes are from there); see also Troßbach, 2008, especially columns 586–587, 593; Emanuelsson, 2009, 157; Peterken, 2013, 170–171, 191, 268.

cally relevant small-scale drainage usually organized by landlords to make the land more suitable for farming reflect a process that was nothing exceptional in Central Europe.[1308] Early examples from the Alps include the drainage carried out in the Upper Inn Valley at Prutz in 1476 and in the bottom of the valley along the middle course of the Adige River (Germ. *Etsch*) in the fifteenth century.[1309] At Piskertschach (Sln. *Piskrče*) in the Jaun Valley (Germ. *Jauntal*, Sln. *Podjuna*) of Carinthia, digging drainage ditches was ordered in the second decade of the seventeenth century in the wetland covering part of the common pasture due to springs there.[1310] Such early drainage projects in the Alps were not limited to valley bottoms. For instance, gradual drainage of wetlands in high mountain pastures in the Villandro (Germ. *Villanders*) area in South Tyrol was ordered in the mid-seventeenth century; many wetlands on impermeable rocks and moraines (have) existed there.[1311] A few larger drainage projects were also carried out early. In the Upper Pinzgau region, three sections of the Salzach were straightened in the late sixteenth century, and in one case work probably continued until the early seventeenth century. Each of these sections (at Bramberg, west of Hollersbach, and east of Mittersill) is several kilometers long. The new, straighter sections of the Salzach were presumably only about 15 to 20 m and about 30 m wide (in the latter case, 43 m together with the banks), and within the new straighter sections the length of the riverbed was significantly reduced (e.g., by 15%, and in one case allegedly even by half).[1312] Hydrological conditions in many former wetlands near the river changed, and so did the grassland species composition.[1313] However, in the aforementioned centuries, drainage was not a general phenomenon in wetlands in the Alps, and "the great majority of European wetlands remained unaffected" at that time.[1314]

Swamp drainage fever started in Central Europe from the mid-eighteenth century onward. It involved large-scale and small-scale drainage of wetlands in the Alps between the eighteenth and especially nineteenth century and the third quarter of the twentieth century. The majority of the wetland areas in the Alps and elsewhere in Central Europe were drained at that time.[1315] Apart from projects in which wetland drainage was the main goal, drainage was also

[1308] McNeill, 2010, 364.
[1309] Wopfner, 1997, 234, 299.
[1310] Zwitter, 2014e, 334; an example of traditional small-scale drainage: Babai, Molnár, & Molnár, 2014, 104.
[1311] Hinterwaldner, 2002, 24–25, 49.
[1312] Rohr, 2007a, 318, 350–353, 365–367.
[1313] Compare Peterken, 2013, 194.
[1314] McNeill, 2010, 364 (quote); see also Rohr, 2007a, 350, 366–367.
[1315] Pils, 1994, 121–122, 151, 157, 165; Wopfner, 1997, 221, 234, 287; Emanuelsson, 2009, 304–310; Peterken, 2013, 262; Poschlod, 2015, 114, 126, 135–143, 146–149, 154, 170–172, 176–177, 209, 223, 230–231; Bätzing, 2015, 88, 94, 96–98, 271; Schmid, 2016; McNeill, 2010, 364; Wöbse, 2022, 29, 31–32, 36, 39; Kuret (ed.), 1989, 263, 271.

one of the consequences of river regulation in the floodplains in broader valleys and basins in the Alps, causing a decrease in groundwater level.[1316] Numerous plant and animal species occurring at wet sites have become endangered or locally extinct due to drainage.[1317] However, agricultural land providing human food or better fodder increased, and this happened in times of quick population growth in the Alps (see Section 3.1)[1318] and in Europe in general.[1319] The risk of malaria, which occurred as far north as Northern Europe, and thus also in the Alps, and some other wetland-related human and animal diseases (see, e.g., Sections 6.1.2.5.1.1, 6.1.2.5.1.2, and 6.4.1 for liver fluke) also decreased.[1320]

6.1.2.5.1 Wetland hay as fodder

6.1.2.5.1.1 Spring mowing and grazing in wetlands

In the literature on grassland history, the importance of wetlands for animal husbandry is usually presented overly pessimistically; they are often reduced to low-quality pastures,[1321] sources of animal bedding,[1322] and sources of low-quality hay. This section presents a specific type of historical wetland use that often provided good fodder.

A major historical advantage of many wetlands as fodder resources was that water temperatures in late winter and early spring are higher than air temperatures. This caused an early onset of the vegetation period in the critical time in early spring when many farmers were often running out of hay (see Figures 60 and 64, and Section 6.2.2.1).[1323] Section 6.1.2.5.1.1 demonstrates the attractiveness of early spring grazing in wetlands based on examples of monocots from Lake Gösselsdorf, comparative examples from what is now Trentino–Alto Adige (South Tyrol) and Switzerland, and selected wetland and aquatic dicots in the Julian Alps, Karawanks (Germ. *Karawanken*, Sln. *Karavanke*), and the Allues (Francoprovençal *Alués*) Valley (French *Les Allues*).

[1316] Mathieu, 1996, 148, 155; Wopfner, 1997, 221, 234, 287; Kauter, 2002, 171; Emanuelsson, 2009, 305, 308–309; Poschlod, 2015, 131, 134–135; Haidvogl, Pont, & Zwitter, 2019, 41–42.

[1317] Pils, 1994, 121–122, 156–157, 161; Aeschimann et al., 2004, vol. 2, 14–15, 1038–1039, 1086–1087, 1096–1097, see also the legend in Aeschimann et al., 2004, vol. 1, inside front cover; Hable & Präsent, 1985, 401; Poptcheva et al., 2009, 108; Peterken, 2013, 209–210, 215, 262–263, 270; Poschlod, 2015, 126–131, 134–143, 146–149; Jaksetič (interview with Kaligarič), 2019, 22; Wöbse, 2022, 31, 34.

[1318] Haidvogl, Pont, & Zwitter, 2019, 41.

[1319] Livi Bacci, 2000; see also McNeill, 2010, 364.

[1320] McNeill, 2010, 364; Haidvogl, Pont, & Zwitter, 2019, 41.

[1321] Kapfer, 2010, 137.

[1322] For example, Peterken, 2013, 17.

[1323] Peterken, 2013, 179–180; Leibundgut & Vonderstrass, 2016a, 125.

Figure 60: The vegetation period of wetland plants starts when snow still covers the surroundings. Note the green plant tissue. Bohinj, March 17th, 2018, photo: Žiga Zwitter.

Although the vegetation period in many wetlands starts early, the growth dynamics and attractiveness of fodder depend on the nutrient status of the wetland and on water temperature. Growth is quicker in nutrient-rich wetlands and slow in nutrient-poor ones, comparable to the difference between moderately fertilized and non-fertilized grasslands in general (see Sections 6.2.2.1 and 6.2.2.2). Concerning water temperature, vegetation development is slow in wetlands that receive water from local cold springs and quicker where they are supplied by warmer water. Thus, where nutrient-poor wetlands are fed by cold water, growth dynamics are very slow. Very low nutrient status or cold water feeding the wetland all the year round favor the dominance of mosses,[1324] which are unattractive to livestock,[1325] and so they were not grazed in early spring. Consequently, only wetlands fed by water that was not too cold and characterized by medium nutrient availability, such as common reed (*Phragmites australis*) sites, or by high nutrient availability, such as butterbur (*Petasites hybridus*) sites,[1326] provided significant fodder quantities for livestock in early spring. Common reed is among the most quickly growing grassland

[1324] Pils, 1994, 112–114, 117–118, 122, 126.
[1325] Author's interview with L. and M. Odar, 2017, in Studor.
[1326] Aeschimann et al., 2004, vol. 2, 524–525, 992–993; Pils, 1994, 112, 239.

and wetland plant species, especially where the nutrient availability is high and the site is wet.[1327]

Occasional floods in some wetlands in winter or early spring, leaving behind muddy deposits on plants, rendered early spring mowing or grazing impossible.[1328]

The hazard of livestock being infected with common liver fluke (*Fasciola hepatica*) was possibly significantly lower during early spring mowing or early spring grazing in wetlands than later. This hazard likely increased during May. Nonetheless, even in late May, the risk of infection might have been lower than in late summer and early autumn, at several sites at least.[1329] If previously infected animals graze in wetlands in early spring, their excrement increases the later liver fluke hazard in the case of grazing in the same wetland. Common liver fluke is a serious and widespread challenge for sheep, cattle, and pigs grazing in wetlands. This parasite is often fatal in sheep, whereas it causes suffering and economic losses in cattle farming without usually being fatal. Less frequently, it causes problems in horses and donkeys.[1330]

The first case study, representative of many productive wetlands mowed or grazed in spring, focuses on the extensive siltation zone around Lake Gösselsdorf (Figures 61 and 62). A combination of a wetland and a lake is clearly proven since the seventeenth century. According to information from 1606, certain sections of the bottom of that lake were not covered by water for a considerable part of the year.[1331] Similarly, this lake was designated as *mosßsee* (*Gösselstarffer mosßsee*) in 1684[1332] and 1694.[1333] Extensive temporarily flooded wetlands at Lake Gösselsdorf were not a novelty at that time because the lake formed behind a very low barrier from fluvial deposits, and so the shallow lake has a wide siltation zone with various vegetation, starting with submerged or floating-leaved plants and up to reeds and wet meadows.[1334] Nonetheless, by the seventeenth century, this lake was a complex socio-natural site. Information from 1613 reveals that the regular temporary reduction of the area of the lake was a combined result of natural and anthropogenic factors because Lake Gösselsdorf also functioned as a mill pond and was an integral part of carp

1327 Pils, 1994, 225–226, 239.

1328 Pfister, 1985a, 48–49, 83.

1329 Novobilský et al., 2014. Gratitude is expressed to the agronomists Martin Škrlep and Jože Verbič for recommending this article.

1330 Ballweber, 2024; Tacer, no date.

1331 Zwitter, 2014e, 342.

1332 SI NŠAL 100, Kapiteljski arhiv Ljubljana (Archives of the Ljubljana Cathedral Chapter), fasc. 103, folder 2: account of the seigneury of Eberndorf (Sln. *Dobrla Vas*), *Einkhommen auß füschwässern.*

1333 AT StiAStP, Grundherrschaft Eberndorf (Handwritings), vol. 186, Gerichtsprotokoll 1694–1698, 28–29 (March 17th, 1694).

1334 Melik, 1954, 502.

Figure 61: Lake Gösselsdorf in southern Carinthia and extensive wetlands (note the light brown color of dry wetland monocots) south and north of it in the early twenty-first century. The area covered by largely treeless wetlands still resembles the one in the era of wetland farming in the mid-twentieth century, one of the main exceptions being spontaneous afforestation in the northeast.[1335] Contour lines are marked in dark brown, the contour interval is 10 m, and the lake is located just below 470 m. In the northwest of the lake, outflow to the north-northwest is seen as a curved line crossing the wetland; this is the only surface outflow from this lake. Adapted from *KAGIS Maps*, Basiskarten, Hintergrund (Basemap), Ortho 2019–2021 & KAGIS Kartendienste, Geländehöhe, Höheninformation, Höhenschichten (10 m).

1335 Compare the digital orthophoto from 1952–1953 in *KAGIS Maps*, Basiskarten, Hintergrund (Basemap), Ortho 1952–1953.

breeding. Consequently, the sluice dividing the lake from the stream flowing from the lake to the north affected not only the water level in the lake, but also the area temporarily covered by water.[1336] As a consequence of the interplay between natural and human factors, wetland plants were mowed at Lake Gösselsdorf by poor locals on March 10th, 1694,[1337] in March 1714,[1338] and on April 11th, 1676.[1339] The plants mowed there are designated in these sources as *Sacher*—which can refer to various wetland plants, such as reeds, sedges, wood club-rush, and so on. Although the word *Sacher* can have a general meaning independent of the wetland plant development phase (also in Carinthia),[1340] a description of the meaning of the word *Sacher* in Matthias Lexer's nineteenth-century dictionary of Carinthian German points precisely to such early mowing of wetland or aquatic plants; namely, *Sacher* is described as the leaves of some wetland plants before stem formation.[1341] Because there were many wetlands in the bottoms of valleys and basins in Carinthia,[1342] this plant development phase was so economically relevant that a special meaning of the word *Sacher* evolved. A parallel with grazing in the wetland at the shore of the same lake is relevant for understanding such early wetland mowing. Based on personal observations, a farmer born before the mid-twentieth century at the isolated Sapotnik (Sln. *Zapotnik*) farm on the shore of Lake Gösselsdorf, who used to tend cattle (including milking cows) grazing in this wetland in spring in her youth, recalled that cattle liked soft young common reed (*Phragmites australis*) very much, until it grew about a meter high; if the reed was younger, it was appreciated even more. She used to keep cattle in this wetland in May. Her traditional ecological knowledge includes the crucial element of the high regeneration ability of common reed in spring; the farmer noticed that cattle consumed accessible organs of young common reed, but this plant regenerated, providing attractive young cattle fodder more than once per spring.[1343] However, the number of grazing events had to be low if the goal was to keep the reeds (see below). This kind of traditional ecological knowledge must have a long history there because the same farm claimed to have grazing rights in the wetland at the same lake in mid-seventeenth century, and by that time

[1336] Zwitter, 2014e, 343–344, 347.
[1337] AT StiAStP, Grundherrschaft Eberndorf (Handwritings), vol. 186, Gerichtsprotokoll 1694–1698, 28–29 (March 17th, 1694).
[1338] AT StiAStP, Grundherrschaft Eberndorf (Handwritings), vol. 189, Gerichtsprotokoll 1711–1717, 145–147 (April 1714).
[1339] AT StiAStP, Grundherrschaft Eberndorf (Handwritings), vol. 183, Gerichtsprotokoll 1675–1683, 29 (April 22nd, 1676).
[1340] Zwitter, 2020, 65.
[1341] Lexer, 1862, column 211; Lexer, *Mittelhochdeutsches Handwörterbuch*: Saher.
[1342] *Arcanum Maps*, Cadastral Maps, Habsburg Empire – Cadastral Maps (XIX. [sic] Century).
[1343] Author's interview with Plesnik, 2021, at Sapotnik (Sln. *Zapotnik*) farm on the shore of Lake Gösselsdorf.

the farmer had already regarded the grazing rights there as an old custom.[1344] A parallel with traditional ecological knowledge about young common reed from the Biebrza Valley in Poland is telling: "[I]n spring cattle do not let it grow because [they] ea[t] everything immediately."[1345] *Phragmites australis* could not, however, tolerate being consumed many times per vegetation season; where this was the case, it disappeared. The impacts of trampling by heavy livestock on sensitive rhizomes are also critical and can suppress the regeneration of common reed if trampling lasts long, and so the grazing time had to be short.[1346] In general, these impacts strengthened with the increase in livestock weight (Section 5). Traditional ecological knowledge of farmers from the shore of Lake Gösselsdorf and the Biebrza Valley corresponds with the results of an analysis announced by the Ministry of Agriculture in 1915, when hay was scarce due to drought and war (see Section 7.2): common reed, if only cut as young as possible, can be regarded as a suitable source of fodder that cattle and horses like fresh and dry. Its sugar content, for instance, is considerable.[1347] This explains the traditional ecological knowledge about young common reed in the Biebrza Valley in Poland: "it is sweet like corn" and cattle like to eat it "because it is sweet."[1348] In contrast, cattle do not eat fully grown common reed.[1349] This is a consequence of the fact that reeds and rushes usually have an internal nutrient cycle, so that sugar is stored in underground parts (the rhizome and stem base) in the winter and pumped into the growing parts of leaves and stems in the spring. Consequently, even *Juncus* leaves are attractive to livestock at that time. Only later on is sugar again stored in underground parts and the upper stems become tough and less attractive.[1350]

The previous paragraph made it clear that there existed some farmed reed beds in the history of the Alps, although in general common reed is much more frequent in the landscape nowadays than it used to be because it has spread in many abandoned wet meadows void of reed as long as they were farmed, and the increased eutrophication of wetlands and fresh waters in general is also favorable for *Phragmites australis*.[1351]

1344 AT StiAStP, Grundherrschaft Eberndorf (Handschriften), vol. 4, Grundbeschreibung 1664/65, 340.

1345 Sucholas et al., 2022, 19 (quote), 36.

1346 Pils, 1994, 223, 239; Davidson et al., 2017; Duncan, 2019; Sucholas et al., 2022, 19.

1347 SI AS 137, Okrajno glavarstvo Radovljica (District Administrative Authority [Germ. *Bezirkshauptmannschaft*] of Radovljica, Germ. *Radmannsdorf*), box 592, folder *Preskrba z živili, prekupčevanje, nastavitev zaupnikov za nadzor zalog, potrebe*: August 28th, 1915, no. 22751. The botanical name *Arundo phragmites* has been updated according to Aeschimann et al., 2004, vol. 2, 992.

1348 Sucholas et al., 2022, 19.

1349 Sucholas et al., 2022, 36.

1350 A special case is the inflorescences of wood club-rush (*Scirpus sylvaticus*), which cattle like to graze on, probably due to their high nutritive value (observation by Leonid Rasran).

1351 Pils, 1994, 120, 133, 136, 223, 236, 239; Sucholas et al., 2022, 2–3, 19, 36, 38.

Figure 62: Various habitats near the shore of the wetland at Lake Gösselsdorf, at many places dominated by tall sedges (*Carex spp.*) or by common reed (*Phragmites australis*) in an environment with a long history of significant human intervention into ecosystems. Apart from the long history of human influences on water level (see Section 6.4.3.2.2), mowing to obtain fodder and animal bedding, and grazing (see also Sections 6.1.2.5.1.2 and 6.1.2.5.2), the impact of fire on habitats of common reed was also significant. From 1902 to 1971, a narrow-gauge steam railway connection crossed this wetland.[1352] Steam locomotives started fires in dry common reed in many years between autumn and spring. Moreover, boys played with fire in dry common reed. According to traditional ecological knowledge, burning made mowing easier because the tussocks burned down, and so the surface became more even, and the vegetation period of wetland monocots was observed to start significantly earlier in the spring due to these fires.[1353] The reasons are a cooler and wetter microclimate below slowly decomposing reed remains,[1354] and so the site became warmer after being burned, and the nutrient release from ash also stimulated reed growth. Abandonment of historical wetland farming was another human factor causing changes in these ecosystems. Photo: Žiga Zwitter, August 25th, 2020.

Even several sedge species (*Carex* spp., commonly summed up as tall sedges, such as *Carex elata* and *Carex acuta*), avoided by cattle when full grown (see Section 6.1.2.5.2), were alternately consumed when young by cattle and not only by horses. Similarly, reed sweet-grass (*Glyceria maxima*) is consumed by cattle and other herbivorous livestock as a delicacy when young, but avoided by cattle when old and hard.[1355]

It is not clear how typical early spring mowing was at Lake Gösselsdorf and which wetland plants were mowed there in spring, when this practice was

[1352] Piko, 1996, 303, 329; Siedler, 1992, 144–147.
[1353] Piko, 1996, 303, 329; author's interview with Plesnik, 2021, at Sapotnik farm on the shore of Lake Gösselsdorf.
[1354] Pils, 1994, 226.
[1355] Sucholas et al., 2022, 18, 23, 36.

recorded in the seventeenth and eighteenth centuries. According to a court settlement from 1645 and later data, some farmers had mowing and grazing rights in the wetland near the lake,[1356] and so apart from later plant uses they most likely mowed or grazed their livestock there in early spring. In the few well-documented cases from 1676, 1694, and 1714, early spring mowing was carried out illegally, although in significant quantities; for instance, a cart of *Sacher* was stored per farm involved in 1676 (one farm) and 1694 (two farms), and seven or eight farms from nearby Gösselsdorf (Sln. *Goselna Vas*) and Sittersdorf (Sln. *Žitara Vas*) participated in mowing in March 1714.[1357]

Similar uses of wetland plants were characteristic of several wetlands; for example, along rivers and lakes at the bottom of broader valleys and basins before the channelization of rivers and drainage of such wetlands (see the introduction to Section 6.1.2.5). From autumn to spring, wetlands along the middle Adige (Germ. *Etsch*) River offered grazing grounds for sheep belonging not only to farmers from the bottom of the Adige Valley, but also to farms located at high elevations and in side valleys, such as the Senales Valley (Ital. *Val Senales*, Germ. *Schnalstal*), Passiria Valley (Ital. *Val Passiria*, Germ. *Passeiertal*), Sarentino Valley (Ital. *Val Sarentino*, Germ. *Sarntal*), Fiemme Valley (Ital. *Val di Fiemme*, Germ. *Fleimstal*), Fassa Valley (Germ. *Fassatal*, Ladin. *Fascia*, Ital. *Val di Fassa*,), and Non Valley (Ital. *Val di Non*, Germ. *Nonstal*). For example, farms in Fondo and Malosco in the upper Non Valley had grazing rights for their sheep in wetlands along the section of the Adige River between Vadena (Germ. *Pfatten*) and Termeno (Germ. *Tramin*) from mid-March until April 24th according to a charter from 1185 (Gregorian date equivalents in terms

[1356] AT StiAStP, Grundherrschaft Eberndorf (Handwritings), vol. 179, Gerichtsprotokoll 1653–1655, 101 (February 12th, 1663), 102 (March 5th & April 20th, 1663); AT StiAStP, Grundherrschaft Eberndorf (Handwritings), vol. 181, Gerichtsprotokoll 1661–1668, 39–40 (December 1st, 1662), 40–41 (June 17th, 1645) and 110–111 (September 8th, 1665); mowing rights are also clearly proven in 1713 (AT StiAStP, Grundherrschaft Eberndorf (Handwritings), vol. 189, Gerichtsprotokoll 1711–1717, 128–129 (1713)), grazing rights in the wetland of Lake Gösselsdorf are also mentioned in the mid-seventeenth century and later (AT StiAStP, Grundherrschaft Eberndorf (Handschriften), vol. 4, Grundbeschreibung 1664/65, 339–340; Zwitter, 2014d, 343; AT StiAStP, Grundherrschaft Eberndorf (Handwritings), vol. 192, Gerichtsprotokoll 1728–1742, 79, 86–87 (1730), 338.

[1357] AT StiAStP, Grundherrschaft Eberndorf (Handwritings), vol. 183, Gerichtsprotokoll 1675–1683, 29 (April 22nd, 1676); AT StiAStP, Grundherrschaft Eberndorf (Handwritings), vol. 186, Gerichtsprotokoll 1694–1698, 28–29 (March 17th, 1694); AT StiAStP, Grundherrschaft Eberndorf (Handwritings), vol. 189, Gerichtsprotokoll 1711–1717, 145–147 (April 1714); localization of the fragmentarily preserved data from March 1714 (these were subjects of the seigneury of Eberndorf from a certain "---storff"; i.e., "---sdorf"; most of the sheet where this entry begins was torn out) is based on the list of all subjects of the seigneury of Eberndorf from 1664/65; its subjects "Lackher" and "Petschnigg," who mowed *Sacher* in 1714, lived in Sittersdorf (Sln. *Žitara Vas*): AT StiAStP, Grundherrschaft Eberndorf (Handschriften), vol. 4, Grundbeschreibung 1664/65, Index iuxta ordinem alphabeti.

of vegetation development: late March–May 1st). Sheep from the Fassa Valley, for instance, walked for about ten hours to reach these pastures (see Figure 4). Several examples of such grazing rights can be proven in the thirteenth century, but winter grazing in these wetlands must have started even much earlier in history. Wetland grazing grounds decreased there significantly in the eighteenth century, when extensive wetland sections were drained; nonetheless, this type of wetland grazing persisted until the twentieth century in some parts of the bottom of the middle Adige Valley. Whereas wetland grazing in early spring was a regular practice in many places, some communities only had the right to exceptional spring grazing in wetlands in the valley bottom if snow made usual grazing elsewhere impossible. Thus, according to sixteenth-century data, in the case of snow, the communities of Aldrans and Rinn on the slopes above Innsbruck had the right to keep their livestock in a wetland section in the valley bottom close to Amras two weeks after April 24th.[1358]

A similar mid-eighteenth-century example refers to the parish of Kerzers west of Bern, some 40 km from the Alps but following the same principle of early spring wetland grazing. Cows whose lactation stopped in winter due to the lack of appropriate fodder started grazing in the extensive wetland in March, and soon they visibly gained weight and provided milk again. Oxen were even more seriously starved in wintertime because more hay was allotted to dairy livestock, and so oxen also gained a lot of weight during a month of wetland grazing.[1359]

The historical relevance of some wetland dicots in early spring is presented with a case study of butterbur (*Petasites*), followed by a more specific example of aconite-leaved buttercup (*Ranunculus aconitifolius*). Butterbur (*Petasites hybridus*) and white butterbur (*Petasites albus*) especially grow in nutrient-rich gumbo soils (i.e., clay dominates the soil texture) on both carbonate and silicate rocks in the colline and montane vegetation belts.[1360] The parallel of a mid-eighteenth-century list of plants in irrigated meadows at Memmingen (west of Munich) before major changes in grassland management there suggests that butterbur (*Petasites hybridus*) used to be frequent; for example, in the immediate vicinity of irrigation ditches, where soils were nutrient rich due to irrigation.[1361] These must have often belonged to butterbur sites (see Section 6.4.4). This is an example of a plant genus that used to be highly relevant for cattle fodder, but its perception by livestock farmers changed thoroughly. In the Middle Ages, some settlements in the Alps were probably named after butterbur, possibly due to its relevance in the livestock diet or simply for its ob-

[1358] Grass, 1948, 108–110; Mathieu, 1996, 148, 155; Wopfner, 1997, 230, 234, 277–278, 287; Bätzing, 2015, 88, 94, 96–98, 271.
[1359] Pfister, 1985a, 38–39.
[1360] Aeschimann et al., 2004, vol. 2, 524–525.
[1361] Kauter, 2002, 141–142, 154.

Figure 63: White butterbur (*Petasites albus*) abounds just below the meadow in the Karawanks north of Jesenice in the right rear part of the photo, which was named after *Petasites* as already recorded in 1609. Note that the grassland plants in the meadow would not yet make cattle grazing possible, whereas white butterbur has already started blooming and its leaves could easily be consumed by grazing livestock. Photo: Žiga Zwitter, May 1st, 2018, at about 1,140 m.

vious large leaves.[1362] Similarly, a meadow in the Karawanks north of Jesenice was named after *Petasites*, as recorded in the early seventeenth century (Figure 63), at that location probably due to its relevance as fodder in spring (Sections 6.2.2.1 and 6.2.2.3).[1363] At present, farmers dislike this genus because the leaves need much more time to dry than grasses, and so there is the risk of mold if hay is already stored when the other plants are dry while butterbur is not dry yet—but, if they wait until the *Petasites* leaves dry, they become so fragile (due to their high protein content and little cellulose) that they fall to pieces, so that only leaf petioles are stored.[1364] However, oral history interviews made it clear that the plant used to regularly be important in the spring for smallholders, and at least in the years following hay harvest failures evidently also for larger farms, for instance, in Bohinj and in the Upper Soča catchment area in the Julian Alps. Namely, *Petasites* leaves grow early (Figure 64) and cows like to eat

[1362] For example, Snoj, 2009, 230–231; compare Makarovič, 1978, 238–239.
[1363] Zwitter, 2020, 68–69.
[1364] Author's interview with Mencinger, 2017, in Stara Fužina.

Figure 64: When a snow blanket still covers the majority of the valley, *Petasites* leaves already develop in aquatic environments. They used to be mowed on a large scale in the spring in the former wetland on this location. Bohinj, March 17th, 2018. Photo: Žiga Zwitter.

them, and so farmers mowed *Petasites* in the spring to prevent starvation of cattle and also not to have *Petasites* leaves in hay.[1365] Entire hand-driven carts of fresh leaves were transported home.[1366] However, a former cheesemaker that made large wheels of Emmental cheese observed that cheese made from the milk of cows fed with *Petasites* in the spring was problematic due to unwanted gas development in cheese.[1367] The former importance of the plant for preventing starvation of cattle in late winter and early spring has disappeared, and butterbur is also no longer used in medicine,[1368] and so *Petasites* is now regarded only as a grassland weed by farmers.

Similarly, in the Allues Valley, early spring grazing usually took place in wet meadows. Aconite-leaved buttercup (*Ranunculus aconitifolius*), widely

1365 Zwitter, 2020; author's interview with Mencinger, 2017, in Stara Fužina. Data from the Upper Soča catchment area: Makarovič, 1978, 238–239.
1366 Author's interview with Boltar, 2018, in Stara Fužina.
1367 Author's interview with A. and M. Taler, 2017, in Studor.
1368 Mlakar, 2015, 316.

present in nutrient-rich wetlands in the Alps in montane and subalpine vegetation belts, was considered acceptable fodder during early spring grazing in the Allues Valley. In the same valley, this species was otherwise viewed as a bad fodder giving a bad taste to cow's milk "[i]f consumed in large enough amounts."[1369] Secondary plant metabolites, which determine the flavor, gradually form and concentrate during the vegetation period, so that the chemical composition of the tissue of this plant species was likely more favorable from the perspective of livestock farming in early spring than later.

6.1.2.5.1.2 Later mowing of wetland hay as fodder

Farmers took time for mowing full-grown or at least relatively well-developed plants in some wetlands in early summer and midsummer, although they had many other tasks to do (see also Sections 6.1.2.1 and 6.1.2.2). This was the first—or, if preceded by early spring mowing or grazing (Section 6.1.2.5.1.1)—second agricultural use of wetland biomass, and the aim was to harvest fodder. The wetlands where animal bedding or building material were obtained were mowed later (see Section 6.1.2.5.2).

Some nutrient-rich marshes were already mowed in the first half of June, as recalled for the mid-twentieth century (Section 6.1.2.2). Haymaking in some nutrient-rich wetlands in June or July has a long historical tradition. Mowing of a wetland common below Bolzano (Germ. *Bozen*) in the Adige (Germ. *Etsch*) catchment was ordered in 1573: this occurred on June 16th[1370] prior to the calendar reform (corresponding to June 25th in today's calendar).[1371] In 1639, summer mowing of wetland and aquatic plants in shallow Lake Gösselsdorf (Germ. *Gösselsdorfer See*, Sln. *Goslinsko jezero*) in southern Carinthia (Germ. *Kärnten*, Sln. *Koroška*) was carried out in July, and partly (allegedly illegally) already in June.[1372]

The influence of plant development on fodder quality stimulated mowing in some marshes in June, as illustrated by the following examples. Dominance of reed sweet-grass (*Glyceria maxima*) was one of the motives for mowing in June. The species is thermophilic, and so it occurs in the colline vegetation belt in some regions of the Eastern Alps. It can abound in some nutrient- and base-rich wetlands.[1373] Reed sweet-grass is a highly productive species that reaches a height of 1.5 meters and was traditionally mowed two times annually without fertilization. In the wider context of wetland hay, reed sweet-grass makes good fodder for horses. Contrary to most wetland plants, its taste is not sour or

1369 Meilleur, 1986, 284, 296–297; Aeschimann et al., 2004, vol. 1, 174–175.
1370 Grass, 1948, 167–168.
1371 The conversion is based on *Julian Day and Civil Date Calculator* and the calculations by the astronomer Tomaž Zwitter.
1372 Zwitter, 2014e, 342.
1373 Aeschimann et al., 2004, vol. 2, 908–909.

hot.[1374] Thus, as long as the quantities of manure available for grasslands were very limited (Section 6.3), it was an advantage for a farm to be entitled to mow a wetland parcel dominated by reed sweet-grass.[1375] Cattle used to wetland hay ate reed sweet-grass as a delicacy as long as it was young and aromatic, but they would not produce much milk on a diet of this fodder type. Cattle used to better hay would find reed sweet-grass acceptable in the absence of better fodder. As it matures, reed sweet-grass becomes harder than many tall sedges, and cattle would reject it at that time.[1376] Thus, the motives for the relatively early timing of first mowing were the later significant decrease in hay quality and the wish to make second mowing possible (see Section 6.1.2.3). Like any wetland hay, there existed the hazard of liver flukes,[1377] often serious in the case of sheep or cattle grazing in wetlands but moderate to negligible in the case of traditional timing of wetland hay uses, as long as grazing in the summer and autumn was commonplace (see Section 6.4.1).

Reed canary-grass (*Phalaris arundinacea*), which occurs in colline and montane vegetation belts in all phytogeographic mesoregions in the Eastern Alps and can abound at nutrient-rich sites with close to neutral soil reaction,[1378] is another example of a wetland species producing high yields of horse fodder twice annually without fertilization. Reed canary-grass hay was relatively good horse fodder if mowed in time; cattle would consume the stalks and leaves as long as they were very young, and later only the leaves. Namely, its stalk gradually becomes exceptionally hard, even harder than that of *Glyceria maxima*. Reed canary-grass's taste is not sour or hot.[1379] In this case also, a relatively early first cutting was motivated by two mowings annually (see Section 6.1.2.3) and higher fodder quality.[1380]

One of the best fodder grasses in wetlands was marsh foxtail (*Alopecurus geniculatus*), producing middling hay quality if compared with the best fodder grasses from mesic sites. Marsh foxtail can abound at moderately wet sites from the colline to subalpine belt, characterized by close to neutral reaction. In some places, it was mowed for the first time in early summer, which allowed mowing of second-crop hay later.[1381]

In early summer, when tussocks produce young leaves and stems, even the quality of hay consisting of some sedge (*Carex* spp.)[1382] and rush (*Juncus* sp.)

[1374] Pils, 1994, 242; Sucholas et al., 2022, 9–10, 14, 16, 23.
[1375] Pils, 1994, 242.
[1376] Sucholas et al., 2022, 9, 18, 23, 35–37.
[1377] Pils, 1994, 242.
[1378] Aeschimann et al., 2004, vol. 2, 984–985.
[1379] Pils, 1994, 242; Sucholas et al., 2022, 19, 36.
[1380] Sucholas et al., 2022, 10, 16, 19, 35.
[1381] Sadar, 1956, 29; Aeschimann et al., 2004, vol. 2, 980–981; Sucholas et al., 2022, 14, 16, 20, 36.
[1382] Sucholas et al., 2022, 18, 36.

species might be not so low (see Section 6.1.2.5.1.1). The stimulation for relatively early mowing dates by the abundance of common reed (*Phragmites australis*) in hay is presented in Section 6.1.2.2.

Another possible motivation for relatively early haymaking in some wetlands, apart from the internal nutrient cycle in many wetland monocots and the option of making second-crop hay, was flood hazard. Thus, some farmers did their best to mow flood-prone wetlands in early summer before a potential flood.[1383] Namely, mud that might be deposited by a flood would spoil the hay and increase the risk of infections in livestock.[1384] If a flood did not visibly spoil the hay, it was likely to cause a longer delay in the timing of fodder mowing than the rain itself.[1385]

Mowing in wetlands in late July or later usually produced low but variable fodder quality at a meadow level.[1386] Several wetlands producing poor fodder were only mowed after better hay had been made in the valley (compare Section 6.1.2.2). For instance, on July 20th, 1679, a small part of the wetland west of Lake Sablatnig/Turner (Germ. *Sablatnigsee,* later *Turnersee*, Sln. *Zablaško jezero*, c. 480 m) was mowed, whereas a major part of this wetland was intended to be mowed later.[1387] Similarly, at 560 m[1388] wetland hay at Lake Pressegg (Germ. *Pressegger See*, Sln. *Pazerško jezero*) in the Gail Valley (Germ. *Gailtal*, Sln. *Ziljska dolina*), Carinthia, was mowed around St. James's Day (July 25th) in the first half of the twentieth century (Figure 65).[1389]

Farmers have learned from the sharpness of plants, animals' eating behavior, and the performance of livestock that the quality of wetland hay cut in midsummer or later was often poor. Awareness of the low quality of hay from many wetlands is proven for almost the entire last millennium.[1390] However, wetland biomass was often used as fodder (Figures 65 and 66) because hay from elsewhere would not suffice for the livestock in winter (see Section 6.2.2.1).

How wetland hay was used depended greatly on the percentage of wetland hay among all hay at the farm level. If farmers had enough hay from meadows that were not wet, they used hay from wet meadows to feed animals that were not milked, were not about to deliver, did not have young offspring, were

[1383] "Košnja," 1864, 199; see also Studen, 2024.
[1384] "Košnja," 1864, 199.
[1385] Compare Sucholas et al., 2022, 14, 37.
[1386] Sadar, 1956, 29, 127–128, 211; Pils, 1994, 116, 138.
[1387] AT StiAStP, Grundherrschaft Eberndorf (Handwritings), vol. 183, Gerichtsprotokoll 1675–1683, 195–197 (August 1st, 1679).
[1388] *KAGIS Maps*, Basiskarten, Hintergrund (Basemap), Franzi [land cadastre] (1822–1828) & *KAGIS Kartendienste*, Geländehöhe, Höheninformation, Höhenschichten (10 m).
[1389] *Slovenski etnografski muzej . . . Jernej Šušteršič / 166 F0006141.*
[1390] Zwitter, 2020, 64–66.

Figure 65: Typically for many wetlands covering significant areas,[1391] "sour" wetland hay at Lake Pressegg in the Gail Valley, Carinthia, was stored (1) in elevated hay barns in the wetland, some of which had already been built before the early nineteenth century,[1392] and (2) in many haystacks,[1393] like the one in the left rear of the photo. A haystack frame with an elevated base as an adaptation to wet conditions stands in the foreground. The elevation of the platform would usually prevent possible floods from reaching the hay.[1394] In late

[1391] For instance, Zwitter, 2014c, 614; Sucholas et al., 2022, 15–17.

[1392] *KAGIS Maps,* Basiskarten, Hintergrund (Basemap), Franzi [land cadastre] (1822–1828).

[1393] *Slovenski etnografski muzej . . . Jernej Šušteršič / 166 F0006141.*

[1394] Compare Sucholas et al., 2022, 17.

autumn, the hay was transported across the frozen wetland. Each large stack, like the one in Figure 65 or the one that could be built on the construction in the center of this photo, contained about 1.2 metric tons of hay, or three cartloads. There were about fifty stacks of "sour" hay in this wetland in the early twentieth century. By 1951, when this photo was taken, their number had decreased to three or four.[1395] Photo: Jernej Šušteršič, August 1951. Slovenian Ethnographic Museum, Documentation: Jernej Šušteršič photo collection: photo 6141 and *Slovenski etnografski muzej . . .* Jernej Šušteršič / 166 F0006141. © Slovenian Ethnographic Museum.

not offspring, and were not planned to be used as draft animals soon. Consequently, wetland hay was usually fed to horses and oxen in intervals when they were not being used, and to cows in late gestation when they were not milked and were not due to deliver soon. Feeding sheep with sour to hot wetland hay resulted in stiffer wool, useable only in loden production. If a farm had little wetland hay and much other hay, even the types of livestock mentioned above were given a mixture of wetland hay and hay from elsewhere. Feeding wetland hay to a gestating cow was a traditional means of stopping the lactation process before new offspring. However, if a farm had a lot of wetland hay, it was used in larger quantities and also given to the livestock types for which it was less appropriate, often mixed with better hay. Consequently, there was a paradox that even the cows of some large farms with a large area of wet meadows were thin and produced little milk. Some other large farms making large quantities of better hay abstained from feeding their livestock low-quality wetland hay and allowed smallholders among fellow villagers to mow and use their wetland hay as fodder, such as in the wetland at the delta of Savica Creek on the western shore of Lake Bohinj (Sln. *Bohinjsko jezero*) in the mid-twentieth century. Smallholders facing a lack of hay also used wetland hay for cows.[1396]

Quantities of stored low-quality (or even lowest-quality) wetland hay increased in dry years when the hay quantities from other meadows was reduced, as demonstrated in the Allues (Francoprovençal *Alués*) Valley (French *Les Allues*). This stimulated farmers to mow more wetland hay for fodder, especially for sheep, goats, horses, and close to grown-up calves and heifers; dairy cows were the last to receive it.[1397] The increase in mowed wetland area in dry summers was certainly a general rule in the Alps, except for places

[1395] *Slovenski etnografski muzej . . . Jernej Šušteršič / 166 F0006141.*

[1396] Author's interviews with Grm, 2017, and with Škantar, 2017, both in Srednja Vas in Bohinj; with Mencinger, 2017, and with Urh, 2017, both in Stara Fužina; with L. and M. Odar, 2017, and with A. and M. Taler, 2017, in Studor. These interviews provide a more diverse picture of uses than the usual statement that sour hay from wet meadows was only used to feed horses (e.g., in Wopfner, 1997, 226, 299; Babai, Molnár, & Molnár, 2014, 56–57). See also Sucholas et al., 2020. On sheep, see Novak, 1970, 379; Pils, 1994, 138, 302.

[1397] Meilleur, 1986, 264, 266–267, 280, 286–287, 295.

where nearly all agriculturally useable wetlands were regularly mowed for hay. Salt was sometimes added to the toughest wetland hay to increase its attractiveness to livestock.[1398]

Even though wetland hay was not only used to feed horses, as demonstrated above, horses were the main consumers of wetland hay. For instance, a probate inventory from 1673 explains that the large Fasching farm east of Feldkirchen in Carinthia used wetland hay to feed horses.[1399] The seigneury of Eberndorf (Sln. *Dobrla Vas*) used hay from Lake Gösselsdorf to feed horses, as recorded in 1710.[1400] The widespread use of wetland hay for horses was a consequence of several factors. First, horses were "believed to have the highest tolerance for different types of fodder"[1401] among all livestock types present throughout the Alps. When the use of a horse's draught power was not approaching, a horse would eat the rest of the hay rejected by cattle or would receive other poor-quality hay, often from a wetland, mixed into other fodder, if possible, such as wheat straw or barley straw and clovers, or low-quality hay mowed too late for cows.[1402] Moreover, horses are much more resistant than cattle or sheep to common liver fluke (*Fasciola hepatica*), an economically highly significant parasite living in many wetlands (see Section 6.4.1).[1403]

Purple moor-grass (*Molinia caerulea*; Figure 66; see also Section 6.1.2.5.2), a species occurring in wet meadows that does not tolerate mowing earlier than in July, was widely used as horse fodder, as demonstrated by data from the foot of the Alps in Upper Austria (Germ. *Oberösterreich*). The nutrient content of such hay was low, and so this was viable for some smaller historical horse breeds.[1404] Tall sedges (*Carex* spp.) were also sometimes used as horse fodder. They are "sour" (probably hot rather than sour), and most species are full of spinules, which can cause injuries in the mouth and elsewhere in the mucosa. In addition, many tall sedge species involve bundles of calcium oxalate crystals, making the plant taste burning hot and potentially causing diarrhea. Lack of calcium can lead to disorders and illnesses in livestock, such as pica, bone fragility, and rickets.[1405] In Bohinj, for instance, horses that received large amounts of hay from tall sedges (*Carex* spp.) on days when they did not

[1398] Meilleur, 1986, 286.
[1399] Dinklage, 1966b, 130–131.
[1400] AT StiAStP, Grundherrschaft Eberndorf (Handwritings), vol. 188, Gerichtsprotokoll 1704–1711, 782 (June 25th, 1710).
[1401] Sucholas et al., 2020, 36.
[1402] Author's interviews with F. Mlakar, 2017, in Bohinjska Češnjica; with Grm, 2017, and Škantar, 2017, both in Srednja Vas in Bohinj; with A. and M. Taler, 2017, and with L. and M. Odar, 2017, both in Studor; and with Mencinger, 2017, and Urh, 2017, both in Stara Fužina.
[1403] Pils, 1994, 138; Ballweber, 2024.
[1404] Pils, 1994, 151.
[1405] Pils, 1994, 138–139, 302.

Figure 66: Mowed plants in a wet montane meadow on Vogar Hill in the Julian Alps, 1,100 m, July 2018. This hay was intended for fodder. The grayish color of the wetland plants does not change when hay dries; consequently, it is less obvious than in usual hay when it can be stored.[1406] A grayish color is common for many grasses growing at base-rich sites in calcareous marshes, like this one. Note purple moor-grass (*Molinia caerulea*, with gray blossoms, on the left) and lady's-mantle (*Alchemilla*) raising the hay quality (see Section 4.1.1.2). Photo: Žiga Zwitter, 2018.

work often suffered from helminths[1407] and diarrhea[1408] (probably caused by helminths and/or hay properties). Yellow gentian (*Gentiana lutea*) rhizome was added to fodder against helminths (Section 8.1.2).

Only exceptional wetlands produced good or middling hay in midsummer. Parallels from Poland and Scandinavia explain that bog-bean (*Menyanthes trifoliata*), which can abound in the colline, montane, or subalpine vegetation belt, at nutrient-poor sites characterized by close to neutral reaction, made excellent, highly fragrant sheep fodder; the species also has a relevant healing impact, the same as large gentian species.[1409] A parallel from Lake Cerknica in

1406 Author's interview with L. Hodnik, 2018, during participant observation of haymaking on Vogar Hill.

1407 Author's interview with A. and M. Taler, 2017, in Studor.

1408 Author's interview with Škantar, 2017, in Srednja Vas in Bohinj. Plant genus (*rudna*) identification follows Tuma, 1935, 251.

1409 Pils, 1994, 122–123, 133, 135, 141–142, Table 23, Photo 63; Emanuelsson, 2009, 281; Sucholas et al., 2022, 13, 26, 35–36; Aeschimann et al., 2004, vol. 2, 54–55.

the northern Dinaric Karst demonstrates that cutting the tops of full-grown common reed (*Phragmites australis*) from boats with sickles, in years when farmers failed to make enough hay elsewhere,[1410] would also make relatively good winter fodder for cattle and horses (see Section 6.1.2.5.1.1).

Some wetland plant species were so unpalatable (even for horses) and at the same time difficult to mow that they were avoided by haymakers in the summer. The example of Lake Gösselsdorf makes it clear that, although much wetland hay was used to feed horses there, one of the rushes, at laest (*Juncus sp.*), was left unmowed.[1411] Some wetland dicots are poisonous: the majority of grassland plants poisonous to livestock grow at wet and moist sites. For example, marsh horsetail (*Equisetum palustre*) is toxic to cattle, not only fresh but also dry.[1412]

6.1.2.5.2 Hay from wetlands and damp meadows as animal bedding

Litter-meadows were wetlands or damp grasslands where the biomass was mainly harvested for animal bedding. They were mowed extremely late, only in late summer when the leaves started turning yellow, or later, in autumn or winter (see 6.1.2.5.2.1), so that most "nutrients have [by that time] retreated from the foliage down to the roots and other low storage organs." Most plant nutrients "are translocated back to rootstocks before the withered herbage is harvested." Thus, despite the absence of fertilization, no soil impoverishment usually occurred in litter meadows over the centuries. Extremely late mowing and moist to wet conditions made litter meadows highly suitable habitats for several animal species threatened or near threatened nowadays, including ground-breeding birds, such as the Eurasian curlew (*Numenius arquata*).[1413]

Focusing on plants, litter-meadows have been variously species rich. There have been great differences in species composition between litter meadows or even within a single litter meadow. Late mowing of grasslands on damp or slightly damp soils in all relevant vegetation belts has made possible, for instance, successful competition of purple moor-grass (*Molinia caerulea*; see Figure 9 in Section 2), which blooms the latest among all grasses and cannot tolerate early mowing or trampling by grazing livestock. Purple moor-grass meadows were often only mowed when its leaves turned orange in the autumn

---

1410 Kebe, 2011, 28, 731.

1411 Author's interview with Plesnik, 2021, at Sapotnik (Sln. *Zapotnik*) farm on the shore of Lake Gösselsdorf; during the interview this monocot was collected in this wetland and later determined by Leonid Rasran.

1412 Sadar, 1956, 11, 126, 128, 133; Pils, 1994, 116.

1413 Pils, 1994, 139–140, 145, 147–148, 152, 170; Ellenberg & Leuschner, 2010, 957–959; Luick, Jones, & Oppermann, 2012, 38; Peterken, 2013, 222 (second quote), 261–263 (first quote from p. 261), 274–278; Poschlod, 2015, 143, 148–153. But see also Pils, 1994, 120.

(a special feature of this grass species) if the biomass was used as animal bedding. Purple moor-grass dominated many late-mowed moist meadows at nutrient-poor sites in the Alps without being sown; however, it was even sown in litter-meadows, at least in the Swiss pre-Alps. Namely, this grass provides high and stable yields of animal bedding without fertilization, but it would make poor fodder (Section 6.1.2.5.1.2).[1414]

Less suitable than purple moor-grass but nonetheless widely used as bedding were tall sedges. They absorb less liquid than purple moor-grass; however, tall-sedge yields are much higher than those of purple moor-grass meadows. At eutrophic sites, the biomass productivity in tall-sedge wetlands is extremely high. They can produce up to ten tons of dry matter per hectare, without additional fertilization, as compared to three to six tons per hectare in purple moor-grass meadows. Contrary to purple moor-grass meadow communities, but similar to reedbeds, tall sedge communities at highly productive eutrophic sites are species poor.[1415]

Some other wetland plant communities were also mowed for livestock bedding, such as common reed (*Phragmites australis*),[1416] rushes (*Juncus* spp.),[1417] or even the communities of some peat bogs.[1418]

Exceptionally, some tall sedges could not be used as animal bedding; an interviewee with personal experience of mowing animal bedding in the wetland at Lake Gösselsdorf (Germ. *Gösselsdorfer See*, Sln. *Goslinsko jezero*) recalled that dry slender-tufted sedge (*Carex acuta*; known as *ostra woč* in the local dialect, literally 'sharp wetland monocot') is so sharp that such bedding could injure the cow's udder (probably only if mowed until autumn; see the presentation of winter mowing in Section 6.1.2.5.2.1), and so this species was avoided when animal bedding was mowed in that wetland.[1419]

Only a few litter-meadows have been preserved to date and are still used in the pre-Alps, but their total area used to be extensive in some parts of the pre-Alps in the nineteenth and first half of the twentieth century.[1420] It was in the nineteenth century that the total area of litter-meadows greatly increased, at least in the western part of the northern pre-Alps. It is already a familiar fact that this area is known for its early specialization in animal husbandry

---

[1414] Pils, 1994, 112, 139–140, 145, 147–148, 152; Ellenberg & Leuschner, 2010, 957; Peterken, 2013, 262, 274–278; Poschlod, 2015, 151; Aeschimann et al., 2004, vol. 2, 994.
[1415] Pils, 1994, 139, 144, 148.
[1416] Pils, 1994, 139–140; Peterken, 2013, 262.
[1417] Luick, Jones, & Oppermann, 2012, 38.
[1418] Peterken, 2013, 262.
[1419] Author's interview with Plesnik, 2021, at Sapotnik farm on the shore of Lake Gösselsdorf; during the interview the species was collected in this wetland and later identified by Leonid Rasran.
[1420] Ellenberg & Leuschner, 2010, 957–959; Poschlod, 2015, 143, 148–153.

and that in the nineteenth century the area of grasslands in general increased there significantly again (see Section 3.1). The investigation of the Loisach–Lake Kochel Marshes (Germ. *Loisach-Kochelsee-Moore*), located just east of Allgäu, made clear that there were no litter-meadows as late as the early nineteenth century. At that time, cereal cultivation was still important there, apart from animal husbandry. Consequently, the area of meadows was limited, and so hay from wet meadows was used as fodder, if only the species composition was at least partly suitable for this. After railway connections started operating in the area in the second half of the nineteenth century, large-scale imports of cereals started. Consequently, former cereal fields were converted into fertilized meadows, partly also in fields for fodder crops. This made it possible to keep much larger herds of livestock, which required more bedding. Possibly also less bedding was taken from forests. Consequently, wet meadows were used as litter-meadows from then on. Litter meadows were not only a landscape feature of the pre-Alps and larger basins in the Alps, but at least a more limited percentage of litter meadows in the landscape also existed at several other places in this mountain range, even in the Limestone Alps where there were local wetlands (on moraines, for instance), such as in the Trenta Valley or Bovec Basin, both in the Upper Soča catchment, as recorded for the first half of the twentieth century. Since the 1960s, the total area of litter-meadows has strongly declined. In Allgäu, for instance, the majority were converted into intensively cultivated drained and fertilized meadows.[1421]

Although the area of litter meadows expanded heavily in the nineteenth century, they are not an invention of late eighteenth- to nineteenth-century changes in grassland management. In fact, some litter meadows have a much longer history, at least in certain regions of the Alps (see also Section 6.1.2.5.2.1).[1422] The list of demesne meadows of the seigneury of Bleiburg (Sln. *Pliberk*) in southern Carinthia from 1570 differentiates between three meadow categories with regard to the abundance of agriculturally undesired wetland plants. The list states that the most appropriate use of hay from a meadow dominated by low-quality wetland plants was livestock bedding.[1423] In the case of the hamlet of Buchhalm (Sln. *Podhom*) in the Jaun Valley (Germ. *Jauntal*, Sln. *Podjuna*), the early nineteenth-century cadastral map, in this case useful for understanding the mid-seventeenth-century situation, reveals that complexes of wet meadows used to be located northwest, west, and southwest of this settlement. Many meadows held by the residents of this hamlet were wet at that time.[1424] In 1664/65, at least one out of four farms in the hamlet used very poor wetland hay as bedding, harvested from the worst of the four meadows

1421 Makarovič, 1978, 239; Poschlod, 2015, 148–153.
1422 Kapfer, 2010, 137–139; Poschlod, Baumann, & Karlík, 2009, 18–20; Peterken, 2013, 262.
1423 Zwitter, 2020, 65.
1424 *Kärnten Atlas*, Darstellung, Kartenhintergrund, Franziszeischer Kataster.

belonging to that farm.[1425] Thus, a more limited area of litter-meadows than from the nineteenth to mid-twentieth century had already existed in the Alps in the sixteenth, seventeenth, and eighteenth centuries, and possibly earlier (see Section 6.1.2.5.2.1). Species composition is one of the reasons for this because not all wetland plants could be used as fodder, not even for horses (see Section 6.1.2.5.1.2),[1426] and so some types of wetland hay inappropriate for fodder were used as bedding. Moreover, not all the rather low-quality hay potentially useable as fodder was always used as fodder, depending on agricultural density and fodder demand in general, but also on the quality of hay from the various parcels where a particular farm had mowing rights.

6.1.2.5.2.1 Winter mowing of animal bedding in wetlands

Winter mowing of weathered monocots and/or horsetails on frozen wetlands and shallow lakes or winter sickle-cutting of common reed (*Phragmites australis*)[1427] was recorded in as culturally diverse areas as Scandinavia,[1428] the Pannonian Basin (proven by archival records already in the early fifteenth century),[1429] Lake Cerknica in the northern Dinaric Karst,[1430] and Lake Gösselsdorf at the foot of the Alps in Carinthia, for instance. At Lake Gösselsdorf, winter mowing in the wetland is clearly documented from the seventeenth[1431] to twentieth century.[1432]

The main aims of winter mowing were to harvest animal bedding or to obtain common reed stalks as building material for roofs, insulation, and plaster lining (see Section 8.2.3). Practicing winter mowing on ice to obtain animal

1425 AT StiAStP, Grundherrschaft Eberndorf (Handschriften), vol. 4, Grundbeschreibung 1664/65, 547–553; Zwitter, 2014d, 343.

1426 See, e.g., Pils, 1994, 131; Sucholas et al., 2020.

1427 For sickle-cutting or mowing common reed to harvest building material, see Jones, 2012, and *Hiss Reet: Die Schilfernte*.

1428 Emanuelsson, 2009, 281.

1429 Kiss, 2020, 12–13.

1430 Kebe, 2011, 732.

1431 Zwitter, 2014e, 342–343; winter mowing is confirmed by AT StiAStP, Grundherrschaft Eberndorf (Handwritings), vol. 181, Gerichtsprotokoll 1661–1668, 40–41 (June 17th, 1645); see also the references to eighteenth-century records lower in this section.

1432 Author's interview with Plesnik, 2021, at Sapotnik (Sln. *Zapotnik*) farm on the shore of Lake Gösselsdorf. Concerning the terminology in this section, note that the verb *mähen* and its derivatives used in the archival materials cited in Section 6.1.2.5.2.1 do not allow any distinction between actual mowing carried out by scythe and cutting with a sickle (*Frühneuhochdeutsches Worterbuch*: mäen [!]). In addition, even knowledge of the implement type would not allow the distinction between mowing animal bedding (Kebe, 2011, 732) and mowing or sickle-cutting building material (*Hiss Reet: Die Schilfernte*) based on archival records from the seventeenth and eighteenth centuries, which are cited in Section 6.1.2.5.2.1. Information from the first half of the twentieth century, cited in Sections 6.1.2.5.2.1 and 8.2.3, explains that biomass was harvested in Lake Gösselsdorf for both types of uses: as animal bedding and common reed as building material.

bedding in such culturally diverse and distant parts of Europe proves that this practice has a long history and also that it used to be more widespread in wetlands in the Alps, especially where areas of deciduous forest close to settlements were insufficient for supplies with quality animal bedding or if farmers were not allowed to rake and store fallen leaves there. Winter mowing had even fewer ecological impacts on the wetland than the autumn mowing presented above in Section 6.1.2.5.2.

Ice was sufficiently stable to carry people mowing and storing wetland monocots "after at least one or two weeks of deep frost," as documented in the Pannonian Basin.[1433] Another crucial factor was that the ice had to be free of snow. Thus, winter mowing of animal bedding and mowing or cutting common reed as building material were not possible every year. Winter mowing took place in December, for instance, as recalled for the twentieth century.[1434] In case of stable, deep ice, livestock transported full carts of harvested biomass on frozen shallow lakes and wetlands.[1435] Suitable conditions for winter mowing were much more frequent in the Little Ice Age than later. From 1500 to 1799, for instance, winters in Central Europe were on average about 1.1 °C colder than from 1961 to 1990, from 1561 to 1600 1.8 °C, from 1591 to 1600 2 °C, and from 1691 to 1700 even 2.6 °C colder than the average from 1961 to 1990. Nonetheless, winter temperatures varied significantly during the Little Ice Age as well. Especially the winters between 1603 and 1618 were extremely variable, with (very) cold weather in 1602/03, 1607/08, 1611/12, 1613/14, 1615/16, and 1617/18, and (very) warm winters in 1603/04, 1606/07, 1608/09, 1612/13, and 1616/17;[1436] this is an example of a subperiod when winter mowing in wetlands was certainly not possible every year.

Preserved data on the wetland at Lake Gösselsdorf in the first half of the eighteenth century demonstrate that the rights to mow *Sacher* (i.e., wetland monocots in any phase of plant development[1437]) there were allocated on December 2nd, 1738; that is, after the seigneury had already stored enough *Sacher* to cover its own economic demand and before farmers started mowing.[1438] It remains unclear whether the farmers could make use of mowing rights that winter because "the 1730s in particular stand out for the absence of even moderately cold winters."[1439] On January 11th, 1710, evidently soon after *Sacher* in

[1433] Kiss, 2020, 12–13.
[1434] Author's interview with Plesnik, 2021, at Sapotnik farm on the shore of Lake Gösselsdorf.
[1435] Kebe, 2011, 732.
[1436] Pfister et al., 2018, 266, 268–269, 273–274, 276–279, 282; Zwitter, 2013.
[1437] Zwitter, 2020, 65; Lexer, *Mittelhochdeutsches Handwörterbuch*: Sacher; Lexer, 1862, column 211.
[1438] AT StiAStP, Grundherrschaft Eberndorf (Handwritings), vol. 192, Gerichtsprotokoll 1728–1742, 421 (December 2nd, 1738).
[1439] Climate history context: Pfister et al., 2018, 273.

Lake Gösselsdorf had been mowed, the seigneury of Eberndorf (Sln. *Dobrla Vas*) complained to the neighboring seigneury of Sonnegg (Sln. *Ženek*) that not only its subjects with the right to mow *Sacher* there, but also additional subjects, had mowed it illegally.[1440] These data make it clear that allocation of *Sacher* mowing rights to several farmers on December 23rd, 1702,[1441] allocation of *Sacher* mowing rights to twenty-two households on December 29th, 1735,[1442] and the fine recorded in December 1723 for illegal *Sacher* mowing at Lake Gösselsdorf[1443] also referred to winter mowing of animal bedding (or harvesting common reed as building material).

Mowing animal bedding on ice was easy; weathered and frozen sedges were so fragile that there was no need to sharpen the scythe, and the quality of animal bedding was much higher than in the case of the same sedges mowed earlier, resulting in much sharper, harder, and less absorbent bedding.[1444] In the Alps, the target plants of winter mowing were mainly some tall sedges. The interviewee with personal experience of winter mowing of a wetland at Lake Gösselsdorf to harvest animal bedding recalled that there was no winter mowing of common reed (*Phragmites australis*), bulrush (*Typha* sp.), slender-tufted sedge (*Carex acuta*), or (soft) rush (*Juncus (effusus)*), all occurring in this wetland,[1445] whereas winter mowing of *Carex acuta* is reported from the northern Dinaric Karst.[1446] By winter time tall sedges became weathered, and so they were a highly esteemed source of animal bedding because they perfectly absorbed the liquid manure, and weathered sedges were not sharp anymore, and so such bedding was comfortable for livestock.[1447] Moreover, later, weathered sedges decomposed quickly, and so such manure was perfect for arable land, not leaving hard residues in the soil that could cause trouble during plowing, unlike low-quality, woody animal bedding from common heather (*Calluna vulgaris*) mowed for this purpose for centuries in Scots pine (*Pinus sylvestris*) forests in parts of the Jaun Valley not far from Lake Gösselsdorf.[1448]

[1440] AT StiAStP, Grundherrschaft Eberndorf (Handwritings), vol. 188, Gerichtsprotokoll 1704–1711, 778 (January 11th, 1710).

[1441] AT StiAStP, Grundherrschaft Eberndorf (Handwritings), vol. 187, Gerichtsprotokoll 1698–1703, 360–362 (December 23rd, 1702).

[1442] AT StiAStP, Grundherrschaft Eberndorf (Handwritings), vol. 192, Gerichtsprotokoll 1728–1742, 337 (December 29th, 1735).

[1443] AT StiAStP, Grundherrschaft Eberndorf (Handwritings), vol. 191, Gerichtsprotokoll (from the 1660s that begins in 1661), 198 (December 1723).

[1444] Kebe, 2011, 732.

[1445] Author's interview with Plesnik, 2021, at Sapotnik farm on the shore of Lake Gösselsdorf.

[1446] Kebe, 2011, 732 (identification of the folk taxon *kravina* as slender-tufted sedge, *Carex acuta*, follows Sadar, 1956, 25 & Aeschimann et al., 2004, vol. 2, 834–835, see the same identification in Martinčič, 2002, 76).

[1447] Kebe, 2011, 732.

[1448] On this type of forest litter see Zwitter, 2014e, 317–322; Paschinger, 1939, 80–83; Mlakar, 2015, 68.

The following microhistory data referring to the wetland at Lake Gösselsdorf support the general statement that weathered high sedges mowed on ice were an esteemed type of animal bedding (and/or common reed was in demand as building material). The availability of tall sedges in that wetland useable as animal bedding (and/or of common reed as building material) was a sufficient motivation for farmers to apply for the winter mowing right,[1449] even though they were expected to deliver a due to the seigneury for winter mowing (typically, a capon per household), as recorded in 1735.[1450] Another proof of the relevance of weathered high sedges mowed in winter as bedding (or common reed harvested as building material) is provided by the information that the seigneury of Eberndorf rewarded those subjects, otherwise heavily burdened by corvée, by giving them priority in the allocation of the right to mow and store *Sacher* in the wetland at Lake Gösselsdorf in those winters when the quantities of *Sacher* available exceeded the demand of the seigneury itself, as documented in 1738.[1451] The significant importance of *Sacher* mowed in winter for local agriculture (and/or as a building material) is also reflected in the fact that people and institutions entitled to winter mowing in this wetland according to early eighteenth-century data included not only those residing close to the lake, but also some distant ones. The former group includes people from the isolated Sapotnik (Sln. *Zapotnik*) farm on the lakeshore, and some residents of Gösselsdorf (Sln. *Goselna Vas*), Sittersdorf (Sln. *Žitara Vas*), Proboj, Eberndorf (Sln. *Dobrla Vas*), Köcking (Sln. *Kokje*), Hart (Sln. *Dobrova*), Rain (Sln. *Breg*), and Kleinzapfen (Sln. *Malčape*); in the latter group, there is a resident of Müllnern (Sln. *Mlinče*), a village four kilometers straight-line distance away, the priest from the village of Globasnitz (Sln. *Globasnica*) more than five kilometers distant, and a resident of Peratschitzen (Sln. *Peračija*), a village six kilometers straight-line distance from this wetland, for instance (see Figure 4). Residents of not all these settlements had permanent rights to mow *Sacher*, and so some did not participate in mowing every year when weather made winter mowing possible; participation in winter mowing depended on the quantities of *Sacher* available, apart from other factors.[1452]

The example of the wetland at Lake Gösselsdorf demonstrates that winter mowing of sedges was practiced longer in history than early spring mowing (see Section 6.1.2.5.1.1). The oldest inhabitants farming this wetland were

1449 AT StiAStP, Grundherrschaft Eberndorf (Handwritings), vol. 192, Gerichtsprotokoll 1728–1742, 337 (December 29th, 1735); and 421 (December 2nd, 1738).

1450 AT StiAStP, Grundherrschaft Eberndorf (Handwritings), vol. 192, Gerichtsprotokoll 1728–1742, 337 (December 29th, 1735).

1451 AT StiAStP, Grundherrschaft Eberndorf (Handwritings), vol. 192, Gerichtsprotokoll 1728–1742, 421 (December 2nd, 1738).

1452 AT StiAStP, Grundherrschaft Eberndorf (Handwritings), vol. 187, Gerichtsprotokoll 1698–1703, 360–362 (December 23rd, 1702); AT StiAStP, Grundherrschaft Eberndorf (Handwritings), vol. 192, Gerichtsprotokoll 1728–1742, 337–338 (December 29th, 1735) and 421 (December 2nd, 1738).

surprised to have heard about the proofs of early spring mowing in archival records during the interview (but they recall spring grazing there), whereas they clearly recalled winter mowing of livestock bedding from personal experience.[1453] Section 6.1.2.5 has already presented the reason for this diversity in the persistence of historical practices: whereas winter mowing easily provided highly esteemed animal bedding, early spring mowing in wetlands was a much more labor-intensive practice (plants were still short, and so the biomass quantities mowed per area were low) that mainly covered the frequent lack of hay in early spring (see also Section 6.2.2.1). This could be replaced by grazing. Moreover, starvation of livestock in early spring stopped being the usual reality many decades before the role of livestock bedding started gradually decreasing as a general phenomenon due to the spread of slatted barn floors or abandonment of livestock farming.

6.1.2.6 Regular mowing carried out less than once annually

Many regularly used meadows in the Alps were not mowed every year. In several places, there were extensive areas of non-fertilized and unirrigated (see Section 6.3.1) remote meadows mowed every second or third year.[1454] For instance, according to detailed information about Tyrol and Vorarlberg from 1839, in many cases non-fertilized (and unirrigated) meadows at higher elevations were mowed every second or third year.[1455]

Some meadows in the Alps were mowed at even longer intervals. At certain places in the Sellrain Valley (Germ. *Sellraintal*), mowing was allegedly only carried out once every six years in the early twentieth century.[1456] This last case must have been exceptional, and the data might refer to specific socio-economic conditions; a gradual decline in mowing the most remote mountain meadows started in the second half of the nineteenth century[1457] (see also the note on abandonment of some of the highest meadows in the first half of the twentieth century in the caption to Figure 11 in Section 3.1). Mowing there was probably more frequent in some other periods of history.[1458]

[1453] Author's interview with Plesnik, 2021, at Sapotnik farm on the shore of Lake Gösselsdorf.

[1454] Grass, 1948, 177, 191; Mathieu, 1996, 144–145; Hubatschek, 1992, 175; Niederer, 1996d, 174; Keller, 2004, 19–20, 27, 30–32, 42, 53, 84, 86; Machatschek, 2007, 149; Bavcon, Ravnjak, & Praprotnik, 2019, 123.

[1455] Mathieu, 1996, 144–145.

[1456] Grass, 1948, 191.

[1457] Grass, 1948, 206.

[1458] A six-year interval between mowings was only possible at high elevations or on soils providing the grassland plants with lowest amounts of nutrients. According to data from the interior of Slovenia, a saw is already needed in some productive places to clear bushes and allow mowing after a meadow has not been mowed for five years (Bavcon, Ravnjak, & Praprotnik, 2019, 23).

Environmental factors linked with wider economic circumstances had an influence on the decision whether to mow a certain meadow every year or every second or third year.[1459] Some of the most relevant environmental factors are presented first, followed by the economic history context.

Nutrient-poor soil and a harsh climate were among the main environmental factors resulting in mowing less often than once a year. Use of meadows characterized by low biomass productivity only once in two or three years was based on the same principle as field rotation systems including fallow,[1460] so that every second year or two out of three years grassland plants decomposed in the meadow and nutrients returned to the soil.[1461] One of the main reasons, often the only one, for mowing every second or third year was the awareness that it was important not to overuse non-fertilized meadows on nutrient-poor soils. Otherwise, the yields would have become too low to be feasible even by the standards of the time.[1462] To quote a farmer from the Lesach Valley (Germ. *Lesachtal*) in Carinthia managing a poorly productive distant mountain meadow: "If you mow this meadow every year, it goes bust."[1463] Namely, due to the soil nutrient availability being too low, not only the grassland species composition would have changed (Sections 6.3.1 and 9), but the height of plants would have also markedly decreased if mowing in such meadows had been too frequent. In some non-fertilized meadows, the plant height decreased so much that it was physically impossible to mow the meadow with a scythe every year.[1464] The increase in meadow productivity due to less-than-annual mowing depended on site conditions, species composition, and the duration of the interval between mowings. The yield was, for instance, about one-third more in a mowing year if a poorly productive meadow was mowed every second year, in comparison with regular annual mowing.[1465] Thus, farmers easily learned from experience that regular breaks were needed between years of mowing in meadows at very nutrient-poor sites.

In some regions, such as Tyrol, meadows that were not mowed every year were divided into narrow belts from top to bottom, and every year only one such belt was mowed in each meadow.[1466] Such uniformity of mowing whole meadows or meadow complexes only every second year was not the rule. Microenvironmental conditions mattered as well: considerable meadow sections with a higher biomass production (often in microrelief depressions) were

[1459] Keller, 2004, 19–20, 27, 30, 32, 53.
[1460] Keller, 2004, 19–20, 27, 30, 32, 53; Leibundgut & Vonderstrass, 2016a, 74.
[1461] Wopfner, 1997, 310–311; Machatschek, 2007, 149.
[1462] Grass, 1948, 177; Hubatschek, 1992, 175; Niederer, 1996d, 174; Keller, 2004, 20, 27.
[1463] Keller, 2004, 20.
[1464] Keller, 2004, 20, 27; Kapfer, 2010, 135.
[1465] Keller, 2004, 19, 84.
[1466] Grass, 1948, 191; Hubatschek, 1992, 172, 175; Keller, 2004, 30.

mowed every year, whereas the prevalent less productive sites within the same meadow were cut every second year or less frequently. Some of the least productive sites were mowed less often than the majority of the meadow, thus even less frequently than every second year.[1467]

Meadows in the Alps at very high elevations were never manured (see Section 6.3.2.1.1). Thus, it is quite understandable that many of the highest meadows in this harsh climate (often exceeding 2,000 m and exceptionally reaching 2,700 m; see Section 3.1) were mowed only every second or third year.[1468] In the Ötz Valley (Germ. *Ötztal*) in Tyrol, where manuring reached higher elevations than typically (Section 6.3.2), non-fertilized meadows above Niederthai that were mowed only every second or third year were located at 1,900 to 2,000 m.[1469] In Carinthia, mowing every second or third year was frequent in meadows at elevations of more than 1,600 or more than 1,800 m, depending on the region.[1470] Similarly, better meadows (or better meadow sections) in subalpine meadows on the Golica Ridge in today's northwestern Slovenia, at elevations from around 1,500 to 1,835 m, were mowed every year, and less productive sites every second or third year, as reported for the first half of the twentieth century.[1471]

However, it was not only many of the highest meadows in the Alps that were mowed no more frequently than every second year. Namely, many meadows at lower elevations were also not fertilized (Section 6.3.1); where the productivity of soils was low, mowing such meadows every year did not make sense. For instance, remote steep montane meadows on the uppermost sections of the sunny southern slopes of Kriška Gora in the Kamnik–Savinja Alps (Sln. *Kamniško-Savinjske Alpe*; the elevations of these meadows are about 1,000 to 1,600 m) were mowed every second year.[1472]

Focusing on social factors with an influence on changing mowing frequency throughout history, periods when economic pressures on meadows decreased (see Sections 3.1, 3.2, and 9) resulted in less frequent mowing in poorly productive meadows, whereas fodder scarcity resulted in more frequent (e.g.,

---

[1467] Keller, 2004, 19.

[1468] Keller, 2004, 19–20, 27, 30, 32.

[1469] Griesser, 1999, 42.

[1470] Keller, 2004, 19–20, 27, 30, 32.

[1471] Torkar Tahir, 2012, 133. Localization: *Arcanum Maps*, Cadastral Maps, Habsburg Empire – Cadastral Maps (XIX. [sic] Century), cadastral municipality of Planina pod Golico (Germ. *Alpen*), 1826; localization: *Environmental Atlas of Slovenia*, Overview Maps (detailed topographical map) & Spatial units, Land cadaster.

[1472] Bavcon, Ravnjak, & Praprotnik, 2019, 123; elevations: *Arcanum Maps*, Cadastral Maps, Habsburg Empire – Cadastral Maps (XIX. [sic] Century), cadastral municipality of Križe (Germ. *Kreutz*), 1826, combined with *Environmental Atlas of Slovenia*, Backgrounds, DEM [Digital elevation model] and *Environmental Atlas of Slovenia*, Spatial Units, Land Cadastre.

annual) mowing in some nutrient-poor grasslands, which resulted in significantly lower yields in mowing years.[1473]

Mowing every two or three years allowed the development of an early stage of ecological succession between mowings, especially if the parcel was excluded from grazing. Consequently, in many places in the Alps, low bushes of green alder (*Alnus alnobetula*) grew among other grassland plants in meadows mowed every second year. Namely, this bush cannot tolerate annual mowing. Within the Eastern Alps, such meadows with green alder were most frequent in the Central Alps, in the Gastein Valley and Rauris Valley in the province of Salzburg, for instance. In contrast, the impacts of bedrocks on soils in the Southern and Northern Limestone Alps made most meadow sites unsuitable for green alder. Because the meadows with green alder were mowed every second year, green alder bushes (reaching up to 60 cm) were mowed with a sturdy short scythe together with other hay because they had not yet become too woody for such mowing. Mixing grassland hay and alder shoots was also practiced near Genoa (Ital. *Genova*) in Liguria. Symbiotic bacteria fixing airborne nitrogen and making it available to plants in soil live in alder root nodules. Thus, the nutritive value of alder leaves was high due to their high nitrogen content. In addition, airborne nitrogen fixation in alder roots also favorably impacted the availability of nutrients for other meadow plants, and this impact of alders on raising soil fertility might have been known to farmers of preindustrial Europe (see Section 3.2.2). Due to the favorable impacts of airborne nitrogen fixation, it can be assumed that the nutritive value of hay from such meadows as a whole was also relatively favorable. No other scrub species has such a strong impact on raising the soil productivity. All this was reflected in the farmers' traditional ecological knowledge recorded during Michael Machatschek's participant observation in the Alps in the twentieth century: in the alders "there is fat" ("ist das Schmalz drin"). Leaves of green alder were also esteemed for their medicinal qualities due to their mineral content. In addition, green alder roots contributed to the stability of soils in meadows on steep slopes more significantly than most other tree and scrub species. Consequently, in some places, another likely reason for regular but not annual mowing was the wish to have green alder in meadows and in hay. Nonetheless, livestock had to become accustomed to eating alder leaves; fine cutting and mixing the leaves of a few tree species, including those that the livestock preferred, made it impossible for the livestock to discard alder leaves from the fodder. Hay with alder leaves was not staple fodder. It was stored separately and used in small quantities.[1474]

[1473] Keller, 2004, 19, 53, 84.

[1474] Aichinger, 1933, 171–179; Schneiter, 1970, 33–39; Machatschek, 2002, 108, 141–142, 171, 329–331, 434 (quote from p. 330); Grove & Rackham, 2003, 183; Machatschek, 2007, 149; Emanuelsson, 2009, 119, 189, 192, 206.

However, depending on site conditions, mowing not taking place every year could also promote other species of early ecological succession that were relatively unproductive from the point of view of animal husbandry, such as several dwarf shrubs (see Sections 6.1.2.1.2 and 6.6.2), and so traditional ecological knowledge of the meadow sites was of utmost importance.

The plant diversity in meadows mowed only every second or third year is often high. Animal diversity also is particularly high there (see Section 6.3.1). Namely, the development of several insects, such as bumblebees and grasshoppers, is only completed at high elevations by late August or early September, and thus slightly later than historical haymaking in many cases (Sections 6.1.2.1 and 6.1.2.2). Butterflies also profit from plant development that is not disturbed by mowing by the end of August. Nonetheless, regular mowing events every couple of years are needed to keep biodiversity so high, and historical mowing every second or third year unintentionally ensured the long-term preservation of such exceptionally biodiverse grassland ecosystems.[1475]

Non-constant use of non-fertilized and unirrigated, regularly mowed meadows was, however, not a general rule in the Alps.[1476] In the eastern Julian Alps, for instance, in the 1830s, detailed information on meadow management—which definitely does not present land use in a more intensive manner than the historical reality—documents mowing in remote meadows every year,[1477] even though in places where the soils were less fertile or gradually too impoverished matgrass abounded at that time (see Section 6.3.1.2). However, despite these data from the eastern Julian Alps, it was not a general rule that mowing at relatively high

[1475] Keller, 2004, 19, 33, 40, 42, 86, 110.

[1476] Wopfner, 1997, 313, 318–319.

[1477] For example, SI AS 176, Franciscejski kataster za Kranjsko (Franciscan Land Cadaster for Carniola), L30, cadastral municipality of Nemški Rovt, cadastral appraisal protocols: s.n., 30/4, 49, and Catastral-Schätzungs-Elaborat, Schätzung des natural Ertrages, § 4: Wiesen; SI AS 176, L56, cadastral municipality of Bohinjska Bistrica, cadastral appraisal protocols: s.n., 56/4, 49, and Catastral-Schätzungs-Elaborat, Schätzung des natural Ertrages, § 4: Wiesen; SI AS 176, L72, cadastral municipality of Gorjuše, cadastral appraisal protocols: s.n., 72/4, 49, and Catastral-Schätzungs-Elaborat, Schätzung des natural Ertrags, § 4: Wiesen; SI AS 176, L107, cadastral municipality of Češnjica v Bohinju [i.e., Bohinjska Češnjica], cadastral appraisal protocols: s.n., 107/4, 49, and Catastral-Schätzungs-Elaborat, Schätzung des natural Ertrages, § 4: Wiesen; SI AS 176, L157, cadastral municipality of Srednja Vas v Bohinju, cadastral appraisal protocols: s.n., 157/4, 49, and Catastral-Schätzungs-Elaborat, Schätzung des natural Ertrages, § 4: Wiesen; SI AS 176, L172, cadastral municipality of Nomenj, cadastral appraisal protocols: s.n., 172/4, 49, and Catastral-Schätzungs-Elaborat, Schaetzung des Natural Ertrages, § 4: Wiesen; SI AS 176, L242, cadastral municipality of Savica, cadastral appraisal protocols: s.n., 242/4, 49, and Catastral-Schätzungs-Elaborat, Schaetzung des Natural Ertrages, § 4: Wiesen; SI AS 176, L284, cadastral municipality of Studor, cadastral appraisal protocols: s.n., 284/4, 49, and Catastral-Schätzungs-Elaborat, Schaetzung des Natural Ertrages, § 4: Wiesen; SI AS 176, L335, cadastral municipality of Bohinjska Bela, cadastral appraisal protocols: s.n., 335/4, 49, and Catastral-Schätzungs-Elaborat, Schaetzung des Natural Ertrages, § 4: Wiesen.

Figure 67: Coppice meadow near the village of Mara in the Carpathians, northwest Romania, soon after the trees or bushes were cut. Photo: Žiga Zwitter, 2015.

elevations in the Limestone Alps of what is now Slovenia took place every year; mention has already been made of montane meadows in the uppermost sections of the sunny southern slopes of Kriška Gora mowed every second year.

So far, this section has focused on meadows not mowed annually that were not regularly overshadowed by trees. Another type of meadow not mowed every year—sometimes at nutrient-poor sites similar to the ones presented so far, sometimes at somewhat more productive non-fertilized sites—was a coppice including a meadow phase, where grassland patches gradually became significantly overshadowed by trees. Apart from grassland plants, a "coppice can contain a large number of broadleaf tree species and is characterized by the multi-stemmed stools. The stems are cut at more or less regular intervals [usually every five to thirty years], after which new stems regenerate from the stools."[1478] Such a wood management system allowed quick regeneration of woody plants and provided high quantities of firewood or small wood; for fences, for example. In addition, it was possible to make hay below the trees and bushes for several years within a coppice cycle (Figure 67), often as long as the grassland was not entirely shaded by woody plants. Due to the many nutrients released from the leaves of trees and bushes stemming from lower soil horizons that are inaccessible to grassland species (e.g., hazel leaves make

[1478] Emanuelsson, 2009, 196–197, 199, 201 (quote)–202; Peterken, 2013, 259.

the soil particularly good for productive meadows; see Section 6.3.1), some coppice meadows remained relatively productive for several years.[1479] Grassland species were again able to develop extensive cover and significant biomass immediately after the woody plants had been cut. Numerous species of coppice grasslands "are equally at home in the open or shade," such as globeflower (*Trollius europaeus*). "[A]cknowledged shade-bearers, such as herb-Paris [*Paris quadrifolia*] and lily-of-the-valley [*Convallaria majalis*], gr[e]w [in coppice grasslands] commonly . . . amongst species with a high greater affinity with unshaded conditions, such as blue moor-grass [*Sesleria caerulea*], dropwort [*Filipendula vulgaris*] and military orchid [*Orchis militaris*]." Due to constant changes in light and shade caused by tree growth and felling, "the sward [was] a mixture of species on the way up and others on the way down."[1480] On the quality of hay from meadows shaded by woody vegetation, see Section 6.6.1.

Coppices were landscape elements in all regions of the Alps; nonetheless, in general, they were more frequent in the south of the mountain chain.[1481] In the Eastern Julian Alps (e.g., on the Jelovica and Pokljuka Plateaus and to the west of the latter), some montane meadows with bushes (hazel, *Corylus avellana*, for instance), such as in the cadastral municipalities of Zgornje Gorje, Gorjuše, Srednja Vas in Bohinj, and Selo pri Bledu in the early nineteenth century,[1482] were most likely coppice meadows. A photograph taken about 1970 documents a likely coppice in the montane meadow complex Za Jamo on the southeasternmost Pokljuka Plateau.[1483] In hazel coppices, however, mowing could take place annually, without interruption, such as in some remote montane meadows in Bohinj where the hazels were cut every five years.[1484] Cutting the hazel at the stump every five to seven years was widespread in the Eastern Alps to avoid too much shade in the grassland, too strong competition for water with grassland plants, and to make sure that the quality of the hazel wood would not decline.[1485]

1479 Emanuelsson, 2009, 157, 201–204, 206–209.

1480 Peterken, 2013, 282 (quotes)–283 (quote); Emanuelsson, 2009, 160, 199, 209–210.

1481 Bätzing, 2015, 266.

1482 SI AS 176, Franciscejski kataster za Kranjsko (Franciscan Land Cadaster for Carniola), L72, cadastral municipality of Gorjuše, cadastral appraisal protocols: s.n., 72/4, 75, and Catastral-Schätzungs-Elaborat, Schätzung des natural Ertrags, § 4: Wiesen, Dritte Classe; SI AS 176, L157, cadastral municipality of Srednja Vas v Bohinju, cadastral appraisal protocols: s.n., 157/4, 75, and Catastral-Schätzungs-Elaborat, Schätzung des natural Ertrages, § 4: Wiesen, Vierte & Fünfte Classe; SI AS 176, L179, cadastral municipality of Zgornje Gorje, cadastral appraisal protocols: s.n., 179/4, 75, and Catastral-Schätzungs-Elaborat, Schaetzung des Natural Ertrages, § 4: Wiesen, Vierte Classe; SI AS 176, L343, cadastral municipality of Selo pri Bledu, cadastral appraisal protocols: s.n., 343/4, 75, and Catastral-Schätzungs-Elaborat, Schaetzung des Natural Ertrages, § 4: Wiesen, Dritte & Vierte Classe.

1483 Šarf, 1974, 137.

1484 Author's interview with Kunavec, 2021, in Stara Fužina, who participated in such a coppice management.

1485 Machatschek, 2002, 308; author's interview with Kunavec, 2021, in Stara Fužina.

Figure 68: The flood-prone grassland on the southern shore of Upper Lake Fusine was not only grazed, but also mowed for a long time. However, according to early seventeenth-century information, mowing did not take place there regularly. The absence of a gravel shore clearly explains why the grassland was so flood prone. Photo: Žiga Zwitter, 2016.

This section dealt with regularly mowed meadows cut less frequently than every year, but there also existed less regularly or less frequently mowed meadows in the Alps. They are the topic of the following section.

6.1.2.7 Sporadically mowed grasslands

In certain periods, the nexus between particular environmental and social conditions led to sporadic mowing in some specific places. As an example, by the 1520s there was a flood-prone meadow at the southern shore of Upper Lake Fusine (Ital. *Lago di Fusine Superiore*, Germ. *Oberer Weißenfelser See*, Sln. *Zgornje Mangartsko jezero*, Friul. *Lât Superiôr di Fusinis*; Figure 68), where a flood sometimes destroyed all the hay. A century later, in a rent-roll (*Urbar*) written between 1617 and 1636, the same grassland was characterized as pasture with the notice that part of this pasture could be mowed,[1486] and so at least for part of the seventeenth century there was an occasionally mowed pasture.

There was a different reason for sporadic mowing in the cadastral municipality of Javornik in the Western Karawanks based on early nineteenth-cen-

[1486] Zwitter, 2020, 80–82.

tury information. There, pastures close to the permanently settled area were significantly better than in the neighboring cadastral municipalities, and a certain share of parcels mapped as pastures in the land cadaster in the early nineteenth century were mowed in very favorable years.[1487] Indeed, a mapped pasture complex at the bottom of the Sava Dolinka Valley and on the lowest slope sections above it in the southwest of this cadastral municipality, on drought-prone soils,[1488] was named *na Snoschettach* (literally, 'in the hay meadows', *Senožeta*), as recorded in the early nineteenth century.[1489] These pastures measured 22.8 hectares, divided into nineteen parcels (average parcel size: 1.2 ha and standard deviation: 0.7 ha; median: 1.0 ha). The toponym, the size of the parcels (it was a complex of contiguous medium-size and small parcels, instead of one large pasture), and the possession rights (not officially a common, but parcels formally in farmers' private possession,[1490] although certainly grazed collectively[1491]) indicate that that meadow complex experienced a long history of (sporadic) mowing. The first use of grassland resources from the Senožeta grasslands was grazing in spring, a practice usual for hay meadows in Central Europe until the eighteenth or nineteenth century (see Section 6.2). In this case, however, spring grazing in the upper section of the colline vegetation belt at the foot of the Alps lasted very long in some years, whereas it ended soon in other years. Namely, spring grazing at Senožeta lasted until Ascension Day,[1492] a movable feast that takes place between April 30th and June 3rd (see Section 6.2.1.2: text and Figure 70). From then on, these meadows were closed for grazing until July 25th, when another grazing season started that lasted until September 29th.[1493] This means that, when the date of Ascension Day was late, the duration of the period between the end of spring grazing and a

1487 SI AS 176, Franciscejski kataster za Kranjsko (Franciscan Land Cadaster for Carniola), L90, cadastral municipality of Javornik, map sheets A01–A07, and cadastral appraisal protocols: Reclamations verhandlungen über den catastral act der steuergemeinde Jauerburg, Gutachten 90/5, December 15th, 1837: ad 8. It is clear that this notice does not refer to mountain pastures, which were dealt with in this report as a separate category (see land-use categories in SI AS 176, L90, cadastral municipality of Javornik, cadastral appraisal protocols: Protocoll sämmtlicher Grund- und Bauparzellen der Gemeinde Jauerburg.

1488 *Environmental Atlas of Slovenia*, Land and Soil, Soil map (1:25,000).

1489 SI AS 176, Franciscejski kataster za Kranjsko (Franciscan Land Cadaster for Carniola), L90, cadastral municipality of Javornik, map sheets A04, A06, and A07.

1490 SI AS 176, Franciscejski kataster za Kranjsko (Franciscan Land Cadaster for Carniola), L90, cadastral municipality of Javornik, Grundparzellen Protokoll, parcel nos. 110–128.

1491 Compare Kapfer, 2010, 136; see also Section 6.2.

1492 SI AS 48, Deželna komisija za odkup in uravnavo zemljiških bremen za Kranjsko (Provincial Commission for Redemption and Regulation of *Profits à Prendre* in Carniola), fasc. 56, 9:1–5, Protokoll, September 3rd, 1859, & Protokoll, July 11th, 1868.

1493 SI AS 48, Deželna komisija za odkup in uravnavo zemljiških bremen za Kranjsko (Provincial Commission for Redemption and Regulation of *Profits à Prendre* in Carniola), fasc. 56, 9:1–5, Protokoll, September 3rd, 1859, & Protokoll, July 11th, 1868.

potential mowing was unusually short at Senožeta in the broader context of hay meadows in Central Europe (usually about nine weeks passed between the end of spring grazing and mowing in eutrophic meadows, and thirteen weeks in oligotrophic grasslands in Central Europe;[1494] see Section 6.1.2.1.1.3). Thus, as a result of the timing of spring grazing and, possibly, drought—the grassland sites at Senožeta were drought-prone—grassland plants did not grow tall enough that farmers decided to mow in mid-July after a late Ascension Day (in late May or early June). In contrast, an early Ascension Day (in late April or early May) made mowing possible there before the re-introduction of grazing in late July. Mowing must have been stimulated by the high quality of hay from Senožeta. Namely, in 1858—thus only one year earlier than the presented grazing regime was documented and only two decades later than the land cadaster noted that these "pastures" were sporadically mowed and they were better than the pastures in the neighboring cadastral municipalities—the botanist Valentin Plemel carried out fieldwork in this very meadow complex. He reported that sainfoin (*Onobrychis viciifolia*; see Figure 164 in Section 10.1.2.2.4) abounded at Senožeta.[1495] Sainfoin is an excellent fodder plant that thrives in some non-fertilized grasslands at relatively dry sites, such as at Senožeta (see Section 6.3.1.1.1). Thus, a specific management regime including highly diverse dates when spring grazing ended, followed by an atypically early start of the following grazing period in the summer, and high cover of a good fodder plant or good fodder plants in the grassland could also stimulate sporadic mowing in some grasslands usually used as pastures.

In contrast to the presented case of Senožeta—where sainfoin abounded, which raised the productivity of the grassland, not only because sainfoin grows vigorously at relatively dry sites, but also because it is a legume, thus symbiotic bacteria in its root nodules add air-borne nitrogen to the soil—long-term soil impoverishment in non-fertilized meadows was likely one of the driving forces that gradually led to the introduction of sporadic mowing and prevalent grazing in some grasslands despite high agricultural density. A wider Central European context makes it clear that it was no exception that plants in some grassland sites on very nutrient-poor soils were only rarely tall enough that farmers decided to mow. Such grasslands were usually grazed, as recorded in the early eighteenth century.[1496]

Sporadic mowing in exceptional pastures was usually carried out either when the yield of grassland biomass was particularly favorable, as in the example of

---

[1494] Janišová et al., 2023, 9, 11.

[1495] In 1858, the botanist Valentin Plemel observed a high abundance of "*Onobrychis sativa* Lmk [Lamarck]" in this grassland complex (Plemel, 1862, 146). "*Onobrychis sativa*" designates sainfoin (*Onobrychis viciifolia*) (compare Aeschimann et al., 2004, vol. 1, 954–955 and Martinčič, 2007c, 317).

[1496] Kapfer, 2010, 135.

Javornik above, or when it was motivated by the scarcity of hay caused by an increased hay demand or by low hay yields on other parcels used by the same farm. Hay scarcity often led to sporadic haymaking in the least favorable parcels where the farm was entitled to mow. In some cases, even such parcels were mowed regularly, due to high agricultural density or specific economic motivation, such as high fodder demand along major roads (see Section 3.1). In the Alps, such least favorable parcels, often mowed irregularly, were usually some wetlands (Section 6.1.2.5.1)[1497] or some grasslands at the highest elevations where mowing was allowed to take place, often very steep, rocky, or barely accessible,[1498] especially where poor hay quality did not stimulate regular mowing (see Section 6.3.1.2). Mowing more wild hay than usual in the case of a hay harvest failure elsewhere was a widespread phenomenon at least after the general decrease in wild haymaking; that is, from the second half of the nineteenth century onward[1499] and in earlier periods of moderate agricultural density and hay demand.

The example of Mount Begunjščica in the Western Karavanks is telling. The section of the rent-roll (*Urbar*) of the seigneury of Radovljica (Germ. *Radmannsdorf*) from 1579, dealing with residents of the villages of Begunje and Zgoša, located at the foot of Begunjščica, documents that those that mowed hay on the slopes of that mountain had to deliver a due in hay.[1500] The highest hay meadows on Mount Begunjščica reached an elevation of at least 1,800 m.[1501] The absence of a statement in the 1579 rent-roll about which farms had mowing rights on that mountain[1502] speaks in favor of an instable area mowed per year. In line with information from the late nineteenth century, hay from Mount Begunjščica was important not only in usual years, because the meadow area at the bottom of the basin was small,[1503] but its availability also significantly increased the resilience of local agriculture regarding the drought hazard. Namely, in the case of an agricultural drought in spring, the hay harvest failed in several meadows in the bottom of the basin and so mountain hay replaced it. This had a direct influence on the mowed area of the high mountain grassland on Begunjščica: in the years of agricultural drought in the valley,

---

1497 Meilleur, 1986, 264, 266–267, 280, 286–287, 295.

1498 Mathieu, 1996, 145.

1499 Grass, 1948, 206.

1500 Zwitter, 2016, 138.

1501 *Arcanum Maps*, Cadastral Maps, Habsburg Empire – Cadastral Maps (XIX. [sic] Century), cadastral municipality of Begunje (Germ. *Vigaun*), 1826; elevations of parcels according to *Environmental Atlas of Slovenia*, Backgrounds, DEM [digital elevation model]. The short even steeper sections of the slope between the mountain meadows and the ridge were used as pastures; the name *Ouzhia Pascha* 'sheep pasture', recorded in the early nineteenth century, reveals that these were not pastures for cattle or horses.

1502 SI AS 173, Imenjska knjiga za Kranjsko (*Gültbuch* for Carniola): Rent-roll (*Urbar*) of the seigneury of Radovljica 1579, 40.

1503 Prešeren, 1933, 98; *Arcanum Maps*, Cadastral Maps, Habsburg Empire – Cadastral Maps (XIX. [sic] Century), cadastral municipality of Begunje (Germ. *Vigaun*), 1826.

the slopes of Begunjščica were mowed up to the rocky uppermost part of the slope.[1504] The data presented from 1579 suggest that local farming might have already functioned in a similar manner in the sixteenth century.

Sporadic mowing had diverse ecological impacts. Although the details depended on the site conditions and on the duration of exclusion of grazing before mowing, a selection of basic impacts of sporadic mowing in pastures can be named. For some general ecological impacts of mowing, see Section 6.1.1.3. In addition, in general, if the plants were allowed to set seeds that ripened by the time of haymaking, which did not happen during usual pasture use, such an interruption of the usual grazing regime was likely to change the species composition to a certain degree, especially if mowing was carried out in a few consecutive years. For instance, some grazing-sensitive plants would temporarily become established in the gaps in the mowed grassland, and the abundance of some heliophilous rosette plants in the pasture would be reduced in a longer absence of grazing; the scythe did not cut rosettes, but a significantly long interval in the shade of taller plants decreased the competition abilities of small-growing heliophilous species.[1505] Another ecological impact, probably not planned as long as pasture management relied on traditional ecological knowledge, but welcome for farmers, was that, because the grassland was left ungrazed longer than usual, sporadic mowing could reduce the hazard of worm infestation in sheep or goats in the pasture after sporadic mowing.[1506]

This section ends with sporadic mowing in pastures, and the next section deals with the opposite case: regular spring grazing in meadows, a topic already addressed at Senožeta near Javornik in this section.

6.2 Spring grazing in meadows

Collective spring grazing in hay meadows was characteristic of Central European grassland management in the colline and the montane vegetation belts. Just like in the rest of this volume, a "meadow" designates here a hay meadow, a parcel that was regularly mowed, in addition to grazing, in line with the terminological explanation in Section 1. Spring grazing in meadows was introduced in Classical Antiquity at the latest and continuously practiced since the Middle Ages, at least. Although it was carried out collectively (with the exception of many isolated farms), it took place in meadows in private possession or, less frequently, in collectively possessed meadows; the total area of meadows that were collective all year round in the Alps in the second millennium was much more limited than meadows in private possession, in particular in the

[1504] Prešeren, 1933, 98.
[1505] Sadar, 1956, 50, 54, 118–119, 131.
[1506] Vidrih, 2005, 119.

second half of the millennium, and it decreased over time. Spring grazing was usually *profit à prendre*; namely, a group of farmers had grazing rights there independent of the ownership and possession rights. Although the hay harvest belonged to the farmer in possession of the meadow, he or she was not the only one (or not at all) entitled to use grazing resources there in spring and autumn. In many places in the Alps, spring grazing in meadows by sheep, cattle, horses, and goats, and in some places also pigs (with rings in their snouts to limit burrowing) and geese, was practiced until between the mid-eighteenth century and the second half of the nineteenth century.[1507] Where it was abandoned prior to the childhood of the oldest living residents, this type of historical meadow use has often already been forgotten.[1508]

In some locations in the lower montane vegetation belt in the Alps, spring grazing still took place in the second half of the nineteenth century or even longer;[1509] in meadows on slopes above the bottom of the Lower Gail Valley (Germ. *Gailtal*, Sln. *Ziljska dolina*) in Carinthia, at Silz in Tyrol, or in several parts of the Ötz Valley (Germ. *Ötztal*: e.g., at Umhausen, Lehn, Längenfeld, and Sölden), for instance; it was practiced at least until the mid-twentieth century, and in several cases longer.[1510] A specific type of early spring grazing in meadows was documented in the Allues Valley (Francoprovençal *Alués*, French *Les Allues*) in the northern French Alps in the first half of the twentieth century. There, spring grazing was carried out individually, on meadows of the livestock owner—it is not clear how long ago such individual management was introduced there. It took place only in a very limited percentage of meadows close to the hamlets and only for a few days. The practice was motivated by the fact that the majority of pastures were quite far away from the settlements in the Allues Valley, from lower settlements at least, and so cattle were first allowed to graze in meadows close to the settlement, when they left the barns after winter, to become fit for grazing again.[1511]

Recently, spring grazing has been restored in a few places, providing interesting insights into its historical ecological impacts.[1512] Spring grazing in

[1507] Novak, 1955; Blaznik, 1970a, 172; Wopfner, 1997, 228–230, 233–234, 242, 264, 269–270, 272–274, 277–278, 289–292, 295, 316–318, 380, 393–394, 403, 474, 477; Kauter, 2002, 9, 109, 116, 129–130, 165; Troßbach, 2008, 600–601; Kapfer, 2010, 133, 136–137, 139–140; Peterken, 2013, 119; Kačičnik Gabrič, 2014, especially pages 8, 12–14, 34, 62, 91–92, 180, 230, 238–240, 276–277, 360–361, 366, 378, 380; Poschlod, 2015, 209; Janišová et al., 2023.

[1508] For example, author's interviews with Mencinger, 2017, in Stara Fužina, and with F. Mlakar, 2017, in Bohinjska Češnjica, both over ninety years old.

[1509] For example, Novak, 1955; Melik, 1950, 119.

[1510] Ilešič, 1931, 141–142; Grass, 1948, 188; Wopfner, 1997, 290–291; Hessenberger, 2022, 147, 149.

[1511] Meilleur, 1986, 284–285, 377–379.

[1512] See Section 6.2.2.

meadows might have been an important factor in the development of certain plant communities.[1513]

Spring grazing was practiced in the great majority of meadows, but not in all of them. Livestock did not graze in spring in intensively managed meadows (at that time represented by only some of the best irrigated meadows and some small grasslands in the immediate vicinity of farmhouses; see Sections 6.1.2.4 and 6.4),[1514] and at some remote meadows where wild hay was made. Namely, distant meadows on mountain slopes were not exceptionally located in places where grazing was impossible due to overly steep slopes (Section 3.1, but see also Section 5), where access for livestock was too hazardous and/or where the absence of surface water nearby made the area unsuitable for (cattle) grazing.[1515] The following examples demonstrate this. Early nineteenth-century records make it clear that haymakers mowing steep meadows at Hudi Konec above Županje Njive north of Kamnik (Germ. *Stein*) in the Kamnik–Savinja Alps (Sln. *Kamniško-Savinjske Alpe*; see slope steepness at Hudi Konec in Section 3.1) wore crampons and that grazing obviously did not take place there because of overly steep slopes.[1516] In some other very steep meadows, the duration of grazing was shorter. In the large steep montane meadow complex of Alniče on the steep slopes of the hill north and northeast of the village of Koroška Bela, for instance, sheep and goat grazing took place only in the autumn, but not in spring, according to mid-nineteenth century data; that is, before the long history of spring grazing in the nearby montane meadows ended.[1517] According to cadastral information from Bohinj in the first half of the nineteenth century, spring grazing was not practiced or at least not economically

---

[1513] Kauter, 2002, 138; Kapfer, 2010, 139.

[1514] Wopfner, 1997, 295; Kapfer, 2010, 136–138 (the example of the large irrigated meadow below the village of Hlebce, where spring grazing took place in a long run, such as presented in Sections 6.2.1.2, 6.4.3.3, and 6.4.5, emphasizes that not all irrigated meadows were excluded from spring grazing).

[1515] Grass, 1948, 146, 163, 183, 206; Novak, 1970, 363; Hubatschek, 1992, 5, 34, 51, 172–173, 178; Pils, 1994, 194; Wopfner, 1997, 317, 330–331; Keller, 2004, 15, 20, 27, 59, 65–66, 137.

[1516] SI AS 176, Franciscejski kataster za Kranjsko (Franciscan Land Cadaster for Carniola), L287, cadastral municipality of Županje Njive, cadastral appraisal protocols: Catastral-Schätzungs-Elaborat, Schätzung des Natural-Ertrages, § 4: Wiesen, and a slightly earlier protocol bound in this source: Protokoll, August 25th, 1830: II. Wiesen.

[1517] SI AS 48, Deželna komisija za odkup in uravnavo zemljiških bremen za Kranjsko (Provincial Commission for Redemption and Regulation of *Profits à Prendre* in Carniola), fasc. 56, 9–4: no. 760, Protocoll, June 28th, 1861, and the list of plots belonging to Alniče. Localization: comparison of *Arcanum Maps*, Cadastral Maps, Habsburg Empire – Cadastral Maps (XIX. [sic] Century), cadastral municipality of Koroška Bela (Germ. *Karnervellach*), 1826, with the georeferenced present-day cadaster (*Environmental Atlas of Slovenia*, Spatial Units, Land Cadastre, & Backgrounds, DEM [digital elevation model]). See also the map *Ledinska imena v občini Jesenice*, 2015, and Section 6.2.2.3 of this volume.

important in high-quality meadows close to villages, but it was a general characteristic of somewhat less productive meadows on forest clearings above the villages (from the vicinity of the villages to the highest meadow plots) and also in remote meadows in parts of the valley bottom that were not permanently populated, such as in Ukanc west of Lake Bohinj (Sln. *Bohinjsko jezero*) and in the Sava Bohinjka Valley east of Nomenj (Figure 69). An exception was the dispersed high permanent settlement of Gorjuše (1,000 m), where spring grazing was also economically important in the meadows between farms but not in intensive ones; there, meadows grazed in spring were only mowed once a year.[1518]

6.2.1 Timing of spring grazing in meadows

The timing of spring grazing was ecologically and economically particularly important (see Section 6.2.2).[1519] First the dates of spring grazing are presented in Section 6.2.1, and then selected ecological and economic impacts of spring grazing in meadows are discussed in Section 6.2.2.

6.2.1.1 Dates of the beginning of spring grazing in meadows

The beginning of spring grazing strongly depended on weather in early spring.[1520] In very many places, spring grazing in meadows started as soon as the snow melted on sunny slopes or the vegetation period started.[1521] This frequent principle is illustrated by a brief selection of examples from several parts of the Alps. The onset of spring grazing as early as soon after the snow had melted where sheep or goats were grazing, or (soon) after the vegetation period started in meadows aimed at cattle, was recorded in the Binn Valley (Germ. *Binntal*) in Wallis (Francoprovençal *Valês*, French *Valais*) in the late eighteenth

---

1518 SI AS 176, Franciscejski kataster za Kranjsko (Franciscan Land Cadaster for Carniola), L30, cadastral municipality of Nemški Rovt, cadastral appraisal protocols: Catastral-Schätzungs-Elaborat, Schätzung des natural Ertrages, § 4: Wiesen; SI AS 176, L56, cadastral municipality of Bohinjska Bistrica, cadastral appraisal protocols: Catastral-Schätzungs-Elaborat, Schätzung des natural Ertrages, § 4: Wiesen; SI AS 176, L72, cadastral municipality of Gorjuše, cadastral appraisal protocols: Catastral-Schätzungs-Elaborat, Schätzung des natural Ertrags, § 4: Wiesen; SI AS 176, L107, cadastral municipality of Češnjica v Bohinju [i.e., Bohinjska Češnjica], cadastral appraisal protocols: Catastral-Schätzungs-Elaborat, Schätzung des natural Ertrages, § 4: Wiesen; SI AS 176, L157, cadastral municipality of Srednja Vas v Bohinju, cadastral appraisal protocols: Catastral-Schätzungs-Elaborat, Schätzung des natural Ertrages, § 4: Wiesen; SI AS 176, L172, cadastral municipality of Nomenj, cadastral appraisal protocols: Catastral-Schätzungs-Elaborat, Schaetzung des Natural Ertrages, § 4: Wiesen; SI AS 176, L284, cadastral municipality of Studor, cadastral appraisal protocols: Catastral-Schätzungs-Elaborat: Schaetzung des Natural Ertrages: § 4: Wiesen.

1519 Peterken, 2013, 121.

1520 Melik, 1950, 118; Novak, 1955, 265, 295; Zenhäusern, 2008, 152, 158–161, 171–174, 186–188, 192–194; compare Janišová et al., 2023, 9.

1521 Wopfner, 1997, 228, 233–234, 269, 378.

and early nineteenth century;[1522] in montane meadows of the western Pokljuka Plateau in the Eastern Julian Alps (Sln. *Julijske Alpe*, Ital. *Alpi Giulie*, Friul. *Alps Juliis*; e.g., in the broader surroundings of Gorjuše, at Ravne, Pokrovec, Goreljek, Jelje, Konjska Dolina, and Zajamniki);[1523] in montane meadows above Hrušica and at the top of the Golica Ridge in the Western Karawanks (Germ. *Karawanken*, Sln. *Karavanke*; on Golica, see Section 6.2.2.3); and in the western Ljubljana Basin in a large meadow below the village of Hlebce,[1524] for instance. Consequently, the variability of spring temperatures mattered a lot.

Historical climatological research for the period between 1500 and 1799, of relevance here because spring grazing in meadows was still commonplace at that time, demonstrates that the average spring temperatures of these three centuries in Central Europe were 0.3 °C lower than the 1961–1990 average. The coldest intervals in those three centuries were from 1687 to 1700 (1.4 °C below the 1961–1990 average), 1740 to 1785 (0.7 °C below the 1961–1990 average), and 1568 to 1600 (0.6 °C below the 1961–1990 average). In the 1680s and 1690s in particular, simultaneously lower May and September temperatures "drastically curtailed the grazing period in the Alps, which led repeatedly to shortages of fodder."[1525] Yearly variations were pronounced; after average spring temperatures cooled down in the 1560s, there were three extremely cold springs (1587, 1596, and 1600), three very warm springs (1571, 1583, and 1599), and nine "average" or "warm" springs (1567, 1574, 1576, 1579, 1581, 1584, 1585, 1591, and 1596) in the period of usually cold springs until the end of the century.[1526] The extreme range of spring temperature variability is illustrated by two opposite anomalies in consecutive years, even though the precise dates of the start of spring grazing in these years are not known. The anomalous early spring of 1614 was evidently extremely cold everywhere in the Eastern Alps because information from Switzerland corresponds with that from the Slovenian Alps. Severe cold lasted from November 1613 to April 1614, followed by cold weather in a considerable part of May that must have acutely reduced the possibilities of early spring grazing in meadows. However, the previous winter (1612/13) was unusually mild.[1527]

The impacts of weather variations on the beginning of spring grazing can be understood based on data from the Binn Valley in Wallis from 1770 to 1812. The majority of meteorological springs within this interval were cold there, and many of them were also wet. Goats opened the spring grazing season in

1522 See below in this section.

1523 Novak, 1955, 265, 295.

1524 SI AS 48, Deželna komisija za odkup in uravnavo zemljiških bremen za Kranjsko (Provincial Commission for Redemption and Regulation of *Profits à Prendre* in Carniola), fasc. 163, 17–19: no. 2277 de 1858, Protocoll, May 10th, 1858.

1525 Pfister et al., 2018, especially pages 274, 277–280 (quote from page 280).

1526 Pfister et al., 2018, 274, 278.

1527 Zwitter, 2013b, 341–344, 348, 362–363, 371–375.

meadows in the valley (at 1,300–1,600 m) after the snow had melted at least on sunny slopes (and possibly some of the earliest plants started sprouting there). The dates when goats started grazing in the Binn Valley are available for thirty years between 1770 and 1812. According to these data, grazing of goats started between February 25th (in 1779) and May 20th (in 1771), on average on April 21st (median: April 27th), but with a standard deviation of nineteen days. In some years, spring grazing had to be interrupted because snow had covered the grasslands again.[1528] This is in line with the finding that, at several places in the Alps and Pre-Alps, spring grazing in meadows started well before Saint George's Day (i.e., before April 23rd or 24th, depending on the bishopric[1529]) before and after the Gregorian calendar reform (thus, see many examples where spring grazing in meadows already ended on Saint George's Day in Section 6.2.1.2).[1530] The example from the Binn Valley clearly demonstrates that such early starts of spring grazing in meadows were not limited to the colline and lower montane vegetation belt but were also characteristic of some settled places at higher elevations. In general, where sheep grazed early in remote grasslands, abundant snow in April, for instance, could kill some of them before the avalanche hazard was mitigated to the degree that farmers could ascend to their sheep.[1531] At some places in the Alps, farmers had the right to move their livestock to lower (wetland) pastures if snow covered those meadows where spring grazing would have usually taken place at that time (see Sections 4.1.1.2 and 6.1.2.5.1.1).

Apart from the weather in the year in question, one must also consider the influence of the following further environmental and socioeconomic factors on the starting date of spring grazing: (1) the quantity and quality of hay stored in the previous year in relation to the number of livestock present, (2) the management of meadows where spring grazing took place, (3) plant nutrient availability in spring-grazed wetland meadows as a matter of natural processes and of broader human impact in the landscape, and (4) local rules or traditions, which are discussed in the next paragraph. Namely, in spring, significant quantities of livestock fodder were available earlier in fertilized and some irrigated meadows and in nutrient-rich or mesotrophic wetlands than in nutrient-poor grasslands or oligotrophic wetlands (see Sections 6.1.2.5.1.1, 6.2.2, 6.3.1, 6.3.2, and 6.4.1). When little hay was stored, when the quality of hay was lower than usual (e.g., because rain during hay-drying time leached nutrients from mowed plants; see Sections 4.1.1.3 and 4.1.1.4), and/or when an early winter required an early start of indoor livestock keeping, so that larger quantities of hay were consumed until spring, farmers were more likely to

[1528] Zenhäusern, 2008, 152, 158–161, 171–174, 186–188, 192–194.
[1529] On regional differences in Saint George's Day (April 23rd or 24th), see Grotefend & Ulrich, 1971, 60.
[1530] See also Zwitter, 2013a, 89–90.
[1531] Wopfner, 1997, 269–270.

open the spring grazing season earlier, even when plant development was still very poor.[1532] On the other hand, in the grasslands grazed intensively late in the previous autumn, there is delayed vegetation development the next spring,[1533] which limited the possibility of early spring grazing.

Rules or traditions also had an impact on the start of spring grazing in meadows. In some cases, a single date of beginning spring grazing in meadows was recorded in history. In Bohinj, for instance, in some locations close to the lowest edge of the montane vegetation belt, spring grazing in meadows started on (or around) Saint George's Day (April 24th) or in early May based on mid-nineteenth-century information.[1534] Saint George's Day was a frequent start of common grazing in spring, not limited to meadows, as demonstrated by data from Slovenian territory (see the comment on the symbolic meaning of this saint's day in Section 6.2.1.2).[1535] Similarly, in the eastern part of the Pokljuka Plateau or at the foot of the Alps there (in the community of Bohinjska Bela), spring grazing in meadows started on May 1st according to a source from 1831.[1536] Clearly, such dates could not be observed when the weather was particularly severe. However, it is not clear whether the contrary option of an earlier grazing start in meadows was allowed there. Variable dates around a certain day (such as Saint George's Day) speak in favor of the interpretation that there was more flexibility. If so, this means that the weather around 1830 (in the community of Bohinjska Bela) or in the mid-nineteenth century (in Bohinj) in combination with socioeconomic factors and tradition usually led to the start of the spring grazing season in meadows on the reported date or around it, rather than this being the earliest allowed date of spring grazing in meadows.

6.2.1.2 Dates when spring grazing in meadows ended

In many Central European meadows in the colline and lower montane vegetation belts, spring grazing lasted no longer than late April or mid-May.[1537] However, in the Alps,[1538] just like in the Carpathians,[1539] later dates when meadows were closed in the spring (i.e., grazing was not allowed to take place there any

1532 Wopfner, 1997, 259; Kauter, 2002, 106, 109, 128; Zenhäusern, 2008, 194–196, 211–215.
1533 Sadar, 1956, 169–170, 194–195.
1534 Novak, 1955, 268, 275, 277–279, 290; conversion of Saint George's Day into the date is based on Grotefend & Ulrich, 1971, 60, and the church historian Lilijana Žnidaršič Golec, personal communication.
1535 Novak, 1970, 348.
1536 SI AS 176, Franciscejski kataster za Kranjsko (Franciscan Land Cadaster for Carniola), L335, cadastral municipality of Bohinjska Bela, cadastral appraisal protocols: s.n., 335/4, 57, 99.
1537 Kapfer, 2010, 133–134, 136.
1538 See below in this section.
1539 Janišová et al., 2023, 9, 12.

more until mowing) were also frequent. Only in some locations was a harsher climate a relevant reason for later spring grazing in meadows, as explained in this section. Namely, in many regions in Europe, "[l]ocal climate and elevation were not always the primary factors influencing farming traditions."[1540] This finding of the research into historical final dates of spring grazing in meadows in the Carpathians is in line with examples from Great Britain in the late eighteenth or early nineteenth century. They show that "[i]n Derbyshire, on the upland–lowland boundary, some [meadows] were shut [i.e. closed for grazing] on April 1st, but others were grazed until mid-May, and in Shropshire they were shut at Candlemas (February 2nd)."[1541] This section demonstrates that the ends of spring grazing in meadows in the Alps were equally diverse if they are looked at from the perspective of climate and vegetation development. Before presenting the dates when spring grazing in meadows ended, knowledge of the historical calendar update and its impact on the agricultural calendar are required, and so they are outlined in the next paragraph.

In 1582, ten days were left out as a result of the calendar reform, which was a challenge for the agricultural calendar relying on saints' days. The reform was implemented in 1582 or within the following two years in many parts of the Alps, but in some places the previous calendar was kept in use much longer; for example, in Wallis until 1655, in many other parts of Switzerland until the eighteenth century, and exceptionally even until the early nineteenth century.[1542] Nikolaus Grass was able to demonstrate that, in some cases, the last date of spring grazing in meadows in Tyrol was directly adjusted to the 1582 calendar reform, so that meadows were closed on a date ten days later.[1543] However, in general, the adjustment of the agricultural calendar to the shift from the Julian to the Gregorian calendar was only partial,[1544] and so the final dates of spring grazing were not adjusted to the Gregorian calendar reform everywhere in the Alps, as presented in this section.

Early ends of spring grazing in meadows in the Alps are now examined. Saint George's Day (April 23rd or 24th) was one of the frequent early dates when meadows were closed for grazing.[1545] Saint George's Day was the Christian version of the celebration of green spring. In the Rosen Valley (Germ. *Rosental*, Sln. *Rož*) in Carinthia, the saying was recorded by ethnologists in the

[1540] Janišová et al., 2023, 9.
[1541] Peterken, 2013, 120.
[1542] Grotefend & Ulrich, 1971, 24–27; Makarovič, 1995, 102; Kočevar, 2014.
[1543] Grass, 1948, 84, 188–189, 196–198, 200. Most probably, the fact that livestock had to leave the meadows at Silz in Tyrol "zechen tag nach sanct Pongrazentag" (i.e., ten days after May 12th, Saint Pancras's Day) according to a source from 1616 is also a consequence of this postponement by ten days due to the calendar reform. The conversion of Saint Pancras's Day into a date is based on Grotefend & Ulrich, 1971, 85.
[1544] Makarovič, 1995, 102, 108.
[1545] See below in this and the following paragraphs.

early twentieth century that "God blesses the grass on Saint George's Day."[1546] Such symbolism, in the Alps often more appropriate before the calendar reform than later (focusing on the times before the recent global warming),[1547] explains why this was a frequent early date when spring grazing in meadows ended. For instance, meadows in the Inn Valley (Germ. *Inntal*) close to Innsbruck (c. 600 m) were closed for grazing from Saint George's Day onward, according to a charter from about 1200 (corresponding to early May in today's calendar). Another example from North Tyrol: since about 1440, spring grazing ended on the same date in those fertilized meadows in the Kauns Valley (Germ. *Kaunertal*, in the Upper Inn catchment; at Feichten, for instance, at an elevation of at least about 1,300 m) where the spring grazing season ended the soonest, thus in a much harsher climate than close to Innsbruck. Nevertheless, spring grazing ended on the same day in both cases presented.[1548] Further meadow complexes where spring grazing ended on St. George's Day include examples from Trentino–Alto Adige (South Tyrol): meadows on slopes above Tesido (Germ. *Taisten*), a village at 1,200 m, according to a source from 1537; meadows close to San Martin de Tor (Germ. *St. Martin in Thurn*, c. 1,150 m);[1549] and those meadows at Prato allo Stelvio (Germ. *Prad am Stilfserjoch*) and Agumes (Germ. *Agums*) (villages at c. 900 m), where the spring grazing season ended soon, in line with an ordinance from 1711, which was based on older regulations and clearly indicated Julian calendar dates.[1550] Spring grazing in meadows also ended on Saint George's Day in many other complexes in the Alps. Heading further east, other examples were certain meadows in private possession between Rateče and Fusine in Valromana (Friul. *Fusinis*, Germ. *Weißenfels*, Sln. *Bela Peč*) according to regulations from 1523 and 1559.[1551] The same date, April 24th, designated the end of spring grazing in a large meadow below the village of Hlebce in the upper Ljubljana Basin (c. 510 m), which was documented in 1579 as an established custom.[1552] In all these cases, spring

---

1546 Kuret, 1989, 253–255, 268 (quote from p. 255).

1547 Compare Kuret, 1989, 261, 265, 268.

1548 Wopfner, 1997, 272–273, 393–394, 477.

1549 Grass, 1948, 189.

1550 Wopfner, 1997, 272.

1551 AT KLA 112, Allgemeine Handschriftenreihe, Ms. 2732, Rent-roll of the seigneury of Fusine (Friul. *Fusinis*, Germ. *Weißenfels*, Sln. *Bela Peč*) 1636, 422–423, 427–430, 433.

1552 Zwitter, 2016, 144; Zwitter, 2018, 172; SI AS 173, Imenjska knjiga za Kranjsko (*Gültbuch* for Carniola): Rent-roll (*Urbar*) of the seigneury of Radovljica, 1579, 384–385. More precise localization is based on the following information: this rent-roll makes it clear that this meadow was located below the village of Hlebce. Moreover, there was still a large meadow (17 ha) consisting of two adjacent parcels divided only by a road in the early nineteenth century. Its toponym *na Knesnek* means 'on the princely land' (SI AS 176, Franciscejski kataster za Kranjsko [Franciscan Land Cadaster for Carniola], L169, cadastral municipality of Nova Vas, map sheet A03, and Protocoll der Grund-Parcellen der Gemeinde, parcel nos. 343, 343½). In fact, the ruler of the historical province of Carniola (Germ. *Krain*, Sln. *Kranjska*) was designated as a prince (Sln. *knez*; Štih, grazing ended on Saint George's Day at least prior to the calendar reform. A parallel with the Carpathians shows that, also in that mountain range, spring grazing ended in some places on Saint George's Day, Julian calendar style.[1553]

The same date, April 24th, still designated the end of spring grazing in the same large meadow below the village of Hlebce in the upper Ljubljana Basin after the calendar reform, until the abolition of spring grazing there in the mid-nineteenth century.[1554] Thus, the duration of spring grazing there was not adjusted to the calendar reform, which would require a postponement of the end of spring grazing until May 4th. The motivation for this must have been the strong tradition of Saint George's Day in the agricultural calendar (see above), possibly in combination with the fact that the reformed calendar made two mowings between the grazing in the same grassland in spring and autumn more certain (but at the same time it reduced the amounts of fodder available there during the spring grazing season). Namely, mid-nineteenth-century sources explain that such an early end of spring grazing there (April 24th after the calendar reform) allowed two mowings in this irrigated meadow area at 500 m above sea level in the bottom of the basin (see Section 6.1.2.3), followed by autumn grazing, which usually started significantly earlier than September 29th.[1555] Where dates of spring grazing in meadows did not adapt to the sixteenth-century calendar reform and grazing was limited to early spring, such as in the case of the large meadow below Hlebce presented in Section 6.2.1.2, the calendar reform made a real difference: where a pre-Gregorian date was

Simoniti, & Vodopivec, 2008, 67) and the seigneury of Radovljica (Germ. *Radmannsdorf*), which this large meadow belonged to in 1579, was at that time formally a princely seigneury. In the 1820s, this meadow still belonged to the seigneury of Radovljica; the owner of both grassland parcels was Vinzenz Thurn, who was at that time one of the two landlords in the seigneury of Radovljica (SI AS 176, L169, cadastral municipality of Nova Vas, Protocoll der Grund-Parcellen der Gemeinde, parcel nos. 343, 343½; Smole, 1982, 407). This was the only meadow under the direct management of the seigneury of Radovljica in the cadastral municipality of Nova Vas in the early nineteenth century (SI AS 176, L169, cadastral municipality of Nova Vas, Alphabetisches Verzeichniss der Grund-Eigenthümer der Gemeinde, no. 37). This meadow is also dealt with in Sections 6.1.2.3, 6.4.3.3, 6.4.5, and 6.6.1.

1553 Janišová et al., 2023, 9, 12. The map in Janišová et al., 2023, 12, shows all dates in Gregorian calendar style, they are all comparable from the perspective of vegetation development (Monika Janišová, personal communication). Thus the dates May 5th or May 6th on this map show the historical end of spring grazing around Saint George's Day, Julian style. In addition, April 23rd or April 24th was also kept in the Gregorian calendar in some cases, building on the older Julian calendar tradition.

1554 SI AS 48, Deželna komisija za odkup in uravnavo zemljiških bremen za Kranjsko (Provincial Commission for Redemption and Regulation of *Profits à Prendre* in Carniola), fasc. 163, 17–19: no. 2277 de 1858, Zur Sitzung am 14. Oct. 1858 and Protocoll, May 10th, 1858.

1555 SI AS 48, Deželna komisija za odkup in uravnavo zemljiških bremen za Kranjsko (Provincial Commission for Redemption and Regulation of *Profits à Prendre* in Carniola), fasc. 163, 17–19: no. 2277 de 1858, Zur Sitzung am 14. Oct. 1858.

also kept after the reform was introduced, spring grazing had to end ten days earlier in terms of vegetation development.

Spring grazing until April 24th after the calendar reform also has parallels elsewhere. In line with a 1792 record, this was the case in the community of Ponteacco (Sln. *Petjag*), located at the southern foot of the Julian Alps, close to the Venetian Plain, thus characterized by relatively mild winters, without long-lasting snow, as stated around 1830.[1556] Very similar was the timing of the end of spring grazing (eight days before April 24th, Julian calendar, in the time of the calendar reform equivalent to April 26th in today's calendar) in those meadows at Prato allo Stelvio and Agumes where the spring grazing season ended the soonest, in line with the aforementioned ordinance from 1711 that was based on older regulations. However, the climate there is harsher than at Ponteacco. Even harsher is the climate of the Ötz Valley, where spring grazing in some meadows at its bottom (at more than 1,000 m) also ended on April 24th or May 1st in the 1930s or 1940s.[1557] Again, a parallel with the Carpathians shows that the same date, St. George's Day, Gregorian style, was respected there in some places.[1558]

Another group of early ends of the spring grazing season consists of early May dates after the calendar reform; they were similar to April 24th prior to the calendar reform in terms of plant development. For instance, spring grazing in meadows ended on May 1st in the relatively harsh climate in parts of the bottom of the Ötz Valley in the 1930s or 1940s, as mentioned above.[1559] May 1st as the end of spring grazing in meadows also has parallels in the eastern Julian Alps or in much lower areas at the foot of the Alps; for example, in part of the low central inner-Alpine Ljubljana Basin (e.g., in the cadastral municipality of Hraše)[1560] and in certain meadows in the community of Podhom,[1561] according to sources from 1831. In the Eastern Julian Alps, spring grazing in meadows sometimes lasted only until May 1st also at higher sites, at elevations comparable with the aforementioned data from the Ötz Valley. According to the rule valid in the community of Bohinjska Češnjica and recorded in 1608, grazing in meadows in Koprivnik or around it (Figure 69; at least about 1,000 m) was prohibited after

1556 Beguš, 2015, 37–38, 90, 95–96; the date conversion of Saint George's Day is based on Grotefend & Ulrich, 1971, 60.

1557 Wopfner, 1997, 272, 291; the chronological definition of the time of Wopfner's writing follows Meixner & Siegl, 2004, 232–233, 235.

1558 Janišová et al., 2023, 9, 12.

1559 Wopfner, 1997, 291; the chronological definition of the time of Wopfner's writing follows Meixner & Siegl, 2004, 232–233, 235.

1560 Geršič, Gabrovec, & Zwitter, 2018, 68, 71.

1561 SI AS 176, Franciscejski kataster za Kranjsko (Franciscan Land Cadaster for Carniola), L22, cadastral municipality of Podhom, cadastral appraisal protocols: s.n., 22/4, 57, 99; p. 89 proves that the information can also refer to the areas used by people from the community of Podhom, not only to the cadastral municipality.

May 1st and until hay was stored there. In the meantime, farmers were not even allowed to graze their livestock at Koprivnik in the meadows they possessed. By 1608 this rule was already regarded as a custom.[1562] However, the duration of spring grazing around Koprivnik was probably prolonged later. Namely, it was recorded in 1831 that spring grazing in that cadastral municipality lasted in all types of meadows (except for meadows at farmyards) until May 25th,[1563] and so this information indicates a shift in spring grazing dates in the vicinity of Koprivnik between the early seventeenth and early nineteenth centuries.

Such late dates for the end of spring grazing in meadows were also relatively common in the Alps and in the Carpathians.[1564] Figure 69 presents a selection of dates when spring grazing in meadows ended in Bohinj in the mid-nineteenth century. They demonstrate that the end of spring grazing in meadows there did not depend on elevation and microclimatic conditions: May 25th was a very common end date of spring grazing in meadows in the valleys and also in higher montane meadows, except for the north of the territory observed, where spring grazing ended on May 12th in the mid-nineteenth century.[1565] The meso- and microclimate of the mapped locations differ significantly. For example, according to a modeled phenological map, considering phenological observations and temperature data from 1965 to 1994, at least at sixty stations around Slovenia, dandelion (*Taraxacum officinale*) started blooming in extensively managed meadows between April 11th and 20th in the majority of locations of historical spring grazing at the bottom of the basin in Figure 69. At the highest mapped locations of historical spring grazing in meadows in the south (such as the Za Malim Vrhom Alp [Sln. *Planina za Malim vrhom*], southwest of Strmne at 1,100 m) and in the north (e.g., at Zgornje Gorjuše, 1,000 m), it only started blooming about three weeks later, in May.[1566] Similarly, phenological modeling for 1971–2000 indicates that first beech (*Fagus sylvatica*) leaves unfold at the end of April at the lowest sites where spring grazing used to last until May 25th, and they unfold about three weeks later in the climatically least favorable sites of spring grazing in meadows that lasted until May 12th or May 25th as mapped in Figure 69.[1567] Despite these significant differences in plant development, historical spring grazing in meadows in all locations listed lasted until May 25th (or May 12th).[1568] Due to climate change, these phases

[1562] SI AS 721, Gospostvo Bled (The Seigneury of Bled, Germ. *Veldes*), Iustitialia, vol. 13, [July 1st, 1608].
[1563] SI AS 176, Franciscejski kataster za Kranjsko (Franciscan Land Cadaster for Carniola), L72, cadastral municipality of Gorjuše, cadastral appraisal protocols: s.n., 72/4, 57, 95, 97, 99.
[1564] On the Carpathians, see Janišová et al., 2023, 9, 12.
[1565] Novak, 1955, 279, 293, 295.
[1566] Zrnec, 1998, 112–113; Zrnec, 2001, 66; Žust, 2016, 26.
[1567] *Environmental Atlas of Slovenia*, Climate, Phenology, Leaf Unfolding of Common Beech, 1971–2000; Žust, 2016, 42–43.
[1568] Novak, 1955.

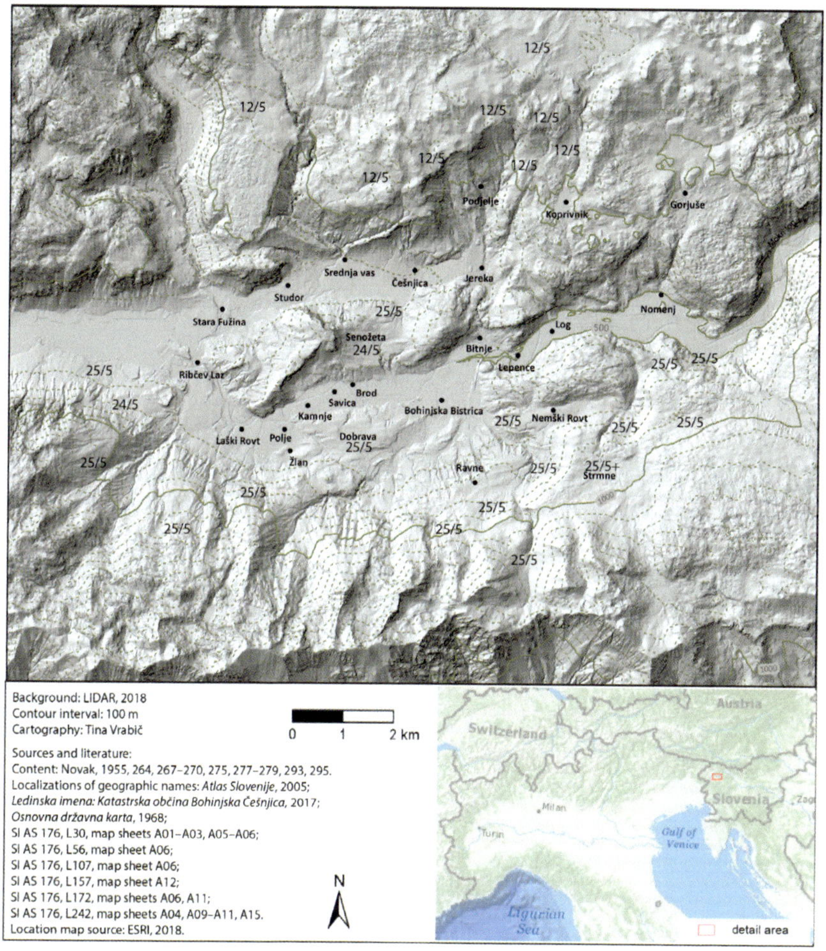

Figure 69: Mid-nineteenth-century dates in May when spring grazing in meadows in Bohinj ended.

of plant development were not the same in times of historical spring grazing in meadows; nevertheless, the data presented here highlight the diversity of microclimates in areas where spring grazing ended at the same time.

Sources from Bohinj prove that May 25th had a long tradition of being the final limit of spring grazing in meadows, also in some places in the northern part of Figure 69, where only May 12th occurs in Figure 69. According to information from 1635, for instance, unfenced spring grazing in meadows on slopes in the lower montane vegetation belt close to the village of Gorjuše (Figure 69, roughly 1,000 m) was allowed to last until May 25th.[1569] The same end

[1569] Novak, 1975, 5–6; for conversion of *Urbani* 'Saint Urban's Day' to the date, see Grotefend & Ulrich, 1971, 104.

of spring grazing there was still the general rule in the cadastral municipality of Gorjuše in the early nineteenth century.[1570]

May 25th as the date when meadows were closed for grazing has parallels elsewhere in the Alps. As early as before the calendar reform, spring grazing at Wieting in Carinthia west of Saualpe, located at 700 m, lasted until May 25th, as reported in 1471 (corresponding to June 2nd in today's calendar).[1571] The selection of May 25th as the end of spring grazing, where it ended late, is logical because in the Julian calendar Saint Urban's Day (May 25th) was considered the first day of summer in terms of the agricultural calendar, as recorded in Primož Trubar's sixteenth-century calendar. In the sixteenth century, May 25th corresponded to June 3rd or June 4th in today's calendar,[1572] and so this date was just after the start of the meteorological summer, as defined today. The tradition of May 25th as the first day of summer was also kept by farmers after the Gregorian calendar reform.[1573] After that reform, spring grazing in meadows ended on May 25th at Gramais in Tyrol (a village at 1,340 m), for instance, as documented in 1690. In contrast to Wieting, May 25th was the result of adjustment to the sixteenth-century calendar reform at Gramais; earlier, spring grazing in meadows there had ended in mid-May. Similarly, spring grazing in meadows at Silz in Tyrol ended on May 22nd in line with a document from 1616, as an obvious adaptation to the calendar reform.[1574]

The end of spring grazing in meadows corresponding with the end of the first third of May (Gregorian calendar), mentioned above in some cases in Bohinj (see Figure 69), also has parallels elsewhere. In a couple of meadows at Zgornje Jezersko (about 900 m) in the contact zone between the central part of the Karawanks and the Kamnik–Savinja Alps, spring grazing lasted until May 12th according to a regulation from 1663 that possibly had deeper historical roots.[1575] Similar examples from South and North Tyrol follow. At Luson (Germ. *Lüsen*, a village at 980 m), spring grazing in higher meadows lasted until May 12th,

1570 SI AS 176, Franciscejski kataster za Kranjsko (Franciscan Land Cadaster for Carniola), L72, cadastral municipality of Gorjuše, cadastral appraisal protocols: s.n., 72/4, 57, 95, 97. In 1860, May 11th was the last permitted day of spring grazing for some of the farms with grazing rights in that cadastral municipality (Novak, 1955, 265), but further sources need to be studied to determine whether this involved a general shift in the end of spring grazing to an earlier date in the mid-nineteenth century or, locally, May 11th had already been the final date of spring grazing in meadows in that area before.

1571 Dinklage, 1966a, 82, 84. The conversion is based on *Julian Day and Civil Date Calculator* and the calculations by the astronomer Tomaž Zwitter.

1572 The conversion for two examples, 1501 and 1581, is based on *Julian Day and Civil Date Calculator* and the calculations by the astronomer Tomaž Zwitter

1573 Makarovič, 1995, 102–104.

1574 Grass, 1948, 188–189.

1575 AT StiAStP, Grundherrschaft Eberndorf (Handwritings), vol. 181, Gerichtsprotokoll (from the 1660s that begins in 1661), 69–70 (September 5th, 1663); for conversion of *Pancratii* 'Saint Pancras's Day' to the date, see Grotefend & Ulrich, 1971, 85.

allegedly for a long period. The end of spring grazing in some meadows at Prato allo Stelvio and Agumes (villages at c. 900 m) was similar because it lasted until May 1st Julian style (a 1711 ordinance, as above, thus at the time of the calendar reform corresponding to May 11th in today's calendar). Similarly, after 1440, spring grazing lasted until May 2nd on the Julian calendar (corresponding to May 11th in today's calendar) in those manured meadows in the Kauns Valley (at least about 1,300 m, at Feichten, for instance) that were not closed for grazing on April 24th (see above). Both dates from the Kauns Valley refer to fertilized meadows, whereas non-fertilized meadows were grazed longer.[1576]

Less frequently, spring grazing in meadows in the Alps ended later than May 25th or even much later. In the Strmne meadow complex, located in the lower montane vegetation belt east of Bohinjska Bistrica (Figure 69), spring grazing lasted later than May 25th in the mid-nineteenth century. Livestock grazed there until the ascent to much higher mountain pastures. Sources from 1681 and 1784 already prove that grazing took place in those meadows later than May 25th.[1577] Grazing in meadows at 700 m at Wieting in Carinthia until May 25th, as reported in 1471 (i.e., the equivalent of June 2nd in today's calendar),[1578] was mentioned above. Spring grazing in some meadows in possession of farmers from Covelano (Germ. *Göflan*) in South Tyrol had to end by June 15th according to a document from 1564, thus prior to the calendar reform (corresponding to June 25th in today's calendar).[1579] The same date, June 15th (Julian calendar), designated the end of spring grazing in non-fertilized meadows at Feichten in the Kauns Valley; then livestock moved to higher mountain pastures, until 1440 (in 1440 corresponding to June 24th in today's calendar); in 1440, the end of spring grazing in meadows there was moved to a date more than two weeks earlier (the end of May in the Julian calendar; i.e., around June 9th in today's calendar). Spring grazing also ended on June 15th in the last meadows used by farmers from Prato allo Stelvio and Agumes (the aforementioned ordinance from 1711 is based on older regulations, thus Julian calendar dates).[1580] At the Villandro Alp (Ital. *Alpe di Villandro*, Germ. *Villanderer Alm*, with the lower edge at about 1,700 m[1581]), northeast of Bolzano (Germ. *Bozen*), mowed meadows covered about a half of the large grassland complex, in line with later data, possibly from the nineteenth century. Grazing in the meadows there started in spring and lasted until July 25th, as reported in the early sixteenth century (in 1523, this corresponded

1576 Wopfner, 1997, 264, 272, 318, 393–394. The date conversion is based on *Julian Day and Civil Date Calculator* and the calculations by the astronomer Tomaž Zwitter.

1577 Novak, 1955, 268, 282; Novak, 1975, 5.

1578 Dinklage, 1966a, 82, 84. Conversion is based on *Julian Day and Civil Date Calculator* and the calculations by the astronomer Tomaž Zwitter.

1579 Grass, 1948, 189. Conversion is based on *Julian Day and Civil Date Calculator* and the calculations by the astronomer Tomaž Zwitter.

1580 Wopfner, 1997, 264, 272–273, 318, 393–394. Conversions are based on *Julian Day and Civil Date Calculator* and the calculations by the astronomer Tomaž Zwitter.

1581 *Villanderer Alm.*

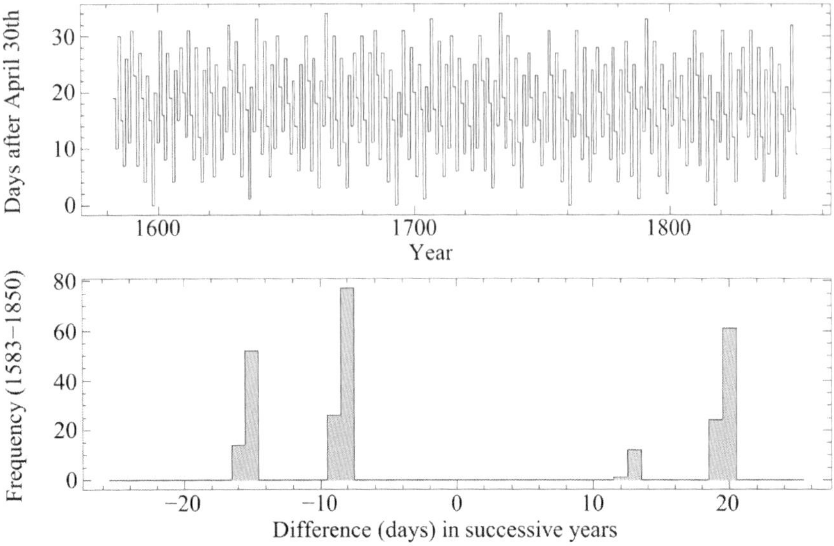

Figure 70: Yearly oscillations of Ascension Day, 1583–1850. The upper graph shows the date of Ascension Day from 1583 to 1850; that is, from the Gregorian calendar reform until roughly the end of spring grazing in meadows in the great majority of the Alps. The lower graph shows the frequency of difference in dates of Ascension Day from one year to another from 1583 to 1850: the highest bar at –8 means that Ascension Day was most frequently (almost eighty times from 1583 to 1850) eight days earlier than in the previous year, and the second-highest bar designating +20 indicates that the second most frequent option (more than sixty times in the same period) was that Ascension Day was twenty days later than in the year before. Data source: Grotefend & Ulrich, 1971, 144–213. Author of the graph: Tomaž Zwitter.

to August 3rd in today's calendar). After the calendar reform in the 1580s, the date was accordingly postponed to August 4th. During this spring-and-summer grazing period, every meadow was grazed at least two times before being closed for grazing. After such long grazing, meadows were still mowed there in the same year; it is no wonder that the 1523 complaint laments that at those elevations only short hay plants grow after July 25th (i.e., August 3rd in today's calendar) and that such long spring and summer grazing in the meadows led to centuries of disputes. Namely, farmers with grazing rights in spring and summer were not the hay owners there. The *profit à prendre* of spring and summer grazing in meadows at the Villandro Alp was abolished in 1823.[1582]

In some cases, the end of spring grazing was determined by a movable feast, and so it varied greatly at the year-to-year level, most often eight or fifteen days earlier, or twenty days later (Figure 70). The persistence of such regulations varied. Spring grazing in meadows at the Siusi Alp (Ital. *Alpe di Siusi*, Germ.

[1582] Grass, 1948, 84, 200–201, 207. Conversion is based on *Julian Day and Civil Date Calculator* and the calculations by the astronomer Tomaž Zwitter.

Seiser Alm, at more than c. 1,800 m) in South Tyrol east of Bolzano (Germ. *Bozen*) lasted until Trinity Sunday, a movable feast between May 17th and June 20th, until 1582 (at the time of the calendar reform corresponding to the dates between May 27th and June 30th in today's calendar). With the calendar reform, the end of spring grazing there was converted into a fixed date, June 1st. The area of meadows on the Siusi Alp was extensive, so that according to a rough estimate about 1,800 carts of (short) hay were reported to have been mowed there (mainly in late August) in about 1600.[1583]

In contrast, there were also complexes of meadows in the Alps where such strong variation in the end of spring grazing determined by a movable feast persisted well into the nineteenth century. These include very common spring grazing in meadows until Ascension Day (a movable feast; Figure 70) in the easternmost Julian Alps—for example, at elevations of 700 to 1,000 m but also higher, and even as low as slightly above 400 m at the foot of the hills, according to information from 1831.[1584] The limit of Ascension Day means that spring grazing in meadows ended between April 30th and June 3rd (Figure 70), depending on the date of Easter.[1585] Even though there are many examples in which Ascension Day still represented the end of spring grazing in the easternmost Julian Alps in the early nineteenth century, in some places in that region this rule was also abolished earlier. In the meadows on the northern slope of the Jelovica Plateau east and south of the village of Kupljenik (c. 400–1,000 m; see Figure 4), there seems to have been an attempt in the mid-seventeenth century to move the end of spring grazing from Ascension Day to the fixed date of May 3rd or 4th, but in the end a compromise was made in 1659, clearly stating that spring grazing in these meadows was allowed to take place there until May 20th regardless of the date of Ascension Day. The fixed date of May 20th was evidently important because the document from 1659 was copied in 1731. The validity of this rule was, however, limited to a section of meadows, whereas in the others the older custom of grazing until the date with a long

---

1583 Grass, 1948, 167, 192, 196–198, 200, 208; Wopfner, 1997, 316–317, 323–324; Grotefend & Ulrich, 1971, 144–213.

1584 These examples include meadows close to the eastern edge of the Pokljuka Plateau, major parts of the Radovna Valley at elevations of about 700 to 1,000 m, meadows in the Poljana Valley, meadows at the southern foot of the Mežakla Plateau and on the southern slopes of the Mežakla Plateau, meadows on the western Jelovica Plateau, and even at elevations of only slightly more than 400 m at the northern foot of the Jelovica Plateau (SI AS 176, Franciscejski kataster za Kranjsko [Franciscan Land Cadaster for Carniola], L210, cadastral municipality of Poljšica, cadastral appraisal protocols: s.n., 210/4, 57, 99; SI AS 176, L310, cadastral municipality of Spodnje Gorje, cadastral appraisal protocols: s.n., 310/4, 57, 97, 99; SI AS 176, L322, cadastral municipality of Višelnica, cadastral appraisal protocols: s.n., 322/4, 57, 97, 99; SI AS 176, L335, cadastral municipality of Bohinjska Bela, cadastral appraisal protocols: s.n., 335/4, 57, 99; SI AS 176, L343, cadastral municipality of Selo pri Bledu, cadastral appraisal protocols: s.n., 343/4, 57, 99; for localization, see also *Osnovna državna karta*, 1967).

1585 Grotefend & Ulrich, 1971, 65, 144–213.

tradition remained respected.[1586] Ascension Day was still documented as the end of spring grazing in many meadows in the same cadastral municipality in the first half of the nineteenth century (1831).[1587]

6.2.2 Ecological and economic implications of spring grazing

Spring grazing in meadows had many important ecological and economic consequences, impacting grassland biocenoses and hay quality and quantity, as explained in this introduction and in subsequent subsections. As usual for grazing, the types of livestock involved and the grazing regime led to significant differences in the impact of this kind of land use on the ecological communities in grasslands.[1588]

Nutrients from the excrement of grazing livestock returned to the meadow in a form available to plants. If the meadow had only been used for haymaking, nutrients would have accumulated in the barn and would have been applied to arable land in many places (see the introduction to Section 6.3 and Section 6.3.2). Excrement was naturally distributed more regularly around the grassland in the case of sheep grazing than in the case of cattle grazing.[1589] On a daily basis, the cow pies of a single bovine weighing 500 kg (thus much more than an average cow in most regions of the Alps almost throughout the study period of this book; see Section 5) cover about 0.8 m^2 of grazed grassland and weigh about 25 kg, and so the manuring impacts of spring and autumn grazing in meadows were not negligible.[1590] This was particularly relevant in most spring- and autumn- grazed meadows that were otherwise not fertilized, and so the manuring impact of grazing additionally increased the species rich-

[1586] SI AS 721, Gospostvo Bled (The Seigneury of Bled), Iustitialia, fasc. 32: dispute between the communities from Kupljenik and Selo over grazing rights on the slopes of Baba Hill (Sln. *Pod Babo*), 1658–1659, 1731: May 30th, 1658, January 21st, 1659, and its copy from 1731; the localization of meadows in clearings below Baba Hill and above the bridge leading to Selo is based on the comparison of SI AS 176, Franciscejski kataster za Kranjsko (Franciscan Land Cadaster for Carniola), L343, cadastral municipality of Selo pri Bledu, map sheets A03–A04, A07–A08 with *Environmental Atlas of Slovenia*, Spatial Units, Land Cadastre and *Environmental Atlas of Slovenia*, Backgrounds, DEM [digital elevation model]. This is also in accordance with the recorded fact that the area belonged to the territorial court (Germ. *Landgericht*) of Radovljica (Germ. *Radmannsdorf*; see the map of territorial courts: Kaspret, 1929).

[1587] SI AS 176, Franciscejski kataster za Kranjsko (Franciscan Land Cadaster for Carniola), L343, cadastral municipality of Selo pri Bledu, cadastral appraisal protocols: s.n., 343/4, 57, 99. This was a general rule in meadows on forest clearings higher than in the valley bottom and in the *V kraju* meadows located south of the Sava Bohinjka River at the northern foot of the Jelovica Plateau (the toponym from the cited nineteenth-century cadaster was compared with *Osnovna državna karta*, 1967).

[1588] Compare Oppermann & Paracchini, 2012, 19.

[1589] Vidrih, 2005, especially, page 94.

[1590] Pils, 1994, 212.

ness of highly species-rich habitats in non-fertilized meadows (Section 6.3.1). Namely, small and dispersed patches of such a prevalently nutrient-poor meadow were actually nutrient-rich (dung and urine secretion sites in times of grazing, where dandelion, *Taraxacum officinale*, occurred, for instance), and crushing and spreading the dung also resulted in a moderate manuring impact in the surroundings of these patches. Crushing and spreading the dung in temporarily grazed but otherwise non-fertilized meadows was done with wooden rakes (not with implements dragged by livestock; compare Section 6.3.2.1.1),[1591] even in (at least some) large meadows in flatland, as reported in the mid-nineteenth century.[1592] It is no coincidence that the material of the rake head was specified in the cited nineteenth-century record; wooden rakes are lighter than rakes with an iron head, and so wooden rakes made possible less labor-intensive work, and at the same time they plucked less living meadow plant matter than iron rakes would when manure was crushed and rubbed into the meadow.[1593] Through the excrement of grazing livestock and spreading dung, the patchwork character of the ecosystem within a meadow increased significantly, ranging from the prevalent non-fertilized sections to fairly limited moderately fertilized sections to small nutrient-rich sites, and each of the three types of sites was inhabited by different species.[1594]

In many meadows, only sheep and goats grazed in spring. They were the first to open the grazing season (see Section 6.2.2.1). The same hay meadows could not be used as cattle pastures later in spring because cattle are so repelled by the smell of sheep excrement that they cannot graze after sheep in the same grassland.[1595] Recent experiences in sheep spring grazing in meadows in Bohinj demonstrate that despite the wet climate, with more than 2,000 mm of annual precipitation between 1981 and 2010,[1596] sheep dung, which stays in meadows after spring grazing at 900 m and 1,200 m, does not fully decom-

---

1591 Author's interview with Kunavec, 2021, in Stara Fužina.

1592 For instance, according to mid-nineteenth-century data, in a large meadow below the village of Hlebce in the upper Ljubljana Basin that for centuries had belonged to the demesne of the seigneury of Radovljica (see further data on this meadow in Section 6.2.1.2), on the first day after the end of spring grazing wooden rakes were used to crush and spread the dung that had accumulated there during the autumn and spring grazing periods (SI AS 48, Deželna komisija za odkup in uravnavo zemljiških bremen za Kranjsko (Provincial Commission for Redemption and Regulation of *Profits à Prendre* in Carniola), fasc. 163, 17–19: no. 2277 de 1858, Protocoll, May 10th, 1858, and Zur Sitzung am 14. Oct. 1858).

1593 Author's interview with Kunavec, 2021, in Stara Fužina and discussion with Jožef Cvetek during participant observation at a traditional spring cleaning with wooden rakes at a meadow near Stara Fužina in early May 2021.

1594 Sadar, 1956, 50; Pils, 1994, 58, 60, 212; Babai, Molnár, & Molnár, 2014, 112; Ivaşcu, Öllerer, & Rákosy, 2016, 43–44; conversation with the botanist Metka Škornik, 2022.

1595 Wopfner, 1997, 378; Janišová et al., 2023, 9, 15.

1596 *Environmental Atlas of Slovenia*, Climate, Precipitation, Mean Annual Corrected Precipitation 1981–2010.

pose by mowing time in July, and consequently it gives the hay an unpleasant scent, in dry weather in particular, and so such hay is usable for feeding sheep, whereas cattle do not like it. These recent data refer to a management system in which sheep dung was not crushed.[1597] In history, when spring grazing was a usual practice, dung was crushed at the very end of the spring grazing season (see Section 6.3.2.1.1).[1598] It remains an open question whether crushing dung leading to quicker dung decomposition made possible the use of such hay for any livestock type, including hay from meadows where sheep had grazed in spring, in years when the weather was not dry, at least, because dry weather hinders dung decomposition.[1599] An observation from Dorset in southern England that "spring grazing improved quality at the expense of quantity: meadows grazed by sheep in winter 'may not cut so great a burden . . . , yet the grasses will be finer, and more free from noxious plants'" suggests that the hay was not spoiled by non-decomposed sheep dung, at least in regions with a humid climate.[1600]

Spring grazing postponed the timing of mowing to allow seed ripening in the regrown vegetation after the meadow was closed for grazing (see Sections 6.1.2.1 and 6.1.2.2).[1601] In the Alps, early dates when spring grazing in meadows ended allowed two mowings (see Section 6.1.2.3) and autumn grazing in moderately fertilized meadows at moderate elevations, such as up to more than 1,300 m in the Kauns Valley, with significant differences within the Alps strongly influenced by climatic conditions. Grazing in meadows lasting longer than until early May (Section 6.2.1.2) or grazing in prevalent non-fertilized meadows (Section 6.3.1) allowed only one mowing, followed by autumn grazing.[1602]

The historical contribution of gaps created by grazing in spring and autumn (Figure 71) to successful reproduction of meadow plant species and thus to the preservation of high species diversity in meadows in the long term is presented in Section 6.5.

Grazing on steep slopes, especially on moist soils by relatively heavy livestock, or grazing in wet meadows resulted in extremely high numbers of gaps in meadow thatch vegetation (Figure 71). The grazed wet meadow surface also became uneven, and a significant percentage of soil was compacted. On un-

[1597] Author's interview with T. Soklič, 2021, in Podjelje, with personal experience in spring sheep grazing in meadows and haymaking there.
[1598] SI AS 48, Deželna komisija za odkup in uravnavo zemljiških bremen za Kranjsko (Provincial Commission for Redemption and Regulation of *Profits à Prendre* in Carniola), fasc. 163, 17–19: no. 2277 de 1858, Zur Sitzung am 14. Oct. 1858.
[1599] On the impacts of weather on dung decomposition, see Vidrih, 2005, 94.
[1600] Peterken, 2013, 121.
[1601] Kapfer, 2010, 136–137; Peterken, 2013, 122; Janišová et al., 2023, 9, 11.
[1602] Wopfner, 1997, 273–274, 295.

Figure 71: Countless gaps in a meadow in the hamlet of Slevo in the Kamnik–Savinja Alps caused by cattle grazing on wet soil (in the autumn). Note the remains of a cow pie in the lower left part of the photo. Even the thatch in meadows on steeper slopes is not as ruined as this one in the case of adequate grazing duration after mowing unless the soil is wet. Photo: Žiga Zwitter, December 25th, 2019.

stable slopes, grazing on wet soils could even trigger mass movements, such as landslides.[1603] Relatively long grazing duration could also lead to livestock trail erosion on steep meadows; numerous paths started to form, and the trail network was denser on steeper slopes and sparser on gentler ones. Lighter animals (e.g., sheep) form trails only on very steep slopes, and cattle also on somewhat gentler ones (but see Section 5 on historical changes in livestock size). This made mowing more difficult and changed site conditions because plants on trails were more exposed to trampling and soil on trails became compacted; this resulted in changed soil water availability as a result of a diverse trampling load. All of this caused more or less pronounced differences in the plant communities within a meadow. As long as livestock were light by present standards (Section 5), these issues were somewhat less serious.[1604] Different types of damage due to spring grazing on damp soil in steep meadows were an important argument leading to the abolition of spring grazing rights in sev-

[1603] Keller, 2004, 59, 89; Vidrih, 2005, 78–80, 90–92, 122; Kačičnik Gabrič, 2014, 173, 366.

[1604] Keller, 2004, 59, 89; Vidrih, 2005, 78–80; Ellenberg & Leuschner, 2010, 35; Bätzing, 2015, 92–93, 104.

eral meadow complexes in the southern part of Bohinj in the mid-nineteenth century, for instance.[1605] Similar is true for autumn grazing. An interviewee from Bohinj recalled a twentieth-century prohibition on autumn grazing in the Pokrovec complex of steep montane meadows on the southern slope of the Pokljuka Plateau in the Eastern Julian Alps due to trail erosion.[1606] These challenges were particularly strong in the case of overgrazing.

Spring grazing also affected wild fauna. Depending on the duration, intensity, and timing of spring grazing, in some cases the practice was likely to reduce an excessive (from farmers' perspective) abundance of animals that came into conflict with farming, such as small rodents (see Section 4.1.3.2); many voles and moles fled from grazed plots of land. The negative impacts of spring grazing on cockchafer larvae in some meadows are also imaginable.[1607] Despite the light weight of sheep in comparison to cattle (Section 5), sheep grazing had the potential to reduce vole populations (e.g., the common vole, *Microtus arvalis*) due to the sheep's sharp hooves.[1608] In addition, at least where spring grazing was intensive and it ended late, it also affected some other animal species; it probably had negative consequences for some ground-breeding birds,[1609] but not all of them. Early spring grazing lasting no later than the end of April, followed by a ban on livestock entering the meadow, was probably compatible with whinchat (*Saxiola rubetra*; see Sections 10.1.1. and 10.1.2.2.5),[1610] for instance, a migratory species breeding in May and June.[1611]

Many other ecological and economic consequences of spring grazing strongly depended on the timing of grazing. They are presented in chronological order within a year in the next two sections.

6.2.2.1 Ecological and economic implications of early spring grazing

In many places, the grazing season began soon after the snow had melted (see Section 6.2.1.1). This was especially the case when the quantity and/or quality of hay stored in the previous year was low (Sections 4.1.1.1 to 4.1.1.4 and 4.1.3) and/or the winter was unusually long, so that many farmers faced a fodder scarcity.[1612] In usual years, the early start of spring grazing was a consequence of the fact that, until the mid-twentieth century, it was not unusual for farmers

1605 Novak, 1955, 270.
1606 Author's interview with F. Mlakar, 2017, in Bohinjska Češnjica.
1607 Vidrih, 2005, 93, 123.
1608 Janišová et al., 2023, 15.
1609 Winter et al., 2014, 531; *Društvo za opazovanje in proučevanje ptic Slovenije*, Kosec; Gooders (adapted by Gregori), 1998, 177; Božič, 2019, 138–139; see also Rasran & Jeromin, 2010, 127, 130; Peterken, 2013, 308–309; Janišová et al., 2023, 15.
1610 Janišová et al., 2023, 16.
1611 Gooders (adapted by Gregori), 1998, 341.
1612 Pfister, 1985a, 39, 45, 47, 51–53, 61–63, 83–86, 91–92, 102–104; Machatschek, 2002, 391.

in many parts of the Alps to keep more livestock than they could adequately feed over the winter.[1613] An important reason leading to excessive numbers of livestock was the very common rule, already documented in previous centuries, that a farmer was allowed to keep as many animals at a summer mountain pasture and in other commons as the farm was able to feed during the winter with its own fodder supplies (but there were exceptions; e.g., areas of long-distance transhumance; see Section 7.1). Consequently, livestock that were malnourished but survived during the winter were well fed during the grazing period.[1614] An additional motive for keeping too many livestock was the consideration that more livestock would produce more manure, which was usually a scarce resource.[1615] Under the conditions of a frequent lack of fodder in late winter, early spring grazing significantly increased quantities of milk; when lactation was interrupted in winter due to insufficient fodder, spring grazing promoted earlier restoration of lactation (see Section 6.1.2.5.1.1). This also means that occasions of cold weather causing a delayed start of the vegetation period sometimes significantly decreased the milk quantities at that time. This was relevant; namely, late winter and early spring were the most frequent seasons of historical scarce protein availability for human nutrition, and so milk was very important at that time of the year—yet another argument supporting an early start of spring grazing.[1616]

1613 Hacquet, 1779, 5–6, 28; Pfister, 1985a, 38–39, 44, 52–53; Novak, 1987, 74, 76; Žontar, 1987, 63; Kauter, 2002, 128; Machatschek, 2002, 271–272, 391; Jurca, 2003, 77; Vogl-Lukasser et al., 2006, 62, 81, 84–85, 88, 189, 229–231; Janišová et al., 2023, 8; author's interviews with G. Gartner, 2017, at the Laz Alp in the Julian Alps, and with L. and M. Odar, 2017, in Studor.

1614 Novak, 1955, 259; Blaznik, 1970b, 156; Novak, 1970, 354, 386; Pfister, 1985a, 29, 51, 63, 92; Viazzo, 1989, 26–27, 112–114, 280, 282–284; Niederer, 1996d, 163, 177, 208; Wopfner, 1997, 221, 266, 429, 431; Hinterwaldner, 2002, 88, 129; Kačičnik Gabrič, 2008, 296; Kačičnik Gabrič, 2014, 109, 154, 214; Bätzing, 2015, 106–107; Orland, 2016, 31–32. Archival records from the Slovenian Alps confirm the statement in quoted publications that farm's winter fodder supplies determined the number of livestock the same farm was allowed to keep in common pastures. This was documented about 1660, for example, in SI AS 721, Gospostvo Bled (The Seigneury of Bled), Iustitialia, fasc. 34: Lawsuit between Lorenz Schwarz and the communities of Ribno, Bodešče, and Koritno over the confiscation of livestock that were grazing in a mountain pasture complex (on the Jelovica Plateau, Eastern Julian Alps), 1660–1661: *Woll Edl Gestrenger Herr bestandinhaber der herrschaft Veldes etc.*, no date. The same rule was respected by inhabitants of the village of Hlebce during early spring grazing in the large Kneżnek meadow (SI AS 48, Deželna komisija za odkup in uravnavo zemljiških bremen za Kranjsko [Provincial Commission for Redemption and Regulation of *Profits à Prendre* in Carniola], fasc. 163, 17–19: no. 2277 de 1858, Zur Sitzung am 14. Oct. 1858; on this meadow, see Section 6.2.1.2). The frequent difficulty of storing enough hay for the winter at farms belonging to the community with its center in Bohinjska Češnjica was documented in 1608 (SI AS 721, Iustitialia, vol. 13, [July 1st, 1608]), in a late phase of a significant peak of the Little Ice Age (see Sections 4.1.1.2, 4.1.1.3, and 4.1.1.4).

1615 Pfister, 1985a, 27, 84, 86, 92, 124, 126; Pavković, 2021, 84, 95, 114.

1616 Pfister, 1985a, 38–39, 43–47, 51–53, 61–64, 85, 90–92, 96, 102–103; Pavković, 2021, 95–101.

The grazing season for goats and sometimes also for sheep started in winter or early spring, when they often first browsed on young scrub, before significant fodder quantities were available in grasslands.[1617] When snow had just melted (see the example from the Binn Valley in the late eighteenth and early nineteenth century in Section 6.2.1.1), the growth of grassland plants was yet to begin. Consequently, only sheep, which graze the closest to the ground among all domestic herbivorous ungulates, and goats, which graze almost as close to the ground as sheep, grazed in some grasslands this early.[1618] Sheep, which played a prominent role in livestock farming in the Alps (Sections 5 and 6.3.1.1) and grazed in many meadows in early spring, graze highly selectively, with a preference for forbs and soft grasses with many leaves, although the spectrum of plants they eat is wide. Sheep usually bite off the plants; if fodder availability in the grazing ground is very low, sheep often uproot their preferred fodder plants. Their trampling is also significant because their hooves are sharp.[1619] Whether sheep and goats found at least a small amount of relatively palatable fodder in grasslands so early in spring or on warm winter days, in addition to the few dry plant remnants from the previous autumn, depended on the grassland species composition, site conditions (see also Section 6.1.2.5.1.1), and grassland management. Indeed, in non-fertilized meadows on steep dry sunny slopes at lower elevations, where snow melts first, which called for starting spring grazing there (if not in wetlands; see Section 6.1.2.5.1.1), some plants often occur or abound that are able to cope with such extreme (cold and often dry) conditions, and their leaves are green throughout the winter. They are capable of carrying out photosynthesis at temperatures slightly above 0 °C on warm winter days when the grassland is free of snow. An example is large thyme (*Thymus pulegioides*).[1620] However, the quantity of such fodder before the general start of the grassland vegetation period is negligible. Thus, sheep and goats often caused damage in meadows this early by consuming all parts of the plant above ground, including regeneration buds, by uprooting some plant species, and through trampling. This was experienced in the northeast part of the Senožeta meadow complex in the middle of the Bohinj Basin in the mid-nineteenth century, for instance. Sheep grazing there started before grassland plants grew considerably, and so they consumed the tender shoots.[1621]

Trampling by horses is the strongest among all herbivorous domestic ungulates, especially if they wear horseshoes. The spectrum of plants eaten by hors-

---

1617 Novak, 2024, 39.
1618 Ellenberg & Leuschner, 2010, 1007; Janišová et al., 2023, 9, 15.
1619 Sadar, 1956, 177; Pils, 1994, 202; Ellenberg & Leuschner, 2010, 1007; Janišová et al., 2023, 15.
1620 Pils, 1994, 83; Wopfner, 1997, 228, 233–234, 378.
1621 Novak, 1955, 270, 278; SI AS 176, Franciscejski kataster za Kranjsko (Franciscan Land Cadaster for Carniola), L107, cadastral municipality of Češnjica v Bohinju [i.e., Bohinjska Češnjica], map sheet A12.

es is the narrowest, and they also bite off plants very close to the ground.[1622] Consequently, the negative impacts of early horse grazing in meadows were significant.[1623] Grazing horses reduce plant species richness in grasslands more strongly than grazing by cattle, sheep, or goats.[1624]

If the grazing season began too early in the spring, when fodder plants were not tall enough (e.g., only 6 to 8 cm tall or less), regeneration and regrowth after such an early shock caused by biting and trampling frequently took several weeks. In the best case, livestock did not bite off the majority of plants close to the ground. Theoretically, 2 to 3 cm should be left, which is not possible in reality, but grazing pressures that are not overly strong contribute to more successful regeneration (see the section on cattle grazing in the next paragraph). Intensive grazing from early spring onward harmed some species that sprout early, such as the useful and highly productive grass cock's-foot (*Dactylis glomerata*).[1625] The negative impacts of early spring grazing were particularly strong if it was preceded by intensive late autumn grazing before the end of the vegetation period in the previous year.[1626] Too early a start of spring grazing was, however not a general phenomenon. For instance, in places where the spring grazing season started on April 23rd/24th or in early May (Section 6.2.1.1), the height of grassland plants often met the aforementioned recommendation of at least 6 to 8 cm in the colline and lower part of the montane vegetation belt.

Cattle can only graze if plants are at least 3 to 5 cm tall, otherwise they cannot take hold of the plants because they tear off the plants using their tongue or lips. The economic consequence of this fact is that cattle grazing in meadows could not start as early as sheep or goat grazing. Moreover, cattle usually tear off the plants relatively high, but they are able to do so as close to the ground as 1.2 to 2 cm if fodder availability is low.[1627] Despite the cattle's inability to graze on plants lower than 3 to 5 cm, moderate cattle grazing started early and livestock, cattle in particular, had to browse on bushes, such as on twigs and nutritive willow catkins (*Salix* spp.);[1628] they announce the start of early spring significantly earlier than temperatures allow beech leaves to unfold.[1629] Herders also cut scrub branches to increase their accessibility for browsing livestock.[1630] Even later in spring, when the first beech leaves fully unfolded,

[1622] Sadar, 1956, 177; Ellenberg & Leuschner, 2010, 1007; Pils, 1994, 201–202; Janišová et al., 2023, 15.
[1623] Novak, 1955, 278.
[1624] Ellenberg & Leuschner, 2010, 1007.
[1625] Sadar, 1956, 169–170, 176, 183–185; Pfister, 1985a, 91; see also Poschlod, 2015, 209.
[1626] Sadar, 1956, 169, 194–195.
[1627] Sadar, 1956, 176, 196; Pils, 1994, 201; Janišová et al., 2023, 9, 15.
[1628] Pfister, 1985a, 39; Machatschek, 2002, 47, 343; Babai, Molnár, & Molnár, 2014, 87–88.
[1629] Crane, 1999, 62; Žust, 2016, 42–43, 45.
[1630] Machatschek, 2002, 47; Babai, Molnár, & Molnár, 2014, 88.

Figure 72: Spring crocus (*Crocus vernus*) at the temporary mountain settlement of Zajamniki in Bohinj, about 1,300 m, April 28th, 2004. In 2004, blooming of spring crocus was later than usual nowadays (see Figure 73). Photo: Tomaž Zwitter.

fodder available for grazing cattle was still scarce in (non-fertilized; see Figure 74) grasslands, thus herders additionally fed them with cut beech branches full of freshly developing leaves;[1631] early in spring the fodder value of tree foliage was lower than later.[1632] A comparison with late twentieth-century information sheds light on the practice of using beech leaves as fodder in early spring: around Koprivnik, at 1,000 m in the Eastern Julian Alps, the first beech leaves on average fully unfolded in mid-May between 1971 and 2000.[1633] Nonetheless, this phenophase usually took place later in previous centuries; see the dimensions of early spring climate change since 1961 in Section 6.2.2.3. Soon after the beech leaves unfolded, grassland plant availability for grazing cattle increased highly significantly. Earlier spring grazing was often combined with significant barn feeding,[1634] which means that the transition from hay to fresh fodder was gradual—a practice good for livestock health.[1635] One of the many examples was a combination of spring grazing that lasted

[1631] Author's interview with L. and M. Odar, 2017, in Studor.

[1632] Machatschek, 2002, 190–191.

[1633] *Environmental Atlas of Slovenia*, Climate, Phenology, Leaf Unfolding of Common Beech, 1971–2000; Žust, 2016, 14, 42–43.

[1634] Wopfner, 1997, 320.

[1635] Compare, e.g., Machatschek, 2002, 220.

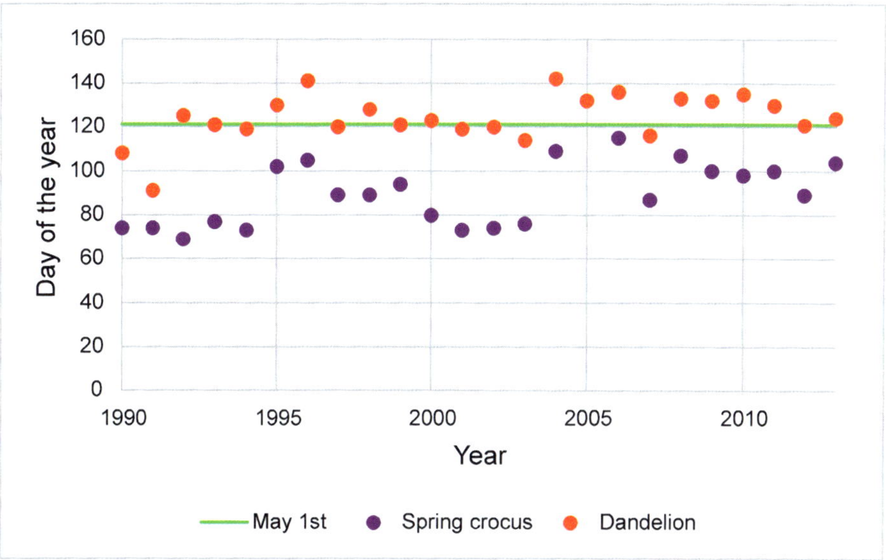

Figure 73: The first flowering day of spring crocus (*Crocus vernus*; i.e., the day when the first blossoms developed to the phase that the observer could see the three stamens and the pistil) and dandelion (*Taraxacum officinale*, i.e. the day when first fully developed inflorescences appeared in extensively managed meadows) in Koprivnik (about 1,000 m), Eastern Julian Alps, according to documented phenological observations (1990–2013).[1636] Source of information: Slovenian Environment Agency: Archives of phenological data, Koprivnik, and Ana Žust, personal communication on locations of observations. Author of the graph: Žiga Zwitter.

from snowmelt until May 12th and barn feeding in meadows at Zajamniki (about 1,250 m) in the Eastern Julian Alps, as recorded in the mid-nineteenth century.[1637]

Apart from the scarcity of available grassland fodder plants that were tall enough to be accessible to cattle, very early cattle grazing faced another challenge. It was carried out when toxic spring crocus (*Crocus vernus*; Figure 72) started blooming (Figure 73) and its blossoms were high enough for cattle to eat.[1638] According to eighteenth-century observations by the natural scientist Balthasar Hacquet, spring crocus was very frequent in grasslands in the lower montane vegetation belt of the province of Carniola (which included the Alps and Pre-Alps in what is now part of northwestern, northern, central,

1636 Designation of phenophases from Žust, 2016, 25–26; following the instructions, the beginning of dandelion (*Taraxacum officinale*) blooming is observed in extensively managed meadows, where, for example, it starts blooming later than in fields, gardens, and roadsides (Zrnec, 1998, 113).

1637 Novak, 1955, 295.

1638 Jurca, 2003, 76–77; Hacquet, 1779, 28, 31–32; species identification based on Praprotnik, 2015, 79.

and eastern Slovenia), where it constituted up to one-third of the grassland plant cover in early spring. Spring crocus was particularly common in grasslands that were nutrient-rich according to the standards of the time[1639] (i.e., characterized by moderate nutrient availability according to present-day criteria; see Section 6.3.2), at mesic sites.[1640] In such grasslands, the vegetation season started earlier than in nutrient-poor ones with similar water availability. In Hacquet's work, published in 1779, March is given as the usual month when spring crocus bloomed at that time at unspecified elevations in the lower montane vegetation belt.[1641] Although this information was likely not based on long-term observations at the same locations, it fits quite well with Figure 73 and the general start of spring crocus blossoming in March and April in these areas nowadays.[1642] If the weather was good, spring crocus only bloomed for a short time.[1643] In the past, livestock were generally able to avoid toxic plants because they were used to grazing. However, starving livestock also ate poisonous plants that they otherwise would have avoided. Livestock poisoning with plants was not rare according to eighteenth-century information.[1644] Namely, starving livestock readily consuming spring crocus was a consequence of the frequent fodder scarcity in early spring (see the beginning of this section) when spring crocus sprouts were tall enough for cattle to eat, while most grassland plants were still inaccessible to them. Animals poisoned with this plant often recovered, but not always. In the eighteenth century, the main study area (Bohinj) was known as the territory where lethal poisonings with spring crocus were most frequent in the entire province of Carniola.[1645]

As presented in Section 6.2.1.2, spring grazing in meadows around Koprivnik in the Eastern Julian Alps was allowed to last until May 1st, according to information from 1608. In 2017, fieldwork was performed around Koprivnik. It proved the great variation in heights of grassland plants in the area observed by the end of April, strongly depending on grassland management. Where grazing pressures had been very strong in late autumn and the grassland had not been fertilized, the growth of plants had just begun (Figure 74); in most meadows, grasses were about 5 to 10 cm tall, whereas in meadows relatively abundantly fertilized with barn manure[1646] grassland plants were already over 25 cm tall. March and April 2017 were warmer than average from 1960

1639 Hacquet, 1779, 27–28, 31.
1640 Aeschimann et al., 2004, vol. 2, 1092–1093.
1641 Hacquet, 1779, 27–29.
1642 Žust, 2016, 25.
1643 Compare Žust, 2016, 25.
1644 Hacquet, 1779, 5–6, 14; Jurca, 2003, 76–77.
1645 Hacquet, 1779, 28, 31–32.
1646 Author's interview with A. Soklič, 2017, in Podjelje.

Figure 74: April 29th, 2017: Growth of plants is just beginning in a non-fertilized sheep pasture (Zgornje Podjelje, about 1,060 m), which had been heavily grazed in late autumn, causing delayed vegetation development in the spring.[1647] Photo: Žiga Zwitter.

to 2017.[1648] Although the average stage of plant development has been changing through history (e.g., due to climate change), there have also always been fluctuations at the year-to-year level. Some were extreme, as demonstrated by another early spring plant phenophase; namely, the start of cherry blooming. In many parts of Switzerland at elevations from 400 to 500 m, cherry trees on average started blooming in mid-April from 1901 to 1960, but based on sixteenth- to early eighteenth-century information they bloomed as early as March in 1530, 1559, 1602, 1607, 1661, and 1702, and as late as in May in

1647 Sadar, 1956, 169–170, 194–195.

1648 The average March temperature at Kredarica (2,513 m) from 1960 to 2017 was –6.5 °C, and at Stara Fužina and Bohinjska Češnjica at the bottom of the valley (548 and 596 m; measurements were carried out at Stara Fužina until 2002 and at Bohinjska Češnjica after that) it was +3.4 °C. The average March temperature in 2017 was –3.0 °C at Kredarica and +7.3 °C at Bohinjska Češnjica. The average April temperature at Kredarica from 1960 to 2017 was –3.9 °C, and at Stara Fužina and Bohinjska Češnjica at the bottom of the valley it was +7.7 °C. The average April temperature in 2017 was –3.4 °C at Kredarica and +9.4 °C at Bohinjska Češnjica (Mekinda-Majaron, 1995, 55, 113; *ARSO, Arhiv – opazovani in merjeni meteorološki podatki po Sloveniji*: Kredarica, Stara Fužina, Bohinjska Češnjica).

Figure 75: April 29th, 2017: A meadow located next to the pasture in Figure 74, but well fertilized with barn manure. Plants in moderately fertilized meadows grow more quickly than in non-fertilized ones. In addition, because plants will grow taller and the grassland will become denser than a non-fertilized meadow, the leaves of dominant plant species are upright to receive more light than the shaded bottom layer,[1649] which makes them more accessible to grazing livestock. A comparison of Figures 74 and 75 makes it clear that fertilized meadows provided significantly more resources for grazing earlier in spring than non-fertilized ones. Quicker growth and the higher position of most leaves in moderately fertilized meadows allowed early spring grazing even by cattle in some places. Photo: Žiga Zwitter.

1542, 1574, 1716, 1729, and 1731.[1650] To give an eighteenth-century example of very early growth of most grassland plants: at Binn in the Binn Valley (Wallis), following the warm winter of 1793/94 and a very warm early spring, on sunny slopes at as much as roughly 1,450 m grassland plants were typically one span tall as early as the beginning of May 1794, and in (fertilized) meadows close to the village grasses reached the middle of the shins at that time.[1651] These data, combined with the 2017 observations in grasslands around Koprivnik presented above, make it clear that, as long as spring grazing around Koprivnik had to end on May 1st, in many years it was not possible to graze cattle in the spring in the majority of meadows, which were not fertilized. However, the quantity of grazing resources in moderately fertilized meadows in warm springs became significant by the end of the spring grazing period (Figures 73–76). Dif-

[1649] Pils, 1994, 162.
[1650] Pfister, 1985a, 82, 84–85.
[1651] Zenhäusern, 2008, 151–152, 158–161, 174–177.

Figure 76: April 29th, 2017: sheep grazing in a meadow in Koprivnik in early spring. Photo: Žiga Zwitter.

ferences in microclimate mattered as well. In this context, it is also important that hummocky meadows (see Sections 3.2 and 4.1.2) used to be characteristic of montane meadow complexes in Bohinj. In their pits, snow melts later than on hummocks,[1652] and so there were many differences in microclimate within a single meadow parcel, which were relevant in terms of grazing in early spring.

Properly timed grazing in early spring, at least by sheep, could improve the quality of hay (resulting in higher protein content) and it did not significantly reduce the quantities of hay mowed later on some parcels,[1653] whereas the decrease in hay quantity was up to 25% in some others, depending on the local plant community and early spring grazing regime.[1654] "In such a management scheme, possible negative impacts of the early biomass removal on species diversity may be compensated by a reduced biomass of competitive grasses in the [hay], leaving more space for less competitive species. Many plant species show an intrinsic ability for compensatory regrowth, which is due to higher relative growth rates early in the season"[1655] and branching of some harmed plants, such as highly productive grasses (more than one stem grows from a

1652 Cvetek, 1971, 69.
1653 Sadar, 1956, 55, 148; Wopfner, 1997, 289; Peterken, 2013, 121; van Eekeren, Iepema, & de Wit, 2014, 665–667; Janišová et al., 2023, 16.
1654 Janišová et al., 2023, 16.
1655 Winter et al., 2014, 530–531; see also Strauss & Agrawal, 1999.

grazed or trampled stem).[1656] Excrement during spring (and autumn) grazing in meadows led to a more favorable nutrient balance in comparison with meadows that are only mowed, as explained in the introduction to Section 6.2.2. Stem branching and nutrient intake from excrement result in denser meadow vegetation as a consequence of properly timed spring grazing if the carrying capacity of the grassland is not exceeded; that is, if the numbers of livestock are well adapted to the available resources and site conditions and grazing does not last too long on the same plot of land.[1657] However, "[s]pring grazing . . . before the mass development of grasses prevents the first flowering cycle, during which larger competitors (both grasses and herbs) are more affected than smaller-growing species, as the former regenerate more from stored reserves than from the remaining leaf area for photosynthesis."[1658]

The dates found in mid-twentieth century agronomic literature for when spring grazing should end in order not to significantly decrease hay quantities roughly correspond with the start of dandelion (*Taraxacum officinale*) blooming in moderately fertilized meadows.[1659] This means about one month after toxic spring crocus (*Crocus vernus*) starts blooming (Figure 73). The start of dandelion blooming is also roughly the upper time limit of the early spring grazing presented in Section 6.2.2.1, whereas later grazing in spring is dealt with in Section 6.2.2.2. Judging from a modeled phenological map showing average dates of dandelion blooming between 1965 and 1994, grazing until April 24th west of Rateče prior to the calendar reform usually did not significantly reduce the quantities of hay mowed there,[1660] unless such early grazing pressures were too strong. The same applies to grazing in meadows around Koprivnik until May 1st after the calendar reform. In the case of Koprivnik, the cited modeled map roughly corresponds with the plant-phenological observations carried exactly at that place from 1990 to 2003 and showing that dandelion often starts blooming there around May 1st or later, with significant differences (Figure 73).[1661] Although the average stage of plant development has changed through history, there have also always been fluctuations at the year-to-year level (see above), thus grazing in meadows around Koprivnik by May 1st in some years ended roughly by the beginning of dandelion blooming. This was in line with optimization of the quantities of hay and grazing resources there, a goal recorded for this area in 1608.[1662]

1656 Strauss & Agrawal, 1999; Wopfner, 1997, 289–290; Babai, Molnár, & Molnár, 2014, 103.

1657 Sadar, 1956, 54–55, 169–209, 211; Babai, Molnár, & Molnár, 2014, 103.

1658 Janišová et al., 2023, 16.

1659 Based on the general explanation by the agronomist Sadar, 1956, 55, 148; the dates given there were compared with the onset of dandelion blooming in Zrnec, 1998, 113; Zrnec, 2001, 66; Žust, 2016, 14.

1660 Based on the general explanation by the agronomist Sadar, 1956, 55, 148; the dates given there were compared with the onset of dandelion blooming in Zrnec, 1998, 113; Zrnec, 2001, 66.

1661 Zrnec, 1998, 113; Zrnec, 2001, 66.

1662 SI AS 721, Gospostvo Bled (The Seigneury of Bled), Iustitialia, vol. 13, [July 1st, 1608].

Spring grazing resulted in comparatively high species richness, resulting from the variety of available ecological niches. From an ecological perspective, "[a] short period of spring grazing (for a few days until the beginning of May) with a higher stocking rate could be beneficial for most semi-dry and mesic hay meadows" (see also Section 6.5).[1663] Indeed, such a grazing regime was frequent in historical early spring grazing. Namely, a parallel with autumn grazing in the same type of plots of land (montane meadows in the large Uskovnica meadow complex) in the mid-twentieth century shows that the entire herd usually grazed in one meadow per day; exceptions were only disproportionately small or large meadows.[1664] This means that at least in large meadow complexes the access of livestock to various meadows during spring grazing was limited in a way that was favorable from an integrated economic-ecological perspective: the herd did not have access to all parts of a large grassland complex over a long time.[1665] This was a general rule as suggested by the parallel with some areas further north in Central Europe where "livestock spent just part of the [spring] grazing days in the hay meadows, probably with a maximum of two to three grazing days per hay meadow or meadow complex."[1666] Some plant species adapted to early spring grazing in meadows by flowering later than in those meadows where mowing represents the first agricultural use. These include, for instance, eyebright (*Euphrasia rostkoviana*), greater yellow-rattle (*Rhinantus alectorolophus*; see Section 4.1.1.1), and small scabious (*Scabiosa columbaria*).[1667]

Nonetheless, not all plants can recover after spring grazing.[1668] In general, early spring grazing favored "late-fruiting plants instead of early-spring flowering plants."[1669] Grazing in early spring suppressed or decreased the abundance of certain undesired plants from the point of view of animal husbandry; for example, sharp tufted hairgrass (*Deschampsia cespitosa*), at least on water-saturated soils.[1670] At the same time, it "may support low-competitive rare species such as [moonwort in non-fertilized meadows] (*Botrychium lunaria*) and [some gentian species] (*Gentinanella* spp.) by reducing competition from fast-growing plants."[1671] As hypothesized, "[s]pecies without the ability to resprout after an early cut, such as orchids, are . . . likely to decrease when their infructescence is repeatedly removed."[1672]

[1663] Janišová et al., 2023, 20.
[1664] Hodnik, 2016, 44.
[1665] Sadar, 1956, 50–51, 147, 168–170, 176–177, 183, 185, 196, 199–202.
[1666] Cv. Janišová et al., 2023, 9
[1667] Janišová et al., 2023, 20.
[1668] Sadar, 1956, 55.
[1669] Peterken, 2013, 392.
[1670] Sadar, 1956, 29, 134.
[1671] Janišová et al., 2023, 19.
[1672] Winter et al., 2014, 531.

Nonetheless, empirical historical data warn against oversimplifications. For instance, in Valentin Plemel's botanical treatise on flora of Carniola published in 1862 it is written that the abundance of elder-flowered orchid (*Dactylorhiza sambucina*, belonging to the family *Orchidaceae*; Figure 77), was very high in montane meadows above Jesenice and Javornik. Its abundance there was the highest he knew within the province of Carniola, even though he also listed some other sites where this species occurred.[1673] Plemel was very familiar with these meadows above Jesenice and Javornik, and archival records make it clear that the long history of spring grazing in these montane meadows was just coming to an end at the time of his observations (see Section 6.2.2.3), and so it is clear that spring grazing did not suppress elder-flowered orchid there. Orchids grow for more than three years and often more than ten years before they blossom for the first time,[1674] and so it is clear that a high abundance of flowering elder-flowered orchids around 1860 reflected the situation in times of spring grazing in that area (Section 6.2.2.3). The most important reason for this was probably the early end of spring grazing (May 1st) combined with the fact that springs were on average significantly colder than they are now, and so spring grazing ended at a very early stage of elder-flowered orchid development (for details and references, see Section 6.2.2.3). In terms of grassland species richness, orchids are nowadays considered "super-positive indicators," reflecting an "exceptionally good state of biodiversity" by present standards.[1675] Adequate historical early spring grazing regimes evidently did not have significant negative impacts on the occurrence and abundance of (some) spring orchids. Indeed, "[s]hort but intensive episodic grazing that disturbs the grass turf is recommended to maintain orchid populations in hay meadows."[1676] Some types of grazing livestock avoid some orchids, such as cattle avoiding green-winged orchid (*Orchis morio*),[1677] which occurs in non-fertilized, relatively dry meadows Alps-wide, especially in colline and montane vegetation belts.[1678] Thus at least some orchids are fostered by cattle grazing.[1679] Goats probably consumed several orchid species.[1680]

[1673] Plemel, 1862, 147; identification of the old botanical name *Orchis sambucina* that Plemel used is based on Aeschimann et al., 2004, vol. 2, 1120–1121. Plemel's description of purple and ocher blossoms also clearly confirms the species identification.

[1674] Pils, 1994, 306.

[1675] Twenty-first-century ecological views on orchids: Šumrada et al., 2021, 5.

[1676] Janišová et al., 2023, 19.

[1677] Janišová et al., 2023, 16.

[1678] Aeschimann et al., 2004, vol. 2, 1126.

[1679] Janišová et al., 2023, 16.

[1680] Author's observation from the Krstenica mountain pasture complex in the Julian Alps: whereas cattle did not touch a small population of pink blossoming *Nigritella* orchids (*Nigritella rubra* and *Nigritella lithopolitanica* occur in the region; Aeschimann et al., 2004, vol. 2, 1116–1117), no inflorescence was left after a flock of goats from another alp passed by.

Figure 77: Elder-flowered orchid (*Dactylorhiza sambucina*, plants with purple and light yellow blossoms in the photo) in a nutrient-poor meadow (see Section 6.3.1) above Javorniški Rovt in the Karawanks, at 1,280 m, in early spring. A high abundance of this species in meadows in this region was observed by Valentin Plemel in the mid-nineteenth century. Nowadays, this species is designated as vulnerable on the red list of endangered species in Slovenia,[1681] and it is, for instance, also endangered in Austria.[1682] Photo: Žiga Zwitter, May 1st, 2018.

[1681] "Pravilnik o uvrstitvi ogroženih rastlinskih in živalskih vrst v rdeči seznam," 2002, 8893, 8899.

[1682] Keller, 2004, 28.

Whereas sheep consume the blossoms of some orchid species,[1683] Plemel's data presented above on the very high abundance of elder-flowered orchid in montane meadows above Jesenice and Javornik in times of spring grazing demonstrate that sheep grazing was as favorable for elder-flowered orchid as cattle grazing. Namely, it is clear that historical spring grazing in the meadows above Javornik lasted no longer than May 1st,[1684] when *Dactylorhiza sambucina* started blossoming in warm springs (Figure 77), so sheep evidently did not represent a threat for this species that abounded there. Spring grazing there only involved sheep and goats.[1685] Because elder-flowered orchid can only abound in non-fertilized grasslands,[1686] but its abundance was still very high despite the livestock excrement, a hypothesis that its symbiotic fungus profits from short-term grazing is worth testing. Namely, like all orchids, *Dactylorhiza sambucina* benefits from mycorrhiza; that is, the symbiotic interactions between plant roots and fungal hyphae (see Section 6.3.1).

On the one hand, Section 6.2.1.2 identified a great variety of dates when meadows that had been grazed in the spring were closed, even under similar environmental conditions. On the other hand, similar to equal dates for different sites were presented (see also the comparison between late final dates of spring grazing in meadows and vegetation development in Section 6.2.2.2). The chronological approach leads to later spring grazing.

6.2.2.2 Ecological and economic implications of late spring grazing

The dates of spring grazing were in general not well adapted to climate and its fluctuations, and partly also not to vegetation development, as presented in Section 6.2.1.2. Nonetheless, later spring grazing sometimes took place in meadows at higher elevations, where the vegetation period starts later (Section 6.2.1.2). Typically, spring grazing lasted long in some non-fertilized meadows at various elevations because the vegetation growth in spring is slower in non-fertilized grasslands than in fertilized grasslands (Section 6.2.2.1).[1687] For instance, when dandelion (*Taraxacum officinale*) starts blooming in moderately fertilized meadows, cowslip (*Primula veris*, which stands out in height in comparison with most other vegetation in non-fertilized meadows at that time, but is detested by cattle), poisonous cypress spurge (*Euphorbia cyparissias*), and hairy violet (*Viola hirta*) blossom in some mesic to relatively dry non-fertilized grasslands. It is only at that time that the general appearance of non-fertilized mesic or dry grasslands changes from brownish to green.[1688] After that

[1683] Norz, 2014, 119.
[1684] See Section 6.2.2.3.
[1685] See Section 6.2.2.3.
[1686] Aeschimann et al., 2004, vol. 2, 1120–1121.
[1687] E.g., Grass, 1948, 188–189, 196–198, 200.
[1688] Pils, 1994, 72, 88, 98, 306 and Table 61, photo 151; Aeschimann et al., 2004, vol. 1, 434, 632.

time, growth was relatively quick and grazing animals could feed on significant amounts of the tender young shoots. The palatability of many plants detested by livestock in later development phases is higher when they are young; thus even matgrass (*Nardus stricta*; see Section 6.3.1.2), later rejected by livestock, was suitable for grazing in May.[1689]

Where spring grazing lasted somewhat longer than presented in Section 6.2.2.1, it was likely to significantly decrease the quantity of hay but increase its quality,[1690] at least at nutrient-poor sites. Where spring grazing lasted even longer, it had very strong negative influences on the quantities of hay harvested there in the summer.[1691] Sources give numerous examples emphasizing the negative economic impacts of spring grazing in meadows that ended too late. Farmers in possession of a couple of meadows at Zgornje Jezersko (in the contact zone between the Karawanks and the Kamnik–Savinja Alps, at about 900 m), where a community of farmers had spring and autumn grazing rights, unsuccessfully tried to move the end of spring grazing from May 12th to April 24th in 1663,[1692] probably to increase the quantities of hay because hay yields are recorded to have been low or, at most, moderate, for the standards of the time.[1693] In another example, this one from Bohinj, it was claimed in 1859 that the traditional practice of spring grazing in only slightly elevated meadows at Dobrava in Bohinj, close to the bottom of the valley and west of the village of Bohinjska Bistrica, lasting until May 25th (see Figure 69), was devastating the quantities of hay.[1694] Similarly, it was recorded in 1831 that, due to spring grazing on the eastern edge of the Pokljuka Plateau in the Eastern Julian Alps, from May 1st to Ascension Day (between April 30th and June 3rd), hay was deprived of "the best grass" (possibly meaning grassland plants producing the most biomass or those with the highest fodder value),[1695] at least when Ascension Day was late.[1696] Section 6.1.2.7 presents an example of grasslands in the upper section of the colline vegetation belt at the foot of the Alps near Javornik (today northwestern Slovenia) grazed until Ascension Day, which were mowed

[1689] Pils, 1994, 304.

[1690] Peterken, 2013, 121; Babai, Molnár, & Molnár, 2014, 103, and the general explanation by the agronomist Sadar, 1956, 55, 148.

[1691] Sadar, 1956, 55, 148; Wopfner, 1997, 273, 393–394; see also Menke, 1992.

[1692] AT StiAStP, Grundherrschaft Eberndorf (Handwritings), vol. 181, Gerichtsprotokoll (from the 1660s that begins in 1661), 69–70 (September 5th, 1663).

[1693] AT StiAStP, Grundherrschaft Eberndorf (Handwritings), vol. 181, Gerichtsprotokoll (from the 1660s that begins in 1661), 70 (September 5th, 1663).

[1694] Novak, 1955, 266–267.

[1695] SI AS 176, Franciscejski kataster za Kranjsko (Franciscan Land Cadaster for Carniola), L335, cadastral municipality of Bohinjska Bela, cadastral appraisal protocols: s.n., 335/4, 57, 99.

[1696] In 1830, the last season of hay harvest before this enquiry was made, Ascension Day was late, on May 20th (Grotefend & Ulrich, 1971, 184–185); that is, already after the modeled usual beginning of dandelion (*Taraxacum officinale*) blooming there between 1965 and 1994 (Zrnec, 1998, 113; Zrnec, 2001, 66).

only in good years—evidently, when Ascension Day was early—and otherwise only grazed.

Despite these facts speaking in favor of an earlier end of spring grazing, the end of spring grazing around Koprivnik in the Eastern Julian Alps was probably postponed by three and a half weeks between 1608 and the early nineteenth century, and at many other sites grazing until late May had a long tradition (Section 6.2.1.2). Namely, such late spring grazing had not only economic negative impacts, but also agroecologically desired ones (see the following paragraphs). Whether negative or positive impacts of late spring grazing in meadows prevailed from the agricultural perspective was site-specific and also depended on past environmental change. Consequently, prolongation of the spring grazing season in meadows (exceptionally, as likely around Koprivnik, or regularly in the case of a movable date designating the end of spring grazing, such as Ascension Day) is also a complex topic, and it could have both negative and positive impacts from the agricultural point of view. It can nonetheless be assumed that Ascension Day as the end of spring grazing in meadows was rooted more in religion than in traditional ecological knowledge. Namely, Ascension Day played a role in spiritual beliefs about the fertility of farmland; small crosses made of blessed wood were placed in meadows on Ascension Day eve, sprinkled with blessed water, and a prayer for a good harvest followed, as documented in part of Lower Styria (Germ. *Untersteiermark*, Sln. *Spodnja Štajerska*) in the first half of the twentieth century.[1697] Thus it seems plausible that farmers stopped grazing their livestock in some meadows on Ascension Day in hope of a good hay harvest. An ethnobotanical parallel with southern Portugal demonstrates the role of religious feasts, even movable feasts dependent on the date of Easter, to cope with a lack of adequate understanding of ecological processes. In southern Portugal, people gathered golden thistle (*Scolymus hispanicus*). This species is considered edible for people only before fibrous tissues develop, which allow stem growth and flowering. Nonetheless, to determine the end of gathering that thistle, people respected the date of Easter.[1698] Thus, due to the lack of a more precise understanding of natural processes, in some cases even very specific stages of plant-phenological development have traditionally been predicted based on feasts depending on some movable religious holy days.

Focusing on the positive impacts of longer spring grazing in meadows in the Alps from the agrarian viewpoint, the dense sward and higher crude protein content of hay have already been mentioned. In moderately manured meadows, shorter grass length caused by spring grazing until the second half of May decreased the hazard of lodging in grass caused by wind in June (see

[1697] Kuret, 1989, 316.
[1698] Barão & Soveral Dias, 2013, 196–197, 203, 205.

Section 6.3.2.2).[1699] Grazing in May suppressed some plant species (and also animal species; see the introduction to Section 6.2.2), some of them non-intentionally,[1700] and some others intentionally, with the aim of suppressing species perceived as weeds or pests in grassland farming. Some meadow "weeds" that are not toxic are normally avoided by livestock because the leaves are too rough (e.g., soft rush, *Juncus effusus*). In the spring, however, they pump sugar into young leaves and become attractive to grazing livestock for a short time. By establishing high grazing pressures in May in a meadow where soft rush abounds, this species undesired in grassland farming could be suppressed in some cases, thus it played a smaller role later in the hay.[1701] Purple moor-grass (*Molinia caerulea*), providing very low-quality fodder when fully developed (see Sections 3.2.2, 6.1.2.5.1.2, and 6.1.2.5.2), could also be suppressed through long spring grazing in some late-mowed, non-fertilized meadows. Namely, purple moor-grass's tussocks can suffer significantly from trampling during grazing in spring, carried out on comparatively wet soil and characterized by a high stocking rate. Similarly, trampling during grazing in May suppressed tor-grass (*Brachypodium pinnatum*; on this species, see also Section 6.3.1.2), whose long runners suffer from intensive grazing pressures. Moreover, long spring grazing was a means of eliminating scrub seedlings from meadows because they are consumed by livestock.[1702] An excessive abundance (from the farmer's viewpoint) of meadow clary (*Salvia pratensis*), hogweed (*Heracleum sphondylium*), cow parsley (*Anthriscus sylvestris*), and some other herb species of moderately manured meadows was also successfully decreased through intensive spring grazing and allowed an increased abundance of productive grasses and white clover (*Trifolium repens*) according to eighteenth-century observations.[1703] Intensive short-term spring grazing in May also suppressed some herbs presenting more serious challenges to grassland farming or prevented their establishment in the meadow, such as the hemiparasite yellow rattle (*Rhinanthus* spp.; see Section 4.1.1.1)[1704] and the toxic autumn crocus (*Colchicum autumnale*), which are common species in various types of moderately used, mesic (or moist to damp in the case of autumn crocus) grasslands in the Alps.[1705]

[1699] Babai, Molnár, & Molnár, 2014, 103.

[1700] Winter et al., 2011, 1777, 1782; Winter et al., 2014, especially 527–531; Peratoner et al., 2014, 418–420.

[1701] Rasran & Jeromin, 2010, 121–123, 126–128, 130–131.

[1702] Janišová et al., 2023, 16.

[1703] Pils, 1994, 179, 211 and Table 37, photo 102; Wopfner, 1997, 289–290; Kauter, 2002, 133, mentioning spring grazing but its timing is not given. We assume that late spring grazing was more efficient in this case; Peterken, 2013, 225. See also Vogl-Lukasser et al., 2006, 70–72, 91.

[1704] Pils, 1994, 211; Wopfner, 1997, 289; Winter et al., 2014, 531.

[1705] Aeschimann et al., 2004, vol. 2, 272–279, 1030–1031; Peterken, 2013, 32–33, 417; Winter et al., 2014, 525.

Autumn crocus prefers sites characterized by close to neutral pH as well as medium nutrient availability.[1706] Seed dispersal is facilitated through traditional dates of hay harvest in late June or later (Sections 6.1.2.1 and 6.1.2.2)[1707] because it is impossible to prevent the dispersal of ripe seeds from the capsule during mowing and shaking of the hay if it is mowed when grasses are ripe.[1708]

Suppression of autumn crocus through spring grazing is relevant from the agricultural perspective because all parts of this plant are poisonous, as well as the dried green herb in hay. Consumption of *Colchicum autumnale* can lead to poisoning of cattle, horses, and sheep, especially if the livestock suffer a fodder scarcity or the animals lack experience with autumn crocus because they are young or come from areas without this species. Well-nourished livestock learned to avoid *Colchicum autumnale* in pastures and also removed it from hay or freshly-mowed fodder by throwing it into their bedding. Consequently, poisonings are rare nowadays; nonetheless, some of them are lethal.[1709] They were more frequent in the past, when livestock were often malnourished during the barn feeding period and in early spring.[1710] Interestingly, some animals suffer after having consumed autumn crocus, whereas some others do not. In the eighteenth century, Balthasar Hacquet found that where there were both autumn crocus and also a lot of common sorrel (*Rumex acetosa*) in hay, livestock were not poisoned. Thus, Hacquet carried out an experiment involving two animals in order to determine whether common sorrel really prevented poisoning with autumn crocus in the spring. The first one was fed a mixture of fresh leaves of autumn crocus (one-quarter) and common sorrel (three-quarters), and the second one only fresh leaves of autumn crocus. The first animal did not show detectable signs of poisoning, whereas the second one was poisoned but recuperated after being fed common sorrel.[1711] The veterinary surgeon Jože Jurca believes that the experiment was valid, and he assumes that the oxaloacetic acid in the common sorrel might have neutralized the basic alkaloids in the autumn crocus, including colchicine.[1712] This was, however, not generally known. Moreover, not only is autumn crocus poisonous and rejected by most livestock unless they are starving (which used to be common in winter; see Section 6.2.2.1), but its fleshy leaves also dry longer than the rest of hay. Thus, in the case of a high abundance of autumn crocus in a meadow, there was a need to manually remove this herb from mowed hay to reduce the

---

[1706] Aeschimann et al., 2004, vol. 2, 1030–1031; Peterken, 2013, 32–33, 417; Winter et al., 2011, 1764, 1781–1782; Winter et al., 2014, 525.

[1707] Winter et al., 2011, 1764, 1782.

[1708] Author's interview with L. and M. Odar, 2017, in Studor.

[1709] Hacquet, 1779, 5–6, 14; Jurca, 2003, 76–77; Sadar, 1956, 135; Winter et al., 2011, 1764, 1774–1775, 1779; Babai, Molnár, & Molnár, 2014, 101; Winter et al., 2014, 525.

[1710] Jurca, 2003, 76–77.

[1711] Hacquet, 1779, 19–21; Jurca, 2003, 78.

[1712] Jurca, 2003, 78, 81.

risk of a mold attack in the hay barn (see Section 6.6.1.1). Therefore, a spring grazing regime decreasing the abundance of autumn crocus was relevant for livestock farming.

In the spring, autumn crocus is particularly sensitive because it produces most of its aboveground biomass as well as the fruit capsule. Moreover, it stores photoassimilates for the entire year in only about three months in the spring and early summer: at that time there is a green plant because the species is characterized not only by winter dormancy, but also by summer dormancy.[1713] During spring grazing, sheep do not consume the toxic autumn crocus herb, but if grazing is intensive they consume (almost) all surrounding plants. Autumn crocus is thus exposed to their trampling and nutrient intake due to the livestock excrement.[1714] If the leaves are destroyed, the plant cannot grow again in the same year, and therefore the destruction of the leaves in May leads to the inability of *Colchicum autumnale* "to store assimilates and [it] has therefore spent its starch reserves on leaf and capsule growth," which usually leads to smaller populations and a decrease in the size of autumn crocus plants. The destruction of the green herb at an early stage (e.g., early in May, when it is about 25 cm tall), is particularly suppressive, probably because the plant has just "mobilized most of its storage reserves for leaf and capsule development, whereas the formation of new reserves from photosynthetic products is just about to start." Nevertheless, the autumn crocus herbs must not be too small to reach the farming-motivated goal of suppression of this species, otherwise less tissue is destroyed.[1715] In the eighteenth century, the botanist Balthasar Hacquet observed that in the lower montane vegetation belt of Carniola leaves of autumn crocus appeared in April or May (Figure 78).[1716] This means that in many years grazing until May 25th or until later possible Ascension Day dates was properly timed to affect the most sensitive phenophase of autumn crocus, even though the impacts on *Colchicum autumnale* populations differed somewhat at the yearly level due to the changing weather and climate if the dates of disturbance (e.g., of spring grazing) were fixed.[1717] Grazing until Ascension Day (see Section 6.2.1.2, text and Figure 70) was likely an interesting compromise involving regular shifts from years when hay quantities were high (early Ascension Day dates) to years when a longer grazing period actively contributed to autumn crocus suppression in some meadow agroecosystems.

The likely postponement of spring grazing around Koprivnik from May 1st to May 25th might have been an attempt to reduce the populations of autumn

[1713] Winter et al., 2011, 1764, 1782; Winter et al., 2014, 525–526.
[1714] Author's interview with T. Soklič, 2017, in Podjelje.
[1715] Winter et al., 2011, 1777, 1782 (quote p. 1782); Winter et al., 2014, especially 527–531 (quote p. 529); Peratoner et al., 2014, 419–420.
[1716] Hacquet, 1779, 12–13.
[1717] Compare Winter et al., 2014, 530.

Figure 78: Autumn crocus in Podjelje, close to Koprivnik, on April 29th, 2017, at the time of dandelion blooming. In the weeks following this plant development stage, autumn crocus grows significantly taller. Photo: Žiga Zwitter.

crocus there. Namely, it was better to harvest less hay with fewer autumn crocus herbs than more hay containing a lot of autumn crocus (see above). As shown, autumn crocus was often not yet in its most sensitive phenophase by May 1st there (Figure 78), whereas it was highly likely that its most sensitive phenophase was affected if grazing there lasted until May 25th.

In the immediate vicinity of Koprivnik, a farmer at Podjelje accidentally established a system similar to the historical one. He believes that sheep grazing in May had a negative impact on autumn crocus in a non-fertilized meadow rich in this species, where May grazing with sheep was introduced in 2015. In line with his observations until 2017, the green herbs of autumn crocus turned brownish due to sheep grazing, and by mowing time in late June or in July autumn crocus was already decomposing instead of competing with other grassland plants (perhaps reflecting an earlier start of summer dormancy or a result of breaking the capsule-carrying stems by trampling, after which they are no longer able to recover). At another farm at Podjelje, the meadows were also full of autumn crocus. The farmer decided to graze his sheep there in the first half of May, for slightly more than ten days, until they had consumed all the edible plants. At that time, the autumn crocus was about 20 cm tall. After spring grazing, grassland plants were left to grow and hay was mowed there once in late June or in July, followed by autumn grazing from about late September until snow, so that all the edible plants were again consumed. Such a regime lasted for two years, and then spring grazing was abandoned. Unlike the management before the start of spring grazing, when these meadows had not been fertilized, in this second case barn manure was also applied to the meadows.[1718] This probably significantly contributed to suppressing autumn crocus; in general, the abundance of the species is higher in sites with medium than with high nutrient availability,[1719] but experiments with the impact of barn manure on autumn crocus have yielded different results.[1720] Stronger fertilization would lead to highly productive nitrophytes such as cock's-foot (*Dactylis glomerata*) gradually outcompeting autumn crocus.[1721] Negative impacts on autumn crocus plants soon became evident, and in approximately four years autumn crocus totally disappeared from those grazed and organically fertilized meadows. Due to sheep grazing in the first half of May, hay was mowed about fourteen days later than in non-grazed meadows. Yields were slightly reduced, but the farmer noted the better hay quality because there were no hard grasses and the hay was softer, and thus more attractive to livestock.[1722]

However, one should not overemphasize the impacts of spring grazing in reducing autumn crocus in meadows. The results were site-specific and they also depended on the plant phenophases in which spring grazing was practiced. Despite widespread spring grazing, autumn crocus was frequent in many somewhat damp meadows in Carniola in the eighteenth century, as proven by the observations by the botanist Balthasar Hacquet. He even thought that Carniola was at that time exceptionally rich in autumn crocus.[1723] Carbonate bedrock prevails in the mountains of the former province of Carniola, belonging to the Southern Limestone Alps, and autumn crocus prefers base-rich soils.[1724] One of the reasons for the high abundance of this species must have been the early end of spring grazing in many meadow complexes (Section 6.2.1.2), such as in Koprivnik as reported in 1608.

The entire complex of factors causing the suppression of autumn crocus through spring grazing deserves further study. Due to the fact that grazing animals avoid autumn crocus, one would expect a selective advantage for this species over fodder plants in the same grassland.[1725] For mechanical suppression through trampling, a high density of grazing animals is necessary, but in this case they have to be moved forward to the next parcel after a comparative-

[1718] Author's interviews with A. Soklič, 2017, and with T. Soklič, 2017, both in Podjelje.
[1719] Aeschimann et al., 2004, vol. 2, 1030.
[1720] Winter et al., 2011, 1772–1774, 1777–1778; Peratoner et al., 2014, 419–420; Sadar, 1956, 135.
[1721] Klapp & Stahlin, 1936; Jung et al., 2011.
[1722] Author's interview with A. Soklič, 2017, in Podjelje.
[1723] Hacquet, 1779, 11, 13–14.
[1724] Compare Pils, 1994, 163; but see also Aeschimann et al., 2004, vol. 2, 1030.
[1725] Compare the general connection between grazing and populations of the plant species poisonous for livestock in Janišová et al., 2023, 16.

ly short time, before they start to starve and eat toxic plants, and so very short time spans of spring grazing are optimal for this method of control of a species considered a weed in grassland farming.

Moreover, it was stated above that spring grazing has the potential to suppress yellow rattles (*Rhinantus* spp.), "annual short-lived [hemiparasitic] species that rely exclusively on generative reproduction" (see Section 4.1.1.1).[1726] Nonetheless, some plant genera, such as yellow rattles, are able to adapt to early grazing within a comparatively short time to avoid negative grazing effects by late-flowering morphs.[1727] Thus, it is not clear how realistic the explanation attempt from around 1780 was that interpreted spring grazing as the main factor leading to the difference between meadows at Laces (Germ. *Latsch*) in the Venosta Valley (Ital. *Val Venosta*, Germ. *Vinschgau*) in South Tyrol, which were grazed in spring and in good agricultural condition, in contrast to non-grazed meadows at Silandro (Germ. *Schlanders*), some six kilometers away in the same valley, where yellow rattle abounded.[1728]

Interestingly, some relatively grazing-sensitive good fodder plants tolerated specific regimes of moderate spring grazing, even if the grazing ended late in many years (see the mid-nineteenth-century example of sainfoin (*Onobrychis viciifolia*) at Senožeta at the foot of the western Karawanks near Jesenice in today's northwestern Slovenia in Section 6.3.1.1.1). The following subsection also focuses on the western Karawanks and deals with the impacts of spring grazing in hay meadows on daffodils.

6.2.2.3 Case study: spring grazing and daffodils

Salisbury's daffodil (*Narcissus poeticus* subsp. *radiiflorus*; Figure 79) grows in the Western Balkans and in the central part of Southern Europe, also including many places in the Alps. This case study deals with the region where the abundance of daffodils in Slovenia is the highest; that is, the southern slopes of the Western Karawanks, where it has been a remarkable and frequent plant in meadows and pastures between Mount Dovška Baba (Germ. *Frauenkogel*) and the Seča complex of mountain pastures,[1729] at elevations of 600 m (large-scale appearance at some places above 800 m) to 1,800 m (see Figure 4). At all these elevations, daffodils currently bloom in May and/or (at higher elevations) in early June.[1730] In the early twentieth century, Julius Kugy, a plant-loving alpinist

---

[1726] Winter et al., 2014, 531; this study refers to the time when green plants of autumn crocus (*Colchicum autumnale*) are about 25 cm tall (Winter et al., 2014, 526).

[1727] For example, Ter Borg, 1972; Poschlod, Baumann, & Karlík, 2009, 16.

[1728] Compare Wopfner, 1997, 289.

[1729] A mountain grazing complex at approximately 1,700 m, north of the village of Moste or southwest of Bärental (Sln. *Rute*).

[1730] Aeschimann et al., 2004, vol. 2, 1082–1083; Praprotnik, 1993a, 215–218; Praprotnik, 1993b, 260; Krepfl, 2012, 12; Praprotnik, 2012, 30.

and mountaineer,[1731] observed them in full blossom in subalpine meadows on Golica between late May and mid-June:[1732] "Masses of white daffodils covered all the mountain meadows, daffodils so beautiful and in such profusion that I have never seen a flower dominate so abundantly, anywhere. Infinite numbers of daffodils, daffodils everywhere . . . . I climbed a real daffodil mountain. And I came through the white daffodil gardens to the top of Golica, one feast of Corpus Christi after the other, year after year."[1733]

The montane and subalpine sites in the study area where daffodils abound belong to non-fertilized and moderately fertilized grasslands. Focusing on moderately fertilized sites, daffodils abound in the vegetation type of yellow oatgrass meadows (see Sections 2 and 6.3.2) that are located mainly at lower elevations. In non-fertilized meadows, daffodils abound in mesic to dry matgrass swards (*Nardetum strictae*; Sections 2 and 6.3.1.2), in acidic grasslands dominated by bents and fescues (*Violion caninae*), and in the subalpine rusty sedge grasslands on carbonate bedrock (*Caricetum ferrugineae carniolicum*).[1734] The majority of these grasslands were created in forest clearings in the Late Middle Ages,[1735] some of them later, probably in the sixteenth century.[1736]

Historical botanical investigations have shown that the first known information on daffodils possibly from this study area is a specimen of *Narcissus poeticus* subsp. *radiiflorus* preserved in the herbarium collection of the botanist Karl Zois (1756–1799). Botany developed greatly in the eighteenth century, and so this eighteenth-century oldest preserved specimen does not mean that daffodils were not growing in the area earlier. Zois specialized in plants of the Alps, and he spent part of his life precisely in the area of the Karawanks where daffodils now grow. He and his gatherers performed botanical fieldwork there. Nonetheless, due to the lack of information on provenance, his specimen

1731 Jevnikar, 2013.

1732 In the volume *Aus dem Leben eines Bergsteigers* (From the Life of a Mountaineer), first published in 1925 (Jevnikar, 2013), Kugy mentioned: "In recent years, I have made it a habit to ascend [Golica] on the feast of Corpus Christi" (Praprotnik, 1993a, 215). The feast of Corpus Christi can take place between May 21st and June 24th. After the First World War and before the publication year of the aforementioned volume by Kugy, the feast of Corpus Christi was celebrated on May 26th, 1921, on May 31st, 1923, on June 3rd, 1920, on June 15th, 1922, and on June 19th in two years, 1919 and 1924 (Grotefend & Ulrich, 1971, 144, 154–155, 164–165, 171, 195, 203, 213).

1733 Praprotnik, 1993a, 215.

1734 Aichinger, 1933; Praprotnik, 1993a, 218; Praprotnik, 1993b, 260–261; Rogelj, 2012, 55–57, 59. The proportion of grasslands where *Caricetum ferrugineae carniolicum* is represented on Golica is erroneously presented by Rogelj (2012, 57) as higher than the actual proportion (Aichinger, 1933, and the biologist Metod Rogelj, personal communication).

1735 Mlinar, 2004, 14, 16, 18; Zwitter, 2013a, 86–87; Mlinar, 2019, 36–37, 39–42, 44–45.

1736 See the presentation of influences of iron extraction and forging on grassland areas above Jesenice in Section 3.1.

Figure 79: Salisbury's daffodil (*Narcissus poeticus* subsp. *radiiflorus*) in a meadow at 1,200 m above Javorniški Rovt in the Karawanks. Photo: Rok Dovjak, May 1st, 2019. © Rok Dovjak.

might have also been collected elsewhere.[1737] In 1793, the Slovenian name *kluzhavenza* was documented as a name for *Narcissus poeticus*.[1738] Ethnobotanical research has shown that this expression has been used for the daffodil in dialects in our study area and its surroundings and that, as far as investigated, other expressions (with few exceptions) have been or were used for this flower in distant dialects,[1739] and so the 1793 information likely comes from the study area. From 1857 until 1875, the botanist Valentin Plemel served as the priest in the parish of Koroška Bela, precisely at the foot of the section of the Karawanks where daffodils occur. In his botanical treatise published in 1862, he noted that daffodils grew so densely on the upper section of the Golica Ridge at that time that one could observe their whiteness from a point two hours distant.[1740] He also documented that at middle elevations daffodils were very abundant (*in größter Menge*; see Figures 79 and 81) in montane meadows above Jesenice and Javornik; that is, again in this study area.[1741] The botanist Alfonz Paulin

1737 Praprotnik, 1993a, 217–218; for more on the biography of Karl Zois, see Praprotnik, 2015, especially 83–85, 93, 109.
1738 Praprotnik, 1993a, 216.
1739 Matičetov, 1976/77, especially 280–281, 285.
1740 Praprotnik, 1993a, 216–217.
1741 Plemel, 1862, 145.

Figure 80: Salisbury's daffodil is sensitive to trampling. Note the absence of blooming white daffodils in the seldom-used path crossing the meadow above the settlement of Javorniški Rovt. The white flowers in the surrounding meadows are daffodils. Photo: Žiga Zwitter, 2018.

(1853–1942) was more precise, stating that daffodils abounded in meadows on the slopes of the Golica Ridge at elevations of 900 to 1,700 m.[1742]

Daffodils are inedible and poisonous to livestock, cattle, sheep, and goats, at least (if cattle nonetheless eat them, their milk becomes unpalatably bitter),[1743] which fostered them in grazed grasslands to a certain degree. Nonetheless, intensive grazing of sheep or cattle in the spring damages *Narcissus poeticus* subsp. *radiiflorus* in a sensitive phenophase—when it starts building (fragile) flowering stems, for instance. These statements given by local interviewees[1744] correspond with field observations of Salisbury's daffodil trampling sensitivity on May 1st, 2018 (Figure 80), after an abnormally warm April leading to very early daffodil blooming[1745] close to the settlement of Javorniški Rovt. All of the

1742 Praprotnik, 1993a, 218.

1743 Praprotnik, 1993a, 217, 219; *Lamb Care Australia*, 2023; *Shrewdnia: Can Goats Eat Daffodils?*, 2024.

1744 Praprotnik, 1993a, 218; Rogelj, 2012, 60; Mlakar, 2018.

1745 On the influences of weather in 2018 on the timing of daffodil blooming, see Mlakar, 2018. The average April temperature at the Kredarica mountain hut (2,513 m) from 1960 to 2018 was –3.8 °C, and at Stara Fužina or Bohinjska Češnjica at the bottom of mechanisms behind this are not entirely clear, but trampling by livestock is definitely an important factor suppressing young daffodils. It is even considered to harm the bulbs.[1746]

Spring grazing was abolished in meadows in the study area around 1860.[1747] This means that *Narcissus poeticus* subsp. *radiiflorus* colonized these meadows in the centuries when spring grazing was carried out there. Historical spring-grazing regimes did not prevent this species from establishing its high abundance there, which has persisted for centuries.[1748] The interesting question of how spring grazing and daffodils coexisted for centuries has not been addressed so far because presentations of traditional meadow management in this area that have already been considered in daffodil publications omit spring grazing. They are based on oral traditions, but data on historical common spring grazing in meadows in this region have already been forgotten.[1749] Consequently, this section presents historical findings about the timing of spring grazing, which contributed to the historical coexistence of daffodils and spring grazing, based on a detailed study of preserved mid-nineteenth-century records produced by the Commission for Redemption and Regulation of *Profits à Prendre* in the Province of Carniola, combined with early nineteenth-century cadastral and mid-nineteenth century botanical data.

Records from the early and mid-nineteenth century make it clear that spring grazing took place at least in some montane meadow complexes rich in daffodils. Attention is first drawn to meadows at middle elevations. In the

the valley in Bohinj it was +7.8 °C. The average April temperature in 2018 was +0.7 °C at the Kredarica mountain hut and +12.0 °C at Bohinjska Češnjica (Mekinda-Majaron, 1995, 55, 113; *ARSO, Arhiv – opazovani in merjeni meteorološki podatki po Sloveniji*: Kredarica, Stara Fužina, Bohinjska Češnjica).

1746 Rogelj, 2012, 60; Mlakar, 2018.

1747 For instance, SI AS 48, Deželna komisija za odkup in uravnavo zemljiških bremen za Kranjsko (Provincial Commission for Redemption and Regulation of *Profits à Prendre* in Carniola), fasc. 56, 9–2, Protocoll, October 17th, 1861, and no. 3253 de 1861, Vergleichs-Urkunde, undated; SI AS 48, fasc. 56, 9–2: Tagebuch über den Verhandlungsact in Sachen der von Anton Scherjov et Valentin Janša von Alpen und mehreren Insassen von Birnbaum angemeldeten Ablösung oder Reguli[!]rung des Weiderechtes der Berechtigten von Birnbaum: Protokoll, June 5th, 1865; SI AS 48, fasc. 56, 9–4: no. 3309 de 1861.

1748 See above in Sections 6.2 and 6.2.2.3, and the following paragraphs.

1749 See the presentation of mid-twentieth-century meadow management there in the geographical paper by Jordan (1945, 89, 92–93) and of the traditional twentieth-century meadow management in the botanical publication by Praprotnik (1993a, 218). It would be erroneous to consider the traditional management presented in these publications, which no longer involved spring grazing in meadows, unchanged in the long run. Even the presentation of historical farming in the article ambitiously titled "Kmetijstvo v Rovtih nekoč in danes" (Farming in Rovti [i.e., Javorniški Rovt and Planina pod Golico in Figure 4] Past and Present; Klinar, 2012), which deals with this area, avoids the study of archival sources from before the twentieth century, and so it omits spring grazing in meadows.

Figure 81: A postcard sent in September 1930—thus taken in spring 1930 at the latest—shows that daffodils abounded in mid-elevation meadows close to the highest permanent settlements above Jesenice. Plemel's and Paulin's data prove that this photograph shows a typical site, not an exceptional one. AT KLA 129, Ansichtskartensammlung, E-Kahlkogel, 6.

westernmost part of the territory of daffodils analyzed, spring grazing in a large complex of montane meadows at middle elevations is clearly documented in the cadastral municipality of Hrušica.[1750] In the central part of the study area, scant data suggest that spring grazing was practiced in some middle-elevation montane meadows in the cadastral municipality of Planina pod Golico.[1751] In

1750 Kačičnik Gabrič, 2016, 506–507 (where documented spring grazing based on this cadaster is mistakenly also mentioned in the cadastral municipality of Jesenice; compare SI AS 176, Franciscejski kataster za Kranjsko [Franciscan Land Cadaster for Carniola], L10, cadastral municipality of Jesenice, cadastral appraisal protocols); SI AS 176, L16, cadastral municipality of Hrušica, cadastral appraisal protocols: Catastral-Schätzungs-Elaborat, Schaetzung des Natural Ertrages, § 4: Wiesen, Dritter Classe.

1751 Only one out of three protocols for the cadastral municipality of Planina pod Golico mentions spring grazing in third-class meadows, located at middle elevations (SI AS 176, Franciscejski kataster za Kranjsko [Franciscan Land Cadaster for Carniola], L3, cadastral municipality of Planina pod Golico, cadastral appraisal protocols: Catastral-Schätzungs-Elaborat, Schaetzung des Natural Ertrages, § 4: Wiesen and two slightly earlier protocols bound in this source: Protocoll, September 14th, 1830, II. Wiesen; Einvernehmungs Protocoll, May 1st, 1832, Wiesen). The same protocol mentions only autumn grazing and no spring grazing in the fourth-class meadows at the top of the mountains of that cadastral municipality, and so a lapse (in the sense that spring grazing was noted for third-class instead of fourth-class meadows) cannot be excluded. Nonetheless, comparisons with the nearby cadastral municipalities given above suggest that spring grazing was likely practiced in third-class meadows there as well.

the eastern part of the microregion studied, spring grazing is clearly documented in the large montane meadow complex of Jezerca north and northeast of the settlement of Javorniški Rovt.[1752] The following paragraphs present more precise data.

In the cadastral municipality of Hrušica, spring grazing in the area relevant for this case study was practiced in the meadows located north and northwest of the village of Hrušica, at elevations of about 800 to 1,200 m.[1753] According to data from the first half of the nineteenth century, the turf there was relatively thin, and the grassland plants relatively short and sharp.[1754] This probably points to a high abundance of matgrass (*Nardus stricta*) in these meadows (see the plant communities where Salisbury's daffodil abounds in the study area at the beginning of Section 6.2.2.3). Some of these meadows (e.g., Rogarjev Rovt at 1,150 m) are currently extremely rich in *Narcissus poeticus* subsp. *radiiflorus*.[1755] Only sheep and goats were involved in spring grazing in these meadows, which started when the snow had melted and lasted no longer than until May 1st, when spring grazing had to end.[1756] In practice, this meant that

The early nineteenth-century cadaster is an incomplete source of data on spring grazing; this practice was not documented in the cadastral municipality of Koroška Bela, where mid-nineteenth century records of the Commission for Redemption and Regulation of *Profits à Prendre* in the Province of Carniola make clear that it was carried out; see the text above on the Jezerca meadow complex.

1752 SI AS 48, Deželna komisija za odkup in uravnavo zemljiških bremen za Kranjsko (Provincial Commission for Redemption and Regulation of *Profits à Prendre* in Carniola), fasc. 56, 9–4: no. 760, Protocoll, June 28th, 1861; see the localization in the second following paragraph.

1753 SI AS 48, Deželna komisija za odkup in uravnavo zemljiških bremen za Kranjsko (Provincial Commission for Redemption and Regulation of *Profits à Prendre* in Carniola), fasc. 56, 9–2: Tagebuch über den Verhandlungsact in Sachen der von Anton Scherjov et Valentin Janša von Alpen und mehreren Insassen von Birnbaum angemeldeten Ablösung oder Reguli[!]rung des Weiderechtes der Berechtigten von Birnbaum: Protokoll, May 13th, 1859, and Protokoll, June 5th, 1865; localization: SI AS 176, Franciscejski kataster za Kranjsko (Franciscan Land Cadaster for Carniola), L16, cadastral municipality of Hrušica, map sheets A04 and A06; the maps are not georeferenced and there is no relief on them, and so the boundaries of historical parcels were localized based on comparisons of the map sheets of the nineteenth-century cadaster with the georeferenced present-day cadaster, on which historical parcels can easily be recognized (*Environmental Atlas of Slovenia*, Spatial Units, Land Cadastre & Backgrounds, DEM [digital elevation model]).

1754 SI AS 176, Franciscejski kataster za Kranjsko (Franciscan Land Cadaster for Carniola), L16, cadastral municipality of Hrušica, cadastral appraisal protocols: Catastral-Schätzungs-Elaborat, Schaetzung des Natural Ertrages, § 4: Wiesen, Dritter Classe.

1755 Late twentieth-century data on the abundance and occurrence of Salisbury's daffodil from Praprotnik, 1993b, 260.

1756 SI AS 48, Deželna komisija za odkup in uravnavo zemljiških bremen za Kranjsko (Provincial Commission for Redemption and Regulation of *Profits à Prendre* in Carniola), fasc. 56, 9–2: Tagebuch über den Verhandlungsact in Sachen der von Anton Scherjov et Valentin Janša von Alpen und mehreren Insassen von Birnbaum angemeldeten around 1830 spring grazing in these same meadows lasted for about three weeks,[1757] and so it started roughly at the end of the first week of April, but the total duration of grazing in a certain meadow was much more limited because of the low fodder availability.[1758]

Similarly, spring grazing in the meadows at Jezerca to the north and northeast of Javorniški Rovt (also including the former Mavre meadows to the north), at elevations of about 930 to 1,260 m, had to end by May 1st.[1759] Due to the fact that fodder availability was so low at such early dates, early sprouting butterbur (*Petasites*) was such an important fodder plant that the large Lepenje meadow was named after it (see also Section 6.1.2.5.1.1).[1760]

It has already been stated that in the mid-nineteenth century the botanist Valentin Plemel, very familiar with the area, observed a high abundance of Salisbury's daffodil in montane meadows at middle elevations in the study area. The archival data presented above make it clear that this was just at the time when the long history of spring grazing in these meadows was coming to an end, and so these data are representative for previous centuries as well. The early date when spring grazing ended (May 1st) in montane meadows at middle elevations together with either only moderate or intensive but short-term grazing pressures (only once or twice per spring per average meadow if grazing usually took place in one meadow daily; see Section 6.2.2.1)[1761] must have contributed to the colonization and long-term preservation of Salisbury's

Ablösung oder Reguli[!]rung des Weiderechtes der Berechtigten von Birnbaum: Protokoll, June 5th, 1865.

1757 SI AS 176, Franciscejski kataster za Kranjsko (Franciscan Land Cadaster for Carniola), L16, cadastral municipality of Hrušica, cadastral appraisal protocols: Catastral–Schätzungs-Elaborat and a slightly earlier protocol bound in this source, that is, Einvernehmungs Protocoll, August 8th, 1831: Wiesen, III[.] Classe.

1758 SI AS 48, Deželna komisija za odkup in uravnavo zemljiških bremen za Kranjsko (Provincial Commission for Redemption and Regulation of *Profits à Prendre* in Carniola), fasc. 56, 9–2: Tagebuch über den Verhandlungsact in Sachen der von Anton Scherjov et Valentin Janša von Alpen und mehreren Insassen von Birnbaum angemeldeten Ablösung oder Reguli[!]rung des Weiderechtes der Berechtigten von Birnbaum: Gutachten, November 4th, 1861.

1759 SI AS 48, Deželna komisija za odkup in uravnavo zemljiških bremen za Kranjsko (Provincial Commission for Redemption and Regulation of *Profits à Prendre* in Carniola), fasc. 56, 9–4: no. 760, Protocoll, June 28th, 1861, and the list of parcels belonging to Jezerca. Localization: comparison of *Arcanum Maps*, Cadastral Maps, Habsburg Empire – Cadastral Maps (XIX. [sic] Century), cadastral municipality of Koroška Bela (Germ. *Karnervellach*), 1826 and SI AS 176, Franciscejski kataster za Kranjsko (Franciscan Land Cadaster for Carniola), L90, cadastral municipality of Javornik, map sheets A02 and A03, with the georeferenced present-day cadaster (*Environmental Atlas of Slovenia*, Spatial Units, Land Cadastre & Backgrounds, DEM [digital elevation model]); see also the map *Ledinska imena v občini Jesenice*, 2015.

1760 Zwitter, 2020, 68–69.

1761 The formulation "either only moderate or intensive but short-term grazing pressures" is a consequence of a lack of detail in the source. In line with the above statement, it is daffodil in montane meadows at middle elevations. Namely, even though daffodils often start blooming in meadows at middle elevations of the case-study area around May 1st now, it is obvious that they usually started blooming later in previous centuries: between 1961 and 2018, global warming cut the average number of days with snow cover in the daffodil meadows close to the highest farms in this area (data from Javorniški Rovt, 940 m) by no less than forty-five days (i.e., by one-third), and the average annual temperature trend there has risen by 1.8 °C in the same period (data from Planina pod Golico, 947 m).[1762]

Apart from the meadows at middle elevations dealt with so far, an extremely high historical abundance of *Narcissus poeticus* subsp. *radiiflorus* was also recorded in meadows at high elevations on the Golica Ridge by the botanist Plemel in the mid-nineteenth century, when spring grazing was just coming to an end there (see above). At least from the sixteenth to mid-twentieth centuries, there used to exist mowed grasslands on the Golica Ridge above the (farming-lowered) forest line. The first precise land-use map from 1826 makes it clear that the lower margin of the belt of wild hay meadows on the south of this ridge was located at 1,500 to 1,590 m, the upper boundary of this meadow belt on the tops of the mountains, including the highest peak, *Velika Golica* (Germ. *Kahlkogel*, 1,835 m).[1763] According to mid-nineteenth-century data, spring grazing in these meadows commenced as soon as the vegetation period started and it lasted until May 25th.[1764] This date, one of the standard days when late spring grazing in meadows ended in the Eastern Alps (see Section 6.2.1.2), suggests that these mid-nineteenth-century data also reflect well the end of spring grazing in earlier centuries, at least since the sixteenth-century calendar reform. It is important to emphasize that grazing was not carried out continuously in these grasslands from the onset of the vegetation period until May 25th; grazing in these meadows distant from any kind of either permanently or temporarily settled buildings did not take place in unfavorable

clear that "the total duration of grazing in a certain meadow was much more limited [than about three weeks] because of the low fodder availability."

1762 Hrvatin & Zorn, 2020, 204–209.

1763 AT-KLA 112, Allgemeine Handschriftenreihe, Ms. 2732, 467–468, 471–472; Janez Mlinar already noted that these are important sources for the mountain pasture complex observed (Mlinar (ed.), 2018b, 163, footnote 139), and he allowed use of his copies of archival records; Praprotnik 1993a, 218; SI AS 176, Franciscejski kataster za Kranjsko (Franciscan Land Cadaster for Carniola), L3, cadastral municipality of Planina pod Golico, map sheets A01 and A03; the maps are not georeferenced and there is no relief on them, so compare the boundaries of parcels in the present-day cadaster (*Environmental Atlas of Slovenia*, Spatial Units, Land Cadastre) for spatial orientation. Very similar conclusions based on observation in the mid-twentieth century are found in Jordan, 1945, 93.

1764 SI AS 48, Deželna komisija za odkup in uravnavo zemljiških bremen za Kranjsko (Provincial Commission for Redemption and Regulation of *Profits à Prendre* in Carniola), fasc. 56, 9–2: Protokoll, June 1st, 1859.

weather,[1765] and the grazing circuit probably also had a major influence on the clearly reported fact that grazing in a certain meadow at the top of the Golica Ridge was carried out only during a minor part of the total formal spring grazing period. All in all, grazing was allegedly only practiced during about a quarter of the formal spring and autumn grazing periods together, as reported for three meadows from this complex.[1766] Spring grazing in these (at many places very steep) meadows involved only sheep and goats.[1767] How can this coexistence of extremely high abundance of Salisbury's daffodil (so that one could observe their whiteness from a point two hours distant, as presented above) and centuries of spring grazing be explained? Spring grazing pressures were either moderate or, if they were intense, discontinuous. Moreover, due to weather variability, the daffodil phenophase during spring grazing was not the same from year to year. Nonetheless, it is clear that in many years grazing at the top of the Golica Ridge until May 25th corresponded with a sensitive phase of daffodil development, at least in some meadows, despite climate change.[1768] Herders probably paid attention to prevent livestock from consuming poi-

[1765] Compare SI AS 48, Deželna komisija za odkup in uravnavo zemljiških bremen za Kranjsko (Provincial Commission for Redemption and Regulation of *Profits à Prendre* in Carniola), fasc. 56, 9–5: Tagebuch über den Verhandlungsact in Sachen der von der Gemeinde Bach angemeldeten Ablösung oder Reguli[!]rung des Weiderechtes an der Alpe Beušca der Berechtigten von Hrašach, Studenčič, Rann, Velben et Hlebic: Protokoll, December 27th, 1871, 3, 4, and Protokoll, January 8th, 1872, 2, and Protokoll, December 20th, 1883, and Schlussverhandlungsprotokoll, July 5th, 1886.

[1766] SI AS 48, Deželna komisija za odkup in uravnavo zemljiških bremen za Kranjsko (Provincial Commission for Redemption and Regulation of *Profits à Prendre* in Carniola), fasc. 56, 9–2: Protokoll, June 1st, 1859. This protocol refers to three (parcel numbers 622, 630, and 634) out of seventeen meadows, altogether forming two meadow complexes at the top of the Golica Ridge. It stands in this protocol that grazing in these meadows was a part of the system involving further meadows as well, evidently the remaining fourteen meadows at the top of this mountain ridge.

[1767] SI AS 48, Deželna komisija za odkup in uravnavo zemljiških bremen za Kranjsko (Provincial Commission for Redemption and Regulation of *Profits à Prendre* in Carniola), fasc. 56, 9–2: Vergleichs-Urkunde, undated. Only sheep and goats were allowed to graze there in the autumn, after hay harvest, when spring grazing had just been abolished. Slope inclinations must have been the reason for this, and so the same types of livestock had also grazed there during spring grazing.

[1768] Daffodils in subalpine meadows of Golica were in full blossom between late May and mid-June as observed in the early twentieth century (see the first paragraph of Section 6.2.2.3). There is obviously a mistake in Zdenka Torkar Tahir's paper, which—with a reference to the testimonies of the locals interviewed in 1981—claims: "In July, when they [i.e., daffodils] had almost been gone for a few weeks at lower elevations, they were just starting to sprout on the slopes of Golica" (Torkar Tahir, 2012, 134). July erroneously stands in this statement instead of June (in Slovenian, there is only the difference of one letter: *junij* 'June' vs. *julij* 'July'), as proven by two independent pieces of evidence. (1) In the early twentieth century, daffodils were in full blossom in the subalpine meadows on Golica in late May to mid-June, as revealed by contemporary evidence (see the first paragraph of Section 6.2.2.3). (2) Torkar Tahir's statement that at lower elevations, in montane meadows, daffodils had by that time "been almost gone for a few weeks" sonous daffodils even though grown-up livestock used to the local environment learned to avoid toxic plants; however, livestock did not avoid poisonous plants when fodder availability was low. This was definitely often the case for historical grazing in May at elevations of 1,500 to 1,800 m. As an adaptation to the hazard of poisonous daffodils, shepherds probably decreased the grazing pressures in a sensitive stage of daffodil development (such as a shorter grazing duration per meadow), which contributed to the extremely high abundance of Salisbury's daffodil despite spring grazing.

Historical spring grazing in meadows was thus much more daffodil-friendly than one would have expected without empirical historical data. In the second half of the twentieth century, new, much more serious hazards to daffodils gradually appeared in the montane meadows in the microregion studied, for the following reasons. First, former mowing in subalpine meadows (up to 1,835 m) on Golica was abandoned; mowing ceased there in the mid-twentieth century—in the last grasslands in 1957. Former meadows were converted into pastures. According to information from the 1990s, intensive grazing of sheep already started there in May, but its environmental impacts are generally not the same as in a different grazing regime and the colder climate of the times when grazing in May and an extremely high abundance of daffodils coexisted. A reintroduction of mowing in almost one hectare of a subalpine meadow on the eastern part of Golica around 2005 (which also means that there is no intensive grazing when daffodils are in a sensitive phenophase) was followed by a significantly increased abundance of daffodils, and so mid-twentieth-century large-scale abandonment of mowing in combination with an altered grazing regime contributed to the decreased abundance and even occurrence of daffodils in grasslands on the upper slope of Golica. Second, lower than that, agricultural use of many montane meadows in the area was entirely abandoned, which has led to spontaneous afforestation and consequently suppression of daffodils by shading. Third, intensive livestock farming increasingly spread, including mowing too early, intensive manuring, and flattening former meadows through the use of heavy machinery; such places are then abundantly treated with mineral fertilizers and used as pastures where daffodils no longer grow. Finally, in a limited area ski slopes have been flattened using heavy machinery. Chemicals from ski slopes have also negatively influenced the daffodil habitats at middle elevations.[1769]

Due to changing environmental conditions, daffodils no longer grow on a large scale in the highest grasslands on the Golica Ridge, which among the

(Torkar Tahir, 2012, 134) actually refers to June (see Section 6.1.2.2) and not July as erroneously stated by the author.

1769 *Arcanum Maps*, Cadastral Maps, Habsburg Empire – Cadastral Maps (XIX. [sic] Century), cadastral municipality of Planina pod Golico (Germ. *Alpen*), 1826; Praprotnik, 1993a, 218–219; Praprotnik, 1993b, 261; Klinar, 2012, 98–100; Krepfl, 2012, 15; Rogelj, 2012, 59.

general public still has the reputation as the best-known site of daffodils in the Slovenian Alps due to the historically high abundance of daffodils at the top of the ridge. In subalpine grasslands of Golica, the high abundance of daffodils is today largely limited to some ravines and close to forest edges. They have disappeared completely from the pasture at the top of *Velika* Golica. An exception is a meadow on the east of the Golica Ridge where mowing was reintroduced, and so daffodils abound there again. Otherwise, today daffodils are more frequent in subalpine pastures on some mountains west of Golica than on this mountain ridge famous for them. In contrast to subalpine pastures on Golica, at middle elevations many grassland parcels with an abundance of daffodils have been preserved in montane meadows in the study area (Figure 79).[1770] A program including agri-environment payments to thirty-five farmers managing 41 ha of grasslands in a daffodil-friendly manner (data from 2022) has been carried out in the area under consideration.[1771]

6.2.3 Abandonment of spring grazing in meadows

There were several, usually interrelated, reasons leading to the general abolition of spring grazing in meadows between the mid-eighteenth century and the second half of the nineteenth century.

It is already known that meadow plants usually belonged to common-pool resources during spring grazing; an exception was many isolated farms with compact holdings and much more individual agriculture than hamlets, villages, and towns (see the introduction to Section 6.2). Not only was spring grazing in meadows communal, but in many cases people with spring grazing rights did not possess any grassland in the same meadow complex. In certain cases, farmers with spring grazing rights lived in different villages than people with possession of meadows.[1772] Consequently, in some cases, abandonment of spring grazing aimed to solve disputes resulting from very complex grazing rights.[1773] Grazing pressures in some commons exceeded their carrying capacity,[1774] which was also the case for some of the meadows grazed in the spring.[1775] Expectations that more individual management would result in higher yields for the owner of the meadow played an important role in the abolition of spring

[1770] Praprotnik, 1993a, 218–219; Praprotnik, 1993b, 260–261; Klinar, 2012, 113; Rogelj, 2012, 59.

[1771] Mlakar, 2018; Mlakar, 2022.

[1772] SI AS 721, Gospostvo Bled (The Seigneury of Bled), Iustitialia, fasc. 34: Lawsuit between Krištof Wuria and the Nomenj community over grazing rights "in der [!] Stiegen" 1613/1614; Grass, 1948, especially 84, 200–201, 207–208; Novak, 1955, 264–265, 267–270, 275, 279, 295; Wopfner, 1997, 264, 273, 393–394; Kačičnik Gabrič, 2014, 348; Geršič, Gabrovec, & Zwitter, 2018, 71.

[1773] Novak, 1955, 279, 310; Kačičnik Gabrič, 2014, 348, 366.

[1774] Blaznik, 1970a, 176; Blaznik, 1970b, 158.

[1775] Novak, 1955, 266–267, 278.

grazing. At roughly the same time, many complexes of common permanent pastures were also divided into individually managed plots of land. Grazing in meadows had more opponents among agricultural improvers (both actual ones and those whose knowledge was too partial to trigger real progress) than grazing in permanent common pastures. Nonetheless, unlike spring grazing, the economic consequences of which were strong in many cases, collective autumn grazing in (individual) meadows has continued in many places.[1776]

One reason for the negative economic impacts of spring grazing in many meadows, mentioned above and contributing to the abandonment of this type of meadow use, was that the historical timing of spring grazing was not always well adjusted to plant development (see Sections 6.2.2.1 and 6.2.2.2). On the one hand, fixed dates, designated by important religious feasts and name days for easy memorization or with a particular symbolic value, did not adequately consider the factors influencing plant development[1777] and ignored the impacts of soil moisture on the suitability of grazing grounds (see the introduction to Section 6.2.2). On the other hand, fixed dates prevented various disputes between individual possessors of meadows and the farmers with grazing rights in the same meadows, which would have theoretically arisen if the timing of spring grazing had been flexible, adapted to ecological conditions, which would at the same time open the floor to a stronger influence of subjective interpretations. Because every grazing day and every bundle of hay often mattered in the context of historically frequent fodder scarcity in early spring, it would have been too hazardous if the possessors of meadows and farmers with grazing rights had tried to find consensus about the economically optimal beginning and end of grazing in meadows every year. Due to the fixed dates of the end of spring grazing, frequent late spring grazing lasting long in May, and lack of attention to erosion risks, in many cases this type of communal use of grassland resources significantly reduced the quantity of the hay harvest and/or rendered mowing of steep meadows more difficult (see Sections 6.2.2 and 6.2.2.2).[1778]

A further reason for abolition of spring grazing in meadows was other major changes in farming that also took place in the late eighteenth century and the nineteenth century. Modernization and intensification of agriculture led to a gradual general decrease in the importance of grazing within animal husbandry, whereas raising animals in barns gradually gained importance, later in the Alps than in their surroundings. This went hand in hand with an increased area of fodder legume cultivation (Section 3.2.1),[1779] which had the

[1776] Novak, 1955; Wopfner, 1997, 278–291; Kapfer, 2010, 133–140.
[1777] Kirchinger, 2016, 113, 115.
[1778] Novak, 1955, 266–267, 270, 278.
[1779] Blaznik, 1970a, 173; Blaznik, 1970b, 156; Novak, 1970, 363–364, 372, 382–383; Kapfer, 2010, 137; Kačičnik Gabrič, 2014, 76–77, 245–255, 329, 335; Poschlod, 2015, 115–125;

potential to reduce the fodder scarcity in late winter and early spring. Another consequence of this modernization of animal farming, relevant for the history of spring grazing, is that the use of some (but certainly not the majority of) meadows intensified at that time; they were then mowed twice rather than only once a year, as a result of higher manure availability (Section 6.3.2). In some cases, the expansion of the area of meadows mowed twice annually was related to the abolition of spring grazing.[1780] All of this made the scarcity of fodder in early spring less frequent, again reducing the need for spring grazing in meadows.

Information from the Julian Alps reveals another factor contributing to the abolition of this type of meadow use. In the mid-twentieth century, people still recalled historical spring grazing in the meadows at Goreljek until about May 12th, corresponding with the mid-nineteenth-century information in Figure 69 (northernmost location). They emphasized that the former transition from grazing in meadows to subsequent grazing in lower-quality pastures was not smooth because livestock did not want to graze in lower-quality pastures due to their previous grazing in meadows.[1781] To a limited degree there existed an easy strategy of coping with this challenge: where spring-grazed meadows were fertilized with barn manure, this was done just after the end of spring grazing, and so the smell helped suppress any interest of livestock in fleeing from pastures back to meadows.[1782]

Not all impulses leading to the modernization of agriculture in the eighteenth and nineteenth centuries have been studied yet.[1783] Political authorities and agricultural societies provided enlightened narratives of better land cultivation, resulting in higher agricultural production and an improved economic situation. In many cases, such considerations were economically appropriate (e.g., the abolition of spring grazing in meadows where this *profit à prendre* caused significant damage to meadow owners and in some other cases where spring grazing was not well adapted to ecological conditions). Nonetheless, in some cases such information was too theoretical or based on observations in environments too different for its implementation at a local level to lead to optimal results.[1784] Some, but not all, intellectuals looked down on farmers' traditional ecological knowledge,[1785] which in reality included much valuable

Grüne, 2016, 27.

1780 Compare Melik, 1950, 119; Kapfer, 2010, 136–139; Grüne, 2016, 27.

1781 Melik, 1950, 119.

1782 Grass, 1948, 188; Wopfner, 1997, 296.

1783 Popplow, 2016, 259.

1784 For example, Wopfner, 1997, 278–291; Vodopivec, 2008; Kačičnik Gabrič, 2014, 14, 17, 77, 173, 239–240, 244–250, 255, 277, 280–282, 285, 307–313, 317–318, 335, 360–361, 366, 376, 378–380; Grüne, 2016, 25–27, 31, 35; Kirchinger, 2016, 107–108, 120, 125; Mahlerwein, 2016, 15–22; Popplow, 2016, 259.

1785 Wopfner, 1997, 271–272, 291; Kačičnik Gabrič, 2014, 71–72, 244–245, 255; Mahlerwein, 2016, 18–21; Kirchinger, 2016.

local long-term experience, even though it was not correct in all cases from an agroecological perspective,[1786] and so a transdisciplinary approach linking farmers and science would be optimal. Underestimation of local knowledge might have also played a role in the locally unjustified abolition of spring grazing, followed by a decrease in hay quality.[1787] See Section 10.1.2.2.6 for a recent potential for the reintroduction of spring sheep grazing in some meadows in the Alps.

It was shown in Section 6.2 that during spring grazing in meadows some nutrients returned to these grasslands in a form available to plants. Another practice achieving this goal, not only at a selection of patches but throughout the meadow area—namely, grassland fertilization—is presented in the next section.

6.3 Fertilizing meadows in the Alps

Non-fertilized hay meadows that received limited quantities of nutrients available to plants through livestock only during grazing in autumn (and in spring earlier in history; see Section 6.2) prevailed in the total meadow area of Central Europe in the vast majority of history until the first half of the twentieth century in most regions.[1788] Figure 82 shows this in the Alps in the sixteenth century. In most of this mountain range, the situation started gradually changing in the second half of the nineteenth century, and fertilized meadow sites have strongly dominated in the landscape since the second half of the twentieth century (see below and Sections 1.2, 10.1.2.1, the introduction to 10.1.2.2, 10.1.2.2.3, 10.1.2.2.4, and 10.2.1).

Fertilizer availability before the late nineteenth century depended on several factors: social and environmental ones. Fertilization of arable land was given precedence over fertilization of meadows, and thus quantities of manure available for meadows were very limited in most of Europe until the changes presented below in the introduction to Section 6.3. The ratio between livestock rearing and arable land in the local farm economy also mattered. These were two examples of human factors influencing the availability of barn manure. An example of environmental factors with a significant impact on manure availability were periods of rainy summers and cold springs or rainy autumns, which were especially frequent in the last third of the sixteenth century and in the 1690s, when quantities of stored hay were often insufficient and its quality

[1786] For traditional ecological knowledge about grasslands in the sixteenth and seventeenth centuries, see Zwitter, 2020.

[1787] E.g., Wopfner, 1997, 289, and the discussion of impacts of spring grazing on hay quality in Section 6.2.2.

[1788] Wopfner, 1997, 292–298; Kapfer, 2010, 136–139; Kačičnik Gabrič, 2016, 506–507; Janišová et al., 2023.

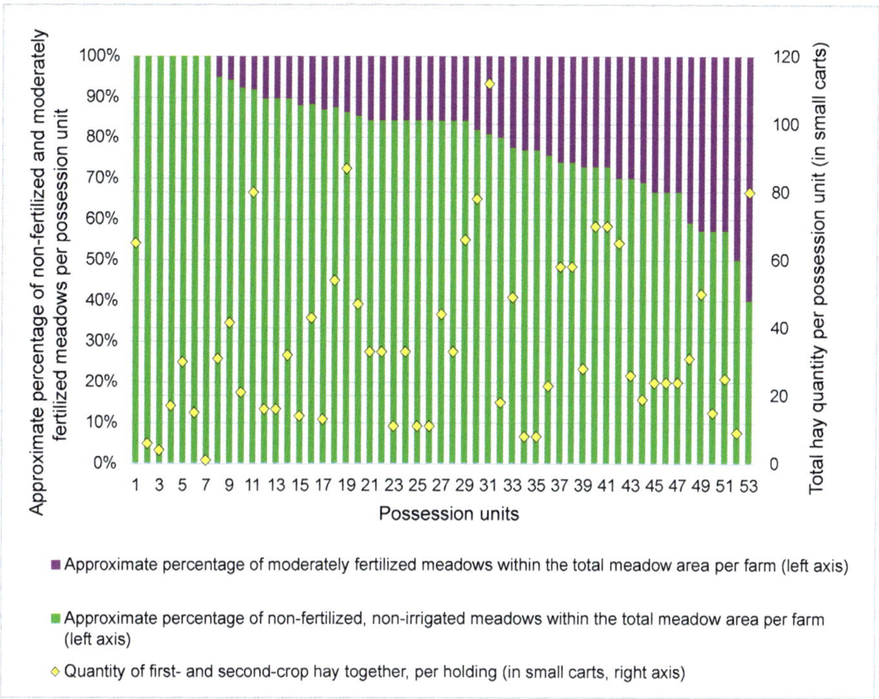

Figure 82: Ratio of fertilized to non-fertilized meadows at fifty-three possession units, each represented by a column. The approximate calculations are based on descriptions of holdings from a 1567 rent-roll. The great majority of these possession units are isolated mountain farms at elevations of around 550 to 1,200 m; rarely, a few hay meadows were located significantly higher than the farm, reaching an elevation of 1,300 m, for instance (see the example of the Anko farm in Section 6.3.0.1). The only holding entirely located higher is a demesne alp with hay meadows at about 1,300 m (i.e., column no. 5; see Section 6.3.0.1). These fifty-three possession units are located in the Central Karawanks (Germ. *Karawanken*, Sln. *Karavanke*) and nearby Kamnik–Savinja Alps (Sln. *Kamniško-Savinjske Alpe*) in the upper Vellach (Sln. *Bela*) and upper Kokra River catchments, from Spodnje Jezersko in the south to Lobnig (Sln. *Lobnik*), the surroundings of (Eisen)Kappel (Sln. *(Železna) Kapla*), and Ebriach (Sln. *Obirsko*) in the north.[1789] Diamonds show the quantity of first- and second-crop hay together per possession unit to give an idea whether a large or small total meadow area belonged to it. See the methodological explanation and comment on this Figure in Sections 6.3.0.1, 6.1.2.3.1, and the caption to Figure 57 in Section 6.1.2.3; namely, Figures 57 and 82 use the same input data. Data source: AT StLA, Innerösterreichische Hofkammer, Sachabteilung, box 69, folder 8: Des ambts Capl bereuttung und beschreibung, 1567. Author of the graph: Žiga Zwitter.

was low at many places (see Section 4.1.1.3), and so numbers of livestock had to be reduced in many cases, resulting in less manure. Roofed hayracks, only used in some regions of the Alps, made a partial difference (Section 4.1.1.3).

[1789] Localization is based on a comparison of this rent-roll with the early nineteenth-century cadaster: *Arcanum Maps*, Cadastral Maps, Habsburg Empire – Cadastral Maps (XIX. [sic] Century).

Nonetheless, over the long run and in many cases, the availability of manure for meadows was greater in the Alps than in their surroundings due to the higher numbers of livestock per spatial unit of fields and meadows together.[1790]

It is clear that quantities of manure available for application to meadows were exceptional in the northernmost Alps from Savoy to Allgäu for several centuries due to early specialization of that region in livestock rearing (see Section 3.1). Similar was the situation in some smaller regions strongly oriented toward raising livestock. In the Eastern Alps, these included relatively small sections at the southern edge of the mountain chain north of some parts of the Venetian Plain (Ital. *Pianura Veneta*), with Venice and other large urban settlements demanding many products of animal husbandry—the Vicentine Alps (Ital. *Prealpi Vicentine*) north of the city of Verona, and the majority of the Alps north of the eastern Venetian Plain (higher sections of the Alps south of the Tagliamento River and the western Julian Alps [Ital. *Alpi Giulie*, Friul. *Alps Juliis*, Sln. *Julijske Alpe*] to the east of it)—as well as the Kitzbühel Alps (Germ. *Kitzbüheler Alpen*) and part of their immediate surroundings, according to data from 1900. Manure availability for meadows was also much more favorable in extensive inner Alpine territories than average in Europe because livestock rearing prevailed over arable land in the inner Alps. At several other places in the Eastern Alps, quantities of available manure were much more limited in general, often insufficient even for arable land.[1791]

More difficult manure transport in the Alps (uphill; one of the main challenges for downhill transport was the long distance between remote grasslands and permanent settlements, and at some places also hazardous track sections) than in lowlands, which was a consequence of landforms, influenced manure availability for meadows in two main and opposite ways. On the one hand, distant meadows at higher elevations with no livestock barn were usually not manured because barn manure was not transported over significant distances

[1790] Pfister, 1985a, 27–28, 84, 86, 92, 104, 119, 124, 126–128; Granda, 1994, 56; Mathieu, 1996, 142–143, 148; Wopfner, 1997, 320; Kauter, 2002, 110–113, 121–122, 124–125, 143–144, 165, 167–170; Kapfer, 2010, 136–139; Leibundgut & Vonderstrass, 2016a, 60, 74, 77, 97, 100, 103; Pavković, 2021, 84, 95, 114; see also Henning, 2014, 338, for at least exceptional early medieval evidence of a manure surplus.

[1791] Mathieu, 1996, 142–143, 148; Granda, 1994, 56; Kauter, 2002, 119–121; Kapfer, 2010, 138; AT KLA 75, Franziszeischer Kataster, cadastral municipality no. 76205, Ebriach (Sln. *Obirsko*), cadastral appraisal protocols: s.n., 47/4, 23–25; AT KLA 75, cadastral municipality no. 76213, Leppen (Sln. *Lepena*), cadastral appraisal protocols: s.n., 248/4, 24–25; AT KLA 75, cadastral municipality no. 76214, Lobnig (Sln. *Lobnik*), cadastral appraisal protocols: s.n., 246/4, 24–25; AT KLA 75, cadastral municipality no. 76218, Remschenig (Sln. *Remšenik*), cadastral appraisal protocols: s.n., 371/4, 24–25; SI AS 178, Franciscejski kataster za Koroško (Franciscan Land Cadaster for Carinthia), K312, cadastral municipality of Zgornje Jezersko, cadastral appraisal protocols: s.n., 312/4, 24–25; SI AS 178, K470, cadastral municipality of Spodnje Jezersko, cadastral appraisal protocols: s.n., 470/4, 24–25.

uphill as long as the transport depended on muscle power.[1792] On the other hand, manure from livestock barns in remote meadows or in their vicinity was often locally applied to some sections of these remote meadows. As a consequence, the general statement in ecological literature that meadows far from the village never received manure[1793] was not a general rule in the Alps and also not in several other parts of Europe (Section 6.3.2.1).

The average quantities of nutrient inputs to meadows have risen tremendously especially since the mid- and late nineteenth century because arable fields in the Alps shrank (see Section 3.1), and so increasingly more fertilizers were available for grasslands. In addition, nitrogen-binding clovers replaced some manure in arable land (see also Section 3.2.1). Moreover, animal husbandry in the Alps has gradually become increasingly oriented toward barn feeding, and fertilizers from barns have often been applied to agricultural land (i.e., increasingly meadows). Since the mid-twentieth century in particular, mineral fertilizers of distant origin were introduced, such as large-scale use of Thomas meal ("a highly potent industrial fertilizer," a byproduct of iron smelting, "rich in phosphates, but also, unfortunately, in heavy metals," widely used especially in the second half of the twentieth century[1794]) in the mid-twentieth century. It needs to be emphasized that the timing and significance of changes in agriculture leading to increased fertilizer availability were locally specific. Thus, mineral fertilizers were already applied to some distant meadows at Tires (Germ. *Tiers*) in South Tyrol in the early twentieth century. Roughly at the same time, cableways allowed fertilizer transport to remote meadows in some places in the Alps and livestock were also capable of transporting significant quantities of mineral fertilizers to distant meadows due to their high concentration, and thus lighter weight, compared to barn manure for a roughly comparable impact on hay quantity. In addition, since the twentieth century, it is not only intentional fertilization and grazing that matter as means of human-induced soil-nutrient intake in permanent meadows: heavier air pollution with nitrates also matters as a general rule, although with a certain level of site-specificity (see Section 10.1.2.2.4).[1795]

Trends in meadow fertilization and their environmental impacts after the mid-twentieth century are not presented here because they are beyond the chronological scope of this volume; they often create grasslands that are no longer species-rich. They are briefly presented in Sections 1.2, 10.1.2.1, the in-

[1792] A sled with a square meter of space for manure can nowadays only be pulled uphill by a horse over a short distance (Babai, Molnár, & Molnár, 2014, 113).

[1793] This statement appears in Luick, Jones, & Oppermann, 2012, 38.

[1794] Groß, 2015, 191 (both quotes), 197.

[1795] Melik, 1950, 114; Pfister, 1985a, 109–116, 119, 122–125, 127–129; Wopfner, 1997, 297, 320–321 (see also Hubatschek, 1992, 173, 224–226); Kauter, 2002, 111–113, 122, 125, 165, 167; Keller, 2004, 22–23; Emanuelsson, 2009, 301; Kapfer, 2010, 137–139; author's interview with Kunavec, 2023, in Stara Fužina.

troduction to 10.1.2.2, in 10.1.2.2.3, 10.1.2.2.4, and 10.2.1 because one needs to take them into account when the relevance of earlier historical data for a more sustainable future is considered.

6.3.0.1 Methodological explanation and comment to calculations of the ratio of non-fertilized to moderately fertilized meadows in the Central Karawanks in 1567

Figure 82 is a result of calculations based on recorded quantities of (1) total first-crop hay (referred to below as h_1, which will be divided into $h_{1,non}$ and $h_{1,fert}$ in our approximate calculations, where $h_{1,non}$ designates first-crop hay from non-fertilized meadows and $h_{1,fert}$ first-crop hay from fertilized meadows), and (2) second-crop hay (referred to as h_2) per possession unit from the 1567 description of holdings in the seigneury of (Eisen)Kappel. All hay quantities were reported in carts; see the explanation in Section 6.1.2.3.1.

The following calculations were used in Figure 82:

1. Non-fertilized, non-irrigated meadows:

In non-fertilized and non-irrigated meadows, there is no second-crop hay (see Sections 6.1.2.3, 6.3.2.2, and 6.4.1), so:

total hay quantity from non-fertilized meadows = $h_{1,non}$ (no h_2 was harvested in non-fertilized meadows)

2. Moderately fertilized meadows (early nineteenth-century data make it clear that irrigation was not relevant in the study area of Figure 82,[1796] *except for the application of liquid manure from livestock barns or the liquid leaving manure piles near the barn, directed to the nearest meadow, and so second-crop hay mowing in such meadows was primarily made possible by fertilizer application; see also Sections 6.1.2.3, 6.3.2.2, and 6.4.1):*

In moderately fertilized meadows, there are first- and second-crop hay:

[1796] AT KLA 75, Franziszeischer Kataster, cadastral municipality no. 76203, (Bad) Vellach (Sln. *Bela*), cadastral appraisal protocols: s.n., 476/4, 50–51; AT KLA 75, cadastral municipality no. 76205, Ebriach (Sln. *Obirsko*), cadastral appraisal protocols: s.n., 47/4, 50–51; AT KLA 75, cadastral municipality no. 76213, Leppen (Sln. *Lepena*), cadastral appraisal protocols: s.n., 248/4, 50–51; AT KLA 75, cadastral municipality no. 76214, Lobnig (Sln. *Lobnik*), cadastral appraisal protocols: s.n., 246/4, 50–51; AT KLA 75, cadastral municipality no. 76218, Remschenig (Sln. *Remšenik*), cadastral appraisal protocols: s.n., 371/4, 50–51; SI AS 178, Franciscejski kataster za Koroško (Franciscan Land Cadaster for Carinthia), K312, cadastral municipality of Zgornje Jezersko, cadastral appraisal protocols: s.n., 312/4, 50–51, and Catastral-Schätzungs-Elaborat, Einleitung, § 10: Cultur des Bodens, & Schätzung des natural ertrages der gemeinde, § 4: Wiesen; SI AS 178, K470, cadastral municipality of Spodnje Jezersko, cadastral appraisal protocols: s.n., 470/4, 50–51, and Catastral-Schätzungs-Elaborat, Einleitung, § 10: Cultur des Bodens.

h_{fert} (total hay yield in moderately fertilized meadows) = $h_{1,fert} + h_2$

3. In productive moderately fertilized meadows, the quantity of second-crop hay is half the quantity of first-crop hay,[1797] so

1797 "Košnja," 1864, 199, reporting a ratio of $h_1:h_2 = 12:7$. Lower figures were also usual in moderately fertilized meadows in some regions, such as 3:1 in mid-twentieth-century Bohinj, and drought would further reduce the second-crop hay yield (author's interview with Kunavec, 2023, in Stara Fužina). Thus, early nineteenth-century cadastral data from the very cadastral municipalities covering the area dealt with in the 1567 rent-roll analyzed in Figure 82 were studied. The values reported in the cadastral appraisal will not be overestimations. In most of the cadastral municipalities later covering the area presented in the 1567 rent-roll, this ratio was close to $h_1:h_2 = 2:1$, and in the others it was between 2:1 and 3:1. Namely, in the cadastral municipalities of Ebriach (Sln. *Obirsko*), Leppen (Sln. *Lepena*), Lobnig (Sln. *Lobnik*), and Remschenig (Sln. *Remšenik*) and in the more productive ones among the fertilized meadows in the municipality of Spodnje Jezersko, this ratio was 8:5, where 5 also includes autumn grazing after second mowing, and so a ratio of close to $h_1:h_2 = 2:1$ can be deduced from these data. A ratio of about $h_1:h_2 = 2:1$ was also characteristic of the more productive among the fertilized meadows in the cadastral municipality of Zgornje Jezersko, where a ratio of $h_1:h_2 = 5:3$ can be calculated and 3 also includes grazing after second mowing; similarly, only slightly less second-crop hay was typically harvested in the more productive of the moderately fertilized meadows in the municipality of Vellach (Sln. *Bela*), allowing calculation of a ratio of $h_1:h_2 = 9:5$, where 5 also includes grazing after second mowing. A ratio of $h_1:h_2$ = (between 2 and 3):1 can be deduced from data on the less productive among the moderately fertilized meadows in the cadastral municipalities of Vellach, Spodnje Jezersko, and Zgornje Jezersko; namely, a ratio of 2:1 can be calculated in all these cases, where 1 also includes grazing after second mowing, and so the amount of second-crop hay was less than half the first-crop hay yield. AT KLA 75, Franziszeischer Kataster, cadastral municipality no. 76203, (Bad) Vellach, cadastral appraisal protocols: Catastral-Schätzungs-Elaborat, Schätzung des natural ertrages der gemeinde, § 4: Wiesen; AT KLA 75, cadastral municipality no. 76205, Ebriach, cadastral appraisal protocols: Catastral-Schätzungs-Elaborat, Schätzung des natural ertrages der gemeinde, § 4: Wiesen; AT KLA 75, cadastral municipality no. 76213, Leppen, cadastral appraisal protocols: Catastral-Schätzungs-Elaborat, Schätzung des natural ertrages der gemeinde, § 4: Wiesen; AT KLA 75, cadastral municipality no. 76214, Lobnig, cadastral appraisal protocols: Catastral-Schätzungs-Elaborat, Schätzung des natural-ertrages der gemeinde, § 4: Wiesen; AT KLA 75, cadastral municipality no. 76218, Remschenig, cadastral appraisal protocols: Catastral-Schätzungs-Elaborat, Schätzung des natural ertrages der gemeinde, § 4: Wiesen; SI AS 178, Franciscejski kataster za Koroško (Franciscan Land Cadaster for Carinthia), K312, cadastral municipality of Zgornje Jezersko, cadastral appraisal protocols: Catastral-Schätzungs-Elaborat, Schätzung des natural ertrages der gemeinde, § 4: Wiesen; SI AS 178, K470, cadastral municipality of Spodnje Jezersko, cadastral appraisal protocols: Catastral-Schätzungs-Elaborat, Schätzung des natural ertrages der Gemeinde, § 4: Wiesen.
Similarly, in 1605, a meadow not far from Koroška Bela (a village at the foot of the Western Karawanks) delivered one wagon of first-crop hay (estimated value: 1 gulden) and four sheets of second-crop hay (0.67 gulden) (SI AS 721, Gospostvo Bled [The Seigneury of Bled, Germ. *Veldes*], Iustitialia, vol. 13, August 21st, 1605). This also suggests that the ratio of $h_1:h_2$ might have been close to 2:1, with first-crop hay constituting slightly more than twice the quantity of second-crop hay. Namely, (1) the nutritive value of second-crop hay is significantly higher than that of first-crop hay (Section

$h_2 = h_{1,\text{fert}} : 2 \Rightarrow \underline{h_{1,\text{fert}} = 2h_2}$

This means that the approximate quantity of $h_{1,non}$ *based on reported input data on* h_1 *and* h_2 *can be calculated as follows*:

$\underline{h_{1,\text{non}} = h_1 - 2h_2}$ (namely, see above: $h_{1,\text{fert}} = 2h_2$)

4. The yield per area unit (first- and second-crop hay together) in moderately fertilized meadows is usually at least twice the yield from non-fertilized meadows,[1798] so

the approximate share of area of moderately fertilized meadows per possession unit can be calculated from the data on the total first-crop and second-crop hay per possession unit as follows:

$(h_{1,\text{fert}} + h_2) / 2{:}((h_{1,\text{fert}} + h_2) / 2 + h_{1,\text{non}}) = 3h_2 / 2{:}(3h_2 / 2 + h_1 - 2h_2) = \underline{3h_2{:}(2h_1 - h_2)}$;

6.1.2.3), thus its price must have been higher than that of first-crop hay, and so the estimated values of 1 gulden and 0.67 gulden are not a good reflection of the ratio between the quantities of first- and second-crop hay. Instead, these data explain that the quantity of second-crop hay was significantly less than two-thirds the quantity of first-crop hay. (2) Assuming that the size of sheets and the wagon were similar to the ones in Figure 58, four hay sheets might have been close to half a wagon: in Figure 58, there are six full sheets of second-crop hay plus the load below the sheets, probably equivalent to an additional two or three sheets. This would mean around nine sheets of first-crop hay (namely, one cart) and four sheets of second-crop hay from that meadow in 1605.

1798 Dietl & Lehmann, 2004, 121, 124–125. Sometimes lower values are reported: 70% more in a moderately fertilized than in a non-fertilized *Agrostis capillaris–Festuca rubra* meadow in the Apuseni Mountains, Romania (Janišová et al., 2023, 18). In the study area of Figure 82, early nineteenth-century data make it clear that yields in moderately fertilized meadows were at least twice the yields in non-fertilized meadows, and often the difference was more pronounced: in less productive moderately fertilized meadows, typical yields were slightly more than twice the typical yields in non-fertilized meadows; yields in more productive moderately fertilized meadows were often close to four times those in typical non-fertilized ones. AT KLA 75, Franziszeischer Kataster, cadastral municipality no. 76203, (Bad) Vellach, cadastral appraisal protocols: Catastral-Schätzungs-Elaborat, Schätzung des natural ertrages der gemeinde, § 4: Wiesen; AT KLA 75, cadastral municipality no. 76205, Ebriach, cadastral appraisal protocols: Catastral-Schätzungs-Elaborat, Schätzung des natural ertrages der gemeinde, § 4: Wiesen; AT KLA 75, cadastral municipality no. 76213, Leppen, cadastral appraisal protocols: Catastral-Schätzungs-Elaborat, Schätzung des natural ertrages der gemeinde, § 4: Wiesen; AT KLA 75, cadastral municipality no. 76214, Lobnig, cadastral appraisal protocols: Catastral-Schätzungs-Elaborat, Schätzung des natural-ertrages der gemeinde, § 4: Wiesen; AT KLA 75, cadastral municipality no. 76218, Remschenig, cadastral appraisal protocols: Catastral-Schätzungs-Elaborat, Schätzung des natural ertrages der gemeinde, § 4: Wiesen; SI AS 178, Franciscejski kataster za Koroško (Franciscan Land Cadaster for Carinthia), K312, cadastral municipality of Zgornje Jezersko, cadastral appraisal protocols: Catastral-Schätzungs-Elaborat, Schätzung des natural ertrages der gemeinde, § 4: Wiesen; SI AS 178, K470, cadastral municipality of Spodnje Jezersko, cadastral appraisal protocols: Catastral-Schätzungs-Elaborat, Schätzung des natural ertrages der gemeinde, § 4: Wiesen.

the same expressed as a percentage, as in Figure 82: $\mathbf{100 \times 3h_2{:}(2h_1 - h_2)}$

>>> approximate *percentage of area of non-fertilized meadows per possession unit:*

$\mathbf{100 \times (1 - (3h_2{:}(2h_1 - h_2)))}$

In reality, the yield in some typical non-fertilized meadows in the study area was very close to half the yield in moderately fertilized meadows, in line with point 4 above. Nonetheless, the comment in the footnote to point 4 explains that less than half the yield in moderately fertilized meadows was characteristic of some non-fertilized meadows, which would lead to an overestimation of the area of moderately fertilized meadows in Figure 82.

The quantity of first-crop hay in moderately fertilized meadows was usually twice the yield of second-crop hay in the same meadow, in line with point 3. Nonetheless, as explained in the footnote to point 3, the first-crop hay yield was more than twice the second-crop hay yield in some moderately fertilized meadows, which would cause an underestimation of the percentage of the moderately fertilized meadow area in Figure 82.

Thus, these positive and negative deviations from the assumptions above partly cancel each other out, which implies that Figure 82 gives the correct general message: non-fertilized meadows strongly prevailed in most sixteenth-century cultural landscapes in the Alps. However, a limited area of moderately fertilized meadows was also typical of almost every farm. In the Central Karawanks in the mid-1560s, moderately fertilized meadows usually covered between about 15% and about 30% of the total meadow area per farm, whereas farms fertilizing more than half of the total meadow area were very rare (Figure 82).[1799] Contextualization of atypically high (the rightmost column in Figure 82) and low values (in the left of the graph) is given in the following paragraphs.

In Figure 82, the rightmost column shows the large Anko farm at the bottom of the basin at Zgornje Jezersko,[1800] at about 900 m, where manure was applied to flat parcels accessible to a livestock-drawn cart of barn manure.[1801]

1799 This conclusion from Figure 82 is in line with Karl Dinklage's data from 1584 on the Kaunz farm at Gariusch (Sln. *Gorjuše*, northeast of Völkermarkt, Carinthia), located at about 730 m (Dinklage, 1966b, 107; the microlocalization is based on the farm name in the early nineteenth-century cadaster: *KAGIS Maps*, Basiskarten, Franziszeischer Kataster & Abfragen, Koordinaten / Höhe). Calculating the percentage of non-fertilized and moderately fertilized meadows with the above equation, 27% of the meadow area was moderately fertilized.

1800 AT StLA, Innerösterreichische Hofkammer, Sachabteilung, box 69, folder 8: Des ambts Capl bereuttung und beschreibung, 1567, 12. Compare Zwitter, 2019, 306.

1801 Compare SI AS 178, Franciscejski kataster za Koroško (Franciscan Land Cadaster for Carinthia), K312, cadastral municipality of Zgornje Jezersko, cadastral appraisal protocols: Catastral-Schätzungs-Elaborat, Protokoll, August 4th, 1831.

In addition, the 1567 rent-roll already documented a large individual summer mountain pasture complex belonging to that isolated farm (i.e., the Anko Alp, Sln. *Ankova planina*);[1802] judging from an early nineteenth-century land-use map, there were meadows next to the individual summer pasture of this mountain farm at about 1,300 m.[1803] Application of locally available manure in that summer alp must have motivated mowing in that grassland complex (see Section 6.3.2.1), which again raised the percentage of moderately fertilized meadows of the Anko farm. Cases of farms fertilizing more than half of all meadows as early as before the nineteenth century could, however, at least for a very limited percentage of farms, also be found elsewhere: another sixteenth-century example from Carinthia (Germ. *Kärnten*, Sln. *Koroška*) was a farm at Gariusch northeast of Völkermarkt, at about 640 m in 1584:[1804] calculating the percentage of non-fertilized and moderately fertilized meadows with the above equation, 69% of the meadow area was moderately fertilized. Although Figure 82 shows low and high percentages of moderately fertilized meadows at small and large farms, sparse data suggest that, focusing on family farms in the Alps, all meadows were moderately fertilized as early as the sixteenth or seventeenth century mainly in some smallholdings, such as a cottage from Ruden (Sln. *Ruda*), at about 450 m, at the foot of the Alps east of Völkermarkt, in 1692,[1805] as a matter of specific decisions of their integrated farm economy.[1806]

An explanation is available for the non-existing or unusually low percentage of moderately fertilized meadows in several possession units in the left of Figure 82. These were not representative of a usual family farm. As regards the first seven columns from the left, all without any fertilized meadows, almost none of them functioned as a family farm, and many of them were additional holdings used by people living elsewhere, which might have caused an absence of locally available barn manure.[1807] Some of these holdings might even have included some fertilized meadows, which were exceptional in the sense

1802 AT StLA, Innerösterreichische Hofkammer, Sachabteilung, box 69, folder 8: Des ambts Capl bereuttung und beschreibung, 1567, 12; see also Zwitter, 2020, 78.

1803 *Arcanum Maps*, Cadastral Maps, Habsburg Empire – Cadastral Maps (XIX. [sic] Century), cadastral municipality of Zgornje Jezersko (Germ. *Oberseeland*), 1826; *Environmental Atlas of Slovenia*, Overview Maps (detailed topographical map) and Spatial units, Land cadaster.

1804 Dinklage, 1966b, 106; the microlocalization is based on toponyms in the early nineteenth-century cadaster: *KAGIS Maps*, Basiskarten, Franziszeischer Kataster & Abfragen, Koordinaten / Höhe.

1805 Calculation based on data in Dinklage, 1966b, 104.

1806 Compare Netting, 1993; Panjek, 2017.

1807 AT StLA, Innerösterreichische Hofkammer, Sachabteilung, box 69, folder 8: Des ambts Capl bereuttung und beschreibung, 1567, 11, 20, 24–27, 32. The reason for the absence of fertilized meadows in the third and sixth columns from the left cannot be explained based on this source. The third column designates a small farm where a mill was an important source of income, and so this was not a typical family farm.

that second-crop hay was not mowed but only grazed—the fifth column, for instance, is probably such an exception. It designates the Govca Alp (Sln. *Govševa planina*; on part of the northern slopes of Mt. Olševa and in its immediate northern surroundings, see Figure 4), the unit of possession with the highest center among all in Figure 82. The Govca Alp extended from about 1,230 m to almost 2,000 m, with the center (the human habitation) at around 1,285 m. The hay meadows belonging to the Govca Alp were located at its lowest margin, and the highest ones at about 1,320 or 1,340 m, as documented in the early nineteenth-century cadaster.[1808] Because the area of arable land at the Govca Alp was relatively small and the numbers of livestock grazing there relatively high,[1809] it is likely that (part of the) mowed area was manured with locally available barn manure, but no second-crop hay was made there due to the exceptionally strong connection between the meadow and pasture management—a potentially long duration of grazing in fertilized meadows would only allow one mowing annually.

[1808] The localization is based on the following: the 1567 rent-roll reports that this was "des Hagen albm an der Erlach" and it stands in the rent-roll's chapter "Remsenig." There were not only pastures and forest, but also some hay meadows and a small area of arable land (AT StLA, Innerösterreichische Hofkammer, Sachabteilung, box 69, folder 8: Des ambts Capl bereuttung und beschreibung, 1567, 25). There was no alp meeting all these criteria in the valley of Remschenig proper; in particular, there was no alp belonging to the seigneury of Hagenegg in the east of Remschenig close to Mt. Olševa or in its northern surroundings ("Erlach" in the 1567 rent-roll) (AT KLA 75, Franziszeischer Kataster, cadastral municipality no.76218, Remschenig, Protocoll sämmtlicher grund- und bau-parzellen der gemeinde, 1827, especially the column "Grundherrschaft;" the cadastral map in *Arcanum Maps*, Cadastral Maps, Habsburg Empire - Cadastral Maps (XIX. [sic] Century), cadastral municipality of Remschenig, 1826).
However, in line with the early nineteenth-century cadaster, the Govca Alp located just east of Remschenig proper met all these criteria: it was located on the northern slope of Mt. Olševa, and in its immediate northern surroundings, there were not only pastures and forest but also hay meadows and arable land, and it belonged to the count "Carl Christallnigg" (AT KLA 75, cadastral municipality no. 76211, Koprein Petzen (Sln. *Koprivna*), Protokoll sämtlicher grund- und bauparzellen der steuergemeinde, c. 1828, and Alphabetisches verzeichniss sämtlicher grundeigenthümer der steuergemeinde, 1828; the cadastral map in *Arcanum Maps*, Cadastral Maps, Habsburg Empire - Cadastral Maps (XIX. [sic] Century), cadastral municipality of Koprein-Petzen, 1827). Indeed, a look at Hagenegg Castle in the cadastral municipality of (Bad) Vellach explains that the count "Karl Christallnig" was at that time in charge of the seigneury of Hagenegg (he was "herrschafts innhaber zu Hagenegg;" AT KLA 75, cadastral municipality no. 76203, (Bad) Vellach, Alphabetisches verzeichniss der grundeigenthümer und ihrer nach sectionen abgetheilten grund-parzellen, c. 1827; *Arcanum Maps*, Cadastral Maps, Habsburg Empire Cadastral Maps (XIX. [sic] Century), cadastral municipality of Bad Vellach, 1826).
Concerning the elevations, compare the cited early nineteenth-century cadastral map with *Environmental Atlas of Slovenia*, Overview Maps (detailed topographical map) and Spatial units, Land cadaster and *Atlas Slovenije*, 2005, 55.

[1809] AT StLA, Innerösterreichische Hofkammer, Sachabteilung, box 69, folder 8: Des ambts Capl bereuttung und beschreibung, 1567, 25.

The situation of several holdings with a very small area of moderately fertilized meadows was also specific. Focusing on the eighth and tenth column from the left, only part of those family farms belonged to the princely seigneury of (Eisen)Kappel that produced this rent-roll;[1810] thus, the farmers might have used further moderately fertilized meadows, not included in Figure 82, if the parts of the holdings belonging to another seigneury included moderately fertilized meadows.

In the cases of the ninth and fourteenth columns from the left, the holding was used by two farm heads (and their families), and the large farm represented by the eleventh column was divided between four farm heads.[1811] Early nineteenth-century information demonstrates that a scarcity of manure was relevant in this study area;[1812] thus, farm divisions probably concentrated the use of fertilizers on the arable land, which possibly explains the low percentages of fertilized meadows in ninth, eleventh, and fourteenth columns in the 1560s.

6.3.1 Non-fertilized meadows[1813]

As long as the quantities of manure available were insufficient from the perspective of farming, and the atmospheric inputs, significant since the second half of the twentieth century, were negligible (see Section 10.1.2.2.4), nutrient cycling within most meadows was very limited. Namely, nutrients were regularly removed from meadows through hay, but most meadows were not fertilized. Consequently, long-term reduction of the nutrient status in the upper soil layer was taking place in several non-fertilized meadows so that nutrient-poor meadows became frequent (see also Section 9).[1814]

Nonetheless, biological processes significantly increased the nutrient availability in many non-fertilized meadows, and so they mitigated the long-term soil impoverishment or they increased the availability of nutrients to plants in another way. (1) The most important biological process of relevance here

1810 AT StLA, Innerösterreichische Hofkammer, Sachabteilung, box 69, folder 8: Des ambts Capl bereuttung und beschreibung, 1567, 9.

1811 AT StLA, Innerösterreichische Hofkammer, Sachabteilung, box 69, folder 8: Des ambts Capl bereuttung und beschreibung, 1567, 25–27, 37.

1812 AT KLA 75, Franziszeischer Kataster, cadastral municipality no. 76213, Leppen, cadastral appraisal protocols: s.n., 248/4, 24–25 (the holding represented by column 14 in Figure 82 is located in the territory of the later cadastral municipality of Leppen); AT KLA 75, cadastral municipality no. 76218, Remschenig, cadastral appraisal protocols: s.n., 371/4, 24–25 (the holdings represented by columns 9 and 11 in Figure 82 are located in the territory of the later cadastral municipality of Remschenig).

1813 Section 6.1.2.5 deals with non-fertilized wet meadows, and this one focuses on moderately moist, mesic, and dry sites. Section 6.4 deals with irrigated non-fertilized meadows, which are not dealt with here due to the regular nutrient intake through irrigation.

1814 Wopfner, 1997, 311–315; Eriksson, Cousins, & Bruun, 2002, 744; Keller, 2004; Emanuelsson, 2009, 340; Peterken, 2013, 221; Babai, Molnár, & Molnár, 2014, 109; Krausmann, Weisz, & Eisenmenger, 2016, 79.

is airborne nitrogen fixation, which significantly contributed to soil nitrogen content at some non-fertilized sites in the case of a high abundance of leguminous plants (e.g., kidney vetch, *Anthyllis vulneraria*, see Figure 10 and the text of Section 2; bird's-foot trefoil, *Lotus* spp., and many others). See Sections 6.3.1.1.1, 6.3.1.2, and 6.3.2.2. (2) Moreover, several plant species in non-fertilized meadows profit from nutrients made available to them by a symbiotic interaction with a fungus (mycorrhiza, interactions between plant roots and fungal hyphae). The majority of perennial grassland species in non-fertilized grasslands, such as gentians (*Gentiana* spp., see Section 8.1.2; *Gentianella* spp.), some grass species (e.g., matgrass, *Nardus stricta*, Figure 30 in Section 4.1.2; heath-grass, *Danthonia decumbens*), all orchids (*Orchidaceae*), mountain arnica (*Arnica montana*, Figure 85), clubmosses (*Lycopodium* spp.), moonwort (*Botrychium lunaria*), and some dwarf shrubs, such as bilberry (*Vaccinium myrtillus*), cowberry (*Vaccinium vitis-idaea*), heather (*Calluna vulgaris*), winter-flowering heath (*Erica carnea*), and alpenrose (*Rhododendron* spp.), are known to establish close mycorrhiza connections and profit from the additional supply of water and soil nutrients through mycorrhiza. (3) In addition, highly relevantly, as a general rule, some plant nutrients regularly returned to non-fertilized meadows through spring and autumn grazing, with the main exceptions of the meadows inaccessible to livestock or located on slopes too steep for grazing (see below and Section 6.2).[1815] (4) Increased grassland productivity through earthworm activities also mattered in some non-fertilized meadows, but not as much as earthworm activities at moderately manured sites (see Section 10.1.2.2.3). (5) In some remote meadow sites mowed only once every two or three years, or two out of three years, grassland plants decomposed in the meadow and nutrients returned to the soil; see Section 6.1.2.6. (6) The accumulation of plant nutrients from meltwater in concave landforms in meadows had a relevant impact on the productivity of non-fertilized meadows (see Section 6.3.1.2). (7) Fallen tree leaves often resulted in a significant amount of nutrients available for plants in meadow sections along forest edges, next to hedges, and below trees growing in meadows. The quantity of hay growing around different tree and shrub species differs; hazel (*Corylus avellana*; see also Section 6.1.2.6) is one of those with the best impacts on hay quality and quantity. Leaves decay in situ from autumn until their partial removal during spring cleaning (see Section 6.6.1). Roots of trees and bushes absorb nutrients more deeply in the soil than grassland plants can, and they are released from decomposing leaves into the top of the soil (Figure 83). Consequently, haymakers knelt below the hedges (hazel hedges, at least), to harvest all the grass on such nutrient-rich sites.[1816] Locally, some further processes significantly mitigated

[1815] Pils, 1994, 75, 90–91, 190, 265–266, and Table 16, Photo 43, and Table 17, Photo 47; compare Babai, Molnár, & Molnár, 2014, 35–36, 43–44.

[1816] Author's interview with Kranjc, 2018, in Stara Fužina; Machatschek, 2002, especially p. 308; Emanuelsson, 2009, 204; Ivaşcu, Öllerer, & Rákosy, 2016, 48.

Figure 83: Hazel hedges (*Corylus avellana*) were sometimes planted or allowed to grow at the top of a meadow on a slope. Their roots often significantly contributed to slope stability and farmers were well aware of this.[1817] Their fallen leaves regularly released a slight amount of plant nutrients from deeper soil that was inaccessible to most grassland plants into meadow topsoil. Fallen leaves accumulated on a much wider area on steep meadows than on the edges of flat meadows. This meadow below fallen leaves is still regularly mowed. Photo: Žiga Zwitter, 2020, south of the Ribnica River at Studor, Bohinj.

the long-term decline in the productivity of several non-fertilized meadows (see a likely example from Mt. Golica in the Karawanks in Section 6.3.1.2).

6.3.1.1 Biodiversity and yields

Nutrient-poor late-mowed meadows, dry calcareous grasslands in particular, are the most species-rich grassland types in the Alps and Central Europe. The biodiversity in extensively managed meadows is significantly higher than in extensively managed pastures. An absence of systematic fertilization significantly contributes to a great diversity of site conditions in non-fertilized meadows. Recent data show that about fifty to sixty-five plant species usually occur on 50 m^2 in nutrient-poor late-mowed meadows, also including most rare species among all meadow types as classified according to fertilization regimes. Spring and autumn grazing in most non-fertilized meadows

[1817] Machatschek, 2017, 274–275, 286.

Figure 84: Carniola lily (*Lilium carniolicum*) in a non-fertilized montane meadow on Vogar Hill, Bohinj, regularly mowed once a year. Photo: Savina Zwitter, June 26th, 2004.

additionally increased species richness, forming small and disjunct patches of nutrient-rich and moderately manured sites, surrounded by the prevalent nutrient-poor grassland (see also Sections 6.2 and 6.3.2). Species richness is usually the highest in moderately dry, sunny meadows where the grass upright brome (*Bromus erectus*, see Section 2) occurs as a character species (e.g., from forty-five to more than seventy flowering plant species per 50 m^2). Upright brome is a dominant species building the community structure especially suitable for high species diversity. The main matrix of the community is built by its small separate tussocks with a comparatively short lifespan of five to ten years, so that various gaps for germination and establishment of subordinated spe-

Figure 85: Mountain arnica (*Arnica montana*, orange blossoms with "disheveled" petals as characteristic for this species) abounding in several sections of a non-fertilized hummocky montane meadow on Vogar Hill, regularly mowed once annually. Photo: Žiga Zwitter, late June 2021.

cies are always available. Grassland communities characterized by rusty sedge (*Carex ferruginea* subsp. *ferruginea*)—abounding at many moist sites on deep soils at higher elevations, in subalpine and alpine vegetation belts, and occurring in most of the Alps—are another example of communities belonging to the most floriferous and colorful meadows. Plant species richness in meadows dominated by red fescue (*Festuca rubra*) and common bent (*Agrostis capillaris*) is somewhat lower than in upright brome communities.[1818] The probable reason is that, especially stoloniferous common bent, if dominant, builds a dense sward and "closes" the establishment gaps for subordinated species in the course of its vegetative propagation.

Many plant species that are now threatened, including some highly aesthetically attractive ones, grow only in nutrient-poor moderately dry and mesic sites. Such plants, occurring in many parts of the Alps, include, for instance, Carthusian pink (*Dianthus carthusianorum* subsp. *carthusianorum*), lilac pink (*Dianthus superbus*), some lilies, such as St. Bruno's lily (*Paradisea liliastrum*), orange lily (*Lilium bulbiferum*) and Carniola lily (*Lilium carniolicum*, Figure 84), bearded bellflower (*Campanula barbata*), gentians, such as yellow gentian (*Gentiana lutea*, Figures 132 and 134 in Sections 8.1.2 and 8.1.2.1),

1818 Pils, 1994, 58, 60; Dietl & Lehmann, 2004, 123–124; Keller, 2004, 33–34; Peterken, 2013, 273; Babai, Molnár, & Molnár, 2014, 35–36, 43, 139–140.

mountain arnica (*Arnica montana*, Figure 85), many species from the orchid family, especially on base-rich or neutral soils on carbonates, such as black vanilla orchid (*Nigritella rhellicani*, Figure 136 in Section 8.1.3.1), pyramidal orchid (*Anacamptis pyramidalis*), fly orchid (*Ophrys insectifera*), early spider orchid (*Ophrys sphegodes* subsp. *sphegodes*), bee orchid (*Ophrys apifera*), late spider-orchid (*Ophrys holosericea* subsp. *holosericea*), round headed orchid (*Traunsteinera globosa*), and elder-flowered orchid (*Dactylorhiza sambucina*; Figure 77 in Section 6.2.2.1, this species nonetheless occurs mainly in acidic soils on several kinds of bedrock); see Section 6.2.2.1 for the impacts of spring grazing on orchids.[1819]

Species-richness in matgrass swards (see Section 2), occurring at acidic nutrient-poor sites, is not as high; this is a characteristic phenomenon of European acidic grasslands or grasslands in extreme site conditions in general (usually fewer than thirty flowering plant species per 50 m^2; often less than twenty-five species per 20 m^2 in acidic extensively managed grasslands in Central Europe). Nonetheless, matgrass swards also are habitats of some rare species, including habitat specialists growing in only a few habitat types (see below in this section and Section 8.1.1). Moreover, mowed matgrass swards are significantly richer in species than grazed matgrass swards in permanent pastures, and grassland plants grow taller in mowed than in grazed matgrass swards.[1820] There are also significant differences between matgrass communities; relatively species-rich and colorful matgrass swards in the subalpine vegetation belt include, for instance, matgrass (*Nardus stricta*; Figure 30 in Section 4.1.2), bearded bellflower (*Campanula barbata*), purple colt's foot (*Homogyne alpina*), black vanilla orchid (*Nigritella rhellicani*), mountain arnica (*Arnica montana*; Figure 85), frog orchid (*Coeloglossum viride*), and Scheuchzer's bellflower (*Campanula scheuchzeri*). Matgrass swards on even more acidic soils in the subalpine vegetation belt are less species-rich and include, for instance, matgrass (*Nardus stricta*), golden cinquefoil (*Potentilla aurea*), mountain arnica (*Arnica montana*), mountain avens (*Geum montanum*), and tormentil (*Potentilla erecta*).[1821]

Similar to plant species richness, animal biodiversity is also the highest in non-fertilized meadows, in those mowed only every second or third year in particular (Section 6.1.2.6). Such an example is the biodiversity-rich complex

[1819] Pils, 1994, 75, 97, 265–266, and Table 16, Photo 43; Keller, 2004, 28, 33–34, 37; Papež Kristanc, 2012, 10; Aeschimann et al., 2004, vol. 1, 352–353, 364–365; Aeschimann et al., 2004, vol. 2, 310–311, 1042–1045, 1114–1115, 1120–1121, 1124–1125, 1136–1145; "Pravilnik o uvrstitvi ogroženih rastlinskih in živalskih vrst v rdeči seznam," 2002, 8893, 8897, 8901, 8903–8904.

[1820] Keller, 2004, 35, 89; Ellenberg & Leuschner, 2010, 953; Babai, Molnár, & Molnár, 2014, 35, 50; Dietl & Lehmann, 2004, 123. Compare Ewald, 2003. In some other parts of the world—for example, in South Africa—the majority of species are acidophilic and acidic sites are the species-rich ones.

[1821] Keller, 2004, 35, 37, 91.

of mowed montane and subalpine meadows at Mussen (1,460–2,040 m) above St. Jakob in the Lesach Valley (Germ. *Lesachtal*) in western Carinthia, managed in a history-inspired way, including mowing every second year. More than one thousand animal species live at Mussen, including 670 butterfly species (221 of them live exclusively in meadows), more than one hundred spider species, twenty-two cicada species, various ground beetles, bumblebees, grasshoppers, ants, birds and some other organisms.[1822] The extremely high number of 1,300 species of butterflies and moths (*Lepidoptera*) was identified in a non-fertilized grassland slope at Fließ in western Tyrol in ten years of fieldwork.[1823]

Historical yields in non-fertilized meadows were tremendously diverse.[1824] This diversity was influenced by natural differences in site conditions, by management diversity (e.g., by significantly different duration of spring grazing in meadows; see Section 6.2), and by further factors impacting species composition. Due to the absence of fertilizer application, the natural diversity of site conditions generally had a significantly stronger impact on the yield diversity in non-fertilized meadows than in fertilized grasslands. However, as a general rule, there is no second-crop hay in non-fertilized and non-irrigated meadows in the Alps, regardless of elevation (see Sections 6.1.2.3, 6.3.2.2, and 6.4.1). Data from the mid-nineteenth to the early twentieth century, with an emphasis on southwestern Germany, give the following annual hay yields in non-fertilized meadows mowed once annually: usually from 1.5 (or 1.2) to 1.8 metric tons/hectare, but in reality ranging from 0.8 to 3.0 tons/hectare.[1825] This roughly corresponds to the present estimates: yields of non-fertilized meadows at climatically favorable damp sites often reach four metric tons of dry substance per hectare at 500 m (3.2 tons when usual 20% losses during the mowing, drying, and storing are subtracted), three metric tons at 1,000 m (2.4 tons), two metric tons at 1,500 m (1.6 ton), and one metric ton at 2,000 m (0.8 ton), but the combination of an (from the perspective of farming) unsustainably high percentage of herbs (significantly more than 50%) and less favorable weather and climate often reduces these figures by 30% to 40%.[1826] At the highest elevations, yields of non-fertilized meadows can be as low as less than 0.4 tons/hectare annually.[1827]

Yields depend strongly on several site factors, not only on elevation. Thus, even in the subalpine belt (Section 2), yields in highly species-rich rusty sedge (*Carex ferruginea* subsp. *ferruginea*) communities on favorable, shady northern slopes in rare cases reached as much (for the standards of non-fertilized meadows) as three metric tons/hectare. These communities are rich in broad-

[1822] Pils, 1994, 60; Keller, 2004, 27–28, 33, 40–42.
[1823] Pils, 1994, 61.
[1824] Compare Wopfner, 1997, 315.
[1825] Kapfer, 2010, 137–138; Wopfner, 1997, 311–315; see also Pils, 1994, 49–50.
[1826] Dietl & Lehmann, 2004, 121; Pils, 1994, 78.
[1827] Mathieu, 1996, 145; Keller, 2004, 84.

leaved herbs that require moist sites and produce significant biomass in spite of elevation. Natural processes at the sites where rusty sedge communities occur result in an impact comparable to moderate fertilization. Namely, these slopes are characterized by moderate to high snow accumulation, and so additional nutrients are washed in by melting water. Probably even more important is the fertilizing impact of some legumes (e.g., honeysuckle, *Hedysarum* spp., and oxytropis, *Oxytropis* spp.). The yields in the subalpine vegetation belt were usually much lower, and they were often significantly lower than three metric tons/hectare even in rusty sedge communities. In Swiss alpine and subalpine evergreen sedge (*Carex sempervirens*) communities growing at drought-prone sites on sunny slopes and on flat but wind-exposed sites, yields were often extremely low, as little as 0.2 to 0.4 metric tons/hectare, and the hay was mostly mowed there only once every two years (see Section 6.1.2.6), according to early twentieth-century data; however, see the high quality of such hay below in this section.[1828] These examples demonstrate that, apart from elevation, there are several other partly interrelated factors significantly influencing yields, such as slope aspect, soil water availability, and species composition.

Considering the slope aspect, average quantities of hay on east- and west-facing slopes are highest in general. Yields in shady meadows (on north-facing slopes) are usually up to a quarter lower than in meadows with similar site conditions except for the less favorable slope aspect.[1829] Nonetheless, there exist plant communities delivering the highest yields on shadier slopes, as learned based on the example of *Carex ferruginea* subsp. *ferruginea* communities. Meadow yields on steep sunny slopes with shallow soils poor in humus content, on limestone or in other frequently dry sites in the Alps, are also lower: drought is usually a more frequent hazard in steep grasslands than in flat ones, on south-facing slopes in particular.[1830] A long root system as one of the plant adaptations to drought and supporting tissue requires so much energy that the aboveground biomass of several species abounding at frequently dry grassland sites grows relatively slowly,[1831] and so hay quantities are low as a plant adaptation to the general site conditions, even if the spring and early summer are not dry in a particular year.

In the case of meadows with short plants (usual, but not ubiquitous in non-fertilized grasslands, at higher elevations in particular), quantities of hay cut by scythe were significantly (e.g., by one-third) higher than in the case of present-day motorized mowing because hand mowing was more thorough and performed closer to the sod.[1832] In addition, several herbs that become fragile

[1828] Pils, 1994, 187, 190–191, 193–198.
[1829] Brugger & Wohlfarter, 1983, 17, 20, 179–180.
[1830] Pils, 1994, 105; Vidrih, 2005, 78.
[1831] Ellenberg & Leuschner, 2010, 916–927, but see also Pils, 1994, 105, 108.
[1832] Babai, Molnár, & Molnár, 2014, 105–106.

when they dry were picked up much more thoroughly than nowadays, as long as haymaking was carried out manually, with wooden rakes, and it was considered unacceptable to step on the hay when it was turned and raked, to reduce herb foliage losses.[1833] However, more systematic hay removal also meant more thorough nutrient removal from non-fertilized meadows, with possible negative long-term impacts on meadow yields at several sites. A historical strategy of coping with this challenge in non-fertilized and non-irrigated meadows was mowing only every second or third year (Section 6.1.2.6).

Non-fertilized meadows where the dominant species was upright brome (*Bromus erectus*, mainly on base-rich, relatively dry sites; see Section 2; the species is not easy to mow; see Section 6.1.1.2), red fescue (*Festuca rubra*, predominately on neutral soils, at mesic sites), and/or common bent (*Agrostis capillaris*, mainly on acidic soils, at mesic sites) were relatively productive (e.g., half the yield of moderately manured meadows). These are usually the highest-yielding grass species of non-fertilized meadows; that is, their yields are middling for twenty-first-century standards. All these species occur (almost) throughout the Alps and can abound up to the subalpine vegetation belt, on silicate (mainly *Agrostis*) or carbonate (mainly *Bromus*) bedrock. By the beginning of the eighteenth century in Central Europe in general and well into the nineteenth century in several regions of the Alps, upright brome, red fescue, and common bent, together with sweet vernal-grass (*Anthoxanthum odoratum*, mainly on acidic soils, in valley bottoms and montane meadows in particular) were grass species characterized by a high occurrence and major economic importance due to the strong predominance of non-fertilized meadows in the landscape. Sweet vernal-grass gives the hay its characteristic flavor. Significantly different were species compositions in economically highly relevant meadows in regions with large areas of non-fertilized meadows in subalpine and/or alpine vegetation belts: high productivity of rusty sedge (*Carex ferruginea* subsp. *ferruginea*) communities in (sub)alpine belt(s) was mentioned above as a relevant example. With the gradual decrease in non-fertilized and moderately fertilized meadow areas in the last couple of centuries, in the last century in particular, the occurrence and abundance of these species has been substantially reduced. Still focusing on grass species, not only those as important for livestock rearing as the ones listed above can abound in non-fertilized meadows. Quaking-grass (*Briza media*), pyramidal hair-grass (*Koeleria pyramidata*), rock fescue (*Festuca stenantha*, on carbonate bedrock), fine-leaved sheep's-fescue (*Festuca filiformis*), small sheep's-fescue (*Festuca airoides*, on silicate bedrock), matgrass (*Nardus stricta*, see below), and many other monocots and dicots also occur or abound in non-fertilized meadows in the Alps.[1834]

[1833] Compare Pfister, 1985a, 42; author's interview with Kunavec, 2021, in Stara Fužina.

[1834] Dietl & Lehmann, 2004, 17, 20, 23, 25, 42, 54; Keller, 2004, especially page 34; see also Kauter, 2002, 20–21, 66–68, 73–74; Aeschimann et al., 2004, vol. 2, especially pages 822–823, 854–855, 860–865, 898–899, 916–917, 954–955, 964–965, 970–971.

Mountain hay is usually shorter than hay from sites in lower elevations.[1835] This is a consequence of several interrelated factors, such as plant adaptations to climate and more intensive solar radiation at higher elevations, other impacts of slope inclination on plant height, common absence of fertilization, and so on. The short length of hay from meadows at higher elevations was historically an important quality as long as the numbers of sheep and goats were high. Namely, hay for sheep and goats had to be short, otherwise they wasted a lot of hay when pulling long plants from the manger and many tall plants fell out, becoming mixed with excrement instead of being consumed.[1836] Some longer hay had to be cut into shorter pieces,[1837] and so additional work was required to feed these animals with longer hay economically. Even though sheep and goats were also fed in the winter with dried tree leaves (Section 8.2.2.1), straw, dried weeds from fields, and feed surrogates such as bark,[1838] short hay did in general play an important role in their diet, as recorded, for example, in the early nineteenth century.[1839] Feeding sheep only with tree leaves (and bark from twigs) throughout the winter had already been considered a result of exceptional scarcity at the farm level in the Meža Valley (Sln. *Mežiška dolina*, Germ. *Mießtal*) in southern Carinthia in 1570.[1840]

In general, the numbers of sheep in the Alps strongly decreased between the early seventeenth century (about 1630) and the late nineteenth or early twentieth century, possibly by half. At the same time, the numbers of cattle increased. The timing of this change and the rates of decrease in the numbers of sheep and increase in the numbers of cows differed regionally. This change was particularly strong in the Northern Alps, whereas Provence (Occitan

[1835] For instance, Keller, 2007, 205.

[1836] Author's interviews with F. Mlakar, 2017, in Bohinjska Češnjica, and with L. and M. Odar, 2017, in Studor.

[1837] Vogl-Lukasser et al., 2006, 65.

[1838] Emanuelsson, 2009; Kačičnik Gabrič, 2009, 35; Kačičnik Gabrič, 2016, 503; Geršič, Gabrovec, & Zwitter, 2018, 69.

[1839] A selection of examples from the southeastern Alps (Bohinj and the Sava Dolinka Valley [Sln. *Zgornjesavska dolina*]) in the early nineteenth century: Kačičnik Gabrič, 2016, 503; SI AS 176, Franciscejski kataster za Kranjsko (Franciscan Land Cadaster for Carniola), L10, cadastral municipality of Jesenice, cadastral appraisal protocols: Catastral-Schätzungs-Elaborat, Einleitung, § 4; SI AS 176, L16, cadastral municipality of Hrušica, cadastral appraisal protocols: Catastral-Schätzungs-Elaborat, Einleitung, § 4; SI AS 176, L30, cadastral municipality of Nemški Rovt, cadastral appraisal protocols: Catastral-Schätzungs-Elaborat, Einleitung, § 4; SI AS 176, L56, cadastral municipality of Bohinjska Bistrica, cadastral appraisal protocols: Catastral-Schätzungs-Elaborat, Einleitung, § 4; SI AS 176, L72, cadastral municipality of Gorjuše, cadastral appraisal protocols: Catastral-Schätzungs-Elaborat, Einleitung, § 4; SI AS 176, L157, cadastral municipality of Srednja Vas v Bohinju, cadastral appraisal protocols: Catastral-Schätzungs-Elaborat, Einleitung, § 4; SI AS 176, L179, cadastral municipality of Zgornje Gorje, cadastral appraisal protocols, Catastral-Schätzungs-Elaborat, Einleitung, § 4.

[1840] Zwitter, 2016, 122–123.

Provença) and some regions of the Limestone Alps in the East of the mountain chain with summer pastures unsuitable for cattle were exceptions even in the southern Alpine context because raising sheep still prevailed there around 1900. An example of such a region with very long-lasting prevalent sheep keeping was the Bovec region in the Upper Soča river catchment in the Julian Alps, with a rich tradition of high-quality cheese export, aimed at meeting the demands of the urban population in the Republic of Venice; the export of quality cheese reached back to the sixteenth century, at least (see Section 8.2.1.1). Many grazing complexes in the karst of the Bovec region were usable only for sheep and goats,[1841] and a significant percentage of wild hay collected in steep and small grasslands at high elevations, partly by sickle because the terrain was unsuitable for a scythe (see Section 6.1.1.2), was characteristic of the region.[1842] Some other regions in the Eastern Alps such as Tyrol[1843] also differed from the general European trend of decline in sheep-rearing in the late nineteenth century. Nonetheless, between the late nineteenth and mid-twentieth century, an important global economic factor that in general significantly influenced the great decline in sheep keeping in the Alps was the introduction of large-scale import of cheaper wool from Australia, South America, New Zealand, and South Africa to Europe, based on steamship transport. "By the 1930s, there were 300 million sheep in the Southern Hemisphere—most of them in the grasslands of Australia, New Zealand, South Africa, and South America." At the same time, imported cotton took part of the former role of wool.[1844] The decline in sheep farming thus made the short length of mountain hay an unwanted characteristic and, where dairy cattle replaced sheep, also made use of hay from several types of non-fertilized meadows less economicaly viable at many farms (Section 6.3.1.2).

Mountain hay also used to be harvested in many non-fertilized grasslands, where it was shorter than the eating behavior of sheep and goats would require, and very short hay was used not only for sheep and goats but also to feed cattle and other livestock. Hay quality (see Section 6.3.1.2) and the lack of more productive meadows importantly influenced the decision to mow many such low-yielding grasslands. In certain places (e.g., in the Ötz Valley, Germ. *Ötztal*), haymakers kept the short plants upright with the help of a broom-like bound bunch of branches, about 20 to 40 cm long and 25 cm wide on the lower side, held in the left hand, to make easier mowing with a short-handled scythe,

[1841] Melik, 1950, 188–193; Gestrin, 1965, 197; Vojvoda, 1969, 28–29, 40; Novak, 1970, 390; Wopfner, 1997, 227–236 (see also Meixner & Siegl, 2004); Mathieu, 2019, 53–55; Stadler-Planzer, Stadler, & Aschwanden, 2015, 100; Stadler-Planzer, Stadler, & Degler-Spengler, 2015, 98, 100.

[1842] Melik, 1950, 115–117, 188.

[1843] The historian Kurt Scharr, personal communication.

[1844] Schöller, 2010, 636; Isenberg, 2017, 145 (quote); Poschlod, 2015, 168–169, 177–179; Jordan, 1945, 59, 97; Melik, 1950, 18, 147–148; Novak, 1970, 390.

adapted to steep slopes and held in the right hand.[1845] This technique also made it possible to harvest the leaves of rosulate plants (i.e., those with a leaf rosette close to the ground), such as rough hawkbit (*Leontodon hispidus*) or golden hawk's-beard (*Crepis aurea*; see Figure 8 in Section 2), both highly desired for fodder quality[1846] but difficult to obtain due to the growth form of the leaves. Such mowing was not introduced only with the high agricultural density of the nineteenth century: *Sichelmähder* 'sickle meadows' were documented in the Late Medieval Venosta Valley (Germ. *Vinschgau*, Ital. *Val Venosta*);[1847] nonetheless, their area fluctuated in history (see Sections 3.1, 3.2.2, and 9). Cutting such short plants on uneven terrain was evidently significantly more labor intensive than, for instance, cutting another short species: alpine cotton-grass (*Eriophorum scheuchzeri*) in the fens of northern Sweden; this grass grows only about 10 cm high and has only a few leaves but livestock liked it very much because it tastes sweet.[1848] This means that haymaking in some remote non-fertilized meadows in the Alps belonged to the most labor-intensive mowing types in a broader European context (see also Section 6.1.1.2). The decision to mow the grassland sites where this practice was so labor-intensive was partly a consequence of the fact that quantities of hay often hardly (if at all) met the demand (see Sections 3.1 and 6.2.2.1).[1849]

Plant properties required that short wild hay from some very nutrient-poor meadows, usually at high elevations, was swept with a broom (e.g., made from alpenrose, *Rhododendron* sp.) instead of being raked, as reported in Tyrol and in the French Alps. The short hay length as the only explanation for hay sweeping in cited historical and ethnological publications[1850] does not fully explain this practice because rakes were also adjusted to short hay from non-fertilized meadows. Data from mid-twentieth-century Carinthia demonstrate such hay rake adaptation; namely, the wooden peg teeth used to be doweled more densely in the rake head than usual nowadays to allow short hay raking:

[1845] Hubatschek, 1992, 172–173; Niederer, 1996d, 174; Wopfner, 1997, 333. The broom-like bunch to make mowing of short plants easier was made of Norway spruce (*Picea abies*), juniper (*Juniperus communis*), alpenrose (*Rhododendron* sp.) branches, or stiff monocots (Hubatschek, 1992, 173; Wopfner, 1997, 333). See the illustration of such a broom-like bound bunch and the short-handled scythe from the Ötz Valley in Griesser, 1999, 42.

[1846] Compare Dietl & Lehmann, 2004, 43.

[1847] Wopfner, 1997, 333.

[1848] Emanuelsson, 2009, 286. We assume that "alpine cotton-grass" in this case designates *Eriophorum scheuchzeri*. *Trichophorum alpinum* is not a possible alternative; namely, *Trichophorum alpinum* never forms dominant patches, whereas *Eriophorum scheuchzeri* can be dominant, which is relevant in the context of haymaking. *Eriophorum scheuchzeri* can cover large areas and provide a significant amount of biomass, whereas *Trichophorum alpinum* only grows in tussocks distant from each other.

[1849] Compare Pfister, 1985a, 38, 52–53.

[1850] Grass, 1948, 182; Hubatschek, 1992, 172; Niederer, 1996d, 174; Wopfner, 1997, 333; see the illustration of alpenrose broom from the Ötz Valley in Griesser, 1999, 42.

whereas the usual space width between the peg teeth was from 1.7 to 2.0 cm in mid-twentieth-century Carinthia, smaller spaces were usual for rakes used in non-fertilized mountain meadows with short hay.[1851] Similarly, data from Bohinj demonstrate that peg teeth used to be doweled more densely in the rake head than usual nowadays to allow short hay raking.[1852] Specific plant tissue characteristics of highly abundant plant species were another factor that made more labor-intensive short hay sweeping instead of raking in meadows at highly nutrient-poor sites most efficient, at least in many cases. The sclerenchyma-rich short and stiff matgrass (*Nardus stricta*, Figure 30 in Section 4.1.2, and see Section 6.3.1.2) leaves fell to the ground when mowed, and they became stuck in the brush-like mowed turf. Consequently, often they could not be thoroughly stored with a rake but only in a more labor-intensive way, with a broom.[1853]

6.3.1.1.1 Historical sowing of fodder legumes in species-rich grasslands

Some legumes were usually sown in species-rich grasslands with barn sweepings (Section 6.5), and it depended on site conditions and further context whether they became established there (see the introduction to Section 4). However, actual fodder legume cultivation also took place in some non-fertilized, species-rich grasslands in Central Europe. The most relevant sown "clover" for the topic of non-fertilized grasslands is sainfoin (*Onobrychis viciifolia*, see Figure 164 in Section 10.1.2.2.4), a species that is either native or an archaeophyte in this part of Europe.[1854] Sainfoin can abound in colline and montane vegetation belts, on base-rich or neutral, nutrient-poor soils, at relatively dry sites. Nowadays, it occurs at many sites Alps-wide.[1855] It is one of the least demanding among all purposefully sown "clovers" on base-rich or neutral soils,[1856] a fact highly relevant for the topic of this book. Sainfoin is also cold-tolerant and less limited by low potassium and phosphorous availability than the common clover species. Because this species grows well in site conditions that are difficult for many plants, it can be highly productively sown not only in fallow fields,[1857] but also in species-rich nutrient-poor (ancient) grasslands, without any ploughing or fertilizing, as demonstrated by recent research in Ghimeş (Hung. *Gyimes*) in the Eastern Carpathians.[1858] Gaps in relatively dry nutrient-poor meadows are

[1851] Moser, 1952, 466, 469–470, 489.
[1852] Author's interview with Kunavec, 2021, in Stara Fužina.
[1853] Pils, 1994, 94.
[1854] Jogan, Bačič, & Strgulc Krajšek, 2012, 170; Küzmič & Šilc, 2017, 208; Bosi et al., 2019, 11; Filatova, Gyulai, & Kirleis, 2019, 66; Axmanová et al., 2021, 5–7.
[1855] Aeschimann et al., 2004, vol. 1, 954–955; Hayot Carbonero et al., 2011, 73.
[1856] Compare Britovšek, 1958, 142–143, 145–146, 149; Karlík & Poschlod, 2009, 334; Poschlod, 2015, 85–86, 120–121; Aeschimann et al., 2004, vol. 1, especially pp. 906–907, 926–929, 948–949, 954–955.
[1857] Karlík & Poschlod, 2009, 324; Hayot Carbonero et al., 2011; Poschlod, 2015, 85–86.
[1858] Babai, Molnár, & Molnár, 2014, 43, 95–96, 139.

niches for sainfoin.[1859] A recent history of cultural landscapes in Central Europe considers sainfoin one of the plant indicators of former arable (or hoe) farming,[1860] which is either erroneous or it is correct only in some regions. Namely, this is only an option; occurrence of sainfoin alone cannot indicate former arable (or hoed) fields with any certainty. This problematic interpretation is based on too quick a generalization[1861] of the empirical results of a historical-ecology investigation of dry calcareous grasslands in the Swabian Jura (Germ. *Schwäbische Alb*)[1862] that compared the occurrence of sainfoin in ancient grasslands and in grasslands on former arable land. Namely, all ancient grasslands studied in the Swabian Jura were pastures; no ancient meadow (as defined in Section 1 of this volume) was included in that research (but in Swabia, sainfoin was also sown in meadows; see below).[1863] This makes a difference because sainfoin is grazing sensitive: "Regrowth is slow, and it is important to allow [sainfoin] enough time to replenish root reserves to maintain its persistence and longevity."[1864] Absence of sainfoin from pastures was, however, not a general rule; see sainfoin occurring in pastures in the Allues Valley (Francoprovençal *Alués*, French *Les Allues*) in the northern French Alps below.

Sainfoin yields are relatively favorable; in the Alps, the plant usually grows 30 to 70 cm tall. If grown in arable land, sainfoin yields about seven to fifteen metric tons of dry matter per hectare;[1865] a small sample of eighteenth-century data reports lower yields, such as a seven- or eight-year annual average of 3.2 to 4.0 metric tons/hectare at moderately favorable sites and moderate farming efforts, and a three-year annual average of more than 5.3 metric tons/hectare,[1866] probably at less suitable sites than the recent data given above. Total hay yields in species-rich non-fertilized meadows rich in sainfoin must have been significantly lower than seven to fifteen metric tons of dry matter per hectare, but still high for non-fertilized meadows at relatively dry sites. Farmers in Ghimeş observed that, with the establishment of a high abundance of sainfoin in (relatively) dry, non-fertilized meadows on shallow soils, mowed once annually, hay yields tripled.[1867] Sainfoin's good yields at non-fertilized sites so dry that they were otherwise not suitable for grasslands considered productive

[1859] Compare Babai, Molnár, & Molnár, 2014, 95–96, 136, 139.
[1860] Poschlod, 2015, 85–86, 121.
[1861] In Poschlod, 2015, 85–86, 121.
[1862] Karlík & Poschlod, 2009.
[1863] Karlík & Poschlod, 2009, 322, 324.
[1864] Compare Hayot Carbonero et al., 2011, quote from p. 75.
[1865] Aeschimann et al., 2004, vol. 1, 954–955; Hayot Carbonero et al., 2011, 73–75.
[1866] Pfister, 1985a, 111. Pfister's calculations of unit conversions are erroneous in this case. Following his statement that one *Morgen* equals 0.3418 ha and that 1 *dz* equals 100 kg or 1.82 *Zentner* à 55 kg (Pfister, 1985a, 110–111), the conversion for 20, 25, and 33 *Zentner* per *Morgen* in metric tons per hectare is 3.2, 4.0, and 5.3 (instead of 3.7, 4.7, and 6.2, as claimed by Pfister).
[1867] Babai, Molnár, & Molnár, 2014, 36, 43, 95–96.

for early modern standards was emphasized in a leaflet from Regensburg from 1717.[1868] Moreover, as recorded in the mid-nineteenth century, sainfoin yields at nutrient-poor dry sites on shallow soils were more reliable than those of most meadow plant species and of other intentionally sown fodder plants.[1869] In the case of appropriate mowing timing, sainfoin provides high-quality hay that was used, for instance, for horses when they performed heavy physical work, or the quality sainfoin hay was given to milk cows and sheep or goats.[1870] Such fodder is nutritive, tasty, and healthy; moreover, contrary to other fodder legumes, such as clovers (*Trifolium* spp.) or alfalfa (*Medicago sativa*), "[r]uminants can safely consume large amounts [of sainfoin] because it does not cause bloat," which can be lethal in the worst cases (see also Section 10.1.2.2.4).[1871]

Sainfoin might have been intentionally sown in southern France in the Late Middle Ages, and its cultivation in the seventeenth century was certainly not a novelty.[1872] According to information in a leaflet published in Regensburg in 1717, sowing sainfoin was believed to have reached the area of Neuchâtel by around 1675, whence the practice spread to Bavaria, Swabia, and further. This leaflet from 1717, promoting sainfoin sowing, claimed that this species was still quite unknown (as an intentionally sown plant) in the surroundings of Regensburg in the early eighteenth century.[1873] Nonetheless, if the 1717 leaflet mentioned sainfoin in the area of Neuchâtel in the late seventeenth century, it is clear that sainfoin spread in the nearby area of Aarberg (northwest of Bern) around 1750. A sainfoin crop is recorded at nearby Nidau around 1760, but sainfoin was not grown in the nearby parish of Kerzers at that time. These data emphasize microregional spatiotemporal differences in the intentional sowing of "clovers."[1874] If sainfoin was indeed grown in the area of Neuchâtel in the seventeenth century as suggested by the leaflet, the pieces of information presented also illustrate historical discontinuities. "Clovers" are sensitive to several pests and diseases. Sainfoin is "relatively free from serious pest and disease problems compared with other [fodder] legumes"; nonetheless, some organisms can have disastrous impacts on it.[1875] Consequently, the observation that vulnerability of "clovers" to pests and diseases hampered a quick early modern spread of "clover" sowing probably partly

---

[1868] Kauter, 2002, 95.
[1869] Poschlod, 2015, 121, 301; compare Hayot Carbonero et al., 2011, 73–76.
[1870] Babai, Molnár, & Molnár, 2014, 43, 96; Hayot Carbonero et al., 2011, 73, 76; Poschlod, 2015, 121.
[1871] Hayot Carbonero et al., 2011, 76–80 (quote from p. 76).
[1872] Ambrosoli, 1993, especially paragraphs 14, 16, and 18; Hayot Carbonero et al., 2011, 73, 76 (the identification of the plant mentioned in the sixteenth century by Hayot Carbonero et al. (2011, 76) is erroneous; see Ambrosoli, 1993, paragraph 16).
[1873] Kauter, 2002, 95.
[1874] Pfister, 1985a, 110.
[1875] Hayot Carbonero et al., 2011, 75–76 (quote from p. 75).

also applied to sainfoin[1876] and contributed to historical discontinuities in its abundance and occurrence.

In Ghimeş, sainfoin was only introduced in the 1940s, and thus farmers still reliably recall how it spread in meadows, and so this is a relevant parallel to understand the history of sainfoin sowing in the Alps. A good half century after the introduction of purposeful sainfoin sowing, the species frequently abounds in montane meadows of highly traditional farmers, on shallow soils on sunny slopes, in non-fertilized meadows too dry for highly productive grasses. Farmers have been sowing sainfoin's seeds in non-fertilized meadows in appropriate site conditions not reached by the species yet. Sainfoin became naturalized, and so it also spreads on its own provided that haymakers keep some plants standing where sainfoin has not abounded yet so that seeds of this long-blooming plant ripen and are dispersed down the meadow slope. Moreover, sainfoin is present in barn sweepings sown in meadows; sweepings abounding in sainfoin seeds are often kept separately in Ghimeş and applied to sunny meadows at dry, non-fertilized sites where seeds of this highly esteemed species are most needed to foster its persistence in the long run (Section 6.5).[1877] Such meadows maintain much of their biodiversity, even though sainfoin causes some changes (see Section 10.1.2.2.4).[1878]

The practice of sowing sainfoin in non-fertilized meadows was carried out in some parts of Central Europe. In Swabia, for instance, sowing sainfoin in (relatively) dry meadows was no rarity according to mid-eighteenth-century data,[1879] at a time of demographic growth in the region.[1880] However, there existed regional diversities in the role of sainfoin; contrary to eighteenth-century data from Swabia, in the territory of Slovenia sainfoin cultivation has never become very common despite its stronger cold, heat, and drought tolerance than alfalfa, as determined by Marjan Britovšek.[1881] Similarly, in the Allues Valley, sainfoin cultivation was limited to periodic experiments in clover fields of some farms, as far as recalled for the early twentieth century, "and escapes occurred into the cultivated field margins and lower hayfields. In its wild form . . . sainfoin was . . . a fairly common component of the higher hayfields and pasturelands."[1882]

It is, however, possible that sainfoin has abounded at more sites in non-fertilized meadows in the Alps since spring grazing was abandoned in the eigh-

[1876] Pfister, 1985a, 109.

[1877] Babai, Molnár, & Molnár, 2014, 31, 33–36, 43, 95–96; compare Hayot Carbonero et al., 2011, 75.

[1878] Babai, Molnár, & Molnár, 2014, 43, 139; Babai & Molnár, 2014, 126–128.

[1879] Kauter, 2002, 94; see also Oppermann, 2012, 225.

[1880] Karlík & Poschlod, 2009, 323.

[1881] Britovšek, 1958, 146.

[1882] Meilleur, 1986, 294.

teenth to nineteenth century, especially where grazing in meadows had previously ended late, such as in the second half of May (Sections 6.2.1.2 and 6.2.2.2). Namely, (1) interest in sowing sainfoin in meadows was high when spring grazing in meadows was abandoned: the peak of "clover" farming in Central Europe was yet to come;[1883] and (2) the species is grazing-sensitive especially during the growth phase in late spring. This might have partly hindered its spread before the abolition of spring grazing. Christine Hayot Carbonero et al. (2011) even claim—citing older literature—that sainfoin "can be used for light grazing, but only in the late autumn, to allow the crop time to replenish root reserves."[1884] Nonetheless, empirical historical evidence warns against oversimplifications. Sainfoin can tolerate some moderate spring grazing regimes. In the Senožeta meadow complex at the foot of the Karawanks near Javornik (Jesenice), in today's northwestern Slovenia, a high abundance of sainfoin was recorded in the mid-nineteenth century in grasslands grazed in spring until Ascension Day; thus, depending on the year, this grazing ended between late April and early June (see Sections 6.1.2.7 and 6.2.1.2). It is clear that grazing in that grassland complex was not very frequent. It was usually carried out two days per week[1885] (and possibly no more than a few times per spring in the same meadow; see Sections 6.2.2.1 and 6.2.2.2).

Sowing sainfoin in species-rich nutrient-poor meadows might have regionally contributed to the fact that sainfoin is nowadays a character species of Mesobrometum; that is, semi-dry nutrient-poor grasslands on calcareous soils (see, e.g., Section 6.1.2.7). Mesobrometum is a widespread unit within Festuco-Brometea; see Section 2 (as the supply of water and nutrients increases, they come closer to the productive agricultural grasslands of Arrhenateretalia; see Section 2). Mesobrometum sites can be relatively highly productive; characteristic are some legumes like sainfoin (*Onobrychis viciifolia*) and other tall-growing herbs such as meadow clary (*Salvia pratensis*—see Sections 10.1.2.2.3 and 10.2.2—it is the dark-purple flowering plant in Figure 166).[1886] Nonetheless, it is not clear how usual sowing sainfoin in non-fertilized meadows was in different regions of the Alps and how this has changed over time in the last few centuries. The relevance of this practice might be underestimated in some "clover" history studies focused very much on "clover" cultivation in

---

[1883] See general trends in "clover" farming history in Britovšek, 1958, 149, and Poschlod, 2015, 119.

[1884] Hayot Carbonero et al., 2011, 75 (quote).

[1885] SI AS 48, Deželna komisija za odkup in uravnavo zemljiških bremen za Kranjsko (Provincial Commission for Redemption and Regulation of *Profits à Prendre* in Carniola), fasc. 56, 9:1–5: Protokoll, July 11th, 1868.

[1886] Poschlod, 2015, 86, 121, mentions that it is a character species of such meadows, but claims contradictory and erroneously that sainfoin sowing was limited to arable fields (see above). On the species' occurrence in Slovenia nowadays, see Martinčič, 2007c, 317; Aeschimann et al., 2004, vol. 1, 954–955; *Notranjski regijski park*: Navadna turška detelja.

arable land,[1887] a topic better documented in historical records than sowing sainfoin in permanent grasslands; leaving some sainfoin plants unmowed in a meadow to allow seed dispersal would be even less likely documented.[1888] In any case, human factors, not only site conditions, had an impact on the decision to sow sainfoin in meadows, as demonstrated by a parallel from the sub-Pannonian hills of Lower Styria (Sln. *Spodnja Štajerska*, Germ. *Untersteiermark*). There, sowing sainfoin was much less common on grassland slopes in vineyards (on base-rich or neutral soils) than it could be, as observed in the mid-nineteenth century,[1889] and similar situations can be assumed in several regions of the Alps.

6.3.1.2 Quality of hay from non-fertilized meadows

Hay quality in non-fertilized meadows was as diverse as species compositions and yields. In general, hay quality in such meadows at low and middle elevations, mowed at times when grass seeds ripen, taken as a whole (grasses and herbs), is usually considered middling to low by present-day standards.[1890] An integrated quality indicator linking the nutritive value, palatability, and health impacts of some grass species frequently abounding there, such as upright brome (*Bromus erectus*), red fescue (*Festuca rubra*), and/or common bent (*Agrostis capillaris*) (see Section 6.3.1.1), is middling.[1891] Nonetheless, differences in species compositions and in site conditions (e.g., higher concentrations of essential oils as a plant adaptation to site conditions in higher elevations) had major influences on fodder quality, on the quality of livestock products, and on the relevance of hay from non-fertilized meadows for livestock's health.[1892] Namely, not only high-quality medicinal herbs but also high-quality fodder plants abound in some non-fertilized meadows. French honeysuckle (*Hedysarum hedysaroides*, Figure 86) is one of the best fodder plants naturally occurring in the Alps; it includes a lot of proteins and fats, and livestock like its taste. This legume occurs in stony meadows on limestone bedrock and abounds only in mesic to moderately moist, mowed meadows in subalpine and alpine belts in several parts of the Alps, in rusty sedge (*Carex ferruginea* subsp. *ferruginea*) communities (mentioned in Section 6.3.1.1 for their relatively high yields and high biodiversity).[1893]

1887 E.g., Britovšek, 1958.

1888 Compare the general comment on tracing sown "clover" species in historical records in Ambrosoli, 1993, paragraph 14.

1889 Slatnar, 2016, 9; see examples of base-rich and neutral soils in the hills of Lower Styria in Repe, 2010, 146–148, and see Debeljak, 2020.

1890 Dietl & Lehmann, 2004, 17, 39, 124.

1891 Dietl & Lehmann, 2004, 20, 39, 42, 54; see also Babai, Molnár, & Molnár, 2014, 35–36.

1892 Hubatschek, 1992, 175; Wopfner, 1997, 311–313; Vogl-Lukasser et al., 2006, 83; Keller, 2007, 206–212.

1893 Keller, 2004, 34, 37; Martinčič, 2007c, 316; Aeschimann et al., 2004, vol. 1, 950–951.

Figure 86: French honeysuckle (*Hedysarum hedysaroides*). Thomé, *Flora von Deutschland, Österreich und der Schweiz,* 1885. *Wikimedia Commons,* File:Illustration Hedysarum hedisaroides0.jpg, public domain.

For centuries and probably millennia, farmers' traditional ecological knowledge has distinguished between the main quality classes of hay, as already demonstrated by archival evidence from the sixteenth and seventeenth centuries.[1894] Due to these quality differences, high-quality mountain hay from some non-fertilized meadows was primarily given to dairy cows or as a high-quality addition to lower-quality hay from the valley. Hard hay from non-fertilized dry calcareous grasslands at lower elevations, with extremely sparse vegetation cover, where the hay contained only a little juice during mowing, was used to feed oxen (not before heavy physical work), other non-dairy adult cattle, sheep, goats, donkeys, and mules.[1895] Even lower-quality hay from matgrass

1894 Zwitter, 2020, 60–71, 81.

1895 Klemenčič, 1959, 43, 115, 123–124, 141, 145; Pils, 1994, 74; Machatschek, 2002, 247, 252; Keller, 2007, 206–212; Bavcon, Ravnjak, & Praprotnik, 2019, 41, 79; Zwitter, 2020, 66.

swards at acidic sites was again used specifically (see below), and wetland hay was used differently (see Section 6.1.2.5). Here, the focus is first on the positive extreme in hay quality from non-fertilized meadows because it is less widely known that hay quality from some non-fertilized meadows has been high.[1896] Presentation of the negative quality extreme of non-fertilized mesic, humid, or dry meadows for animal husbandry follows later in this section.

Historical non-fertilized mountain meadows providing high-quality first-crop hay for the standards of the time were no exception (see also a holistic discussion of the merits of hay from species-rich grasslands in the introduction to Section 10.1.2.2, and Subsections 10.1.2.2.2, 10.1.2.2.3, 10.1.2.2.4, and 10.1.2.2.7), as proven by farmers' historical information from as culturally diverse and spatially distant places as the Alps of the province of Salzburg, from Carinthia, on the southern slopes of the western Karawanks, from Bohinj in the Julian Alps, and from Ghimeş in the Eastern Carpathians.[1897] Animal husbandry played a major role in the historical farm economy in these regions, and farmers living at such different and distant places would not independently severely overestimate the quality of hay from the best non-fertilized mountain meadows. They knew how much hay of the main different local quality classes their livestock needed, and they observed the key impacts of different hay quality classes on livestock's performance, such as on draft power, and on milk quantity, taste, and, roughly, its fat content.[1898]

It is scientifically proven that hay from high elevations contains higher contents of essential oils, making the fodder more healthful and the milk more fragrant. Moreover, although grassland yields decline in quantitative terms with rising elevation in the case of similar other site conditions and management, plants "process greater amounts of energy."[1899] The species-rich blue moor-grass (*Sesleria caerulea*) and evergreen sedge (*Carex sempervirens*) communities, for instance, which are widespread in the Alps in the subalpine and alpine vegetation belt, including edelweiss (*Leontopodium alpinum*, Figure 135 in Section 8.1.3.1), trumpet gentian (*Gentiana clusii*, Figure 168 in Section 10.2.2), and many other grasses and herbs, were mowed in many relatively dry neutral to base-rich and moderately nutrient-poor stony meadows at steep sites where haymakers' lives were not exceptionally at risk (Section 3.1) and hay quantities were extremely low (see Section 6.3.1.1). Nonetheless, fodder quality was very

---

[1896] Keller, 2007, 207–208, 214.

[1897] Hubatschek, 1992, 175; Keller, 2004, 27, 31–33, 43, 47, 58, 62, 64, 67, 71–72, 75, 137–138, 142; Keller, 2007, 206–209; Babai, Molnár, & Molnár, 2014, 15–20, 35–38, 42–44, 110, 112, 136, author's interviews with Mencinger, 2017, in Stara Fužina, and with L. and M. Odar, 2017, in Studor. On southern slopes of the western Karawanks, see the example of Golica below in this section.

[1898] For instance, Keller, 2007, 206–208; Zwitter, 2020, 60–69. Concerning draft power, see the example of the hay from Golica below in this section.

[1899] Brugger & Wohlfarter, 1983, 54; Pils, 1994, 221; Orland, 2016, 27 (quote).

high despite the hardness of such hay; its crude fiber content is about twice as high as in fertilized lowland meadows. The high quality of the hay from the mentioned meadows in high elevations was the result of plants rich in biologically active substances, such as vitamins and minerals.[1900] Such hay also included a lot of fat, which helps understand the high milk fat content.[1901] The crude fat content in hay from non-fertilized *Carex sempervirens* communities mowed in the (sub)alpine belt is almost four times the maximum fat content in hay from intensive lowland meadows (maximum in the context of intensive meadow standards around 1970), as proven by a case study. The crude protein content in hay from subalpine and alpine *Carex sempervirens* communities, however, is significantly lower, only about one-third of the average in fertilized lowland meadows. Another quality advantage of mountain hay from appropriate sites is demonstrated by the following example: in *Carex sempervirens* stands on neutral to base-rich, mesic soils in the lower section of alpine belt, also rich in rusty sedge (*Carex ferruginea* subsp. *ferruginea*), Austrian fescue (*Festuca norica*), and herbs, plants grow sparsely and reach a relatively low height (not low for the alpine vegetation belt), but the flavor of hay from such meadows was excellent and many such meadows were mowed in the Alps.[1902] Similarly, quantities of hay in meadows dominated by alpine curly sedge (*Carex curvula* subsp. *curvula*) were very low, but the fodder quality from these communities was high; such meadows were frequent on nutrient-poor, acidic soils on silicate bedrock, mainly in the alpine vegetation belt.[1903]

Grassland plant protein content is also reported to increase at high elevations in some cases.[1904] The protein content was usually relatively high in meadows where some protein-rich herbs abounded; that is, at some mesic to moderately moist sites, such as in some concave slope landforms characterized by moderately longer persistence of snow cover and moderate soil humidity in the summer, or on some shady slopes where water availability in the soil was comparable. Ecological processes (and grazing; see Section 6.2) maintained middling nutrient availability at such sites over the long term, without fertilizer application. The example of highly nutritive French honeysuckle (*Hedysarum hedysaroides*), occurring at such sites, was presented above; further high-quality fodder plants frequent at such sites in subalpine and alpine vegetation belts include alpine spicknel (*Ligusticum mutellina*, mainly on silicate bedrock; see

[1900] Pils, 1994, 196; Keller, 2004, 34; Brugger & Wohlfarter, 1983, 54, 186; Aeschimann et al., 2004, vol. 2, 16–17, 452–453, 820–821, 902–903, 1153.

[1901] A comparison with summer mountain pastures is telling: “it is a scientifically proven fact that Alpine milk [in some plant communities] contains between 15 percent and 30 percent more fat than in the valley” (Orland, 2016, 27).

[1902] Pils, 1994, 190, 193, 196, 221; Aeschimann et al., 2004, vol. 2, 820–823, 850–851; Orland, 2016, 27.

[1903] Brugger & Wohlfarter, 1983, 186, 188; Keller, 2004, 36; Aeschimann et al., 2004, vol. 2, 796–797.

[1904] Brugger & Wohlfarter, 1983, 177; Orland, 2016, 27.

Section 4.1.1.2), and some Lady's mantles (*Alchemilla* spp.; see Section 4.1.1.2), for instance. In such stands, the hay crude fiber content was significantly lower than at drier sites (e.g., lower than in *Carex sempervirens* communities).[1905] These data explain why milk cows at some farms received hay from shady sites in the mountains, whereas hay from sunny meadows in the mountains served as fodder for non-milking cattle, according to one of the mountain farmers interviewed in Carinthia by Erika Keller.[1906] High fodder quality on northern slopes was, however, not a general rule; the hay quality on northern slopes was often lower than on slopes with different aspects.[1907]

The previous paragraphs contribute explaining traditional ecological knowledge in old proverbs from the Alps of Carinthia, claiming that hay from mountain meadows is significantly better than hay from the valley, and so cattle need much less of this kind of hay to satisfy their hunger.[1908] The following four paragraphs provide more detailed insight into the impact of the species composition of mountain meadows on the high nutritive value of hay, based on the example of the Golica Ridge in the western Karawanks in the mid-nineteenth century and first half of the twentieth century.

Sunny subalpine meadows on south-facing slopes at the top of the Golica Ridge were known for their quality hay. It is no wonder that selling subalpine hay from this mountain ridge commenced early, before 1616. Due to the high quality of the hay, these efforts were justified, even though some farms entitled to mow there were located quite far away in the valley (Section 3.1). Around sixty tons of hay or even more were made there at elevations of around 1,500 to 1,835 m in the first half of the twentieth century, in those years when farmers mowed the entire area where mowing was allowed.[1909] The hay from Golica

1905 Brugger & Wohlfarter, 1983, 20; Wopfner, 1997, 333; Aeschimann et al., 2004, vol. 1, 786–791, 950–951, 1126–1127; Aeschimann et al., 2004, vol. 2, 820–821; compare Martinčič, 2007d, 403. See also the presentation of Lady's mantle sites lower in this section, where montane meadows are dealt with.

1906 Keller, 2004, 66.

1907 Brugger & Wohlfarter, 1983, 20, 54, 180; Wopfner, 1997, 311; Ivaşcu, Öllerer, & Rákosy, 2016, 42.

1908 Keller, 2004, 27, 31–33, 43, 62, 64, 67, 71, 75, 137–138, 142; Keller, 2007, 207–208.

1909 Smolej (2008, 15), who participated in such large-scale mowing on Golica in the mid-twentieth century, mentioned more than a hundred carts of hay made in subalpine grasslands on this mountain ridge when they mowed all the subalpine area that could be mowed. Torkar Tahir (2012, 133) mentions up to a hundred carts of hay, each of them weighting 500 to 600 kg, based on notes taken in 1975 during interviews with locals that most likely participated in haymaking on Golica until the mid-twentieth century. Klinar (2012, 98) reports a quantity of up to 100 tons of hay. It is plausible that the conversion of loads of hay to tons is erroneous in Klinar's publication. Localization of these meadows: *Arcanum Maps*, Cadastral Maps, Habsburg Empire – Cadastral Maps (XIX. [sic] Century), cadastral municipality of Planina pod Golico (Germ. *Alpen*), 1826; localization: *Environmental Atlas of Slovenia*, Overview Maps (detailed topographical map) and Spatial units, Land cadaster.

was rich in broad-leaved and other herbs[1910] and considered the best fodder available in the meadows where farms from the village of Planina pod Golico were entitled to mow. They believed that non-fertilized subalpine meadow sites on Golica increased the amount of wood their oxen were able to haul because this high-quality hay gave the oxen more energy (or they recuperated sooner after the effort) than hay from elsewhere.[1911] This has parallels at other places in the Alps: farmers hauling cargo with horses over the Katschberg Pass from Carinthia to the province of Salzburg saved hay from particular meadows for this purpose.[1912]

The meadows on Golica have received significant botanical attention since the eighteenth century.[1913] Relevant information not only on plant occurrence but also selected data on plant abundance on Golica is available in botanical publications from the nineteenth century and first half of the twentieth century, when mowing regularly took place there. This makes it possible to understand some of the main reasons for the high hay quality and provides some insights into the species richness in these meadows before the 1860s, when mowing on Golica was preceded by spring grazing in hay meadows, and also later, from the first half of the twentieth century, when mowing still took place on Golica.

Valentin Plemel's mid-nineteenth-century botanical observations in meadows on Golica refer to a period when spring grazing was still practiced in these meadows until May 25th (Section 6.2.2.3). A comparison of his data and the pedological map explains that these meadows on prevalent rendzinas on limestone and dolomite[1914] were species rich (*pflanzenreich*)[1915] despite the late regular end grazing in spring. The protein-rich legume wood vetch (*Vicia sylvatica*) was frequent (*häufig*) in these meadows at that time,[1916] and the legume French honeysuckle (*Hedysarum hedysaroides*), one of the best fodder plants naturally occurring in the Alps (Figure 86), also grew there[1917] despite spring

1910 Klinar, 2012, 100; Torkar Tahir, 2012, 133.
1911 Smolej, 2008, 15; Torkar Tahir, 2012, 133.
1912 Keller, 2004, 30.
1913 Praprotnik, 2012.
1914 *Environmental Atlas of Slovenia*, Spatial Units, Cadastral Areas and Land and soil, Soil map (1:25,000).
1915 Plemel, 1862, 146.
1916 Plemel, 1862, 164.
1917 Plemel, 1862, 138. Plemel used the Latin name *Hedysarum obscurum* L. *Hedysarum obscurum* can designate a few *Hedysarum* species, among which only *Hedysarum hedysaroides* occurs in the Slovenian Alps (Aeschimann et al., 2004, vol. 1, 950–953, 1159). Among the subspecies of *Hedysarum hedysaroides* only *Hedysarum hedysaroides* subsp. *hedysaroides* occurs in the Karawanks where Golica is located (Martinčič, 2007c, 316), and so this subspecies is mentioned in the text above.

grazing. Among other species, Salisbury's daffodil (*Narcissus poeticus* subsp. *radiiflorus*) was very abundant there (Figure 79 in Section 6.2.2.3), which possibly had an indirect positive impact on the hay quality in these meadows: the daffodils did not impact the quality of the hay directly because their leaves and stems decomposed before haymaking (Section 6.2.2.3). However, this means that nitrogen from the biomass of decomposing daffodils regularly returned to the soil before haymaking, unlike most other plant biomass from these meadows that was harvested by farmers in the summer or grazed in spring and autumn. Because the abundance of daffodils in these meadows was very high (see the following paragraph and the beginning of Section 6.2.2.3), such nitrogen cycling possibly contributed to the maintenance of meadow productivity and the related preservation of the high quality of the hay from these non-fertilized meadows in the long run. Moonwort (*Botrychium lunaria*),[1918] Dinaric alpine thistle (*Carduus carduelis*),[1919] and bearded bellflower (*Campanula barbata*; its fodder value is nothing special)[1920] were all frequent in these meadows, pink scorzonera (*Scorzonera rosea*) was (relatively) frequent,[1921] and single-flowered cat's ear (*Hypochaeris uniflora*)[1922] and Alpine aster (*Aster alpinus*)[1923] were also observed there.[1924] Plemel focused on a selection of plant species, and he did not determine the plant communities; nonetheless, his cited data suggest that the particularly high hay quality was not only the consequence of changes in meadow species composition after the abolition of spring grazing in mowed grasslands here around 1860 (see Section 6.2.2.3). Some good fodder plants occurred in meadows on Golica and some of them even abounded in the hay from there already in the long centuries when mowing there was preceded by spring grazing. In addition to the species composition, the sunny subalpine sites on this mountain ridge, resulting in significantly higher concentrations

[1918] Plemel, 1862, 124.

[1919] Plemel, 1862, 126. Plemel used the Latin name *Carduus arctioides* Willd., a synonym for *Carduus carduelis* (*Global Biodiversity Information Facility*: *Carduus arctioides* Willd.).

[1920] Plemel, 1862, 125; on the fodder value, see Brugger & Wohlfarter, 1983, 26.

[1921] Plemel mentioned *Scorzonera purpurea* L. on Golica (1862, 157). Nonetheless, *Scorzonera purpurea* does not occur in the Slovenian Alps (Aeschimann et al., 2004, vol. 2, 640–641), but *Scorzonera rosea* used to be called *Scorzonera purpurea* subsp *rosea*. Around 1930, when these meadows were still mowed, the botanist Erwin Aichinger documented *Scorzonera rosea* in meadows on Golica (Aichinger, 1933, 128). *Scorzonera rosea* still occurs in the Karawanks (Wraber, 2007c, 697), and so this one is mentioned in the text above.

[1922] Plemel, 1862, 140.

[1923] Plemel, 1862, 123.

[1924] Green alder (*Alnus alnobetula*) also grew on Golica at that time (Plemel, 1862, 121). If it had grown in those meadows mowed once every two or three years, this scrub would also have contributed to the high quality of the hay (see Section 6.1.2.6). Nonetheless, the subalpine meadow sites on Golica, located on sunny slopes on carbonate bedrock, were not suitable for this species. Green alder grows on the northern slopes of Golica, not in the grasslands on the south (Aichinger, 1933, 171–179).

of essential oils in hay than in the biomass from lower elevations, must have contributed significantly to the high hay quality.

Around 1930, after spring grazing in meadows that lasted until May 25th had already been abolished on Golica, but still in the period of regular mowing in subalpine meadows, the plant sociologist Erwin Aichinger studied meadows and pastures[1925] there. He was the first to apply the Central European phytocoenological method in Slovenian territory.[1926] He determined three plant sociological units in these grasslands. The results of his study of one of these units—*Caricetum ferruginei carniolicum*, the plant sociological unit named after rusty sedge (*Carex ferruginea*)[1927]—refer to relevant sections of hay meadows on Golica. This plant sociological unit occurs on southern slopes in meadows in concave landforms on relatively gentle slope sections (15° to 20°), on soil with a basic to slightly acidic pH. It is species rich, with 47 plant species on up to 50 m^2 around 1930, as the average of the three meadow sites that Aichinger studied on Golica.[1928] A presentation of the species that established a high abundance in *Caricetum ferruginei carniolicum* meadows on Golica as found around 1930 follows.[1929] The abundance of the broad-leaved herb mountain hawksbeard (*Crepis bocconei*[1930]), which could make quality hay,[1931] was (very) high, but it varied. This species covered a quarter to half of one site, on

[1925] On historical land use types in the grasslands on Golica, see *Arcanum Maps*, Cadastral Maps, Habsburg Empire - Cadastral Maps (XIX. [sic] Century), cadastral municipality of Planina pod Golico (Germ. *Alpen*), 1826; SI AS 176, Franciscejski kataster za Kranjsko (Franciscan Land Cadaster for Carniola), L16, cadastral municipality of Hrušica, map sheet A01. See also the land use at Jekelj in Section 3.1.

[1926] For a very general notice on Aichinger's research in this study area, see Praprotnik, 1993a, 218, and Praprotnik, 2012, 30.

[1927] In contrast to *Caricetum ferruginei carniolicum* studied in meadows, Aichinger mentioned that he studied the grassland plant association on Golica's steeper slopes at a site that was "impoverished through heavy grazing" (Aichinger, 1933, 121–125, 127, quote from p. 123), and so his results about other grasslands on Golica are not relevant for the hay meadows dealt with here; they deal with a pasture.

[1928] Aichinger, 1933, 127–132. The sites studied by Aichinger were all located in meadows and not in pastures (compare Aichinger, 1933, 129, with *Arcanum Maps*, Cadastral Maps, Habsburg Empire - Cadastral Maps (XIX. [sic] Century), cadastral municipality of Planina pod Golico (Germ. *Alpen*), 1826; to determine the locations, this early nineteenth-century cadastral map was compared with the georeferenced present cadaster in *Environmental Atlas of Slovenia*, Backgrounds, Overview maps (detailed topographical maps) and Spatial units, Land cadaster).

[1929] Aichinger, 1933, 11, 128–129.

[1930] Aichinger mentioned *Crepis montana*, which is a synonym for *Crepis bocconei* (Aeschimann et al., 2004, vol. 2, 672–673).

[1931] Its relative golden hawk's beard (*Crepis aurea*) is a good fodder plant (Brugger & Wohlfarter, 1983, 25). An integrated quality indicator linking the nutritive value, palatability, and health impacts of another relative, rough hawk's beard (*Crepis biennis*), can be high, middling, or low (Dietl & Lehmann, 2004, 39, 43). Because *Crepis bocconei* abounds at significantly higher elevations than rough hawk's beard, the nutritive value of *Crepis bocconei* at the sunny subalpine sites of Golica was likely high, especially when another site it still abounded, covering between a twentieth and a quarter of the site, and at the third location studied it was abundant but covered a low percentage of the meadow. The protein-rich legume yellow mountain vetchling (*Lathyrus ochraceus* subsp. *ochraceus*[1932]), rusty sedge (*Carex ferruginea*), and great masterwort (*Astrantia major*) varied from very abundant, covering between a twentieth and a quarter of the meadow site, to abundant but covering a low percentage of the meadow surface at all three sites. At one site, alpine saw-wort (*Serratula tinctoria* var. *alpina*[1933]) covered a quarter to half of the area. Evergreen sedge (*Carex sempervirens*) and common bent (*Agrostis capillaris*[1934])—the fodder value of the first one is nothing special, and the latter makes middling-quality hay[1935]—covered between a twentieth and a quarter of one of the three meadow sites. Salisbury's daffodil (*Narcissus poeticus* subsp. *radiiflorus*), which possibly had an indirect positive impact on hay quality (see above), was very abundant and covered between a twentieth and a quarter of one meadow site; it was abundant but with low cover at another site, and rare at a third one.[1936] Wood vetch (*Vicia sylvatica*), documented in these meadows as frequent in the mid-nineteenth century (see above), still occurred there around 1930. At one of the sites analyzed it was frequent (and so it is not possible to determine whether the abolition of spring grazing had any significant impact on the representation of this species) but nonetheless covered less than a twentieth of the area. Pink scorzonera (*Scorzonera rosea*), reported by Plemel as (relatively) frequent, occurred at all three sites analyzed around 1930 but its abundance was low or very low, and moonwort (*Botrychium lunaria*), mentioned by Plemel, was documented at only one of the three sites analyzed around 1930, and its abundance was low or very low.[1937] All in all, similar reasons can be recognized for the high hay quality around 1930 as in the mid-nineteenth century. It should, however, be emphasized that, on sunny slopes and prevalent rendzina on carbonates of Golica, Aichinger's data refer to moist sites in relief depressions along which meltwater runs, and to gentle slope sections. The plant association on steeper slopes that prevailed in meadows on Golica were not studied by Aichinger.[1938]

---

hay was mowed with a scythe, thus close to the ground, and turned and collected manually with wooden implements, and so the losses of fragile leaves were moderate.

1932 Aichinger mentioned *Lathyrus laevigatus*, which is a synonym for *Lathyrus ochraceus* subsp. *ochraceus* (Aeschimann et al., 2004, vol. 1, 880–881).

1933 Aichinger (1933, 128) mentioned *Serratula macrocephala*, which is a synonym for *Serratula tinctoria* subsp. *monticola* (Aeschimann et al., 2004, vol. 2, 592–593).

1934 Aichinger (1933, 129) mentioned *Agrostis tenuis*, which is a synonym for *Agrostis capillaris* (Aeschimann et al., 2004, vol. 2, 970–971).

1935 On evergreen sedge's fodder value, see Brugger & Wohlfarter, 1983, 24, and on common bent's fodder value, see Dietl & Lehmann, 2004, 39, 42.

1936 Aichinger, 1933, 128 (concerning the species identification, see also Section 6.2.2.3).

1937 Aichinger, 1933, 11, 129.

1938 The statement that most meadow sites on Golica were too steep for *Caricetum ferruginei carniolicum* as presented by Aichinger is based on a comparison of Aichinger, 1933,

High nutritive value was not the only advantage of mountain hay from many sites; the positive impacts of hay on livestock's health also mattered a great deal. In the province of Salzburg, farmers in Lungau describing hay from Zoponitzen (in the Upper Mur catchment, on the slopes of Marislwand and Kreuzhöhe close to Hintermuhr; these meadows were located on slopes facing southeast and southwest, and they reached 2,500 m) emphasized that they considered hay from higher elevations significantly better and healthier, as recorded in the mid-twentieth century.[1939] According to another piece of traditional ecological knowledge from Carinthia, the higher one ascends, the healthier the fodder.[1940] This type of traditional ecological knowledge was widespread in the Alps: in the mid-twentieth century, old farmers reported to ethnologists in the Trenta Valley (the Limestone Alps of the Upper Soča catchment: the Julian Alps) that the hay they cut with scythes and sickles in remote meadows high in the mountains, up to four hours' walk from their farms (the highest hay meadows in this microregion exceeded the altitude of 2000 m – see Sections 3.1 and 6.1.1.2), was considerably healthier for their livestock due to the significant abundance of medicinally active forbs in it, such as wormwood (*Artemisia* spp.), mountain arnica (*Arnica montana*), and masterwort (*Peucedanum ostruthium*).[1941] Masterwort was a multifunctional healing plant (called *vse dobro,* literally 'all good')[1942] abounding only at high elevations, where it was mowed in some mesic grasslands at rocky, nutrient-rich sites in subalpine and alpine vegetation belts.[1943] Farmers in Carinthia and East Tyrol have been using mountain hay from non-fertilized meadows against cattle diarrhea and lack of fertility, and regularly to promote general wellbeing and strengthen livestock's immunity; interviewees in East Tyrol have additionally been feeding cattle mountain hay from non-fertilized meadows to increase their appetite because livestock like it so much.[1944] High hay quality and favorable impacts on livestock health (see the introduction to Section 10.1.2.2, and Sections 10.1.2.2.2, 10.1.2.2.3, 10.1.2.2.4, and 10.1.2.2.7) are still among the main reasons for the limited areas of preserved haymaking in non-fertilized meadows at high elevations in the mountains of Carinthia now, and some types of beneficial health impacts of hay from

130–131, with *Arcanum Maps*, Cadastral Maps, Habsburg Empire – Cadastral Maps (XIX. [sic] Century), cadastral municipality of Planina pod Golico (Germ. *Alpen*), 1826, and calculations of average slope gradient based on *Environmental Atlas of Slovenia*, Overview Maps (detailed topographical map) and Spatial units, Land cadaster.

1939 Hubatschek, 1992, 175; localization: *Arcanum Maps*, Cadastral Maps, Habsburg Empire – Cadastral Maps (XIX. [sic] Century), cadastral municipality of Hintermuhr, 1829, and *Elevation Map*.

1940 Keller, 2007, 208–209.

1941 Makarovič, 1978, 238. For the meaning of the dialect name *jaščarica*, see Mlakar, 2015, 306, and Martinčič, 2007d, 407.

1942 Mlakar, 2015, 306, 350–351.

1943 Martinčič, 2007d, 407; Aeschimann et al., 2004, vol. 1, 1136–1137.

1944 Keller, 2004, 31–32, 43, 47, 58, 62, 75, 137, 142; Vogl-Lukasser et al., 2006, 55–57, 67–69, 72, 83, 87–90, 195, 234, 252.

appropriate plant species compositions at high elevations have been scientifically proven,[1945] not only the high nutritive value of such hay, as explained above.

The impacts of high elevations on plant development are much more evident in the subalpine and alpine vegetation belts, where short plants prevail, than in the montane belt with usual middling grassland vegetation size even in many non-fertilized meadows unless site conditions are extreme (see Figure 85, showing vegetation in a non-fertilized montane meadow).[1946] Nonetheless, although average fodder qualities in agriculturally favorable non-fertilized meadows were often highest in subalpine and alpine vegetation belts, several non-fertilized meadows at middle and higher elevations in the montane belt also provided better fodder than meadows at significantly lower elevations. Favorable health impacts of hay from appropriate species compositions at high elevations were demonstrated above in the context of the hay from the subalpine and alpine vegetation belts. They are also evidently significant in several grasslands with appropriate species compositions in the montane belt. Traditional Csángó farmers in Ghimeş (Eastern Carpathians), for instance, emphasize the significantly higher share of medicinally active plants in hay from mostly non-fertilized montane meadows at higher elevations relevant for livestock's health than in moderately manured meadows close to farms.[1947] Some of the data on the health impacts of hay from mountain grasslands in the Alps presented above in the context of (sub)alpine meadows were also relevant in montane meadows, and old farmers emphasize the favorable impacts that the milk of livestock largely fed with such hay had on human health.[1948]

Moreover, experiences of traditional Csángó farmers show that, during the winter period of stable-foddering, milk is tastier and has a higher fat content if cows are fed with hay from more species-rich non-fertilized mountain meadows with a higher leaf content than it would be if the cows consumed relatively good hay from moderately manured meadows closer to the farms, where yields are higher. These farmers belong to communities with the best-preserved traditional agriculture in Europe today; they farm in an area where the mountains do not exceed the upper limit of the montane vegetation belt.[1949] Experience-based knowledge from the montane belt in various parts of the Alps, such as Bohinj in the Eastern Julian Alps and Tyrol, emphasizes that the same volume unit of montane hay (from sunny meadows at least) weighs much more than hay from the valley bottoms, as was easily perceived when hay was carried in sheets (Figure 87). This helps understand the high fodder value of

[1945] Brugger & Wohlfarter, 1983, 54, 180; Keller, 2004, 27, 31–32, 43, 62, 64, 67, 72, 75, 137–138, 142; Keller, 2007, 206–209.
[1946] Pils, 1994, especially p. 187.
[1947] Babai, Molnár, & Molnár, 2014, 35–38, 40, 42–44, 112.
[1948] E.g., author's interview with Milena Sodja, 2017, in Srednja Vas.
[1949] Babai, Molnár, & Molnár, 2014, 15–20, 35–38, 40, 42–44, 95–96, 110, 112, 136; Babai & Molnár, 2013.

Figure 87: Carrying heavy loads of hay from the meadow to the hay barn or haystack was usually a task for men in historical patterns of work division in the Alps. Nonetheless, in some regions such as parts of the Upper Soča catchment, where this photo was taken, women frequently did it, according to mid-twentieth century data, probably in the absence of men.[1950] The size and weight of loads of hay carried by women were no smaller than those carried by men.[1951] The hay was packed in a linen sheet based on traditional knowledge about how to transport as much as possible in one sheet; that is, the hay was not simply piled on the sheet, but several layers of hay were neatly spread on the sheet with a rake.[1952] Hay was carried in a way that most of the weight from the load pressed on the carrier's shoulders, neck, and head; some made sure that the majority of the hay's weight pressed on the shoulders to make this hard work bearable, as in Figure 87, whereas others carried more on their heads.[1953] The load in this figure is large; it must have weighed much more than the woman carrying it.[1954] Such large loads of hay were typical in many cases.[1955] Significantly smaller hay sheets for smaller loads of hay were also used in some regions or on some farms in the Alps.[1956] Carrying the loads of hay allowed the carrier to notice significant differences in the weight of a load of hay from different meadows. Storing first-crop hay at Vrsnik in the Upper Soča catchment, July 1952. Photo: Boris Orel. Slovenian Ethnographic Museum, Documentation, Fieldwork campaign 8, Trenta – Bovec (1952): photo 8/138 and *Slovenski etnografski muzej* . . . Trenta – Bovec / 140 F0000008/138. © Slovenian Ethnographic Museum.

[1950] Matičetov, 1948, 21; Smerdel, 1994, 61, 106–108.
[1951] Makarovič, 1978, 148; see also Figure 87 in this volume.
[1952] Šarf, 1974, 140.
[1953] Explanation of the carrying strategy: Hodnik, 2016, 77; author's interviews with J. Kavčič, 2021, in Dolina above Tržič, and with Kunavec, 2023, in Stara Fužina.
[1954] Namely, even a much smaller hay load would weigh about 60 to 70 kilograms (Šarf, 1974, 140–141).
[1955] See, e.g., Melik, 1950, 116; Cevc, 1992, 84; Novak, 2024, 23, 97.
[1956] Šarf, 1974, 141; Novak, 2024, 23.

such hay.[1957] In the case of appropriate species compositions in montane meadows, cows even licked the hay dust from the manger—an exceptional behavior demonstrating how tasty such hay was.[1958] A farmer in Carinthia, interviewed by Erika Keller, emphasized that the milk quality from extensively managed mountain meadows is higher but its quantity is lower.[1959] Similarly, a farmer from Stara Fužina in Bohinj claimed that cattle need less such hay to satisfy their hunger, but they produce less milk than if fed with the hay from moderately fertilized meadows in the valley.[1960]

Still focusing on high-quality non-fertilized meadows, a historical strategy of increasing hay digestibility, which had an impact on milk quantity, needs to be presented. Namely, mid-twentieth-century information from several places in the Eastern Alps, such as East Tyrol and Bohinj, probably reflecting a longer history, reveals that a limited percentage of the total daily fodder (e.g., quality mountain hay[1961] in temporary settlements in montane meadow complexes inhabited in spring and autumn) for dairy cows was scalded with water. Hay was, for instance, filled in a wooden trough, boiling water was poured over it so that the fodder softened, it was allowed to cool down and slightly salted, and cows liked this type of fodder very much. This practice evidently increased the digestibility of fodder, and so it raised the milk quantities a bit, as observed independently at several places.[1962] Only a limited percentage of the daily fodder was scalded, and so the question of loss of some biologically active substances was not an issue because the rest of the hay and/or grazing resources also came from species-rich grasslands. This hay-saving practice was abandoned when mineral fertilizers were introduced, and so more hay was available.[1963]

Just as in the subalpine and alpine belts dealt with above, these generalized and simplified data on high-quality montane hay require further contextualization due to habitat diversity in non-fertilized meadows.[1964] According to the traditional ecological knowledge from several parts of Slovenia, several sunny, species-rich and (close to) mesic meadows at higher elevations, on soils on

1957 Wopfner, 1997, 311; author's interview with A. Hribar, 2017, in Bohinjska Češnjica.
1958 Author's interview with F. Mlakar, 2017, in Bohinjska Češnjica.
1959 Keller, 2004, 67.
1960 Author's interview with Kunavec, 2023, in Stara Fužina.
1961 At several places in the Alps, limited quantities of other dry fodder types also were scalded. For instance, dry clover was used instead of mountain hay; similar to mountain hay, an increased amount of milk was observed after the cows consumed scalded clover (Bavcon, Ravnjak, & Praprotnik, 2019, 79). Other ingredients were added to scalded fodder at many farms, ranging from those with favorable impacts on livestock's performance, such as barn sweepings (see Section 6.5) to those with very limited nutritive value, even if scalded, such as cereal straw (compare Vogl-Lukasser et al., 2006, 65–66).
1962 Hodnik, 2016, 44; Bavcon, Ravnjak, & Praprotnik, 2019, 79; see also Vogl-Lukasser et al., 2006, 65–66; author's interview with Kunavec, 2023, in Stara Fužina.
1963 Author's interview with Kunavec, 2023, in Stara Fužina.
1964 Keller, 2004, 43, 66.

limestone bedrock, provide particularly good hay (see Section 10.1.2.2.4),[1965] and the importance of some sunny sites has been emphasized elsewhere in the Alps.[1966] For instance, farmers from Stara Fužina and Studor in Bohinj that were adults by the mid-twentieth century agreed that hay from some sunny, mesic meadows on Vogar Hill and at the Uskovnica grassland complex (located at about 1,000 to more than 1,200 m, some 500 or 600 m higher than the villages) was the best hay from montane meadows belonging to these villages because cows needed less such hay to satisfy their hunger and to provide much milk by historical standards. Moreover, they emphasized that cows, whose winter diet was largely based on hay from Vogar Hill or Uskovnica, were fat by the standards of the time. They also noted that hay from these montane meadows was much better than hay from shadier montane meadows; for example, from the Voje Valley (at lower elevations and with a north–south orientation, thus missing some morning and afternoon sun), at Blatca north of Studor Hill, or at Zagradec on north-facing slopes above the easternmost part of Lake Bohinj (Sln. *Bohinjsko jezero*).[1967]

Comparison with a soil map shows that at Vogar Hill and Uskovnica there are (1) rendzinas on limestone and dolomite, which retain very little water available to plants, and (2) rendzinas on carbonate moraines and few eutric cambisols, which store little water available to plants, and thus relatively sunny (but hummocky, thus highly diverse; see Sections 3.2 and 4.1.2, text and Figures 15 and 30) and relatively dry (actually mesic because the climate of the area is wet, with about 2,600 m of annual precipitation from 1971 to 2000[1968]) meadows, with often close to a neutral or slightly alkaline soil pH, providing high-quality montane hay. The soils at Vogar Hill and Uskovnica on the one hand, and in the Voje Valley and at most of Zagradec on the other, are similar, and so the slope aspect causing shadier sites and lower elevation must have played an important role in limiting the hay quality at Voje and Zagradec.[1969] Present agronomy agrees that the quality of hay from sunny sites on south-facing slopes that are neither too dry nor wet is usually high, and that livestock like such hay very much.[1970] Nonetheless, just like in (sub)alpine belts, not all

---

[1965] Bavcon, Ravnjak, & Praprotnik, 2019, 81, 131; Zwitter, 2020, 66.

[1966] Wopfner, 1997, 311.

[1967] Zwitter, 2020, 66; author's interviews with Mencinger, 2017, in Stara Fužina, and with L. and M. Odar, 2017, in Studor.

[1968] *Environmental Atlas of Slovenia*, Climate, Precipitation, Mean annual corrected precipitation 1971–2000.

[1969] This is a result of the comparison of a soil map (*Environmental Atlas of Slovenia*, Spatial Units, Cadastral Areas & Land and soil, Soil map (1:25.000)) with meadow areas in the early nineteenth-century cadaster (*Arcanum Maps*, Cadastral Maps, Habsburg Empire – Cadastral Maps (XIX. [sic] Century), cadastral municipalities of Savica [Germ. *Saviz*], Srednja Vas v Bohinju [Germ. *Mitterdorf*], and Studor [Germ. *Studorf*], all 1826), and the oral-history interviews cited above.

[1970] Brugger & Wohlfarter, 1983, 20, 54, 180.

non-fertilized sunny meadows provided high-quality hay, depending on site conditions, management, and species composition (see also the introduction to Section 10.1.2.2 and Section 10.1.2.2.4).

At Blatca, a different geology leading to prevalent dystric cambisols retaining medium quantities of water available to plants, which in the wet climate of the region often meant moist soil, evidently contributed to lower hay quality, together with shadier sites.[1971] Present agronomy agrees that fodder quality on north-facing slopes is often significantly lower than on slopes facing other directions.[1972] However, just like in the (sub)alpine belt, hay quality from some shady non-fertilized mountain meadows in Ghimeş, rich in Lady's mantles (*Alchemilla* spp.), clovers (*Trifolium* spp.), and several other herbs and grasses, is also considered very high if there are appropriate species compositions; the usual dominant grasses in these meadows prefer acidic to neutral soils.[1973] Agronomy confirms the validity of this piece of traditional ecological knowledge; readers are familiar with the high quality of Lady's mantles as fodder (Section 4.1.1.2), and clovers also provide quality hay and bind airborne nitrogen to the soil, making nutrient availability to other grassland plants higher.[1974]

However, many non-fertilized sites did not provide comparably good grassland fodder, and some plants were gladly consumed by some livestock types or individual animals of the same livestock species but avoided or less readily consumed by others. Hay from relatively dry to mesic sites in the montane or colline belt in some places (see the introduction to Section 4) abounded in *Peucedanum* spp.[1975] Such hay has a strong flavor. It was good for sheep and goats and highly esteemed in the context of livestock's health (see above), whereas cattle were to a certain degree repelled by the smell of *Peucedanum*. Thus, if used for cattle, such hay had to be mixed with other hay.[1976]

Extremely poor hay quality in non-fertilized meadows from the viewpoint of animal husbandry was typical for hay from matgrass (*Nardus stricta*, Fig-

---

1971 This is a result of the comparison of a soil map (*Environmental Atlas of Slovenia*, Spatial Units, Cadastral Areas & Land and soil, Soil map (1:25.000)) with meadow areas in the early nineteenth-century cadaster (*Arcanum Maps*, Cadastral Maps, Habsburg Empire – Cadastral Maps (XIX. [sic] Century), cadastral municipality of Studor [Germ. *Studorf*], 1826), and the oral-history interviews cited above.

1972 Brugger & Wohlfarter, 1983, 20, 54, 180.

1973 Babai, Molnár, & Molnár, 2014, 36, 43–44, 49, 110, and a comparison of basic ecological requirements with data on the usual dominant species in these meadows: Babai, Molnár, & Molnár, 2014, 35–36; see also a note on geology in Babai & Molnár, 2013, 3.

1974 Dietl & Lehmann, 2004, 10, 19, 36–37, 39, 41–43, 79–81, 88–90; Pfister, 1985a, 110–111, 113.

1975 Aeschimann et al., 2004, vol. 1, 1130–1137.

1976 Author's interviews with Kunavec, 2023, in Stara Fužina, and with A. Soklič, 2017, at Podjelje; folk taxon identification: Tuma, 1935, 251.

ure 30 in Section 4.1.2) swards (see also Section 2). Matgrass abounds in many nutrient-poor and acidic meadows and pastures in the subalpine and alpine vegetation belt in all regions of the Alps, and less frequently in the montane and even colline vegetation belt,[1977] where it used to be more frequent as long as nutrient-poor grasslands were usual. Thick tussocks of short and tough sclerenchyma-rich matgrass presented a special challenge to haymakers; only experienced haymakers were able to mow matgrass systematically, and even they had to invest more energy in mowing matgrass than in usual mowing, as reported in Tyrol.[1978] Similarly, in the Allues Valley, Savoy, matgrass was the most difficult to mow among all mowed grass species.[1979] It was a general rule that mowing by scythe was easier at morning dew than later;[1980] however, this rule was more important in the case of short vegetation (if first-crop hay at non-fertilized sites is considered), and it was crucial in the case of matgrass due to a combined impact of its short length and tissue structure. Even experienced haymakers could only mow matgrass in the morning (until about 9 am); later it stuck to the scythe blade. Thus, mowing started particularly early, at 2 or 3 am, to mow significant matgrass areas, and the scythe blade had to be particularly sharp.[1981] In some places in the Alps, this led to an adaptation of the mowing direction to mow matgrass somewhat more smoothly (see Section 6.1.1.2). Not only mowing matgrass hay, but also storing it, was particularly labor intensive, as explained in Section 6.3.1.1. To make matters worse, matgrass has been recognized as low-quality fodder, and in the mid-twentieth century low-quality hay from matgrass-dominated mountain meadows was stored separately (by many farms) and used to feed non-milking adult cattle or horses when they did not work, as reported in Bohinj.[1982] Farmers used such hay as fodder because quantities of other fodder at the farm level were insufficient (see Section 6.2.2.1).[1983] The unwantedness of matgrass resulted in its frequent recognition as a separate folk taxon. Traditional ecological knowledge of tough matgrass is reflected in several folk names, such as *Borst* and *Bierstling* (*Borste* 'bristle', *Bürste* 'brush') in Tyrol, *ščetina* (i.e., 'bristle') in some parts of Slovenian territory, *Hirschhaar* ('deer hair') in northeastern Tyrol,[1984] and *vásna* (i.e., *lasna*, 'hair-like grass') at Dolina above Tržič in the Karawanks.[1985]

[1977] Sadar, 1956, 24, 126–128, 134, 140; Meilleur, 1986, 267, 287, 306; Pils, 1994, 78, 90; Aeschimann et al., 2004, vol. 2, 996–997.
[1978] Pils, 1994, 94; Wopfner, 1997, 312. Similar report from the Eastern Carpathians: Babai, Molnár, & Molnár, 2014, 50, 120.
[1979] Meilleur, 1986, 291, 293–294, 296, 306.
[1980] Iuga, 2016, 73.
[1981] Pils, 1994, 94; Keller, 2004, 15, 67.
[1982] Author's interviews with Kunavec, 2017, in Stara Fužina; with Minka Sodja, 2017, in Jereka; and with A. Soklič, 2018, in Podjelje; the species collected and confirmed by Soklič was identified by Leonid Rasran.
[1983] Author's interview with J. Kavčič, 2021, in Dolina.
[1984] Wopfner, 1997, 312; Dakskobler, 2023, 8.
[1985] Author's interview with J. Kavčič, 2021, in Dolina.

A high or significantly increased abundance of matgrass in meadows was either a consequence of natural soil properties, or, frequently, of regular annual mowing of non-fertilized meadows, gradually leading to soil impoverishment, and thus suppressing species that are more productive from the viewpoint of livestock farming. Low matgrass height under historical conditions of frequent hay scarcity promoted mowing matgrass swards very close to the sod, which made soil impoverishment an even more serious issue in the long run.[1986] In the eastern Julian Alps, matgrass is now frequent in some meadows in the meadow complexes on acidic soils at Konjska Dolina, Jelje, near Podjelje, at Pokrovec, and around Koprivnik, for instance. Today, the high abundance of matgrass starts there slightly above the elevations of the highest permanent settlements.[1987] Early nineteenth-century information proves the existence of nutrient-poor montane meadows with hard and sharp vegetation in the same cadastral municipality where Konjska Dolina, Jelje, Podjelje, and Pokrovec are located,[1988] pointing to the long-term preservation of limited sections of larger historical areas of nutrient-poor montane meadows abounding in hard plants such as matgrass.

It is relevant to know that nutrient-poor (matgrass) swards were so frequent in the area in the early nineteenth century that they were presented in a land cadaster as the typical vegetation in many montane meadows. Namely, in the nineteenth-century, corralling sheep resulting in moderately manured meadows and suppressing matgrass (Section 6.3.2.2) is documented in montane meadows on the edge of the Pokljuka Plateau,[1989] where Konjska Dolina, Jelje, Podjelje, and Pokrovec are located. In the early nineteenth century, when nutrient-poor (matgrass) swards were recorded in these montane meadow complexes, rearing sheep was still an important branch of livestock farming in the area: farmers managing the meadow complexes at Konjska Dolina, Jelje, Podjelje, and Pokrovec reared significant numbers of sheep; big farms were estimated to have kept about fifteen sheep per farm (and other livestock), and this was certainly not an overestimate.[1990] The data presented above explain that, despite some importance of rearing sheep, the numbers of sheep grazing in meadows in part of spring (Section 6.2) and autumn did not suffice for corralling to sup-

1986 Wopfner, 1997, 312; Keller, 2004, 29, 35, 66, 89.

1987 Dakskobler et al., 2019, 18, 26; author's interview with A. Soklič, 2017, in Zgornje Podjelje. On the status of a temporary versus permanent settlement at Pokrovec in the early twentieth century, see Melik, 1927, 76–77.

1988 SI AS 176, Franciscejski kataster za Kranjsko (Franciscan Land Cadaster for Carniola), L107, cadastral municipality of Češnjica v Bohinju [i.e., Bohinjska Češnjica], cadastral appraisal protocols: Catastral-Schätzungs-Elaborat, Einleitung, §11: "wegen der harten spissigen grasnarbe da solche weniger nahrungsstoff enthält . . . ."

1989 Novak, 1970, 357; Šarf, 1974, 143.

1990 Žontar, 1987, 63; SI AS 176, Franciscejski kataster za Kranjsko (Franciscan Land Cadaster for Carniola), L107, cadastral municipality of Češnjica v Bohinju [i.e., Bohinjska Češnjica], cadastral appraisal protocols: Catastral-Schätzungs-Elaborat, Einleitung, §4.

press matgrass in more than selected sections of montane meadows. The existence of meadows with a high abundance of matgrass despite the importance of rearing sheep and the practice of corralling (see Sections 6.3.2.1 and 6.3.2.2) is considered representative of several regions of the Alps. Similarly, significant areas of matgrass swards suggest that wood was not burned for the specific purpose of ash production for its application to matgrass swards. Household ash was usually applied to relatively nearby meadows,[1991] but its quantities were insufficient to convert matgrass swards in remote meadows on an area-wide basis into meadows more productive from the perspective of farming (Section 6.3.2.1).

Despite all the challenges that matgrass posed to farmers, a limited percentage of matgrass swards in the total meadow area where a farm was entitled to mowing strengthened the resilience of farm economies in the Alps in times of limited long-distance road transport of hay (Section 3.1), following the general principle in agriculture that "the more elements that are integrated into a farming system, the more resilient it becomes."[1992] Namely, hay from matgrass is less sensitive to mold attacks; mold does not infest the hay if matgrass is stored earlier than entirely dry.[1993] In the case of unsuitable weather during mowing times—forcing farmers to store largely but not completely dry hay (if a roofed hayrack was not at hand), which led to mold attacks in hay from certain species compositions, even though some coping strategies mitigated the mold hazard (Section 4.1.1.3)—low-quality hay from matgrass swards was a welcome substitute for some higher-quality hay more prone to mold infestation. Livestock's performance was highly significantly worse due to a diet based on matgrass swards, but the number of animals that had to be sold or slaughtered was reduced.[1994] The situation was faced in which low-quality hay strengthened the general resilience of local farming. Section 9 provides an example of another low-quality kind of hay that was common in nutrient-poor grasslands and that strengthened the resilience of local farming in a manner similar to matgrass. Tor-grass (*Brachypodium pinnatum*; see Section 2: text and Figure 10) also historically strengthened the resilience of the farm economy. It abounds in some non-fertilized, relatively dry meadows, mainly on base-rich soils. Livestock usually do not like tor-grass, but when mowing is delayed rain leaches significantly more nutrients from grasses in moderately manured meadows with a generally higher feed value (such as false oat-grass, *Arrhenatherum elatius*,

[1991] Author's interview with Boltar, 2018, in Stara Fužina (household ash application to a lower meadow at Zavaze); compare Janišová et al., 2023, 15.
[1992] Compare *Transforming Food and Agriculture to Achieve the SDGs*, 2018, 51.
[1993] Author's interview with R. Sodja, 2022, in Stara Fužina, with personal experience in harvesting matgrass at Ši above Srednja Vas in Bohinj.
[1994] Compare the general explanation of impacts of lack of fodder and worse fodder quality on livestock's performance in Pfister, 1985a, 38, 52–53.

and yellow oat-grass, *Trisetum flavescens*, Section 6.3.2.2) than from tor-grass. In this case, livestock preferred hay from tor-grass, as observed by traditional Csángó farmers in the Eastern Carpathians and possibly also by historical farmers in the Alps. Tor-grass was also difficult to mow; it tends to escape the scythe, and so additional efforts were needed to mow it.[1995]

Many nutrient-poor grasslands were highly economically relevant despite the low hay yields, as explained in this section, (1) due to the high percentage of such grasslands in the landscape, (2) because of the medicinal qualities of the hay from several non-fertilized sites and high nutritive value of the hay from specific non-fertilized sites, and (3) in order to increase the resilience of the farm economy, which depended on local fodder resources (see also Sections 6.1.2.1 and 6.1.2.2). (4) Another aspect that used to make some nutrient-poor grasslands (particularly) economically attractive was the high abundance of specific gathered plant species in some non-fertilized grasslands. In many places in the easternmost Alps, for centuries this was especially the case for frequent grasslands on nutrient-poor, acidic soils at mesic sites, on silicate bedrock, at elevations of about 1,800 to 2,800 m, dominated by matgrass (*Nardus stricta*) or alpine curly sedge (*Carex curvula* subsp. *curvula*). They were economically highly relevant because plant communities harbored some specific target species, even though livestock did not like the fodder from matgrass swards, and quantities of hay or grazing resources in alpine curly sedge communities were very low (but the fodder quality in alpine curly sedge communities is high; see above in this section).[1996] Namely, *Valeriana celtica* subsp. *norica*, a highly esteemed and economically highly relevant organism with a long history of human gathering and long-distance trade, occurred and at places even abounded in matgrass and alpine curly sedge grasslands (see Section 8.1.1).[1997] Matgrass swards in montane, subalpine, and alpine vegetation belts can also abound in dotted-flowered gentian (*Gentiana punctata*, see Section 8.1.2), mountain arnica (*Arnica montana*, see Section 8.1.2), bilberry (*Vaccinium myrtillus*), and cowberry (*Vaccinium vitis-idaea*),[1998] offering significantly important gathering opportunities (see Sections 6.6.2.2 and 8.2.1.1). Calcareous non-fertilized grasslands at high elevations made possible economically relevant gathering of edelweiss (*Leontopodium alpinum*, Section 8.1.3.1) and yellow gentian (*Gentiana lutea*, Section 8.1.2), for instance.

1995 Babai, Molnár, & Molnár, 2014, 38–39; Aeschimann et al., 2004, vol. 2, 926–927.

1996 On the agricultural relevance of alpine curly sedge communities and the frequency of both communities in the Alps, see Brugger & Wohlfarter, 1983, 186, 188; Keller, 2004, 36.

1997 Gentner, 1932.

1998 Babai, Molnár, & Molnár, 2014, 35; Aeschimann et al., 2004, vol. 1, 624–627.

6.3.2 Moderately fertilized meadows

Despite the scarcity of manure in many parts of the Alps until the nineteenth and twentieth centuries, a limited but significantly important percentage of the total meadow area in the Alps had already been regularly fertilized in the long run before these changes (see the introduction to Section 6.3 and Section 6.3.0.1). In general, there were two main provenances of fertilizers for meadows before the large-scale use of mineral fertilizers: locally produced ones and manure transported to meadows from higher mountain pastures (see more in Section 6.3.2.1).[1999] Before the increased availability of fertilizers, moderately fertilized meadows in the Alps usually included the few most productive meadows close to the farms (see also Section 6.4.1) and many (sections of) distant meadows located between the summer mountain pasture complex and the farms, or, less frequently, in the belt of high summer mountain pastures (see Section 6.3.2.1). Where field rotation systems included mowed grassland fallows lasting for several years, these also functioned as moderately manured meadows (see Section 3.2.1).[2000] This was a specific type because these meadows were not fertilized in the grassland phase or just before it; soil nutrient availability in grassland fallows, comparable to that in moderately manured meadows, was an ecological legacy from field fertilization. The rest of agricultural nutrient intake to fallow grasslands was excrement in times of grazing, just like in most other meadows.

6.3.2.1 Usual fertilizer sources for meadows until the mid-twentieth century

Livestock excrement and barn manure were by far the most common fertilizers until the mid-twentieth century, and so they are dealt with first. Excrement was deposited in meadows during grazing in spring (Section 6.2) and autumn, and some dung from pastures, at least from extremely nutrient-rich sites around mountain cottages, was also transferred to nearby meadows in some cases. Regularly, local manure was applied to meadows from remote barns, where animals were kept overnight, mainly in the meadow complexes distant from farms and arable land and located in the montane vegetation belt. Some human excrement from people working there was also in the manure pile,[2001] which was not exceptional in a wider European context.[2002] In the immediate vicinity of farms, nutrient-rich water, leaching nutrients from the manure pile in the farmyard, or—especially since the nineteenth century—moderate quantities of largely

[1999] Compare Grass, 1948, 147–158; Wopfner, 1997, 295–296; Emanuelsson, 2009, 158; see also Peterken, 2013, 170–172, 191, 268.

[2000] Bätzing, 2015, 66–67.

[2001] Novak, 1955, 300–301; Šarf, 1974, 143–144; Niederer, 1996d, 157, 160–161; Wopfner, 1997, 296; Hodnik, 2016, 8, 30, 37–39, 44, 48. On human excrement: observations at Zajamniki in the Eastern Julian Alps; this was not exceptional; compare Kauter, 2002, 122–124.

[2002] Troßbach, 2008, 589–590.

urine-based liquid manure (unless absorbed by animal bedding) were also often applied in a nearby meadow (see Section 6.4.1). Thus, the few most productive meadows close to the farms were fertilized in many places, such as in the cadastral municipality of Bohinjska Češnjica in the Eastern Julian Alps, where they were fertilized with liquid manure in the 1830s.[2003] A selection of further historical sources of fertilizer for meadows is briefly presented at the end of this section.

Manure was applied to a selection of meadows in the Alps throughout the study period. A late medieval example is transport of livestock feces to montane meadows through irrigation channels from a mountain pasture complex in the Silandro Valley (Germ. *Schlandraun*, Ital. *Val di Silandro*) in the Venosta Valley (Germ. *Vinschgau*, Ital. *Val Venosta*), which was mentioned as an option in 1442, when two former farms were converted into a common mountain pasture complex.[2004]

Manure for meadows frequently came from livestock barns located in remote meadow complexes. Historical records of mountain meadows equipped with livestock barns[2005] suggest that such meadows were at least partly fertilized with local barn manure. In such barns, animals were kept overnight during spring and autumn grazing in these meadows. In addition to grazing in meadows, a pasture (as the only land-use category) was located nearby in some cases. This means that some nutrients consumed and synthesized by livestock in meadows or pastures during the day accumulated in barns, far distant from the farm, in the form of excrement. In addition, when hoarfrost formed at night during spring and autumn livestock keeping in remote barns, animals were kept inside longer in the morning until the hoarfrost melted; in the barn, they were mainly fed with hay, tree leaves (Section 8.2.2.1), and/or cereal straw where locally available. In the same manner, livestock were kept indoors in remote barns all day long when there was snow cover outside during the period when livestock were kept there. Such typical practices of keeping livestock in distant barns increased the quantities of manure available far from the farm.[2006] In many parts of the Alps, it was usual for sheep, goats, and cattle to even be kept in the winter in such remote barns located below high summer moun-

[2003] SI AS 176, Franciscejski kataster za Kranjsko (Franciscan Land Cadaster for Carniola), L107, cadastral municipality of Češnjica v Bohinju [i.e., Bohinjska Češnjica], cadastral appraisal protocols: Catastral-Schätzungs-Elaborat, Einleitung, §10, Cultur des bodens, & Schätzung des natural Ertrages: § 4: Wiesen. See also Kauter, 2002, 124–125, 167.

[2004] Grass, 1948, 149.

[2005] E.g., in the Sava Dolinka Valley of what is now northwest Slovenia such barns in possession of people from the villages of Kranjska Gora, Srednji Vrh, and Jesenice were recorded in 1636 (Mlinar (ed.), 2018b, 125, 134, 147).

[2006] Melik, 1950, 127; Novak, 1955, 293, 295, 300–301; Niederer, 1996d, 157, 159–162, 177; Wopfner, 1997, 319–320, 380, 382, 403; Hinterwaldner, 2002, 179; Hodnik, 2016, 7, 30, 35, 37, 39, 42, 44, 48, 76–77; author's interviews with Gašperin, 2017, and with Mencinger, 2017, both in Stara Fužina; with Grm, 2017, and with Milena Sodja, 2017, both in Srednja Vas in Bohinj.

tain pastures but above permanent settlements (although this was not a general rule in the Alps). In those regions of the Alps where the migratory character of rearing livestock was particularly pronounced, so that it included migration with ruminants even at the peak of winter, a herder stayed with livestock in the mountains (or a worker visited such places daily from the farm) until the hay from that meadow as well as leaf hay from trees, and straw in the case of adjacent distant mountain fields, had been consumed, instead of transporting fodder from remote meadows to the farm. When the fodder was consumed at a remote barn, migration with livestock to the farm at lower elevations or to the next barn, up to two hours distant, followed, often in deep snow. In some cases, the majority of the livestock were only kept at the farm one to two months a year; they spent the rest of the time in distant meadows (and pastures) with barns and in higher summer pastures. At exceptional places where the annual number of livestock migrations was very high, an individual large farm was in possession of fourteen to sixteen distant livestock barns (in some places even up to twenty, such as at Mayrhofen in the Ziller Valley (Germ. *Zillertal*) in Tyrol and its surroundings in the mid-nineteenth century), and a smaller farm up to five; this was the case in the community of Gerlos in the Ziller Valley. On some farms, goats stayed in the same remote barn throughout winter. Regions where a livestock rearing system was common that involved ruminants being kept in distant barns in the winter include, for example, parts of southern Allgäu according to eighteenth-century data, Wallis (Francoprovençal *Valês*, French *Valais*), Graubünden (Ital. *Grigioni*; Romansh *Grishun*), Glarus, and Vorarlberg, parts of Tyrol (the Kitzbühel Alps [Germ. *Kitzbüheler Alpen*] and Tux Alps [Germ. *Tuxer Alpen*] in particular), some parts of the Alps in what is now northern Italy (e.g., Alto Adige—i.e., a part of South Tyrol—and neighboring areas), the Bovec region in the Upper Soča catchment, and some parts of Styria before the major changes in farming in the twentieth century.[2007] Fragmentary data prove that the system of keeping livestock in remote barns in winter has a long history. For instance, a remote livestock barn at Beilstein southwest of Obergurgl in the Ötz Valley was radiocarbon dated to about 1665.[2008] Furthermore, the livestock of some farms was kept in distant barns during a significant part of the winter in some other regions of the Alps, where this was not as common as in the regions mentioned above, or the duration of staying with livestock in the mountains in winter was shorter than in the cases presented above. For instance, livestock were kept by some farms in remote barns in the montane belt at some places

[2007] Novak, 1955, 293, 295; Novak, 1970, 349, 357, 363; Hubatschek, 1992, 40, 84–85; "Bildteil," 1996, 410–411; Niederer, 1996c, 53, 63; Niederer, 1996d, 157, 159–162, 177; Wopfner, 1997, 247–248, 306, 380, 382–383, 395, 403–404; Keller, 2004, 21, 120; Jäger, 2006, 57; Zenhäusern, 2008, 190; Hodnik, 2016, 7, 39, 73; Kravanja & Komac, 2023, 19; Novak, 2024, 39, 42, 259; author's interview with Škantar, 2017, in Srednja Vas in Bohinj. At many other places, all the hay used to be transported from mountain meadows to the farms (Grass, 1948, 213).

[2008] Patzelt, 1996, 66.

in the Julian Alps as proven in the sixteenth,[2009] seventeenth or eighteenth,[2010] mid-nineteenth,[2011] and mid-twentieth century.[2012] Similar was the situation at some places in the Western Karawanks. On January 2nd, 1777, for instance, in deep snow, the sheep of farmers from the settlement of Srednji Vrh (950 m) in the Karawanks were kept in barns at Hudi Hlevi at approximately 1,350 to 1,420 m, more than an hour's walk from the farms; a livestock barn had already been mentioned in preserved records in one of the meadows in that same meadow complex in the seventeenth century.[2013] Livestock rearing in remote

2009 SI AS 721, Gospostvo Bled (The Seigneury of Bled), Iustitialia, fasc. 34: Lawsuit between Krištof Wuria and the Nomenj community over grazing rights "in der [!] Stiegen" 1613/1614; localization of these remote meadows at the bottom of the unpopulated Sava Bohinjka Valley between Nomenj and Bohinjska Bela is based on a comparison with toponyms on early nineteenth-century cadastral maps (SI AS 176, Franciscejski kataster za Kranjsko [Franciscan Land Cadaster for Carniola], L172, cadastral municipality of Nomenj, map sheets A02 and A07). It is clear that the man keeping goats in a remote livestock barn until the local hay was consumed died in the late sixteenth century. Keeping goats is documented there in at least one sixteenth-century winter. Several data support the hypothesis that much hay in Bohinj was transported home from remote meadows and not consumed in the mountains in winter, as illustrated by the following selection of examples. (1) The barn dealt with in the previous paragraph was no longer used this way later (SI AS 721, Iustitialia, fasc. 34: Lawsuit between Krištof Wuria and the community of Nomenj over grazing rights "in der [!] Stiegen" 1613/1614). (2) The exceptionally snowy winter in early 1608 (Zwitter, 2013b, 341, 343, 362, 369; Zwitter, 2020, 72–73) made the maintenance of a passable connection on the slope linking the villages of Gorjuše and Nomenj a vital issue. According to the court settlement between these two villages from February 1608, a passable connection had to be established after snowfalls, also in deep snow. It seems that people from Nomenj were particularly interested in the passability of that route in deep snow. Namely, they were the ones that had to make their way up the hill first—a clause that was only valid in deep snow. Cargo types transported there on snow are not mentioned in this court settlement (SI AS 721, Iustitialia, vol. 13, February 11th, 1608), but hay sled transport from the Pokljuka Plateau to Nomenj was most likely one of the relevant purposes this connection served on snowy winter days, and so hay from the mountains was transported to the farms in the valley. (3) In the mid-twentieth century, only a few farms in Bohinj kept their livestock in the barns at higher elevations until Christmas; the livestock of most farmers returned to the farms in autumn (Hodnik, 2016, 7, 38–39).

2010 The documented grazing of goats in the winter in Ukanc (SI AS 721, Gospostvo Bled [The Seigneury of Bled], Iustitialia, fasc. 33, Lawsuit between the communities with centers in Savica and Studor over grazing rights at Lake Bohinj and in Ukanc, 1655–1666: Discurs über die geman [!] Vkaniz, no date) can be explained only as a supplementary source of fodder because these goats must have been kept in barns for a great deal of the winter in the cold and wet climate of the region (compare Emanuelsson, 2009, 29, 122, 139, 155, 158; Bätzing, 2015, 37; *Environmental Atlas of Slovenia*, Climate, Precipitation, Mean annual corrected precipitation 1971–2000 and Mean annual corrected precipitation 1981–2010).

2011 Novak, 1955, 293, 295; Novak, 1970, 357, 363; Kravanja & Komac, 2023, 19.

2012 Melik, 1950, 178–179; Novak, 1970, 349; Novak, 1978, 57; Kačičnik Gabrič, 2014, 106; Hodnik, 2016, 7, 39; Kravanja & Komac, 2023, 19; Novak, 2024, 39, 42, 259.

2013 Hlebanja & Hlebanja, 2006, 24–25, 54, 69. Localization: *Arcanum Maps*, Cadastral Maps, Habsburg Empire – Cadastral Maps (XIX. [sic] Century), cadastral municipality barns in winter thus had a long history in several regions of the Alps, though its relative role within the entire livestock rearing system differed significantly between some of these regions, and it might to a certain degree have also changed over time before its abandonment in the second half of the twentieth century.

The data presented make it clear that non-negligible quantities of manure were produced and stored in remote livestock barns adjacent to meadows. Even though manure was primarily applied to arable fields before the changes briefly presented in the introduction to Section 6.3, manure was usually applied to the nearby meadows if there were no arable fields in the vicinity of remote livestock barns. For example, this sufficed to fertilize a section of the remote montane meadow every year or once every several years where quantities of manure were low, so that it was only taken from the barn every third or fourth year. Thus at least a limited part of the meadow close to the livestock barn (often on the slope below the barn) was manured.[2014] Some farmers, however, purposefully used significantly more animal bedding in remote barns than required by livestock with the aim of producing more manure, and at such places the entire montane meadow, even if large (2 ha, for instance; ten cows of a small breed were kept in the remote barn in this case) was (very moderately) fertilized, such as in some parts of Konjska Dolina (around 1,300 m or a bit higher) in the Eastern Julian Alps[2015] (see also seventeenth-century data on the Voje Valley in Bohinj in Section 6.1.2.3). To explain the relevance of a high amount of bedding used in such remote barns: tree leaves, the most frequent animal bedding type in several regions of the Alps with deciduous forest, added significantly more nutrients to the manure than cereal straw, which was not available as bedding in many remote meadows.[2016] In Bohinj, hazel (*Corylus avellana*) leaves were believed to

of Gozd (Germ. *Wald*), 1820, and orientation based on comparison with *Environmental Atlas of Slovenia*, Spatial Units, Land Cadaster and *Environmental Atlas of Slovenia*, Backgrounds, DEM [Digital elevation model].

2014 Šarf, 1974, 143–144; Wopfner, 1997, 248, 319–320, 323, 381, 403, 440; Hubatschek, 1992, 84–85; Niederer, 1996d, 157, 161; Kauter, 2002, 145; Keller, 2004, 21; Hodnik, 2016, 35, 42, 44, 76; author's interviews with Gašperin, 2017, with Mencinger, 2017, and with Urh, 2017, all in Stara Fužina; with Grm, 2017, and with Milena Sodja, 2017, both in Srednja Vas in Bohinj; with J. Smukavec, 2017, in Podjelje; with A. and M. Taler, 2017, in Studor.

2015 Recollections from Konjska Dolina: author's interview with A. Soklič, 2017, in Podjelje.

2016 Machatschek, 2002, 142, 259–271, 275, 281–282, 284–289; Kačičnik Gabrič, 2016, 504; author's interviews with Kunavec, 2017, and with Gašperin, 2017, both in Stara Fužina. SI AS 176, Franciscejski kataster za Kranjsko (Franciscan Land Cadaster for Carniola), L22, cadastral municipality of Podhom, cadastral appraisal protocols: s.n., 22/4, 19, 83; SI AS 176, L30, cadastral municipality of Nemški Rovt, cadastral appraisal protocols: s.n., 30/4, 19; SI AS 176, L35, cadastral municipality of Blejska Dobrava, cadastral appraisal protocols: s.n., 35/4, 19, 83; SI AS 176, L56, cadastral municipality of Bohinjska Bistrica, cadastral appraisal protocols: s.n., 56/4, 19; SI AS 176, L72, cadastral municipality of Gorjuše, cadastral appraisal protocols: s.n., 72/4, 18–19, 83; SI AS 176, L107, cadastral municipality of Češnjica v Bohinju, cadastral appraisal protocols: s.n., 107/4, 19, 83; SI AS 176, L157, cadastral municipality of Srednja Vas v Bohinju, cadastral appraisal protocols:

make particularly good manure, but they were not the most widely used bedding because the availability of hazel leaves was limited, and so beech (*Fagus sylvatica*) leaves, for instance, were widely used.[2017]

Indeed, manure production to maintain the productivity of meadows was sometimes one of the main motives for feeding livestock in remote barns outside the grazing period, apart from the difficult hay transport to the valley.[2018] Thus, representatives of villages in the cadastral municipality of Gorjuše, on the Pokljuka Plateau in the eastern Julian Alps in the immediate vicinity of Konjska Dolina, mentioned in the 1830s that they were manuring their montane meadows.[2019] In another example located nearby, in the mid-nineteenth century—namely, at Zajamniki (about 1,250 m)—one of the motives for feeding locally harvested hay in these remote livestock barns in early spring and late autumn was to keep the meadows productive; apart from barn manure, livestock excrement from grazing in meadows in spring (Section 6.2) and autumn also mattered in this context.[2020] Fertilizing remote meadows with the manure from the local barn was not unique of the Alps. For instance, in similar circumstances of remote livestock barns with no or very few arable fields in the surroundings, meadows also were manured at many seasonally inhabited settlements in the Carpathians[2021] and in Scandinavia.[2022]

So far remote barns were given special attention because most manure in several regions of the Alps that was applied to meadows (and not to arable land)

---

s.n., 157/4, 19, 83; SI AS 176, L172, cadastral municipality of Nomenj, cadastral appraisal protocols: s.n., 172/4, 19; SI AS 176, L179, cadastral municipality of Zgornje Gorje, cadastral appraisal protocols: s.n., 179/4, 19; SI AS 176, L210, cadastral municipality of Poljšica, cadastral appraisal protocols: s.n., 210/4, 19, 83; SI AS 176, L227, cadastral municipality of Rečica, cadastral appraisal protocols: s.n., 227/4, 19; SI AS 176, L242, cadastral municipality of Savica, cadastral appraisal protocols: s.n., 242/4, 19; SI AS 176, L284, cadastral municipality of Studor, cadastral appraisal protocols: s.n., 284/4, 19, 83; SI AS 176, L310, cadastral municipality of Spodnje Gorje, cadastral appraisal protocols: s.n., 310/4, 19, 73, 83; SI AS 176, L322, cadastral municipality of Višelnica, cadastral appraisal protocols: s.n., 322/4, 19; SI AS 176, L335, cadastral municipality of Bohinjska Bela, cadastral appraisal protocols: s.n., 335/4, 19, 83; SI AS 176, L343, cadastral municipality of Selo pri Bledu, cadastral appraisal protocols: s.n., 343/4, 19.

2017 Author's interviews with Kunavec, 2017, and with Gašperin, 2017, both in Stara Fužina. Beech leaves were also typical animal bedding in the early nineteenth century, as recorded in SI AS 176, Franciscejski kataster za Kranjsko (Franciscan Land Cadaster for Carniola), L30, cadastral municipality of Nemški Rovt, cadastral appraisal protocols: s.n., 30/4, 19; SI AS 176, L56, cadastral municipality of Bohinjska Bistrica, cadastral appraisal protocols: s.n., 56/4, 19; SI AS 176, L242, cadastral municipality of Savica, cadastral appraisal protocols: s.n., 242/4, 19.

2018 Janišová et al., 2023, 13, 15; Novak, 1955, e.g., p. 295.

2019 SI AS 176, Franciscejski kataster za Kranjsko (Franciscan Land Cadaster for Carniola), L72, cadastral municipality of Gorjuše, cadastral appraisal protocols: s.n., 72/4, 51.

2020 Novak, 1955, 295.

2021 Janišová et al., 2023, 9–10, 13, 15.

2022 Emanuelsson, 2009, 158.

originated there, but attention is now turned to the practice of corralling sheep due to the most significant agriculturally desired ecological impacts of this manner of livestock keeping in meadows, as explained in Section 6.3.2.2. In many places, sheep were corralled in meadows to spend the night in pens for the sake of manure. Data from the nineteenth and early twentieth century reveal that corralling sheep in meadows was usual in several parts of the Alps, such as Tyrol,[2023] the Julian Alps, and the Western Karawanks;[2024] it is also documented in the Carpathians, where evidence reaches back to the thirteenth century.[2025] Distant locations where corralling sheep in meadows was usual in the nineteenth century suggest that the practice is old and that it used to be widespread in the Alps and beyond when sheep still belonged to the most frequent livestock type there (see Section 6.3.1.1). Corralling sheep in meadows is documented, for instance, in the Upper Inn Valley by a contract from 1842, but in this case its roots reach at least to the eighteenth century. According to this document, the sheep of farmers from Fiss had to be corralled overnight in a specified complex of meadows during the autumn grazing period after the return from the summer mountain grazing complex until weather conditions required the sheep to be kept in barns at night. Similarly, in the Upper Ötz Valley, sheep from the Rofen mountain pasture had to be kept overnight in mountain meadows after mowing.[2026] Information from the Julian Alps makes it clear that the fences of such corrals were relocated within the meadow or the meadow complex after a certain section was sufficiently fertilized.[2027] Data from Tyrol demonstrate that sheep were corralled at a different location every night, at least in some cases.[2028]

In some other places, montane meadows were long fertilized using manure transported there from a higher mountain pasture. Manure from the Langetsberg complex of mountain pastures in the Kauns Valley (Germ. *Kaunertal*, in the Upper Inn catchment), for instance, had to be regularly applied to meadows below these pastures; in return, the farmers that held the meadows there had to allow the livestock from these pastures to retreat to those meadows in the case of untimely snow in higher pastures. This rule was valid until 1935. Exceptionally, there were long wooden pipes to transport manure in liquid form from higher pastures across steep slopes to meadows, such as at the Kagring Alp (Germ. *Kagringalm*) northeast of Kitzbühel (on the southwestern slopes of Kitzbüheler Horn), where pipes about two kilometers long were maintained for this purpose, at least between the late eighteenth and twentieth century. It is probably no coincidence to find such an exceptional form of manure transport precisely in the Kitzbühel Alps, an area where agriculture was exceptionally

---

2023 Hubatschek, 1992, 175.
2024 Novak, 1942, 92; Novak, 1970, 357; Šarf, 1974, 143.
2025 Babai, Molnár, & Molnár, 2014, 50, 88, 112, 120–121; Janišová et al., 2023, 11, 13.
2026 Grass, 1948, 156–158; Hubatschek, 1992, 175.
2027 Novak, 1970, 357; Šarf, 1974, 143.
2028 Hubatschek, 1992, 175.

strongly oriented toward cattle husbandry and meadow farming, comparable to the larger area of the northern Alps from Savoy to Allgäu. However, although manure from summer mountain pastures was applied to lower mountain meadows in many cases, this was not a general practice—in some cases it was applied to the mountain pasture itself, and so removal of dung from common pastures was prohibited in some places, such as at Untermieming in the Upper Inn Valley in 1618. Manure from some other common pastures was transported to farms, or it was not used at all.[2029]

Ash and soot were also applied to meadows to a certain degree.[2030] Application of ash to meadows has a long history; it was already recommended by Roman agricultural authors, who also had an influence later.[2031] Application of ash to meadows was usually recommended in agricultural literature, at least until the nineteenth century. It was seen as a fertilizer on the one hand, and on the other hand as a means of moss suppression (effective in certain conditions, but not always).[2032] Farmers' experience-based knowledge in various regions of Central Europe, such as the Julian Alps and Upper Austria (Germ. *Oberösterreich*), also emphasizes that the application of ash every year significantly reduces the abundance of matgrass (*Nardus stricta*; see Section 6.3.1.2).[2033] Experience from Upper Austria emphasizes that ash stimulates the growth of legumes (*Fabaceae*), which is a consequence of the high potassium content in ash. Potassium as a plant nutrient usually limits the suitability of non-fertilized sites for many plant species, legumes in particular, because a lack of nitrogen at the site is not an obstacle for them due to the airborne nitrogen fixation going on in their root nodules (see the introduction to Section 6.3.1). Consequently, nitrogen binding in legumes' root nodules stimulates the growth of agriculturally more desired grasses.[2034]

Nutrient-rich mud from depressions in forests was also used as a fertilizer,[2035] and mud washed from the roads and containing nutrients played a similar role. For example, it was collected in pits where the gutters adjacent to the roads in the villages and outside of settlements ended or the gutters crossed these spots so that the water came to rest there and deposited its fine, nutrient-rich load. Such depos-

[2029] Grass, 1948, 120–121, 147–156 (localization of the Kagring Alp: *Arcanum Maps*, Cadastral Maps, Habsburg Empire – Cadastral Maps (XIX. [sic] Century), cadastral municipality of Kitzbühel, 1855), 158, 208–213; Wopfner, 1997, 267, 385–386; Ledinek, 1999, section Odgon in paša; Keller, 2004, 20, 58, 66; Novak, 2024, 96; contextualization of the farming type in the Kitzbühel Alps is based on Emmanuel de Martonne's map of farming types in the Alps around 1900 reprinted in Mathieu, 1996, 142–143.

[2030] Pfister, 1985a, 27; Pils, 1994, 24–25; Kauter, 2002, 120–122; Torkar Tahir, 2012, 136.

[2031] See, e.g., Winiwarter, 1999; Zwitter, 2020, 73–76.

[2032] Kauter, 2002, 105–106, 120, 122, 125, 131; Babai & Molnár, 2014, 126; Zwitter, 2020.

[2033] Pils, 1994, 95, and experiences at Ši above Srednja Vas in Bohinj (author's interview with R. Sodja, 2022, in Stara Fužina).

[2034] Pils, 1994, 95.

[2035] Leibundgut & Vonderstrass, 2016a, 109.

its were excavated, and in many cases they were applied to meadows. Wastewater from households was also applied to a nearby meadow in many cases.[2036]

A significant source of nutrients in many grasslands in the Alps was also proper irrigation (see Section 6.4).

6.3.2.1.1 Application of barn manure to meadows

Focusing on fertilization frequency and intensity before the general increase in fertilizer availability in mid-twentieth century, information referring to moderately manured meadows of traditional rural Csángós communities in Ghimeş in the Eastern Carpathians demonstrates that moderately manured meadows in the montane vegetation belt were manured once every two to three years, for instance. These data from Ghimeş correspond with the most usual frequency of manuring in distant montane meadows in the Alps of Carinthia (e.g., at elevations of 1,400 to 1,700 m).[2037] The manuring frequency in those meadows that were fertilized once every three years would correspond with the manuring frequency in arable land under a three-field rotation system that was common in several parts of the Alps in much of the last millennium by the eighteenth century.[2038] Because most meadows were not manured more frequently than arable land, this means that the same meadow section was normally manured no more than every third year. More frequent was fertilization in some meadows close to the farms,[2039] such as those where the water leaching nutrients from barn manure was applied, in meadows adjacent to some remote livestock barns,[2040] and in a higher percentage of meadows in the regions or farm economies (nearly) specialized in livestock rearing (see Sections 3.1 and 10.1.2.1) or with particularly extensive swidden cultivation where the ash in fields possibly allowed saving more manure for meadows. In contrast to these cases of frequent fertilizing early in history, manure was applied to some meadows less frequently than presented so far; for example, every sixth year, even in meadows relatively close to farms, such as at Marebbe (Ladin *Mareo*; Germ. *Enneberg*) in southeastern Tyrol in the first half of the nineteenth century.[2041] An increased specialization toward animal farming in the Alps in general since the nineteenth century made possible an increased frequency of fertilizer ap-

[2036] Pfister, 1985a, 27; Meilleur, 1986, 109; Granda, 1994, 56; Leibundgut & Vonderstrass, 2016a, 109; Kauter, 2002, 121–122; Renes et al., 2020, 117; author's interviews with Gašperin, 2017, in Stara Fužina, with F. Mlakar, 2017, in Bohinjska Češnjica, and with L. and M. Odar, 2017, in Studor. See also Peterken, 2013, 140.

[2037] Keller, 2004, 29, 91; Babai, Molnár, & Molnár, 2014, 17–20, 24, 37, 89, and Babai & Molnár, 2013, 3.

[2038] Pfister, 1985a, 26–28, 106–116; Grafenauer, 1970c, 228–229, 236, 240–245; but see Mathieu, 1996, 147–148, 156, 159.

[2039] E.g., Kačičnik Gabrič, 2016, 506; compare Janišová et al., 2023, 15.

[2040] Compare Kačičnik Gabrič, 2016, 506.

[2041] Wopfner, 1997, 292–293; the chronological definition of what "a century ago" means in Wopfner's text follows Meixner & Siegl, 2004, 232–233, 235.

Figure 88: Oxen rubbing barn manure into a meadow. Padaun, Vals Valley (Germ. *Valsertal*), Tyrol, 1950. Photo: Erika Hubatschek. Hubatschek, 1992, 135. © Irmtraud Hubatschek.

plication, and fertilization changed thoroughly with the general accessibility of mineral fertilizers starting in the mid-twentieth century (see Sections 1.2, 10.1.2.1, the introduction to 10.1.2.2, and Section 10.2.1).

Barn manure was piled in meadows in small heaps in late autumn, winter, or spring, crushed and distributed around the meadow with a pitchfork (only on very steep slopes barehanded; see below) in autumn, in winter when it was piled on snow (e.g., in December at around 1,800 to 1,900 m at Stabele in the Ötz Valley), or in spring, together with the dung deposited in meadows during autumn and spring grazing. As long as spring grazing took place in meadows (Section 6.2), barn manure could not be piled in small heaps and distributed in meadows earlier than the end of spring grazing. Since the abolition of spring grazing (Section 6.2.3), some farmers have fertilized their meadows in autumn or winter; manure spread in the autumn decomposes by the time it is finely crushed in the meadow in spring. Others have maintained the habit of spring manuring, one of the reasons being late autumn grazing in some meadows, so that some farmers do not have time for manure application before snow.[2042]

Manure crushed and distributed around the meadows in autumn or spring was finely rubbed into the meadow in spring. Until the mid-twentieth centu-

[2042] For instance, Wopfner, 1997, 296–298; Hubatschek, 1992, 134–135; Makarovič, 2005, 109, 124, 298; Babai, Molnár, & Molnár, 2014, 89, 113; Hessenberger, 2022, 154, 158 (concerning the localization of Stabele, see Section 6.1.2.3); author's interviews with Kunavec, 2021, in Stara Fužina, and with M. Gartner, 2022, in Studor.

ry, this was usually done by draft livestock (Figure 88), except for very small parcels or steep slopes, where manure was rubbed into the meadows in a more labor-intensive manner, with rakes.[2043] Depending on the region and farm, rubbing the manure into meadows with draft livestock was done using variously shaped implements with slightly diverse ecological impacts. Evidence from the early twentieth century demonstrates that farmers in the Eastern Alps most often used one of the following implements dragged by livestock all over a fertilized meadow: (1) a two-to-four-meter-long bunch of branches (e.g., hazel, *Corylus avellana*; Norway spruce, *Picea abies*; or larch, *Larix decidua*) or thorny bushes (barberry, *Berberis* spp.; blackthorn, *Prunus spinosa*; or buckthorn, *Rhamnus* spp.) placed parallel next to each other in the direction of movement and dragged by livestock; the branches or bushes were fixed and tied to livestock at the thicker end of the wood, but weighted down at the bushy final section, which was often tied to a log placed on the bushy bunch to weight it down and to keep the form of the bushy bunch rubbing the manure flatter; alternatively, a Norway spruce (*Picea abies*) tree top or tops of a few spruce trees with many thick branches (weighted down or not) were dragged by livestock; (2) a wickerwork from branches or branches and bushes interwoven in a ladder-like frame or in a field harrow placed upside down, so that the harrow acted only as a weight and frame to secure the branches; (3) two or more bunches of branches or bushes placed horizontally, transversally to the movement direction, fixed one behind the other and often weighted down; (4) bunches of branches (e.g., willow, *Salix* sp.; hazel, *Corylus avellana*; and alder, *Alnus* sp.) fixed vertically in a wooden frame (a wooden harrow with teeth or without them), acting as a brush (Figure 88); (5) branches interwoven between the field harrow's horizontal bars, so that only a few centimeters of the harrow teeth were lower than the branches; this way, the harrow teeth together with the branches contributed to manure-rubbing, and the branches prevented significant damage to meadow turf by the harrow teeth.[2044] Various types of these simple meadow harrows evidently have a long history. For instance, a meadow harrow, evidently a durable specialized implement serving this purpose only (not just branches or a spruce top), was documented in a probate inventory of the large Fasching farm east of Feldkirchen in Carinthia in 1673.[2045] Thus, for

[2043] Wopfner, 1997, 297–298; Makarovič, 2005, 109, 124, 298; Babai, Molnár, & Molnár, 2014, 89; author's interviews with Kunavec, 2017, in Stara Fužina, with A. and M. Taler, 2017, in Studor.

[2044] Šarf, 1955, 89; Dosedla, 1971; Hubatschek, 1992, 135; Wopfner, 1997, 298; Makarovič, 2005, 124, 293, 298; author's interviews with Gašperin, 2017, and with Kunavec, 2017, both in Stara Fužina, and with A. and M. Taler, 2017, in Studor; see also the same purpose of use of blackthorn branches, spruce branches and spruce tops (or young spruce trees), more sophisticated meadow harrows, and a combination of branches and a harrow in the Carpathians in Babai, Molnár, & Molnár, 2014, 89; Ivaşcu, Öllerer, & Rákosy, 2016, 44; and Janišová et al., 2023, 15.

[2045] Note that Dosedla's statement that in some places regular meadow cultivation was only introduced in the nineteenth century or at the transition from the nineteenth to twencenturies, diverse implements for rubbing manure into meadows coexisted in the Alps, from very plain ones (a spruce top or weighted-down branches) to more sophisticated harrows. Apart from cultural transfer, one of the factors influencing the equipment available was general economic conditions: more sophisticated meadow harrows were more likely available at larger farms than at small holdings, as a parallel from the mid-twentieth century suggests.[2046] This hypothesis finds partial confirmation in the meadow harrow in the inventory of the large Fasching farm presented above. The longevity of use of some very simple implements until the mid-twentieth century suggests that a skilled farmer was able to make relatively good use of them.

The ecological impacts of the various implements used to rub the manure into the meadow differed somewhat. Rubbing the manure into steep meadows with a rake allowed action best adapted to local site specifics within the meadow; a farmer seeing a patch of remnants of dead plant would press the rake more heavily to remove them at the same time as finely crushing and distributing the manure, but this practice was most labor intensive. Branches placed horizontally acted most uniformly if weighted down by a field harrow (types 2 and 5 above) and least uniformly if the pressure on the branches was less equally distributed, such as if they were weighted down by a few large stones (a variety from type 1 above).[2047]

The main aim of rubbing the manure into meadows was further finely crushing the manure lumps in spring so that the manure fully decomposed by mowing time. Otherwise, remaining dung particles would be crushed during haymaking, especially during raking, and so the hay would stink of manure, and livestock—cattle in particular—would not eat such hay.[2048] Sheep manure is particularly compact; it had to mature and compost for more than a year before being applied to meadows.[2049] At least as long as spring grazing (sheep grazing in particular) took place in meadows, and to a lesser degree until mowing was carried out by scythe (closer to the turf than with a motorized mower), rubbing the manure (probably done at the same time as rubbing the more compact dung from spring grazing *in situ*) had to be thorough to prevent hay from stinking (see also Section 6.2.2).[2050] A rule documented in Tyrol says: "The nutrient-poorer the meadow and the higher its elevation, the finer the manure has to be crushed in order

tieth century (Dosedla, 1971, 2) is erroneous; the reference he cited does not prove this. Compare Dinklage, 1966b, 130–131: *Wiesenschlapfe* is mentioned in the inventory from 1673. *Schlapfen* means *schleppen* 'to drag' (Lexer, 1862, 219), and *Wiesenschleppe* means 'meadow harrow' nowadays.

2046 Impacts of the holding size: Šarf, 1955, 89.

2047 Dosedla, 1971, 4–6.

2048 Šarf, 1955, 89; Wopfner, 1997, 297–298; Babai, Molnár, & Molnár, 2014, 89; author's interview with T. Soklič, in Spodnje Podjelje, 2021.

2049 Author's interview with Kunavec, 2023, in Stara Fužina.

2050 Compare the author's interview with Kunavec, 2023, in Stara Fužina.

... not to appear in the hay."[2051] Crushing manure finer at higher elevations was adapted to the impacts of climate on the decomposition of barn manure. A farmer in Carinthia that tried to fertilize his meadow with sheep manure at 1,700 m in the twentieth century reported that his attempt at gaining an economic benefit failed because the manure did not decompose quickly enough.[2052]

This means that there has always existed an upper limit of meadows fertilized with barn manure in the Alps, located higher than the great majority of meadows, but nonetheless lower than the highest meadows in many parts of the Alps (see Section 3.1). Twentieth-century data from Carinthia, where meadows used to reach an elevation of at least 2,400 m, demonstrate that the highest manured meadows were located at about 1,700 to 1,800 m.[2053] In the Ötz Valley in Tyrol, the highest moderately manured meadows significantly exceeded an elevation of 1,900 m, and most likely they reached more than 2,000 m.[2054] In the Ötz Valley, the upper line of historical cereal cultivation also reached an extremely high elevation; it exceeded 1,700 m.[2055] This means that, similar to the upper line of cereal cultivation, the upper forest line (see Section 3.1),[2056] and the upper line of second-crop hay mowing in the Alps (Section 6.1.2.3), the upper limit where manuring meadows made sense varied within the mountain range as a result of interrelated environmental and human factors. Differences in climate are an example of a relevant environmental factor.[2057] One of the main human factors concerns the spatial distribution of the barns where livestock were kept. Namely, as long as transport up the hill depended on muscle power, farmers would never transport barn manure, which is heavy cargo, up a hill over more than short distances (see the introduction to Section 6.3). Thus, barn manure could only be applied to remote meadows if it was available locally or where it was transported from higher elevations.

Slope inclination (in terms of steep slopes) was not an insurmountable challenge for barn manure application unless it involved upslope transport, and so many steep meadows below animal barns or on the slopes close to them were

2051 Wopfner, 1997, 297–298.

2052 Keller, 2004, 58.

2053 Keller, 2004, 13, 27–28, 32, 58.

2054 Wopfner, 1997, 325; localization of manured meadows close to livestock barns at Kaiserberg northeast of Kaisers and at Kleble (this meadow complex is mapped in Figure 4) east of Sölden in the inner Ötz Valley is based on a comparison of *Arcanum Maps*, Cadastral Maps, Habsburg Empire – Cadastral Maps (XIX. [sic] Century), cadastral municipality of Sölden, 1856) with *tiris: Tiroler Rauminformationssystem*, Hintergrund: Basiskarte Beschriftung, & Basisthemen: Kataster.

2055 Scharr, 2001, 22; *Arcanum Maps*, Cadastral Maps, Habsburg Empire – Cadastral Maps (XIX. [sic] Century), cadastral municipality of Sölden, 1856.

2056 Bätzing, 2015, 40, 54; Gams, 1960 and Zwitter, 2015d.

2057 E.g., general considerations on environmental diversity in the Alps in Viazzo, 1989, 16–19. Concerning the possible impact of historical climate change on the upper line of meadow manuring, compare Pfister, 1985a, especially page 127: the dynamics of manure decomposition depended on the weather and climate.

Figure 89: A species-rich meadow at Vržnica, moderately fertilized with matured barn manure, south of the village of Studor, Bohinj.[2058] Photo: Žiga Zwitter, April 9th, 2021.

also manured in the history. On some extremely steep slopes, manure piles were distributed barehanded, instead of by fork, to cope with gravity more efficiently, and wooden boards were placed on the lower meadow margin to catch the manure that rolled down the slope to prevent losses of this valuable resource.[2059] In 1950, Erika Hubatschek's fieldwork at Perfall (in the Sellrain Valley [Germ. *Sellraintal*] west of Innsbruck) documented that on meadows with a slope inclination of 40° barn manure was first finely crushed and then transported down the slope where small dung particles were distributed by being thrown by hand; larger pieces of dung would have rolled down the slope.[2060]

6.3.2.2 Ecological and economic impacts of fertilizer application

Moderate manuring (Figure 89) has resulted in a high abundance of species that are highly relevant as livestock fodder, and the general nutritive value of hay is higher than at most non-fertilized sites. Farmers' awareness of the high hay quality from such meadows (in line with contemporaneous standards) in the southeast of the Alps is proven at least since the early seventeenth century.[2061] Moderate application of matured barn manure still maintained habitats

[2058] The intensity of fertilization in distant meadows in the Voje Valley in 1978 was similar; see Novak, 2024, photos on p. 106.
[2059] Wopfner, 1997, 298.
[2060] Hubatschek, 1992, 136.
[2061] Zwitter, 2020, 66.

of many herbs relevant for livestock health and wellbeing, and for the quality of livestock products (Section 10.1.2.2.3). This management practice raised hay quantities in comparison with non-fertilized grasslands; quantities of first-crop hay increased, and manuring usually also made an additional mowing possible (see Section 6.1.2.3), allowing the harvest of particularly nutritive second-crop hay (see below). In addition to the direct impacts of manure, ecological processes in moderately manured meadows led to further indirect economically favorable impacts meeting the standards of ecological intensification (Section 10.1.2.2.3). As a consequence of the direct and indirect impacts of manure application, grassland turf became denser and plants grew taller. Moderate manuring also had further economic advantages: it added organic matter to the soil and soil nitrogen availability has no longer limited plant growth; consequently the limiting effects of further factors (e.g., hazards of drought and of temperatures that are too low) have decreased in comparison with non-fertilized grasslands.[2062] However, higher nutrient availability made grasses of moderately manured meadows (also including mowed grassland fallows at the beginning of the grassland stage; see Section 3.2.1), which grew taller (e.g., 1 m tall) than at non-fertilized sites, much more prone to lodging before first-crop hay mowing than grasses at non-fertilized sites.[2063] This occasionally increased the labor intensity of first-crop hay mowing. Taller plants and thicker plant growth in moderately manured meadows also required more time and (an) additional first-crop hay turning during drying.[2064] Often, the hay from moderately fertilized meadows was placed on stacks to dry or in a hayrack, especially in the case of unstable weather (see Section 4.1.1.3). Thus, longer plants required additional work to decrease the risk or extent of damage posed by rain in comparison with non-fertilized meadows.

To understand the main economic impacts, an insight into the main ecological impacts of fertilizer application is needed before the economic impact can be presented in greater detail. Thus, this section starts with the ecological impacts of moderate manuring, and the main general economic impacts of fertilization follow.

From an ecological perspective, moderate historical fertilization has significantly impacted meadow species composition through influences on site characteristics,[2065] but also through direct dispersal of undigested seeds from manure, such as white clover (*Trifolium repens*) seed: "60 tons of sheep dung may contain

[2062] Pils, 1994, 171, 304; Peterken, 2013, 222; Sadar, 1956, 55, 57–58, 66–67, 105–107; Šarf, 1974, 134; Babai, Molnár, & Molnár, 2014, 87, 89; Ivaşcu, Öllerer, & Rákosy, 2016, especially p. 43; Janišová et al., 2023, 18. See also data below in this section and compare Section 6.3.1.

[2063] Babai, Molnár, & Molnár, 2014, 36, 109–110; Bavcon, Ravnjak, & Praprotnik, 2019, 14, 16–17.

[2064] Keller, 2004, 15.

[2065] Peterken, 2013, 222; Sadar, 1956, 55–58, 66–67, 105–107.

as many as 825,000 seeds."[2066] Moderate manuring changed the species composition and, except for some grasslands in extreme site conditions, it reduced plant species richness in comparison with non-fertilized meadows. Several species, such as mountain arnica (*Arnica montana*; see also the text of Section 6.3.1.1 and Figure 85), cannot tolerate moderate manuring, and so their absence is an indicator of fertilizer application in many cases where they frequently occur (or occurred) in the surrounding nutrient-poor grasslands characterized by similar site conditions except for fertilization. Some other plants, such as meadow clary (*Salvia pratensis*; Figure 166 in Section 10.2.2), occur in both non-fertilized and moderately manured grasslands. Yet others do not occur in nutrient-poor, non-fertilized grasslands, but they can abound or even dominate in moderately manured ones, such as yellow oat-grass (*Trisetum flavescens*).[2067]

Despite the general decrease in species numbers as a consequence of manure application, plant species richness remained relatively high if moderate quantities of matured barn manure were applied (Figure 89), which was most often the case until the mid-twentieth century as the time dealt with in this volume. Thirty-five to fifty plant species often occur on 50 m^2 in such meadows recently[2068] (see also Sections 1.2, 2, 10.1.2, and 10.2).

Farmers usually selected only naturally more productive, mesic meadows for manure application, if socioeconomic conditions allowed this, to make optimal economic use of limited quantities of the available manure. Namely, higher quantities of manure would have been needed for a high abundance of good fodder grasses of moderately manured meadows at dry sites; moreover, in dry conditions yields of these grasses would be lower than at mesic sites. Another case was wet meadows, where a lack of soil aeration hindered the growth of grasses even in the case of moderate manuring, and so they were also usually only manured where more suitable sites were not available to the same farmer.[2069] Similarly, the same amount of manure applied to highly acidic sites would not result in comparable hay quantity and quality than manuring sites with close to neutral pH.

At moderate sites most frequently selected for fertilizer application as long as amounts of fertilizers were scarce, false oat-grass (*Arrhenatherum elatius*, mainly on neutral soils, at mesic or relatively dry sites; Figure 90), yellow oat-grass (*Trisetum flavescens*, mainly on neutral to basic soils), common bent (*Agrostis capillaris*, mainly on acidic soils, at mesic sites), red fescue (*Festuca rubra* agg.,

2066 Janišová et al., 2023, 18.

2067 Sadar, 1956, 55, 142; Kauter, 2002, 19, 110; Peterken, 2013, 269–270; Aeschimann et al., 2004, vol. 2, 164–165, 528–529; Babai, Molnár, & Molnár, 2014, 40; Dakskobler et al., 2019, 58.

2068 Dietl & Lehmann, 2004, 124–125; Babai, Molnár, & Molnár, 2014, 35–36; Machatschek, 2015, 33–34.

2069 Pils, 1994, 153–154, 165; compare Sucholas et al., 2022.

Figure 90: False oat-grass (*Arrhenatherum elatius*). Photo: Leonid Rasran.

predominately on neutral soils, at mesic sites), meadow fescue (*Festuca pratensis*, mainly on neutral soils, at mesic sites), crested dog's-tail (*Cynosurus cristatus*, mainly on neutral and slightly acidic soils), and perennial rye-grass (*Lolium perenne*, predominately on neutral soils, at mesic sites) are the most important dominant grasses for animal husbandry. They occur on silicate or carbonate bedrock throughout the Alps. Most of these species can abound in all vegetation belts relevant for moderately manured meadows, whereas false oat-grass can grow as a dominant species only in the colline and montane vegetation belts (up to about 800 m). At higher elevations, its niche is mainly filled by yellow oat-grass and/or common bent. The abundance of perennial rye-grass also is low at high elevations.[2070]

A general occurrence and abundance of false oat-grass, yellow oat-grass, or downy oat-grass (*Helictotrichon pubescens*) in many moderately manured meadows in Central Europe is proven as early as before the agricultural reforms of the eighteenth century (thus before the abolition of spring grazing in meadows; see Section 6.2) and in the nineteenth century as well, based on Dirk Kauter's analyses. In addition to the grasses listed in the previous paragraph, downy oat-grass, a species that can abound in moderately dry, moderately manured meadows in colline and montane vegetation belts, in sites on carbonate or silicate bedrock, seems to have relatively frequently belonged to the dominant species in moderately manured meadows (see Section 2). This was likely related to the fact that, although grazing can represent a challenge for both species mentioned, downy oat-grass usually tolerates it more successfully. Moreover, within moderately manured meadows, downy oat-grass prefers

[2070] Compare Pils, 1994, 164, 179, 182; Kauter, 2002, 16–17; Dietl & Lehmann, 2004, 18, 20, 23–25, 37, 54; Babai, Molnár, & Molnár, 2014, 35–38; Aeschimann et al., 2004, vol. 2, 840–841, 854–855, 872–873, 894–895, 950–951, 956–957, 970–971.

less-fertilized sites than its competitor false oat-grass.[2071] Similarly, several other plant species, such as large thyme (*Thymus pulegioides*), highly relevant for the scent of milk and dairy products (see Section 10.1.2.2.2), or fragrant orchid (*Gymnadenia conopsea*; see Figure 10 in Section 2) benefitted from very moderate fertilizing with matured barn manure (ten tons of manure per hectare once in two or three years), whereas they became extinct in the meadow in the case of annual application of the same manure quantity,[2072] which nowadays still belongs to moderate meadow fertilization in the colline belt and lowest section of the montane belt, whereas fertilization at higher elevations must be less frequent to attain a similar grassland status.[2073]

False oat-grass (Figure 90) also deserves historical attention. It is a species of great importance for animal husbandry, which benefited much from moderate manuring. The species is an archaeophyte in northwestern Europe[2074] and an archaeophyte or native species in the Alps.[2075] A reliable botanical record from 1719 proves that at that time false oat-grass was already the dominant species in meadows around Zürich—in the immediate vicinity of the northwesternmost Alps, where agriculture specialized in animal husbandry early. In 1770, the natural scientist Albrecht von Haller emphasized that the high abundance of false oat-grass in Swiss meadows was not a consequence of sowing, which is in line with the aforementioned long history of occurrence of this species in the landscape. Similarly, mid-eighteenth-century information from the surroundings of Memmingen, prior to major changes in grassland management there, suggests that manured false oat-grass meadows were not exceptional at that time. False oat-grass is usually considered very sensitive to grazing, spring grazing in particular. Kauter emphasizes that it is not yet clear how this grass could grow in spring-grazed meadows (see also Section 6.2); nonetheless, there is mid-eighteenth-century evidence that false oat-grass also grew in manured spring-grazed meadows. Local regulations on grazing in the context of other environmental parameters, such as climate and grazing timing, duration, and intensity, probably importantly influenced the possibilities for survival of false oat-grass in spring-grazed manured meadows (compare Section 6.2.2). Nonetheless, it is clear that the abundance of false oat-grass meadows in the valleys in the Alps increased significantly from the eighteenth to mid-twentieth centuries. There are two main reasons for this change. First, there were changes in grassland management, such as wider application of fertilizers to meadows, abolition of spring grazing in meadows, and sowing of false oat-grass seeds. Second, there was an enormous increase in the total area of meadows in Alpine valleys

---

2071 Kauter, 2002, 20, 60–65, 73–74; Dietl & Lehmann, 2004, 20; Aeschimann et al., 2004, vol. 2, 946–947, 950–951.

2072 Janišová et al., 2023, 18.

2073 Dietl & Lehmann, 2004, 125.

2074 Pfitzenmeyer, 1962; Roehrs et al., 2013.

2075 Akeret & Geith-Chauviere, 2011.

Figure 91: A species-rich moderately fertilized meadow east of the village of Srednja Vas at the bottom of the valley in Bohinj (at about 590 m), abounding in false oat-grass (*Arrhenatherum elatius*), crested dog's-tail (*Cynosurus cristatus*), cock's-foot (*Dactylis glomerata*), Kentucky bluegrass (*Poa pratensis* agg.), goat's beard (yellow blossoms, *Tragopon pratensis* subsp. *orientalis*), field scabious (purple blossoms, *Knautia arvensis*), and oxeye daisy (white blossoms, *Leucanthemum vulgare*). Note the hayrack behind, often used for drying hay from moderately fertilized sites in the last centuries, and in this region some small farms still use them (see the text above and Section 4.1.1.3). The entire plant of goat's beard is edible.[2076] The "sweet, succulent interior part of the stem" was eaten raw prior to flowering in several parts of the Alps in history, such as in the Allues Valley,[2077] and young goat's beard was a frequent raw children's snack on their way to school, as reported in the Eastern Alps;[2078] children's consumption of the tasty interior of goat's beard stem was evidently widespread; it is also attested in the Eastern Carpathians.[2079] Photo: Žiga Zwitter, May 30th, 2020.

in lower elevations, where there had previously been arable fields (Section 3.1). Namely, false oat-grass could not abound at high elevations (see above),[2080] where so many meadows had previously been located.

In moderately manured meadows, the grasses mentioned in the previous three paragraphs have been joined by several other grasses (such as sweet vernal-grass, *Anthoxanthum odoratum*, black bent, *Agrostis gigantea*, smooth

[2076] Seliškar & Wraber, 1986, 150–151, Machatschek, 2017, 119–121; see also its traditional uses in the White Carpathians in Pawera et al., 2017, 659, 666.

[2077] Meilleur, 1982, 168–170; Meilleur, 1986, 283–284, 290.

[2078] Machatschek, 2017, 120.

[2079] Babai, Molnár, & Molnár, 2014, 40.

[2080] Kauter, 2002, 7, 17–20, 26–28, 31–42, 45–47, 51–60, 73–74, 89, 100, 103, 105, 108–109, 118–119, 139–141, 143, 155, 160, 162–163–165, 165–168; Kapfer, 2010, 138–140; Peterken, 2013, 271.

meadow-grass, *Poa pratensis*, and cock's foot, *Dactylis glomerata* subsp. *glomerata*; mainly or exclusively at higher elevations: Alpine meadow-grass, *Poa alpina,* Alpine vernal grass, *Anthoxantum alpinum*, violet fescue, *Festuca violacea*, and alpine timothy, *Phleum rhaeticum*). Moreover, many herb species occur in moderately manured meadows. A selection of those that often abound there follows: goat's-beard (*Tragopon pratensis*; Figure 91), red clover (*Trifolium pratense*; Figure 166 in Section 10.2.2), oxeye daisy (*Leucanthemum vulgare*; Figure 166), meadow buttercup (*Ranunculus acris* subsp. *acris*), upright hedge-bedstraw (*Galium album* subsp. *album*), yellow-rattle (*Rhinanthus minor*; Figure 17 in Section 4.1.1.1), caraway (*Carum carvi*), rough hawkbit (*Leontodon hispidus*), spreading bellflower (*Campanula patula*), rough hawk's-beard (*Crepis biennis*), autumn crocus (*Colchicum autumnale*; Figure 78 in Section 6.2.2.2 and Figure 126 in Section 8), meadow crane's-bill (*Geranium pratense*); mainly at relatively dry sites: meadow clary (*Salvia pratensis*; Figure 166), field scabious (*Knautia arvensis*; Figure 91), wild carrot (*Daucus carota*), ribwort plantain (*Plantago lanceolata*); mainly or only at higher elevations: long-leaf widow flower (*Knautia longifolia*), greater burnet-saxifrage (*Pimpinella major*), alpine spicknel (*Ligusticum mutellina*; see Section 4.1.1.2), alpine pasture clover (*Trifolium thalii*), imperforate St. John's wort (*Hypericum maculatum*), golden hawk's beard (*Crepis aurea*; Figure 8 in Section 2), great masterwort (*Astrantia major*, in very late-mowed meadows); mainly at moist sites: globe-flower (*Trollius europaeus*), leafy lousewort (*Pedicularis foliosa*), intermediate, smooth, and other subalpine lady's mantles (*Alchemilla xanthochlora, A. glabra*, and *A. fissa*); at wet sites also broad-leaved marsh orchid (*Dactylorhiza majalis*) and leatherleaf lady's mantle (*Alchemilla coriacea*). See further examples of herbs in moderately manured meadows in Sections 10.1.2.2.3 and 10.2.2, and Figure 3 in Section 1.2).[2081]

Despite the prevalence of moderate site conditions in fertilized meadows, the nexus between environmental and socioeconomic conditions sometimes led to manure application at agriculturally less favorable sites. Some dry grasslands were manured in history, where this was fostered by the socioeconomic conditions: a parallel from the Swabian Jura makes it clear that some dry calcareous grasslands there were manured annually, "although only 'little and short grass'" was harvested there, as reported in early nineteenth century.[2082] Acidic sites were also sometimes manured, such as some remote meadows where manure was

[2081] Pils, 1994, 179–180, 182, and Table 44, Photo 117; Machatschek, 2015, 35–36; Dietl & Lehmann, 2004, 20, 25, 37, 43, 54; Keller, 2004, 35; Babai & Molnár, 2013, 18; Babai, Molnár, & Molnár, 2014, 35–40; Aeschimann et al., 2004, vol. 1, 126–127, 160–161, 404–405, 788–789, 916–917, 928–929, 1054–1055, 1072–1073, 1088–1089, 1124–1127, 1152–1153; Aeschimann et al., 2004, vol. 2, 164–165, 174–175, 260–261, 272–273, 306–307, 356–357, 410–411, 416–417, 504–505, 630–631, 646–647, 670–671, 674–675, 850–851, 880–881, 890–895, 926–927, 946–947, 964–965, 970–971, 978–979, 1122–1123, 1030–1031, 1122–1123.

[2082] Kauter, 2002, quote from p. 122; for general ecological conditions in the Swabian Jura, see Karlík & Poschlod, 2009.

available and it would be difficult to transport it elsewhere (see Section 6.3.2.1), and so it was applied to poorly productive sites, such as acidic matgrass swards, to convert them to agriculturally more favorable grasslands not meeting the standard of more productive moderately fertilized meadows.[2083] Regular fertilization more or less neutralized the soil pH at formerly acidic sites, and so species such as the following often started occurring commonly in former matgrass swards due to frequent moderate manuring in (lower sections of) the subalpine vegetation belt: long-leaf widow flower (*Knautia longifolia*), greater burnet-saxifrage (*Pimpinella major*), alpine spicknel (*Ligusticum mutellina*; see Section 4.1.1.2), imperforate St. John's wort (*Hypericum maculatum*), Lady's mantles (*Alchemilla* spp.), Chewing's fescue (*Festuca nigrescens*), Alpine vernal grass (*Anthoxantum alpinum*), alpine timothy (*Phleum rhaeticum*), Alpine meadow-grass (*Poa alpina*), and golden hawk's beard (*Crepis aurea*; Figure 8 in Section 2).[2084] See also Section 6.3.2.1 on the impacts of ash application on matgrass swards. At extreme sites, such as former matgrass swards or other nutrient-poor soils on silicate bedrock, historical moderate fertilization or irrigation with water with appropriate characteristics (Section 6.4.1) made the mowing process significantly easier and somewhat less labor-intensive; namely, mowing on fertilized sites was easier than mowing hard plants growing on silicate-rich soils and other matgrass swards (see also Section 6.1.1.2).[2085]

More frequent than barn manure application to extreme sites was probably corralling sheep there (Section 6.3.2.1), a practice with a particularly strong ecological impact. The parallel from Ghimeş in the Eastern Carpathians, where autumn sheep corralling in montane meadows is still well recalled and is still practiced in mountain pastures, provides crucial ecological insights. It suggests that the pen's fence in a meadow was moved after no more than three or four nights. Three or four nights and an appropriate stocking density (see below for a rough estimate) provided enough manure for the next three or four years. Even in the case of nutrient-poor grassland as the starting point, keeping sheep in the same pen for significantly more than a week would decrease the species richness and provide so many nutrients that nitrophilous flora (such as nettles, *Urtica dioica*; large-leaved docks: broad-leaved dock, *Rumex obtusifolius*, and Alpine dock, *Rumex alpinus*; and nitrophilous thistles, *Cirsium* spp., burdock, *Arctium* spp., etc.) would later abound at the site of the corral instead of a higher cover of meadow plants more desired from the viewpoint of livestock farming. Weather also mattered: in the case of rain, corrals had to be moved sooner—every second or third day, for instance. The duration of corrals in the same meadow section also depended on species composition: matgrass swards made it possible

2083 Compare Keller, 2004, 35; see also Pils, 1994, 95.

2084 Keller, 2004, 35; Janišová et al., 2023, 17; Aeschimann et al., 2004, vol. 1, 404–405, 786–791, 1088–1089, 1126–1127; Aeschimann et al., 2004, vol. 2, 410–411, 670–671, 852–853, 890–891, 964–965, 978–979.

2085 Machatschek, 2007, 149–150.

to keep sheep in corrals longer in the same location than other non-fertilized grasslands.[2086] A study from the Slovak Carpathians recommends "a two-day corralling schedule in the spring with a load of 7.5 m^2 per cow/ox or 1.5 m^2 per sheep" (current livestock size; compare Section 5 for historical sizes) to achieve the highest yields and "a better representation of forage species [in a grassland that is still species-rich], as well as improvement in soil's agrochemical and biological properties" in some extensively managed nutrient-poor meadows.[2087] The combination of fertilizing and trampling effects of corralling (including regular movement of fences) was able to suppress agriculturally undesired mosses, as recommended by agricultural handbooks from the sixteenth and seventeenth centuries, matgrass (*Nardus stricta*; see Section 6.3.1.2), *Vaccinium* spp. (this genus includes, e.g., bilberry, *Vaccinium myrtillus*, and cowberry, *Vaccinium vitis-idaea*), and sedges (*Carex* spp.), and limit the spread of bracken (*Pteridium aquilinum*) in meadows. Sheep were crucial to attaining these goals; cattle would not suppress matgrass significantly,[2088] two of the reasons being the sharpness of sheep hooves and the higher overall raw-humus mineralization (i.e., the biochemical dissolution of organic matter,[2089] making the nutrients available to plants) in the case of corralling sheep instead of corralling cattle.[2090] Appropriate corralling of sheep at nutrient-poor acidic sites promoted some better fodder plants in a relatively species-rich grassland, including excellent white clover (*Trifolium repens*) and meadow-grass (*Poa* spp.).[2091]

Similar to plant species diversity, animal species diversity decreases with moderate manuring because many animals depend on non-fertilized sites (see Sections 1.2 and 10.1.2.2.4). However, due to the moderate level of fertilization, animal diversity remains relatively high. Moderate manuring has much fewer negative impacts on meadow fauna than strong fertilization. A few animal species, from the farming perspective most notably earthworms, are more abundant in moderately manured than in non-fertilized meadows (see Section 10.1.2.2.3).[2092]

Due to the direct and indirect impacts of manuring, including earthworm activities, an additional mowing, denser turf, and higher plants, yields of moderately manured meadows are often twice those of non-fertilized meadows at similar sites (see also Section 6.3.0.1). At climatically favorable damp sites where yields are highest, they often reach eight metric tons of dry substance per hec-

[2086] Babai & Molnár, 2014, 127; Babai, Molnár, & Molnár, 2014, 88, 112, 120–121; see also Janišová et al., 2023, 17, and Aeschimann et al., 2004, vol. 2, 582–591.
[2087] Janišová et al., 2023, 17.
[2088] Kauter, 2002, 131, 191 (compare Ellenberg & Leuschner, 2010, 33); Babai & Molnár, 2014, 126–128 and the electronic appendix; Babai, Molnár, & Molnár, 2014, 50, 88, 120–121; Janišová et al., 2023, 17.
[2089] *Geografski terminološki slovar*: Mineralizacija.
[2090] Janišová et al., 2023, 17.
[2091] Janišová et al., 2023, 17; Dietl & Lehmann, 2004, 39, 42–43.
[2092] Machatschek, 2015, 34.

tare at 500 m (6.4 tons when usual 20% losses during the mowing, drying, and storing are subtracted), six tons at 1,000 m (4.8 tons), four tons at 1,500 m (3.2 tons), and two tons at 2,000 m (1.6 tons) according to recent data, but the combination of an unsustainably high percentage of herbs (much more than 50%) and less favorable weather and climate often reduces these figures by 30% to 40%.[2093] These data match relatively well with historical information from southwestern Germany in Franz Häfener's agricultural handbook published in 1847: average annual yields of first-crop and second-crop hay from moderately manured meadows (where two mowings were followed by autumn grazing) were 3.6 to 4.6 metric tons per hectare; 5.5 to 6.8 tons/hectare was considered very high for such meadows, and up to 9.1 tons/hectare was yielded by irrigated meadows (Section 6.4).[2094] This relative correspondence and somewhat lower average yields in the nineteenth century are not surprising. Namely, in the context of downy oat-grass (*Helictotrichon pubescens*), it was stated above that many fertilized meadows were manured even more moderately than false oat-grass meadows, and so the yields were somewhat lower. Such moderate manuring is also in line with the quantity of second-crop hay being well below half the quantity of first-crop hay in some cases (see Section 6.3.0.1). At the Kaunz farm at Gariusch (northeast of Völkermarkt, Carinthia), located at about 730 m, the quantity of second-crop hay in 1584 was reportedly one-third of the quantity of first-crop hay from the same meadows mowed twice annually.[2095] This is in line with mid-twentieth century observations from Bohinj; the quantity of second-crop hay at some farms was often one-third of the quantity of first-crop hay from the same meadows.[2096] Because the intensity of fertilizer application rose in the second half of the twentieth century, the quantity of second-crop hay at the farm level in Bohinj, at some farms still managing relatively species-rich, moderately fertilized meadows, has risen to half the yield of first-crop hay at 550 m nowadays.[2097] Such meadow productivity, however, can already be found in older records in many cases (see early nineteenth-century cadastral data from the Central Karawanks and their surroundings in Section 6.3.0.1). In line with a record from 1584, the quantity of second-crop hay in a meadow or meadows belonging to a smallholding (i.e., a cottage) at Ruden (about 450 m, at the foot of the Alps east of Völkermarkt) was reportedly half the amount of first-crop hay there, and the same ratio was reported for another smallholding in the same village in 1692.[2098] Thus, in several cases the manuring intensity in meadows that were fertilized seems to have long

2093 Šarf, 1974, 143; Dietl & Lehmann, 2004, 16, 40, 44, 107, 121; Babai, Molnár, & Molnár, 2014, 36.

2094 Kapfer, 2010, 137–138.

2095 Dinklage, 1966b, 107; microlocalization is based on the farm name in the early nineteenth-century cadaster: *KAGIS Maps*, Basiskarten, Franziszeischer Kataster & Abfragen, Koordinaten / Höhe.

2096 Author's interview with Kunavec, 2023, in Stara Fužina.

2097 Author's interview with M. Gartner, 2022, in Studor.

2098 Dinklage, 1966b, 104.

been somewhat weaker or comparable to the recent intensity in most moderately manured meadows that still exist.

Yields of the grasses of moderately fertilized meadows that are most important for animal husbandry (false oat-grass, *Arrhenatherum elatius*; yellow oat-grass, *Trisetum flavescens*; and meadow fescue, *Festuca pratensis*) are middling by twenty-first-century standards. Net energy value for lactation is significantly higher in the case of first-crop hay mowed when grass seeds ripen and in second-crop hay from moderately manured meadows dominated by yellow oat-grass (mainly at higher elevations, about 5.75 MJ/kg dry substance in first-crop and second-crop hay —that is comparable to net energy value for lactation of the significantly less biodiverse hay from third and fourth grassland use in some highly intensively managed, mowed and grazed grasslands used five times annually, but less than in highly intensive meadows mowed five times annually) than in the case of hay from moderately manured meadows dominated by false oat-grass (about 4.9 MJ/kg dry substance in first-crop hay mowed when grass seeds ripen, that is comparable to some good non-fertilized meadows in similar elevations; net energy for lactation in false oat-grass meadows is significantly higher in second-crop hay: 5.4 MJ/kg dry substance, and more than 6 MJ/kg dry substance during the following autumn grazing). A general integrated quality indicator linking the nutritive value, palatability, and health impacts of most grasses abounding in such meadows is relatively high (grade 2 on a five-grade scale, where 1 is excellent; in the case of dominant red fescue or common bent, it is middling).[2099]

An additional mowing in many fertilized meadows was mentioned several times in this section. The fact that moderate manuring made possible a second mowing is confirmed by various sources such as a description of Tyrol and Vorarlberg from 1839 claiming that, unlike manured meadows, non-fertilized meadows at the bottom of the valleys were only mowed once a year,[2100] by oral history interviews in the core study area in the eastern Julian Alps,[2101] and by ethnological publications.[2102] Oral history interviews in Bohinj referring to the

[2099] Dietl & Lehmann, 2004, 18, 39, 42, 122, 125–126.

[2100] Mathieu, 1996, 144–145.

[2101] Hodnik, 2016, 35, 42, 44, 76; Babai, Molnár, & Molnár, 2014, 87, 89; author's interviews with Gašperin, 2017, with Kunavec, 2017, with Mencinger, 2017, and with Urh, 2017, all in Stara Fužina; with Grm, 2017 and with Milena Sodja, 2017, both in Srednja Vas in Bohinj; with J. Smukavec, 2017, in Podjelje; with A. and M. Taler, 2017, in Studor; and with I. Mlakar, 2020, at the Zajamniki Alp. It remains an open question whether the information from the early nineteenth-century cadaster that the best meadows close to farms at Gorjuše (about 1,000 m), which were mowed twice a year, were mainly non-fertilized and unirrigated (SI AS 176, Franciscejski kataster za Kranjsko [Franciscan Land Cadaster for Carniola], L72, cadastral municipality of Gorjuše, cadastral appraisal protocols: Catastral-Schätzungs-Elaborat, Schätzung des natural Ertrags, § 4: Wiesen) is reliable, or whether these meadows were in fact also usually fertilized.

[2102] E.g., Babai & Molnár, 2014, 125; see also Kauter, 2002, 145.

first half of the twentieth century proved that moderate manuring also made possible second-crop hay mowing in distant montane meadows in September (see Section 6.1.2.3); first-crop hay was mowed there in July; see Section 6.1.2.2. This information refers not only to those montane meadows at lower elevations, but also to meadows on Vogar Hill (at an elevation of about 1,100 m; i.e., about 550 m higher than the village), at Uskovnica (at about 1200 m; i.e., about 600 m higher than the village), and at Zajamniki (about 1,250 m), for instance.[2103] Similarly, the manured and irrigated meadow in the lower section of the Stift Alp (Germ. *Stiftsalm*, about 1,200 m) southeast of Hall in Tyrol was mowed twice according to data from 1714.[2104]

Manured meadows at higher elevations, at many places higher than about 1,300 m, nonetheless allowed only one mowing (but the elevation of the upper limit of second-crop hay mowing varied significantly across space and time; see Section 6.1.2.3). Timing of spring and autumn grazing (Section 6.2) also allowed only one mowing between in some meadows at more limited elevations.[2105] However, where spring grazing ended early, it was usually followed by two mowings and autumn grazing in moderately manured meadows, up to a certain elevation reaching well into the montane vegetation belt (see also Section 6.2.2).[2106] This can be understood based on a comparison of Figures 74 and 75 in Section 6.2.2.1, showing ecological consequences of opposing types of impacts of land use on plant development in neighboring non-fertilized and organically fertilized grasslands at around 1,050 to 1,100 m in early spring.

Another practice that often allowed second-crop hay mowing was meadow irrigation, presented in the next section.

6.4 Grassland irrigation

Traditional grassland irrigation (Figures 92, 93, 94, 95, 97, and 98) was widespread in the Alps over the long run, until the late nineteenth century and, in many cases, the mid-twentieth century. Since the Late Middle Ages, at least, basic irrigation techniques were practiced in the majority of this mountain chain, but the intensity of irrigation and relevant details about how it was carried out differed significantly. Irrigation practices and the rate of their adjustment to local environmental conditions may have changed over the course of history.[2107]

[2103] Hodnik, 2016, 35, 42, 44, 76; author's interviews with Gašperin, 2017, and with Mencinger, 2017, both in Stara Fužina; and with I. Mlakar, 2020, at Zajamniki.

[2104] Wopfner, 1997, 445–446; localization of the meadow in the lower section of Stift Alp: *Arcanum Maps*, Cadastral Maps, Habsburg Empire – Cadastral Maps (XIX. [sic] Century), cadastral municipality of Tulfes, 1856); the Stift Alp was created from two former farms: Grass, 1948, 209.

[2105] Wopfner, 1997, 381.

[2106] Wopfner, 1997, 288–290, 295–296.

[2107] Emanuelsson, 2009, 281–283, 285–287; Wopfner, 1997, 348–367; Leibundgut & Vonderstrass, 2016a, b; Mathieu, 2019, 53.

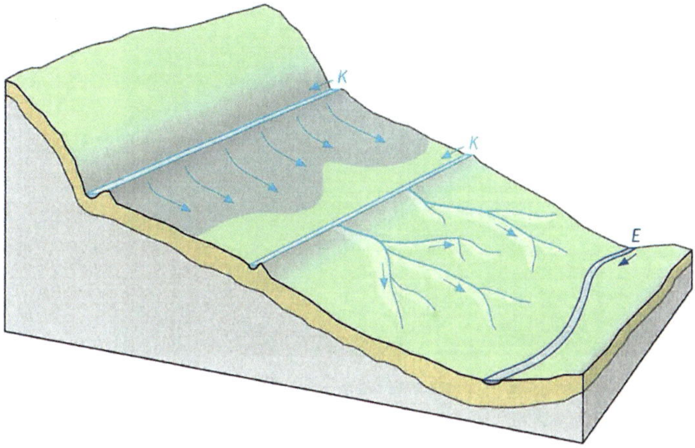

Figure 92: Model of usual historical irrigation on slopes, from the channels labeled K. Leibundgut & Vonderstrass, 2016a, 170. © Ingeborg Vonderstrass.

The oldest known archaeological finds prove irrigation in the Alps already in the Bronze Age and early Iron Age—for example, in the Venosta Valley (Germ. *Vinschgau*, Ital. *Val Venosta*, now in Alto Adige / South Tyrol), in the Ötz Valley (at Obergurgl and Vent, in North Tyrol), and in Dauphiné. In Classical Antiquity, grassland irrigation must have also played a role in the Alps, but preserved records make it clear that in the second millennium AD, at least, grassland irrigation in the Alps was widespread and its economic impact in grassland farming and farming in general was significant in several parts of the mountain chain (see Section 6.4.1). In the first half of the second millennium AD, many new irrigation systems were built and older ones extended. In the following four centuries, major changes seem to have been relatively rare because maintenance and in some cases also extension of existing systems was in the foreground. The next important phase of construction of irrigation systems followed in the nineteenth century, and in Briançonnais in the eighteenth century. Significant increases in agricultural density in agrarian societies often stimulated more labor-intensive farming. It is thus no coincidence that the infrastructure allowing grassland irrigation as a labor-intensive practice (Section 6.4.2) aimed at higher hay quantities and quality (Section 6.4.1) was usually built during periods of strong demographic growth in the Alps, such as the twelfth and thirteenth centuries, and the eighteenth and especially nineteenth century. The population of the Alps as delimited now by the Alpine Convention increased from about 5.8 million in 1800 to 8.4 million in 1900, with more pronounced population growth in the Western Alps (see Section 3.1). Moreover, the population growth characteristic for nineteenth-century Europe in general was even stronger, which is relevant for the market for animal husbandry products from the Alps (Section 3.1), and animal husbandry in many areas of this mountain range was influenced by labor-intensive

Figure 93: Irrigated meadows at Obergurgl, Ötz Valley (Germ. *Ötztal*), Tyrol, at 2,300 m. These were high elevations for irrigated meadows in the Alps. Note the wooden gutters directing the water from the creek to the irrigation channels. In the early nineteenth century, the village of Obergurgl, located at an extreme elevation of 1,930 m, had to face particularly high agricultural density in the context of local environmental conditions. Consequently, a total ban on marriages was introduced in 1831 and lasted until 1850, an extreme measure in the broader Alpine context, to prevent the overuse of local environmental resources,[2108] despite the option of emigration. Thus high agricultural density stimulated further development or at least maintenance of historical labor-intensive arrangements, such as the complex meadow irrigation system at 2,300 m. However, irrigation at such elevations at Obergurgl was not an exception in the wider context of the Alps. In the southern part of the Tauern mountains, for instance, several meadow complexes in the lower section of the alpine vegetation belt used to be irrigated.[2109] Photo: Erika Hubatschek, 1946. Hubatschek, 1992, 120. © Irmtraud Hubatschek.

grassland irrigation. In some smaller regions, such as Wallis (Francoprovençal *Valês*, French *Valais*; Figure 94) and the Venosta Valley, two areas with historically widespread grassland irrigation, new irrigation systems were also built in the fifteenth and/or sixteenth centuries, again during times of population growth.[2110]

[2108] Bätzing, 2015, 122–124, 419.

[2109] Pils, 1994, 193.

[2110] Leibundgut & Vonderstrass, 2016a, 53–55, 58–59, 61–73, 78, 165, 216; Leibundgut & Vonderstrass, 2016b, 89, 99–100, 102–104, 113, 117–119, 124, 127, 132, 152–153, 160–162, 179; Mathieu, 1996, 145, 156; Patzelt, 1996, 65; Wopfner, 1997, 352; Scharr,

Figure 94: In the relatively dry parts of Wallis, many main channels providing water for grassland irrigation constructed in Late Middle Ages involved a section or sections passing along a cliff in a wooden gutter. This was placed on a wooden support that was fastened to a cliff with wooden beams set horizontally in holes hewn into the cliff. Outside of Wallis, sections of irrigation channels passing along cliffs were a rarity in the Alps; some were located especially in the Venosta Valley and in Briançonnais, two regions with a comparably dry climate, like the relatively dry parts of Wallis. Maintenance work in such channel sections was dangerous, as seen in this photograph; activities in a hanging gutter passing along a cliff sometimes claimed lives. This photo, taken around 1920, shows spring maintenance in such a section of a main irrigation channel—the watertightness of the gutter is being improved with enormous quantities of tree leaves and mud (or clay[2111]). Irrigation channel at Savièse (Fr. *Bisse de Savièse*), Upper Rhône Valley in Wallis. A channel at this site provided irrigation water for half a millennium, until the mid-1930s.[2112] Recently, the Savièse Irrigation Channel was partly restored (Section 10.2.2).[2113] Photo: (probably) Charles Paris. "Bildteil," 1996, 428, 515. © Charles Paris, Mediathek Wallis – Martigny.

2001, 20–21, 23; Krausmann, Weisz, & Eisenmenger, 2016, 76; Mathieu, 2019, 49, 53, 124–125, 155; and Bätzing, 2015, 71.

2111 Compare Wopfner, 1997, 356.

2112 "Bildteil," 1996, 428; Wopfner, 1997, 356–357 and Figure 47; Leibundgut & Vonderstrass, 2016a, 158–159, 164–166; Leibundgut & Vonderstrass, 2016b, 92–97, 100, 102–103, 106–108, 126–128, 149–151 (on precipitation, see also Bätzing, 2015, 37).

2113 Leibundgut & Vonderstrass, 2016b, 106–108; *Bisse du Torrent Neuf/Savièse*.

By the early twentieth century, the total length of big irrigation channels in the Alps was several thousand kilometers. As a general rule, irrigation was not carried out in grasslands on limestone due to the challenges posed by karst, and it was also absent from some other places. About 1935, there were 380 irrigation channels more than 800 m long in the Venosta Valley alone with a total length of about a thousand kilometers; the total maximum historical length of all irrigation channels of all sizes in the same valley was estimated at about ten thousand kilometers. Before the introduction of sprinkling irrigation in the mid-twentieth century, about 35 m^3/s of irrigation water was used in the Venosta Valley, including side valleys, in July and August, which corresponded to roughly half of the Adige (Germ. *Etsch*) River's discharge at Tel (Germ. *Töll*). About 100 km^2 of agricultural land was irrigated in the Venosta Valley in the mid-1930s. Grassland irrigation, especially meadow irrigation, played a major role there; nonetheless, in contrast to many other regions in the Alps, grasslands were not the only land-use types irrigated in the Venosta Valley.[2114]

Since the decline in traditional irrigation in the Alps that started in the late nineteenth century and its rapid reduction since the mid-twentieth century, as in most of Europe, only very few traditional irrigation systems still function in grasslands in the Alps. The great majority of traditional grassland irrigation systems have been abandoned or replaced by sprinkling or drip irrigation, mineral fertilizers, or pesticides (see the functions of historical irrigation in Section 6.4.1). Whereas the historical irrigation dealt with in this section was multifunctional, modern irrigation techniques are used almost exclusively to mitigate drought, but they are much less labor-intensive, distribute the water more evenly over the meadow, and demand about 90% less water than traditional irrigation. This, however, means that the threat of soil salinization in relatively dry agroecosystems increased with the transition from the historical irrigation technique, which is presented here, to sprinkler irrigation. Mineral fertilizers and pesticides, which replaced some functions of historical irrigation, cause pollution in agroecosystems.[2115]

6.4.1 Functions of historical grassland irrigation

Historical grassland irrigation in the Alps was a labor-intensive practice, but in general it was sustainable, where it was well adapted to the local environmental conditions. Grassland irrigation included transport of water based

2114 Wopfner, 1997, 366–367; Leibundgut & Vonderstrass, 2016a, 251; Leibundgut & Vonderstrass, 2016b, 89, 151; see also Section 6.4.1.

2115 Niederer, 1996a, 124; Wopfner, 1997, 353–356 (the chronological definition of the time of Wopfner's writing follows Meixner & Siegl, 2004, 232–233, 235); Leibundgut & Vonderstrass, 2016a, 91–92, 96–97, 108, 202; Leibundgut & Vonderstrass, 2016b, 105–106, 110, 120, 153–159, 161, 174, 182; Emanuelsson, 2009, 304; Peterken, 2013, 182; Mathieu, 2019, 125; Renes et al., 2020, 111, 116. See also Pintar, 2003, 23–24.

on sunshine and gravity. The results were higher hay quantity, more stable hay yields at a year-to-year level, and in many cases also higher hay quality than in non-irrigated meadows (see below in this section). At the micro- and mesoregional level, the impacts of irrigation on hay quantities were diverse. At the mesoregional scale, the increase in meadow yield due to irrigation in the French Hautes-Alpes was estimated at 50% on average in the nineteenth century.[2116] Studies in the Black Forest (Germ. *Schwarzwald*) found twice the amount of dry substance in annual yield from irrigated meadows compared to non-watered ones.[2117] Locally, yields were sometimes increased even more significantly,[2118] reaching four times as much hay in irrigated than in non-irrigated meadows at Umhausen in the Ötz Valley in the case of a dry summer.[2119] Irrigated meadows in some microregions of the relatively dry Alto Adige (South Tyrol) usually yielded five times the quantity of hay compared to non-irrigated ones,[2120] but a factor of three was probably more usual than five even in the relatively dry parts of that region; the price of a parcel in the Venosta Valley decreased by two-thirds if sold without irrigation rights in the 1930s or 1940s.[2121] Exceptionally and locally, differences in yields between non-irrigated and watered meadows were even larger than a factor of five.[2122] More hay and better hay quality in the long run usually meant more livestock kept locally (but see the trade in hay in Section 3.1), and thus more manure, and so grassland irrigation also contributed to higher yields in arable land.[2123]

Historical grassland irrigation in the Alps had several important and interrelated functions (see also Section 6.4.3.3):

(1) In inner-Alpine dry valleys and basins—especially the Rhône Valley in Wallis, the Aosta Valley (Ital. *Valle d'Aosta*, Francoprovençal *Val d'Aoûta*, Fr. *Vallée d'Aoste*), the Vorderrhein Valley, the Venosta Valley, and the Upper Inn Valley (Figure 4), where precipitation during the vegetation period of grassland plants is low—irrigation was the crucial method of drought mitigation in grassland agroecosystems providing significantly more hay than plant commu-

[2116] Pintar, 2003, 23; Leibundgut & Vonderstrass, 2016a, especially pp. 59, 77, 96–97, 99–101, 126, 128, 131–132, 143, 147–149, 152; Leibundgut & Vonderstrass, 2016b, especially pp. 120, 128, 170, 181–182, 189; Kuret (ed.), 1987, 199; Hubatschek, 1992, 120; Pils, 1994, 94; Kauter, 2002, 124; Peterken, 2013, 179–180; Poschlod, 2015, 145.

[2117] Leibundgut & Vonderstrass, 2016a, 126.

[2118] E.g., Poschlod, 2015, 145; Leibundgut & Vonderstrass, 2016a, 131; see also Emanuelsson, 2009, 285–286 (with an evident lapse in units of measurement, where the yield of "one to three kilos dry substance per square kilometer" is given on p. 286).

[2119] Wopfner, 1997, 350.

[2120] Leibundgut & Vonderstrass, 2016a, 131.

[2121] Wopfner, 1997, 350; the chronological definition of the time of Wopfner's writing follows Meixner & Siegl, 2004, 232–233, 235.

[2122] Even a yield increase by a factor of five to eight is mentioned in extreme cases (Mathieu, 1996, 146). See also Leibundgut & Vonderstrass, 2016a, 131.

[2123] Leibundgut & Vonderstrass, 2016a, 59–60, 64, 74, 77, 96–97, 100.

Figure 95: The lower section of the sunny slope above the bottom of the Venosta Valley (Germ. *Vinschgau*) at Sluderno (Germ. *Schluderns*), where a dry steppe and irrigated (and fertilized?[2124]) grasslands are divided by the irrigation channel in the middle of the photo, surrounded by trees. Leibundgut & Vonderstrass, 2016b, 150. © Ingeborg Vonderstrass.

nities adapted to more frequently dry, non-irrigated sites (Figure 95). In some of the dry parts of the Alps, grassland irrigation was widespread in the range of elevations from valley bottoms to high meadows and pastures at elevations of more than 2,000 m. Frequent agricultural drought in the summer, during an early stage of growth of the valuable second-crop hay (Section 6.1.2.3), must have also co-influenced the building of many grassland irrigation systems in the Alps, not limited to the valleys mentioned above.[2125] Historical weather variability and climate change also played a role in some cases: for instance, a permit for construction of a new irrigation channel in the Ötz Valley, from Niederthai to Umhausen, to mitigate the challenge of insufficient irrigation water availability was issued in 1654, during a relatively dry climate.[2126]

In some places, such as in the Allues Valley (Francoprovençal *Alués*, French *Les Allues*) in the Western Alps (with about 900 mm of annual precipitation in the mid-twentieth century, concentrated in spring and autumn, and thus subject to agricultural summer droughts), there was no irrigation infrastructure and no irrigation was carried out in usual years as part of traditional agriculture of the early twentieth century. Nonetheless, in the case of severe drought,

[2124] Irrigation alone was indeed enough in this landscape to cause the differences between at least the less intensively green meadows below the irrigation channel and the dry grassland above the channel (Wopfner, 1997, 348–351, 358; Leibundgut & Vonderstrass, 2016a, 99–100); see the precipitation data in Section 6.4.3.1.

[2125] Leibundgut & Vonderstrass, 2016a, 47, 96, 98–100; Leibundgut & Vonderstrass, 2016b, 88, 94–95, 149–150, 165–166, 181; Pintar, 2003, 23; Pintar, 2005, 332, 335–336.

[2126] Jäger, 2006, 55.

when quantities of first-crop hay were very low, "water might be crudely diverted from small watercourses onto nearby hayfields" during the growth of the second-crop hay in the summer to raise its quality and quantity. It is noteworthy that regular irrigation with usual irrigation infrastructure existed in the neighboring communities, which probably inspired such irregular irrigation in the Allues Valley.[2127]

Even though the grassland irrigation systems in areas of the Alps with a relatively dry climate were often particularly complex in the wider context of the Alps, the following paragraphs explain that drought mitigation was only one of the goals of traditional irrigation, and in damper sites a subordinate one.[2128]

(2) Locally adapted irrigation led to a more favorable nutrient balance and contributed to further aspects of soil fertility from the agricultural point of view, through deposition of different types of plant nutrients that the water carried as external dissolved matter and suspended particles. A scarcity of manure (see Section 6.3) made irrigation an important source of nutrients for grasslands. The impacts were strong where water was rich in organic matter or minerals. Concerning organic matter, the natural fertilizing impact of irrigation was the strongest if its timing in the cold part of the year allowed it to make use of organic material that had accumulated in the autumn and reached the water (see the example from Kematen below). Focusing on minerals, the water rising below glaciers, with varying but high amounts of dissolved minerals and particles held in suspension (known as glacial milk)—also containing some highly fertile organic matter from glacial meltwater—was particularly suitable for increasing soil fertility. The Late Medieval innovation of long water channels—that crossed significant elevation differences and even passed along cliffs (in Wallis, in particular) to provide irrigation water richer in plant nutrients than local creeks usually—enabled the application of glacial milk to the meadows often located a few hours' walk lower. However, even nutrient-poor irrigation water (e.g., from streams on old igneous and metamorphic bedrock or sandstone) had visible albeit (very) limited fertilizing effects, also depending on the season and thus on the amount of organic matter, and on irrigation details. From the perspective of grassland farming, irrigation also improved some other soil properties, such as soil structure, texture, and pH value at acidic sites. Thus, irrigation influenced chemical processes and stimulated the activity of soil organisms, soil oxygenation, and so on. Running—not standing—water was obligatory to attain many of these goals.[2129]

[2127] Meilleur, 1986, 16, 269, 280 (quote from p. 269).

[2128] Netting, 1974; Emanuelsson, 2009, 285; Leibundgut & Vonderstrass, 2016a, especially pp. 49, 60, 100.

[2129] Leibundgut & Vonderstrass, 2016a, 47, 49, 60, 74, 77, 92, 96–97, 99–104, 106–108, 116–122, 126, 128–129, 147–148, 154; Leibundgut & Vonderstrass, 2016b, 88–89, 91,

Irrigation led to changes in grassland vegetation. This grassland management practice could have varying impacts on species composition, depending on site conditions, irrigation water characteristics, and irrigation practices. In general, its impacts were strongest in the case of a nutrient-poor grassland site irrigated with water carrying relatively many nutrients. For instance, irrigation with glacial milk, rich in calcium and containing organic matter from the glacier, was able to suppress matgrass (*Nardus stricta*; see Section 6.3.1.2). Glacial milk was also available just at the right time: after snowmelt, water from melting glaciers was available during the intensive growth of grassland plants, when the climate in many inner Alpine valleys is dry. Even if some farmers in various historical periods were not aware of the relevance of nutrient intake through irrigation, the appearance of matgrass likely stimulated the decision for irrigation, although the site was often not dry in reality. Namely, the tissue of matgrass stems is stiff (Section 6.3.1.2), its color is pallid green, its dead leaves decompose very slowly, and its growth dynamics are slow, and so it looks as though the site were dry. In reality, matgrass often abounds at mesic (sometimes also wet, moist, or relatively dry) sites. Similar might have been the motivation for some farmers to irrigate most nutrient-poor meadow types because many of them appear as though they were dry, and this way the fodder was often improved by plant nutrient intake through irrigation. In some cases, irrigation was a measure making possible conversion of some former poor sheep pastures into productive meadows (Figure 95). Observations in the Black Forest proved that winter irrigation of grasslands on a slope led to transition from a sparse yellow oatgrass meadow to a nutrient-richer meadow with cow parsley (*Anthriscus sylvestris*). Nutrient input through irrigation water was, however, usually not so significant that irrigation alone would turn a nutrient-poor grassland into a moderately fertilized one; nonetheless, its impact on species composition and grassland productivity was economically relevant in many cases (see above and below in this section). In several places, this effect of irrigation was replaced by an increase in atmospheric nitrogen inputs to grasslands in the second half of the twentieth century (see Section 10.1.2.2.4). Despite the usually too limited nutrient intake through irrigation to convert nutrient-poor grasslands into moderately fertilized ones, irrigation indeed created moderately fertilized meadows at some sites: winter irrigation at Kematen in Tyrol (610 m, in the bottom of the Inn Valley, Germ. *Inntal*), between November 11th and the end of February (a rare irrigation timing in the context of North Tyrol west of Innsbruck, but grassland irrigation in the cold part of the year used to be frequent in the hills of Central Europe), as prac-

98, 105, 150–151, 165–166, 169–170, 181–182, 186–189; Pils, 1994, 94, 193; Niederer, 1996a, 123; Wopfner, 1997, 259, 296, 350, 353; Keller, 2004, 20, 30; Emanuelsson, 2009, 281–282; Peterken, 2013, 180. Gratitude is expressed to the microbiologist and high mountain ecologist Kurt Hanselmann for his explanation concerning the organic matter in glacial meltwater.

ticed, for example, in the mid-twentieth century, resulted in ice a meter thick in some depressions. When it melted, a layer of largely organic mud up 0.1 m thick formed. These meadows were mowed up to three times a year, and so the fertilizing impact was very strong in this case. The specific irrigation timing was relevant to attain such a strong fertilizing impact. Namely, the water used at Kematen transported a lot of organic material from fallen leaves and similar biomass reaching the waters from the autumn onward.[2130]

Focusing on the impacts of irrigation in wet meadows on plant species composition, this practice could decrease the abundance of sedges and raise the abundance of grasses and forbs in some cases, resulting in higher-quality wetland hay that changed from "sour" (see Section 6.1.2.5.1.2) to middling quality. Between the intervals of irrigation, the same channels functioned as drainage ditches. In parts of the Black Forest, grasslands with high water availability on the lower parts of slopes and in valley bottoms turned from low-productivity small sedge communities and purple moor-grass (litter-) meadows (see Sections 2, 6.1.2.5.1.2, and 6.1.2.5.2) to more productive and nutrient-richer marsh marigold meadows. For example, irrigation promoted tufted grass (*Holcus lanatus*), sweet vernal-grass (*Anthoxanthum odoratum*), nitrophilous common bistort (*Bistorta officinalis*), aconite-leaved buttercup (*Ranunculus aconitifolius*), and marsh marigold (*Caltha palustris*). The nutritive value of these grasses is middling (sweet vernal-grass) or middling to low (tufted grass), whereas that of marsh marigold is very low, but the overall fodder quality was still significantly better than that from small sedges and purple moorgrass. At least in other parts of Europe, such as Sweden, irrigation of peat bogs with running water transformed them into fens: peat decomposed and slightly more favorable plants from the point of view of animal husbandry became established, such as sedges (*Carex* spp.) and purple moor-grass (*Molinia caerulea*).[2131]

To reach desired goals, the practices of irrigation had to be appropriate. Gaps in traditional ecological knowledge in some cases resulted in irrigation systems directing water that was too cold to the meadows. Water that is too cold is usually a step in the wrong direction if the goal is to provide better fodder. The temperature of irrigation water was a frequent challenge especially when using meltwater in spring or meltwater from glaciers in the summer, unless the grassland was distant enough from the melting place that the running water warmed up sufficiently. Some irrigation systems coped with this issue with

[2130] Pils, 1994, 49–53, 75, 90–91, 94, 97–99, 129–130, 290–291 and Table 13, Photo 34; Wopfner, 1997, 350–351; Ellenberg & Leuschner, 2010, 1042; Poschlod, 2015, 145; Leibundgut & Vonderstrass, 2016a, 47, 96, 99, 101, 126, 128–129, 143; Leibundgut & Vonderstrass, 2016b, 165, 169–170.

[2131] Emanuelsson, 2009, 282; Ellenberg & Leuschner, 2010, 1042–1043; Poschlod, 2015, 145; Leibundgut & Vonderstrass, 2016a, 128–130; Dietl & Lehmann, 2004, 39, 42.

ponds allowing the water to warm up before being directed to meadows. Such small ponds were frequent in the Upper Inn Valley, for instance. Apart from water that was too cold being directed to grasslands, another example of an erroneous historical irrigation practice in some parts of the Alps was a duration of irrigation that was too long. Excessive irrigation occurred in some places for generations, and an important consequence was leaching of much nitrogen. The threat of excessive irrigation was especially relevant in landscapes of isolated farms, where irrigation was carried out individually, not communally, and so water rights were not temporally limited (see Section 6.4.3.1). Improper irrigation could have negative impacts on hay quality. In addition to nutrient leaching, a mid-nineteenth-century record from Switzerland documents an unpleasant mossy smell of hay from irrigated meadows. Irrigation practices and their suitability in the context of local environmental conditions might have changed over the course of history.[2132] Nonetheless, the fact that grassland irrigation as a highly labor-intensive practice (Sections 6.4.2 and 6.4.3.1) was carried out in so many places and for so many centuries proves that the way it was practiced usually led to favorable results.[2133]

In some places, additional nutrients (e.g., in the form of the liquid leaving a farmyard manure pile, urine-based liquid manure from the barn, nutrient-rich drainage from roads, or sewage) were added to irrigation water (see also Section 6.3.2.1), which was applied to a limited area of meadows.[2134] A source from 1612 referring to the hamlet of Zgornje Laze, northwest of Bled (in the Eastern Julian Alps, Sln. *Julijske Alpe*), for instance, documents meadow irrigation with water, which, when it rained, allegedly also leached many nutrients from (manure at) a farmyard.[2135] Even though the court record does not prove that this defendant's statement was correct, it reveals the awareness of seventeenth-century farmers that irrigation with nutrient-rich water would make much sense. This is not surprising. A system directing a mixture of water and manure to agricultural land is attested at Rüeggisberg in the Pre-Alps south of Bern in 1572.[2136] Similarly, mid-eighteenth-century data from the Swiss northern Alps explain that directing the nutrient-rich liquid leaving farmyard manure piles into water for meadow irrigation was a frequent farmers' practice.[2137]

[2132] Leibundgut & Vonderstrass, 2016a, 49, 81, 100–101, 108, 118, 121, 130, 132, 143, 147–149, 152; Leibundgut & Vonderstrass, 2016b, 153, 186–189; Kauter, 2002, 124; Pils, 1994, 94; Grießer, 1999, 54; Wopfner, 1997, 350, 353, 359–360. See also Section 6.4.3.1.

[2133] Pils, 1994, 298–299; Leibundgut & Vonderstrass, 2016a; Leibundgut & Vonderstrass, 2016b; Mathieu, 2019, 53; Renes et al., 2020, 108, 110.

[2134] Leibundgut & Vonderstrass, 2016a, 49, 100, 104–105, 128; Kauter, 2002, 122–124; Emanuelsson, 2009, 285; Peterken, 2013, 174.

[2135] SI AS 721, Gospostvo Bled (The Seigneury of Bled, Germ. *Veldes*), Iustitialia, vol. 14, July 31st, 1612; localization is based on SI AS 721, Iustitialia, vol. 12, October 20th, 1608, and SI AS 721, Iustitialia, vol. 14, [June 16th,] 1611.

[2136] Kauter, 2002, 124.

[2137] Kauter, 2002, 121; Pfister, 1985a, 27, 110.

Data from the Central Karawanks (Germ. *Karawanken*, Sln. *Karavanke*) in the early nineteenth century—such as in Lobnig (Sln. *Lobnik*), Leppen (Sln. *Lepena*), and Remschenigg (Sln. *Remšenik*)— demonstrate that liquid manure from livestock barns or the liquid leaving manure piles near the barn was frequently directed to the nearest meadow, and such irrigation made it possible to mow second-crop hay (see Sections 6.1.2.3 and 6.3).[2138] In some places in the Alps where the slope inclination was sufficient but at the same time the landslide hazard was not serious, irrigation water was also used to distribute piles of manure, which had been brought to a meadow, over the grassland surface. In meadows close to a grazing complex[2139] in the Defereggen Valley (Germ. *Defereggental*) in East Tyrol, for instance, barn manure from the alp was placed in small irrigation ditches and distributed around the meadows through irrigation, as observed in the nineteenth or early twentieth century. Alternatively, an irrigation ditch was directed through a pile of farmyard manure near the barn in some summer grazing complexes.[2140] The option of transport of livestock feces from an alp in the Venosta Valley to montane meadows through irrigation channels was recorded as early as in 1442 (Section 6.3.2.1).

(3) Irrigation in the winter and early spring in the hills and lower valleys and basins led to an earlier beginning of the vegetation period, which—together with nutrient intake from irrigation water—allowed an additional hay harvest per year, or at least guaranteed an otherwise uncertain second harvest at the elevations where two mowings were possible if the nutrient and water availability sufficed (Section 6.1.2.3). This was relevant not only in terms of higher fodder quantities, but also because of the higher nutritive value of second-crop hay than that of first-crop hay (Section 6.1.2.3). Alternatively, an irrigation-induced earlier start of the vegetation period could allow an earlier start of grazing (Section 6.2.2.1), and so it helped overcome the frequent lack of fodder in early spring. Carefully timed irrigation in spring also limited frost damage after the beginning of the vegetation period.[2141]

2138 AT KLA 75, Franziszeischer Kataster, cadastral municipality no. 76213, Leppen (Sln. *Lepena*), cadastral appraisal protocols: Catastral-Schätzungs-Elaborat, Protocoll, June 15th, 1831 [or 1832]; AT KLA 75, cadastral municipality no. 76214, Lobnig (Sln. *Lobnik*), cadastral appraisal protocols: s.n., 246/4, 48–49, and Catastral-Schätzungs-Elaborat, Protocoll, August 31st, 1831, and Einvernehmungs-Protocoll, October 11th, 1832; AT KLA 75, cadastral municipality no. 76218, Remschenig Sln. *Remšenik*), cadastral appraisal protocols: Catastral-Schätzungs-Elaborat, Protocoll, June 13th, 1831.

2139 The location is not clear. According to Hermann Wopfner, this grazing complex was *Jöiser Alm* (Wopfner, 1997, 321), which does not exist, in line with the explanation from the municipality of St. Jakob in the Defereggen Valley (Hannes Erlsbacher, personal communication). Either the alp was abandoned long ago, or Wopfner's toponym is erroneous.

2140 Wopfner, 1997, 321; Leibundgut & Vonderstrass, 2016a, 49, 63, 105; Leibundgut & Vonderstrass, 2016b, 115, 153, 165, 170.

2141 Pils, 1994, 94, 193; Peterken, 2013, 173, 179–180; Leibundgut & Vonderstrass, 2016a, 49, 96, 99, 123, 125–128; Renes et al., 2020, 109.

(4) Less regular but also important functions of irrigation included suppression of animals that came in conflict with grassland farming, such as voles, moles, and cockchafers. Irrigation when cockchafers were flying prevented them from laying eggs in that grassland, a fully reliable strategy. Alternatively, later irrigation drowned cockchafer larvae or drove them to the surface, where they became easy prey and suffered from sun exposure (see Section 4.1.3.1). Many voles and moles were drowned in their burrows through irrigation (see Sections 4.1.3.2 and 6.6.1), but some escaped to microelevated relief or vegetation. Christian Leibundgut's personal experience proved that birds of prey became used to this irrigation purpose and learned that meadow irrigation quickly produced easy prey for them.[2142]

Nonetheless, grassland irrigation in some places and historical periods also had a contrary impact: it helped some animal species come into conflict with grassland farming. Namely, the livestock parasite common liver fluke (*Fasciola hepatica*), which is frequently lethal for sheep, whereas it causes suffering in cattle and economic losses in cattle farming without usually being lethal, finds suitable habitats in irrigation channels and can be spread around a grassland through irrigation.[2143] Common liver fluke's development stage of metacercariae attached to grasses and hazardous for livestock can survive in hay "for a short time, depending on temperature and humidity during drying and storage; this can be from two to three weeks."[2144] Thus, in historical farming, when most livestock were fully dependent on grazing during first- and second-crop haymaking (Sections 6.1.2.1, 6.1.2.2, and 6.1.2.3) and usually for a long period afterward, common liver fluke metacercariae in hay from irrigated meadows usually perished before the hay was consumed. The main exceptions were (1) working animals, such as oxen, which received some freshly mowed grassland fodder; (2) ill livestock fed in the barn and potentially more sensitive to liver fluke due to general weakness; and (3) the feeding regime in higher montane meadow complexes, where livestock were kept in the autumn; there, second-crop haymaking was soon followed by limited barn-feeding (Section 6.3.2.1), and so the risk of potential active common liver fluke metacercariae consumption with hay from irrigated meadows was higher. (4) Similarly, unusually early snow in the autumn would cause an increased risk of liver fluke infestation from second-crop hay from irrigated meadows at lower elevations because second-crop hay consumption in such a case started soon after mowing. In general, cold autumn weather might have also prolonged the survival of common liver fluke metacercariae

---

[2142] Leibundgut & Vonderstrass, 2016a, 92, 96–97, 127, 143, 149; Leibundgut & Vonderstrass, 2016b, 153, 158, 165; Wopfner, 1997, 350; Peterken, 2013, 175; Renes et al., 2020, 109.

[2143] Leibundgut & Vonderstrass, 2016a, 149; Ballweber, 2024; *Veterina Novo mesto:* Fasciola hepatica – *veliki metljaj.*

[2144] Ballweber, 2024.

in second-crop hay in a cold barn for some additional weeks.[2145] All in all, the risk of common liver fluke reaching livestock hosts through hay from irrigated meadows was moderate in many places; during the time it was consumed, most hay from irrigated meadows was safe. Grazing in irrigated hay meadows was probably less safe from liver flukes.[2146] Additional spread of common liver fluke metacercariae in hay meadows that were grazed after the last mowing, before the start of autumn grazing, was prevented through rules such as never irrigating in the autumn, as claimed by old farmers in the Ötz Valley interviewed in the first half of the twentieth century.[2147] Concerning irrigated permanent pastures—where the risk of common liver fluke infestation was generally higher than in the case of hay consumption—irrigating grazing ground was rarer than irrigating hay meadows in many regions of the Alps.[2148] In addition, the timing of pasture irrigation mattered. It was limited to spring before the beginning of the grazing season and autumn after the end of grazing, as reported in summer pastures in North Tyrol in the first half of the twentieth century.[2149] This posed a certain level of liver fluke hazard for grazing livestock, and the level of this hazard also depended on environmental conditions in the pasture and the details of pasture management.[2150] Nonetheless, it can be assumed that such irrigation timing resulted in conditions that were too dry for infesting livestock with liver fluke[2151] during a significant part of the grazing season in many pastures, unless the livestock had access to the vegetation along water channels.

Continuing with the goals of historical grassland irrigation, (5) adapted irrigation was a means of suppressing some plants considered weeds from the agricultural perspective, such as mosses through winter irrigation. Autumn crocus (*Colchicumm autumnale*) is another plant unwanted by farming (see Sections 6.2.2.2 and 6.6.1.1). It can be supported through grassland irrigation because it likes damp and relatively nutrient-rich soils, but intensive irrigation in the spring and autumn suppresses it. Moreover, site-adapted, carefully timed winter and spring irrigation also decreased the risk of evident damage in grasslands posed by fungi. Namely, irrigation shortened

[2145] A specific combination of high humidity and cold temperatures would allow the survival of common liver fluke metacercariae for up to four months (Ballweber, 2024). The humidity in a hay barn was not high, and so a cold autumn resulted in a more limited expansion of the survival of common liver fluke metacercariae.

[2146] Compare Ballweber, 2024; Novobilský et al., 2014 (gratitude is expressed to the agronomists Martin Škrlep and Jože Verbič for recommending this article).

[2147] Compare Wopfner, 1997, 350; the chronological definition of the time of Wopfner's writing follows Meixner & Siegl, 2004, 232–233, 235.

[2148] E.g., rare irrigation in summer mountain pastures in South Tyrol: Leibundgut & Vonderstrass, 2016b, 153.

[2149] Leibundgut & Vonderstrass, 2016b, 165.

[2150] Compare Novobilský et al., 2014.

[2151] Ballweber, 2024.

the duration of snow cover, thus preventing fungal attacks below persistent snow.[2152]

In addition, (6) irrigation on slopes with channels one above the other almost following contour lines helped stabilize slopes because in the case of excessive precipitation the channels also functioned as drainage ditches (Figure 92). At the same time, the channels mitigated peak discharges in the main stream.[2153]

Meadows in the Pohorje Hills (in the Central Alps, today northeast Slovenia) and at the foot of the Pohorje Hills illustrate how differences in the main goals of irrigation influenced its various timing. Information from about 1900 reveals that in several locations in the Pohorje Hills—where igneous and metamorphic rocks prevail and influx of nutrients was among the most important objectives of meadow irrigation—meadows were mainly irrigated in the spring and autumn. However, summer irrigation (e.g., as a means of drought prevention) also played a role there.[2154] To a certain degree, this is compatible with information from 1812, referring to the northern Pohorje Hills above Fala, which designated March as the (most important) month for meadow irrigation.[2155] Such early irrigation timing in the montane vegetation belt in the Alps suggests that an early start of the vegetation period was one of the goals of this practice.

On the other hand, in the milder climate at the foot of the Pohorje Hills, an early start of the vegetation period evidently was not among the main goals of meadow irrigation. There, summer irrigation was more important than in the hills. In meadows on coarse fluvial sediments at the eastern foot of the Pohorje Hills (Figure 96), drought prevention and an influx of nutrients were the main objectives of irrigation. According to information from the early twentieth century, the irrigation period there lasted from Saint George's Day (April 24th, and so it started more than a month later than higher in the hills due to a different irrigation goal) to Michaelmas (September 29th). The irrigation dates were different again in less drought-prone meadows, located on clays at the eastern foot of the Pohorje Hills. These were irrigated only during the summer drought: from Saint John's Day (June 24th) to the Nativity of Mary (September 8th).[2156] At the same time, these examples prove the great importance of fixed feasts in the agricultural calendar.[2157]

2152 Leibundgut & Vonderstrass, 2016a, 97, 126–127, 130–131; Leibundgut & Vonderstrass, 2016b, 165; Kauter, 2002, 107, 109; Theimer, 2000.
2153 Leibundgut & Vonderstrass, 2016a, 49, 96, 132–134; Leibundgut & Vonderstrass, 2016b, 91, 129.
2154 Koprivnik, 1923, 77; Ložar, 1944, 147.
2155 Kuret (ed.), 1987, 208.
2156 Baš, 1937, 327, 329, 331–333.
2157 See also examples of fixed feasts in Wopfner, 1997, 349.

Figure 96: Razvanje Creek (Sln. *Razvanjski potok*) is one of the watercourses that used to provide water for meadow irrigation at the eastern foot of the Pohorje Hills in northeastern Slovenia (see no. 22 in Figure 103). The preserved southern bank of Razvanje Creek currently rises a couple of meters above the surroundings. Photo: Žiga Zwitter, 2020.

The environmental impacts of irrigation exceeded the goals set by grassland farming. It impacted the groundwater level. The discharge in some springs below the irrigated grasslands increased significantly due to irrigation, whereas discharges in some rivers and creeks, and the water levels of some lakes decreased markedly or slightly but significantly from the ecosystem perspective (see the introduction to Section 6.4, and Sections 6.4.3.2.1. and 6.4.3.2.2). In some cases, irrigation led to the creation and preservation of small-scale wetlands and moist sites in depressions, contributing to the small-scale diversity of ecosystems in the landscape as an example of vegetation change not desired by agriculture. Impacts of irrigation on wild fauna in the soil and on grassland surfaces were significant and diverse, ranging from killing fauna through drowning or driving some animal species elsewhere to attracting some others to these habitats. For instance, some irrigated grasslands were important resting sites for migratory birds, and animals preferring occasionally wet habitats were fostered.[2158]

[2158] Emanuelsson, 2009, 287; Leibundgut & Vonderstrass, 2016a, 96, 137–139, 143, 223; Leibundgut & Vonderstrass, 2016b, 150–151, 186, 188–189.

Figure 97: A side channel is blocked with a stone plate that directs the water to a meadow. The next plate is prepared to block the channel later. Photo: Odilo Schmid in Wallis.[2159] Leibundgut & Vonderstrass, 2016a, 172. © Odilo Schmid.

6.4.2 Labor intensity of traditional grassland irrigation

Grassland irrigation used to be very labor intensive. For instance, sluices had to be opened and closed (Section 6.4.3), and numerous barriers—for example, wooden boards and stones (Figure 97), excavated pieces of turf, and at least later also sheets of iron (Figure 98), and so on—directing water to small grassland sections had to be relocated again and again. The very time-consuming fine distribution of water in meadows was carried out with an implement resembling an axe and hoe (Figure 98), and the entire process of irrigation had to be (permanently) monitored so that the entire grassland parcel was irrigated and that irrigation on slopes did not lead to erosion or mass movement.[2160] Monitoring was somewhat more challenging in the darkness of the night, in torch- or lanternlight, and moonlight. Thus, nighttime irrigation (see Section 6.4.3.1) was not only more tiring but also more likely to be less efficient (see below in this section) than daytime irrigation. Thus, higher water availability could limit the duration of irrigation to daytime. Meadow irrigation in Bur-

2159 Leibundgut & Vonderstrass, 2016b, 103.

2160 Netting, 1974, 72–73; Wopfner, 1997, 354, 360, 363–365, 367, and Figures 48, 49; Emanuelsson, 2009, 282, 285; Peterken, 2013, 174, 176, 182; Leibundgut & Vonderstrass, 2016a, especially 147, 152, 170–175; Leibundgut & Vonderstrass, 2016b, 104, 115, 140, 177, 181.

Figure 98: Fine-scale irrigation water distribution in a meadow with iron plates and an axe-and-hoe like implement—the tool used for this purpose throughout the Alps.[2161] Photo: Giovanni Vauterin in the Aosta Valley. © Giovanni Vauterin.

graviato (Germ. *Burggrafenamt*) in South Tyrol, with water from the Adige, the Passirio (Germ. *Passer*), and the Valsura (Germ. *Falschauer*), and characterized by high discharges in main irrigation channels (0.3–3 m^3/s), was only carried out in daytime. In the Venosta Valley, located nearby, where discharge in the main irrigation channels was 0.05 to 0.3 m^3/s, irrigation took place day and night.[2162] See also the note on the relation between water availability per grassland area and labor intensity of irrigation in Section 6.4.3.1.

[2161] Leibundgut & Vonderstrass, 2016a, especially p. 175; Leibundgut & Vonderstrass, 2016b, especially p. 140.

[2162] Compare Wopfner, 1997, 348–351, 358, 360, 363, 366.

At Törbel in Wallis, located in a relatively dry section of the Upper Rhône Valley, a "larger-than-average landowner might have water rights requiring irrigation of some plot almost every day of the week. To actual watering time of one to four hours, must be added a round trip of up to an hour and a half," as observed around 1970.[2163] According to early twentieth-century computation from the Visp Valley (Germ. *Vispertal*) in Wallis, where grassland irrigation was highly labor intensive, a certain farmer keeping four adult cattle spent about 80 working days in the summer in meadow irrigation; that is, at least twice as much as in mowing, hay drying, and storing taken together.[2164] One should also consider that irrigation during second-crop hay growth (Section 6.1.2.3), which was a general rule in irrigated landscapes where drought hazard was relevant, fell "at the height of the summer agricultural season, when time is at a premium."[2165] All this, together with the irrigation infrastructure maintenance outlined in the next paragraph, explains why it was said in the dry Wallis in the sixteenth century "that water cost the people more than wine."[2166]

Irrigation channels and the rest of irrigation infrastructure, such as sluices, had to be regularly maintained. The main channels were often a couple of kilometers long, with the maximum length of a single main channel in the Alps of 32 km (in the Nendaz Valley [Fr. *Val de Nendaz*], a side valley of the Upper Rhône Valley in Wallis, south of Savièse that is mapped in Figure 4); it functioned from the second half of the nineteenth century until 1964. In some regions, several main lines built in the Middle Ages and later started at high elevations, sometimes holding water from close to 2,000 m or more, very exceptionally from more than 3,000 m. In addition, the numerous side channels had to be kept in good condition. Maintenance involved removal of deposits and repair of damage in the channel sections caused through water erosion, avalanches, slope processes (such as rockfalls, landslides, mudflows, etc.), and/or fallen trees; it also decreased the losses of water into the ground (Figure 94).[2167] Channels were cleared of vegetation,[2168] and damage to the sides of channels caused by trampling during grazing in meadows (Section 6.2) was repaired.[2169] Thorough adaptation of irrigation systems had to be carried out after changes in the courses of river and creek beds during flood events[2170] and

2163 Netting, 1974, 67–68 (quote from p. 68).
2164 Netting, 1974, 68; Mathieu, 1996, 145–146, 159.
2165 Netting, 1974, 68–69 (quote from p. 69).
2166 Mathieu, 2019, 53 (quote); see also Mathieu, 1996, 145–146, 159.
2167 Netting, 1974, 69, 72–73; Wopfner, 1997, 353–354, 356–360, 366; Emanuelsson, 2009, 285; Peterken, 2013, 178; Leibundgut & Vonderstrass, 2016a, especially pp. 48, 83, 86, 155–175; Leibundgut & Vonderstrass, 2016b, 89, 93, 96–97, 99–121, 127–131, 134–140, 142, 144–147, 149–151, 153–163, 167–173, 175–177, 179–185.
2168 Peterken, 2013, 178, 307; Zwitter, 2014e, 346.
2169 Wopfner, 1997, 355; Emanuelsson, 2009, 285; Peterken, 2013, 178.
2170 Zwitter, 2014e, 346–348.

after strong earthquakes. For instance, in 1855, two earthquakes took place in Wallis (on July 25th and 26th), both with a magnitude of about 6. The epicenters were located in the immediate vicinity of the village of Törbel, where complex grassland irrigation systems existed. More than twenty-one springs near Törbel dried up due to the 1855 earthquakes, one new spring appeared, and numerous springs carried more water than previously.[2171] Similar were the impacts of these earthquakes on grassland irrigation systems near other villages in the vicinity,[2172] which required adaptations of the irrigation systems.

There were significant differences in the labor intensity of grassland irrigation within the Alps and also at the same location in time as a consequence of factors such as historical expansion of irrigation systems, an increasing percentage of time invested in walking to the meadows and back because the parcels belonging to a farm gradually became increasingly dispersed where partible inheritance prevailed, and so on.[2173]

In general, additional labor input through grassland irrigation increased yields (see Section 6.4.1). A saying from the Upper Engadin (Romansh *Engiadina*), locally subject to serious temporary agricultural drought, goes: "The more water channels, the more hay!"[2174] Nonetheless, transgressing a certain boundary of labor intensity in farming led to declining labor efficiency. There was "an overall tendency" in historical agrarian societies "to increase land-use efficiency (output per unit of area) at the expense of labour efficiency." Consequently, "in the long run, the material/energy output per capita reache[d] a limit or even start[ed] to decline."[2175] Thus historical grassland irrigation systems in general did make possible higher agricultural densities in the Alps than non-irrigated grassland management, but, when irrigation workload transgressed a certain boundary, more irrigation works contributed to "declining per capita wealth."[2176]

In many European regions, a high percentage of non-agricultural income in the integrated farm economy decreased the amount of labor invested in local farming.[2177] Thus, where non-agricultural sources of income played a major role in the integrated peasant economy, labor intensity was a limiting factor for historical irrigation. Consequently, intense engagement in traffic and road

---

2171 Fritsche et al., 2010, 2–3, 10–17; the irrigation systems of Törbel were studied by Robert Netting (1974), who failed to find this information.

2172 For instance, numerous springs involved in grassland irrigation dried out at Visperterminen at that time (Niederer, 1996c, 88).

2173 Netting, 1974; Mathieu, 1996, 145.

2174 Leibundgut & Vonderstrass, 2016b, 143.

2175 Krausmann, Weisz, & Eisenmenger, 2016, 77, 79 (quotes from p. 79).

2176 Compare Wopfner, 1997, 367 and Krausmann, Weisz, & Eisenmenger, 2016, 77 (quote).

2177 Compare Panjek, 2015b, 27–28, 107–108; on the concept of the integrated peasant economy, see Panjek, 2017.

maintenance could be one of the factors hindering the development of grassland irrigation. This seems to be (one of) the reason(s) why in Graubünden (Ital. *Grigioni*; Romansh *Grishun*), where grassland irrigation used to be widespread, the areas around the passes Splügen (Germ. *Splügenpass*, Ital. *Passo dello Spluga*), San Bernardino (Germ: *Bernhardinpass*, Ital. *Passo del San Bernardino*), Septimer (Germ. *Septimerpass*, Ital. *Passo del Settimo*), and Julier (Germ. *Julierpass*, Ital. *Passo del Giulia*) were exceptions, so that many grassland patches potentially suitable for irrigation remained non-irrigated there.[2178]

Nonetheless, the that past reality was usually more complex to explain the absence of grassland irrigation by one factor alone. Labor intensity was likely one of the factors leading to the absence of grassland irrigation in the village of Alagna (Walser Germ. *Im Land*; northeast of Turin) as an example of those neither exceptional nor frequent settlements in the southern Alps "from the Venetian lands over Lombardy into the Piedmont" characterized by large-scale economically profitable temporary male emigration lasting throughout the vegetation period. Considerable flows of temporary migrations of craftsmen from this village started in the fifteenth or sixteenth century at the latest, and "to judge from the exceedingly pronounced seasonality of births, it would seem that at least from the early seventeenth century agro-pastoralism was in Alagna as strongly subordinated to emigration as it was in the nineteenth century." Strong flows of temporary emigration from Alagna lasted until the 1930s, interrupted only by short intervals, usually in times of crises. This means that a large percentage of men were absent during the time of most farmwork. Consequently, women, children, the elderly, and a limited number of hired workers had many more tasks to do in agriculture than in most farm economies in the Alps. Unlike the relatively close Aosta Valley, where grassland irrigation was usual, farmers from Alagna did not irrigate the meadows. Nonetheless, despite the lack of male workforce, not all agricultural work in Alagna was optimized from the labor perspective: tending cattle in high summer mountain pastures and carrying out dairy operations in summer alps in Alagna were individual tasks, requiring a much larger workforce than communal alp management would have employed; individual alp management continued there from the late medieval permanent settlement foundation onward, and so cultural history and traditions could also impact economic history in the long run. Another social factor that might have contributed to the absence of grassland irrigation in Alagna was the involvement of this village in long-distance cattle movements, and so not so much hay was needed because many cows from Alagna overwintered in lower parts of the Sesia Valley (Ital. *Valsesia*) or even in the distant flatlands of Piedmont and Lombardy—a practice documented as "an old custom" in 1840s. Finally, the climate of Alagna is humid (with about

---

[2178] Leibundgut & Vonderstrass, 2016b, 141; compare Viazzo, 1989, 36–38, 142–143.

1,500 mm of annual precipitation at Alagna and more in the mountains; there is usually no water deficit, in contrast to the nearby Aosta Valley—these data refer to the last period of significant irrigation in the Alps before sprinkling irrigation; i.e., the 1920s to 1950s). Therefore, drought hazard as one of the main motivations for large-scale irrigation usually did not represent an issue there,[2179] whereas other functions of traditional irrigation did not lead to its introduction in the complex socioeconomic and environmental reality of this village.

6.4.3 Social factors limiting access to irrigation water

6.4.3.1 Irrigation cycle

In compact villages where parcels belonging to the same farm are scattered around the village territory, grassland irrigation required the cooperation of many fellow villagers, and so irrigation of a single parcel was limited to a time-span lasting from several hours (exceptionally only half an hour or less) to somewhat more than a day.[2180] The irrigation sequence of a larger number of parcels formed an irrigation cycle, which lasted for at least several days or one week, but often two or three weeks, and exceptionally four; the number of weeks covered by one irrigation cycle was not always an integer.[2181] The duration of an irrigation cycle depended, for example, on the grassland area, climate, soil properties, and water availability, as well as on social factors, and so it could differ significantly not only in neighboring villages, but also in different irrigation systems within a village where more than one irrigation system existed. At Törbel in Wallis, the cycle of the Augstborderin historical irrigation system—that is, the largest out of the three irrigation systems of the village, each providing water to different meadow complexes—lasted for twenty-one days, whereas the irrigation cycle of the smaller Felderin irrigation system at

[2179] Viazzo, 1989, 103–108, 110–117, 125, 127–135 (second quote from page 135), 146–152, 167–169, 171, 231, 247, 249–253, 255, 260–261, 279–280, 293–294; Mathieu, 2019, 59 (first quote), 61, 63. Precipitation and humidity: Pinna, 1989a; Pinna, 1989b.

[2180] Netting, 1974; Wopfner, 1997, 352, 362–366; Leibundgut & Vonderstrass, 2016a, 79–80, 82–86; Leibundgut & Vonderstrass, 2016b, 100, 103–104, 108, 152, 179, 189; Baš, 1937, 332–333. In a small irrigation system at Klopein (Sln. *Klopinj*) in Carinthia (Germ. *Kärnten*, Sln. *Koroška*), each of the four farms involved was entitled to an irrigation time slot of forty-two hours, and so the irrigation cycle covered one week, in line with the regulation from 1718. It is clear that every farmer involved had the irrigation right in a clearly specified block of forty-two hours (such as from Sunday at 6 am to midnight on Monday night, and so on), but it is not clear whether each farmer involved irrigated one or more meadow parcels located in the same meadow complex within those forty-two hours (AT StiAStP, Grundherrschaft Eberndorf (Handwritings), vol. 190, Gerichtsprotokoll 1718–1720, 82–83 (October 11th, 1718)).

[2181] Wopfner, 1997, 362–365; Leibundgut & Vonderstrass, 2016a, 79, 81–86; Leibundgut & Vonderstrass, 2016b, 100, 103–104, 152, 179; Baš, 1937, 332–333.

Törbel (irrigating 19 ha) lasted for sixteen days, as recorded around 1970. At the eastern foot of the Pohorje Hills (now northeastern Slovenia), the irrigation cycle lasted for fourteen days in the village of Razvanje, ten days in the smaller neighboring village of Bohova, and one week in the even smaller village of Slivnica, all in the 1930s.[2182]

The number of repetitions of an irrigation cycle per year depended on the soil, climate, weather and water availability, and vegetation (see Section 6.4.1), and also on human factors, such as locally usual farming and legal customs, both rooted in tradition.[2183] An example of the influence of local traditions on irrigation timing is that meadows were not irrigated immediately after first-crop hay mowing in some parts of Central Europe, such as the Langete Valley (Germ. *Langetetal*) northeast of Bern, although drought was a relevant hazard at that time in many years, to prevent the water from carrying away the seeds shed during haymaking, which maintained meadow productivity in the long run (Sections 6.5, 10.1.2.2.3, and 10.1.2.2.4).[2184] Another example of the influence of traditions on irrigation timing were the first and last day of the year when grassland irrigation was an option; these were usually determined by a fixed saint's day or another fixed day of the week close to a specified saint's day. Thus, in the 1930s, villages at the eastern foot of the Pohorje Hills irrigated their drought-prone meadows on coarser sediments from Saint George's Day (April 24th) to Michaelmas (September 29th), whereas less drought-prone meadows on clays were irrigated in the same region during summer drought, from Saint John's Day (June 24th) to the Nativity of Mary (September 8th).[2185] Meadow irrigation in Törbel's Felderin system lasted from the first Saturday in April to early September, as recorded around 1970.[2186]

In the relatively dry Ötz Valley,[2187] old farmers interviewed in the first half of the twentieth century claimed that one irrigation was needed during first-crop hay growth, and two more in times of second-crop hay growth.[2188] At a similar historical time, meadows in the dry Venosta Valley were irrigated five to twelve times annually;[2189] of course not during haymaking (but see an example of irrigation in an inner-Alpine dry valley during the blossoming of many tall meadow flowers such as oxeye daisy (*Leucanthemum vulgare*), thus in a relatively late phase of plant development, in Figure 98). At sunny sites at lower elevations of the central Venosta Valley, irrigation cycles were repeated several

2182 Baš, 1937, 332–333; Netting, 1974.
2183 Wopfner, 1997, 349, 363.
2184 Leibundgut & Vonderstrass, 2016a, 148–149.
2185 Baš, 1937, 332.
2186 Netting, 1974, 70.
2187 Bätzing, 2015, 37.
2188 Wopfner, 1997, 350; the chronological definition of the time of Wopfner's writing follows Meixner & Siegl, 2004, 232–233, 235.
2189 Leibundgut & Vonderstrass, 2016b, 153.

times from (the first half of) March until the end of October, as observed in the first half of the twentieth century. Thus at Silandro (Germ. *Schlanders*, 485 mm precipitation annually, 320 mm in the vegetation period, in the mid-twentieth century), the irrigation cycle was repeated up to ten times per year, so that each typical single plot of land involved was irrigated up to ten times annually. In spring, irrigation on drought-prone sunny slopes began earlier than on shady sites, sooner at lower elevations and later at higher ones, so that, in contrast to Silandro, irrigation at Resia (Germ. *Reschen*, about 330 mm of precipitation in the vegetation period and twice as much annually in the mid-twentieth century) at 1,500 m in the uppermost Venosta Valley was only carried out between mid-June and the end of August.[2190]

The dates of the beginning and end of irrigation underwent some changes over time or they were spatially more diverse than the general picture outlined above: at the bottom of the eastern Venosta Valley, irrigation rights using water from the Church Creek (Ital. *Rio della Chiesa*, Germ. *Kirchbach*) at Naturno (Germ. *Naturns*, 483 mm precipitation annually, 320 mm in the vegetation period, in the mid-twentieth century) were only divided from Saint George's Day (there meaning April 25th) to Saint Vitus's Day (June 15th), as documented in the second half of the fourteenth century.[2191] These were times of possibly very low agricultural density after the Black Death[2192] (see Sections 3.1, 3.2.2, and 6.4.3.2), which might have reduced the duration of labor-intensive irrigation. Compare the start of irrigation in the first half of March in this region in the first half of the twentieth century, as presented above—at that time, the agricultural density in the microregion was certainly incomparably higher than in the second half of the fourteenth century.[2193] Not only the duration of the entire annual irrigation but also the complexity of many irrigation systems must have increased over time with the rising agricultural density during the early modern period and the nineteenth century, when many farms in compact villages were split.

Understanding historical irrigation cycles based on archival records is no easy task: the documentation is often basic because it was aimed to be used by people with specific knowledge about the complex irrigation sequence.[2194] The example from the archival source dealing with the irrigation cycle at Lasa (Germ. *Laas*) in the Venosta Valley (Germ. *Vinschgau*) using water from Alliz Creek (Germ. *Allitzbach*) in Table 1 demonstrates this.

2190 Wopfner, 1997, 349, 363 (the chronological definition of the time of Wopfner's writing follows Meixner & Siegl, 2004, 232–233, 235); Leibundgut & Vonderstrass, 2016b, especially p. 149.
2191 Wopfner, 1997, 349; Leibundgut & Vonderstrass, 2016b, 149.
2192 Fornasin, 2017, 64–66; Izdebski et al., 2022.
2193 Compare Livi Bacci, 2000.
2194 Compare Wopfner, 1997, 364, and Leibundgut & Vonderstrass, 2016b, 179.

Table 1: Water rights allocation for the fifteenth irrigation day in the third out of eight annual irrigation cycles from spring until autumn using water from Alliz Creek (Germ. *Allitzbach*) near Lasa (Germ. *Laas*) in the Venosta Valley (Germ. *Vinschgau*) as recorded in 1949. The "Line number" column has been added to facilitate orientation in the table. Data source: Lechthaler family archive, Lasa, notebook Aufschrebung [!] für den Allitzbach im Jahre 1949, page starting with "15. III. Rod."

Line number	Farm owner	Number of irrigation time slots (*Weilen*)	Share of total irrigation water
1	Rosa Holze[r]	1	¼
2	Franz Alber	½	¼
3	Mathies Reich	2½	¼
4	widow Telfser	2, *Tagweilen*	¼
5	Johann Marx	2	¼
6	Kaspar Telser	4	¼
7	Josef Wallnöfer	3	¼
8	Anton Wellenzohn	3	¼
9	Jakob Lechthaler	2	¼
10	Herman Niederfriniger	2	¼
11	Josef Reich	2	¼
Nachmittag			
12	Johann Wellenzohn	6	⅓
13	Anton Gruber	2	⅙
14	Martin Metz	1	⅙
15	Matascher	1½	⅙
16	Franz Gemasmer	1½	⅙
17	Barbara Lechthaler	1½	⅓
18	Niederholzer	4½	⅓
19*	Johann Gruber	6	⅙

* Line 19 was added later in a different hand in different ink, but Johann Gruber was evidently added because he was originally missing as a mistake. This is proven by the total duration of the irrigation cycle including that entry (12 *Weilen* using all the irrigation water available: 12 is the sum of every duration of the irrigation time multiplied by the share of water used) and the comparison with the fifteenth irrigation day in the other irrigation cycles in this system, which also include Johann Gruber's irrigation slot as documented in the original script and ink of this source.

Other sections of this source make it clear that the order of the farms in Table 1 does not linearly follow the time of the day, but there are entities of one to four farms (not evident in Table 1) irrigating at the same time, as clearly documented at least for the second, third, fourth, and fifth out of the eight irrigation cycles in the irrigation system analyzed. Each such entity consisting of one to four farms was entitled to irrigate its parcels for six out of the twelve *Weilen* (twelve *Weilen* lasted for 24 hours).[2195] Thus, the farms in lines one to four formed one such entity, meaning that the morning irrigation time (starting around 3 am and lasting until 11:30 am) of one group of farms begins in line 1 and ends in line 4; it begins again in line 5 and ends in line 6, begins in line 7 and ends in line 8, and begins in line 9 and ends in line 11. Each of these groups of farms entitled to irrigate at the same time used a quarter of the irrigation water available in this case (see the column "Share of total irrigation water" in Table 1). Note that *Nachmittag* in the source does not only refer to the afternoon, but designates a longer part of the irrigation day, including most of the night, until the beginning of the morning irrigation time. In this case, the farms entitled to irrigation in the first out of the two unequally long parts of the 24-hour irrigation day in the irrigation cycles designated by an odd number (such as the third one presented here) had the irrigation right in the other part of the day during the irrigation cycles designated by even numbers, and so the farm's irrigation timing in the same parcel on the same consecutive day of the irrigation cycle shifted from one irrigation cycle to the other. In the fourth line, the word *Tagweilen* (i.e., irrigation slots during the daylight) designates that fact that in the mid-twentieth century on the fifteenth day of irrigation cycles the widow Telfser was consistently entitled to irrigation as the last one within the group of farms in lines one to four, when she was entitled to irrigation in the part of the day ending in late morning—and she was the first one starting the irrigation in the middle of the day when it was her turn to irrigate in the part of the day that started at 11:30 and lasted until the night on the fifteenth day of an irrigation cycle. This is clearly proven by further evidence in this source. Due to the widow Telfser's privileged irrigation timing (always in the daylight, in late morning, or around noon and in early afternoon), the irrigation slots of the farms in lines one to three in Table 1 had to be more frequently scheduled during the nighttime (in early morning or between late evening and around 3 am), in the second to fifth irrigation cycle at least[2196] (see also the beginning of Section 6.4.2).

[2195] Concerning the usual division of 24 hours into 12 *Weilen*, see Wopfner, 1997, 363.

[2196] Lechthaler family archive, Lasa (Germ. *Laas*), notebook Aufschrebung [!] für den Allitzbach im Jahre 1949, fifteenth day of every irrigation cycle; time orientation on the third sheet before the sheet designated by the letter C at the beginning of the notebook; and the unbound sheet starting with the words "15. Rod" in the end of this notebook. The page starting with "15. Rod" provides specific comments relevant to understand the irrigation rights of Mathies Reich (whose name stands in Table 1, line 3) on the fifteenth day of the irrigation cycles; "15. Rod" does not designate a fifteenth irrigation cycle (this is confirmed by (1) the fact that the irrigation cycle was not repeated so many times annually in this case, and (2) the comparison with the content of the tables dealing with the fifteenth day of irrigation cycles in the same source).

The complexity of this irrigation system at Lasa was overlooked in Hermann Wopfner's study.[2197] He probably collected the data based on interviews with participating farmers, without paying enough attention to how well the interviewees knew the functioning of the entire irrigation system, far exceeding their own irrigation rights.[2198]

The data presented on the irrigation system at Lasa shows that irrigation timing was constant in some irrigation systems, and so some parcels would always have their irrigation right in the same part of the day or night,[2199] whereas it moved regularly in some others,[2200] such as that mentioned above in the case of Alliz Creek, so that irrigation during the night and daytime was in general more fairly distributed over the long run, with possible exceptions, such as the widow mentioned above (see Section 6.4.2). Netting's detailed study of the irrigation at Törbel demonstrated that the start of the irrigation cycle of the Augstborderin irrigation system moved by six hours every time, "so that watering times for an individual rotate[d] regularly through the 24-hour period."

2197 Wopfner, 1997, 363.

2198 Namely, the details corrected in this footnote are not given in the section of the village bylaw cited by Wopfner. Wopfner was known for his close contacts with farmers and extensive fieldwork (compare Meixner & Siegl, 2004, 228–229, 235), which suggests that he collected his data during his talks with farmers. The main author of this volume's preliminary analysis of the division of archivally documented irrigation water rights from Alliz Creek near Lasa written down in 1949 identified the following fundamental mistake in Wopfner's publication. His statement "Four authorised users have a share in each *Weile*; each of them receives a quarter of the *Bach* ('stream'), i.e., of the water flowing along the main channel. In this way, 24 authorized users are satisfied on one day" (Wopfner, 1997, 363) is much too simplified and erroneous. Indeed, many irrigation slots in this system allocated a quarter of the water from the main channel to one authorized user, but this was not a general rule: the right to use a half, a third, a sixth, or an eighth of the water from the main channel was also nothing special there (see Table 1) and, rarely, a different share of the total water in the main channel was allocated to one user. In addition, most irrigation slots lasted for more than one *Weile* (see Table 1) and a few users irrigated at the same time because each one was only entitled to a specified share of the total water from the main channel. In Table 1, there are 24 *Weilen* × ¼ water from the main channel + 12 *Weilen* × ⅓ water from the main channel + 12 *Weilen* × ⅙ water from the main channel. Consequently, 19 farm owners were allowed to make use of the irrigation water on that day that consisted of 12 *Weilen* altogether, but a few farmers were irrigating at the same time. As a general rule, the number of farms irrigating their parcels in this irrigation system per day was significantly less than 24 (which is the figure given by Wopfner) (Lechthaler family archive, Lasa (Germ. *Laas*), notebook Aufschrebung [!] für den Allitzbach im Jahre 1949).

2199 For instance, at the eastern foot of the Pohorje Hills as recorded in the 1930s (Baš, 1937, 332). At Klopein in Carinthia, the forty-two-hour irrigation timeslots were constant, and so some farmers' irrigation right was more daytime centered and that of some fellow villagers more nighttime centered according to the 1718 ruling (AT StiAStP, Grundherrschaft Eberndorf (Handwritings), vol. 190, Gerichtsprotokoll 1718–1720, 82–83 (October 11th, 1718)).

2200 Leibundgut & Vonderstrass, 2016b; e.g., p. 179; see also Lechthaler family archive, Lasa (Germ. *Laas*), notebook Aufschrebung [!] für den Allitzbach im Jahre 1949.

Augstborderin's irrigation sequence was spatially logical, with contiguous parcels irrigated successively, as recorded around 1970.[2201] The irrigation sequence in Törbel's Felderin was much more complex than that of Augstborderin, despite Felderin's smaller size: it included many regular changes in the irrigation sequence and in the duration of irrigation in the same meadow from one irrigation cycle to another in the same year and sometimes also between years. Often, noncontiguous parcels were irrigated successively in Felderin, which caused water losses because the dry channels had to first become wet.

The quantities of irrigation water needed and the water available per spatial unit strongly depend on the microclimate and the weather, soil, vegetation, and human factors.[2202] There were huge differences in the rate of (in)equality in the distribution of water rights in different villages but also within some villages. At Wenns (in the Pitze Valley [Germ. *Pitztal*], Tyrol), every 650 m^2 of meadows had the right to an hour of irrigation within an irrigation cycle in 1930s or 1940s, if Hermann Wopfner's information is reliable. Similarly, at the same time, at Prato allo Stelvio (Germ. *Prad am Stilfser Joch*) in the Venosta Valley, every meadow area that would be mowed by one haymaker in a day (Wopfner's estimate in this case: c. 2,300 to 2,600 m^2) had an irrigation right lasting for four hours within the same irrigation cycle, which involved 203 land parcels. In contrast, according to a regulation for the community of Lana—with water from the Foiana River (Ital. *Rio Foiana*, Germ. *Völlanerbach*) or the Brandis River (Ital. *Rio Brandis*, Germ. *Brandisbach*) from 1720, every meadow area that would be mowed by one haymaker in a day (Wopfner's estimate in this case: c. 2,900 m^2) had the right to only two hours of irrigation.[2203] Grassland irrigation rights at Slivnica east of Pohorje on a selected day (24 hours) in the 1930s included time slots per farm lasting from half an hour to eight hours (average 2.4 hours, median 2.0 hours); due to the weekly irrigation cycle, only some households had irrigation rights on the day presented here.[2204] Usually, absence of relevant pieces of information—such as the discharge in the channels, grassland site microenvironmental conditions, and plant species compositions—limits our understanding of the (in)equalities presented and results in less telling comparisons.

The chronological orientation in uniformly defined hours, such as in the examples above from Prato allo Stelvio, Wenns, and Slivnica, makes it clear that the water right allocations presented were in fact not unaltered for many centuries, but instead the results of modernization of local irrigation systems when they were adapted to the mechanical clock, probably in recent centu-

---

[2201] Netting, 1974, 69, see also p. 73.

[2202] Netting, 1974; Wopfner, 1997, 349–352, 355–357, 359, 362–367; Leibundgut & Vonderstrass, 2016a, 130; Leibundgut & Vonderstrass, 2016b, 181, 186–189.

[2203] Wopfner, 1997, 352, 362–363; the chronological definition of Wopfner's data at Wenns and Prato allo Stelvio follows Meixner & Siegl, 2004, 232–233, 235.

[2204] Baš, 1937, 332.

ries.[2205] Törbel's Felderin irrigation system, as analyzed by Robert McNetting around 1970, still involved centuries-old elements, such as delimitation of irrigation time slots by movable natural time markers—mainly shadow or the rays of the Sun reaching a certain landscape structure instead of fixed hours. The same type of chronological orientation persisted until the twentieth century in some other regions as well, such as the Venosta Valley. The irrigation right for a single parcel (in three irrigation systems) at Törbel usually lasted from about an hour to four hours, but in some cases from 20 minutes to six hours, as recorded around 1970. Netting's preliminary calculation of water directed to individual parcels per unit of land and time in the Felderin irrigation system showed "gross inequalities in the amount of water available per unit of land." "[S]ome plots ha[d] a minute of water for 3.7 square meters of area, whereas the same amount of time [had to] serve for an average of almost 50 square meters in other meadows."[2206] Shorter duration of an irrigation right per unit of land usually meant more frequent redirections of waterflows within the meadow by the farmer carrying out irrigation to water the entire parcel in the time slot available, and so such irrigation was even more labor-intensive than usual.[2207] Concerning such severe inequalities in irrigation time allocated per unit of meadow space, it can be assumed that human factors had a more decisive influence on these inequalities in some cases than discharge differences within the same irrigation system and the diversity of grassland site microenvironmental conditions.[2208]

Some irrigation systems were flexible enough to allow allocation of additional irrigation water if another farmer or legal entity did not use its irrigation rights alone, such as in Törbel's Felderin around 1970.[2209] In some other places this was forbidden, and the question whether this was allowed or not changed over time in some cases.[2210] More usual was seemingly the right to use one's own irrigation time slot in another meadow that needed water more urgently than the one whose turn it would otherwise be.[2211]

Another limiting impact of the irrigation cycle on water availability for grasslands was relevant in some minor parts of the Alps with a relatively dry climate, and it was rooted in the broader economic history context. There, ir-

[2205] Compare the history of measuring time in Makarovič, 1995, 64–67, 69–80, 119–126, and Wopfner, 1997, 364.

[2206] Netting, 1974, 68–72, 73 (both quotes), 74; compare Wopfner, 1997, 364; a general statement concerning the temporal orientation in the frame of historical irrigation systems in the Venosta Valley: Leibundgut & Vonderstrass, 2016b, 152. An example is the documentation of the rights to use irrigation water from Alliz Creek near Lasa in the Venosta Valley from 1949 (Lechthaler family archive, Lasa (Germ. *Laas*), notebook Aufschrebung [!] für den Allitzbach im Jahre 1949); this source uses movable natural time markers but at the same time adds the conversion to fixed hours in many cases.

[2207] Leibundgut & Vonderstrass, 2016b, 182.

[2208] Compare Netting, 1974, and Wopfner, 1997, 359.

[2209] Netting, 1974, 72–73.

[2210] Wopfner, 1997, 364.

[2211] Netting, 1974, 71–72; Wopfner, 1997, 364.

rigation of arable fields in spring was very important, and so at some drought-prone sites it had priority over grassland irrigation. This was the case in the Venosta Valley, as reported for the nineteenth century and first half of the twentieth century; nonetheless, arable land required much less water per unit of land than grasslands, and so this limitation was only relevant in a minor part of the grassland vegetation period. Irrigation doubled the cereal yields in the Venosta Valley. That region had a tradition of growing more cereals than necessary to cover the local demand. Cereals were sold from the Venosta Valley for a long time, until the drop in prices of long-distance cereal transport in the mid-nineteenth century due to rail transport, and so it is understandable that economically highly relevant cereals were prioritized in the irrigation cycle in spring. Similarly, irrigation of cereals had priority over grasslands at Stanz in North Tyrol, another relatively dry place, as reported for the time before the end of cereal cultivation in the twentieth century.[2212] When there was a need to water arable land and meadows, fields were irrigated in the day and meadows at night in some of these irrigation systems (compare Section 6.4.2).[2213]

The absence of an agreed-upon irrigation cycle was a rarity in regions of Central Europe where irrigation was a frequent practice. Nonetheless, locally such cases also posed human limitations on access to irrigation water. An example of a complex situation of this kind was the irrigation of meadows on drought-prone coarse alluvial deposits at villages further east from the eastern foot of the Pohorje Hills (such as from Brezula to Stražgonjca), where water availability was far from sufficient, as recorded in the 1930s. There was disorder and competition for irrigation water, during the day or at night, following every rain after the first-crop hay harvest in June. It depended on how quick, plucky, and persistent a person was, whether the water for the meadow managed to be secured, and only those that succeeded also mowed the valuable second-crop hay (see Section 6.1.2.3) in those meadows. Such grasslands were guarded by up to five people when irrigation was in progress so that nobody would direct the water away from the parcel.[2214]

Apart from the irrigation cycle, another human factor that limited the availability of irrigation water per spatial unit in some cases was rooted in cultural history. In line with not fully standardized perceptions of prohibited and allowed farmwork on Sundays and other holy days (see Section 4.1.1.4), in some places irrigation was considered prohibited on these days. Although more drought-prone sites were to a certain degree more likely to be irrigated on Sundays and other holy days, such a deterministic explanation that can be found in the literature[2215] is too simplified, and the interplay between environmental and human

[2212] Wopfner, 1997, 351–352, 363, 367; Leibundgut & Vonderstrass, 2016a, 59, 131; Leibundgut & Vonderstrass, 2016b, 178–181, 189; precipitation map: Bätzing, 2015, 37.
[2213] Wopfner, 1997, 363.
[2214] Baš, 1937, 327, 333.
[2215] Wopfner, 1997, 349, providing only the deterministic explanation.

factors mattered in this context. Thus, the following example from the relatively dry Venosta Valley[2216] also emphasizes the importance of the cultural history factor dealt with in this paragraph. At Glorenza (Germ. *Glurns*, only 270 mm precipitation in the vegetation period, 442 mm annually in mid-twentieth century), with two Sunday Christ paintings nearby, at Tubre (Germ. *Taufers*) and Tarces (Germ. *Tartsch*)—which might have contributed to the wide range of farmwork considered prohibited on holy days (Section 4.1.1.4)—water had to be kept in the creek on Sundays, and irrigation of the fields in *Glurnserfeld* was prohibited at that time, as written down in 1489. To publicly announce an infringement of this rule, it was optional to ring the large church bell.[2217] Similarly, prohibition of irrigation on Sundays and other holy days was documented in Lagundo (Germ. *Algund*) near Merano (Germ. *Meran*) in 1648.[2218]

Nonetheless, the prohibition of irrigation on Sundays was not a general rule, not even when the standards set by the Council of Trent came into force, and in specific cases there were formal ecclesiastical exemptions (compare Section 4.1.1.4). To give but a few contrasting examples: the 1718 division of water rights in a small meadow irrigation system among four farmers from the village of Klopein (Sln. *Klopinj*) in the Jaun Valley (Germ. *Jauntal*, Sln. *Podjuna*), Carinthia, partitioned a week into four equal time slots, each covering 42 hours; each time slot belonged to one farmer. It is clear that irrigation there was also carried out on Sundays, even though this division of irrigation rights was the result of a decision made at the patrimonial court of an ecclesiastical seigneury, the seigneury of Eberndorf (Sln. *Dobrla Vas*),[2219] at that time in the hands of Jesuits.[2220] In part of the dry Visp Valley in Wallis, there was special ecclesiastical permission for irrigating on Sundays, which would otherwise have been prohibited.[2221] This exemption is likely reflected in Törbel's Felderin irrigation system in the same valley. There, the church had "rights to all water on Sundays" and individual farmers purchased one-hour Sunday irrigation time slots from the church, as noted around 1970,[2222] but this rule reflects the situation before the Second Vatican Council, which was followed by a general abolition of the ban on any farmwork on Sundays and other holy days in Roman Catholicism (compare Section 4.1.1.4). The question whether this was a later adaptation of potentially irrigation-free Sundays in Felderin earlier in history remains open. Even in the Venosta Valley—examples of prohibition of

---

[2216] Compare Wopfner, 1997, 348–351, 358.

[2217] Grass, 1950, 168–169; Wildhaber, 1956, 30; precipitation data from Leibundgut & Vonderstrass, 2016b, 149.

[2218] Grass, 1950, 168.

[2219] AT StiAStP, Grundherrschaft Eberndorf (Handwritings), vol. 190, Gerichtsprotokoll 1718–1720, 82–83 (October 11th, 1718).

[2220] For a brief overview of history of the seigneury of Eberndorf, see *Archives Portal Europe: Eberndorf, Augustiner-Chorherrenstift (Urkunden)*.

[2221] Niederer, 1996d, 179.

[2222] Netting, 1974, 68–70, 72.

irrigation on Sundays and other holy days from the wider region documented in the fifteenth and seventeenth century were presented above—some communities practiced irrigation on Sundays in the first half of the twentieth century and somewhat earlier, (one of) the principle motives being the dry climate.[2223] Partly similarly to Törbel's Felderin, the example of the irrigation system using water from Alliz Creek near Lasa in the Venosta Valley, documented in 1949, proves a specific allotment of irrigation rights on Sundays, differing from the other days of the week,[2224] which again raises the likelihood that people using that system abstained from irrigation on Sundays for some time in earlier history. Individual irrigation at isolated mountain farms in the Pohorje Hills and communal irrigation systems at the eastern foot of the Pohorje Hills also operated on Sundays, as documented in the 1930s.[2225] Nonetheless, similar to Törbel's Felderin, Sunday was an exception at the eastern foot of the Pohorje Hills: due to the extensive irrigation in the Pohorje Hills on Sundays, the water quantity that reached the foot of the hills was usually so limited that the farmer entitled to meadow irrigation there on those days had an irrigation right lasting for the entire day (in line with the rule from the village of Bohova).[2226] It is explained in Section 6.4.3.2.1 that such a concentration of meadow irrigation in the Pohorje Hills on Sundays was closely connected with the spread of sawmills there, mainly in the nineteenth century. The question how irrigation at the eastern foot of the Pohorje Hills functioned before the sawmills became so numerous in the hills remains open. It is most likely that irrigation in the Pohorje Hills had been more evenly distributed throughout the week. Thus, similar to Törbel's Felderin, irrigation on Sundays might have been considered prohibited in earlier centuries, as suggested by the exceptional Sunday irrigation order in the early twentieth-century irrigation cycles at the eastern foot of the Pohorje Hills. See also the example of Sunday irrigation at Podkoren in the Sava Dolinka Valley (Sln. *Zgornjesavska dolina*) in the late nineteenth century in Section 6.4.3.2.1.

6.4.3.2 Water uses for other activities

In many places, grassland irrigation was part of a broader system of water uses, and so access to water for irrigation of grassland parcels was limited not only as a consequence of natural water availability and the phase of the irrigation cycle,[2227] but also due to other types of water uses. These often included the use of water power or fish breeding. Rarely, grassland irrigation was limited by

2223 Wopfner, 1997, 349.
2224 Lechthaler family archive, Lasa (Germ. *Laas*), notebook Aufschrebung [!] für den Allitzbach im Jahre 1949.
2225 Baš, 1937, 332–333.
2226 Baš, 1937, 332.
2227 Wopfner, 1997, 351–352; Leibundgut & Vonderstrass, 2016a, especially 78–87, 143–146, and examples from various parts of the Alps in Leibundgut & Vonderstrass, 2016b.

other purposes, such as securing water as a preventive measure to put out fires. Nocturnal irrigation was prohibited to ensure water for potential firefighting at Wenns in the Pitze Valley, according to a record from the second half of the fourteenth century.[2228] Nonetheless, the historical context of this rule from Wenns is specific—it comes from the period before the significant early-modern and nineteenth-century increases in agricultural density (Section 3.1) and labor intensity in agriculture (Section 4.1.1.4). Possibly, this rule from Wenns even stems from a period of an exceptionally low agricultural density, if the microregion was seriously affected by the Black Death (Section 3.2.2), and so it is no wonder that the interests of potential firefighting did not more generally limit access to irrigation water in the Alps later, especially not at places where water availability was higher than at Wenns.

6.4.3.2.1 Use of waterpower and irrigation

Characteristically, access to water was usually more limited for grassland irrigation than for crafts (or industry).[2229] For instance, according to an agreement from the early 1880s, the water supply for a meadow immediately south of the Alpine village of Podkoren (today in northwesternmost Slovenia), which had already had a certain irrigation tradition by that time, was unlimited, provided that discharges were sufficient for a grain mill and a sawmill below the confluence of the uppermost flow of the Sava Dolinka River and its tributary, Krotnjek Creek, to fully operate despite the irrigation. However, whenever watercourses were low, irrigation of the meadow was only allowed on Sundays and other holy days, when the grain mill and the sawmill did not operate. Thus, irrigation was allowed to start on Saturday evenings and on evenings before holidays, and to continue until the morning of the first working day for craftsmen. However, when the mills already started operating during the night after a Sunday or holiday, irrigation had to stop the previous evening.[2230]

Exceptions are also known. In the moderately humid climate of Pontresina in Engadin, for instance, milling had to stop during low water availability to allow irrigation higher in the catchment according to a 1587 ordinance; however, this regulation was changed in favor of mills in 1692.[2231] Another case is meadow irrigation at the eastern foot of the Pohorje Hills, where milling was partly hampered and partly rendered impossible when meadows were irrigated during the growth of high-quality second-crop hay (see Section 6.1.2.3); that

---

2228 Wopfner, 1997, 352; Leibundgut & Vonderstrass, 2016a, 49, 81, 96.

2229 Wopfner, 1997, 362; Leibundgut & Vonderstrass, 2016a, 49, 77, 81; Leibundgut & Vonderstrass, 2016b, 143.

2230 SI AS 137, Okrajno glavarstvo Radovljica (District Administrative Authority [Germ. *Bezirkshauptmannschaft*] of Radovljica, Germ. *Radmannsdorf*), box 121, archival unit 568: May 5th, 1880; May 16th, 1880; June 18th, 1880; October 27th, 1882.

2231 Leibundgut & Vonderstrass, 2016b, 143; on precipitation in Pontresina, see Bätzing, 2015, 37.

is, during the summer drought, according to early twentieth-century information. The situation upstream on the same creeks, in the area of mountain farms on impermeable rocks of the eastern Pohorje Hills, was different: there, grasslands were only irrigated on Sundays, so that sawmills and grain mills could operate six days a week, as recorded in the 1930s.[2232] This information refers to a time when an exceptionally high density of sawmills and the strong role of trade in lumber were characteristic for the economy of the Pohorje Hills. After the major increase in wood trade, in the nineteenth century in particular, the number of sawmills there reached its peak in the late nineteenth century and early twentieth century. The great majority of sawmills were owned by farmers because every medium and large isolated farm in this area usually had a water-powered sawmill,[2233] and so the priority of wood processing over grassland irrigation in this case is understandable in the context of the role of wood processing in the regional farm economy.

6.4.3.2.2 Fish breeding and irrigation

Fish breeding had priority over irrigation using water from a fish pond or lake in certain cases. This was not limited to small water bodies, which could dry out due to historical grassland irrigation in the Alps or where the water temperature or oxygenation in a small remaining waterbody in the summer could become an issue for fish, as explained in this section. The case of Lake Klopein (Germ. *Klopeiner See*, Sln. *Klopinjsko jezero*) in Carinthia is instructive. This lake is almost two kilometers long and at its maximum almost one kilometer wide, covering 1.1 km^2. It is a relatively deep body of water, reaching a depth of 48 m.[2234] Over a long period, fish breeding there had an impact on water availability for grassland irrigation. Fishermen were allowed to partly block the outflow from this lake at any time of the year according to a regulation from 1708, which limited the possibilities of meadow irrigation by inhabitants of Seelach (Sln. *Selo*).[2235] The ruling from 1719 introduced an important new feature: fishermen were only allowed to partly block the outflow from this lake from February 24th to May 12th due to fish spawning time.[2236] In the rest of

[2232] Baš, 1937, 332–333.

[2233] Cimperšek, 2014, 368–369, 375–382.

[2234] For the lake's size and depth: *Tiefenkarte des Klopeiner See*; Höbaus, Sommer, & Reinthaler, 2023.

[2235] AT StiAStP, Grundherrschaft Eberndorf (Handwritings), vol. 189, Gerichtsprotokoll 1711–1717, 13 (October 2nd, 1708); AT StiAStP, Grundherrschaft Eberndorf (Handwritings), vol. 191, Gerichtsprotokoll 1721–1728, 96 (July 18th, 1722), 101 (September 2nd, 1722).

[2236] AT StiAStP, Grundherrschaft Eberndorf (Handwritings), vol. 190, Gerichtsprotokoll 1718–1720, 218 (November 8th, 1719); the conversion of Saint Mathias's Day and Saint Pancras's Day into dates is based on Grotenfend & Ulrich, 1971, 78, 85; AT StiAStP, Grundherrschaft Eberndorf (Handwritings), vol. 191, Gerichtsprotokoll 1721–1728, 101 (September 2nd, 1722), 102 (September 9th, 1722); AT StiAStP, Grundherrschaft the year, the outflow had to remain free to allow meadow irrigation, to provide a regular water supply for the inhabitants of Seelach, and to operate a mill (or mills) further downstream.[2237] Thus, from 1719 onward, fish breeding still limited the grassland irrigation there, but only for two and a half months annually, including no more than the beginning of the period of the quickest growth of most grassland plants. An interdisciplinary approach linking fish ecology and economic history explains why fishery posed limits for irrigation exactly at that time of the year. Fish species and genera of economic importance in this lake at that time included pike (*Esox lucius*),[2238] carp (*Cyprinus carpio*),[2239] whitefishes (*Coregonus* sp.),[2240] and a few others.[2241] Pike and carp were the most valued fish species in Lake Klopein according to data from 1724.[2242] Whitefish were so abundant and economically relevant in Lake Klopein that in 1684 this lake was designated as *Clapeinner reinankhen see* (*Reinanken* means 'whitefish', thus "Klopein Whitefish Lake") in the account of the seigneury of Eberndorf, which received income from fishing there.[2243] Whitefish spawn earlier than the

Eberndorf (Handwritings), vol. 191, Gerichtsprotokoll 1721–1728, 409–410 (April 26th, 1728), 410 (April 27th and April 29th, 1728); AT StiAStP, Grundherrschaft Eberndorf (Handwritings), vol. 192, Gerichtsprotokoll 1728–1742, 11 (December 2nd, 1728), 242 (February 26th, 1733).

2237 AT StiAStP, Grundherrschaft Eberndorf (Handwritings), vol. 190, Gerichtsprotokoll 1718–1720, 218 (November 8th, 1719); AT StiAStP, Grundherrschaft Eberndorf (Handwritings), vol. 191, Gerichtsprotokoll 1721–1728, 96 (July 18th, 1722), 97 (July 19th, 1722), 98 (July 29th, 1722), 101 (September 2nd, 1722); AT StiAStP, Grundherrschaft Eberndorf (Handwritings), vol. 192, Gerichtsprotokoll 1728–1742, 11 (December 2nd, 1728), 242 (February 26th, 1733).

2238 AT StiAStP, Grundherrschaft Eberndorf (Handwritings), vol. 191, Gerichtsprotokoll 1721–1728, 221 (May 26th, 1724).

2239 AT StiAStP, Grundherrschaft Eberndorf (Handwritings), vol. 183, Gerichtsprotokoll 1675–1683, 148 (July 24th, 1678); AT StiAStP, Grundherrschaft Eberndorf (Handwritings), vol. 191, Gerichtsprotokoll 1721–1728, 221 (May 26th, 1724).

2240 AT StiAStP, Grundherrschaft Eberndorf (Handwritings), vol. 187, Gerichtsprotokoll 1698–1703, 123 (January 5th, 1700), 127 (January 23rd, 1700), 133 (March 3rd, 1700), 151–152 (March 8th, 1700), 157–162 (March 13th, 1700), 216 (December 29th, 1700); AT StiAStP, Grundherrschaft Eberndorf (Handwritings), vol. 191, Gerichtsprotokoll 1721–1728, 221 (May 26th, 1724).

2241 Another cyprinid (*alten*) spawning in late spring 1719 is recorded in the sources studied (AT StiAStP, Grundherrschaft Eberndorf (Handwritings), vol. 190, Gerichtsprotokoll 1718–1720, 172–173 (June 15th, 1719), 172–174 (June 17th, 1719)); compare *Frühneuhochdeutsches Wörterbuch*, 1989, columns 741 and 866). Moreover, *persten*, possibly perch, are also documented there in the early eighteenth century (AT StiAStP, Grundherrschaft Eberndorf (Handwritings), vol. 191, Gerichtsprotokoll 1721–1728, 221 (May 26th, 1724)).

2242 AT StiAStP, Grundherrschaft Eberndorf (Handwritings), vol. 191, Gerichtsprotokoll 1721–1728, 221 (May 26th, 1724).

2243 SI NŠAL 100, Kapiteljski arhiv Ljubljana (Archives of the Ljubljana Cathedral Chapter), fasc. 103, folder 2: account of the seigneury of Eberndorf, *Einkhommen auß füschwässern*. On the historical economic relevance of this genus, see also Höbaus, Sommer, & Reinthaler, 2023.

partial lake's outflow blockade introduced in 1719. Their spawning takes place between November and January; it starts when the water temperature reaches 6 °C,[2244] and so whitefish spawning was never affected by grassland irrigation. In addition to the spawning timing differing from meadow irrigation, whitefish eggs are released into open water, they are fertilized in the water, and then they sink to sand or pebbles on the bottom of the lake to hatch; however, some spawn closer to the shore and the eggs become attached to aquatic plants.[2245] Such reproduction and rare visits to shallow water during their lifetime[2246] make whitefish largely independent from the narrow belt of shallow water in Lake Klopein (see below), potentially drying out as a consequence of irrigation. In contrast to whitefish, carp spawning does depend on shallow water, but they spawn later (see the paragraph on Lake Gösselsdorf, Germ. *Gösselsdorfer See*, Sln. *Goslinsko jezero*, below in this section), and so the 1719 regulation was evidently largely oriented toward pike fishing. Namely, pike spawn in early spring (between late February and May, corresponding to the permitted historical partial blocking of the lake from February 24th to May 12th introduced in 1719), when the water temperature reaches 8 to 10 °C.[2247] Pike "[s]pawning takes place in shallow, calm waters with rich vegetation along the banks." Pike "can also spawn on flooded grassland. The most suitable substrata for laying eggs are the leaves and stems of aquatic plants."[2248] About one month is needed for the development of young pike to a stage allowing their migration to other parts of the lake. Consequently, a drop in water level in a month following spawning causes large numbers of young pike to die.[2249] Lake Klopein is relatively deep, and the shallows most relevant for pike spawning are found only in a narrow belt close to the shoreline.[2250] This makes pike eggs and pike fry particularly vulnerable to a decrease in water level, which explains why the outflow from the lake was usually partly blocked at that time, placing social limits on grassland irrigation and corresponding fairly well with the sensitive phase of pike development. This makes it clear why such a limitation on irrigation was needed despite the fact that water availability in this lake was never an issue, but the limited habitats of the shallows were sensitive. Contrary to these findings, carp, which spawn in similar environments as pike but later (see the

---

2244 *Institut für Seenforschung (KIS): Reinanke (Coregonus sp. "Kärnten").*

2245 Höbaus, Sommer, & Reinthaler, 2023; *Institut für Seenforschung (KIS): Reinanke (Coregonus sp. "Kärnten").*

2246 Höbaus, Sommer, & Reinthaler, 2013.

2247 Govedič, 2004, 140. This also corresponds relatively well with the fact that the closed season for pike in Lake Klopein now lasts from January 1st to April 30th (*Gemeinde St. Kanzian am Klopeiner See: Beiblatt zur Fischereiausübungsberechtigung Klopeiner See*); its aim is to prevent fishing during the entire spawning time (Povž & Sket, 1990, 55–56).

2248 Both quotes from Govedič, 2004, 140.

2249 Vovk, 1978, 279–280, 292.

2250 *Tiefenkarte des Klopeiner See.*

following paragraph), were evidently not protected by the 1719 ruling; irrigation had priority over carp breeding in Lake Klopein.

The link between dynamic environmental conditions and human economic uses resulted in some water-use systems involving fish breeding and irrigation, which were too complex to allow fixed irrigation rights to evolve. According to a ruling from the early seventeenth century, peasants that irrigated meadows in their possession with water from the creek flowing north from shallow lake Lake Gösselsdorf in the Jaun Valley of Carinthia had to ask for permission every time they wanted to irrigate their meadows to ensure that irrigation would not affect the interests of both milling and carp breeding.[2251] Carp spawning starts when the temperature reaches 18 °C and is interrupted if the water temperature falls below 17 °C. Today, carp spawning in standing waters, a process that can take from a couple of days to a couple of weeks or even longer, usually takes place in May and/or June.[2252] In 1728, carp spawned in Lake Gösselsdorf before June 12th,[2253] whereas in 1681 carp spawning took place (or was resumed?) there in July.[2254] Carp eggs are fixed to flooded grassland plant leaves and stems in shallow water (25–50 cm deep) on flooded meadows and start hatching after approximately three to four days.[2255] After hatching, carp larvae hang vertically on the blades of grass for about two additional days and are able to start actively swimming away from the shallow spawning ground five to ten days after the eggs are laid, depending on the water temperature.[2256] Moreover, carp prefer to spawn during a rising water level.[2257] These are some reasons why the carp population was particularly sensitive to a drop in water level in spawning time and shortly thereafter; that is, some reasons to ban grassland irrigation with water released from a pond, when carp were about to spawn. The absence of fixed dates of permitted irrigation using water from the shallow lake serving as a carp pond is quite understandable from the viewpoint of the seigneury with fishing rights: it is an adaptation to spring (and summer) temperature variability leading to diverse carp spawning dates, as briefly outlined above. However, from the farmers' point of view, this means that the fishery caused limited access to irrigation water at a time when grassland irrigation was relevant in some years. Nonetheless, the historical data presented above suggest that carp spawning timing there often fully allowed irrigation

---

2251 Zwitter, 2014e, 343–344.

2252 Povž & Sket, 1990, 47, 116–117; Balon, 1995; Govedič, 2004, 138, 148, 150; *Gradivo za strokovni izpit za ribogojca*, 2009, 46, 63–65; *Carp Reproduction*, 2007.

2253 AT StiAStP, Grundherrschaft Eberndorf (Handwritings), vol. 191, Gerichtsprotokoll 1721–1728, 417 (June 12th, 1728).

2254 AT StiAStP, Grundherrschaft Eberndorf (Handwritings), vol. 183, Gerichtsprotokoll 1675–1683, 252–253 (November 8th, 1681).

2255 Povž & Sket, 1990, 116; Balon, 1995, 18, 21; Govedič, 2004, 138–139; *Gradivo za strokovni izpit za ribogojca*, 2009, 45–46, 63–65.

2256 Balon, 1995, 21; *Gradivo za strokovni izpit za ribogojca*, 2009, 65.

2257 *Carp Reproduction*, 2007.

during the most important time of grassland plant growth in May. Namely, quick grass growth requires a sufficient but not too high air temperature (e.g., 13–14 °C for very quick growth) and adequate soil water availability,[2258] and such conditions were obviously already over in many years by the time of carp spawning because the water had to reach 18 °C. In years of late carp spawning, the prohibition on irrigation was likely relevant, for instance, relatively soon after first-crop hay mowing, when irrigation would mitigate drought impacts on the valuable second-crop hay.

6.4.3.3 The specific social position of power

The impact of the specific social position of power on access to irrigation water was implicitly already partly discussed in Section 6.4.3.1, where inequalities in irrigation timing per unit of area were presented. Moreover, it was presented explicitly in Section 6.4.3.2.2: a specific position of power allowed the seigneuries to prioritize fish breeding over meadow irrigation. This section presents the impact of a specific position of power on the presence or absence of grassland irrigation in an area where limited water quantities were often an issue. It also discusses diverse motives for irrigation in this case.

In some cases, a specific social position of power made grassland irrigation possible where one would not expect it, at least not in its actual extent, based on hydrological conditions. An ordinance from 1579 for the upper Ljubljana Basin, for instance, referred to irrigation of a large demesne meadow below the village of Hlebce located on soils that retain low amounts of water available to plants. Irrigation allowed valuable second-crop mowing there (Section 6.1.2.3). Blatnica Creek, which provided irrigation water for this meadow, passed by about one kilometer of meadows and arable fields on soils in an alluvial plain characterized by relatively low amounts of water available to plants after it exited the wetland on its way through the plain to the demesne meadow.[2259] The stream crossed the land of various seigneuries, but despite the drought hazard no other seigneury and no subject was allowed to use this water for irrigation to ensure that as much of the water as possible would reach the distant meadow of the seigneury of Radovljica (Germ. *Radmannsdorf*) below Hlebce, according to the ruling from 1579.[2260] Two and a half centuries later, in the early nineteenth century, this was still a demesne meadow of the same seigneury

---

[2258] For general information on the timing of grassland plant growth, see Pfister, 1985a, 39–40, 47, 62.

[2259] *Arcanum Maps*, Cadastral Maps, Habsburg Empire – Cadastral Maps (XIX. [sic] Century), cadastral municipalities of Begunje (Germ. *Vigaun*), 1826, Doslovče (Germ. *Doschlovitsch*), 1826, Hraše (Germ. *Hraschasch* [!], 1826), and Nova Vas (Germ. *Neudorf*), 1826; *Environmental Atlas of Slovenia*: Spatial Units, Cadastral Areas & Land and Soil, Soil Map (1:25,000); see also Humerca, 2020, 18–21. Localization of this demesne meadow is explained in Section 6.2.1.2.

[2260] Zwitter, 2018, 172.

and it was still partly temporarily irrigated.[2261] In 1858, a more precise specification of rights of water uses was recorded than in 1579, and to a certain extent this also reflects earlier history because these rights were described in 1858 as an old custom: on the one hand, inhabitants of Hlebce were allowed to use unlimited quantities of water from Blatnica Creek for their household demand (it is reasonable that the seigneury allowed such use because there was no other surface water in the immediate vicinity[2262]). On the other hand, farmers' rights to irrigate their meadows were limited to flood events when only water spilling over the banks of Blatnica Creek was allowed to be used for irrigation in an unlimited manner; the rest had to reach the demesne meadow. The only other legal regular use of this water in the context of irrigation of meadows belonging to farms was direction of water into irrigation channels on one occasion annually; that is, when farmers were cleaning their irrigation ditches as part of spring cleaning (see Section 6.6.1).[2263] Nonetheless, the right to irrigation during flood events must have been important because the water crossing the plain during minor floods was full of fertile particles held in suspension and dissolved substances, and so its fertilizing impact was significant. Thus, farmers maintained their irrigation ditches, although this was quite a labor-intensive activity. For this reason, grassland irrigation during minor floods used to be usual at some places in Central Europe.[2264]

On the one hand, this example demonstrates the role of the specific position of power of the seigneury in grassland irrigation; on the other hand, it illustrates that grassland irrigation often targeted several factors at the same time (drought hazard mitigation and the fertilizing impact in the case of this demesne meadow), and that the motive of irrigation, which seems most obvious based on environmental conditions (drought hazard mitigation due to the low amounts of water available to plants), was not always the actual aim of irrigation in the complex socio-natural reality. In spite of the prohibition

[2261] SI AS 176, Franciscejski kataster za Kranjsko (Franciscan Land Cadaster for Carniola), L169, cadastral municipality of Nova Vas, cadastral appraisal protocols: Protocoll sämmtlicher Grund- und Bau-Parzellen der Gemeinde, parcel nos. 343a, 343b, 343½, and Catastral-Schätzungs-Elaborat, Einleitung, § 10 and Schätzung des Natural Ertrages, § 4: Wiesen. This information makes it clear that part of this large grassland was classified as a first-class meadow and the rest as a second-class meadow. It was a common characteristic of at least some first-class class meadows in this cadastral municipality that they were irrigated. The absence of irrigation in second-class meadows was the main criterion dividing first- from second-class meadows, leading to yields up to a quarter lower in second-class meadows.

[2262] On historical hydrological situation there, see Humerca, 2020, 18–19.

[2263] SI AS 48, Deželna komisija za odkup in uravnavo zemljiških bremen za Kranjsko (Provincial Commission for Redemption and Regulation of *Profits à Prendre* in Carniola), fasc. 163, 17–19: no. 2277 de 1858, Zur Sitzung am 14. Oct. 1858 and Protocoll, May 10th, 1858.

[2264] Compare the statement concerning the Swiss Plateau (Germ. *Schweizer Mittelland*, Fr. *Plateau suisse*, Ital. *Altopiano svizzero*) in Leibundgut & Vonderstrass, 2016a, 102.

on irrigation for farmers during agricultural drought events, they nonetheless made significant use of the fertilizing impact of irrigation water during floods immediately following rainy weather.[2265]

6.4.4 Landscape legacies of historical grassland irrigation

Among the legacies of historical grassland irrigation, their impact on soils is mentioned first. Several centuries of irrigation caused deposition of fine mineral and organic matter, which could lead to the formation of locally specific but significantly deep and highly fertile soils rich in humus (see Section 6.4.1). From the point of view of an agrarian society, the importance of this process was the greatest on poor soils. Good irrigation converted them into much nutrient-richer but occasionally dry soils. At least since the sixteenth century, in some places irrigation was also practiced in wetlands in depressions during high discharges, when the water carried a high load. This process formed sufficiently thick layers of sediments to create better agricultural grasslands; numerous cases are known from the French Alps in particular.[2266]

A second type of legacy left by historical irrigation is geomorphic, such as some preserved (or restored) irrigation channels or networks of such channels passing through grasslands. They are often several kilometers long. Especially where irrigation channels were located on microrelief ridges, which was very common for several centuries, irrigation caused significant relief alteration. It left clear convex uneven geomorphic legacies in formerly irrigated grasslands, formed through centuries of deposition of fine particles (Figures 99 and 100; fractions larger than fine sand did not exit the small irrigation channels). The depth of deposited layers decreased with increasing distance from exits of water from channels into grasslands. The total depth of deposits of irrigation varies from centimeters to meters. Some ridges with irrigation channels gradually grew due to the material deposited. In addition, material dug from some larger channels was thrown to their sides during maintenance work to strengthen the channel banks, and so farmers also built the ridges intentionally. In some cases (e.g., at the eastern foot of the Pohorje Hills), the banks of the main irrigation channels used for centuries were raised a few meters above the surrounding landscape (Figures 96 and 101). Numerous examples of uneven relief as a consequence of former grassland irrigation can still be found in the Alps, where they have not been flattened more recently (Figures 96, 99, 100, 101, and 102). However, to a certain degree, historical farmers already removed the deposits at the side of irrigation channels when the system was still in use if the lifted

2265 Irrigation was usually not carried out when it was raining (compare Wopfner, 1997, 363–364, 367).

2266 Leibundgut & Vonderstrass, 2016a, 47, 96, 111–114, 116–122, 128; Leibundgut & Vonderstrass, 2016b, 98, 170, 174, 181, 186–189.

Figure 99: Diverse microrelief in the Müstair Valley (Romansh *Val Müstair*), eastern Switzerland, formed by centuries of grassland irrigation. Irrigation channels were located at the top of small ridges, and water arrived from the channel in the middle of the photo; see the lower section of Figure 92. Leibundgut & Vonderstrass, 2016a, 113. © Archives of Stiftung Landschaftsschutz Schweiz.

relief was not useful for irrigation; the accumulated soil was spread around the meadow.[2267]

A third type of legacy is hydrological and hydrotechnical. Water still flows along some preserved or restored former irrigation channels, and some other elements of water infrastructure, such as some sluices, are still there. The preservation of some old main channels in the Alps has been the result of their recognized attractiveness; spectacular constructions such as the hanging channels passing along cliffs in Wallis (Figure 94) have themselves been important elements of tangible cultural heritage. Thus, some main channels have been maintained or restored even if they no longer serve traditional irrigation (see Section 10.2.2). Only very exceptionally, part of the larger grassland irrigation system deeply rooted in long historical tradition still operates without being

[2267] Baš, 1937, especially pp. 325–326, 328, 331 and Figures 1 and 2; Wopfner, 1997, 356, 359; Leibundgut & Vonderstrass, 2016a 47–48, 92, 111–113, 120, 159, 170–175, 179–181, 211–212, 234; Leibundgut & Vonderstrass, 2016b, 98–99, 110, 133–134, 138, 140, 143–147, 151, 155, 158–159, 163, 170–171, 174; Renes et al., 2020, 109, 123, 125–126.

Figure 100: Despite today's intensive industrialized agriculture, some alluvial fans[2268] as long-term legacies of former grassland irrigation have been preserved at Razvanje Creek (see no. 22 in Figure 103). They formed at places where water was long directed over the raised creek bed. Note that the people in dark jackets are standing on alluvial fans, and the woman in the blue jacket is in the depression between them. Photo: Žiga Zwitter, 2020.

converted to sprinkler irrigation (such as in small parts of Briançonnais; in eastern Wallis at Ausserberg, Mund, Birgisch, and Bitsch; 424 ha of agricultural land altogether in the Aosta Valley at Etroubles (Fr. *Étroubles*, Francoprovençal *Ètrobles*), Gignod (Francoprovençal *Gignôd*), and Aosta (Fr. *Aoste*, Francoprovençal *Aoûta*); 500 ha in the upper Venosta Valley at Malles Venosta (Germ. *Mals*), Sluderno (Germ. *Schluderns*), and Mazia (Germ. *Matsch*)) or was restored in its functionality. However, preserved open channels often provide water for modern sprinkling irrigation instead of construction of pipes.[2269]

A fourth type of legacy concerns biodiversity. Losses of water at the bottom and sides of a channel create much damper habitats than naturally, especially where they cross relatively dry environments, but also elsewhere. This impact

[2268] Already mentioned by Baš, 1937, 331.

[2269] Wopfner, 1997, 356–360 and Figure 47; Leibundgut & Vonderstrass, 2016a, 47–48, 92, 158–159, 162–175, 179–181, 211–212, 219–237, 251; Leibundgut & Vonderstrass, 2016b, 89, 96–97, 99–110, 112–121, 127–131, 134–140, 142, 144, 146–147, 149–163, 167–173, 175–177, 180–185; Renes et al., 2020, 109, 123, 125–126.

Figure 101: Removal of legacies of historical grassland irrigation has been going on in landscapes of intensive industrialized agriculture. Razvanje Creek (a creek in an artificial channel[2270]) east of the village of Razvanje at the foot of the Central Alps of Pohorje was used for grassland irrigation for several centuries. Its higher southern bank has been preserved, whereas its northern bank was recently flattened. Photo: Žiga Zwitter, 2020.

is particularly visible along the channel sides and, where the channel passes along a cliff, on the slope below, and so the biocenoses there differ significantly from those in the surroundings (note the lush vegetation on the cliff below the channel compared to the (largely) barren rock above the channel in Figure 94; see also the example of butterbur, *Petasites hybridus*, in Section 6.1.2.5.1.1).[2271] Wetter conditions are also evident in abandoned but not flattened irrigation ditches or their remnants that are not filled with water any more, as well as in abandoned irrigation ponds.[2272] Wetland or damp grassland habitats have often formed and have been kept at such sites until the ecological succession

[2270] Baš, 1937, 330.

[2271] Leibundgut & Vonderstrass, 2016a, 135–136, 212; Leibundgut & Vonderstrass, 2016b, 150; Wopfner, 1997, 351; Renes et al., 2020, 116.

[2272] Wopfner, 1997, 359–360; Leibundgut & Vonderstrass, 2016a, 168–169, 212, 247; Leibundgut & Vonderstrass, 2016b, 93, 108, 175, 180.

Figure 102: Historical maps[2273] and an article by the ethnologist Franjo Baš (published in 1937, when the irrigation system was still in use)[2274] make it possible to understand why this hedge and the convex landform beneath, located east of the village of Bohova (east of the Pohorje Hills; see no. 22 in Figure 103), are present in the landscape, currently characterized by intensive industrialized agriculture. The convex landform is the preserved bank of a historical irrigation channel, whereas the other bank and the channel itself were removed. Photo: Žiga Zwitter, 2020.

progresses further or land-use intensification destroys the site. For example, tall-growing and easily recognizable meadowsweet (*Filipendula ulmaria*, Figure 119 in Section 7.2), a species preferring damp and nutrient-rich sites with a neutral reaction, from the colline to subalpine vegetation belt, often shows the course of partly preserved but abandoned irrigation channels in the land scape.[2275] In some cases, legacies of historical irrigation can also fragmentarily be recognized in the species composition elsewhere in formerly irrigated grasslands because some enduring species that were favored by historical irrigation have survived there.[2276]

[2273] *Arcanum Maps*, Maps of Europe, Europe in the XIX. [sic] century (with the Third Military Survey), Habsburg Empire (1869–1887) – Third Military Survey (1:25,000).
[2274] Baš, 1937.
[2275] Peterken, 2013, 28, 80; Renes et al., 2020, 116; Aeschimann et al., 2004, vol. 1, 730–731.
[2276] Leibundgut & Vonderstrass, 2016a, 212.

6.4.5 The spatial distribution of historical grassland irrigation in the Alps, pre-Alps, and foot of the Alps in present-day Slovenia

Irrigation of farmed grasslands is not currently practiced in the Alps of Slovenia,[2277] with very rare exceptions of preserved traditional irrigation at a few isolated mountain farms.[2278] Two different starting points for this historical investigation are offered by a map by Christian Leibundgut and Ingeborg Vonderstrass, presenting historical irrigation systems in Europe according to the 2015 state of knowledge (based on studies taken into account in their extensive international synthesis on traditional grassland irrigation) and a publication by the Slovenian geographer, historian, and ethnologist Franjo Baš. According to the synthesis, no irrigation system in the Alpine and pre-Alpine parts of Slovenia was known in international literature, whereas in 1937 Baš published the general remark that meadow irrigation was commonplace in Slovenian territory at that time.[2279] The main impression based on the synthesis by Leibundgut and Vonderstrass is correct; namely, not much was known about the spatial distribution of historical grassland irrigation systems in the territory observed. Consequently, the majority of locations on the map (Figure 103) are new findings based on the study of archival records, and only a few areas of grassland irrigation in the Alps of what is now Slovenia are known in previous literature. Figure 103 and Table 2 offer a clear message: irrigation of a limited share of grasslands per village was common in the Slovenian Alps and pre-Alps, and at the foot of hills. However, the Slovenian Alps did not belong to the hotspots of grassland irrigation in a broader Alpine context. Complex communal meadow irrigation cycles have been proven so far only in the villages at the eastern foot of the Pohorje Hills, which had irrigation systems functioning seven days per week, 24 hours per day, and in some villages there were clearly defined hours of irrigation for individual meadows (Section 6.4.3.1).[2280]

Note that the map (Figure 103) is not a result of systematic research, and so many more sites and information from older and more recent periods can be added in the future. However, early-nineteenth-century cadastral data for Bohinj were studied systematically, proving that meadow irrigation was virtually absent from this Alpine basin (see numbers 2 and 3 in Figure 103), which is characterized by a very humid climate (2,000–2,600 mm in settlement areas between 1981 and 2010)[2281] and torrential waters. Similarly, despite the many

[2277] Pintar, 2003, 26.
[2278] Gabrič et al., 2019, 113.
[2279] Baš, 1937, 326–327, 332; Leibundgut & Vonderstrass, 2016a, 53–55, 72–73; maps of historical grassland irrigation in Central Europe are also presented and discussed in Renes et al., 2020, 110–111, see also pp. 120–121.
[2280] Baš, 1937, 332–333.
[2281] *Environmental Atlas of Slovenia*, Climate, Precipitation, Mean annual corrected precipitation 1981–2010.

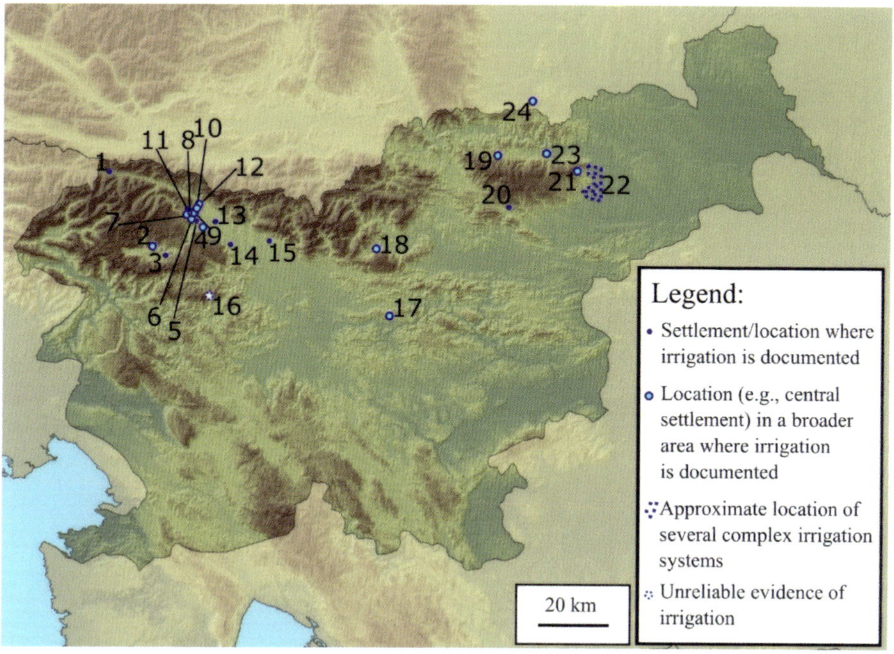

Figure 103: Selected locations of historical meadow irrigation in the Alps, pre-Alps, and foot of the Alps in what is now Slovenia. An explanation of numbers on the map and references is given in Table 2. Cartography: Žiga Zwitter. Background: Mladenović, 2012.

functions of historical irrigation (Section 6.4.1), only a single example of an irrigated meadow is known from Podkoren (no. 1 in Figure 103; the area received 1,800 to 2,000 mm of annual precipitation from 1981 to 2010).[2282]

Table 2: Explanation of information on historical meadow irrigation in Figure 103.

No. on map (Fig. 103)	**Location**	**Data on irrigation**
1	Podkoren	Irrigation of a single meadow south of the village was documented in the late nineteenth century (see Section 6.4.3.2.1); irrigation there already had a certain tradition by that time.[2283]

[2282] *Environmental Atlas of Slovenia*, Climate, Precipitation, Mean annual corrected precipitation 1981–2010.

[2283] SI AS 137, Okrajno glavarstvo Radovljica (District Administrative Authority [Germ. *Bezirkshauptmannschaft*] of Radovljica), box 121, archival unit 568: May 5th, 1880; May 16th, 1880; June 18th, 1880; October 27th, 1882: the meadow was called *Ledina*; it was located below the village and upstream from the mill and sawmill southeast of the village. It belonged to the meadow complex called *Ledine* south of the settlement (*Kulturportal der Flur- und Hausnamen*, Maps; *Arcanum Maps*, Cadastral Maps, Habsburg

No. on map (Fig. 103)	Location	Data on irrigation
2	Srednja Vas cadastral municipality	Information from the early nineteenth century: meadows were divided into five classes, and some of the first-class meadows were irrigated[2284] (it is likely that no more than directing the liquid manure flowing from manure piles to the nearest meadows is meant).[2285]
3	Nemški Rovt	Information from the early nineteenth century: meadows were divided into three classes; some of the first-class meadows were irrigated, and channels were directed there from the village[2286] (to direct nutrient-rich water there, such as that from manure piles; see Section 6.4.1).
4	Selo pri Bledu cadastral municipality	Two meadows were irrigated in the early nineteenth century (due to the fact that farmers tended to present their farming in a dark light in order not to pay high taxes, some more meadows might have been irrigated, but definitely not many because a more objective report states that not even first-class meadows were irrigated, and thus irrigation was an exception even for first-class meadows there).[2287]
5	Rečica	A dispute over an oak board in an irrigation channel next to meadows close to Rečica was documented in 1608,[2288] probably due to meadow irrigation. The irrigation of some meadows near Rečica Creek between the village of Rečica and Lake Bled (Sln. *Blejsko jezero*) is proven in the early nineteenth century.[2289]

Empire – Cadastral Maps (XIX. [sic] Century), cadastral municipality of Podkoren (Germ. *Wurzen*), 1826.

2284 SI AS 176, Franciscejski kataster za Kranjsko (Franciscan Land Cadaster for Carniola), L157, cadastral municipality of Srednja Vas v Bohinju, cadastral appraisal protocols: Catastral-Schätzungs-Elaborat, Schätzung des Natural Ertrages: § 4: Wiesen.

2285 See also SI AS 176, Franciscejski kataster za Kranjsko (Franciscan Land Cadaster for Carniola), L157, cadastral municipality of Srednja Vas v Bohinju, cadastral appraisal protocols: s.n., 157/4, 51.

2286 SI AS 176, Franciscejski kataster za Kranjsko (Franciscan Land Cadaster for Carniola), L30, cadastral municipality of Nemški Rovt, cadastral appraisal protocols: s.n., 30/4, 51, and Catastral-Schätzungs-Elaborat, Schätzung des Natural Ertrages: § 4: Wiesen.

2287 SI AS 176, Franciscejski kataster za Kranjsko (Franciscan Land Cadaster for Carniola), L343, cadastral municipality of Selo pri Bledu, cadastral appraisal protocols: s.n., 343/4, 51, and Catastral-Schätzungs-Elaborat, Schaetzung des Natural Ertrages: § 4: Wiesen.

2288 SI AS 721, Gospostvo Bled (The Seigneury of Bled), Iustitialia, vol. 13, April 11th, 1608; localization based on AS 721, Iustitialia, vol. 12, February 15th, 1605, and March 28th, 1605; SI AS 721, Iustitialia, vol. 14, October 28th, 1612.

2289 SI AS 176, Franciscejski kataster za Kranjsko (Franciscan Land Cadaster for Carniola), L227, cadastral municipality of Rečica, cadastral appraisal protocols: s.n., 227/4, 51.

No. on map (Fig. 103)	Location	Data on irrigation
6	Poljšica cadastral municipality	Some (of the best) meadows were irrigated according to information from the early nineteenth century.[2290]
7	Zgornje Gorje cadastral municipality	Some of the best meadows were irrigated according to information from the early nineteenth century.[2291]
8	Višelnica cadastral municipality	Some of the best meadows were irrigated according to information from the early nineteenth century.[2292]
9	Spodnje Gorje cadastral municipality	Information from the early nineteenth century: meadows were divided into four classes, and at least three first-class meadows in the plain close to the village in the southeast part of the cadastral municipality were irrigated.[2293]
10	Podhom cadastral municipality	Information from the early nineteenth century: some (of the best) meadows were irrigated.[2294]
11	Zgornje Laze	Meadow irrigation documented in 1612, but the record makes it clear that it had already been practiced earlier.[2295]
12	Blejska Dobrava cadastral municipality	Only water from a little stream was used for meadow irrigation in the early nineteenth century.[2296]

2290 SI AS 176, Franciscejski kataster za Kranjsko (Franciscan Land Cadaster for Carniola), L210, cadastral municipality of Poljšica, cadastral appraisal protocols: s.n., 210/4, 3, 51.

2291 SI AS 176, Franciscejski kataster za Kranjsko (Franciscan Land Cadaster for Carniola), L179, cadastral municipality of Zgornje Gorje, cadastral appraisal protocols: s.n., 179/4, 3, 51, 97.

2292 SI AS 176, Franciscejski kataster za Kranjsko (Franciscan Land Cadaster for Carniola), L322, cadastral municipality of Višelnica, cadastral appraisal protocols: s.n., 322/4, 3, 51, 97.

2293 SI AS 176, Franciscejski kataster za Kranjsko (Franciscan Land Cadaster for Carniola), L310, cadastral municipality of Spodnje Gorje, cadastral appraisal protocols: s.n., 310/4, 51, and Catastral-Schätzungs-Elaborat, Schätzung des Natural Ertrages: § 4: Wiesen.

2294 SI AS 176, Franciscejski kataster za Kranjsko (Franciscan Land Cadaster for Carniola), L22, cadastral municipality of Podhom, cadastral appraisal protocols: s.n., 22/4, 51.

2295 SI AS 721, Gospostvo Bled (The Seigneury of Bled), Iustitialia, vol. 14, June 14th and July 31st, 1612, localization based on SI AS 721, Iustitialia, vol. 12, October 20th, 1608, and SI AS 721, Iustitialia, vol. 14, [June 16th] 1611.

2296 SI AS 176, Franciscejski kataster za Kranjsko (Franciscan Land Cadaster for Carniola), L35, cadastral municipality of Blejska Dobrava, cadastral appraisal protocols: s.n., 35/4, 3, 51.

No. on map (Fig. 103)	Location	Data on irrigation
13	Meadow southwest of Hlebce (today part of this meadow is used as an airport[2297]), only during flood events also meadows upstream	Irrigation of the large demesne meadow below the village of Hlebce was recorded in 1579[2298] and still in the nineteenth century, when a much more limited but relevant irrigation of other meadows located upstream is also proven (Section 6.4.3.3).[2299]
14	Southeast of Brezovica near Kropa	Irrigation of a meadow at Lipnica Creek documented in 1861.[2300]
15	Čadovlje	Meadow irrigation recorded in 1657.[2301]
16	Hills in the Poljane Sora (Sln. *Poljanska Sora*) catchment	Irrigation of a meadow in this area was mentioned in the novel *Cvetje v jeseni* (Autumn Blossoms) by the writer Ivan Tavčar, who spent his early childhood there in the mid-nineteenth century as a farmer's son;[2302] a phytogeographical analysis of this work proved that the plant species in this novel correspond with reality.[2303]
17	Vače cadastral municipality	Meadows along creeks and roads and close to houses were irrigated according to information from the 1830s.[2304]
18	Major parts of the Savinja and Dreta river catchments above Ljubno, Okonina, and Zgornje Kraše (central settlement: Gornji Grad, Germ. *Oberburg*)	Some dry meadows were irrigated according to information from 1843.[2305]

2297 See the explanation of localization in Section 6.2.1.2.

2298 Zwitter, 2016, 144.

2299 SI AS 176, Franciscejski kataster za Kranjsko (Franciscan Land Cadaster for Carniola), L169, cadastral municipality of Nova Vas, cadastral appraisal protocols: Protocoll sämmtlicher Grund- und Bau-Parzellen der Gemeinde, parcel no. 343a, and Catastral-Schätzungs-Elaborat: Einleitung, § 10 and Schätzung des Natural Ertrages, § 4: Wiesen; SI AS 48, Deželna komisija za odkup in uravnavo zemljiških bremen za Kranjsko (Provincial Commission for Redemption and Regulation of *Profits à Prendre* in Carniola), fasc. 163, 17–19: no. 2277 de 1858, Zur Sitzung am 14. Oct. 1858 and Protocoll, May 10th, 1858.

2300 SI AS 137, Okrajno glavarstvo Radovljica (District Administrative Authority [Germ. *Bezirkshauptmannschaft*] of Radovljica), box 39, archival unit 158, April 29th, 1861.

2301 SI AS 721, Gospostvo Bled (The Seigneury of Bled), Iustitialia, vol. 21, April 26th, 1657; localization: *Slovenska historična topografija*, Paleonimi: "Forst," proving not only the same village name but also the same seigneury.

2302 Ložar, 1944, 147; on Tavčar, see Boršnik, 2013.

2303 Mestnik, 2018.

2304 Vodopivec, 2011, 508–509.

2305 Kuret (ed.), 1989, 263, 271.

No. on map (Fig. 103)	Location	Data on irrigation
19	Ribniško Pohorje (e.g., Hudi Kot)	Meadow irrigation documented in the mid-twentieth century: where there was a creek with appropriate slope near a meadow, farmers dug a small ditch more or less horizontally, approximately 100 meters long (e.g., in the middle of the meadow or at another suitable location) to allow meadow irrigation.[2306]
20	Vitanjsko Skomarje	Meadow irrigation documented in 1840 (unspecified location in the broader hilly territory) and in the early twentieth century at Vitanjsko Skomarje. In this microregion, exceptional meadows are still irrigated in a traditional manner.[2307]
21	Easternmost Pohorje Plateau; e.g., in the catchment of Hoče Creek (Sln. *Hočki potok*)	Meadow irrigation clearly documented in the early twentieth century[2308] and in the broader territory already in 1843.[2309]
22	Villages at the eastern foot of the Pohorje Plateau	Meadow irrigation documented in 1843[2310] and in the early twentieth century, but landscape legacies (Figures 96, 100, and 101) reveal that grassland irrigation there had a history at least several centuries long.[2311]
23	Broader surroundings of Fala (Germ. *Faal*), including slopes of the Pohorje Hills	An either irrigated or drained meadow was documented at Ruše in the 1620s.[2312] Evidence from 1812 makes it clear that irrigation of meadows in hilly parts of the Fala district (Germ. *Bezirksobrigkeit*; e.g., on the slopes of the Pohorje Hills) was widespread, but to a certain degree irrigation was still spreading there at that time: the recent conversion of pastures into meadows by an exemplary progressive farmer through creation of an irrigation system was documented in 1812.[2313]
24	Arnfels (Sln. *Arnež*) district (*Bezirksobrigkeit*), also including Kapla in the Kozjak Hills in what is now Slovenia)	Meadows were irrigated where this was possible (information from 1811); it is not entirely clear whether this information also refers to the district section that is in Slovenia today.[2314]

[2306] The ethnologist Marija Makarovič's interview with G. Miklavc, born at Hudi Kot (Makarovič, 2016).
[2307] Gabrič et al., 2019, 46, 113.
[2308] Baš, 1937, 328, 332.
[2309] Kuret (ed.), 1989, 325, 327.
[2310] Kuret (ed.), 1989, 325, 327.
[2311] Baš, 1937.
[2312] Zwitter, 2016, 144.
[2313] Kuret (ed.), 1987, 195–196, 199, 208–209.
[2314] Kuret (ed.), 1985, 58–60.

6.5 Care for long-term renewal of meadows

Over the long term, until the mid-twentieth century, farmers' strategies for purposeful influence on the renewal of species-rich non-fertilized and moderately fertilized meadows relied on two main principles: (1) mowing dates allowing seed ripening and dispersal when grasses and herbs were mowed, shaken out, turned, and raked in situ on the ground (see Sections 4.1.1.3, 6.1.2.1, and 6.1.2.2), and (2) sowing seeds of meadow plant species.[2315] To quote a knowledgeable interviewee from the oldest generation from Stara Fužina in Bohinj in the Julian Alps: "Old people said that grass has to ripen so that seeds fall to the ground [during haymaking] to keep [species-rich] meadows productive for the future."[2316] Ripe seeds, every second year at least, are a prerequisite for both principles of productive species-rich meadows maintenance.

Information from Bohinj, referring to the mid-twentieth century, demonstrates that some farms mowed their meadows in the valley bottom in a different order every year, starting in early to mid-June (depending on the weather), when grasses were only blooming. The hay harvest in the valley bottom continued until July, when seeds had already fallen to the ground.[2317] A similar (possibly history-inspired; see below) practice of alternating early, middle, and late mowing dates was observed in the late twentieth century in the Lesach Valley (Germ. *Lesachtal*) in western Carinthia (Germ. *Kärnten*).[2318] Deliberate alternations between middle and late mowing dates have also been documented in the Carpathians (in Ghimeş, Hung. Gyimes).[2319] The wide spatial range of alternating mowing dates suggests that this practice was not exceptional in some productive meadows close to farms. At higher elevations in the Alps, first-crop hay mowing then typically continued until August or early September, when the last montane, subalpine, or alpine meadows were mowed (see Sections 6.1.2.1 and 6.1.2.2); namely, mowing by scythe was highly time-consuming, and work in arable land also took up farmers' time in the summer. Thus, by the time when most meadows located significantly higher than permanent settlements were mowed, the seeds were already ripe there every year, not only every second year, and so the alternation between earlier and later mowing dates was not practiced there.[2320]

There were two main goals of regular alternations between middle and late mowing dates in productive meadows close to farms: (1) the long-term preservation of productive species-rich meadows and (2) hay quality as high as the

2315 Papež Kristanc, 2012, 16; Bavcon, Ravnjak, & Praprotnik, 2019, 25, 83, 101, 171, 173.
2316 Author's interviews with Kranjc, 2018, in Stara Fužina.
2317 Author's interview with Kunavec, 2017, in Stara Fužina.
2318 Machatschek, 2015, 32, 39–40.
2319 Babai & Molnár, 2014, 126, 128; Babai, Molnár & Molnár, 2014, 107–108.
2320 E.g., author's interviews with Kunavec, 2017, in Stara Fužina, and with F. Mlakar, 2017, in Bohinjska Češnjica. See also Keller, 2004, 60–61, 139.

long-term preservation of productive moderately manured meadows allowed it. The practice was adjusted to the fact that tussock grasses dominating in moderately manured meadows and providing relatively high quality hay can tolerate mowing before the first seeds ripen only for a few years; their seeds need to ripen (by the time of haymaking or when the hay dries on the ground) every second year at least, otherwise yields would strongly decline. Namely, unlike most grass species in recent intensive meadows, tussock grasses dominating in productive moderately manured meadows do not develop vegetative buds from which new tillers would develop; new tillers of tussock grasses develop only from seeds. The lifespan of individual tussocks of many important fodder grasses (such as false oat-grass, *Arrhenatherum elatius*) is restricted to five to ten years, so that a significant number of them need to be regularly replaced by seedlings. In years of middle mowing dates when most grasses were only blossoming during mowing time, the nutritive value of hay was higher because the majority of seeds and more nutrients remained in the hay and the crude fiber content was also lower than in years of late mowing. In Ghimeş, where mowing close to farms at elevations of no less than about 900 m started after July 22nd by the mid-twentieth century, when grasses started turning yellow (the quantities of fertilizer available were scant and mowing was preceded by spring grazing, and so grassland plant seeds were ripe late; see Sections 6.1.2.1, 6.1.2.2, 6.2, and 6.3), a bit of hay was nonetheless mowed before June 24th. Older persons believed that the medicinal value of such hay with medicinal herbs in bloom was particularly high.[2321]

It is not clear how old the practice of alternating middle and late mowing dates is, but it seems that the first carts of hay have a long history of being stored earlier than seeds of the majority of grasses fully ripened, even though most hay was stored only when seeds were ripe. The fact that nutritive value is higher if hay is cut before the seeds of most plants ripen was already known to the Roman agricultural writer Lucius Junius Moderatus Columella in the first century AD. This information was copied by later agricultural writers (e.g., in the sixteenth to seventeenth centuries, and partly in the eighteenth century), and Columella's work was also known in parts of the Alps (see Section 6.1.2.4). The agricultural handbook by Olivier de Serres published about 1600 (this research used the 1603 edition) is relevant in the context of alternations between middle and late mowing dates. De Serres's *Théâtre d'agriculture* combines his three decades of observations, inquiries, and experiments carried out at Pradel—located at the eastern foot of the Massif Central, thus close to the French Alps—with knowledge acquired from publications of ancient and contemporaneous

[2321] Bekker et al., 2000; Dierschke & Briemle, 2002; Dietl & Lehmann, 2004, 20, 24, 66, 83, 106, 118; Schwienbacher, Marcante, & Erschbamer, 2010; Babai & Molnár, 2014, 126, 128; Babai, Molnár & Molnár, 2014, 36, 107–110 (elevations from Babai & Molnár, 2013, 3); Machatschek, 2015, 32, 39–40. On the lifespan of individual tussocks of many important fodder grasses, see Lauenroth & Adler, 2008; Karrer, unpublished.

authors. 18 or 19 seventeenth-century French reprints definitely made *Théâtre d'agriculture* accessible to some landlords, in the Western Alps in particular.[2322] On the one hand, there is the idea in de Serres's publication that mowing before seeds ripen does not allow the regeneration of meadows (species-rich, as usual at that time), and so seeds need to ripen at least when the grassland species composition starts worsening from the viewpoint of farming. On the other hand, de Serres was aware of the fact that mowing only when seeds ripen results in less palatable hay, which is also less favorable for milk production.[2323] Such considerations must have stimulated alternating middle and late mowing dates. Not too far from the Alps, mowing in May in the Upper Rhine Valley in the sixteenth century and the start of mowing at Orbe (in the Jura Mountains) in late May in the mid-eighteenth century[2324] point to the start of mowing of grasses that were not yet ripe. In Bohinj, yearly alternations between middle and late mowing dates in the valley bottom might have already been usual for a minor meadow area in the early nineteenth century at the latest. Namely, at that time it was documented that hay mowing there usually started in mid-June,[2325] which must have usually corresponded with the blossoming time of most grasses in the cool to cold 1810s to 1830s.[2326]

Another indication of the great attention paid to the reproduction of species-rich meadows in history was sowing seeds of meadow plants, usually by spreading barn-sweepings rich in seeds that fell from hay. This practice was already known in Roman Antiquity. It is not clear how widespread the sowing of

2322 Beutler, 1973, 1280, 1282–1286, 1288–1295, 1299–1301; localization of Pradel is based on *Olivier de Serres: Le domaine du Pradel.* On the reception of agricultural handbooks in the Alps in the seventeenth century, see also Zwitter, 2020, 55–57, 73–76, 82.

2323 Kauter, 2002, 127; de Serres, 1603, 242. Gratitude is expressed to the French language specialist Primož Vitez for his help in translating early modern French.

2324 Kauter, 2002, 106–107, 127.

2325 SI AS 176, Franciscejski kataster za Kranjsko (Franciscan Land Cadaster for Carniola), L284, cadastral municipality of Studor, cadastral appraisal protocols: Catastral-Schätzungs-Elaborat, Einleitung, §10. This cadaster has many strengths, but also some weaknesses. One of the weaknesses is possible generalizations by commissioners in charge of cadaster production as a consequence of their routine and extensive workload (Bauer, 2017, especially pp. 244–245). Even though a similar date of starting mowing (after mid-June) was recorded, for instance, in the cadastral municipality of Hraše near Radovljica, in the somewhat milder climate of the central upper Ljubljana Basin (SI AS 176, L80, cadastral municipality of Hraše pri Radovljici, cadastral appraisal protocols: Catastral-Schätzungs-Elaborat, Einleitung, §10), the data on the start of mowing in the cadastral municipality of Studor do not seem to be a construction of the cadastral commission. Namely, in the cadastral municipality of Nemški Rovt, located at higher elevations in Bohinj, a later start of mowing (around the end of June) was recorded (SI AS 176, L30, cadastral municipality of Nemški Rovt, cadastral appraisal protocols: Catastral-Schätzungs-Elaborat, Einleitung, §10), and so we are not dealing with a simple copying of the same start of mowing in all cadastral municipalities in Bohinj.

2326 Climate history context: Brönnimann, White, & Slonosky, 2018, 312–315, Pfister et al., 2018, 268–269, 274, and Pfister & White, 2018, 551–557.

barn sweepings was in Central Europe in the Middle Ages, but limited reliable information makes it clear that this was practiced. Barn sweepings were relatively frequently sown in Northern Europe and England in the eighteenth century (and most probably earlier), and this was practiced in at least some parts of Central and Eastern Europe as well.[2327] It was also frequent in the Alps.[2328] The investigation of barn sweepings of traditional Csángó farmers in Ghimeş by Dániel Babai, Ábel Molnár, and Zsolt Molnár found up to twenty thousand seeds of monocots and dicots in one liter of barn sweepings. The quantities of barn sweepings collected per farm were quite significant.[2329] At least some farmers in the Alps paid attention to the weather; they sowed barn sweepings in wet weather so that the seeds germinated soon and were less hampered by the remaining litter, which quickly decomposed. Paying attention to moist soil and rainy weather in the days when barn sweepings were sown was evidently widespread: farmers in Bohinj[2330] and Csángós in Ghimeş respect(ed) this rule.[2331] Indeed, humid to damp soil is one of the important preconditions for successful establishment of sown seeds (see also the introduction to Section 4).[2332] Significantly more plants from barn sweepings became established if the meadow section was also manured (see also Section 6.3). The remnants of plants in barn sweepings themselves have a slight manuring impact.[2333] Barn-sweepings as sowing material promoting regeneration of meadows have the important advantage of quite an appropriate proportion between seeds and other plant remnants, maintaining humidity and providing some protection for germinating seedlings, but not hampering them through a litter layer that is too thick.

Barn sweepings were usually sown and had a significant impact on meadow vegetation where the history of a grassland was (locally) discontinued due to natural processes, agricultural uses, and other human interventions in the landscape, or due to the interplay of social and natural factors. Larger areas where barn sweepings were sown (if available in sufficient quantities) include, for instance, many grassland fallows and swidden fields after the cereal phase in the rotation cycle or in its final stage (Section 3.2.1), as well as meadow sections covered by fine flood sediments.[2334] They were also sown at many very small sites where the hills of meadow ants, molehills, or a lot of moss had been removed during spring cleaning (see also Section 6.6.1). Moreover, they

[2327] Kauter, 2002, 102–103, 105–106; Emanuelsson, 2009, 160; Peterken, 2013, 170–172, 186–187, 191, 232; Babai, Molnár, & Molnár, 2014, 92–94.
[2328] Wopfner, 1997, 294; Bavcon, Ravnjak, & Praprotnik, 2019, 83, 171.
[2329] Babai, Molnár, & Molnár, 2014, 92–94.
[2330] Author's interview with L. and M. Odar, 2017, in Studor.
[2331] Babai, Molnár, & Molnár, 2014, 93.
[2332] Sadar, 1956, 48.
[2333] Babai, Molnár, & Molnár, 2014, 92–93.
[2334] Author's interview with Kunavec, 2023, in Stara Fužina.

were applied to other gaps, such as turf considerably damaged by grazing (e.g., through trampling; see Section 6.2) or by cart transport in the meadow.[2335]

Sowing barn sweepings mainly in gaps in meadows relatively close to farms but also in some distant higher meadows seems representative of many places.[2336] Oral history interviews in Bohinj reveal that, in the mid-twentieth century, some barn sweepings were usually sown in gaps in meadow vegetation in the valley, and in some cases also in lower montane meadows. Only some farmers also transported barn-sweepings from the valley (or from the vicinity of farms) to low-quality parts of montane meadows (located higher than farms). Only a limited area of better-accessible and more productive montane meadows was mowed early enough that most of the seeds were not shed in the meadow during haymaking (see Sections 6.1.2.1 and 6.1.2.2), and thus a significant quantity of hay seed fell off in the hay barn, next to it when a cart was loaded with hay from the barn, or at the foot of a hayrack (see Section 4.1.1.3); such seeds were often applied to that meadow. Due to the late mowing dates, the seeds in most of the distant montane meadows did not reach the local hay shed.[2337] Nonetheless, the first mowing in manured sections of distant meadows mowed twice annually took place early enough that such hay left some germinable seeds in hay barns (Sections 6.1.2.1, 6.1.2.2, and 6.3.2),[2338] and the seeds of these species originating from higher elevations were also more likely to become established in the moderately fertilized meadows in the valley where the hay was transported than most plant species from non-fertilized meadows. Sowing barn sweepings mainly in gaps in meadows relatively close to farms but also in some distant higher meadows is also recorded in Ghimeş, where barn sweepings are usually sown in better, moderately manured meadow sites.[2339]

Sowing barn sweepings contributed not only to the regeneration of meadow ecosystems, but to a certain degree also to their historical dynamics (see also Section 4).[2340] Namely, the majority of farmers in many regions of the Alps

---

[2335] Babai, Molnár, & Molnár, 2014, 93; Ivaşcu, Öllerer, & Rákosy, 2016, 43, 46, 49.

[2336] Babai & Molnár, 2014, 125–126, 128; Babai, Molnár, & Molnár, 2014, 36, 92–93; an important recent exception is sowing barn sweepings rich in sainfoin (*Onobrychis viciifolia*) seeds, an economically highly relevant species introduced to Ghimeş only in the mid-twentieth century, in montane meadows in nutrient-poor sites; see Section 6.3.1.1.1 (Babai, Molnár, & Molnár, 2014, 95–96).

[2337] Author's interviews with Gašperin, 2017, with Kunavec, 2017 and 2023, and with Mencinger, 2017, all in Stara Fužina; with Grm, 2017, and with Milena Sodja, 2017, in Srednja Vas in Bohinj; with F. Mlakar, 2017, in Bohinjska Češnjica; with L. and M. Odar, 2017, in Studor; and with J. Smukavec, 2017, in Podjelje. See also Ivaşcu, Öllerer, & Rákosy, 2016, 46.

[2338] Author's interviews with Kunavec, 2017, in Stara Fužina; with F. Mlakar, 2017, in Bohinjska Češnjica; and with Milena Sodja, 2017, in Srednja Vas in Bohinj; compare also Hodnik, 2016, 42, 77, 79.

[2339] Babai & Molnár, 2014, 125–126, 128; Babai, Molnár, & Molnár, 2014, 36, 92–93.

[2340] Bavcon, Ravnjak, & Praprotnik, 2019, 171.

transported a lot of hay from montane (or higher) meadows to their farms at lower elevations[2341] at those places where keeping livestock in remote barns in late autumn and winter was not usual (see Section 6.3.2.1). Even though the majority of seeds fell off in mountain meadows, seeds from some late-flowering or long-flowering species—in addition to those from manured sites—were stored with the hay in some meadow complexes. Consequently, farmers sowed (predominately at lower elevations) a mixture of highly local seeds and seeds from meadows at different site conditions (e.g., from grasslands at higher elevations). Parallels with Ghimeş, where sowing barn sweepings by Csángó farmers is still common today, demonstrate that the role of barn sweepings in the long-term maintenance of productive species-rich meadows close to farms is significant, but that some species from higher elevations can become established in moderately manured meadows in montane (or colline) vegetation belts; for example, cross-leaved gentian (*Gentiana cruciata*) from non-fertilized meadows, white false-hellebore (*Veratrum album*) in meadows in the montane vegetation belt, and also Alpine dock (*Rumex alpinus*) from nutrient-rich sites, and so on. Such germinable seed transport in Ghimeş from higher meadows to those closer to farms has been possible at least since the second half of the twentieth century, with mowing one month earlier (nonetheless, in a warmer climate and usually no longer preceded by spring grazing; see Section 6.2.2) than in the early twentieth century.[2342] Characteristically, the seeds of all the plant species mentioned introduced to meadows in Ghimeş at lower elevations through sown barn sweepings ripen late,[2343] and so barn sweepings include the seeds of these plants after the seeds of many other species have already fallen to the ground in higher meadows during haymaking in the case of late mowing. On average, the barn-sweepings-driven dynamics in grassland ecosystems presented must have been most limited in those regions in the Alps where keeping livestock in distant barns at remote meadows in autumn and winter was more usual than hay transport down to the farm (see Section 6.3.2.1), and

[2341] E.g., Melik, 1950, 112–114, Hubatschek, 1992, 196–201, and Hodnik, 2016, 77–78; Bavcon, Ravnjak, & Praprotnik, 2019, 155, 157, 161, 163, 165, 165, 167, 169, 171, 173. For older historical proofs of hay transport to farms, see, e.g., the large-scale hay transports from the Siusi Alp (Ital. *Alpe di Siusi*, Germ. *Seiser Alm*) in South Tyrol around 1600 in Grass, 1948, p. 208 (the chronological orientation is based on *Deutsche Bibliographie: Wolkenstein, Marx Sittich von*). The court settlement from 1608 dividing tasks in making the route in Bohinj between the villages of Nomenj (about 500 m) and Gorjuše (roughly 1,000 m, with access to further montane meadow complexes) passable for sleds in winter in deep snow (SI AS 721, Gospostvo Bled [The Seigneury of Bled, Germ. *Veldes*], Iustitialia, vol. 13, February 11th, 1608) was definitely highly important for montane hay (and other items) transported to the valley.

[2342] Babai & Molnár, 2014, 124–126, 128; Babai, Molnár, & Molnár, 2014, 36, 92–94, 103 (and Aeschimann et al., 2004, vol. 1, 382–383; Aeschimann et al., 2004, vol. 2, 16–17, 1026–1027).

[2343] Aeschimann et al., 2004, vol. 1, 382–383; Aeschimann et al., 2004, vol. 2, 16–17, 1026–1027.

so barn sweepings must have very often been sown in the very meadow of their origin or consumed by the livestock locally (see the following paragraph).

The quantities of barn sweepings sown in the framework of care for long-term regeneration or restoration of productive meadows were significant, which is clear proof of the high attention farmers paid to maintaining and restoring meadow productivity. Namely, farmers were also eager to use barn sweepings as a highly esteemed fodder ingredient. The nutritive value of barn sweepings is high. Consequently, barn sweepings were a usual source of calorific fodder before potatoes (as pig fodder), cereals, and industrially processed concentrated feedstuff (e.g., for cattle and swine) became easily available and mineral fertilizers increased the quantity of relatively good hay from moderately fertilized sites. Seeds rich in starch and also leaf fragments mattered; depending on the meadow species composition, barn sweepings were often rich in fragile but nutritious clover leaves, including proteins. In the case of mowing by scythe, barn sweepings have been a more suitable source of fodder than in the case of motorized mowing because the latter results in a significantly higher soil content (from anthills and molehills) in barn sweepings, as explained by independent evidence from as distant places as East Tyrol and Ghimeş. Barn sweepings were sifted before being used as fodder to remove stones and some other impurities (but not fine soil), at some places in the Alps at least, such as East Tyrol. The use of some barn sweepings as fodder must have a long tradition and it must have been generally present in the Alps, as suggested by twentieth-century data from as distant and culturally diverse areas as examples from the Western and Eastern Alps and traditional Csángó farmers in Ghimeş. Barn sweepings were slightly salted, mixed with a bit of wheat bran or barley bran, soaked in boiled water to swell significantly during the night, and used on the following day (examples from Bohinj in the mid-twentieth century), soaked in a bit of hot water and mixed with wheat bran or flax residues (in East Tyrol), and/or with several gathered organisms (see Section 8.2.2), such as boiled Iceland moss (*Cetraria islandica*), nettles (*Urtica dioica*), Alpine dock (*Rumex alpinus*), thistles (*Cirsium* sp.), *Chenopodium* species such as good King Henry (*Chenopodium bonus-henricus*), and/or other mowed young vegetation of nutrient-rich sites around cottages (see Section 6.1.2.4) and/or juniper (*Juniperus communis*; see Section 6.6.2.2). Alternatively, barn sweepings were infused in (cold) water (e.g., in the Allues Valley [Francoprovençal *Alués*, French *Les Allues*] in the Western Alps and the Lesach Valley in Carinthia in the Eastern Alps) and some of the additions mentioned above were admixed. These mixtures were confirmed in several places, such as summer pastures in Austria and/or Switzerland. Such fodder was widely used for delivering cows, milk cows, ailing livestock, swine fattening, or horses, for instance. Barn sweepings were also a source of chicken fodder (e.g., steamed and mixed with cereals to stimulate increased egg laying in East Tyrol), for example, and in much smaller quantities barn sweepings were also used to cure ill livestock. Csángós still using barn sweepings as fodder emphasize that attention has to be paid to the absence of poisonous plant species; consumption of white false-hellebore (*Veratrum album*) in barn sweepings could kill a pig.[2344]

The high and diverse demand for barn sweepings as presented in the preceding paragraphs made barn sweepings for meadow regeneration a scarce commodity. This helps explain some farmers' striving to avoid unnecessary and easily avoidable types of meadow turf damage: although court proceedings against driving logs across montane meadows when grasslands were not frozen[2345] might have been motivated by a variety of reasons, avoiding unnecessary damage in grasslands in the context of the scarcity of barn sweepings and the related additional workload of meadow maintenance must have been one of them. Similarly, information from Vals in Graubünden (Ital. *Grigioni*; Romansh *Grishun*) suggests that attention was paid to minimizing erosion when driving through steep or partly moist mowed meadows when gathering the hay. The information here comes from the early twenty-first century, but similar behavior can be expected in times of ox- or horse-driven carts on those steep meadows from which the hay was not transported by sleds in winter: "farmers have to make precise decisions about whether and how to drive a hay truck on a particular slope in order to minimize erosion. Conditions can vary from day to day, depending on the level of moisture in the soil."[2346]

Barn sweepings were not the only source of local grassland seeds on all farms in the Alps. An interviewee in Bohinj, over ninety years old, recalled having been sent many times to the front of a haymaker in his childhood to collect seeds by rubbing grasses from which seeds could be collected in this way and to throw them onto the ground to promote grass seed dispersal. He was not supposed to collect the seeds of dicot herbs. He performed this task in the entire meadow in the Senožeta meadow complex in Bohinj (Figure 53 in Section 6.1.2.2). Haymakers mowed parallel to each other, but there was a bit of a distance between the length of the strip mowed by every haymaker in the same meadow, and so it was possible to collect seeds in front of those haymakers that were not the quickest to mow in the team without trampling the grassland plants because the neighboring haymaker had already progressed a bit further. Another interviewee in Bohinj recalled gathering seeds of grasses in front of a haymaker to sow them elsewhere.[2347] These data prove the par-

[2344] Meilleur, 1986, 282; Machatschek, 2006, 133–134, 142; Vogl-Lukasser et al., 2006, 66, 72–73, 76–77, 82, 85, 89–90, 145–146, 195, 212, 223, 227, 233, 237; Keller, 2007, 208; Babai & Molnár, 2014, 126; Babai, Molnár, & Molnár, 2014, 92–94, 105; Pfister, 1985a, 110–111, 113; author's interviews with A. Kavčič, 2017, at the Laz Alp in Bohinj; with Škantar, 2017, in Srednja Vas in Bohinj; and with Kunavec, 2023, in Stara Fužina.

[2345] An early seventeenth-century example from Bohinj: SI AS 721, Gospostvo Bled (The Seigneury of Bled), Iustitialia, vol. 11, January 14th, 1604, and January 29th, 1604.

[2346] von Glasenapp & Thornton, 2011, 775.

[2347] Author's interviews with Boltar, 2018, in Stara Fužina; with F. Mlakar, 2017 and 2018, in Bohinjska Češnjica; and with C. Mlakar, 2020, at the Zajamniki Alp.

ticularly high attention that some farmers paid to the regeneration of meadows and maintenance of meadow productivity, at least in the mid-twentieth century.

Generative reproduction is crucial for species-rich meadows (see Sections 10.1.2.2.3 and 10.1.2.2.4). Meadow plants also regenerate with new shoots growing from underground plant organs, as observed at the beginning of the eighteenth century at the latest.[2348] However, only some grassland species can actually effectively regenerate from the belowground organs over the long term. Today it is known that only species-poor plant communities of intensively managed meadows can successfully reproduce (almost) exclusively vegetatively in the long run without declining yields (Sections 10.1.2.2.3 and 10.1.2.2.4). Seeds and rhizomes are highly relevant for grassland regeneration, but this paragraph shows that their relative importance has changed over the course of time. In the early twenty-first century, there are often only a few places where meadow vegetation is not dense (caused by harrowing, wheels of mechanization, mice, voles, moles, shrews, rabbits, wild boars, and badgers, and rarely by extreme drought, dry freezes, and exceptionally long-lasting puddles resulting in drowned plants), and some more such places can be found in pastures and autumn-grazed meadows, where additional gaps are caused by livestock trampling and feces. The seeds of grasses and forbs can become established in such gaps if the site conditions are appropriate; vegetative regrowth also can take place in the gaps. Due to the scarcity of appropriate gaps in many meadows at present, and early mowing dates, vegetative reproduction is crucial for most plant species occurring in intensive species-poor meadows nowadays, whereas other plants have disappeared from there.[2349] Much more widespread and longer-lasting historical grazing in meadows, and higher grazing densities in pastures, created more gaps than usual nowadays. Before the late eighteenth to mid-nineteenth century, this longer grazing duration creating more gaps in meadows included not only autumn grazing but also spring grazing (see Section 6.2).[2350] This impact was so relevant that in many cases it more than merely compensated for the grazing-posed disturbance: "although target species such as [mountain arnica] *Arnica montana* [see Figure 85 in Section 6.3.1.1] and [viper's-grass] (*Scorzonera humilis*) seemed to be sensitive to disturbances . . . [,] for their long-term generative reproduction open soil patches are necessary," and so research in the eastern Ore Mountains (Germ. *Erzgebirge*) in Saxony (Germ. *Saxen*) "recommended extensive grazing with sheep or cattle together with late mowing in nutrient-poor meadows."[2351] Gaps

---

2348 Kauter, 2002, 127.

2349 Adapted from Dietl & Lehmann, 2004, 57–58; Ellenberg & Leuschner, 2010, 1018–1020; Peterken, 2013, 223–224, 392–393; Babai & Molnár, 2014, 126; Janišová et al., 2023, 18. See also Section 10.1.2.1.

2350 Janišová et al., 2023, 9, 11–13, 16, 18–19.

2351 Janišová et al., 2023, 19.

caused by the scythe during mowing very close to the soil (see Sections 6.1.1.1 and 6.1.1.3), which are created in species-rich meadows exactly when seeds are ripe, also mattered. The role of generative reproduction of species-rich meadows is currently just as high as it has been throughout history. Present grassland management regimes often create less favorable circumstances for successful plant generative reproduction, even in preserved species-rich meadows, than historical management regimes used to.

6.6 Spring cleaning in grasslands and other removal of plants undesired in livestock farming

6.6.1 Cleaning hay meadows in spring and later

The main goals of meadow cleaning were (1) to keep these areas productive for animal husbandry in the long run; (2) to prevent negative impacts on hay quality due to impurities, poisonous plants, some herbs that take longer to dry than most meadow vegetation and are thus likely to cause mold infestation in hay, or plants rejected by livestock (Section 6.6.1.1); and (3) to alleviate the hard work of highly time-consuming mowing with a scythe in the summer (Sections 6.1.1.1, 6.1.1.2, 6.1.2.1, and 6.1.2.2).[2352] These aspects were so important that not only making hay but also spring cleaning are listed as annual management practices that farms were allowed to carry out individually in montane meadows at (or near) Poljana Valley (at the eastern foot of the Mežakla Plateau, in the eastern Julian Alps, Sln. *Julijske Alpe*) according to a record by the court of the seigneury of Bled (Germ. *Veldes*) from 1611.[2353]

Cleaning, regularly carried out in spring, included the removal of woody vegetation that spreads to meadows from forest edges[2354] and of broken tree branches and cones that fell onto meadows since the previous haymaking and would impede mowing (Figures 104 and 107). Moreover, spring cleaning involved clearing water-deposited wood, sand, and gravel from flood-prone meadows, and removal of rocks deposited by avalanches (Figure 105) and also trees and parts of them in avalanche-prone meadows. Removal of other stones that reached the meadow surface since the previous mowing was also part of spring cleaning. Even small stones from molehills had to be removed; they were exposed on the surface while leveling molehills, which was itself also part of spring cleaning. Otherwise, even small stones would ruin the scythe blade and require additional peening (see Sections 6.1.1.1 and 6.1.1.3). Not

2352 Adapted from Dosedla, 1971, 1, 3, 11–14; Wopfner, 1997, 298–299, 318–319; Kauter, 2002, 131–132; Peterken, 2013, 140, 187, 259, 282; Babai, Molnár, & Molnár, 2014, 90–91, 112; Ivaşcu, Öllerer, & Rákosy, 2016, 43.

2353 SI AS 721, Gospostvo Bled (The Seigneury of Bled, Germ. *Veldes*), Iustitialia, vol. 14, June 16th, 1611.

2354 Šarf, 1974, 138; Wopfner, 1997, 318–319; Kauter, 2002, 131.

Figure 104: Farmers and tourists during spring cleaning in a larch meadow at Gschwandtkopf close to Reith bei Seefeld in Tyrol around 1960. First, large branches were removed to be used as firewood. Then the twigs and cones were raked up with a densely toothed rake and carried in linen sheets to a site where the removed biomass was burned[2355] or it was kept for animal bedding (see below). Some larch twigs were used as tinder.[2356] Note the further full sheets of biomass at the feet of the woman in the central photo section and the raked organic material waiting to be removed, which the arrows point to. Norz, 2014, 21. © Christian Norz.

only molehills, but also vole runways were leveled during spring cleaning. Furthermore, the hills of meadow ants were removed, otherwise numerous and considerable hummocks would develop from them (Figure 106; but see the different case of much rarer hills of red forest ants on meadow edges in Section 10.1.1). Spring cleaning also involved crushing the remnants of livestock dung deposited in meadows during autumn and spring grazing (see also Section 6.2.2; thus, the weather had to be wet, otherwise old cow pies, for instance, could not be crushed with a rake) and removal of hard livestock bedding remnants that arrived in fertilized meadows with manure. The bedding remnants were removed after (most of) the manure rubbed into the meadow (Section 6.3.2.1.1) was leached into the soil by rain. Near trees and bushes, nutrients were released into meadows from fallen leaves, and so these were only partly

2355 Norz, 2014, 19–21.
2356 Machatschek, 2002, 406–407; Machatschek, 2015, 52.

Figure 105: Removal of rocks deposited by an avalanche from a meadow in a gentle part of the avalanche track. The Oberberg Valley (Germ. *Oberbergtal*), a side valley of the Stubai Valley (Germ. *Stubaital*), near Innsbruck in Tyrol, 1945. Photo: Erika Hubatschek. Hubatschek, 1992, 132. © Irmtraud Hubatschek.

Figure 106: Hills of meadow ants (small hummocks in the photo marked by yellow arrows) in a steep, non-fertilized meadow section that was mowed in 2021 but not in 2022, when this photo was taken. Keeping hills of meadow ants in late-mowed meadows, especially those mowed once every second or third year, would significantly increase the time needed for mowing because the density of anthills gradually increases, and so it is understandable why farmers typically removed them.[2357] Photo: Žiga Zwitter, October 31st, 2022.

[2357] Cvetek, 1971, 73–74.

removed during spring cleaning, at least from tree species such as hazel, whose leaves favorably affect meadow productivity (Section 6.3.1). Action was nonetheless needed because thick leaf accumulations would locally suppress many grassland plants. Piles of leaves that had already largely decomposed by the time of spring cleaning were only distributed more evenly to release the remaining nutrients.[2358] Spring-cleaned leaves and twigs were burned in the Alps in some cases, and otherwise they were deposited in depressions to decompose there, or the leaves were stored for animal bedding.[2359] Moreover, irrigation channels in irrigated meadows were cleaned in spring (see Section 6.4).

The types of materials and plants removed during spring cleaning as documented in the previous paragraph for the mid-twentieth century have not changed much in comparison with previous centuries. Descriptions of spring cleaning in the Eastern Julian Alps and their surroundings, and at Spodnje Jezersko in the Kamnik–Savinja Alps, all from the early and mid-nineteenth century, clearly record at least some of the following tasks: crushing the livestock dung deposited during grazing in meadows,[2360] leveling molehills, removing anthills, fallen tree branches, (some) fallen leaves, and unwanted bushes and plants harmful for livestock (or for general hay quality; see Section 6.6.1.1), clearing channels, and removing (sand and larger fractions of) flood deposits[2361] and other stones of considerable size (see Section 6.1.1.3). Some

---

[2358] Author's interviews with Gašperin, 2017, with Mencinger, 2017, with Urh, 2017, with Kunavec, 2017 and 2021, with Kranjc, 2018, all in Stara Fužina; with L. and M. Odar, 2017, in Studor; with Grm, 2017, with Milena Sodja, 2017, both at Srednja Vas in Bohinj; with F. Mlakar, 2017, in Bohinjska Češnjica; and with J. Smukavec, 2017, in Podjelje; participant observation at a traditional spring cleaning with wooden rakes at a meadow near Stara Fužina in early May 2021; SI AS 176, Franciscejski kataster za Kranjsko (Franciscan Land Cadaster for Carniola), L35, cadastral municipality of Blejska Dobrava, cadastral appraisal protocols: s.n., 35/4, 59; Dosedla, 1971, 1, 3, 11–14; Pils, 1994, 253 and Table 74, photo 179; Wopfner, 1997, 298–299, 318; Kauter, 2002, 131; Machatschek, 2002, 262, 399–401, 413; Emanuelsson, 2009, 208, Peterken 2013, 259, 282, and Babai, Molnár, & Molnár, 2014, 90–91, 112.

[2359] Author's interviews with Gašperin, 2017 and with Kunavec, 2017, both in Stara Fužina; with L. and M. Odar, 2017, in Studor; with Šetina, 2018, in Srednja Vas in Bohinj (data refer to the farm at Holy Spirit Church south of Lake Bohinj, Sln. *Bohinjsko jezero*). Compare Machatschek, 2002, 264–265.

[2360] SI AS 48, Deželna komisija za odkup in uravnavo zemljiških bremen za Kranjsko (Provincial Commission for Redemption and Regulation of *Profits à Prendre* in Carniola), fasc. 163, 17–19: no. 2277 de 1858, Zur Sitzung am 14. Oct. 1858 and Protocoll, May 10th, 1858.

[2361] SI AS 176, Franciscejski kataster za Kranjsko (Franciscan Land Cadaster for Carniola), L30, cadastral municipality of Nemški Rovt, cadastral appraisal protocols: s.n., 30/4, 59; SI AS 176, L35, cadastral municipality of Blejska Dobrava, cadastral appraisal protocols: s.n., 35/4, 59; SI AS 176, L107, cadastral municipality of Češnjica v Bohinju (i.e., Bohinjska Češnjica), cadastral appraisal protocols: s.n., 107/4, 50–51; SI AS 176, L157, cadastral municipality of Srednja Vas v Bohinju, cadastral appraisal protocols: s.n., 157/4, 50–51; SI AS 176, L227, cadastral municipality of Rečica, cadastral appraisal protocols: s.n., 227/4, 59; SI AS 176, L284, cadastral municipality of Studor, cadasfurther tasks are hidden behind the general term "spring cleaning" in the records cited in this paragraph.

Most of the tasks listed were carried out manually (Figures 104, 105, 109, and 112), in non-fertilized meadows in particular. For most types of tasks requiring raking, a wooden rake was used because it did not destroy the turf and is easier to handle; an iron rake is too aggressive for most spring cleaning.[2362] The seigneury of Radovljica (Germ. *Radmannsdorf*), for instance, required farmers to crush cow pies from the grazing period and spread molehills in its large demesne meadow at Hlebce in the upper Ljubljana Basin with wooden rakes, as recorded in the mid-nineteenth century.[2363] Thus, they focused only on the target microsites and did not use draft livestock to harrow the grassland although the meadow was large and located on flat land. One reason for this must have been irrigation in that meadow, so that meadow harrowing would have damaged the network of irrigation channels (Section 6.4: text, and Figures 93, 97, 98, and 99) and additional trampling by relatively heavy animals (but see Section 5) would have caused undesired soil compaction at water-saturated sites in spring if irrigation had already started by the time of spring cleaning.

Alternatively, where spring cleaning was linked to rubbing manure into the meadow (Section 6.3.2.1.1) or where a lot of moss was to be removed, livestock were used to do some of the spring-cleaning tasks, as explained by data from unirrigated meadows from the first half of the twentieth century.[2364] Namely, meadow harrowing in spring with various kinds of implements, such as spruce tops, weighted-down branches, branches interwoven between a field harrow's horizontal bars, or a field harrow placed upside down, as presented in Section 6.3.2.1.1, not only rubbed the manure into the soil, but also crushed the earthen hills of meadow ants, molehills, and vole runways, and removed some moss.[2365] Specific types of implements were sometimes used to uproot moss in wet meadows, such as rakes with an iron head[2366] or a field harrow

tral appraisal protocols: s.n., 284/4, 50–51; SI AS 178, K470, cadastral municipality of Spodnje Jezersko, cadastral appraisal protocols: s.n., 470/4, 50–51, and Catastral-Schätzungs-Elaborat, Einleitung, § 10: Cultur des bodens; SI AS 48, Deželna komisija za odkup in uravnavo zemljiških bremen za Kranjsko (Provincial Commission for Redemption and Regulation of *Profits à Prendre* in Carniola), fasc. 163, 17–19: no. 2277 de 1858, Zur Sitzung am 14. Oct. 1858 and Protocoll, May 10th, 1858.

2362 Author's interview with J. Cvetek during manual spring cleaning near Stara Fužina in 2021.

2363 SI AS 48, Deželna komisija za odkup in uravnavo zemljiških bremen za Kranjsko (Provincial Commission for Redemption and Regulation of *Profits à Prendre* in Carniola), fasc. 163, 17–19: no. 2277 de 1858, Zur Sitzung am 14. Oct. 1858.

2364 Šarf, 1955, 89; Novak, 1970, 362; author's interviews with Gašperin, 2017, and with Kunavec, 2017, both in Stara Fužina, and with A. and M. Taler, 2017, in Studor.

2365 Šarf, 1955, 89; Dosedla, 1971, 1, 3, 11–14; author's interview with Kunavec, 2017, in Stara Fužina.

2366 Novak, 1970, 362; Babai, Molnár, & Molnár, 2014, 104.

with wooden teeth or worn-out iron teeth, with branches (placed below the teeth to reduce their tearing impact) or without such branches.[2367] Alternatively, fire or manure (see also 6.3.2) might have also been used as a means of moss suppression, as suggested by a parallel from Ghimeş (Hung. Gyimes) in the Eastern Carpathians.[2368] Another strategy of moss suppression during spring cleaning by some farmers in the mid-twentieth century was application of quicklime.[2369] Removal of moss was, however, not generally widespread. Some farmers in the mid-twentieth century left moss in the meadow. They mowed the sites where moss abounded at a greater distance from the soil and discarded mowed moss.[2370]

As long as hay meadows were grazed in spring (Section 6.2), spring cleaning followed after the end of spring grazing, and so it was never carried out very early (see Section 6.2.1.2). According to the mid-nineteenth-century description of spring cleaning in the large seigneury of Radovljica's meadow below the village of Hlebce (see Sections 6.2.1.2, 6.4.3.3, and 6.4.5), this was performed on the day when spring grazing ended, on April 24th, or on the following day, so that the manure left in the meadow during spring grazing was spread contemporaneously with leveling molehills and other spring cleaning.[2371] April 24th was one of the earliest likely dates for spring cleaning before the abolition of spring grazing in hay meadows. Such an early timing of spring cleaning was relevant only where spring grazing ended that early (Section 6.2.1.2). Meadows grazed until May 25th, for instance, would be cleaned in late May.[2372] In typical meadows in farmers' possession, the (earliest allowed) timing of spring cleaning was determined collectively, with the end of collective spring grazing in the same meadow complex.[2373] Thus, the (earliest possible) dates of spring cleaning could only be determined more individually (1) in a few meadows exempt from spring grazing in the regions of compact villages and (2) in meadows in possession of some isolated farms with compact holdings and more individually organized agriculture (Section 6.2).

2367 Dosedla, 1971, 1, 3–4.

2368 Pils, 1994, 251; Babai, Molnár, & Molnár, 2014, 104; Ivaşcu, Öllerer, & Rákosy, 2016, 49.

2369 Author's interviews with Kunavec, 2017, in Stara Fužina, and with F. Mlakar, 2017, in Bohinjska Češnjica.

2370 Author's interviews with L. and M. Odar, 2017, in Studor, and with Kunavec, 2023, in Stara Fužina.

2371 SI AS 48, Deželna komisija za odkup in uravnavo zemljiških bremen za Kranjsko (Provincial Commission for Redemption and Regulation of *Profits à Prendre* in Carniola), fasc. 163, 17–19: no. 2277 de 1858, Protocoll, May 10th, 1858, and Zur Sitzung am 14. Oct. 1858.

2372 See the next paragraph.

2373 For instance, spring grazing at Poljane east of the Mežakla Plateau had to be carried out "zu re[c]hter weil vnd zeit," whereas earlier in spring those meadows functioned as a common pasture (SI AS 721, Gospostvo Bled [The Seigneury of Bled], Iustitialia, vol. 14, June 16th, 1611).

The abolition of spring grazing (Section 6.2.3) allowed the individualization of timing spring cleaning. Some farmers still respected dates with a tradition several centuries long, at least as the final deadline for spring cleaning. In the first half of the twentieth century, residents of the isolated farm close to Holy Spirit Church at Lake Bohinj, at the base of the shady north-facing slope, respected the oral tradition that their last meadow in the remotest shady depression in the valley bottom (thus snow melted there relatively late) at an elevation of about 550 m had to be spring cleaned by May 25th, Saint Urban's Day. Consequently, that meadow was called *Urbanova konta*[2374] (literally, 'Urban's Hollow'). The date (May 25th) corresponds with the historical end of spring grazing there (see Figure 69 in Section 6.2.1.2), which had ceased long before that date was still seen as the final deadline for spring cleaning, a late date after the abolition of grazing in May. On the other hand, most farmers have moved their spring cleaning forward since the abolition of spring grazing, at least in the meadows where spring grazing had lasted long into May earlier in history (Section 6.2.1.2). In some cases, this led to very early spring cleaning. An interviewee in Stara Fužina recalled an elderly villager that commented that some fellow villagers raked their meadows too early in the spring, when the turf could easily be broken and raked away. She recommended later spring cleaning, after the grass had already started growing[2375] (note the significant height of grassland plants during spring cleaning in Figure 104[2376]). This advice was well founded because raking or harrowing (sections of) a meadow too early in the spring was likely to remove some patches of grassland vegetation where plants were actually in good condition but seemed to be dead because they had not sprouted yet.[2377]

Spring cleaning of meadows in the Alps can be traced in historical records at least since the sixteenth century, but for obvious reasons presented above it can be assumed to have also been practiced previously. Sixteenth-century examples from the Julian Alps and their surroundings include not only spring cleaning of the large irrigated seigneury of Radovljica's demesne meadow below the village of Hlebce in the upper Ljubljana Basin according to a regulation from 1579,[2378] but also remote meadows in the unpopulated bottom of the Sava Bohinjka Valley between the villages of Nomenj and Bohinjska Bela, in farmers' possession.[2379] Spring cleaning was performed there too, even though such remote meadows were managed in a less labor-intensive manner than

[2374] Author's interview with Šetina, 2018, in Srednja Vas in Bohinj.
[2375] Author's interview with Kranjc, 2018, in Stara Fužina.
[2376] The forester Christian Norz, personal communication.
[2377] *Rake matted grass.*
[2378] Zwitter, 2016, 144; localization of this meadow is explained in Section 6.2.1.2.
[2379] SI AS 721, Gospostvo Bled (The Seigneury of Bled), Iustitialia, fasc. 34: Lawsuit between Krištof Wuria and the community of Nomenj over grazing rights "in der [!] Stiegen" 1613/1614, 25 February 1614.

irrigated ones (such as the 1579 example at Hlebce). Spring grazing in remote montane meadows is not surprising in the context of early nineteenth-century data documenting it in such meadows as a general phenomenon.[2380] Moreover, vegetative and generative reproduction of trees at forest edges required regular spring cleaning in remote montane meadows to prevent the reduction in the meadow area, and further processes presented at the beginning of this section would pose obstacles to mowing in remote montane meadows unless they were spring cleaned. Late nineteenth-century information proves that even those sections of remote meadows located at the very edge of the forest used to be neatly spring cleaned, both in the Slovenian Alps and in Tyrol,[2381] and thus this was a general phenomenon in the Eastern Alps, at least at times of high agricultural density (see Section 3.1).

The historical labor intensity of spring cleaning was high not only during the maximum historical agricultural density in the Alps in the nineteenth century (see Section 3.1), but also earlier. For instance, in the sixteenth century, spring cleaning (together with other early spring activities[2382]) in what was probably a large remote meadow[2383] that belonged to a farm took as long as

2380 SI AS 176, Franciscejski kataster za Kranjsko (Franciscan Land Cadaster for Carniola), L179, cadastral municipality of Zgornje Gorje, cadastral appraisal protocols: s.n., 179/4, 59; SI AS 176, L210, cadastral municipality of Poljšica, cadastral appraisal protocols: s.n., 210/4, 59.

2381 Šarf, 1974, 138; Wopfner, 1997, 298–299, 318–320.

2382 The following underlined clauses suggest that the main sixteenth-century activity carried out in that meadow in spring and lasting for around three days was spring cleaning: "alß zeug vor zwainzig jaren, beÿ dem Tschube selligen, fünf jar lang gedient, ist man frÿelings zeit, mit den rossen, auf ein dreÿ tag hinein, *so man die gereüter geraumbt, oder was daselbst gearbeith hat*, vnd soliche rosß nur in den gereüthen gehen oder wäidnen lassen"; or "alß der zeugin schweher Leonhart Tschube sellig, der vor 30 jaren gestorben, *die gereüter in frielling geraumbt*, hat man die rosß mitgenomben, vnd ein dreÿ tag daselbst waidnen lassen" (SI AS 721, Gospostvo Bled [The Seigneury of Bled], Iustitialia, fasc. 34: Lawsuit between Krištof Wuria and the Nomenj community over grazing rights "in der [!] Stiegen" 1613/1614, February 25th, 1614; emphasis by Ž. Z.).
Further tasks performed in the more or less three days mentioned above possibly involve maintenance work on fences (SI AS 176, Franciscejski kataster za Kranjsko [Franciscan Land Cadaster for Carniola], L179, cadastral municipality of Zgornje Gorje, cadastral appraisal protocols: s.n., 179/4, 59; SI AS 176, L210, cadastral municipality of Poljšica, cadastral appraisal protocols: s.n., 210/4, 59). In addition, early twentieth-century data from the same microregion suggest that the firewood needed at the time of mowing was likely made during spring cleaning (Šarf, 1974, 138).

2383 The plural (*gereüter*) in the citations from 1614 in the previous footnote should be understood in the manner that, during the general spring cleaning in remote meadows, in this meadow members of that family spent around three days. Namely, the court recorded the following question by the farmer in possession of this meadow: "Ob der zeug nit selber gesehen, oder von andern gehört, das [!] meine vorfordern, vnd weliche[,] das gereüth [singular!] innen gehabt …" (SI AS 721, Gospostvo Bled [The Seigneury of Bled], Iustitialia, fasc. 34: Lawsuit between Krištof Wuria and the community of Nomenj over grazing rights "in der [!] Stiegen" 1613/1614, February 25th, 1614).

about three days if performed by two or more people.[2384] The activity of spring cleaning and its duration were not a matter of the dispute in that court record, and so they are likely reliable,[2385] in particular because the quoted testimonies by two witnesses match concerning the duration of spring cleaning. Focusing on early nineteenth-century data on the labor intensity of meadow spring cleaning in the eastern Julian Alps, they suggest significant variations in the amount of time typically invested in spring cleaning and fence repair in different meadows: one day of two people's work, one day of four people's work, one day of eight people's work,[2386] or even more (around one day of ten people's work),[2387] and in one case spring cleaning alone required three people (in one day, probably).[2388] All these pieces of information refer to one *Joch* (0.57 ha) of a montane meadow. These variations are most likely partly a result of different environmental conditions in meadows requiring a different workload during spring cleaning, and a different state of the fence also resulted in huge differ-

[2384] The first testimony cited above, in the context of the duration of spring cleaning, claims that the witness served as a farmhand at Tschube's farm for five years, about two decades earlier (SI AS 721, Gospostvo Bled [The Seigneury of Bled], Iustitialia, fasc. 34: Lawsuit between Krištof Wuria and the community of Nomenj over grazing rights "in der [!] Stiegen" 1613/1614, February 25th, 1614). The fact that a farmhand participated in that spring cleaning in the late sixteenth century suggests that more than one family member, a couple of people or more, were involved. The other testimony cited in the context of the sixteenth-century duration of spring cleaning, claiming that Lenart Tschube spring cleaned that meadow, does not mean that he did it alone, only he was mentioned by his relative because he had been the head of the farm during that spring cleaning.

[2385] Compare Fuchs & Schulze, 2002, 9.

[2386] SI AS 176, Franciscejski kataster za Kranjsko (Franciscan Land Cadaster for Carniola), L22, cadastral municipality of Podhom, cadastral appraisal protocols: s.n., 22/4, 59; SI AS 176, L30, cadastral municipality of Nemški Rovt, cadastral appraisal protocols: s.n., 30/4, 59; SI AS 176, L179, cadastral municipality of Zgornje Gorje, cadastral appraisal protocols: s.n., 179/4, 59; SI AS 176, L210, cadastral municipality of Poljšica, cadastral appraisal protocols: s.n., 210/4, 59; SI AS 176, L227, cadastral municipality of Rečica, cadastral appraisal protocols: s.n., 227/4, 59.

[2387] The community of Blejska Dobrava reported 4 fl 20 kr costs of spring cleaning per *Joch* (0.57 ha) (involving removal of flood deposits and anthills, channel clearing, and other spring cleaning) and fence repair (SI AS 176, Franciscejski kataster za Kranjsko [Franciscan Land Cadaster for Carniola], L35, cadastral municipality of Blejska Dobrava, cadastral appraisal protocols: s.n., 35/4, 59). In the community of Zgornje Gorje, the costs of two workers cleaning a meadow in spring, a female and a male, were reportedly 36 kr per day, and in the community of Poljšica 50 kr (SI AS 176, L179, cadastral municipality of Zgornje Gorje, cadastral appraisal protocols: s.n., 179/4, 59; SI AS 176, L210, cadastral municipality of Poljšica, cadastral appraisal protocols: s.n., 210/4, 59). Because spring cleaning in the community of Blejska Dobrava reportedly involved removal of flood deposits, relatively heavy work can be assumed and so the costs were likely relatively high. Reckoning 50 kr per a pair of workers per day, 4 fl 20 kr would mean about 10 workers' tasks for one day (1 fl = 60 kr, and so ten workers paid 50 kr per pair would cost 4 fl 10 kr).

[2388] SI AS 176, Franciscejski kataster za Kranjsko (Franciscan Land Cadaster for Carniola), L227, cadastral municipality of Rečica, cadastral appraisal protocols: s.n., 227/4, 59.

ences in the time needed for its repair in different years. Focusing on spring cleaning, a higher abundance of trees and bushes, molehills, anthills, plants that were unwanted from the farmers' perspective, movable stones on the surface (see Section 6.1.1.3), coarse flood deposits in a meadow, and the existence of irrigation (see Section 6.4) or drainage channels—all these rendered spring cleaning more labor intensive. A far distant meadow location (some of the early nineteenth-century data presented above refer to meadows a two- or three-hour walk from the village) required more time for the round trip to the meadow and back, and so less time was available for spring cleaning the same day. Nonetheless, two people sufficed to spring clean one *Joch* of some meadows allegedly three hours distant from the village, as recorded in the cadastral municipality of Poljšica.[2389] The ratio of 1:4 (or around 1:5) between the least demanding and most demanding spring cleaning and fence repair might partly also reflect differences in the thoroughness of spring cleaning at different farms; for instance, some would likely remove green shoots of autumn crocus (*Colchicum autumnale*), and some others did not (see Section 6.6.1.1). However, the vast differences in the figures recorded might also reflect inequalities in the degree of reliability of farmers' responses concerning their workload because the data were collected for taxation purposes, in greater detail than ever before. Thus, the responses might have ranged from relatively trustworthy (one day's work for two people or somewhat more appears plausible if these data are compared with more partial twentieth-century information presented below, where fence repair is not included) to possibly exaggerated or exceptional (the highest figures given).

In the second half of the twentieth century (in the eastern Julian Alps by the 1970s), the labor intensity of spring cleaning in remote montane meadows decreased significantly. In many places, less labor-intensive cleaning still included removal (i.e., raking) of fallen tree leaves and thorough removal of branches, completed in a large farm meadow (possibly around 2.5 hectares) by three to four people in one day, for example. Forest edges where bushes and trees spread to the meadow most intensively had no longer been mowed by that time, and so the labor-intensive spring cleaning at such places is not included in one day's work for three to four people. In some other remote meadows that were still mowed in the 1970s, spring cleaning had already been completely abandoned by that time.[2390] At present, the most widespread method of meadow spring cleaning is harrowing by tractor, performed at better accessible sites. Raking has become exceptional.[2391] Nonetheless, examples from Carinthia (Germ. *Kärnten*, Sln. *Koroška*) demonstrate that manual spring cleaning

2389 SI AS 176, Franciscejski kataster za Kranjsko (Franciscan Land Cadaster for Carniola), L210, cadastral municipality of Poljšica, cadastral appraisal protocols: s.n., 210/4, 59.
2390 Šarf, 1974, 138, 143–144; Wopfner, 1997, 318, 320.
2391 Personal observations in Bohinj.

Figure 107: The amount of biomass removed in spring from larch meadows was high. This former larch hay meadow at Gschwandtkopf close to Reith bei Seefeld in Tyrol is now used as a cattle pasture. Note that not only fallen branches and twigs, but also the dense layer of larch needles was removed during the cleaning of this grassland. Photo: Christian Norz. Norz, 2014, 21. © Christian Norz.

still takes places where it is not avoidable, such as in the very limited area of steep meadows nowadays mowed with a scythe.[2392]

Spring cleaning was usually particularly intensive where grasslands in light forests were mowed (Figures 104 and 107). Although the majority of fallen branches were removed during spring cleaning, this type of work had to be repeated in such meadows in the case of later weather with heavy winds. In addition, in front of a haymaker, children had to use a rake to remove the last branches that had fallen during intensive grassland plant growth.[2393] Due to shade, the quantity of hay from light forests per spatial unit and also the quality of such forest hay was lower in many places than from open grasslands under similar environmental conditions. Many grasses and herbs abounding in the shadow of trees are "non-nutritious, distasteful or poisonous."[2394] Nonetheless, in light deciduous forests "light-demanding edible plants also grow." Most tree species obtain water and nutrients mainly at soil depths inaccessible to meadow plants, and a significant amount of these nutrients is released into meadow soil from fallen leaves, increasing the meadow productivity. Moreover, on

[2392] Keller, 2004, especially pages 55–56, 68–69, 72, 84–85.
[2393] Norz, 2014, 20–22.
[2394] Sadar, 1956, 14; Grove & Rackham, 2003, 191 (quote); Peterken, 2013, 249.

sunny slopes, the shade of trees decreased evaporation from soil, and so trees created damper grassland sites, which were favorable for biomass productivity on some sunny slopes.[2395]

A typical example of meadows in light forests requiring labor-intensive spring cleaning were those on sunny slopes with European larch trees (*Larix decidua*). They used to be frequent at nutrient-poor mesic sites in montane and subalpine vegetation belts (Section 2).[2396] A moderate area of larch meadows within the total meadow area per village was relatively frequent in some parts of Wallis (Francoprovençal *Valês*, French *Valais*), Graubünden (Ital. *Grigioni*; Romansh *Grishun*), south, northwest, and east Tyrol, and Upper Carinthia.[2397] One could also find them in the Alps of Upper Austria,[2398] in Piedmont,[2399] and in the Slovenian Alps,[2400] for instance. In northwest Tyrol (up to and including the district of Innsbruck in the east), there were around 90 km^2 of larch meadows and pastures with larch trees together in 1948, and the maximum historical area of larch meadows and pastures with this tree species there was larger than that.[2401] Namely, larches have been an esteemed source of durable timber. Moreover, starting in the first half of the seventeenth century at the latest, larch resin or turpentine was exported, the medicinally esteemed agarikon (*Laricifomes officinalis*)—a fungus growing on larch trees—was a matter of

2395 Grove & Rackham, 2003, 191 (quote); Brus, 2012, 46; Norz, 2014, 18–19, 124–126, 130; Machatschek, 2015, 44, 52, 60.

2396 Aeschimann et al., 2004, vol. 1, 102–103; Brus, 2012, 46; Norz, 2014, 11, 19. See also the examples from Dolina above Tržič and Rakovec below, both located in the montane vegetation belt.

2397 E.g., Norz, 2014, especially pp. 3, 6–7.

2398 Machatschek, 2015, 52.

2399 Viazzo, 1989, 111.

2400 E.g., a complex of montane wooded hay meadows named Macesnovc ("Mesesnouz"; named after the Slovenian common noun *macesen* 'larch') was mapped north of Dolina above Tržič in the early nineteenth century (*Arcanum Maps*, Cadastral Maps, Habsburg Empire – Cadastral Maps (XIX. [sic] Century), cadastral municipality of Sv. Katarina (Germ. *St. Katarina*), 1826). Mowing was abandoned in that meadow complex by the mid-twentieth century. Namely, it not only required labor-intensive spring cleaning, but it was also so difficult to access that bundles of hay were thrown over a short steep slope section; see Section 3.1 (author's interview with J. Kavčič, 2021, in Dolina). Another example: in the Pohorje Hills, a larch meadow existed west of the hamlet of Rakovec at upper Hudinja; in the 1820s, there was still a pasture (*Arcanum Maps*, Cadastral Maps, Habsburg Empire – Cadastral Maps (XIX. [sic] Century), cadastral municipality of Hudinja (Germ. *Hudina*), 1824). A high population density in this remote (proto)industrial mountain settlement in the nineteenth century that still depended on livestock-powered transport (compare Gabrič et al., 2019, 48–51, 54, 113) probably resulted in conversion of the pasture into a hay meadow, which is clearly recalled for the twentieth century (author's interview with Lamot, 2019, in Rakovec, who used to farm that larch meadow). Gratitude is expressed to Gorazd Fric from the Municipality of Vitanje, who recalled that the present-day larch pasture used to be a larch meadow, which was a stimulus for the interview.

2401 Norz, 2014, 29.

Figure 108: A non-fertilized larch meadow at Gschwandtkopf close to Reith bei Seefeld in Tyrol mowed once annually. Kidney vetch (*Anthyllis vulneraria*), arnica (*Arnica montana*), round-headed rampion (*Phyteuma orbiculare*; i.e., the blue flowers in the right), mountain clover (*Trifolium montanum*), red clover (*Trifolium pratense*), and goldilocks (*Ranunculus auricomus*) belong to the many species occurring there. Haymaking in such a meadow has to be preceded by labor-intensive spring cleaning.[2402] Photo: Christian Norz. Norz, 2014, 13. © Christian Norz.

trade, and in the eighteenth and nineteenth centuries, larch seeds from the Alps were sold to regions outside of this mountain range, and so larches were the basis for some further sources of income.[2403] The crown of a larch tree is thin, the Sun's rays partly penetrate through it, and larch forests are sparse because the species is heliophilous.[2404] In addition, only a limited number of saplings were allowed to grow in larch meadows; the others were removed by scythe during mowing to prevent the grassland from being too shaded.[2405] Thus, there was sufficient undergrowth for meadows and fodder quality was better than in denser forests (Figure 108). Differences in light conditions at shadier and sunnier sites, differences in water availability due to the impact of tree crowns, and local differences in soil and relief led to patchworks of grassland plant communities within wooded meadows. Meadow species dominate more open sections of larch meadows,

---

[2402] Norz, 2014, 20–21.

[2403] Valentinitsch, 1989, 88–92; Niederer, 1996d, 192; Brus, 2012, 47; Tappeiner, Borsdorf, & Bahn, 2013, 517; Norz, 2014, 18–19, 24.

[2404] Brus, 2012, 46; Norz, 2014, 18.

[2405] Norz, 2014, 24, 110.

and forest species shadier sites.[2406] In some places in the Alps, fallen larch needles were raked in late autumn and used as animal bedding.[2407] Use of needles raked during spring cleaning as livestock bedding was also possible. Partly decomposed needles (and twigs) absorbed urine better than recently fallen ones. Alternatively, raked larch needles that had already largely decomposed were further composted and used directly as fertilizer, such as in some places in the Venosta Valley (Germ. *Vinschgau*, Ital. *Val Venosta*).[2408] The partial removal of needles mitigated meadow soil acidification and prevented the suppression of some grassland plants through the dense layer of larch needles, which decomposes slowly, but this also reduced the amount of nutrients released into the soil.[2409]

Larch meadows have a long historical tradition. They were, for instance, mentioned in the Stanz Valley (Germ. *Stanzertal*) and Paznaun Valley (Germ. *Paznauntal*) in the northwestern part of Tyrol in 1459. In another example, when the community of Telfs, a village west of Innsbruck, received a permit to create new meadows on Zimmerberg Hill in 1631, farmers were ordered to maintain the larch trees there. Thus, many larch meadows most likely originated in the medieval settlement process, whereas many others are a result of later increases in agricultural density (Section 3.1) and other historical socio-economic processes. The rise in the systematic nature of land records in the eighteenth century resulted in more data on larch meadows.[2410] Some larch meadows were created later: when a forest section at Mieders, a settlement located south of Innsbruck, was divided in the early nineteenth century to create meadows, larch trees were ordered to be planted in the forest clearings.[2411] However, soon thereafter, starting in the mid-nineteenth century, a significant decrease in the total area of larch meadows followed in some regions; many were converted into pastures, at least in the Upper Inn Valley.[2412] Small areas of larch meadows have been mowed to date (Figure 108).[2413]

Larch meadows were not the only relatively densely wooded meadow type in the Alps. Coppices are dealt with in Section 6.1.2.6. Moreover, a rent-roll (*Urbar*) of the seigneury of Bleiburg (Sln. *Pliberk*) in Carinthia from 1570 documents mowing about two to four carts of hay per farm in years of good forest hay harvest by some farmers from the villages of Replach (Sln. *Replje*) and Mittlern (Sln. *Metlova*) in the Jaun Valley (Germ. *Jauntal*, Sln. *Podjuna*). This was performed in sparse (e.g., birch, *Betula pendula*, and Scots pine, *Pinus sylvestris*) forests. This rent-roll emphasized that yields of forest hay varied

[2406] Ellenberg & Leuschner, 2010, 998; Norz, 2014, 13–14.
[2407] Viazzo, 1989, 111; Machatschek, 2002, 397; Machatschek, 2015, 52.
[2408] Machatschek, 2002, 385–386, 405–408.
[2409] Machatschek, 2002, 407–408; Norz, 2014, 123–126.
[2410] Grass, 1948, 187; Norz, 2014, 16–17, 19.
[2411] Wopfner, 1997, 295–296.
[2412] Norz, 2014, 28.
[2413] Machatschek, 2015, 52.

significantly from year to year.[2414] Spring cleaning there must have also been relatively intensive.

In sum, spring cleaning in meadows had relevant economic and ecological impacts. The economic impacts are briefly presented at the beginning of this section. The historical ecological impacts of spring cleaning differed between sites. They involved the removal or redistribution of target biomass and inorganic material, such as molehills, hills of meadow ants, flood-deposits, and so on. All this had significant impacts on site conditions if one considers the cumulative impacts in the long run in particular, and also direct impacts on plant and animal species living there. Examples of meadow agroecosystems in which the impacts of spring cleaning were ecologically particularly relevant involve sites that would become largely covered by sand, pebbles, or rocks deposited by floods or slope processes, unless spring-cleaned, and sites where false hellebore (*Veratrum* spp.; see Section 6.6.1.1) would dominate unless spring cleaning was carried out (see Section 3.2.2).[2415] Where undesired plants, such as moss, were removed by being uprooted, spring cleaning also created gaps, used as regeneration niches by meadow species, whose seeds fell off during haymaking.

6.6.1.1 Removal of herbs considered harmful, useless, or having a negative impact on hay quality

Removal of some plants undesired from the perspective of livestock farming in spring has a long history. For instance, records from the 1830s referring to the cadastral municipalities of Studor, Srednja Vas, and Češnjica in western and northern Bohinj report that some meadows, ranging from the bottom of the valley to montane level, were cleaned of harmful plants.[2416] Several plant species removed from non-fertilized and moderately fertilized meadows are presented in this section. Nonetheless, they occur at various sites, and so one forb species or no more than a few species were removed from a usual meadow.

The forbs removed from meadows, as reported in Bohinj in the 1830s, surely included false hellebore (*Veratrum* sp.; Figure 16 in Section 3.2.2) in montane meadows. In the eighteenth century, the botanist Balthasar Hacquet observed that white false hellebore (*Veratrum album*) and black false hellebore (*Veratrum nigrum*) grew in many meadows in the Pre-Alps and Alps of what is now

2414 Zwitter, 2016, 122.

2415 Pils, 1994, Table 58, photo 145.

2416 SI AS 176, Franciscejski kataster za Kranjsko (Franciscan Land Cadaster for Carniola), L107, cadastral municipality of Češnjica v Bohinju (i.e., Bohinjska Češnjica), cadastral appraisal protocols: s.n., 107/4, 50–51; SI AS 176, L157, cadastral municipality of Srednja Vas v Bohinju, cadastral appraisal protocols: s.n., 157/4, 50–51; SI AS 176, L284, cadastral municipality of Studor, cadastral appraisal protocols: s.n., 284/4, 50–51.

northwest and northern Slovenia.[2417] At that time, spring grazing in meadows, a practice suppressing some species, was still usual (see Section 6.2). Thus, there is firm evidence that false hellebore had also been an important challenge earlier in history, when spring grazing was practiced in meadows. Hacquet noted many farmers' claim that both false hellebore species were more dangerous to their livestock than autumn crocus (*Colchicum autumnale*); in numerous cases, cattle perished due to poisoning with false hellebores.[2418] False hellebore can be lethal to grazing calves and heifers not familiar with this plant; adult cattle used to such grasslands avoid it, provided that other fodder is available.[2419] The frequent poisoning in the eighteenth century, recorded by Hacquet, must be a consequence of fodder scarcity that was often experienced by cattle while grazing in meadows in early spring (Section 6.2.2.1). False hellebores grow in spring much more quickly than the majority of grassland plants,[2420] and so cattle can potentially consume them earlier than most other grassland vegetation. Thus, starving cattle would consume false hellebores in spring. Sheep are less likely to avoid false hellebores systematically where other fodder is available, and for some sheep the consequent poisoning is lethal.[2421] The symptoms of livestock poisoning (severe diarrhea, vomiting, and foaming at the mouth) are evident, and so they must have also been observed earlier in history. Early to mid-twentieth-century information demonstrates that farmers usually mowed false hellebores in the spring when the grasses were still low, when false hellebore had already reached a significant size.[2422] Historical removal of false hellebore by scythe in May or June was evidently widespread in the Alps because it was identified at such distant and culturally diverse places as the Julian Alps in the Eastern Alps[2423] and the Allues Valley (Francoprovençal *Alués*, French *Les Allues*) in the northern French Alps[2424] and by Csángó farmers in Ghimeş in the Eastern Carpathians.[2425] If cut in spring, false hellebore does not sprout again in the same year (or the shoot is very weak), and so the hay was free of this poisonous species.[2426] Further reasons for its removal are that, by the time of his-

2417 Hacquet, 1779, 40–41, 43 (plant species identification by Praprotnik, 2015, 80).
2418 Hacquet, 1779, 43 (plant species identification by Praprotnik, 2015, 80).
2419 Author's interviews with Mencinger, 2017, in Stara Fužina, and with C. Mlakar, 2017, at the Zajamniki Alp.
2420 Author's interviews with Gašperin, 2017, with Kunavec, 2017, and with Mencinger, 2017, all in Stara Fužina; with L. and M. Odar, 2017, in Studor; with Grm, 2017, in Srednja Vas; with Minka Sodja, 2017, in Jereka; and with J. Smukavec, 2017, in Podjelje.
2421 Babai, Molnár, & Molnár, 2014, 97–98.
2422 Hacquet, 1779, 43–44; Babai, Molnár, & Molnár, 2014, 97–98; author's interviews with Gašperin, 2017, and with Kunavec, 2017, both in Stara Fužina; with L. and M. Odar, 2017, in Studor; with Grm, 2017, in Srednja Vas; and with Minka Sodja, 2017, in Jereka.
2423 The interviews are cited above.
2424 Meilleur, 1986, 270, 306.
2425 Babai, Molnár, & Molnár, 2014, 97–98.
2426 Babai, Molnár, & Molnár, 2014, 97–98; author's interviews with Gašperin, 2017, in Stara Fužina, and with F. Mlakar, 2017, in Bohinjska Češnjica.

torical haymaking, stalks of false hellebore became hard and difficult to mow, and that, if its seeds had been allowed to ripen, the abundance of false hellebore in grasslands would have increased quickly and strongly (see Section 3.2.2).[2427] Removal by scythe in spring prevents strong false hellebore plants from being in hay, but it does not lead to a total disappearance of false hellebores, which tend to continuously spread to grasslands from forest edges.[2428] Thus, from the farmers' perspective, at white false hellebore sites there was always enough of the species for limited use as an insecticide. Its "roots were smashed and macerated in water" and this solution was used externally in autumn "to scrub down bovines before their reentry into the stables for over-wintering"; for instance, in the Allues Valley in the first half of the twentieth century.[2429] Information from Ghimeş explains that cattle washed with white false hellebore brew as a means of lice suppression were tied in a manner that they could not lick the poisonous substance from their hair.[2430] This kind of white false hellebore use was widespread in the Alps, not only in the Allues Valley, but also at many other places, such as Hudi Kot in the northern Pohorje Hills, in northeastern Slovenia, by the mid-twentieth century,[2431] and possibly in East Tyrol.[2432]

Yellow gentian (*Gentiana lutea*) is another plant species that used to be suppressed in hay meadows in some places in the Alps in some historical periods if its occurrence and abundance significantly exceeded economically motivated ones while the demand for livestock fodder was (relatively) high. In such cases, yellow gentian was removed from hay meadows and, optionally, some pastures, with a scythe, but farmers let it grow at least in lower-quality pastures for the purpose of gathering (see Sections 6.6.2 and 8.1.2). In the Allues Valley, for instance, yellow gentian and false hellebore were usually removed from meadows through maintenance mowing in June (hay was made in the same meadows in July and August) during the relatively high agricultural density of the first half of the twentieth century. Like false hellebore, yellow gentian is difficult to mow during haymaking due to its hard stalks, and it is rejected by livestock. Maintenance mowing in June in several successive years could indeed eradicate yellow gentian[2433] because it was repeatedly unable to store enough nutrients.

Removal of butterbur (*Petasites* spp.) leaves from meadows in spring with a scythe, aiming at producing fresh fodder and at the same time reducing the content of butterbur in hay mowed later, is presented in Section 6.1.2.5.1.1.

2427 Meilleur, 1986, 270, 306; Babai, Molnár, & Molnár, 2014, 97–98.
2428 Babai, Molnár, & Molnár, 2014, 97–98.
2429 Meilleur, 1986, 306 (both quotes).
2430 Babai, Molnár, & Molnár, 2014, 98.
2431 Interview by Marija Makarovič, with Uran, born at Hudi Kot (Makarovič, 2016).
2432 Vogl-Lukasser et al., 2006, 190, 239.
2433 Meilleur, 1986, 30–31, 122, 270, 292, 306, 382; relevant insight into the decreased intensity of agricultural use of mountain pastures in that area by the first half of the twentieth century is also given on p. 374.

Abundance of docks (*Rumex* spp.) in manured meadows was decreased by digging out the roots with a pick in spring, when the dock leaves grow in grasslands sooner than the majority of plants, and so they were easy to identify. This was practiced in hay meadows, at various elevations, also including the mowed surroundings of cottages in mountain summer settlements, unless docks were used there (see Sections 8.2.1.1, 8.2.2, and 8.2.3). Alternatively, as documented in the first half of the twentieth century at least, some farmers in various regions of the Alps used a so-called *Ampferstecher*—a special sharp narrow iron implement, making possible easy penetration deep into the soil by stepping on the iron crossbar on its handle—to dig out dock roots. This reduced the workload and saved many other plants.[2434] Uprooting the docks with a pick or this special implement was a successful but labor-intensive method,[2435] adapted to the fact that there are root buds on the upper sections of dock's taproot. In the case of broad-leaved dock (*Rumex obtusifolius*), for instance, the root buds extend about 12 cm deep into the soil; if the dock's taproot is removed more shallowly than up to the end of the section with buds, the herb can resprout.[2436]

Removal of at least some autumn crocus (*Colchicum autumnale*) leaves in moderately fertilized and non-fertilized mesic meadows was carried out in spring by some farmers, as recalled for the mid-twentieth century (Figure 109). Some others removed many green shoots of autumn crocus from mowed hay in the meadow during haymaking—a particularly unpleasant task in the summer heat due to the high abundance of the species in some meadows. Namely, many domestic animals do not consume autumn crocus's green shoots. Much more important than rejection of the species by some livestock, which would allow simply discarding rejected *Colchicum autumnale* from the manger, is the impact of autumn crocus's green shoots on the general hay quality. Namely, it takes much longer to dry than most other grassland plants, and so a significant proportion of *Colchicum autumnale* leaves in hay can result in mold infestation. Another reason for the removal of autumn crocus leaves was the fact that not all cattle and possibly horses avoided them in hay, and so some farmers claimed that such hay caused health issues (see Section 6.2.2.2).[2437]

2434 Novak, 1978, 58; author's interviews with Urh, 2017, with Kunavec, 2017, and with Kranjc, 2018, all in Stara Fužina; see an illustration in Hinterwaldner, 2002, 218.

2435 Rasran & Jeromin, 125, 129.

2436 Dietl & Lehmann, 2004, 48–49.

2437 Winter et al., 2011, 1774, 1777–1778; author's interviews with Gašperin, 2017, with Kunavec, 2017, with Mencinger, 2017, and with Urh, 2017, all in Stara Fužina; with Grm, 2017, and with Milena Sodja, 2017, both in Srednja Vas; with Minka Sodja, 2017, in Jereka; with A. and M. Taler, 2017, and with L. and M. Odar, 2017, both in Studor. Similar information comes from the Eastern Carpathians: Babai, Molnár, & Molnár, 2014, 101.

Figure 109: Manual spring cleaning of meadows has become exceptional: legacies of pulling green shoots of autumn crocus in a remote montane meadow in the Javornica meadow complex, west of Podjelje in the Julian Alps, to reduce its abundance later, during haymaking. This extensively managed meadow at 1,100 m is mowed in July. Photo: Žiga Zwitter, June 3rd, 2017.

Another example of a species historically weeded in meadows is Alpine eryngo (*Eryngium alpinum*, Figure 110), which can abound in some regions of the Eastern and Western Alps, at mesic base- and nutrient-rich sites within the subalpine vegetation belt (Section 2).[2438] Today, it is a rare species, fully protected in Austrian Carinthia and on the Red List of endangered species in Austria as a whole.[2439] Likewise, it is a potentially endangered species in Slovenia.[2440] However, it used to abound on the southern slopes of the Golica Ridge in the Karawanks (Sln. *Karavanke*, Germ. *Karawanken*) as recorded in the second half of the eighteenth century. In the nineteenth century, the abundance of Alpine eryngo on the slopes of Golica decreased a lot. It was still gathered there by nineteenth-century botanists; for example, in 1874. By the early twentieth century, the species had almost disappeared from these mountain meadows. An important reason for this major decrease in the abundance of Alpine

[2438] Aeschimann et al., 2004, vol. 1, 1074–1075.
[2439] Keller, 2004, 27.
[2440] "Pravilnik o uvrstitvi ogroženih rastlinskih in živalskih vrst v rdeči seznam," 2002, 8893, 8900.

Figure 110: Alpine eryngo (*Eryngium alpinum*) on Mt. Lisec in the Julian Alps, close to Mt. Črna Prst. Photo: Tomaž Zwitter, 2021.

eryngo was uprooting of this thorny plant to improve hay in remote montane meadows, and this weeding was stimulated by the fact that it provided marketable specimens: tourists bought Alpine eryngo for decorative reasons, as an ecologically relevant phenomenon (on the high interest of tourists of that time in plants from the Alps and the fashionable gardens of Alpine plants built then all around Europe, see Section 8.1.3.1). *Eryngium alpinum* almost certainly became extinct in these grasslands.[2441] Currently, this endangered species occurs in the Slovenian Alps mainly in the surroundings of Mt. Črna Prst in the Julian Alps. An early twentieth-century botanical notice makes it clear that Alpine eryngo was also gathered by herders there and sold to tourists, but the long-

[2441] Praprotnik, 2012, 34; Praprotnik, 2015, 89, 157, 163–164, 243; Bavcon, Ravnjak, & Praprotnik, 2019, 113; Dakskobler et al., 2019, 110–111; Peternel, 2022; compare field eryngo (*Eryngium camperstre*) in Pils, 1994, 203.

term ecological impacts of gathering *Eryngium alpinum* (and, possibly, other adverse management impacts) in this area have been incomparably lower than those on the slopes of Mt. Golica in the Karawanks.[2442]

Few other meadow herbs were also likely to be removed during spring cleaning. An example is bracken (*Pteridium aquilinum*), which can dominate at some nutrient-poor mesic sites unless its abundance is limited by grassland management.[2443] Bracken is poisonous to some livestock types, at least.[2444] Nonetheless, scant twentieth-century evidence from East Tyrol explains that, at some farms, limited quantities of bracken were perceived as useful; sheep liked to eat it and some farmers fed their sheep with fresh or dry bracken for some days as a means they believed to suppress worms.[2445] In this context, regional to local diversity in the attractiveness of bracken for sheep is worth mentioning, ranging from rejection to consumption and poisoning that can cause severe health issues in the long run.[2446] Further plants partly removed during spring cleaning in meadows include some dwarf shrubs, shrubs, and saplings of tree species unwanted from the perspective of livestock farming, especially in meadows not mowed every year (see Sections 6.1.2.6. and 6.1.2.7).[2447] Suppression of moss was mentioned at the beginning of Section 6.6.1.

Spring cleaning was not the only means of removal of plant species that farmers wanted to exclude from fodder. The abundance of some non-desired species was reduced through grazing in hay meadows in spring (Section 6.2.2.2). Some other species were discarded during haymaking, as long as hay was shaken out and turned manually (see Section 9).[2448] Examples included autumn crocus (see above) and dodder (*Cuscuta* spp.) which occurs in meadows at various elevations, mainly in nutrient-poor sites.[2449] Hay plants bound together with dodder (Figure 111) used to be regularly removed from hay by many farmers in Bohinj, for instance, because dodder was considered to cause problems with digestion and diarrhea if cattle ate it in hay.[2450] Discarding the hay plants that dodder climbs on was probably more widespread in the Alps because some dodder species occur in many regions in the Alps,[2451] and

[2442] Dakskobler et al., 2019, 109–112, 144.

[2443] Pils, 1994, 209; Babai, Molnár, & Molnár, 2014, 97, 100; Ivaşcu, Öllerer, & Rákosy, 2016, 46–47, 49; Aeschimann et al., 2004, vol. 1, 70–71.

[2444] Pils, 1994, 203, 305.

[2445] Vogl-Lukasser et al., 2006, 177–178.

[2446] Pils, 1994, 305.

[2447] Wopfner, 1997, 318–319; see also Ivaşcu, Öllerer, & Rákosy, 2016, 47–48.

[2448] Author's interview with J. Smukavec, 2017, in Podjelje.

[2449] Aeschimann et al., 2004, vol. 2, 50–51.

[2450] Author's interviews with Kranjc, 2018, and with Kunavec, 2017, both in Stara Fužina. Removal of dodder from hay is also recommended in official guidelines by the Chamber of Agriculture and Forestry of Slovenia to promote best practices in healthy fodder production (*Smernice dobre prakse* ..., 2007, 33).

[2451] Aeschimann et al., 2004, vol. 2, 48–51.

Figure 111: Dodder (*Cuscuta spp.*;[2452] i.e., the pink stalks and inflorescences), occurs here and there in non-fertilized meadows on Vogar Hill in Bohinj. Photo: Žiga Zwitter, July 2018, during haymaking.

dodder's negative impacts on livestock health are also reported in some other regions of the world, such as in parts of northern India. Similar poisoning symptoms are reported there with a different dodder species (*Cuscuta reflexa*): vomiting, abdominal pain, purgation, uneasiness and anorexia, and also abortion in gestating livestock.[2453] As a parasite, dodder suppresses other plants and continues to obtain liquids from them even after being cut.[2454] For this reason, it needs longer to dry than the majority of grassland plants, which is another reason to remove it from hay.

Spiny restharrow (*Ononis spinosa*) was also removed from mowed biomass by some farmers during haymaking (see Section 9), as were some docks (*Rumex* spp.) to prevent their seed dispersal.[2455] For the same reason, some sand leek (*Allium scorodoprasum*) that was not uprooted (see below) was removed from mowed hay because it produces a lot of seed, but the plant is rejected by some livestock.[2456] On discarding mowed moss because livestock would not consume

[2452] Thanks to the botanist Andrej Podobnik, who identified the genus in the photo.
[2453] Bhatia, 2014, 209, 213–215.
[2454] Orloff et al., 1989.
[2455] Author's interview with Urh, 2017 and 2022, in Stara Fužina.
[2456] Author's interviews with Kunavec, 2021 and 2023, in Stara Fužina.

it,[2457] see Section 6.6.1. A more detailed study of removal of unwanted plant species from hay when it was shaken out in the Allues Valley documented that discarding focused there on the following plants: (1) hogweed (*Heracleum sphondylium*), which has an inedible stalk; (2) sermontain (*Laserpitium latifolium*), a species that is difficult to mow during haymaking (Section 6.1.2.2) and its stalks are discarded by cattle, and so children often uprooted and removed it in front of haymakers. A parallel from the Eastern Carpathians suggests that its leaves might be used as hay, whereas the stalks were discarded during haymaking; (3) highly thorny thistles such as spiniest thistle (*Cirsium spinosissimum*) and masked thistle (*Carduus personata*) were removed to make sure that their prickles would not cause "injuries to the mouths of livestock"; (4) horsetails (*Equisetum* spp.) were discarded because they were rejected by some livestock types;[2458] marsh horsetail (*Equisetum palustre*) is poisonous to cattle fresh and dry (see Section 6.1.2.5.1.2); and (5) white false hellebore (*Veratrum album*) and yellow gentian (*Gentiana lutea*) were removed during haymaking where they had not been cut during maintenance mowing in spring.[2459] Removal of some further species such as thorny thistles from the genus *Cirsium* can be assumed as general characteristics in those meadows in the Alps where they occur in significant numbers, as suggested by equal evidence at a place as culturally different and spatially distant from the Allues Valley as Ghimeş in the Eastern Carpathians (but see the specific historical use of some thistle species in Sections 8.2.1.1 and 8.2.2).[2460]

In specific cases, plant tissue properties allowed significantly easier uprooting during haymaking than before trampling became an issue in hay meadows in spring. Uprooting by pulling the specimens in front of the scythe was particularly relevant if the main aim of such weeding was to reduce the plant's abundance in grazing resources in meadows in the autumn. Namely, if the main motive was to reduce the species's impact on milk quality during autumn grazing, it was not enough to remove the mowed specimens of a species able to resprout after haymaking. This was the case of leeks, which were partially removed from some grasslands in general[2461] and also in Bohinj. Namely, milk from livestock that had grazed in a meadow in autumn, where young wild leek (about 10 cm tall) that resprouted after mowing abounded in grazing resources, stank of garlic. In contrast to autumn grazing, the presence of some leek in hay would not be a challenge for milk flavor because many dairy cows rejected dry and full-grown, relatively hard leek stalks, as experienced in Bohinj,[2462] where

[2457] Author's interview with L. and M. Odar, 2017, in Studor.
[2458] Meilleur, 1986, 273, 297–298, 305 (quote from p. 305); on sermontain in the Carpathians: Babai, Molnár, & Molnár, 2014, 97, 102.
[2459] Meilleur, 1986, 273, 292, 306.
[2460] Babai, Molnár, & Molnár, 2014, 52.
[2461] E.g., Peterken, 2013, 187.
[2462] Hodnik, 2016, 73; author's interviews with Urh, 2017, with Kranjc, 2018, and with Kunavec, 2023, all in Stara Fužina.

hay stocks were sufficient and leek's abundance in hay was moderate, and so cows were not starving by having rejected the leek.[2463] These data most probably at least refer to sand leek (*Allium scorodoprasum*) because this species occurs in many semi-dry moderately-manured meadows in Bohinj.[2464] Moreover, sand leek primarily grows in the colline vegetation belt,[2465] and according to the interviewees leek weeding was performed in the valley.[2466] A haymaker working with a scythe would easily uproot sand leek by hand without any tool because its stalk became harder by the time of haymaking (Section 6.1.2.2), and so the plant was uprooted with its bulb, without tearing the stalk.[2467] The experience of stinking milk reported by several interviewees for the mid-twentieth century[2468] proves that this type of weeding was not (always) thorough, and so from the ecological perspective it merely reduced the leek's abundance, but it did not fully remove the species. In the case of a high leek abundance, farmers did not remove it from grasslands and used such hay for non-milk livestock, to feed horses[2469] and those oxen willing to consume it; some oxen even liked it very much.[2470]

Since mowing, spreading, and turning hay have been motorized and farmers also lack time for such labor-intensive practices, fewer plants have been removed. Some farmers and veterinarians claim that the abandonment of removing poisonous plants has led to more livestock diseases in some cases if the animals do not reject the species hazardous to their health, but this is not a general phenomenon, and poisoning with some plants, such as autumn crocus, is not easily recognizable.[2471]

Changed species composition due to an intensified meadow management has brought new challenges, such as an increase in the abundance of parsnip (*Pastinaca sativa*) in meadows heavily fertilized with liquid manure. In the valleys of the Julian Alps, this species is often removed manually during a careful walk through the meadow, trampling as little meadow plants as possible, before second-crop hay is mowed.[2472] On the other hand, the abundance and occurrence of some toxic species in hay meadows, such as autumn crocus, has decreased due to more intensive management[2473] at easily accessible sites.

2463 Compare Erjavec & Peršič, 2005, 796.
2464 Dakskobler et al., 2019, 65.
2465 Aeschimann et al., 2004, vol. 2, 1070–1071.
2466 Hodnik, 2016, 73; author's interview with Kranjc, 2018, in Stara Fužina.
2467 Author's interview with Kunavec, 2021, in Stara Fužina.
2468 Hodnik, 2016, 73; author's interviews with Urh, 2017, with Kranjc, 2018, and with Kunavec, 2021 and 2023, all in Stara Fužina.
2469 Erjavec & Peršič, 2005, 796.
2470 Author's interview with Kunavec, 2021, in Stara Fužina.
2471 Winter et al., 2011, 1774–1775, 1777–1778; author's interview with Grm, 2017, in Srednja Vas.
2472 Author's interviews with Mencinger, 2017, in Stara Fužina, and with F. Mlakar, 2017, in Bohinjska Češnjica; and observations in meadows.
2473 Author's interview with Rozman, 2020, in Srednja Vas.

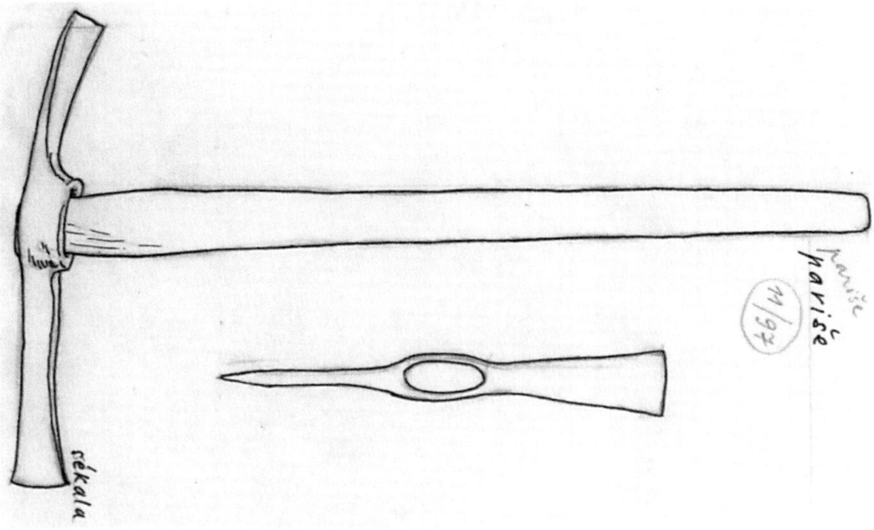

Figure 112: A pick for bush uprooting. Zakriž above Cerkno, pre-Alpine hills of western Slovenia, documented in 1954. Drawn by Ivan Romih. Slovenian Ethnographic Museum, Documentation, Fieldwork campaign 11, Cerkno Region (1954): drawing 11/100 & *Slovenski etnografski muzej* . . . Cerkljansko / 235 R0000011/100. © Slovenian Ethnographic Museum.

6.6.2 Pasture clearing

Pasture clearing[2474] usually involved decreasing the abundance of plant species considered hazardous or useless as livestock fodder. Livestock's selective grazing behavior, in interaction with site conditions and species composition, required cleaning in pastures to prevent a long-term decrease in their carrying capacity.[2475] In stony pastures, some stones lying on the surface were also removed on a regular basis and piled to slightly increase the productive grazing area or, in the case of continuous stone accumulation, to keep the pasture productive in the long run. In flood- or avalanche-prone pastures, coarse deposits were removed in periods of high fodder demand, where their gradual accumulation was considered to significantly reduce the grassland's carrying capacity in the long run.[2476]

In pastures below the tree line (Section 2), many bushes and some young trees not consumed by livestock were uprooted (Figure 112) or cut on a more or less yearly basis to keep the grasslands productive. In many places,

2474 On work organization, see Niederer, 1996b, 342, 344; Niederer, 1996c, 23, 28, 31, 33, 40, 42–47, 50, 67, 69–71, 79–80, 84–86; Niederer, 1996d, 163.

2475 Pils, 1994, especially p. 203; Wopfner, 1997, 267, 397–398, 449–450; Babai, Molnár, & Molnár, 2014, 116–119; Kačičnik Gabrič, 2014, 175; Zwitter, 2014e, 333–334.

2476 "Bildteil," 1996, 439; Niederer, 1996c, 31, 40–42, 55, 69, 79–80; Niederer, 1996d, 163; Wopfner, 1997, 449; Hodnik, 2016, 61, 87; Novak, 2024, 96.

Figure 113: Thistles (lower right part of the figure) grow in many pastures unless they are cleaned or consumed by horses. Photo: Žiga Zwitter, 2016, at the foot of the Sinjajevina Plateau, Montenegro.

regular spring cleaning of pastures included burning of (cut) bushes, as documented, for example, in the Konjska Dolina complex of mountain pastures in the Julian Alps in 1605,[2477] in Carinthia according to nineteenth-century data,[2478] and in the Allues Valley in the early twentieth century.[2479] Goat grazing also suppressed some bushes and trees in pastures that were avoided by other livestock.[2480] A specific example, where clearing of grazing land was at least partly motivated by strengthening the resilience of local animal husbandry in terms of flood and drought risk, is a court decision from the northern Jaun Valley (Germ. *Jauntal*; Sln. *Podjuna*) in Carinthia in 1615. In line with this, in order to maintain or restore quality grassy pastures near Piskertschach (Sln. *Piskrče*), thornbush, pine (*Pinus* spp.), silver fir (*Abies alba*), Norway spruce (*Picea abies*), and thistle(s) had to be pulled out or

---

[2477] SI AS 721, Gospostvo Bled (The Seigneury of Bled), Iustitialia, vol. 13, [May 21st] 1605; author's interview with Mencinger, 2017, in Stara Fužina; on burning shrubs in pastures in the Eastern Carpathians: Babai, Molnár, & Molnár, 2014, 116, 118.

[2478] Kačičnik Gabrič, 2014, 175.

[2479] Meilleur, 1986, 347.

[2480] Author's interview with Kunavec, 2017, in Stara Fužina; Machatschek, 2017, 94.

removed.[2481] Some thistles are nitrophilous,[2482] and so they tend to grow from cowpies in pastures. Rejected by livestock, thistle blossoms are relevant nectar sources for insects in many pastures (Figure 113).[2483] Often, some thistle species left standing by other domestic animals were consumed by grazing horses.[2484]

Reduction of the shading impact of lower conifer branches in grasslands was not limited to cutting the trees. In many places, farmers would only cut the lower conifer branches. In line with the regulation on the Hölzelstal Alp (Germ. *Hölzelstalalm*) and Flach Alp (Germ. *Flachalm*) grazing complexes in Tyrol northeast of Innsbruck from 1627, branches, mainly of conifers (occurrence of conifers is mentioned in those pastures, and the regulation mentions trees whose lower branches decreased the fodder availability in the pasture or prevented access to grazing sites), were allowed to be cut as high as reachable standing on the ground.[2485] In this manner, the trees provided shade for grazing livestock and allowed wood production. Some valued tree species were not allowed to be removed from some pastures in any stage of development, such as Swiss stone pine (*Pinus cembra*) in the Hölzelstal Alp and Flach Alp according to the 1627 regulation.[2486] Swiss stone pine is one of the tree species reaching the highest elevations among all trees in the Alps. Dwarf mountain pine (*Pinus mugo*) dominates at such sites in many carbonate regions of the Alps. Cutting dwarf mountain pine has often been practiced in mountain pastures above the actual forest line, which had been lowered by clearing, grazing and pasture maintenance (Section 3.1).[2487] On occasions when dwarf mountain pine was cleared to expand summer pastures on sites outside the usual grazing territory, the outcome of clearing dwarf mountain pine was uncertain (Section 3.2.2).

Still focusing on woody plants, a specific case was pastures that followed swidden field-crop cultivation in the same parcels (Section 3.2.1). In such cases, pasture's afforestation was allowed to proceed, and farmers did not remove woody species to produce biomass needed for the next burning. Namely, too little biomass (for burning and fencing) required leaving some parcel sections unburned, which made them unsuitable for cereal growth, and the quality of subsequent pasture there was low. Similarly, forest undergrowth, such as ferns and bramble (*Rubus* sp.), were allowed grow in swidden pastures, at least where bushes had already densely covered the grassland.[2488] Succession was

[2481] Zwitter, 2014e, 333–334.
[2482] Aeschimann et al., 2004, vol. 2, 558–565, 570–593, 604–605, 618–621, 648–651.
[2483] Pils, 1994, 169, 204, 221, and Table 14, photo 37; Table 60, photos 149 and 150.
[2484] Babai, Molnár, & Molnár, 2014, 117.
[2485] Wopfner, 1997, 398.
[2486] Wopfner, 1997, 398.
[2487] For instance, Wopfner, 1997, 397–398, 449–450.
[2488] Predan, 1956b (and *Slovar slovenskega knjižnega jezika*: ostrogovina); Vršnik, 1962, 22–24 (see the biography of this author in Kumer, 2005); Schneiter, 1970, 80–81, 86.

intensive, especially because some tree and scrub species (such as birch, *Betula pendula*; alder, *Alnus* spp.; and hazel, *Corylus avellana*) re-sprouted from the roots after being cut and burned (see Section 3.2.1), and fern (mainly bracken, *Pteridium aquilinum*) and bramble occurred with significant abundance in some swidden fields as soon as in the first year of cereal cultivation that followed late-spring to early-summer spring cutting and burning. During the stage of cereal cultivation, at least, fern and bracken were weeded.[2489] Wood cutting followed by burning were the most intensive phases of swidden plot clearing.[2490]

Turning attention to the removal of herbs from pastures in general, some species that were removed from meadows, such as white false hellebore (*Veratrum album*; Section 6.6.1.1), were suppressed in pastures in the same manner as in mowed grasslands.[2491] This also applies to yellow gentian (*Gentiana lutea*), a species that is avoided by livestock because of its bitterness. It was suppressed in some better pastures in some periods and regions of the Alps, where the abundance of yellow gentian in coarser pastures was considered sufficient for gathering rhizomes in the long run, while the demand for livestock fodder was (relatively) high (see Sections 6.6.1.1 and 8.1.2), such as in the Allues Valley in the first half of the twentieth century.[2492]

Where the leaves of large-leaved dock species (*Rumex* spp.) were not used (Sections 8.2.1.1, 8.2.2, and 8.2.3), suppression of Alpine dock (*Rumex alpinus*) or broad-leaved dock (*Rumex obtusifolius*) in pastures in the mid-twentieth century, before the widespread use of herbicides, usually involved removal of above-ground organs, which is still practiced regularly in some mountain pastures nowadays. This is less labor intensive but also less efficient than digging the taproot, a more usual approach in hay meadows (Section 6.6.1.1). Most frequently, large-leaved docks were mowed in pastures before their seeds were set. Because these grasslands were otherwise grazed, trampling was not an issue, in contrast to hay meadows before mowing. Thus, pastures allowed more frequent mowing of dock than meadows; dock resprouts, grows, and sets seeds more quickly than most grassland plants (Section 6.6.1.1).[2493] Mid-twentieth-century data from Bohinj—specifically, the Konjščica and Laz mountain pasture complexes—demonstrate that the first Alpine dock mowing was carried out before the livestock arrived at these high summer alps. Migration to Konjščica took place very late by today's standards (on July 12th as

[2489] Predan, 1956b, 25, 27–28; Vršnik, 1962, 21–22; Schneiter, 1970, 40–42, 81–82.
[2490] Predan, 1956b; Vršnik, 1962, 21–22, 24; Schneiter, 1970, 81–86.
[2491] Meilleur, 1986, 292, 306; Babai, Molnár, & Molnár, 2014, 97–98, 117.
[2492] Meilleur, 1986, 292, 382; Pils, 1994, 203.
[2493] Author's interviews with Gašperin, 2017, with Kranjc, 2018, with Kunavec, 2017, and with Mencinger, 2017, all in Stara Fužina; with Grm, 2017, in Srednja Vas; with L. and M. Odar, 2017, in Studor; with J. Smukavec, 2017, in Podjelje; and with Milena Sodja, 2017, in Jereka.

a fixed date;[2494] probably not with a long historical tradition[2495]), and so there was a need to cut Alpine dock before the arrival of livestock, at least in the mid-twentieth century. Frequently repeated mowing has some limited success in suppressing Alpine dock or at least preventing its further spread. Most importantly, timely mowing prevented dock from setting seeds.[2496] To reach the desired goal, such dock removal had to take into account the ability of docks cut too late after flowering (but before their seeds ripen) and left in situ to produce ripe seeds in the course of post-ripening. Docks' seed production is enormous, and docks can also become established in relatively dense grasslands; no significant gaps are required.[2497]

The contemporary history of Alpine dock suppression demonstrates the dynamics of traditional ecological knowledge, and it emphasizes that, apart from the strong points presented in several sections of this volume, traditional ecological knowledge also has its weaknesses. In the Laz high summer pasture complex, in the eastern Julian Alps, there has been a lot of Alpine dock (Figure 114). In August 2001, an elderly herder and cheesemaker was visited there by an acquaintance that had not been to Laz for half a century. The acquaintance advised the herder how to get rid of Alpine dock, and she considered the advice so important that she recorded it in her notebook kept in the cottage at Laz. According to this advice, Alpine dock should be mowed on June 18th by 1 pm or on July 24th. In this case, even dock's roots should dry out.[2498] This example of traditional ecological knowledge clearly demonstrates the strong desire to eradicate Alpine dock. Emphasizing the role of plant gathering at auspicious moments (in this case a clearly defined part of the determined day) was a frequent magical component of traditional ecological knowledge systems in

2494 Hodnik, 2016, 37.

2495 Around 1830, the grazing period at Konjščica probably started earlier, on June 24th (as reported in three protocols) or in early July (i.e., about nine weeks before September 8th), which still means that the grazing period started more than half a week earlier than July 12th; there was no reason for reporting too early a start of the grazing period in these types of sources (SI AS 176, Franciscejski kataster za Kranjsko [Franciscan Land Cadaster for Carniola], L157, cadastral municipality of Srednja Vas v Bohinju, cadastral appraisal protocols: s.n., 157/4, 91, and Catastral-Schätzungs-Elaborat, Schätzung des natural ertrages: § 6: Alpen, and two slightly earlier protocols bound in this source: Protocoll, September 16th, 1830, IV. Alpen; Einvernehmungs Protocoll, May 21st, 1832, Alpen, einzige klasse).

2496 Author's interviews with C. Mlakar, 2018, at the Zajamniki Alp in Bohinj (experience of suppression by scythe in a pasture in the vicinity of a cottage at Zajamniki); with F. Mlakar, 2017, in Bohinjska Češnjica; with A. Kavčič, 2017, at the Laz Alp; with J. Smukavec, 2017, in Podjelje; with Milena Sodja, 2017, in Jereka; and with Urh, 2022, in Stara Fužina. See also Rasran & Jeromin, 2010, 124–125, 129–130.

2497 Rasran & Jeromin, 124–125, 129–130.

2498 *Bundrov stan* summer cottage in the Laz Alp in the Julian Alps: notebook starting with "Planšarska staja v Lazu," 38. Thanks to the ethnologist Saša Roškar for the information that this notebook exists there.

Figure 114: Alpine dock in the Laz high summer pasture complex in the Julian Alps at the time of the mowing experiment on June 18th, 2018. Photo: Žiga Zwitter.

Europe.[2499] The rule did not seem likely to be a general truth, and so an experiment was carried out in the Laz mountain pasture in 2018. On the morning of June 18th, some Alpine dock was mowed there using a scythe, and a visit to the pasture three weeks later confirmed that the plants had resprouted.[2500]

A few forb species were uprooted in pastures. See the example of cat's ear (*Hypochaeris* sp.) in Sections 8.2.1.1, 8.2.1.2, and 8.2.2, linking (1) removal of a species that was unwanted from the perspective of grazing and (2) edible plants gathering for human consumption and feeding to pigs. The period since the eighteenth century also saw political incentives to suppress a small number of poisonous herb species, such as deadly nightshade (*Atropa bella-donna*), highly toxic for people, in late eighteenth-century Lower Austria[2501]—a measure that in reality probably had a limited influence on some shady grassland edges.

2499 Compare Brisebarre, 1996, 84.
2500 Thanks to the temporary mountain settlement personnel, Lucija and Gregor Gartner, keeping livestock in the mountain pasture at that time, for their kind support of the experiment.
2501 Lang, 2020, 117–123, 127–128; Maček, 1979, 307.

6.6.2.1 Pasture burning

Fire was or has been used in pasture maintenance in many world regions, including several mountain ranges in Europe, such as the Alps, Carpathians, Apennines, and Pyrenees.[2502] In certain periods and parts of the Alps, pastures were greatly overgrown by dwarf shrubs or monocots unproductive from the point of view of animal husbandry. Such pastures were burned in many cases to improve the quality of grazing resources.

Burning matgrass swards (see Sections 2 and 6.3.1.2) is presented first because it was practiced in several regions of the Alps, as demonstrated by selected examples. In southern Ticino, grasslands on sunny slopes, characterized by a plant community abounding in matgrass (*Nardus stricta*) and heather (*Calluna vulgaris*), used to be occasionally burned in the winter, as documented in the late nineteenth century or first half of the twentieth century. Some years of higher-quality pasture followed.[2503] The same vegetation type was burned in some parts of Tyrol, such as in the Ötz Valley (Germ. *Ötztal*).[2504] In Carinthia, the practice of burning summer mountain matgrass pasture sections dominated by dwarf shrubs, such as heather, has been preserved in the Kreuzeck Group (farms in the Upper Drava [Germ. *Drau*] Valley) and in the southern Nock Mountains (Germ. *Nockberge*), which is probably a small remnant of the formerly much larger territory of pasture burning.[2505] In eastern Styria (Germ. *Steiermark*), burning matgrass swards was reported in the first half of the twentieth century at Teich Alp (Germ. *Teichalm*) during fence repair.[2506]

Another example of burning to improve the availability and quality of grazing resources periodically took place in extensive stands of alpenrose (*Rhododendron ferrugineum*), juniper (*Juniperus communis* subsp. *alpina*), bilberry (*Vaccinium myrtillus*), bog bilberry (*Vaccinium uliginosum*), and cowberry (*Vaccinium vitis-ideae*), which overgrew summer pastures in some parts of the Alps, such as in the Allues Valley in the early twentieth century.[2507]

Other grassland species that used to be intentionally suppressed in the Alps by fire, as reported in the first half of the twentieth century, include, for instance, one or several stiff fescues (*Festuca* spp.) in Sottoceneri, Ticino, and at some places between southwest Carinthia, Sciliar Massif (Germ. *Schlern*, east of Bolzano, Germ. *Bozen*), and Lake Garda (Ital. *Lago di Garda*). Their sharp adult leaves caused injuries from the snout to the stomach of live-

[2502] Schneiter, 1970, especially pp. 44–47, 131–133; Grove & Rackham, 2003, 230.
[2503] Schneiter, 1970, 44, 131.
[2504] Machatschek & Kurz, 2006, 113.
[2505] Kerschbaumer, 2007, 141–143, 145–148.
[2506] Schneiter, 1970, 44.
[2507] Meilleur, 1986, 342, 347, 350, 354, 356, 374, 380, 396, 404; alpenrose removal from pastures in East Tyrol: Vogl-Lukasser et al., 2006, 91.

stock[2508] (or livestock cut their mouths, at least, on these grasses). Only in early spring were these fescues suitable for (sheep) grazing.[2509]

Experimental research into the pros and cons of traditional pasture burning in the Alps has shown that if the plants farmers want to eradicate, such as heather and other dwarf shrubs, are burned, but humus (except for raw humus) is preserved, this leads to a change in species composition, creating somewhat more productive pastures and more palatable fodder. For instance, matgrass swards that became dominated by dwarf shrubs were converted by skilled burning into matgrass swards with a higher share of more attractive grasses and forbs, which make more nutritious and palatable fodder. Similarly, skilled burning of a thick plant cover dominated by alpenrose and juniper (see Section 2) resulted in an increase in the total abundance and number of herbaceous plant species more relevant as fodder plants (another possible aim of burning was increased blueberry production; see below in this section). This is a consequence of changed competition abilities due to different fire tolerance, altered nutrient availability, and higher pH value. For instance, dwarf shrubs, which have managed to compete with the help of wood formation, lose their competitive abilities. Burning destroys older woody parts of *Calluna* bushes and provides gaps for the establishment of young plants. If heather grows there again, its shoots are also eaten by livestock as long as the wood proportion is low. Focusing on burning's impact on nutrient availability, on the one hand, some nutrients released by burning were lost in smoke and some others were leached later. On the other hand, in high mountain pastures used in the summer, where organic material decomposes slowly due to the mountain climate, many nutrients were quickly released and became available to plants through burning. However, nutrient intake through burning has not been so strong that it would make possible the establishment of species requiring nutrient-rich soils. Periodic burning carried out once every several years by experienced farmers has preserved relatively productive pastures in the long run. At some site types where species unproductive from the perspective of livestock farming would otherwise colonize the pasture in large numbers, they were successfully suppressed by burning. Periodic burning was needed because unwanted species gradually became abundant again in the case of a similar grazing regime.[2510]

The burning temperature and the depth reached by fire mattered greatly. These factors are mutually related and depend, for example, on the duration of fire and quantities of burned biomass. This depends not only on the quantities of biomass available, but also on the frozen or unfrozen state of the soil; the

[2508] Schneiter, 1970, 131.
[2509] Meilleur, 1986, 296.
[2510] Meilleur, 1986, 342, 347, 350, 354, 356, 374, 380, 396, 404; Machatschek & Kurz, 2006, 109–116; Kerschbaumer, 2007, 141–145, 147; see also Måren et al., 2010.

dry status of the soil and the entire dead vegetation layer, or the high humidity of soil and the lower layer of dead vegetation but dry higher layer of dead vegetation; burning practices; slope gradient; wind; and so on. The temperature and depth of burning, together with the timing of burning, influenced the opportunities for the survival of seeds, of structures allowing vegetative reproduction, and of underground organs, and this had a decisive impact on the survival or loss of much grassland fauna (see Section 10.1.2.2.6) and on the preservation or burning of humus. Overly deep burning can promote the reestablishment of heather, for example. To suppress dwarf shrubs and achieve pasture improvement from the perspective of livestock farming, fire temperatures should be low, the duration of burning should be short, and burning should be superficial, so that much soil organic matter and many diaspores are preserved. To avoid negative consequences, burning should never be carried out on shallow soils and on slopes with a gradient of more than about 50% because steeper slopes cause burning that is too deep. The right time for burning dwarf shrubs is snow-free intervals in the winter. Namely, in frozen soils, the grassland plants aimed to be fostered have stored their energy in underground organs.[2511] Indeed, traditional ecological knowledge included some of these data: early seventeenth-century information on the Karst (Sln. *Kras*, Ital. *Carso*) Plateau, about 40 km south of the Julian Alps, documents burning of old grassland plants at the transition from winter to spring, during windy weather,[2512] probably fulfilling the criteria of relatively low temperatures and shallow burning depth. Stable wind direction was a prerequisite for burning precisely defined plots without endangering surrounding landscape structures (including buildings and other farm facilities). An additional function of skilled burning at the transition from winter to spring was that burning dry grassland plants "triggered the growth of new vegetation with a higher nutritional value" than the withered old biomass.[2513] Winter burning in Ticino has already been mentioned.

Contrary to the findings presented, unskilled grassland burning could have many negative consequences. It could promote long-term soil degradation and dwarf shrubs.[2514] Nonetheless, in specific socioeconomic contexts, it was the main aim of burning of pasture sections to promote dwarf shrubs with highly esteemed berries. In the Allues Valley, for instance, after the numbers of livestock had decreased in the nineteenth century, which caused lesser demand for productive grazing areas, some pasture sections where bilberry (*Vaccinium myrtillus*) occurred were periodically burned in the early twentieth century.

[2511] Pils, 1994, 248–254, 274–275; Machatschek & Kurz, 2006, 110, 112; Kerschbaumer, 2007, 143–146; see also Måren et al., 2010; Davies et al., 2016.
[2512] Zwitter, 2015a, 463.
[2513] Lauk & Erb, 2016, 343.
[2514] Kerschbaumer, 2007, 143, 145; Måren et al., 2010; Schneiter, 1970, 133–134.

The aim was to increase the quantities of blueberries gathered, and thus to increase the abundance or at least to extend the duration of bilberry's successful competition. It is not clear whether some fairly limited sections of pastures had already previously been "sacrificed" for the benefit of gathering, or whether this practice was an innovation from the period of decreased demand for productive pastures and decreased grazing pressures (increased abundance of some *Vaccinium* species such as bilberry is a frequent early stage of ecological succession in mountain pastures).[2515] The history of options for preserving blueberry products suggests that the quantities of blueberries gathered likely increased in several places in the Alps from the early modern period to nineteenth century or the first half of the twentieth century, potentially promoting the type of pasture management presented here. Namely, in contrast to cowberries (*Vaccinium vitis-idaea*), which also occur in pastures—and which can be preserved with sugar or without it, not merely in the form of brandy[2516]—blueberries were long preserved as a dietary product only where processed into (or soaked in) brandy, with a tradition dating back a few centuries (compare Section 8.1.2). Gathering blueberries for brandy production was a widespread business in some regions of the Alps, such as the Lower Inn Valley in Tyrol, as reported in the mid-nineteenth century. There, traditional ecological knowledge emphasized a higher amount of brandy obtained from blueberries gathered at high elevations in the mountains than at lower sites, which made such berries particularly esteemed. Burning was reportedly carried out at some blueberry gathering sites there.[2517] In regions of blueberry brandy production, it played a role in the general rise of alcoholism in the Alps in the nineteenth century.[2518] The other manner of usual preservation of blueberries since the twentieth century—namely, jam production—was not practiced by the agrarian population in previous centuries. Widespread jam making is closely related to a nineteenth-century transition from imported sugar cane to sugar beets grown in Central Europe in sugar production in this macroregion, which caused a significant reduction in sugar prices. Since the early twentieth century, when sugar consumption rose significantly, conditions have become more favorable for preservation of large quantities of blueberries in regions where blueberry spirit production had not been widespread before,[2519] and this may have also influenced the management of selected pasture sections in the same manner as the increase in blueberry spirit production from the early modern period to the nineteenth century.

[2515] Meilleur, 1986, 374, 396, 404.
[2516] von Hörmann, 1877, 152, 167; Babai, Molnár, & Molnár, 2014, 51.
[2517] von Hörmann, 1877, 152, 164–167.
[2518] von Hörmann, 1877, 153.
[2519] von Hörmann, 1877, 152–153; Łuczaj et al., 2012, 362; see also Abbet et al., 2014, 628; compare Crane, 1999, 493–494, and Sandgruber, 1989, 105, 107–109.

6.6.2.2 Economic relevance of some species suppressed in some pastures

Some species reducing the carrying capacity of pastures were nonetheless economically relevant, such as the example of yellow gentian (*Gentiana lutea*) presented above (see also Sections 6.6.1.1, 6.6.2, and 8.1.2). Moreover, perceptions of economically desired and non-desired grassland species have changed over time, as demonstrated by large-leaved docks (Section 8.2.2) and bilberry (Section 6.6.2.1).

Another case is heather (*Calluna vulgaris*). It is a species unwanted from the perspective of grazing (Section 6.6.2.1), but its importance in autumn bee pasture probably rendered the suppression of heather less attractive in some places, at least prior to the sixteenth century. Namely, when heather blossoms, bees can forage on only a few other plants (including winter savory, *Satureja montana*). Bees foraging on heather had significant impacts on honey production, and this could extend the season of making honey by more than a month. The relevance of bee products was significant. In some regions of the Alps, honey was an export product; for example, in seventeenth-century Carniola.[2520] Seventeenth-century information refers to the time after the general spread of buckwheat as stubble crop (by the sixteenth century) that often offered excellent bee foraging sites in the autumn in many places at the foot of the Alps where late summer and early autumn weather is usually mild enough for frost-sensitive buckwheat. The general role of beekeeping increased when buckwheat spread.[2521] Since then, the perception of heather most likely became more locally specific and its economic appreciation more limited to elevations too high for buckwheat and places where it did not abound in vast areas, and so the motif of clearing it to improve a pasture's carrying capacity was less relevant. There is every reason to assume that farmers became more motivated to suppress heather where buckwheat replaced its former role in beekeeping.

Juniper (*Juniperus communis*) is another example of a shrub in pastures that served special functions and was only removed in some places or its partial removal could become a matter of dispute.[2522] Juniper's ability to spread around the pasture[2523] required its (partial) clearing in many pastures in history. Where there were enough rock piles in grasslands, juniper could grow

---

[2520] Mihelič, 1970, 400–401, 403; on the importance of heather for beekeeping elsewhere in Europe: Crane, 1999, 229, 232–233, 235, 239–240, 243–244, 249, 350, 352–353.

[2521] Zupan, 1934, 96–97; Mihelič, 1970, 400; Valenčič, 1970a, 255, 257; Zwitter, 2014a, 221; Crane, 1999, 232, 244, 249.

[2522] Machatschek, 2015, 278; author's interviews with Gašperin, 2017, with Kunavec, 2017, and with Mencinger, 2017, all in Stara Fužina; with L. and M. Odar, 2017, and with A. and M. Taler, 2017, both in Studor; with Minka Sodja, 2017, in Jereka; and with V. Smukavec, 2017, in Spodnje Podjelje.

[2523] Babai, Molnár, & Molnár, 2014, 116–118.

there (see Section 6.1.1.3); during periods of high agricultural density, it was often not allowed to grow at sites with soils suitable for grasslands.[2524] Juniper's multipurpose use is presented in the following paragraphs.

Juniper was widely used in folk practice for humans and veterinary medicine.[2525] Its uses in the Alps are demonstrated largely based on examples from East Tyrol in the mid-twentieth century, observations from Tyrol in general in the nineteenth century, and some data from the Slovenian Alps from the early modern period onward. In East Tyrol, juniper was slowly burned in cattle barns to produce smoke as a preventive measure against viral diseases (flu) or to heal cattle colds, pulmonary issues in cattle or horses (consumption of ground dry juniper (needles or berries?) or finely cut fresh needles were alternatives), or diarrhea. Juniper in fact contributes to disinfection with its smoke.[2526] In some parts of the Eastern Alps, juniper twigs were burned to disinfect buildings, barrels, and wooden buckets, including butter churns.[2527] Juniper and pine wood were also distilled and used in human and veterinary medicine, as revealed by nineteenth-century data from the Upper Savinja Valley of Lower Styria.[2528] In mid-twentieth-century East Tyrol, fresh juniper needles were finely cut and added to fodder for horses against colic. Juniper needles and berries were used to promote livestock's appetite—a good practice because juniper berries do increase appetite. A tisane mixture with juniper berries was administered against diarrhea in cattle, and a tisane mixture including juniper berries or a juniper brandy against cattle rumen or stomach issues (probably effectively in many cases),[2529] a practice with a clearly longer tradition because the use of juniper berries against livestock stomach issues was documented in nineteenth-century Tyrol.[2530]

Similarly, juniper berries or brandy made from them was applied against stomachache in humans, as reported in Tyrol in the nineteenth century. In the same area and period, farmers believed that prophylactically chewing juniper berries on an empty stomach in the morning strengthened immunity.[2531] A concoction from juniper berries macerated in spirit was taken against cold and flu, as was recalled in the Allues Valley for the first half of the twentieth century.[2532] Oil collected at the top of the juniper-berry brandy was adminis-

2524 Machatschek, 2002, 392–394; Machatschek, 2015, 279.

2525 Machatschek, 2015, 273, 276, 278–280, 282–283, 285–287, 289; Mlakar, 2015, 74–75, 285, 303, 349–350.

2526 Vogl-Lukasser et al., 2006, 41, 53, 71, 83, 166–168, 189, 266.

2527 Machatschek, 2015, 282.

2528 Mlakar, 2015, 62.

2529 Vogl-Lukasser et al., 2006, 41, 53, 71, 166–168, 174, 189, 266.

2530 von Hörmann, 1877, 168.

2531 von Hörmann, 1877, 168.

2532 Meilleur, 1986, 347; see also Abbet et al., 2014, Appendix C for some other medical uses, but it is not clear whether they are old (see Abbet et al., 2014, page 625).

tered, for instance, against intestinal worms, as proven in the pre-Alpine hills around Cerkno and Idrija, and on the southern slopes of the Pohorje Hills in the mid-twentieth century.[2533] The use of juniper berries for producing spirits and essential oil has a long tradition: it started by the seventeenth century, as attested for the province of Carniola, including the Alps in what is now north-west Slovenia. In seventeenth-century Carniola, production of essential oil from juniper berries was partly market-oriented; this oil was an export product.[2534] Nineteenth-century data from Tyrol help explain this: the essential oil from juniper berries was very expensive.[2535]

The use of juniper berries in the human diet also has a long tradition, and history knew some types of uses not usual nowadays, such as juniper berries soaked in water for two to three weeks to produce a refreshing drink, according to seventeenth-century information referring to the province of Carniola.[2536] Moreover, juniper has been used as a spice. Berries have been added to sauerkraut, to meat soups and some other dishes in some parts of the Alps, like in Tyrol and Alto Adige (South Tyrol), unlike several other regions in Central Europe where pepper has been used instead of them in recent centuries. Juniper berries and twigs were added to the brine in which meat was treated before smoking.[2537] Juniper twigs were used in beer brewing in Carinthia in the eighteenth and nineteenth centuries, for instance, as a filter that gave the beer a specific aroma.[2538]

Juniper was also used as fodder, at least in certain parts of the Alps. The Gößnitz Valley (Germ. *Gößnitztal*) in the High Tauern of Carinthia and several places in East Tyrol and Wallis are just some examples. Young juniper twigs were pruned (e.g., with a sickle), and needles and berries were collected from woody branches. Shoot tips, needles, bark, and berries were finely cut or beaten. Boiling water was poured over them, and this was mixed, e.g., with nettles, thistles, and, optionally, thyme, bran or barn sweepings (see Sections 6.5 and 8.2.2), and salt, and used now and then or regularly as an esteemed fodder that included trace elements and was rich in minerals, thus reducing the quantities of salt given to livestock, as reported from Carinthia. At least cattle like juniper berries very much. The fodder mixture with juniper presented here was believed to significantly increase milk fat content. This was relevant because butter was one of the main products from summer alps. In some regions and summer mountain pasture complexes, where cheese was not made in the Alps

---

2533 Sosič, 2000, 239–240, 247, 249.
2534 Zwitter, 2014d, 619.
2535 von Hörmann, 1877, 168.
2536 Zwitter, 2014d, 619.
2537 Machatschek, 2015, 276, 279–280, 284, 288; Lea Lechthaler, personal communication; main author's personal experience in Tyrol.
2538 Dinklage, 1966b, 142–145.

or its taste differed from that demanded by the market, clarified or salted butter was the main commercial dairy product. Juniper turned butter yellow (East Tyrol), stimulated animal digestion, and rendered fur finer and softer. Juniper was also a source of fodder at some high summer pastures in cases of untimely summer snow, and a source of winter fodder in times of scarcity.[2539] Longer or excessive use of such fodder would not be recommendable because too much juniper (berries, at least) could cause kidney problems,[2540] whereas moderate consumption of juniper berries and shoots stimulated kidney and intestinal function.[2541]

Moreover, juniper was applied in magic. As a fodder ingredient, it was believed to protect livestock against lightning strikes, and burnt juniper was considered to drive away evil forces, and so it was used for centuries as incense linking magic and religion.[2542] Another example of magic is that juniper was worn on the hat against skin inflammation, as reported in the south of the Eastern Alps in the eighteenth century (see Section 8.1.3.1).

Juniper was not the only thorny bush valued for its fruits; blackthorn fruit (*Prunus spinosa*) and rose hips (*Rosa* spp.), for instance, were also gathered by the local population according to early twentieth-century information, a more general trend presented here based on data from Bohinj.[2543] Gathering wild fruits must have been one of the reasons for the observation that "neglected" pastures with bushes were frequent in nineteenth-century Carniola.[2544] Several types of "weeds" in pastures, such as the aforementioned bushes, contributed significantly to the patchwork vegetation structure in extensively managed pastures, which increased the diversity of animal species, in contrast to recent intensively managed uniform pastures.[2545] Thorny bushes in some extensively managed pastures provided shelter for some tree species palatable to livestock. When a young tree overgrew the bush, it was tall enough that browsing cattle would not destroy it. This allowed long-term regeneration of wooded pastures with trees in some places, where livestock would otherwise have thoroughly consumed tree saplings.[2546]

[2539] Niederer, 1996d, 190; Machatschek, 2002, 112, 331, 392–394; Machatschek, 2006, 134; Vogl-Lukasser et al., 2006, 41, 53, 78, 83–84, 266; Machatschek, 2015, 57, 278–280; Novak, 2024, 54, 180–181, 183–184, 196, 198–199, 210.

[2540] Vogl-Lukasser et al., 2006, 168.

[2541] Machatschek, 2002, 220.

[2542] Vogl-Lukasser et al., 2006, 83; Zwitter, 2014d, 619; Machatschek, 2015, 278, 287; Mlakar, 2015, 281–282, 437, 452.

[2543] Author's interviews with Mencinger, 2017, in Stara Fužina, and with Minka Sodja, 2017, in Jereka.

[2544] Compare Kačičnik Gabrič, 2014, 236–237.

[2545] Pils, 1994, 204, 208.

[2546] Rotherham, 2013, 81; Samojlik & Kuijper, 2013, 145–146, 148–150; Vera, 2013, 108–116.

Section 6 explored various ecological and economic roles of species-rich grassland management. Livestock's role in the migration of plant species is presented in Section 7.

7 Introduction of plant species to grasslands in the Alps through human activities in the context of economic history, political history, and the history of botany and ornamental plants

This section makes it clear that there has existed a potential for plant introduction to some grasslands in the Alps as a consequence of human activities throughout the study period and beyond. The likelihood for plant introduction nonetheless varied significantly in the course of history, as did the distance over which diaspores of some introduced plants were carried. In general, the likelihood for introduction increased in the case of several types of discontinuities, such as grassland area expansion, increased volumes or new directions of traffic and trade flows, some changes in agriculture, increased intercultural connections, and/or developments of modern science (botany). In other words, as we move closer to the present, more alien plant species have been introduced each century, since the waning Middle Ages at least; however, as we approach recent history, increasingly clearer data are also available about this process. Local site conditions and processes in affected grasslands also had a crucial impact. Namely, only part of the introduced species became established and spread in new environments. Many alien species could not grow there or died out comparatively quickly, whereas a few alien species even became invasive.[2547]

Adaptation to a new environment takes years, decades, or even centuries.[2548] A study from Slovenia demonstrates that most of the naturalized species known to have been introduced since 1750 only became naturalized in the twentieth century. Also, most of those that became invasive turned invasive in the twentieth century; the majority of them only became a serious threat for biodiversity in the Alps in the last decades of the twentieth century, when they spread significantly around rural areas.[2549] Nowadays, invasive alien species are one of the five core drivers of biodiversity loss globally, but there are significant

[2547] Emanuelsson, 2009, 160; Jogan, Bačič, & Strgulc Krajšek, 2012, 162–164, 167–168, 179; Šilc et al., 2012; Küzmič & Šilc, 2017; Axmanová et al., 2021, 3, 7.

[2548] Jogan, Bačič, & Strgulc Krajšek, 2012, 164; Kus Veenvliet, 2012, 106; Küzmič & Šilc, 2017, 203, 205.

[2549] Jogan, Bačič, & Strgulc Krajšek, 2012, 163–165.

regional differences,[2550] and the Alps are not one of the particularly heavily affected regions in the broader European context,[2551] above the middle of the montane vegetation belt in particular.[2552]

The focus in this section is on a selection of plants introduced to Europe or the Alps from the Middle Ages or earlier to the end of the First World War, with (later) durable establishment and persistence in grasslands in the Alps,[2553] and on a selection of human-related introduction pathways of species that later became naturalized in grasslands in the Alps. The introduction of new species occurred in several main ways, including unintentional (such as through farming or transportation) or intentional ones (planting).

Farming "substantially contributed to the spread of the majority of grassland species."[2554] For instance, some species of fodder legumes introduced as fodder crops in Early Modern Period (Sections 3.2.1 and 6.3.1.1.1) became naturalized in grasslands in the Eastern Alps. Bastard alfalfa (*Medicago × varia*) is an example of a neophyte in moderately manured meadows of Central Europe that can abound in species-rich sites of former arable fields characterized by middling plant nutrient availability and a basic to neutral soil reaction (Section 3.1).[2555]

First, Section 7.1 focuses on the plant introductions through long-term and medium-term livestock migrations in grasslands, with a possible example at the species level. This section also explores the history of species introduction in the context of a nexus between farming and transport for purposes of the usual non-agrarian economy. Second, the nexus between farming and transport is again dealt with in the following section, but in a different context. There, examples of plant species of polemochorous origin (i.e., "alien plants appearing outside the limits of their natural range as a result of military operations"[2556]) demonstrate that major conflicts over political boundaries also led to the transgression of ecosystem boundaries and resulted in long-term legacies in species composition due to exceptional, intensive, short-term livestock-related impacts.

Finally, the examples of annual fleabane (*Erigeron annuus*), Japanese and giant knotweed (*Reynoutria japonica* and *Reynoutria sachalinensis*), Himala-

---

[2550] Šilc et al., 2012, 720; *The Future of Nature and Business*, 2020, 21.

[2551] Axmanová et al., 2021.

[2552] Jogan, Bačič, & Strgulc Krajšek, 2012, 174–175; Šilc et al., 2012, 725–728; Zelnik, 2012a, 27, 35; Zelnik, 2012b, 62.

[2553] For comparison with introduced species naturalized in British grasslands, see Peterken, 2013, 210–215.

[2554] Ellenberg & Leuschner, 2010, 1019 (quote); Luick, Jones, & Oppermann, 2012, 32, 36; Peterken, 2013, 203.

[2555] Karlík & Poschlod, 2009; Poschlod, 2015, 85–86; Aeschimann et al., 2004, vol. 1, 906–907.

[2556] Reshetnikova, 2016, 62.

yan touch-me-not (*Impatiens glandulifera*), and coneflower (*Rudbeckia laciniata*) in Section 7.3 demonstrate that the seemingly innocent introduction of species for ornamental or botanical purposes in certain cases resulted not only in naturalization but also in invasiveness of these species or their hybrids in grasslands in the Alps. Section 7.2 also contributes to this issue because some species introduced through military activities were perceived as aesthetically appealing and were transplanted to further habitats. In the last centuries, more alien species that became naturalized were introduced as decorative species than as agricultural crops. The manners of introducing plants presented in this section were relevant, but this volume does not aim to offer a complete overview of all possible pathways of grassland plant introduction through human activities within the study period.[2557]

7.1 Species introduction through medium-term and long-term livestock migration

This section deals first with plant species introduction through historical sheep migration and continues with the movement of horses and cattle. It focuses on the introduction of extra-Alpine species to the Alps. In addition, livestock movement within the Alps also had an influence on the occurrence and abundance of grassland plants.

Sheep transhumance in Europe was characteristic mainly of the southeast, south, and west of the continent and to a lesser degree of what is now southern Germany.[2558] Many routes of sheep transhumance between winter pastures in lowlands outside the Alps, where environmental conditions made grazing in the cold part of the year possible, and more suitable summer pastures in mountains of the Alps had a long history. Some sections of the Alps may have already been involved in transhumance in prehistory, but the high and late medieval increase in population density (Section 3.1) must have resulted in many new routes of livestock migration. Transhumant sheep routes linked the southwestern Alps with the coastal areas of Languedoc and Provence (Occitan *Provença*): transhumance between the latter and the Durance (Occitan *Durença*) region is proven since the thirteenth century; it is documented between the Hauts Plateaux of the Vercors (Occitan *Vercòrs*) in the Alps and the Crau region in the Rhône delta, more than 100 km away in a straight line, from the fourteenth century on, when it was already well established there. Transhumance linked the Po Plain (Ital. *Pianura Padana*) with the side valleys of the Aosta Valley (Ital. *Valle d'Aosta*, Francoprovençal *Val d'Aoûta,* Fr. *Vallée d'Aoste*) or with the Swiss Engadin (Romansh *Engiadina*) region (see Figure 4). South Tyrol was also involved in such long-distance transhumance: before

[2557] See, for instance, Ellenberg & Leuschner, 2010, 76–77; Jogan, Bačič, & Strgulc Krajšek, 2012, 165, 168, 176; Kus Veenvliet, 2012, 121.

[2558] Schöller, 2010, 634.

1525 sheep from Lombardy (Ital. *Lombardia*) were grazing in the pastures of the community of Planol (Germ. *Planeil*) in the uppermost Venosta Valley (Ital. *Val Venosta*, Germ. *Vinschgau*) close to the Ötz Valley Alps (Germ. *Ötztaler Alpen*), and thousands of sheep from what is now northern Italy grazed on mountain pastures of the inner Ulten Valley (Ital. *Val d'Ultimo*, Germ. *Ultental*) in the sixteenth century. Moreover, thousands of sheep from Folgaria (Germ. *Vielgereuth*, above Trento, Germ. *Trient*) and neighboring southernmost Tyrol were still grazing in the plain of Lombardy in winters around the mid-nineteenth century, as a continuation of an older tradition. Further east, the Venetian Plain (Ital. *Pianura Veneta*) was linked by transhumant sheep movement with the Alps of the Gail Valley (Germ. *Gailtal*, Sln. *Ziljska dolina*) in Carinthia (Germ. *Kärnten*, Sln. *Koroška*) (Figure 4). These selected examples reaching from the Western to the Eastern Alps suggest that historical sheep transhumance linking the Alps and their southern surroundings was an important phenomenon. The economic and ecological significance of these movements is demonstrated quantitatively in the next two paragraphs.[2559]

According to mid-nineteenth-century information from southeastern France, about 400,000 sheep were involved in transhumance linking extra-Alpine territories and the Alps there. About 41,000 sheep (the number includes some goats) involved in transhumant movement were estimated to have been grazing solely on the Hauts Plateaux of the Vercors in the eighteenth century. The numbers of sheep involved in long-distance transhumant movement have varied over time (see also Section 6.3.1.1). The nineteenth century saw a significant increase in numbers of sheep from Provence grazing in the Alps of Briançonnais, for instance. Namely, in the Alps of what is now southeasternmost France, the prevalence of sheep and goats over cattle had a tradition. However, the nineteenth century saw a decline in local extensive sheep husbandry in Briançonnais and related decreased uses of upland grasslands by local farmers due to several reasons, the introduction of intensive dairy cattle breeding being one of them. Consequently, more upland pastures were available for transhumant sheep from Provence. In contrast to this case of increased numbers of sheep involved in transhumant movement in the nineteenth century, early modern significant intensification of agriculture in the plains of what is now northern Italy, which was carried out early in a broader European context, caused a decrease in winter pastures there. This led to a significant drop in numbers of sheep involved in transhumance between the Alps and these plains in the eighteenth century, for instance.[2560]

2559 Viazzo, 1989, especially pp. 125–126; Mathieu, 1996, 150, 160; Niederer, 1996d, 150–151, 153–155; Wopfner, 1997, 230, 232, 474; Emanuelsson, 2009, 143, 154, 161; Véron, 2009, 165–166; Mathieu, 2019, 36–37, 51, 55.

2560 Mathieu, 1996, 139, 142–143, 145, 150, 160; Niederer, 1996d, 151; Véron, 2009, 166; Mathieu, 2019, 50–51, 55; Carrer, Walsh, & Mocci, 2020, 74, 79 (gratitude is expressed to the archaeologist Jana Horvat for recommending this publication).

Transhumance made possible effective epizoochorous (on the outside of livestock) and endozoochorous (via ingestion by livestock) "transport of germinable diaspores of many grassland species," characterized by low or medium diaspore mass (below 2 mg) over long distances. For some plant species, it was the only means of long-distance seed transport. A single sheep can carry thousands of diaspores of several dozen species in its fleece, and a study from sub-Mediterranean Mount Vremščica in southwestern Slovenia proved that diaspores of "nearly 50% of the species found in the aboveground vegetation of the pastures on Mt. Vremščica recorded in previous studies were also found in the [sheep] fur." Similarly, repeated fleece research on a test sheep that was "allowed to move around for three months between patches of limestone grassland on the [Swabian Jura, Germ.:] Schwäbische Alb … revealed an extraordinary total of 8,511 diaspores from 85 plant species, together with hundreds of grasshoppers, a few spiders, beetles and snails, and three lizards." This suggests that a couple of billion diaspores were moved annually in the mid-nineteenth century in the fleece of about 400,000 sheep involved in transhumance linking the Alps and extra-Alpine territories in southeastern France only (taking into account that these sheep were smaller than they are now; see Section 5). Transhumance was a means of exchange of species not only among pastures; it also included other grazed land such as meadows and fallow fields grazed after mowing in late summer and autumn, when sheep were full of seeds from elsewhere. Considering epizoochorous seed diaspore transport in fur, seeds do not attach to sheep only during walking, grazing, and lying; most importantly, sheep also wallow.[2561] Sheep's grazing behavior in general promotes grassland species-richness. Creation of gaps in grassland vegetation through sheep's ability to bite the plants the closest to the soil among all historically usual domestic ungulates (see also Section 6.2.2.1)[2562] was one of the factors allowing establishment of plants introduced to the grasslands via zoochory.

Transhumance was not the only driver of long-distance dispersal of diaspores through livestock in the Alps.[2563] Sheep were also sold over long distances; people from the Passiria Valley (Ital. *Val Passiria*, Germ. *Passeiertal*) and Sarentino Valley (Germ. *Sarntal*) in South Tyrol, for instance, bought significant numbers of sheep (wethers) in historical southern Hungary and the Balkans in spring, kept them in their summer pastures, and sold them to France in the autumn, according to data from the first and second half of the nineteenth century, thus not limited to the times when railway connections

[2561] Poschlod, 2008, 138; Emanuelsson, 2009, 151–152, 154, 258; Véron, 2009, 165; Ellenberg & Leuschner, 2010, 45–47, 1019–1020; Hejcman et al., 2013, 355 (first quote), 356; Peterken, 2013, 232–233 (third quote from both pages); Babai & Molnár, 2014, 128; Bätzing, 2015, 61, 277–278; Poschlod, 2015, 179–181; Kaligarič, Brecl, & Škornik, 2016 (second quote from p. 182).

[2562] Ellenberg & Leuschner, 2010, 1007.

[2563] Emanuelsson, 2009, 151–154; Kaligarič, Brecl, & Škornik, 2016, 183.

made such long-distance movements much easier. Similarly, farmers from the Ötz Valley (Germ. *Ötztal*) and Ziller Valley (*Zillertal*) bought thousands of sheep in the same areas of Southeastern Europe in spring and kept them in summer mountain pastures until autumn.[2564]

In addition to sheep, long-distance movement of horses was a further important dispersal vector[2565] relevant not only for grasslands near permanent settlements and traffic routes but also for some remote mountain pastures because pack-horses grazed there as well.[2566] One example of species possibly linked to horse grazing is wild Illyrian gladiolus (*Gladiolus illyricus*, Figure 115); this link has never been established in previous literature, and so a novel interdisciplinary environmental history synthesis follows, linking plant ecology and economic history. This (sub-)Mediterranean species occurs today at one single species-rich purple moor-grass meadow consisting of a patchwork of wet/damp and drier sites, northwest of the village of Oberschütt (Sln. *Rogaje*) in the lower Gail Valley. The meadow is located at the bottom of the valley, close to the lower edge of the thermal belt of a southern mountain slope. Wild Illyrian gladiolus is a botanical peculiarity that colors this meadow pink in June and grows only there within the entire territory of present-day Austria.[2567] Historically, nutrient-poor hay from this meadow was either used to feed horses (see Section 6.1.2.5.1.2)[2568] or this was a litter-meadow (see Section 6.1.2.5.2).[2569] The area of wet meadows in valley bottoms in this part of Carinthia was exceptionally high in the broader Eastern Alpine context, according to data from the early 1930s at least;[2570] by that time, large-scale drainage had already strongly reduced the area of wet meadows in many valleys in the Alps (see the introduction to Section 6.1.2.5). Mid-twentieth-century information

2564 Wopfner, 1997, 230, 232–233, 473–474; the chronological definition of the time of Wopfner's writing follows Meixner & Siegl, 2004, 232–233, 235.

2565 Emanuelsson, 2009, 151, 153.

2566 Grazing of packhorses in high mountain pastures is documented, for example, on Mount Velika Golica (Germ. *Kahlkogel*) in the Western Karawanks (Germ. *Karawanken*, Sln. *Karavanke*) and in its surroundings according to information from 1606 (AT-KLA 112, Allgemeine Handschriftenreihe, Ms. 2435, 148; AT-KLA 112, Ms. 2732, 471) and in some high mountain pastures in the Julian Alps (Kos, 1960, 131 mentions 150 to two hundred horses on a high mountain pasture in the Julian Alps in the sixteenth century).

2567 Melzer, 1965, 183; Friess, 2001, 361–362, 380–381; "Schwendeaktion Gladiolenwiese," 2003, 2; Aeschminann et al., 2004, vol. 2, 1094–1095; information about the site conditions of this meadow is also summarized on the local information sign titled *Gladiolenwiese*.

2568 Even though the information comes from a tourist information sign (*Oberschütt: Gladiolenwiese – Weinitzen*), information about this and similar meadows was gathered within the framework of the LIFE project "Schütt-Dobratsch" ("Ein erfolgreiches Projekt ...," 2005, 1–2). Both of these purposes of use of wetland hay also have clear parallels elsewhere in the Alps (see Section 6.1.2.5.1.2).

2569 Friess, 2001, 362.

2570 Ilešič, 1931, 135.

reveals that the meadow where wild Illyrian gladiolus occurs now was historically not the only meadow with this species near Oberschütt. At that time, observations proved that *Gladiolus illyricus* also grew in the nearby Canale Valley (Friul. *Val Cjanâl*, Germ. *Kanaltal*, Sln. *Kanalska dolina*) in northeastern Italy. Some other southern plant species also occur and even locally abound in the Lower Gail Valley as well as in the Canale Valley, at least in (relatively) wet meadows.[2571] The fact that trees recently started colonizing sections of the damp meadow with wild Illyrian gladiolus at the northern end of the village of Oberschütt[2572] demonstrates that *Gladiolus illyricus* only colonized this grassland in the Alps after forest was cleared there during the human colonization process; the site was not affected by the nearby 1348 rockfalls and the related floods (Section 4.2).[2573] Agriculture in the area has long been well connected with long- and medium-distance trade. Important traffic flows linking northeastern Italy and central Noricum through the Canale Valley and the Lower Gail Valley are known to have already existed in Roman Antiquity.[2574] The traffic connection through the Canale Valley saw positive and negative developments of traffic flows. Traffic in this area gradually gained importance again in the later Middle Ages, especially after the thirteenth century, when the former bridleway was converted into a road passable by carts. Later, the state of the road and traffic flows were still dynamic there, but usually this was a traffic connection of major importance. In the Late Middle Ages, thriving horse-powered transport provided by the inhabitants of the Lower Gail Valley through the Canale Valley is clearly documented as early as 1381. At that time, inhabitants of villages around Arnoldstein (Sln. *Podklošter*), where Oberschütt is located, were particularly active in this transport by pack horses and horse-driven wagons linking Carinthia and Italian urban settlements. Gradually, villagers from several parts of the Lower Gail Valley became highly active in cargo transport linking Italian markets with the territory of Salzburg through the Canale Valley and further across the Tauern mountains. People from the Lower Gail Valley were highly active in these directions of cargo transport in the sixteenth and seventeenth centuries, for instance.[2575] Horses were the main consumers of hay from (partly) wet meadows, where (sub-)Mediterranean *Gladiolus illyricus* occurs in the Gail Valley. Moreover, horses also grazed in meadows in the spring and autumn (see also Sections 6.1.2.5.1.2, 6.2, and 9). The great importance of historical grazing in damp or wet meadows for horse breeding in the Gail Valley is clearly documented,[2576] and this was also usual in the broader context of the Alps.[2577] Consequently, the wild Illyrian gladiolus might

2571 Melzer, 1965, 183–184; Aeschminann et al., 2004, vol. 2, 1094.
2572 Friess, 2001, 361–362; "Schwendeaktion Gladiolenwiese," 2003, 2.
2573 See the map of the 1348 rockfalls in Zorn, 2002, 13.
2574 Šašel, 1955, 73, 79; Kosi, 1998, 14.
2575 Kosi, 1998, 136, 139–142, 144, 229–230, 245–246; Oitzl, 2021, 427, 429.
2576 Ilešič, 1931, 135, 142; Ložar, 1944, 146; Novak, 1970, 381.
2577 Wopfner, 1997, 226.

Figure 115: Wild Illyrian gladiolus (*Gladiolus illyricus*) in the meadow at Oberschütt in the Gail Valley, 2017. *Wikimedia Commons*, File:Illyrische Siegwurz (Gladiolus illyricus).jpg.

have been brought there by horses traveling to the Alps of the Gail Valley from sub-Mediterranean environments through the Canale Valley, or from the Canale Valley if the species was first established there. Diaspores of wild Illyrian gladiolus are of medium size and weight, and they are smooth and structureless, without specific morphological adaptations to any dispersal vector.[2578] The round form of the diaspore would speak in favor of hoof-zoochory, for example, and endozoochory by horses might also have been possible because they have less efficient and somewhat less destructive digestion than ruminants, at least concerning seeds of some grassland species.[2579] Even though no reliable information about long-distance dispersal pathways of gladiolus exists in the

[2578] Kaligarič, Brecl, & Škornik, 2016, 179–180 and *Supplemental data*.

[2579] Emanuelsson, 2009, 151, 153; Kaligarič, Brecl, & Škornik, 2016, 177; Milotić & Hoffmann, 2016b; *Gladiolus illyricus MHNT.BOT.2009.17.12.jpg*; but see also Milotić & Hoffmann, 2016a.

literature and targeted sowing by people cannot be fully excluded for such an esthetically attractive plant, the occurrence of wild Illyrian gladiolus in the partly wet and damp meadow(s) at Oberschütt is likely a long-term legacy of historical horse-powered transport, horse grazing, and/or hay transport. The effect of trampling in grazed grasslands by horses wearing horseshoes was the strongest among all herbivorous domestic ungulates,[2580] and damp or wet meadows were soft. Consequently, one can be certain about the existence of patches in meadows in the lower Gail Valley where vegetation was damaged through horse grazing, creating gaps that allowed the successful establishment of seeds that could not become established in a dense grassland (see Sections 6.2.2 and 6.5).

Cattle also moved in large numbers over long distances, not only between grazing areas in various elevation belts within the Alps. In some parts of the mountain chain, many cattle moved over long distances between pastures in the Alps in the warm part of the year and the surrounding plains out of the Alps in winter; for instance, between Alagna (Walser German *Im Land*) and its summer pastures at high elevations in the upper Sesia Valley (Ital. *Valsesia*, Walser German *Tseschrutol*) northeast of Turin (Ital. *Torino*) on the one hand, and the plains of Piedmont or Lombardy on the other. These migrations were recorded in 1845 "as 'an old custom.'"[2581] Apart from regular grazing cycles within the Alps or linking the Alps and their surroundings, and transport activities, selling domestic animals on distant markets was another reason for medium- and long-distance cattle movement, as already mentioned in the context of sheep. Apart from trade in cattle from the Alps to their broader surroundings, huge numbers of oxen were driven along some major traffic connections leading through a few valleys in the Alps and over a few Alpine passes to Italian markets from the Late Middle Ages until the railway network was established in the nineteenth century. The role of individual routes changed significantly during the course of history.[2582] Particularly important was the trade in Hungarian oxen from the Carpathian Basin to Italy or Central Europe; from the late fourteenth to the sixteenth century at least, part of this trade crossed the Alps.[2583] In the late fourteenth and early fifteenth centuries, one out of two routes for Hungarian cattle toward Cologne (Germ. *Köln*) led through Ptuj (Germ. *Pettau*) via Villach to Augsburg, including two alternatives for crossing the Alps: across the Brenner Pass (Germ. *Brennerpass*, Ital. *Passo del Brennero*) or Tauern. At that time, long-distance cattle trade was still of little importance and precarious; trade in Hungarian oxen decreased in the second quarter of the fifteenth century. In the late fifteenth century, there were fundamental changes in the long-distance cattle trade in Europe. After that time,

[2580] Ellenberg & Leuschner, 2010, 1007; Sadar, 1956, 177.
[2581] Viazzo, 1989, 110–116 (quote from p. 113).
[2582] Blanchard, 1986; Mathieu, 2019, 63–64; Winiwarter & Bork, 2019, 60–61.
[2583] Blanchard, 1986; Kosi, 1998, 229–230; Bartosiewicz et al., 2018, 129.

its scale increased heavily,[2584] and so a significant share of the 13,000 oxen annually provided by a company in Villach from Hungary and from what is now Slovenia and southeastern Austria to the Venetian market walked through the Canale Valley between 1512 and 1517. This was the peak of the importance of the road through the Canale Valley for Hungarian cattle sold to Venice; the 1530s saw a major long-term drop in the role of this inner-Alpine route of Hungarian oxen trade.[2585] Such oxen movements also had the potential to contribute to long-distance migration of grassland plant species.[2586] The way such long-distance traffic was carried out mattered in this context, as demonstrated by Hungarian oxen. In the fourteenth to mid-sixteenth century, the times mentioned above, these oxen walked for 15 to 20 km daily. A long rest followed after no more than 450 km, usually in the western Carpathian Basin, from autumn to spring, aimed at fattening, otherwise the oxen would have lost too much weight and their price would sink too low.[2587] This means that Hungarian oxen crossing the Alps in the late fourteenth to early sixteenth century were a short-to-medium-distance dispersal vector for germinable diaspores. They were unlikely to introduce plants to the Alps directly from the places of origin of many oxen in the central Carpathian Basin.

7.2 Polemochores: species introduced to the Julian Alps as a long-term environmental legacy of the First World War, with an emphasis on the Bohinj Basin

It is clear that the long-term environmental consequences of wars included the introduction of some grassland species, and some of them later became naturalized in those ecosystems.[2588] Here, this issue is addressed by focusing on environmental legacies of the First World War. This was not the first war leading to polemochorous plant introduction.[2589] Nonetheless, we focus on the First World War because of the unprecedented combination of the scale of this war, of extreme numbers of livestock involved in this war (see below in this section), and of the well-established railway network in several world regions,[2590] making possible much easier and cheaper long-distance inland cargo (e.g.,

[2584] Blanchard, 1986.

[2585] Blanchard, 1986, 434; Kosi, 1998, 229–230.

[2586] Emanuelsson, 2009, 151.

[2587] Blanchard, 1986, especially pp. 432, 434, 449, 451–452.

[2588] See, for example, Chauvel et al., 2006; Poschlod, 2015, 102–104; Reshetnikova, 2016. Poschlod and Reshetnikova deal with examples of polemochores from other territories, not with plants in the Alps, and with various wars, but not with the First World War.

[2589] See some earlier examples and potential earlier examples in Poschlod, 2015, 102–104.

[2590] A very general overview of the global history of railroads, with major shortcomings: Westwood, 2009; for a precise chronology of railway constructions and locations of railway routes close to the area studied in Section 7.2 see Černe, 2001, 161, and *Slovenski zgodovinski atlas*, 2011, 153.

hay) transport than before the steam engine (Section 3.1). Consequently, the likelihood of polemochorous species introduction to grasslands as a side effect of the First World War must have far exceeded any earlier war in European history, despite the difficulties of the war economy. The environmental history of this war has received some scholarly attention.[2591] John R. McNeill's overviews of environmental history research worldwide[2592] and Richard P. Tucker's overview of the environmental historiography of wars[2593] explain that military environmental history is a very young research field that, except for few early studies,[2594] has only been developing in the twenty-first century. This makes it possible to understand why polemochorous plant introduction still has not been fully incorporated into it. Thus, it is still absent from Tait Keller's recent introduction to the environmental history of the First World War.[2595] Nonetheless, plant dispersal during this war has yielded some findings, such as the polemochorous introduction of ragweed (*Ambrosia artemisiifolia*) to a few places in western France and potentially also southern France.[2596] Tone Wraber[2597] and Igor Dakskobler et al.[2598] studied polemochorous plant introduction on the Isonzo (Sln. *Soča*) Front; they focused on the botanical viewpoint, but the humanities have shown no interest in their findings so far.[2599] Moreover, no known study provides the complex historical context of polemochorous plant introduction, even though the authors do provide some basic historical information. This section contributes to the understanding of polemochorous introduction by also focusing on the complexity of the economic and environmental history context of military livestock feed provision.

During the First World War, the numbers of animals involved—horses in particular—and the diversity of their uses reached a marked peak,[2600] even though that war represented an important phase in the transition from biomass energy to fossil fuels. Coal was highly relevant[2601] and "[t]he First World War was the first military conflict where oil played a defining role"; nonetheless, oil

2591 For an overview, see Keller, 2018; see also Pirc & Budkovič, 1996; Lavrenčič, 1998; Souvent & Pirc, 2001; Budkovič, 2001, 118–120; Keller, 2009, especially pp. 263–265, 267; Koncilija, 2009; Budkovič, 2010; Štular, 2011, 394–396, 403–405, 409–411, 413–414, 418–419, 421, 427; Saunders et al., 2013, 49–52, 54, 58–63; Mlekuž, Košir, & Črešnar, 2016; Breg Valjavec, Zorn, & Ribeiro, 2018.

2592 McNeill, 2003, 17; McNeill, 2010, 358, 372.

2593 Tucker, 2015.

2594 E.g., Pirc & Budkovič, 1996; Lavrenčič, 1998; see also McNeill, 2010, 358, 372.

2595 Keller, 2018.

2596 Chauvel et al., 2006.

2597 Wraber, 1960, 35; Wraber, 1964, 101; Wraber, 2005.

2598 Dakskobler et al., 2019, 16–17, 24, 33, 54, 60.

2599 Even a research monograph chapter on environmental impacts of the First World War in Bohinj (Budkovič, 2001, 118–120) and a recent article on First World War memory landscapes and heritage in the Julian Alps (Repič, 2021) ignore their results.

2600 Koncilija, 2009, 80; see also Keller, 2018, 8–9; Uekötter, 2018, 291.

2601 Keller, 2018, 5, 10; see also Stergar, 2019, 17–20, 30.

had not yet prevailed in quantitative terms.[2602] On average, about 400,000 horses were still needed for every million men in the First World War. Especially in the mountains, there were many strategically relevant places inaccessible to the machinery of the time, with the main exception of cableway networks, and so animals (mainly horses, mules, and oxen) carrying and drawing heavy cargo were particularly relevant there.[2603] Large-scale use of animals impacted the grassland plant species composition, offering examples of polemochores.[2604] Nine such examples from the broader area of the Isonzo Front in the Julian Alps are presented here, eight from the Bohinj Basin, and one from the Trenta Valley. The most attention is paid to the introduction of yellow oxeye (*Telekia speciosa*; Figure 119) to Bohinj because the evidence is clearest in this case.

In late June 1915, major fighting started on the Isonzo Front, and it lasted until October 1917. The 90 km long Isonzo Front was the easternmost section of the Italian Front, the front line hundreds of kilometers long between Austria-Hungary and Italy, extending from the Swiss border to the Adriatic northwest of Trieste. Nonetheless, along the Italian Front, all Italian attempts at military invasions of Habsburg territory took place precisely on the Isonzo Front, raising its significance within the entire front between the Swiss border and the northern Adriatic. The Isonzo Front crossed the Julian Alps.[2605] Bohinj was a transitional territory between the areas of military operations and the hinterland. It was an inhabited area in Austria-Hungary at the foot of the mountains very close to this front and accessible by train but out of reach of Italian artillery from the other side of the front line. Bohinj was crucial for provisions for the Austro-Hungarian army on the section of the Isonzo Front between Mount Batognica and Mount Sleme, and to a lesser degree up to Mount Mrzli Vrh (Figure 116). Victuals, weapons, and especially at the beginning also wooden planks for buildings in the mountains, for instance, were delivered to that front line section through the study area. From the closest railway station at Bohinjska Bistrica, cargo had to be transported about 15 km to the foot of the mountains in Ukanc, west of Lake Bohinj (Sln. *Bohinjsko jezero*; Figure 116). Until late autumn 1915, this was mainly carried out by several hundred wagons drawn by animals (horses, oxen, or dogs), and later by a more energy-efficient horse-drawn railway until electrification in 1917. From Ukanc to the mountains, cargo was carried by horses (the small but strong and persistent Bosnian horse breed was very important) or people until the cableway network started

2602 Uekötter, 2018, 286 (quote), see also p. 287, and Keller, 2018, 10.
2603 Lavrenčič, 1998, 33; Koncilija, 2009, 80–87; Keller, 2009, 261; Stergar, 2019, 17.
2604 Wraber, 2005, 4–8; Chauvel et al., 2006, 669, 671–672; Dakskobler et al., 2019, 16–17, 24, 33, 54, 60.
2605 Leisering, 1998, 74; Guštin, 2005a, 131–137; Štih, Simoniti, & Vodopivec, 2008, 344–345; *Slovenski zgodovinski atlas*, 2011, 156–157; Budkovič, 2001, 25; Mlekuž, Košir, & Črešnar, 2016, 130; Svoljšak & Antoličič, 2018, especially pp. 106–107, 198–199, 161–215; Komac, 2022.

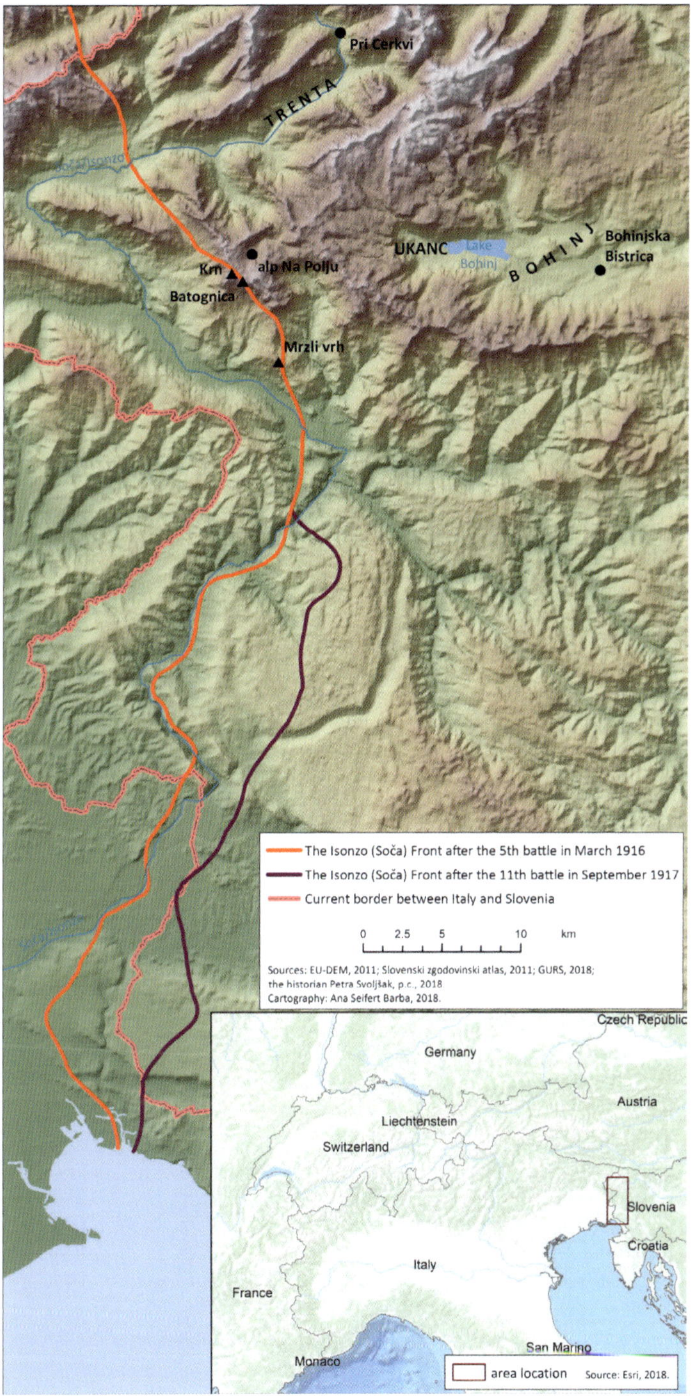

Figure 116: The Isonzo (Sln. *Soča*) Front and locations discussed in the context of polemochores. Adapted from the general map of the Isonzo Front in *Slovenski zgodovinski atlas*, 2011, 156.

operating. The first cableway linking Ukanc and the mountains was built in the autumn of 1915, and with the cableway network extension it reached the immediate hinterland of this front section in the autumn of 1916. The enormous accumulation of military horses (partly local or regional and partly from beyond the region) in Ukanc, especially until late autumn 1915, and to a lesser degree until 1917, significantly impacted the grassland environments there,[2606] as explained in this section.

Looking more closely at the military draft livestock, horses greatly prevailed over oxen. Economic considerations were one of the main reasons for this. Namely, during the fodder scarcity of mid-October 1917, the supreme Austro-Hungarian military command stated that a draft ox required 15 kg of coarse fodder (*Rauhfutter*, i.e., hay, straw, reeds—see Section 6.1.2.5.1.1—which could partly be replaced with tree leaf-hay; see Section 8.2.2.1)[2607] daily, and an additional 2 kg of "hard feed" (*Hartfutter*, i.e. field crops) on working days. A well-nourished horse, on the other hand, required only 4 kg of coarse fodder and up to 3 kg of hard feed daily. Horses were thus more economical for the army in most cases. Moreover, draft oxen were incomparably more sensitive to inclement weather than horses. Draft oxen also required more water than horses: fifty liters of water per ox daily represented an insurmountable challenge to the widespread military use of oxen in mountain karst of the Southern Limestone Alps (see Section 4.1.1.2). In addition, an ox as a ruminant requires about four hours of rest daily for chewing the cud to remain healthy; a horse is not a ruminant. Moreover, a likely outbreak of foot-and-mouth disease would require slaughtering a significant number of oxen, which would upset the availability of draft power. As the war progressed, further factors were in favor of the prevalent horses, such as more widespread availability of horse than ox harnesses, and more common knowledge of how to shoe and treat horses properly than oxen. All in all, military oxen were economically more rational than military horses only under specific local conditions in the First World War.[2608]

[2606] Budkovič, 2001, 26–27, 30–79, 119, as well as "Karte 2" and "Karte 3" between pp. 212 and 213; Budkovič, 1999, 34 (the relevant section of a sentence is missing in the translation in Budkovič, 2001); Habjan, 2014, 69; Svoljšak & Antoličič, 2018, archival photographs of military transports between pp. 208 and 209; on characteristics of the Bosnian horse breed, see Novak, 1970, 382. At a regional level, forced purchases of horses for the Austrian army were carried out; for example, in distant Bela Krajina, the southeasternmost part of what is now Slovenia (Dražumerič, 1988, 89–90).

[2607] AT-OeStA/KA FA NFA HHK HFK/HGK Kdo Südwestfront II, box 621: K. u. k. Armeeoberkommando, no. 100 071, July 18th, 1917, Darstellung der materiellen Lage der Armee im Felde: Beilage 1, Die Verpflegung der Armee im Felde.

[2608] AT-OeStA/KA FA NFA HHK HFK/HGK Kdo Südwestfront II, box 587: no. 154 234, October 15th, 1917; AT-OeStA/KA FA NFA HHK HFK/HGK Kdo Südwestfront II, box 621: K.u.k. 5. Armeekommando (Qu. Abt.), no. 52 500/1917: B. Ressortbeiträge: Verpflegs-Situation am 1. August [1917].

Focusing on grassland fodder, only part of the hay was of local origin. For instance, in September 1915, a message from the military authorities was sent to two municipalities covering the entire Bohinj stating that, due to the extremely high military demand for hay there, local supplies of hay in Bohinj could not be denied for military purposes.[2609] Again, in the spring of 1916, the military took large amounts (from the local perspective) of hay from the farms and barns in the montane vegetation belt in the northern part of Bohinj.[2610] However, local hay supplies were evidently insufficient—a highly significant fact for polemochorous species dispersal.

Mesoregional hay supplies reached Bohinj from the upper Ljubljana Basin, for instance: from the Municipality of Bled (Germ. *Veldes*), where soldiers from Bohinj purchased hay daily, as recorded in September 1915;[2611] from municipalities in the vicinity of Radovljica (Germ. *Radmannsdorf*), where all first- and second-crop hay not necessary for the local livestock was purchased for the Fifteenth Austro-Hungarian Corps according to a record dated September 19th, 1915; and also from the Municipality of Mošnje, where about seven wagons of hay for the Fifteenth Corps were purchased between September 11th and 28th, 1915.[2612] Mesoregional supplies almost certainly also included hay of more distant origin. For instance, when the facility rearing livestock to be slaughtered for the military, located in Jesenice (in the Alps not far from Bohinj), was looking for 100 metric tons of hay in September 1915, it was advised to try purchasing it in the surroundings of Ljubljana (Germ. *Laibach*) or in southeastern Carniola (Germ. *Krain*, Sln. *Kranjska*; today southeastern Slovenia), where the record stated that important quantities of hay might still be available at that time.[2613] Indeed, forced purchases of hay transported to railway stations in the southeasternmost part of present-day Slovenia (Bela Krajina) for the Austrian army are documented at least in the autumn of 1915.[2614] Hay prices in the territories neighboring Croatia were relatively favorable even in times of fodder scarcity in the Isonzo Army in the summer of

---

2609 SI AS 137, Okrajno glavarstvo Radovljica (District Administrative Authority (Germ. *Bezirkshauptmannschaft*) of Radovljica, Germ. *Radmannsdorf*), box 591, September 21st, 1915, Feldpostamt no. 325.

2610 Budkovič, 2001, 147; with an unclear translation; the source says that hay was collected at those places (Budkovič, 1999, 144).

2611 SI AS 137, Okrajno glavarstvo Radovljica (District Administrative Authority of Radovljica), box 591, September 27th, 1915, no. 2569.

2612 SI AS 137, Okrajno glavarstvo Radovljica (District Administrative Authority of Radovljica), box 591, September 19th, 1915, no. 25220 and September 28th, 1915, no. 1071. Several thousand soldiers belonging to the Fifteenth Corps gathered in Bohinj in the autumn of 1915, and provisions for the northern part of the territory of the Fifteenth Corps were delivered to the section of the Soča Front between Mount Batognica and the peak Mrzli Vrh through Bohinj (Budkovič, 2001, 27).

2613 SI AS 137, Okrajno glavarstvo Radovljica (District Administrative Authority of Radovljica), box 591, September 17th, 1915, no. 567 and September 19th, 1915, no. 25220.

2614 Dražumerič, 1988, 89–91, 99.

1917.[2615] Yellow oxeye (*Telekia speciosa*), a highly relevant species for this case study, is considered to have reached those regions (e.g., southernmost Lower Carniola, Sln. *Dolenjska*) before the First World War,[2616] after having spread throughout Southeastern Europe.[2617] In 1916, the province of Carniola alone provided hundreds of thousands of hundredweights of hay and straw for Austro-Hungarian military horses, and in the summer of 1917 feed for horses was being plundered by the military in southern and western Carniola. According to possibly exaggerated data, 20% of the usual fodder harvest in Carniola was taken for the military. Fodder supplies from Carniola were disproportionally higher than from its neighboring provinces.[2618] In 1916 and 1917, regional hay supplies from the province of Carniola, which Bohinj belonged to, must have faced the challenge of a hay harvest failure caused in both years by drought together with the absence of men serving as soldiers. The labor of absent male farmers was only partly replaced by women, the elderly, children, refugees, prisoners of war, and soldiers. In addition, fodder production was reduced by a lack of draft animals and manure. Instead of being able to sell superfluous hay to the military, farmers experienced a lack of fodder for their own livestock.[2619] These data should be understood in the broader context of the large-scale subsistence crisis in extensive areas of the Alps (in several territories of the Central Powers, and even in neutral Switzerland, at least) from 1916 to 1918, caused by the war coinciding with impacts of a severe Little Ice Age–type weather anomaly on agriculture.[2620] No wonder that the hay failure in 1917, for instance, was more than a mesoregional phenomenon.[2621]

The long winter of 1916/17 and serious drought in late spring and early summer 1917 were widespread phenomena; their interaction with the war economy impeded forage imports and adequate harvest at a macro- and mesoregional level. This nexus between environmental and human challenges threatened fodder supplies for the Austro-Hungarian armies on different European front lines. In some parts of western Carinthia, the coarse fodder yield

[2615] AT-OeStA/KA FA NFA HHK HFK/HGK Kdo Südwestfront II, box 621: K.u.k. 5. Armeekommando (Qu. Abt.), no. 52 500/1917: A. Zusammenfassende Darstellung der materiellen Lage, Juli 1917.

[2616] Melzer, 1965, 181.

[2617] Hayek, 1931, 610; thanks to botanist Jernej Jogan for the recommendation and a scan of the relevant section of this reference.

[2618] Guštin, 2005b, 73–74.

[2619] Dražumerič, 1988, 91, 97–98, 100–101; Fischer, 2005, 149; Studen & Kresal, 2005, 150–151, 153; Habjan, 2014, 67–68; Langthaler, 2018, 39–44, 60–61; Komac, 2022, 51–53, 76–88, 90, 92–105, 108–112, 147, 151–152, 154, 165. In some parts of Carniola, such as Bela Krajina, farmers even had to buy fodder for their own livestock elsewhere, or they used fodder surrogates such as straw from thatched roofs due to the lack of fodder on the market (Dražumerič, 1988, 91, 97–98, 100–101).

[2620] Pfister, Camenisch, & Dobrovolný, 2018, 126.

[2621] AT-OeStA/KA FA NFA HHK HFK/HGK Kdo Südwestfront II, box 586: Die Pferdekrise in der Armee, June 30th, 1917, 2.

in 1917 was only half of what it had been in 1916, as reported after the end of the haymaking season, on October 1st, 1917.[2622] Similarly, in distant Poland, occupied by the Central Powers, a 50% oat and hay harvest failure in 1917 was predicted in a secret report by the Austro-Hungarian army supreme command from mid-July 1917.[2623] Thus, limited hay quantities might have been supplied to the vicinity of the Isonzo Front in the Julian Alps from outside the region and its usual hinterland,[2624] but, in any case, they were insufficient.

During the First World War, supplies of the Central Powers, including Austria-Hungary, were more vulnerable than those of the Allies.[2625] Malnutrition of military livestock is discussed here because it is relevant for grassland history. In 1916, the poor condition of military oxen in the Bohinj Basin is documented due to a scarcity of appropriate relatively energy-rich fodder such as bran.[2626] Subsequent fodder scarcity in Bohinj can be better understood based on information on the broader context. In 1917, fodder scarcity became a more serious

[2622] AT-OeStA/KA FA NFA HHK HFK/HGK Kdo Südwestfront II, box 621: K.u.k. Quartiermeisterabteilung des 10. A.K., no. 4200, October 13th, 1917: A. Allgemeine Darstellung mit Subbeilagen A1, A2, A3: Allgemeiner Bericht zur materiellen Situation vom 1. Oktober 1917; and the map: Beilage A1, Situation der Armeekörper – Marschformationen – und Trains, 1. Oktober 1917.

[2623] AT-OeStA/KA FA NFA HHK HFK/HGK Kdo Südwestfront II, box 621: K. u. k. Armeeoberkommando, no. 100 071, July 18th, 1917, Darstellung der materiellen Lage der Armee im Felde: Beilage 1, Die Verpflegung der Armee im Felde, and Beilage 2, Wirtschaftliche Verwertung der okkupierten Gebiete für die Monarchie. Compare Pfister, Camenisch, & Dobrovolný, 2018, 126; Uekötter, 2018, 286, 288.

[2624] The railway network was crucial for long-distance land transport. Before the First World War, the Austrian part of the Monarchy had been importing enormous quantities of agricultural products from the Hungarian part of the Monarchy, and such long-distance transport, although at a smaller scale, continued during the war. Furthermore, especially until early 1916, imports of agricultural products from neutral states to the Monarchy were extensive. For example, by summer 1916 agricultural products were imported from (still) neutral Romania, but quantities were low; even lower quantities of agricultural products were again imported from the occupied parts of Romanian territory from late 1916 onward (Himmelreich, 2001, 10, 14–15; Brückmüller & Stekl, 1998, 67; Leisering, 1998, 74; Langthaler, 2018, 44–50, 60); an Austro-Hungarian army supreme command report from mid-July 1917, emphasized the high economic relevance of Romania, nonetheless, not in terms of exports of coarse fodder from Romania to the Monarchy (AT-OeStA/KA FA NFA HHK HFK/HGK Kdo Südwestfront II, box 621: K. u. k. Armeeoberkommando, no. 100 071, 18 July, 1917, Darstellung der materiellen Lage der Armee im Felde: Beilage 1, Die Verpflegung der Armee im Felde, and Beilage 2, Wirtschaftliche Verwertung der okkupierten Gebiete für die Monarchie). It remains an open question whether, due to the exceptionally high demand for hay in the Bohinj Basin, hay was also transported there from far distant areas. Genetic analyses could potentially add to the understanding of the provenance of yellow oxeye in Bohinj (advice by Gertrud Haidvogl; see also Chauvel et al., 2006, 672). For long-distance hay transports during the First World War—from the UK to its ally France, thus, by the opposite site, the Allies—see, for example, Peterken, 2013, 145–146.

[2625] Keller, 2018, 2–3, 7–8, 12.

[2626] AT-OeStA/KA FA NFA HHK AK/AGKdo, 5. Armee (HGK Boroevic), box 1108: K. und k. 15. Korpskommando, Op. Nr. 504/1.

issue than earlier.[2627] Starting in February 1917, a fodder crisis was recorded on the Austro-Hungarian side of the Italian Front.[2628] In June 1917, the official military command's report on the provisions for Austria-Hungary's Fifth Army (from June 1917 onward called the Isonzo Army, the Austro-Hungarian military formation at the Isonzo Front south of the alp Na Polju, Sln. *Planina na Polju*, north of Mt. Batognica[2629]) claimed that coarse fodder quantities were sufficient, but its quality was fully inadequate: almost no hay but low-quality straw was provided, and grazing resources were fully insufficient for such large numbers of horses. In addition, the quantity and quality of hard feed was low (probably bran, as well as ground corncobs and only a small amount of oats), and so horses were almost no longer able to perform draft service.[2630] The same report type for July 1917 emphasized that a fodder crisis was causing a transport crisis there. Namely, in the second half of July 1917, the fodder crisis worsened as the supplies with coarse fodder declined significantly; in the last days of July, only around a quarter of the coarse fodder demanded was provided. It was impossible to obtain sufficient coarse fodder quantities in the province. Despite the significant reduction in military horse numbers, the fodder crisis worsened in the last days of July.[2631] In August, supplies of coarse fodder were completely inadequate in terms of quantity and quality; namely, only straw was provided. At that time, the horses of the Austro-Hungarian Isonzo Army were able to transport only greatly reduced cargo quantities over short distances, despite a further reduction in horse numbers.[2632] In September 1917, coarse fodder supplies to the Second Isonzo Army, operating in the north of the former Isonzo Army, were still extremely low and qualitatively inappropriate (mainly low-quality straw), whereas provisions of hard feed were adequate.[2633] In October 1917, supplies of coarse fodder still represented

[2627] Budkovič, 2001, 38.

[2628] See later in this section.

[2629] AT-OeStA/KA FA NFA HHK HFK/HGK Kdo Südwestfront II, box 621: K.u.k. Heeresgruppenkommando G.O. v. Boroević (Qu. Abt.), no. 59 400: A. Zusammenfassende Darstellung der materiellen Lage, August 1917, Beilage A3, Subbeilage A1, map Lage der H. G. GO. von Boroević am 15. Sept. 1917, and general information in Guštin, 2005a, 131, 136–137, Sedmak, 1998a, 160–167, and Sedmak, 1998b, 171.

[2630] AT-OeStA/KA FA NFA HHK HFK/HGK Kdo Südwestfront II, box 621: Kmdo. der k.u.k. Isonzo-Armee (Q. Abt.), no. 46 500: A. Zusammenfassende Darstellung der materiellen Lage, Juni 1917; AT-OeStA/KA FA NFA HHK HFK/HGK Kdo Südwestfront II, box 586: Die Pferdekrise in der Armee, June 30th, 1917.

[2631] AT-OeStA/KA FA NFA HHK HFK/HGK Kdo Südwestfront II, box 621: K.u.k. 5. Armeekommando (Qu. Abt.), no. 52 500/1917: A. Zusammenfassende Darstellung der materiellen Lage, Juli 1917.

[2632] AT-OeStA/KA FA NFA HHK HFK/HGK Kdo Südwestfront II, box 621: K.u.k. Heeresgruppenkommando G.O. v. Boroević (Qu. Abt.), no. 59 400: A. Zusammenfassende Darstellung der materiellen Lage, August 1917.

[2633] AT-OeStA/KA FA NFA HHK HFK/HGK Kdo Südwestfront II, box 621: Kommando der k.u.k. 2. Isonzo-Armee (Q. Abt.), no. 500: A. Zusammenfassende Darstellung der materiellen Lage, September 1917.

a greater challenge than supplies with hard feed for Austro-Hungarian armies on various front lines, including the Italian Front.[2634] The fodder scarcity faced by the Austro-Hungarian Isonzo Army (the First and Second Isonzo Army) in the summer and autumn of 1917 was particularly severe, more challenging than in other military formations of the Habsburg Monarchy on the Italian Front, as shown in Figures 117 and 118.

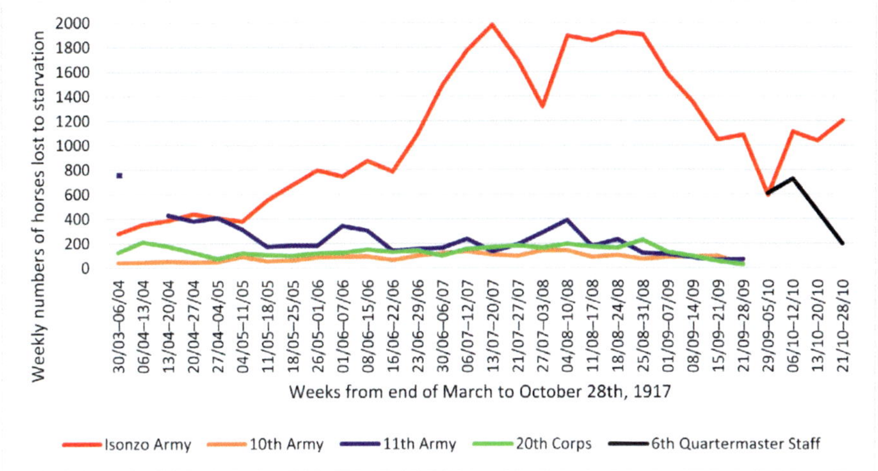

Figure 117: Weekly numbers of horses lost to starvation on the Austro-Hungarian side on various sections of the Italian Front from the end of March to late October 1917, according to official military statistics. These horses reportedly died of malnutrition or were slaughtered out of necessity due to malnutrition. Exact dates are given in reports referring to the time from April 27th onward; the first four sources cited only state that the data refer to the previous week, whereas later reports sometimes refer to seven or eight days, and so the dates limiting the first four weeks are approximate. A short mountainous section of the territory of the Isonzo Army was supplied through Bohinj in the manner presented above. Divisions of the territory of the former Fifth Army (i.e., the Isonzo Army) in late summer and autumn 1917 are not reflected in this graph; the data on different army supply areas, which developed in the territory of the former Isonzo Army (the First and Second Isonzo Armies, and a Second Isonzo Army branch (*Expositur*) that covered the northern territories of the former Isonzo Army, and so it included the area supplied through Bohinj[2635]) were summed under the label of the Isonzo Army in this graph until September 28th, 1917, to make the data comparable with earlier information. The Tenth Austro-Hungarian Army operated in the upper Soča River catchment,[2636] in northwesternmost Upper Carniola, and

2634 AT-OeStA/KA FA NFA HHK HFK/HGK Kdo Südwestfront II, box 587: no. 154 234, October 15th, 1917.

2635 AT-OeStA/KA FA NFA HHK HFK/HGK Kdo Südwestfront II, box 621: K.u.k. Heeresgruppenkommando G.O. v. Boroević (Qu. Abt.), no. 59 400: A. Zusammenfassende Darstellung der materiellen Lage, August 1917, Beilage A3, map Situation vom 1. September 1917, and Subbeilage A1, map Lage der H. G. GO. von Boroević am 15. Sept. 1917; AT-OeStA/KA FA NFA HHK HFK/HGK Kdo Südwestfront II, box 621: Kommando der k.u.k. 2. Isonzo-Armee (Q. Abt.), no. 500: A. Zusammenfassende Darstellung der materiellen Lage, September 1917.

2636 Komac, 2022, 138.

in Carinthia.[2637] The Twentieth Austro-Hungarian Corps and the Eleventh Austro-Hungarian Army were both located further west along the southern Austro-Hungarian border of South Tyrol;[2638] data on horse losses to starvation in the Eleventh Army in the second week of April 1917 (roughly from April 6th to April 13th) are not reported in the sources analyzed, hence the space between the blue dot in the first week presented and the blue line starting in the third week presented in this graph. In late September and early October 1917, during the military reorganization before the last Battle of the Isonzo, the new Sixth Quartermaster Staff (*Quartiermeisterabteilung*) was placed in charge of material supplies to and from various front sections and territories in the hinterland, at least in some areas where material supply had previously been organized by the Tenth Army command and the *Expositur* of the Second Isonzo Army, including Bohinj. Concerning the Tenth Army, the Sixth Quartermaster Staff was placed in charge of material supplies involving eastern Carinthia, the Upper Soča catchment, and northwesternmost Upper Carniola.[2639] The graph ends in the week when the Italian Front moved from the Soča catchment to the west, and so further data are irrelevant for this case study. Data sources for the graph: AT-OeStA/KA FA NFA HHK HFK/HGK Kdo Südwestfront II, box 587: no. 2331 (April 8th, 1917), no. 3086 (April [16th], 1917), no. 3726 (April 23rd, 1917), no. 4377 (April 30th, 1917), no. 5075 (May 6th, 1917), no. 5780 (May 13th, 1917), no. 6403 (May 20th, 1917), no. 7164 (May 27th, 1917), no. 7954 (June 3rd, 1917), no. 8786 (June 10th, 1917), no. 9533 (June 17th, 1917), no. 10313 (June 26th, 1917), no. 11113 (July 1st, 1917), no. 11882 (July 8th, 1917), no. 12477 (July 15th, 1917), no. 13140 (July 22nd, 1917), no. 13867 (July 29th, 1917), no. 14523 (August 5th, 1917), no. 15279 (August [13th], 1917), no. 15970 (August 19th, 1917), no. 16854 (August 26th, 1917), no. 17625 (September 2nd, 1917), no. 18535 (September 9th, 1917), no. 19640 (September 16th, 1917), no. 20891 (September 23rd, 1917), no. 21949 (September 30th, 1917), no. 23363 (October 8th, 1917), no. 24784 (October 15th, 1917), no. 26152 (October 21st, 1917), no. 27184 (October 28th, 1917). Author of the graph: Žiga Zwitter.

The fodder scarcity thus resulted in a horrific death toll among the military horses: 31,502 horses died of malnutrition or were slaughtered out of necessity due to malnutrition from February 1st to September 28th, 1917, in the territory of the Austro-Hungarian Fifth Army (and later military formations on the Isonzo Army's former territory), according to official military statistics. The data referring to the Isonzo Army are presented by the red line in Figures 117 and 118.[2640] This means that on average 5.5 horses died only for these reasons

[2637] AT-OeStA/KA FA NFA HHK HFK/HGK Kdo Südwestfront II, box 621: K.u.k. Quartiermeisterabteilung des 10. A.K., no. 3570, August 12th, 1917: map Situation der Armeekörper und Marschformationen am 1. August 1917.

[2638] A., 1967, map supplement between pp. 396 and 397.

[2639] AT-OeStA/KA FA NFA HHK HFK/HGK Kdo Südwestfront II, box 621: K.u.k. Quartiermeisterabteilung des 10. A.K., no. 4200, October 13th, 1917: A. Allgemeine Darstellung mit Subbeilagen A1, A2, A3: Allgemeiner Bericht zur materiellen Situation vom 1. Oktober 1917; and the map: Beilage A1, Situation der Armeekörper – Marschformationen – und Trains, 1. Oktober 1917; AT-OeStA/KA FA NFA HHK HFK/HGK Kdo Südwestfront II, box 621: Kommando der k.u.k. 2. Isonzo-Armee (Q. Abt.), no. 500: A. Zusammenfassende Darstellung der materiellen Lage, September 1917. Compare Svoljšak & Antoličič, 2018, 209–210, and Komac, 2022, 138, 213–214.

[2640] Calculated from data in AT-OeStA/KA FA NFA HHK HFK/HGK Kdo Südwestfront II, box 587: no. 20891 (September 23rd, 1917), no. 21949 (September 30th, 1917).

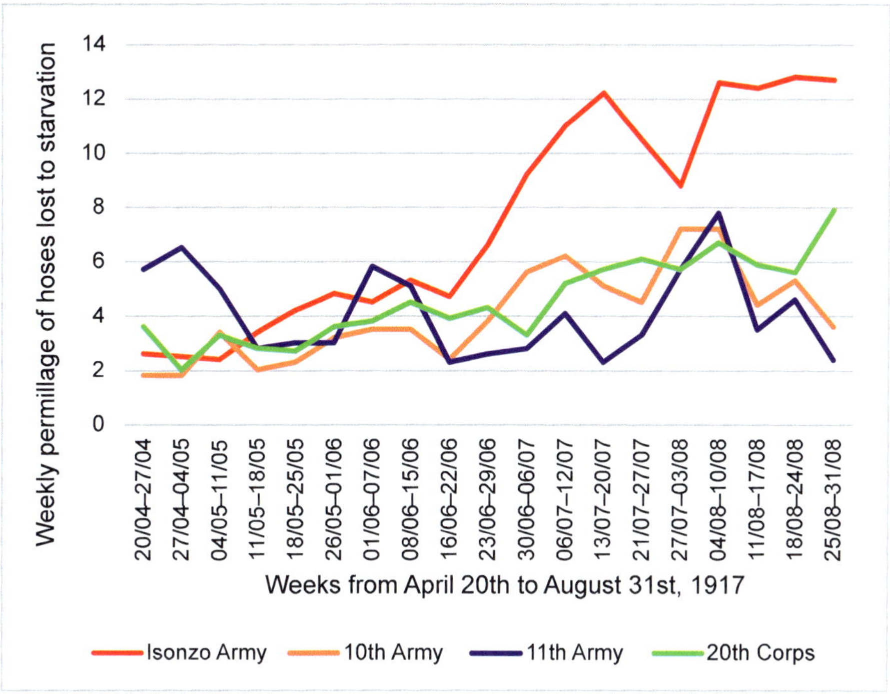

Figure 118: Weekly permillage of horses that died of starvation or were slaughtered out of necessity due to malnutrition on the Austro-Hungarian side of the Italian Front from the last decade of April to end of August 1917, according to official military statistics. As in the previous figure, the dates limiting the first week presented are approximate. The numbers of horses lost to starvation (presented in Figure 117) in every week were divided by the number of horses kept by the same military formation on the first day of the month (these data were rounded to the nearest thousand in the cited records, which makes sense because the number was not constant throughout the month). The central military official received updated numbers of horses kept by every military formation on the first day of the month with a significant time lag of more than a month and a week, and so at the beginning of the month he still calculated the permillage using the obsolete dividend from the beginning of the previous month. Archival data made possible recalculations based on more contemporaneous data, and so the number of horses kept by the same military formation on the first day of the same month serves as the dividend in Figure 118. In the weeks crossing the month divide, the number of horses on the first day of the month included in the interval analyzed serves as the dividend. Data sources: AT-OeStA/KA FA NFA HHK HFK/HGK Kdo Südwestfront II, box 587: no. 3726 (April 23rd, 1917), no. 4377 (April 30th, 1917), no. 5075 (May 6th, 1917), no. 5780 (May 13th, 1917), no. 6403 (May 20th, 1917), no. 7164 (May 27th, 1917), no. 7954 (June 3rd, 1917), no. 8786 (June 10th, 1917), no. 9533 (June 17th, 1917), no. 10313 (June 26th, 1917), no. 11113 (July 1st, 1917), no. 11882 (July 8th, 1917), no. 12477 (July 15th, 1917), no. 13140 (July 22nd, 1917), no. 13867 (July 29th, 1917), no. 14523 (August 5th, 1917), no. 15279 (August [13th], 1917), no. 15970 (August 19th, 1917), no. 16854 (August 26th, 1917), no. 17625 (September 2nd, 1917). Author of the graph: Žiga Zwitter.

in that area every hour for 240 days, with a weekly maximum of 11.4 horses per hour.[2641] These 31,502 horses correspond to 64% of all military horses that reportedly died of malnutrition or were slaughtered out of necessity due to malnutrition on the Austro-Hungarian side of the entire Italian Front from the Adriatic to the Swiss border between February 1st and September 28th, 1917; in other words, the Fifth Army and later military formations on its former territory faced the death of more than three times (318%) the number of starving horses that perished or were reportedly slaughtered out of necessity due to malnutrition in the Austro-Hungarian Eleventh Army, the military formation in South Tyrol that lost the second-largest number of starving horses on the Italian Front in the same interval.[2642] These statistics might be slightly exaggerated because some of the slaughtered horses were possibly not as malnourished as recorded, but were slaughtered to be eaten by undernourished soldiers.[2643] Nonetheless, as noted in the official military command report on the provisions for Austria-Hungary's Isonzo Army in July 1917, meat was one of the rare food types available for soldiers in sufficient quantities,[2644] and so the number of horses slaughtered illegally to feed the soldiers was not high in this territory at that time. Hence, the statistical data presented explain that malnutrition of horses was an urgent issue on the Isonzo Front. Horses dying on roads of starvation and exhaustion were also noticed by civilians.[2645] Due to malnutrition, many military mares miscarried or foals soon perished because the undernourished mares did not have milk.[2646]

2641 Calculated from AT-OeStA/KA FA NFA HHK HFK/HGK Kdo Südwestfront II, box 587: no. 16854 (August 26th, 1917).

2642 Calculated from AT-OeStA/KA FA NFA HHK HFK/HGK Kdo Südwestfront II, box 587: no. 20891 (September 23rd, 1917), no. 21949 (September 30th, 1917).

2643 Horse meat consumption gradually became more acceptable in Europe during the nineteenth century; in the previous centuries, it was limited to drastic food shortage events in most places (Thomas, 1987, 53–55, 115–116). Due to the scarcity of food, the meat of worn-out military horses was also eaten by civilians during the First World War. Civilians in the countryside close to the front line faced much more severe food shortages than the average rural population. The Austro-Hungarian Army on the Isonzo Front also often faced scarcity of some food types. According to information from 1918, when the front line crossed northeastern Italy, and so it was no longer located in the Alps of the Soča River catchment, even many cats and dogs, as well as carrion, were eaten by starving Austro-Hungarian soldiers (Godina Golija, 2011, 89, 93; Habjan, 2014, 68; Svoljšak & Antoličič, 2018, 216, 300–308; Stergar, 2015, 31, 33, 35–46; Stergar, 2019, especially pp. 24–30; Komac, 2022, 232–237).

2644 AT-OeStA/KA FA NFA HHK HFK/HGK Kdo Südwestfront II, box 621: K.u.k. 5. Armeekommando (Qu. Abt.), no. 52 500/1917: A. Zusammenfassende Darstellung der materiellen Lage, Juli 1917.

2645 Verginella, 2010, 91.

2646 AT-OeStA/KA FA NFA HHK HFK/HGK Kdo Südwestfront II, box 586: Referat zur Studie "Die Pferdekrise in der Armee," July 2nd, 1917, 5; see also the general remarks on effects of malnutrition on the physical performance of animals in Langthaler, 2018, 43.

A comparison of Figure 117 with the monthly sums of horses that perished not only through starvation and were slaughtered out of necessity due to malnutrition but also died as a result of enemy action in the Austro-Hungarian Isonzo Army from May to August 1917 demonstrates that starvation and malnutrition took around 85% of horses that died of all these reasons in June and August, around 90% in May, and roughly all horses that perished for these reasons in "peaceful" July 1917.[2647] In contrast to July, May and June 1917 experienced the bloody Tenth Battle of the Isonzo, lasting from May 12th to June 5th, with some 43,300 soldiers killed altogether (7,300 on the Austro-Hungarian side) and some 141,000 soldiers wounded (45,000 Austro-Hungarian), and the bloody Eleventh Battle of the Isonzo took place in August (from August 17th to September 15th), again taking tens of thousands of human lives on both sides of the front,[2648] and many horses. Focusing on military horses, the data presented in this paragraph demonstrate that war-related starvation was a tremendously more significant killer of horses in the Isonzo Army from May to August 1917 than the warfare itself.

Severe malnutrition of livestock along the Isonzo Front south of Mount Krn, usually the severest within the entire Italian Front, the high importance of grassland plants in the diet of military livestock, and insufficiency of grazing resources all clearly point to the significance of overgrazing. Turf devastation through overgrazing and increased trampling in meadows in Ukanc by military horses is documented in a forestry chronicle for 1915/16.[2649] Establishment of yellow oxeye (*Telekia speciosa*) in Ukanc[2650] was by far most likely made possible by unprecedented numbers of horses there. The introduction was supported by a combination of the following factors: (1) the devastation of turf in meadows, providing gaps in meadow vegetation functioning as niches for the new species;[2651] (2) ripe seeds transferred with the hay to feed horses,

---

[2647] Calculations based on comparisons of data presented in Figures 117 and 118 with the following reports: AT-OeStA/KA FA NFA HHK HFK/HGK Kdo Südwestfront II, box 621: Kmdo. der k.u.k. Isonzo-Armee (Q. Abt.), no. 46 500: A. Zusammenfassende Darstellung der materiellen Lage, Juni 1917; AT-OeStA/KA FA NFA HHK HFK/HGK Kdo Südwestfront II, box 621: K.u.k. 5. Armeekommando (Qu. Abt.), no. 52 500/1917: A. Zusammenfassende Darstellung der materiellen Lage, Juli 1917; AT-OeStA/KA FA NFA HHK HFK/HGK Kdo Südwestfront II, box 621: K.u.k. Heeresgruppenkommando G.O. v. Boroević (Qu. Abt.), no. 59 400: A. Zusammenfassende Darstellung der materiellen Lage, August 1917. Statistics on horses that died of starvation or were slaughtered out of necessity due to malnutrition are given at a weekly level as seen in Figures 117 and 118. Consequently, statistics for the weeks crossing the divide between two months were split proportionally between the two months in the above calculations of percentages of horses that died of starvation.

[2648] Svoljšak & Antoličič, 2018, 206–208.

[2649] Budkovič, 2001, 119, 140.

[2650] Mainly based on Wraber, 1960, 35, Wraber, 1964, 101, and Wraber, 2005, 4. Wraber was, however, not aware of the recorded devastation in meadows.

[2651] Compare Zelnik, 2012a, 26; Axmanová et al., 2021, 12.

Figure 119: Yellow oxeye (*Telekia speciosa*) in a tall-forb community of an abandoned half-shady damp meadow[2652] in Ukanc close to the mouth of the Savica River (July 7th, 2018). Blossoming meadowsweet (*Filipendula ulmaria*, the white blooms in the left rear part of the photo) and cabbage thistle (*Cirsium oleraceum*, pale-green forb in the right rear part of the photo) demonstrate that this wet meadow is abandoned.[2653] Photo: Žiga Zwitter.

which was not only of local origin, but, as presented above, also came from the broader surrounding region or even from beyond it; the severe scarcity of coarse fodder makes it possible to understand the provision of hay rich in yellow oxeye from distant regions; and/or (3) seed input through zoochory (e.g., small Bosnian horses might have traveled long distances to reach Bohinj from the southeast; nonetheless, it is not clear where they arrived from to reach Bohinj; namely, the Bosnian horse breed was, for instance, also kept in the province of Carniola according to early modern data, in its southeasternmost territories in particular);[2654] or even (4) imports of animal bedding.[2655] Locals

[2652] The cadastral map from around 1820 makes it clear that the parcel used to be a meadow (SI AS 176, Franciscejski kataster za Kranjsko [Franciscan Land Cadaster for Carniola], L284, cadastral municipality of Studor, map sheet A12; for orientation, see *Environmental Atlas of Slovenia*, Spatial Units, Land Cadastre).

[2653] Pils, 1994, 158, 239, 243, Tafel 72, Photos 174–175; Ellenberg & Leuschner, 2010, 1031; Peterken, 2013, 275.

[2654] On the Bosnian horse in Slovenian territory, see Novak, 1970, 382.

[2655] For example, according to a report dated June 3rd, 1916, 231 hundredweights of animal bedding were sent from the area around Domžale in northeastern Carniola to the sta-

Figure 120: Yellow oxeye close to the forest edge in a former wooded meadow[2656] used as a horse pasture in 2018. Photo: Žiga Zwitter, July 7th, 2018.

in Bohinj observed that *Telekia speciosa* grew at first only in the vicinity of the valley station of the First World War cableway in Ukanc,[2657] thus not at the camp of the Russian prisoners of war,[2658] which excludes any kind of intentional species introduction.[2659] Yellow oxeye is a perennial plant. Today it grows in the Alps in the colline and montane vegetation belts, in alkaline to neutral soils. It likes places rich in nutrients, but it also grows in mesotrophic sites that are damp or at least mesic.[2660] This species originates in Asia Minor

---

tion in Bohinj (AT-OeStA/KA FA NFA HHK AK/AGKdo, 5. Armee (HGK Boroević), box 1108: Situation des Etappentraingruppenkommandos 2/15, June 3rd, 1916).

2656 SI AS 176, Franciscejski kataster za Kranjsko (Franciscan Land Cadaster for Carniola), L284, cadastral municipality of Studor, map sheet A11; for orientation, see *Environmental Atlas of Slovenia*, Spatial Units, Land Cadastre.

2657 Wraber, 1964, 101.

2658 Budkovič, 2001, 53–54, 82–84, 115, and "Karte 3" between pp. 212 and 213; for localization see also *Arcanum Maps*, Maps of Europe, Europe in the XIX. [sic] century (with the Third Military Survey), Habsburg Empire (1869–1887) – Third Military Survey (1:25,000).

2659 *Telekia* looks much like *Inula*, a well-known medicinal plant in Russia (Grodzins'kyj (ed.), 1989; Sokolov, 1993), and so the possibility that Russian soldiers mixed them up and introduced yellow oxeye intentionally could not be entirely excluded, had the location of its first recorded site in the microregion not spoken against this hypothesis.

2660 Aeschimann et al., 2004, vol. 1, 22, 34–36; Aeschimann et al., 2004, vol. 2, 466.

or the Caucasus; however, from a global perspective it currently grows mainly in the mountains of Southeastern Europe. It also occurs at some places in the Alps, from Austria and Slovenia in the east to France in the west.[2661] Locals in Bohinj observed that the plant started growing there during the First World War and that it spread around Ukanc as the species became naturalized. Its initial spread was nonetheless additionally supported by people because locals transplanted it to another site in Ukanc for decoration, and then yellow oxeye spread to the surroundings.[2662] Such a remarkable plant would not have been unnoticed if it had grown there earlier.[2663] In 2019, an almost 104-year-old interviewee, born in the nearest village during the First World War, recalled the local dialect name *ruska roža* (literally, 'Russian flower') for yellow oxeye[2664] because there was a camp of Russian prisoners of war in Ukanc during the First World War.[2665] Botanists noted the presence of yellow oxeye in Ukanc for the first time about 1930, and it still grows there, at least in abandoned damp meadows partly shaded by trees and bushes, in riparian zones, along forest edges, and also in grasslands (Figures 119 and 120).[2666]

Further spread of yellow oxeye in Bohinj has not been limited to Ukanc, and this was again related to human activity: for instance, it was planted for decorative purposes at a shed in a montane meadow on Vogar Hill, where it still grows[2667] and from which it has started slowly spreading to the surroundings (Figure 121). This is another typical manner of introduction of yellow oxeye to several places in Europe. Namely, in Central Europe *Telekia speciosa* was cultivated in some gardens, possibly since the eighteenth century.[2668] In this way, it was also introduced to at least one site in the Eastern Alps and it became naturalized there before the First World War. Namely, yellow oxeye was planted for decoration at the mountain health resort in Bad Vellach (Sln. *Bela*), below the Seeberg Pass (Germ. *Seebergsattel*, Sln. *Jezerski vrh*), in the Karawanks of Carinthia. This type of introduction of *Telekia speciosa* there was proven by Helmut Melzer's botanical fieldwork in the mid-twentieth cen-

2661 Aeschimann et al., 2004, vol. 1, 28; Aeschimann et al., 2004, vol. 2, 466; *Global Biodiversity Information Facility: Telekia speciosa* (Schreb.) Baumg. Designation of the main region of global occurrence of yellow oxeye is based on Aeschimann et al., 2004, vol. 2; in addition, note that *Global Biodiversity Information Facility* lists many sites in Central, Eastern, Northern, and northwestern Europe.

2662 Wraber, 1960, 35; Wraber, 1964, 101; Wraber, 2005, 4.

2663 Melzer, 1965, 182; see also general remarks in Babai & Molnár, 2013, 10, 18.

2664 Author's interview with A. Hodnik, 2019, in Srednja Vas in Bohinj.

2665 Budkovič, 2001, 82–84, 115–117.

2666 Waber, 2007b, 652 and field observations in many but not all existing grasslands in Ukanc by the author on July 7th, 2018, after hay had already mostly been harvested in those meadows that are still mowed.

2667 Author's interviews with M. Hodnik, 2018, in Stara Fužina, and with A. Hodnik, 2019, in Srednja Vas in Bohinj.

2668 Poschlod, 2015, 186–187.

Figure 121: Yellow oxeye started spreading slowly on Vogar Hill; the photo shows it in an abandoned meadow section (August 30th, 2019). It colonized some forest edges there and started occurring in the middle of a meadow mowed every year in July. Photo: Žiga Zwitter, 2019.

tury. He determined that yellow oxeye abounded in the vicinity of this institution; it was still frequent some kilometers down the valley, it became rare at lower elevations, and it was totally absent from the valley above the health resort and from side valleys. Wild growth of this conspicuous species in the Vellach Valley was first recorded by a botanist in 1907.[2669] A comparative perspective reveals that in some other places yellow oxeye was introduced to the Eastern Alps in Austria even by being sown at damp sites in forest clear-cuts to serve as food for wildlife there.[2670] Polemochorous dispersal was thus an important pathway, but not the only one, of introducing yellow oxeye to the Eastern Alps.

As emphasized at the beginning of Section 7, the environmental consequences of the introduction of some species of polemochorous origin were not the strongest soon after the introduction and successful naturalization of the plant. In the case of yellow oxeye, this aesthetically attractive organism can potentially become invasive. Yellow oxeye has been recognized as a potentially

[2669] Melzer, 1965, 182.
[2670] Melzer, 1965, 182.

invasive species in some European countries, including Austria, Germany,[2671] and Slovenia: it is the twenty-fifth most common invasive alien plant species in Slovenia, the most frequent one in the upper montane and subalpine vegetation belts, where very few invasive species occur, but less common at lower elevations than many other invasive species.[2672] This leads to the identification of a likely delayed and long-term negative environmental issue caused by the First World War in the grasslands of Bohinj. Namely, in contemporary history, *Telekia speciosa* started abounding in some abandoned partly shaded wet meadows and in some riparian areas along the Savica River in Ukanc as well as along some forest edges northwest of Vogar Hill and in Ukanc. A century after its polemochorous introduction, the species continues to spread around Bohinj.[2673] Nonetheless, yellow oxeye is believed to represent no more than a "minor threat to biodiversity" in Slovenia;[2674] it is not so invasive that ecosystems would require its general eradication, but its populations have to be monitored.[2675]

Yellow oxeye is not the only plant species that became naturalized in the Julian Alps during the First World War. Wig knapweed (*Centaurea phrygia* subsp. *pseudophrygia*), not known to grow elsewhere in the Julian Alps, was collected in the late 1950s in a forest clearing above Ukanc below the First World War cableway, in the vicinity of its station, and was also possibly a legacy of the Isonzo Front.[2676] Moreover, at the level of Bohinj, occurrence of some other species is known only in one single dry meadow in Ukanc; for instance, rock centaury (*Centaurea rupestris*), red dianthus (*Dianthus sanguineus*), Illyrian scabious (*Knautia illyrica*), Liburnian plantain (*Plantago argentea* subsp. *liburnica*), holosteum plantain (*Plantago holosteum*), and hairy salsify (*Scorzonera villosa*); all these species do not occur in the great majority of the Alps in general, but they are character species of sub-Mediterranean dry grasslands on limestone or flysch. Due to the First World War activities in Ukanc presented above, they might also be long-term environmental legacies of this war. This is not as clear[2677] as in the case of the much more conspicuous yellow oxeye. Another likely influence of the First World War on flora close to the Isonzo Front is meadow saxifrage (*Saxifraga granulata*) in the Trenta Valley, close to

2671 Pergl et al., 2020; examples of lacking ecological attention to yellow oxeye: Melzer, 1965, 182 (yellow oxeye sowing at a damp site in forest clearcuts), and Pergl et al., 2020.

2672 Zelnik, 2012a, 27–31; Zelnik, 2012b, 63; Jogan, Bačič, & Strgulc Krajšek, 2012, 174–175.

2673 Author's observations on Vogar Hill. A small isolated population at the cart track at Zavaze on the western slopes of Rudnica might be either a consequence of natural or anthropogenic seed transport.

2674 Zelnik, 2012a, 29–31 (quote from p. 29).

2675 Essl, Rabitsch, & Breuss, 2002; *Bundesamt für Naturschutz: Neobiota.de: Gebietsfremde und invasive Arten in Deutschland*; compare Jogan, Bačič, & Strgulc Krajšek, 2012, 177–178.

2676 Wraber, 2005, 7–8.

2677 Dakskobler et al., 2019, 16–17, 24, 33, 54, 60; Aeschimann, 2004, vol. 1, 354–355; Aeschimann, 2004, vol. 2, 172–173, 176–177, 416–417, 598–599, 642–643.

the hamlet of Pri Cerkvi (Figure 116). This species, which does not grow in the surroundings, was first discovered in the Trenta Valley in 1961—specifically, in a meadow where hay for the Austro-Hungarian army had been stored during the First World War. In 1961, countless specimens of meadow saxifrage grew in that meadow, but its distribution in the Trenta Valley was strictly limited. Meadow saxifrage still grows in the Trenta Valley, but meanwhile its abundance has become very low.[2678]

Large-scale environmental impacts of the First World War in the Alps were particularly concentrated relatively close to the front line, where massive use of weapons directly and indirectly affected wildlife (e.g., many animals were killed by gas and other weapons; noise and crowding with people caused several animal species to leave, but the front line attracted others, such as rats, lice, and organisms feeding on carrion). Large-scale use of barbed wire—a late-nineteenth century invention, thus a novelty in this part of the Alps—also impacted habitat conditions. Some animal species were severely over-hunted or over-fished close to the front line, such as marble trout (*Salmo marmoratus*) in the Soča River. Projectiles and attacks using buried explosives caused large-scale devastation of buildings, many species, and soil, including deeper soil horizons. Where soils in karst environments of the Julian Alps were shallow, they were completely removed at affected sites. Use of firearms (artillery) and explosions of ammunition magazines caused long-term varied and patchy geochemical (soil) pollution, at many sites currently exceeding the maximum allowed concentrations of lead, rarely of copper, and exceptionally also of mercury, zinc, and antimony. Shelling and manmade or artificially reshaped caverns significantly transformed the morphology of bedrock. Soils and bedrock were also affected by lines or networks of trenches, and by building platforms and other construction, such as new traffic connections. There were concentrations of garbage, human excrement, and other water pollution in sensitive aquatic environments, and there were remains of dead animals and soldiers waiting to be buried after the end of combat, whereas many others were buried during the war. Many artefacts and geomorphic legacies in less accessible areas of hard limestone or dolomite have been left in situ in the long run, despite later removal of huge quantities of metal artifacts and munitions, removal of some building platforms, artificial backfilling of trenches and shell craters at better accessible sites close to the former front line, filling of depressions by natural processes, such as erosion and deposition, changed landscape functioning and appearance due to ecological succession, and human efforts to restore agricultural land. On steep slopes with soft stone, trenches and bombarded areas turned into erosion hotspots.[2679]

[2678] Wraber, 1964, 101; Wraber, 2005, 5–6.

[2679] Pirc & Budkovič, 1996; Lavrenčič, 1998, 33, 35–36; Sedmak, 1998b, 171; Budkovič, 2001, 47–79, 119–120; Keller, 2009, 263–265, 267; Koncilija, 2009, 80, 91, 93–94; Budkovič, 2010, 61–62; Štular, 2011, 394–396, 403–405, 409–411, 413–414, 418–419, 421,

However, as learned from these botanical examples, the First World War left long-term environmental impacts not only along the front line, but also in the hinterland and the transitional territories between the areas of military operations and the hinterland, such as Bohinj. Environmental impacts of the First World War in transitional territories toward the front hinterlands, in the hinterlands, and in far distant areas in terms of environmental impacts of large-scale mining (e.g., for mercury) and ore processing, industrial and traffic pollution, oil drilling, large-scale logging throughout Europe and overseas, impacts of war on diseases, the global food economy (see also Section 11), and, focusing on the territories closer to the front line, excessive use of natural resources in some places (such as overfishing in Lake Bohinj), devastation by aerial bombing in the hinterland, and construction (e.g., traffic infrastructure, building platforms, and hydropower plants in the transitional territories toward the hinterlands) have been addressed in historiography.[2680] This study aims to raise historians' interest in polemochorous plant introduction as a long-term ecological legacy of a fundamental economic history topic of the First World War: military livestock provision. In addition to historical and contemporary ecology, these plants are ethically highly relevant (Section 10.3).

7.3 Examples of plant introduction to gardens for ornamental and botanical purposes followed by naturalization in grasslands

Later naturalization of plants that were first introduced to gardens is a significantly relevant pathway of grassland plant introduction. In particular, it is one of the most usual ways of non-native plant introduction to (relatively) nutrient-rich flood-prone wet and moist grasslands along rivers and streams, disturbed by hydrological processes. Other means of species introduction, such as hydrotechnical works on river banks or disposal of organic garbage from

427; Saunders et al., 2013, 49–52, 54, 58–60, 62; Tucker, 2015, 329; Mlekuž, Košir, & Črešnar, 2016, 132–143; Isenberg, 2017, 145; Breg Valjavec, Zorn, & Ribeiro, 2018; Keller, 2018, 1–2; Svoljšak & Antoličič, 2018, 241–246 and an archival photograph of soldiers doing laundry at the highest small and sensitive mountain lake below Mount Krn between pp. 208 and 209; Uekötter, 2018, 281–285, 289, 295; Drole, 2019, 200, 209–210; Košir, 2019, 181–189; Košir et al., 2019, 93–94, 97, 99–103, 105; Štepec, 2019, 226, 228; Fikfak & Jezernik, 2021; Hazler, 2021; Jezernik, 2021, 122; Kravanja, 2021; Repič, 2021; Komac, 2022, 67, 111, 133–136, 139, 145, 196–197, 212, 216–229, 251, 256–257. Thanks to Marko Viduka, Tourism Bohinj, for information on the existence of First World War horse graveyards in Bohinj.

[2680] E.g., Coates, 2004, 415; Tucker, 2015, 329; Keller, 2018, 2, 5–13; Uekötter, 2018, 281–282, 287–289; Komac, 2022, 136–137, 139–140, 144, 198, 220, 228–229; Torkar & Zwitter, 2015; Svoljšak & Antoličič, 2018, 310–312, 314–315; Repič, 2021, 186; in the study area especially Budkovič, 2001, 31, 36–79, 118–120.

gardens, are also relevant there. Such riparian and flood-prone sites belong to the habitat types where neophytes are particularly frequent.[2681] For comparative reasons, plant introduction from gardens was already briefly addressed in Section 7.2 as one of the means of yellow oxeye introduction to Central Europe and the Alps. A selection of further examples is given here, all introduced to Europe between the seventeenth and nineteenth centuries.

Daisy fleabane (*Erigeron annuus*; Figures 122 and 123), a species native to North America and one of the most frequent neophytes in the Alps and Europe today—the most common invasive neophyte in Slovenia, the most common invasive neophyte in grasslands in the same country and in Austria, and one of the most common invasive plant species in (semi)natural habitats in Slovenia—was first introduced to Europe in the seventeenth century, and further independent introductions of this species to Europe followed. Daisy fleabane prefers nutrient-rich and neutral soils, but it also occurs at sites with medium nutrient availability, on alkaline to acidic soils, from dry to wet. In the eighteenth century, Carl Linnaeus still recorded that this forb only grew in botanical gardens in Europe, but significant emigration of *Erigeron annuus* from gardens and, consequently, its high abundance near Altona (at Hamburg) was already observed at that time. In the province of Carniola (now northwestern, central, southeastern, and southern Slovenia), for instance, *Erigeron annuus* became naturalized and frequent in some (damp and) wet late-mowed meadows between the late eighteenth and mid- nineteenth centuries. Currently, the species grows in the colline vegetation belt and up to the middle of the montane belt Alps-wide.[2682] The highest limit where it grows has probably increased in the last decades. The spread of *Erigeron annuus* from roadsides and ruderal sites to grasslands, mainly at elevations up to 1,000 m, has been observed recently, about four centuries after its introduction to Europe (e.g., in the Swiss Alps, and in the Julian Alps).[2683] The massive spread of *Erigeron annuus* to hay meadows, which only happened in the twenty-first century, as demonstrated by data from Slovenia, is closely related to the recent adaptation of the species: its lifetime changed from being an annual to a perennial, thanks

[2681] Ellenberg & Leuschner, 2010, 1067; Jogan, Bačič, & Strgulc Krajšek, 2012, 169, 174; Šilc et al., 2012, 726; Zelnik, 2012a, 26, 30–36; Küzmič & Šilc, 2017, 202; Axmanová et al., 2021, 7, 10–11, 14; see also Urbančič-Zemljič & Škerlavaj, 1999, and Jogan, 2006, 437–439, for instance.

[2682] Deschmann, 1868, 571 (see also Praprotnik, 2015, 266); the author was the leading natural scientist in Carniola in the second half of the nineteenth century (Praprotnik, 2015, 263–264); Frey, 2003; Wraber, 2007b, 646; Trtikova, Edwards, & Güsewell, 2010, 557; Jogan, Bačič, & Strgulc Krajšek, 2012, 173; Zelnik, 2012a, 28–29, 31, 34–36 (see also Jogan, Bačič, & Strgulc Krajšek, 2012, 173–174, 178); Poschlod, 2015, 186–187; Küzmič & Šilc, 2017, 202; Axmanová et al., 2021, 7–8; Aeschimann et al., 2004, vol. 2, 434–437; *Global Biodiversity Information Facility: Erigeron annuus* (L.) Pers.

[2683] Trtikova, Edwards, & Güsewell, 2010, 557, 560; author's interview with Kunavec, 2021, in Stara Fužina.

Figure 122: Flowering *Erigeron annuus* along a stream in a semi-natural grassland close to the bottom of a valley (565 m). The belt of the tall forb community along the stream is not mowed every year, but this invasive species also grows in regularly mowed meadows. Bohinj Basin, August 8th, 2018. Annual fleabane is one of the most common introduced species in Triglav National Park.[2684] Photo: Žiga Zwitter.

to its recently developed forms with a tuber-like stem base, allowing it to tolerate even frequent mowing.[2685]

Introduction of Japanese knotweed (*Reynoutria japonica*) from East Asia to Central Europe in the 1820s for ornamental purposes has resulted in major environmental problems. For more than a century, people planted it intentionally, in gardens as decoration and as autumn forage for bees, for instance. However, the plant turned out to be highly aggressive, making use of unintentional spread by people (e.g., in soil transport and on vehicles) and suppressing the previous vegetation. Its abundance and occurrence have been increasing quickly,[2686] reaching the colline and montane vegetation belts Alps-wide by the early twenty-first century,[2687] as one of the most common invasive alien

[2684] Papež Kristanc, 2012, 45.
[2685] Jogan, Bačič, & Strgulc Krajšek, 2012, 173–174, 178.
[2686] Urbančič-Zemljič & Škerlavaj, 1999; Jogan, 2006, 439; Poschlod, 2015, 187.
[2687] Aeschimann et al., 2004, vol. 1, 376; Jogan, 2006, 440; *Global Biodiversity Information Facility: Reynoutria japonica* Houtt.

Figure 123: One of the flowering plants from the previous figure. Photo: Žiga Zwitter.

neophytes.[2688] In the Julian Alps, the species occurs at elevations as high as 1,343 m.[2689] It has also been colonizing grasslands. In what is now Slovenia, for instance, the species was first recorded in 1908, at least, and it reached many locations in the Alps by 1980.[2690] Its increased spread in the last decades can be linked to the fact that, besides the original Japanese knotweed, a hybrid between *Reynoutria japonica* and giant knotweed (*Reynoutria sachalinensis*, first introduced to Central Europe in 1863), so-called Bohemian knotweed (*Reynoutria* × *bohemica*, first described 1983 in Bohemia),[2691] appears to play an important role. This hybrid knotweed arose in Europe, is well adapted to regional environmental conditions, and is even more aggressive than the parental species, replacing it on various sites. Nowadays a significant part of the European population of "Japanese knotweed" is most likely Bohemian knotweed.[2692] Apart from riparian sites and floodplains, it also occurs elsewhere in

[2688] Zelnik, 2012a, 28–31, 34, 36.
[2689] Jogan, Bačič, & Strgulc Krajšek, 2012, 174.
[2690] Strgar, 1981; Strgar, 1982; Urbančič-Zemljič & Škerlavaj, 1999, 85–87.
[2691] *Bundesamt für Naturschutz: Neobiota.de: Gebietsfremde und invasive Arten in Deutschland; The World Flora Online:* Reynoutria × bohemica *Chrtek & Chrtková*; see also Poschlod, 2015, 187.
[2692] Bailey, Child, & Conolly, 1996.

Figure 124: Himalayan touch-me-not (*Impatiens glandulifera*, purple blossoming tall herb) on the bank of the Ribnica River, Bohinj. Photo: Žiga Zwitter, August 11th, 2022.

pastures and meadows, and its rhizome system tolerates frequent mowing for several years.[2693]

Wet meadows have been colonized by another aggressive invasive species introduced to Europe for decorative purposes, Himalayan touch-me-not (*Impatiens glandulifera*, Figure 124), which spread from gardens and parks, and was also promoted by beekeepers. The species is native to the Himalayas and eastern India. It was first introduced to Europe in the first half of the nineteenth century, and to Central Europe in 1839. Naturalized Himalayan touch-me-not reached the Alps of Switzerland and Austria in the early twentieth century. In western Styria (Germ. *Steiermark*), for instance, it was only sporadically observed to grow wild in the first half of the twentieth century, but it started to abound in some (probably abandoned) wet meadows at Frojach, Pux, Lind, Scheifling, and St. Lambrecht, for instance, by the 1980s. At that time, many invasive alien neophytes also spread around Slovenia, where *Impatiens glandulifera* is now one of the most common invasive neophytes that widely colonizes suitable seminatural habitats. Himalayan touch-me-not occurs and abounds in nutrient-rich, damp, abandoned or extensively managed grasslands in the

[2693] Jogan, Bačič, & Strgulc Krajšek, 2012, 176; Zelnik, 2012a, 34, 36; Zelnik, 2012b, 62.

Figure 125: Coneflower at a forest edge in the Voje Valley in Bohinj. Photo: Žiga Zwitter, September 2nd, 2019.

colline vegetation belt Alps-wide, where it tends to suppress naturally present vegetation.[2694]

Coneflower (*Rudbeckia laciniata*, Figure 125) was introduced from North America to Europe in the early seventeenth century as an ornamental plant, and its naturalized occurrence was first recorded in Europe in the late eighteenth century. For instance, in the mid-nineteenth century it was observed to have just become naturalized in what is now central Slovenia, and it was predicted that a significant increase in the occurrence of this species would follow.[2695] By 2000, it grew wild in the colline vegetation belt in major parts of

2694 Hable & Präsent, 1985, 391; Frajman, 2003; Aeschimann et al., 2004, vol. 1, 1068; Ellenberg & Leuschner, 2010, 1067; Jogan, Bačič, & Strgulc Krajšek, 2012, 163, 165, 168, 176; Zelnik, 2012a, 28–31, 34, 36; Machatschek, 2015, 185–186; Poschlod, 2015, 187; Küzmič & Šilc, 2017, 202; *Global Biodiversity Information Facility*: *Impatiens glandulifera* Royle.

2695 Deschmann, 1868, 571; Praprotnik, 2015, 266; Wraber, 2000, 82; Wraber, 2007b, 654.

the Alps.[2696] In Slovenia, for instance, it is one of the most common invasive neophytes, threatening several seminatural habitat types, including wetlands and damp grassland sites, often in riparian zones and on forest edges, preferably on nutrient-rich soils.[2697]

2696 Aeschimann et al., 2004, vol. 2, 470; Wraber, 2000, 83; Wraber, 2007b, 654; *Global Biodiversity Information Facility: Rudbeckia laciniata* L.

2697 Wraber, 2007b, 654; Zelnik, 2012a, 28–31, 34–36; Zelnik, 2012b, 62; Küzmič & Šilc, 2017, 202; Deschmann, 1868, 571; Aeschimann et al., 2004, vol. 2, 470.

8 Gathering of grassland plants and lichens

This section explains that gathering of grassland plants and lichens has had a long tradition in the Alps—not only for local uses (Section 8.2), but also as a commodity to be sold at distant markets, for centuries even as distant as the eastern Mediterranean. Consequently, broader socioeconomic conditions and cultural history flows significantly influenced gathering pressures on grassland ecosystems in the Alps (see Section 8.1). This is demonstrated based on the presentation of the gathering history of selected species and its main ecological impacts.

The history of gathering was closely related to historical knowledge about grassland species and their characteristics. A historical perspective reveals that knowledge of traditional uses of wild plants was rich and dynamic.[2698] Historical perceptions of grassland plants were based on their qualities from the human perspective, such as their importance in animal husbandry and beekeeping, as human food or seasoning, as fibers, dye, and packaging (the large leaves of certain species), in folk medicine and body hygiene, in magic and in poisons, as decoration, in pest management, and as a source of other raw materials.[2699] Agricultural activities and gathering enabled farmers to learn many characteristics of the agro-ecosystems in parcels of land they were using. The details of traditional plant knowledge differed between fellow villagers due to the different micro-environmental conditions and species compositions in plots of land they were using, unequal motivation for learning, different levels of creativity, different thinking abilities, and further social factors. Farmers' traditional plant knowledge was not static; it evolved under changing environmental and social conditions and it was embedded into the broader cultural milieu. Traditional ecological knowledge was partly of local origin, was experience-based, and ran in families and local communities, but it was also subject to external influences of trans-regional cultural transfer, including learning from distant farmers (e.g., based on long-distance migrations of farmhands and maidservants because their work on a distant farm made participant observation possible). In some cases, traditional ecological knowledge was possibly also influenced by agricultural and medical literature; the influ-

[2698] E.g., Meilleur, 1986; Babai, Molnár, & Molnár, 2014; Zwitter, 2020.
[2699] Meilleur, 1982; Meilleur, 1986; Brisebarre, 1996; Thomas, 1987, 52, 65, 71–76, 90, 224; Vogl-Lukasser et al., 2006; Christanell et al., 2010, 62, 67–68, 70; Babai, Molnár, & Molnár, 2014; Mlakar, 2015; Poschlod, 2015; Sõukand & Pieroni, 2016; Zwitter, 2020.

Figure 126: In autumn, autumn crocus (*Colchicum autumnale*) blossoms without any leaves. The plant stores energy for blossoming in spring, when there appears only a green herb, without blossoms. Seeds from the blossom will ripen next summer. The seeds were gathered (Section 8.1.2). Photo: Žiga Zwitter, 2022.

ence of volumes on ecological knowledge transfer gradually strengthened after the spread of the printing press. Some published content had already reached many farmers through seigneuries or priests before literacy generally spread in the countryside (see Section 6.1.2.4, 6.5, and 8.2.2). The history of the spread of literacy in the Alps, gradually opening the way for quicker transformations of traditional ecological knowledge, was complex, with great regional and local diversity—including, for instance, proofs of high literacy rates in some regions of the French Alps starting in the fifteenth or seventeenth century at least, and great diversity in the state of education in the southern Pennine Alps (Ital. *Alpi Pennine*, French *Alpes Pennines*, Germ. *Walliser Alpen*) in the mid-nineteenth century. The spread of literacy was more or less completed in all regions of the Alps by 1900, at least considering younger generations.[2700]

Folk grassland plant taxonomy did not correspond with the Linnean one. Folk taxa sometimes designated a level equal to a species of scientific botany,

[2700] Thomas, 1987, 72; Zwitter, 2020; Babai, Molnár, & Molnár, 2014; Vogl-Lukasser et al., 2006; Sucholas et al., 2022; on the history of literacy in the Alps, see Viazzo, 1989, 138–142; see also Section 8.2.2 on the history of literacy in the Slovenian Alps.

Figure 127: Blossoming stemless thistle (*Carlina acaulis*) in a non-fertilized meadow on Vogar Hill in Bohinj, which was mowed in July. Most local stemless thistle specimens blossomed earlier. The species used to be gathered as a vegetable (Section 8.2.1.1). Photo: Žiga Zwitter, August 14th, 2022.

often they corresponded with a (few species belonging to the same) genus, sometimes a single folk taxon included species from more than one genus of Linnean taxonomy, and exceptionally different stages of development of a single species formed two separate folk taxa;[2701] for example, the autumn flower (Figure 126) and the green spring shoot (Figure 78 in Section 6.2.2.2) of autumn crocus (*Colchicum autumnale*) in many places,[2702] or what was highly likely the leaf rosette of mountain arnica (*Arnica montana*) in spring and its later phenophase according to data from the Allues (Francoprovençal *Alués*) Valley (French *Les Allues*).[2703]

Although grassland observations by ordinary people were often correct, a lack of understanding of natural processes sometimes led to wrong interpretations.[2704] The examples of two methods of weather forecasting based on grass-

2701 Meilleur, 1986; Babai & Molnár, 2013; Babai, Molnár, & Molnár, 2014; Sucholas et al., 2022, 17–34, 36.
2702 Meilleur, 1986, 292, 301; Winter et al., 2011, 1780; Babai, Molnár, & Molnár, 2014, 101.
2703 Meilleur, 1986, 290, 393.
2704 Thomas, 1987, 74–77; Scheurer, 2010, especially pages 72–83; Zwitter, 2020.

land plant observations can demonstrate this, even though it does not involve gathering. Stemless thistle (*Carlina acaulis*, Figure 127) was one of the widely observed grassland plants in weather forecasting in as distant and culturally diverse places as the Allues Valley on the one hand, and the Alps of Carinthia in Austria and of present-day Slovenia on the other hand. Namely, the inflorescence of *Carlina acaulis* closes in humid air, predicting rain, whereas it is fully open in dry weather,[2705] thus the Carinthian name *Sunnrose* (literally, 'sunflower').[2706] The occurrence and abundance of stemless thistle in the Alps used to be higher than they are now, not only because the species occurs in non-fertilized grasslands (Section 6.3.1) or, less abundantly, in some moderately manured grasslands, but also because mowing helped *Carlina acaulis* compete in grassland ecosystems (Section 6.1.1.3).[2707] The weather forecasting method based on stemless thistle was one of the pieces of traditional ecological knowledge extremely important to avoid mowing in rainy weather, which would decrease hay quality (see Section 4.1.1.3, and an example of a grassland-based weather forecast involving gathering in Section 8.2.3).[2708] The high abundance of *Carlina acaulis* in (relatively) dry to mesic grasslands at all meadow elevations in the great majority of the Alps—preferably in neutral, but also in base-rich and acidic soils, definitely not heavily manured—and its long blossoming timing from June to September both supported the usefulness of this species for weather forecasting.[2709] Unlike this well-founded practice, prediction of the depth of snow in the next winter based on heights of stalks of mature yellow gentian (*Gentiana lutea*) plants, as proven in the Allues Valley in the first half of the twentieth century,[2710] belongs to unfounded beliefs based on accidental observations. Despite its groundlessness, the practice of snow prediction based on the height of certain tall grassland plant species seems to have been more widespread because it has a parallel among traditional Csángó farmers in Ghimeş (Hung. *Gyimes*) in the Eastern Carpathians. They believe that, if stalks of white false-hellebore (*Veratrum album*) grow high and do not dry, they predict a lot of snow next winter.[2711] Such unfounded beliefs also have parallels in observations of animal behavior in the Alps; for example, in Wallis (Francoprovençal *Valês*, French *Valais*), using the height of the flight of a kestrel (belonging to the genus *Falco*) in early winter: a bird flying close to the ground when looking for rodents was considered to forecast the height of the snow.[2712]

2705 Meilleur, 1982, 169; Meilleur, 1986, 305; Machatschek, 1997a, 24; Ledinek, 1999, section Vreme; Mlakar, 2015, 460–463; Dakskobler et al., 2019, 59; Novak, 2024, 219.
2706 Machatschek, 1997a, 24; in Carinthian German, the word *Rose* has a broader meaning, not limited to a rose (Lexer, 1862, column 210).
2707 Aeschimann et al., 2004, vol. 2, 560–561; Pils, 1994, 46.
2708 Machatschek, 1997a.
2709 Aeschimann et al., 2004, vol. 2, 560–561; Pils, 1994, 46.
2710 Meilleur, 1986, 293.
2711 Babai, Molnár, & Molnár, 2014, 98.
2712 Scheurer, 2010, 80.

For millennia, gathered plants and products from them belonged to the most usual types of folk and learned medicines.[2713] Use of medicinal plants in folk medicine was particularly tricky not only because of the limited knowledge of the advantages and disadvantages of using plants to cure people and livestock (issues such as limited knowledge about the impacts of plant substances on humans and animals, and fluctuating concentrations of substances in plants during the year, from year to year, and between populations),[2714] but also because of the limited ability to diagnose diseases. Several types of detected symptoms (e.g., diarrhea) could be a consequence of several different health issues. Many plant uses in folk medicine were nonetheless appropriate, but this was not a general rule. The appropriateness of several historical wild plant uses in folk medicine and veterinary medicine has still not been tested to date; that is, about a century after the beginning of exceptionally dynamic times of traditional plant knowledge transformations and erosion.[2715] Folk veterinary medicine very often used the same substances to treat livestock as to treat people in the case of similar symptoms.[2716] Scientific botany was predominately motivated by medicine until the seventeenth century;[2717] nonetheless, botanical knowledge of learned medicine evolved gradually. For instance, it was during the Renaissance in the sixteenth century, when much fieldwork was inspired by rediscovered texts from classical antiquity (including sketchy and brief plant descriptions calling for reidentification of the plants described), that the highly toxic plant named *Aconitum* by the ancient Dioscorides was identified correctly as leopard's bane (*Doronicum*). Previously, this plant genus had mistakenly been used internally in learned medicine. Stimulated by Dioscorides, sixteenth-century botanists carried out experiments and found out that leopard's bane was too poisonous for (internal) pharmaceutical use—a discovery that saved many lives. Many substances were erroneously used in learned medicine well into the eighteenth century.[2718]

Plant gathering performed in the Alps by even the oldest local inhabitants at present is only partly based on old traditions. Although the selection of gathered plants and their uses have always been dynamic to a certain degree, it is clear that the magnitude of changes taking place since the late nineteenth century or the first half of the twentieth century (and not only since the

[2713] For instance, Kos, 2015, 392–396.

[2714] For example, Kos, 2015, 392; Sevindik et al., 2017; Kabusch, 2020, 47, 71; we thank the pharmacist Matej Pavli for drawing our attention to the pharmaceutical challenges posed by unstable contents of substances in plants.

[2715] Compare Brisebarre, 1996; Vogl-Lukasser et al., 2006; Babai, Molnár, & Molnár, 2014, especially page 99; Kos, 2015, 392–396; Mlakar, 2015, especially pages 16, 276–296, 332–333; Ramšak, 2017, 1072–1073; see also the final paragraphs of this chapter.

[2716] Vogl-Lukasser et al., 2006, 103, 126, 212, 245.

[2717] Thomas, 1987, 66–67, 73.

[2718] Palmer, 1985, 149–156; Kos, 2015, 392–396.

mid-twentieth century) is higher than in previous centuries.[2719] This section of the book focuses on the history of gathering before the recent high-magnitude changes briefly outlined in the next two paragraphs.

Many plants that experienced a long history of gathering are not gathered any more. The reasons for this are, for instance, consequences of (1) a decrease in the role of gathering in the human diet and in medicine due to quicker, easier, and increased transport of goods and people (e.g., plant gathering in small quantities along the way was usual, when people walked much more over short and longer distances, and so gatherers knew the sites along many paths, including distant places in case of species of specific interest), increased monetary orientation of the economy, progress in formal education, a higher living standard, developments in medicine, availability of commercial, chemically synthesized medicines and perfumes, increased availability and accessibility of physicians, and health insurance becoming common; (2) changes in farming, such as increased diversity of locally grown vegetables in gardens and fields; disappearance of herders spending whole days in grasslands with livestock thus having plenty of time for gathering; the end of the frequent lack of hay in spring promoting specific early spring gathering (see Section 6.1.2.5.1.1); a totally different diet for pigs; progress in veterinary medicine; and increased accessibility and availability of veterinarians; (3) substitution of several raw materials from grasslands with industrial products from other kinds of raw materials; (4) the end of magical practices; (5) changes in ecosystems leading to the rarity of some sought-after species, and environmental conservation measures; and (6) significant losses of historical knowledge about useful plants and about (un)sustainable gathering.[2720]

In contrast, at some farms more species of grassland plants have been gathered recently than in the mid-twentieth century[2721] and at many others the selection of species gathered has changed significantly:[2722] stronger intercultural encounters than in previous centuries have led to local use of some organisms historically not gathered by locals, but recently gathered for various purposes of local use, or only for market (for the external market or for local tourists).[2723]

2719 Meilleur, 1982; Sosič, 2000, 214; Vogl-Lukasser et al., 2006; Abbet et al., 2014, 625, 627–630, 632; Mlakar, 2015.

2720 Novak, 1978, 58; Maček, 1979, 305–306 (and Valenčič, 1970a); Meilleur, 1982; Sosič, 2000, 214; Vogl-Lukasser et al., 2006; Christanell et al., 2010, 62, 64–71; Łuczaj et al., 2012, 362, 364–365; Abbet et al., 2014, 625; Mlakar, 2015; Poschlod, 2015; Ramšak, 2017, 1072–1073; Rogelja Caf & Ledinek Lozej, 2021, especially pp. 67–68, 75–76; author's interviews with Boltar, 2018, with Kunavec, 2021, and with Mencinger, 2017, all in Stara Fužina.

2721 Author's interview with L. and M. Odar, 2017, in Studor.

2722 E.g., Meilleur, 1982.

2723 Meilleur, 1982; Meilleur, 1986; Łuczaj et al., 2012, 364–365; Abbet et al., 2014, 632; Sõukand & Pieroni, 2016, 39.

These processes were influenced by various media, such as herbal manuals and newspaper articles. Cousin's, Kneipp's, Künzle's (published in 1915), Treben's, and Couplan's publications were influential in Wallis and in many other places in the Alps where they were linguistically accessible;[2724] the translation of Richard Willfort's handbook *Health through Medicinal Herbs* has, for instance, been widely used in recent decades in the Julian Alps (Sln. *Julijske Alpe*, Ital. *Alpi Giulie*).[2725]

This section of the book is exceptional because plants of different levels of nutrient availability are mentioned, including plants of "nutrient-rich" sites. It is important to emphasize that optimal nutrient availability in this section is given according to ecological standards that differ from the standards of present-day agronomy: low and medium nutrient availability in this text match with low nutrient availability according to recent standards in agronomy, and nutrient-rich sites in this text can refer not only to nutrient-rich sites according to present agronomy standards, but also to medium nutrient availability (e.g., moderately manured grasslands).[2726]

8.1 Plants from the Alps reaching distant markets

This section deals with long-term and short-term wild plant gathering in the Alps in the last five hundred years. Although the demand of distant markets and thus also the impacts of broader economic flows on grassland ecosystems in the Alps were specific for plants dealt with in this section of the book, many of these species were also used locally and regionally. The first subsection (8.1.1) deals with spikenard valerian, a species from the Eastern Alps whose long-term large-scale use in distant markets is most clearly proven in written

[2724] Makarovič, 1982, 394–395, 397; Scheidegger, 2008, 33–34; Abbet et al., 2014, 632; see also Christanell et al., 2010, 62. Treben's volume has also had a significant influence on wild herb knowledge by highly traditional Csángó farmers in the Eastern Carpathians (Babai, Molnár, & Molnár, 2014, 40).

[2725] During the author's interview with the ninety-year-old herder A. Kavčič in the high temporary settlement of Laz, located in a complex of summer mountain pastures in the Julian Alps (2017), the interviewee responded that she used this book to gather medicinal plants; she kept the Slovenian translation of this book in her summer mountain cottage.

[2726] For instance, false oat-grass (*Arrhenatherum elatius*), a characteristic species of moderately manured sites according to the standards of agronomy, is characterized as a plant preferring nutrient-rich sites in ecological literature, but, in fact, this species does not at all tolerate intensive fertilization according to present-day standards of intensive fertilization in agronomy. Cat's ear (*Hypochaeris radicata*) prefers medium nutrient availability according to ecological standards, corresponding to low nutrient availability according to the standards of today's agronomy (compare the ecological perspective in Aeschimann et al., 2004, vol. 2, 628–629, 950–951 with the views of agronomy in Dietl & Lehmann, 2004, 12–13, 20, 23, 25, 37–38, 43, 53–54, 66, 106–108, 117, 118, 125; for identification of *Wiesen-Ferkelkraut*, see also *Info Flora 2004: Hypochaeris radicata* L.).

sources. The aims of spikenard valerian use have linked human pleasure and medicinal qualities. The next sections present a selection of species administered in medicine and veterinary medicine (8.1.2) or gathered to evoke human aesthetic pleasure or gourmet delight (8.1.3). The last subsection (8.1.4) provides an exceptional historical example of locally significant ecological impacts of gathering by the international scientific community of botanists.

8.1.1 Spikenard valerian: an item of ex-regional and trans-continental trade

Spikenard valerian (*Valeriana celtica*) gathering is documented in classical antiquity and from the Late Middle Ages to the twentieth (twenty-first) century. There are two subspecies of *Valeriana celtica*, both endemic to a few regions in the Alps. Most esteemed by people was *Valeriana celtica* subsp. *norica*, an endemic occurring only in the Alps of Styria, Carinthia, East Tyrol, Salzburg, and Upper Austria; it has possibly gone extinct in Trentino–Alto Adige (South Tyrol),[2727] but the evidence is not clear.[2728] All of Section 8.1.1 deals with this most esteemed subspecies, with the exception of a separate paragraph at the end of Section 8.1.1.1.2, where the history of the other subspecies, *Valeriana celtica* subsp. *celtica*, is briefly addressed.

Valeriana celtica subsp. *norica* (from here on simply called *Valeriana celtica* or spikenard valerian) is a perennial plant 3 to 15 cm tall. Considering its small size, the species has an exceptionally strongly developed root system, with a taproot in the center. Spikenard valerian occurs only in subalpine and alpine vegetation belts, at about 1,800 to 2,800 m. Its abundance in meadows and pastures can be high only on nutrient-poor, acidic soils with medium water availability, on silicate bedrock. Nonetheless, it also occurs in the eastern part of the Northern Limestone Alps, but only on carbonate-poor soils.[2729] On suitable sites of its occurrence, the abundance of spikenard valerian is usually very high.[2730] Farmers tended not to cultivate species that were characterized by sufficiently high natural abundance.[2731] Recent attempts at *Valeriana celtica* cultivation for commercial purposes have failed because such gardening was too work-intensive and the roots did not develop as well at the elevation of highest farms (1,600 m) as they did higher up in the mountains.[2732] These data

2727 Gentner, 1932; Wendelberger, 1969, 110; Aeschimann et al., 2004, vol. 2, 400–401; Kačičnik Gabrič, 2014, 189.

2728 Vierhapper, 1925, 246–248; thanks to the environmental historian Verena Winiwarter for recommending this paper.

2729 Vierhapper, 1925; Gentner, 1932, 63; Wendelberger, 1969, 110; Novak, Franz, & Novak, 1998, 637; Kabusch, 2020, 9, 11–12; Aeschimann et al., 2004, vol. 2, 400–401.

2730 Novak, 1999/2000, 113.

2731 Compare Meilleur, 1986, especially pages 157, 167.

2732 Novak, 1999/2000, 109–110; Kabusch, 2020, 56–57, 59, 68–69, 71, 74–75.

Figure 128: Spikenard valerian (*Valeriana celtica* subsp. *norica*), photo Christian Berg. © Ch. Berg.

demonstrate that gathering in high summer pastures and in the highest meadows (see Section 3.1) was the only source of spikenard valerian in the past.[2733]

8.1.1.1 The long history of spikenard valerian trade

This section first gives an overview of chronologically structured data on spikenard valerian gathering and trade in the Eastern Alps (8.1.1.1.1). Then it focuses on distant markets in a selection of the main regions of spikenard valerian use (8.1.1.1.2). This second part is also needed because knowledge about the demand significantly contributes to the understanding of historical ecological gathering impacts, discussed in Section 8.1.1.2.

8.1.1.1.1 Spikenard valerian in the economy in gathering areas

In the first century AD, Plinius the Elder noted that valerian grew in the provinces of Noricum and Pannonia. This probably referred to *Valeriana celtica*, occurring at many places in what used to be Noricum. Plinius's contemporary, the famous physician Dioscorides, documented gathering and selling of ex-

[2733] Nonetheless, a specimen from the Styrian Alps growing in a garden in Vienna for several years in the sixteenth century is documented (Gentner, 1932, 68–69). A single plant or a couple of plants growing for several years is not the same as maintaining a stand characterized by high abundance of harvestable specimens.

pensive *Nardos keltike* / *Nardus celtica* in "Istria"—in fact, spikenard valerian does not grow there, but it is likely that the *Valeriana celtica* gathered in Noricum was exported through the ports of Istria. In classical antiquity, it was used in medicine, for body care, and as a spice.[2734]

The Late Middle Ages saw a significant general increase in long-distance trade flows, and also the emergence of institutions recording these flows much more thoroughly than earlier. Since that time, historians can again trace spikenard valerian trade. In the Late Middle Ages, the Modern Period, and later, spikenard valerian was an esteemed source of perfume and a healing plant used in human and veterinary medicine as well as body care, it was placed in wardrobes to give a pleasant aroma to clothes and against moths, and it was used in magic as a means of protection against evil spirits. The most important center of trade in endemic *Valeriana celtica* from the mid-fifteenth to mid-sixteenth century at least was Judenburg,[2735] a town in the upper Mur Valley, with sites abounding in *Valeriana celtica* in its hinterland. Judenburg was also the most important commerce center in (Upper) Styria at that time. It was located on the main traffic route linking Venice (Ital. *Venezia*) as a major commercial power in the Mediterranean, and Vienna (Germ. *Wien*) through Villach (see Section 7.1). In addition, Judenburg was well connected to Salzburg and Nuremberg (Germ. *Nürnberg*).[2736] Strong trade connections between Judenburg and Venice are clearly documented in the Late Middle Ages and at the beginning of the modern period. *Valeriana celtica* roots (note the strong root system; introduction to Section 8.1.1) were a very important type of merchandise, exchanged for luxury items such as sweet wines, citrus fruits, Asian spices, glassware, and so on, arriving in Judenburg via Venice or from Venice. Trade flows changed over time, in terms of quantities and directions. Sometimes, as in most of the first half of the sixteenth century (in contrast to the fifteenth century), the spikenard valerian trade heading north, to Nuremberg in particular, exceeded the traffic flows to Venice. Nuremberg was strongly connected, among other cities, to the flourishing North Sea ports (Antwerp [Dutch *Antwerpen*, French *Anvers*], in the sixteenth century; some trade connections between Judenburg and Antwerp are proven at that time) that were well embedded into long-distance international trade.[2737]

In 1460, the town of Judenburg was granted the exclusive right to trade in *Valeriana celtica* in the provinces of Upper and Lower Austria (Germ. *Ober-*, *Niederösterreich*), Styria (Germ. *Steiermark*, Sln. *Štajerska*), Carinthia (Germ.

[2734] Gentner, 1932, 63–65, 69, 71, 73–74; Petkovšek, 1966, 300–301.

[2735] Gentner, 1932, 71–73; Tremel, 1947, 96–97, 117–118, 144; Andritsch, 1989, 31; Mlacović, 2023. See also paragraphs on trade in the Middle East below.

[2736] Tremel, 1947, 95, 103, 119, 143–144, 149; Teibenbacher, 1989, 97; "Die Wirtschaft Mittel- und Westeuropas im Zeitalter des Frühkapitalismus (um 1550)," 1998, 43; Oitzl, 2021, 424.

[2737] Tremel, 1947, 95–97, 107, 109, 119, 144, 148–157, 160–161, 164.

Kärnten, Sln. *Koroška*), Carniola (Germ. *Krain*, Sln. *Kranjska*), Gorizia (Germ. *Görz*, Sln. *Goriška*), and the territory of Trieste, also including, for instance, spikenard valerian intended for export to the Republic of Venice. The town paid an annual leasehold rent for this privilege. Uprooting, transport, and trade in *Valeriana celtica* in these provinces (covering most of the territory where this endemic species occurred) was only allowed with permission from Judenburg and a threat of confiscation of illegal stocks of spikenard valerian in these provinces was announced. In practice, the town of Judenburg leased this privilege to one of its merchants. Numerous peasant families and other inhabitants in areas where *Valeriana celtica* occurred uprooted it and provided it to the leaseholder. Large-scale uprooting took place in the Seetal Alps (Germ. *Seetaler Alpen*, in the immediate hinterland of Judenburg, see Figure 4) at least, where the most important *Valeriana celtica* gathering sites were located in the first half of the sixteenth century. At the same time, commercial gathering of this plant took place outside the Seetal Alps as well: the evidence from Judenburg proves legal supplies from Feldkirchen in Carinthia (mountains above Villach were mentioned as spikenard valerian gathering sites in the mid-sixteenth century by the Italian botanist Pietro Andrea Mattioli, who spent part of his life in the Eastern Alps and also had fieldwork done by his colleagues[2738]), from Rottenmann in Styria, and from the province of Salzburg (from Lungau, for instance). Gathering and selling *Valeriana celtica* was a significant element of the integrated peasant economy in the region. The example of the years 1542 and 1543 demonstrates that in both years together 5,072 kg (9,057 pounds, 0.56 kg each[2739]) of *Valeriana celtica* were delivered to the leaseholder. A single supplier provided from 1 kg to 156 kg, and there were more than sixty suppliers annually.[2740] A single Judenburg burgher and spikenard valerian merchant, Toman Trittenprein, owned ninety packhorse loads (between 12.6 and 15.3 metric tons[2741]) of *Valeriana celtica* stored in Venice (forty loads), Salzburg (thirty loads), and Nuremberg (twenty loads) when he died in 1523.[2742] It is important to note that it is not clear whether these quantities refer to spikenard valerian roots cleared of soil or not; unless soil was removed, it could represent up to 80% of the weight.[2743] Nonetheless, in line with the legal clauses such as the provision in the 1547 trade contract that the spikenard valerian had to be "good, dry, and appropriate,"[2744] the soil was most likely removed.

[2738] On Mattioli's method of work, see Palmer, 1985, 151–156, and Petkovšek, 1966.

[2739] The conversion is based on Baravalle, 1929, 47, 50–51, 76–78, and Baravalle, 1931, 190.

[2740] Patz, 1946, 3; Tremel, 1947, 97, 117–118; Gentner, 1932, 63; Petkovšek, 1966, 300; Wendelberger, 1969, 111; Andritsch, 1989, 31; Valentinitsch, 1989, 86.

[2741] Tremel, 1947, 121; Kosi, 1998, 136–137.

[2742] Tremel, 1947, 119; Andritsch, 1989, 31.

[2743] Compare Gentner, 1932, 72.

[2744] AT StLA, Stadt Murau, box 85, folder 2201, Speikhandel 1547. This manuscript was found by the doctoral student Tadej Pavković during his livestock history research, and he kindly provided the main author with a scan.

In the 1530s quantities of spikenard valerian from the Eastern Alps, traded under the control of Judenburg, declined as a consequence of several interacting factors; according to the historian Ferdinand Tremel, excessive uprooting of spikenard valerian, possibly not allowing the stands to regenerate, was one of them. Another important factor was smuggling; the monopoly in spikenard valerian trade was not always respected, as documented in the 1530s. At that time, smuggling *Valeriana celtica* to Venice is known to have been performed by burghers from Völkermarkt and Feldkirchen, for instance, and illegal direct trade in spikenard valerian was carried out by some landlords as owners of grasslands where *Valeriana celtica* occurred.[2745]

In 1542 and 1543, a Viennese pound (0.56 kg) of freshly gathered (not yet dried) spikenard valerian cost from 1.6 to 1.75 gulden in Judenburg.[2746] A nearly contemporary comparison with a bulk purchase of figs in Judenburg demonstrates that one hundred Judenburg pounds of figs (most probably about 56 kg[2747]) cost no more than three gulden in 1535,[2748] and so local and fresh *Valeriana celtica* cost about fifty to sixty times the value of figs from the Mediterranean transported to Judenburg in the Alps per a similar unit of weight. In 1547, the price of a pound of Asian pepper (an expensive spice widely used in upper-class cuisine of the time[2749]) in Salzburg was forty-eight kreuzer (thus 0.8 of a gulden),[2750] and so it cost about half the price of fresh *Valeriana celtica*. Further trade in spikenard valerian was also profitable: based on data from 1542 and 1543, the rate of the profit in trade in spikenard valerian for the town of Judenburg was more than 20%. It is no wonder that historical investigation of one of the reputed families of Judenburg's burghers, the Körbler family, revealed that this family acquired its economic power in the early sixteenth century mainly thanks to the privilege of trade in spikenard valerian.[2751] At a no less prominent position than on the main square of the Alpine town of Oberwölz, west of Judenburg (see Figure 4), there is the *Speikerhaus*, where large quantities of spikenard valerian roots used to be dried and stored.[2752]

Judenburg's lease on the spikenard valerian trade monopoly lasted from 1460 until 1560, as stated by the historian Johann Andritsch.[2753] Nonetheless, it

---

[2745] Tremel, 1947, 105–108, 119–120, 159.

[2746] Tremel, 1947, 118: one gulden and five or six schillings; the conversion of schillings into gulden is based on Vilfan, 1986, 401, 403, 405, 407, 410.

[2747] Baravalle, 1929, 47, 50–51, 76–78, and Baravalle, 1931, 190.

[2748] Tremel, 1947, 132.

[2749] Sandgruber, 1989, 103; see also Lane, 1973, 70.

[2750] Tremel, 1947, 133.

[2751] Tremel, 1947, 107–109, 117–119.

[2752] Wendelberger, 1969, 111; Hable & Präsent, 1985, 411; for confirmation that this house is still there, see *Equadrat engineering GmbH: Sanierung Speikerhaus Oberwölz.*

[2753] Andritsch, 1989, 31.

is clear that Hanns Seÿdlinger, a burgher from Murau in Upper Styria, leased the princely monopoly in spikenard valerian trade in 1547 (he obliged himself to sell all the spikenard valerian "so er … in dem bestanndt des speyckh von Romischen Khön[iglichen] M[aies]t[ä]t hatt vnd bekhummen wierdt" to a merchant from Salzburg),[2754] and therefore either the century of Judenburg's monopoly lease on this trade was not uninterrupted, or Judenburg did not always lease this privilege to one of its merchants, and so burghers from some other urban settlements such as Murau were also eligible. After 1560, trade in *Valeriana celtica* was carried out for some time by merchants from Carinthia.[2755] Successive spikenard valerian trade monopoly leaseholders or monopoly commanders for Styria and Carinthia—burghers and noblemen—are again known continuously from 1622 (or 1630, at least) to 1744 and again in 1763–1787; in the mid-eighteenth century not covered by the data presented so far, the monopoly in the spikenard valerian trade from the Habsburg Monarchy was in the hands of Venetian merchants. In 1787, the more than three-centuries-long era of the princely monopoly in trade in *Valeriana celtica* ended. Trade in spikenard valerian was not one of those trade monopolies delivering very high rents to people close to the Inner Austrian prince or to Inner Austrian officials (Styria and Carinthia, where there have been many spikenard valerian sites, belonged to Inner Austria). Nonetheless, in general, princely trade monopoly leaseholders were very knowledgeable about market conditions, and their business was lucrative.[2756] Thus, long continuous interest in being in charge of the Styrian and Carinthian spikenard valerian trade monopoly lease seems to suggest that market-oriented (intensive?) gathering of *Valeriana celtica* in fact had a long history in modern times, not limited to the sixteenth century and second half of the eighteenth century. Nonetheless, the quantities of gathered and exported spikenard valerian changed over time; they declined in the early eighteenth century; as the historian Helfried Valentinitsch claimed, long-term overexploitation was a (likely) reason for this.[2757] According to studies of exports from Styria in the eighteenth century, especially from Maria Theresa's reign (1740–1780), thus in a period when international traffic flows were rising significantly,[2758] *Valeriana celtica* was exported in relatively significant quantities as a highly esteemed luxury item. Of course, its market value among total Styrian exports was far from comparable to that

[2754] AT StLA, Stadt Murau, box 85, folder 2201, Speikhandel 1547. Thanks to Tadej Pavković, who found this document.
[2755] Andritsch, 1989, 31.
[2756] Valentinitsch, 1989, 88–93; Cova, 2000, 421.
[2757] Valentinitsch, 1989, 91. The main author wanted to check in the archives whether this is Valentinitsch's own attempt at an explanation or if this argument is given in the cited archival sources; it turned out that at least one of the signatures cited by Valentinitsch is erroneous, and the source(s) designated by the other signature could not be found in the archival storage.
[2758] Teibenbacher, 1989, 100–101; Cova, 2000.

of mass-export merchandise, such as iron, cereals, or cattle.[2759] In line with the general history of maritime trade in the northern Adriatic, the Habsburg port of Trieste gained importance in spikenard valerian export in the eighteenth century. Competition from Venetian merchants declined after the 1771 diplomatic dispute: the Trieste company Pincherle und Abendana accused the Venetian trader Bernardi of having persuaded the merchants in Alexandria not to buy spikenard valerian roots from the Trieste company. Diplomatic intervention by the Habsburg vice-consul in Venice and an appeal to the Venetian consul in Trieste led to an order by the Venetian authorities that Bernardi had to allow the Habsburg merchants' *Valeriana celtica* trade in Alexandria, not to complicate the republic's relations with the Habsburg Monarchy.[2760]

In the late nineteenth and early twentieth century, the example from Mount Preber on the border between the provinces of Styria and Salzburg demonstrates that, if gathering took place in grasslands that did not belong to the farm in question, the right to (commercially) gather spikenard valerian was formalized as *profit à prendre*. As such, it was also entered into the land register, at least in some cases. In this case, an unlimited and exclusive right to gather *Valeriana celtica* belonging to four farms in a specified territory was documented in a land register in 1888.[2761] This shows that spikenard valerian gathering was still considered a relevant economic activity in the late nineteenth century.[2762] Roughly at that time, a father and two sons working as herders in a summer mountain pasture complex above Krakaudorf in Upper Styria were recalled to have bought a cow with their spikenard valerian earnings from one gathering season.[2763] Data on *Valeriana celtica* in North African markets in the second half of the nineteenth century and in 1918 (see the next paragraph) also confirm the profitability of this business.[2764] For instance, in 1904, the district administrative authority (Germ. *Bezirkshauptmannschaft*) of Murau reported that full railroad wagons of *Valeriana celtica* were exported at that time, and in March 1918 a Viennese newspaper was advertising 25 metric tons of dry spikenard valerian packed in five hundred barrels readily available to be sold (exported; see below).[2765] About 1930, one of the main centers of *Valeriana celtica* gathering was the Metnitz Valley in Carinthia. In the nearby Nock Mountains (Germ. *Nockberge*), where there used to be extensive spikenard valerian sites, gatherers coming to the area from distant places are remembered, and gathering was a particularly important economic activity for the poor. During that economic crisis, some mountain meadows in the Nock Mountains were so

2759 Ebner, 1958, 84–87; Cova, 2000, especially pp. 406, 410, 420–421.
2760 Cova, 2000.
2761 Wendelberger, 1969, 112–113.
2762 Compare Hable & Präsent, 1985, 412.
2763 Novak, 1999/2000, 109.
2764 Gentner, 1932, 72–73.
2765 Patz, 1946, 3; Novak, Franz, & Novak, 1998, 637, 640; Hilpert, 2014, 6.

extensively dug up by *Valeriana celtica* gatherers that farmers could not mow there.[2766] At the same time, an important spikenard valerian trade center in Styria was Oberwölz (the *Speikerhaus* there, with a longer history, was mentioned above). At that time, the entire trade in spikenard valerian from the Eastern Alps was estimated at one to two metric tons annually, with offers ranging from a kilogram to more than 600 kg (not all roots included in this weight were cleaned of soil), but with a decreasing trend in the entire weight of gathered roots. Decreased demand was one of the main reasons for this;[2767] in some locations, in no more than a small proportion of the relevant habitats, overexploitation of wild-growing populations might have also played a role (see the presentation of ecological impacts of spikenard valerian gathering in Section 8.1.1.2). (Large-scale) uprooting and international trade in *Valeriana celtica* ceased in the second quarter of the twentieth century; protection of this endemic species led to its end (see below). Thus, commercial spikenard valerian gathering lasted more or less continuously for half a millennium at least, since the Late Middle Ages, including, as it seems, only short-term thorough breaks during some severe crises in international trade, but it experienced significant fluctuations over time (on such historical fluctuations in commercial grassland plant gathering, see also Section 8.1.2).[2768]

8.1.1.1.2 Distant markets for spikenard valerian

Focusing on distant regions where *Valeriana celtica* from the Alps was used, Levantine trade deserves special attention. In the twelfth and thirteenth centuries, trade flows between the Levant and Europe increased significantly. In most of the Late Middle Ages and much of the Modern Period, trade flows between the Levant and Central Europe were bilateral and lucrative, although with serious changes over time. According to well-known data, within this trade, the Alps provided items such as some metals and metal products (see Section 3.1), and some wooden objects demanded in the Levant. Expensive merchandise sought after due to its scent or taste—spices, for instance—was largely transported westward in European Levantine trade.[2769] However, this section demonstrates that, with spikenard valerian, grasslands in the Alps offered an expensive item sought after in the Levant for its scent. It was trans-

---

[2766] Keller, 2004, 21.

[2767] Gentner, 1932, 72.

[2768] The end of the history of spikenard valerian gathering was dated based on Gentner, 1932; Patz, 1946, 3; Wendelberger, 1969, 113; and Novak, Franz, & Novak, 1998, 637. See also some general remarks on the history of Styrian international trade in Andritsch, 1989 and in Teibenbacher, 1989.

[2769] Lane, 1973, 60–61, 64, 73, 79, 200, 292–294, 297–299, 338–339, 386, 388–389, 401–402, 407–408, 416, 418–420, 424–425, and pages given in other footnotes of this section; Tremel, 1947, 144; Ebner, 1958, 84–88; Cova, 2000, 407; Pedani, 2009; Oitzl, 2021, 422, 427–428.

ported in Levantine trade in the opposite direction than usual for aromatic goods, in quantities that were significant though not comparable with the trade in spices in the opposite direction, at least in the great majority of the interval covering the waning Middle Ages and the Early Modern Period. Nonetheless, this kind of eastward Levantine trade was relevant in terms of economic, cultural, and environmental history.

For several centuries, much of the long-distance trade in *Valeriana celtica* took place in Egypt.[2770] Because Venice was the most important late-medieval maritime center of long-distance spikenard valerian trade, the explanation is relevant that Venetian ship convoys with precious trade items regularly and periodically sailed to the eastern Mediterranean, with Alexandria (a large port firmly connected with Cairo) and (initially, but less regularly) Beirut belonging to their usual stops from the mid- and especially late fourteenth century to the late fifteenth century, building on earlier historical experiences. These voyages were absent only in exceptional times. In the fifteenth century, Venetian merchants dominated European trade in spices in Alexandria (an important destination for expensive merchandise exported through Venice from the Holy Roman Empire, and so it is no wonder that spikenard valerian was aboard) and in Syria[2771] (at that time a European term designating territories east of the Mediterranean between the Syrian Desert and Taurus Mountains).[2772] Egypt, which had belonged to the Mamluk Empire in the Late Middle Ages,[2773] was conquered by the Ottoman Empire in 1517, which in general did not represent a radical turning point in its history.[2774] It did, however, contribute to a gradual recovery of European Levantine trade, in Egypt in particular, which had declined seriously at the very beginning of the sixteenth century in the context of the new Portuguese maritime trade route linking the Indian Ocean and Europe south of Africa. The Portuguese were now faced with the much more formidable military power of the Ottoman Empire exactly at a time when it was quickly strengthening its navy and developing into a major maritime power.[2775] Just before the Ottoman conquest of Egypt, the Venetians "again became the leading Westerners trading at Alexandria[.]"[2776] The aforementioned data on spikenard valerian stocks owned by the Judenburg merchant Toman Trittenprein in 1523—that documents his large spikenard valerian stock in Venice—should be understood in this context; but the main buyer of spikenard valerian in the 1510s and 1520s was Caspar Nuczl, a merchant from Nurem-

[2770] Gentner, 1932, 72–73; Cova, 2000, 421.
[2771] Lane, 1973, 69–72, 82, 125–126, 130–132, 186, 197, 199, 286–290, 338–339, 341, 349, 351–352.
[2772] Masters, 2009g.
[2773] Masters, 2009e, 169.
[2774] Masters, 2009f, 204.
[2775] Lane, 1973, 285, 290–292, 351–352, quote from p. 292; Ágoston, 2010, 113–114.
[2776] Lane, 1973, 290, 351, quote from p. 290.

berg.[2777] Nonetheless, Trittenprein's probate inventory demonstrates that the direction to Venice was also relevant in trade in *Valeriana celtica*, in times of restored Venetian trade at Alexandria, at least. Egypt became the wealthiest Ottoman province, providing about a third of the empire's revenues. Cairo, the Egyptian capital—the richest and second-largest city in the Ottoman Empire with a population of 100,000 well before the Ottoman conquest and over 200,000 by the late eighteenth century—was a major center of international trade (with the East, like India and Yemen, with Constantinople,[2778] with Sudan, non-Ottoman Europe, etc.), crossing the boundaries of states and continents. Ottoman visitors were surprised by the diversity of merchandise available in Cairo;[2779] *Valeriana celtica* from the Eastern Alps was a part thereof. According to mid-sixteenth-century data by the botanist Pietro Andrea Mattioli, *Valeriana celtica* from the Alps of Carinthia and Styria was exported to Egypt and Syria;[2780] these data chronologically correspond to a restored strong position of Venice in Levantine trade.[2781] Syria, which like Egypt belonged to the Mamluk Empire in the Late Middle Ages, was conquered by the Ottoman Empire in 1516, with Aleppo becoming the third-largest city in the Ottoman Empire. It was one of its main flourishing commercial centers, building on its pre-Ottoman tradition of inter-continental trade linking regional, Indian, Persian, and European merchandise, mainly involving Venetian, French, and English merchants on the European side in the sixteenth century.[2782] When European Levantine trade gradually recovered after the Ottoman conquest of Syria and Egypt, Venetians for some time became "far and away the leading Western merchants in Aleppo."[2783] This is also reflected in Mattioli's data presented on the spikenard valerian trade.

A record from 1766 documents that large quantities of *Valeriana celtica* were still sold to Alexandria in Egypt, and other places in North Africa. At that time, merchants from Alexandria paid about 15,000 gulden annually for

[2777] Compare Tremel, 1947, 151. Tremel did not place Trittenprein's data into the broader Levantine context.

[2778] The common belief that this city was renamed after the Ottoman conquest is erroneous: the Ottomans continued using the name *Kostantiniyye*, the Arabic form of *Constantinople*, to designate the city and its suburbs (as proven in documents and on coins), and western sources also continued using the name *Constantinople* in Ottoman times. In parallel, the name *Istanbul* (which evolved from the Greek phrase *eis tin polin* 'to the city') was used by the common people and in official language, initially only to designate the space within the city walls, thus excluding even important suburbs. This corresponds with the use of the western name *Stamboul* in the early modern period (Eldem, 2009, 286; *Note on Transliteration and Spelling*, 2009, xxiii–xxiv).

[2779] Faroqhi, 2002, 358–360, 362; Ágoston, 2009, 593; Masters, 2009c; Masters, 2009d; Masters, 2009f, 204; Calic, 2019, 81.

[2780] Gentner, 1932, 67; Petkovšek, 1966, 298–300, 302–304.

[2781] Lane, 1973, 285, 292.

[2782] Faroqhi, 2002, 358–362; Masters, 2009a; Masters, 2009e.

[2783] Lane, 1973, 285, 292, 304, quote from p. 304.

Valeriana celtica.[2784] Part of the Egyptian spikenard valerian trade was of a transit character. Possible historical trade further east has not been examined,[2785] but it is clear that some spikenard valerian that crossed Egypt was heading south, to Sudan, at least; as recorded around 1800 and in the second half of the nineteenth century, for example.[2786] Peter Rosegger's novel *Die Schriften des Waldschulmeisters* (The Papers of the Forest School Master, 1875) provides indirect evidence for the nineteenth-century Levantine spikenard valerian trade. Namely, in this novel, a *Valeriana celtica* gatherer sings a song mentioning the use of *Valeriana celtica* in the Ottoman Empire. Rosegger (born 1843) grew up in an area located relatively close to *Valeriana celtica* gathering sites, in a poverty-stricken mountain-farm family.[2787] The aforementioned entire railroad wagons of spikenard valerian roots sold from the district administrative authority of Murau around 1904 were exported to the Orient, and in 1918 trade in spikenard valerian on the drug and herb market in Cairo was still recorded. Egypt was probably still the main consumption area of *Valeriana celtica* from the Eastern Alps between the First and Second World War.[2788]

Contrary to the culture in the territories of sites of spikenard valerian gathering in the Eastern Alps,[2789] Islamic tradition in the Middle East required that Muslims maintain a high level of bodily hygiene.[2790] In the mid-sixteenth century, the botanist Mattioli, who was also well informed about the eastern Mediterranean, where Renaissance botanists were looking for plants mentioned by natural scientists (Dioscorides) in classical antiquity, stated that in Egypt and Syria *Valeriana celtica* was frequently used in baths.[2791] In the Ottoman Empire, "[b]alneotherapy remained in practice, basically unaltered, well into the twentieth century" and "traditional herbal and animal pharmacological formulae remained in circulation even after ready-made drugs from Europe

2784 Ebner, 1958, 84–85, 87. In Carniola, this price roughly corresponded to earnings from trade in somewhat less than fifty metric tons of locally produced iron nails around 1780 (Šorn, 1984, 145).

2785 Existing routes of large-scale trade and cultural history suggest that some *Valeriana celtica* might have been sold further east, in the Late Middle Ages and modern period, for instance (compare Lane, 1973, 70–73, 131, 286). Nonetheless, another valerian with an esteemed aromatic essential oil similar to spikenard valerian—namely, Indian nard (*Nardostachys jatamansi*), occurring in mountains such as the Himalayas, and at some places in the mountains of East Asia—was relevant in Asian markets (Gentner, 1932, 64; Kabusch, 2020, 68; *Global Biodiversity Information Facility*: *Nardostachys jatamansi* (D.Don) DC.).

2786 Gentner, 1932, 73; chronological determination of Schweinfurth's data is based on *Encyclopaedia Britannica*: Georg August Schweinfurth.

2787 Gentner, 1932, 70; on Rosegger, see also Wagner, 2005; on spikenard valerian sites, see Kabusch, 2020, 24–25.

2788 Gentner, 1932, 72–73; Patz, 1946, 3.

2789 Remec, 2015; see also Wopfner, 1997, 461–462, and Šarf, 1974, 139, 142.

2790 Masters, 2009b.

2791 Gentner, 1932, 67; on Mattioli, see Palmer, 1985.

began to be imported into the Ottoman market in the mid-nineteenth century."[2792] In the second half of the sixteenth century, another early botanist, Charles de l'Ecluse (Clusius), noted that spikenard valerian was allegedly often used in (possibly the Ottoman part of) Hungary for washing hair.[2793] Knowing such distant areas of historical *Valeriana celtica* use and its administration in body care, it becomes understandable that Rosegger's novel mentioned above cites the use of *Valeriana celtica* as perfume by women.[2794]

Less esteemed, but nonetheless gathered and traded, was the other subspecies, *Valeriana celtica* subsp. *celtica*, endemic only in the Western Alps of what are now France, northwestern Italy, and southwestern Switzerland, mainly in the Graian Alps (Ital. *Alpi Graie*, French *Alpes grées*). Based on observations published in 1932, its abundance in the Western Alps, and even in the Graian Alps, was significantly lower than that of *Valeriana celtica* subsp. *norica* in the Eastern Alps.[2795] Trade in what was most likely *Valeriana celtica* subsp. *celtica* in Liguria was mentioned by Dioscorides in the first century AD. Similarly, trade in this subspecies in Genoa (Ital. *Genova*) was documented in the sixteenth century, for instance.[2796] Its aroma and chemical composition differ significantly from the subspecies *norica*.[2797] Apart from preferences of the buyers' side, dependent on aroma, political history probably also had an influence on which *Valeriana celtica* subspecies was available as a luxury item in distant markets. Dynamic power relations between Venice (exporting *Valeriana celtica* subsp. *norica*) and Genoa (with *Valeriana celtica* subsp. *celtica* sites in its hinterland) in the eastern Mediterranean[2798] are an example of such impacts.

One or both subspecies of *Valeriana celtica* from the Alps were also used in many places in Europe, as documented in the mid-sixteenth century by Valerius Cordus, the author of the first pharmacopoeia north of the Alps, or by the Swiss naturalist and polymath Konrad Gesner, who edited Cordus's publication.[2799] Local uses of spikenard valerian (*Valeriana celtica* subsp. *norica*) proven more recently include placing it in wardrobes to give a nice scent to clothes and against moths, its application in magic as a means of protection against evil spirits, and its use in medicine: against fever, nervous ailments, cramps, toothache, some heart issues (in Carinthia, for instance), and stomachache in the lower abdomen (Upper Styria, as recorded in the nineteenth

[2792] Günergun & Etker, 2009 (quotes from pp. 357, 359).
[2793] Gentner, 1932, 68.
[2794] Gentner, 1932, 70.
[2795] Gentner, 1932, 67–68, 73–74; Aeschimann et al., 2004, vol. 2, 400–401.
[2796] Gentner, 1932, 64, 67–68.
[2797] Novak, Franz, & Novak, 1998, 638–639.
[2798] On these political history aspects, see, for instance, Lane, 1973.
[2799] Gentner, 1932, 67; on Cordus, see Ziegenspeck, 1957; on Gesner, see Fueter, 1964.

century). The species was also used locally to treat livestock;[2800] a tisane from its above-ground parts was administered to cattle and sheep for stomach problems (in East Tyrol, at least).[2801] Administration of spikenard valerian or another valerian (*Valeriana* spp.) against some livestock health issues has a long tradition. It was not limited to the territories of spikenard valerian sites, in the demesne stables of seigneuries, at least. For instance, the seigneury of Gornja Radgona (Germ. *Oberradkersburg*), in a sub-Pannonian environment several dozen kilometers from the Alps and spikenard valerian sites,[2802] purchased a certain type of valerian to cure its livestock in 1667/68.[2803]

8.1.1.2 Historical ecological impacts of gathering spikenard valerian

Early twentieth-century data make clear that spikenard valerian gatherers in the Eastern Alps used a tool called a *Speikkramperl*, similar to a short-handled pickaxe, with a three-toothed curved fork instead of a broad flat pickaxe blade on the one side and a pointed curved end on the other side, enabling the gatherer to take longer root sections from the soil (Figure 129).[2804] The *Speikkramperl* made it possible not only to dig single spikenard valerian roots, but also remove entire belts of turf. The latter gathering practice, carried out in places with a high abundance of *Valeriana celtica*, led to soil erosion in some places. This was particularly problematic in some locations where spikenard valerian occurred in the Northern Limestone Alps; for example, at Schießeck and Hoher Zinken west of Judenburg. There, limestone bedrock is exposed on the surface instead of the former vegetation cover due to such historical turf removal.[2805] Historical large-scale gathering, mainly oriented toward long-distance trade, seems to have decreased the abundance (and occurrence) of *Valeriana celtica* several times. Historians assume that this was the case in the sixteenth and eighteenth centuries, as presented above. This finds confirmation in observations carried out by the botanist Heinrich Hoppe, who visited the Alps of Salzburg and Carinthia. In 1806, he stated that some stands were overharvested, so that *Valeriana celtica* became extremely rare there. In addition, gathering of the early twentieth century, for instance, was also excessive in some places. Farmers from the Nock Mountains reported that spikenard valerian gatherers dug out the vegetation of their high mountain meadows to such a degree that they could not mow there. Thus, around 1930, as spikenard valerian gathering was decreasing significantly, the botanist and agriculturalist Georg Gentner stated that *Valeriana celtica* was at that time slowly recolonizing sev-

[2800] Keller, 2004, 21; Scheidegger, 2008, 11; see also Makarovič, 1982, 396–397, and Novak, 1999/2000, 109.
[2801] Vogl-Lukasser et al., 2006, 188; see also Novak, 1999/2000, 109.
[2802] Mlinarič, 1979, 11–12.
[2803] Pavković, in progress.
[2804] Wendelberger, 1969, 110–112; Novak, 1999/2000, 108–109; Kabusch, 2020, 16.
[2805] Wendelberger, 1969, 111, 113; Hable & Präsent, 1985, 412.

Figure 129: A *Speik-kramperl* in the Wölz Tauern. Photo: Ewald Gabardi, 2013. © Ewald Gabardi.

eral sites where excessive gathering in the past had suppressed it. Although Gentner's statement is correct in the case of excessive historical gathering at some sites, ecological arguments warn against over-simplified conclusions regarding ecological impacts of gathering spikenard valerian in older history. Indeed, despite the large-scale gathering, *Valeriana celtica* was probably not driven to extinction more than locally. Namely, Gentner also reported that, despite the long history of strong gathering pressures, he could identify sites in the Eastern Alps where hundreds of thousands of specimens of *Valeriana celtica* occurred in his time.[2806] Similarly, in the mid-twentieth century the botanist Gustav Wendelberger observed a high abundance of *Valeriana celtica* in some places, where historical records make clear that it had been collected on a large scale, such as in the Seetal Alps above Judenburg and in the Rottenmann and Wölz Tauern (Germ. *Rottenmanner* and *Wölzer Tauern*, both belong to the Lower Tauern).[2807]

The reason for such somewhat surprising observations is that gathering *Valeriana celtica* can be sustainable[2808] because spikenard valerian can regen-

[2806] Gentner, 1932, 72, 75; Patz, 1946, 3; Keller, 2004, 21.
[2807] Wendelberger, 1969, 113.
[2808] Gentner, 1932, 69; Wendelberger, 1969, 113.

erate well if the gathering pressure is moderate (i.e., 25% to 50% of spikenard valerian per plot is gathered at once) and gathering plots are not constant, so that moderate gathering does not take place more frequently than once every three years in the same place. This was proven by a four-year experiment in the 1990s;[2809] the long-term monitoring of moderately gathered spikenard valerian populations in the following decades has so far confirmed the positive impacts of moderate gathering on *Valeriana celtica* populations.[2810] Namely, spikenard valerian is a pioneer plant species, able to relatively quickly colonize gaps in grasslands created through moderate gathering pressures. Gathering *Valeriana celtica* roots unintentionally removes nearby plants, and spikenard valerian uses such sites as regeneration niches in the following couple of years.[2811] Gentner claims that traditional ecological knowledge (almost certainly) included the following information to allow long-term use of this species perceived as a "natural resource": after a year of gathering, this practice was banned in the same plot for some years to allow regeneration of spikenard valerian stands.[2812] Significant sections of *Valeriana celtica* roots remain stuck in the soil during the gathering process (Figure 130) and plants sprout from such root fragments. This allows regeneration in the case of moderate gathering pressure.[2813] Another important fact contributing to the survival of the species despite its long history of intensive root gathering can be identified based on the opus of the sixteenth-century botanist Charles de l'Ecluse, who presented the species based on the knowledge acquired in the years he spent in Vienna. He documented that *Valeriana celtica* was gathered by local inhabitants in the Eastern Alps in late August and early September, when its leaves started turning yellow and its scent was the strongest.[2814] The coincidence of the approaching end of photosynthetic activities (leaves turning yellow) and the highest attractiveness for people due to the strongest pleasant smell resulted in a long history of gathering exactly at the time when the species had already stored high quantities of nutrients in its roots, significantly contributing to its ability to resprout from some root fragments left in situ. Moreover, the species usually blooms in July and August,[2815] so that gathering in late August and September allowed seed dispersal. In the province of Salzburg, at least, it

[2809] Novak, 1999/2000, 111–112.

[2810] Kabusch, 2020, 46–47, 64–65, 71.

[2811] Novak, 1999/2000, 113.

[2812] Gentner, 1932, 69.

[2813] Wendelberger, 1969, 113.

[2814] Gentner, 1932, 68. These details suggest that de l'Ecluse was better informed than his contemporary botanist Joachim Kammermeister (Camerarius), claiming that *Valeriana celtica* was collected in July and August, before its above-ground herb decomposed in autumn. However, even the latter report confirms the explanation above because *Valeriana celtica* leaves start becoming yellowish as early as late July (compare Gentner, 1932, 63, 71).

[2815] Aeschimann et al., 2004, vol. 2, 400.

Figure 130: Gathered spikenard valerian (*Valeriana celtica* subsp. *norica*) in the Wölz Tauern, August 2013. Note that fine roots stayed in the soil in many cases, that the first leaves had started turning yellow by the gathering time, and that some seeds look ripe. Photo: Ewald Gabardi. © Ewald Gabardi.

was not allowed to uproot *Valeriana celtica* before its seeds ripened, as recorded in the first years of the nineteenth century.[2816] All of this helps understand the initially surprising observation of farmers from the Wölz Tauern reported to the botanist Helmut Melzer in the mid-twentieth century; namely, that the abundance of *Valeriana celtica* was much higher in the early twentieth century (that is, in the era of large-scale gathering of spikenard valerian) than in the mid-twentieth century, after the end of the long history of intensive gathering of this species there.[2817] This corresponds to the results of a botanical experiment proving higher abundance in the case of moderate spikenard valerian root gathering pressures (e.g., removal of 50% of the stand) than in the case of absence of gathering (but a significant decrease in abundance in the case of excessive gathering).[2818] As a pioneer species, *Valeriana celtica* quickly colonizes gaps in grassland vegetation, whereas its competition ability in dense

[2816] Patz, 1946, 3.
[2817] Wendelberger, 1969, 113.
[2818] Novak, 1999/2000, 111–112; Keller, 2004, 21.

grasslands is low.[2819] Spikenard valerian also uses natural gaps in grasslands for recruitment, which are rather frequent in the alpine vegetation belt,[2820] and those created by grazing livestock, but moderate human disturbance through gathering increased the suitability of the site for this species.[2821]

In the early twentieth century, *Valeriana celtica* was legally protected in most of the provinces where it occurs, except for Carinthia and East Tyrol. In Styria, for instance, this happened in 1915. Nonetheless, not only rather small-scale local gathering to treat livestock but also commercial gathering was still allowed until the 1930s; for commercial purposes, the gatherers had to receive permission in advance not only from the parcel owner but also from the political authorities. In Styria, such permission referred to a clearly specified site and was valid in a specified year; permission to gather spikenard valerian roots in the same parcel could only be obtained every third year. This measure indeed had the potential to lead to regeneration of the stands. As observed around 1930, the political authorities paid attention to the economic crisis, so that it was easy for the locals to obtain such permission. In 1936 and 1940, protection of the species at the state level followed, and commercial *Valeriana celtica* gathering was fully banned. According to data from the late 1960s, the species was strictly protected in the provinces of Salzburg and Upper Austria, trade in this plant was prohibited in Styria and Tyrol, and spikenard valerian was partially protected in Carinthia. Relevant for this species, partial protection in that province meant a ban on uprooting the plant, as explained in the law from 1988, with exceptions possible in cases of research, education, or matters of public interest.[2822] Such a regulation allowed a revival of small-scale and local spikenard valerian gathering in Carinthia since the late twentieth century[2823] (Section 10.2.3).

8.1.2 Large gentians and further examples of traded medicinal plants

Gentians have long belonged to the most popular healing plants in the Alps, for both local and commercial use.[2824] The underground parts of a few gentian species locally abounding at mesic to relatively dry sites were gathered (Figures 131 and 132). In alkaline soils on carbonate bedrock, yellow gentian (*Gentiana lutea*), generally the largest gentian in the Alps, locally abounds in montane and subalpine vegetation belts, so that it was the most frequently

2819 Wendelberger, 1969, 113; Novak, Franz, & Novak, 1998, 637; Hilpert, 2014, 8–9.
2820 Pils, 1994, 198–200.
2821 Novak, 1999/2000, 110–114.
2822 Gentner, 1932, 72; Wendelberger, 1969, 113; Novak, 1999/2000, 113; Hilpert, 2014, 6–8.
2823 Kabusch, 2020, 4, 31–35, 38, 56, 59, 61, 68, 71–72.
2824 von Hörmann, 1877, 149; Jäger, 2007a, 222–224; Niederer, 1996d, 191, 205; Sosič, 2000, 235, 249; Mlakar, 2015, 308–309, 348, 350, 428, 432.

gathered large gentian species. Danube gentian (*Gentiana pannonica*), another quite large gentian species that can abound in montane and subalpine belts, in grasslands with close to neutral soil pH, on carbonate or slightly acidic silicate bedrock, was also uprooted. At acidic sites on silicate bedrock, the abundance of dotted-flowered gentian (*Gentiana punctata*, mainly in subalpine and alpine belts) and purple gentian (*Gentiana purpurea*, largely in the subalpine belt) was locally high, and so these quite large species were gathered. Purple gentian usually replaced the role of yellow or dotted-flowered gentian for people in the areas where these two species were less common, as proven in the Anniviers Valley (French *Val d' Anniviers*) in Wallis (Francoprovençal *Valês*, French *Valais*). Small gentian species with large blue blossoms (trumpet gentian, *Gentiana clusii*, can abound in subalpine and alpine grasslands, at alkaline sites on carbonate bedrock, see Figure 168 in Section 10.2.2; stemless gentian, *Gentiana acaulis*, a plant of similar size, abounds at some acidic sites on silicate bedrock) were also gathered by being uprooted. In addition, centauries (*Centaurium* spp.), which belong to the Gentianaceae family and contain partly similar substances as gentians, were esteemed healing plants mixed with gentians—for instance, in brandy production, as reported from Tyrol in the nineteenth century.[2825]

Focusing on older historical data from the Alps, in 1560, the seigneury of Gornji Grad (Germ. *Oberburg*) paid for gentian collected locally in the Kamnik–Savinja Alps. In 1607, gentian was recorded in a probate inventory of a farmer in Koroška Bela, a village at the foot of the Western Karavanks close to Jesenice.[2826] In North Tyrol as well, gathering gentian rhizomes can be proven at least since the mid-sixteenth century. This activity was so important that limitations on gathering gentian and gentian spirit production there reach as far back as the second half of the seventeenth century.[2827] Namely, where brandy was made from gentian rhizomes (in Tyrol and some other regions of the Alps), and not just a bit of gentian was macerated in another spirit type, enormous quantities of the rhizomes were needed. About 150 to 200 kg of below-ground parts of yellow gentian were necessary to produce four or five liters of the spirit, according to data from the Allues (Francoprovençal *Alués*) Valley (French *Les*

[2825] von Hörmann, 1877, 146, 150; Niederer, 1996d, 191, 205; Abbet et al., 2014, especially p. 631; Juen senior, 2014, 56–57; Aeschimann et al., 2004, vol. 2, 6–13, 16–19. The list of large gentians gathered above is not fully exhaustive. Further large gentians gathered surely included hybrids (Wraber, 1969, 287, mentions hybrids between yellow gentian (*Gentiana lutea* subsp. *lutea*) and Danube gentian (*Gentiana pannonica*), and between *Gentiana lutea* subsp. *symphyandra* and Danube gentian on the Komna Plateau in the Julian Alps, where gentian rhizome gathering took place; author's interview with Kranjc, 2020, in Stara Fužina). Compare the chemical composition of centaury and yellow gentian in Galle Toplak, 2015, 295–296, 303–304, 440.

[2826] Zwitter, 2020, 67–68.

[2827] Jäger, 2007a, 222–223.

Figure 131: Gathering dotted-flowered gentian (*Gentiana punctata*) at Galtür; see Section 10.2.3. Photo: Herbert Raffalt. © Municipality of Galtür.

Allues) in the first half of the twentieth century.[2828] About seven to nine liters of dotted-flowered gentian (*Gentiana punctata*) brandy are made from the same quantity of below-ground parts at Galtür now, according to some data;[2829] note that the volume of brandy produced from a certain quantity of gentian is considered a trade secret,[2830] and so significantly higher spirit quantities (seven liters from 100 kg of dotted-flowered gentian rhizomes) can also be found in the literature on Galtür.[2831] However, in addition to inaccurate data, different quantities of the spirit produced from the same amount of gentian roots were likely also impacted by variations in the production process. Gentian spirit production has evolved over time,[2832] which might have had an impact on efficiency in terms of the quantity of the rhizomes needed per volume of spirit.

[2828] Meilleur, 1986, 292.
[2829] Juen senior, 2014, 57.
[2830] Helmut Pöll, personal communication.
[2831] *Austrian Commission for UNESCO:* Immaterielles Kulturerbe: Bräuche, Wissen, Handwerkstechniken: Wissen um die Standorte, das Ernten und das Verarbeiten des Punktierten Enzians – Bewerbungsformular, 2013.
[2832] E.g., von Hörmann, 1877, 148–149.

Figure 132: Gathered dotted-flowered gentian rhizomes; see Section 10.2.3. Photo: Herbert Raffalt. © Municipality of Galtür.

Figure 133: A pickaxe adapted specifically for gathering gentian rhizomes in the mountains: the tool without a broad head part was lighter and easier to handle in rocky sites.[2833] This museum exhibit is hanging on chains. Similar are implements for uprooting large gentians on some nineteenth-century illustrations showing a tool with an iron head bent somewhat more than this one, and without any element of the other head.[2834] Museum Alpinarium Galtür, photo Žiga Zwitter. © Alpinarium Galtür.

The following lines make clear that we are dealing with relatively early spirit distillation in the presented case of gentian brandy in Tyrol (for another early example, juniper-berry brandy, see Section 6.6.2.2). In general, the spirit production procedure including repeated distillation was invented in the Middle Ages, but at first brandy was only used for medicinal purposes in Europe. From the fifteenth to nineteenth century, brandy consumption in Europe increased heavily. The first known data on farmers as brandy producers come from late-sixteenth-century France. In the seventeenth century, selling brandy in taverns on Sundays and other holy days before the end of the mass was prohibited in Anras, east Tyrol, which suggests that brandy had already

[2833] Helmut Pöll, Alpinarium Galtür, personal communication, 2022.
[2834] Jäger, 2007a, 223–224.

been drunk there for pleasure by that time. Similarly, in 1642, in the hills of the Natisone (Sln. *Nadiža*) River catchment at the southern foot of the Julian Alps, more than just exceptional brandy consumption by local farmers was indicated. In the second half of the seventeenth or first years of the eighteenth century, proposal(s) to introduce renting out the princely monopoly on the brandy trade in Inner Austria (i.e., just across the state border from the Natisone catchment) failed because the central authorities believed that this would have impacted too high a percentage of the population. In general, nonetheless, brandy consumption in rural societies of Central Europe increased significantly in the eighteenth and nineteenth centuries.[2835] The relatively early gentian spirit distillation at the beginning of the Early Modern Period is understandable because this brandy has medicinal qualities (see below), but it also resulted in drinking for pleasure, especially because the bitter taste fades with the ageing of gentian brandy or because people addicted to alcohol did not mind drinking bitter brandy for pleasure.[2836]

Large gentian rhizomes and gentian products were also items of medium- or long-distance trade.[2837] In Tyrol, women and men from the Ziller Valley (Germ. *Zillertal*) were particularly well-known gentian gatherers. In the eighteenth century at the latest, people from the Ziller Valley were no longer gathering gentian in their local environments only; they also started gathering gentian rhizomes on a large scale in distant mountain pastures in Tyrol, Bavaria, and Carinthia, and according to mid-nineteenth-century information also in Swabia and Styria. In many places, gathering gentian rhizomes was a source of income not only for gatherers and brandy makers, often building part of the integrated peasant economy, but to a much lesser extent also for the owners of pastures where this activity took place (and for the state in the case of taxes on brandy). This was not the only example of yellow-gentian rhizome gathering transgressing the boundaries of village commons. Smallholders from Ganda (Germ. *Gand*) in the Martello Valley (Ital. *Val Martello*, Germ. *Martelltal*), South Tyrol, for instance, crossed the glaciers in the westernmost Martello

2835 von Hörmann, 1877, 149–150, 152–153; Konersmann, 2005, 404–405; Priesner, 2010, 372–373; Grass, 1950, 169–170; Valentinitsch, 1989, 90, 92; Spode, 2005, 395–396; Beguš, 2015, 80, 189–190; Pfister, 1985a, 121.

2836 von Hörmann, 1877, 149–150, 152–153, and knowledge of an alcoholic in the Julian Alps that would gladly drink brandy infused with mountain arnica.

2837 Mlakar, 2015, 309. For instance, the seigneury of Gornja Radgona (Germ. *Oberradkersburg*), situated in a sub-Pannonian environment several dozen kilometers away from the Alps (Mlinarič, 1979, 11–12), bought 13.5 kg (24 pounds) of gentian for its livestock in 1667 and again more than 12 kg (22 pounds) for its bovines and donkeys in 1669 (Pavković, 2021, 110). Use of significant quantities of gentian rhizomes for livestock there suggests that one of the gentians from the Alps with large rhizomes might have been bought on these occasions for consumption outside from the Alps, although this is not the only option (for more about the uses of yellow gentian in animal husbandry, see below).

Valley to gather gentian rhizomes in the Italian mountains; this activity was interrupted by significant early-nineteenth-century glacier advances starting in the 1810s.[2838]

In the second half of the nineteenth century and the first half of the twentieth century, commercial gathering of a limited selection of species for pharmaceutical enterprises gained importance in many regions of the Alps (not everywhere, though; see examples of large-scale earlier spikenard valerian and large gentian gathering above). This was the era of the beginnings of the pharmaceutical industry and its further developments. In the times dealt with in this volume, the industry was still processing large quantities of gathered wild plants.[2839] Yellow gentian rhizomes and mountain arnica (*Arnica montana*) blooms belonged to the most common items of commercial gathering for pharmaceutical purposes. Oral history interviews carried out in Bohinj in the Julian Alps have proved that yellow gentian was the only (or almost the only) grassland plant species that was uprooted for market on a large scale, although some other montane, subalpine, and alpine plants—including blossoms of mountain arnica, blossoms of wormwood (*Artemisia* sp.), and rhizomes of spignel (*Meum athamanticum*)—were also gathered there commercially, apart from their local uses in the mid-twentieth century,[2840] and the autumn crocus (*Colchicum autumnale*) seed trade was also highly economically relevant (see below).

The following contextualization of this mid-twentieth-century information on yellow gentian gathering in Bohinj demonstrates that broader economic conditions importantly influenced the pressures of gatherers on grassland ecosystems. In this respect, this case study is representative of many parts of the Alps at that time. Moreover, many sites in this mountain chain must to a certain degree have already faced similar increases in grassland gathering stimulated by broader economic conditions some decades or centuries earlier. From shortly after the Second World War until the early 1950s, the population of Slovenia and large parts of the Alps faced major economic difficulties and a low living standard. This was a combined result of global and internal developments, especially after 1929, including the long economic crisis (in Yugoslavia from 1930 onward), the Second World War,[2841] and postwar economic condi-

---

2838 Niederer, 1996d, 191; Jäger, 2006, 30 (chronology of glacier advances supplemented according to Pfister et al., 2018, 268–269); Jäger, 2007a, 222–224; von Hörmann, 1877, 151.

2839 Scheidegger, 2008, 33–34; Friedrich & Müller-Jahncke, 2009, 1083–1084; Mlakar, 2015, 285, 293, 296.

2840 Author's interviews with Boltar, 2018, with Kranjc, 2018, with Kunavec, 2017, and with Mencinger, 2017 and 2020—all in Stara Fužina, and with Grm, 2017, in Srednja Vas in Bohinj.

2841 The general connection between broader economic trends and gathering pressures was inspired by Panjek, 2015b, 108. During the Second World War, gathering wild food for local consumption increased (see Section 8.2.1.1). Moreover, some regions of the Alps, such as East Tyrol, introduced compulsory deliveries of some wild medicinal herbs tions. Weather anomalies such as the Central European summer droughts of 1945, 1947, 1949, 1950 (well known for its severity in Slovenia, for instance), and 1954 added to the tremendous postwar suffering.[2842] In the Slovenian Alps, farmers, except for the smallest, had to sell significant quantities of their crops and a certain number of livestock to the state at a limited price. Industrial products were too expensive in the context of the prices of agricultural products.[2843] Similar in part to Slovenia, in Austria the Second World War and the subsequent crisis were times of chaotic conditions, hunger, and general scarcity.[2844] Consequently, pressures on natural resources rose until the early 1950s; for example, in forests[2845] and in grasslands. Under these circumstances of major economic difficulties, gathering medicinal grassland herbs was particularly relevant in hilly and mountainous areas at places fairly distant from railroad and bus connections, where commuting to industrial centers was not yet usual.[2846] In the late 1940s and in the 1950s, gathered medicinal plants were one of the export items from Slovenia; nonetheless, this was of limited importance in the broader context of export trade. Medicinal plants also played a minor role in export from Slovenia to Western Europe[2847] when the Yugoslav economy gradually opened to the West starting in 1948.[2848] Gathering the rhizomes of yellow gentian (*Gentiana lutea*) was extensive and it was a significant source of income for many families. Apart from the well-known medicinally

---

such as ribwort plantain (*Plantago lanceolata*), coltsfoot (*Tussilago farfara*), and yarrow (*Achillea millefolium*) gathered by women and children for military hospital supplies (Vogl-Lukasser et al., 2006, 51).

2842 Štih, Simoniti, & Vodopivec, 2008, 448–449, 454–455, 457, 461–462, 467, 470, 474; Čepič, 2005a, 890; Čepič, 2005b, 893–899; Lazarević, 2005, 473–479; Brönnimann, White, & Slonosky, 2018, 316–317.

2843 Čepič, 2005a, 891; Čepič, 2005b, 893, 897–899; historian Zdenko Čepič, personal communication. In addition, cited literature and communication explain that from 1948 on those farmers that sold an additional part of their crops to the state (which facilitated supplies of agricultural crops to non-farmers) could buy industrial products more cheaply if such commodities were available.

2844 Beer, 1989, 207–211; Mathieu, 2019, 133–134, 136–137.

2845 Prinčič, 1996, 166.

2846 SI AS 2164, Zadružništvo na Slovenskem, 1922–1962 (Cooperatives in Slovenian Territory, 1922–1962), box 3, unit 445, Report on Fieldwork Promoting Medicinal Plant Gathering, June 25th–26th, 1948. This source presents differences in the role of medicinal plant gathering in local economies in the surroundings of Kamnik.

2847 SI AS 241, Ministrstvo za izvoz in uvod Ljudske republike Slovenije, 1945–1951 (Ministry of Export and Import of the People's Republic of Slovenia, 1945–1951), box 1, unit 16, Record of the Meeting, January 5th, 1948; SI AS 1131, Republiški sekretariat za trgovino Socialistične republike Slovenije, 1962–1965 (Secretariat for Trade of the Socialist Republic of Slovenia, 1962–1965), box 15, unit 623, "Gosad Ljubljana": Issues of Medicinal Herb and Mushroom Export from the Point of View of Our Foreign Trade Development, 1963, 1; SI AS 2164, Zadružništvo na Slovenskem, 1922–1962 (Cooperatives in Slovenian Territory, 1922–1962), box 3, unit 451, To All District Agricultural Cooperative Associations!, May 4th, 1948.

2848 Lazarević, 2022, 105–106, 108, 110.

active substances and the related market demand, the aforementioned summer droughts must have also motivated increased gathering of those grassland organisms whose gathered parts (e.g., the rhizomes) were not severely affected by weather anomalies. For instance, in the years after the Second World War a single smallholder family from the village of Stara Fužina once dug nine full bags of rhizomes of yellow gentian in one single complex of mountain meadows (Trstje, at about 1,375 m, below Mount Tosc), and they pulled this heavy load placed on tree branches down to the valley to sell them. Gathering of this species was widespread. The fact that gentian rhizomes were dug in autumn (usually in September or early October), when farmers had more time, additionally stimulated this activity.[2849] Similarly, in the 1950s, children in the Upper Soča river basin, from the village of Drežnica, for example, collected medicinal herbs, mainly high in the mountains during the summer holidays; they sold them to buy their notebooks for the next school year.[2850] In these specific times of economic crisis, similar gathering was carried out at several places in the Alps. In East Tyrol (in the mountain village of Kartitsch and in its surroundings, for instance), "[t]he district pharmacy was provided with medicinal herbs (e.g., *Arnica montana*) gathered in the region. This was an important economic activity during the Second World War and in postwar times for poor people."[2851] Some other grassland organisms were also gathered commercially in significant quantities in some parts of the Alps. In the Möll Valley (Germ. *Mölltal*) in Carinthia, for instance, Iceland moss (*Cetraria islandica*; see more in Sections 8.2.1.3 and 8.2.2) belonged to commercially gathered organisms in the mid-twentieth century.[2852]

On the one hand, the pharmaceutical industry stimulated gathering of a narrow selection of healing organisms that had already been gathered and used earlier, such as large gentians, mountain arnica, or Iceland moss (see also Section 8.2.1.3). On the other hand, it demanded enormous amounts of target parts of a few grassland species that had not been gathered on a large scale before. Seeds of autumn crocus (*Colchicum autumnale*; Figure 126 and Figure 78 in Section 6.2.2.2) are a characteristic example. Earlier, only very limited uses of autumn crocus in vernacular medicine were known in the Alps, such as its external application for suppressing head lice, and very limited internal uses, mainly against gout; the latter autumn crocus use easily led to lethal poisonings in cases of overdose.[2853] The example of the Slovenian Alps demonstrates that selling autumn crocus seeds to the pharmaceutical industry flourished for long decades in the twentieth century. Experience-based knowledge from Bohinj

[2849] Author's interviews with Boltar, 2018, and with Mencinger, 2017, both in Stara Fužina; see also Meilleur, 1986, 382; Abbet et al., 2014, 631.
[2850] Kurinčič, 2000, 135; Štepec, 2019, 228; Bavcon, Ravnjak, & Praprotnik, 2019, 123.
[2851] Christanell et al., 2010, 65–66 (quote from page 66).
[2852] Machatschek, 1997b, 114–115.
[2853] Mlakar, 2015, 333, 367.

explains that seed capsules were particularly large in wet (damp?) meadows with medium nutrient availability (such as the wetland complex of Blato below Srednja Vas), and the abundance of the species there was high. Gathering took place in the first half of June; that is, before mowing. In the second half of the 1970s, one of the many autumn crocus seed capsule gatherers (an exceptionally industrious gatherer) collected so many seeds in one spring in the meadows around Srednja Vas that the dry seed weight amounted to 13 kg. This adult gatherer collected seed capsules for about three hours a day for two weeks. Additional work was needed to remove the seeds from the capsules and dry them in the air. With the income from 13 kg of dry autumn crocus seeds sold to a pharmaceutical enterprise, she bought a bicycle, a tent for two people, and a food processor. Gathering autumn crocus seed capsules was also a source of income for children. All in all, gathering created an ambivalent attitude toward autumn crocus in the region. On the one hand, it was a significant source of income for some locals. On the other hand, a negative attitude toward the plant persisted with regard to haymaking and animal husbandry, and this was even exacerbated because in several cases the seed gatherers (i.e., not only close relatives of meadow owners) trampled grassland plants in the meadows before mowing.[2854]

Most healing plants gathered commercially were also used locally, to treat people and livestock, as demonstrated in this paragraph by data mainly from the nineteenth and early twentieth century. When scientific medicine and pharmacies were gaining importance in the countryside in the nineteenth and twentieth century, traditional folk medicine was declining, but its relevance was still significant during the times dealt with in this volume.[2855] The following uses of large gentian rhizomes in the Alps in the nineteenth and early twentieth century partly reflect earlier historical practices. The use of gentian rhizomes in the Slovenian Alps to treat some livestock health issues is documented since the seventeenth century, at least.[2856] In Bohinj, for instance, grated rhizome was given to horses against helminths as documented for the twentieth century; some owners of draft horses used the same substance as a regular addition to fodder (in small quantities) to strengthen horses' immunity. Data from the Trenta Valley (also in the Julian Alps) suggest that gentian rhizome was used again human worms as well. Use of *Gentiana lutea* rhizomes or their products against worms at as distant and culturally diverse places as in the Julian Alps and in the Allues Valley (against worms in livestock and humans) proves that this was the case in several parts of the Alps and also

[2854] Author's interviews with Rozman, 2020, and with Grm, 2017, both in Srednja Vas in Bohinj; with Kunavec, 2017, with Mencinger, 2017, and with Kranjc, 2020, all in Stara Fužina. Compare ecological requirements in Aeschimann et al., 2004, vol. 2, 1030–1031.

[2855] Mlakar, 2015, 292–296, 347, 463.

[2856] Pečnik, 2024, 57.

that this practice in fact has a long history, even though it might have evolved independently. In the Julian Alps (data from Bohinj and the Upper Soča River catchment), grated yellow gentian rhizome was also given to cows (mixed with salt) if they were not in heat when expected (it is not clear whether this is an example of the ancient doctrine of signatures, based on the rhizome's shape—see Section 8.2.1.3.1 and Figure 132—or actually effective), and to sick cows—for instance, to promote their appetite. Similarly, in East Tyrol, yellow gentian spirit (diluted) or tisane was often administered against (cattle and most likely also human) problems with the rumen/stomach or bowels, and the use of gentian rhizome against indigestion in livestock is documented in Bohinj. Local veterinarians in Tyrol, interviewed by Brigitte Vogl-Lukasser et al., agree with such a traditional use of yellow gentian tisane because yellow gentian in fact promotes secretion of juices in the pancreas and stomach, stimulates peristalsis in bowels, and so on. Data from East Tyrol demonstrate that yellow gentian was believed to be helpful against colic or flatulence (in the latter case, only partly supported by local veterinarians, who prefer uses of plants such as caraway, *Carum carvi* (see 8.2.1.2), or scented mayweed, *Matricaria chamomilla*, against flatulence), and to promote appetite (again supported by science because yellow gentian promotes saliva production). Similarly, in the Allues Valley and in the Julian Alps (in Bohinj and in the Upper Soča River catchment) an alcoholic drink with yellow gentian rhizome was also used against human indigestion (belching and stomach problems), and the rhizome of large gentians was also known as a human appetizer. The use of gentian brandy as human appetizer is also known in Wallis. In Bohinj, dried and cut yellow gentian rhizome was alternatively infused in cold water and left in a warm place on the oven, and this very bitter liquid was drunk in very small quantities against stomach problems or sore throat. Similarly, in Wallis (e.g., in the Valley of Entremont [French *Val d' Entremont*]) and in Tyrol, gentian spirit was drunk after greasy meals (farmers in the Alps—where areas suitable for cereal cultivation were relatively limited on many farms, but much more land was appropriate for animal husbandry—often ate greasy food, one of the motives being to save home-grown or purchased flour through the calories from fat;[2857] nonetheless this belief is considered erroneous in medicine because gentian spirit after a meal increases indigestion rather than easing it[2858]), in Wallis, at least, also to cure some other health issues, mainly of the gastrointestinal system, but not against pulmonary problems. Contrary to this, more regionally limited uses of *Gentiana lutea* included its administration against pulmonary diseases or diseases of the respiratory system more generally (coughing in general, pneumonia, and allegedly tuberculosis) in the Upper Soča River catchment.[2859] Nonetheless, the

[2857] Wopfner, 1997, 16–19, 189, 375, 459–460.

[2858] Galle Toplak, 2015, 296.

[2859] Bohinj: Novak, 2024, 221, and author's interviews with Boltar, 2018, with Gašperin, 2017, with Kunavec, 2017 and 2021—all in Stara Fužina—and with C. Mlakar, 2017, use of gentian against diseases of the respiratory system was not exceptional—in nineteenth-century Tyrol, application of gentian brandy against coughing and phlegm is reported.[2860] An example without a clear parallel in the cited literature was the use of gentian rhizome, in combination with an unspecified mineral, as an abortifacient by farmers, which was mentioned in the Eastern Julian Alps in the late seventeenth century, but in the documented case this was inefficient.[2861] In nineteenth-century Tyrol, gentian brandy was also used against backache, against what was considered joint pain due to rheumatism, and even as "a universal medicine against all afflictions and diseases,"[2862] which most clearly demonstrates the limits of folk medicine.

Data from Slovenia, representative of many regions in the Alps, show that a general decline in medicinal grassland herb gathering and export was observed by the early 1960s. People were no longer so interested in gathering commercial herbs because the general economic situation in the countryside had improved in the 1950s; in addition, agricultural cooperatives involved in selling herbs were more interested in fostering agriculture than gathering.[2863] Nonetheless, gathering in the Alps was still significant, as the example of autumn crocus has already demonstrated. For comparative reasons, quantitative information mainly referring to Slovenian territories south of the Alps is given, even though these data are not representative of the Slovenian Alps. Despite the aforementioned decline in herb gathering, a regional newspaper reported that the pharmaceutical enterprise Krka was still able to buy about 117 metric tons of herbs, roots, dry fungi, bark, and so on from well over 2,100 gatherers from Slovenia in 1969 (the figure refers to the period from January 1st until November 15th). Because interest in commercial gathering had already decreased by the late 1960s, gathering was stimulated by a lottery allocating lots to those that gathered an amount of plants or fungi that reached a minimum prescribed price when they were sold to Krka from May to October, the first prize being a car, with further prizes included a washing machine, a

---

at the Zajamniki Alp; Upper Soča River catchment: Sosič, 2000, 233, 235, 247; East Tyrol: Vogl-Lukasser et al., 2006, 41, 125, 129, 137–139, 174, 234, 266 (and also 103, 126, 212, 245); see also Galle Toplak, 2015, 296; Tyrol in general: von Hörmann, 1877, 149; Wopfner, 1997, 375; Wallis: Abbet et al., 2014, 626, 628–631 and Appendix C; the Allues Valley: Meilleur, 1986, 292, 382.

2860 von Hörmann, 1877, 149.

2861 Pečnik, 2024, 51–59.

2862 von Hörmann, 1877, 149.

2863 SI AS 1126, Državni sekretariat za blagovni promet Ljudske republike Slovenije, 1958–1965 (State Secretariat for Trade of the People's Republic of Slovenia, 1958–1965), box 32, unit 2294, Permission for Purchases of Medicinal Herbs and Forest Fruits, June 6th, 1960; SI AS 1131, Republiški sekretariat za trgovino Socialistične republike Slovenije, 1962–1965 (Secretariat for Trade of the Socialist Republic of Slovenia, 1962–1965), box 15, unit 623, "Gosad Ljubljana": Issues of Medicinal Herb and Mushroom Export from the Point of View of Our Foreign Trade Development, 1963, 2.

moped, a sewing machine, a chainsaw, a television, and a cauldron for making brandy. These plants and fungi were gathered in almost all regions of Slovenia; nonetheless, the great majority of gathering took place south of the Alps, in southeastern Slovenia, where the factory is also located (in Novo Mesto). At that time, gathered plants and fungi were reportedly the third most important export item of the Krka factory.[2864] During the following half century, this type of gathering decreased significantly.

8.1.2.1 Historical ecological impacts of gathering and farming on populations of large gentian species

The presentation of the ecological impacts of human activities in this section focuses on large gentians as widely gathered plants occurring throughout the Alps. This is a particularly interesting subject because gathering in this case meant uprooting, which represented a certain challenge, not really similar to but also not fully different from the case of spikenard valerian (Section 8.1.1). A long-term historical perspective reveals that human activities had significant and opposite impacts on the occurrence and abundance of gentians.

(1) Extensive expansion of grasslands on former forest grounds in the Alps, in the Neolithic and later (see Sections 2 and 3 for the creation of grasslands in the last millennium) created new suitable sites, so that the occurrence and abundance of large gentians in the landscape increased significantly. In addition, grazing livestock perceive gentians as unpalatable and avoided them, and so their selective grazing behavior facilitates competition of gentians in pasture ecosystems (Figure 134). Moreover, with the only exception of dotted-flowered gentian, occurrence of none of the large gentian species can be high in the alpine vegetation belt; that is, above the potential natural climatic forest line, and even *Gentiana punctata* reaches high abundances mainly in the subalpine belt, as explained in the introduction to Section 8.1.2. Consequently, all large gentians profited from the creation of extensive grassland areas in upper montane and subalpine zones, including anthropogenically caused lowering of the forest line. Thus, pollen analyses at some places in the Alps, such as the Ötz Valley (Germ. *Ötztal*), document a relative increase in gentian pollen as early as the Neolithic.[2865] This process was also relevant in later centuries at many sites, unless gathering and removal significantly decreased gentian's occurrence and abundance, as explained in paragraphs (3) and (4).

(2) Historical gentian rhizome gathering was sustainable in several cases, and so it functioned in the long run. Nonetheless, gathering pressures changed over the course of history; in some places sustainable and unsustainable gath-

[2864] "Sreča je izbrala Krkine nagrajence," 1969, 20.
[2865] Bortenschlager, 1999.

Figure 134: Relatively high abundance of one of the large gentian species (*Gentiana* sp., large green leaves) in a summer mountain pasture in a forest clearing in the Julian Alps, slightly below 1,600 m. Note that gentian leaves were not grazed. Photo: Žiga Zwitter, August 2017. These specimens did not blossom that year.

ering coexisted, and in some others history saw a transition from sustainable to unsustainable gathering or alternation between the two. Traditional farmers in Bohinj that participated in large-scale yellow gentian gathering in the mid-twentieth century, as presented above, observed that unless gathering was excessive (i.e., only the thick part of the rhizome was gathered and the rest left in the soil) yellow gentian re-sprouted from parts left in the soil. Based on their knowledge, informed gathering could even increase the abundance of this species.[2866] This kind of knowledge existed in several parts of the Alps; Ludwig von Hörmann emphasized in his 1877 publication on Tyrol that large gentians regenerate if the gatherer cuts the underground part in a manner such that only the thick part of the rhizome is gathered, but regeneration takes several years.[2867] Gentian regeneration from roots remaining in the soil after moderate gathering is indeed possible (at least if soil openings created during the harvest

[2866] The identity of this interviewee from Stara Fužina is kept confidential; she is one of the interviewees given in the final list of references, born in the first half of the twentieth century.

[2867] von Hörmann, 1877, 151.

are closed),[2868] and, indeed, this is a long process. Namely, gentian roots grow very slowly in the first years (with significant differences within different yellow gentian populations and between them), and so some vegetation periods are needed for a harvestable rhizome.[2869] At some places in the Alps, gentian rhizome gathering was carried out such that after a year of gathering this activity was banned for some years in the same parcel to allow a certain degree of restoration of the gathered populations of large gentians.[2870] Timing the gathering also mattered. Dotted-flowered (*Gentiana punctata*) and purple gentian (*Gentiana purpurea*) bloom between July and September, whereas yellow gentian (*Gentiana lutea*) blossoms slightly earlier, between June and August.[2871] Gathering in autumn thus allowed not only vegetative propagation from roots remaining in situ, but also dispersal of ripe seeds. Mid-twentieth-century data from Bohinj, and probably roughly from the same time from the Valley of Entremont and the Allues Valley, mention gathering in autumn, in September or early October.[2872] In several parts of the Alps, such as the Allues Valley or Bohinj, strong deagrarianization shortly after the Second World War greatly decreased the number of people at temporary summer mountain settlements (the elderly that only helped others, and herders keeping non-milking livestock had previously had much time for gentian gathering). Thus, the example of Bohinj demonstrates that, even though yellow gentian rhizome gathering still mattered in the rural economy of the region at that time, the many tasks to be performed by a limited workforce did not allow farmers to spend time gathering earlier than in autumn.[2873] This was important from the ecological perspective to allow seed dispersal. At Galtür, Tyrol, gathering inspired by historical experience (and not simply following history), with late gathering dates, is nowadays recognized as a good, sustainable practice, acceptable from the point of view of result-oriented environmental conservation (see Section 10.2.3). However, late gathering dates were not the only gathering timing earlier in history, as presented in paragraph (3).

(3) Excessive gathering of gentian rhizomes had adverse ecological consequences. Gentian gathering pressures were the strongest in past centuries, during times of strong demographic growth and rising agricultural density in the eighteenth and nineteenth centuries at least (see Section 3.1). At that time,

[2868] *Austrian Commission for UNESCO*: Immaterielles Kulturerbe: Bräuche, Wissen, Handwerkstechniken: Wissen um die Standorte, das Ernten und das Verarbeiten des Punktierten Enzians & Bewerbungsformular, 2013.

[2869] Franz & Fritz, 1978, 308–309, 311.

[2870] Gentner, 1932, 69.

[2871] Aeschimann et al., 2004, vol. 2, 10–13.

[2872] Author's interviews with Boltar, 2018, and with Mencinger, 2017, both in Stara Fužina; Meilleur, 1986, 382; Abbet et al., 2014, 631.

[2873] Author's interviews with Boltar, 2018, in Stara Fužina; Novak, 1978, 55; Meilleur, 1986, 30–31, 378, 383, 386–388; Wopfner, 1997, 375.

gathering pressures were particularly strong in places (the Ziller Valley is a well-known case) where gentian gathering was not only carried out as a supplementary economic activity, but a few people—usually widows without any other significant source of income, young maidservants, junior forestry workers, charcoal burners, and so on—specialized or nearly specialized in gentian gathering. They spent much of summer and early autumn in the mountains gathering on a large scale, with an early start. Some of them made a good living through the gentian business.[2874] People gathered such quantities of underground gentian parts and stored them in the mountains in the summer and early autumn that they had to carry them gradually down to the valley in late autumn, even though the rhizomes had dried in the mountains and dry rhizomes weigh only a fifth of the weight of fresh ones. In the mid-nineteenth century, a lucky forager would gather about 110 kg (two hundredweights) of gentian rhizomes per day in the richest sites of the Stubai Valley, where gentians were still outstandingly abundant at that time, or around 30 kg a day in some other parts of the same valley. Indeed, historical overexploitation of populations of large gentians was noticed in Tyrol in the mid-nineteenth century. Many gatherers did not cut the underground part below the thick rhizome, and so they gathered excessively. Populations of gentians gathered unsustainably on a large scale were decimated by the mid-nineteenth century in some valleys, such as the Lech Valley (Germ. *Lechtal*) in northern Tyrol and the Martello Valley in today's Trentino–Alto Adige (South Tyrol). Gentian populations in the Ziller Valley that had been so famous for large-scale gentian gathering were smaller and smaller year after year.[2875]

Certainly, people from the Ziller Valley were not the only ones gathering unsustainably in several cases, as demonstrated by the example from the southeastern Alps, the territory that some of the sixteenth- and seventeenth-century data mentioned above refer to, and places out of reach of gathering from the Ziller Valley, as presented above. In the mid-nineteenth century, the botanist Valentin Plemel observed (objectively?) that by that time gathering had significantly decreased the number of yellow gentian specimens in the Julian Alps above the Bohinj Basin.[2876] As presented above, large-scale gentian gathering in Bohinj had not ended by that time. Currently, yellow gentian is a protected species,[2877] designated as vulnerable on the red list of endangered species in Slovenia, Austria, and Germany. Large-scale mid-twentieth-century gathering, which was highly likely partly excessive, must have mattered in this respect.[2878]

2874 Jäger, 2007a, 222–224; von Hörmann, 1877, 150–151.

2875 von Hörmann, 1877, 146–148, 150–152.

2876 Praprotnik, 2015, 232, 235, 241–242.

2877 "Uredba o zavarovanih prosto živečih rastlinskih vrstah", 2004, 5939.

2878 "Pravilnik o uvrstitvi ogroženih rastlinskih in živalskih vrst v rdeči seznam," 2002, 8893, 8901.

Where gathering was excessive, populations of yellow gentian became small and isolated; Kéry, Matthies, and Spillmann have proved that this additionally led to a significant decrease in yellow gentian's (*Gentian lutea*) fecundity—an additional challenge for regeneration of stands.[2879] Too early a start to the season for large-scale gathering also mattered. Whereas the history-inspired sustainable gentian uprooting at Galtür nowadays takes place in October, this was not the only historical timing for gathering gentian rhizomes. At several places in the Alps, this activity started as early as summer and continued until (early) autumn in previous centuries.[2880] In Galtür, for instance, dotted-flowered gentian (*Gentiana punctata*) rhizomes were dug not only after the second-crop hay harvest (see Section 6.1.2.3), but also in late July, after the first-crop hay harvest, according to data from the first half of the twentieth century (see Section 6.1.2.2).[2881] Readers are already familiar with later blossoming of dotted-flowered and purple gentian compared to yellow gentian. Data from the Central Alps of Galtür, where the two late-flowering gentian species occur,[2882] demonstrate that, looking at Alps as a whole, not only gathering yellow gentian that blossoms slightly earlier but also gathering later flowering species started as early as summer. Such early gathering had negative ecological impacts in terms of decreased generative propagation, which could have been avoided if gathering had been carried out later.

(4) Moreover, despite the advantages of large gentians as esteemed healing plants and, consequently, also a source of income, yellow gentian (*Gentiana lutea*), at least, was intentionally suppressed as a grassland "weed" in meadows and better pastures in some parts of the Alps, such as the Allues Valley in the early twentieth century, to increase quantities of fodder appropriate for livestock (Sections 6.6.1.1 and 6.6.2). Yellow gentian gathering there focused on lower-quality pastures, where the species was not suppressed.[2883]

(5) After the end of the study period of this book, abandonment of mowing and grazing at extensive gentian sites, ongoing ecological succession (Sections 2 and 9), and, less frequently, grassland farming intensification (Sections 1.2 and 10.1.2.1) carried out in a limited area of grasslands at elevations of gentian sites have become some of the main triggers for the decrease in occurrence and abundance of large gentian species in the Alps.

[2879] Kéry, Matthies, & Spillmann, 2000.
[2880] Jäger, 2007a, 224; Niederer, 1996d, 191.
[2881] *Austrian Commission for UNESCO:* Immaterielles Kulturerbe: Bräuche, Wissen, Handwerkstechniken: Wissen um die Standorte, das Ernten und das Verarbeiten des Punktierten Enzians – Bewerbungsformular, 2013.
[2882] Juen senior, 2014, 56.
[2883] Meilleur, 1986, 270, 292.

8.1.3 Plants from the Alps for human pleasure: from decoration to Asian tea surrogates

8.1.3.1 Discontinuity in aesthetic perceptions: the example of edelweiss

In the course of history, the perception of the aesthetic qualities of certain plant species occurring in the Alps has changed greatly; this is particularly marked in the case of edelweiss (*Leontopodium alpinum*). Prior to the nineteenth century, this species was esteemed among the local population of many parts of the Alps, but mainly—at least in some places exclusively—for its medicinal qualities. Most widespread was its use against stomach problems and against intestinal diseases, such as intestinal colic and diarrhea (thus the old German name *Bauchwehblüml* 'stomachache wort'),[2884] but its folk uses included treating some other health problems as well. In the Julian Alps, for instance, edelweiss tisane was administered against fever and to treat some livestock illnesses according to data from mid-twentieth century, meaning that edelweiss was gathered in considerable quantities (a big bag per farm, for instance) as a medicinal herb.[2885] Information from nineteenth- and early twentieth-century Tyrol demonstrates that some farmers used edelweiss for human and cattle stomachache or diarrhea, to treat human pulmonary tuberculosis and (eighteenth-century information) also in magic—as incense in exorcising evil spirits.[2886] Similarly, in Entremont in Wallis, edelweiss was administered against cough and bronchitis, apart from its application against diarrhea.[2887] The role of edelweiss varied regionally: the examples above have demonstrated its prominent role in folk medicine in some parts of the Alps; in contrast, it was not perceived as one of the most important plants of folk medicine in Switzerland,[2888] even though there are many edelweiss sites in the Swiss Alps.[2889]

Until the nineteenth century, the aesthetic role of edelweiss was subordinate, as far as is known, in several regions of the Alps, at least. The Swiss ethnologist Arnold Niederer, who carried out research in some regions of the Alps, determined that "mountain farmers had not really liked edelweiss, before they read edelweiss-novels [i.e., not before the second half of the nineteenth century; see below]. For them, it had been the *Bauchwehblume* ['stomachache wort'] . . . ."[2890] Similarly, in 1881, a Swiss mountaineering journal article

[2884] Niederer, 1996d, 191–192; Scheidegger, 2008, 9, 11, 13, 58; Novak, 2024, 221.

[2885] Bohinj: Ledinek, 1999, sections Prehranjevanje, Prosti čas, and Nezgode in bolezni; author's interviews with Kunavec, 2017, in Stara Fužina, and with C. Mlakar, 2017, at the Zajamniki Alp; the Upper Soča catchment area: Sosič, 2000, 234–236 (additionally documenting its use against "male health issues").

[2886] Vogl-Lukasser et al., 2006, 178, 182, 234; Scheidegger, 2008, 9, 12–13.

[2887] Abbet et al., 2014, Appendix C.

[2888] Scheidegger, 2008, 9, 12–13.

[2889] For instance, a selection of sites is recorded and mapped in *Global Biodiversity Information Facility: Leontopodium alpinum* Cass.

[2890] Niederer, 1991d, 191–192.

Figure 135: Edelweiss in a subalpine grassland, mountain ridge east of Mount Vogel in the Julian Alps. Note that not every edelweiss inflorescence appears as appealing as the ones in this picture, and personal perceptions also differ. Thus, in 1881, the American writer Mark Twain, stimulated by the thriving edelweiss cult, noted his opinion that "the ugly Swiss favorite called Edelweiss . . . is not attractive, and it is not white. The fuzzy blossom is the color of bad cigar ashes, and appears to be made of a cheap quality of gray plush."[2891] Photo: Mojca Osvald, July 27th, 2016. © Mojca Osvald.

criticizing the newly developed edelweiss cult stated that, two decades earlier, edelweiss had been ignored by the inhabitants of Swiss Alps, or it at least had not received any more attention from them than other wild herbs. The German name *Edelweiss*, reflecting the appreciation of this species' inflorescence's appearance, was first published in the Eastern Alps, as determined by Tobias Scheidegger, in 1785, in an interview by the naturalist Karl Ehrenbert von Moll with a farmer from the Ziller Valley. *Edelweiss* is believed to be the expression actually used in the Ziller Valley at that time (and not just the writer's or editor's construct) because there are more proofs of this term designating this plant species in the broader region of the High Tauern (Germ. *Hohe Tauern*) as early as the late eighteenth to early nineteenth century. However, several (other) folk names for edelweiss were also used in the (Eastern) Alps at that time, not praising the inflorescence's appearance (see above).[2892] A preliminary result

[2891] Scheidegger, 2008, 26 (quote), see also pp. 32 and 41.
[2892] Scheidegger, 2008, 5–37, 48–50, 53–60.

from the southeast Alps, a territory where edelweiss gathering later became particularly extensive in some regions (see below), identified a description of a mountaineer that visited the Kamnik–Savinja Alps (now in northern Slovenia) in 1795. He wore a hat with a bouquet of common juniper (*Juniperus communis*) fixed to it, not edelweiss. However, this was reportedly a means of magical protection against skin inflammation,[2893] and so the influence of aesthetic perceptions on plant selection is unclear in this case.

The generally changed aesthetic perception of edelweiss was influenced by the changed attitude of the foreign upper class, and gradually also the middle class, to the Alps in the eighteenth and, more generally, nineteenth century. At that time Alpine summer tourism developed, which experienced a marked increase starting in the late eighteenth century—and particularly after the mid-nineteenth century, when railway connections made trips to the Alps much quicker and cheaper. Since the late nineteenth and early twentieth century, the times of flourishing *Belle Époque* Alpine tourism, a couple of cogwheel railway routes, mainly in Switzerland, have even made possible an effortless ascent to high valleys, passes, and viewing summits at usual elevations of edelweiss sites. However, in general, more comfort in everyday life made some exertion on vacations attractive to the middle class, and in their view it was particularly worth making the effort in a natural environment. In the second half of the nineteenth century, the activities of mountaineering clubs in towns and cities created the edelweiss cult. This increased interest in mountains and in edelweiss as a wildflower happened during times when a significant part of the bourgeoisie, dissatisfied with some negative impacts of industrialization and strong increase in urbanization, established movements aiming at (re-)introduction of a lifestyle "closer to nature." With the developments presented above, the Alps gradually became a European "central model" for a "beautiful landscape", perceived, among other aspects, "as a kind of a large garden." After the Napoleonic wars, nineteenth-century Alpine tourism corresponded with the *Biedermeier* style and its later impacts—*Biedermeier* is famous for its further developments in bouquet design from fresh or dry flowers, including edelweiss. In the nineteenth century, there was a general change in aesthetic perception, recognizing edelweiss as the most—or at least one of the most—beautiful plants of the Alps. Thus, the novel *Edelweiss* (1861) by one of the most popular German writers of the time, Berthold Auerbach, is likely the first prose work attributing a prominent role to edelweiss. At the same time, Auerbach's novel is likely the first piece of literary art presenting a dry edelweiss as a keepsake. Influenced by broader historical developments, edelweiss gradually also became perceived among the local population as the most (or one of the most) beautiful wild plant(s)—a position previously reserved for alpenrose (*Rhododendron* sp.), for instance. The motif of edelweiss gradu

[2893] Baš, 1992, 222.

Figure 136: Painting of a bouquet of mountain flowers by the local folk artist Angela Hodnik (born 1915) on the exterior wall of a cottage in a montane meadow on Vogar Hill (1,100 m; i.e., 550 m above the village) in Bohinj, on a trail leading to well-known mountain hiking destinations in the eastern Julian Alps. The central position and large size give edelweiss the most prominent role in this work of art: the brown spots in the center are the middle of edelweiss's inflorescence; note also the edelweiss's white involucral leaves not obvious against the pale background. The roles of gentian (*Gentiana clusii* or *Gentiana acaulis*), alpenrose (*Rhododendron ferrugineum* or *Rhododendron hirsutum*), and black vanilla orchid (*Nigritella rhellicani*) are not as prominent. The motif is likely a result of the changed aesthetic perception of this plant by the local population. It was painted around 2000, but the motif is firmly rooted in twentieth-century tradition.[2894] Photo: Savina Zwitter, 2018.

ally found its way into folk art; for example, as embroidery on some folk costumes (starting in the late nineteenth century), as burnt decoration on wooden objects, mainly spoons,[2895] and in painting (e.g., in wall painting, Figure 136,

[2894] Thanks to Marjeta Hodnik, who showed this painting and identified her great-aunt as the artist when the author of the text helped her family turn hay in this meadow in 2018. She also explained when the painting was made.

[2895] Meilleur, 1982, 167; Meilleur, 1986, 389, 395; Thomas, 1987, 260–268, 281–284; Elias, 1994, 405–407; Niederer, 1996d, 191; Hanhart & Sonderegger, 1998, 55–56; Ferle, 2004, 33; Hachtmann, 2007, 30–33, 36–38, 59–62, 71–76, 86–88, 94–97, 100–105, 109–112, 114–116, 170; Scheidegger, 2008, especially 5–6, 9–37 (fourth quote from p. 28), 48–50, 53–60; Westwood, 2009, 136–141 (in addition: *Le tramway du Mont-Blanc, 110 ans d'histoire*, 2021); Bätzing, 2015, 138, 172–173, 175, 183, 237–239, 244; Mlakar, 2015, 390–392; Mathieu, 2019, 94–96, 99–105, 108–109 (second and third quotes from and painted decoration on wooden objects, such as souvenir plates, Figure 137). The recollection of a farmer born in Bohinj in the late 1930s that she used to find dry edelweiss specimens as bookmarks in old prayer books—thus deeply pious predecessors were using them for a special purpose and at the same time enabling long preservation of herbs[2896]—might also be a reflection of the changed attitude toward edelweiss. Namely, literacy only spread generally in this region in the late nineteenth century (see Section 8.2.2), and so prayer books were not frequently kept by farmers earlier. As shown in the following paragraphs, edelweiss became an esthetically highly esteemed plant more than Europe-wide by the late nineteenth century, and so it is no surprise that the same manner of expressing respect toward it by keeping a specimen or two in a prayer book also became a habit of traditional Csángó farmers in Ghimeş (Hung. *Gyimes*) in the Eastern Carpathians, as revealed by twentieth-century data.[2897] This parallel suggests that edelweiss specimens were kept in prayer books in many parts of the Alps, not only in the main study area, since the nineteenth century at the latest.[2898]

Locals and tourists picked large numbers of trendy edelweiss. Concerning the tourists—both men and women—looking for plants, identifying and gathering them was a popular hobby of the bourgeoisie in the last decades of the nineteenth century, rooted in Early Modern dealings with natural history "for pleasure, curiosity and emotional satisfaction." This was another stimulus for edelweiss gathering, apart from the trendiness of this species. The locals sold the vast majority of specimens gathered: as souvenirs to Alpine visitors (flowers and bouquets) and to merchants exporting edelweiss, many to distant markets for decorative purposes—as wall decoration, for instance. Local inhabitants also commercially uprooted specimens to be planted in fashionable rock gardens of Alpine plants that were built all around Europe from the 1870s onward.[2899] Obviously, an Alps-like rock garden in the lowland often belonged to an English-style garden that "opted for increasingly natural forms."[2900] In the Western World, domestic flower gardening predominately had an aesthetic motivation, and in particular the repertoire of species grown in small-scale gardens expanded enormously during the Modern Period—a process that the English cultural and environmental historian Kate Thomas called a "gardening revolution"[2901] or, better, a thorough gardening evolution, to take into account

p. 109), 111, 113, 120, 148–149; Winiwarter & Bork, 2019, 130–131 (first quote from p. 130); on Auerbach, see Martini, 1953.

2896 Author's interview with Kranjc, 2020, in Stara Fužina. On nineteenth-century prayer books, see Polajnar, 2008, 12.

2897 Babai, Molnár, & Molnár, 2014, 85.

2898 Compare Scheidegger, 2008, 25.

2899 Thomas, 1987, 281–284 (quote from p. 281); Scheidegger, 2008, 23–25, 28–29, 48–49; Niederer, 1996d, 191; Peternel, 2022.

2900 Thomas, 1987, 262–268 (quote from p. 262); Scheidegger, 2008, 29.

2901 Thomas, 1987, 224, 226, 228–233, 239.

its gradual nature. In some parts of Europe, England in particular, it was in a mature stage by the nineteenth century,[2902] when edelweiss became fashionable, as explained above; in several parts of continental Europe, decorative gardening with a rich species repertoire was becoming widespread just at that time. Selling exotic ornamental plants "first from southern Europe and the Levant, then from [the] America[s], the West Indies and South Africa" to distant places in Europe was a widespread phenomenon,[2903] and so edelweiss from the Alps became part of this. However, taking into account the European locations of the gathering sites, edelweiss was a latecomer to this business, which had much to do with the general chronology of Alpine tourism developments and the related spread of the appreciation of the Alps (see above). The demand for edelweiss for rock gardens in the Western World nonetheless became high in the second half of the nineteenth century and in the early twentieth century not only in regions where flower gardening was only spreading highly significantly at that time but also in those with a rich tradition in this activity. Namely, species selection in decorative gardening has been subject to the strong impacts of ever-changing fashion preoccupied by rarity and novelty.[2904] Edelweiss from the Alps (and possibly from elsewhere because this species also occurs in several other places in Eurasia, such as the Carpathian Mountains and Pyrenees) was sold, for instance, in England, at distant places in France and Germany, in the United States, and closer to the Alps, as in lower parts of Switzerland and other Alpine countries in the last third of the nineteenth century. British, French, German, American, and Swiss tourists were all well represented in the Swiss Alps in the late nineteenth century, when destinations in Switzerland attracted two-thirds of all tourists in the Alps; the popularity of Alpine tourism definitely stimulated the demand for edelweiss in the mentioned countries.[2905]

As explained in Section 3.1, the second half of the nineteenth century, the time when the edelweiss cult evolved, probably saw the highest number of people ever making a living in the Alps largely from farming, and so the edelweiss business was a highly welcome source of income. Most commercial edelweiss gathering took place in the southeastern Alps, known for its high abundance of beautiful specimens (see caption to Figure 135). Namely, *Leontopodium alpinum* occurs most frequently in base-rich sites on carbonate bedrock; it avoids acidic sites on silicate bedrock. Consequently, edelweiss gathering in the Central Alps was not comparatively extensive. Nonetheless, there are quite a few edelweiss sites almost Alps-wide, and so the southeastern part of the mountain chain was only the main gathering area. Apart from small local merchants, large enterprises were involved in the edelweiss business in

2902 Thomas, 1987, 224, 226, 228–229, 239.
2903 Thomas, 1987, 226, 229, 239.
2904 Compare the general presentation of impacts of fashion on the history of flower gardening in Thomas, 1987, 231–233.
2905 Scheidegger, 2008, 23, 29, 41; compare Thomas, 1987, 260–261.

the Alps. For instance, the Viennese company Michael Frank am Neubau reportedly had about three million freshly gathered edelweiss blooms in stock at a certain point in the early twentieth century. Vienna is located at the foot of the Northern Limestone Alps, and at that time it was well connected with the southeastern Alps through the railroad network. Similar enterprises existed in Switzerland as well. For example, one in Lucerne (Germ. *Luzern*) offered a 10% discount if one purchased five hundred specimens, a 20% discount for one thousand specimens at the same time, and a greater discount if one bought even more; allegedly, the enterprise sold edelweiss imported from other countries. Or, once per week a market of uprooted rare Alpine plants was held in Geneva (French: *Genève*) in the early 1880s. Even though marketable edelweiss garden cultivation appeared in some regions of the Alps in the second half of the nineteenth century, so that a significant number of exported blossoming edelweiss specimens were not gathered wild, gathering was still highly relevant. Namely, herders and hunters in the Alps knew edelweiss sites, and they were satisfied with a low purchase price. To offer only one concrete example from a microregion in the southeastern (limestone) Alps where the edelweiss business was particularly lucrative: in the village of Spodnji Log below Mount Mangrt in the Julian Alps, the edelweiss trade was not only regional, but it was export oriented. Some houses in Spodnji Log were designated edelweiss depots in the late nineteenth and/or early twentieth century. Merchants involved in this business purchased edelweiss from the inhabitants of the region, mainly youth, who collected the specimens in the mountains. Market-oriented edelweiss gathering was also a frequent activity for adult women in some places in the southeastern Alps, constituting a relevant source of their income. The case of a shoemaker from Spodnji Log is instructive because it helps understand the role of edelweiss gathering despite the option of edelweiss cultivation. He first grew marketable edelweiss in his garden in the valley. However, having realized that herders and hunters were willing to provide him with a full backpack of edelweiss gathered at high elevations if he repaired a pair of their shoes in return, he switched to gathered specimens in his business oriented toward long- or medium-distance trade.[2906]

Some other grassland plants from the Alps were also gathered and sold at markets for ornamental purposes, among others the martagon lily (*Lilium martagon*) and the orchid lady's slipper (*Cypripedium calceolus*, the herb with the largest flower of all European wild orchids), according to data from the Allues Valley.[2907] A selection of examples from the Slovenian Alps includes purple

[2906] "Edelweissexport," 1893 (thanks to the historian Marija Mojca Peternel for her recommendation of this newspaper article); Meilleur, 1982, 167; Meilleur, 1986, 389, 395; Niederer, 1996d, 191; Keller, 2004, 34; Scheidegger, 2008, 28–29, 31, 49; Brückmüller & Stekl, 1998, 67 (this railroad network map is not always precise, so see also Černe, 2001, 161); Aeschimann et al., 2004, vol. 2, 452–453.

[2907] Meilleur, 1982, 167; Meilleur, 1986, 389, 395; compare Dakskobler et al., 2019, 89.

Figure 137: Apart from real, gathered edelweiss, arts and crafts have also played a significant role in edelweiss souvenir manufacturing in the Alps, being another type of edelweiss-related income for local inhabitants:[2908] here, a bouquet from the Alps with edelweiss (*Leontopodium alpinum*), gentian (*Gentiana clusii* or *Gentiana acaulis*), and alpenrose (*Rhododendron ferrugineum* or *Rhododendron hirsutum*) as a wooden souvenir plate decoration. These three blossoming plants are typical for souvenirs.[2909] Note the typical position of edelweiss in the center (just like in Figure 136). This souvenir was bought at Obergurgl, Tyrol (a village at 1,930 m), by tourists from Slovenia around the mid-twentieth century or earlier. Zwitter family collection. Photo: Tomaž Zwitter.

cyclamen (*Cyclamen purpurascens*)[2910] and Salisbury's daffodil (*Narcissus poeticus* subsp. *radiiflorus*); the latter used to be sold in Yugoslavia on the markets in Ljubljana and as far away as the capital Belgrade, some 600 km distant, by railway employees and their children living in the valley below the grasslands

[2908] Scheidegger, 2008, 25.
[2909] Scheidegger, 2008, 25.
[2910] Bavcon, Ravnjak, & Praprotnik, 2019, 123.

in the Western Karawanks where this plant species abounded[2911] (see Section 6.2.2.3). Nonetheless, contrary to other grassland flowers gathered in the Alps and sold for decoration, edelweiss, which keeps its unchanged appearance if dried properly, was particularly suitable as ornamental plant, also making possible long-distance trade and decorative use for several years.[2912]

Thus, the aforementioned shift in aesthetic perception had important ecological consequences because it resulted in large-scale edelweiss gathering. It is clear that the abundance of edelweiss locally or regionally markedly decreased due to the large-scale gathering. It is, however, not clear how endangered it actually became regionally in the Alps at that time. One should consider that edelweiss habitats were anthropogenically much expanded in history, mainly when woody vegetation, such as dwarf mountain-pine (*Pinus mugo*) and trees, were cleared in the subalpine vegetation belt to create pastures and non-manured subalpine meadows lower than the potential climatic forest line (Sections 2 and 3.1), especially at sites on carbonate bedrock. This means that edelweiss habitats in the Alps were much more extensive during flourishing Alpine summer tourism, which stimulated edelweiss gathering in the second half of the nineteenth and first half of the twentieth century, than they are now. Namely, shrubs and trees have recolonized many subalpine grasslands with a high abundance of edelweiss since agricultural uses of these grasslands have ceased (Sections 2 and 9).[2913]

It was exactly during large-scale edelweiss gathering that the idea of nature protection arose in the industrial society of European towns and cities, and it quickly spread around the continent starting about 1880 or 1890. Step by step, this influenced the history of edelweiss gathering. Exceptionally, mountaineering clubs at that time suggested that their members stop picking and buying edelweiss and promote such thought among their acquaintances, such as the quickly growing German and Austrian Alpine Club in 1874; that is, only five years after the German (and twelve years after the Austrian) Alpine Club was founded, and one year after their unification. At the legislative level, uprooting edelweiss was first prohibited in 1878 in the Swiss canton of Obwalden. At that

---

[2911] Praprotnik, 1993a, 219; Torkar Tahir, 2012, 134.

[2912] Scheidegger, 2008, 24–25, 28–29. In 1910, the botanist Ernst Moriz Kronfeld described the drying practice, ensuring that dry edelweiss for bouquets kept its form and color, as follows: edelweiss specimens were set on a wooden board or cardboard, with stems placed in holes in the board and inflorescences lying on the board. The involucral leaves of the inflorescences were nicely arranged, and then the inflorescences were covered with cotton wool and a thin layer of sand. The edelweiss then dried on a sunny window or outside in a sunny place for six to eight days (Scheidegger, 2008, 29). A slightly different practice was used by local farmers in Bohinj in the twentieth century: the inflorescence was cautiously set and wrapped in a piece of paper, with the stem projecting though a hole in the paper, and the flowers dried whole hung upside down (the identity of this interviewee from Stara Fužina has been kept confidential).

[2913] See the following footnote.

time, the main motive for this measure was not complex ecosystem considerations. The outcomes of the romantic edelweiss cult belong to the main motives for this edelweiss protection act; protection of nature "from humans and for humans" has played a prominent role in the history of nature conservation in Europe. From the commercial viewpoint, this meant interest in the long-term availability of edelweiss as a "resource", and so gathering had to be sustainable. Moreover, the emerging edelweiss cult was not driven solely by aesthetic perceptions; a rich plant symbolism appeared, such as edelweiss as an allegory of romantically perceived nature in the Alps, edelweiss as a symbol of the Alps during rising modern nationalisms, in which the Alps played a significant role (more in Austria than in Switzerland), and, last but not least, edelweiss as a token of love (possibly only since the times of the edelweiss cult, in some parts of the Alps, at least). All these had a certain impact on early legislative edelweiss protection, again with significant differences between countries; nationalistic connotations of plant protection were more widespread among alpine hikers in Austria and Germany than in Switzerland, for instance. Protective acts elsewhere in the Alps followed gradually, with significantly diverse chronology; for example, in Carniola in 1898 (in this case at the same time as protection of Blagay's daphne, *Daphne blagayana*, a plant with a certain market demand, although not comparable to that of edelweiss, and without general symbolism as rich as that of edelweiss). However, gathering wild plants in the mountains, including edelweiss, picking its above-ground parts, and partly also uprooting it, stimulated by demand from buyers, remained part of the integrated peasant economy in many places in the Alps (although with significant differences within the mountain chain) as a supplementary economic activity, providing a source of income to the local population until the mid-twentieth century.[2914] In the Alps of the Kingdom of Yugoslavia, for instance, protected plant species were sometimes illegally gathered on a large scale, not only by farmers but also by vacationers.[2915]

8.1.3.2 Plants evoking gourmet delight: the example of Swiss tea

The market for tea mixtures consisting of plants gathered in the Alps developed in the eighteenth and nineteenth centuries, when Asian tea gradually became a popular drink of the nobility and bourgeoisie in Europe. Asian tea was much more expensive than coffee at that time, making trade in Asian tea substitutes a lucrative business.[2916] Consequently, tea substitutes became relevant export products in some regions of the Alps. Judging from an unfinished document

[2914] Scheidegger, 2008, especially pp. 6, 11, 14–22, 28–50, 55, 77 (and Niederer, 1996d, 191–192); Bätzing, 2015, 99, 237–239, 244; Praprotnik, 2015, 168–172; Mathieu, 2019, 100–101, 106–109; Kupper, 2021, 130, 134, 138, 144, 173; Aschwanden et al., 2022, 219–222, 224; Hasenöhrl & Groß, 2022, 193 (quote), 199; Kupper & Wöbse, 2022, 5; Peternel, 2022; Novak, 2024, 248–249. These references refer to two paragraphs, this one and the previous one.

[2915] Erhatič Širnik, 2011, 171, 183.

[2916] Sandgruber, 1989, 106; Abbet et al., 2014, 630.

presenting part of the commercial economy of the canton of Bern in 1780s that exceeded local consumption, two types of tea mixtures from the canton of Bern were sold in significant quantities in Basel, Strasbourg (Germ. *Straßburg*), and the Holy Roman Empire (in Central Europe), at least: Swiss tea substituting for expensive Asian teas in Europe, and another medicinal tisane from gathered wild plants.[2917] Based on nineteenth-century data, Swiss tea consisted of coltsfoot (*Tussilago farfara*), wild thyme (*Thymus serpyllum*), claries or common sage (*Salvia* sp.), meadowsweet (*Filipendula ulmaria*), milfoil/yarrow/sneezewort (*Achillea* sp.), wild marjoram (*Origanum vulgare*), hyssop (*Hyssopus officinalis*), betony (*Stachys officinalis*), wall germander (*Teucrium chamaedrys*), rosemary (*Rosmarinus officinalis*, not gathered wild in the Swiss Alps[2918]), mountain everlasting (*Gnaphalium dioicum*), glossy scabious (*Scabiosa lucida*), woodruff (*Asperula odorata*), ground-ivy (*Glechoma hederacea*), mountain arnica (*Arnica montana*), and alpine marguerite (*Tanacetum alpinum*), and either mountain avens (*Dryas octopetala*) or heath speedwell (*Veronica officinalis*) was often added.[2919] An increase in the export of Swiss tea aimed at consumers elsewhere in Europe during the eighteenth and nineteenth centuries means that gathering pressures on ecosystems in the Alps rose at that time.

8.1.4 Plant gathering for scientific botany

The development of scientific botany exceptionally also had significant ecological impacts triggered by gathering pressures. Botanists collected specimens to transplant wild flora to botanical gardens and for herbarium collections. These also included published collections of dry plants, such as *Flora Germanica exsiccata* (1830–1845) with 2,600 sheets, one of the first botanical collections of dry plants available on the market, or *Flora exsiccata Austro-Hungarica* (1881–1913), with four thousand species, available in one hundred copies, each of them with original dry plants. An exceptional example from the Western Karavanks demonstrates the ecological impacts of botanical gathering. According to the observation by the botanist Alfonz Paulin from 1906, botanical gathering of the rare yellow Zois' violet (*Viola calcarata* subsp. *zoysii*) on Mount Stol (Germ. *Hochstuhl*) jeopardized it. Zois' violet had by then been gathered there for more than a century by Slovenian and foreign botanists.[2920] In general, *Viola zoysii* is now on the red list of endangered species in Slovenia, designated as a rare species.[2921]

[2917] Flückiger & Radeff, 2000, 5, 8–11, 13–14, 19–20, 31, 39; Abbet et al., 2014, 630.
[2918] Aeschimann et al., 2004, vol. 2, 158.
[2919] Abbet et al., 2014, 630.
[2920] Praprotnik, 2015, especially pages 14–16, 19–20, 97–98, 108–110, 119, 135, 137–140, 142–145, 157–158, 160–163, 168, 172–173, 179, 181–182, 185–186, 189, 195–202, 204, 208, 210–211, 215–231, 247, 267, 276, 278, 280, 282, 285, 294, 297, 302–308, 310–324, 359–361.
[2921] "Pravilnik o uvrstitvi ogroženih rastlinskih in živalskih vrst v rdeči seznam," 2002, 8893, 8910.

8.2 Gathering for local consumption

This section presents plant gathering for the human diet and medicine (8.2.1), fodder plants (8.2.2), and other historical raw materials from grasslands (8.2.3). The selection of gathered plant species has never been influenced solely by species composition in local grasslands. Cultural differences in space and time have always had an influence, and further factors have impacted the selection of gathered wild plants.[2922] Consequently, this section also deals with historical similarities and diversities of gathering history at several places in the Alps and with changes in the selection of gathered plants in the long run.

Section 8.2 deals with grassland plants largely gathered for local consumption. Some of them were nonetheless also gathered commercially, such as dandelion as a vegetable (Section 8.2.1) or several healing organisms (Section 8.2.1.3), but local uses of the plants presented in this chapter prevailed, or at least a high percentage of the gathered plants were aimed at local use.

Comparisons with plant gathering in the Eastern Carpathians, populated by rural societies that are currently among the most traditional in European mountains,[2923] partly also with traditional gathering in the Western Carpathians and in Durmitor (Montenegro), are drawn where these parallels can improve the understanding of gathering history in the Alps.

8.2.1 Plants gathered for the human diet or medicine

8.2.1.1 Gathering wild food for the human diet

The average diversity of vegetables in gardens and fields in most of the times dealt with in this volume used to be significantly lower than today. In general, the deeper into the history of the last millennium one looks, the more this was the case.[2924] Some plant species were historically gathered because farmers perceived it a waste of arable land to sow these species that typically abounded in some grasslands or to plant their cultivated substitutes that found their ways to farm gardens later, leading to abandonment of gathering wild parallels. In the Allues Valley, for instance, this applies to two species that had formerly been significantly economically important: Good King Henry (*Chenopodium bonus-henricus*, a typical "culture accompanying" wild species for nutrient-rich relatively dry montane and subalpine grassland sites near cottages in several regions of the Alps, which in the twentieth century was replaced by cultivated spinach, which had historically been considered to occupy too much space in

[2922] Christanell et al., 2010; Abbet et al., 2014, 627–632; Sõukand & Pieroni, 2016, especially pages 17–18, 38–39.
[2923] E.g., Emanuelsson, 2009; Babai, Molnár, & Molnár, 2014.
[2924] Valenčič, 1970a; Maček, 1979, 305–306.

the garden in comparison to the low quantity of cooked leaves), and lamb's lettuce (*Valerianella locusta*, cultivated specimens from the same genus later found their way into the gardens, or the abundance of lamb's lettuce in gardens increased at least). Both species have no longer been gathered in the Allues Valley since the mid-twentieth century.[2925] Wild fruits (including those from grasslands) mattered as well. This section explains that even a limited amount of food ingredients replacing cereals was gathered in grasslands in the Alps in some cases. Quantities of food from farmland, vegetables in particular, also were insufficient on some occasions. Quantitative food demand and food diversity both increased the historical relevance of gathered wild plants in human diet.[2926] The chemical composition of wild varieties can differ significantly from the cultivated ones and between populations, which makes some wild greens less reliable concerning their taste,[2927] but due to this feature gathered wild food can also have nutritional advantages (Section 10.2.3).

The selection of grassland plants with significant importance in local cuisines differed regionally only to a limited degree, less than locally gathered medicinal herbs.[2928] Gathering grassland vegetables was particularly important in spring. The main aim of usual early spring gathering was to diversify the diet after the rather monotonous winter dishes.[2929] Brien Meilleur's investigation of the Allues Valley in the first half of the twentieth century, representative of many high settlements in the Alps, demonstrates that gathering in grasslands for the human diet started in mid-April and was frequent in May and June. It started in the area of permanent settlements in the valley (1,000–1,520 m; the climate of the valley is relatively dry, with about 900 mm of annual precipitation, concentrated in spring and autumn, with a snow blanket in the habitation zone until April), and it lasted as long as trampling was not considered a problem for hay production. Later, it continued around mid-elevation temporary settlements (e.g., at 1,300 m and somewhat higher), where livestock arrived in early or mid-May for two to four weeks.[2930] In the Allues Valley, plants such as dandelion (*Taraxacum officinale* and most probably also some other species from the genus *Taraxacum*; mainly from nutrient-rich sites), nettles (especially *Urtica dioica*, at nutrient-rich sites), Good King Henry (*Chenopodium bonus-henricus*, like nettles belonging to nitrophilous flora in excessively nutrient-rich sites around cottages in mid- and high-elevation pastures), common

[2925] Meilleur, 1982, especially pages 168–170, 172; Meilleur, 1986, 4–5, 30–31, 157, 167, 185–186; Oppermann & Paracchini, 2012, 18 (quote); Abbet et al., 2014, 632; Aeschimann et al., 2004, vol. 1, 238–239.

[2926] Maček, 1979, 305; Meilleur, 1982; Łuczaj et al., 2012.

[2927] Compare Sevindik et al., 2017.

[2928] Łuczaj et al., 2012, 367; Abbet et al., 2014, 627–629, 632; Sõukand & Pieroni, 2016, especially pages 17–18, 38–39.

[2929] Meilleur, 1982, 169; Meilleur, 1986, 283, 303; Abbet et al., 2014, 627–630; Łuczaj et al., 2012, 361; Łuczaj et al., 2015, 339.

[2930] Meilleur, 1982, 167, 169; Meilleur, 1986, 15–21, 31, 33, 41–42, 283, 289, 377–382, 402.

bistort (*Polygonum bistorta*, mainly at nutrient-rich damp sites in montane and subalpine vegetation belts), lamb's lettuce (*Valerianella locusta*, mainly at nutrient-rich, mesic sites in colline and lower montane belts, in sunny grasslands full of gaps), chicory (*Cichorium intybus*; mainly at relatively dry sites with medium nutrient availability in colline and montane belts), and brooklime (*Veronica beccabunga*, early start of the vegetation season in wet and nutrient-rich meadows) had significant economic importance in the human diet in spring, even though they covered only a minor part of the entire calorie intake. They were usually served as salad (e.g., dandelion, lamb's lettuce, small amounts of brooklime) or in soups (e.g., nettles, Good King Henry, common bistort, small amounts of brooklime).[2931] Some of the plant species listed were typical wild vegetables of significant importance in the Alps in general, such as nettles (also in many other places in the Alps, such as Wallis, East Tyrol, southwest Styria, and the Slovenian Alps), and dandelion (also in, e.g., southwest Styria, East Tyrol, Wallis [Francoprovençal *Valês*, French *Valais*], the Aosta Valley [Ital. *Valle d'Aosta*, Francoprovençal *Val d'Aoûta*, Fr. *Vallée d'Aoste*], Piedmont [Ital. *Piemonte*], and the Slovenian Alps), only that their manner of preparation differed somewhat.[2932]

It is obvious that plants gathered in early spring usually grow in nutrient-rich or at least moderately manured sites; nutrients promote a quick start of the vegetation period, and so these plants were available early in the season. However, the preference for nutrient-rich sites also means that the occurrence and abundance of these species (dandelions, for instance) in meadows were significantly lower than they are now, as excellently observed by old farmers in Bohinj.[2933] The wide territorial distribution of gathering of these species also proves that they have a very long history of human consumption. The case of Good King Henry must have been similar: although the cited references document its use in the human diet in the Alps only in the Allues Valley and in Wallis (in both cases, this was one of the most frequently gathered wild greens),[2934] a comparative perspective reveals that Good King Henry was an important wild vegetable elsewhere; for example in Durmitor (in the Dinaric Alps of Montenegro)[2935] and in the Carpathians close to the present Roma-

[2931] Meilleur, 1982, 168–170, 172, 174; Meilleur, 1986, 178, 184–185, 228, 260, 267, 283, 298–299, 303, 375–376, 381, 402; Abbet et al., 2014, 627–631; Łuczaj et al., 2015, 339; site conditions from Aeschimann et al., 2004, vol. 1, 372–373 and vol. 2, 224–225, 388–389, 620–621, 658–665, and from Pils, 1994, 81, 103.

[2932] Vogl-Lukasser et al., 2006, 50; Christanell et al., 2010, 62–63; Schunko & Vogl, 2010, 2, 10, 12; Abbet et al., 2014, 627–632; Mlakar, 2015, 130–131; author's interview with Gašperin, 2017, in Stara Fužina.

[2933] Author's interview with Kranjc, 2018, in Stara Fužina.

[2934] Wallis: Abbet et al., 2014, 627–628, 630–632; the Allues Valley: see above.

[2935] The art historian Tatjana Koprivica (Historical Institute of Montenegro, experience-based knowledge because her family originates from Durmitor), personal communication, 2016 and 2021.

nian–Ukrainian border.[2936] Consequently, there is every reason to assume that the use of Good King Henry in the Alps was more widespread in previous centuries and that origins of its use in human nutrition are ancient, even though people might have started using it independently in several places. Even Alpine dock (*Rumex alpinus*, abounding in nutrient-rich sites around cottages in relatively high-elevation summer pastures in the subalpine vegetation belt (Section 2) in particular; see Figure 114 in Section 6.6.2) was historically eaten by people. Clear evidence for this comes from the Western Alps. In the Allues Valley it was still eaten in the first half of the twentieth century: its boiled "'meaty' leaf petioles were often [or sometimes, at least] eaten" in grazing complexes at middle and high elevations, where Alpine dock abounds at over-manured sites around the cottages.[2937] The "stems" (probably leaf petioles) of Alpine dock were also eaten as vegetables in Wallis (in the Valleys of Entrement and Anniviers), for instance, and common sorrel (*Rumex acetosa*) was eaten in salads there.[2938] In the Allues Valley, occasional chewing of small-leaved docks and sorrels (belonging to *Rumex* sp.) in summer was widespread among children and in some families, and leaves were also added to soups in small quantities.[2939] Similarly, leaves of some *Rumex* species such as patience dock (*Rumex patientia*), common sorrel (*Rumex acetosa*), and thyrse sorrel (*Rumex thyrsiflorus*) were used as boiled greens in the Carpathians.[2940]

These are not the only species used as significantly important wild vegetables in the history of the Alps. At several places in the Italian Alps (e.g., in the surroundings of Bergamo), in Wallis, and in the Tolmin region (in the Julian Alps in the Soča River catchment), for instance, cooked young leaves of bladder campion (*Silene vulgaris*, growing best at nutrient-poor sites; Figure 139) have been consumed and highly esteemed in the human diet.[2941] Evidence from Spain shows that bladder campion was even intentionally promoted (and spread) by local inhabitants,[2942] and so this might have been practiced at some places in the Alps as well, but definite proof is missing. The leaves of ribwort plantain (*Plantago lanceolata*) and hoary plantain (*Plantago media*) were eaten in salads and as cooked vegetables in parts of Wallis, such as the Valley of Entremont.[2943] Hoary plantain abounds in nutrient-poor and (too) frequently used pastures (the largest leaf rosettes in Figure 74, Section 6.2.2.1), and rib-

---

[2936] Sõukand & Pieroni, 2016, 23.

[2937] Meilleur, 1982, 168–172; Meilleur, 1986, 375–377, 379, 381, 397 (quote); Aeschimann et al., 2004, vol. 1, 382. Meilleur's text (1986, 397) allows the interpretation that (young?) leaves of Alpine dock also were consumed by people in soups.

[2938] Abbet et al., 2014, 628–629, 631.

[2939] Meilleur, 1982, 168–172; Meilleur, 1986, 303–304.

[2940] Łuczaj et al., 2015, 342–344; Sõukand & Pieroni, 2016, 29, 35.

[2941] Machatschek, 2017, especially pages 142–150; Abbet et al., 2014, 627; Mlakar, 2015, 131; site conditions from Aeschimann et al., 2004, vol. 1, 332–333.

[2942] Łuczaj et al., 2012, 362–363.

[2943] Abbet et al., 2014, 628.

Figure 138: In some parts of the Alps, young leaves of bladder campion (*Silene vulgaris*) have belonged to esteemed wild greens. The photo shows a fully developed plant. Photo: Žiga Zwitter, early September 2022.

wort plantain abounds in extensively used meadows with medium nutrient availability, both of them on relatively dry sites in colline and montane vegetation belts in all regions of the Alps,[2944] so that grasslands characterized by their occurrence and high abundance used to be frequent. Plantains were consumed as vegetables in some other places in Europe as well, such as greater plantain (*Plantago major*) in the Carpathians of Ukraine.[2945] Young flower buds (and possibly young leaf stems) of hogweed (*Heracleum sphondylium*) were eaten as cooked vegetables, in parts of Wallis, at least. The aerial parts of cuckoo-flower (*Cardamine pratensis*) were consumed in salads in parts of Wallis, according to data from the nineteenth and twentieth centuries.[2946] In the Alps of Lower Styria and Carinthia, for instance, the leaves and also roots of cat's ear (*Hypochaeris* sp., especially *Hypochaeris radicata*, a species growing best at mesic sites in colline and montane belts, characterized by medium nutrient availability, such as (too) frequently used pastures without additional fertilization) were widely consumed in significant quantities, mainly in salads, but also as cooked vegetables. Cat's ear leaf rosette is an evergreen so that not only fresh

[2944] Dietl & Lehmann, 2004, 12–13, 37; Aeschimann et al., 2004, vol. 2, 174–175.
[2945] Sõukand & Pieroni, 2016, 38.
[2946] Abbet et al., 2014, 627–628, 631.

young leaves but also leaves from the previous year were eaten in spring.[2947] The ten bunches of *wurzl salat* bought by the seigneury of Gornji Grad (Germ. *Oberburg*) in the Kamnik–Savinja Alps of Lower Styria during Lent, on February 26th, 1553 (corresponding to March 7th in today's calendar; average spring temperatures in 1500–1567 were very close to the 1961–1990 average) might have consisted of cat's ear because of the early availability of its palatable leaves and roots. Not only the early date but also the fact that all other entries in this seigneury's account section designate gathered organisms (mushrooms)[2948] speaks in favor of the hypothesis that these were gathered wild plants; moreover, as just mentioned, more recent gathering of cat's ear in the Alps of Lower Styria is well documented. Parts of the Eastern Alps are not the only territories where consumption of cat's ear is proven: a parallel from Romania[2949] suggests that gathering cat's ear might actually have historically been more widespread in the Alps as well. In some parts of the Alps, such as the Paznaun Valley (Germ. *Paznauntal*) in Tyrol, young shoots of rosebay willowherb (*Epilobium angustifolium*) were consumed as (cooked) vegetables. Young shoots of further plants such as common comfrey (*Symphytum officinale*), bulbous comfrey (*Symphytum bulbosum*), great burnet (*Sanguisorba officinalis*), and so on were consumed as vegetables, probably in the Alps.[2950] At some places where watercress (*Nasturtium officinale*) grew in aquatic environments, it was also gathered as an early spring vegetable and eaten in salads (proven, for instance, at Stara Fužina in Bohinj, where a wetland used to be located close to the village).[2951] The same type of use is known in the Valley of Entremont in Wallis,[2952] for instance. Due to its aquatic character, the attitude toward this species was not uniform: mid-twentieth-century data from Bohinj demonstrate that, although it was consumed at some farms in significant quantities, some fellow villagers were afraid of possible health issues (impurities or parasites) and did not eat the watercress at all.[2953] The spicy flavor of watercress might have also contributed to such a diversity of attitudes.

2947 Machatschek, 2017, 176–178, 180–189; site conditions also from Aeschimann et al., 2004, vol. 2, 628–629 and Dietl & Lehmann, 2004, 12–13, 37.

2948 SI NŠAL 20, Gornji Grad Avstrija (Gornji Grad Austria), box 24, seigneury's account by Marx Textor, 1553: "Ops [sic!, i.e. Obst] vnnd salatt"; allocation to Lent of that year is based on Grotefend & Ulrich, 1971, 166; conversion is based on *Julian Day and Civil Date Calculator* and the calculations by the astronomer Tomaž Zwitter; climate history context from Pfister et al., 2018, 274, 277, 279.

2949 Machatschek, 2017, 177.

2950 Machatschek, 2017, especially pages 115–117, 154, 159–162, 370.

2951 Author's interviews with M. Hodnik, 2018, Markeš, 2020, Kunavec, 2021, and Kranjc, 2020, all in Stara Fužina.

2952 Abbet et al., 2014, 627, 629–630.

2953 Author's interviews with M. Hodnik, 2018, Markeš, 2020, Kunavec, 2021 and Kranjc, 2020, all in Stara Fužina. One of these interviewees recalled that she once joined an experienced fellow villager for watercress gathering in the mid-twentieth century, but her father, a smallholder, ordered her to discard all she had collected.

Summer gathering in grasslands for the human diet was more limited than spring gathering (but see Section 8.2.2 for summer gathering of large amounts of wild fodder plants). It had to focus on pastures to avoid trampling in meadows, where gathering could largely only start again at mowing time, except for sections that could be reached from paths. In summer, some gathering of plants continued at overly nutrient-rich sites around alpine cottages (such as Good King Henry). Note that these were specific grassland environments: grasses abounding at such over-fertilized sites frequently trampled by people and livestock are resistant to trampling, and so trampling in the course of gathering was not an issue there. Moreover, such nutrient-rich meadow sites were mowed several times during summer in many cases (Section 6.1.2.4), and so herbs gathered there for human consumption were significantly more tender and more suitable for human food than old plants would have been. Another type of gathering characteristic for summer targeted flower receptacles.[2954] The flowers of some thistles, for instance, were consumed: flower receptacles of stemless thistle (*Carlina acaulis*, Figure 127; a species promoted by mowing that used to abound in many grasslands: Section 6.1.1.3 and the introduction to Section 8) were eaten as a summer raw snack in many parts of the Alps, as clearly documented, for example, in Wallis in the nineteenth and twentieth century,[2955] and in the Allues Valley.[2956] The same use is practiced by spatially distant traditional farmers in Ghimeş (Hung. Gyimes) in the Eastern Carpathians.[2957] Children spent a lot of time in grasslands as herders. This was one of the main occasions to eat grassland plants as raw snacks; plants consumed raw by children usually tasted sweet (e.g., goat's beard; see Section 6.3.2.2) or sour (see, e.g., docks and sorrels above); plant snacks tasting sour were sought, for instance, by thirsty people. Raw plant snacks were occasionally still eaten by adults, but the selection of plants for raw snacks usually differed from the list of plants used as vegetables in local cuisine.[2958] A parallel from the White Carpathians (now the Czech Republic) demonstrates that gathering flower receptacles of stemless thistle was substantial in certain cases even though they were consumed only as raw snacks (like artichokes).[2959] Moreover, in parts of the Alps, such as Wallis, the flowers (receptacles) of stemless thistle (*Carlina acaulis*, clear evidence from the nineteenth and twentieth century) and spiniest thistle (*Cirsium spinosissimum*, evidence from the twentieth century) were nonetheless also cooked as vegetables.[2960] The limited occurrence of some species within the Alps was one of the factors causing their spatially limited consumption. An example is stemless thistle, *Carlina acanthifolia*; in the Alps

2954 Meilleur, 1982, 169–170; Meilleur, 1986, especially pages 283, 289, 382.
2955 Abbet et al., 2014, 627, 632.
2956 Meilleur, 1982, 168–170; Meilleur, 1986, 305, 402.
2957 Babai, Molnár, & Molnár, 2014, 46.
2958 Meilleur, 1982, 168–172; Łuczaj et al., 2012, 362; Abbet et al., 2014, 627–632.
2959 Pawera et al., 2017, 666; Abbet et al., 2014, 629.
2960 Abbet et al., 2014, 627, 629.

it occurs almost only in the western part of the mountain chain.[2961] Its inflorescences were consumed within this territory not only as a snack but also as a boiled vegetable, as recorded in the nineteenth century. Another reason for its limited consumption area must have been its low edibility rating.[2962]

Gathering of wild fruits in grasslands and at dwarf shrub sites also started in summer and continued in autumn. This activity focused on berries for local consumption or as a source of income, in particular for limited women's and children's income. The majority of berries that have been usually gathered in grasslands, such as cowberries (*Vaccinium vitis-idaea*), blueberries (*Vaccinium myrtillus*; see also Sections 2, 6.3.1, and 6.6.2.1), and strawberries (*Fragaria vesca*), are well known,[2963] and so this paragraph focuses on a less well-known but characteristic example. Ripe fruits of stone bramble (*Rubus saxatilis*), which is widespread in the Alps,[2964] used to be collected for food during mowing in many mountain meadows and also in pastures. It grows at forest margins, around rocks and next to small structures within grasslands. Stone bramble was used in many places, including such distant and culturally diverse areas as the Julian Alps[2965] and the Allues Valley. Nonetheless, data from the first half of the twentieth century make clear that this was not a species of significant economic importance and it is usually not gathered any more. In this respect, stone bramble is a typical case; namely, gathering of wild plants without significant dietary importance in the first half of the twentieth century ceased in later decades, with very few exceptions (in the Allues Valley, for instance, only the sweet wild strawberry, *Fragaria vesca*).[2966] At present, ripe stone bramble fruits can easily be found along the most frequented mountain trails because visitors are not aware of their edibility. Knowledge of this species is also no longer widely present in local communities.[2967] Abandonment of historical uses is a characteristic driving force for losses of knowledge about grassland plants because this knowledge was largely motivated by historical plant use.[2968]

2961 Aeschimann et al., 2004, vol. 2, 560–561; *Global Biodiversity Information Facility*: *Carlina acanthifolia* All.

2962 Abbet et al., 2014, 630; *Plants for a Future*: *Carlina acanthifolia* All.

2963 Meilleur, 1982, 169–170; Meilleur, 1986, especially page 382; Vogl-Lukasser et al., 2006, 50, 53; author's interviews with C. Mlakar, 2015, in Bohinjska Češnjica; with Kunavec, 2017 and 2020, and with Mencinger, 2017, all in Stara Fužina; with L. and M. Odar, 2017, and with A. and M. Taler, 2017, both in Studor.

2964 Aeschimann et al., 2004, vol. 1, 732–733.

2965 Author's interviews with Boltar, 2018, with Kranjc, 2018, with Mencinger, 2017, all in Stara Fužina, with C. Mlakar, 2017, at the Zajamniki Alp, with L. and M. Odar, 2017, in Studor, and with Minka Sodja, 2017, in Jereka.

2966 Meilleur, 1982, 165–166, 168–172.

2967 Author's field discussions with locals—young adults, and children—close to stone bramble sites in Bohinj, in the summers of 2020–2022: in the Laz Alp and during a walk to the Krstenica Alp.

2968 Sucholas et al., 2022, 37.

Figure 139: Stone bramble. Photo: Žiga Zwitter, August 30th, 2019.

Some economically relevant gathered plant uses in the human diet were more persistent in the twentieth century, whereas further historical uses of the same species were abandoned earlier. A good example is nettle (see its persistent use as vegetable above in this section). Forgotten use includes its application in cheese production. Within the last millennium, cheesemaking from fresh milk using rennet gained much importance and spread in extensive areas of the Alps from the sixteenth century onward. It gradually replaced older cheesemaking techniques from sour milk.[2969] Apart from or combined with widely used animal rennet from boiled and dried stomachs of calves, kids, or lambs, in which the digestive enzymes acted as a rennet,[2970] a few plants also became used as (an addition to) rennet in some places.[2971] A mid-twentieth-century ethnological study by Vilko Novak referring to the nineteenth or early twentieth century was able to document that nettles were used in rennet production in parts of the Southeastern Alps, in the high summer pastures to

[2969] Pfister, 1985a, 56, 96–97; Niederer, 1996d, 177–178; Kauter, 2002, 117–118; Orland, 2016, 36–37, 51; Bätzing, 2015, 72–75.

[2970] Novak, 1942, 94; Niederer, 1996d, 171; Ledinek, 1999, section Predelava mleka; Orland, 2016, 37, 51.

[2971] Orland, 2016, 51; Mlakar, 2015, 477; the latter documents such a use of lady's bedstraw (*Galium verum*) in Slovenian territory, but it is not clear if this information comes from the Alps.

the south and southwest of Rateče, at least. They were mixed with the smoked relevant part of the stomachs of young calves and lambs and then stored until this mixture was used as rennet.[2972] Addition of nettles to rennet is usually not recorded, and the role of nettles was not explained by Novak. A comparative approach reveals that the selection of nettles was well-considered: a recent experiment carried out in the Iberian Peninsula and inspired by the traditional Basque use of nettles instead of a strainer to remove impurities from milk proved that fresh nettle (*Urtica dioica*) and probably also small nettle (*Urtica urens*) can be used as rennet even alone.[2973] (Salted) nettle rennet makes not only fresh cheese, but also semi-hard cheese; a combination of nettles with animal rennet allowed harder cheese production. It is possible to store nettle rennet, namely, dry nettles also work,[2974] and so nettles in dried animal–herbal rennet at Rateče were indeed active. The traditional use of nettles in cheese-making—practiced in at least part of the Julian Alps—evidently aimed to extend the limited available substance from livestock stomachs. Because the use of rennet only became widespread in cheese production in the Alps some centuries before the nineteenth century, the addition of nettle rennet was likely invented in the modern period. In the northern Julian Alps, where such a sophisticated historical use of nettles is documented in the nineteenth and/or early twentieth century, market-oriented cheese production had by that time already had a long tradition. Tasty sheep cheese (or cheese from a mixture of sheep and cattle milk) had belonged to the items of long-distance trade from that part of the Julian Alps at least since the seventeenth century. In the late seventeenth century, cheese from the Sava Dolinka Valley (Sln. *Zgornjesavska dolina*) was sold to distant places in the Holy Roman Empire as a kind of parmesan.[2975] High-quality cheese production in the region must have started earlier than at that time. Namely, as recorded in 1636,[2976] people from a couple of villages from this region (Kranjska Gora and Podkoren; that is, the village next to Rateče as the settlement mentioned above in the context of adding nettle to rennet) kept livestock in summer pastures and produced cheese in the distant Upper Soča catchment, on the other, southern side of the mountain range, in part of the Trenta Valley and at higher elevations, most likely at

2972 Novak, 1942, 92–94. The information evidently refers to the nineteenth or early twentieth century because at the time when this investigation was carried out sheep were no longer milked there; cheese production lasted there until 1912 (Novak, 1942, 92, 95).

2973 Fiol et al., 2016, 19–23.

2974 Wilde, 2013; *Cultures for Health: How to Make Nettle Rennet for Cheesemaking.*

2975 Mlinar, 2018b, 48; Valvasor, 1689, volume 1, book II, 135, 137. In some microregions of the Julian Alps, with a history of cheese trade reaching back much longer than the introduction of Swiss cheese recipes in the second half of the nineteenth century, like Bohinj, sheep and cattle milk were mixed in cheese production, as recorded in the early nineteenth century (Novak, 1987, 73, 76–78; Žontar, 1987, 63).

2976 This source, written in 1636, reflects conditions in a certain year between 1617 and 1636 (Mlinar, 2018a, 27–28).

Kranjska Alp (Sln. *Kranjska planina*) mountain pasture complex on the southern slopes of Mount Prisojnik. A parallel from 1498 suggests that sheep (or sheep and goat) cheese was made there.[2977] Farmers from the Capitanate of Bovec kept dairy livestock and made cheese in the other half of the same grazing complex, probably a neighboring mountain pasture complex, as recorded in 1636.[2978] From the perspective of people from Kranjska Gora and Podkoren, this grazing complex was far from their home farms, and it belonged to the same seigneury as the neighboring grazing complex used in summer by people from the Capitanate of Bovec. Farmers from Kranjska Gora, Podkoren, and the Bovec region had to deliver dues in cheese from these mountain pastures in the Upper Soča catchment to the same seigneury, the seigneury of Fusine (Friul. *Fusinis*, Germ. *Weißenfels*, Sln. *Bela Peč*). This seigneury, as the recipient of some cheese, was certainly interested in high cheese quality, which was probably one of the stimulations for knowledge exchange about dairying in the uppermost Soča catchment. In addition, likely contacts with temporary mountain settlement personnel from the Capitanate of Bovec in the nearby mountain pasture complex possibly contributed to this kind of knowledge transfer. Namely, the cheese exported from the Sava Dolinka Valley in the late seventeenth century was called *Flitscher Käse* in German[2979] (i.e., 'Bovec cheese'), named after a region of the summer mountain pastures dealt with above, in the upper Soča river catchment, not in the Sava river catchment. People from the Capitanate of Bovec already had a tradition of market-oriented cheese production by that time; buyers from the Republic of Venice visited the Bovec region to buy cheese there, and farmers also exported it to the Republic of Venice and to the provinces of Carniola and Carinthia, as recorded in the mid-sixteenth century.[2980] These data support the statement above that rennet cheese from the area was a high-quality product, as was characteristic for Alpine regions exporting cheese at that time; not everywhere in the Alps was rennet cheese good enough for long-distance trade.[2981] Adding nettle to rennet was thus practiced (or even invented) in a region of the Julian Alps with a rich and old history of high-quality rennet-cheese production. Nonetheless, it is

[2977] Mlinar, 2018b, 49–50; Mlinar (ed.), 2018a, 80–81; Mlinar (ed.), 2018b, 120, 129. A loaf of at least one type of cheese produced there in the early seventeenth century weighed about 3 kg (Mlinar, 2018b, 49–50; Mlinar (ed.), 2018b, 120, 129).

[2978] Mlinar (ed.), 2018b, 120: a loaf of at least one type of cheese that farmers from the Capitanate of Bovec made there was larger (it weighed about 17 kg, as mentioned in 1636) than the cheese made in the Upper Soča catchment from milk of livestock owned by farmers from the upper Sava Dolinka Valley. This means that seventeenth-century cheesemaking procedures to deliver dues from summer pastures in the uppermost Soča Valley were not entirely uniform.

[2979] See the quote from Valvasor in Mlinar, 2018b, 48; our interpretation on dairying knowledge exchange is original.

[2980] Gestrin, 1965, 197; on the more recent history of Bovec cheese, see Ledinek Lozej, 2021.

[2981] Bätzing, 2015, 73.

not clear whether nettles replaced part of the animal rennet in the production of such an aged cheese[2982] as Bovec cheese, sold as a kind of parmesan at distant places in the late seventeenth century. The addition of nettles to rennet was also not generally widespread in the southeastern Alps, at least not in times when sheep were not milked anymore in major parts of the Julian Alps (see Section 6.3.1.1). An experienced traditional farmer, herbalist, and cheesemaker (processing cattle milk) from Stara Fužina in Bohinj, born in the 1940s, for instance, stated that plants were not added to rennet there in times she could remember and she had never heard of it, although she had learned a lot of traditional local herbalist knowledge, mainly from her grandfather.[2983] Not even an organic Bovec sheep-cheese producer from the Upper Soča catchment, born in the mid-twentieth century, recalled the practice.[2984] Nettles were not the only grassland plants historically used as (an additive to) rennet in some regions of the Alps. Some parts of the southern Alps had a historical tradition of rennet production from thistle blossoms.[2985] Unlike nettles working with any usual milk type, rennet from the flower heads of (some) thistles (of spear thistle, *Cirsium vulgare*, at least, which is widespread in the Alps) can only be used with sheep or goat milk in the production of cheeses that are not aged for more than two months, otherwise the cheese may develop a bitter flavor.[2986]

The regular role of gathered grassland plants in the human diet was even more important in previous centuries than in the early twentieth century.[2987] Independent sources from the seventeenth and early nineteenth century, for instance, emphasize that the role of gathering in the farm economy of the hills of the Natisone (Sln. *Nadiža*) river catchment, in the southern Julian Alps, was significant. Gathering wild "herbs" and "wild grass" was documented there. The local economy was largely based on animal husbandry, trade in livestock, dairy products and fat, wood, charcoal and some fruits, whereas locally grown cereals only covered a minor part of the annual demand. Cereals were regularly bought in the Friulian Plain, and the overall economic conditions of an average family were rather difficult.[2988] Even though "grass" could in general refer to many non-woody species of grasses and/or herbs, it is highly likely that gathering the seeds of hairy finger-grass (*Digitaria sanguinalis*), growing

2982 According to the website of *Cultures for Health*, providing a recipe for nettle rennet, "cheese made with vegetable rennet may develop a bitter flavor if aged for a long period of time (over two months)" (*Cultures for Health: How to Make Nettle Rennet for Cheesemaking*).

2983 Author's interviews with Kunavec, 2021 and 2023, in Stara Fužina.

2984 Her identity is kept condifential.

2985 Orland, 2016, 51.

2986 *Cultures for Health: How to Make Nettle Rennet for Cheesemaking*; *Cultures for Health: How to Make Thistle Rennet for Cheesemaking*; Aeschimann et al., 2004, vol. 2, 582–583.

2987 Meilleur, 1982, 169, 174; Abbet et al., 2014; Łuczaj et al., 2012.

2988 Beguš, 2015, 78–126, 150–152, 157–173, 189–192, quotes from pages 88 and 112.

at nutrient-rich sites in the colline and montane vegetation belts,[2989] was in fact usual there. Namely, in the sixteenth century, the botanist Pietro Andrea Mattioli spent part of his life in the province of Gorizia (Germ. *Görz*, Sln. *Goriška*), and he also visited Carniola. He documented that in the provinces of Gorizia, and Carniola, thus in the immediate vicinity of the Natisone catchment, people gathered the seeds of this grass species to make porridge from them.[2990] The immediate territorial vicinity, the fact that farmers from the Natisone catchment had trade contacts with people from the province of Gorizia,[2991] the lack of local cereals, and overall tough economic conditions speak in favor of this hypothesis. The seeds of hairy finger-grass are easy to store and it is possible to make fine white flour from them.[2992]

The economic importance of gathering (especially in spring or to create food stocks for winter) was even higher in the case of food scarcity (e.g., years with cereal stocks running out in spring), hunger, and famine, either in times of general scarcity or in terms of economic difficulties of some families.[2993] In the Valley of Illiez in Wallis (French *Val d'Illiez*), for instance, "some families fought against each other to collect *Rumex alpinus*" during the Second World War.[2994] Selected examples of wild grassland species exceptionally eaten as foods during times of scarcity in the Alps include starch-rich bulbs of martagon lily (*Lilium martagon*; easy to store dried, no harmful impacts on human health are known) and possibly the starchy tubers of spotted orchid (*Dactylorhiza maculata*) and broad-leaved marsh orchid (*Dactylorhiza majalis*), all these in a part of the Western Alps, at least.[2995] In parts of the Eastern Alps of Austria, for instance, infructescences, and roots and oily seeds of common mallow (*Malva sylvestris*), probably also of dwarf mallow (*Malva neglecta*), are recalled to have been added to bread dough in the severe economic crisis following the Second World War.[2996] This might have been a more general habit in the Alps; both mallow species mentioned occur in nutrient-rich, not too moist grasslands in colline and montane vegetation belts in all regions of the Alps.[2997] In the Allues Valley, for instance, some interviewees recalled consumption of common mallow, dwarf mallow, and hollyhock mallow (*Malva alcea*) seeds as

[2989] Site conditions from Aeschimann et al., 2004, vol. 2, 1008.

[2990] Petkovšek, 1966, 298–304: the botanist Viktor Petkovšek determined that in this case Mattioli was dealing with *Digitaria sanguinalis*.

[2991] On such trade, see Beguš, 2015, 158, and Zwitter, 2014b, 44–46.

[2992] *Plants for a Future*: *Digitaria sanguinalis* (L.)Scop.

[2993] Meilleur, 1982, 167, 169, 174; Pfister, 1985a, 60, 119; Meilleur, 1986, 283, 289; Machatschek, 1997b, 215–216; Jäger, 2006, 13; Łuczaj et al., 2012, 360–361; Abbet et al., 2014, 633; Slavin, 2018, 503–504, 508.

[2994] Abbet et al., 2014, 631.

[2995] Meilleur, 1986, 181, 299–300, 356; *Plants for a Future*: *Lilium martagon* L., *Dactylorhiza maculata* L.

[2996] Machatschek, 2017, 220–227.

[2997] Aeschimann et al., 2004, vol. 1, 410–413; Martinčič, 2007e, 470–474.

snacks in their childhood in the first half of the twentieth century, possibly a remnant of the former more general use of these species in times of scarcity. A parallel from the White Carpathians in the Czech Republic also includes consumption of immature dwarf mallow fruits.[2998]

All plants with a long history of significant annual human consumption lasting until the early twentieth century or longer "appear devoid of acute toxicity. However, it must be critically noticed that some species contain secondary metabolites of toxicological relevance."[2999] Among the plants dealt with in this volume, those from the genera *Rumex*, *Chenopodium*, possibly also *Polygonum*, are "rich in oxalic acid, which may lead [and which in fact led] to hypocalcemia and formation of calcium oxalate crystals in kidneys,"[3000] in cases of immoderate consumption or if consumed by people prone to several acid-related health issues. These included arthritis, for instance,[3001] particularly annoying for people performing much heavy manual physical work, such as farmers in the timeframe dealt with in this book. At certain places in the Alps, Michael Machatschek's fieldwork was able to document an example of historical coping with oxalic acid in Alpine dock, a substance that does not decompose by cooking only, and carbonates naturally dissolved in water were usually insufficient to neutralize it. Consequently, a bit of lime was added to the (first) water in which Alpine dock was cooked. The reaction of neutralization led to deposition of the harmful calcium oxalate at the bottom of the pot, and this deposit was discarded, together with the water.[3002] It is not known how old this practice was and how widely it was applied. Hogweed (*Heracleum sphondylium*, Figure 142) gathered in open, sunny meadows "may cause photosensitization and possess[es] carcinogenic properties."[3003] It can be assumed that consumption of these plants in small quantities was harmless, whereas regular consumption in significant quantities possibly led to health issues, especially

---

2998 Pawera et al., 2017, 661.

2999 Abbet et al., 2014, 632 (quote); for proofs of the harmlessness of the plants mentioned in this section but not dealt with by Abbet et al., namely of cuckoo-flower and cat's ear, and of the general acceptability of young rosebay willowherb, young shoots of common comfrey, and great burnet for the human diet (not to be eaten in excessive quantities, or in the case of specific illnesses or pregnancy), see *Plants for a Future*: *Cardamine pratensis* L., *Dactylorhiza maculata* L., *Digitaria sanguinalis* (L.) Scop., *Epilobium angustifolium* L., *Hypochoeris radicata* L., *Lilium martagon* L., *Polygonum bistorta* L., *Sanguisorba officinalis* L., *Veronica beccabunga* L., and Machatschek, 2017, 142–150, 154, 158–161, 220–227.

3000 Maček, 1979, 306; Abbet et al., 2014, 632 (quote).

3001 *Plants for a Future*: *Polygonum bistorta* L.

3002 Machatschek, 2017, 161–162, 370.

3003 Abbet et al., 2014, 632. Phytophotodermatitis caused by hogweed from open, sunny meadows might be a problem, but hogweed (*Heracleum sphondylium*) is much less harmful than the well-known phototoxic species giant hogweed (*Heracleum mantegazzianum*), which arrived in Central Europe in the second half of the twentieth century, later than the time span dealt with in this book.

if tricks of proper treatment, like the one presented for Alpine dock, were not applied. Harmful impacts must have been aggravated in times of hunger or famine, when undernourished, thus particularly sensitive, populations consumed large quantities of such plants. Health issues triggered by some wild plant species consumed in large quantities added to general compromising of the immune system in times of malnourishment and increased susceptibility to other diseases.[3004] Historical overdosing might have led to total exclusion of these species from the diet in some parts of the Alps before the time the oldest living interviewees could recall, in the course of developments of modern medicine. An exception is occasional fresh *Rumex* snacks in rather low quantities.[3005] Potentially toxic are also saponins in bladder campion (*Silene vulgaris*); nonetheless, they are "broken down by thorough cooking" and the body tends to absorb them very poorly, so that consumption of moderate quantities of *Silene vulgaris* is harmless.[3006] Several grassland plants that were gathered as vegetables in early spring would later become less palatable or unpalatable to people, but their young shoots contain less if any bitter or poisonous (including psychoactive) substances and hard fibers.[3007]

In the last centuries, the ecological impacts of large-scale and regular gathering of above-ground parts of wild food plants were comparatively limited, in general. Overharvesting was only an exceptional phenomenon, as a parallel from Poland suggests. Namely, large-scale gathering could only work in the long term if species were common in the landscape[3008] and if their regeneration abilities were high (see the example of gathering thistles for pig forage in Section 8.2.2). In certain cases, like some cat's ears (*Hypochaeris* sp.), large-scale gathering including below-ground parts helped reduce the high abundance of plants perceived negatively from the viewpoint of animal husbandry. Namely, cat's ears are trample-resistant rosette plants; animal-husbandry-related management of grasslands fertilized no more than moderately promotes the occurrence and abundance of cat's ears: some cat's ear species (such as *Hypochaeris radicata*) have leaf rosettes tight to the ground that can usually be neither mowed nor grazed, and so they profited from historical farming. Consequently, the abundance of cat's ears in some grasslands has been very high (in some moderately manured, mesic grasslands in colline and montane vegetation belts in the case of *Hypochaeris radicata*). Under these circumstances, these species appear to be competitive and their rosette leaves hamper the

[3004] For general considerations about the impacts of malnourishment and famine food on the immune system (without a focus on wild food species and their health impacts), see Slavin, 2018, 504.

[3005] On sorrel snacks, see, e.g., Meilleur, 1982, 168–170; Meilleur, 1986, 303–304, 401; Abbet et al., 2014, 632.

[3006] *Plants for a Future*: *Silene vulgaris* (Moench.) Garcke.

[3007] Machatschek, 2017, 114, 184–185; *Plants for a Future*: *Epilobium angustifolium* L.

[3008] Łuczaj et al., 2012, 363.

recruitment of other meadow species, reducing the yield of the plant community. Early spring gathering of cat's ear with roots or autumn gathering of cat's ear roots for human consumption, gathering of cat's ear roots as a coffee surrogate, as well as cat's ear gathering for feeding pigs also fulfilled the role of removal of species unproductive from the viewpoint of grazing animal husbandry (Sections 6.6.2, 8.2.1.1, and 8.2.1.2), and so to the general agricultural value of the grassland site.[3009]

Also vulnerable are plants with single tubers (e.g., *Dactylorhiza* spp.; see the paragraph on gathering in times of scarcity), where tuber gathering for food could lead to significant reduction of populations. Contrary to the usual wild food from above-ground organs, in some cases large-scale gathering of medicinal plants as well as large-scale gathering for decorative purposes had significant negative ecological consequences (see Section 8.1).[3010]

8.2.1.2 Gathering spices, staple tisanes, and coffee and tobacco surrogates

Grasslands in the Alps made possible gathering of good and even highly esteemed tisanes (see also Section 8.1.3.2) and spices used not only in staple dishes, but also in some festive ones, such as stuffed pasta (*Schlipfkrapfen*) spiced with nettles in East Tyrol according to data from the early twentieth century.[3011] Moreover, less esteemed coffee and tobacco surrogates were gathered there that nonetheless played a role in the daily life of ordinary people during a shorter historical period. Selected examples shed light on both similarities and diversities in these aspects of the history of the Alps and on changes over time, triggered by either cultural history developments or changes in grassland ecosystems.

Caraway (*Carum carvi*) seeds used to be widely collected in the Alps, in moderately manured, late-mowed meadows on soils with a neutral pH, in mesic sites, at various elevations, from the colline to subalpine vegetation belt (Sections 2 and 6.3.2.2).[3012] According to information referring to some parts of Central Europe from the seventeenth to nineteenth centuries, caraway was a common species in meadows with enough nutrients (well-manured for the standards of the time, corresponding to moderately manured meadows now)

[3009] Pils, 1994, Table 11, Photo 31; Machatschek, 2017, 176–188; on site conditions, see also Aeschimann et al., 2004, vol. 2, 626–629.

[3010] Compare Łuczaj et al., 2012, 363.

[3011] Christanell et al., 2010, 62.

[3012] Author's interviews in Bohinj with Boltar, 2018, in Stara Fužina; with Grm, 2017, and with Milena Sodja, 2017, both in Srednja Vas; with Minka Sodja, 2017, in Jereka; with C. Mlakar, 2017, at the Zajamniki Alp; and with L. and M. Odar, 2017, in Studor. Meilleur, 1986, 290; Abbet et al., 2014, 627–628 and Appendix C; Torkar Tahir, 2012, 141; Mlakar, 2015, 204, 206, 270, 299, 338–339, 348; Bavcon, Ravnjak, & Praprotnik, 2019, 123; Aeschimann et al., 2004, vol. 1, 1124.

at mesic sites.[3013] One interviewee in Bohinj recalled that she collected a full linen sheet of caraway seeds in a distant but manured montane meadow at Uskovnica (see Section 6.3.2.1).[3014] Caraway seed was used locally as a spice and as a human or veterinary medicine (most widespread were uses against flatulence and indigestion, proven, for instance, in the Allues Valley, Wallis, East Tyrol, the Slovenian Alps, and the Eastern Carpathians).[3015] Moreover, regionally sold caraway seeds must have been a source of income for some women. In the sixteenth century, caraway was, for instance, a spice traded in larger quantities than anise, imported pepper, bay leaves, mace, and ginger in Judenburg, the main commerce center of (Upper) Styria of the time, if one may judge from recorded toll payments.[3016] Caraway still grows in the Alps,[3017] but twentieth- and twenty-first-century landscape eutrophication (intensified fertilizing), mowing at times of an earlier plant development stage at many sites (Sections 1.2, 10.1.2.1, introduction to 10.1.2.2, and Section 10.2.1), and afforestation in many others (Section 9) significantly decreased its abundance and occurrence. As a comparatively short-lived (usually biennial) plant, caraway depends on regeneration gaps in meadows, which also disappeared from many grasslands with increasing sward density due to increased fertilization. Abandonment of grassland use and consecutive afforestation also shrank caraway habitats.[3018]

Beyond caraway as an example of a widely used meadow spice in the Alps, there also existed significant regional diversities in the selection of gathered wild spices, as demonstrated by the example of meadow clary (*Salvia pratensis*; Figure 166 in Section 10.2.2). It is another widespread species in moderately manured meadows in the Alps.[3019] Ethnological information shows that in some parts of this mountain chain, such as in the Sattnitz (Sln. *Gure*) Hills in southern Carinthia, meadow clary sprouts were used as a spice (e.g., in soups), at least in the nineteenth and early twentieth centuries.[3020] However, use of meadow clary was spatially limited; oral history interviews in Bohinj, for in-

---

3013 Kauter, 2002, 156–158.

3014 Author's interviews with Boltar, 2018, in Stara Fužina; with Grm, 2017, and with Milena Sodja, 2017, both in Srednja Vas; with Minka Sodja, 2017, in Jereka; with L. and M. Odar, 2017, in Studor; and with C. Mlakar, 2017, at the Zajamniki Alp.

3015 Meilleur, 1986, 290; Vogl-Lukasser et al., 2006, 50, 184–185; Abbet et al., 2014, 627–628 and Appendix C; Babai, Molnár, & Molnár, 2014, 42; Mlakar, 2015, 204, 206, 270, 299, 338–339, 348.

3016 Tremel, 1947, 97, 99, 103, 132. In the late sixteenth century (data from 1592), quantities of pepper in these statistics almost reached the quantities of caraway (Tremel, 1947, 132), which should be understood in the context of sinking pepper prices, leading to an increased demand for pepper (Sandgruber, 1989, 103).

3017 Aeschimann et al., 2004, vol. 1, 1124; Kauter, 2002, 157–158.

3018 Author's interview with Boltar, 2018, in Stara Fužina; Christanell et al., 2010, 64.

3019 Aeschimann et al., 2004, vol. 2, 164–165; Papež Kristanc, 2012, 9.

3020 Angerer, 1987, 5–7, 13, 15.

stance, show that the plant was not used in the households of interviewees,[3021] although meadow clary has been frequent in many moderately manured meadows there.[3022] Similarly, use of meadow clary as a spice in recent centuries was also not recorded in the Valleys of Entremont, Anniviers, Illiez, and Hérens in Wallis, where meadow clary has only been used as a medicinal plant,[3023] or in the Allues Valley, where its cultivated relative, common sage (*Salvia officinalis*), was also only used as a medicinal herb and not as a spice.[3024] Apart from cultural differences (use of the genus *Salvia* in the Alps as a spice is more spatially limited than as a medicine), another factor that must have limited the role of meadow clary in the human diet in recent centuries is significantly higher qualities of *Salvia officinalis* as a spice or medicine; *Salvia pratensis* contains much smaller amounts of essential oils than common sage.[3025]

The selection of plants for staple tisanes was particularly diverse. In Wallis in the mid-twentieth century, "little consensus was found between informants with regard to the species to be used" as a tisane,[3026] which is characteristic of many parts of the Alps. Members of various households from the same village were used to different types of tisanes as an everyday beverage, even with respect to plants that are widespread in the region. In the mid-twentieth century, some families in Bohinj, for instance, frequently drank tisane from lady's-mantle (*Alchemilla*), thyme (*Thymus* sp.), and St. John's wort (*Hypericum* sp.), whereas some other families did not gather these plants at all, but instead collected bilberry leaves, wild strawberry leaves, and elderflower blooms, for example.[3027] Partly, this is a question of different tastes among different people, but also of the time available to women for gathering plants in certain meadows or pastures in different parts of the year. In addition to local differences, there have also existed regional differences in gathering widespread plants. A comparison of the aforementioned selection of wild plants for tisanes in Bohinj with Lower and Central Wallis and with the Allues Valley demonstrates that some of these plants were widely used for tisanes in many places in the Alps (lady's-mantle in Bohinj, Wallis, and the Allues Valley; thyme in Bohinj and Wallis, and in the Allues Valley mainly as a medicinal tisane). Use of some others that were frequently used in tisanes in Bohinj, as mentioned above, was

3021 For instance, author's interviews with Boltar, 2018, and with Mencinger, 2018, both in Stara Fužina.

3022 Author's interviews with Boltar, 2018, in Stara Fužina, and with C. Mlakar, 2017, at the Zajamniki Alp; Papež Kristanc, 2012, 9; personal observations of grasslands in Bohinj.

3023 Abbet et al., 2014, 628 and Appendix C; absence of use of meadow clary as a spice in Table 2 suggests that *Salvia* added to sirups in the valley of Anniviers is another species.

3024 Meilleur, 1986, 159, 187.

3025 Dakskobler et al., 2019, 53.

3026 Abbet et al., 2014, 627–628.

3027 Author's interviews with Kunavec, 2017, and with Urh, 2017, both in Stara Fužina, and with C. Mlakar, 2017, at the Zajamniki Alp. Similar data are given in Mlakar, 2015, 268–270, 298–299 and Novak, 2024, 221.

more regionally limited, practiced only in one or two out of four valleys analyzed in Wallis, and only for specific purposes in the Allues Valley: bilberry leaves (used in the Allues Valley as a medicinal infusion), St. John's wort (in the Allues Valley only as an eye wash), and elderflower blooms (used only as medicine in the Allues Valley). Wild strawberry leaves, frequently used in Bohinj, are not known to have been a tisane ingredient in the Valleys of Entremont, Anniviers, Illiez, and Hérens in Wallis or in the Allues Valley,[3028] even though all these plants occur in all regions of the Alps.[3029]

Broader trends in the cultural history of beverages also had an influence on the dynamic history of gathering grassland plants. In the nineteenth (or early twentieth) century, when coffee also became a staple in the diet of farmers, craftsmen, and industry workers, the roots of some grassland plants such as cat's ear (*Hypochaeris* sp.) were gathered in significant quantities as a coffee surrogate until the mid-twentieth century.[3030]

In the modern period, when smoking spread in Europe,[3031] this led to specific roles of some (grassland) plants, such as the use of what were most probably mountain arnica leaves (*Arnica montana*) as a tobacco surrogate, proven in the late modern period in the Allues Valley.[3032] The roots of sweet vernal-grass (*Anthoxanthum odoratum*) are another example of a historical tobacco surrogate from grasslands in some other places in the Alps.[3033]

8.2.1.3 Gathering medicinal herbs and grassland lichens

Through uses of healing plants and lichens, grassland history impacted human health and wellbeing. Some of these organisms could even decisively influence human recovery or death. Examples of healing plants gathered in the Alps and aimed not only at local and regional markets, but also at ex-regional markets, were presented in Sections 8.1.1 and 8.1.2, and local medicinal uses of edelweiss (*Leontopodium alpinum*) were briefly addressed in Section 8.1.3.1. Building on general considerations of historical medicinal herb uses, given in the introduction to Section 8, this section focuses on organisms that were historically widely gathered for medicinal purposes, emphasizes the regional diversity in plant gathering, and links the history of grassland plant gathering

[3028] Meilleur, 1986, 300, 302, 348, 359, 396, 400–401; Abbet et al., 2014, 627–629.
[3029] Aeschimann et al., 2004, vol. 1, 396–407, 626–627, 782–785; Aeschimann et al., 2004, vol. 2, 378–379.
[3030] Sandgruber, 1989, 105–106; Mlakar, 2015, 259–262; Machatschek, 2017, 185; see also Poschlod, 2015, 105–106. Chicory (*Cichorium intybus*) was often grown as a field crop and not collected wild as a coffee surrogate (Abbet et al., 2014, 627, 630–631; Mlakar, 2015, 191–192, 261–262; Poschlod, 2015, 105).
[3031] Sandgruber, 1989, 109–111.
[3032] Meilleur, 1986, 283, 290, 393.
[3033] Machatschek, 2017, 199.

with the history of morality. A selection of case studies is presented. The text starts with uses of Iceland moss (*Cetraria islandica*) in folk medicine as an example of a lichen historically widely gathered in grasslands in the Alps for healing purposes. A general comparison of gathering medicinal plants in two microregions in the southeastern part of the Alps, but discussed in a wider context, follows. The final part of this section places gathering of grassland plants used as abortifacients and aphrodisiacs or aphrodisiac ingredients in a broader context of the history of sexual morals to make it possible to understand the historical demand for such plants. This is a topic most readers of this volume would not usually relate to grassland history, and so it receives much attention in the second half of this section.

Some historically gathered species were replaced by cultivated substitutes, which found their ways to farm gardens later; this note from the section on wild food plants (8.2.1.1) is also partly relevant in the context of medicinal herbs. One example is wild horse mint (*Mentha longifolia*), which had been used in several regions of the Alps, such as the Allues Valley, as a medicinal plant against indigestion, and it was most frequently replaced there by cultivated mints in the early twentieth century.[3034] Nonetheless, for a variety of reasons, a greater diversity of medicinal plants than herbs used as vegetables were still gathered in the mid-twentieth century, and many of them are also gathered today (e.g., Section 10.2.3).

Iceland moss (*Cetraria islandica*; Figure 140) is an example of a grassland organism with a long history of medicinal uses in several parts of the Alps, such as the Alps of Slovenia, Tyrol, and Wallis.[3035] Large quantities were collected and used in the Alps (see also Section 8.2.2 for Iceland moss as fodder), and some *Cetraria islandica* was also sold to merchants trading in healing plants and lichens.[3036] In the Julian Alps, for instance, gathering Iceland moss by locals was clearly documented in the mid-eighteenth century, at least for medicinal purposes,[3037] and it can be assumed that at least some of the healing uses of this lichen built on older historical traditions. In the eighteenth century, the naturalist Balthasar Hacquet observed that (some) shepherds in Bohinj used boiled Iceland moss as medicine against what they believed to be sheep poisonings by one of the yellow flowering aconite species (*Aconitum* sp.), probably with one of the subspecies of wolf's bane (*Aconitum lycoctonum*).[3038] Furthermore, healing effects of Iceland moss on respiratory

---

[3034] Meilleur, 1986, 300.

[3035] East Tyrol: Vogl-Lukasser et al., 2006, 41, 71, 73, 82, 125, 129, 145–147, 174, 234, 238, 266; Christanell et al., 2010, 58, 64, 69; Wallis: Abbet et al. 2014, 630–631 and Appendix C; references to Iceland moss uses in the Slovenian Alps are given in the following footnotes.

[3036] Machatschek, 1997b, 114.

[3037] Wraber, 1969, 270, 288.

[3038] Hacquet, 1779, 39; plant identification: Praprotnik, 2015, 80.

Figure 140: Iceland moss (*Cetraria islandica*; light green and brown lichen in the center of the picture) in the Ovčarija summer pasture complex (about 1,650 m) in the Julian Alps. Photo: Žiga Zwitter, October 25th, 2020, after the first autumn snow blanket had melted.

diseases of livestock were widely recognized in the Alps.[3039] This means that it was one of the medicines applied against coughing in cattle, in East Tyrol, for instance.[3040] Similarly, mid-twentieth-century data from Bohinj reveal that Iceland moss was used for treating pigs if they were coughing.[3041] Moreover, in the same area this lichen was boiled and given to pigs to promote appetite.[3042] This kind of use was not limited to pigs; Iceland moss was also administered to cattle, mainly to livestock that needed extra energy, to promote the appetite of breeding bulls, or to fatten bulls in East Tyrol, for instance.[3043] According to data from the same area and from the Slovenian Alps, this lichen was used to treat cattle and pigs with rumen/stomach and bowel issues, including diarrhea; a handful of Iceland moss mixed with charcoal and ash was applied against

3039 Machatschek, 1997b, 114; Machatschek, 2006, 138; Vogl-Lukasser et al., 2006, 41, 71, 73, 125, 129, 145–147, 234, 238, 266; Mlakar, 2015, 352; Novak, 2024, 221.

3040 Vogl-Lukasser et al., 2006, 146; Mlakar, 2015, 352.

3041 Author's interview with Kunavec, 2017, in Stara Fužina.

3042 Author's interview with A. Kavčič, 2017, at the Laz Alp in the Julian Alps; Novak, 2024, 221; Mlakar, 2015, 352.

3043 Vogl-Lukasser et al., 2006, 146–147, 234.

diarrhea in pigs. Iceland moss was also administered for problems during and after calving (in East Tyrol, at least).[3044] Moreover, it was used in human medicine: against human respiratory (and some other health) problems in many parts of the Alps—for example, as an antitussive against bronchitis and asthma in several parts of Wallis, such as the Valleys of Entrement and Illiez,[3045] against asthma on mountain farms in the Karawanks above Jesenice[3046]—or as a tisane against cold, and against stomach problems in Bohinj,[3047] and against pneumonia, bronchitis, and what was believed to be tuberculosis in (parts of) the Slovenian Alps. Powder from this lichen was applied to external livestock wounds to promote healing, in some places in the Slovenian Alps at least.[3048] Iceland moss's antibiotic effect and ability to cure some respiratory diseases have been scientifically confirmed. Its applications against certain rumen/stomach and bowel problems of cattle, pigs, or people and to promote appetite are also understood by (veterinary) medicine, and the same applies to application of powder from this lichen to promote healing of external wounds. The effectiveness of folk medicine's administration of Iceland moss for problems during and after calving is not clear to veterinarians.[3049]

Regional diversities in gathering healing plants are demonstrated based on data on the Slovenian Alps from 1948. By that time, gathering commercially relevant medicinal herbs had already had a tradition in the District (Sln. *okraj*) of Jesenice, whereas it was by and large considered a novelty in the District of Dravograd (in northern Slovenia); among medicinal herbs considered commercially relevant, only gathering of mountain arnica (*Arnica montana*, Figure 85 in Section 6.3.1.1) and centaury (common or lesser; *Centaurium erythraea* or *pulchellum*) had had a tradition in the District of Dravograd.[3050] Thus, the selection of healing herbs usually gathered in these two microregions of the Alps differed significantly at that time.

Focusing on the plant species mentioned, a wider mid-twentieth-century context demonstrates that mountain arnica was one of the most usual medicinal herbs in many, but not all, regions in the Alps (see also Section 8.1.2). It is no wonder that it was gathered in both districts studied. Namely, moun-

---

[3044] Vogl-Lukasser et al., 2006, 82, 146–147, 174; Mlakar, 2015, 352; Novak, 2024, 221.

[3045] Abbet et al., 2014, 631, and Appendix C; Mlakar, 2015, 306, 432.

[3046] Torkar Tahir, 2012, 141.

[3047] Novak, 2024, 221; author's interviews with Grm, 2017, in Srednja Vas in Bohinj, and with Mencinger, 2017, in Stara Fužina.

[3048] Mlakar, 2015, 306, 352.

[3049] Vogl-Lukasser et al., 2006, 82, 147–148; Galle Toplak, 2015, 179–181.

[3050] SI AS 2164, Zadružništvo na Slovenskem, 1922–1962 (Cooperatives in Slovenian Territory, 1922–1962), box 2, unit 287, Report on Medicinal Herb Gathering at the Level of District Business Associations, March 26th, 1948. *Tavžentroža* designates the genus *Centaurium*, but only *C. erythraea* and *C. pulchellum* grow in the Slovenian Alps (Aeschimann et al., 2004, vol. 2, 6–9; Wraber, 2007a, 506).

tain arnica was frequently collected in such distant and culturally diverse regions as the Alps of Slovenia, Carinthia, East Tyrol, the Allues Valley, and some other places in the northern Alps (see also Section 8.1.2), but not in the Valleys of Entremont, Anniviers, Illiez, and Hérens in Wallis. The historical occurrence and abundance of mountain arnica used to be much higher than it is nowadays because it is a species of nutrient-poor grasslands: it used to be "quite common" in montane meadows on nutrient-poor, acidic soils, and it was even more frequent at higher elevations, in non-fertilized subalpine and alpine meadows and pastures (see Section 6.3.1). Most widespread was the application of mountain arnica for external human and livestock wound care in as distant regions as the Alps of Slovenia, East Tyrol, and the Allues Valley; a good historical practice because mountain arnica in fact includes substances with antiseptic and anti-inflammatory effects. At several places such as in mid-twentieth-century Strojna above Dravograd, in the Pohorje Hills, in the Upper Soča River catchment and in nineteenth-century Wallis, mountain arnica was also used internally in human medicine, and in the Slovenian Alps at least against stomach problems. This was, however, a poor historical practice because arnica is toxic, unsuitable for this kind of internal use. Similarly, historical internal use of arnica against flatulence is proven in treating livestock in East Tyrol.[3051]

The use of centaury, also mentioned in the District of Dravograd in 1948, seems to have been more spatially limited,[3052] but it was a usual medicinal plant in folk medicine in the Alps of Slovenia at that time.[3053] Nonetheless, nineteenth-century information from Tyrol emphasizes that farmers were generally aware of the partial similarity of substances in centaury—which belongs to

3051 The Alps in Slovenia in general: Mlakar, 2015, 236, 300–301, 348–349, 432; the Upper Soča River catchment in the Julian Alps and Cerkno Hills (Sln. *Cerkljansko hribovje*): Sosič, 2000, 233–235, 240, 246; Pohorje: Sosič, 2000, 249; Carinthia: Keller, 2004, 21 and Strojna in Carinthia: Makarovič, 1982, 396; East Tyrol: Vogl-Lukasser et al., 2006, 41, 129–132, 175, 177, 237, 266 (and also 103, 126, 212, 245) and Christanell et al., 2010, 58, 60, 63, 66, 69; the Allues Valley and some other places in the Northern Alps: Meilleur, 1986, 290, 393. Medical uses of *Arnica montana* were not mentioned by the interviewees in four valleys of Central and Lower Wallis (Abbet et al., 2014, 630–632 and Appendix C). The real health impacts of mountain arnica blooms on the human body and livestock are summarized from Galle Toplak, 2015, 62–64 and Vogl-Lukasser et al., 2006, 131–132; the note on historical occurrence and abundance is based on Keller, 2004, 43, Bavcon, Ravnjak, & Praprotnik, 2019, 19 (quote), 111, 123, Dakskobler et al., 2019, 21–22, 26, 58, and Aeschimann et al., 2004, vol. 2, 528–529.

3052 Even though *Centaurium erythraea* occurs in all regions of the Alps as defined by *Flora Alpina* and *Centaurium pulchellum* occurs in almost all regions of the Alps (Aeschimann et al., 2004, vol. 2, 6–9), no species from this genus was documented to have been used in the Allues Valley (Meilleur, 1986), in the valleys of Illiez, Entrement, Hérens, and Anniviers in Central and Lower Wallis (Abbet et al., 2014), or in East Tyrol (Vogl-Lukasser et al., 2006; Christanell et al., 2010) at that time.

3053 See the references in the second following footnote.

the *Gentianaceae* family—and gentians, so they used centaury together with gentians.[3054] Consequently, research might have overlooked the historical uses of centaury hidden behind the folk taxon of gentian. Centaury was administered against several diverse health issues, such as stomach problems (a good practice because this plant is effective in the case of stomach issues related to a lack of gastric juice), to purify blood, and so on.[3055]

The cited notice from the District of Dravograd in 1948 certainly does not mean that mountain arnica and centaury were the only healing plants gathered there at that time. Quite the contrary, the notice from Dravograd reflects a situation still usual in the majority of the countryside in the late 1940s: diversity of medicinal plants gathered for local use was significantly higher than diversity of plants gathered for pharmaceutical enterprises (see Section 8.1.2). Ethnological investigations documented that, even though the number of plant species gathered by an average household at Strojna above Dravograd in the mid-twentieth century was relatively low, in addition to the two aforementioned plant genera farmers also gathered the following healing herbs and fruits in grasslands: yarrows (*Achillea* spp.), juniper berries (*Juniperus communis*), probably lady's mantle (*Alchemilla* spp.) and St. John's wort (*Hypericum* sp.), and so on.[3056] However, these were already times of a great erosion of traditional gathering and knowledge about wild herbs. Historical herb knowledge and gathering by ordinary people had already been dynamic before; nonetheless the large-scale erosion they underwent in the second half of the nineteenth century and in the twentieth century is unprecedented in the broader frame of the past millennium (for reasons, see the introduction to Section 8). In general, this large-scale loss of historical knowledge became significant later in remoter regions, such as isolated farms in the hills above Dravograd, than in areas with significantly more frequent contacts of the population with (larger) urban settlements.[3057] The next subsection takes us back to the times of rich historical plant knowledge.

8.2.1.3.1 Historical grassland abortifacients and aphrodisiacs

The strict sexual discipline prescribed by religion was followed by some and broken by others from all social strata, especially concerning premarital sex. In the countryside, for instance, in the long run, many peasant couples practiced

---

[3054] von Hörmann, 1877, 150; compare the chemical composition of centaury and yellow gentian in Galle Toplak, 2015, 295–296, 303–304, 440.

[3055] Examples from the Alps of Slovenia in the mid-twentieth century include areas as distant as the Upper Soča River catchment, the hills around Cerkno and Idrija, and the Pohorje Hills (Sosič, 2000, 232–233, 235–236, 238–239, 246–247, 249); on the real health impacts of centaury on the human body, see Galle Toplak, 2015, 303–304.

[3056] Makarovič, 1982, 395–396.

[3057] Compare Mlakar, 2015, 285, 292–296, 347, 463; see also Makarovič, 1982, 385–412, 423–426.

premarital sex, as clearly recorded since the sixteenth century. As long as the couples were faithful to each other, the lovers complied with other rules of unmarried adolescents and young adults in the village, and their activities did not result in out-of-wedlock children, community members in general tolerated such love without considerable condemnation. Despite the violation of fundamental Church rules and moral ideals, sex was perceived as an informal betrothal. Investigations from Slovenian territories demonstrate that such relatively relaxed and publicly known premarital behavior was usually abandoned in the mid-eighteenth century. Then, "general tabooization" of sexuality and the related "cult of shyness" spread throughout society and persisted until the twentieth century; however, the level of rigidity differed among rural communities. The new moral standards resulted not only in an actual decrease in premarital sexuality, but also in more thoroughly hidden actions of premarital love and their eventual consequences, and in new forms and thresholds of uneasiness for young adults, women in particular, because the society of the time was more tolerant toward men than women violating the norms of sexual behavior.[3058] Since the nineteenth century and possibly earlier, *coitus interruptus* was widespread in premarital sex, among the agricultural population, for instance, as stated in Slovenian territory before the First World War and found by studies in several regions of Europe. However, this contraception method is unreliable.[3059] An eventual conception was often followed by a quick marriage.[3060] If this was impossible, either abortion or out-of-wedlock children followed. Tolerance toward out-of-wedlock children before the twentieth century varied within the Alps across space and time, which is reflected in highly diverse percentages of such children. Very high extramarital fertility ("up to two-thirds of all births in some districts") and remarkable tolerance toward out-of-wedlock children were characteristic for extensive areas in the eastern part of the Eastern Alps in the nineteenth century, characterized by the specific settlement type of prevalent isolated farms: only the couple in charge of the farm was typically married, and several unmarried farmhands and maidservants also lived there. Inheritance was impartible, and in several cases and time periods population density was lower than required by the environmental carrying capacity. However, such regions of high levels of extramarital fertility and tolerance toward out-of-wedlock children were an exception in the broader Alpine context; very low percentages of such children and cultural prejudice

[3058] Polajnar, 2008, especially pp. 5–93; Studen, 2011, 33–42; Kos, 2015, 52–53, 77–84, 271–273, 275–277, 279–280, 282–294, 296–301, 307–311, 313–318, 322, 329–334, 337–338, 341, 344, 346, 348–352, 355–368, 388–391, 403–404, 415, 424, 432–440; Kos, 2016, 309–310, 350–351, 353–354, 358–364, 371, 378–381, 436 (quotes from p. 350). Compare Makarovič, 1982, 446–447.

[3059] Labouvie, 2006, 263; Kos, 2015, 55–56; Cergol Paradiž, 2022, 85, 121–122; Polajnar, 2008, 74.

[3060] Kos, 2015, 285, 289, 358, 362–363, 365–366, 388, 390–391, 434.

against them were characteristic for several regions and subperiods,[3061] and so lower historical rates of deliberate abortion can be assumed for isolated farms, at least in some regions and periods.

The nobility went thorough periods of even more relaxed sexual behavior than farmers. There was a peak in conjugal infidelity among the nobility in the seventeenth century, despite ecclesiastical and state measures to raise moral standards. The notion of the acceptable sexual behavior of a noble wife at that time differed significantly from that of her husband; in some cases, the lady—unlike her husband, who enjoyed more freedom—could socially acceptably cheat on the husband as long as this was not publicly known and the reputation of the family was maintained.[3062] Such notions must have also led to abortions. Again, this was not a general rule; part of the nobility complied with high moral standards.[3063]

In the long run, until the twentieth century, the most frequent reasons for deliberate abortion were socioeconomic hardships, usually combined with sex outside of marriage, often premarital, without a realistic option for the marriage by the couple for social or economic reasons after the conception was detected.[3064] In the long run, before the twentieth century, deliberate abortion was possibly not one of the most common strategies of limiting the number of conceptions: long periods of late marriage and abstinence from sexual intercourse by a significant percentage of childless unmarried women and men roughly in several subperiods from the sixteenth to mid-twentieth century belonged to the main strategies of deliberate family limitation in the Alps.[3065]

Deliberate abortion is a phenomenon that can be traced "in all societies, cultures, and epochs." Christianity rejected sexuality without the aim of marital reproduction; nonetheless, evolution of this kind of knowledge prior to the nineteenth century needs to be understood in the context of the history of discussions on the fetus and the soul. Medieval ecclesiastical law, following Aristotle's theory, considered the soul to enter a male fetus forty days after conception, and a female fetus eighty days after conception. In the modern period, the question when a fetus becomes a living child was contested in theo-

---

3061 Makarovič, 1982, 85, 88–89, 458, 460, 468; Viazzo, 1989; Bätzing, 2015, 122–124; Mathieu, 2019, 66–67, quote from p. 66; Cergol Paradiž, 2022, 157–158; Kos, 2015, 288–289; Kos, 2016, 380–381.

3062 Kos, 2015, 85–86, 277–281, 300, 351, 375–380, 390, 416–418, 425; Kos, 2016, 309–310, 312, 314–315, 321, 323, 335–336, 344–348, 378; but see Štuhec, 1985, 135.

3063 Kos, 2015, 85, 292–294.

3064 Polajnar, 2008, 5–93; Studen, 2011, 33–42; Kos, 2015, 285–286, 290, 356–357, 362, 367, 420; Kos, 2016, 364–371; Cergol Paradiž, 2022, 50–51, 61, 77, 81, 90, 94–95, 104, 113, 116, 118, 138, 145, 147, 150–169, 172–173.

3065 Viazzo, 1989; Bätzing, 2015, 122–124, 419; Mathieu, 2019, 66–67; on the dynamic limits of the integrated farm economy, see also Panjek, 2017; Zwitter, 2015d; and Mathieu, 2019, 59–61, 63.

logical, juridical, and medical discussions until the nineteenth century. In 1869, an adaptation of ecclesiastical law accepted the notion that the soul enters the fetus at the time of conception. On the one hand, this means that an early deliberate abortion was long seen as less of a moral issue than in the late nineteenth century. On the other hand, the penalties for abortion have slowly and gradually become more lenient but also reached an increased percentage of cases. Capital punishment was threatened in the case of an abortion at the stage when the fetus was already considered a living child in the Holy Roman Empire, from the sixteenth (1532) to the eighteenth century, unless the rules for territorial courts (*Landgericht*) in the region were different. However, it was difficult for physicians and midwives of the time to determine whether a fetus was older or younger than forty or eighty days, and sometimes also whether the abortion was spontaneous or deliberate. These issues and the rarity of cases in which information on this topic reached the courts limited the percentage of convictions and resulted in less rigid punishment in several cases than one would presume from the penal code. Since the second half of the seventeenth century, the fetus was considered alive by the Habsburg court system in the eastern part of the Alps only from the middle of the pregnancy, when the mother was able to feel its movements. After 1743, hidden childbirths were forbidden in Austria, one of the aims being to reduce abortion; women that concealed their pregnancy and gave birth to an out-of-wedlock child without asking anyone for help, if the child did not survive, had already been suspected by courts earlier. Since the mid-eighteenth century, long periods of forced labor or imprisonment for deliberate abortion are proven in practice, in some cases at least, even though the penal code would still allow capital punishment. Joseph II's penal code (1787) officially ended capital punishment for abortion in the Habsburg Monarchy; it was replaced by lengthy imprisonment, from a year to five years, but less for unmarried women. In addition, earlier than in many other states in Europe, this code stopped distinguishing between a "nonliving" and living stage of fetus development. Similar were some regulations of the 1803 penal code that remained in force in most of what is now Slovenia until 1929 and longer in Austria; in reality, the punishments were lighter, such as a month in jail for the couple and more for those that induced an abortion, as found by Ana Cergol Paradiž's study of court records from the 1920s. Most abortion cases were difficult to trace and prove for the authorities, and thus the number of such prosecutions was low in the first half of the nineteenth century. Thus, the 1852 penal code, otherwise largely following the 1803 rules, required doctors, midwives, pharmacists, and medical examiners to report suspected abortions to the authorities, to be informed at least of cases in which a deliberate abortion (attempt) caused complications. Indeed, in the second half of the nineteenth century and the first half of the twentieth century, the numbers of people convicted of abortion increased for several reasons, such as the definition of the fetus as a living being starting at conception, and enhancement of the state's insights into citizens' lives. The 1929 Yugoslav penal code was relatively liberal concerning abortion in a broader European context; the practice was still illegal, and the people involved were threatened with imprisonment until the second half of the twentieth century; however, after 1929 the couples and a potential third person inducing the abortion were in reality usually not imprisoned any longer if the partner did not play a more active role than helping in the process, and if the third person inducing the abortion did so free of charge and had not been sentenced for similar activities before (a criterion often not fulfilled in court records). Contrary to these principles, involvement of formally educated specialists—physicians, midwives, or pharmacists—in abortion (unless the mother's health status did not allow a safe continuation of pregnancy) was still considered a crime after 1929.[3066]

Socioeconomic developments after the end of the *ancien régime* added new motives for deliberate family limitation. Its demographic impacts (deliberately low marital fertility) highly likely became usual in several western parts of the Western Alps by the first half of the nineteenth century, and significantly later in eastern parts of the Western Alps and in the Eastern Alps (the case of the low birth rate in the Swiss canton of Glarus in the 1880s significantly impacted by deliberately low marital fertility was an early exception in a regional context).[3067] Gradually, the demographic transition spread throughout the Alps and reached Slovenian territories in the late nineteenth century: the birth rate in Slovenian territories started a slow but persistent decline in the 1890s, which led to birth rates in the Drava Banate (Sln. *Dravska banovina*; it covered most of the present-day Slovenia, except for much of its western part that belonged to Italy) of 28.7‰ in 1921–1930, and 23.4‰ in 1931–1940. In the interwar period, birth rates in the Drava Banate were significantly lower than in Italy, but higher than in Austria. One of the factors reducing the birth rate was the increasing practice of deliberate abortion also among married couples: according to a medical assessment from the Drava Banate in the 1930s, abortions accounted for at least a quarter and up to 40% of the total number of births and abortions; this figure includes not only deliberate abortions but also spontaneous ones. Deliberate abortion was motivated by many factors, as outlined above, and aggravated by the specifics of the interwar period, such as (1) the economic crisis in the 1930s, promoting abortion not only among nonmarried people, as most usual before, but also in an increased percentage of families living in poverty and unable to feed more children; (2) general life-

[3066] Labouvie, 2006, 261–263; Brinkschulte, 2010, 962–964 (quote from column 962); Kos, 2015, 297–298; Kos, 2016, 367–370; Cergol Paradiž, 2022, 14–23, 27–29, 44, 72, 132–141, 144–145, 153, 195–199; Crane, 1999, 537; Pečnik, 2024, 51–58. On the impact of Constitutio Criminalis Carolina (1532) and the rules for territorial courts, and subsequent developments in criminal law until the nineteenth century, see Kambič, 2004a, 172–178 (and more extensively in the Slovenian version: Kambič, 2004b, 152–154, 157–159).

[3067] Viazzo, 1989, 91–93, 198–210, 244–246, 292–293.

style modernization, involving a desire for a higher standard of living in times of a general rise of consumer society and increased intellectual engagement of women; and (3) a general lack of fairly reliable and cheap contraception until the mid-twentieth century.[3068]

In the long run, until the legalization of abortion in the second half of the twentieth century, most cases of abortion were undetected by authorities. They were a "public secret," legally strictly prohibited, but part of reality and relatively acceptable for many in private life. This is also reflected in the words that several children had already suffered the fate of abortion, as allegedly said by an extramarital partner, a married man (cheating on his wife was of course not socially acceptable in this specific case), to his pregnant lover in the seigneury of Bled (Germ. *Veldes*) in the late seventeenth century. Sometimes abortion gave rise to rumors, but these were usually not reported to the authorities unless stimulated by specific circumstances, such as other conflicts between those involved and people that learned about the event.[3069]

Abortion was triggered by diverse practices, based on chemical (often plant-based), physical (e.g., heavy physical work,[3070] interventions in the uterus), or thermal impacts.[3071] In line with the topic of this volume, this paragraph focuses on the role of grassland plants. European pharmacopeias, herbals, and midwives' handbooks from the sixteenth and seventeenth centuries recommend more than a hundred drugs to cause abortion, often tisanes, some of them also effective according to today's state of knowledge, and some of them life-threatening for the pregnant woman.[3072] Between the First and Second World War, the better off afforded deliberate abortion illegally induced by a formally educated physician or midwife; the majority of people practicing this kind of deed depended on illegal abortion induced by the pregnant woman herself, relatives or neighbors, or experienced persons without a formal education, which significantly increased the risk for the pregnant women. Many of those lacking formal education continued using traditional abortifacients, such as tisanes or parts of a plant to physically intervene in the uterus: in no less than a quarter out of the eighty-three abortion cases dealt with at the court in Celje, thus covering part of Lower Styria (now part of northeastern Slovenia), in the interwar period, hellebore (most likely Christmas rose, *Helleborus niger*; see Figure 167 in Sec-

[3068] Cergol Paradiž, 2022, especially pp. 7–9, 29, 43–125, 137, 143, 150–154, 157, 168, 172–173, 190.

[3069] Kos, 2016, 368; Cergol Paradiž, 2022, especially pp. 11–12, 43–125, 143, 149, 167–168, 170–172, 195–199; Pečnik, 2024, 51–58.

[3070] Compare Kos, 2016, 370, footnote 114.

[3071] An example of a thermal practice: sitting at a very hot place, as documented in the hills or Dinaric plateaus west or south of Idrija, now in western Slovenia, in ethnological field notes from 1959 (Sosič, 2000, 246–247). Examples of other types of practices are cited below.

[3072] Brinkschulte, 2010, 962–963; Kos, 2015, 56, 298; Kos, 2016, 368; Mlakar, 2015, 339.

tion 10.2.2, at least) rhizome was used for this purpose, as found by Ana Cergol Paradiž.[3073] Christmas rose rhizomes are highly toxic, and an overdose can be lethal for people.[3074] The frequent use of hellebore rhizome to intervene in the uterus in cases dealt with by the Celje court might be a consequence of the riskiness of this abortifacient, especially if applied by people lacking a formal education. Namely, the court was usually only informed of illegal abortions when the health of the woman was seriously threatened by complications caused by the illegal operation,[3075] and so hellebore rhizome use might have been overrepresented in the cases dealt with by the court compared to the actual frequency of the use of this abortifacient. Another example involving a grassland plant was the use of a sharpened branch of barberry (*Berberis vulgaris*), a frequent pasture scrub, to penetrate the uterus[3076]—a method exposing the pregnant woman to the risk of severe (and potentially lethal) physical injuries.[3077]

Folk medicine also used some substances without a real abortive impact in the hope of triggering an abortion,[3078] and there were contradictions in some plant uses caused by the typical challenge of limited knowledge of plant impacts on the human body (see the introduction to Section 8). Such a contradiction can be demonstrated by the use of roseroot (*Rhodiola rosea*) in the Alps. A very knowledgeable interviewee from Bohinj—where the abundance of roseroot was high in some mesic subalpine and alpine scree slopes, in rock fissures, and among dwarf shrubs, according to mid-nineteenth century botanical observations[3079]—explained that a tisane from its root was used to trigger abortion in that area.[3080] In contrast to this practice, around 1600, farmers in Tyrol considered roseroot rhizome boiled in milk to promote conception if couples suffered from what they considered a lack of feminine fecundity.[3081] Similarly, yellow gentian (*Gentiana lutea*) rhizome was often used as an aphrodisiac for livestock (Section 8.1.2), whereas a seventeenth-century failed attempt of the use of gentian (*Gentiana* sp.) rhizome as an ingredient of what farmers hoped to be an abortifacient is also known (Section 8.1.2).

3073 Cergol Paradiž, 2022, 30–31, 66–67, 80–81, 87, 94, 128, 136, 145, 153–154, 163–164, 175–189, 198.

3074 Willfort, 1971, 390.

3075 Compare Cergol Paradiž, 2022, 127–128, 143–148, 189.

3076 Compare Mlakar, 2015, 339–340, naming further grassland species historically used as abortifacients, but it is not clear whether the data refer to the Alps.

3077 Brinkschulte, 2010, 963.

3078 E.g., Cergol Paradiž, 2022, 138.

3079 Plemel, 1862, 154; the note on site conditions takes into account Aeschimann et al., 2004, vol. 1, 688, and Martinčič, 2007b, 237.

3080 Author's interviews in 2021 and 2022. The identity of this interviewee from Stara Fužina is kept confidential. Species identification is clear. The interviewee named the folk taxon *naštrbec*. Her designation of the site characteristics for *naštrbec* corresponds with her plant identification in the photo published in Aeschimann, 2004, vol. 1, 689.

3081 Machatschek, 2015, 164–165; the chronological orientation is based on *Deutsche Bibliographie: Wolkenstein, Marx Sittich von.*

In contrast to the previous paragraphs were the desired impacts of grassland plants used as an aphrodisiac or aphrodisiac ingredients. Focusing on family history, the most important use of an aphrodisiac was to cure occasional and treatable (male) impotence, or what was considered a lack of female fecundity. Impotence was a health issue that severely negatively impacted self-image and social success. It was perceived a shame for a man, in particular because this was not a taboo until the mid-eighteenth or nineteenth century, and so this reason for the absence of children of a married couple was not always a secret restricted to their private life but became publicly known in some cases. An eventual attempt at premarital sexual intercourse showing a boy's inability to achieve an erection for the girl would in many cases make their marriage impossible. Similarly to abortifacients, several historically used aphrodisiacs were ineffective and some threatened human health,[3082] and some others might have had some impact. Orchid tubers belonged to the usual aphrodisiac ingredients, with a long tradition reaching back to the Late Middle Ages at least.[3083] Historical administration of some grassland orchids (e.g., three-toothed orchid, *Orchis tridentata*) as an aphrodisiac in the Eastern Alps is proven by examples from the province of Carniola in the seventeenth century and the nineteenth century (in the Kamnik–Savinja Alps in the surroundings of Preddvor (now northern Slovenia), for instance). In the seventeenth century, pharmacists in Carniola regularly stocked orchid tubers. Thus, because non-fertilized grasslands strongly prevailed in the Alps (Section 6.3), these were market items aimed at aphrodisiac production; their use was not limited to the local population in the Alps. It is nonetheless assumed that orchids were mainly gathered for local consumption or for the micro- or mesoregional market because orchid species of interest as aphrodisiacs also occurred and abounded elsewhere in non-fertilized grasslands in lowlands and hills; they are neither endemic in the Alps nor do they grow only high in the mountains. In the seventeenth century, some girls were said to take with them such (magically treated) orchids, so that "boys would run after them and quench their impure fire," as a presentation from Carniola, aimed at amusing readers, puts it. Again, a nineteenth-century interview collecting folk plant knowledge at the foot of the Kamnik–Savinja Alps recorded the opinion that boys would run after a girl that had a plant from the *Orchis* genus with her. The old Slovenian vernacular name of this folk taxon, documented in the seventeenth and nineteenth centuries (*ceptec*—i.e., *cepetec*—still means 'sexual lust or agitation' in some dialects) also points to this kind of historical use.[3084] Such a plant name, later considered

[3082] Kos, 2015, 282–291, 294, 392–396, 401, 420–426, 430–433, 441–442.

[3083] Kos, 2015, 393–394.

[3084] Wraber, 1977–1978 (quotes from pages 70 and 71; Wraber refers to *Die Ehre Deß Hertzogthums Crain*, 1689, and to nineteenth-century correspondence with the botanist Wilhelm Voss); Kos, 2015, 393–394; Wraber et al., 1989–1990, 344 give a clear seventeenth-century proof for species identification, whereas Mlakar (2015, 468) erroindecent, is no exception in a wider European context. Bowdlerization of such plant names by means of their replacement with deliberately invented polite names complying with changed moral standards is known in some European languages in the second half of the eighteenth century and in the nineteenth century;[3085] in Slovenian, the expression *ceptec* was still used for orchids at the very end of the nineteenth century[3086] and later.[3087] Less telling but pointing in the same direction is the German equivalent of *ceptec*—*Knabenkraut*, literally, 'boy herb'[3088]—suggesting a more widespread use of orchids as an aphrodisiac in Central Europe and the Alps, not only in the southeasternmost territories of this mountain chain. Clear seventeenth- and nineteenth-century data from Slovenian territory, the very explicit old Czech plant name,[3089] and the use of underground parts of grassland orchids as an aphrodisiac in distant and culturally diverse Ghimeş in the Eastern Carpathians[3090] confirm that the German word *Knabenkraut* reflects its historical uses, not only the shape of tubers. Namely, as far as the genus *Orchis* is considered, this type of use was likely inspired mainly by the shape of its pair of tubers (the current tuber and the one from the previous year) resembling testicles.[3091] Numerous other parallels of historical treatment per analogiam (including the notion that parts of a plant have an impact on human organs of an analogous shape) were identified in early learned and in folk medicine[3092] (another likely example is the historical use of yellow gentian rhizome as an aphrodisiac for cows; see Section 8.1.2). This is known as the doctrine of signatures, the belief that a plant's "colour, shape and texture were designed to give some external indication" of the human use of that organism.[3093] The doctrine of signatures was not limited to flora, but also applied in historical healing uses of wild fauna.[3094]

---

neously writes about *Cypripedium calceolus*; see also *Slovar slovenskega knjižnega jezika*: *cepetec*. On habitats, see Aeschimann et al., 2004, vol. 2, 1112, 1124, 1128 (orchids mentioned in the paragraph above and clearly used as an aphrodisiac in history); on their historical abundance, see Jaksetič (interview with Kaligarič), 2019, 21–22.

3085 Thomas, 1987, 85.

3086 Jaksetič (interview with Kaligarič), 2019, 21.

3087 *Slovar slovenskega knjižnega jezika*: ceptec; see also Wraber, 1977–1978, 70–71.

3088 Kos, 2015, 394.

3089 Wraber, 1977–1978, 70.

3090 Babai, Molnár, & Molnár, 2014, 45.

3091 Wraber, 1977–1978.

3092 Brisebarre, 1996, especially pages 80–85; Babai, Molnár, & Molnár, 2014, 47; Mlakar, 2015, 287.

3093 Thomas, 1987, 84.

3094 Parallels from the historical use of wild fauna in vernacular medicine in the Alps following the doctrine of signatures include consumption of marmot (an animal well known for its hibernation in winter, and tolerating humid, cold below-ground conditions well) against insomnia and rheumatism, and consumption of ibex or chamois (well known for their balance on steep slopes) against vertigo (Scheurer, 2010, 78–79), or consumption of the testicles or penis of animals, such as a ram, goat, bull, stag or boar, against impotence (Kos, 2015, 393–394, 396).

Contrary to folk medicine, scientific medicine increasingly rejected the doctrine of signatures from the late seventeenth century onward.[3095] Future analyses will need to demonstrate whether the doctrine of signatures was the only motive for the use of some grassland orchids as (an ingredient of) an aphrodisiac in this case. Experiences with psychoactive impacts of black vanilla orchid (*Nigritella rhellicani*), another species from the Orchidaceae family, are relevant in this context. Some households in the Alps made a tisane including *Nigritella* orchids in the mid-twentieth century. This was recalled by several farmers in Carinthia managing non-fertilized subalpine and alpine meadows (Section 6.3.1), where (nowadays rare) *Nigritella* orchids used to occur frequently,[3096] and by only a very few interviewees from Bohinj. An experienced interviewee emphasized that, although the tisane from the dry blossoms of this species (now protected mainly due to historical pressures of gathering for decorative purposes and habitat loss due to the abandonment of farming in grasslands below the upper line of dwarf mountain pine) has a delicious scent, like chocolate, locals paid attention not to use more than one blossom per cup of tisane to keep their thoughts sober; the interviewee claimed that people would be intoxicated by black vanilla orchid tisane if more than one blossom per cup were used.[3097] Similarly, a study in Ghimeş found that people that consumed underground parts of fragrant orchid (*Gymnadenia conopsea*) or spotted orchid (*Dactylorhiza maculata*) started feeling intoxicated; they used this kind of grassland aphrodisiac sex-specifically, depending on the shape of the plant's underground parts.[3098]

8.2.2 From esteemed fodder plants to "nasty weeds" and plants or grassland lichens without recognized fodder value in animal husbandry

This section demonstrates that historical grassland gathering for the diet of livestock much exceeded the framework of the hay harvest. Several organisms, such as Alpine dock (*Rumex alpinus*), goosefoot species (*Chenopodium* sp.), nettles (*Urtica* sp.) and dead-nettles (*Lamium* sp.), thistles (*Cirsium* sp.), hogweed (*Heracleum sphondylium*), Iceland moss (*Cetraria islandica*), juniper (*Juniperus communis*; Section 6.6.2.2) and some others, including leaves and needles of several tree species (Section 8.2.2.1), had a long history of large-scale and purposeful gathering for pigs, cattle, horses, donkeys, sheep

[3095] Thomas, 1987, 84.
[3096] Keller, 2004, 21; Aeschimann et al., 2004, vol. 2, 1114–1119.
[3097] Use as an aphrodisiac was nonetheless not known to the interviewee, who showed remarkable trust and honesty during the interview, as confirmed by the response to the question about abortifacients (the identity of this interviewee from Stara Fužina born in the first half of the twentieth century is kept confidential).
[3098] Babai, Molnár, & Molnár, 2014, 45.

and goats, chickens, and some other domestic animals. This gathering type was either motivated by specific goals, or it was simply a hay substitute. Such specific goals were to provide more calorific fodder in the phase of livestock growth (e.g., for chickens or pigs) or fattening (for pigs, bulls, and oxen, for instance); to attain higher milk fat content than with hay or grazing alone; to gather feed with medicinal qualities;[3099] to compensate for a lack of minerals in hay (Section 8.2.2.1); to achieve specific dairy product qualities, such as color (see also Sections 6.6.2.2 and 8.2.2.1) or flavor; and so on. Among all large domestic animals, the diet of pigs in the Alps changed most significantly. Before the introduction of fodder plants and potatoes, the diet of pigs was largely based on grassland plants and lichens, other grazing (including forest grazing, highly significant in occasional autumns offering abundant pannage in oak or beech forests[3100]), dairy and cereal farming side products, such as whey, buttermilk,[3101] bran,[3102] barn sweepings (Section 6.5), stored acorns, and so on.[3103] Even though gathering certain grassland organisms as pig fodder continued until the mid-twentieth century in some parts of the Alps at least, the following presentation makes it clear that the development of agriculture with increased importance of fodder crops starting in the eighteenth and nineteenth centuries had an impact on gathering wild plants even in high summer mountain pastures.

The leaves of Alpine dock (*Rumex alpinus*, Figure 114 in Section 6.6.2), belonging to typical nitrophilous vegetation of highly nutrient-rich damp sites around cottages in middle and high-elevation pastures, mainly in the subalpine vegetation belt (Section 2), were gathered as fodder in many parts of the Alps (e.g., in the Allues (Francoprovençal *Alués*) Valley (French *Les Allues*), Switzerland, Vorarlberg, East Tyrol, and Bohinj), until the 1960s or earlier.[3104] Dock leaves are usually rich in proteins and minerals. Nonetheless, the remark on the adverse health impacts of oxalic acid in the case of immoderate human dock consumption (Section 8.2.1.1) is also relevant for animal husbandry, and (mature) dock leaves include some other substances that can have problematic impacts on the health of livestock. Although moderate quantities of boiled Alpine dock fodder were harmless, overdosing as a consequence of large-scale use of such fodder could lead to serious livestock health issues.[3105] Some leaves

[3099] E.g., Machatschek, 2006; Vogl-Lukasser et al., 2006, especially pages 76–85, 91, 145–146.
[3100] Szabó, 2013.
[3101] Niederer 1996d, 171, 173, 190; Ledinek, 1999, sections Predelava mleka, Prosti čas, Sirarstvo, and Živina; Machatschek, 2006.
[3102] Author's interview with Kunavec, 2017, in Stara Fužina.
[3103] Machatschek, 2006, 131; Zwitter, 2013a, 90; Zwitter, 2016, 124.
[3104] Meilleur, 1986, 375–377, 379, 381, 386, 388, 397, 402; Machatschek, 2006, 133–137; Vogl-Lukasser et al., 2006, 76–77; Aeschimann et al., 2004, vol. 1, 382–383. The state of evidence from Bohinj is presented below.
[3105] Dietl & Lehmann, 2004, 49; Machatschek, 2006, 137.

were used immediately after harvest, and so they were cut, and boiling (second) whey (i.e., the residue after butter, cheese, and albumin cottage cheese production from the same milk after rennet was introduced to cheese production; i.e., from the sixteenth century onward; see Section 8.2.1.1) was poured over them. This was a frequent pig fodder (the basis of pigs' diet in several places) in many temporary summer mountain settlements (except for some of the highest ones in the Alps located too high for Alpine dock to abound there) and also in some temporary settlements at medium elevations inhabited in spring and autumn. Several other locally collected plants, such as Good King Henry (*Chenopodium bonus-henricus*; see further plant and lichen ingredients of this fodder type in Section 6.5) could be added. Boiled Alpine dock with Iceland moss (*Cetraria islandica*) and barn sweepings or bran was not only a pig fodder but also a horse fodder. In addition, as examples from Switzerland and Vorarlberg demonstrate, large quantities of Alpine dock leaves were stored for winter. They were gathered in late June or early July, and again in late August or early September. The leaves were either harvested manually, so that the white underground leaf base was also collected, or reaped with sickles or mowed. They were either soured in fresh form with a bit of salt like cabbage, or they were boiled and strained, placed in boxes, the air was pressed out, and then the product was stored in a pressed form to sour. Both ways of making this kind of soured fodder were observed in Tyrol in the nineteenth century, for instance. The result was a kind of silage. Cooking and fermentation rendered the plants more easily digestible and their nutrients better accessible to pigs, also making pig fattening possible. In addition, cooking reduced limited impurities as a consequence of the fact that dock leaves were collected around cottages, so they were not entirely free of livestock urine. Moreover, cooking reduced the bitterness that gradually develops with maturation of Alpine dock and Good King Henry herbs. Sour fodder from Alpine dock was either used in summer (especially in late summer, when the availability of good fodder plants in grasslands decreased significantly) or it was transported from the mountain pastures and used between late autumn and spring. Similarly to fodder from freshly gathered Alpine dock, other plants were sometimes added to this silage, such as nettles, thistles (e.g., spiniest thistle, *Cirsium spinosissimum*, Figure 143), and Good King Henry (*Chenopodium bonus-henricus*). Spicing the dish with mint (e.g., horse mint, *Mentha longifolia*) was optional, possibly a relic from the time when Alpine dock was also eaten by people (see Section 8.2.1.1). Pigs liked this diet type very much.[3106] Bacon from pigs fed with such an Alpine-dock-based diet was considered significantly tastier and more durable than bacon from pigs fattened with slops. The other option of storing Alpine dock for winter fodder was to dry the leaves and use them crushed and

---

[3106] Meilleur, 1986, 381, 386, 388, 397, 402; Machatschek, 2006, 133–137 (and Novak, 1978, 58); Vogl-Lukasser et al., 2006, 76; Aeschimann et al., 2004, vol. 1, 382–383.

boiled. Drying in the air makes Alpine dock leaves fragrant, and so not only pigs, but also cattle liked them.[3107]

Families of timber workers that migrated from the Alps to the Giant Mountains (Cz. *Krkonoše*, Germ. *Riesengebirge*) in the Sudetes in the sixteenth century took Alpine dock with them as a source of fodder, as a vegetable and medicinal plant.[3108] Similar Alpine dock introductions also happened within the Alps. In the Zajavornik Alp on the Pokljuka Plateau in the eastern Julian Alps, for instance, Alpine dock did not occur, and so it was intentionally introduced from the Lipanca Alp, grown in gardens near cottages at Zajavornik, and cooked as fodder for pigs.[3109] Alpine dock cultivation in gardens in summer alps was nothing special in the broader context of the Alps. Plants such as Alpine dock, nettle (*Urtica dioica*), small nettle (*Urtica urens*), and various goosefoot species (*Chenopodium* spp.) were once so important that in several summer alps they were cultivated (in times of high agricultural density, at least): such places were manured and weeded, so that other grassland plants were removed.[3110]

In the course of modern history, cultivation or at least use of Alpine dock, nettle, small nettle, and various goosefoot species in pigs' diet was most likely (partly) replaced by fertilized and fenced gardens in the vicinity of every typical cottage in high summer mountain pastures to grow turnips as pig fodder, and salad and in some cases potatoes for people working in the temporary settlement and in mountain pastures. Such gardens were a usual feature for high summer temporary settlements in Bohinj, for instance.[3111] This text focuses on fodder plants; such turnip gardens existed at Zajavornik (1,280 m) on the Pokljuka Plateau,[3112] Konjščica (1,450 m, half surrounded by mountains),[3113] Krstenica (1,655 m)[3114] and the Na Kraju Alp (at 1,500 m) on the Komna Plateau,[3115] and so on. These gardens in sunny sites are clearly proven as a typical feature in the first half of the twentieth century, at several places still shortly after the Second World War, and exceptionally turnip cultivation in summer alps

3107 Machatschek, 2006, 135–137; Vogl-Lukasser et al., 2006, 76–77.

3108 Hejcman et al., 2013, 356; for Alpine dock uses in folk medicine, see Meilleur, 1986, 397, and in folk veterinary medicine see Vogl-Lukasser et al., 2006, 77.

3109 Novak, 2024, 105. The Zajavornik Alp is named Javornik on some maps, such as in *Atlas Slovenije*, 2005, map sheet 75 (compare the location of Zajavornik in Novak, 2024, inside back cover of the volume).

3110 Machatschek, 2006, 137.

3111 Novak, 1978, 58; Novak, 2024, 105–106; the connection with the earlier, wilder type of mountain pasture gardens is novel.

3112 Novak, 2024, 105.

3113 Hodnik, 2016, 8, 37, 59, 81, citing her grandmother, who used to work at Konjščica when there were turnip gardens.

3114 Author's interview with C. Mlakar, 2017, at the Zajamniki Alp; when she worked at the Krstenica Alp in the mid-twentieth century, there were turnip gardens.

3115 Novak, 2024, 105–106.

persisted longer.[3116] Fodder turnip cultivation there must have been a relatively recent phenomenon. Turnip was most likely not introduced to summer grazing complexes before the nineteenth century.[3117] A similar process might have taken place in some other parts of the Alps. It should be emphasized, however, that turnips were only sown at high summer cottages in late May, in early or mid-June, meaning, for instance, two to three weeks before livestock arrived there.[3118] Consequently, turnips were only available for pigs in the later part of the grazing season high in the mountains. On the one hand, these locally grown turnips helped feed the pigs in times of scarcer fodder availability in the alps at high elevations in late summer. Namely, fewer grazing resources in late summer and progress in gestation in cows caused a decrease in milk quantities over the course of summer. Consequently, the amount of whey available for pigs fell short of the demand in late summer, and so turnips were welcome at that time.[3119] On the other hand, late turnip ripening still retained the significance of gathering of several grassland organisms for pig fodder as an addition to grazing during much of summer, despite the existence of turnip gardens, as explained lower in this section.

There are differences in the chronology of the end of large-scale use of Alpine dock as pig fodder between various regions in the Alps. In Bohinj, it was mostly abandoned relatively soon. Consequently, unlike some other regions of the Eastern and Western Alps given at the beginning of the first paragraph on Alpine dock as a fodder plant, no interviewee recalled the use of Alpine dock in Bohinj from personal experience. However, an interviewee born in the early 1930s remembered the use of leaves of another or some other dock species (most probably including or limited to broad-leaved dock, *Rumex obtusifolius*, and/or curled dock, *Rumex crispus*, both occurring in arable land in the area[3120]) as pig fodder. These dock leaves were collected in fields and good meadows at the bottom of the valley, at an elevation of less than 600 m.[3121] As proven in East Tyrol, large-leaved docks such as Alpine dock and broad-leaved dock were used the same way,[3122] and so this finding is also relevant for *Rumex alpinus*. Moreover, historical use of Alpine dock as a fodder plant in the Zajavornik Alp (and, by analogy, at Lipanca) was mentioned above, and oral tradition still recalls that Alpine dock with boiling second whey poured over

---

[3116] Novak, 1978, 58; Hodnik, 2016, 8, 37, 59, 81; Novak, 2024, 105–106; author's interview with C. Mlakar, 2017, at the Zajamniki Alp.

[3117] General history of fodder crops, including the turnip; for example, in Valenčič, 1970a, especially 260, 262; compare Meilleur, 1986, 128.

[3118] Novak, 1978, 58; Hodnik, 2016, 81.

[3119] On the impact of changing fodder availability in high mountain pastures over the course of summer on pigs' diet, see Novak, 1978, 58; Novak, 2024, 105.

[3120] Vreš & Jogan, 2007, 216–219; compare Meilleur, 1986, 135.

[3121] Author's interview with Šetina, 2018, in Srednja Vas; data refer to the isolated farm at Holy Spirit Church south of Lake Bohinj (Sln. *Bohinjsko jezero*).

[3122] Vogl-Lukasser et al., 2006, 76.

it was historically used as pig fodder in the Laz Alp in the eastern Julian Alps (Figure 114).[3123]

Alpine dock demonstrates the shift in perception from an esteemed fodder plant to a weed.[3124] Most interviewees in Bohinj considered docks noxious weeds according to the mid-twentieth-century standards of their youth. They observed that cattle consumed full-grown Alpine dock only after the plants had been affected by hoarfrost[3125] or if the plants had been mowed and withered away or dried.[3126] Traditionally, they tried to limit and decrease the occurrence of Alpine dock in mountain pastures and other docks in manured meadows through the methods presented in Sections 6.6.1.1 and 6.6.2. As explained there, removal of the taproot relatively deeply below the ground is needed to prevent Alpine dock from resprouting, which explains that large-scale historical gathering by cutting its aboveground parts or by pulling the leaves from the soil a couple of times per year was easily sustainable.

Nettles (*Urtica dioica* and *Urtica urens*; Figure 141) were widely gathered as pig fodder (used boiled) and cattle fodder: dried nettle bunches were added to hay. Where cattle were accustomed to fresh young nettles, they were eager to consume them fresh in spring. Nettles were an addition to other fodder types mentioned elsewhere in this section and in Section 6.5. They were said to promote lactation, according to twentieth-century traditional ecological knowledge originating from an unknown time and recorded at some places in

3123 Information from the ethnologist and geographer Lucija Gartner from Bohinj, nowadays tending livestock and producing dairy products at Laz. The high abundance of Alpine dock in significant sections of summer mountain pastures close to cottages in Bohinj in the photos taken in the 1920s (see reproductions of the photos taken at Laz and Vogel in Novak, 2024, 45, 175) and in the mid-twentieth century (at Zajavornik, Pri Jezeru, Viševnik (the alp, not the mountain with the same name), Dedno Polje, Laz, and Konjščica, for instance: see the reproductions in Melik, 1950, 19, Figure 6; Hodnik, 2016, 40, Figure 13; Novak, 2024, 115, 125, 136, 242; author's interview with G. Gartner, 2017, at the Laz Alp, and photos exhibited on walls in the cheese dairy in Laz, taken by Marijan Lipovšek in the 1960s) partly points to the same direction of historical uses. Nonetheless, the photos from the mid-twentieth century should not be romantically interpreted in the sense that they present the landscape in times of maximum historical use of natural resources in summer alps. The workforce in summer mountain alps in Bohinj had already declined highly significantly by the 1950s and 1960s, as compared to times before the Second World War (about a half or two-thirds fewer personnel: Novak, 1978, 55). Consequently, removal of species perceived unwanted from the viewpoint of animal husbandry (Section 6.6) could not be as thorough in the 1950s and 1960s as it had been, for instance, in times of an exceptionally high agrarian density in the region in the late nineteenth century (Section 3.1). However, the text above suggests that quite a lot of Alpine dock was used in Bohinj as a fodder plant in the late nineteenth century.

3124 Vogl-Lukasser et al., 2006, 76.

3125 Author's interviews with L. and M. Odar, 2017, in Studor, and with Kunavec, 2017, in Stara Fužina.

3126 Vogl-Lukasser et al., 2006, 76; author's interview with L. Gartner, 2018, in Studor.

Figure 141: Nettle (*Urtica dioica*) abounds in a strictly limited section of a seminatural meadow. At this site, a pile of barn manure used to be stored in the meadow unusually long. Stands that colonize overly manured sites "remain stable for long periods" without additional manuring, so that the species composition becomes a legacy of the land-use history. Even more stable than overly manured sites in meadows, where nutrients are gradually removed with hay if the farmer abstains from fertilizing such a meadow section, are extremely nutrient-rich sites in pastures: around cottages and at other usual livestock resting sites in mountain pastures where excrement accumulates, above the potential forest line in particular. Namely, nettles, Alpine dock (*Rumex alpinus*), aconite (*Aconitum* spp.) or spiniest thistle (*Cirsium spinosissimum*) growing at such sites are avoided by grazing livestock, so that nutrients usually return from the decomposed plant remains to the soil. Because nettles, Alpine dock, and spiniest thistle are not used any more, nutrients are not removed and the site remains eutrophic for a long time. The plant nutrient availability at spatially strictly limited livestock resting sites in pastures and around cottages is extremely high, usually as high as or even higher than in most heavily manured lowland meadows. Extremely nutrient-rich sites populated by characteristic plant stands remain in the landscape for several decades, presumably for centuries, even in the case of abandonment of livestock grazing, unless they are heavily grazed by wild fauna, such as red deer. When pasture complexes are abandoned, in several cases such extremely nutrient-rich livestock resting sites have been the most stable landscape components in abandoned grasslands for more than half a century, for instance.[3127] Photo: Žiga Zwitter, November 2020.

[3127] Meilleur, 1986, 375–376, 381, 383, 387–388, 397, 402; Austad & Hauge, 1989; Pils, 1994, 220; Erschbamer, Virtanen, & Nagy, 2003, 386; Ellenberg & Leuschner, 2010, 37.

the Alps, such as East Tyrol and part of Slovenian territory.[3128] Moreover, according to traditional ecological knowledge from summer mountain pastures in East Tyrol, a mixture of nettles, juniper (see Section 6.6.2.2), and thistles was given to cows, in addition to grazing, the whole summer long, and in this case the milk fat content was considered significantly higher.[3129] Nettles also were relevant in historical livestock treatment.[3130] These pieces of historical knowledge deserve scientific evaluation in the future.[3131] They were not generally widespread throughout the Alps. A couple of old interviewees from Bohinj, for instance, confirmed that nettles have not been used there in the manner mentioned in times they could recall. The entire attitude of local people toward this plant genus is largely shaped by the stinging felt when it is touched.[3132] Similarly, fieldwork in Bohinj showed that some other old farmers in that area leave nettles unstored during the hay harvest.

Another species that was widely collected for pig fodder, especially in nutrient-rich meadows and pastures or in grasslands with medium nutrient availability, at elevations ranging from permanent settlements to the highest summer settlements, is hogweed (*Heracleum sphondylium*; Figure 142). It was particularly important as pig fodder in spring, when it grows much more quickly than grasses, so that leaves of hogweed were already available at the time the grass could not yet be mowed, not even in grassy gardens (Section 6.1.2.4). In spring at least, pigs consumed them fresh. Full baskets of hogweed leaves were also regularly gathered later, especially by those that did not have a sufficient grassy garden to feed pigs, and in some summer mountain pastures. Due to changes in pig rearing, hogweed is perceived only as a useless grassland plant because cattle do not consume its tall stacks in hay.[3133] In some places it

[3128] Makarovič, 1978, 239; Vogl-Lukasser et al., 2006, 77–78; Mlakar, 2015, 351; author's interview with Kunavec, 2021, in Stara Fužina; gathering nettles to feed pigs is still practiced by traditional Csángó farmers in the Eastern Carpathians (Babai, Molnár, & Molnár, 2014, 75).

[3129] Vogl-Lukasser et al., 2006, 78.

[3130] See examples of an invigorating drink for weakened livestock from nettles, silver fir bark, and hot water in the Allues Valley (Meilleur, 1986, 178), a practice of nettle seed application to strengthen livestock's organism after an illness in East Tyrol, administration of dry nettles there against cattle and sheep diarrhea (Vogl-Lukasser et al., 2006, 181), and use of nettles to treat cramps in pigs in the hilly surroundings of Idrija (Sosič, 2000, 247).

[3131] A locally limited example of a similar type of traditional ecological knowledge, which is actually an unfounded belief, was the saying from Stara Fužina that the herd of cattle that arrived first at the watering place (in a millrace in the village) on Easter would give the creamiest milk. The old informant, a local farmer, considered it uncertain whether this corresponded to reality or not (the identity of the interviewee is kept confidential).

[3132] Author's interviews with Kunavec, 2021, in Stara Fužina, and with I. Mlakar, 2021, at the Zajamniki Alp.

[3133] Author's interviews with Gašperin, 2017, with M. Hodnik, 2017, with Kranjc, 2018, with Kunavec, 2017, and with Mencinger, 2017, all in Stara Fužina; with C. Mlakar,

Figure 142: Hogweed (*Heracleum sphondylium*) below the Krstenica summer pasture complex. Photo: Žiga Zwitter, 2017.

has been uprooted from meadows to keep it out of second-crop hay.[3134] Such a highly labor-intensive suppression method proves the high regeneration abilities of hogweed, making its historical large-scale gathering sustainable.

Thistles have been considered noxious weeds by some agronomists; for example, in the early nineteenth century[3135] and still in the late twentieth century.[3136] Nonetheless, some thistles with spines that are not too hard are rich in proteins and were also used as a source of fodder for cattle and pigs. One of the expressions for spiniest thistle (*Cirsium spinosissimum*, Figure 143; in spite of the name, this thistle has comparatively soft leaves, much less spiny than, e.g., spear thistle, *Cirsium vulgare*) in East Tyrol is *Rahmdistel* (literally, 'cream-

2017, at the Zajamniki Alp; and with Minka Sodja, 2017, in Jereka; these interviews and the plants that C. Mlakar and M. Markeš—both with personal experience in gathering *brš*—showed me at the village of Podjelje and in the Zajamniki mountain pasture complex also made it clear that the dialect name *brš* (Hodnik, 2016, ix, 36, 60), previously identified only at the level of the genus *Heracleum* (Tuma, 1935, 251), refers at least to *Heracleum sphondylium*; Aeschimann et al., 2004, vol. 1, 1140–1141.

3134 Author's interviews with T. Hodnik, 2017, and with Kunavec, 2017, both in Stara Fužina.

3135 Peterken, 2013, 187.

3136 An example: Brugger & Wohlfarter, 1983, 203.

Figure 143: The occurrence of spiniest thistle (*Cirsium spinosissimum*) in pastures at Jezerce in the Julian Alps can illustrate significant historical changes in the occurrence and abundance of some species, even ones abounding at certain sites in a certain period, although grazing was not generally abandoned in the area. In the mid-nineteenth century, the botanist Valentin Plemel observed that the abundance of spiniest thistle around the Jezerce mountain grazing complex was very high.[3137] Five and a half hours of fieldwork in the great majority of pastures around Jezerce on July 18th, 2020, carried out for the purpose of this book, made it clear that this species rarely occurs there nowadays; it grows only at very few sites and even there in low numbers. One of the rare specimens in a pasture south of Jezerce, at an elevation of 1,740 m, is seen in the photo. The great decline in abundance of spiniest thistle there is most probably a result of several factors (see also the introduction to Section 4). Some of them are related to grazing: (1) in some years after Plemel's visit, Jezerce functioned throughout the summer as a pasture complex for horses only,[3138] and horses consume thistles in pastures,[3139] and so their impact was likely significant; (2) the temporary settlement at Jezerce was displaced to (the existing settlement of) Krstenica—this displacement was completed after the time of Plemel's observations (around 1900 or in the early twentieth century); and (3) later, numbers of livestock grazing in the entire complex of Krstenica (also including Jezerce) decreased.[3140] The second and third factors presented must have gradually resulted in smaller areas of nutrient-rich sites above Jezerce. As emphasized in the caption to Figure 141, reduction of nutrient availability is a very slow process. Photo: Žiga Zwitter, July 18th, 2020.

[3137] Plemel, 1862, 130: "sehr häufig in der Wocheiner Alpen bei Jezerce."

[3138] Novak, 2024, 66.

[3139] Babai, Molnár, & Molnár, 2014, 117, 147.

[3140] On the history of grazing there in the late nineteenth and twentieth century, see Ledinek, 1999, sections Kuharjev Joža, Planinsko naselje, Podoba in struktura planine, Sirarstvo, Stanovi, and Voda, and Novak, 2024, 66; *Arcanum Maps*, Maps of Europe, Europe in the XIX. [sic] century (with the Third Military Survey), Habsburg Empire (1869–1887) – Third Military Survey (1:25,000).

thistle'). Namely, by the mid-twentieth century, this species and some other thistles were gathered by herders in high mountain summer pastures, finely cut, mixed with bran—addition of nettles (*Urtica dioica* and *Urtica urens*), thyme (*Thymus* sp.), and juniper (*Juniperus communis*) was optional—and given to milk cows fresh as an addition to grazing. This fodder was believed to increase milk fat content significantly.[3141] Spiniest thistle often abounds at highly nutrient-rich damp grassland sites characterized by close to neutral pH, in the alpine or subalpine vegetation belt (Section 2), in soils on carbonate or silicate bedrock. Shallow soils suffice. The alpine belt is too high for a high abundance of Alpine dock (*Rumex alpinus*) or tall herbs frequently abounding at similar sites in the subalpine belt, whereas the upper limit of spiniest thistle's occurrence is far above 2,000 m (for instance, about 2,460 m in southern Germany or its surroundings). Spiniest thistle is an exceptional species because its biomass production is very high, despite the harsh environmental conditions at such high elevations. The height of this herb can exceed one meter. No other species produces a comparable amount of biomass annually at such elevations,[3142] and so farmers were lucky that, with some labor investment, their livestock could make good use of it. High biomass production and high abundance at appropriate sites also mean that not much time was needed to gather significant quantities, another factor that made the species attractive for gatherers.

In some parts of the Alps, such as Bohinj, use of thistles seems to have been limited to pigs' diet, at least in the mid-twentieth century. Use of thistles in the Alps as pig fodder characteristically involved fresh leaves, not dried ones; they were cut and boiling liquid (whey, where available) was poured over them, improving their digestibility; if possible, bran also was added.[3143] Some soft and possibly most pleasant-tasting thistle species such as cabbage thistle (*Cirsium oleraceum*) and Carniolic thistle (*Cirsium carniolicum*) were consumed fresh by pigs.[3144] Pigs particularly liked the sprouts of some thistle species,[3145] but, considering the species dealt with here, they were also eager to consume leaves of older specimens in summer. Moreover, gathering probably caused regrowth of young leaves. In the Julian Alps, for instance, the leaves of yellow thistle (*Cirsium erisithales*; Figure 144), cabbage thistle (*Cirsium oleraceum*), and Carniolic thistle (*Cirsium carniolicum*) were collected on a large scale as pig fodder, at all elevations where they occur and abound. The abundance of cabbage thistle is high at some damp sites in colline and montane vegetation

[3141] Vogl-Lukasser et al., 2006, 78–79, 84.
[3142] Pils, 1994, 191, 199–200, 217, 324 and Table 53, photo 136; Aeschimann et al., 2004, vol. 2, 586–587.
[3143] Author's interview with C. Mlakar, 2017, at the Zajamniki Alp; Machatschek, 2006, 131, 133–135, 137.
[3144] Author's interviews with M. Hodnik, 2018, and with Kunavec, 2017, both in Stara Fužina.
[3145] Machatschek, 2006, 134, 137, 139.

belts in base- and nutrient-rich soils, whereas yellow thistle (mesic sites, base-rich soils, medium nutrient availability) and Carniolic thistle (damp sites, base- and nutrient-rich soils) abound higher, in the montane and subalpine belts. Not the entire plant, but particularly the large leaves of the basal rosette growing close to the ground were gathered. There are no hard spines on the leaves of these three species, contrary to many others in the genus *Cirsium*, and so they are easily eaten by livestock even when fully developed (but see the example of spiniest thistle above, edible for livestock despite the spines). Herders from those temporary mountain settlements where these thistles grow—Konjščica and Krstenica in the Julian Alps, for instance—used textile bags about 1.5 m long and 80 cm wide to gather thistle leaves. Thistles were commonly collected together with hogweed if both plants grew in the vicinity; otherwise, collection of yellow thistle (and/or Carniolic thistle at these elevations) alone was also an option. Apart from the size of bags, the fact that bags full of leaves were rolled down the slope on the way back to cottages where the terrain made rolling possible also proves that gathering took place on a large scale. If the bags had been light, they would have been carried. Such a quantity sufficed for several days.[3146] As long as the apical bud of the rosette remains intact, thistles possess a high regeneration ability,[3147] and so our observation, focused on yellow thistle, proved that it is very common in areas where its leaves used to be collected on a large scale (Figure 144). Thus, the practice of gathering thistle leaves on a large scale appears to have been sustainable. Large-scale gathering of yellow (or Carniolic) thistle and hogweed leaves evidently did not end when turnip gardens were introduced to the same summer mountain pastures (clearly proven, for instance, for Krstenica and Konjščica).[3148] The late availability of ripe turnips, as explained above, must have contributed to this. It is nonetheless noteworthy that large-scale use of Alpine dock as fodder had already been abandoned in Bohinj by the mid-twentieth century, whereas large-scale thistle and hogweed gathering still took place there at that time. This means that, even though herders had to mow Alpine dock to prevent its spread across too extensive areas of grasslands providing high-quality fodder (Section 6.6.2),

[3146] Tuma, 1935, 251; Hodnik, 2016, ix, 36, 60; author's interviews with C. Mlakar, 2017, at the Zajamniki Alp; with Minka Sodja, 2017, in Jereka; and with M. Hodnik, 2021, in Stara Fužina: the interview with Mlakar made it clear that the hitherto undeciphered dialect names *przof* (Ledinek, 1999, sections Prosti čas, and Živina), *przofc* (Novak, 1978, 58), *frzofc* (Hodnik, 2016, ix, 60), and *prezofc* (author's interview with Minka Sodja, 2017, at Jereka) designate yellow thistle (*Cirsium erisithales*), at least: the interviewee described the plant and stated that its leaves used to be collected on a large scale at the eastern foot of Mount Ogradi, belonging to the Krstenica mountain pasture complex; I visited the place and collected plants there that to a certain degree corresponded to the description. These were brought to the interviewee with personal experience of historical *przof* gathering, and she was asked to identify the right one. Site conditions: Aeschimann et al., 2004, vol. 2, 584–587.

[3147] Machatschek, 2006, 139.

[3148] Hodnik, 2016, ix, 36, 60; author's interview with C. Mlakar, 2017, at the Zajamniki Alp.

Figure 144: Yellow thistle (the plant with dissected leaves) still abounds at a location at the eastern foot of Mount Ogradi, where herders from the temporary settlement of Krstenica used to collect its lower leaves on a large scale. Photo: Žiga Zwitter, 2017.

they nonetheless reserved additional time for thistles (and hogweed) gathering instead of extensively using the dock leaves as a side product of maintenance mowing. It is likely that negative earlier historical experiences with excessive use of Alpine dock leaves in terms of impacts on pigs' health might have been a reason for this. A lack of documented use of thistles in research literature does not prove that they did not represent a significant part of pigs' diet in the area. The importance of gathering is frequently underestimated in economic history and economic geography literature.[3149] However, it is also true that appropriate

[3149] For instance, Anton Melik's mid-twentieth-century volume on mountain pastures and adjacent temporary settlements in the Julian Alps ignores thistles. It states, imprecisely, that in temporary mountain settlements pigs were fed with dairy side-products, and, better (in another chapter), that dairy side-products represented the basis of pigs' diet in temporary mountain settlements (Melik, 1950, 123, 157). Our interviews and the literature cited above prove that thistles and hogweed were used at some temporary summer settlements dealt with by Melik at the time of his work. Similarly, Vilko Novak's synthesis of the history of animal husbandry in the Alps in Slovenian territory only states that in temporary settlements adjacent to mountain pastures pigs grazing there consumed whey as well, whereas gathering is not mentioned (Novak, 1970, 366, 392), and Bogo Grafenauer's short synthesis of the history of gathering in Slovenian thistle species were not available in all microregions of the Alps. The abundance of almost no species from the genera *Cirsium* and *Carduus* can be high on acidic soils on silicate bedrock,[3150] for instance, and suitable fodder thistles can be locally absent in generally more suitable sites.[3151]

Highly relevant for domestic animals and gathered on a large scale in grasslands at high elevations was also the lichen Iceland moss (*Cetraria islandica*; Figure 140). It was collected in the mountain summer pastures and mountain meadows, not only for medicinal purposes (see Section 8.2.1.3). Namely, at several places in the Alps, such as in the Julian Alps, Carinthia, East Tyrol, the Enns Valley, and the Lower Tauern (Germ. *Niedere Tauern*), Iceland moss was an important source of calorific cooked fodder for pigs, thus the name *Saugraupe* (*Sau* means 'sow') in the Enns Valley and the Lower Tauern. In some regions of the Alps it was also used to feed cattle (e.g., milk cows or to fatten bulls), sheep, goats, and donkeys.[3152] According to folk tales from the Ötz Valley (Germ. *Ötztal*) and the Swiss canton of St. Gallen, the "milk-richness" of some complexes of mountain pastures was related to Iceland moss and some other organisms (see also Section 4.1.1.2).[3153] Iceland moss was either gathered manually or raked; in the latter case, rakes of a specific shape appropriate for Iceland moss harvest were used (thus, this task was so relevant that it resulted in a specific implement at some regions) – narrow rakes were adjusted to the uneven relief in pastures.[3154] Gathering was carried out in humid weather (in heavy dew, in rain, or thereafter), when this lichen is heavily swollen, easy to see, easy to remove by raking, and all the biomass gathered is clean, which would not be the case in dry weather.[3155] Namely, when the relative (air) humidity is low, lichens dry rapidly[3156] and reach a state of anabiosis, so that

territory does not document grassland plant gathering to feed pigs (Grafenauer, 1970b, 465–466), even though this was one of the most large-scale and widespread gathering types targeting plant species selected in a well-considered manner.

3150 Aeschimann et al., 2004, vol. 2, 570–591.

3151 Herders observed that no suitable thistles were available for pigs near the high summer mountain settlement of Ovčarija, in the Julian Alps, for instance (author's interview with Kunavec, 2021, in Stara Fužina), even though site conditions are similar in many aspects in some nearby mountain pasture complexes where suitable thistles locally abound (*Environmental Atlas of Slovenia*, Land and Soil, Soil Map, 1:25,000).

3152 Machatschek, 1997b, 114; Keller, 2004, 13–14, 21; Machatschek, 2006, 133, 138–139, 142; Vogl-Lukasser et al., 2006, 82, 146; Mlakar, 2015, 432; Novak, 2024, 221; author's interviews with Gašperin, 2017, and with Mencinger, 2017, both in Stara Fužina.

3153 Machatschek, 1997b, 112.

3154 Moser, 1952, 468, 473; Machatschek, 1997b, 114; Keller, 2004, 13–14, 21; manual gathering in Bohinj: author's interviews with Kunavec, 2021, in Stara Fužina, and with I. Mlakar, 2021, in Bohinjska Češnjica, both with personal experience in gathering Iceland moss.

3155 Machatschek, 1997b, 111–112, 114; Keller, 2004, 13–14, 21; Vogl-Lukasser et al., 2006, 82–83.

3156 Matthews, 1993.

drought causes no harm to them, regardless of its duration.[3157] Iceland moss was gathered in summer (appropriately humid weather was more frequent in late summer) and in autumn, until October.[3158] Gathering dates in the second half of August and in September must have been usual where this lichen was collected in summer pastures and used fresh.[3159] Iceland moss gathering for winter fodder in remote meadows took place in autumn, when farmers had more time; at Saureggen in the Nock Mountains (Germ. *Nockberge*), Carinthia, it was collected on a large scale by raking its sites in high mountain meadows in October.[3160] Drying made it easy to store stocks of this lichen for winter;[3161] substances in the tissue make dry Iceland moss resistant to mold and insect attacks, and so it was a reliable source of stored fodder.[3162]

Iceland moss gathering for winter fodder in the Alps seems to have had a long history, at least in some of the regions where it occurs, but its use as fodder seems to have increased during the First World War. In the mid-nineteenth century, for instance, the Tyrolian natural scientist Adolf Trientl mentioned that Iceland moss can be used as excellent fodder.[3163] However, an investigation in Bohinj in the Julian Alps showed that in the mid-twentieth century this manner of using Iceland moss was not as common there as its role in folk medicine and folk veterinary medicine (Section 8.2.1.3). Archival evidence suggests, but does not firmly prove, that these differences in use—only as medicine or also as fodder—might have (partly) been a consequence of the First World War economy. In times of general scarcity during that war (Section 7.2), economic use of wild plants and lichens was stimulated,[3164] reviving earlier uses of some wild organisms[3165] or introducing the use of organisms that had previously not been used that way in these regions. In late 1915, the government of the province of Carniola, in accordance with the guidelines of the Ministry of Agriculture, sent recommendations promoting the use of lichens as feed surrogates to the district administrative offices of the province, including the district authority (Germ. *Bezirkshauptmannschaft*) of Radovljica (Germ. *Radmannsdorf*), which Bohinj belonged to. The district administrative authority forwarded these feed recommendations to the lowest level of territo-

---

3157 Pils, 1994, 105.
3158 Machatschek, 1997b, 114; Keller, 2004, 14, 21; author's interviews with Kunavec, 2017, and with Gašperin, 2017, both in Stara Fužina.
3159 For instance, Machatschek, 1997b, 114; author's interview with Gašperin, 2017, in Stara Fužina.
3160 Keller, 2004, 13–14, 21.
3161 Machatschek, 1997b, 114; Keller, 2004, 21, Machatschek, 2006, 138.
3162 SI AS 137, Okrajno glavarstvo Radovljica (District Administrative Authority [Germ. *Bezirkshauptmannschaft*] of Radovljica, Germ. *Radmannsdorf*), box 591, November 10th, 1915, no. 30866.
3163 Machatschek, 1997b, 112; on Trientl's biography: Hofinger & Bergmann, 2015.
3164 Himmelreich, 2001, 17; Studen & Kresal, 2005, 150–153; Fischer, 2005, 148–150.
3165 Poschlod, 2015, 105–113.

rial administrative units, to municipalities, and requested that they announce them publicly.[3166] In addition, these recommendations were sent to the editorial boards of both daily newspapers in Slovenian that were most relevant for the province of Carniola, the Catholic *Slovenec* and the liberal *Slovenski narod*.[3167] Indeed, in the first half of January 1916, an identical short summary of these guidelines was published in *Slovenec*[3168] and in *Slovenski narod*,[3169] relevant media for some village intellectuals and farmers because literacy in this part of the Alps rose rapidly and reading culture spread quickly in the late nineteenth century.[3170] In both ways, through proclamations and newspapers, these (summarized) recommendations became accessible to several farmers in the area. Proposed feed surrogates included Iceland moss due to its high calorific value. According to these guidelines, *Cetraria islandica* could replace a certain percentage of potatoes. If used for feeding pigs, guidelines recommended cooking Iceland moss in whey or buttermilk, or, less optimally, using it fresh and raw, again with whey or buttermilk. The addition of boiled potatoes to such feed was recommended, as was a gradual introduction of lichen fodder, meaning small quantities first and more later when the pigs got accustomed to it.[3171] These considerations are correct: Iceland moss is rich in carbohydrates

3166 SI AS 137, Okrajno glavarstvo Radovljica (District Administrative Authority of Radovljica), box 591, November 20th, 1915, no. 31141, and November 10th, 1915, no. 30866.

3167 SI AS 137, Okrajno glavarstvo Radovljica (District Administrative Authority of Radovljica), box 591, Notica za Slovenca & Slovenski Narod, s.d., following document no. 30866, November 10th, 1915; for a general history of Slovenian newspapers in the region, see Štih, Simoniti, & Vodopivec, 2008, 329; Balkovec, 2013, 197; Šlebinger, 1937, especially 13–14, 19.

3168 "Lišaji kot živinska poklaja," 1916, 5

3169 "Lišaji kot živinska krma," 1916, 4.

3170 Vodopivec, 2005, 61–62; Štih, Simoniti, & Vodopivec, 2008, 326–329. Before the outbreak of the First World War, 85% to 89% of the Slovenian population of the province of Carniola was literate; almost all children attended school from 1869 onward (Vodopivec, 2005, 61–62). At least intellectuals in villages, such as an elementary school teacher living in a room at a farm, subscribed to one of the daily newspapers, and some of them also allowed the farmer they were staying with and the neighbor to read the newspaper daily. Information from newspapers could also spread further orally, including information announced by priests. Moreover, newspapers were also likely available at inns, where farmers read them. This is how it also worked in this area; for example, in the village of Stara Fužina, between the First and Second World War. In the interwar period at the latest, *Slovenec* was read at least at one of those farms at Stara Fužina that used Iceland moss as pig fodder in the mid-twentieth century. This farmer, born in the late nineteenth century or in the first years of the twentieth century, borrowed it from a teacher living next door (author's interview with Mencinger, 2017, in Stara Fužina).

3171 SI AS 137, Okrajno glavarstvo Radovljica (District Administrative Authority of Radovljica), box 591, November 10th, 1915, no. 30866, and December 21st, 1915, no. 35377. The German and Slovenian versions differ in one important aspect: the German version includes 'buttermilk' and the Slovenian version 'skimmed milk' in the same place. It is clear that buttermilk was much more easily available for pigs than skimmed milk (starch, about 60% of this lichen), which is exactly the type of a macronutrient otherwise available to pigs only in small quantities in nutrient-poor sections of summer mountain pastures. Moreover, Iceland moss becomes available in several mountain pastures in the humid weather of late summer, when the availability of good plant fodder in pastures declines significantly. Thus, from the economic viewpoint, Iceland moss was very important.[3172] Farmers from several regions of the Alps, such as Bohinj and Carinthia, that fed pigs with boiled Iceland moss in the mid-twentieth century noted this high nutritive value, claiming that boiled Iceland moss was "fatty" or "oily."[3173] In the pastures, livestock could graze on swollen *Cetraria islandica* in rainy weather when better-tasting fodder was no longer available. Namely, fresh Iceland moss is bitter, but, when herders poured boiling liquid on gathered *Cetraria islandica*, the bitterness faded. Moreover, in many places farmers discarded the first hot water, which became bitter in a minute after Iceland moss was added, and only used as fodder the second, much less bitter, liquid and the boiled lichen, characterized by a slight smell of mushrooms. Thus, if Iceland moss was properly prepared, pigs liked it very much. They could also be fattened using Iceland moss, although the increase in weight was not comparable to that of animals fed with potatoes or cereals.[3174] This was again relatively favorable as far as animal husbandry in summer mountain pastures is considered; the main goal of feeding pigs in high summer mountain pastures was that they grew significantly.[3175] Farmers did not want to fatten them too much there, otherwise the risk was too great that pigs would die on their long way to a lower grazing complex or back home in autumn due to too much physical strain. A farmer belonging to the old generation recalled that she provided her mother keeping their livestock in the Ovčarija mountain pasture complex in the third quarter of the twentieth century with bran, but the mother did not use it in order not to fatten the pigs too much for their long way on foot from Ovčarija (about 1,700 m) to Grintojca (about 1,200 m), which they had to make in one day: about 6 km aerial distance and probably close to 10 km in reality, including descents and ascents.[3176] Boiled Iceland moss was often mixed with Alpine dock (*Rumex alpinus*), and/or nettles (*Urtica dioica*, *Urtica urens*) and/or flax

in times of scarcity, and so this term is used in this text. The newspaper summaries only mention lichens and do not name *Cetraria islandica* specifically ("Lišaji kot živinska krma," 1916, 4; "Lišaji kot živinska poklaja," 1916, 5)—but Iceland moss is one of the lichens recorded in the first archival document cited in this footnote.

3172 Machatschek, 1997b, 114; Machatschek, 2006, 138–139; compare Novak, 1978, 58.

3173 Keller, 2004, 13, 21; author's interviews with Gašperin, 2017, in Stara Fužina, and with C. Mlakar, 2020, at the Zajamniki Alp.

3174 Machatschek, 1997b, 111–112, 114; Machatschek, 2006, 133, 138–139; author's interview with Gašperin, 2017, in Stara Fužina.

3175 Machatschek, 2006, 131. In cases of inappropriate fodder, pigs would not grow much, they would remain small; compare Doll, 2003, 153.

3176 Author's interview with Kunavec, 2021, in Stara Fužina.

seeds and/or bran or barley or barn sweepings and other usual gathered fodder plants (see above in this section and Section 6.5).[3177]

Reports on the regularity of use of Iceland moss as fodder in the mid-twentieth century differ. In many parts of the Alps, large quantities of this lichen were consumed by livestock. At Saureggen in the Nock Mountains, an enormous quantity of one cart[3178] of Iceland moss, collected in meadows and pastures, was stored per average farm; the lichen was intended for feeding pigs there, and it represented a significant component of the total winter pig fodder at that time.[3179] Similarly, it was reported in Bohinj that at least two weeks had to pass between the end of feeding a pig with Iceland moss and slaughter because otherwise the meat was bitter.[3180] This clearly demonstrates that quantities of Iceland moss fed to pigs at some farms in Bohinj were high. Data from East Tyrol in the (mid-)twentieth century provide a more diversified picture. They range from regular daily fodder mixture including Iceland moss as pig fodder at a summer mountain settlement to an example of a farmer (considered exceptional), who offered some Iceland moss to her 'super cow' every morning.[3181] The occurrence and abundance of Iceland moss, and thus the time people took to gather this lichen, differed significantly. The question whether feeding pigs (apparently most usual) or cattle also mattered. In cases of overdose, Iceland moss could have negative health impacts.[3182]

Historical large-scale Iceland moss gathering seems to have been sustainable in some cases. At Saureggen, for instance, where enormous quantities of Iceland moss were harvested, *Cetraria islandica* populations were observed to have strongly declined after the abandonment of gathering in the 1960s, and so adapted sustainable gathering might have had a positive impact on the abundance of this lichen.[3183] Late gathering dates (e.g., in October) might have ecologically mattered somewhat in this respect, but more important must have been raking itself. Namely, Iceland moss rarely reproduces generatively; most usual is reproduction "by means of thallus fragmentation or the dispersal of isidia[,]"[3184] hence raking most probably fostered vegetative reproduction. Haymaking with a scythe most likely also had a favorable impact on Iceland moss reproduction in high meadows: a record from Carinthia explains that

3177 Vogl-Lukasser et al., 2006, 82, 145–146.

3178 A cart is a variable unit; based on a rough estimate from mountain farms in Carinthia, 300 to 340 kg was one of the possible weights of a load of hay transported by one cart (calculated from data in Keller, 2004, 30).

3179 Keller, 2004, 13–14, 21.

3180 Author's interview with A. Kavčič, 2017, at the Laz Alp in the Julian Alps.

3181 Vogl-Lukasser et al., 2006, 82.

3182 On the impacts of overdosing, see Vogl-Lukasser et al., 2006, 169, and Galle Toplak, 2015, 181.

3183 Keller, 2004, 14, 21.

3184 Matthews, 1993. In addition to isidia, soredia are also mentioned by Matthews, but they are not typical for Iceland moss.

meadows rich in Iceland moss could only be mowed in moist weather or afterward, when grassland plants and lichens were wet, otherwise the haymaker could not cope with the scythe blade slipping and getting stuck in the lichen.[3185] This means that Iceland moss was to a certain degree mowed together with grassland plants, and so mowing, spreading, and turning the hay resulted in dispersal of Iceland moss talus fragments. Nonetheless, the growth rates of lichens, even as large as Iceland moss, are incomparably lower than those of higher plants of meadows and pastures, so that regeneration periods need to last for several years or even decades to make Iceland moss gathering sustainable.[3186] Iceland moss thallus grows, for instance, 0.5 cm annually.[3187] This means that large-scale historical gathering could not take place every year at the same place, if it was sustainable. According to the traditional ecological knowledge of an experienced herder from Stara Fužina, born in the first half of the twentieth century, with personal experience in Iceland moss gathering for pig fodder in the Laz mountain pasture complex, people say that it takes seven years (to be taken approximately because seven is one of the usual numbers in folk sayings) before gathered Iceland moss regenerates.[3188] Considering meadow sites, *Cetraria islandica* must have profited from longer intermowing intervals, such as mowing only once every second or third year, which was frequent in the history of remote, non-fertilized meadows at high elevations in several regions of the Alps (Section 6.1.2.6). A parallel with *Cetraria islandica* harvest in Iceland is telling: there, Iceland moss colonies "are harvested every three years to allow the lichen to grow to a 'profitable size.'"[3189] Thus, mowing every third year (see Section 6.1.2.6) was likely quite compatible with gathering *Cetraria islandica* in the Alps. In addition, the abovementioned fact that it was impossible to mow grassland stands rich in Iceland moss in dry weather most likely led to some meadow sections rich in this lichen left unmowed in many years when the weather was too dry for mowing there. This, again, decreased the frequency of Iceland moss removal.

In many parts of the mountain chain where this lichen occurs, large-scale historical Iceland moss gathering is hardly imaginable today. Data from several regions in the Alps such as the Julian Alps and various places in the Alps of Carinthia demonstrate that Iceland moss became much rarer than it had been in the mid-twentieth century, even in still-managed grasslands in nutrient-poor sites.[3190] The decreasing abundance of this lichen is probably a result

[3185] Keller, 2004, 14–15.
[3186] Zarabska-Bożejewicz, Studzińska-Sroka, & Fałtynowicz, 2015.
[3187] Matthews, 1993.
[3188] Author's interview with A. Kavčič, 2017, at the Laz Alp.
[3189] Data for Iceland from Matthews, 1993.
[3190] Julian Alps: author's interviews with Grm, 2017, in Srednja Vas in Bohinj; with A. Kavčič, 2017, at the Laz Alp; and with I. Mlakar, 2020, at the Zajamniki Alp. Iceland moss used to be gathered at both mentioned alps. Carinthia: Keller, 2004, 21; Tyrol: Machatschek, 1997b, 114–115.

of the interplay of several factors. Potentially excessive gathering (especially where performed manually, not by raking) in some regions during the economic crisis in the mid-twentieth century (see Section 8.1.2), the end of large-scale but not too frequent Iceland moss gathering for fodder using rakes that had probably fostered Iceland moss in some other regions, nutrient intake from the polluted atmosphere (see Section 10.1.2.2.4), and in some cases mineral fertilizer application even in high mountain pastures[3191] seem to be four of them.

A few other grassland plants were also gathered for pigs, and there were probably some regional differences in the selection of plants gathered for this purpose, impacted not only by the abundance and occurrence of plants but also by cultural factors. In the limestone Alps of the Bovec region of the Upper Soča catchment (Julian Alps), where there were a lot of meadows in the colline and montane vegetation belts at relatively dry, moderately fertilized to non-fertilized, base-rich to neutral soils in which meadow clary (*Salvia pratensis*) frequently abounds, this species was often used as one of the ingredients of pig fodder, as reported for the first half of the twentieth century. In the context of the meadow vegetation at such sites, meadow clary is one of the forbs with relatively large leaves, which must have been one of the motivations for gathering for pigs. Moreover, farmers there sometimes gathered chicory (*Cichorium intybus*), and they even uprooted wild carrot (*Daucus carota*) for pigs, as reported for the first half of the twentieth century;[3192] gathering wild carrots in meadows required detailed traditional ecological knowledge, so that some extremely toxic species such as hemlock (*Conium maculatum*) were not mistaken for carrots.[3193]

Alpine spicknel (*Ligusticum mutellina*; see also Section 4.1.1.2), a species that pigs liked very much, was uprooted and added fresh or dried to fodder in at least some parts of the Alps.[3194] Further species that used to be gathered for pigs include, for instance, cat's ears (*Hypochaeris* sp.), such as young leaves of *Hypochaeris uniflora*, growing in nutrient-poor high summer pastures with medium water availability. This is yet another example of a gathered plant that used to be scalded and some bran was added to make pig fodder. In this way, its bitterness was reduced and pigs liked it, whereas the plant was too hairy to be eaten by people.[3195]

Fodder for hens gathered in grasslands, except for mowed grassland plants,[3196] also included targeted gathering, such as nettles (*Urtica dioica*,

[3191] E.g., personal observation of a large bag of mineral fertilizers at one of the high summer alps in Triglav National Park.
[3192] Makarovič, 1978, 239; Aeschimann et al., 2004, vol. 2, 164.
[3193] Compare Janovsky (interview with Cortese), 2014.
[3194] Machatschek, 2006, 139; Vogl-Lukasser et al., 2006, 91.
[3195] Machatschek, 2017, 176, 179, 182, 184–185.
[3196] Author's interview with Kunavec, 2021, in Stara Fužina.

Urtica urens), dead-nettles (*Lamium* sp.), common sorrel (*Rumex acetosa*), and comfrey (*Symphytum officinale*), for instance, as demonstrated by twentieth-century examples from East Tyrol.[3197] Nettles were widely used as hen fodder, also confirmed by traditional Csángó farmers in the Eastern Carpathians.[3198] Gathering hen fodder included some further herbs, such as watercress (*Nasturtium officinale*), available in wetlands very early in spring.[3199]

8.2.2.1 Gathering tree-leaf fodder

Use of tree leaves as livestock fodder (Figures 145, 147, and 148) was a usual practice in most regions of animal husbandry in Europe, including the Alps, with a rich tradition that must have been rooted in prehistory. Within the last millennium, it played a significant role from the Middle Ages to mid-twentieth century. Feeding livestock on tree leaves decreased significantly in the nineteenth century and virtually disappeared in the twentieth century.[3200] Storing tree leaves for fodder lay in the contact zone between gathering and cultivation,[3201] as explained in this section.

This book deals with the history of grasslands, with the main emphasis on meadows, thus harvesting tree leaves in pollard-meadows is presented in this section, but tree-leaf fodder gathering also took place in many pastures and in some forests.[3202] Trees in meadows and pastures take advantage of greater availability of light than in a forest; the quantity of leaves growing on freestanding trees far exceeds the leaf mass of trees in a closed-canopy forest at a similar site.[3203] Use of leaves from trees in meadows and at meadow edges increased the total fodder yield from the meadow. A tree in good condition pruned every second year provided more than five times as much fodder as the grassland below its crown, as a rough estimate; in practice, there existed significant variability. Nutrients built into leaves were largely taken from deeper soil, inaccessible to grassland plants; consequently, by and large, trees did not decrease soil nutrient availability for grassland plants. The quantity of tree leaves was much more stable than the quantity of hay from the surrounding meadow plants.

3197 Vogl-Lukasser et al., 2006, 77, 83, 85.

3198 Babai, Molnár, & Molnár, 2014, 75.

3199 Author's interview with Kunavec, 2021, in Stara Fužina.

3200 SI AS 137, Okrajno glavarstvo Radovljica (District Administrative Authority of Radovljica), box 592, folder *Preskrba z živili, prekupčevanje, nastavitev zaupnikov za nadzor zalog, potrebe*: August 28th, 1915, no. 22751; Machatschek, 2002; Kotnik, 2002/2003, 62; Emanuelsson, 2009, 155–157, 176, 196–201, 204–207; Peterken, 2013, 257–260; Niederer, 1996d, 189–190; Zwitter, 2013a, 91; Zwitter, 2014e, 321, 347; Zwitter, 2015d, 150; Hartel, Craioveanu, & Réti, 2016, especially pp. 56, 61–62; Zwitter, 2016, 122–123.

3201 Vogl-Lukasser et al., 2006, 54.

3202 Niederer, 1996d, 189–190; Machatschek, 2002, 37, 129; Emanuelsson, 2009, 156–157, 196, 200–201.

3203 Machatschek, 2002, 41.

Figure 145: Linden-tree pruning at Zakraj in the pre-Alpine hills in the Cerkno region, August[3204] 1954. Note the man in the tree. In this case, linden trees were pruned every third year. Photo: Boris Orel. Slovenian Ethnographic Museum, Documentation: Fieldwork campaign 11: Cerkno region (1954): photo 11/53 and *Slovenski etnografski muzej* . . . Cerkljansko / 34 F0000011/053. © Slovenian Ethnographic Museum.

Tree-leaf fodder, for, instance, mitigated the impacts of drought hazard on total grassland harvest. Namely, frequent leaf-fodder tree species—common ash (*Fraxinus excelsior*) and linden (*Tilia* sp.), for instance—access water deeper in the soil than grassland plants, and the shade of trees mitigated the drought hazard in the surrounding grassland.[3205] Tree roots in grasslands on potentially unstable slopes reduced the hazard of mass movements (Section 4.2).

3204 This photo was taken between July 30th and August 31st, 1954; that is, at an unknown time during that ethnological campaign. The date July 30th in the online database of the museum (*Slovenski etnografski muzej* . . . Cerkljansko / 34 F0000011/053), which is most unlikely (see below the usual pruning dates and their impact on trees), is fictitious. The e-system requires an exact date, and so the museum enters the first day of the campaign when a precise date is not known (the ethnologist Miha Špiček, personal communication).

3205 Machatschek, 2002, 34, 44, 48–49, 59, 73–75, 282, 417, 425, 495; Sever, 2005, 9–11; Emanuelsson, 2009, 198, 204; Peterken, 2013, 102; author's interview with Kranjc, 2020, in Stara Fužina. Nonetheless, common ash is known for its particularly high water demand (Sever, 2005, 11), and so it is drought sensitive (Brus, 2012, 366).

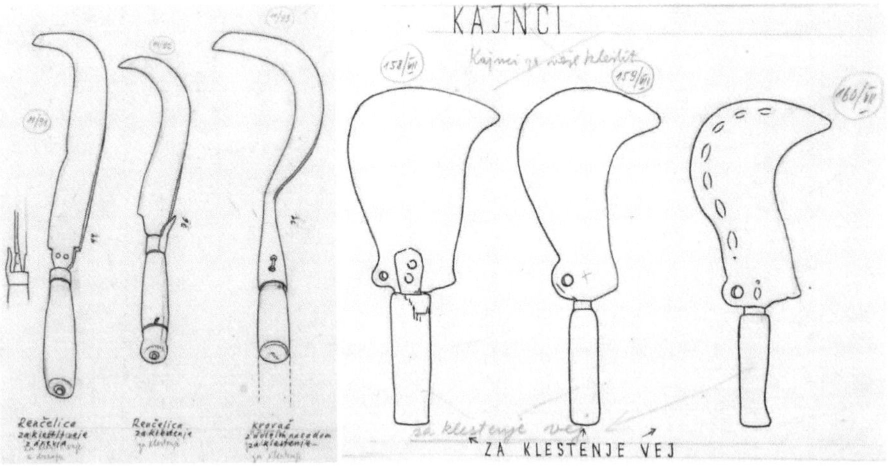

Figure 146: Implements for tree pruning for leaf-fodder gathering as documented at two sites in the Slovenian Alps and pre-Alps in the 1950s. Prapetno Brdo in the Cerkno region (left), and Kred in the Kobarid region (right). Such pruning implements could function as multi-purpose tools not limited to leaf-fodder gathering. The implements on the right allow cutting thicker branches than those on the left, which are suitable for separating small branches from the thicker ones, as explained by data from the Upper Savinja Valley (Sln. *Zgornja Savinjska dolina*) in the Alps of northern Slovenia; there, implements resembling those on the right and those on the left were usual.[3206] Many of these tool forms are ancient, archaeologically documented in Southeast Europe since late prehistory and classical antiquity. It is not clear how widespread their use was in the subsequent centuries and millennia,[3207] although they seem to have been widely used in the timeframe dealt with in this book. A tool that is likely a pruning implement resembling the first one from the left in this figure can be seen in the Sunday Christ wall painting (see Section 4.1.1.4) at Ormalingen near Basel, created in the second half of the fourteenth century.[3208] A pruning implement partly resembling the second one on the right (however, with a longer blade) can be seen in the 1557 Sunday Christ wall painting at St Roch's church in Tesero in Alto Adige (South Tyrol).[3209] Ordinary people were expected to recognize the implements to understand the message of the Sunday Christ painting. Although such paintings themselves do not provide reliable insights in the forms of locally used implements—because the artist might have been inspired by painted examples from distant places or have observed farm implements at distant places where he might have spent a significant part of his life—the connection with ethnologically documented implements in this figure demonstrates that implements of similar shapes were indeed used in the southeastern Alps in previous centuries.[3210] Both drawings presented in this figure are the result of ethnological fieldwork. Slovenian Ethnographic Museum, Documentation, Fieldwork campaign 7, Kobarid (1951): drawing 7/161 and *Slovenski etnografski muzej* . . . Kobarid / 176 R0000007/161; Fieldwork campaign 11, Cerkno Region (1954): drawing 11/91 and *Slovenski etnografski muzej* … Cerkljansko / 29 R0000011/91. Drawn by Ivan Romih © Slovenian Ethnographic Museum.

[3206] Predan, 1956b, 25–26.
[3207] Henning, 1987, 93–96 and Table 35, Table 36.
[3208] Wildhaber, 1956, 9–10 and Table 2, Figure 9.
[3209] Wildhaber, 1956, 16–17 and Table 7, Figure 21.
[3210] Compare Wildhaber, 1956, 25.

Figure 147: The crown shapes of many historically pollarded trees in meadows and in hedges at meadow edges show clear legacies of historical pruning[3211] (see Figures 147 and 148). Look at the trunk above the lower part of the crown and note the sharp transition from the main stem to a multibranched crown where this common ash (*Fraxinus excelsior*) used to be pruned in contemporary history. Montane meadow on Vogar Hill, at 1,035 m; the closest village of Stara Fužina lies at 550 m Photo: Žiga Zwitter, May 24th, 2020.

The list of tree species pruned for leaf fodder differed significantly in the Alps, depending on site conditions and local habits. Differences in tree species occurrence and abundance, site conditions, tree management, and leaf-fodder management significantly impacted the quality of leaf fodder. The impacts of site conditions were so significant that they led to diverse local quality orders

[3211] Compare Machatschek, 2002; Emanuelsson, 2009, 155–157, 176, 196–201, 204–207.

of the trees providing leaf fodder and to diverse quantities of leaf yields from comparably managed trees of the same species.[3212] Ashes (*Fraxinus* sp., in the Alps most frequently common ash, *Fraxinus excelsior*), elms (*Ulmus* sp., especially wych elm, *Ulmus glabra*, and large-leaved elm, *Ulmus laevis*), lindens (*Tilia* sp.), and mainly in harsher climates sycamore (*Acer pseudoplatanus*) belonged to most usual leaf-fodder trees, but it depended on the region and livestock type which leaf-fodder tree was the most highly esteemed.[3213] Species and genera, such as common ash, elm (wych elm, *Ulmus glabra*, at least), linden (*Tilia* sp.), and sycamore (*Acer pseudoplatanus*) were also planted in meadows and meadow edges; it was not simply a matter of allowing some naturally occurring trees to grow.[3214] In general, ashes or eventually elms were in first place among leaf-fodder trees in Central Europe, both providing tasty fodder that was easy to digest. One of the advantages of some elm species was their ability to keep relatively many nutrients in their leaves late in the season; their leaves also stay green very long in autumn. Linden leaves are not as tasty as ash and elm leaves—their taste is rather neutral—but the leaves are easy to chew and digest, and they were widely used.[3215]

Common ash (*Fraxinus excelsior*) is a heliophilous species,[3216] and so it grows well in meadows. Mid-nineteenth-century data from Upper Austria[3217] and early-nineteenth-century records from the Julian Alps and Western Karawanks (Sln. *Karavanke*, Germ. *Karawanken*)[3218] make clear that common ash

[3212] Machatschek, 2002, 59–60, 65–66, 104, 235, 294, 439; Emanuelsson, 2009, 198–199; Hartel, Craioveanu, & Réti, 2016, 61. See also the impacts of pruning on leaf quality and quantity, and comments on the significance of processes during leaf drying on fodder quality later in this section.

[3213] Machatschek, 2002, 105–107, 109, 213–214, 239, 291–293, 299, 300; Machatschek, 2015, 57, 59; Melik, 1950, 117; Niederer, 1996d, 190; Emanuelsson, 2009, 198–199, 201; Šmid Hribar & Lisec, 2011, 175–176, 178; Klinar, 2012, 101; Peterken, 2013, 260; Geršič, Gabrovec, & Zwitter, 2018, 71, 73.

[3214] Machatschek, 2002, 25, 128, 299, 304–305; Emanuelsson, 2009, 199; Šmid Hribar & Lisec, 2011, 176, 178; and the testimony of an old farmer that planted linden trees for leaf fodder at the Kogelnik farm in Planina (author's interview with an anonymous farmer, 2011, Planina north of Ljubno ob Savinji in the Kamnik–Savinja Alps of northern Slovenia).

[3215] Machatschek, 2002, 108–109, 213–214, 239, 291–294, 299–300, 312, 456.

[3216] Sever, 2005, 3, 11.

[3217] Pils, 1994, 23, 283.

[3218] Kačičnik Gabrič, 2016, 503, 510; SI AS 176, Franciscejski kataster za Kranjsko (Franciscan Land Cadaster for Carniola), L10, cadastral municipality of Jesenice, cadastral appraisal protocols: Catastral-Schätzungs-Elaborat, Einleitung, § 4, and Schätzung des natural ertrages, § Wiesen mit Holznutzen in the document dated July 10th, 1833, standing in front of § 8, and Protokoll dated August 9th, 1830, § VIII; SI AS 176, L30, cadastral municipality of Nemški Rovt, cadastral appraisal protocols: Catastral-Schätzungs-Elaborat, Einleitung, § 4; SI AS 176, L56, cadastral municipality of Bohinjska Bistrica, cadastral appraisal protocols: Catastral-Schätzungs-Elaborat, Einleitung, § 4; SI AS 176, L72, cadastral municipality of Gorjuše, cadastral appraisal protocols:

belonged to the most frequent trees pruned for fodder in pollard-meadows. These data clearly also reflect conditions since the eighteenth century, and to a more limited degree since the sixteenth and seventeenth century, because a common ash tree can live as long as two or three centuries, and pruning adapted to site conditions increases the longevity of the specimen.[3219] It is no wonder that ashes were mentioned as relevant trees providing leaf fodder in Zell (Sln. *Sele*) in the Central Karawanks in 1697.[3220] Common ash was still one of the most usual pruned trees to obtain leaf fodder from in the Julian Alps and Western Karawanks in the mid-twentieth century.[3221] At that time, the same preferred role of common ash is recorded in meadows in the mountains of Carinthia;[3222] common ash was one of the most usual trees pruned for leaf winter fodder in East Tyrol[3223] and in the Allues Valley.[3224] These data from several places make it clear that common ash used to be one of the most esteemed tree-leaf fodder species in the Alps in the last centuries. Ashes do not grow in highest meadow complexes in the Alps, with about 1,300 m currently representing the upper limit of common ash (*Fraxinus excelsior*) occurrence at many places; in some regions of the Alps, the species exceptionally reaches elevations of up to 1,600 m.[3225] Leaf yields close to the upper limit were significantly lower than at lower elevations. Common ash trees could nonetheless be pruned there every second year, and a higher number of pruned common ash trees could compensate for lower yields. It is noteworthy that site conditions significantly impacted not only yield quantity, but also the quality of ash-tree leaf fodder, and so some other tree species—such as small-leaved elm, (*Ulmus minor*), broad-leaved linden (*Tilia platyphyllos*), small-leaved linden (*Tilia cordata*), sycamore (*Acer pseudoplatanus*), and some others, in certain cases even

Catastral-Schätzungs-Elaborat, Einleitung, § 4; SI AS 176, L90, cadastral municipality of Javornik, cadastral appraisal protocols: Catastral-Schätzungs-Elaborat, Einleitung, § 4, and Schätzung des natural ertrages, § Wiesen mit Holznutzen in the document dated July 12th, 1833, standing in front of § 9, and Protocoll dated August 31st, 1830, § VIII; SI AS 176, L102, cadastral municipality of Koroška Bela, cadastral appraisal protocols: Catastral-Schätzungs-Elaborat, Schätzung des Natural Bruto Ertrages, § Wiesen mit Holznutzen in the document dated July 13th, 1833, standing in front of § 8; SI AS 176, L157, cadastral municipality of Srednja Vas v Bohinju, cadastral appraisal protocols: Catastral-Schätzungs-Elaborat, Einleitung, § 4; SI AS 176, L179, cadastral municipality of Zgornje Gorje, cadastral appraisal protocols: Catastral-Schätzungs-Elaborat, Einleitung, § 4.

3219 Sever, 2005, 10; Emanuelsson, 2009, 198; Machatschek, 2002, 478; Grove & Rackham, 2003, 194; Machatschek, 2015, 57.

3220 Kotnik, 2002/2003, 62.

3221 Melik, 1950, 117; Sever, 2010, supplements A, B, and C; Šmid Hribar & Lisec, 2011, 175–176, 178; Hodnik, 2016, 73; author's observations of legacies of historical pruning in tree crown forms in meadows in Bohinj, in the valley and in montane meadows.

3222 Keller, 2004, 13.

3223 Vogl-Lukasser et al., 2006, 54, 65, 71, 79, 85, 248.

3224 Meilleur, 1986, 324, 348–349.

3225 Sever, 2005, 9.

oak (*Quercus* sp.)—provided nutrient-richer leaf fodder than common ash at sites barely suitable for common ash.[3226] Oaks were the usual leaf-fodder trees in some regions. At elevations too high for common ash and at some other sites not allowing the growth of high-quality ash leaves, pruning for fodder included sycamore (*Acer pseudoplatanus*), birch (*Betula*), green alder (*Alnus alnobetula*; in general, alder causes constipation in sheep if consumed in large quantities. During the First World War, alder, oak, and hazel leaves were promoted for use in sheep farming as fodder additions to grassland fodder that was too moist, probably to suppress parasites; see also Section 6.1.2.6). Leaves of several other tree and bush species also were pruned and used as fodder (for instance, see Section 6.6.2.2 on juniper, *Juniperus communis*).[3227]

Due to regular pruning, leaf-fodder quality was significantly higher (a less bitter and less sharp taste, and more tender leaves) than from trees that were not regularly pruned. Other pruning impacts included (1) quicker shoot growth than on non-pruned specimens belonging to the same tree or bush species; young branches of pruned trees could grow as much as 1.5 to 2 meters in two years, (2) larger leaves than in the case of trees that were not regularly pruned, and (3) many more leaves on the same branch length of pruned deciduous trees than on non-pruned ones.[3228]

Pruning in the Alps to obtain winter fodder usually took place at two times of the year, in late spring and in late summer: in June—not earlier because more nutrients accumulate in leaves after the frost hazard is over—or in (late) August or September. According to nineteenth-century data, pruning in June was widespread in the Alps, but pruning in late summer was also usual. The phase of the Moon was respected for timing pruning in the same way as in grassland hay harvest (Section 6.1.2.1.2); pruning was carried out during a waning Moon, at some places at least, such as in the Malta Valley (Germ. *Maltatal*) in Carinthia or in East Tyrol. Pruning in June usually resulted in the highest mineral content, favorable protein content, and tastiest tree-leaf fodder. The leaves of oaks—for example, sessile oak (*Quercus petrea*), turkey oak (*Quercus cerris*), common oak (*Quercus robur*), and white oak (*Querus pubescens*)—and some other tree species such as birch (*Betula* sp.), hornbeam (*Carpinus betulus*) and poplars—aspen (*Populus tremula*) and black poplar (*Populus nigra* subsp. *nigra*)—could only be used as appropriate fodder if trees were pruned regularly and pruning was carried out when the leaves were rel-

[3226] Machatschek, 2002, 76–83, 294, 304–305.

[3227] Machatschek, 2002, 90, 104–110, 112, 291–334, 370, 391–393, 512–513; SI AS 137, Okrajno glavarstvo Radovljica (District Administrative Authority of Radovljica), box 592, folder *Preskrba z živili, prekupčevanje, nastavitev zaupnikov za nadzor zalog, potrebe*: August 28th, 1915, no. 22751; Melik, 1950, 117; Meilleur, 1986, 324, 355; Vogl-Lukasser et al., 2006, 85; Klinar, 2012, 101; Zwitter, 2013a, 91; Machatschek, 2015, 57, 59; Geršič, Gabrovec, & Zwitter, 2018, 71, 73.

[3228] Machatschek, 2002, 102, 138, 147–148, 165–167, 193–194, 219, 257, 312, 329, 355, 358.

atively young; for instance, the palatability of birch leaves pruned later than in June was too low. As a general rule, bitterness faded during the drying of leaves cut green. Oak leaves, for instance, had to be dried; consumption of fresh oak leaves could trigger health issues. The later general pruning timing, in (late) August or September, usually resulted in higher quantity of the leaf yield, but its quality was no longer as high as in June. Linden species are an exception; their mild taste has already been mentioned, and so lindens were pruned in August or September (Figure 145) to attain at least a somewhat stronger taste and allegedly also higher nutritive value than in June. As a general rule, pruning in July was less frequent, in the case of skilled farmers limited to specific species not allowing a later harvest due to an increasing tan content. Pruning in July was highly likely to jeopardize the tree in winter. Namely, in the case of pruning in June, shoots that regrew were strong enough by the first frosts; pruning in late August or September usually resulted in no leaf and branch regrowth in the same year, again allowing smooth overwintering of the tree; pruning in July would result in regrowth still sensitive to frosts.[3229]

Tree pruning frequency to obtain leaf fodder depended on site conditions and the demand for leaf fodder. Pruning annually or every second or third year resulted in higher feed quality than in the case of less frequent pruning. Common ash trees were pruned every year at certain sites; in this way, the fodder quality was the highest. However, in many environments in the Alps (for instance, with drier soils and/or a colder climate), common ash trees could not cope with annual pruning, and so they were pruned every second year. It was possible to gather leaves from the tree by hand or with a narrow rake in years when pruning did not take place, or to use fallen leaves as lower-quality fodder or bedding. Regrowth after pruning in June also allowed gathering leaves from the tree crown in autumn or using leaves fallen off in autumn.[3230]

Properly treated recently fallen leaves of ashes, elms, sycamore, alders, apple (*Malus domestica*), and wild cherry (*Prunus avium*), for instance, were still considered (relatively) good fodder, even though their nutritive value was lower than of pruned and dried green leaves. Leaves became increasingly bitter and less digestible in late autumn, and their nutrient content decreased. If the tree was pruned in June, the quality of regrown young, recently fallen leaves in autumn was high.[3231]

[3229] Machatschek, 2002, 88, 108–110, 135, 191–197, 232–238, 312, 322; Vogl-Lukasser et al., 2006, 79; SI AS 137, Okrajno glavarstvo Radovljica (District Administrative Authority of Radovljica), box 592, folder *Preskrba z živili, prekupčevanje, nastavitev zaupnikov za nadzor zalog, potrebe*: August 28th, 1915, no. 22751.

[3230] Machatschek, 2002, 18, 42, 76, 80, 86, 130, 132, 135–138, 143, 148, 167, 194, 236–239.

[3231] Machatschek, 2002, 108, 138, 142, 191–197, 200, 232–238, 271–275, 278–279, 334, 336–338; Vogl-Lukasser et al., 2006, 79–80, 85; Emanuelsson, 2009, 196; author's interviews with Gašperin, 2017, in Stara Fužina, and with Šetina, 2018, in Srednja Vas in Bohinj. See also the text on leaf fermentation at the end of Section 8.2.2.1.

Figure 148: An old regularly pruned tree at a sheep farm at Javorniški Rovt in the Western Karavanks, resprouting in spring after it was pruned in the previous late summer. Note that the branches were not sawed off at the trunk but cut a bit further away, a prerequisite for successful pollard-tree management.[3232] Photo: Rok Dovjak, May 1st, 2019.

Many farmers recall experiences with their livestock proving the high relevance of a limited percentage of high-quality tree leaves in the total fodder for the organism of the livestock in preindustrial agriculture. For instance, a farmer in Bohinj remembered that when their farm used fallen apple tree (*Malus domestica*) leaves (possibly also leaves of wild trees in the meadow, such as common ash, *Fraxinus excelsior*) raked in a meadow as animal bedding, their

[3232] Compare Machatschek, 2002.

cows consumed half of these leaves thrown to the barn floor and used only the remaining half as bedding.[3233] A farmer from the Bergamo Alps (Ital. *Alpi Orobie*) recalled that a cow fell ill in the barn and veterinarians could not diagnose the illness. The cow lost much weight and its milk yield fell significantly; on the one hand, she was lethargic and stood up with difficulty, but on the other hand she would also bolt, and so on. The cow managed to escape from the barn and broke several fences while running to a stream bank, where she fed greedily on elder (*Sambucus* sp.), birch (*Betula* sp.), and alder (*Alnus* sp.). In a couple of days, the cow recovered, regained her appetite, and gave milk again.[3234] At present, farmers in East Tyrol using fallen common ash (*Fraxinus excelsior*) leaves as sheep bedding observe that their sheep are eager to consume large quantities of this "bedding" if it is stored in good condition.[3235] Livestock also consume leaves, fine branches, and young bark from trees and bushes when they can freely select their feed in pastures; such a fodder selection is not limited to sheep and goats, and cattle also prefer tree leaves in some cases.[3236] The main reason behind these given examples of animal feeding behavior is that tree leaves contain several minerals (largely) absent from other historical livestock fodder types, such as hay and cereal straw. Minerals from leaves have a positive impact on livestock health and performance. Due to the high mineral content in tree leaves, (some) farmers increased the percentage of tree leaves in fodder after the livestock had performed heavy physical activity. Again, due to the mineral content in tree leaves representing a significant percentage of all fodder, (some) farmers reduced the quantities of salt provided to their livestock.[3237] This was relevant because salt used to be a relatively expensive commodity: in seventeenth-century Carniola (including the southeasternmost Alps), its price equaled the price of quality cereals, and salt cost 1.5 times as much as low-quality cereal grains.[3238] Nonetheless, minerals from tree leaves only replaced a certain quantity of salt, and farmers combined both.[3239]

[3233] Author's interview with Šetina, 2018, in Srednja Vas. The interviewee mentioned the leaves of fruit trees. Another farmer from Stara Fužina emphasized that, among the fruit trees growing in this Alpine basin, cattle preferred apple leaves (pear leaves, for instance, were rejected; author's interview with Gašperin, 2018, in Stara Fužina). In addition, in 2022, the author of this text visited the farm in question, where the interviewee's daughter confirmed that there were apple trees around the farmstead at that time.

[3234] Machatschek, 2002, 302–303.

[3235] Vogl-Lukasser et al., 2006, 79–80.

[3236] Machatschek, 2002, 27–28, 245.

[3237] Machatschek, 2002, 18, 28, 189, 192, 205, 207, 209–210, 214, 218, 224, 226, 232–233, 392–393, 411; Niederer, 1996d, 190; Machatschek, 2015, 84; Machatschek, 2017, 93, 347.

[3238] Zwitter, 2014d, 672.

[3239] Machatschek, 2002, 226; e.g., a seventeenth-century farmer provided salt to his cattle twice a week (Pavkovič, 2021, 101–104) in times when one can assume a significant role of tree-leaf fodder.

Proper tree management, appropriate site conditions, favored tree species—for example, ash (*Fraxinus* sp.) or elm (*Ulmus* sp.)—and proper management of harvested tree leaves led to fodder quality as high as in some meadows, but consisting of different substances. Dry leaves were used in cold months, whereas freshly pruned ones were used as an addition to early spring grazing (Sections 6.2.1.2 and 6.2.2.1); when grazing resources in late spring, summer, or autumn for any livestock type were scarce (e.g., because of drought); or to increase feed diversity. Nutrients in high-quality tree-leaf fodder exceeded medium-quality grassland hay. Farmers observed that the addition of high-quality tree leaves to poor-quality (or even middling-quality) grassland hay resulted in more and higher-fat milk. Beta carotene from leaves colored butter more intensively yellow, and livestock were healthier than if fed on grassland and straw fodder alone. Cereal straw was a usual but particularly nutrient-poor fodder ingredient in preindustrial agriculture, so that simultaneous use of (second-crop) hay, straw, tree leaves and, usually since the eighteenth or early nineteenth century, clover, compensated for poor straw and some poor-quality grassland fodder. Similarly, where sheep received poor-quality wetland hay, it was important to add appropriate tree leaves (see the above note on parasite suppression). The quantities of tree leaves consumed by any type of dairy livestock during the lactation period had to be moderate, otherwise there existed the risk of bitter milk and difficulties in cheese production. This is reflected in testimonies in the use of tree-leaf fodder for non-dairy cattle. However, dairy cows, milked sheep, and dairy goats also received small quantities of tree leaves in many cases for its favorable impacts on milk quality and quantity. Dry tree leaves were more recommendable than fresh ones because the latter could have a negative impact on milk taste. Livestock type mattered. Sheep and goats particularly liked tree leaves (high-quality ash leaves were a desert for them), and they also consumed young twigs and bark. Tree leaves could easily represent the majority of their fodder. Sheep consume only the tastier parts, and so fodder from properly dried leafy branches from the right tree species cut in the right part of the year was as good for them as middling to high-quality grassland hay (tree leaves often reach the feed value of alfalfa, *Medicago sativa*; the twigs are less nutritive). Sheep and goats safely received entire bundles of leaved branches to consume the edible parts. Where enough quality ash leaves were available, sheep and goats received them regularly. Elsewhere, sheep, goats, horses, mules, and donkeys often received lower-quality tree-leaf fodder, whereas better tree-leaf fodder was saved for dairy cattle or pigs. Where sheep and goats were important dairy animals, high-quality leaves were usually part of their feed. In contrast to sheep, cattle had to get used to tree leaves being added to hay, optimally at a young age. For cattle accustomed to such a diet, the quality of leafy tree branches resembled some low-quality to middling grassland hay. In contrast to sheep, only leaves were used in cattle feed or the twigs were finely cut, crushed, and often scalded or fermented and mixed with other fodder types. Feeding cattle with fallen apple-tree leaves was easy and cattle liked them very much. Cattle, horses, and pigs liked an appropriate percentage of tree leaves in fodder. For these livestock types, tree leaves represented a minor percentage of the fodder. Cattle rarely received more than a third of their feed in tree leaves; only calves liked high-quality tree leaves more than grassland fodder. Fresh or dry tree (e.g., ash) leaves were (at least in combination with other fodder types, such as hay and/or cereals) used as a relevant ingredient in fodder aimed at fattening the cattle, such as calves, bulls, and oxen. Fresh leaves were highly esteemed in this context. Large quantities of crushed dry tree leaves—mainly ash (*Fraxinus* sp.), elm (*Ulmus* sp.), or linden, (*Tilia* sp.)—were also used for fattening pigs, alongside other fodder types (e.g., Section 8.2.2). In several cases, dry leaves were crushed and boiling water was poured over them to reduce bitterness before feeding, as in feeding pigs; leaves were also sometimes scalded for cattle, although cattle also consume dry tree leaves without any treatment if the quality is high. At Bohinjska Bela in the Julian Alps, the use of a mixture of dry fruit-tree leaves and cereal chaff, scalded to increase the digestibility, is documented as one of the better types of winter fodder for cattle in the nineteenth century.[3240]

This section has explained that use of tree leaves as fodder required more specific knowledge than using grassland hay, including (1) leaves of which tree species are most suitable, (2) what a proper pruning timing for the species was to make good fodder and not to jeopardize the tree, (3) what the optimal pruning frequency in the local microenvironment was, and (4) the awareness that pruning a regularly pruned tree resulted in better fodder than pruning trees usually not pruned. Moreover, skilled tree-leaf fodder management respected (5) the knowledge of how to shape the tree's crown through pruning to allow highly productive management of the tree species in the long run, and also (6) knowledge of appropriate ways to dry and store leaves. Namely, the aim was to provide a drying environment that was neither too dry and airy nor too

[3240] SI AS 137, Okrajno glavarstvo Radovljica (District Administrative Authority of Radovljica), box 592, folder *Preskrba z živili, prekupčevanje, nastavitev zaupnikov za nadzor zalog, potrebe*: August 28th, 1915, no. 22751; Machatschek, 2002, 74–75, 78–81, 138, 189, 192–193, 202–205, 207, 209–210, 212–214, 216–218, 220, 224–227, 232–237, 245, 247–248, 250–251, 256–258, 276–277, 284, 392–394, 410–412, 430, 437, 440, 444, 446, 449, 497, 499, 504, 512–513; Novak, 1970, 363, 373, 383, 389; Valenčič, 1970a, 262–263, 265–268, 270; Meilleur, 1986, 278, 324, 348–349; Niederer, 1996d, 189–190; Makarovič, 2005, 302, 365, 380, 570; Vogl-Lukasser et al., 2006, 79–80, 85; Emanuelsson, 2009, 155–157, 196–201; Klinar, 2012, 101; Machatschek, 2015, 84; Hodnik, 2016, 73; Machatschek, 2017, 92–93, 174, 347; author's interviews with Gašperin, 2017, and with Urh, 2017, both in Stara Fužina. Information from Bohinjska Bela: Šarf, 1974, 134. Many thanks to Matevž Gartner, a cheesemaker and farmer that studied pharmacy, for his explanation that beta carotene from fodder colors butter yellow. The main author's observation of cattle browsing dry leaves from cut maple branches in a pasture at Grießen in the province of Salzburg in late September 2024, where the availability of grassland fodder was low.

moist or too closed so that the leaves would dry relatively slowly, including appropriate fermentation, increasing tree-leaf fodder quality, but at the same time preventing a damaging mold attack.[3241] In practice, farmers' traditional ecological knowledge differed significantly,[3242] and so not every farmer knew the facts briefly outlined in this section well enough to be able to produce high quality tree-leaf fodder for the livestock.

Feeding tree leaves to livestock accustomed to consuming some tree-leaf fodder was not a sign of poverty or scarcity, as explained in this section, unless (1) the percentage of tree leaves in the total fodder was excessive, (2) scarcity led to pruning trees or bushes not regularly pruned, or to pruning in a part of the year when leaf-fodder quality was low, such as birch (*Betula* sp.) pruning later than in spring or oak (*Quercus* sp.) pruning in autumn, or (3) a lack of fodder required pruning species of trees or bushes providing less appropriate leaf fodder, such as rowan (*Sorbus aucuparia*) or even European silver fir (*Abies alba*).[3243]

The decrease in and virtual abandonment of tree-leaf fodder uses in the nineteenth and twentieth centuries were a combination of several interrelated factors. They include (1) campaigns by agricultural experts against tree-leaf fodder based on lack of traditional ecological knowledge on the part of experts; the result was inadequate experiments showing lower quality of analyzed tree-leaf fodder than obtainable by skilled farmers; (2) lack of traditional ecological knowledge about tree and tree-leaf fodder management among some farmers resulting in poor quality of tree-leaf fodder and milk; (3) labor intensity (a bit more than a quarter of an hour to prune a medium-size pollarded tree for an experienced farmer, plus later ripping of leaves and small twigs from larger shoots); (4) higher grassland hay availability when former arable land was converted into meadows and a significant increase in moderately fertilized meadow sites resulted in higher grassland yields (see Sections 3.1, 6.3, and 10.1.2.1); similar were impacts of the introduction of mineral fertilizers to grasslands in the twentieth century, and so farmers ended the use of tree-leaf hay in many places; a significant decrease in livestock numbers at many traditional farms also contributed to higher grassland hay availability, and meadow hay is easier to make than tree-leaf fodder; (5) widespread use of field crops in livestock diet (see Section 10.1.2.1, the introduction to 10.1.2.2, and Section 10.2.1); (6) a strong general decrease in keeping goats and sheep (see Sections 6.3.1.1 and 9); these animals were the main consumers of tree leaves; (7) introduction of

[3241] Machatschek, 2002; Niederer, 1996d, 190; Emanuelsson, 2009, 155–157, 196–199; SI AS 137, Okrajno glavarstvo Radovljica (District Administrative Authority of Radovljica), box 592, folder *Preskrba z živili, prekupčevanje, nastavitev zaupnikov za nadzor zalog, potrebe*: August 28th, 1915, no. 22751.

[3242] Zwitter, 2020.

[3243] Niederer, 1996d, 190; Machatschek, 2002, 42–43, 78–79, 109–110, 370, 391, 398–399; Vogl-Lukasser et al., 2006, 85; Emanuelsson, 2009, 196, 198, 201, 204.

chemically synthesized mineral fodder additives; (8) agricultural lobbies promoting industrial fodder products; and (9) absence of suitable sites for optimal tree leaf quality on some farms.[3244]

8.2.3 Other historical raw materials from grasslands

Several plants occurring in grasslands were historically used as highly diverse raw materials.[3245] Selected examples from the Alps demonstrate this. Several of these types of grassland plant raw material uses have a long history, starting much sooner than the beginning of the second millennium as the start of the study period of this book. Times of introduction of several historical uses of plant raw materials must have differed significantly across time and space. Nonetheless, not all the examples of grassland plant uses as raw materials presented in this section were ancient anywhere in the Alps, as demonstrated most clearly by grass used for cleaning smoking pipes.

Some tenacious plants from damp, wet, and mesic sites (wetlands used to be much more frequent and extensive than they are nowadays; see the introduction to Section 6.1.2.5) were used as valued raw materials. Quaking grass sedge (*Carex brizoides*), growing mainly on nutrient-poor mesic sites in the colline and montane vegetation belt, was highly relevant.[3246] Places dominated by this species were more limited in many locations than fieldwork nowadays suggests. Namely, except for periods of farming abandonment, this sedge species used to abound in open-canopy forests in wet, acidic sites, and on the edges of marshes. Only farming abandonment led to extensive areas of species-poor stands strongly dominated by *Carex brizoides* at sites of former wet meadows, where mowing had significantly decreased the abundance of this sedge species.[3247] Quaking grass sedge provides fine (soft) but tough material, if harvested at the right time: at the foot of the Alps in Upper Austria from late July to mid-August.[3248] Rain capes were produced from long leaves of quaking grass sedge fixed on a birch rod, as documented in Styria in the eighteenth century and until the nineteenth century—in the Pohorje Hills, for instance.[3249] The same sedge species was used to weave mats, cords, ropes, and carrying bands, to produce baskets and buckets (impregnated buckets from quaking grass sedge were used by firefighters in some places; such buckets were light

[3244] Adapted, based on Machatschek, 2002; Emanuelsson, 2009, 157, 198; Vidrih, 2005, 10, 150, 153, 155–156; Hartel, Craioveanu, & Réti, 2016, 56, 59; and Zwitter, 2020.

[3245] For data on site conditions preferred by plants mentioned in this section, see Aeschimann et al., 2004, vol. 1, 48–49, 382–383; Aeschimann et al., 2004, vol. 2, 10–11, 190–191, 732–747, 786–787, 964–965, 1020–1023.

[3246] Aeschimann et al., 2004, vol. 2, 786–787; Martinčič, 2007f, 817; Machatschek, 2017, 249–250.

[3247] Pils, 1994, 161, 234, 236, 240, 242, 246 and Table 69, Figure 169.

[3248] Machatschek, 2017, 247, 252–253, 258; Pils, 1994, 234, 242.

[3249] Mlakar, 2015, 65; Kačičnik Gabrič, 2014, 100; Machatschek, 2017, 244.

Figure 149: A tool that was used for combing sedges and flax by two people, at the Pri Tinčku farm in Stara Fužina in Bohinj. Several wetland parcels belonged to this farm. I thank Tine Ceklin for guiding me around the artifacts he has been keeping. Photo: Žiga Zwitter, 2020.

and not fragile), to produce shoes made up of quaking grass sedge braids tied together by being sewn up (as in Attergau, Upper Austria), and to fill mattresses, pillows, horse collars, and saddles, for instance. Due to its multiple uses, in some places quaking grass sedge was also a trade item that farmers sold to craftsmen.[3250] When sedges were used as raw material in the handicrafts presented above, monocots had to be durable, and so any non-sedge plant species and fragile parts were removed in situ. However, quaking grass sedge tends to build quite pure, monodominant stands, so that very little extraneous material was harvested. Moreover, sedges were treated with a comb (Figure 149).[3251] In addition, in some places, such as Bohinj, beating of combed fibers (of another or some other tall sedge species) with a wooden bat followed.[3252] In some other places, quaking grass sedge was scalded before being used as a raw material.[3253] Another example of monocots historically gathered for weaving is bulrush (*Typha* sp.),[3254] a plant of wet sites.[3255]

3250 Machatschek, 2017, 242–263; Mlakar, 2015, 65.
3251 Machatschek, 2017, 254–257.
3252 Author's interview with Ceklin, 2020, in Stara Fužina.
3253 Machatschek, 2017, 250, see also pages 259–260.
3254 Machatschek, 2017, 263; see also Poschlod, 2015, 105–109.
3255 Aeschimann et al., 2004, vol. 2, 1020–1023.

In the 1720s and 1730s, at least three cases of farmers weaving baskets from unspecified tenacious monocots (possibly sedges or bulrush) were clearly documented in partial evidence from the Jaun Valley (Germ. *Jauntal*, Sln. *Podjuna*), Carinthia. Weavers delivered the seigneury of Eberndorf (Sln. *Dobrla Vas*) annual dues of four to twelve baskets for permission to harvest plants for basket weaving at wet or damp sites near the shoreline of a pond (or ponds).[3256] Some monocots from wetlands or mesic sites were also used in beekeeping, such as in coiled-work skeps (inverted baskets) made of tough long-stemmed monocots. They were the usual type of beehives in apiculture in a few parts of the Alps during most of the last millennium.[3257]

The above presentation of the use of sedges already dealt with monocots as human bedding. Not only sedges (*Carex* sp.), but also rushes (*Juncus* sp.) were used as human bedding, in the Allues Valley, for instance,[3258] as was some fine, non-wetland hay from the surroundings of huts in summer grazing complexes where wetland hay or other less esteemed material for bedding was not at hand.[3259] In addition, some men "wore" rushes (*Juncus* sp.) instead of socks, as documented in the Allues Valley; "they were said to ensure against blisters and to keep the feet much cooler in summer than socks."[3260] Use of grassland and wetland plants in human functional dress (not focusing on flowers as decoration) was not limited to monocots. Some people wore the large leaf of some dicots, butterbur (*Petasites* sp.) at least, growing in moist to wet meadows, simply turned upside down and placed on top of the head (over the headscarf in the case of women) to cover the head and shoulders, as a means of protection against sun or rain, as documented in the mid-twentieth century.[3261] This was a summer habit not only because protection against the sun's rays was most needed in that season, but also because butterbur leaves reached sufficient size

3256 AT StiAStP, Grundherrschaft Eberndorf (Handwritings), book 191, Gerichtsprotokoll 1721–1728, 307 (March 2nd, 1726); AT StiAStP, Grundherrschaft Eberndorf (Handwritings), book 192, Gerichtsprotokoll 1728–1742, 156 (April 12th, 1731). Gathering (precisely or not called "mowing" and "cutting" by sources; pulling the plants up would also be imaginable; compare Machatschek, 2017, 242, 247, 252–254) was carried out *am . . . teicht*, not *im . . . teicht*. Weavers resided in Lauchenholz (Sln. *Gluhi Les*), Oberburg (Sln. *Podgrad*), and Unterburg (Sln. *Spodnji Podgrad*).

3257 Crane, 1999, especially pages 238, 241–247, 251–252, 256, 352, 389, 398. Crane also explains that wild monocots were not the only species used in coiled-work skep production; use of cereal straw was even more frequent and considered better, at least by some beekeepers.

3258 Meilleur, 1986, 295, 388.

3259 In the grazing complexes of Krstenica and Viševnik in the Julian Alps, for instance, people slept on quality hay in the mid-twentieth century (author's interviews with G. Gartner, 2017, at the Laz Alp, and with C. Mlakar, 2015, at the Krstenica Alp, both in the eastern Julian Alps).

3260 Meilleur, 1986, 295 (quote), 388.

3261 Mlakar, 2015, 408; on site conditions, see Aeschimann et al., 2004, vol. 2, 524–525, and the author's observations of butterbur abounding in a wet to aquatic environment.

at that time. Even the Latin name of the genus *Petasites* (from *petasus*, an ancient broad-brimmed hat) implies that it was used in this way.

Further uses of wetland plants as raw materials included common reed (*Phragmites australis*) instead of rope in fruit tree care, in the Jaun Valley, for instance.[3262] Bulrush (*Typha* sp.) leaves were used to seal wooden barrels, as proven by twentieth-century data recalled at the Sapotnik (Sln. *Zapotnik*) farm in the Jaun Valley, on the shore of Lake Gösselsdorf (Germ. *Gösselsdorfer See*, Sln. *Goslinsko jezero*), where bulrush occurs. This farm used many such barrels for cider.[3263] This tiny piece of micro-history evidence certainly echoes a more general phenomenon with a longer history because the suitability of bulrush leaves—which are elastic and robust—for sealing barrels was emphasized in Georg Rudolf Böhmer's late eighteenth-century volume on plant raw materials. In some places in Central Europe, such as the wine-growing areas of Lower Carniola, which included part of the pre-Alpine Posavsko Hills (Sln. *Posavsko hribovje*), bulrush was even planted for this reason, as reported in the nineteenth century.[3264]

The use of tough, durable monocots in the Alps for several of the purposes listed above must have been widespread in previous centuries, even though it was not uniform. This hypothesis is confirmed by scattered evidence from distant parts of the world where such uses most likely developed independently,[3265] as demonstrated here by a short selection of examples. More than five thousand years ago, Ötzi wore shoes partly fixed with grass wickerwork and filled with grasses. Analysis of a small sample of the grassy shoe filling showed the presence of tor-grass (*Brachypodium pinnatum*), matgrass (*Nardus stricta*), and some others. This filling was possibly at least partly collected in subalpine grasslands to thermally insulate the shoes. His cape was made of tor-grass in combination with some other grasses (probably including purple moorgrass, *Molinia caerulea*) and linden bast fiber string, suggesting that this piece of clothing was produced at the valley bottom.[3266] In early modern Europe, Zedler's *Universal Lexicon* documents the use of baskets from wetland plants as packaging for dry figs imported from Provence (Occitan *Provença*).[3267] Not all types of plant uses presented in the previous paragraphs were practiced

[3262] Author's interview with Plesnik, 2021, at Sapotnik (Sln. *Zapotnik*) farm on the shore of Lake Gösselsdorf. Common reed abounds in the immediate vicinity; moreover, a large orchard used to grow at this farmstead.
[3263] Author's interview with Plesnik, 2021, at Sapotnik farm on the shore of Lake Gösselsdorf.
[3264] Poschlod, 2015, 108, 255; Plemel, 1862, 162–163.
[3265] E.g., in the Inca Empire, basket and rope weaving from hard monocots is proven before the Spanish conquest (Jesenšek, 2020, 27).
[3266] Acs, Wilhalm, & Oeggl, 2005; Machatschek, 2017, 243.
[3267] *Universal-Lexicon*, 1731–1754, Supplement 4, column 1184 (headword "Cabas, oder Cabat").

Figure 150: Plaster lining from common reed (*Phragmites australis*) and wire. Slovenski Javornik (Jesenice), most probably built in the early twentieth century. Photo: Rok Humerca, 2021. © Rok Humerca.

everywhere; some interviewees in Bohinj recalled the use of mowed wetland hay (sedge(s) up to about 60 cm long, much less fragile than grass) as filling in mattresses, and wetland hay as human bedding in temporarily inhabited huts at some lower meadows in forest clearings (in Ukanc, for instance), but no shoes from sedge in the mid-twentieth century.[3268]

Different from sedges and bulrush were uses of stems of another wetland monocot, common reed (*Phragmites australis*), as building material. Its stems tied parallel with a wire were widely used as plaster lining[3269] binding the wooden wall with the lime plaster, supporting the plaster, and providing building insulation.[3270] Such plaster lining can still be seen in old houses at the

[3268] Author's interviews with C. Mlakar, 2020, at the Zajamniki Alp; with Ceklin, 2020, in Stara Fužina; and with Mencinger, 2020, in Stara Fužina. Compare Tuma, 1935, 251.

[3269] Makarovič, 1978, 241; Humerca, 2020, 18; Humerca, 2019; the archaeologist Rok Humerca, personal communication; author's interview with Plesnik, 2021, at Sapotnik farm on the shore of Lake Gösselsdorf.

[3270] Rok Humerca, personal communication, and author's interview with Plesnik, 2021, at Sapotnik farm.

foot of the Alps in the western Ljubljana Basin, for instance. Plaster lining from common reed was in use for millennia, as proven, for instance, by archaeological excavations from prehistory, and by excavations of the *villa rustica* at Rodine near Radovljica from Roman antiquity. Plaster linings from common reed were constructed until the mid-twentieth century.[3271]

Higher-quality hay that would otherwise provide good fodder was also used as insulation for specific purposes. This way, fresh hay made possible cooking at some places in the Alps, by some women in Bohinj, for instance, according to mid-twentieth-century data.[3272] A full large earthen pot was heated up at a fireplace, covered, and cautiously placed into recently made, dry hay in a hay barn for a couple of hours. The hay was probably also pressed around the pot to provide better insulation. This cooking technique made it possible to cook dishes requiring as long a cooking time as dry meat. This way, lunch in remote montane meadows was ready in a way that saved both firewood and the time of the cook, who had more time for spreading, turning and raking hay. Hay from the previous year would not make such cooking possible, traditional ecological knowledge claims.[3273] Indeed, agronomy explains that the quality of hay that dried on the ground often gradually slightly declines in the barn because the moisture content in the hay that dried on the ground relatively frequently stays a little bit above the threshold of circa 13% when "the living conditions for bacteria and fungi deteriorate to such an extent that they can no longer work."[3274] Only some women practiced this manner of cooking[3275] because the cook had to pay careful attention not to set fire to the hay barn.[3276] "[H]ay boxes for slow cooking"[3277] were likely inspired by such a cooking technique if it is in fact an old practice.

The previous paragraph ended with a reference to daily life in the haymaking season. In times of haymaking and reaping, the success of local weather forecasting was of utmost importance. Before the progress in meteorology and large-scale spread of mass media broadcasting scientific weather forecasts in the countryside in the twentieth century, local weather forecasting partly relied on grassland plants. Weather forecasting based on stemless thistle has already been mentioned in the introduction to Section 8. Less widespread was another plant-based weather forecasting method, involving sweet vernal-grass

[3271] Humerca, 2020, 18; Humerca, 2019; Rok Humerca, personal communication.

[3272] This manner of cooking was practiced in the montane meadows at Za Robam above Uskovnica in the mid-twentieth century, for instance (Hodnik, 2016, 80–81).

[3273] Hodnik, 2016, 80–81.

[3274] *Traditional Hay Milk Farming* ..., [2023], 66–67 (quote from p. 66); Verbič & Verbič, 2014–2020, 27, 29.

[3275] Author's interview with C. Mlakar, 2020, at the Zajamniki Alp, stating that her family did not cook this way.

[3276] Hodnik, 2016, 81.

[3277] Peterken, 2013, 381.

(*Anthoxanthum odoratum*) as a raw material. Sweet vernal-grass is a species abounding in moderately manured and not too frequently cut mesic meadows and pastures, on rather acidic soils. It occurs in all regions of the Alps, on silicate bedrock, on carbonate moraines or, rarely, on carbonate bedrock, in valley bottoms and montane meadows in particular.[3278] Sweet vernal-grass is by far the most aromatic grass species, giving hay its special scent. It was collected by some farmers between April and July and they used to braid it. In some regions of the Alps, braids were hung in entrance halls of farms for decorative purposes and to forecast weather. Namely, the scent of braids was stronger when the humidity level was high, announcing rain, which was crucial for planning haymaking (see Section 6.1.2.1.2).[3279] In the mid-twentieth century, weather forecasting using sweet vernal-grass was only practiced in some parts of the Alps, not in Bohinj, for instance.[3280] Weather forecasting using stemless thistle and sweet vernal-grass are, similar to some other locally adapted ways of weather forecasting (based on specific cloud locations, directions of cloud movements, or animal behavior, for instance[3281]) likely older than some weather forecasting practices that we call "traditional" nowadays but that are not locally adapted. Namely, it is possible that some unfounded predictions of future weather claiming that weather observations on a particular date can forecast weather in the future were broadly accepted by farmers later, only in the eighteenth and nineteenth centuries.[3282]

Some dicotyledons also were used in handicraft as raw materials, not only for decoration (on decoration, see Section 8.1.3.1), as demonstrated by selected examples. Hard but not woody flower stalks of great mullein (*Verbascum thapsus*) were used as lamp wicks in the Allues Valley, for instance.[3283] The Slovenian vernacular name *papeževa sveča* 'pope's candle' designating *Verbascum*[3284] and the German name *Königskerze* 'king's candle'[3285] suggest that great mullein might have been used in a similar way in several parts of the Alps, unless the name was inspired solely by the blossoming plant appearance, abounding in small yellow blooms on top of a high straight stem (like another Slovenian vernacular name, *svečnik* 'candlestick', for the same species[3286]). Another example of a dicotyledonous grassland plant as raw material deals with plant dyes before the spread of chemically synthesized dyes in the second

3278 Aeschimann et al., 2004, vol. 2, 964–965; Machatschek, 2017, 191–195.

3279 Machatschek, 2017, 190–191, 194–195, 199.

3280 Author's interviews with C. Mlakar, 2020, at the Zajamniki Alp, and with Kranjc, 2020, in Stara Fužina.

3281 For instance, Ledinek, 1999, chapter Vreme; Novak, 2024, 219; author's interview with Kranjc, 2020, in Stara Fužina.

3282 Landsteiner, 1993, 49–50, 56–59, 62.

3283 Meilleur, 1986, 170.

3284 Mlakar, 2015, 310–311.

3285 Aeschimann et al., vol. 2, 190–191.

3286 Mlakar, 2015, 310–311.

half of the nineteenth century. Until then, plants were one of the usual dye sources. They were grown in fields, and some others were gathered in Europe or imported from other continents. Dyer's greenweed (*Genista tinctoria*) was one of the plants used in historical yellow dyeing in Europe, and it was usual to gather wild specimens.[3287] Dyer's greenweed occurs in most of the Alps in nutrient-poor mesic to dry meadows and forest margins, and it can abound in colline and montane vegetation belts.[3288] The general importance of the species in dyeing and its abundance at many places in the Alps suggest that it was likely gathered in some regions of the Alps as well.

Several grassland plants played practical roles in housekeeping. Staple food processing included the historically common use of interrupted clubmoss (*Lycopodium annotinum*), a species of forest undergrowth but also of shady grassland edges in nutrient-poor, predominantly acidic, mesic or moist sites. The species occurs in almost all regions of the Alps in the montane and subalpine vegetation belts.[3289] Despite frequent gathering, a high abundance of interrupted clubmoss was locally observed in some places when use of this moss was ending;[3290] nonetheless, it is now a protected species in Slovenia,[3291] for instance. Long interrupted clubmoss was used at many farms to sweep away the ash before bread was placed in a brick oven (Figure 151).[3292] Moreover, the same moss species was used to strain milk because interrupted clubmoss made possible highly efficient removal of even fine particles that fell into milk during milking (Figure 152). Use of interrupted clubmoss to strain milk was common in major parts of the Julian Alps, for instance, in both the Sava and the Soča catchment areas. The same plants were reused many times because they were always washed.[3293] This type of strainer was not used in all regions of the Alps, at least not comparably long; in Tyrol, for instance, woven milk strainers from thread braided from cattle hair had a tradition.[3294] Continuing in the sphere of food processing, one frequent dairy product was butter, and its transport required vessels or wrapping material. Plant leaves were one of the usual wrapping materials before the widespread use of paper and, more recently, plastics.

[3287] Engel, 2006a; Engel, 2006b; Troßbach, 2006.
[3288] Aeschimann et al., 2004, vol. 1, 826–827; Martinčič, 2007c, 296.
[3289] Aeschimann et al., 2004, vol. 1, 48–49.
[3290] Author's interview with L. and M. Odar, 2017, in Studor.
[3291] Martinčič, 2007a, 85.
[3292] Author's interviews with Mencinger, 2017, and with Urh, 2017, both in Stara Fužina; and with L. and M. Odar, 2017, in Studor.
[3293] Alič, 2011 (interpretation by the director of the Planika Dairy with personal experience in milk straining through interrupted clubmoss); Ledinek, 1999, section Predelava mleka; Pintar, 2012, 427; Hodnik, 2016, 34. Author's interviews with Gašperin, 2017, with Kunavec, 2017, and with Mencinger, 2017, all in Stara Fužina; with G. Gartner, 2017, at the Laz Alp; with L. and M. Odar, 2017, and with A. and M. Taler, 2017, both in Studor.
[3294] Wopfner, 1997, 503.

Figure 151: Broom made of interrupted clubmoss for sweeping a brick oven before bread baking. Jenkova Kasarna Museum at Zgornje Jezersko in the Karawanks. Photo: Žiga Zwitter, 2020.

The English plant name *butterbur* (i.e., *Petasites* sp.) reveals that some plants with large leaves were used to wrap butter. In several parts of the Alps, such as the spatially distant and culturally diverse Allues Valley and the Julian Alps, the historical use of the large leathery leaves of Alpine dock (*Rumex alpinus*, Figure 114 in Section 6.6.2) to wrap butter for transport from the high summer pastures to the valley has been proven. In Bohinj, for instance, elongated rolls of butter, weighing from one to two kilograms, were made in temporary mountain settlements and transported to the valley wrapped in Alpine dock leaves.[3295] This species abounds in the immediate vicinity of cottages in moun-

[3295] Meilleur, 1986, 397; Dolžan Eržen, 2024, 304; Novak, 2024, 198–199, 217.

Figure 152: Milk strainer using interrupted clubmoss. Cheese Museum of the Planika Dairy (Mlekarna Planika Kobarid), © Planika Dairy. Photo: Žiga Zwitter, 2017. Alternatively and probably as a general rule earlier, the vessel of the strainer was wooden, made of staves.[3296]

tain pastures, thus one could gather its leaves quickly. Large leaves of a few plant species, like large-leaved docks[3297] and probably butterbur in summer and autumn, were also used to wrap several other items before transportation, such as fish—like marble trout (*Salmo marmoratus*) and eel (*Anguilla anguilla*)—as reported in the region of Idrija in the mid-eighteenth century.[3298] This simple use of gathered plant leaves was probably practiced more or less continuously from prehistory[3299] until the twentieth century.

3296 Dolžan Eržen, 2024, 297.

3297 Meilleur, 1986, 397.

3298 Someone carried a few trouts and eels in late July 1758 "in laubwerck eingewicklter" (SI AS 736, The seigneury of Idrija, box 1, unit 1: Gerichts prothocoll buch de anno 1754 biß 1759, 444–448, quote from p. 446). The matter of the dispute was whether the fish were stolen, not the wrapping material, thus this information can be considered reliable. The source does not document which plant species was gathered to wrap the fish in this case. Tree leaves can be excluded being too small to wrap a trout or an eel. Butterbur abounds at many riparian sites and by late July (the time of the year when this kind of use of leaves was documented in 1758) its leaves are large. The above identification of fish species in the Idrijca River follows Torkar & Zwitter, 2015, 387.

3299 Butterbur leaves, for instance, were found in Early Iron Age waste deposits in the Hallstatt region (Uzunoglu-Obenaus, 2008). The purpose of their use is not clear. None-

The domain of grassland plants in housekeeping was not limited to food wrapping and processing. Nettle (*Urtica dioica*) is an example of a plant that was used in some households to produce cleaning fluid.[3300] Thorough late-spring washing of the exteriors of entirely wooden farmhouses by brushing them with a cleaning fluid including nettle was documented in the Brixen Valley (Germ. *Brixental*) and its surroundings (above the Lower Inn Valley in Tyrol) in the first half of the twentieth century. The aim was not only to clean the house exterior, but also to protect the wood against organisms perceived as pests.[3301] Continuing in the sphere of cleaning, winter-flowering heath (*Erica carnea*) was collected (pulled from the soil with part of the root to obtain material about 30 cm long; this length was needed) in heathlands in pastures to make brooms from heath biomass, neatly arranged in about three bunches fastened together, intertwined, and weighed down to obtain a flat shape, with a shaft added. Such brooms were very useful in livestock barns, and in cottages in temporary settlements, for instance. They were produced locally at all elevations, also including summer settlements in the mountains.[3302] Brooms were also made out of some other dwarf shrubs occurring in non-manured grasslands; see the example of alpenrose (*Rhododendron* sp.) brooms for sweeping hay in remote non-fertilized meadows at high elevations in several regions of the Alps in Section 6.3.1.1. In some parts of Central Europe, brooms were made out of purple moor-grass (*Molinia caerulea*; see Sections 3.2.2, 6.1.2.5.1.2, and 6.1.2.5.2), thus its Swiss name *Besenried* (*Besen* means 'broom'). When pipe smoking spread around Europe in the modern period, purple moor-grass was used to clean the inside of the pipes.[3303]

Grassland plants were also used as toys, and some of them as a raw material for toys, such as a whistle from hogweed (*Heracleum sphondylium*) stalk or a musical instrument similar to a kazoo "from the easily separated multiple dermis stalk and a piece of membrane removed from the lower leaf of an immature" yellow gentian plant (*Gentiana lutea*) proven in the Allues Valley in the first half of the twentieth century.[3304] Some girls made "hairdresses" from matgrass (*Nardus stricta*), at a small holding at Dolina above Tržič in the Karawanks, for instance.[3305] In eighteenth-century Carniola, the botanist Balthasar

---

theless, this proves that the leaves of this genus were gathered on purpose at that time already, and so the use of such leaves as wrapping material by the same communities is likely.

3300 This was not the only kind of use of nettles as technical raw material. See, for instance, also Poschlod, 2015, 14, 105–107, 109–112; Fiol et al., 2016, 19.

3301 Hubatschek, 1992, 147.

3302 Novak, 1978, 59; author's interviews with Kunavec, 2017 and 2023, and with Kranjc, 2018, all in Stara Fužina.

3303 Pils, 1994, 151; compare Dinklage, 1966b, 114, 137.

3304 Meilleur, 1982, 169; Meilleur, 1986, 293.

3305 Author's interview with J. Kavčič, 2021, at Dolina; the grass specimen taken was identified by Leonid Rasran.

Hacquet observed that it was usual for children to play with autumn crocus (*Colchicum autumnale*) seed capsules in summer. It is easy to imagine how attractive such natural rattles were. This was, however, a hazardous toy due to the high toxicity of this species for people. Nonetheless, severe poisonings were very rare because the disgusting hot taste usually prevented children from consuming the seeds. This was, however, not always the case: a paper by Hacquet published in 1779 documented that two children from the village of Mače near Preddvor, at the foot of the Kamnik–Savinja Alps, now northern Slovenia, died after consuming autumn crocus seeds around 1770.[3306]

[3306] Hacquet, 1779, 11–14.

9 Abandonment of grasslands in the Alps: a case study with an emphasis on the role of plant species composition

The long-term prevalent increase in grassland area was discussed in Section 3.1. Section 3.2.2 emphasized that forest regrowth in grasslands is not only a phenomenon of recent centuries. Nonetheless, the scale and longevity of the large-scale afforestation of grasslands followed by uninterrupted forest cover until today was unprecedented earlier in the second millennium as the study period of this volume. This afforestation has largely been spontaneous in most cases, and only a minor percentage of afforested land has resulted from plantations. The "forest transition," defined as the historical change "from net deforestation to net [af]forestation," was already characteristic for the Alps in the mid-nineteenth century or in its second half, soon after the beginning of the long-term decrease in agricultural density, and afforestation has since then been particularly strong in the southern and easternmost part of the mountain chain. In general, afforestation rates in the Alps have been significantly higher in the twentieth century than in the second half of the nineteenth century.[3307] The complex of driving forces leading to land-use and land-cover changes from grasslands to forests usually consists of interacting social and environmental factors.[3308] Two examples of social factors from the eighteenth and nineteenth centuries are the following. First, mainly since the late eighteenth century and the nineteenth century, cultivation of fodder crops has covered part of the fodder demand in the Alps, which rendered some former pastures superfluous; although this process was less characteristic for the Alps than for lowlands, some sections of former open pastures in the Alps were afforested in the nineteenth century.[3309] Second, in the second half of the nineteenth century, the importance of wood as a marketable commodity increased due to industrialization and urbanization. Since that time, in some regions in the eastern part of the Eastern Alps—namely, in parts of eastern Carinthia (Germ. *Kärnten*, Sln. *Koroška*), Styria (Germ. *Steiermark*, Sln. *Štajerska*), and Lower and Upper Austria (Germ. *Nieder-, Oberösterreich*)—a significant percentage of all farms

[3307] Mathieu, 2019, 125, 138 (quote); Bätzing, 2015, 156, 158–159, 274, 276, 314, 322.
[3308] Bürgi, Hersperger, & Schneeberger, 2004.
[3309] Vojvoda, 1965, 26; Grafenauer, 1970c, 243–245; Valenčič, 1970a, 260, 262, 266–267; Poschlod, 2015, 120–123.

were bought by non-agrarian owners. Centuries-old agricultural land was afforested to become a source of marketable wood, and game also played a role in this process. In the second half of the nineteenth century, this happened, for example, with many isolated farms with compact holdings in the Meža (Germ. *Mieß*) Valley and the Vellach (Sln. *Bela*) Valley in southern Carinthia. Chronologically, this process corresponded with a crisis of the farm economy there.[3310] There exist many other differently timed examples of this process in the Alps, such as forest planted in many mountain meadows in Austrian Carinthia in the 1950s, during times of favorable wood prices.[3311]

Over the last century and a half, losses and transformations of species-rich grasslands in the Alps due to abandonment or intensification of land use became commonplace, leading (1) to earlier and more frequent mowing, and higher nutrient availability, or (2) to gradual afforestation as result of ecological succession (see Sections 1.2, 6.2.2.3, 6.3, 8.1.2.1, 8.1.3.1, 10.1.2.1, the introduction to 10.1.2.2, and Section 10.2.1).[3312] Abandonment of agricultural land and afforestation on the one hand, and land-use intensification on the other, are complex topics, belonging to the most common twentieth-century processes of land-use change all over Europe.[3313] However, until recently, the majority of studies of land-use change in the last century and a half focused on changes in the spatial distribution of land-use categories. Even though problem-oriented research discussing the driving forces behind these changes has also had a decades-old tradition, it has gained much more attention recently.[3314] Consequently, this section presents a case study, adding fresh insight based on grassland history, improving the understanding of land-use and land-cover change from grasslands at nutrient-poor sites to forests.

Unfavorable grassland (or hay) quality from the point of view of animal husbandry,[3315] slopes that were too steep or too rocky, or (especially in the limestone Alps) meadows too hummocky for usual mowers, together with social

[3310] Blaznik, 1970a, 175, 179; Grafenauer, 1970a, 635–637; Bätzing, 2015, 276.

[3311] Keller, 2004, 23, 30.

[3312] For various aspects relevant for the Alps, see, for example, Melik, 1950; Novak, 1970, 391; Wopfner, 1997, 311, 320–321; Matouch, Traxler, & Grass, 2000; Kauter, 2002, 1, 17; Erschbamer, Virtanen, & Nagy, 2003, especially pages 380–385, 388–389; Dietl & Lehmann, 2004; Keller, 2004; Nicolussi et al., 2005; Petek, 2005; Emanuelsson, 2009; Kapfer, 2010, 138–139; Wunderli, 2010; Schirpke et al., 2013; Tappeiner, Borsdorf, & Bahn, 2013; Gingrich et al., 2013; Groß, 2015; Bürgi et al., 2017; Gabrovec & Kumer, 2019; Groß, 2019; Carrer, Walsh, & Mocci, 2020. See also Babai & Molnár, 2014, 123–124, 129; Eriksson, Cousins, & Bruun, 2002; Poschlod, Baumann, & Karlík, 2009, 19–20; Bätzing, 2015, 152–163, 265–276, 279, 281–283, 287–289, 314, 322, and pages cited in Sections 1.2 and 10; Poschlod, 2015.

[3313] See, for example, Bürgi et al., 2017, especially pp. 2104, 2108, and Section 1.2 in this publication.

[3314] See the syntheses by Bürgi, Hersperger, & Schneeberger, 2004, and Bürgi et al., 2017.

[3315] Vojvoda, 1965, 27.

Figure 153: Blooming spiny restharrow (pink) in the remnants of the Zavaze meadow complex. In the centuries studied in this book, meadow edges close to cart tracks often belonged to nutrient-richer (moderately fertilized) meadow sections, mainly due to the nutrients leached from the dung of livestock traveling along the cart track. Currently, the situation is usually the opposite: meadow margins belong to the nutrient-poorer meadow sections because fertilizer application is less intensive on the meadow edges and because machinery replaced the livestock on the cart tracks. Photo: Žiga Zwitter, June 24th, 2018.

factors—for instance, deagrarianization and transformations of agriculture and transport—are some of the frequent reasons for abandonment of meadows in the Alps.[3316] Grassland abandonment is often associated with the intensification of grassland use elsewhere (e.g., through fertilization) as a result of a shift from agriculturally less favorable to more promising sites. This study contributes to the knowledge of the influence of local plant species composition on the abandonment of grasslands, which is rarely taken into account in studies of land-use and land-cover change. The study of autumn crocus (*Colchicum autumnale*) in Austria by Winter, Penker, and Kriechbaum emphasizes that "unpalatable species may indirectly threat[en] biodiversity, as they could promote land abandonment, which clearly reduces plant species richness in the

[3316] Vojvoda, 1965; Odar, 1993; Petek, 2005; Bürgi et al. 2004; Bürgi et al., 2017; Bavcon, Ravnjak, & Praprotnik, 2019, 21–22, 31, 33, 125, 127.

[longterm]."[3317] This case study focuses on the influence of the abundance of spiny restharrow (*Ononis spinosa*, Figure 153) as one of the intrinsic environmental driving forces that, together with other intrinsic and extrinsic environmental and social driving forces, led to the abandonment of several meadows in the study area of Zavaze on the slopes of Rudnica Hill in the Bohinj Basin, at elevations of about 550 to 900 meters.[3318] The broader importance of this study in the Central European context is explained by the fact that spiny restharrow was a common plant species in comparatively nutrient-poor and mesic to dry meadows and pastures, at slightly acidic to slightly alkaline sites, as explained by Dirk Kauter's analysis of grasslands in southwestern Central Europe in the seventeenth to nineteenth centuries.[3319] Spiny restharrow has been no rarity in remote montane meadows still existing in Slovenia.[3320]

Based on the first precise land-use map from the early nineteenth century, the case study area was a large complex of montane meadows and a few pastures (Figure 154) on moraine deposits. Rendzinas prevailed, but to a lesser degree eutric cambisols also developed there.[3321] Accessibility of the area was not problematic: the distance from the closest meadows to the villages of Stara Fužina and Studor, where the great majority of the owners of these parcels lived,[3322] is a ten-minute walk, and a forty-five–minute walk from the remotest one. Meadows were usually accessible with an ox or horse and sled, and later at least with small tractors. The context of other distant meadows in the territory of the same two villages reveals that meadows at Zavaze were not the most difficult to access and to gather in the hay.[3323] Consequently, many remote meadows in the same cadastral municipality had already become afforested by 1960, when the great majority of meadows at Zavaze were still forestless (see Figure 156), even though by that time some meadows had been abandoned at Zavaze as well.[3324]

The meadows at Zavaze experienced long-term deterioration of soil nutrient availability. On the one hand, nutrients were removed with hay every year

---

3317 Winter et al., 2011, 1782.

3318 The theoretical background was summarized from Bürgi, Hersperger, & Schneeberger, 2004, 858–862, 865–866. As is characteristic for non-fertilized meadows, species compositions were strongly dependent on natural site conditions and extensive management, and so they could change over relatively short distances (Section 6.3.1). In line with this, spiny restharrow was frequent in many meadows at Zavaze but not everywhere (author's interviews with Gašperin, 2017, with Mencinger, 2017, with Urh, 2017 and 2022, and with Kunavec, 2023, all in Stara Fužina; with L. and M. Odar, 2017, and with A. and M. Taler, 2017, both in Studor).

3319 Kauter, 2002, 150–151,154, 156–158, 221 (Table 26).

3320 Bavcon, Ravnjak, & Praprotnik, 2019, 19.

3321 *Environmental Atlas of Slovenia*, Land and Soil, Soil map (1:25,000).

3322 Novak, 1955, 274.

3323 Vojvoda, 1965, 27, 61, 137; author's interview with Kunavec, 2017, in Stara Fužina.

3324 Vojvoda, 1965, 20, 27.

Figure 154

and, on the other hand, input of nutrients was limited. At many sites, hay from Zavaze was hard and it included a lot of spiny restharrow (*Ononis spinosa*).[3325] Nonetheless, leguminous spiny restharrow was itself important for meadow ecosystems due to fixation of nitrogen from the air by bacteria in its root nodules.[3326] Another source of nutrients was animal excrement. Spring grazing in meadows there (see Section 6.2) was practiced until the mid-nineteenth century. After that, the grazing period in the meadows studied was reduced to the autumn.[3327] Autumn grazing was documented there in the early eighteenth century,[3328] but the practice was definitely not introduced there only at that time. Many meadows in this complex, at least higher ones, were mowed in August, and some of them only in mid-August (Section 6.1.2.2).[3329] Consequently, grazing in these meadows lasted from September until snow.[3330] In the spring,

3325 Author's interviews with Gašperin, 2017, with Mencinger, 2017, and with Urh, 2017, all in Stara Fužina; with L. and M. Odar, 2017, and with A. and M. Taler, 2017, in Studor.

3326 Emanuelsson, 2009, 38.

3327 Novak, 1955, 274.

3328 SI AS 721, Gospostvo Bled (The Seigneury of Bled, Germ. *Veldes*), Iustitialia, fasc. 25: Lawsuits between the seigneuries of Bled and Radovljica IV, Over the confiscation of livestock in the "Rudenca" common 1716–1722: Hoh [!] vnd wollgebohrner Herr Graff, besonders hohgeehrtester Herr Herr, December 3rd, 1716, Haubtweÿsung, 1717, and Gegnweÿsung [!], 1718.

3329 Author's interviews with Gašperin, 2017; with Kunavec, 2017; and with Kranjc, 2018—all in Stara Fužina.

3330 Novak, 1955, 274.

cow pies were crushed and distributed (Section 6.2.2). However, at least from 1860 onward, but possibly already earlier, it was prohibited to keep livestock at Zavaze overnight; animals had to be driven to graze at Zavaze from the farms in the valley every day.[3331] A century later, in the mid-twentieth century and in its second half, still no livestock were kept there overnight.[3332] Consequently, unlike many other complexes of distant meadows in the surroundings, there was no locally produced barn manure available (see also Section 6.3), and quantities of livestock excrement were limited at least since the 1860s because livestock grazed in these meadows only in the autumn. In the mid-twentieth century, some smallholders fertilized the meadows in the lower part of the Zavaze complex with household ash, making possible mowing of second-crop hay there as well (Section 6.1.2.3),[3333] and only those that could afford it started using Thomas meal at Zavaze,[3334] also in the remotest meadows.[3335] Thomas meal is an industrial mineral fertilizer rich in phosphates but also in heavy metals.[3336] In contrast to barnyard manure (Section 6.3.2.1), mineral fertilizers could be feasibly transported uphill even by livestock because they are more concentrated, and so the load was lighter, making possible the application of many nutrients in remote meadows. The goal was to improve hay quality and quantity—for instance, to harvest better hay in what had been a late-mowed purple-moorgrass (*Molinia caerulea*) meadow.[3337]

When asked about the characteristics of hay from Zavaze, the majority of interviewees emphasized that hay from non-fertilized meadows in the study area was of low quality, principally because of the abundance of spiny restharrow. Some farmers removed this species when manually shaking the mowed hay and turning the hay while it was drying on the ground. Such refined hay was also useful for cattle; namely, at least some farmers would never feed hay including spiny restharrow to cattle because it was said that its thorns could cause wounds and infection in cattle's muzzles. Other farmers did not remove spiny restharrow and used such hard and hard-thorny hay for sheep, goats,

3331 Novak, 1955, 274.

3332 Author's interviews with Gašperin, 2017, in Stara Fužina.

3333 Author's interview with Boltar, 2018, in Stara Fužina.

3334 Author's interviews with Boltar, 2018, and with Gašperin, 2017, both in Stara Fužina; and with L. and M. Odar, 2017, in Studor.

3335 Author's interview with Kunavec, 2023, in Stara Fužina.

3336 Groß, 2015, 191.

3337 Author's interview with Kunavec, 2023, in Stara Fužina: the interviewee, a meadow owner, recalled a high historical abundance of a grass with black blossoms, making poor-quality hay resembling the low-quality wetland hay, in that mesic to moist meadow at the top of Zavaze, on acidic soils (common heather, *Calluna vulgaris*, was recalled to have occurred there, blossoming in autumn), unless a moderate quantity of fertilizers (e.g., Thomas meal) was applied once every couple of years. Purple moor-grass is the only dominant "black blossoming" plant from the families of *Poaceae*, *Cyperaceae*, and *Juncaceae* occurring in the Julian Alps (Sln. *Julijske Alpe*, Ital. *Alpi Giulie*) and abounding at mesic to moist, acidic sites (Aeschimann et al., 2004, vol. 2, 732–1017).

Figure 155

Figure 156

and horses. A high content of spiny restharrow in hay was also unpleasant from the human perspective:[3338] the species was harder to mow (increasing the labor intensity of mowing and more work with sharpening and peening scythes);[3339] moreover, the thorns of spiny restharrow caused injuries to people when hay was carried in sheets on the back and shoulders to hay barns (Figure 87 in Section 6.3.1.2) or in carts and sleds, and when such hay was brought to livestock in the winter, but spiny restharrow was not removed from the hay for the aforementioned livestock types that could safely consume it because spiny restharrow's abundance was too high.[3340]

Although the interviewees agreed that hay from the Zavaze meadows complex including a lot of spiny restharrow was of low quality (e.g., third class), they emphasized that hay from Zavaze also had its merits for all livestock under unusual conditions. Namely, such hay dried quickly and it was not prone to mold infestation. Some owners of meadows at Zavaze also owned a meadow in the Voje Valley. Hay from there was usually considered better (second class) than from Zavaze for the majority of livestock. Farmers link the suboptimal quality of the majority of hay from the Voje Valley (thus not first class) with less sunshine there due to the north–south orientation of the valley (see also Section 6.3.1.2). In addition, hay from Voje was highly prone to mold. Consequently, in the case of humid weather during the hay harvest, third-class hay from Zavaze had to play the usual role of the second-class hay from the lower montane meadows in the Voje Valley.[3341] A parallel can be drawn with tor-grass (*Brachypodium pinnatum*) and even matgrass (*Nardus stricta*); see Section 6.3.1.2.[3342]

In many places, abandonment or intensification of agriculture in former nutrient-poor meadows was related to the decrease in sheep farming,[3343] which is also one of the important factors for spontaneous afforestation in the study area of Zavaze. Between the early nineteenth century and the aerial photo taken in 1956 (Figure 155), there were only minor changes in terms of spontaneous afforestation as a consequence of specific socioeconomic and/or environmental conditions (see red arrows in Figure 155). Forest regrowth significantly

---

3338 Author's interviews with Kunavec, 2017, with Gašperin, 2017, with Mencinger, 2017, and with Urh, 2017—all in Stara Fužina; with G. Gartner, 2017, at the Laz Alp in the Julian Alps; with L. and M. Odar, 2017, in Studor; with V. Smukavec, 2017, in Spodnje Podjelje; with A. and M. Taler, 2017, in Studor.

3339 Bavcon, Ravnjak, & Praprotnik, 2019, 19.

3340 Author's interviews with Gašperin, 2017, and with Kunavec, 2017, both in Stara Fužina; and with L. and M. Odar, 2017, in Studor.

3341 Author's interviews with Boltar, 2018, with Gašperin, 2017, with Kranjc, 2018, with Mencinger, 2017, and with Urh, 2017—all in Stara Fužina, the village closest to the Voje Valley.

3342 Babai, Molnár, & Molnár, 2014, 39; Aeschimann et al., 2004, vol. 2, 926–927.

3343 Poschlod, 2015, 177.

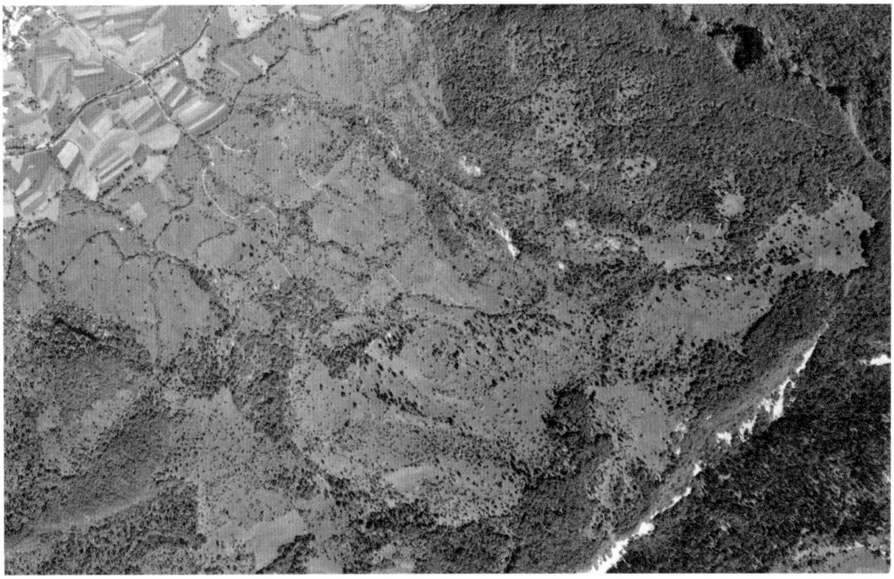

Figure 157

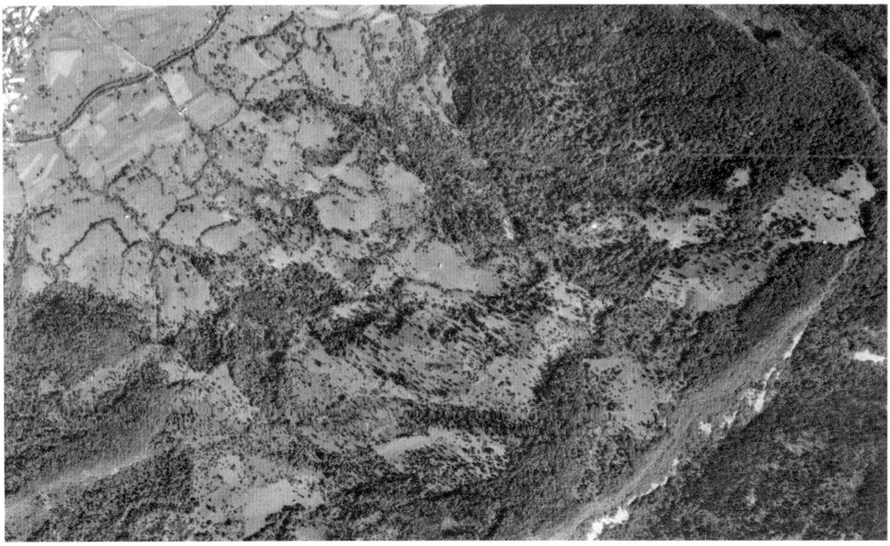

Figure 158

increased especially between 1972 (Figure 157) and 1982 (Figure 158), and in every following decade, as proven by the aerial photos taken in 1992 (Figure 159), 2006 (Figure 160), and 2015/16 (Figure 161). The decrease in sheep and goat farming in the two nearest villages (Stara Fužina and Studor), where the great majority of owners of these meadows lived, was an important factor, but not the only one influencing this afforestation. A decrease in sheep farming in Bohinj in the late nineteenth or early twentieth century,[3344] stimulated by the development of the global wool trade (Section 6.3.1.1), did not lead to any significant abandonment of meadows at Zavaze because the total number of livestock was high.[3345] On the one hand, very high agricultural density (see Section 3.1) must have increased the use of low-quality hay from Zavaze in cattle farming in the late nineteenth and early twentieth century. On the other hand, many farms at that time still (or again; see below) kept some sheep and also other types of livestock able to make relatively good use of the hay from Zavaze, such as goats and horses. In 1960, the numbers of sheep and goats together in the entire Upper Bohinj Valley (669), also including further villages to the east, were by and large comparable with the 1900 census data (653); but the structure had changed because the number of sheep in Bohinj almost doubled between 1910 and 1960, whereas goats had almost disappeared. However, the half century between 1910 and 1960 saw great variations; it was not characterized by a constant trend. The numbers of sheep and goats strongly increased in part of the interwar period, and the scarcity of wool immediately following the Second World War led to another short-term increase in sheep numbers. This was followed by a large decline in sheep and goat farming between 1955 and 1960, when the number of these animals in the entire Upper Bohinj Valley decreased from 1,177 to 653.[3346] Figure 162 shows the continuation of the immense decline in raising sheep between 1960 and 1981 in the two villages most relevant for this case study (Stara Fužina and Studor).

The number of horses in the whole valley had already decreased by a third between 1910 (187) and 1960 (122);[3347] that is, before the first tractors. Namely, motorized road transport gradually gained so much importance in the study area that by 1965 horses were typically used only in farmwork and for the transport of privately owned wood. The rest of road cargo traffic had been motorized by that time and bus lines operated, which all belonged to the main reasons for the decreased numbers of horses and oxen.[3348] The introduction of tractors in Bohinj starting in the 1970s,[3349] along with the continued increase

[3344] On this decrease in sheep numbers, see Novak, 2024, 26.

[3345] See the discussion of livestock numbers in Bohinj in Vojvoda 1965, 141–147, especially p. 142, and see Figure 162 in this volume.

[3346] Vojvoda 1965, 141–145, and Table 18; Novak, 2024, 26.

[3347] Vojvoda 1965, Table 18.

[3348] Vojvoda 1965, 141–143, 145–146.

[3349] The timing of the local introduction of tractors follows Novak, 2024, 26.

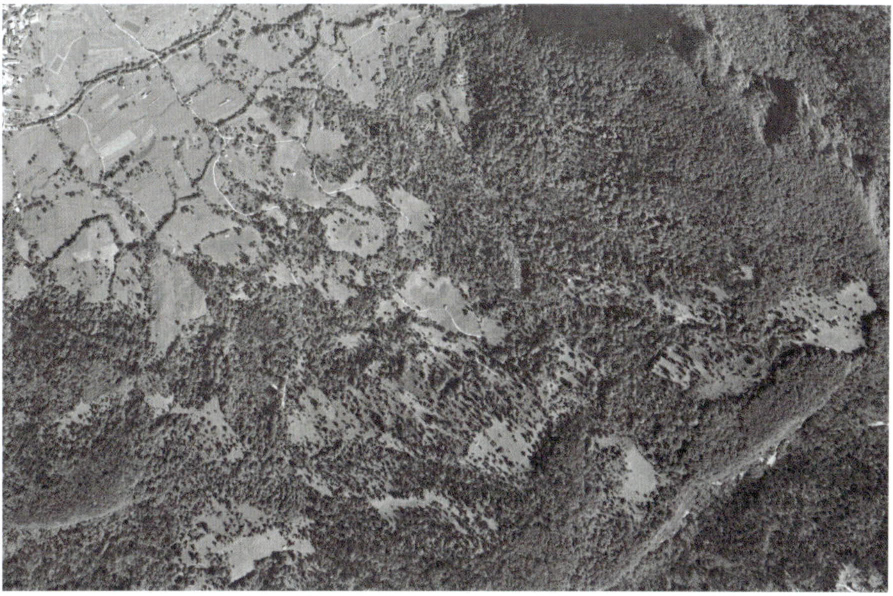

Figure 159

Figure 160

in other motorized road traffic, belonged to the main reasons for the major decrease in the number of horses at Stara Fužina and Studor in the 1980s evident in Figure 162. The motorization of farm transport in the study area belonged to the late stage of the broad process that largely took place in Central Europe from the 1950s to the mid-1970s but continued mainly until the 1980s.[3350] However, Figure 162 explains that horses only represented a low share of all hay consumers.

By 1960, the number of cattle in the entire valley had just begun to decrease significantly, and so the great decline in numbers of cattle in Stara Fužina and Studor in the periods 1960–1971, 1971–1981, and 1981–1991 and the later continuous decline in small family farms (see Section 1.2) also had an impact on the afforestation of Zavaze.

The statistics presented are not entirely comparable;[3351] nonetheless, they clearly reflect the general trend of decline in animal husbandry, which influenced the afforestation of Zavaze—with a time lag because ecological succession is a gradual process. Due to the fact that about fifteen years (as a rough estimation) is usually needed for a young forest to overgrow unused montane

[3350] Ehlert et al., 2012, 478, 480; Poschlod, 2015, 152.

[3351] At least the following three reasons might cause mismatches in the comparability of data:
First, the key reference date in the 1900 and 1910 censuses was December 31st (Teibenbacher, Kramer, & Göderle, 2012) and January 15th in 1955, whereas the data from 1960 refer to May (Lavrič, 1960, 7), and March 31st was the key reference date in 1971, 1981, and 1991 (Vertot & Ilić, 2001, 6); livestock numbers cannot be presumed to be constant throughout the year (Lavrič, 1960, 7; Novak, 1970, 373).
Second, published statistics calculate unweighted sums of young and adult livestock and they do not allow a presentation of different age categories of every livestock type for each of the years in Figure 162, and so the data within each column are not homogeneous. For instance, whereas the results of the 1960 census include different categories of cattle, sheep are only listed as "sheep" and horses only as "horses," and so it is assumed that all ages and both sexes of sheep and horses were summed up in these categories in an unweighted manner because it is evident that the sum of cattle was calculated this way. It is assumed that the methodology of synthesizing the data for the extensive agricultural census of 1960 was also followed in the synthesis of the basic livestock census in 1971, listing only "cattle," "sheep," and "horses," without any classes by age or sex.
Third, in 1960 mules were added to the broader category of "horses", but they were also listed separately (Vojvoda 1965, Table 19), and so they are not included in the graph (Figure 162). In 1981 at the latest, the census questionnaires left out hinnies and mules; in 1991 they were added to the questionnaire again (Ilić et al., 1994, 16–17). Syntheses of results of the 1971 census and of the 1981 census do not list hinnies and mules, and it is not clear whether they were added to the sum of horses, as in 1960, or whether they were left out; this does not significantly affect the results because mules (and hinnies) only represented 10% of all "horses" in Stara Fužina and Studor in 1960, and it can be safely assumed that the numbers of hinnies and mules decreased later because they were disappearing from the countryside.

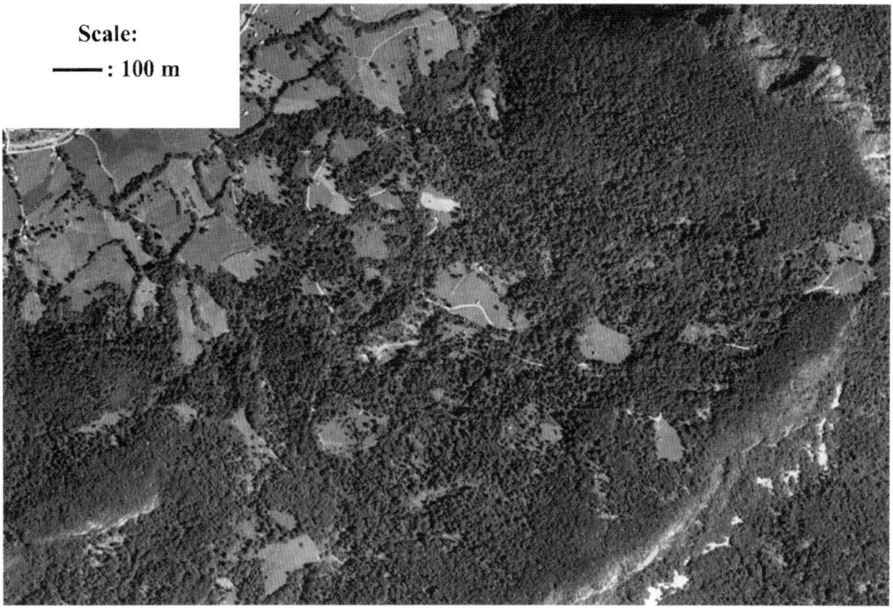

Figures 154–161: Spontaneous afforestation of the former Zavaze complex of montane grasslands between the early nineteenth century and 2015. Figure 154: land-use categories about 1820. Figure 155: aerial photo from 1956; red arrows point to a selection of areas where afforestation progressed significantly between 1820s and 1956. Figure 156: aerial photo from 1964. Figure 157: aerial photo from 1972. Figure 158: aerial photo from 1982. Figure 159: aerial photo from 1992. Figure 160: digital orthophoto from 2006. Figure 161: digital orthophoto from 2015/16. To make the comparison with Figure 154 easier, cadastral parcels are added to Figure 160. Note that both cart tracks to Zavaze from the early nineteenth-century map can be clearly recognized, making comparisons between the early nineteenth-century land use and land use in 2006 easy; the eastern boundary of the cadastral municipality (red line in 2006 = edge of the early nineteenth-century map) is also helpful. Sources: Figure 154: SI AS 176, Franciscejski kataster za Kranjsko (Franciscan Land Cadaster for Carniola), L284, cadastral municipality of Studor, map sheet A16; Figures 155–159: GURS, aerial photos, cadastral municipality of Studor: years 1956, 1964, 1972, 1982, 1992; Figure 160: *Environmental Atlas of Slovenia*, Backgrounds, Dof 2006 & *Environmental Atlas of Slovenia*, Spatial units, Land Cadastre; Figure 161: *Environmental Atlas of Slovenia*, Backgrounds, Dof 2016.

meadows in the interior of Slovenia[3352] (long delays are possible in case of specific species compositions, such as a thick interconnected litter cover of some dominant grasses like east Alpine violet fescue, *Festuca paniculata*, not allowing the penetration of trees and scrub through it, but the meadows at Zavaze are too moist and located at elevations that are too low for the dominance of this grass species),[3353] significant afforestation starting in the 1970s can be

[3352] General ecological succession dynamics in the interior of Slovenia: Bavcon, Ravnjak, & Praprotnik, 2019, 23; see also Norz, 2014, 132–133.

[3353] Keller, 2004, 23, 87; Aeschimann et al., 2004, vol. 2, 838–839; on different succession dynamics at some other sites in the Alps, see also Norz, 2014, 132–133 and Bätzing,

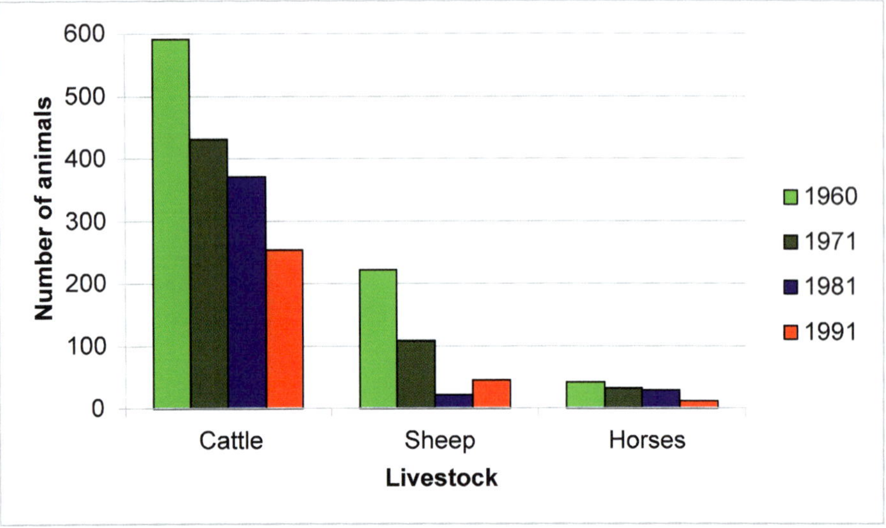

Figure 162: Numbers of selected livestock in the adjacent villages of Stara Fužina and Studor, where the great majority of the owners of (former) meadows at Zavaze live (1960, 1971, 1981, 1991). Sources of data: Vojvoda 1965, Tables 19m and 19n; Prvi podatki …, 1971, 218; AKOG FF UL, SR Slovenija, Popis 1981, …, Part 3, Radovljica, 8; AKOG FF UL, Republika Slovenija – popis 1991, …, 180. Twenty-first-century livestock numbers are not as relevant for this case study to be added to the graph because spontaneous afforestation takes time, and so the 1991 data and the trend in 1981–1991 are more relevant to interpret the 2015/16 digital orthophoto than more recent livestock numbers, in particular (1) because there is no trend of creating new meadows on forest clearings in the study area of Zavaze and (2) because the strength of the connection between the grassland area and livestock numbers has weakened due to the increased use of industrial fodder additions (Section 10.1.2.1 and the introduction to 10.1.2.2). Author of the graph: Žiga Zwitter.

linked with the significant decline in animal husbandry in the 1950s (in sheep rearing in particular) and later. This decrease in livestock numbers is closely connected with further socioeconomic factors promoting afforestation there, such as the decrease in available workforce in farming and the smaller role of farming in the local economy. Many remote meadows providing low-quality hay, and also some montane meadows with high-quality hay that were not easy to access, were abandoned. Increasingly more former arable land close to settlements was converted into meadows, which all decreased the role of distant meadows on slopes, in particular as this corresponded with a major increase in barnyard manure availability for meadows (Section 6.3) and wide access to mineral fertilizers, and so farmers preferred to moderately fertilize the meadows in the valley and abandon the more remote ones. Land owner-

2015, 265–268, but the numbers of years that spontaneous afforestation takes in the case of the quickest forest regrowth given by Bätzing are too high (see Bavcon, Ravnjak, & Praprotnik, 2019, 23).

ship in combination with the socioeconomic situation of the owner played an important role: around 1960, fieldwork in Bohinj proved that many meadows where haymaking was very labor-intensive due to landforms and where the quality of hay was low were still neatly mowed, whereas numerous better ones in the immediate vicinity were already abandoned.[3354] Despite this individuality, the interrelated social and environmental driving forces, among them also the high abundance of spiny restharrow, led to abandonment of the majority of the meadows at Zavaze largely starting in the mid-twentieth century.

Due to the strong decrease in farming, the former positive role of hay that had a high content of spiny restharrow (reducing the climate-induced vulnerability of local animal husbandry) also decreased. In former times, this hay was used as an emergency fodder, allowing the survival of the livestock in case of hay scarcity. In line with present standards, additional hay or silage is purchased from elsewhere if the local harvest is insufficient.[3355] Due to the decreased importance of farming, scarcity of hay is no longer a regional phenomenon, and feeding milk cattle with such low-quality hay that they will survive but lose a lot of weight in the winter is no longer acceptable.

[3354] Vojvoda, 1965, 22–24, 26–28, 33–35, 37, 48, 54–55, 58, 61, 135 136, 141–146 and Table 18; author's interview with Urh, 2022, in Stara Fužina—the interviewee experienced the socioeconomic changes presented; Petek, 2005, 78–79, 88–89, 92–97, 131, 167–171, 177–178, 191–192; compare Wunderli, 2010, 28; Gingrich et al., 2013, 307, 309–310; Poschlod, 2015, 125.

[3355] Author's interview with Kunavec, 2017, in Stara Fužina.

10 Applicability of grassland history in the twenty-first century, with an emphasis on meadows

This chapter demonstrates that some aspects of the historical ecology and environmental history of species-rich grasslands are relevant for today's farming and grassland ecology, for some other on-farm activities, and for tourism in rural areas in general. It also makes it clear that environmental historians can become part of the transdisciplinary conversation aimed at better policy relevant for grassland farming. Furthermore, historical legacies in some grasslands are educative for human ethical reasons.

10.1 Farming and the ecology of species-rich grasslands

10.1.1 Farmers' historical attitudes toward grassland environments: a brief insight up to the mid-twentieth century

Historically, grassland management in all regions of the Alps mainly focused on agricultural production and mitigation of the risks of natural hazards.[3356] Nonetheless, this economy-centered management was influenced by farmers' traditional ecological knowledge, which was sufficiently refined that they were interested in the maintenance of many species-rich grassland agroecosystems: these ecosystems often provided healthy species-rich fodder from several grasses and herbs (Sections 2, 6.3, and 10.1.2) as well as species gathered for use for humans (Sections 8.1, 8.2.1, and 8.2.3) or livestock (Section 8.2.2). Farmers' evaluation of fodder quality followed an integrated approach linking livestock's performance, the evident or assumed impacts of fodder on animal health, fodder taste as reflected in livestock's eating behavior, and impacts of grassland plants on some milk and meat properties, such as their taste and milk fat content (Sections 4.1.1.2, 6.3.1, 6.3.2, 6.6.2.2, the introduction to Section 8, Sections, 8.1.2, 8.2.1.3, 8.2.2, and Section 10.1.2).[3357] In the periods dealt with in this volume, from the Middle Ages to the mid-twentieth century, such farming changed the ecosystems significantly (Sections 2, 3, 5, 6, 7, and 8.2.2).

[3356] von Glasenapp & Thornton, 2011, 780 (quote); Babai & Molnár, 2014, especially pp. 127–129; Orland, 2016, 23.

[3357] Compare Babai, Molnár, & Molnár, 2014, especially pp. 6, 15–20, 24–26, 31–58, 135–136.

However, such grassland management fulfilled several key objectives of what is now designated as land-sharing in environmental conservation: usually, it helped create high levels of biodiversity and maintained patchworks of different and dynamic biocenoses in the landscape.[3358]

From the Late Middle Ages to the mid-twentieth century, agricultural density in the Alps was very high in comparison with many other mountain ranges in Europe and beyond; in absolute terms, it was particularly high in the nineteenth century (Section 3.1). This means that, for a mountain range system, species-rich grassland agroecosystems covered an exceptionally high percentage of the total area in the Alps. High biodiversity (the diversity of ecosystems, species, and genetic diversity) was a general feature of historical agricultural landscapes in the Alps, but not all grassland ecosystems in historical landscapes maintained a comparable level of species-richness for centuries (see the gradual transformation of some non-fertilized meadows to less species-rich matgrass swards at extreme sites in Sections 6.3.1.2 and 9).

Nonetheless, the primary historical economic interest in grassland ecosystems does not mean that until the mid-twentieth century farming-motivated grassland management as a general rule cared for the environment only as much as motivated by economic considerations. Compassion for some animals also played a role, from the mid-nineteenth century onward in particular. Namely, Christianity made possible highly ambivalent views on the appropriate human attitude toward the animal world, including grassland fauna. On the one hand, there was the widely accepted notion among church officials and their congregations, including farmers, "that the world had been created for man's sake and that other species were meant to be subordinate to his wishes and needs." In the *ancien régime*, economic circumstances, stimulated by the aforementioned anthropocentric views inspired by religion, resulted in looking at nature "as a reservoir of resources." On the other hand, there was the view that the world is God's creation, and so humans should not interfere in it in a way that causes unneeded suffering of other living beings: a "doctrine of human stewardship and responsibility for God's creatures." In the following passage from the Bible, from the laws in addition to the Decalogue in the Old Testament's Book of Deuteronomy, benevolent conduct toward wild animals and the view of nature as a reservoir of resources are combined in what is claimed to be the God's words: "If you happen to come upon a bird's nest along the way, in any tree or on the ground, with young ones or eggs, and the mother sitting on the young or on the eggs, you shall not take the mother with the young; you shall certainly let the mother go, but the young you may take for

[3358] E.g., Kauter, 2002; von Glasenapp & Thornton, 2011, 777; Babai, Molnár, & Molnár, 2014; Bätzing, 2015, 277–280, 289, 358, 376–377; Bavcon, Ravnjak, & Praprotnik, 2019; Grass, Batáry, & Tscharntke, 2021. Compare: McNeill, 2003, 15–16, 34; *The Challenge of Feeding the World Sustainably*, 2021, 13–14; and Willett et al., 2019, 481.

yourself, in order that it may be well with you and that you may prolong your days." In general, the ambivalence between a world created for man's sake and avoidance of unnecessary suffering in God's creation was strengthened by selective readings of the Bible concerning human attitudes toward nature, which were not uncommon even among prominent early modern theologians,[3359] let alone village priests.

The cruelty of indifference was widespread in human attitudes toward wild fauna until the eighteenth and nineteenth centuries because "[f]or most persons, the beasts were outside the terms of moral reference." In addition, there was some cruelty of vindictiveness: many boys amused themselves "in the pursuit and torture of living creatures," raiding birds' nests, destroying eggs and birds, catching birds and putting their eyes out, and so on. Hence, even killing animals considered non-noxious and economically irrelevant simply for pleasure was no exception, apart from the many economically or otherwise (e.g., medicinally) motivated actions against wild fauna.[3360] These observations concerning the pursuit and torture of economically irrelevant birds by boys in England[3361] correspond with those from Slovenian territory,[3362] and so they reflect broad European behavioral patterns, also characteristic of the Alps.

This was, however, not the only attitude toward wild animals at that time; avoidance of animal suffering unnecessary for humans was also quite present. Thus, to a certain degree, compassion of many farmers in the past with wild animals that were considered useless or not significantly important for humans played a role, provided that species that were esteemed or at least considered non-noxious and non-detested (for practical reasons or prejudice) are discussed, and that this did not entail significant economic losses. Even though many insects, in general, were particularly detested animals, just like reptiles and amphibians,[3363] superstitious beliefs led to highly positive attitudes toward some species. Ladybugs (Coccinellidae)—an insect family not void of economic importance because it is part of biological control of organisms that come in conflict with farming (suppression of herbivorous hemipterans)—are an example of insects long protected "from wanton injury" by superstition in

[3359] Thomas, 1987, 17–25, 29–30, 49–50, 152–158, 162, 165–167, 242–243, 278 (first and third quotes from pp. 17 and 24); Jaritz & Winiwarter, 1997, 93 (second quote); Erhatič Širnik, 2011, 6, 21–23, 360 (the English translation of the fourth quote follows *Bible: YouVersion, New American Standard Bible*, 1995, Deuteronomy 22, lines 6–7). For a general comment on the Book of Deuteronomy, see Brekelmans, 1968, column 328; de Fraine, 1968, especially columns 577–578.

[3360] Concerning the economically and medicinally motivated actions against wild fauna, see, e.g., Scheurer, 2010; Erhatič Širnik, 2011.

[3361] Thomas, 1987, 147–148, 275 (quote from p. 147).

[3362] Erhatič Širnik, 2011, 15–16, 24–26, 31, 33, 38, 94, 100, 102, 121–122, 127, 132, 135, 139, 144, 205, 207, 354; Kebe, 2011, 201.

[3363] Thomas, 1987, 53, 57–58, 70, 74, 150–158, 162, 164, 173, 191, 273–275, 278; Scheurer, 2010, 67–75, 82–83; Erhatič Širnik, 2011, 191.

some parts of Britain and in as distant and culturally different areas as Slovenian territory (here as a good luck omen, bringing misfortune to a person that kills it),[3364] and thus also in many other places in Europe. Thus, cultural history, such as superstition and Christian symbolism, also mattered, especially if in line with economic interests or use of wild fauna in medicine or their believed contribution to human health. The saying "Whoever scares the swallows away from the house (farm) also scares away the luck" or "Woe betides the person that scares a swallow away from the house (farm)," documented at least since 1849, was probably already well known before the eighteenth and nineteenth century. Namely, in line with the legend of the swallow that alleviated Christ's suffering on the cross, it was thus an esteemed animal. In addition, swallows (such as the barn swallow, *Hirundo rustica*) as insectivorous birds are economically welcome on farms, with the possible exception of bee farming. Although symbolic value protected some animal species such as swallows in their habitats, a specific positive symbolic value could result in sparing the life of a wild animal but at the same time foster hunting of such species and keeping them in captivity: the red crossbill (*Loxia curvirostra*), said to use its specifically shaped bill to remove the nails holding Christ to the cross, was believed to protect a family that kept it in a cage against misfortune, black magic, and diseases in general, as reported from Tyrol in the nineteenth century, where this bird was widely kept in farmhouses, likely following an older historical tradition. Although positive symbolism resulted in a positive attitude toward such species, negative symbolism fostered suppression of some organisms even if they actually did not pose any harm to people.[3365]

In the eighteenth and nineteenth centuries, the broader human civilizing process continued, as explained by the German sociologist, philosopher, and cultural historian Norbert Elias. He defined the civilizing process as long-term changes in the perception of socially acceptable behavior, leading to a shift in the threshold of discomfort in the course of history. This made many things that had been acceptable everyday practice centuries ago unacceptable as a result of the continuing civilizing process.[3366] As part of this process that changed moral standards, compassion with many wild animals increasingly spread among intellectuals in large parts of Europe, mainly from the eighteenth and nineteenth centuries onward, during urbanization and industrialization. The human attitude toward animals underwent "a whole complex of changes which, by the later eighteenth century, had helped to overthrow many established assumptions and to create new sensibilities of a kind which have gained

[3364] Thomas, 1987, 76–78 (quote from p. 76); Cibic, 2011b, 414–415.

[3365] Compare Erhatič Širnik, 2011, 6–7 (see also p. 121), 23–24, 40, 42, 53–54, 105, 113, 118–119, 125, 129, 135, 137, 139, 143, 191, 233–234, 279, 356–358, 363; von Hörmann, 1877, 214–215; Scheurer, 2010, 68–71, 83.

[3366] Elias, 1993; Elias, 1994. On the relevance of this publication in the context of the early twenty-first-century state of research, see Cvirn, 2001, 424–425, 433.

in intensity ever since."[3367] "[T]he area of moral concern had been widened to include many living beings [animals] which had been traditionally regarded as hateful or noxious"[3368] and to foster the positive feelings toward those that had already received some appreciation earlier.

In the context of the new sensibilities toward wild fauna, songbirds received the most attention because most of them have been perceived as lovely, gentle animals, and their singing entertaining or soothing. Birds' care of their offspring has been admired by people, as was their stimulation of farmers to wake up early and carry out farmwork on time, in line with nineteenth-century standards (see Sections 4.1.1.4 and 6.3.1.2). Ecological control of organisms perceived as pests in farming (such as some arthropods and mice) by many songbirds and some other bird species, in orchards, vineyards, arable land, and forests, mattered a great deal in the process that changed the attitude toward many wild bird species, among the agricultural population in particular. This was inspired by more widespread knowledge about the economic role of insectivorous birds in agroecosystems. Namely, there had been major gaps in traditional ecological knowledge concerning the biology of wild fauna, despite some surprisingly detailed insights, hence the significant nineteenth-century increase in formal education mattered in this context.[3369]

Thus, the nineteenth-century widening of moral concern in the attitude toward wild fauna was partly motivated by the shifting threshold of discomfort and partly economically. Focusing on the economic aspect in the southeastern Alps, following a request by the Styrian agricultural society, early decrees limiting bird hunting were published in Styria in 1826 and 1839, and the latter re-enacted in 1842. The 1839 decree banned the hunting of "small meadow and forest birds" until September 1st from 1839 to 1841, and it also banned the destruction of bird nests and stealing their eggs and chicks to reduce the damage caused in orchards by caterpillars. The main goal was to take economic advantage of the predation of caterpillars by insectivorous birds, which were esteemed in farming until the widespread use of pesticides in orchards, cereal and vegetable fields, and gardens. The use of pesticides increased in importance from the late nineteenth century onward, and it has become widespread especially since the mid-twentieth century. However, the 1839 Styrian decree included a general ban on hunting small meadow and forest birds until September 1st, not limited to insectivorous species—this general nature of the prohibition on hunting small meadow and forest birds in Styria, not limited

[3367] Thomas, 1987, 69–70, 149–151, 163–171, 173–188, 190–191, 243, 273, 275, 278–280, 287 (quote from p. 243); see also Erhatič Širnik, 2011, 15–22, 26.

[3368] Thomas, 1987, 173.

[3369] Erhatič Širnik, 2011, especially pp. 17–19, 31, 83, 91–93, 97–98, 100–103, 109–110, 116–117, 120–121, 124, 126, 128, 130, 132, 135, 137, 139–142, 144–147, 197, 203, 207, 354, 358, 361; Praprotnik, 1849, 126–127; Scheurer, 2010, 67–78, 80–83; Wöbse, 2022, 25, 40. See also the following footnote.

to the species considered beneficial in farming, was atypical for the Habsburg Alps. The ban on stealing eggs and chicks called for higher environmental ethics than required by the above quote from the Old Testament, but it was in line with the Christian notion that living beings are God's creation and should not suffer unnecessarily. Children were considered the main actors destroying nests, and stealing the eggs and chicks, and so the 1839 and 1842 Styrian decrees ordered teachers and catechists to educate pupils "about the cruelty and unacceptability of bird hunting and, through the awakening of 'better moral feelings,' to discourage them from such acts to 'stifle the brutal pleasure and joy and the cruel fun of bird hunting, tormenting, and killing animals.'" The 1839 decree emphasized that some teachers and catechists had already taught their pupils in the manner evoking new sensibilities toward wild birds earlier, based on their own initiative. With a few exceptions, the rigorous limits on hunting small meadow and forest birds were also kept in the 1868 Styrian law concerning bird hunting. The 1909 Styrian law then reduced this ban to the protection of bird species beneficial for farming in line with the 1902 International Convention for the Protection of Birds.[3370]

The developments in the neighboring provinces of Carinthia and Carniola were relatively similar, but legislative restrictions for bird hunting there were not as extensive as in Styria, in Carniola in particular. The regulation in Carniola was rooted in older bans on hunting small meadow and forest birds during their nesting period from March until late August, and on stealing their eggs and chicks. The 1838 decree in Carniola was stimulated by "excessive breeding of caterpillars" that caused damage in orchards, which was allegedly the consequence of reduced populations of small songbirds. Thus, with the 1838 decree, hunting of "beneficial" birds was banned in Carniola from March until August, and stealing their eggs and chicks was prohibited. Here also, teachers and catechists were required to educate children in this respect, and this continued later. In the following decades, the actual success of the legislative regulations presented in this paragraph and the previous one was limited at first.[3371]

Characteristically, new standards of environmental ethics first appeared among the intellectuals and, in addition to the measures presented, their thoughts also resulted in edifying publications "designed to excite in children 'a benevolent conduct to the brute creation,'" such as poems lamenting "the an-

[3370] Erhatič Širnik, 2011, 15–22, 26, 30–39, 46–47, 49–51, 59–65, 94, 99, 101–103, 110, 120, 133–142, 144–145, 147, 151, 154, 157–158, 174, 183, 189, 192, 194, 198, 201, 204, 211, 213, 218–219, 234–235, 239–240, 244, 246, 255–256, 262–264, 276–277, 281, 283–284, 296, 299, 301, 305, 310, 339, 354–355, 357, 359–363 (first quote from p. 15, second quote from p. 16). The governorate (*Gubernium*) in Graz, mentioned by Erhatič Širnik in 1826 and 1839, was in charge of Styria as a whole at that time (Žontar et al., 1988, 35). Wöbse, 2022, 23–24.

[3371] Erhatič Širnik, 2011, 15–17, 26, 30–36, 39, 40–45, 63, 65, 69, 94, 100, 102, 135, 142, 151, 157, 355, 359, 362.

guish of the mother bird whose eggs were stolen by marauding schoolboys or whose offspring were shot by heartless sportsmen."[3372] "[P]oetic attacks on cruelty to wild birds of every kind . . . had an incalculable effect upon middle-class sensibilities."[3373] This was a wide European phenomenon, not limited to the middle class, but also affecting farmers in general, when general school attendance spread, as demonstrated by the following example. In Slovenian territory, the influential teacher's journal *Učiteljski tovar(i)š*, with about five hundred subscribers in 1863, half of them teachers, published the poem "Tožba" (The Lament) in 1864, in which a tit, in her conversation with a teacher, laments the destruction of her inhabited nest that was thrown to the ground by male pupils. This pupils' action killed all the tit's chicks. The poem ends with: "Who leaves chicks not in peace / his heart attacks malice."[3374] This poem was rewritten by the belletrist Fran Levstik and published as "Siničja tožba" (The Tit's Lament) in a monthly pupils' journal in 1877. Levstik concludes: "Who kills the chicks in their nest / cold-hearted person manifest."[3375] Levstik's variant of this poem has been widely used in primary-school and family education ever since;[3376] for instance, the text and melody were published in a songbook for elementary schools and families in 1878.[3377] This happened in an area where almost all children in the countryside attended school from 1869 onward, and so literacy increased rapidly and reading soon also became a popular pastime activity in ordinary farmers' families.[3378]

It is no coincidence that a tit was selected as one of the main protagonists in these poems. Tits—which were previously, and sometimes illegally later, also hunted in the nesting period and consumed by people or taken captive as caged birds for human pleasure—belonged to the bird species most esteemed in farming because it became generally known by the mid-nineteenth century that they feed on caterpillars, which come into conflict with farming in orchards. Unlike some other songbird species, tits did not cause any notable harm in farming, with the possible exception of consuming some bees. In addition, tits were among those songbirds whose sound was considered particularly pleasant. Tits were on the list of permanently protected birds in Carniola according to the 1870 bird protection law, with possible legal exceptions only when this was considered safe from the farming perspective and then only with special permission that referred to the period from September until Janu-

[3372] Thomas, 1987, 149 (quote), 150, 280 (quote); Erhatič Širnik, 2011, 18–23, 123, 128–129.
[3373] Thomas, 1987, 150, 280.
[3374] Literally: whoever does not leave chicks in peace fosters malice in his heart.
[3375] Literally: whoever hunts for chicks in their nest is not a good-hearted person; the translation of *priden* is based on Pleteršnik, 2006, 300.
[3376] Mrak, 2012; on *Učiteljski tovariš*, see Dolenc, 2000, 11. Erhatič Širnik (2011, 127) states that this poem was taught for almost a century; nonetheless, in the mid-1990s, the main author of this volume still learned it in elementary school.
[3377] Mrak, 2012.
[3378] Štih, Simoniti, & Vodopivec, 2008, 326–329.

ary and could not be obtained for commercial hunting. Later, when the poem The Tit's Lament was still taught in schools, tits were protected the whole year round with the Yugoslav legislation in the Drava Banate from the 1930s and with new Slovenian legislation from 1946. Thus, the prominent role of a tit in the poems "The Lament" and "The Tit's Lament" aiming to promote benevolent behavior among primary school pupils toward songbirds is not surprising, taking into account the fact that biological knowledge in the school curriculum in the countryside of the second half of the nineteenth century was anthropocentric, dominated by information relevant in farming and forest management. In addition, the selection of a tit in the influential poem "The Tit's Lament" was in line with the role of school in spreading knowledge about legislatively protected wild animal species.[3379]

This was part of the more general but gradual civilizing process. Cruelty toward animals out of carelessness, indifference, or vindictiveness decreased significantly[3380] with regard to wild animals not considered game, noxious, or detested for practical reasons or prejudice inherited from beliefs from former times. When a haymaker came across a nest of bumblebees in a montane meadow, their honey was a welcome delicacy as long as sweets were a luxury in farmers' diet, thus until the second half of the twentieth century. Nonetheless, several farmers in the mid-twentieth century also cared for the welfare of bumblebees, drank only a bit of honey through a stalk of grass, and intentionally covered the nest so that it would not be obvious prey for a fox in a meadow mowed close to the sod, they believed. Some others believe that they ate so much honey when a bumblebee nest was found in a meadow in their childhood that the bumblebees perished.[3381]

This means that there were still relevant limitations on changes in the general attitude toward wild fauna. Several bird species still played a role in the human diet throughout the nineteenth century because there was high agricultural density, frequent poverty, and a widespread modest everyday diet in the countryside, and so smallholders and landless countryside dwellers also ate some songbirds and other birds or sold them to the inhabitants of some urban settlements. For instance, birds as small as sparrows (*Passer* sp.), common chaffinch (*Fringilla coelebs*), and buntings (*Emberiza* sp.) were part of the

[3379] Compare Erhatič Širnik, 2011, 15–17, 20–24, 32–33, 40–43, 53–54, 88, 94, 102–103, 108–110, 113–124, 126, 131, 133–137, 139, 143, 149, 165–166, 170, 174, 192, 198, 200, 213, 233–234, 264, 275, 279, 283–284, 291, 299, 301, 310; Ložar, 1944, 110; Kuhar, 1966, 8.

[3380] Cf. Thomas, 1987, 148, Erhatič Širnik, 2011, 15–22, 26, 31, 34, 46–48, 59–62, 64–65, 69, 83–95, 97–102, 105, 108–114, 121–122, 126–131, 134–135, 147–148, 151, 157, 170, 193–194, 196–197, 203–207, 210–211, 213–214, 236, 246, 263, 291, 354, 361; see also Elias, 1994, 405–407.

[3381] Author's interviews with Boltar, 2018, with Kunavec, 2023, and with Mencinger, 2018, all in Stara Fužina; and with Milena Sodja, 2017, in Srednja Vas in Bohinj.

human diet in many places and consumed in porridge in some regions such as the northwestern Dinaric Karst of today's central Slovenia – despite the full protection (with possible exceptions) of common chaffinch as an insect-feeding species in Carniola in line with the 1870 law. In the late nineteenth century and partly later, hunting small birds for human consumption was more common in northern Italy and the Austro-Hungarian provinces of Tyrol and Görz (Ital. *Gorizia*, Sln. *Goriška*) than in most other regions of the Eastern Alps. Nonetheless, people ate songbirds in gradually decreasing quantities since the late nineteenth century, with a mesoregionally, microregionally, and locally diverse decline dynamic, under the influence of several factors such as the interplay between continuation of the civilizing process, decreased agricultural density with large-scale emigration overseas, and so on. Songbird consumption was fostered in some places in Europe by food shortages as late as the first half of the twentieth century despite some legislative prohibitions.[3382]

In addition to the role of birds in the human diet, another factor limiting changes in the human attitude toward them was the frequent keeping of some native songbird species in captivity for pleasure not only in urban settlements but also in the countryside, on many farms as reported in the nineteenth century in some regions of the Alps such as Tyrol and the Slovenian Alps. The habit gradually declined in part, and native caged birds were partly replaced by exotic species that were bred for this purpose, and thus no longer repeatedly captured.[3383] Bird hunting involving legal practices was also still considered acceptable recreation by many people, and some birds taken captive were used as lures during bird hunting. Some species were still hunted in the late nineteenth century for feathers as hat decoration.[3384]

Concerning bird species that were considered noxious, even destroying their nests and eggs, and killing chicks, was legal and widespread in the nineteenth and early twentieth century. Some teachers at that time praised or rewarded pupils that destroyed nests, chicks, and adult birds belonging to species considered noxious. However, tormenting any kind of animal, including ones

[3382] Erhatič Širnik, 2011, 7, 14, 18–19, 21, 24–26, 30–33, 36–41, 46–50, 55–56, 59–60, 62–67, 86–88, 90, 92, 94, 100–105, 108–109, 112, 116–118, 121–122, 126, 129, 135, 139–140, 143, 148, 150, 157, 166, 169–171, 174, 179, 183, 189, 191, 198–199, 208–209, 245–246, 251, 263–264, 293, 301–302, 310, 343, 345, 354–356, 358–359, 362, see also pp. 51, 54, 63, 135, 137, 139, 266–267; Ložar, 1944, 110–111; Kuhar, 1966, 3, 8, 10, 13–14; Makarovič, 1978, 225–227; Scheurer, 2010, 67–75, 82–83; Wöbse, 2022, 26–29, 38, 40. On small birds consumed in porridge and on local diversity of the end of small bird consumption in the twentieth century: Slovenian Ethnographic Museum, Documentation, Fieldwork campaign 21, Škocjan (1964): notebook no. 3, 55–57.

[3383] von Hörmann, 1877, 214–215; Erhatič Širnik, 2011, 7, 14, 22, 24–26, 30–33, 36, 38–39, 47, 55, 62–64, 100, 105, 108–110, 113–116, 119, 122, 126–127, 129–131, 134, 139–140, 142, 149, 198–200, 203–204, 208, 286, 290, 354–355, 363.

[3384] Erhatič Širnik, 2011, 7, 10, 33, 51, 100, 105, 108–109, 119, 140, 147–148, 151, 354–355, 358, 363; Wöbse, 2022, 26–27.

considered noxious, was rejected.[3385] Moreover, some other teachers in the nineteenth century already encouraged benevolent conduct also toward those (song)bird species feeding on grain or fruit, thus causing considerable harm in farming.[3386] The program of the Slovenian Hunting Association, founded in the first decade of the twentieth century, adopted modern principles present among educated hunters of Austria-Hungary, and thus their formal program already paid much attention to nature protection. The 1917 volume on owls by a member of this association, the jurist and ornithologist Janko Ponebšek, one of the pioneers of environmental protection in Slovenian territory, stated that hunters should ethically care for the preservation of rare birds of prey in the landscape, even if they cause some damage because they are "natural monuments" – a notion the majority of Slovenian hunters had not yet agreed with at that time.[3387]

Despite these nineteenth-century limitations on the widespread attitude toward wild fauna, some of which were still relevant in part of the twentieth century, cases contrasting with the killing of chicks mentioned above became more frequent from the nineteenth century onward, and many followed higher ethical standards in their conduct toward some wild bird species than demanded by the quote from the Old Testament above. When the nests of a little bird, light brown on the lower side and dark brown on the upper side,[3388] were found in a moist lower montane meadow at Zagradec, southeast of Lake Bohinj (the region had previously belonged to the province of Carniola), as the only acceptable behavior in the mid-twentieth century haymakers would leave the nest untouched and keep standing a bit of grass in its immediate vicinity, and they would also pay attention that the vegetation would not be touched when spreading, turning, and raking the hay.[3389] This bird was probably the whinchat (*Saxicola rubetra*), a ground-breeding species with these colors, known to nest in moist meadows in the vicinity.[3390] The whinchat feeds on arthropods, and so it was perceived as beneficial in farming and protected in Carniola by the 1870 law and also by the 1930s legislation in the Drava Banate, which Bohinj belonged to in the Kingdom of Yugoslavia, and with the new 1946 Slovenian legislation, in all cases in the same manner as tits[3391] (see above). Thus, the

[3385] Erhatič Širnik, 2011, 18, 24, 40, 47–49, 60–67, 87, 100–104, 108, 116–117, 121, 129, 143, 150, 169–170, 208–209, 355, 358–359.

[3386] E.g., Erhatič Širnik, 2011, 111–112, 121, 127, 361, 363 (see also pp. 49, 124, 199–201, 205, 208).

[3387] Erhatič Širnik, 2011, 95–99, 133, 138, 143–144, 150, 157, 164, 183–184, 189–190, 193–194, 209, 250–251, 258–259, 266–267, 361; see Ponebšek's biography in Andrejka, 2013.

[3388] Author's interview with Kunavec, 2023, in Stara Fužina.

[3389] Author's interview with Kunavec, 2023, in Stara Fužina.

[3390] On the whinchat's diet and maps of whinchat breeding distribution in late twentieth and early twenty-first century, see Tome, 2019, 426–427.

[3391] On the whinchat in the Carniolan 1870 bird protection law and the Drava Banate hunting legislation from the 1930s, see Erhatič Širnik, 2011, 42, 165–166, 233–234.

aforementioned haymakers' behavior was likely primarily founded on personal moral standards or those usual for local farmers—namely, the likelihood that possible destruction of the nests of a little bird in a remote montane meadow usually not crossed by mountain hikers would be noticed by anyone else than the haymakers was low—but also supported by legislation.

Such haymaking behavior, rarely investigated, was evidently more widespread but not general, starting at least in the second half of the nineteenth century and lasting until the introduction of livestock-powered mowers and the motorization of mowing[3392] (see also Sections 6.1.1.1 and 10.1.2.2.5). A parallel from a large complex of periodically wet meadows at Lake Cerknica in the Dinaric Karst south of the Alps demonstrates that at least some haymakers intentionally avoided corncrake (*Crex crex*) nests as long as mowing was carried out by scythe. When a haymaker detected an inhabited nest, a child was instructed to cut a tree branch and mark the location of the nest to make sure that haymaking would not disturb the mother bird on the eggs or taking care of her chicks, according to mid-twentieth-century information.[3393] As a bird species stimulated by extensive agriculture, and thus historically abundant in many places, corncrakes were considered edible and were consumed. Corncrakes belonged to game long covered by hunting legislation, meaning that hunting rights were not comparably widespread as in the case of songbirds.[3394] The following main reasons for the humane attitude toward this species in the Cerknica Polje[3395] in the mid-twentieth century can be identified:

(1) The Old Testament–based benevolent attitude toward a mother bird found in her nest strengthened by more generally present compassion with wild fauna than before the nineteenth century due to progress of the civilizing process. This may have also been a stimulation to keep the eggs or chicks in the nest.

(2) Awareness that abstinence from hunting during the nesting period is particularly relevant for sustaining bird populations was widespread. It must

---

3392 When the warm spring of 1936 announced unusually early mowing, the Slovenian hunters' journal *Lovec* warned against the "heartless" destruction of many nests of gray partridge (*Perdix perdix*) and common pheasant (*Phasianus colchicus*) through mowing, and encouraged hunters to stimulate the farmers with a payment to preserve the nests and show them to the hunters, who would take the eggs out and have them hatched by chicken and turkey hens to be released into the wild later (Erhatič Širnik, 2011, 187). The destruction of nests was probably more frequent where livestock-powered mowers were used at that time (see Section 6.1.1.1) than where mowing was done with a scythe. On seeing through the plants in the case of mowing with a scythe, see Iuga, 2016, 73.

3393 Kebe, 2011, 204. The man that recalled this was born in 1942.

3394 Erhatič Širnik, 2011, 7, 9–10, 71–72, 150; *They Used to Eat Corncrakes*, 1997; Russell, 2010.

3395 On karst poljes, as geomorphic features, see Bonacci, 2004.

have belonged to traditional ecological knowledge. Farmers most probably learned this based on their observations of natural processes; otherwise a general official hunting ban on wetland fowl stimulated farmers to think of the vulnerability of birds at that time. Namely, the official closed season for wetland fowl hunting was introduced in the province of Carniola in 1711 from March 1st to June 24th and shifted to March 12th to July 1st in 1754. Later, it was abandoned,[3396] but it was reintroduced in 1874 from mid-April until the end of June[3397] and kept the same with the 1929 regulation.[3398] The new Slovenian 1946 legislation introduced the closed season for corncrake hunting from December 1st until August 15th or 31st.[3399] These regulations could not remain unnoticed in the villages on intermittent karst Lake Cerknica, a hotspot of waterfowl diversity. In the mid-twentieth century, after the 1946 regulation, mowing took place during the closed season of corncrake hunting. In addition, the farmer most likely did not have hunting rights for corncrake in any part of the year.

(3) Apart from the morally sensitive nesting period, corncrakes might still not be highly suitable for human consumption during mowing time if mowing in June is considered. Namely, they are a migratory species, exhausted and shrunken in spring from the long flight from southern and eastern Africa; later, they gain weight significantly, which makes them more suitable for human consumption.[3400]

(4) Corncrakes were never among the hunted birds highly relevant for the human diet in the Cerknica Polje, one reason surely being the difficulty in hunting them because fat corncrakes hide effectively and run nimbly through the vegetation,[3401] whereas more easily caught waterfowl, ducks in particular, abounded on Lake Cerknica and were also extensively hunted for food, apart from significant quantities of fish (pike and tench) caught by local farmers or simply collected as the intermittent lake receded.[3402] Thus, the local availability of other kinds of wild animal protein in the human diet was considerable.

All these circumstances likely helped create a sensitive attitude toward an edible bird species whose size would meet the usual standards for human consumption.

3396 Umek, 1970, 480, 488, 492; compare Thomas, 1987, 276–277.

3397 Erhatič Širnik, 2011, 69–70, 73.

3398 Erhatič Širnik, 2011, 160–161, 165; "Uredba, s katero se spreminjajo dosedanji predpisi . . . ," 1929.

3399 Erhatič Širnik, 2011, 232–233, 235, 237–238.

3400 Cf. *They Used to Eat Corncrakes*, 1997; *Društvo za opazovanje in proučevanje ptic Slovenije: Kosec.*

3401 *They Used to Eat Corncrakes*, 1997.

3402 Ložar, 1944, 108–110, 112–115; Zwitter, 2014d, 634–636, 639–640, 655, 660, 672; Kebe, 2011, 24, 195–198; Zwitter, [2021].

Figure 163: An anthill of red forest ants below a spruce tree at a species-rich meadow margin was left untouched during spring cleaning. This meadow at the foot of the Alps is still spring cleaned manually, with a rake; the steep marginal sections are still mowed with a scythe (author's fieldwork observations). Although the farmer cut down the shrubs to prevent the spread of woody vegetation into the meadow (stumps appear in the photo in front of the anthill, to the left), he left the anthill standing. Southeast of Stara Fužina in Bohinj. Photo: Žiga Zwitter, October 31st, 2022.

Similarly to the positive attitude towards some bird species, as a general rule, during meadow spring cleaning (Section 6.6.1) farmers in the Alps did not remove the few large non-earthen anthills from needles and branches in marginal sections of many grasslands (in a meadow-forest transition zone or below coniferous trees in meadows) made by red forest ants (mainly *Formica* spp.[3403]), as revealed by mid-twentieth century data referring to Bohinj. Exceptions were destruction of some anthills of red forest ants, which was usual if they were in serious conflict with people, such as where farmed beehives were not protected from them and numerous ants found their way to the honey.[3404] The widespread toleration of red forest ant habitations was contrary to numerous small earthen anthills of grassland ants that were systematically removed during meadow spring cleaning and in times of mowing because they would otherwise significantly increase the labor intensity of mowing by scythe in the

[3403] On species occurring in the Slovenian Alps, see Titovšek, 1981.
[3404] Compare Zalokar, 1854, 173, 186.

long run (Section 6.6.1). However, in the case of non-earthen anthills of red forest ants, bushes growing next to or through such an anthill were often cut to keep the meadow open (Figure 163), fallen tree branches next to the anthill were removed, and at mowing time grassland plants were cut around the anthill, but the anthill was kept standing as an expression of sympathy for the work the ants had invested in building their nest.[3405] In fact, the vitality of the anthill may have benefited from the aforementioned cutting of bushes growing through the anthill because this prevented the site from becoming increasingly shady; in general, ants prefer sunny sites provided that a protective tree above the anthill is not felled[3406]—a criterion met in Figure 163 because the protective coniferous tree is still there. Showing respect for the work of ants, proverbially diligent animals, by avoiding non-earthen anthills was explicitly mentioned by interviewees.[3407] A father of a smallholder farm family at Stara Fužina in Bohinj is recalled to have ordered his children before the mid-twentieth century not to remove the anthills of red forest ants, claiming that children should imagine how they would feel if someone demolished their own house.[3408] From the cultural history perspective, such a positive attitude toward those ant species that did not cause serious economic challenges is most likely rooted in a combination of the Bible[3409] and practical nature observations. These hard-working social insects building impressive nests were interpreted as moral examples.[3410] In addition, people working in meadows barefoot readily avoided the anthills of red forest ants, not to be bitten by the irritated insects.[3411]

Sparse older data suggest that many non-earthen anthills were also kept intact by people earlier in history or that human interference with them was moderate, and thus a positive attitude toward red forest ants was not only a result of new nineteenth-century sensibilities. For instance, red forest ants were allowed to build their anthill (the "amaßhauffen, so mann Mraulische nent," "amaßhauffen Mraulische genant"; i.e., the 'anthill known as Anthill'[3412]) at a

[3405] Author's interviews with Mencinger, 2017, with Kunavec, 2017, and with Urh, 2017 all in Stara Fužina; with L. and M. Odar, 2017, and with A. and M. Taler, 2017, both in Studor; with Škantar, 2017, in Srednja Vas in Bohinj; with F. Mlakar, 2017, in Bohinjska Češnjica; and with J. Smukavec, 2017, in Podjelje.

[3406] *Die wichtigsten Grundsätze im Ameisenschutz.*

[3407] Author's interviews with Mencinger, 2017, and with Urh, 2017, both in Stara Fužina; with L. and M. Odar, 2017, in Studor; with Škantar, 2017, in Srednja Vas in Bohinj. On ants' proverbial diligence, see Cibic, 2011a, 340.

[3408] Author's interview with Kranjc, 2018, in Stara Fužina.

[3409] *What Does the Bible Say about Ants?*

[3410] Thomas, 1987, 19, 61, 63–65, 125–126.

[3411] Author's interview with Kranjc, 2018, in Stara Fužina.

[3412] SI NŠAL 20, Gornji Grad Avstrija (Gornji Grad Austria), box 97, Various cases concerning the subjects of the seigneury of Gornji Grad (Germ. *Oberburg*), sixteenth–nineteenth century: 1588, Am S: Andreas tag . . . The toponym in the German source quotes is in Slovenian; *mravljišče* means 'anthill'.

crossroads in the montane vegetation belt, at the boundary of the compact holding of the isolated Planinšek farm (Figure 4) southwest of Luče in the Upper Savinja Valley (Sln. *Zgornja Savinjska dolina*) in the sixteenth century or earlier. It must have been large because it was a prominent landscape figure (the largest anthills in the Slovenian Alps reach a volume of about 1.5 m^3, more than 3.0 m^3 or even about 6.5 m^3, depending on the *Formica* ant species, environmental conditions, and further factors[3413]), and so it took time to evolve and it was not destroyed in the meantime. That anthill is mentioned in four court testimonies collected in 1588.[3414] The archival source makes it clear that by 1588 that particular anthill had already been there for roughly a decade, or for several decades, at least.[3415] However, it is likely that the anthill appeared there later than the crossroads, which highly likely formed after the medieval establishment of mountain farms in the region (Section 3.1), unless there had been no dense forest before that crossroads came into being. Namely, due to the crossroads, that place was likely less shaded; red forest ants tend to select not fully shady sites for their nests, and their colonies are more vital (the anthill was large!) at no more than partly shady to relatively open sites, making habitat conditions more suitable for red forest ants.[3416] That anthill became a boundary marker of the isolated Planinšek mountain farm.[3417] The very desig-

[3413] Titovšek, 1981, 90–97, 104–107.

[3414] SI NŠAL 20, Gornji Grad Avstrija (Gornji Grad Austria), box 97, Various cases concerning the subjects of the seigneury of Gornji Grad (Germ. *Oberburg*), sixteenth–nineteenth century: 1588, Am S: Andreas tag . . .

[3415] Namely, on the one hand the brother of the head of the Planinšek farm at the time mentioned that anthill as a boundary marker, suggesting that the anthill was (still) there, at least not long ago. On the other hand, further data in this archival record prove that that anthill was not a new landscape feature in the late 1580s. One of the witnesses stated that at an occasion earlier in his life, when they were hunting together with the grandfather of the 1588 head of the Planinšek farm, they arrived at that anthill, which must have happened quite some time previous to that, probably more than a decade earlier. Similarly, a couple of witnesses that mentioned that anthill in their testimonies referred to a woman named Reich that had participated in an earlier inspection of the boundaries of the Planinšek farm, and that earlier boundary inspection had taken place when one of the witnesses involved in the 1588 hearing had only been about ten years old; Reich herself had died before the 1588 hearing. Moreover, one of the witnesses (i.e., the head of the nearby Palčič farm), had allegedly never heard of any earlier inspection of that boundary. All these pieces of information point to a similar chronological orientation as the hunting episode mentioned higher in this footnote (SI NŠAL 20, Gornji Grad Avstrija [Gornji Grad Austria], box 97, Various cases concerning the subjects of the seigneury of Gornji Grad (Germ. *Oberburg*), sixteenth to nineteenth century: 1588, Am S: Andreas tag . . .).

[3416] Compare the ecological requirements of red forest ants presented in Titovšek, 1981, 85, 90–97.

[3417] For instance, the grandfather of the 1588 head of the Planinšek farm was claimed in the 1588 hearing to have earlier said to his hunting companion that that anthill was a boundary marker, and that his possession ended there. Similarly, the brother of the 1588 head of the Planinšek farm named that anthill as a boundary marker, and the nation as a boundary marker in the sixteenth century means that it was perceived as a rather durable structure by the locals of the time, and that they did not perceive it significantly threatened either by careless cruelty toward the ants (thus, it would most likely be considered immoral to destroy an anthill of red forest ants for fun) or by ant larvae and pupae gatherers (see below). People were believed to leave the large anthill standing, even though it was located at a crossroads of connections to mountains, and so several people from the microregion passed by on their way to mountain pastures and forests. The role of the boundary marker protected that anthill from those aware of this special status; but the recorded court testimonies from 1588 make it clear that only some inhabitants of farms in the area were aware of Planinšek's boundaries. Others could not know that that anthill had that special status, but they were not perceived as a threat to the anthill as a boundary marker.

This would likely not be the case later, from the nineteenth century until the second half of the twentieth century, in certain regions of the Alps, at least. Namely, some rural inhabitants caused damage to anthills of red forest ants when they collected ant larvae and pupae to make incense for their own use, for local and regional trade;[3418] to sell dried ant larvae and pupae as the most esteemed type of fodder for caged birds and pheasants, as recorded since the mid-nineteenth century; to use them in folk medicine,[3419] as in a wound-healing cream, including further substances, such as Norway spruce (*Picea abies*) resin;[3420] or as a trade item sold to pharmaceutical and other enterprises. Commercial ant larvae and pupae gathering must have increased with the general increase in trade flows and the developments of pharmaceutical industry in the nineteenth and twentieth centuries (Section 8.1.2). It became a flourishing partly export-oriented business in some regions of the Alps, such as the Pohorje Hills in the southeast of the mountain chain, at least from the first half of the nineteenth century to the second half of the twentieth century.[3421] Similarly, gathering ant larvae and pupae was a flourishing business in the microregion around the mountain village of Seefeld west of Innsbruck in Tyrol, as reported in the mid-nineteenth century, and they were sold in Innsbruck and Munich (Germ. *München*). The price was relatively high or high, depending on the time and region. Thus, female gatherers from Seefeld wore nice clothes and

old woman Reich was claimed by other witnesses to have mentioned that anthill as a boundary marker (SI NŠAL 20, Gornji Grad Avstrija [Gornji Grad Austria], box 97, Various cases concerning the subjects of the seigneury of Gornji Grad (Germ. Oberburg), sixteenth to nineteenth century: 1588, Am S: Andreas tag . . .).

3418 Ramšak, 2017, 1065; author's interviews with Milena Sodja, 2017, in Srednja Vas in Bohinj; with Kranjc, 2018, and with Mencinger, 2018, both at Stara Fužina; and with C. Mlakar, 2017, at the Zajamniki Alp.

3419 von Hörmann, 1877, 170; Ramšak, 2017, 1065–1068.

3420 Author's interviews with Milena Sodja, 2017, in Srednja Vas, and with C. Mlakar, 2017, at the Zajamniki Alp.

3421 Kužnik, 2002; Ramšak, 2017, 1065–1068.

jewelry on Sundays in the mid-nineteenth century.[3422] In the mid-twentieth century, 50 kg of dried ant larvae and pupae, collected in the Pohorje Hills by an experienced gatherer or a pair of gatherers in two successful weeks, sufficed to buy two pairs of oxen (in the 1950s) or a self-loading hay wagon (in the 1970s), but they gathered about four times this quantity in the entire gathering season.[3423]

These examples, the ones from the Pohorje Hills in particular, demonstrate how especially some poor people but also some relatively well-off people[3424] not only violated general ethical principles but also took the risk of being imprisoned for two weeks or paying a fine (because the activity was illegal at least since the early twentieth century due to the acknowledged ecosystem role of red forest ants)[3425] to either attain an equitable living standard or even to become well-off. Some of them tried to maintain the traditional, sustainable, economically stimulated positive attitude toward ants, and so they tried to help the ants repair the anthills they had taken the larvae or pupae from, and they believed not to have gathered too frequently to have threatened the population of the anthill.[3426] Red forest ants are indeed capable of repairing occasional serious damage to an anthill and the population does not perish from infrequent gathering of larvae and pupae, just as it recovers after occasional anthill predation by bears, wild boars, badgers, and some other wild vertebrates feeding on the protein-rich ant larvae and pupae.[3427] Some other humans, however, did not care and gathered excessively. All in all, large anthills became rare in several regions of large-scale ant larvae and pupae gathering.[3428]

This example demonstrates that, contrary to the trend toward more humane behavior toward wild animals since the nineteenth century, specific economic incentives, such as an exceptionally favorable price of the trade item or poverty, could cause worsened dealings with an animal species that had previously enjoyed special moral favor and that was still treated similarly favorably in some other regions of the Alps (see the example of the prevalent attitude toward red forest ants in Bohinj at the beginning of the narrative about the

3422 von Hörmann, 1877, 169–173; he claimed that “ant eggs” were gathered, which is in line with the typical designation of the items gathered by ordinary people but entomologically erroneous; they gathered ant larvae and pupae (compare Ramšak, 2017, 1065).

3423 Kužnik, 2002, 43, 48.

3424 See the note on the economic background of ant (larvae and) pupae gatherers in Kužnik, 2002, 41–42.

3425 Kužnik, 2002, 47–50; Marošek, 2007; Ramšak, 2017, 1065–1066.

3426 Kužnik, 2002, 41, 45, 52–53.

3427 Titovšek, 1981, 80; Kužnik, 2002, 53; Ramšak, 2017, 1066; and author’s field observations: the anthill in Figure 163 was damaged by wild fauna in winter and repaired in spring.

3428 Kužnik, 2002, 53; Marošek, 2007; author’s interview with Mencinger, 2018, in Stara Fužina.

ants) or even in the same region by those that did not get involved in the ant larvae and pupae business.

In the nineteenth century, the time also dealt with in the previous paragraphs, the British economist John Stuart Mill was thinking about some issues that belong to the core challenges of sustainable species-rich grassland management nowadays. Mill was an English classical economist, generally "stressing *laissez-faire* ideas and free competition."[3429] Concerning the social position of producers, however, he "could not rest satisfied with the distribution that condemned the laboring classes to a cramped and wretched existence."[3430] However, Mill was not only one of the leading economists of the time—he was an intellectual with a multidisciplinary background, also educated in botany and "an enthusiastic botanist all his life."[3431] In *Principles of Political Economy* (1848), he emphasized that there is not much satisfaction "in contemplating the world with nothing left to the spontaneous activity of nature; with every foot of land brought into cultivation, which is capable of growing food for human beings; every flowery waste or [semi]natural pasture ploughed up, all quadrupeds or birds which are not domesticated for man's use exterminated as his rivals for food, every hedgerow or superfluous tree rooted out, and scarcely a place left where a wild shrub or flower could grow without being eradicated as a weed in the name of improved agriculture."[3432]

Unfortunately, despite the significant efforts carried out in recent decades, the developments in the western world in the next almost two centuries have not taken the challenge of solving the contradictions between the social, economic, and environmental goals in Mill's standpoints seriously enough.

10.1.2 A thoroughly changed environmental footprint requires new environmental ethics

10.1.2.1 Twenty-first-century livestock rearing and grassland management, embedded in grassland farming-environment interactions since the nineteenth century

Before addressing the present-day relevance of the contents of this volume, readers need to become familiar with fundamental historical changes in global material and energy flows caused by humanity since the nineteenth century, and their impacts on agriculture and grassland environments. Analyses of historical global annual human use of materials (the total weight of biomass, fossil fuels, industrial minerals and ores, and construction minerals used) and ener-

[3429] *Britannica Academic: Classical Economics.*
[3430] *Britannica Academic: John Stuart Mill.*
[3431] *Britannica Academic: John Stuart Mill.*
[3432] Thomas, 1987, 268–269.

gy conversion (from biomass and coal, and since the (mid-)twentieth century also oil and natural gas, hydropower, and nuclear heat, all in joules) from 1850 to the first decade of the twenty-first century demonstrate that use of materials rose by a factor of nineteen and energy conversion increased by a factor of fourteen in these 160 years, with a rapid rise in the global use of materials and energy conversion since the 1950s. This fundamental change is a consequence of population growth and of multiplication of per capita consumption.[3433] During the world economy's Great Acceleration since the mid-twentieth-century, the human population has more than tripled at the planetary level; "the average person has become 4.4 times richer and lives 25 years longer than in 1950."[3434] However, these processes have also brought about unprecedented environmental loading.

Major changes in agriculture have belonged to this rapid global rise in material use and rapid increase in energy flows since the 1950s. These developments led to a rupture in agricultural history.[3435] In the course of previous developments raising productivity in agriculture, also presented in this book (especially Section 3.2.1, the introduction to Section 6.3, Sections 6.3.2 and 6.4)—such as the marked nineteenth-century increase in farming productivity—external inputs to agriculture remained low (e.g., available fertilizers were still limited to the products of the local agricultural system and local environmental resources, such as irrigation water depositing some nutrients). Farming development since the mid-twentieth century has been fundamentally different. The so-called Green Revolution "led to unprecedented changes in productivity, production and the intensity of land use. It changed the scale of human domination of terrestrial ecosystems and the quality and quantity of sustainability problems related to agriculture." It caused "a fundamental restructuring of the agricultural production system at all scales . . . [w]ithin only a few decades." Agriculture has been changing "from being the central activity for providing society with primary energy"—a role it played for millennia—"to an energy-consuming provider of food and raw material" at major environmental costs.[3436] "Humans have become dominating drivers of [environmental] change, and food production is the largest source of environmental degradation and has the greatest effect on the Earth system."[3437]

Less than a quarter of the global population currently lives in a typical industrial regime of material use and energy conversion, also characteristic for all countries with territories in the Alps. This regime of material use and ener-

[3433] Krausmann et al., 2016; Krausmann, Weisz, & Eisenmenger, 2016, 64–65, 67, 80–88.
[3434] *The Future of Nature and Business*, 2020, 4.
[3435] On the concept of rupture in history, see Dolar, 2022.
[3436] Ehlert et al., 2012, 473; McKittrick, 2015; Krausmann, 2016 (quotes from pp. 434, 440, and 443–444); Grass, Batáry, and Tscharntke, 2021.
[3437] Willett et al., 2019 (quote from p. 461); see also *The Future of Nature and Business*, 2020, especially pp. 22, 30.

gy conversion "is . . . problematic as neither its reliance on exhaustible resources nor its huge outflows [wastes and emissions], which exceed the sink capacity of the earth system, allow for long-term existence."[3438] Thus, the present industrial regime "by definition, lacks the potential for sustainability." It is not only waste- and emission-related environmental problems that are growing; our unsustainable use of materials and energy conversion will also have direct negative impacts on the wellbeing of future human societies: "[T]he abundance experienced [on average by present human societies] relative to the previous agrarian regime may reach limits as well."[3439] In order to reach a sustainable industrial regime, the industrial world needs "to find ways to drastically reduce the throughput of materials and energy."[3440] "A sustainability transition may well require similarly fundamental changes in society as the historical agrarian-industrial transition."[3441] Namely, a Great Reset in human-environment attitudes, and, consequently, in the economy and cultural values, is needed.[3442] "A successful coping with the relationship between an expanding social and a limited natural system has, for the foreseeable future, to rely . . . on a new ethics towards nature."[3443] These are not only conclusions by biologists, social ecologists, geographers and economic historians cited so far. They are shared by open-minded economists, as emphasized in the World Economic Forum's publication *The Future of Nature and Business* (*New Nature Economy Report II*, 2020)[3444] and by open-minded farmers.[3445] This Great Reset is a highly complex issue because it requires strong cooperation between highly diverse actors and sectors, many of them without a tradition in biodiversity thought, to tackle "the indirect forces that underlie the drivers of nature loss—such as global trade, production and consumption patterns, and governance mechanisms and the values and behaviors of society."[3446]

Two major indicators of farming's environmental impacts—global use of industrially fixed nitrogen, and phosphorous soil intake though fertilization, both highly concentrated in regions of intensive farming—nowadays exceed the Earth's planetary boundaries; that is, the safe operating space allowing humanity to maintain the Earth system in its Holocene-like state. The rate of species extinctions, again under significant impact of intensive farming, by

[3438] Krausmann et al., 2016, 206–209; Krausmann, Weisz, & Eisenmenger, 2016, 65 (quote), 87; see also McKittrick, 2015, 423–425.

[3439] Krausmann, Weisz, & Eisenmenger, 2016, 87–88, quotes from both pages.

[3440] Bätzing, 2015, 399; Krausmann et al., 2016, 209–210; Krausmann, Weisz, & Eisenmenger, 2016, 88 (quote).

[3441] Krausmann et al., 2016, 210.

[3442] Bätzing, 2015, 399; *The Future of Nature and Business*, 2020.

[3443] Gustafsson, 1997, 362.

[3444] *The Future of Nature and Business*, 2020.

[3445] E.g., *Global Agenda for Sustainable Livestock: Progress Report 2021*, 2022, 24–25; author's interview with Orešnik, 2023, at the Laz Alp in the Julian Alps.

[3446] *The Future of Nature and Business*, 2020, 19 (quote), 94–97.

far exceeds the planetary boundary as well.[3447] "[H]abitat loss and fragmentation, particularly through human appropriation of land for food production, is the greatest driver of biodiversity loss." "We have entered the sixth mass species extinction on Earth, losing species at a rate 100–1000 times greater than Holocene rates. This loss is measured through rates of species extinction, local changes in community composition, declines in population abundance, and reduced biodiversity intactness [see Section 1.2]. This species extinction is an increasing threat to the Earth system and global food security and has the potential to fundamentally undermine our ability to sustainably feed the global population by 2050."[3448] Namely, "[t]he diversity and richness of all living organisms on land and in water is necessary for the stability of ecosystems, and productivity and resilience of food production systems." It provides pollination and natural control of organisms perceived in agriculture as pests (such as control through insect predators), for instance. "This functional value of biodiversity is often poorly understood and hugely underevaluated."[3449]

The most efficient mechanism of preserving enough pollinating insects is to keep enough species-rich grasslands with high and long-lasting flowering intensity in the landscape. Wild pollinating insects are far more efficient pollinators than farmed bees. The pollinating impacts of a single wild bee can equal the impacts of a hundred farmed bees. For wild bees, seminatural grasslands "are key habitats for nesting and over-wintering," and data from Germany show that 41% of the 560 native wild bee species are endangered. Bumblebees nesting in meadows and hedgerows are active in temperatures too low for honeybees; in contrast to honeybees, bumblebees also carry out pollination in rainy and windy weather, frequent during fruit tree blossoming in the Alps. Moreover, a bumblebee transfers twice as much pollen per blossom visit as a farmed bee and pollinates up to four times as many blossoms per unit of time as a farmed bee. A bumblebee shakes the blossom heavily, which increases the pollination success and is a prerequisite for successful pollination of some plant species, such as bilberry (blueberry, *Vaccinium myrtillus*) and tomato (*Lycopersicon esculentum*). A strawberry (*Fragaria* sp.) blossom is pollinated in fifteen visits by farmed bees or only five visits by bumblebees. But bumblebees' development cycle is much longer than that of honeybees, and so they depend on continuous availability of feed resources from spring to late summer or autumn (see Section 4.1.1.1.1). In quantitative terms, about one-third of our food depends on pollination, and intake of some vitamins in our bodies is limited to food that needs insect pollination.[3450] Pollination and biological "pest" con-

[3447] McKittrick, 2015, 415–416; Steffen et al., 2015; see also Willett et al., 2019, especially pp. 466–468, 472; and *The Future of Nature and Business*, 2020, 19.

[3448] Willett et al., 2019, especially pp. 467–469 (quotes from p. 467).

[3449] Peterken, 2013, 380; Willett et al., 2019, 467 (quote), 481; *The Challenge of Feeding the World Sustainably*, 2021, 8, 12; see also *The Future of Nature and Business*, 2020, 19.

[3450] Bevk, 2018; Grass, Batáry, & Tscharntke, 2021, 266, 279, 282 (quote from p. 282).

trol are two of the aspects challenging the "belief that there is an unavoidable trade-off between agricultural production and [environmental] conservation objectives . . . deeply rooted in both agronomic and naturalist frameworks."[3451] Recent interdisciplinary agronomic-ecological research has emphasized that "[t]he recent dramatic loss of biodiversity in agricultural systems in Europe and other continents . . . requires urgent changes" in environment management within agriculture.[3452] "The number of species that can become extinct while maintaining the ability to feed humanity remains unknown [and contributions of various species are unequal], but each additional species lost represents a fundamental reduction in resilience and capacity to respond to environmental change[s]." An extinction is preceded by a reduction in the number and size of populations of the species and by local extinctions. Even in landscapes of intensive agriculture, such as some basins and valleys in the colline vegetation belt and the lower montane belt in the Alps, "[i]ntegrating a minimum of 10% of ecologically conserved land at fine scales (< 1 km^2) into agricultural systems enables habitat connectivity, which is essential for species survival, and access to the services biodiversity provides to support food production"[3453] (see Sections 1.2, 10.1.2.2, 10.2.1, and 11).

Moreover, for a variety of reasons, today's diets are causing major health issues for a significant percentage of the global population, ranging from insufficient to excessively nutrient-dense. Some diets that are at the same time unhealthy (significantly reducing life expectancy) and environmentally challenging play an important role in much of Central Europe,[3454] and thus in the Alps as well.

"In the future, the global food system could head in quite different directions. One possibility is that environmental degradation, increased inequity and injustice, exclusive reliance on global distribution networks, and homogenization of energy-dense diets will destabilize food systems. . . . The result would be a food system marked by low diversity, continued high levels of waste, reliance on external inputs, inequities, and a lack of agency. An alternate future is for the global food system to become more equitable and just, more respectful of cultural and gender issues, more biodiverse through agroecological management, less wasteful, and more food secure."[3455] Failing to implement fundamental changes in our habits, thus following the business-as-usual scenario for food during global human population growth and "moderate socio-economic development" means that "food production could increase greenhouse-gas

[3451] Magda et al., 2015, 1054.
[3452] Šumrada, Kmecl, & Erjavec, 2021 (quote from p. 12).
[3453] Willett et al., 2019, 467–470, 481 (quotes from pp. 467 and 469).
[3454] McKittrick, 2015, 411; Willett et al., 2019, 449–460; *The Challenge of Feeding the World Sustainably*, 2021, 18.
[3455] *The Challenge of Feeding the World Sustainably*, 2021, 34.

emissions, cropland use, freshwater use, and nitrogen and phosphorous application by 50–90% from 2010 to 2050" as projected by an independent multidisciplinary team of thirty-seven scientists from seventeen countries in 2019, using the latest data.[3456] In order to feed 9.2 billion people in the world sustainably in 2050, "a mindset shift is required" from inequitable productionism in which "the world needs to produce more food to feed a growing [human] population" "to the one in where we must produce foods that are more nutritious and more sustainable," contributing to environmental stability, and we need to lose and discard much less food.[3457] Polarized standpoints in scientific discourses analyzing farming and its environmental footprint, in which one pole focuses far too much only on the economy and feeding the world, and the other pole only on the environment, are still a significant phenomenon in academia, and they are limiting the time, finances, and efforts invested in well-considered integrated approaches toward sustainability.[3458]

Long-term trends in material use and energy conversion since the nineteenth century, as outlined in the first paragraph of Section 10.1.2.1, are clearly also reflected in the history of grasslands in Central Europe. Nineteenth-century farming intensification in the Alps, which depended on plant nutrient sources originating within the farm system, largely resulted in significant increases in relatively species-rich meadows moderately fertilized with farmyard manure (see the introduction to Section 6.3 and Section 6.3.2) and in limited areas of clover fields (Section 3.2.1). It was only in territories that were most specialized in animal husbandry that it also caused a significant increase in medium-intensive meadows harboring hardly any rare species. In contrast to these nineteenth-century developments, grassland farming intensification and abandonment since the mid-twentieth century has caused major increases in all intensive meadow types and significant decreases in all less intensive meadow management regimes, including non-fertilized, moderately fertilized, and medium-intensive meadows.[3459] The introduction to this book (Section 1.2) explained that species-rich grasslands in the Alps have been better preserved than in most of Central, Western, and Northern Europe, but their loss due to farming intensification at better accessible sites and abandonment elsewhere has also been a large-scale process in the Alps,[3460] and the situation in this

[3456] Willett et al., 2019, 452, 471–472, 475–476 (quote from p. 471). On the relevance of the publication Willett et al., 2019, see *The Future of Nature and Business*, 2020, 10, 26, 30, 48, and *The Challenge of Feeding the World Sustainably*, 2021, 18–19.

[3457] Willett et al., 2019; *The Challenge of Feeding the World Sustainably*, 2021 (second quote from p. 30); *[United Nations] Food Systems Summit 2021 . . .*, 2021, 24 (first and third quotes).

[3458] Debuschewitz & Sanders, 2022; Magda et al., 2015; Šumrada et al., 2021, 10.

[3459] Pfister, 1985a, especially pp. 128–129 (but see Luick, Jones, & Oppermann, 2012, 39); Dietl & Lehmann, 2004, 20, 25, 124–126.

[3460] Dietl & Lehmann, 2004; Bätzing, 2015, 277–280, 314, 322; Bavcon, Ravnjak, & Praprotnik, 2019; and Sections 1.2 and 9 of this volume.

mountain range system has been worsening quickly. Previous small-scale grassland patchiness and diversity in the landscape is being replaced by large-scale homogenization or abandonment. Moreover, England, for instance, lost 95% of species-rich meadows from 1950 to 1990; 97% percent of species-rich meadows and pastures disappeared from the English and Welsh lowlands between the 1930s and mid-1980s, and even more have been lost since (Section 1.2). The British ecologist George Peterken argues that with the virtual disappearance of species-rich meadows "something of value has been lost, not just an attractive and species-rich habitat, a part of the culture and traditions of the country, but also a source of happiness, inspiration and awareness of the environment."[3461]

The livestock sector "perhaps, exemplifies the challenges towards sustainability [e.g., the economy, human society and its culture, greenhouse-gas emissions, animal welfare, water pollution, biodiversity stewardship, or depletion] like no other sub-sector in agriculture."[3462] For instance, animal farming generates the most greenhouse gases in agriculture, and greenhouse-gas emissions from animal production will approximately double by 2050 in a business-as-usual scenario, with beef production increasing by about 70% (instead of substantially decreasing to reach a food regime allowing sustainable feeding of 9.2 billion people) and milk production rising by about 60% instead of only a minor increase at the global level to feed the world sustainably in 2050. In a business-as-usual scenario, total greenhouse-gas emissions from agriculture will reach almost twice the limit of the safe operating zone allowing humanity to maintain the Earth system in its Holocene-like state.[3463] Business as usual thus cannot continue.[3464] In 2021, the Global Agenda for Sustainable Livestock, which was initiated by the United Nations Food and Agriculture Organization, carried out a dialog drawing participants worldwide. "The diverse livestock sector actors who joined the dialogue all acknowledge that, as with all sectors, change towards better future food systems must be embraced. What that change looks like [in the livestock sector] is incredibly diverse because of the different ways that animals and products are raised, managed and consumed as well as their multiple roles."[3465] In "wealthier countries and segments of the population across the world" there is a need to switch to healthier, plant-richer diets, strongly reducing average red meat consumption to feed the world

3461 Peterken, 2013, 12.

3462 *Towards Sustainable Livestock*, 2014, 6–9 (quote from p. 9); Willett et al., 2019; *The Challenge of Feeding the World Sustainably*, 2021, 8–9.

3463 Willett et al., 2019, 471–472, 475–476; McKittrick, 2015, 424–425.

3464 *The Challenge of Feeding the World Sustainably*, 2021, 1, 3.

3465 *Global Agenda for Sustainable Livestock: Progress Report 2021*, 2022, 24–25 (quote from p. 25); Willett et al., 2019, especially p. 471; *[United Nations] Food Systems Summit 2021 . . .*, 2021, 75–76. On the Global Agenda for Sustainable Livestock, see *Transforming Food and Agriculture to Achieve the SDGs*, 2018, 55.

sustainably in 2050.[3466] No increase in the role of dairy products in the diet in Europe and Central Asia is needed to sustainably feed the projected population in 2050, unless required by complex intercontinental trade.[3467] There is much space for reduced food waste: in the early twenty-first century, between one-tenth and one-quarter of all milk ("or equivalent in dairy products") is lost or wasted globally, with significant regional differences.[3468] Following such a general sustainable development path allows a great deal of dietary diversity within healthy and sustainable nutrition and offers much space for moderate consumption of dairy products and a more limited amount of meat that are nutrient-dense, and have positive and negative impacts on the environment, with high biodiversity in (relatively) extensively managed grasslands as the most outstanding positive environmental impact of rearing livestock in the Alps.[3469] Nonetheless, feeding livestock largely, or to a limited extent at least, with fodder from species-rich meadows and grazing livestock in species-rich pastures has become very rare in Europe and is becoming increasingly less frequent even in the Alps (Section 1.2). If the ethical basis of humanity, combining environmental ethics and care for the wellbeing of future human generations, is strong enough to lead to a great global sustainable transformation of food systems until 2050, livestock rearing can maintain its high importance within transformed farming in the Alps due to the natural constraints on agriculture[3470] and in order to maintain species-rich grasslands, as explained in the next paragraph.

Namely, shared goals need to be encouraged by the great global food system transformation; nonetheless, "different priorities and pathways need to be tailored to different contexts."[3471] Biodiversity loss related to food production is one of the tremendous challenges of the world nowadays. Species-rich habitats are important local or regional specifics, deserving special attention. The animals, plants, and other organisms of these high-nature-value agroecosystems have co-evolved with farming and have adapted to farming for centuries and

3466 Willett et al., 2019; *[United Nations] Food Systems Summit 2021 . . .*, 2021, 75 (quote); Leitgeb et al., 2023, 188.

3467 Willett et al., 2019, 456, 460, 475, 483; *The Challenge of Feeding the World Sustainably*, 2021, 29, 33; McKittrick, 2015, 423–425.

3468 *Synthesis—Livestock and the Sustainable Development Goals*, 2016, 9; Leitgeb et al., 2023, 188.

3469 Wilkes, Solymosi, & Tennigkeit, 2012, especially pp. 2–3, 8–10, 19–20, 39–42 (and Hörtenhuber et al., 2010, 316, 321); Willett et al., 2019; *[United Nations] Food Systems Summit 2021 . . .*, 2021, 26, 75, 78; *Towards Sustainable Livestock*, 2014, 2, 6–8; *Synthesis—Livestock and the Sustainable Development Goals*, 2016, 3, 6–7.

3470 Compare Bätzing, 2015, especially page 381; Hörtenhuber et al., 2010 (thanks to the sustainable resource-use specialist Fridolin Krausmann for the recommendation of this article); von Glasenapp & Thornton, 2011; Potočnik Slavič et al., 2016, especially pp. 14, 71, 106.

3471 Willett et al., 2019, especially pp. 482–483 (quote from p. 483); Debuschewitz & Sanders, 2022, 5–8.

millennia. The potential natural habitats of some of these species have already been almost entirely destroyed through human activities in the last centuries, and so such organisms depend on seminatural habitats maintained by appropriate farming (see the examples in Sections 1.2 and 10.1.2.2.5). It is our ethical responsibility, grounded in legislation,[3472] not to cause the extinction of such organisms, which would be a result of making their long-term ecological adaptations to humans counterproductive through an excessive spatial scale of farming intensification or abandonment—they cannot tolerate any of these land-use changes. Consequently, in general, minimizing agricultural intensification in or expansion into existing species-rich sites is one of the priorities of development toward sustainable food systems.[3473] Species-rich grasslands belong to the most threatened habitat types worldwide, and they are "among the most endangered ecosystems in the European Union"[3474] (see Section 1.2). Seminatural meadows mowed once or twice a year are even somewhat more species-rich than species-rich pastures at similar sites.[3475]

Given the high ecological and economic relevance of species-rich grasslands and their quick disappearance, as explained in Section 1.2, it is astonishing that, contrary to forests, any specific mention of species-rich grasslands is still absent from works as relevant in policymaking and in the economic turn to biodiversity considerations as the World Economic Forum's publication *The Future of Nature and Business* (2020).[3476] In general, focus on greenhouse-gas emissions is overrepresented in studies of the environmental footprints of agriculture and human diets, leading to a lack of integrated studies taking into account diverse core spheres, including biodiversity and further aspects.[3477] Thus even a perspective focused on greenhouse gas emissions emphasizes that cessation of low-input livestock farming (designating low external energy and material inputs, such as low quantities of industrial animal feed, mineral fertilizers, herbicides, and pesticides) in the mountains would "have tremendous socio-economic and high ecological costs," and that its favorable ethical aspects in terms of livestock treatment also are highly relevant in the context of sustainability.[3478]

The data presented in Section 10.1.1 show that the preindustrial grassland management was primarily economy-driven, but the management at that time kept agroecosystems species-rich, so that high grassland biodiversity

[3472] On the legislation, see, for instance, Erhatič Širnik, 2011, 315–319, 359–360.
[3473] Pils, 1994; Oppermann & Paracchini, 2012; Willett et al., 2019, 472, 474–475, 481–482; Boetzl et al., 2021; Grass, Batáry, & Tscharntke, 2021, especially pp. 260–261.
[3474] Babai, Molnár, & Molnár, 2014, especially pp. 35–47, 133–143; Kaligarič et al., 2019; Willett et al., 2019, 468–469; Šumrada et al., 2021, 1 (quote); Šumrada et al., 2022.
[3475] Fischer et al., 2008, 151–153; Babai, Molnár, & Molnár, 2014, 139–140.
[3476] Compare *The Future of Nature and Business*, 2020.
[3477] Willett et al., 2019, 470.
[3478] Hörtenhuber et al., 2010 (quote from p. 316).

was largely a side-product.[3479] It has already been stated that several grassland plants were highly esteemed, and that preindustrial farming in the modified solar energy system enabled farmers to care for a selection of animal species they would otherwise have suppressed if they had only cared about economics. Humane attitudes toward nonhuman living beings (animals) or sustainable plant gathering were, however, sometimes challenged by an inequitable economic position or greed (Sections 8 and 10.1.1). As part of the civilizing process, farmers' humane attitude toward other species strengthened starting in the nineteenth century (Section 10.1.1), only to take a fully opposite direction in the second half of the twentieth century with the industrial intensification of farming (see also the introduction to 10.1.2.2., and Sections 10.1.2.2.5 and 10.2.1).

Modernized, industrialized grassland management, "bound as it was to industrial notions of efficiency and productivity . . . foreclosed the exploration of possible alternatives in much of the twentieth century. There were roads not traveled as the modern [industrialized] system of food production was being developed."[3480] This path of farming modernization in the second half of the twentieth century stimulated the kind of agronomic grassland research that largely ignored extensively managed grasslands. Consequently, there are still significant knowledge gaps in the agronomic understanding of species-rich grasslands due to the great diversity of such agroecosystems (Section 2) and the related advantages and drawbacks of their agricultural uses, as emphasized by recent inter- and transdisciplinary research, also involving agronomy.[3481] Species-rich grasslands have their pros and cons, and they cannot be regarded by modern livestock farmers in the twenty-first century as no more than landscape legacies of underdeveloped historical agriculture (see Section 1.2 and all of Sections 10.1.2.2, 10.2, and 11). The path of industrialized farming development has also been burdened (1) by agricultural lobbies, sometimes favoring economic decisions that cause unnecessary pollution or decrease livestock's wellbeing without any actual economic benefit for the farmer because not only yields but also production costs rise, as emphasized by agronomists,[3482] and (2) by lacunae in twentieth-century agronomic knowledge and insufficient availability of information to farmers about the complexity of the environmental impacts of agricultural practices (see, e.g., the critical evaluation of meadow management currently promoted among hay milk and hay meat farmers in Slovenia in Section 10.2.1).

[3479] See also Oppermann & Paracchini, 2012, 20–22.

[3480] Compare McKittrick, 2015, 427.

[3481] Dietl & Lehmann, 2004; Willi, 2004; Hubert, Meuret, & Bonnemaire, 2008; Magda et al., 2015.

[3482] Vidrih, 2005, 10, 150, 153, 155–156; and all of Section 10.1.2 in this volume, see also Section 10.2.1.

"As concerns about the cost and future availability of inputs continue to grow and as the limits to the current system become more apparent, notions of what is possible are beginning to shift." Biodiversity has in recent decades started being intentionally fostered in a very low percentage of farmland. "In wealthy countries, consumers are demanding food that is produced differently";[3483] "behaving responsibly [in business] is no longer just a moral imperative. There is now a strong business case for more sustainable, nature-positive models,"[3484] as emphasized in the World Economic Forum's report *The Future of Nature and Business*. There are, however, significant differences in the awareness of the recent biodiversity loss between the citizens of different Alpine countries,[3485] and further raising the public awareness of the environmental footprint of our food, including the mass extinction going on in an increasing percentage of grasslands, is of utmost relevance everywhere in Europe and in all Alpine countries to strengthen the demand for nature-positive business models:[3486] "a strong incentive" for high-nature-value farmland "from society at large" is still lacking.[3487] Nonetheless, "a growing number of processors, retailers, and consumers are interested in the environmental footprint of their purchase or consumption."[3488] *ARGE Heumilch Österreich* (the agricultural cooperative for hay milk in Austria), based on significant efforts, manages to serve the market niche of dairy products produced more sustainably.[3489] Industrialized, intensive livestock farming may soon face serious challenges that will strongly raise the prices of products from extensive rearing, such as the question of how long an average human being in the Western world will still find it morally acceptable to eat meat and dairy products from livestock reared in unethical conditions;[3490] or: "If subtherapeutic antibiotics in animal feed ultimately proved to have highly undesirable effects on the health of humans or the environment—a very real possibility—the entire industrial system of animal production, and thus the supply of relatively inexpensive meat, would be in jeopardy."[3491]

The history of human attitudes toward grassland environments shows that after the generally improved human attitude to nature since the nineteenth century was interrupted by twentieth- and twenty-first-century farming in-

3483 Oppermann & Paracchini, 2012, 22–23; McKittrick, 2015, 427 (both quotes).
3484 *The Future of Nature and Business*, 2020, 97.
3485 A mere look at the range of products of organic farming in supermarkets in Austria and Slovenia confirms this.
3486 Oppermann et al., 2012a, 466; Oppermann & Paracchini, 2012, Riedel, Walter, & Herzog, 2012, especially p. 432.
3487 Oppermann, 2012, 231–232 (quote from p. 231).
3488 Schader et al., 2014, 20–21 (quote from both pages); Cigale & Lampič, 2023, 252–253.
3489 *Traditional Hay Milk Farming . . .*, [2023].
3490 Compare *They Used to Eat Corncrakes*, 1997; but see Leitgeb et al., 2023, 176.
3491 McKittrick, 2015, 425–426.

tensification, a large-scale—though, currently impossibly full[3492]—transition to agri-environmental management is a logical path to go, to continue the civilizing process (Sections 10.1.1 and 11). A historical comparison shows that large-scale and heavy grassland management intensification in significant areas of still-existing species-rich sites (non-fertilized and moderately fertilized) converting them into green grassy plains—where no or only a couple of flowering dicotyledonous species occur—is similarly extremely anthropocentric as the message of an article published in the journal of the Swiss mountaineering club in 1881. The article claimed that it would be an act of charity to organize elimination of edelweiss in the mountains, at certain steep sites difficult to access, to prevent fearless youth from hazardous climbing to gather edelweiss.[3493] These were times of large-scale, mostly commercial edelweiss gathering (Section 8.1.3.1). The anthropocentrism in that mountaineering journal article is now perceived as absurd by the scientific community and general public alike, whereas the increasing anthropocentrism in grassland management is often still not reflected on. Nowadays, when "the number of overnourished people surpassed, for the first time in history, the number of undernourished people,"[3494] grassland species compositions are indicators of human society's environmental ethics, its (1) compassion with, (2) careless suppression of, or (3) excessive exploitation of nonhuman living beings, grassland organisms, and livestock. Human society's environmental ethics, and not only farmers', because the global food system and the unprecedented environmental footprint of food production require changes at several levels, from trade to politics and human values (see above in Section 10.1.2.1 and Section 11) to make possible an equitable standard for farmers managing species-rich grasslands to de-anthropocentrize, or at least stop anthropocentrizing, recent grassland management.

Seminatural grassland ecosystems have evolved under the influence of historical grassland management. Thus, maintaining high biodiversity in species-rich grasslands requires management (Sections 2 and 6). In quantitative terms, yields in seminatural grasslands are middling to low, depending on site conditions, species composition, and management type, and work input is often higher than in less biodiverse meadows.[3495] This means that it would be unfair to expect farmers to follow high standards of environmental ethics in their grassland management if this unethically pushes them into a precarious economic situation. Consequently, economic compensation for lower fodder quantities and an often higher workload—properly high budgetary payments designed in a well-considered manner—and fair prices of high-nature-value

[3492] Compare McKittrick, 2015; Willett et al., 2019; *The Future of Nature and Business*, 2020; *The Challenge of Feeding the World Sustainably*, 2021.
[3493] On edelweiss, see Scheidegger, 2008, 21–22.
[3494] McKittrick, 2015, 411.
[3495] Oppermann & Paracchini, 2012, 17–18.

farm products are essential for the equitable position of farmers managing grasslands in an environmentally friendly manner (see Sections 6.3.1 and 6.3.2, below in this section, and Sections 10.1.2.2 and 10.2.1). Only in this way can significant areas of species-rich grasslands be preserved in the landscape. In the late twentieth and early twenty-first century, non-farming "[f]ood consumers spen[d] a smaller proportion of their incomes on food than ever before," but this has to change to allow for sustainable food production.[3496]

Focusing on agri-environmental payments, historical knowledge about suitable meadow management and its impact on grassland ecosystems, and history-inspired innovations can best be applied in result-based or hybrid schemes (combining a result-based and management-based approach) of agri-environmental payments. Namely, in result-based schemes "payments are directly related to the achievement of environmental outcomes" such as actual plant species diversity or the "breeding success of farmland birds" (as opposed to rigid management-based schemes, including, for instance, fixed dates of allowed mowing to start, taking into account neither the diversity of site conditions nor year-to-year weather variations). Result-based agri-environmental schemes, if carefully designed and carried out, have recently proven to usually be most efficient in terms of both farmers' interest in participation (because these schemes only include necessary losses and avoid unnecessary economic losses in the farm economy) and in the state of involved target species-rich grassland ecosystems, where the plan is "to preserve already established farming practices and traditional systems."[3497] Such novel integrated agroecological approaches, including well-considered agri-environmental payment schemes equitable for farmers, can break history-rooted obstacles such as the "farmer's strong self-perception as a 'producer'" rather than a producer and a landscape manager.[3498] Similarly, based on such an approach, open-minded farmers can establish fruitful cooperation with ecologists, despite the history of much of the twentieth century, when nature conservationists and farmers "have traditionally been in continual conflict."[3499] "In designing the future [European Common Agricultural Policy], decision-makers should . . . design income support schemes that are at least neutral, if not actively promoting farmland

[3496] Compare Pils, 1994; Oppermann & Paracchini, 2012, 22; McKittrick, 2015, quote from p. 411; Favilli, 2016, 118; Herzon et al., 2018; Magda et al., 2015; *Results-Based Nature Conservation Plan,* [2019]; Debeljak et al., 2020; *The Future of Nature and Business,* 2020, 29; Šumrada et al., 2021; Šumrada, Kmecl, & Erjavec, 2021; Šumrada et al., 2022; Šumrada et al., 2024.

[3497] Herzon et al., 2018; Magda et al., 2015; Kaligarič et al., 2019; *Results-Based Nature Conservation Plan,* [2019]; Debeljak et al., 2020; Šumrada et al., 2021 (first quote from p. 11, third from p. 2); Šumrada, Kmecl, & Erjavec, 2021; Vintar Mally et al., 2021, 169; Šumrada et al., 2022 (second quote from p. 2).

[3498] von Glasenapp & Thornton, 2011, 780; author's interview with M. Gartner, 2022, in Stara Fužina.

[3499] Magda et al., 2015, 1054 (quote); see also Pils, 1994.

biodiversity," as determined by an interdisciplinary team of agronomists and ecologists.[3500] Well-designed result-based schemes include site-specific agro-ecological training for farmers individually or in small groups, adjusted to environmental conditions on their farms, and adding further pieces to the puzzles of professional competence in farmers' minds.[3501] "[I]mproved knowledge transfer was . . . recognized as necessary to change farmers' attitude towards biodiversity conservation."[3502] In some cases, there is mutual knowledge exchange, in which farmers learn from ecologists and ecologists learn local (traditional) ecological and agricultural knowledge from farmers.[3503] Knowledge gain is also the basis of recently developed agri-environmental payment schemes in several regions in the Alps (in Switzerland, Austria, and parts of France and Germany, at least) involving farmers looking for target species in their own meadows; that is, for typical species in grasslands moderately fertilized with mature farmyard manure or non-fertilized grasslands.[3504]

Spending some time in grasslands observing various organisms can strengthen farmers' interest in caring for wild living beings and their compassion with them. A reasonable amount of such fieldwork carried out by a farmer in close contact with a flowering grassland is a step in the right direction. Fieldwork as a learning inspiration in cases of the best-designed result-based agri-environmental payment schemes leads to broader knowledge gain among farmers, clearly exceeding the limits of historical traditional ecological knowledge as farmers become familiar with state-of-the-art agroecological information relevant for their grasslands. For instance, in Austria, 143 farms participated in a result-based agri-environmental payment scheme in 2013–2020. When they entered that program, the farmers' environmental consciousness had already been significantly above average because the great majority of them had previously been involved in conventional management-based agri-environmental payments. A survey among these farmers at a mature stage of the novel payment scheme determined that 93% of the respondents (strongly) agreed with the statement that the desire "to understand how cultivation affects plants and animals" was an important reason to participate in the scheme. For 96%, the wish "to learn more about plants and animals" was a similarly relevant reason, and 85% (strongly) appreciated the tailored, farm-specific ecological advice they received. To quote a farmer from Lower Austria: "Working with goals in mind is different from having to fulfill mandatory measures. It allows

3500 Šumrada, Kmecl, & Erjavec, 2021, 12.
3501 *Results-Based Nature Conservation Plan,* [2019]; Šumrada et al., 2021; Šumrada et al., 2022.
3502 Šumrada et al., 2021, 9 (quote); see also Vintar Mally et al., 2021, 158–169.
3503 *Results-Based Nature Conservation Plan,* [2019], 17.
3504 Keller, 2004, 56–57, 60–61, 65, 68–69, 84; Vogl-Lukasser et al., 2006; Keenleyside & Oppermann, 2009, 300–301; Bätzing, 2015, 152–163, 245–260, 268–276, 282, 287–289, 295–299, 301–302, 304, 333–362, 365–366, 377–378, 395–397, 399.

you to learn. When you observe the objectives in an area and make a note of what you are doing, then you really start thinking about it." A mountain farmer at Nauders, in the Alps of Tyrol, specialized in extensive livestock rearing with hay fodder, stated: "Since we've been involved in [the Austrian result-based nature conservation plan], we have become more aware of the plant populations and have been especially surprised by the large number of different orchid species. It makes me proud to have so many gems in my meadows. The hard work [the family manages 3.5 ha of hay meadows involved in the agri-environmental payment scheme at elevations of 1,600 to 1,900 m, the majority of them difficult to reach] becomes easier when you understand why preserving these meadows is so important. I would like to pass this knowledge on to my children so that they too can take the responsibility for our land." Another example concerns wild fauna: a farmer from Mönichwald in the Alps of Styria, involved in the scheme, explained: "On my [result-based nature conservation plan] areas there are wart biters, field crickets[,] and also the lesser-known long-horned grasshoppers. Once you get interested in these animals, you cannot let go. It is fascinating to observe which habitats they populate and how the populations develop. I deliberately mow some areas later or leave them untouched over . . . winter. Grasshoppers need the stalks in which their eggs can pass the winter. I also have old fruit trees on my land, which I leave standing as breeding trees for the hoopoe. And in the surrounding hedges I support the red-backed shrike, which now has several territories on my farm. I find it important for nature to be given sufficient space on my agricultural land." The same survey of 143 farms in Austria determined that 93% of respondents (strongly) agreed with the statement that they "'feel valued as farmers,' especially because they are given the competence to manage their nature-conservation areas themselves."[3505] It is also a matter of fairness in farmers' eyes that they are trusted with monitoring the land they are managing, combined with monitoring by an external specialist no more than once in a multi-annual payment period. Monitoring by a biologist every year would not only require additional funding but would also make the farmers feel untrusted and a notion would arise that biologists are earning money from land the farmers make an effort to manage in an environmentally friendly manner.[3506]

The pride felt by some farmers who deliberately preserve biodiversity is highly relevant because farming has never been just a profession but a way of life. Consequently, it is important for the well-being of farmers that farming achieves more than just economic goals. A high degree of motorization introduced into farming since the second half of the twentieth century saves farmers time, increases the labor productivity (which usually does not result in more free time for farmers, due to the reduced workforce availability and

[3505] *Results-Based Nature Conservation Plan*, [2019], 6 (third quote), 10–11 (first, second, and sixth quote from p. 11), 12 (fifth quote), 13 (fourth quote), 15, 17–19.
[3506] Author's interview with Orešnik, 2023, at the Laz Alp.

changes in the economy and standard of living), and facilitates farm work. Nonetheless, the high degree of motorization also has side effects for the farmers in terms of the abandonment of many social activities characteristic for historical farm life that included numerous tasks performed jointly by (future) partners, family members, more distant relatives, and neighbors. One of the best examples of this was haymaking, when long-lasting hard work was often accompanied by jokes, flirting, and games played by young adults, "often full of erotic implications."[3507] Due to the decline of happiness based on social activities that used to accompany farm work, a new aspect of happiness based on pride resulting from landscape impacts of environmentally responsible farming may (depending on the knowledge and personality of the farmer) be a welcome component in the life of a modern farmer. Namely, the lives of some economically thriving farmers are too economically dominated to be sustainable from a human psychological perspective.[3508]

Farmers, however, are a highly heterogeneous group in terms of their attitudes toward the environment, economic ambitions, and social structure. In addition, environmental conservation activities in farmland differ a lot, ranging from labor-intensive farming practices to those less demanding than conventional farming, and so farmers' care for the environment often does not involve both groups of activities. A detailed study of a small sample of twenty-three carefully selected farms (nine of them dairy farms mainly managing grasslands) and their agroecosystems in eight study areas in Austria, three of them in the Alps, identified the following farm style types with respect to their economic ambitions and environmental impacts (numbers 1 to 4 are the prevalent farming styles): (1) yield optimizer (a large highly profitable farm targeting maximal yields and global market, farm mechanization is modern, less productive farmland is usually abandoned, the attitude toward nature conservation is negative); (2) innovator (typically young or nowadays middle generation, working on specific market niches of high quality products, often in combination with agritourism and/or seminars, which frequently poses a high work load, the farm generates above-average profit and targets local, microregional, (and mesoregional) markets, less productive farmland is cultivated in some cases and abandoned in the others, the attitude toward nature conservation is indifferent to positive; farming is often seen not only as a profession but also as a vocation); (3) traditionalist (a small farm, often in the mountains, usually managed by an elderly farmer or, sometimes, by a part-time farmer; the farming style generates (relatively) low income, less pro-

[3507] Kuret, 1989, especially pp. 438–451; Makarovič, 1982; Smerdel, 1994, especially pp. 105–114 (quote from p. 107); Makarovič, 2005.

[3508] On the impoverishment of social life among farmers in Bohinj with the decline of traditional farming: author's interviews with two anonymized interviewees in Stara Fužina, belonging to the oldest generation. See also the interview with organic farmer and cheesemaker Toni Kukenberger in Kozorog Blatnik, 2025.

ductive farmland requiring labor-intensive farming methods is managed, thus high-nature-value farmland is often maintained as a side result of traditional farming practices, but the farmers' general attitude toward nature conservation is basically negative); (4) support optimizer (farm size is above average, an extensively managed farm generates low yields to maximize governmental agri-environmental support, which also plays a certain role in other farm styles, and support optimizers minimize expenditure; low market surplus is generated and the income is below average; the attitude toward nature conservation is positive as long as the payments meet the expectations; nonetheless, support optimizers avoid labor-intensive farm practices, thus less productive areas are abandoned); (5) production-oriented part-time farmer (struggling with time scarcity, which fosters investment in modern machinery, less productive farmland is abandoned, the farm is small and generates a low income, the attitude toward nature conservation is typically indifferent to negative: the loss of biodiversity through farming intensification is often "regretted but accepted" for the sake of income in this farming style); (6) idealist (a humble person, the farmer and the family are usually characterized by a very modest lifestyle because personal ideals such as environmental ethics are above economic considerations; less productive farmland is also cultivated through labor-intensive methods, the attitude toward nature conservation is highly positive, and nature-friendly farming is a source of personal self-realization, of pride and happiness; a farm is typically small and produces a low income, the earnings per hour can be well below the legal minimal wage in other economic sectors; agriculture is often a leisure activity); (7) forced farmer (farming involuntarily to earn a living; generates a low income, less productive areas are abandoned, shows no interest in nature conservation); and (8) social farmer (care for family members is above average, thus the farmer lacks the time for labor-intensive farming, abandons farmland that requires such management and invests in machinery, the attitude toward environmental conservation differs, the income is often low). Among the eight farming styles identified, four are crucial in the maintenance of high-nature-value agroecosystems (numbers 2, 3, 4, and 6), but among them only some innovative farmers manage to generate a high income in a nature-friendly manner. Shifts in farming styles are possible. Any kind of classification in types requires simplifications (thus, not only the "pure types" presented above exist).[3509] Moreover, the main author's knowledge of farming in Slovenian Alps, including participative observation, adds that a larger sample would also demonstrate some different combinations. Traditional farmers, for instance, are a diverse group, and some of them also do care for the environment. In this context, they are far from idealists but also from yield optimizers—many among their ancestors also cared for the environment to a certain degree, and history saw differing attitudes toward the

[3509] Schmitzberger et al., 2005 (quote from p. 280); see also Beaufoy, Keenleyside, & Oppermann, 2012, 534, and Suske, Gantner, & Gattermaier, 2012, 121–124.

environment by farmers at the same time, as presented in Section 10.1.1. This is also partly reflected in a certain degree of present-day diversity of traditional farmers' attitudes toward the environment.

The results presented from Austria emphasize that each of the four farming styles relevant for preserving high-nature-value farmland has different requirements to be perceived as viable. On the one hand, farms with very high incomes often "maintain markedly less biodiversity on their farmland" than their less affluent colleagues (see also Section 10.2.1); thus, at least upward from a certain level of farm income, the decision whether to care for biodiversity or not is often more a matter of the farmer's mentality and knowledge of the environmental challenges the world is facing than of her or his actual economic ability to also integrate biodiversity into the farming style, at least in a limited but significant part of the farmland. On the other hand, economically weak farms with high levels of biodiversity often face economic difficulties.[3510]

Agri-environmental policies differ substantially within the Alps,[3511] ranging all the way from Switzerland to most of the Alps of Slovenia. In Switzerland, agri-environmental payments are "among the highest in the world,"[3512] and, in addition to the general system of agri-environmental payments, "[f]armers can qualify for bonus payments for E[cological] C[ompensation] A[rea] meadows if certain target species are present" (see also the Swiss system of biodiversity points for farmers in Section 10.2.1). Almost half of all farms in the Alpine canton of Graubünden (Ital. *Grigioni*, Romansh *Grishun*) are certified organic, as reported in 2012.[3513]

The Alps of Slovenia deserve special attention due to the high percentage of species-rich grasslands in the landscape, also including highly species-rich agroecosystems in the colline and lowest section of the montane vegetation belt, until the late twentieth century, and partly still in the twenty-first century, which is of utmost relevance because limiting biodiversity preservation in the Alps to high elevations would not halt the mass extinction (see Section 1.2). Nonetheless, management intensification in such grasslands is an ongoing process in Slovenia (Section 1.2). In Slovenia, the agri-environmental payments themselves are mostly insufficient to foster high-nature-value farming, and their design is not optimal. A combination of different types of budgetary payments available for the same farm, partly to the same parcels, improves this situation significantly, but the challenge is that several other budgetary payment types relevant in this context can also be obtained for (more) intensive farming delivering higher yields, while destroying biodiversity, which of-

[3510] Schmitzberger et al., 2005 (quote from p. 284).
[3511] Compare Beaufoy, Keenleyside, & Oppermann, 2012, 528–529.
[3512] McKittrick, 2015, 423.
[3513] Riedel, Walter, & Herzog, 2012, 421, 423 (quote), 429.

ten puts high-nature-value farming in a precarious position.[3514] This is a wider European challenge.[3515] Farmers in the Slovenian Alps have still not been stimulated to look for target plant and animal species in their meadows because results-based agri-environmental payments for flowering meadows have not been introduced yet,[3516] despite the preparatory work done in this direction.[3517] One reason is probably that "neither the monitoring authorities nor the public agricultural advisory service [are] currently adequately equipped with the knowledge required to verify plant indicators [the basis of eligibility of a meadow for flowering meadows schemes] and offer advice to farmers."[3518] This is a wider European challenge, and Common Agricultural Policy funding is available for this.[3519] Due to the incomparably weaker tradition of integrated agroecological approaches in Slovenia in comparison to Austria, Germany, or Switzerland (see also Section 10.2.1), there is also still no agroecological championship in the Slovenian Alps evaluating species-rich meadows from an integrated, agronomic and ecological perspective (and not only from the ecological perspective or only from the aesthetic[3520] perspective, which are all less attractive for most farmers), whereby a very important person at the national level, such as a minister, would congratulate the winners, and every participating farmer would receive detailed, site-specific, integrated agronomic and ecological feedback on participating species-rich grasslands from an in-

[3514] Šumrada et al., 2024, 2, 10–11; interviews with two open-minded, anonymized representatives of mountain farmers from the Slovenian Alps and pre-Alps, in 2022 and 2023, one belonging to the young generation and the other to the middle generation; the rural geographer Irma Potočnik Slavič, 2024, personal communication.

[3515] E.g., Oppermann, 2012, 232; Poux & Ramain, 2012, 218–219; Redman, 2012, 499, 501; Suske, Gantner, & Gattermaier, 2012, 126; Trisorio, Borlizzi, & Povellato, 2012, 271–272.

[3516] Interviews with two open-minded, anonymized representatives of mountain farmers from the Slovenian Alps and pre-Alps, in 2022 and 2023, one belonging to the young generation and the other to the middle generation.

[3517] Jaksetič (interview with Kaligarič), 2019, 22; Debeljak et al., 2020.

[3518] Šumrada et al., 2021, 9 (quote), 11.

[3519] Oppermann, 2012, 232; Poux & Ramain, 2012, 218; Redman, 2012, 499; Riedel, Walter, & Herzog, 2012, 432; Trisorio, Borlizzi, & Povellato, 2012, 273; Pe'er et al., 2017, 526.

[3520] Since 2020, a local competition for the most aesthetically appealing species-rich meadows has been carried out by the Gorje Tourism Association near Bled (see the villages of Spodnje Gorje and Zgornje Gorje in Figure 4) at the foot of the eastern Julian Alps (Bregant, 2023, 24–25). This is a relevant competition type, especially because the organizer is a tourism association, and they also do their best to spread news about species-rich meadows among the general public, cooperate with botanists, and recognize the winning farmers (Bregant, 2023, 24–25). However, if the goal were to gain more farmers' attention and raise more farmers' awareness of the relevance of species-rich meadows, most farmers would need to "see clear and immediate benefits" in taking their time and learning about species-rich meadows (compare Redman, 2012, 501). Thus, an agroecological competition organized in a well-considered manner and evaluating species-rich, flowering meadows from the vantage point of reconciliation between agronomy and ecology is a different competition type and raises more knowledge among potentially interested farmers.

terdisciplinary jury of specialists striving at reconciliation between agronomy and nature conservation (see Sections 10.1.2.2.3 and 10.2.1). An open-minded representative of mountain farmers involved in talks with the national authorities lacks integrated agri-environmental approaches in Slovenia instead of polarized economy-centered versus polarized ecology-centered perspectives.[3521]

Framing successful agri-environmental schemes for seminatural grasslands requires not only bridging the divide between environmental and agricultural knowledge looking for possible synergies and responsible and sustainable tradeoffs, but also a lot of inclusive transdisciplinary work with farmers and studies of the impact of existing farming and existing agri-environment payments on agroecosystems. Thus, the contribution of social sciences and humanities is also to be taken seriously.[3522] Recent years saw highly positive developments in raising farmers' knowledge of species-rich grassland ecosystems, modern, transdisciplinary research as a background for more farmer-equitable and better designed agri-environment payment schemes in some parts of Slovenia, promotion of management and even re-creation of seminatural grasslands, attempts at integrating such meadows into economically viable farming, and raising the awareness of the importance of species-rich grasslands among the general public.[3523] Nonetheless, the cited publications make it clear that, focusing on the Alps as the study area of this volume, these highly fruitful efforts concerning species-rich meadows and framed as mutual knowledge exchange rarely took place there. This happened in a few Natura 2000 microregions and exceptionally in other microregions where there are also many meadow sites that are biodiversity hotspots, but those areas have not been included in Natura 2000, such as the many daffodil meadow sites in the western Karawanks[3524] (see Section 6.2.2.3). Thus, extensive modern transdisciplinary activities with farmers concerning species-rich meadows have so far been carried out only in a very small percentage of Slovenian territory, not representative of the Slovenian Alps, due to regional diversity in environmental conditions, the socioeconomic structure of farms, farming styles, and farmers' mentality within Slovenia. The activities briefly mentioned provided valuable insights and form a solid background for further transdisciplinary work

[3521] Author's interview with an anonymized representative of mountain farmers from the Slovenian Alps and pre-Alps in 2023.

[3522] Dietl & Lehmann, 2004; Schmitzberger et al., 2005; Erhatič Širnik, 2011, especially p. 364; Magda et al., 2015; Broggi, 2016, 9; Svadlenak-Gomez & Badura, 2016; *Results-Based Nature Conservation Plan*, [2019], and the publications cited in the following footnote.

[3523] See Debeljak et al., 2020; Šumrada et al., 2021; Vintar Mally et al., 2021; Šumrada et al., 2022; Šumrada & Erjavec, 2023; Šumrada et al., 2024. See also the presentation of the new agri-environment measure aimed at the preservation of dry calcareous grasslands in the Dinaric Karst of western Slovenia (thus non-fertilized meadows in the Limestone Alps are not eligible): *Suhi kraški travniki in pašniki*, 2023.

[3524] Krepfl, 2012; Klinar, 2012, 98.

aiming to halt the rapid decline in species-rich meadows in Slovenia, which requires substantial further efforts. This could pave the way toward good economic uses of species-rich meadows in Slovenian Alps in the near future following high ethical standards (see further subsections of Section 10.1.2 and Section 10.2)—but, to make such a scenario possible, significant areas of existing non-fertilized and moderately manured meadows need to be preserved now, in all vegetation belts of the Alps. Preserving high-nature-value grasslands is significantly more efficient and cheaper than restoring them later.[3525]

The regions where the transdisciplinary activities presented above involving extensive and inclusive knowledge exchange with farms in Slovenia about species-rich meadows have taken place so far have also not yet been representative of all Natura 2000 regions in Slovenia with seminatural grassland sites of European relevance. Focusing on the Slovenian Alps, Natura 2000 and other ecologically valuable sites in most of the Limestone Alps and most of the pre-Alps are among those insufficiently involved in extensive recent transdisciplinary work with farmers linking the agroecology of seminatural meadows and the farm economy in Slovenia. One of the main incentives for agroecological activities concerning plants and animals of seminatural meadows and the farm economy in the Alps and pre-Alps of Slovenia was the transdisciplinary project *Life to Grasslands*, with project areas in the Pohorje Hills (in the Central Alps of northeastern Slovenia) and in the Kum microregion in the pre-Alps of eastern Slovenia.[3526] There are also a few other initiatives in the Slovenian Alps, such as the one dealing with daffodil meadows in the Western Karawanks.[3527] The regions where, despite current efforts, more transdisciplinary and trans-sectoral work is needed to reach a comparable level of agri-environmental policy in Slovenian Alps as in Austria, and where more transdisciplinary and trans-sectoral efforts should be invested in a fair market for the products of high-nature-value farming, include even Triglav National Park,[3528] the only national park in Slovenia, primarily aimed to preserve "ecosystems and natural processes, diversity of habitat types, animal and plant species, and the quality and diversity of landscapes." Thus, the continuity of high-nature-value farming, which is in a fragile state and in decline (see Section 1.2), is crucial for this protected area.[3529] There are studies on, conservation efforts in, and restoration of some seminatural grasslands, including some knowledge exchange with farmers, financial incentives, and other cooperation between

3525 Beaufoy, Oppermann, & Paracchini, 2012, 27; see also Section 1.2.

3526 Debeljak et al., 2020.

3527 Krepfl, 2012; Klinar, 2012, 98.

3528 See, e.g., the basic level of recent research on the uptake of agri-environment payments in Triglav National Park (Bertoncelj & Travnikar, 2024) in comparison to the sophisticated research as a background for future agri-environment payments in a few other areas in Slovenia mentioned above. Bertoncelj and Travnikar (2024), for instance, have not yet carried out interviews with farmers.

3529 Bertoncelj & Travnikar, 2024, 81.

nature conservation and farmers.[3530] Despite current efforts, an insufficient level of transdisciplinary and trans-sectoral agroecological activities concerning economically and socially viable high-nature-value farming in seminatural grasslands in the protected area that could help create a more equitable and just agri-environmental policy in Slovenia is reflected in the relatively low general participation of farmers from the national park in the Common Agricultural Policy's agri-environment measures[3531] (see also the caption to Figure 15 in Section 3.2). Extensive, inclusive, transdisciplinary and trans-sectoral, integrated agroecological work (also focusing on the functional role of biodiversity), one of the pillars of modern agri-environment policy in agroecosystems requiring land sharing to preserve biodiversity in all vegetation belts in the Alps,[3532] clearly needs more efforts in the national park and the Slovenian Alps in general for the long-term success of biodiversity preservation in a significant percentage of species-rich meadows.

10.1.2.2 History-inspired best practices in agroecological grassland management and uses

This section explores historical data, not aiming to turn "rural areas into living museums for privileged urban visitors,"[3533] but to explore possible contributions of the history of species-rich grasslands to sustainable development that includes sustainable innovation. The United Nations Food Systems Summit 2021 emphasized the importance of valuing historical, traditional knowledge "when introducing innovations," of "taking the best out of both traditions and innovations."[3534] Combinations of the most recent understanding of ecosystems, older—including historical (traditional)—smart management types and techniques, and "judicious, even sparing, use of industrial inputs" appears to be the most promising guarantee for "long-term success and sustainability" in agriculture, taking agroecological approaches seriously.[3535] Indeed, in order not to follow the English example in grasslands in the Alps below the forest line to a high degree, it seems necessary to work on innovative compromises combining the very much different worlds of historical agrarian societies and of recent industry-rooted service-oriented society that, from the present perspective, will be economically viable, socially equitable, and ecologically sound, thus sustainable. "[E]ach of them [the worlds of agrarian and ser-

[3530] See, for instance, the publication Papež Kristanc, 2012; the recent project *Life for Seeds*, such as the report *Seed Gathering and Other Project Activities in Triglav National Park*, 2024; *Triglavski narodni park: Kmetijstvo*; *Triglavski narodni park: Razpis za kmetijstvo in podeželje na območju TNP 2024*.

[3531] Compare Bertoncelj & Travnikar, 2024, 76–82.

[3532] Broggi, 2016, 9; Svadlenak-Gomez & Badura, 2016; Grass, Batáry, & Tscharntke, 2021.

[3533] Ledinek Lozej, 2021, 155; compare Erhatič Širnik, 2011, 308.

[3534] *[United Nations] Food Systems Summit 2021 . . .*, 2021, 34 (both quotes), 72–73.

[3535] McKittrick, 2015, 427 (quotes); compare *The Future of Nature and Business*, 2020, 30.

vice-oriented societies] has specific qualities indispensable for [a sustainable] future of the Alps . . . ."[3536]

A historical insight is needed to understand the recent potential for history-inspired innovation properly. The contemporary historical context explains that, as a general rule, appreciation of the traditional economy in the Alps was the lowest in the second half of the 1970s. Later, there were gradual major changes in the attitude toward traditional practices and arrangements. Since the 1980s, appreciation of traditions has been gradually rising, and interest in traditional practices and arrangements again became characteristic Alps-wide by the first decade of the twenty-first century. This opened the field for creative history-inspired innovations linking historical practices and arrangements (in our case, species-rich meadows and pastures) with a modern way of life in a sustainable manner—economically viable, socially equitable, ecologically sound, and offering locally or regionally specific products or services.[3537] In practice, innovative actions linking traditions and modernity in agriculture are relatively rare.[3538] Namely, in contemporary history, the perception of modernity in the farm economy is mostly based on thorough breaks with traditions; it focuses on the economy and society, but usually this type of "modern" approach causes large-scale environmental degradation.[3539] Only gradually, this view has started being significantly challenged by well-considered agroecological approaches.[3540]

One of the main general lessons that modern agriculture can learn from history is that a thorough consideration of energy and material flows caused by every decision at the farm level and the entire environmental footprint of every decision at the farm and beyond (see Section 10.1.2.1) should become one of the fundamental principles underlying decisions by modern farmers, just as important as economic and social considerations. This thorough consideration should not be limited to the question of the farm's direct, local

3536 Bätzing, 2015, 240–241, 244, 294–299, 301 (quote), 302, 304, 333, 357–362, 376–387, 391–392, 396–397, 399.

3537 Bätzing, 2015, 240–241, 244, 277–280, 294–299, 301–302, 304, 333, 361–362, 376–387, 391–392, 396–397, 399; Leibundgut & Vonderstrass, 2016a, especially pages 220–221, 226–233; *Oberlin College and Conservatory: About OES: What is Sustainability*.

3538 Bätzing, 2015, 298–299, 301–302, 304, 357–362, 391, 396–397. See the example of a young farmer recommending history-inspired innovative thought in the movie presenting the Gartner Vacation Farm at Studor, Bohinj ([Gartner & Uhan], 2020). This movie was recorded as part of the Horizon 2020 NEWBIE network facilitating "the development and dissemination of new business models, including new entry models, to the full range of new entrants [to farming]—from successors to complete newcomers to the agricultural sector" (*New Entrant netWork: Business Models for Innovation, Entrepreneurship and Resilience in European Agriculture*).

3539 Ehlert et al., 2012; Bätzing, 2015, 279–280, 289, 298–299, 301, 358–361, 377–378.

3540 Bätzing, 2015, 240–241, 244, 277–280, 298–299; Magda et al., 2015; *Results-Based Nature Conservation Plan*, [2019]; Šumrada et al., 2021; Grass, Batáry, & Tscharntke, 2021; Tscharntke et al., 2021; Šumrada et al., 2022.

environmental impacts, but should also include, for instance, the environmental impacts of production and long-distance transport of concentrated feedstuff (Sections 10.1.2.1 and 10.1.2.2.4) and of possible mineral fertilizers, which require an enormous amount of energy to be synthesized (Section 10.1.2.2.4), and so their environmental impacts are not limited to the territory where they are applied.[3541] Moreover, mineral fertilizer application raises not only the questions of soil and water pollution, but also direct impacts on meadow fauna. Mineral fertilizers are so concentrated that when amphibians become contaminated with the fertilizer granules—which often happens when they migrate across fertilized farmland—this leads to alkali burns on the epidermis, and so it causes pain and suffering.[3542] Moreover, considerations should not neglect medicines and veterinary costs that could be avoided through better decisions about healthier, diverse fodder and a higher quality of life for livestock, and so on.

Yet, farmers are only informed about "[t]he environmental impacts of modern agriculture . . . to a certain extent,"[3543] which has to change—what, for instance, is the environmental footprint of the concentrated feeds they use? For example, does the farm contribute to tropical rainforest clearing for soybean plantations because soy of such origin is extensively used in the conventional European livestock sector (see Section 10.1.2.2.4)? History promotes serious thinking about the possibilities of greater reliance on local material and energy resources in farming. Paying greater attention to more complex sustainability in farming, linking the economy, human society, and the rest of the environment, can contribute its pieces to the broader puzzle of developing sustainable farming, also involving market conditions, measures of agricultural politics, and consumers' values. Some intensive farms in the Alps will become more profitable if they reduce environmental loading through large investments in concentrated feed and mineral fertilizers and get rid of their agricultural lobby-dominated mentality.[3544] On some other farms, some more environmentally friendly decisions can prove economically and socially acceptable, even if not always the most cost-effective, and some of them can bring farmers and their families more happiness than a slightly less labor-intensive (or a slightly more profitable) but less ethical action or management type.[3545] The very essence of farming is work with living beings, which can bring happiness but

[3541] See e.g., Leitgeb et al., 2023.

[3542] Ehlert et al., 2012, 474.

[3543] On farmers' limited knowledge about the total environmental footprints of the farm economy, see McKittrick, 2015, 425. Compare the recently developed Sustainability Monitoring and Assessment RouTine farm tool: Leitgeb et al., 2023, 176.

[3544] Vidrih, 2005, 10, 150, 153, 155–156; author's interview with an anonymized knowledgeable representative of farmers from the Slovenian Alps, 2022.

[3545] Compare Schmitzberger et al., 2005, 276, 279; Suske, Gantner, & Gattermaier, 2012, 126; see also Sections 10.1.2.1 and 10.2.1.

also frustration. Selected examples of history-inspired good practices in grassland farming relevant today are given below; just like this volume as a whole, the examples focus more on meadows than on pastures. Some other historical practices and arrangements presented in this volume may also be inspiring, but most of those not presented here are too labor-intensive to be applicable nowadays.

Before focusing on examples of history-inspired good practices and examples of history-inspired innovation in species-rich grassland management, it must be emphasized that local and regional site conditions and ecosystem properties, as well as the human society contexts, matter, and so "[s]olutions that claim to be 'best practices' [in certain places] lead to inappropriate one-size-fits-all interventions and scaling strategies that are not only inappropriate in some contexts but can bring about harm through negative consequences; diversity needs to be considered."[3546] Considering historical contributions to twenty-first-century agroecological thought, this nonetheless does not mean that only local history matters. Namely, cultural transfer and knowledge transfer were more limited in the past than today; moreover, further social practices might have locally prevented implementation of some good practices. "[R]esearch on farmland biodiversity and [respective] policy instruments" in Europe mostly focused on areas in the west and north of the continent, "which are dominated by intensively farmed landscapes," often also in areas of intensive agriculture in Central Europe. This means that in the Eastern Alps, for instance, these findings are relevant in (relatively) intensive farming systems, whereas "the suitability of applying the same approaches" in areas of more extensive agriculture in Europe, such as in several meadows in the southeasternmost Alps, "has . . . frequently been called into question."[3547] Consequently, more context-specific approaches are more suitable in the Eastern Alps and some other parts of this mountain range.[3548] As stated by an interdisciplinary team of agronomists and biologists in 2021, "future research should primarily address the development of sustainable farming models for [high nature value] grasslands that could achieve both nature conservation objectives and adequate farm incomes" and "[t]hese new approaches might need to be quite innovative."[3549] Similarly, the interdisciplinary agroecological cooperation of the renowned grassland agroecologist Walter Dietl and agronomist Josef Lehmann working in the Alps came to the following conclusion: "In recent times, dubious natural science-technology and economy views often promoted [grassland] farming forms" heavily threatening plants, animals, further com-

3546 *[United Nations] Food Systems Summit 2021 . . .*, 2021, 20–22, 26, 28–29 (quote from p. 20).

3547 Bätzing, 2015, especially pp. 277–280; Grass, Batáry, & Tscharntke, 2021, 265; Šumrada et al., 2021, 2; Šumrada, Kmecl, & Erjavec, 2021, 12 (quotes).

3548 See *Results-Based Nature Conservation Plan*, [2019].

3549 Šumrada et al., 2021, 11 (first quote), 12 (second quote).

ponents of the environment, people, and human culture. Consequently, Dietl and Lehmann recommend taking "the historical farmer's tasks more seriously again, through naturally, economically, and socially durable [land-]use forms," which calls for innovative approaches.[3550] Some historical good practices thus have the potential to add pieces to the puzzle of developing such locally adapted history-inspired agroecological approaches, offering further potential for history-inspired innovation.

10.1.2.2.1 Farmers observe (flowering) grasslands thoroughly

History makes clear that precise observations of meadows several times during the vegetation season significantly increase farmers' understanding of ecology of the grasslands they manage,[3551] and this strengthens farmers' interest in grassland organisms. "Manually cutting the grass, then scattering, turning, raking and finally collecting it [allowed and rarely still] allows a close inspection of the land each year."[3552] Historical grassland observations during walks to sites of farmwork also mattered. In addition, largely work-free Sundays and other holidays (Section 4.1.1.4) allowed the farmers to take more time: these were usual occasions for relaxed walks by the agricultural population, when many farmers observed the landscape, including the blossoming fields and hay meadows—not only from an economic perspective, focusing on what the harvest would bring, and to feel pride in the work diligently done, but also from an aesthetic point of view.[3553] To quote a traditional farmer from Ghimeş (Hung. *Gymes*) in the Eastern Carpathians: "honestly said: I need to stop and have a look [at the blooming highly species-rich meadows in June and July], I cannot tell how beautiful the colors of these flowers are, one is more beautiful than the other."[3554] Similar are farmers' testimonies from the Alps of Carinthia: one of the important motives for traditional labor-intensive mountain meadow management nowadays is to maintain "the meadow with its flowers etc." or "work in a [traditional] mountain meadow [much more labor intensive than in the valley, including more contacts of the farmer with the species-rich grassland ecosystem] is simply . . . beautiful."[3555] The examples of historical avoidance of birds' nests and large anthills (Section 10.1.1) have already demonstrated that historical knowledge of grassland organisms also stimulated farmers' compassion with certain wild animals.

[3550] Dietl & Lehmann, 2004, 128 and presentation of authors on the back cover.
[3551] On historical experience-based grassland knowledge in the mountains in Europe at middle latitudes, see Meilleur, 1986; von Glasenapp & Thornton, 2011, 774–776; Babai, Molnár, & Molnár, 2014; and Zwitter, 2020.
[3552] von Glasenapp & Thornton, 2011, 774 (quote); see also Zwitter, 2020.
[3553] Compare Štucin, 2011, 71; Meilleur, 1986, 389, 395.
[3554] Babai, Molnár, & Molnár, 2014, 26 (quote), 35–37, 40, 42–46, 57, 135, 138, 141, 143–144.
[3555] Keller, 2004, 71–72, 137 (quote), 142 (quote).

"An older farmer explained that his father would have cared for his land much more than he does, and that he sees the same difference between himself and his son,"[3556] as recently recorded in the village of Vals in Graubünden, Switzerland. This is a usual reality. A recent investigation comparing grassland plant knowledge by old and young farmers in the Biebrza Valley in Poland led to the assumption "that direct contact with plants during haymaking . . . and pasturing . . . livestock . . . are crucial factors in knowledge development. The more distanced the farmers are (literally—by machines), the poorer their [experience-based] knowledge."[3557] Thus, erosion of farmers' experience-based grassland knowledge is mainly a side effect of factors making modernized farming socially equitable and economically viable under the conditions of limited workforce, limited time, much stronger long-distance trade flows, and unfairly low prices of some farm products, such as milk, but also a result of the general trends in industrial society's mentality developments. Consequently, mowing, spreading and turning the hay, raking, and gathering-in the (bailed) fodder have recently become entirely motorized on many farms in the Alps, in flat meadows or on slopes not too steep for heavy machinery. They are performed quickly, without immediate contact with the grassland. In addition, walking to the parcels is replaced by motorized driving, which is too quick to allow (thorough) observations of grasslands on the way, and the noise (and cabin) of the vehicle do not make it possible to listen to grassland insects and birds—or to notice the silence caused by their sudden disappearance as a result of strong intensification. Changes in grassland ecosystems caused by intensification in the grassland management regime are not as precisely noted by young farmers anymore compared to historical and often less thorough changes that used to be noted by their predecessors.

Consequently, equitably high result-based agri-environmental payments requiring that farmers walk through the grassland (diagonally, not to focus only on the less intensively managed sections at the margins of the parcel) before mowing, with a checklist of target species, to check which ones occur not only in the marginal areas of the parcel but also in the center of the cross-section the farmer walked through, which is the basis for the payment (Section 10.1.2.1),[3558] are a step in the right direction and an innovation inspired by history.

10.1.2.2.2 Farmers smell the fodder

"I remember when we were making hay in the Voje Valley [in a montane meadow mowed twice a year] and we slept there [in hay in a barn], oh, how pleasant it smelled! And grasshoppers were jumping; now it seems to me that

[3556] von Glasenapp & Thornton, 2011, 779.
[3557] Sucholas et al., 2022, 37.
[3558] Compare Keenleyside & Oppermann, 2009, 300–301; Peterken, 2013, 285.

this is largely all gone."[3559] – A knowledgeable old farmer from Stara Fužina, Bohinj.

Smelling the hay used to be widespread in the Alps, not limited to sleeping in hay. "A farmer may smell the grass to judge differences in herb content,"[3560] a history-inspired practice recorded by research in Vals, Graubünden. Smelling the hay in species-rich grasslands was motivated by several purposes that are also relevant today. First, the smells of grasses and herbs in non-fertilized and moderately manured meadows matter because cattle prefer aromatic fodder.[3561] Second, milk based on fodder that includes aromatic plants from several types of species-rich grasslands is the basis for the highest-quality livestock products.[3562] Sweet vernal-grass (*Anthoxanthum odoratum*, mainly on acidic soils, in valley bottoms and montane meadows in particular) usually gives the hay its characteristic flavor, but it occurs only in non-fertilized and moderately fertilized grasslands (Sections 6.3.1, 6.3.2, 10.1.2.2.3, and 10.1.2.2.4). Thymes (*Thymus* spp.) in hay, abounding in some (relatively) dry non-fertilized or very moderately fertilized meadows,[3563] are examples of herbs giving the milk a particularly excellent scent, as traditional ecological knowledge in Bohinj has it.[3564] At the same time, thymes promote appetite among cattle and contribute to livestock's immunity, according to traditional ecological knowledge from East Tyrol.[3565]

The impact of pleasant-smelling herbs on diary products' quality is illustrated by the following historical example. Between the two world wars, butter from the settlements of Gorje and Poče, the highest villages on the southern slopes of Mount Porezen (1,630 m) in the southern Prealps of the Julian Alps (Sln. *Julijske Alpe*), was highly esteemed on the market of the port city of Trieste, some 60 km away by air. Cattle from these villages received a lot of highly fragrant fodder from extensively managed, species-rich, steep montane hay

3559 Author's interview with Kranjc, 2018, in Stara Fužina. Indeed, the species composition in the great majority of meadows in that valley, which is a Natura 2000 site, has changed significantly due to management intensification (the biologist Peter Skoberne, personal communication; *Environmental Atlas of Slovenia*, Nature, Natura 2000).

3560 von Glasenapp & Thornton, 2011, 774.

3561 Pfister, 1985a, 42; Vidrih, 2005, 139.

3562 Dietl & Lehmann, 2004, 91, 94–96; Bätzing, 2015, 381; Machatschek, 2015, 25, 28, 33–34, 37–40; Magda et al., 2015, 1060; Machatschek, 2017, 20, 23, 25, 72–73, 82 (with a romanticized presentation of the economic history of Damüls; compare Gross & Winiwarter, 2015, 492–498 and general remarks on the economic history of villages in the Alps at high elevations in Viazzo, 1989); Bavcon, Ravnjak, & Praprotnik, 2019, 81, 83, 131, 207.

3563 Aeschimann et al., 2004, vol. 2, 148–155.

3564 Author's interview with C. Mlakar, about 2018, an experienced farmer and cheesemaker born in 1930s, visited in the Zajamniki complex of montane meadows and pastures in Bohinj. Mlakar commented on the quality of their hay from a non-fertilized meadow at Zajamniki.

3565 Vogl-Lukasser et al., 2006, 84, 189.

meadows on the south-, east-, and west-facing slopes of Mount Porezen and the mountains nearby, starting at about 650 to 1,000 m and reaching 1,630 m.[3566] The impact of a significant percentage of fragrant herbs, such as thymes, in hay on the excellent aroma of butter is well known.[3567] Another late-twentieth-century example: a farmer in Bohinj whose cattle in the summer mountain pastures fully depended on grazing resources emphasized that, if the cattle received industrially processed fodder as an addition to grazing, the milk and cheese flavor would thoroughly change, and that the cheese taste from the mountain pasture cannot be attained in the valley following the same milk-processing procedure.[3568] Concerning dairy products, smells of grassland plants not only directly impact their aroma and taste; secondary metabolites from plants also favorably influence the microorganisms in cheese; moreover, milk based on fodder from appropriate species-rich grasslands "is rich in protein, [those] fatty acids [beneficial to human health], minerals and antioxidants, which is also healthier for humans."[3569]

The meat flavor from specific types of species-rich grasslands is also best.[3570] In this context, history is enlightening: whereas this aspect is now neglected by most consumers, it used to be taken into account in the early modern market economy, as demonstrated by a parallel from the northwestern Dinaric Karst, located south of the southeastern Alps. Data from 1780 demonstrate that lambs from the Karst (Sln. *Kras*, Ital. *Carso*) Plateau were sold in large numbers in Venice at a particularly high price, which was a consequence of the fodder quality from dry calcareous grasslands.[3571] Although species-rich grasslands were usual at that time, dry calcareous grasslands on the Karst Plateau provided particularly aromatic fodder in comparison to most places located closer to Venice. In addition, aromatic plants positively affect the quality of honey (the aroma, taste, and proportions between sugars depend on nectar sources) and other bee products.[3572]

[3566] Štucin, 2011, 70; *Archivio di Stato di Trieste*, Catasto franceschino, Mappe del Catasto franceschino, Distretto di Tolmino, Comune di Goriano di Circhina [Sln. *Gorje*] . . . (1822 – sec. XIX primo quarto): sheets I, II, III, IV, V (localization: *Environmental Atlas of Slovenia*, Layers, Spatial units, Land Cadastre and *Environmental Atlas of Slovenia*, Backgrounds, Overview maps). See also Section 3.1, which presents twentieth-century long-distance trade in such highly fragrant hay from this microregion.

[3567] Peterken, 2013, 187.

[3568] Ledinek, 1999, chapters "Krma in mlečnost" and "Predelava mleka."

[3569] Dietl & Lehmann, 2004, 91, 94–96; Vidrih, 2005, 164–165; Bätzing, 2015, 381; Machatschek, 2015, 25, 28, 33–34, 37–40; Magda et al., 2015, 1060; Machatschek, 2017, 20, 23, 25, 72–73, 82; Bavcon, Ravnjak, & Praprotnik, 2019, 81 (quote), 83, 131, 207.

[3570] Dietl & Lehmann, 2004, 91, 94–96; Vidrih, 2005, 164–165; Bätzing, 2015, 381; Machatschek, 2015, 25, 28, 33–34, 37–40; Magda et al., 2015, 1060; Machatschek, 2017, 20, 23, 25, 72–73, 82; Bavcon, Ravnjak, & Praprotnik, 2019, 81, 83, 131, 207.

[3571] Panjek, 2015b, 29; Zwitter et al., 2024, 447.

[3572] Crane, 1999, especially page 502; Bavcon, Ravnjak, & Praprotnik, 2019, 207.

10.1.2.2.3 Moderate fertilization and proper mowing timing: an agroecological strategy deserving much attention

Moderate fertilization with composted farmyard manure and mowing when grass seeds are ripening, every second year at least, in a significant percentage of all grasslands in the landscape, can nowadays often be perceived as a good compromise between (1) economic interests, (2) environmental considerations, and (3) animal health and wellbeing (see also Sections 1.2, the rest of 10.1, and 10.2), if supported by appropriate agri-environmental payments and product prices.

Nowadays, inspired by long-term evidence, interdisciplinary cooperation between agronomy and grassland ecology has recognized relatively species-rich, moderately manured meadows[3573] (Section 6.3.2) managed in a well-considered way as a good agroecological solution at some sites without special constraints for grassland farming. Apart from the practice of moderate fertilization, such history-inspired management involves two mowings (and grazing) or three mowings (without grazing) annually in lower valleys in the Alps.[3574] In many cases, but not all, optimal timing of the first mowing in such meadows can follow another history-inspired good practice—namely, the annual alternation between middle mowing dates (when grasses blossom) and late mowing (with ripe grass seeds, often in mid-June at lower elevations, and in late June at higher elevations), making sure that many ripe seeds are dispersed every second year, which is the only way of preventing the decline in productivity of these grassland species compositions in the long run, as even the dominant grass species in moderately fertilized meadows depend on generative reproduction. Nonetheless, alternating mowing dates increases the fodder quality every second year (see Section 6.5).[3575] Yields in such meadows would decline significantly in the case of earlier mowing, such as regular annual mowing dates when grasses blossom; these species compositions cannot tol-

[3573] This means about 1 kg of matured (!) farmyard manure per m^2 annually in lower valleys and the same quantity once every two to three years at higher elevations (see Sections 6.3.2, 10.1.2.1, and the introduction to Section 10.1.2.2). Only in exceptional autumns, up to 20 m^3/ha of highly diluted slurry or liquid manure can be added (Dietl & Lehmann, 2004, 13, 24, 54, 102, 105–111, 118, 125). In practice, moderate annual application of some mild, highly diluted, and properly managed slurries from farm systems largely feeding livestock with grassland fodder cut later than at conventional silage stage, with no more than moderate use of concentrated feed, and without significant mineral fertilizer additions, applied in a well-considered way, also works (Dietl & Lehmann, 2004, 67, 69, 110; Machatschek, 2015, 37; author's interview with an anonymized farmer from Bohinj, 2022).

[3574] Dietl & Lehmann, 2004, 9–11, 20, 23–25, 37, 39, 42–45, 53–54, 91, 106–108, 118, 121, 123, 125, 128; Machatschek, 2015, 25–41; Magda et al. 2015.

[3575] The management type presented above is, for instance, not a good compromise in moist or wet meadows with ground-nesting birds, which could not tolerate early mowing every second year (compare Kaligarič et al., 2019, 128).

erate more intensive management. It is easier to maintain the productivity of moderately manured meadows in the long run than that of medium-intensive meadows with less stable species compositions, and thus less reliable yields.[3576] Even the impact of droughts on the yields in many moderately fertilized meadows is moderate: false oat-grass (*Arrhenatherum elatius*), a highly relevant fodder grass in moderately manured meadows, is remarkably drought resistant (Section 4.1.1.1), which is relevant in the context of today's climate change.

A historically feared, economically challenging quick spread of hemiparasitic yellow rattle (*Rhinantus* sp.) in moderately fertilized or non-fertilized, mesic sites (Section 4.1.1.1) can be followed by an abrupt crash of its population with adapted management in one spring or summer. Namely, yellow rattles are usually well adapted to the time of regular mowing. They blossom and produce their seeds before mowing time, or they grow and blossom after first mowing. They are annuals, depend on seed dispersal, and lack dormant seeds, and so destruction of their flowers in one spring or summer (e.g., through continuous grazing instead of mowing) suffices to overcome this challenge of the farm economy.[3577]

"[E]cological processes play a much more important role [in moderately fertilized meadows] than in intensive [grasslands]." Moderately manured meadows follow the principles of ecological intensification in terms of "increasing production by making use of ecological processes, making it possible to minimize the use of chemical or energy inputs . . . by replacing . . . some of the nitrogen applied (fertilizer, manure)" through ecological processes. Namely, the impacts of activities by earthworm species (such as *Aporrectodea rosea*, *Aporrectodea caliginosa*, *Lumb ricus terrestris*, and *Eisenia fetida*) highly significantly increase grassland productivity. Earthworm excrement includes, for instance, eleven times more potassium, seven times more phosphorus, five times more nitrogen, three times more magnesium, and twice more calcium than the soil they live in. Earthworms are also actively engaged in manure decomposition. The abundance of earthworms is usually highest in moderately manured grasslands; these animals seem to be attracted by higher biomass production in these meadows than in non-fertilized ones, whereas intensive fertilization (with slurry) would kill earthworms.[3578] Another aspect of eco-

[3576] Dietl & Lehmann, 2004, 10, 24, 54, 66, 69, 83, 91, 106–107, 118, 123, 125; Machatschek, 2015, 32, 35–36, and 40, stating that alternations between early (before grasses blossom), middle (when grasses are blossoming), and late (after grasses have blossomed) first-crop hay mowing dates in the same meadow also work in moderately manured meadows (for how long?), which is contrary to Dietl & Lehmann (2004, 83, 91, 106, 118, 125), who emphasize that seeds need to ripen there every second year, at least.

[3577] Peterken, 2013, 226; and expertise by Leonid Rasran.

[3578] The transition from farmyard manure to slurry that kills earthworms also kills cockchafer larvae, which in some years can devastate the yield in moderately fertilized meadows (author's interview with an anonymized farmers' representative from Bohinj, 2022). It is worth seriously considering whether the mitigation of the cockchafer hazlogical intensification in moderately manured grasslands is soil nitrogen input through a significant abundance of leguminous species, with nitrogen fixation taking place in their root nodules. This process replaces some fertilizers.[3579] In addition, grass species abounding in moderately manured meadows are usually able to make use of many nutrients from the soil despite the moderate soil nutrient availability.[3580] Well-considered and history-inspired management of species-rich moderately manured meadows makes possible long-term preservation of soil qualities,[3581] and, in contrast to highly intensive meadow management, it causes no significant (ground)water pollution.[3582]

At sites without special constraints on grassland farming, moderately manured seminatural meadows can—depending on farming style, livestock type, farm size, non-farm rural economy (see Section 10.2), and further socio-economic parameters—be suitable arrangements "embedding biodiversity conservation within the vision of good farming" to contribute to sustainable development.[3583] In the case of well-considered management of moderately manured meadows, "[b]iodiversity may be seen as beneficial or interesting without the farmer recognizing it as an element driving farm production or wishing to organize the entire forage system as a function of this diversity."[3584]

This is clearly recognized at the French Flowering Meadows competition (involving meadows and pastures as defined in Section 1 of this volume), a championship of farmers' professional excellence "which rewards the best agroecological balance found in species-rich grasslands . . . managed by livestock farmers." Since 2014, Flowering Meadows is part of the prestigious

---

ard in farming through spatial dispersal of meadows used by the same farm, including those avoided by cockchafers, such as meadows at high elevations (many such meadows in mesic to moderately dry sites have several merits from the perspective of livestock health and wellbeing, milk taste, and some produce relatively nutritious hay; see Sections 6.3.1 and 6.3.2), relevantly mitigates the cockchafer hazard in the farm economy, which can be serious every third or fourth year, or significantly less frequently (Section 4.1.3.1). The increasing concentration of all meadows that are managed by a certain farm at low elevations, common in today's context of low and decreasing agricultural density in the Alps, is favorable from the perspective of the economic use of farmers' time and fuel, but this increases the risk of cockchafers coming into serious conflict with grassland farming in species-rich meadows where the microenvironmental conditions are favorable for them. For references cited in the text above, see the following footnote.

3579 Magda et al., 2015, 1063 (quote); Pils, 1994, 170–171, 212–213; Vidrih, 2005, 86–87; Papež Kristanc, 2012, 41; Kremen, 2015, 67–68; author's interview with L. and M. Odar, 2017, in Studor, noting that, if one turns an old cow pie in a grazed grassland upside-down, one will find earthworms, among other animals.

3580 Dietl & Lehmann, 2004, 36.

3581 Dietl & Lehmann, 2004, 45; Machatschek, 2015, 34, 37, 41.

3582 Dietl & Lehmann, 2004, 99–100, 105, 116–117.

3583 Magda et al., 2015 (quote from p. 1054); Dietl & Lehmann, 2004, 10–11, 39, 44–45, 53, 93–95, 99–102, 105, 116–118, 121–129; Machatschek, 2015, 25–41.

3584 Magda et al., 2015 (quote from pp. 1062–1063); see also Machatschek, 2015, 25–41.

French General Agriculture Competition. Work on the Flowering Meadows competition underwent "a collective learning process in which the reconciliation between agricultural production and biodiversity conservation [was] rethought in terms of synergistic relationships between agricultural and ecological functioning. This reconciliation involve[d] the integration of criteria other than productivity in terms of high-nitrogen biomass into evaluations of forage production performance and the consideration of biodiversity conservation through its functional role within the grassland ecosystem, rather than simply by noting the presence or abundance of species." Only preserved species-rich grasslands are eligible. Agronomic value evaluation follows more complex indicators than the conventional one. Apart from productivity and nutritive value, the jury evaluates the "ability to maintain the production of large amounts of high-quality forage over long periods of time" and "ability to stimulate animal ingestion during grazing and benefits to animal health." The French agroecological grassland competition had to face "resistance of many experts [in agronomy and biology] to changes in their conventional approaches." "Conversely, many agronomists and ecologists intensively embraced the method, with an agricultural and ecological vision of the assessment process." Following the agroecological preselection of applications, "[m]ost of the meadows [and pastures] assessed by agricultural and ecological experts obtained high scores for both agronomical and ecological values . . . . The results obtained thus demonstrated that species richness could be combined with viable fodder production, thereby challenging the notion of an inevitable trade-off between the goals of farmers and conservationists."[3585]

Moderately manured meadows provide medium quantities of diverse fodder, and this diversity is appreciated because almost all highly esteemed grassland fodder plants sooner or later cause (serious) negative health impacts in the case of an unbalanced, one-sided diet. About thirty-five to fifty flowering plant species usually occur on 50 m^2 of a history-inspired moderately manured meadow, including some five to fifteen rare species. Aboveground biomass cut in a suitably managed moderately fertilized meadow consists of about 50% to 70% grasses (they impact fodder quantities most significantly, and their sugar content and energy value are generally high for grassland plants) and 30% to 50% herbs (their protein content is often higher than in grasses; clovers in particular contain plenty of proteins), including nitrogen-fixing legumes, such as clovers, and other forbs, with non-legume forbs building 10% to 30% of the total hay meadow yield. The content of minerals in herbs that are relevant for livestock is usually higher than in grasses. Moderately manured meadows are also habitats of diverse animal species. Flowering herbs (usually almost fully absent in intensively managed meadows) still provide crucial feed resources for a plethora of wild grassland organisms (even though not as many as

[3585] Magda et al., 2015 (quotes one and two from p. 1054, quotes three and four from p. 1056, quote five from p. 1054, quote six from p. 1059, quote seven from p. 1055).

in non-fertilized meadows), such as butterflies, honeybees and bumblebees, beetles, spiders, and grasshoppers, with further impacts on ecosystems along the trophic chain. Species-richness in moderately fertilized meadows is significantly higher than in more intensive meadows, and yield quantities are lower, but see the note on yield qualities in moderately manured meadows below. Species diversity in moderately fertilized meadows is significantly lower than in non-fertilized ones, but yields of non-fertilized meadows are only about half the yields of moderately manured ones (Section 6.3.1.1) and, in contrast to moderately fertilized grasslands, non-fertilized and non-irrigated ones do not provide any high quality second-crop hay (Sections 6.1.2.3, 6.3.2.2, and 6.4.1).[3586] Several herbs from non-fertilized and moderately manured meadows are "particularly tasty, rich in nutrients and minerals" if the abundance of such plants in a grassland is moderate.[3587] In contrast to highly intensive grasslands, species compositions in moderately manured meadows still differ significantly depending on natural site conditions because human interventions are moderate enough. Thus, communities at base-rich sites differ significantly from those at acidic sites (provided that no liming is practiced), and landscapes of moderately manured meadows preserve their identity (Section 2).[3588]

The quality (linking nutritive value, tastiness, and health impacts) of first-crop hay from traditional moderately manured meadows, mowed when the seeds ripen, is middling according to twenty-first-century standards, but the quality of second-crop hay is high (Section 6.3.2.2). Namely, provided that the first mowing takes place at the right time (just late enough to allow seed dispersal), herbs' carbohydrate content is relatively high in species-rich grasslands, and crude fiber content is not excessive. Ruminants need considerable quantities of roughage and horses need even more of it. The fodder value of many herbs decreases more slowly than that of grasses in the case of later mowing.[3589] Regular combinations of first-crop hay, second-crop hay from moderately fertilized grasslands, and moderate quantities of industrial feed additions can work well with some cattle breeds kept in the Alps nowadays, but not with those delivering the highest milk quantities (contributing more heavily to the mass extinction in meadows). Even medium-intensive dairy farms, which still keep breeds that can feed mainly on grassland fodder, can highly suitably use largely first-crop hay from moderately manured meadows in the late gestation phase (two months before calving) and to feed close to grown-up calves and heifers; this is favorable for cattle health and wellbeing, and it is ecologically and economically sound, and so even medium-intensive dairy farms can keep

[3586] Pils, 1994, 171–174; Dietl & Lehmann, 2004, 10, 16, 19, 39, 42–45, 107, 118, 121, 123–125; Vidrih, 2005, 139; Machatschek, 2015, 29, 33–34; Verbič & Verbič, 2014–2020, 20; Bevk, 2018.
[3587] Pils, 1994, 173.
[3588] Pils, 1994, 176.
[3589] Pils, 1994, 173; Dietl & Lehmann, 2004, 118, 122, 125; Ehlert et al., 2012, 479; Machatschek, 2015, 30, 38.

a significant percentage of their meadows moderately fertilized if the farm size, market conditions, budgetary payments, and the farmer's attitude to the environment make this possible. The same fodder types are also appropriate for non-milking cows, sheep, goats, and horses.[3590]

Focusing on the links between grassland biodiversity and the livestock product market, the main result of the French Flowering Meadows competition is the finding that grassland plant species compositions significantly impact milk, cheese, and meat quality; nowadays, consumers are becoming aware of the relevance of species-rich grasslands for the quality of livestock products.[3591] Thus it is easy to understand that an analysis of the local winners of agroecological pre-selections for the French Flowering Meadows competition reveals "an overrepresentation of farmers aiming to produce high-quality products (such as *appellations d'origine* in particular) and organic farmers, together with farmers selling their products through means other than the mainstream food system (direct sales, for example)."[3592] Similar are experiences elsewhere in the Alps, such as in Slovenia: successful farm economies largely based on added value generated through milk processing (high-quality cheese production), provided that a sufficient grassland area is available, are more likely to profitably manage a large proportion of moderately manured history-inspired species-rich meadows than farm economies largely dependent on selling raw milk.[3593]

In addition, aromatic and other substances in species-rich meadow fodder have a significant positive impact on livestock health. Livestock in extensive agriculture are "more resistant to various diseases and parasites." Animals feeding in such species-rich grasslands or on fodder from them suffer less frequently from infertility, mastitis, problems with claws, displaced abomasum, rumen acidosis, udder edema, and so on than livestock in intensive animal husbandry.[3594] A moderate proportion of tannin-containing plants in moderately manured meadows limits "the need for veterinary antiparasite agents,"[3595] which is even more relevant because nematodes are becoming resistant to drugs and antiparasitic medicines. These medicines also make it impossible to use the milk for human consumption for a considerable time[3596] (see also the text on sainfoin in Section 10.1.2.2.4). Several herbs are rich in minerals and active substances (e.g., vitamins), and consequently they promote appetite and have further favorable impacts on livestock health and wellbeing. Thus, the

3590 Dietl & Lehmann, 2004, 11, 91–96, 101, 122, 124–128; Pils, 1994, 222.
3591 Magda et al., 2015, 1060; *Europarc federation:* Étude de cas: Concours des Prés Fleuris en France.
3592 Magda et al., 2015, 1063.
3593 For instance, the author's interview with M. Gartner, 2022, in Studor.
3594 Dietl & Lehmann, 2004, 10, 39, 71, 91, 93–94, 96; Machatschek, 2015, 30, 33–34.
3595 Magda et al., 2015, 1063.
3596 Hayot Carbonero et al., 2011, 78–79.

qualities of many plants in species-rich grasslands that used to be considered "weeds"—namely, "worthless space takers" or as "low-yielding indicators of low nutrient availability" in conventional agronomy and intensive grassland management—have been reevaluated from the perspective of livestock farming in recent decades.[3597] The list of "[p]lants reputed to be of high aromatic value or good for the health of livestock" and often underestimated as fodder plants include diverse species abounding in many moderately manured and non-manured meadows. Consequently, fodder from moderately manured and non-fertilized meadows provides a much greater variety of substances relevant for livestock health, which are included in herbs such as hawkbits (*Leontodon* spp.), caraway (*Carum carvi*), sages (*Salvia* spp.), hawkbeards (*Crepis* spp.), bedstraws and cleavers (*Galium* spp.), small docks and sorrels (belonging to *Rumex* sp.), yarrows (*Achillea* spp.), thymes (*Thymus* spp.), savories (*Satureja* spp. and *Acinos* spp.), lavenders (*Lavandula* spp.), harebells and bellflowers (*Campanula* spp.), burnets (*Sanguisorba* spp.), hawkweeds (*Hieracium* spp.), fennels (*Foeniculum* spp.), angelicas (*Angelica* spp.), rampions (*Phyteuma* spp.), and mountain arnica (*Arnica montana*). An integrated quality indicator linking nutritive value, tastiness to livestock, and health impacts is also high for some of these species, such as hawkbits (but in quantitative terms their yields are low), and in many cases for caraway (also providing relatively high yield quantities), yarrows (*Achillea* sp., providing middling yield quantities), small docks and sorrels (belonging to *Rumex* sp. and providing relatively low yield quantities), and hawkbeards (but in quantitative terms their yields are low).[3598] On the other hand, several plant species considered worthless or poisonous do not cause any (serious) harm if consumed in small quantities.[3599] In contrast to species-rich meadows, highly aromatic herbs particularly favorable for livestock health in intensively manured meadows are usually limited to yarrows (yarrow, *Achillea millefolium*; meadow yarrow, *Achillea pratensis*).[3600] Medicinal qualities of hay from extensively managed meadows in the Alps are often so high that they significantly reduce the farmer's costs for veterinary medicine.[3601]

Despite the medicinal qualities, animal wellbeing, possible higher prices for better-tasting products, and lower farm investment costs, an additional

[3597] Pils, 1994, 171–173; Dietl & Lehmann, 2004, 10, 39; Vidrih, 2005, 139; Peterken, 2013, 190, 381; Machatschek, 2015, 25, 28, 30, 33–34, 37–38, 40; Magda et al., 2015; Machatschek, 2017, 20, 25, 71–73, 77; Bavcon, Ravnjak, & Praprotnik, 2019, 83.

[3598] Magda et al., 2015, 1057; Aeschimann et al., 2004, vol. 1, 378–391, 756–759, 1104–1105, 1124–1125, 1128–1129; Aeschimann et al., 2004, vol. 2, 138–141, 148–155, 160–167, 304–343, 350–375, 484–497, 528–529, 628–635, 668–705; Pils, 1994, 173; Dietl & Lehmann, 2004, 39, 43.

[3599] Pils, 1994, 171–172.

[3600] Pils, 1994, 184; Dietl & Lehmann, 2004, 43; Magda et al., 2015, 1057; Aeschimann et al., 2004, vol. 2, 492.

[3601] Machatschek, 2015, 30.

economic motivation through equitable agri-environmental payments (and broader economic conditions, such as through a potentially fair market for high-nature-value farm products and further trans-sectoral cooperation) is needed to preserve moderately manured species-rich meadows because the yield quantities and nutritive value of such hay are limited. In the flatland of lower valley bottoms in the Alps, moderately manured meadows "are often more [prone to] intensification than mountain grassland and therefore tend to have a greater need for action"[3602] (see also Section 1.2). In Slovenia, for instance, their area has recently declined highly significantly in many regions where they were still one of the most usual meadow types two decades ago.[3603] A rapid decrease in moderately manured meadows on an area-wide basis has been the prevalent phenomenon in areas without significant constraints on grassland farming, without well-considered agri-environmental payments, and without a fair market for the products of high-nature value farming (Sections 1.2, 10.1.2.1, and 10.2.1). In many cases, this is a result of real economic challenges for the farm economy, at farms selling raw milk and keeping highly productive cattle breeds in particular. However, in several other cases, the uncertain future for moderately manured meadows is related to a lack of knowledge (e.g., introduction of mowing dates that are too early in moderately fertilized meadows gradually decreases the meadow productivity) and/or with uncritical acceptance of one-sided advice too strongly focused on yield quantities (instead of quality) and the lack of an integrated economic view linking yields and costs of medicines and veterinarians, concentrated feed additions, fertilizers, fuel, mechanization, and so on. This kind of one-sided thought is susceptible to agricultural lobbies securing the market for several unneeded and unsustainable products, contrary to the achievement of sustainable development goals. Such intensive farms sometimes make less profit than a farm with less intensively managed and thus more biodiverse meadows would.[3604]

10.1.2.2.4 Absence of fertilization and proper mowing timing: maintaining biodiversity hotspots at sites with specific natural constraints on farming

During the millennia of evolution of farmed grassland organisms in the Alps up to the nineteenth or early twentieth century, non-fertilized meadows—grazed during part of the year; see Section 6.2—covered a larger area than fertilized meadows (see the introduction to Section 6.3 and Section 6.3.1). As a consequence of prevalent non-fertilized sites over this long evolutionary time, many plant and animal species cannot tolerate fertilization.[3605] Many non-fer-

3602 Magda et al., 2015; *Results-Based Nature Conservation Plan*, [2019], 8 (quote).
3603 Jaksetič (interview with Kaligarič), 2019, 22.
3604 Vidrih, 2005, especially pages 10, 150, 153, 155–156; Jones & Poux, 2012, 451, 454–455, 457; Paracchini & Oppermann, 2012, 447, 449–450; Machatschek, 2015, 32; author's interview with an anonymized knowledgeable farmer from Bohinj, 2022.
3605 Jaksetič (interview with Kaligarič), 2019, 21.

tilized meadows have been biodiversity hotspots, the most species-rich habitat types in Central Europe (see the introduction to Section 1 and Section 6.3.1.1). Thus, from the ecological perspective, non-fertilized meadows are extremely relevant local specifics, and many plant and animal species are specialists only occurring in non-fertilized grasslands.[3606] Many endemic species occur in and depend on non-fertilized meadows. Species-rich grasslands have the highest rate "of vascular plant endemism" of all farmland habitat types in Europe and second highest among all habitat types (including non-farmed habitats). Species-rich grasslands' rate of vascular plant endemism is higher than in the forests as potential natural vegetation.[3607]

Despite the long history of natural and human-induced environmental changes during the last millennium, the extinction of numerous species is a recent phenomenon that mainly started in the twentieth century (Section 10.1.2.1).[3608] Currently, nutrient-poor meadows belong to the most threatened vegetation types: among others, alpine grasslands on limestone (Flora-Fauna-Habitat Directive type 6170), boreo-alpine grasslands on silicates (Flora-Fauna-Habitat Directive type 6150), and mountain matgrass meadows (Flora-Fauna-Habitat Directive type 6230). They are threatened by two opposite trends in agriculture—abandonment and intensification (see Sections 1.2, 2, 9, 10.1.2.1, the introduction to 10.1.2.2, and Section 10.2.1)—and by enormous nitrogen oxide emissions from farming, traffic, and industry: a significant amount of pollutants have been blown into the atmosphere and have come down with rain (in Europe on average up to 20 to 30 kg per hectare of nitrogen compounds yearly).[3609] Since the second half of the twentieth century, eutrophication from the atmosphere has been one of the most important threats to biocenoses at nutrient-poor sites.[3610] Orographic lifting and related cooling of air masses make precipitation in the Alps more frequent, and they render landscapes at high elevations in those sections of the Alps that are usually reached by airmasses from densely populated lowlands particularly prone to this type of pollution.[3611] As long as the nutrient level remains limiting, a broad spectrum of life strategies and species with different adaptations are characteristic for the site. In grassland communities of the alpine vegetation belt (Section 2), cessation of nitrogen limitation due to the airborne pollution causes the same effects on plants that are known from various other grasslands: tall-growing nitrophilous forbs and grasses become dominant, whereas subordinate species disappear due to increased light competition (Section 2).[3612] Nonetheless, the

[3606] Pils, 1994.
[3607] Luick, Jones, & Oppermann, 2012, 32; Cremene et al., 2005; see also Section 2.
[3608] Poschlod, 2015, 248.
[3609] *Stickstoff – zuviel des Guten*, no date; Emanuelsson, 2009, 340; Peterken, 2013, 244; Poschlod, 2015, 182–183; Broggi, 2016, 8–9; Willett et al., 2019, 465–467.
[3610] Pils, 1994, 60, 70–71; Kohli, 2011; Peterken, 2013, 244–245; Poschlod, 2015, 182–183.
[3611] Bätzing, 2015, 270.
[3612] Bakker, 1989; Poptcheva et al., 2009.

percentage of rare species in the Alps is still high compared to several surrounding regions,[3613] but declining (see Sections 1.2 and 2).

Due to the limited mobility of many organisms, it is highly relevant for a healthy cultural landscape of the Alps that non-fertilized grasslands be located in a considerable area of grassland sites with natural constraints on farming in all elevation belts where farmed grasslands locally occur, not only in summer mountain pastures. Adequate connectivity of such habitats in the landscape (including hay meadows where biodiversity is higher than in pastures) that makes it possible for species or populations to "move among landscape elements in a mosaic of habitat types," is of utmost importance (see Sections 1.2 and 6.3.1.1).[3614] Many endangered species of butterflies, grasshoppers, orchids, pasque flowers (*Pulsatilla* spp.), and carnations (*Dianthus* spp.) went extinct or are becoming extinct with the disappearance of non-fertilized meadows from European cultural landscapes in the last half century.[3615] Data from Upper Austria in the second half of the twentieth century demonstrate that caterpillars of more than a quarter of all butterfly species and a third of all grasshopper species occurring in that province depend on dry non-fertilized meadows at low elevations. Thus, maintaining managed non-fertilized meadows and pastures at high elevations alone does not suffice. Many insect species (grasshoppers, for instance) need high temperatures and sunlight available in the sparse vegetation of non-fertilized meadows, and some of them are specialists that find food only in non-fertilized meadows.[3616]

Keeping non-fertilized meadows productive in the long run allows mowing times when fodder is still usable for livestock according to present standards if used in a well-considered manner; focusing on plants, such a meadow can be mowed as soon as haymaking allows sufficient dispersal of ripe seeds. This nonetheless means that non-fertilized meadows need to be mowed later than moderately manured ones, usually about two weeks later, to maintain their agricultural productivity and species-richness in the long run—a history-inspired practice (Sections 6.1.2.1 and 6.1.2.2)—due to the slower vegetation development at non-fertilized sites.[3617] Several farms in the Alps mowing significantly diverse meadow sites with respect to fertilization intensity have maintained the diversity of mowing dates,[3618] whereas others have switched to uniform mowing dates, which are easier from the viewpoint of the farmer's schedule but shortsighted if the manuring intensity, microclimate and/or

[3613] Jogan, Bačič, & Strgulc Krajšek, 2012, 172.
[3614] For instance, Fischer et al., 2008, 154 ; Scheurer, 2016, 86 (quote).
[3615] Grass, Batáry, & Tscharntke, 2021, 262; Jaksetič (interview with Kaligarič), 2019, 22 (and Aeschimann et al., 2004, vol. 1, 144–151, 350–365).
[3616] Pils, 1994, 61–64.
[3617] Pils, 1994, 273–274; Dietl & Lehmann, 2004, 54, 124–125.
[3618] For instance, von Glasenapp & Thornton, 2011, 773–774, 778; *Results-Based Nature Conservation Plan*, [2019], 13.

(in some cases) species composition differ significantly. For many non-fertilized meadows, mowing as early as in moderately fertilized sites is too early to keep them productive in the long run. Productivity maintenance in non-manured meadows depends on partial seed dispersal during haymaking (Sections 6.1.2.1, 6.1.2.2, and 6.5). Mowing all the parcels in a day or two also decreases the resilience of the farm economy in the case of untimely rain because precipitation negatively affects the hay cut in all parcels. These arguments speak in favor of the history-inspired good practice of a certain level of diversity of mowing timing in meadows managed by the same farm if the intensity of meadow management and/or microclimate or elevations differ significantly. "One farmer had completely changed her approach to mowing since participating in [the Austrian result-based agri-environmental scheme]. She used to mow everything at once. Now she starts earlier, but mows in small sections because she can take her time. She begins with the more nutrient-rich meadows and cuts the more calcareous ones [at nutrient-poor sites] towards the end."[3619] This is in line with the historical fundamental principle of mowing timing in preindustrial agriculture (Sections 6.1.2.1 and 6.1.2.2).

Among all meadow types classified according to fertilization, the fodder quality in non-fertilized meadows depends most on natural site conditions, and thus the fodder quality from such meadows is most diverse (Section 6.3.1.2). Historical livestock breeds and a diversity of livestock types, including a highly significant percentage of animals other than dairy cows (such as sheep, goats, non-dairy cattle, horses, and in some regions mules, hinnies, and donkeys), a high percentage of agricultural population, thus incomparably more workforce available in farming, and different market conditions made use of hay from highly diverse sites of non-fertilized meadows more easily economically viable than today (Sections 5 and 6.3.1). Whereas cattle need a certain percentage of raw fiber, horses' digestion requires even more raw fiber for the animals to stay healthy, and so even the hay from non-fertilized meadows cut too late for cattle can be suitable for feeding horses when they do not perform heavy physical work. A strong reduction in extensive farming involving livestock other than dairy cattle, the development of intensive cattle breeds, and heavily increased trade flows made matgrass swards (Sections 2, 6.3.1.1, and 6.3.1.2), which provided very low-quality fodder but nonetheless strengthened the resilience of the farm economy in preindustrial farming, almost useless in modern farming. Due to the same processes, non-fertilized calcareous meadows (Sections 2 and 6.3.1) became unattractive in many farming systems, in particular in mass-production-oriented intensive farming.[3620]

However, there are (partly history-inspired) farming systems that make the economic use of non-fertilized calcareous grasslands viable, but they "are rela-

[3619] *Results-Based Nature Conservation Plan*, [2019], 16.

[3620] Pils, 1994; Karlík & Poschlod, 2009; Ehlert et al., 2012, 479; Oppermann & Paracchini, 2012, 22.

tively poorly researched" because they differ from the main trends in farming (see also an innovative grazing circuit in non-fertilized grasslands inspired by traditional ecological knowledge and improving agronomic knowledge in Section 10.1.2.2.6). Their productivity per farm area is low, but they largely represent land sharing in the context of biodiversity preservation in areas with natural constraints on farming.[3621] A large farm size, location in a karst area with natural constraints on farming, and appropriate livestock types can make the use of hay from non-fertilized, dry calcareous grasslands, in combination with budgetary payments, highly profitable. The example of the Karst Plateau in the northwestern Dinaric Karst, south of the southeastern Alps, shows that many large cattle farms (managing, for instance, some 45 to 50 ha of hay meadows and 70 ha of pastures per farm, all in karst areas facing natural constraints on farming, at low elevations) that rear around forty suckler cows and eight heifers of appropriate breeds can economically perform strongly if their livestock feed only on hay from extensively managed meadows—most of them non-fertilized dry calcareous grasslands—and grazing resources from such pastures, without any purchased concentrated feed. Similarly, a sheep farm of comparable size but adding some concentrated feed to the prevalent fodder from dry calcareous grasslands and fertilizing a small percentage of the meadow area is highly profitable there. A smaller cattle farm (still large for Slovenia but highly extensively managed) using 32 ha of non-fertilized dry calcareous grasslands and rearing ten suckler cows and two heifers is, however, often economically unattractive on the Karst Plateau, unless an innovative farm income integration (combining farming and tourism, for instance) improves the economic situation. Horse rearing in particular also generates market demand for the hay from non-fertilized dry calcareous grasslands.[3622]

Where socio-economic conditions differ significantly from those presented in the previous paragraph and natural constraints on farming are not so general, significant areas of the most biodiverse meadows at non-fertilized sites—often habitats of fifty to sixty-five flowering plant species in 50 m^2, including fifteen to twenty rare plant species, and many animal species but the lowest yield quantities (Section 6.3.1.1)[3623]—can be sustainably maintained at selected sites where natural constraints "significantly limit the potential for agricultural intensification." The very low yields must be balanced by well-considered, properly high budgetary payments, and fair prices of high-nature-value farm products (Section 10.2.1), equitable to farmers, thus significantly higher than for moderately manured grasslands. There are nonetheless some economic motives for knowledgeable farmers to maintain a limited percentage of history-inspired non-fertilized meadows at the farm level, in terms of expected favorable impacts on livestock health and wellbeing, and product quality and

[3621] Šumrada et al., 2024 (quote from p. 21).
[3622] Šumrada et al., 2024, 5–6, 8–17, 19–22.
[3623] Dietl & Lehmann, 2004, 121, 124.

other ecosystem services relevant in farming (Section 10.1.2.1).[3624] Based on appropriate payments and prices, open-minded farmers could nowadays appreciate the non-fertilized, high-nature-value seminatural meadows on steep slopes, for instance, where mowing tall grasses would be difficult.[3625]

Closely related to such natural constraints on farming intensification, steep meadows are often extraordinarily species rich. Preventing their abandonment receives much agroecological attention and related ecological and botanical education for farmers in some Alpine countries such as Switzerland,[3626] whereas they are still being abandoned on a large scale in the Alps of Slovenia.[3627] In this case, agri-environmental measures should take seriously traditional farmers' message (recorded around 1980, thus it was certainly not influenced by the desire for agri-environmental payments) that one cannot work safely on steep slopes hastily and frenetically, as grassland management is frequently performed on industrially modernized farms,[3628] and so an equitable income from agri-environmental payments should allow farmers to work deliberately on selected steep slopes where economic considerations can more easily allow the existence of non-fertilized meadows. Appropriate payments could also allow for manually turning hay and raking it on selected steep slopes; turning and raking hay manually causes significantly smaller losses of fragile but valuable herbs (high protein content; Section 10.1.2.2.3; see also Section 6.3.1.2) than carrying out these tasks with tractors.[3629] It has already been explained that, depending on an individual's personality, haymaking outside of a tractor cabin in a steep non-fertilized meadow that is filled with pleasant smells and full of animals, and where there are still some blossoming herbs, can, despite the hard work, be a source of happiness for farmers and their families, even though the most colorful stage of the meadow is over by a sustainable mowing time (Sections 10.1.2.1 and 10.1.2.2.1). This is followed by satisfaction in the livestock barn, where the farmer sees how gratefully the livestock consume the high-quality, aromatic hay from the carefully selected non-fertilized meadows (see Section 6.3.1.2: cows even lick the hay dust from the manger) and thinks about the beneficial impacts of that hay on animal health and wellbeing, and on the quality of the farm's products.

In terms of product quality, the impacts of certain species compositions in non-fertilized meadows are significantly more favorable than in moderately fertilized ones. The excellent aroma of products given to dairy products by thymes (*Thymus* spp.) and their impact on promoting livestock appetite

---

3624 Pils, 1994, 68; Oppermann & Paracchini, 2012, 22; Šumrada et al., 2022, 8–9 (quote).
3625 *The Challenge of Feeding the World Sustainably*, 2021, 11; author's interview with M. Gartner, 2022, in Stara Fužina.
3626 Oppermann et al., 2012a, 470.
3627 Bertoncelj & Travnikar, 2024, 78–80, 82.
3628 Makarovič, 1982, 33–34, 457, and authors' observations in Bohinj.
3629 Machatschek, 2015, 29.

have already been mentioned (Section 10.1.2.2.2). Thymes can only abound in non-fertilized or very moderately fertilized meadows (less than usual moderate fertilizing).[3630] Focusing on the relevance of non-fertilized meadows for livestock wellbeing, their eating habits are instructive. On medium-intensive dairy farms, hay from productive non-fertilized meadows is suitable for feeding dairy cows during late gestation, and for feeding grown-up heifers; it is also suitable for non-dairy cows, sheep, and horses.[3631] Moreover, sheep and goats are eager to feed on fodder rich in crude fiber after they have consumed plants in an early development stage, such as early cut fodder from intensive meadows. Fodder rich in crude fiber from non-fertilized meadows in this case has a positive impact on ruminant digestion and wellbeing. Even matgrass sward (see Section 6.3.1) is an optional combination with an intensively managed pasture in sheep rearing. Cattle also do their best to combine grazing in an intensively managed pasture with grazing in a neighboring species-rich grassland (they show a stronger preference for the neighboring pasture than usual) if it is reachable below the fence, despite the high availability of nutritive fodder in the intensive pasture where the cattle are kept. Thus, at several farms that manage part of their grasslands intensively, hay from non-fertilized meadows is a welcome addition to the monotonous diet from intensively managed meadows to limit the financial and environmental costs of using industrial feed additions, to reduce the farm expenses for veterinary medicine, and this is an expression of responsible care for grassland biodiversity.[3632]

All in all, well-thought-out farming taking into account local and regional specifics in non-fertilized meadows and the farm economy usually requires tradeoffs[3633] unless the natural constraints on farming are very general, as in the case from the Karst Plateau presented above. An example of such tradeoffs is patchworks of grasslands in a farm system managed highly diversely and designed in a well-considered way, ranging from non-fertilized through moderately manured to intensive at the farm level. In such a system, proper proportions between areas of these grassland types matter from an integrated agroecological perspective; thus a considerable area of non-fertilized meadows is also ethically highly relevant.[3634]

The data presented so far could encourage the maintenance or restoration of some non-fertilized meadows. Because of the diversity of fodder quality in non-fertilized meadows (Section 6.3.1.2), historical traditional ecological knowledge about the exact meadows where non-fertilized sites provide(d) good fodder (fragrant fodder and tasty milk; livestock liked the fodder very much; relatively nutritious fodder, livestock needed less such fodder to satisfy their

[3630] Aeschimann et al., 2004, vol. 2, 148–155.
[3631] Dietl & Lehmann, 2004, 11, 91–96, 101, 122, 124–128.
[3632] Pils, 1994, 68, 71, 174.
[3633] Willett et al., 2019, especially pp. 482–483; Debuschewitz & Sanders, 2022, 5–9.
[3634] Dietl & Lehmann, 2004, 9–11, 111, 116–129.

hunger; positive impacts on livestock's immunity) can be highly relevant for farmers that raise appropriate cattle breeds and are willing to manage some non-fertilized meadows in a history-inspired way to contribute to the preservation of biodiversity. Such traditional ecological knowledge matters most in the context of the potential reintroduction of agricultural use of the grassland or when a farmer is looking for a meadow used so far by somebody else. On Vogar Hill above Stara Fužina in Bohinj, for instance, where high hay quality in sunny, mesic meadows was mentioned in Section 6.3.1.2, only some hay currently mowed is transported to the valley, and the rest is left in piles in meadows to decompose there[3635] or used for nature conservation purposes, as a source of seeds sown in non-fertilized meadows elsewhere in Triglav National Park.[3636] This would probably not be the case if traditional ecological knowledge about hay quality were more widely applied. In contrast to Vogar, it is not recommendable from a combined agro-ecological perspective to restore several historical sites of non-fertilized meadows in most of Zavaze near Stara Fužina, even if this largely abandoned meadow complex is closer to the village— unless the hay is intended for non-demanding livestock, such as horses—because historical experience proved that the hay quality will be low there if meadows are non-fertilized (Section 9). Elsewhere in the Alps, where a significant percentage of meadows used to be located at higher elevations, in the subalpine and alpine vegetation belts, the hay quality was often even significantly higher, as explained in Section 6.3.1.2. This is one of the main motivations for the continued use of the hay from such distant meadows at rare farms in the Alps until today.[3637] Exceptionally, the high hay quality stimulated a reintroduction of mowing at such distant sites.[3638] The historical method of improving the digestibility of hay from non-fertilized meadows by scalding part of it (not all of it, in order to retain enough biologically active substances) is worth testing.

The relatively low digestibility of winter fodder from several species-rich (especially non-fertilized) grasslands increases the amount of greenhouse gas emissions. Nonetheless, factors mitigating total greenhouse gas emissions per liter of milk include the following. (1) There is a prominent role of grazing in the farm economy. (2) The absence or low application of nitrogen mineral fertilizers in species-rich meadows matters because industrial conversion of ammonia from nitrogen gas requires a lot of energy, and so it is a relevant source of greenhouse gas emissions. (3) There is an absence of tillage. (4) A higher number of lactations per cow due to a healthier diet reduces the relative duration of heifer rearing within the cow's lifetime when the animal produces greenhouse gasses without providing milk. (5) Such fodder is suitable for

[3635] Author's observations in recent years.
[3636] *Seed Gathering and Other Project Activities in Triglav National Park*, 2024.
[3637] Keller, 2007, 205–209.
[3638] E.g., in a large subalpine meadow in the east of the Golica Ridge (Klinar, 2012, 98, 100; Torkar Tahir, 2012, 133).

breeds requiring low to moderate quantities of feed concentrates. In general, concentrated feed additions to fodder from seminatural grasslands can reduce greenhouse-gas emissions because the digestibility of fodder increases, provided that such feed is produced sustainably and also not transported over distances that are too long. However, concentrated feeds often include soybean meal; soybeans for livestock reared in Europe are often imported from South America (unless prohibited, such as at hay milk farms in Austria; see Section 10.2.1), where European livestock rearing contributes significantly to destruction of savannahs or rainforest clearing for soybean plantations, at the same time causing the destruction of species-rich habitats and greenhouse gas emissions. In addition, such fodder produces greenhouse gas emissions related to long-distance traffic. Feed concentrates cause environmental loading in grasslands in the Alps through local application of animal excrement, including nutrients from extra-Alpine grains and beans.[3639] (6) Animal housing at farms managing species-rich meadows is often litter-based; straw-based (and possibly also sawdust-based) systems usually cause significantly lower greenhouse gas emissions than slatted barn floors and slurry.[3640]

There is some space for ecological intensification in some non-fertilized meadows, and it can be history-inspired, as demonstrated by the example of the legume sainfoin (*Onobrychis viciifolia*; see Figure 164 and Section 6.3.1.1.1). The total area sown with fodder legumes in Central Europe reached its peak in the second half of the nineteenth century and in the early twentieth century.[3641] Agronomic appreciation of legumes decreased a lot with the industrial intensification of farming in the sense of enormous quantities of mineral fertilizers in most of Europe since the mid-twentieth century. "This trend is now changing, and pressure to reduce energy consumption and environmental pollution and to improve agricultural sustainability is driving lower input agronomy . . . , and farmers are reconsidering the use of forage legumes."[3642] Ecological intensification in some non-fertilized meadows at relatively dry, sunny, base-rich or neutral sites on shallow soils could follow the historical good practice of sowing sainfoin (*Onobrychis viciifolia*) seeds without ploughing or fertilizing the grassland—a practice that was carried out at some places in Central Europe in the last centuries due to the far increased fodder quantity and nutritive value, and fairly reliable yields. Sainfoin colonizes the gaps frequent in such dry sites (Section 6.3.1.1.1). Yields of species-rich non-fertilized meadows where sainfoin abounds are extremely high for non-fertilized, rather dry meadows on shallow soils. They can, for instance, triple in comparison

[3639] Dietl & Lehmann, 2004, 93; Hörtenhuber et al., 2010; Steffen et al., 2015; Willett et al., 2019, 462–463, 465–466, 469, 472; *The Future of Nature and Business*, 2020, 41.

[3640] Hörtenhuber et al., 2010, 317, 325 and personal communication with Stefan Hörtenhuber, a specialist in environmental science related to livestock production systems.

[3641] Britovšek, 1958, 149; Hayot Carbonero et al., 2011, 72–73; Poschlod, 2015, 119–120.

[3642] Hayot Carbonero et al., 2011, 70 (quote), 76.

Figure 164: Sainfoin (*Onobrychis viciifolia*). Thomé, *Flora von Deutschland, Österreich und der Schweiz*, cv. Hayot Carbonero et al., 2011, 72.

with the yields before sainfoin.[3643] Despite the significant manuring impact of airborne nitrogen fixation through symbiotic bacteria in sainfoin root nodules (Section 6.3.1.1.1),[3644] such non-fertilized meadows are still highly species-rich, as demonstrated in Ghimeş in the Carpathians.[3645] This corresponds with the observation from arable land that sainfoin "is usually considered to be a non-aggressive crop."[3646] This special feature is partly due to the stem architecture of sainfoin, dependent on the primary taproot throughout the entire lifespan and building a few upright stems but no vegetative propagation or lateral movements. Consequently, sainfoin does not produce a closed canopy outcompeting other herbaceous species. Sainfoin's populations in species-rich

[3643] Compare Babai, Molnár, & Molnár, 2014, 95–96, 139; and Hayot Carbonero et al., 2011.
[3644] Hayot Carbonero et al., 2011, 73.
[3645] Babai, Molnár, & Molnár, 2014, 43, 139.
[3646] Hayot Carbonero et al., 2011, 74 (quote).

meadows are maintained by keeping some specimens standing during mowing and, if needed, by sowing sainfoin-rich barn sweepings (Sections 6.3.1.1.1 and 6.5). Sowing sainfoin in non-fertilized species-rich meadows at appropriate sites increases the resilience of the farm economy in comparison with a sainfoin monoculture in an arable field;[3647] namely, even though sainfoin is less sensitive to pests and diseases than other intentionally sown "clovers," pests and diseases can nonetheless cause a sainfoin harvest failure.[3648] Several other species also occur and abound in non-fertilized meadows, and so, even if there is a sainfoin harvest failure, its economic consequences will be mitigated. A species-rich meadow rich in sainfoin provides nutritive, tasty, and healthy fodder. It should be cut late enough to allow seed dispersal of many species but early enough that the sainfoin stalks have not dried before mowing. Contrary to other sown "clovers," sainfoin does not cause bloat, and its tannins can potentially reduce gastrointestinal "worm burdens in ruminants." This is relevant: when antiparasitic medicines are used, milk and meat are not suitable for human consumption for a long time, and so the use of such industrially produced medicines can be a significant burden for the farm economy. In addition, intestinal parasites are becoming increasingly resistant to medicines. Such issues can be largely avoided with fodder from several species-rich non-fertilized and moderately manured meadows, containing a sufficient amount of tannins. Sainfoin is only one example of such an herb, and there are many others frequently occurring in extensively managed grasslands (see Section 10.1.2.2.3). Sainfoin is also a rich pollen and nectar source.[3649] Nonetheless, the specifics of the local biocenosis matter, and so the decision to sow sainfoin in non-fertilized species-rich meadows should be preceded by careful agroecological consideration.[3650]

10.1.2.2.5 Farmers avoid limited meadow sections with nests of ground-breeding birds at mowing

Cruelty toward grassland animals from carelessness or indifference, which decreased so much in the nineteenth century (Section 10.1.1), has risen again, mainly since the mid-twentieth century. Focusing on birds, widespread motorization of haymaking, the ever-increasing speed of meadow management, and the recent prolongation of work in meadows after nightfall on some farms when animals are asleep, thus fewer of them retreat, belong to the main processes that mattered in this context. A farmer driving a mower in a moist meadow often does not even notice that this activity is killing birds in their nests. In the absence of appropriate agri-environmental payments and appropriate prices, such

[3647] Compare Pfister, 1985a, 109–110; Babai, Molnár, & Molnár, 2014, 95–96.
[3648] Hayot Carbonero et al., 2011, 75–76.
[3649] Hayot Carbonero et al., 2011, 71–72, 76–80 (quote from p. 78); Babai, Molnár, & Molnár, 2014, 43, 95–96, 139; Slatnar, 2016, 10.
[3650] For instance, Kus Veenvliet, 2012, 143, 148–150, and Section 6.3.1 in this volume.

a pattern is largely promoted through the limited workforce and globalized economy. Historical examples of avoidance of corncrake (*Crex crex*) and whinchat (*Saxicola rubetra*) nests, where mowing took place before the chicks were ready to escape (Section 10.1.1), call for history-inspired innovation. The whinchat, which used to be a character species of extensively managed moist grasslands in most of Europe, and the corncrake, which also used to be widespread, are nowadays threatened ground-breeding species in Europe. Their numbers have been quickly declining, mainly due to early mowing, which destroys nests and chicks, and other habitat changes related to grassland management intensification, such as the decline in availability of arthropods that whinchats feed on (Section 1.2). The period when chicks are sensitive to mowing depends on the bird species and elevation; whinchats are sensitive until mid-July in the montane zone of the Upper Inn Valley bottom; in the Julian Alps, the nesting period at 1,800 m starts at least a month later than in the lowlands, where mowing earlier than mid-June destroys 90% of the nests; whinchat chicks still do not escape from a mower for at least 10 days after they leave the nest. The whinchat is one of the most endangered ground-nesting meadow birds, with quickly declining populations (its decline was 50% to 90% in several European countries in 1995–2015).[3651] Thus, many local and regional whinchat populations in countries such as Austria (where high-nature-value farmland covers almost a fifth of the entire country but there is profound regional diversity in the percentage of high-nature-value farmland) and Germany became extinct in recent decades, and many existing whinchat populations are isolated,[3652] which is a challenge for their long-term survival (Section 1.2). Whinchat have also disappeared from several valleys in the Austrian Alps.[3653] Similarly, whinchat populations in the Slovenian Alps are small.[3654] This reflects the impacts of changes in farming on an animal species that depends very heavily on farming "because its natural primary habitat is now almost non-existent." All this is even more alarming because the whinchat population breeding in Europe "accounts for 75–94% of the world population." In the Alps, sustainable whinchat populations require significant areas of species-rich grasslands (including non-fertilized or moderately manured meadows, in part late mowed) in the landscape.[3655]

The corncrake is so severely jeopardized that it will probably disappear from Slovenia unless the threatening factors decrease.[3656] Namely, the number

---

3651 Peterken, 2013, 308, 311; *Repaljščica – življenje na robu izumrtja*, 2015; Božič, 2019, 138–139; *Results-Based Nature Conservation Plan*, [2019], 9; Tome, 2019, 426–427; *European Commission, Environment: Wild Birds: Threatened Bird Species in Annex I*; *European Commission, Environment: Wild Birds: Corncrake*.

3652 Oppermann et al., 2012b, 106; Schmitzberger et al., 2005, 275.

3653 *Results-Based Nature Conservation Plan*, [2019], 9.

3654 Tome, 2019, 426.

3655 Oppermann et al., 2012b, 106–107 (quotes from p. 106).

3656 "Pravilnik o uvrstitvi ogroženih rastlinskih in živalskih vrst v rdeči seznam," 2002, 8893, 8920.

of singing male corncrakes in Slovenia decreased by 30% to 50% during the early twenty-first century alone; this is in contrast to the estimated general trend of stable populations in Europe, and it is related to the strong grassland history transition in Slovenia at present (Section 1.2).[3657]

The innovative practice in the recently developed result-based agri-environmental payment scheme in Austria resembles a history-inspired innovation. It aims "to reestablish the necessary conditions for the whinchat to return" to a meadow area in the Upper Inn valley that was historically inhabited by this species. Mowing performed in times that are usual nowadays kills the birds. However, based on site-specific agroecological (ornithological) training dealing with the whinchat and, optionally, an inspection of the meadow on foot just before mowing using an infrared sensor that detects the nests, the farmer abstains from mowing (until mid-July) only in those limited meadow sections where breeding (potentially) takes place. In this way, suitable nesting sites are saved and protected by standing grassland plants, and only the quality of hay from limited meadow sections declines. This practice has a parallel in the action usual for many farmers mowing with a scythe earlier in history (Section 10.1.1). It resembles the approach of some hunters after the introduction of motorized mowing, such as some Slovenian hunters in the 1980s, who were encouraged by the magazine *Lovec* (The Hunter) to use a trained bird dog to identify bird nests before motorized mowing to save at least part of the severely declining populations of gray partridge (*Perdix perdix*). This is another ground-breeding bird species that had been relatively common in Europe until the nineteenth or mid-twentieth century and nested in low basin and valley bottoms in the Alps, but its populations in Europe declined by 94% from 1980 to 2017. Focusing on meadows, gray partridge nests suffered large-scale destruction by motorized mowers, but the species also requires relatively extensively managed arable land.[3658]

The recent result-based practice to create whinchat-friendly habitats in Austria presented above raises the economic relevance of such meadows in comparison to a conventional nature conservation approach requiring late mowing in entire meadows. In addition, an educated farmer concerned about whinchats does not level irregularities in grassland terrain used for nesting. The practice of avoiding nests during haymaking has become more widespread: many farmers involved in the Austrian result-based agri-environmental scheme "look out for . . . nests of meadow breeders and mow around them."[3659]

[3657] Božič, 2019, 138–139.

[3658] The information on hunters' approach to save gray partridge is from Koren, 1982, 206, and Erhatič Širnik, 2011, 255–256, 304, 307; see also Section 10.1.1 in this volume. On the swiftly progressing extinction of the partridge in Europe since 1980 and its habitat requirements, see Wöbse, 2022, 40, and Šumrada & Trilar, 2019, 66–67.

[3659] *Results-Based Nature Conservation Plan*, [2019], 9 (first quote), 16 (second quote); Ehlert et al., 2012, 478.

Developments in precision farming may soon make possible the use of infrared detection systems on mowing machines to identify bird nests and other vertebrates hidden in the meadow, such as roe deer fawns, in time to allow farmers to avoid them. This will make redundant the aforementioned grassland check on foot by a nature-friendly farmer before mowing. Together with the optional actual spread of efficient modern cutter-bar mowers that are already available on the market, replacing the widespread use of rotary mowers, this has the potential to immensely decrease cruelty toward wild fauna from carelessness or indifference during motorized mowing. Namely, rotary mowers kill and injure about twice the number of amphibians and grasshoppers in comparison to cutter-bar mowers.[3660]

10.1.2.2.6 Mechanical, other grazing-related, and thermal suppression of plants unwanted in animal husbandry

Historical suppression of unwanted plant species through uprooting or maintenance mowing in spring has received agroecological attention to avoid the use of herbicides and keep the grassland productive from the viewpoint of animal husbandry. "[M]anual, mechanical and heat measures" to control the plant species perceived as weeds are highly relevant in organic grassland farming and in an increasingly smaller area of conventional farming that usually gives priority to chemical suppression of the plants perceived as weeds.[3661] The plants suppressed are poisonous to livestock, or they are large and abundant but of hardly any use for domestic animals; the fleshy leaves of some of them would increase the risk of mold in hay (Section 6.6.1.1). Currently, suppression of particular plant species in grasslands is only carried out on some farms and usually less thoroughly than in historical times of an abundant work force. Abandonment of many grasslands at higher elevations led to the end of the removal of false hellebore species (*Veratrum* spp.) there, whereas intensification in the management of existing meadows reduced the abundance and occurrence of some species formerly (partially) removed, such as autumn crocus in moist meadows.[3662] However, intensification also brought new challenges, such as an increased abundance of parsnip (*Pastinaca sativa*) and large-leaved sorrels in meadows heavily fertilized with liquid manure.[3663] Thus, on some farms, removal of mowed plants when spreading freshly mowed hay and during hay turning (Section 6.6.1.1) was partly replaced by a history-inspired practice: a careful walk through a meadow before mowing, taking long strides to trample as little grass as possible to remove much of the parsnip (*Pastinaca sativa*) be-

[3660] Ehlert et al., 2012, 475–478.
[3661] Rahmann, 2011, 191 (quote); see also Tscharntke et al., 2021, 920.
[3662] Author's interview with Rozman, 2020, in Srednja Vas in Bohinj.
[3663] Author's interviews with Mencinger, 2017, in Stara Fužina; and with F. Mlakar, 2017, in Bohinjska Češnjica.

fore motorized mowing. In the valleys of the Julian Alps, this species is partly removed by some farmers before second-crop hay is mowed.[3664]

Where removal of unwanted plants follows historical or history-inspired labor-intensive methods, it is carried out with traditional or modern implements, such as a scythe or a filament mower for maintenance mowing. Traditional ecological knowledge is relevant, for instance, in explaining to young farmers and other new entrants to farming the differences in labor intensity to cope with different plant species. Coping with false-hellebore (*Veratrum* sp.; see Sections 3.2.2, 6.6.1, and 6.6.2) by scythe in meadows or pastures, for instance, is less labor intensive than removing Alpine dock (*Rumex alpinus*) with a scythe. Namely, after the mowing in spring, false-hellebore does not resprout in the same year (or the regrown herb is very weak, it does not blossom). In contrast, Alpine dock regularly resprouts and not only its first shoot blossoms and sets seeds, and so mowing Alpine dock, a species unattractive to cattle and abounding in nutrient-rich sites around mountain barns and water troughs, needs to be repeated the same year or replaced by even more labor-intensive uprooting with a dock fork or similar implement, whereas one mowing suffices to prevent the seed shed of poisonous false-hellebore, whose abundance would increase heavily in the case of seed dispersal (see Sections 3.2.2, 6.6.1, and 6.6.2).[3665] Another piece of information from traditional ecological knowledge is how deeply some plant species need to be uprooted to hopefully prevent them from sprouting again (see the example of broad-leaved dock, *Rumex obtusifolius*, in Section 6.6.1.1). This is fully relevant in agroecological grassland management.[3666]

On flat terrain and at accessible sites, however, maintenance mowing is motorized, using tractor mowers. In this case, the historical practice of maintenance mowing affecting a particular plant species can be an inspiration for history-inspired innovation. As part of a result-based environmental conservation scheme in Austria, a farmer in Fieberbrunn, Tyrol, performs maintenance mowing in May such that he "adjusts the height of the motor mower so that it does not damage the orchids, but it does cut the veratrum." The farm introduced this practice in 2016 and, by the time of the report published in May 2019, the family could "clearly see how the veratrum population [was] in decline while the orchid population [was] recovering."[3667]

[3664] Author's observations in Bohinj.

[3665] The traditional ecological knowledge presented in Sections 6.6.1 and 8.2.2 was confirmed by the author's participant observation at the Laz alpine pasture complex in the Julian Alps in the summer of 2021.

[3666] Compare Dietl & Lehmann, 2004, 31, 37, 45–49, 54–56, 60–61, 63, 65, 71, 85, 106–107, 125–126.

[3667] *Results-Based Nature Conservation Plan*, [2019], 12; datation of the report: *Ergebnisorientierter Naturschutzplan*, 2019, 2.

Another historical practice of suppression of a grassland species unwanted in animal husbandry is cautiously timed sheep grazing in meadows where hay is used for sheep, if characterized by appropriate site conditions (not too steep, not wet), at the right time in spring (Section 6.2). Spring sheep grazing can be a particularly relevant means of suppressing some poisonous meadow plants, such as autumn crocus (*Colchicum autumnale*) in moderately fertilized meadows. Hay from such meadows is, however, only useable to feed sheep (Section 6.2.2). Apart from application of spring grazing to suppress autumn crocus in moderately fertilized meadows at mesic sites as presented in Section 6.2.2.2, occasional spring grazing lasting for five to seven days at the onset of dandelion blossoming (i.e., when cock's-foot, *Dactylis glomerata*, is about 10 cm tall), a history-inspired practice (Section 6.2), has also been recognized as an appropriate means of reducing an unsustainably high percentage of herbs (more than 30% to 50%) in medium-intensive meadows. Moreover, the history-inspired practice of regular spring grazing is a sustainable means of quick suppression of cow parsley (*Anthriscus sylvestris*), an unwanted species abounding in some intensive, nutrient-rich, mesic silage meadows Alps-wide in all vegetation belts relevant for meadows; similarly, parsnip (*Pastinaca sativa*) is highly sensitive to trampling in late spring.[3668]

Traditional ecological knowledge about livestock's grazing behavior[3669] can also contribute to grazing circuits designed innovatively and in a transdisciplinary manner serving several agroecological purposes, including suppression of some plant species and fostering some others. Modern research in pastures in Mediterranean southern France "subject to scrub encroachment" aimed to "use a herd or flock [of sheep or goats] . . . in order to boost herd or flock production, control scrub development [and thus reduce the risk of forest fires], maintain biodiversity and promote the habitat of particular populations of plants or animals in need of protection." The scrub controlled in this manner is woody vegetation in holm oak (*Quercus ilex*) and downy oak (*Quercus pubescens*) coppices (see Section 6.1.2.6), and broom (*Genista* spp.) heathland. The results emphasized "the importance of flock management patterns and therefore . . . the farmer's role," not only the different grazing behavior of diverse livestock species and breeds. Namely, contrary to the premises in conventional agronomy, the concept of palatability as part of the total feed value "is actually unable to take into account the relative nature of food preferences which [grazing] animals express" in pastures: actually, "no individual plant or fenced grazing area possesses an intrinsic [palatability] that can be defined in the absolute and mapped. They all possess a relative . . . [palatability], linked to the way the shepherd [or goatherd] stimulates the feeding motivation of [the] flock."

[3668] Pils, 1994, 179, 211 and Table 37, Photo 102; Dietl & Lehmann, 2004, 13, 16, 20, 24–25, 28, 31, 37–38, 42–43, 54, 56–58, 72, 117–118, 121–125; Aeschimann et al., 2004, vol. 1, 1080–1081; Vogl-Lukasser et al., 2006, 70–72, 91.

[3669] Compare seventeenth-century data in Zwitter, 2020, 69–71.

Namely, the palatability of grazing resources depends not only on their intrinsic, objectively defined palatability, but also on circumstantial palatability, depending on feed availability, synergistic effects of successive grazing phases, and other factors, such as individual animal grazing behavior. This was learned from herders emphasizing that "'the everyday challenge is to offer the flock the feeding resources within a sequence that constantly revives the appetite', a standpoint . . . which animal nutritionists were quite unable to deal with" in the late twentieth century. In addition, conventional agronomy fully underestimated the feed value of the scrub abounding there (see also Section 8.2.2.1). Taking into account traditional and modern multidisciplinary research-based knowledge, a daily grazing circuit was developed, which stimulates fodder intake in the target area to be cleared of highly flammable scrub that offers middling fodder abundance characterized by middling instantaneous palatability. The vegetation in this target zone "would be insufficiently attractive to the herd without a particular organization of the grazing circuit." To suppress the scrub, foster biodiversity, and reduce the fire risk, the herder first takes the flock to a pasture section where the relative abundance of grazing resources is high but their instantaneous palatability is low. Then the herder moves the flock to the target area to be cleared of scrub, where the livestock consumes a high percentage of its daily fodder intake because its palatability is middling, thus higher than in the pasture section where the grazing circuit started. Later, the sheep or goats are allowed to graze in an area where fodder is delicious but its availability is very low. From there, they move again to the target zone to be cleared of scrub where fodder attractiveness is middling, thus lower than at the previous site, but the fodder availability is higher, and they again consume much of the target scrub. Then, the herder allows the flock to have dessert in a patch where there is a lot of fodder and its quality is excellent. After half a day, the flock returns to the barn; a different, carefully planned grazing circuit follows in the afternoon to pursue the same agroecological goal. In this manner, it is possible to "organize the daily feeding patterns so as to increase the overall palatability of resources having low intrinsic palatability by boosting their circumstantial palatability." In some cases—depending on species compositions—an advanced grazing circuit can "interrupt or limit the reproductive cycle of over-invasive shrubs through browsing of a series of 'target plant parts,' which moreover correspond to the 'functional feeds' of animals," such as goats "feeding on rangeland for more than 6 months in the year with no hay supplementation" and providing 700 l of milk each. In the absence of herders, these findings are also applicable in paddock design.[3670]

[3670] Meuret, 1997 (quotes from pp. 54 (tenth quote), 56 (eighth quote)); Hubert, Meuret, & Bonnemaire, 2008 (quotes from pp. 105 (first quote), 104–105 (second quote), 110 (third to fifth quotes), 110 (sixth quote), 111 (seventh quote), 113 (nineth quote)); Sauvant, Baumont, & Faverdin, 1996, 2786–2787. See also Sulzer, 2005. On palatability as part of feed value, see Dietl & Lehmann, 2004, 39.

Another history-inspired practice, well-considered winter grassland burning, can have favorable impacts in some habitats. It aims to reduce the abundance of dwarf shrubs unwanted from the perspective of livestock, such as common heather (*Calluna vulgaris*), and to decrease the abundance of matgrass (*Nardus stricta*) and some fescues (*Festuca* spp.) too hard for grazing (Section 6.6.2.1). The positive impacts of such well-considered winter burning are not limited to the quality and availability of fodder (grasses and forbs), but they also involve preservation and fostering of some rare species, and so this can also be an agroecological good practice. The seemingly universal destructive effect of a well-considered fire is deceptive, but of course the fire actually kills some plant and animal species, fosters some others, and impedes the development of yet others. Well-considered winter burning follows the wind direction when there are uphill winds, which keeps the fire temperature relatively low. It is carried out on moist or frozen soil, never on steep slopes, and never on shallow soil. It does not cause most of the humus to burn, and it saves the diaspores and roots of many grasses and forbs. It can have a favorable impact on populations of rare species, such as orchids (such as elder-flowered orchid, *Dactylorhiza sambucina*; fragrant orchid, *Gymnadenia conopsea*; and common twayblade, *Listera ovata*), gentians (chiltern gentian, *Gentiana germanica*; fringed gentian, *Gentiana ciliata*), and mountain arnica (*Arnica montana*). Concerning the impacts of such burning on grassland fauna, much is still not understood, but many grassland animals that remain in the soil and in below-ground plant parts, such as several grasshopper and butterfly species, survive, whereas the burning kills those overwintering in above-ground plant remnants, such as wild bees and some butterfly species.[3671] In contrast, burning performed in spring or autumn, when the soil is not frozen, leads to significant damage to the majority of plant and animal species and a rapid reduction of biodiversity.[3672]

10.1.2.2.7 Making specific use of plants from grasslands in livestock farming based on sophisticated plant knowledge

Traditional ecological knowledge about fodder plants was refined enough to properly recognize the high fodder value of some herbs from species-rich grasslands, not only of grasses, as proven by scientific testing (Section 4.1.1.2); traditional ecological knowledge about the qualities of several species still deserves scientific proof or rejection (Section 8.2.2). Most applicable to the present conditions with a limited workforce is knowledge about potential fodder plants that do not require gathering and processing but simply meadow and pasture management that fosters the abundance of the desired species. Exam-

[3671] Pils, 1994, 248–254, 274–275; Machatschek & Kurz, 2006, 109–116; Kerschbaumer, 2007.
[3672] E.g., Fynn, Morris, & Edwards, 2004.

ples of lady's mantles (*Alchemilla* spp.) and alpine spicknel (*Ligusticum mutellina*), the first providing as high-quality fodder as yellow oat-grass (*Trisetum flavescens*), alpine spicknel even better, are given in Section 4.1.1.2. Thus, it is a history-inspired practice to do maintenance mowing in the summer pastures to remove the plants overgrowing pasture sections rich in lady's mantle, such as alpine dock (*Rumex alpinus*), false hellebore (*Veratrum* spp.), and thorny thistles (*Cirsium* spp.).[3673] Feeding dried nettles is a method believed to increase lactation based on traditional ecological knowledge (see details in Section 8.2.2); it is worth testing in extensive farming whether it indeed has a potential to reduce the use of concentrated feedstuff. Nettles on the edges of a meadow are currently left unmowed by some extensive farmers, or they are mowed but left to decompose *in situ* by those that nowadays consider them low-quality fodder.[3674]

In line with the general industrialized developments in agriculture in the twentieth century,[3675] the relevance of properly treated (!) tree-leaf fodder from appropriately managed trees (Section 8.2.2.1) has not been fully investigated in agronomy to date. Nonetheless, tree-leaf fodder is appreciated not only by some elderly farmers but also by some history-inspired innovative young farmers seeking to practice agriculture largely based on local resources, with low investment cost and in an environmentally friendly manner, and also trying to reduce veterinary costs this way.[3676] One could, for instance, study whether some other feed additions could be reduced if the cows receive fallen apple leaves stored in good condition, how much they should receive, and what the impacts of such feeding are on milk; cows like apple leaves very much, and it does not take the farmer significant time to gather them (Section 8.2.2.1). The high nutritive value of properly treated tree leaves and their easy application in sheep farming deserve further research attention, as demonstrated by traditional ecological knowledge and historical research. Further studies should explore whether the minerals that livestock (sheep and goats in particular) receive from properly treated leaves from properly managed trees[3677] can be so relevant nowadays that it makes sense to recreate some pollarded fodder trees to reduce industrial feed additions and the use of medicines, to decrease environmental and financial costs, and to improve animals' wellbeing, at least for sheep and goat farms. If so, it makes sense to focus on the historically most appreciated fodder-tree species in the area (Section 8.2.2.1), but the species selection should also be adapted to present-day environmental reality. It is not

3673 Author's observations in the Julian Alps.
3674 Participant observation in haymaking in a moderately manured meadow at Stara Fužina in 2021, and interview with the farmer on that occasion, whose identity is kept confidential.
3675 Compare McKittrick, 2015.
3676 Machatschek, 2002, especially pages 189, 193, 204, 213, 447, 499.
3677 See Section 8.2.2.1 and Machatschek, 2002.

worth recreating common ash (*Fraxinus excelsior*) fodder trees, which belong to the historically most widespread leaf-fodder trees in the Alps, producing the highest-quality fodder at suitable sites (Section 8.2.2.1). Namely, this species is currently severely threatened in Europe by several diseases, most notably *Chalara* ash dieback, caused by the invasive fungus *Hymenoscyphus fraxineus*, introduced to Europe from East Asia in the 1990s. Wounds on branches during pollarding would support the spread of this pathogen. European native ash species are highly sensitive to this fungal attack: the leaves are severely affected, and the trees languish and gradually die.[3678] Some possible alternatives are offered in Section 8.2.2.1.

Another farming aspect is innovative history-inspired veterinary treatments including grassland plants. This still represents a research niche due to the limited scientific understanding of some historical uses of medicinal plants.[3679] Most grassland plants historically used in veterinary medicine can abound only at nutrient-poor sites or at sites with medium plant nutrient availability, which again calls for species-rich grasslands. An example of a traditional plant use that recently resulted in innovative history-inspired veterinary treatments is the practice of placing huge bouquets in livestock barns to prevent or cure some livestock diseases. This has been usual in many French regions since at least the twelfth century, and traditional farmers there still use such bouquets against several livestock diseases that have dermatological symptoms. In the late twentieth century, dissatisfaction with the effectiveness of modern medicines to treat contagious ecthyma—"[a] scabby viral infection that manifests itself mainly on the muzzle of lambs"—and the limited availability of medicines led to reconsideration, modification, and reintroduction of medicinal bouquets in sheepfolds (and to other applications of the same medicinal plants) following modern intellectual principles embedded in the modern cultural milieu by modern sheep farmers and zootechnicians. The results of therapeutically "rubbing the lambs' mouths . . . with a handful" of herb-Robert (*Geranium robertianum*) against contagious ecthyma combined with hanging herb-Robert bouquets in the sheepfold, and the results of prophylactically hanging herb-Robert bouquets in sheepfolds as low as possible (close to the sheep), against the same disease, were so encouraging that some farmers entirely abandoned ecthyma vaccination.[3680] Herb-Robert was widely used in Europe as a healing plant against some health issues with dermatological symptoms. In line with the data on the uses considered tradi-

3678 *Uprava Republike Slovenije za varno hrano, veterinarstvo in varstvo rastlin: Kdo je krivec, da se jeseni že poleti sušijo?*, 2020.

3679 Brisebarre, 1996; Vogl-Lukasser et al., 2006, give examples of several historical plant therapies applied in vernacular veterinary medicine in the Alps of East Tyrol for which no scientific evaluation of effectiveness is available.

3680 Brisebarre, 1996 (first quote from p. 78, second quote from p. 79, third quote from p. 88).

tional in the twentieth century, it was applied in such distant and culturally diverse milieus as in parts of France and Slovenia, and thus also in many other places.[3681]

Apart from these types of history-inspired treatments, a thorough evaluation of some poorly understood historical uses of (grassland) plants in veterinary medicine that were in fact appropriate (which is not the case for all historical medicinal plant uses) could still serve as indirect inspiration in developing new medicines.

10.2 Grassland history and the twenty-first-century sustainable rural non-farm economy: gathering, processing and marketing farm products, and tourism in rural areas

The rural non-farm economy comprises "all those non-agricultural activities which generate [on-farm or off-farm] income to rural households," involving "activities closely linked to farming and the food chain, and those not part of that chain."[3682] Focusing on the on-farm economy, integration of diverse activities—"traditional ones and those providing products targeting the specifics of recent market demand—is based on the innovativeness of farm household members. It strengthens the resilience of a farm."[3683] Processing farm products and farm tourism are currently among the most important on-farm activities of the rural non-farm economy associated with farming.[3684] This section explains that grassland history can provide some pieces in the puzzle of data relevant for processing and marketing farm products, farm tourism and tourism in rural areas in general, and plant gathering, with further implications for the general human attitude toward species-rich grassland environments that needs to change (Section 10.1.2.1).

10.2.1 Marketing and processing farm products in light of high-nature-value history-inspired grassland habitats

Ensuring "that consumers, regulators and investors are able to make informed decisions that, in turn, reinforce responsible production" is one of the key mechanisms for fostering sustainable production.[3685] In general, "[t]he environ-

---

[3681] Brisebarre, 1996, 81, 88, 90; Mlakar, 2015, 309.

[3682] Davis, 2004, 3.

[3683] Potočnik Slavič et al., 2016, 12 (quote); see especially also pages 33, 62, 70–71, 75–80, 84–85, 108–110, 113, 123–126, 129–130, 136; Keenleyside & Oppermann, 2009, 295–302.

[3684] Potočnik Slavič et al., 2016, especially pp. 39–44, 61–85.

[3685] *The Future of Nature and Business*, 2020, 31 (quote), 41.

mental impacts of modern agriculture are . . . distant from the consumers of food."[3686] "The increasingly complex and anonymous links between decisions of production and decisions of consumption, as a consequence of increasing specialization, make it more and more difficult for consumers on the market to signal adequate environmental preferences to producers in various parts of the world."[3687] The World Economic Forum's report *The Future of Nature and Business* emphasizes that "[m]ore comprehensive product labelling requirements can be instituted to encourage consumers to purchase . . . less environmentally harmful foods."[3688] "[T]he trend for greater transparency and accountability demanded by consumers, regulators and investors continues to rise."[3689] As a managing director of the company Speick Natural Cosmetics (see Section 10.2.3) formulated it: "Earlier, people bought the products because they smelled pleasant"; nowadays, they buy them because they can check their background.[3690] The same trend is gaining importance in food production (Sections 10.1.2.1 and 10.1.2.2.3), with the environmental footprint and livestock's wellbeing gaining in importance. "In wealthy countries, consumers are [increasingly] demanding food that is produced differently";[3691] moreover, they are gradually becoming aware of the relevance of species-rich grasslands for the quality of livestock products and the environment (Sections 10.1.2.1, 10.1.2.2.2, 10.1.2.2.3, and 10.1.2.2.4).

There is currently a considerable lack of clarity in labeling dairy products. This contributes to the vanishing of species-rich grasslands from landscapes in the Alps and elsewhere. Unclear and often misleading product labeling from the biodiversity perspective, together with often insufficiently targeted agri-environmental policy measures, significantly decreases the efficient use of public funds aimed at fostering sustainability in farming,[3692] as demonstrated in the following paragraphs, especially because the evidence on the environmental footprint of agriculture at the European level largely relies on "farm numbers under specific commitments" rather than data on the actual state of agroecosystems.[3693] In the absence of fair market options or specific initiatives targeting niche approaches—which the prevention of extinction of many species in European landscapes cannot ethically depend on (Section 1.2)—the current agri-environment payments make economically viable use of hay

[3686] McKittrick, 2015, 425.
[3687] Gustafsson, 1997, 359.
[3688] *The Future of Nature and Business*, 2020, 39 (quote), 47.
[3689] *The Future of Nature and Business*, 2020, 22.
[3690] Kabusch, 2020, 74.
[3691] McKittrick, 2015, 427.
[3692] Adapted from the general consideration that well-considered, socioeconomically viable high-nature-value farming increases the efficiency of public funds use aimed at sustainability goals (Lomba et al., 2019, 41) and the review of agri-environmental policy measures and their environmental impacts (Pe'er et al., 2017; Boetzl et al., 2021).
[3693] Grass, Batáry, & Tscharntke, 2021, 285.

or grazing resources from several types of seminatural grasslands difficult on many farms[3694] (see below in this section).

One of the fundamental current market challenges is that the labels such as "organic," "hay milk," "Traditional Speciality Guaranteed," "from [a certain] family farm," "from milk from [a certain] national park," "fully produced in [a certain well-known] Alpine valley" (including "protected designation of origin" and "protected geographical indication"), or "Selected Quality—Slovenia" are likely to give the impression that these products are made from the milk or meat of livestock that grazed in flower-filled pastures or whose diet was based on hay from species-rich meadows. This corresponds with reality only in some cases. Moreover, some officially registered labels, even several European Union certifications, designate different environmental standards in different countries, and also in different Alpine countries, as demonstrated by the example of hay milk, which has been gaining importance in recent decades.

There are major differences between Alpine countries concerning the use of the European hay milk label. Grassland biodiversity is seriously addressed in the association *ARGE Heumilch Österreich*'s hay milk standard (Figure 165), which is related by the fact that Austria is one of the EU member states with the highest uptake of agri-environmental payments per average hectare of farmland.[3695] In 2023, *Traditional Hay Milk Farming in the Austrian Alpine Arc*—represented by the association *ARGE Heumilch Österreich*, which links around 6,500 certified Austrian hay farms as well as around sixty hay-milk processors[3696]—became the first Food and Agriculture Organization of the United Nations' registered Globally Important Agricultural Heritage System in German-speaking countries.[3697] From about 6,500 hay farms involved, 38% are organic. The hay farms are "located in the mountainous area along the main ridge of the Alps, as well as in the lakeland region with a small structure at its foothills"; 88% of farms certified in this manner are located in mountain areas. About 6,500 certified hay farms produce around 15% of the total milk in Austria. It is healthier for humans than silage milk due to the much higher amount of unsaturated omega-3 fatty acids, which are vital for the human body, where they foster the metabolism and immune system. Hay milk also has evident advantages in cheesemaking in terms of a biodiversity-related more complex dairy product taste and because the challenge of Clostridia spores—which result in a bitter taste of semi-hard and hard cheeses unless the milk is subject to specific treatment—is incomparably lower in hay milk than in silage milk. In 2021 "about 47,000 tons of dairy products made from T[raditional] S[peciality] G[uaranteed] hay milk with a turnover of more than 147,000,000 euros

[3694] Compare Grass, Batáry, & Tscharntke, 2021, 265, 284–285.
[3695] Redman, 2012, 496–497.
[3696] *Traditional Hay Milk Farming . . .*, [2023], 10, 43–44, 79, 129.
[3697] *Food and Agriculture Organization of the United Nations*, GIAHS around the world; Trettler, 2024.

were sold in the Austrian retail trade" and more than 60% of Austrian hay milk products are exported, mainly to Germany.[3698]

Permanent grasslands are the land-use type *ARGE Heumilch Österreich* focuses on. Nearly a thousand plant species occur in the meadows of around 6,500 certified Austrian hay farms, the great majority of them located in the Alps. All certified *ARGE Heumilch Österreich* hay milk farms are obliged to preserve some biodiversity areas (5% of the farmland at the level of every hay milk farm until 2022, 7% since then, and the goal is 9% by 2028[3699]—in comparison to less than 4% of biodiversity enhancement areas in the farmland of most conventional and organic farms in Austria as found by a detailed study of 245 organic farms from 2017 to 2019 and interviews with experts on conventional farming in Austria[3700]). This means that, in addition to full compliance with the *Österreichisches Programm für umweltgerechte Landwirtschaft* (the Austrian Agri-Environmental Program in the framework of the Common European Agricultural Policy) measure Hay Management, *ARGE Heumilch Österreich* hay milk farms are obliged to participate with at least 7% of the farmland in the Austrian Agri-Environmental Program's measure *Umweltgerechte und biodiversitätsfördernde Bewirtschaftung* (Environmentally Sound and Biodiversity-Promoting Management). Some Austrian hay farms also use non-fertilized summer mountain pastures—the criterion of 7% biodiversity areas has to be fulfilled in addition to mountain pastures,[3701] which is in line with the utmost need to preserve enough biodiversity areas at all elevations in the Alps, including extensively managed grasslands at middle and lower elevations, where biodiversity is higher than at high elevations and where many organisms live that cannot withstand the harsher environmental conditions at high elevations (Section 1.2).

Many certified mountain hay milk farms in Austria manage a larger percentage of their farmland in a biodiversity-friendly manner, and some land-use types that do not count as part of the 7% of biodiversity areas, such as summer mountain pastures, are also typically species-rich. Thus, 28% of the total farmland of 6,500 hay milk farms is high-nature-value farmland (in comparison to 13% on silage farms).[3702] Namely, many mountain hay farms in

[3698] *Traditional Hay Milk Farming . . .*, [2023], 6, 8, 16–17, 28–30, 33–34, 40, 44, 52, 54, 79, 113, 122, 124, 132 (first quote from p. 8, second quote from p.17).

[3699] *Traditional Hay Milk Farming . . .*, [2023], 7, 11, 16, 20, 44, 48, 55, 92–115, 121–122, 124, 131, 138–156.

[3700] Leitgeb et al., 2023, 176–177, 186.

[3701] Heidi Trettler from *ARGE Heumilch*, personal communication; on the role of summer mountain pastures on certified Austrian hay milk farms, see *Traditional Hay Milk Farming . . .*, [2023], 4–9, 31, 38–40, 47, 51–57, 74–76, 79–83, 89–90, 93, 95, 98–101, 103, 105–108, 110–111, 113–120.

[3702] *Traditional Hay Milk Farming . . .*, [2023], 37, 49, 52, 55. A question not addressed here is whether the criteria for high nature value farmland are as strict as those for the biodiversity areas above or if the higher percentage is in part a consequence of slightly

Austria also participate in several other agri-environmental measures of the Austrian Agri-Environmental Program, in addition to the formally required 7% of biodiversity areas and the Hay Management measure. For instance, 88% of certified Austrian hay milk farms participate in the measure banning the application of mineral fertilizers and synthetic pesticides in grasslands, 83% participate in the measure for pastures, some participate in the measure for alp farming, some participate in the measure for mowing distant and extensively managed mountain hay meadows, and so on.[3703] Characteristically, hay farms mow smaller areas at the same time than silage farms, which is favorable for biodiversity, and Austrian hay farms mow "many areas only when a variety of grasses and herbs are at full seed maturity."[3704] This is crucial for the long-term preservation of species-rich meadows. Such hay is usually used to feed cows during late gestation or to feed calves and heifers that are close to full-grown,[3705] when less calorific fodder is an advantage for the livestock and the environment (see Sections 10.1.2.2.3 and 10.1.2.2.4).

Such a considerable percentage of high-nature-value farmland (largely history-inspired species-rich grasslands) and the activities that the farms involved carry out to deliberately maintain high-nature-value farmland, make it understandable from the viewpoint of consumers that the Austrian hay milk label is connected with the European label *Traditional Speciality Guaranteed*.[3706] Consumers' interest in biodiversity is clearly expressed (1) in the most frequent answer of interviewed consumers in Austria that their primary motivation for buying hay milk products is "natural feed with grasses, herbs, and hay," (2) in the high market demand for cheese from summer mountain pastures, and (3) in the fact that the demand for organic hay milk on markets in Austria and Germany "currently exceeds the supply." *ARGE Heumilch Österreich* has carried out many activities to increase farmers' knowledge about biodiversity, its functional relevance, such as the favorable impact on the taste of dairy products, and ways of preserving biodiversity.[3707]

Nonetheless, to cope with the challenges of economic sustainability on such a large number of involved dairy farms in Austria, more intensive management is permitted in the majority of meadows belonging to certified hay milk farms than in line with historical traditions before the mid-twentieth century[3708] (see

---

looser criteria for high-nature-value farmland than for biodiversity areas. In any case, 28% of the farmland used by hay milk farms is of high nature value.

[3703] *Traditional Hay Milk Farming . . .*, [2023], 55; Heidi Trettler from *ARGE Heumilch*, personal communication.

[3704] *Traditional Hay Milk Farming . . .*, [2023], 7, 19–20, 35, 37, 48–51, 55, 63, 121 (quote from p. 7); Hartmann et al., 2023, separate pagination, 1, 3, 7–9.

[3705] *Traditional Hay Milk Farming . . .*, [2023], 63.

[3706] *Traditional Hay Milk Farming . . .*, [2023], 10, 29, 80.

[3707] *Traditional Hay Milk Farming . . .*, [2023], 16–18, 34, 40, 44, 54, 113, 117, 124, 127, 129, 131–132 (quote from p. 127).

[3708] *Traditional Hay Milk Farming . . .*, [2023], 48, 121, 123, 128.

Section 6.3). Most meadows belonging to Austrian hay milk farmers are relatively species rich, moderately fertilized, or medium intensive; in medium-intensive meadows, only very few rare plant species occur. At least every second year moderately fertilized and medium-intensive meadows cannot be cut earlier than the first seeds ripen to remain productive in the long run because most plant species (including productive grasses) in these meadows depend on generative reproduction, made possible through partial seed dispersal during haymaking when seeds are ripe.[3709] Certified Austrian hay farms usually mow their meadows two to four times annually, which is usually less frequently than comparable silage farms. This is also reflected in less intensive grassland fertilization: 88% of all *ARGE Heumilch Österreich community*'s hay farms decided on a formal ban on mineral fertilizer and synthetic pesticide application in grasslands. However, they are generally allowed to use farm slurry (a mixture of feces, urine, and water),[3710] which, unless heavily diluted, is usually a significantly more aggressive fertilizer than the traditional, matured farmyard manure mainly because the nitrogen in slurry is predominately bound in ammonium (NH_4-N) instead of in oxidized forms as in matured farmyard manure, and ammonium is incomparably more aggressive than matured farmyard manure and thus not tolerated by the roots of most plants from species-rich meadows and many grassland animals.[3711] Nonetheless, the slurry on most Austrian hay farms, which do not use mineral fertilizers, is significantly milder than the slurry on many intensive conventional silage farms that mix mineral fertilizers with the slurry.[3712] Moderately intensive to relatively intensive meadow fertilization is reflected in the fact that many certified Austrian hay farms mow most of their grasslands three or four times annually, including the first hay cut as early as May and near the farms at high elevations, higher than the lower section of the montane vegetation belt, in the first half of June. All this means that the first mowing usually takes place one to four weeks later than on comparable silage farms; this also means that the first mowing takes place later on hay farms than on many organic silage farms. In this manner and by making use of roofed hay driers, they produce high-quality hay and add only a limited amount of purchased protein feed, thus reaching economically viable milk yields even though they follow the formal *ARGE Heumilch Österreich* hay

[3709] Compare Dietl & Lehmann, 2004, 13, 20, 24–25, 42–43, 51, 54, 60–61, 66, 83, 105, 109, 111, 116–118, 120–121, 124–125.

[3710] *Traditional Hay Milk Farming* . . . , [2023], 19, 37, 55, 63–64, 74, 79, 121, 135, and personal communication with Stefan Hörtenhuber, an Austrian specialist in environmental science related to livestock production systems.

[3711] Dietl & Lehmann, 2004, 9, 13, 20, 24–25, 28, 37–38, 42–43, 52 56, 69, 102–103, 105–111, 116–126. If a farm maintains the tradition of farmyard manure and organic liquid manure (i.e., urine and a little feces), organic liquid manure also is aggressive for the same reason as slurry, but the quantity of organic liquid manure is only half the quantity of slurry (Dietl & Lehmann, 2004, 13, 101).

[3712] Compare Machatschek, 2015, 37; author's interview with an anonymized farmers' representative from the Slovenian Alps, 2022.

milk standard requiring that the roughage portion in the total fodder must be at least 75% annually, and at least 85% on organic hay farms. Namely, growing calorific fodder plants on arable land usually burdens the environment even more than intensive meadow management. Certain high-protein feeds usual in conventional farming, such as urea, genetically modified organisms, and fodder imported into Europe from other continents (see the introduction to Section 10.1.2.2), are entirely banned in *ARGE Heumilch Österreich* hay milk for environmental reasons.[3713]

Thus, the intensity of grassland farming at more productive sites in the Alps—such as the bottoms of wider valleys and basins and on gentle slopes in colline vegetation belt and the lowest section of montane belt—is often lower in hay farming than in silage farming, in organic farming in particular, resulting in higher biodiversity on hay farms. In contrast, in meadows at high elevations and on steep slopes, less favorable conditions for farming are often reflected in generally slightly lower farming intensity, such as atypically late mowing and rather moderate fertilization not only on hay farms, but also on silage farms. Nonetheless, the management in most meadows at hay farms and silage farms is significantly more intensive than optimal for biodiversity, in particular on average conventional hay and silage farms. Consequently, in steep meadows at high elevations, the differences in the diversity of habitats and species diversity between hay farming and silage farming are often insignificant in the Austrian Alps, whereas the differences in species diversity between organic and conventional farms are significant there.[3714]

The tradeoffs presented in terms of meadows of various use intensity on Austrian hay farms, ranging from non-fertilized to the prevalent medium intensive, to allow the preservation of limited areas of biodiversity hotspots, demonstrate how difficult it is nowadays to establish a system involving a large number of dairy farms in agri-environment schemes in a socially equitable manner targeting a specific market niche. The example of *ARGE Heumilch Österreich* emphasizes that a small but relevant and increasing percentage of farmland can be maintained or recreated in a highly biodiversity-friendly manner (such as non-fertilized grasslands) by a significant percentage of all farms based on strong engagement of, increased environmental education for, and fruitful cooperation by various types of stakeholders, such as farmers, farm advisory services, dairy product processors, the dairy product market, politics, and consumers.[3715] The biodiversity enhancement areas at Austrian

3713 *Traditional Hay Milk Farming . . .* , [2023], 9–11, 20, 28, 35–37, 48–49, 54, 63–74, 79, 116, 123, 125, 134–135; Dietl & Lehmann, 2004, 99–100; Hartmann et al., 2023, separate pagination, 3.

3714 Leitgeb et al., 2023, 177, 181–182. See also the previous two paragraphs.

3715 *Traditional Hay Milk Farming . . .* , [2023], 18, 35, 40, 43, 79–80, 121, 123, 126–129, 131–133.

Figure 165: Some farmers in the Alps and various types of stakeholders involved do their best to ethically and proudly combine biodiversity inspired by traditional farming and a viable farm economy. The figure shows two signs on a door of a hay barn on a hay farm that has been practicing organic farming since 1993 in the village of Grießen (at an elevation of 960 m), near the border of the provinces of Salzburg and Tyrol. The text on the board next to the blossoming meadow plants, butterflies, and grasshopper says: "Our space for species richness. Hay milk. Simply perfect" and in the photo on right the text says "Biodiversity label. We live biodiversity." There is fantastic flavor of aromatic hay near the barn and the regularly used, extensively managed seminatural pasture on sunny slopes just above the village is highly species rich. Many species that could not tolerate fertilizer application, such as stemless thistle (*Carlina acaulis*) and a couple of small gentian species (*Gentiana* spp.) occur there. Preserved geodiversity (the pasture is hummocky) is yet another factor that keeps the biodiversity so high. For several reasons, such farming is becoming increasingly rare near permanent settlements in several regions of the Alps, such as in some regions of this mountain range in neighboring Slovenia. Photo: Žiga Zwitter, September 2024.

hay farms gradually approach the minimum threshold of 10% to 20% (roughly; the figure depends on the ecological and biophysical context in the landscape, including the targeted species) of all farmland at a microregional level that, if the connectivity between habitats is sufficient, probably makes it possible to sustain biodiversity in farmed landscapes in the long run.[3716] "While hay milk farmers were considered retrograde just a few decades ago, they are now pioneers of sustainable agriculture";[3717] they typically belong to the innovator farmer type (see Section 10.1.2.1). Nonetheless, alongside efforts concerning the preservation of biodiversity, specific efforts are being carried out in Austria

[3716] Oppermann & Hoffmann, 2012, 94–96; Kremen, 2015, 68, Grass, Batáry, & Tscharntke, 2021, 277–287; Tscharntke et al., 2022, 923, 925–926, 928; Leitgeb et al., 2023, 186.
[3717] *Traditional Hay Milk Farming* . . . , [2023], 80. On serious negative aspects of silage, such as the limited availability of proteins in silage, requiring higher concentrated feed additions and causing significantly heavier environmental pollution in meadows than quality hay, see Dietl & Lehmann, 2004, 67, 69, 94, 99–100. However, see also Verbič & Verbič, 2014–2020, 30, and Kuhar & Simeonov, 2019.

concerning the social and economic pillars of sustainability. Namely, a recent survey of 284 Austrian hay milk farmers showed that—despite the improved economic situation based on budgetary payments, technological advance in haymaking, and a higher price for certified Austrian hay milk in comparison to silage milk, made possible by the high market demand and specific marketing efforts—"there is still a competitive disadvantage [for hay farms that are rather small by today's standards, with an average herd of seventeen cows] compared to the large and powerful silo farms" (yield optimizers; see Section 10.1.2.1). More environmentally friendly farming on hay farms results in a higher workload and lower milk yield. Whereas about 95% of hay farmers are very satisfied or rather satisfied with their choice of profession and more than 90% with the possibility to learn something new, only about two-thirds of them are very satisfied or rather satisfied with the existing level of stress and with the milk price, 60% of them are very satisfied or rather satisfied with the number of work hours per day, and only about 45% of them are very satisfied or rather satisfied with the recognition by society.[3718]

In neighboring Slovenia, biodiversity in any percentage of involved farmland does not belong to the eligibility criteria for certified hay meat or hay milk.[3719] In Slovenia, the position of conventional agriculture is stronger than in Austria (which is partly rooted in the complexities of several aspects of history[3720]), the number of organic farms in Slovenia is low (less than 5% of all agricultural holdings that manage 11% of the total farmland, mainly grasslands) and declining.[3721] In Slovenia, "organic products often fail to achieve a market premium and are sold at a price comparable to conventional products,"[3722] which is in contrast to the market for organic farm products in several other Central European countries, including some Alpine countries.[3723] Intentional agroecological approaches—not just side products of declining traditional farming—have not yet been comparably widely present in Slovenian landscapes in comparison to Austria.[3724] Although biodiversity in any percentage of involved grasslands does not belong to the eligibility criteria for certified

[3718] *Traditional Hay Milk Farming* . . . , [2023], 18, 35, 40, 43, 79–80, 121, 123, 126–129, 131–133 (quote from p. 35).

[3719] Kolarek Novšek, 2024.

[3720] Compare Kupper, 2021, 152–153, 156–163, 172–173, 175–185; Wöbse, 2022, 31; Palmblad, 2024, 667; Erhatič Širnik, 2011, especially pp. 219–220, 226–230, 236–237, 241–242, 248–252, 264, 266–267, 273, 276–277, 278, 305, 310–311, 315, 329, 334, 338, 343, 353, 359; Polajnar Horvat, Smrekar, & Zorn, 2017, 23–31; Baldock & Petersen, 2012, 537.

[3721] Cigale & Lampič, 2023, 257, 263; Bertoncelj & Travnikar, 2024, 81; gratitude is also expressed to the geographer Irma Potočnik Slavič for her information.

[3722] Šumrada et al., 2024, 13 (quote), and the rural geographer Irma Potočnik Slavič, 2024, personal communication.

[3723] E.g., Baldock & Petersen, 2012, 539; Tscharntke et al., 2022, 920; *Traditional Hay Milk Farming* . . ., [2023], 40, 124.

[3724] E.g., Redman, 2012, 496.

hay meat or hay milk in Slovenia, registered Slovenian hay farmers receive the European Common Agricultural Policy's agri-environment-climate payment for natural resources, in line with the operation Hay Farming in the frame of the European Common Agricultural Policy's 2023–2027 Strategic Plan for Slovenia.[3725] One reason for the difference between the Austrian and Slovenian environmental standards for hay milk is the different main goal currently pursued by the operation Hay Farming in the frame of the European Common Agricultural Policy's 2023–2027 Strategic Plan for Slovenia; namely, to reduce fodder maize cultivation on arable land that poses high environmental loading. Thus, in contrast to Austria, grasslands in field rotation (not permanent grasslands) currently belong to the main target areas of this agri-environment operation in Slovenia.[3726] Whereas decreasing fodder maize cultivation is a relevant goal in some regions of Slovenia[3727] (its relevance in the Alps is limited because an average farm in the Slovenian Alps does not grow fodder maize;[3728] the need for regionally tailored agri-environmental measures has been emphasized in Central Europe for decades[3729]) and hay in general has some environmental advantages in comparison to silage,[3730] grasslands in field rotation replacing maize in the landscapes of intensive farming are not species rich. The challenge is that the program for hay farming in Slovenia formally also aims to "preserv[e] the species diversity of permanent grassland and the mosaic character of the landscape,"[3731] but so far this has not resulted in any formal eligibility criteria concerning grassland biodiversity.[3732] Thus, the temptation for Slovenian farmers not to consider biodiversity while entering the Slovenian hay milk or hay meat scheme is high, even though a limited percentage of hay milk and especially hay meat farming in Slovenia will likely contribute to sustainable grassland management because Slovenia currently still "hosts one of the largest relative areas of H[igh] N[ature] V[alue] farmland in Europe."[3733] However, only in about one quarter of the grassland habitat types of European importance as defined in the Habitats Directive, the conservation status is favorable in Slovenia,[3734] and other types of species-rich meadows are also in decline (Section 1.2), thus preserving high-nature-value farming requires strong efforts. With the rapid decrease in the number of traditional farmers

3725 This type of agri-environment-climate payment was presented by Kolarek Novšek, 2024.
3726 Kolarek Novšek, 2024.
3727 Compare, for instance, *Traditional Hay Milk Farming* . . . , [2023], 30, 37.
3728 Main author's field-work based knowledge.
3729 E.g., Schmitzberger et al., 2005, 287; Redman, 2012, 501.
3730 Dietl & Lehmann, 2004, especially pp. 67, 69, 94, 99, 101.
3731 Kolarek Novšek, 2024.
3732 Kolarek Novšek, 2024.
3733 Šumrada et al., 2024, 3.
3734 Petkovšek, 2020b, see sections Definition and Charts; the Methodology section is not translated on this website, so see the Slovenian version: Petkovšek, 2020a, Metodologija.

(Sections 1.2 and 10.1.2.1), the ecological state of grasslands in Slovenia has generally been rapidly worsening: the populations of the bird species that are farmland specialists declined in Slovenia by 40.8% from 2008 to 2018, largely due to grassland management intensification (Section 1.2).

One of the Slovenian results of the 2014–2020 European Innovation Partnership "Agricultural Productivity and Sustainability" project Hay Milk and Meat is the hay production handbook *Pridelovanje kakovostnega sena in uporaba sena v prehrani goved* (The Production of Quality Hay and Use of Hay for Cattle).[3735] The handbook is a contribution to the economy of intensive hay farming. It is problematic that the project aimed to foster not only productivity but also sustainability in grassland farming results in a handbook for farmers that omits agroecological topics and deals only with the economy of hay farming, neglecting the environmental pillar of farming sustainability.[3736] It is our suggestion to make this manual more interdisciplinary and in line with the transdisciplinary goals of sustainability by adding agroecological topics to the existing valuable information on intensive hay meadow management in the same publication. Namely, many cattle farmers that find this handbook on the website for Slovenian hay farming[3737] will probably not look for additional literature dealing with the agroecology of hay farming, based on their own initiative—with likely exceptions only among idealist farmers (which are rare) and some innovator farmers (Section 10.1.2.1). An agroecological addition is relevant because this handbook informs interested hay farmers about economically optimal hay meadow management, and so it has likely been impacting real grassland management changes (see Section 1.2). A farmer currently managing high-nature-value grasslands and reading the current version of this handbook will be stimulated to intensify grassland management and will not at all be stimulated to also seriously consider the impact of intensification of grassland management on grassland biodiversity and the functional role of biodiversity in farming. Namely, the information that large-scale intensification in grassland management is currently causing mass extinction in grassland agroecosystems in Slovenia (see Section 1.2) is missing from the handbook.[3738] In addition, due to the lack of agroecological content, this handbook does not present the functional role of biodiversity.[3739] Although the quantity of hay from

3735 Verbič & Verbič, 2014–2020.

3736 Compare the Austrian approach to the sustainability of hay farming in *Traditional Hay Milk Farming . . .*, [2023].

3737 *Seneno*, Gradiva.

3738 The results of a study of the impact of fertilization on grassland plant species richness in one single meadow in the northwestern Dinaric Karst (Verbič & Verbič, 2014–2020, 19) are not representative of the prevalent process of strong management intensification causing mass extinction in Slovenian grasslands in many areas without specific constraints on farming (see Section 1.2).

3739 Exceptions are a general statement that forbs are richer in minerals relevant for livestock than grasses and clovers (Verbič & Verbič, 2014–2020, 20) and the desire for hay seminatural meadows is by definition significantly lower than that from intensively managed meadows,[3740] the quality of first-crop hay and second-crop hay from many extensively managed, seminatural, moderately fertilized meadows at higher elevations (higher than the lower edge of the montane belt: yellow oat grass meadows) can actually be very high by current Slovenian standards (see Section 6.3.2.2).[3741] This information would be relevant because certified hay farmers currently keep dozens of square kilometers of permanent grassland in the territory of Alpine Convention in Slovenia.[3742] Moreover, based on better knowledge of grassland agroecosystems, some specific non-fertilized, species-rich mountain and other meadows, such as those where sainfoin or other relevant legumes abound, also deliver highly nutritious, protein-rich hay (Sections 6.3.1.2 and 10.1.2.2.4). All of this is missing from the handbook, which simplistically links the high nutritious value of hay only with intensive meadow management in a conventional agronomy manner. There is a risk that the general statement in this handbook that hay mowed late enough to allow seed dispersal relevant for natural meadow regeneration is "suitable only as animal bedding"[3743] will be interpreted by some farmers as also referring to seminatural meadows. Such meadows can only be kept productive and species rich in the long run if some seeds are regularly dispersed during late haymaking.[3744] Only in another section does this handbook state that late mowing makes the worst hay if combined with intensive fertilization,[3745] and so it was probably not the author's intention to suggest regular mowing in non-fertilized and moderately fertilized meadows earlier than allowing some seed shed during haymaking.

Thus, due to the missing agroecological part, the functional role of biodiversity in seminatural grasslands—which specialized conventional agronomy research since the mid-twentieth century, excessively focused on production, "has rendered obsolete from its own fragmented vantage point"[3746] (see Sections 1.2, 10.1.2.1, 10.1.2.2, and 11)—is still absent from the handbook presented. In this context, the following additions are suggested: (1) a farmer-friendly presentation of seminatural meadow plant communities deliver-

---

to have a nice flavor (Verbič & Verbič, 2014–2020, 27), but the explanation that the best hay flavor results from species-rich grasslands (see Section 10.1.2.2.2) is missing.

3740 Dietl & Lehmann, 2004, 121.

3741 Compare Verbič & Verbič, 2014–2020, 29, and Dietl & Lehmann, 2004, 122. See also Section 6.3.2.2.

3742 Spatial representation of the farms involved in the project for hay farming in the Slovenian Alps and a map of the territorial distribution of certified hay farms was shown in Kolarek Novšek, 2024.

3743 Verbič & Verbič, 2014–2020, 15.

3744 Dietl & Lehmann, 2004, 54, 66, 116–117, 121–125.

3745 Compare Verbič & Verbič, 2014–2020, 22.

3746 The general idea and the quote are from Palmblad, 2024, p. 674; its application in the agroecological context is new.

ing good fodder, with photos of easily recognizable indicator plant species; (2) positive impacts of hay from several non-fertilized and moderately fertilized meadows on livestock's health, especially in the case of hay from high elevations, which has the potential to reduce costs for medicines and veterinarians (such as the decreased likelihood of gastrointestinal parasite infestation in livestock due to the tannins in species-rich hay; the related application of medicines prevents the use of milk for a long time); (3) the positive impact of hay from some extensively managed species-rich meadows, such as non-fertilized grasslands rich in thymes (*Thymus* spp.) on dairy product taste; (4) the relevance of flowering meadows in providing bee forage; (5) the presentation of ecosystem services in arable land and orchards provided by organisms depending on species-rich grasslands such as pollination of insect-pollinated crops and fruit trees, and natural control of organisms perceived as pests in farming without the pollution caused by pesticides; and (6) the explanation of the relevance of ethical and responsible, well-considered care for biodiversity in the context of the recent farming's environmental footprint, as much as the farm economy and other relevant agronomic considerations allow[3747] (see Sections 1.2, 6.3.1, 6.3.2, 10.1.1, 10.1.2.1, 10.1.2.2.2, 10.1.2.2.3, and 10.1.2.2.4).

There is a significant difference between this Slovenian handbook and the grassland management handbook by the Swiss grassland agroecologist Walter Dietl and agronomist Josef Lehmann, published by the *Österreichischer Agrarverlag* (Austrian Farming Press), which emphasizes the high relevance of a combination of both (1) intensively managed meadows and (2) extensively managed, non-fertilized, and moderately fertilized meadows inspired by long-term history that are mowed late enough so that seed dispersal during haymaking allows natural grassland renewal and ethical grassland farming. Dietl and Lehmann also emphasize the functional role of hay from such meadows in livestock farming. Dietl and Lehmann emphasize that an adequate percentage of meadows of all intensity types (meaning that the percentage of extensively managed ones is not negligible), from non-fertilized to intensive, at the farm level "takes into account the natural and cultural foundations of a living cultural landscape and allows sustainable, site-appropriate, and species-appropriate meadow management" that can be economically viable on many farms even though this means that yields are not the highest.[3748] In line with Dietl and Lehmann, up to 25% of all hay at the farm level can be made in meadows mowed late enough to allow natural regeneration and highly ethical farming in species-rich seminatural grasslands, if used in a well-considered manner on

[3747] Compare the consideration of such a thought in the presentation of the new agri-environment measure aimed at the preservation of dry calcareous grasslands in the Dinaric Karst of western Slovenia (thus non-fertilized meadows in the Limestone Alps are not eligible): *Suhi kraški travniki in pašniki*, 2023.

[3748] Dietl & Lehmann, 2004; see the entire volume (the quote is from p. 117).

many dairy farms targeting an annual milk yield of five to six tons of milk per cow and also raising offspring.[3749]

Of course, biodiversity considerations cannot drive all grassland management. However, the manual *The Production of Quality Hay and Use of Hay for Cattle* makes it clear that the estimated difference in the annual milk yield between (1) a milk cow fed the best hay type currently produced in Slovenia, supplemented by a moderate addition of feed concentrate, and (2) a milk cow fed the least nutritive hay type currently produced in Slovenia,[3750] supplemented by a moderate addition of feed concentrate, is a bit more than a quarter less milk based on the least nutritious hay compared to the most nutritious hay.[3751] This difference is significant but not immense. Extensive grassland management also requires a larger area for the same yield quantity, but several types of expenses for purchased inputs such as fuel, fertilizers, and medicines are lower (Sections 6.3.1, 6.3.2, 10.1.2.2.3, and 10.1.2.2.4). This means that keeping predominately non-fertilized grasslands on most of a farm would be neither realistic nor economically acceptable for most dairy farms. However, this also means that an innovative approach combining meadows of different intensity types, from non-fertilized to intensive within a farm, thus combining a viable farm economy and ethical dealing with wild living beings in agroecosystems,[3752] would result in only a moderate decrease in milk yield that could be compensated by suitable agri-environment compensation payments if more effort were invested in this direction, and the hay milk and hay meat operation could contribute to this.

This leads to the challenge of hay milk and hay meat marketing in Slovenia. The Slovenian document *Certificiranje senenega mesa in mleka* (Certification of Hay Meat and Milk) states: "As both producers and consumers become increasingly aware of the importance of food quality and concern for the environment, feed or food production and food processing are consequently returning to traditional methods of livestock rearing or the manner in which our ancestors reared their livestock."[3753] Despite this statement, the hay content of Slovenian hay milk and hay meat farms is not required to have anything in common with the hay from history-inspired, species-rich meadows in terms of its species composition. In contradiction with this, Slovenian farmers certified as hay milk or hay meat producers use the label of the European food quality scheme "Traditional Speciality Guaranteed,"[3754] aimed at a product that

---

[3749] Compare Dietl & Lehmann, 2004, 126–127 and Verbič & Verbič, 2014–2020, 22–23, 30–34.

[3750] Compare Verbič & Verbič, 2014–2020, 29, and Dietl & Lehmann, 2004, 122.

[3751] Verbič & Verbič, 2014–2020, 29, 32.

[3752] Compare Verbič & Verbič, 2014–2020, 29, and Dietl & Lehmann, 2004, especially pp. 11, 13, 20, 23–27, 42–45, 53–54, 66, 91–94, 96, 99–102, 105–106, 108–111, 116–129.

[3753] Kolar, [2019 or 2020], 1.

[3754] Kolar, [2019 or 2020], 2.

"is produced from traditional raw materials, has a traditional composition or traditional recipe."[3755] Thus, there is a high threat of large-scale misleading of environmentally conscious buyers not aware of marketing tricks. Some buyers are likely to connect the label "Traditional Speciality Guaranteed" with traditional hay ingredients, traditional haymaking dates, traditional numbers of hay cuttings, traditional fertilizer quantities, and flora and fauna of traditional, colorful species-rich meadows—the meadows of their youth for a typical representative of the older and middle generation[3756] and for some members of the younger generation (Section 1.2). It is up to hay farmers in Slovenia to decide whether (part of) the hay in fact originates from species-rich (moderately fertilized or non-fertilized) meadows. It is irresponsible to allow the sale of products as "Traditional Speciality Guaranteed" if the only thing that is obliged to be traditional is strongly prevalent roughage, largely from dry or fresh grassland feed, in addition to the prohibition of some feed types of industrialized farming.[3757] Namely, "Traditional Speciality Guaranteed" hay milk is a label that, during the recent mass extinction in grasslands, designates higher biodiversity standards in neighboring Austria and is likely to evoke romantic illusions about the state of agroecosystems in the buyer's mind in most cases.

This is an example of how some sustainability funds and marketing strategies nowadays partly result in greenwashing and what common sense would call a partial economic abuse of the nexus between the concepts of traditionality and sustainability, which aggravates fair prices for products of farmers actually managing high-nature-value farmland—which some hay farmers do, but many hay farmers do not. Lack of interdisciplinary, transdisciplinary, and trans-sectoral cooperation in Slovenia[3758] results in Traditional Quality Guaranteed hay milk and hay meat standards that have every potential to contribute more strongly than actually needed to the mass extinction in Slovenian grasslands. Due to the recent period of mass extinction occurring in Slovenian grasslands, agroecological reconciliation (requiring some tradeoffs from both sides) making it possible to reach the formal but not seriously addressed goal to "preserve the species diversity of permanent grassland and the mosaic character of the landscape" in a limited percentage of farmland for every farm participating in the hay milk or hay meat scheme should ethically stand closer to the implementation of the operation Hay Farming in the frame of the European Common Agricultural Policy's 2023–2027 Strategic Plan for Slovenia. This matters because permanent grasslands cover 56% of the total agricultural land in Slovenia.[3759] There are good examples of agroecological practices abroad

---

[3755] Kolarek Novšek, 2024.

[3756] Compare Jaksetič (interview with Kaligarič), 2019.

[3757] On the feeding standards in Slovenian hay farming, see Kuhar & Simeonov, 2019.

[3758] Gratitude is expressed to the geographer Irma Potočnik Slavič for conversation concerning this issue.

[3759] The percentage of permanent grassland within the total agricultural land was presented by Kolarek Novšek, 2024.

striving for reconciliation between agronomy and nature conservation in grasslands, which, however, require significantly more interdisciplinary, transdisciplinary, and trans-sectoral efforts than the implementation of the recent Slovenian hay meat and hay milk criteria. It is high time to invest immense efforts in more sustainable tradeoffs between the economy, ecology, and social equity in grassland farming in Slovenia than in the current hay milk and hay meat schemes in the country. Optional examples of good practices are the reconciliation between agronomy and ecology in the French Flowering Meadows competition—that is, part of the main French agricultural competition looking at biodiversity from a functional perspective[3760]—and the *ARGE Heumilch Österreich* hay milk standards[3761] (see above).

Similar to hay milk, an organic livestock farm is not by definition a synonym for species-rich grasslands and especially not for small-scale diversity of grassland ecosystems, but it is true that biodiversity and the state of the environment in general (e.g., preservation of soil quality) are usually much better in farmland on organic farms than on conventional farms. Certainly, meadow management on organic farms is more strictly regulated than on conventional farms, and it maintains high-nature-value agroecosystems in some cases thanks to farmers' initiatives and decisions.[3762] However, this is not a general rule. Recent findings on dairy farms in the Austrian Alps with grasslands mostly located on steep slopes at high elevations demonstrate that species diversity in the grasslands of organic hay milk farms is significantly higher than on conventional hay milk farms, and species diversity in the grasslands of organic silage milk farms is significantly higher than on conventional silage milk farms. The achievement of the sustainability goal of species diversity is, however, limited: the goal achievement score for species diversity on average reaches around 70% for the organic hay milk and silage milk farms analyzed in the Austrian Alps, in comparison to around 50% for conventional hay milk and silage farms mostly located on steep slopes at high elevations. Another biodiversity component, the diversity of grassland agroecosystems—which is strongly influenced by "the proportion and connectivity of semi natural habitats," which are characterized by small-scale heterogeneity—is also higher on organic farms than on conventional hay and silage dairy farms in the Austrian Alps mostly managing farmland on steep slopes at high elevations. However, in the case of habitat diversity, the difference between organic and conventional farms is not statistically significant, and the achievement score for the sustainability goal of ecosystem diversity averages 61% for organic hay milk farms and 67% for organic silage milk farms in the Austrian Alps (in comparison to

[3760] Magda et al., 2015.
[3761] *Traditional Hay Milk Farming . . . ,* [2023].
[3762] Pils, 1994, 67, 274, 276, and Table 51, photo 131; Dietl & Lehmann, 2004; Jones & Poux, 2012, 457; Oppermann, 2012, 223–224; Machatschek, 2015, 28, 33; Machatschek, 2017, 91–92; Tscharntke et al., 2021 (quote from p. 921); Leitgeb et al., 2023.

50% and 55% for conventional farms), and so the average ecosystem diversity goal achievement scores for organic farms are lower than in the case of species diversity. Namely, the decision whether also seminatural habitats or only more intensively farmed land is maintained on organic farms "depend[s] largely on the commitment of individual farmers" and not on organic farming's legal regulations. Thus, most organic and conventional farms in the Austrian Alps keep less than 4% of the privately owned farmland in a highly biodiversity-friendly seminatural state. The percentage of such seminatural grasslands on organic silage farms located at high elevations and managing steep meadows in the Austrian Alps is higher, but it is still below 10%.[3763] Thus, species richness, and especially the diversity of seminatural habitats, are not guaranteed by the organic farming certificate: grassy plains that are intensively managed, relatively heavily organically fertilized, and cut early are allowed. "[T]argeted on-farm measures to restore biodiversity are not mandatory in any organic certification scheme,"[3764] and organic farming has recently become increasingly intensified in many cases, which partly resembles the trend in conventional farming even though the environmental impacts of the recently intensified conventional and organic farming are not comparable.[3765]

The specification of "Selected Quality—Slovenia" for certified milk and dairy products also does not include any requirement on the state of grassland agroecosystems,[3766] and so selling milk from intensive grassland monocultures certified this way is legal.

Focusing on designations of product origin, it is only optional that the product specifications ensure maintenance of high-nature-value farmland; this is not a general requirement for products with certified origin in Europe.[3767] Because industrial intensification in farming that causes severe biodiversity loss is the prevalent process in managed farmland in Europe (Section 1.2), it is probably becoming increasingly less likely in general that dairy and meat products with a formally designated origin support high-nature-value grasslands, with the exception of products whose certification formally requires this. There are usually some quite intensive meadows in Alpine valleys that products come from according to their specification (Section 1.2), and it is often allowed for such products with a designated origin to support intensive meadow management, thus contributing to the mass extinction in grasslands. Moreover,

[3763] Leitgeb et al., 2023, 176–179, 181–182, 186.
[3764] Pils, 1994, 67, 274, 276, and Table 51, photo 131; Dietl & Lehmann, 2004; Jones & Poux, 2012, 457; Oppermann, 2012, 223–224; Machatschek, 2015, 28, 33; Machatschek, 2017, 91–92; Tscharntke et al., 2021 (quote from p. 921); Leitgeb et al., 2023; *Traditional Hay Milk Farming . . . ,* [2023]; author's interview with an anonymized farmers' representative from the Slovenian Alps, 2022.
[3765] Tscharntke et al., 2021, 920–923; Leitgeb et al., 2023.
[3766] *Specifikacija: Izbrana kakovost: mleko in mlečni izdelki*, 2019.
[3767] Paracchini & Capitani, 2012, 520–522.

contrary to "protected designation of origin," which at least "indicates that all stages of production and processing take place in the designated geographical area," "protected geographical indication" only "demonstrates that at least one step of production and processing takes place in the area specified in the designation,"[3768] which often does not allow any kind of consumer idea about the grasslands or other fodder sources a dairy or meat product might be based on.

A general shift in the policy for product certifications is urgently needed at the level of the European Union and its member states from the simplified question of the area of product origin, which in general creates a lot of space for unsustainable farming—greatly contributing to the mass extinction currently occurring in European cultural landscapes—to protected designations of the link between the environmental impact of farming (where biodiversity has to be addressed) and the area of product origin.

In addition, pictures of blossoming grasslands are allowed to be placed on any product regardless of the actual status of the grasslands the product depends on. They are often abused because in the case of grasslands on product stickers or the package and in advertising, high-nature-value grasslands are usually presented to attract buyers, even if the product has nothing in common with high-nature-value agroecosystems.[3769] Namely, in general commercial labeling, outside formal product certification, many marketing lies and labels evoking romantic illusions about the grasslands that the products are based on are allowed in Alpine countries, with a varying degree of toleration in legislation in different countries, such as somewhat better informed consumers in Switzerland than in Slovenia.[3770] Thus, the reality is often different from what the consumer hopes. Even a conscious consumer usually cannot know what the percentage of grassland plants in the total livestock diet was, and what the grasslands yielding the fodder consumed by the livestock that produced the milk are like (species-rich or intensive grassy plains), unless one knows the farmer personally.

It is often hard to say whether this lack of clarity about the actual status of the agroecosystems that the product is based on is an intentional strategy to attract customers and evoke their illusions or a consequence of a lack of the farmer's, cheesemaker's, or the enterprise's staff's knowledge about the relevance of species-rich grasslands, especially in countries where school curricula have until recently avoided this topic and where there are still no appropriate

---

3768 On the European Union's certification of products "with a guaranteed or protected place of origin," see Ledinek Lozej, 2021, 158–159.

3769 Compare Jones & Poux, 2012, 456–457; Erhatič Širnik, 2011, 357.

3770 Compare Bätzing, 2015, 392, and the discussion at the International Conference on Mountain Cheeses at the 12th annual meeting of the Farmhouse and Artisan Cheese & Dairy Producers European Network, Bohinjska Bistrica, 2022, with broad international participation.

results-based agri-environmental measures fostering farmers' education about species-rich grasslands. Nonetheless, absence of information on the role of species-rich grasslands in livestock's diet on the labels of industrial dairy products, and in some cases also artisanal or homemade ones, is often not a mere coincidence, but a consequence of the fact that the grasslands that the products are based on are no longer species rich: they are more than moderately fertilized and cut too soon to be species rich. It is not legally binding for the producers of dairy and meat products to make any information on the state of grassland agroecosystems available to consumers, and so some intensive farmers in the Alps silently contribute to illusions in customers' minds because the legislation does not stimulate them to think creatively about which of the grasslands they manage could economically feasibly be kept species rich. A legislative regulation forcing the industry, artisans, and farmers producing dairy products to specify the role of species-rich grasslands in their livestock's diet would do much good for preservation of an ecologically and economically sound, sustainably high percentage of species-rich grasslands in all vegetation belts of the Alps—and would at the same time foster appropriate prices for products from species-rich (moderately manured or non-fertilized) grasslands, making the role of a farmer managing species-rich grasslands more easily economically viable and socially equitable. Unless such legislation is introduced, people are unlikely to become aware of the paucity of dairies processing milk based on species-rich grasslands, and so customers are not willing to pay equitably high prices for such products. They live in the illusion that they eat products from such grasslands at a cheaper price, but in fact they buy products made with milk from intensively managed grasslands. This way, the idealized, "imagined delights of rural life," a sentimental topic with a long and diverse history reaching back uninterruptedly to the early modern period,[3771] keeps being produced and circulated among city (and country) dwellers. The actual status of the farm, the status of the agroecosystems the products are based on, the issues of animal health and wellbeing, and the healthiness of the products for people are concealed (Sections 10.1.2.2.2 to 10.1.2.2.7). Such partiality of information currently available "reflect[s] the unease" generated by the environmental footprint of economic development in times of the Great Acceleration (Section 10.1.2.1),[3772] and it demonstrates that many agroecological goals of the European Common Agricultural Policy[3773] cannot be achieved without fundamental changes.

Yet, as observed in Europe at least since the eighteenth century, "[t]he man who live[s] permanently in 'romantic' countryside . . . tend[s] to regard it in a very different light from that in which it [is] seen by the . . . tourist who [comes]

3771 Compare Thomas, 1987, 245 (quote), 248–254.
3772 Inspired by Thomas, 1987, 253–254 (quote from p. 254), and Elias, 1993, 161–163.
3773 E.g., Iuga, Iancu, & Stroe, 2016; Ledinek Lozej, 2021, 155, 163.

only on a brief visit."[3774] "H[igh] N[ature] V[alue] farmers clearly deserve our respect. Many of them have carried on in spite of low incomes year after year, putting in long working hours under tough conditions with little recognition of what they contribute to society and the environment."[3775] There is still insufficient general social appreciation and respect for the farmers providing many public goods and ecosystem services through their management of high-nature-value agroecosystems, which is related to inadequate knowledge about the ecological footprint of today's society and its food production among the general public (Section 10.1.2.1). In addition, despite compensation payments, the total income of farmers managing high-nature-value farmland is in general lower than that of their peers practicing intensive farming: "many of the remaining H[igh] N[ature] V[alue] farmlands are currently losing socioeconomic viability." "The competition is unfair," which has not only ecological and economic impacts but also leads to the lower social prestige of such farmers in the eyes of the farmers' peers.[3776]

Good examples making high-nature-value farmlands socioeconomically viable in Europe are usually either (1) located in specific social, economic, and environmental conditions—large areas of farmland used per farm at sites with significant natural constraints on farming and receiving budgetary payments[3777] (see Section 10.1.2.2.4), such as extensive non-fertilized mountain summer pastures where they are still grazed by livestock in the Alps, or (2) smaller high-nature-value farms focused on high product quality, making their economy viable through significant farm income diversification. Good examples of the second type that involve many farms in a certain region require time-consuming, profound, and innovative trans-sectoral and transdisciplinary engagement by farmers and other relevant economic sectors, such as marketing and tourism, an interdisciplinary group of researchers, politicians, and other stakeholders, looking at high-nature-value farmland as a development opportunity and not as a constraint.[3778] A frequent issue is overly small farms managing high-nature-value farmland, and the related challenge is how to increase the farm size or how to diversify the farm income in a manner that makes a larger or more affluent farm (for instance, processing additional raw materials from other farms) not only economically and socially sustainable,

[3774] Thomas, 1987, 264–265 (quote from p. 264).
[3775] Baldock & Petersen, 2012, 538.
[3776] Baldock & Petersen, 2012, 537–538; Beaufoy, Keenleyside, & Oppermann, 2012, 525–526, 534; Birrer & Oppermann, 2012, 190; Jones & Poux, 2012, 451–457 (second quote from p. 457); Paracchini & Oppermann, 2012, 446–450; Redman, 2012, 501; Lomba et al., 2019, 37–39 (first quote from p. 37); Debeljak et al., 2020; Tscharntke et al., 2021, 925; Šumrada et al., 2024, 2.
[3777] Compare Oppermann, 2012, 231; Šumrada et al., 2024.
[3778] Lomba et al., 2019, 40–41; see also Oppermann, 2012, 231; Redman, 2012, 500; Trisorio, Borlizzi, & Povellato, 2012, 271.

but also environmentally sustainable—a challenge that is not generally subject to serious consideration and solved.[3779]

Innovative solutions—enabling the consumer to know the high-nature-value agroecosystems the products are based on—are highly relevant, in particular because such farms also have the potential for educating farm visitors about their role in biodiversity preservation. However, such innovator farms will never preserve a sufficient percentage of high-nature-farmland countrywide. To halt the mass extinction in grasslands, more than a combination of niche approaches is needed, which calls for (1) changes in the education of producers and consumers, paving the way for possible changes in their behavior, (2) further strengthening of a well-targeted agri-environmental policy, and (3) raising the standards of environmental ethics in marketing, which requires changes in legislation[3780] (see Sections 1.2, 10.1, 10.2, and 11).

Thus, making conditions that at the same time allow an equitable position of a farm family managing species-rich grasslands, and make it possible for the consumer to know which food is based on species-rich and which on intensive grasslands, should become a key priority for policy in the regions where moderately fertilized and non-fertilized grasslands still cover a significant percentage of agricultural land. They are often quickly disappearing just now (Section 1.2). It was perceived as "morally healing" for middle-class tourists from distant lands, England in particular, to visit the attractive species-rich ecosystems in the Alps in the late eighteenth and the nineteenth century,[3781] and it can be morally healing for well-informed consumers nowadays to know that they support species-rich grasslands through their eating habits. Similarly, it can be morally healing for a farmer or cheesemaker to be able to frankly and proudly talk about environmentally and socially responsible[3782] grassland management with farm visitors instead of trying to conceal undesired practices so as not to face an inequitable economic position despite the hard work on the farm, or, in some cases, due to overly ambitious economic goals.[3783]

Nowadays, in general there is "a strong business case for more sustainable, nature-positive models"[3784] in business in the developed world or this business case has at least been rising, thus such a highly ethical economic activity

[3779] Suske, Gantner, & Gattermaier, 2012, 124; Trisorio, Borlizzi, & Povellato, 2012, 271; Šumrada et al., 2024, 6, 9, 11, 14–17, 20–21.

[3780] Adapted from Baldock & Petersen, 2012; Oppermann, 2012, 231–232; Redman, 2012, 501; Trisorio, Borlizzi, & Povellato, 2012, 271–273; Plassmann, 2016b, 33, 36.

[3781] Thomas, 1987, 260–268 (quote from p. 260).

[3782] Compare Dietl & Lehmann, 2004; the use of the concept of responsibility is inspired by the handbook for farmers concerning the management of non-fertilized, dry calcareous grasslands in the Dinaric Karst of western Slovenia (*Suhi kraški travniki in pašniki*, 2023, 1).

[3783] See the presentation of the farm style types with respect to their economic ambitions and environmental impacts in Section 10.1.2.1.

[3784] Quote from *The Future of Nature and Business*, 2020, 97.

is worth being seriously considered when framing or transforming the business model for the following decades (Sections 10.1.2.1 and 10.1.2.2). Many farmhouse cheesemakers in the Alps, for instance, are aware of the fact that sustainability is one of the key and highly complex challenges they need to face.[3785] Nonetheless, to be able to embrace species-rich grasslands as an economically viable business background allowing an equitable position of the producers targeting this market niche, like other livestock farmers, they need fair agri-environmental payments and a fair market for species-rich products, ranging from equitable prices[3786] (the fact that average food consumers that are not farmers spend a lower percentage of their income on food than ever before needs to change in the course of the Great Reset of the human–environment attitude and the related change in the system of human values; Section 10.1.2.1) to regulations allowing consumers to easily know what comes from species-rich grasslands and what does not (compare Section 10.1.2.1).

Thus, returning to the existing "babel of labels," research calls for product denominations deconstructing "stereotyped representations, while creating new forms of conviviality."[3787] For instance, the *Schweizerische Vereinigung integriert produzierender Bauern und Bäuerinnen* (The Swiss Association of Integrated Farming; i.e., *IP-Suisse*) "to which about one third of Swiss farmers belong, has developed a point-ranking system in co-operation with the Swiss Ornithological Station" (see the impact of farming intensity on farmland bird species in Section 1.2). "Farmers have to reach a minimum threshold of 'biodiversity points' in order to sell their products under the *IP-Suisse* label and obtain the associated premium."[3788]

In general, creating a new collective brand designating products that per definition support the preservation of species-rich grasslands—thus the raw materials are based exclusively or to a significant extent on seminatural grasslands at the level of every farm involved—requires a lot of enthusiasm, time, and effort.[3789] Moreover, just adding another label—likely consisting of a couple of blooming grassland flowers, and a butterfly or a bee from history-in-

[3785] This is proven, for instance, by the welcome speech by Frédéric Blanchard, the president of the Farmhouse and Artisan Cheese & Dairy Producers European Network, at the International Conference on Mountain Cheeses at this European network's twelfth annual meeting, Bohinjska Bistrica, 2022; by the program of that conference (*International Conference on Mountain Cheeses and XII Annual Meeting of FACEnetwork*); and by the program of lecture cycle to celebrate the twenty-fifth anniversary of the Association of Small Slovenian Cheesemakers (*Združenje kmečkih sirarjev Slovenije*).

[3786] Author's interviews with anonymized representatives of farmers and farmhouse cheesemakers from the Slovenian Alps and Pre-Alps in 2022 and 2023.

[3787] Ledinek Lozej, 2021, 164.

[3788] Riedel, Walter, & Herzog, 2012, 432 (both quotes).

[3789] See the examples of transdisciplinary efforts to initiate the collective brand From Haloze Meadow Slopes (ZRSVN, *Annex C.7 – Detailed Description . . .*, [2020]) and of hay milk marketing in Austria: *Traditional Hay Milk Farming . . .*, [2023]. See also Šumrada et al., 2024, 20.

spired meadows—to so many existing labels on the product stickers would probably not be successful at all because it might go largely unnoticed:[3790] the potential of labels is "often overestimated" unless embedded in a broader set of activities aimed at the same goals.[3791]

Thus, it is one of the key priorities of agricultural, environmental, food trade policies, and the education systems to create shared goals making possible ethical and fair conditions for farmers managing a significant percentage of species-rich grasslands linking natural and cultural heritage to avoid evoking illusions about rural life and agroecosystem management that have contributed to the swift losses of species-rich grasslands in Europe since the mid-twentieth century (Section 1.2). Alpine countries with such a rich history of livestock products based on species-rich grasslands (Section 3.1), and of tourism with cultural landscapes and biodiversity as significantly important attractions (Sections 8.1.3.1 and 10.2.2), should connect and act accordingly, in the framework of the Alpine Convention.[3792] These and further kinds of actions, such as further developments of grassland championships that already have a tradition in Austria, Germany, France, and Switzerland, with specialists forming interdisciplinary juries evaluating species-rich grassland candidates from a combined agricultural (livestock farming in terms of large mammals and apiculture) and ecological perspective. Optimally, the jury visits every meadow joined by the owner, and their conversation about the specifics of every grassland is a unique opportunity for knowledge exchange. A very prominent person, often a minister, awards the grant to the winning farmer in a major public show and the event receives significant public media attention, which all helps spread awareness that preservation of a network of species-rich grasslands below the forest line depends on farming and that consumers can indirectly co-create species-rich grassland ecosystems through their eating habits. This can also encourage "public appreciation of H[igh]N[ature]V[alue] farmers' work and public support for preserving and maintaining HNV grassland[s]" as a basis for the public appreciation of appropriately high prices and of agri-environmental payments and for raising the social prestige of farmers acting against the recent mass extinction by managing high-nature-value farmland. This "give[s] farmers pride in their work," particularly important because the workload in history-inspired (of course largely motorized) management regimes is somewhat heavier and yields are middling or low. In this way, decisions based on high ethical standards, in line with the required reset

[3790] Author's interview with the ethnologist, geographer, cheesemaker, and farmer L. Gartner, 2022, strongly engaged in development of the brand From Bohinj; and the ethnologist Špela Ledinek Lozej, personal communication.

[3791] Compare Ledinek Lozej, 2021, especially pp. 155 and 163 (quote from p. 155). In the case of the brand From Haloze Meadow Slopes, see the presentation of accompanying activities in Debeljak et al., 2020.

[3792] On the Alpine Convention, see Bätzing, 2015, 390–392.

in the human attitude toward the rest of nature (Section 10.1.2.1), can foster the ecological, economic, and social pillars of sustainable development.[3793]

This section focused on trade in products from history-inspired species-rich grasslands. Grassland history also includes topics relevant for processing dairy and meat products. The impacts of fodder from species-rich grasslands on product quality were outlined in Sections 10.1.2.2.2 to 10.1.2.2.4. Considering milk processing, the forgotten technique of addition of nettles (*Urtica* spp.) to rennet is worth being more widely evaluated in the production of certain fresh and semi-hard cheese types from species-rich grasslands, which do not mature for longer than two months (Section 8.2.1.1). Similarly, using rennet from thistle flower heads, from spear thistle (*Cirsium vulgare*) at least, in the production of sheep and goat cheeses that are not aged for more than two months, is worth testing (Section 8.2.1.1). These historical practices have the potential to inspire innovative cheese recipes. An example of a relevant historical use of plants from grasslands in meat processing is the use of juniper (*Juniperus communis*) berries and twigs in the brine in which meat is treated before smoking (Section 6.6.2.2).

10.2.2 History-inspired species-rich grasslands as a tourist attraction

A look into the history of tourism in the Western Alps reveals that the relative uniformity of enclosures in the English lowland countryside embedded in English intellectual and cultural history was one of the push factors for the English middle class to be so well represented among the summer tourists in the species-rich (agro)ecosystems of the Alps in the eighteenth and nineteenth centuries.[3794] Until the 1880s the British, for whom the landscape they lived in was exceptionally uniform (nonetheless, still significantly more biodiverse than since the mid-twentieth century; see Section 1.2), were the most numerous tourists in Switzerland,[3795] the country in the Alps visited by most foreign tourists in this mountain range until the 1920s.[3796] Similarly, in the early twenty-first century, an attractive landscape often crucially influences people's decision to take vacations in rural areas, as revealed by a recent survey including 440 interviewees in Slovenia, a country where "the annual number of

[3793] Adapted from Keller, 2004; Schmitzberger et al., 2005, 276, 279; Keller, 2007, 214; Keenleyside & Oppermann, 2009, 295–302 (first quote from page 302, second one from page 297); Baldock & Petersen, 2012, 537; Beaufoy, Keenleyside, & Oppermann, 2012, 533; Jones & Poux, 2012, 457; Oppermann et al., 2012a, 466–472; Redman, 2012, 498, 501; Suske, Gantner, & Gattermaier, 2012, 126; Lomba et al., 2019, 40; and Dietl & Lehmann, 2004.

[3794] Thomas, 1987, 260–268, 281–284.

[3795] Scheidegger, 2008, 23.

[3796] Hachtmann, 2007, 86, 133, see also pp. 134–137, 140; Hasenöhrl & Groß, 2022, 188.

Figure 166: A species-rich and spectacularly colorful moderately manured meadow on a sunny slope in northern Bohinj in late spring. Archives of Tourism Bohinj, photo: Mojca Odar. © Tourism Bohinj.

overnight stays on tourist farms increased from 114,739 to 271,894" between 2010 and 2018 and then also stayed similar from 2019 to 2021 "despite the COVID-19 pandemic." An attractive landscape was considered the most important factor influencing the choice of which vacation farm to stay at (relevant for 47% of interviewees) as expressed by 110 interviewees with personal experience in staying at a vacation farm overnight. An attractive landscape was designated as the most important of eleven factors given on a questionnaire—that is, significantly more important than the general attractiveness of the area, the hospitality of the farm family, positive personal experience during previous visits or recommendations by friends or acquaintances, favorable prices, the presence of other tourism attractions nearby, and opportunities for several free-time activities on the farm or in its vicinity, to name only those factors with an important influence on the selection of the vacation farm as determined by this survey.[3797] Moreover, the same survey revealed that 61% of those 330 interviewees that had never stayed at a vacation farm overnight consider an attractive landscape the most important factor influencing their selection of the tourist destination, meaning that attractiveness of the land-

[3797] Cigale & Lampič, 2023, 254 (both quotes); Potočnik Slavič et al., 2016, 74–77, 84; access to the farm and cleanliness of the household and its equipment were not considered important by the interviewees because they are usually unproblematic.

scape is considered far more important than the desire to see something new, to meet friends and make new friends and acquaintances, the diversity of tourist opportunities, accessibility of the destination, or opportunities to engage in another preferred activity.[3798]

The perception of an attractive landscape is subjective; nonetheless, several landscape characteristics are usually perceived to be attractive.[3799] The basic message from the long-term links outlined above between tourism history and the history of cultural landscapes is that increasing uniformity in the local cultural landscape and increasing land-use intensity (Section 1.2) make other less uniform and thus more picturesque cultural landscapes, closer to nature, particularly appealing. In general, "communities of [high] plant species diversity are considered aesthetically more valuable by the public than those with low diversity,"[3800] and seminatural vegetation in high-nature-value agroecosystems is often more aesthetically appealing than the potential natural vegetation.[3801] "Many Europeans recognise that H[igh] N[ature] V[alue] landscapes are as much part of Europe's patrimony as the architecture of our most celebrated cities."[3802]

Non-fertilized and moderately manured meadows managed in a history-inspired manner (especially Sections 10.1.2.2.3, 10.1.2.2.4, and 10.1.2.2.6) are usually colorful.[3803] Flowers were so frequent in grasslands in previous centuries that in legal terminology the word *Blume* 'flower' designated the yield in a grassland (meadow or pasture);[3804] in a lawsuit record from the eastern Julian Alps in 1614, taking "the flowers" and taking "the grass" in a meadow were used synonymously.[3805] However, not all extensively managed meadows have been "wonderfully floriferous."[3806] Moderately manured meadows, in particular those at relatively dry sites abounding in purple meadow clary (*Salvia pratensis*), always accompanied by yellow and often white and pink blossoming herbs, are generally perceived as particularly attractive (Figure 166). Namely, a high abundance of species developing relatively large, intensely colored blossoms or inflorescences makes such meadows particularly colorful, so that

---

3798 Potočnik Slavič et al., 2016, 74, 80, 84.
3799 Potočnik Slavič et al., 2016, 80.
3800 Fischer et al., 2008, 148 (quote), 154; see also Jones & Poux, 2012, 457; Oppermann & Paracchini, 2012, 21; Riedel, Walter, & Herzog, 2012, 432; Suske, Gantner, & Gattermaier, 2012, 125; Lomba et al., 2019, 37–38.
3801 Grass, Batáry, & Tscharntke, 2021, 255.
3802 Baldock & Petersen, 2012, 537 (quote), 538.
3803 Peterken, 2013, 285; Machatschek, 2015, 26–27.
3804 *Frühneuhochdeutsches Wörterbuch*: Blume.
3805 The phrase "die rosen [Sln. *rože* 'flowers'] oder pluemben [i.e., *Blumen* 'flowers'] abgenomben" was used as a synonym of "das graß abgenomben" (SI AS 721, Gospostvo Bled [The Seigneury of Bled, Germ. *Veldes*], Iustitialia, fasc. 34: Lawsuit between Krištof Wuria and the Nomenj community over grazing rights "in der [!] Stiegen" 1613/1614).
3806 Peterken, 2013, 285 (quote); Machatschek, 2015, 26–27.

Figure 167: When snow still covers shady meadows in low valley bottoms, Christmas rose (*Helleborus niger* subsp. *niger*), one of the first sources of bee forage (note the bee in the photo), is in full blossom in extensively managed sunny pastures, at the forest edges of extensively managed sunny meadows, and in sunny, light forests on carbonates as high as the montane vegetation belt. Its blossoms resist night frost and short-term snow blanketing. A sunny meadow edge on the west-facing slope at Zavaze in Bohinj at 670 m, February 23rd, 2022. Photo: Žiga Zwitter.

this meadow type meets the typical expectations of a colorful meadow in late spring; these expectations developed in times of frequent moderately manured meadows from the late nineteenth century to the second half of the twentieth century (Section 6.3.2). Many non-fertilized meadows are also comparably colorful in late spring and in summer, and they are habitats of many rare and endangered species; non-fertilized meadows at calcareous sites are the habitats richest in wild orchids (Section 6.3.1).[3807] Sunny, dry, non-fertilized, or moderately manured meadows are characterized by a great seasonally diverse variety

[3807] Pils, 1994, 168, 176, 179, 187, 267 and Table 3, Photo 7, Table 8, Photo 21, Table 10, Photos 27 and 28, Table 17, Photo 48, Table 35, Photo 96, Table 38, Photo 103, Table 41, Photo 110, Table 42, Photos 112 and 113, Table 43, Photo 114, Table 44, Photo 116, Table 50, Photo 129.

Figure 168: A subalpine pasture rich in trumpet gentian (*Gentiana clusii*) on Mount Ogradi in the Julian Alps, June 25th, 2019. The long-term preservation of this agroecosystem depends on grazing; dwarf mountain pine (*Pinus mugo*), seen in the upper part of the photo, would otherwise overgrow the site. Photo: Eva Petkovšek. © Eva Petkovšek.

of colors, whereas shady and more moist meadows managed this way can also be colorful, though they are usually not as colorful as drier sunny sites.[3808]

The following lines present only a selection of species highly relevant in eco-tourism for aesthetic reasons because their abundance in many non-fertilized (see also Sections 6.3.1, 10.1.2.2.2, and 10.1.2.2.4) or moderately manured grasslands managed in a history-inspired manner (see also Sections 6.3.2 and 10.1.2.2.3) is so high that they can color entire meadows and pastures or significant sections of them. In late winter and early spring, the highly attractive blossoms of Christmas rose (*Helleborus niger* subsp. *niger*, Figure 167) give a white color, with a scarlet shade, to sections of some moderately manured or non-fertilized relatively dry grasslands on carbonates. Christmas rose occurs in many places in the Eastern Alps and in a very limited area of the Western Alps. Another species blossoming in early spring is spring crocus (*Crocus vernus*; Figure 72 in Section 6.2.2.1), which gives a purple color to some moderately fertilized mesic grasslands in the east of the Eastern Alps and in the south of the Western Alps. Slightly later in spring, non-fertilized and moderately manured meadows in some sections of the Alps are white with Salisbury's daffodil (*Narcissus poeticus* subsp. *radiiflorus*; Figures 79 and 81 in Section 6.2.2.3). At the same time, some non-manured dry meadows on carbonates in many parts

[3808] Bavcon, Ravnjak, & Praprotnik, 2019, 17, 91, 93, 131; a couple of examples of colorful flowers in moist and shady meadows: Aeschimann et al., 2004, vol. 2, 1086–1087; Dakskobler et al., 2019, 92.

of the Alps become blue when common ball flower (*Globularia bisnagarica*) blossoms, and purple and yellow blossoms of elder-flowered orchid (*Dactylorhiza sambucina*; Figure 77 in Section 6.2.2.1) abound in some non-fertilized relatively dry grasslands at that time. In late spring and early summer, globe-flower (*Trollius europaeus*) colors some moderately manured or non-fertilized grasslands on moist soils yellow, whereas relatively dry non-fertilized or moderately manured meadows become violet with meadow clary (*Salvia pratensis* subsp. *pratensis*; Figure 166). At the same time, many moderately manured relatively dry meadows in the Alps abound in yellow blossoms of goat's-beard (*Tragopon pratensis* subsp. *orientalis*; Figure 91 in Section 6.3.2.2), and moderately fertilized meadows on moist soils become pink with ragged-robin (*Silene flos-cuculi*) flowers (Figure 3 in Section 1.2). Still in late spring or early summer, some non-fertilized meadows on moist soils on carbonate bedrock in several regions of the Alps abound in pink marsh gladiolus (*Gladiolus palustris*, looking equally appealing as its close relative in Figure 115, Section 7.1). Late spring and early summer are the seasons richest in varied colors of blossoming herbs in meadows moderately fertilized with mature farmyard manure and in non-fertilized ones (Figure 1 in Section 1, Figures 2 and 3 in Section 1.2, Figure 10 in Section 2, Figure 17 in Section 4.1.1.1, Figures 84 and 85 in Section 6.3.1.1, Figure 91 in Section 6.3.2.2, Figure 108 in Section 6.6.1, and Figure 166). In late summer, many non-fertilized or moderately manured meadows in the Alps at lower elevations, which were mowed in July or August and are located on mesic, base-rich soils, were colored purple by European cyclamen (*Cyclamen purpurascens*) soon after mowing (see Sections 6.1.2.1.1.2 and 8.1.3.1). Alps-wide, many moderately manured mesic late-mowed meadows turn purple in late summer and early autumn, when autumn crocus (*Colchicum autumnale*) blossoms. In reality, these flowers are joined by a plethora of other flowering herbs in meadows and pastures, in late spring and early summer in particular, including many highly aesthetically appealing species, such as lilies, orchids, gentians, and many others (Figure 168; Figures 6, 8, and 10 in Section 2; Figure 77 in Section 6.2.2.1; Figure 84 in Section 6.3.1.1).[3809]

Colorful meadows are generally perceived as beautiful and harmonious. They are highly aesthetically appealing (see the admiration of daffodil meadows at the beginning of Section 6.2.2.3 and of blossoming meadows in early summer in Section 10.1.2.2.1; in addition to flowers and butterflies, songbirds depending on food from seminatural meadows can often be observed on hedges and trees in seminatural meadows and at meadow edges, and several song-

[3809] Aeschimann et al., 2004, vol. 1, 122–123, 126–127, 324–325, 660–661; Aeschimann et al., 2004, vol. 2, 164–165, 280–281, 646–647, 1030–1031, 1082–1083, 1092–1093, 1096–1097, 1120–1121; Peterken, 2013, 285–287, 399; Bavcon, Ravnjak, & Praprotnik, 2019, especially pages 23, 25, 43, 47–51, 53–56, 59, 61–63, 65–67, 69, 76–77, 89, 91, 93, 101, 106–107, 173; Dakskobler et al., 2019; Praprotnik, 1993a, 215–218; Praprotnik, 1993b, 260–261.

birds are generally perceived as beautiful, lovely, gentle animals; see Section 10.1.1). Species-rich meadows are also psychologically refreshing (e.g., due to the melodies of songbirds (Section 10.1.1) and delightfully smelling flowers such as great masterwort, *Astrantia major*, that can abound in late-mowed, moderately fertilized meadows in montane and subalpine belts, although its habitats in the Alps have been shrinking) and educational at the same time. All this is a consequence of the rich diversity of their flowers, grasses, butterflies, honeybees, bumblebees, beetles, including bioluminescent fireflies on evenings in late spring and early summer, and other organisms.[3810] As the awareness of the environmental footprint of our society increases, the aforementioned positive perception of meadow birds will likely increase because they are also indicators of a good state of preservation of seminatural habitats characterized by high biodiversity.[3811]

Such positive feelings were widely present among the local inhabitants in history (see the examples of a traditional farmer impressed by the beauty of grassland flowers in the Carpathians in Section 10.1.2.2.1, and of positive perception of songbirds in Section 10.1.1). The traditions of aesthetically appealing and nice-smelling bouquets and garlands of local grassland flowers in the Alps (e.g., as decoration for churches, shrines, and homes)[3812] and of blossoms of meadow flowers strewn on the feast of Corpus Christi (a movable holiday celebrated since the late Middle Ages[3813] between May 21st and June 24th,[3814] when most meadow flowers were in full blossom) in front of every house in some villages of the Alps[3815] or around the altars erected outside to be visited by the Corpus Christi procession[3816] confirm that positive perceptions of meadow flowers have been widespread among the local population. Similar feelings already attracted large numbers of summer tourists in the Alps during

[3810] Pils, 1994 (e.g., p. 182 and Aeschimann et al., 2004, vol. 1, 1072–1073); Keller, 2004, 46–47, 69; Keenleyside & Oppermann, 2009, 296, 299–300, 302; Erhatič Širnik, 2011, 305, 361 (and Section 10.1.1 in this volume); Jones & Poux, 2012, 457; Peterken, 2013, 285–287, 375, 381–390, 399, 404–405; Bavcon, Ravnjak, & Praprotnik, 2019, 17, 23, 25, 45, 91, 93, 131, 207, and pictures on pages 20–21, 47, 50, 59, 63, 98–99, 102–103, 105–106, 184–185, 190, 194–195, 197, 204–205; Dakskobler et al., 2019; Debeljak et al., 2020; Vintar Mally et al., 2021, 160; Bregant, 2023. Author's observation of the shrinking of great masterwort habitats in Bohinj in the last three decades and the general trends presented in Section 1.2.

[3811] Erhatič Širnik, 2011, 354; Wöbse, 2022, 40.

[3812] E.g., Kuret, 1989, 367–368, 370–371; Vogl-Lukasser et al., 2006, 51, 54, 84, 222–226 (dealing with magical-religious use of a bouquet consisting of grassland flowers and other components); Bavcon, Ravnjak, & Praprotnik, 2019, 123; Novak, 2024, 217.

[3813] Kuret, 1989, 358.

[3814] Grotefend & Ulrich, 1971.

[3815] Such as at Plavški Rovt in the Karawanks above Jesenice as documented for the first half of the twentieth century (Ravnik, 2008, 39).

[3816] Such as in the pre-Alpine hills in what is now eastern Friuli-Venetia Giulia as documented in mid-nineteenth century (Kuret, 1989, 362, 367–368, 371).

the widespread interest in botany by the bourgeoisie in the late nineteenth century (Section 8.1.3.1).

After the relative decrease in tourists' interest in biodiversity in the second half of the twentieth century,[3817] the large-scale environmental change caused by the Green Revolution (Section 10.1.2.1) has fostered ecotourism again. Nowadays, "[e]cotourism is the fastest-growing market in tourism" globally.[3818] Thus, the popularity of meadow eco-tourism has been rising among the general public, largely due to the increasing disappearance of species-rich grasslands from European landscapes (Section 1.2)[3819] and thanks to local tourism development offices, national parks, and other stakeholders, recognizing the attractiveness of meadows and pastures managed in a traditional or history-inspired way. An example is the Bohinj International Wild Flower Festival, which took place for the eighteenth time in 2024,[3820] with the accompanying volume *Bohinj in Bloom: The Floral Richness of Bohinj*[3821] published in English and Slovenian by the local tourism development office. Diversity, which is characteristic of non-fertilized and moderately manured species-rich grasslands, is highly relevant in raising tourists' interest in the region because species compositions and colors of blossoming meadows differ from region to region, from farm to farm, and between different meadow parcels on the same farm,[3822] and so local and regional specifics contribute to the attractiveness of the landscape.

Contrary to the above good example from Bohinj—where closer cooperation between tourism workers and farmers in species-rich, flowering meadows is nevertheless still a desideratum[3823]—it is astonishing how little environmental understanding sometimes serves as the background in evaluating landscape attractiveness in tourism. The chapter "Tourism in the Alps—A Nature and Biodiversity Perspective," published in 2016 in the German Federal Ministry for the Environment, Nature Conservation, Building and Nuclear Safety's volume *Alpine Nature 2030: Creating [Ecological] Connectivity for Generations to Come*, and written by a member of the German Federal Nature Conservation Agency's Nature Protection and Society Division, states: "The touristic and landscape attractiveness can be assessed through an evaluation combining the degree of fragmentation, forest cover, energy of the relief (for example the difference in altitude between the highest and lowest point in a[n] individual section of the map), the existence of water courses and coastlines, as well as

[3817] Hasenöhrl & Groß, 2022, 198.
[3818] *The Future of Nature and Business*, 2020, 44, 97.
[3819] Praprotnik 1993a, 215–217, 219; Bavcon, Ravnjak, & Praprotnik, 2019, 25, 97, 99, 101, 133, 207.
[3820] *18th International Wild Flower Festival*; *19th International Wild Flower Festival: 24. May – 9. June 2024*.
[3821] Dakskobler et al., 2019.
[3822] Inspired by Pils, 1994.
[3823] Vilman, 2022.

the number of overnight stays."[3824] Such overly simplified assessments of the tourism attractiveness of a landscape, focusing only on a selection of landscape attractions while ignoring others,[3825] are counterproductive and can negatively impact the recent evolution of actual tourism landscape attractiveness in regions where species-rich grasslands are still well represented, as well as the recent evolution of the state of grassland agroecosystems. It is astonishing that such simplified views are still published in volumes of the cited type, given that (1) the environment "started to receive greater attention in tourism planning" in the mid-1960s, when the environmental impacts of large-scale tourism became widespread and obvious, (2) that environmental thought has gained so much importance in Europe ever since, (3) that Alpine countries have been committed to sustainable tourism through the Alpine Convention since the 1990s,[3826] and (4) that the role of traditional twentieth-century winter tourism has been declining at low elevations in the Alps due to climate change, posing serious challenges of restructuring the economy in many winter tourism centers in the Alps;[3827] this calls for more attention to be paid in rural development plans to other types of tourism niches, including ecotourism.

As a general rule, tourists in rural areas long for less touristy,[3828] "pristine," and scenic landscapes.[3829] Species-rich meadows contribute to the impression that the entire local economy is not driven by tourism, but at the same time these agroecosystems are appealing for tourists in the recent times of rising ecotourism. Such grasslands are important not only for aesthetic reasons but also due to their contribution to healthy landscapes providing ecosystem services that sustain agriculture, such as pollination and biological pest control without additional environmental pollution (see Sections 1.2 and 10.1.2). The related pride and reputation for all involved in such a highly ethical economy also matter.[3830] "Numerous potential synergies between biodiversity conservation, spatial planning, tourism[,] and agriculture are currently under-utilised" in the Alps.[3831]

[3824] Engels, 2016, 100.

[3825] Compare, for instance, *Traditional Hay Milk Farming . . .*, [2023], 47, 88, 99, 103, 105–106, 112, 114, 120, 122.

[3826] Compare Kupper, 2021, 172–185; Hasenöhrl & Groß, 2022, 186, 197–207 (quote from p. 197); Kupper & Wöbse, 2022, 5; see also Aschwanden et al., 2022, 234–235. The Alpine Convention was signed in 1991; Slovenia signed it in 1993 and ratified it in 1995 (Erhatič Širnik, 2011, 319). However, "not all Alpine countries have ratified all of the Alpine Convention Protocols. For example, neither Switzerland nor the European Union ratified the Protocol on Conservation of Nature and the Countryside" (Svadlenak-Gomez, 2016, 14).

[3827] Bucher, 2023.

[3828] On this general preference of twenty-first century tourists, see Hachtmann, 2007, 170.

[3829] Hasenöhrl & Groß, 2022, 185.

[3830] Compare Schmitzberger et al., 2005, 276, 279; Paracchini & Oppermann, 2012, 449; Hasenöhrl & Groß, 2022, 205.

[3831] Plassmann, 2016a, 24 (quote); see also Plassmann, 2016b, especially p. 33.

In contrast to species-rich meadows, city dwellers longing for "nature" do not appreciate walks along monotonous intensively managed, green grasslands if they are aware of the recent mass extinction occurring in grasslands in Europe. Because sustainability education is gaining importance, the monotonous sight of grass and high abundance of yellow dandelion blossoms in intensively managed meadows in spring will probably relatively soon make guests think more of fertilizers and "nitrates in groundwater than of crickets chirping and the spicy smell of a colorful summer meadow."[3832] They will likely feel increasingly uncomfortable in such a landscape.[3833] The history-rooted biodiversity of farmed grasslands disappears if intensive fertilizing and early mowing are introduced (Section 1.2). Thus, some regions in the Alps with an economy largely based on rural tourism have gradually been losing important components of their attractiveness.[3834] The example from the contemporary history of the Siusi Alp (Ital. *Alpe di Siusi* in Alto Adige, Germ. *Seiser Alm* in South Tyrol; see Sections 6.1.2.1.1.3, 6.2.1.2, and 6.5), where a botanical investigation of land-use-intensification impacts focused on aesthetic consequences, is telling. The Siusi Alp is the largest contiguous complex of actively used distant mountain meadows and mountain pastures in Europe (51.5 km^2), located at an elevation of 1,700 to 2,100 m, where a long management history has helped create colorful meadows, predominately at nutrient-poor sites. In three decades, from the 1950s to the 1980s, intensive manuring (every year with farmyard manure—i.e., more than moderate manuring at such elevations—and/or with slurry and/or mineral fertilizers) was introduced in roughly a third of the study area at the Siusi Alp. This was the main reason for the decrease in the total number of attractive summer flowers (conspicuous, well-known, and beautiful species) by about 50,000,000 to 100,000,000, including at least 6,000,000 specimens of trumpet gentian (*Gentiana acaulis*), despite the limited area of intensified manuring. On average, about 60% of all attractive summer flower species disappeared from intensively manured meadows at the Siusi Alp from the 1950s to the 1980s, and the meadow colors in intensified meadows became increasingly monotonous. In rather extreme cases, 85% of all attractive summer flower species disappeared, and in extremes of leveled, intensively manured and sown meadows, there were no longer any attractive flowers in the Siusi Alp in the 1980s. Even moderate fertilization sufficed to strongly reduce the abundance of or entirely suppress some species highly esteemed for their appearance and scent, such as black vanilla orchid (*Nigritella rhellicani*). However, contrary to intensively managed meadows, moderately fertilized grasslands have remained relatively species rich.[3835]

[3832] Inspired by Pils, 1994, in particular p. 67 (quote) and Table 49, Photo 127.

[3833] Compare Oppermann & Paracchini, 2012, 21.

[3834] Dietl & Lehmann, 2004, 9, 13, 24–25, 37–38, 42–43, 53–56, 67, 69, 94, 99–100, 102–103, 105–111, 118, 121–125; Keenleyside & Oppermann, 2009, 296, 303; Peterken, 2013, 285, 392; Bavcon, Ravnjak, & Praprotnik, 2019, 25, 43.

[3835] Grabherr, Kusstatscher, & Mair, 1985, 271–274, 276–285, 288–289.

Thus, in those regions of the Alps where significant areas of farmed species-rich grasslands still exist, it makes sense for tourism development to include high-nature-value grassland farming. This matters because Western Europe, including most of the Alps, is "the most important hotspot of world tourism," Alpine countries "today host about 25 per cent of the world's tourism,"[3836] biodiversity is significantly better preserved in the Alps than in most of the rest of Western Europe but nonetheless in a serious decline, and species-rich meadows are by far the most species-rich and threatened habitat types in Central Europe (Section 1.2). This creates a particular type of tourism niche, with a potential to foster sustainable development in this mountain chain in a partly history-inspired manner, but motivated largely by novel high standards of environmental ethics. The attractiveness of species-rich meadows is not likely to decline in the broader context of strong homogenization of most farmed landscapes in Central, Northern, and Western Europe, continuing mass extinctions, and rising ecological awareness, and so this niche is reliable and relevant in the long run. However, preserving and making use of this highly ethical niche requires more action and closer trans-sectoral cooperation (Sections 1.2, 10.1.2, 10.2, 10.3, and 11): if the tourism economy wants species-rich grasslands, most farmers expect tourism to provide significant financial support for the farm economy to allow such grassland management in the long run[3837] unless another type of appropriate agri-environmental scheme is available.

In this manner, the tourism market is a part of the general "failure of markets to recognize and remunerate farmers for the ecosystem services that [high-nature-value] farmlands deliver to society."[3838] Namely, "farm tourism has a relatively modest importance in the context of overall tourism in each country";[3839] thus, in many regions of the Alps where vacation farms are not the most frequent accommodation type and food provider for tourists, farmers are not the ones to benefit most from the rising biodiversity-related tourism.

Recently, "Slovenia has managed to create an image of a sustainable tourist destination at the international level," it "ranks 8th out of 140 countries for which this data is available" in line with the World Economic Forum's assessment.[3840] The content of Section 10 as a whole emphasizes that well focused and significantly closer economic cooperation between high-nature-value farming (thus, not any kind of farming) and tourism is needed to preserve this status in the following decades—a topic that has in general received a "rather limited scope of efforts" in Slovenia.[3841]

3836 Hasenöhrl & Groß, 2022, 186 (first quote), 201 (second quote).
3837 Author's interview with an anonymized farmer's representative in the Julian Alps, 2022; see also Mlakar, 2018 and Mlakar, 2022.
3838 Lomba et al., 2019, 37 (quote), 40.
3839 Cigale & Lampič, 2023, 252.
3840 Cigale & Lampič, 2023, 253–254 (quote from p. 253).
3841 Compare Cigale & Lampič, 2023, 254.

An insufficient level of such trans-sectoral cooperation is deleterious not only for biodiversity, but also for tourism. Namely, per definition, mass tourism—which is characteristic of many regions in the Alps and even nowadays often follows after an eventual initial stage of ecotourism—thoroughly transforms local economies and landscapes, which threatens tourism's "very foundations through its environmental impacts" unless well-considered tourism and local economic development prevents this. In many parts of the Alps, tourism has changed the landscapes thoroughly,[3842] resulting, among other aspects, in the loss of biodiversity and traditional ecological knowledge, largely because mass tourism fostered large-scale abandonment of high-nature-value farming.[3843]

Thus, the intersection of the contemporary histories of farming and tourism emphasizes that both modern tourism and farming development can destroy or, under certain conditions, maintain (or foster the maintenance of) grassland biodiversity. Hence, it is a failure to apply the same tourism marketing strategies in the regions of the Alps still abounding in species-rich grasslands as in countries where species-rich grasslands are almost gone (Section 1.2). This happens by promoting such regions in the Alps as "green" destinations.[3844] Rural areas in Europe where species-rich grasslands are almost non-existent (Section 1.2) cannot advertise themselves in another appealing color than monotonous green, whereas regions with a significant percentage of species-rich grasslands in the Alps should not be promoted as "green" destinations. "Green" is not always a synonym for environmentally friendly, and in flowering grasslands this is not the case at all. Green is also the color of the Green Revolution and its highly problematic environmental impacts (Section 10.1.2.1), and so advertising a region as a "green" destination blurs any distinction between a depressing landscape of highly intensively managed green grassy plains, characterized by low levels of biodiversity on the one hand, and a landscape where there is also a significant percentage of species-rich agroecosystems in the total agricultural land, with their sounds, smells, and the fantastic feeling of being in a landscape full of life, on the other. Through this kind of sensory experience, species-rich grasslands can foster modern rural tourism, where the attraction involves much more than rural nostalgia.[3845] Instead of being "green," landscapes abounding in species-rich grasslands should be advertised as "COLOR-

---

[3842] Hasenöhrl & Groß, 2022, especially pp. 185–186, 189, 192, 199–205 (quote from p. 185).

[3843] Svadlenak-Gomez & Badura, 2016, 89. Numerous examples can be observed during visits to sites of mass tourism in the Alps. For one of the many examples dealt with the literature, see Meilleur, 1986.

[3844] For instance, *Slovenia Green.*

[3845] Due to the sensory experience presented and the educated visitors' ethical wish to support high-nature-value agroecosystems in the contexts of the recent environmental crisis and the rise in environmental awareness, it is not possible to agree with the simplistic notion that modern agro-tourism is motivated simply by rural nostalgia, which is claimed by Hasenöhrl & Groß, 2022, 204.

FUL" or "GREEN AND COLORFUL" biodiverse destinations to link the most recent trends in global tourism development (see below) with attempts to foster sustainability thinking and implementation in the local economy.

Gaining more economic profit from existing species-rich grasslands in the farm economy can make use of the fact that tourists are also attracted by food and beverages, and other products based on species-rich grasslands, without additional environmental loading by more intensive agriculture (Sections 1.2, 10.1.2.1, 10.1.2.2, 10.2.1, and 10.2.3). High-nature-value farming—and not any kind of farming—contributes to the preservation of the attractive cultural landscape, and short food-supply chains avoid transport-related emissions.[3846] Such product selections can include local dairy, meat products, and honey based on high-nature-value grasslands as a source of fodder and bee forage, food from gathered wild greens, tisanes and other products based on ingredients gathered in high-nature-value grasslands, handicrafts from wool of sheep raised with fodder from species-rich grasslands, fruit products from species-rich meadow orchards, and further products, such as aromatic hay baths or sleeping in aromatic hay from species-rich meadows.[3847]

As a relevant and highly ethical contribution to the general raising of public awareness of the role of high-nature-value grasslands in the landscape, some innovative farms can offer visitors guided tours around their history-rooted species-rich grasslands, combining historical (traditional) local, modern agricultural, and to a certain degree botanical and ecological knowledge, which is currently rarely practiced. Even if not trained in biology, farmers can usually recognize many plant species.[3848] During such tours, more farmers could teach visitors about the dependence of plants on specific grassland management regimes,[3849] spring and summer tours could include gathering some herbs edible or usable as tisanes at that stage of development, and, depending on the interest of the participants, farmers could, for instance, present historical grassland management and its environmental impacts (local or literature-based), present further historical plant uses, or focus on livestock's attitude toward the plants observed or their relevance for wild fauna. It will be important for farmers to include some locally specific information to increase the attractiveness of such tours. A further niche contributing to all the aspects mentioned above is adding scientific botanical, ecological, or environmental history knowledge

[3846] Compare Section 1.2 and Potočnik Slavič et al., 2016, 62, 70–71, 75–77, 80, 84–85, 108–110, 113, 124, 126, 129; Keenleyside & Oppermann, 2009, 295–302; Cigale & Lampič, 2023, 252, 257, 260–261.

[3847] Keller, 2007, 212–213; Łuczaj et al., 2012; ZRSVN, *Annex C.7 – Detailed Description . . .*, [2020].

[3848] Schmitzberger et al., 2005, 279–280; Keenleyside & Oppermann, 2009, 295–297, 300–301; Suske, Gantner, & Gattermaier, 2012, 121; *Traditional Hay Milk Farming . . .*, [2023], 42.

[3849] Babai & Molnár, 2016, 162.

(including explanations of interactions between plants and animals, impacts of historical grassland management on grassland ecology as given in this book, etc.) to such tours guided by farmers, but only a limited number of farmers can act as such guides. Namely, additional ecological education takes time, and personal motivation is needed.[3850] Organized ecological and environmental history courses, guided tours, and lectures for farmers willing to carry out such tours themselves can be helpful: "Depending on the personality and background of the farmer, [she or] he may be more or less eager to acquire new knowledge of plant species or differences in biodiversity. Some farmers are rather indifferent . . . while others embrace the biodiversity initiative" (Section 10.1.2.1). It will make most sense to offer such courses for farmers—as guides in a transdisciplinary manner, so that ecologists or environmental historians will also learn from the farmers' traditional ecological knowledge.[3851]

This section focused on the relevance of species-rich seminatural grasslands with a rich historical tradition for tourism. Apart from species-rich meadows themselves, and products based on them, meadow history offers some further tourist attractions at some places. An example is the exciting routes of some former grassland irrigation systems and some maintained or restored and touristically adapted old main grassland irrigation channels in the Alps, which are only fragments of the entire historical systems. These spectacular constructions (Figure 94 in Section 6.4) are important elements of tangible cultural heritage, preserved or restored for their recognized attractiveness; the great majority of them no longer serve traditional irrigation. Nonetheless, such preserved open channels in some places provide water for modern sprinkling irrigation instead of construction of pipes.[3852]

In special cases, the history of grasslands in the Alps enables tourism developers to foster education on general human ethical principles, which exceed environmental ethics, and promote peace and mutual respect based on grassland plants (see Section 10.3).

10.2.3 History-inspired gathering in grasslands

Section 8 presents many examples of historical gathering to obtain food, livestock fodder (see also Section 10.1.2.2.7), medicines, and technical raw materials. Several historical plant uses can still be relevant today, but there is a need to carefully check the appropriateness of historical uses, in particular in the

[3850] Adapted from Keenleyside & Oppermann, 2009, 297, 301; Babai & Molnar, 2014.

[3851] von Glasenapp & Thornton, 2011 (quote from p. 777); Birrer & Oppermann, 2012, 485–488; Peterken, 2013, 371–373, 375, 385–391.

[3852] Leibundgut & Vonderstrass, 2016a, 47–48, 92, 158–159, 211–212, 225–233; Leibundgut & Vonderstrass, 2016b, 96–97, 101, 108–110, 120–121, 138, 147, 151, 155–158, 160, 183–185.

case of plant-based medicines. In the field of food ingredients, gathering in grasslands can nowadays make small contributions to a healthy and environmentally friendly diet. Gathered wild food usually contains more vitamins C and E, and more polyunsaturated fatty acids than similar farmed crops, which speaks in favor of some gathered food.[3853] The taste of wild greens can be less constant than the taste of cultivars.[3854] Apart from the gatherers' (either locals' or visitors') household demand, gathered grassland herbs and fruits or seeds are attractive food ingredients at some local vacation farms and restaurants.[3855] A few species, such as dandelion (*Taraxacum* spp.), are relevant as marketable wild greens, which are sold, for instance, at markets in towns and cities.[3856]

Any type of gathering has to pay attention to the question of its sustainability. Historically, especially important food and livestock fodder ingredients were usually gathered in large quantities, regularly, and long-term, and so most of the wild greens historically gathered for these purposes regenerate well. Attention has to be paid to the impacts of environmental changes on plants, especially the impacts of changes in land use and land-use intensity. Thus, particular attention has to be paid to the question of sustainability of gathering species abounding only in non-fertilized grasslands; if a certain organism was gathered on a large scale in such grasslands when they prevailed in the landscape (Section 6.3), one cannot just assume that the same species can be gathered on a large scale nowadays, when non-fertilized grasslands are increasingly rare. Climate change and pollution also matter (see Section 10.1.2.2.4). Thus, historical information is a useful inspiration for gathering, but it deserves careful observations of gathering impacts to ensure that the practice is sustainable.

More challenging than in most historical food plants presented in this volume are sustainability issues in the case of gathering, especially uprooting, some healing plants. However, concerns of conventional nature conservation were sometimes right, whereas in some other cases a gathering prohibition in conventional nature conservation was either unnecessarily rigid, or, exceptionally, even counterproductive, as found by recent reconsideration of protection measures referring to a couple of plant species. Consequently, based on findings of history-inspired experiments, innovative nature conservation is being developed, considering specifics at the species level. The evolution of such innovative nature conservation involves testing traditional ecological knowledge, with long-term monitoring of gathered populations that also fo-

---

3853 E.g., Abbet et al., 2014, 624.

3854 Compare Sevindik et al., 2017.

3855 Christanell et al., 2010, 71; Schunko & Vogl, 2010, 2; Łuczaj et al., 2012, 364–367; *15th International Wild Flower Festival: 28 May – 13 June 2021*: Floral Cuisine.

3856 For instance, gathered dandelions are widely available at the organic market in Ljubljana.

cuses on impacts of recent environmental changes on the populations. In the case of certain species, such a nature conservation concept can allow carefully designed, moderate, informed gathering. An extreme example of what counterproductive conventional nature conservation ignoring traditional ecological knowledge can be like is spikenard valerian (*Valeriana celtica* subsp. *norica*), whose populations declined significantly due to a general gathering prohibition (see Section 8.1.1).[3857] It came as a surprise for nature conservation in the 1990s that moderate uprooting, based on traditional ecological knowledge, does not reduce the abundance of spikenard valerian, but increases it in the long run.[3858] Long-term monitoring of moderately gathered populations has so far confirmed these results.[3859] Building on a rich historical tradition several centuries long (Section 8.1.1), two families of mountain farmers at Saureggen in the Nock Mountains (Germ. *Nockberge*) again obtained the right to gather *Valeriana celtica* subsp. *norica* in a well-informed, sustainable manner in 1997, and they have been gathering the species ever since. On 80 hectares of a clearly designated, ungrazed area, both families together are allowed to gather an amount of spikenard valerian allegedly resulting in about 54 kg of dried marketable roots annually, between 15 August and 8 September. The license was first issued for two years, then for five, and later for ten. At the same time, long-term monitoring of spikenard valerian populations at the gathering sites has been carried out. The species has profited from moderate gathering and has so far also proved resistant to climate change and nutrient intake through airborne pollution. The gathered roots are sold to the enterprise Speick Natural Cosmetics in Leinfelden-Echterdingen close to Stuttgart, which has a long tradition of fine cosmetic products containing *Valeriana celtica*. Further tourism products build on the spikenard valerian story in the region, such as spikenard valerian tours and spikenard valerian massages using products from Speick Natural Cosmetics, including a bit of locally gathered *Valeriana celtica*.[3860] Apart from the reconsideration of nature conservation, this example also shows that fluctuating concentrations of substances in plants during the year, from year to year, and between populations (see the introduction to Section 8) can be coped with even in industry if there is goodwill, at least in cases of certain species.[3861]

Similar was a contribution of tested traditional ecological knowledge in the permit for sustainable gentian gathering at Galtür, Tyrol (see also Section 8.1.2). Knowledge inspired by historical experience about dotted-flowered

---

[3857] Gentner, 1932, 69; Wendelberger, 1969, 113; Novak, 1999/2000, 110–114; Kabusch, 2020, 44–49, 56, 59, 64–65, 71.

[3858] Novak, 1999/2000, 109–114.

[3859] Kabusch, 2020, 46–47, 64–65, 71.

[3860] Kabusch, 2020, 4, 35, 38–41, 46–47, 56, 59–61, 63–64, 67–72, 75–76; Hilpert, 2014, 8, 12–13; *SPEICK Natural Cosmetics: An Exuberant Success Story.*

[3861] Kabusch, 2020, 47, 71.

gentian (*Gentiana punctata*) sites, gathering, and rhizome brandy production in Galtür has been recognized and listed on the Austrian Commission for UNESCO's list of intangible cultural heritage since 2013. Gentian rhizome gathering there passed through a difficult phase in the late twentieth and early twenty-first century because this activity was prohibited by nature conservation in the early 1990s. Following several experiments demonstrating that the stands of dotted-flowered gentian have not declined due to the current local gathering practice, gathering was legalized again in 2006, on the condition that the permission is approved every five years, and that gathered populations are monitored every five years. Gentian rhizomes are currently gathered there in a sustainable manner, allowing long-term restoration of populations, even though dotted-flowered gentian grows even more slowly than yellow gentian (*Gentiana lutea*). *Gentiana punctata*'s rhizomes are smaller than those of *Gentiana lutea*; however, the concentrations of the bitter substances, most important from the medicinal viewpoint, are higher in *Gentiana punctata*'s rhizomes than in *Gentiana lutea*'s. The present sustainable dotted-flowered gentian gathering regime in Galtür includes sortition-based selection of up to thirteen families from the entire community entitled to gather gentian in a particular year. They gather up to 1,300 kg of gentian rhizomes (100 kg per family) annually at specified sites characterized by sufficient abundance in *Gentiana punctata*, at elevations of 2,000 m and above. This sustainable gathering is only allowed to take place in autumn; it is carried out in October.[3862] See also general considerations on challenges for sustainable gentian rhizome gathering in Section 8.1.2.1.

10.3 Grassland plants as mementos of wars

Plants of polemochorous origin (see Section 7.2) often occur in grassland environments far from sites where these species grow naturally. Furthermore, some plants of polemochorous origin are aesthetically attractive. Due to both of these facts, species of polemochorous origin are sometimes merely considered botanical peculiarities. Consequently, a recent foreign visitor's report designated yellow oxeye (*Telekia speciosa*) in Bohinj (see Section 7.2) only as one of the "more exciting finds," together with a couple of other botanical attractions in the area.[3863]

A historical perspective can add a crucial ethical component: the occurrence of a polemochorous species offers an underexploited potential for people to learn about the traces of wars in the flora of grasslands they visit. Every

[3862] Franz & Fritz, 1978, 309–310; Juen senior, 2014, 56–57; *Austrian Commission for UNESCO: Immaterielles Kulturerbe: Bräuche, Wissen, Handwerkstechniken: Wissen um die Standorte, das Ernten und das Verarbeiten des Punktierten Enzians* and *Bewerbungsformular*, 2013, on the same website.

[3863] An example of such a report: Harmes & Tout, 2015, 2.

time they observe polemochorous plants, awareness of polemochorous plant origins and of the great suffering that resulted in this introduction (Section 7.2) can stimulate people that know about this manner of grassland species introduction to recall war-related casualties; the daily distress and traumatic experiences of soldiers, civilians, and animals, including long-term psychological and physical consequences; and the decay of values in wartime, and also to think of the environmental consequences of war and the historically experienced psychosis influencing decisions detrimental to humankind[3864] and against the logic of environmental issues in wartime. About eight to ten million soldiers and more than six million civilians (a rough estimate: seven to ten million) died in the First World War; in addition, millions went missing and tens of millions (an estimate: twenty-one million soldiers alone) were wounded. The First World War was one of "the deadliest conflicts in human history." Hundreds of thousands of soldiers of several nationalities and many civilians died along the Isonzo Front alone;[3865] limiting attention to half of the particularly difficult year 1917, tens of thousands of horses died there simply of malnutrition or were slaughtered out of necessity due to malnutrition (Section 7.2). The total number of horses that died in the First World War as a whole is roughly estimated at up to eight million,[3866] making this war the most ferocious slaughterhouse of horses the world had ever seen. The sheer mass of the flesh and bones of the horses that died in the First World War was a multiple of that of the human casualties.[3867]

[3864] Whereas interpretation of polemochorous plants as war mementos is original, the main ethical issues of the First World War were adapted from histories of the Isonzo Front, of life in Slovenian territory during the First World War in general, of Slovenians on distant battlefields of that war, and of Slovenian refugees abroad (Svojšak & Antoličič, 2018; Komac, 2022; Saunders et al., 2013, 49–52, 56, 59–63; Mlekuž, Košir, & Črešnar, 2016; Košir, 2019, 184–186, 188–189; Košir et al., 2019, especially pp. 91–93, 96–98, 106; Štepec, 2019, 214–217, 219–220, 223, 227), from general studies of memories of the First World War (Grdina & Luthar, 2010, 14–17; Rauchensteiner, 2010, 58, 63; Verginella, 2010, 89–97; Jezernik, 2021), from research on food shortages during this war (Godina Golija, 2011, 85–96; Stergar, 2019), from studies on First World War casualties (Svoljšak, 1999; Blaj Hribar et al., 2019), from publications on destroyed and severely damaged settlements along the Isonzo Front, on refugees from the Soča River catchment and the postwar return of many of them home: their buildings were very often demolished, damaged, and/or plundered, and they often found their agricultural lands and forests devastated or damaged—an additional serious threat was unexploded ammunition (*Slovenski zgodovinski atlas*, 2011, 158; Svoljšak, 2019)—and from general considerations on the optimal goals of teaching the general public about wars in history (Vodopivec, 1991, 120), studies of First World War heritage and its presentation, and commemorative events (Testen & Koren, 2015; Bajuk Senčar, 2021; Fikfak & Jezernik, 2021; Hazler, 2021; Kravanja, 2021; Repič, 2021).

[3865] Svoljšak, 1999; Keller, 2018, 3 (quote); Blaj Hribar et al., 2019; Fikfak & Jezernik, 2021, 11, 31; Jezernik, 2021, 139–140; Hazler, 2021, especially pp. 52 and 63; *The Isonzo Front: Register of the Killed Soldiers.*

[3866] Koncilija, 2009, 80–81.

[3867] Lavrenčič, 1998, 31; see also Koncilija, 2009, 80–81.

The interpretation of grassland species as living mementos of wars is not possible everywhere, but, where easily recognizable plants of polemochorous origin occur, the addition of such an interpretation will be ethically extremely important. Namely, people tend to forget about more distressing memories and to retain pleasant memories.[3868] Even though the main message of wars in late-modern and contemporary history has in general been taken seriously as "[t]he Alps have perhaps never been as peaceful at a transnational and transregional level as they are today,"[3869] the disappearance of knowledge of the historical suffering experienced during wars and later on due to wars' consequences is worrying: ignorance makes it easier for such suffering to occur again. Due to the lack of knowledge, the image of a historical war can even degenerate from a major catastrophe to a spectacular event.[3870] Spreading information about the complexity of daily life during historical wars greatly contributes to one of the main goals of history education: namely, making possible comparative critical thought based on knowledge and in-depth understanding of past processes, structures, and events. The goal of history education is not to generate "loyal patriots and uncritical members of a nation or a state." Even though history education must be ideology-free, it must warn against intolerance, aggression, and exclusivity.[3871] There are several highly relevant media for appropriate history education for the general public, such as school classes, public history (e.g., cosmopolitan and inclusive museums, open-air museums, and documentary films), historiography, published war diaries and memoirs, cosmopolitan and inclusive commemorative events, countless landscape monuments (e.g., cemeteries, ossuaries, chapels, and other preserved war-related buildings, and evident geomorphic and other archaeological legacies of the war), and other works of art (e.g., war *belles-lettres*, paintings, sculptures, music, etc.),[3872] but there is an unexploited potential for polemochorous plants to add their piece to this puzzle.

Observations of naturalized polemochorous plants might be a minor contribution in comparison to the aforementioned media of war history education, but their advantage in raising the awareness of the senselessness of wars based on historical knowledge is that one finds polemochorous plants in grasslands, places not usually visited with the purpose of learning history. Conse-

[3868] See general considerations on relevance of war remembrance in Verginella, 2010, 88; see also Drole, 2019, Košir, 2019, and Štepec, 2019.

[3869] Mathieu, 2019, 157.

[3870] Rauchensteiner, 2010, 58; see also Elias, 1993, 263–283; Jerše, 2017, 247; Košir et al., 2019, 104–106.

[3871] Vodopivec, 1991, 119–121 (quote from p. 121).

[3872] Rauchensteiner, 2010, 56, 62, 68–70; Saunders et al., 2013, 52–56, 61–62; Mlekuž, Košir, & Črešnar, 2016, 143; Svoljšak & Antoličič, 2018, 9–11, 31, 216, 252, 314–315, 379–384; Košir, 2019, 186–195; Košir et al., 2019, 93–94, 98, 100–101, 104–106; Drole, 2019, 200–209, 212; Kozorog, 2019, 69, 72; Štepec, 2019, 217–218, 220–222, 224–226, 228–231; Bajuk Senčar, 2021; Fikfak & Jezernik, 2021; Hazler, 2021; Kravanja, 2021; Jezernik, 2021; Repič, 2021, 187–191; Komac, 2022, 179–181, 260–263.

quently, these plants can be an unexpected medium for people to keep remembering what they have learned about ethical aspects of the historical war, or polemochorous plants can become a stimulus to strengthen their knowledge about this. To attain this goal, plants of polemochorous origin need to be presented to the general public (yellow oxeye in Bohinj, for instance, is easy to notice and identify; Section 7.2), and this information should be linked with a stimulus to think about the ethical components of wars in history in a way that polemochorous plants become considered living mementos of great suffering by living organisms and of great disruptions to the environment caused by historical mistakes by diverse actors, such as politicians, military commanders, diplomats, other members of the state apparatus, other public figures, and crowds of people.[3873] Even though visitor information must be linked to locally occurring polemochorous plants, the information should not be limited to local war history; instead, it has to link the local, regional, and global war history context in a well-balanced way.[3874]

Various information for visitors at sites where plants of polemochorous origin occur could become useful in this respect. For instance, there are many yellow oxeye plants at the start of the trail leading to Savica Falls in Ukanc—a popular tourist attraction and a very busy place in the First World War (Section 7.2)—and so the starting point of this path can be considered an ideal place to add this information in the future. The location is already included in the Bohinj section of the Walk of Peace trail, linking the Julian Alps (Sln. *Julijske Alpe*, Ital. *Alpi Giulie*) and the Adriatic, with the main emphasis on presentation of the Isonzo Front heritage in the sense of evoking the broader message of peace and mutual respect.[3875] Additional yellow oxeye sites can be added to the Walk of Peace trail, such as the abandoned wet meadow where it abounds, located on a trail to the beach of the Savica delta on Lake Bohinj (Sln. *Bohinjsko jezero*; in Figure 119, Section 7.2). Through the inclusion of yellow oxeye, the existing endeavor of the Walk of Peace trail with a strong emphasis on cultural heritage, but linking cultural and natural heritage,[3876] could be enriched by another element in the contact zone between natural and cultural heritage.

Nonetheless, the history of polemochorous species is as complex as history usually is. It was learned in Section 7.2 that yellow oxeye is not of polemochorous origin in all parts of the Alps where it occurs. This also has to be clearly explained to the general public.

[3873] General considerations on those responsible for the First World War are summarized from Svoljšak & Antoličič, 2018; Jezernik, 2021; Cergol Paradiž, 2022, 94; and Komac, 2022, especially pp. 42, 70–71, 176.

[3874] Bajuk Senčar, 2021; Kravanja, 2021.

[3875] On the Walk of Peace trail, see Testen & Koren, 2015; *The Walk of Peace, Bohinj*, 2015; Kravanja, 2021; Fikfak & Jezernik, 2021; *Walk of Peace—Map* (many thanks to the historian Petra Svoljšak for this reference).

[3876] *Walk of Peace—Map*.

11 Conclusions

Publications on long-term environmental history and historical ecology have so far paid unequal attention to various topics about farmed grasslands in the Alps—a land-use category that has played a major role in this mountain range. This volume is a result of interdisciplinary cooperation between a historian and geographer, and a botanist. Thus, the topics presented include (1) the history of biodiversity of meadow agroecosystems, the economic relevance of this biodiversity in terms of farming and gathering, and mutual interactions between grassland management and biodiversity, (2) environmental changes in grassland environments and their relevance in the farm economy, and (3) long-term impacts of exceptional historical events on grassland species composition. The topics dealt with in this volume are significantly different from many interdisciplinary publications on steppe history due to major differences between zonal steppes and most grasslands in the Alps in terms of site conditions and ecology, grassland management, and uses (Sections 1.1, 1.3, 2, 4, 6, and 8).[3877]

The strong emphasis on links between the biology of plants in species-rich meadows and grassland history distinguishes this volume from most long-term agricultural environmental history volumes primarily focused on European mountains in the second millennium and written in the traditional historical manner, by a historian as the single author. Examples from recent historiography are John Robert McNeill's *The Mountains of the Mediterranean World: An Environmental History*[3878] and presentation of agricultural environmental history topics in Jon Mathieu's *The Alps: An Environmental History*.[3879] These works do establish some links between history and grassland environments, but not as many as their publication titles might suggest.

Similarly, several other disciplines, such as anthropology, ethnology, ethnoecology and ethnobotany, botany, geography, and paleoecological studies, contributed to the research topic, although their contributions to long-term insights into the second millennium are limited. Some volumes by botanists, for instance, include historical data relevant for explaining plant occurrence and

[3877] Compare Isenberg, 2017.
[3878] McNeill, 2002; see also Butzer, 1996.
[3879] Mathieu, 2019.

abundance, and present traditional uses of plant sociological units, but they usually do not provide much information from archival records. Historical information in botanical publications is mostly limited to the twentieth century as recalled by interviewees or observed by authors, or to research on historical sources (also older archival ones) created by earlier botanists, whereas other types of historical data with the potential to explain the historical context are usually not studied in botanical publications.[3880] Examples from the Eastern Alps are Gerhard Pils's *Die Wiesen Oberösterreichs: Eine Naturgeschichte des oberösterreichischen Grünlandes unter besonderer Berücksichtigung von Naturschutzaspekten* (The Meadows of Upper Austria: A Natural History of Upper Austrian Grasslands with a Special Emphasis on Nature Conservation Aspects)[3881] and Nada Praprotnik's *Botaniki, njihovo delo in herbarijske zbirke praprotnic in semenk v Prirodoslovnem muzeju Slovenije* (The Botanists, Their Work, and Herbarium Collections of Vascular Plants in the Slovenian Museum of Natural History).[3882] Ethnobotanical research in the Alps contributed fundamental results, such as Brien Adrien Meilleur's *Alluetain Ethnoecology and Traditional Economy: The Procurement and Production of Plant Resources in the Northern French Alps*[3883] and *Lokales Erfahrungswissen über Pflanzenarten aus Wildsammlung mit Verwendung in der Fütterung und als Hausmittel in der Volksheilkunde bei landwirtschaftlichen Nutztieren in Osttirol* (Local Experience-Based Knowledge about Gathered Wild Plant Species Used as Fodder and as Home Remedies in Folk Livestock Medicine in East Tyrol) by Brigitte Vogl-Lukasser et al.[3884] Nonetheless, their scope and their main chronological and spatial focus are narrower than those of this volume.

Bridging "the divide between the humanities and the natural sciences" by contributing to the understanding of the complexities of the real world[3885] through links between long-term history and grassland ecology in the Alps is not a novelty.[3886] Such studies, however, focus on more specific topics, and so they also only cover part of the contents of this volume. Tobias Scheidegger's *Mythos Edelweiss: zur Kulturgeschichte eines alpinen Symbols* (The Edelweiss Myth: a Contribution to the Cultural History of an Alpine Symbol)[3887] is a complex historiography publication, dealing with an exceptional species from species-rich grasslands in the Alps. Agricultural plant scientist Dirk Kauter's volume *"Sauergras" und "Wegbreit"? Die Entwicklung der Wiesen in Mitteleuropa zwischen 1500 und 1900* ("Sour Grass" and "Plantain"? The Development of

[3880] On historical botanical research, see also Popplow, 2017, 201–202.
[3881] Pils, 1994.
[3882] Praprotnik, 2015.
[3883] Meilleur, 1986.
[3884] Vogl-Lukasser et al., 2006.
[3885] Compare Izdebski, 2022, 12–18 (quote from p. 13).
[3886] An example of an article written by a historian and a botanist: Stuber & Lienhard, 2007.
[3887] Scheidegger, 2008.

Meadows in Central Europe between 1500 and 1900)[3888] closely links long-term agricultural history and grassland ecology in part of the Alps. Kauter studied many historical agricultural and botanical publications and a few early agronomic and botanical manuscripts and herbaria. However, that volume uses a narrower selection of types of historical sources than this one, and it focuses on a narrower selection of grassland-history topics within a slightly shorter timeframe than this volume. Concerning the Alps, Kauter deals with a region in the northwest of the mountain range that saw a specific course of agricultural history, in several aspects not representative of most of the Eastern Alps dealt with in this volume (see Sections 3.1 and 6.3).

This volume focuses on a variety of types of archival records since the Late Middle Ages, a selection of historical publications, pictures, maps, oral history interviews, historical artefacts, and fieldwork (Section 1.5), presented in an interdisciplinary and somewhat transdisciplinary context, embedded in multidisciplinary literature. Due to the limited attention that grassland history has received in Central Europe, it was possible to include findings not only from historical records not studied so far but also from types of archival sources that have been fully or largely neglected in grassland environmental history research to date, such as court minutes (especially Sections 3.2.2, 4.1.1.3, 4.2, 6.1.2.1, 6.1.2.3, 6.1.2.5, 6.2.1.2, 6.2.2.1, 6.2.2.2, 6.3.2.1, 6.4.1, 6.4.3.1, 6.4.3.2, 6.4.5, 6.6.1, 6.6.2, 10.1.1, and 10.2.2), Sunday Christ paintings (Sections 4.1.1.4, 6.1.2.1.2, and 6.4.3.1), data on tithe in beehives (Section 4.1.1.1.1), and archives of military commands (Section 7.2), which all proved rewarding. Interdisciplinary historical and ecological interpretations of records from the First World War, when major nineteenth-century transformations of agriculture were still in progress and the crisis led to the restoration of some past practices of coping with a lack of resources (supported by expert reports ordered by political authorities), proved relevant to explain some earlier historical practices that are otherwise difficult to understand. One example is early spring mowing in wetlands (Section 6.1.2.5.1.1).

Typically for the agrarian sociometabolic regime and the first phase of the industrial sociometabolic regime (see the introduction to Section 1) worldwide, farmers in the Alps "altered their environments to suit their preferences within the limits of their technologies and populations."[3889] The results were species-rich agroecosystems. Medieval highland colonization, subsequent settlement growth, and the wider socioeconomic history context resulted in the vast expansion and long-term maintenance of species-rich seminatural grasslands in forest clearings in the Alps. The agricultural density in the Alps from the Late Middle Ages onward was high compared to many other mountain

[3888] Kauter, 2002.

[3889] Compare the general presentations in Krausmann, Weisz, & Eisenmenger, 2016, 67, 70–79; McNeill, 2010, 361–362 (quote from p. 361).

ranges in Europe, and it increased over the long run until the nineteenth century, as a general trend. Consequently, the area of seminatural grasslands in the Alps was particularly extensive in the long run. By far, most hay was mowed in meadows in forest clearings in montane vegetation belt and somewhat less in colline and subalpine vegetation belts. Seminatural meadows are the most species-rich habitat types in Central Europe. Many species reach their highest abundance precisely in seminatural grasslands in forest clearings. Between the Late Middle Ages and the nineteenth century, non-fertilized meadows prevailed; moderately fertilized meadows were also found in typical landscapes of the Alps, and some regions had many irrigated grasslands. The thorough high and late medieval increase in species-rich meadow area that also continued at a lesser pace during the following centuries, together with other changes in the cultural landscape of low-energy-input preindustrial farming, also ensured high connectivity between species-rich seminatural grassland agroecosystems. This matters because the mobility of some organisms living there, such as specialized butterflies, is limited (Sections 1, 2, 3, 6, 8.1.2.1, and 10.1).

The volume provides insights into historical seminatural grasslands as dynamic "socio-natural sites."[3890] It also discusses a certain degree of Alpine farmers' ingenuity or adaptability[3891] in the grassland-related aspects of making a living based on subsistence and market-oriented integrated farm economies in patchwork landscapes, where grassland agroecosystems had a multifunctional role in people's daily lives. Seminatural grasslands and wetlands provided several types of ecosystem services. Most obviously, they provided livestock fodder. In some wetlands, making animal bedding was also significant. Farmyard manure based largely on grassland fodder and animal bedding (the latter in some places from wet meadows) crucially contributed to the long-term maintenance of soil fertility on arable land. Grassland fallow decreased the abundance of some plants considered weeds on arable land. In addition, seminatural grasslands were often sites of foraging that was motivated by a wide range of goals: gathering food, livestock fodder, medicinal herbs and other tisanes, raw materials, and flowers for decoration are focused on in this volume. Moreover, species-rich grasslands, characterized by late but diverse mowing dates, provided bee pasture and habitats for wild insects, including more efficient pollinators than farmed bees, and birds. These insects were highly relevant in the pollination of some farm crops, under specific weather conditions in particular, and some insects and birds contributed to the biological control of organisms considered pests on arable land and in orchards (Sections 1, 3, 4, 6, 8, and 10.1). Farming in the Alps has a long history of being economically connected with numerous places in Central Europe and beyond. Thus, long-term general economic and environmental history trends, together with regional and local spe-

[3890] On this concept, see Winiwarter et al., 2013, 105–108.
[3891] Compare Isenberg, 2017, 135, 137.

cifics, influenced meadows and pastures in the Alps. Grasslands in the Alps also had specific impacts on wider economic history (Sections 3, 5, and 8).

The high diversity of grassland site conditions in the Alpine macroregion is a consequence of a complex interplay between diverse natural environmental conditions within the mountain range and management. To a certain degree, differences in species compositions among grasslands in the Alps are related to the diversity of site conditions. Dissimilar sites and species compositions not only between vegetation belts but also within them resulted in major differences in yields and fodder quality, and in the time and energy that had to be invested in mowing by scythe and in hay drying. Most of these differences were notable between different non-fertilized grasslands. Moreover, weather could have a significant impact on the quality and quantity of hay from the same meadow from year to year. Not only the self-evident weather during vigorous plant growth and haymaking mattered. Extreme cold in early spring, in interaction with further aspects of environmental dynamics, not only had a significant impact on the composition of plant species in hay and grazing resources, but all this often also resulted in significantly reduced hay quantity and quality. The occasional abundance (or adequate development phase) of some animals or fungi that came into serious conflict with grassland farming could also affect hay quantity on a yearly basis. The internal dynamics of grassland agroecosystems, which function in a chaotic but not random manner, sometimes added to the diversity in fodder quality and quantity; however, the impacts of these kinds of environmental dynamics were often limited due to the high number of species present (Sections 2, 4, 6, and 10.1).

Focusing on the impacts of management on site conditions and species composition, several management practices were relevant, such as hay removal, livestock grazing not only in pastures but also in hay meadows in spring and autumn, sowing of barn sweepings, meadow spring cleaning and pasture clearing, and possible fertilization or irrigation (Sections 6.1, 6.2, 6.3, 6.4, 6.5, and 6.6). Interdisciplinary perspectives reveal that many grassland management practices proven by historical records had clear ecological consequences, but most of them are usually not documented in archival data. Nonetheless, these ecological impacts mattered in farming; for example, in terms of the differing hay quality and/or quantity as a consequence of a changed species composition and plant nutrient and/or water availability. Recent publications claim that historical grassland management types in the colline and lower mountain vegetation belts of Central Europe were astonishingly uniform in their essential features within generally similar farming systems, despite all the fine-scale agricultural differentiation,[3892] and also that "[t]he correspondence

[3892] Kapfer, 2010, 134. This publication by Kapfer is an overview of the main types of historical grassland management; historical ecological interpretations are poorly represented in it.

between pre-industrial meadow use in Sweden and the current [traditional] use [in the Carpathians] in Romania is striking."[3893] Numerous parallels with the Carpathians drawn in this volume lead us to the finding that grassland management in the Alps in the waning agrarian regime and the first phase of the industrial sociometabolic regime was in many aspects very similar to the management recently documented among traditional farmers in the Carpathians in the colline and montane belts. It is acknowledged that such comparative approaches significantly improve the understanding of traditional species-rich grassland management. For instance, meadows temporarily used as arable land in field rotation used to be common in Sweden and the general impact of such a management regime "on vegetation succession can be studied in Romania even if the species assemblages are somewhat different."[3894]

Nonetheless, this book also emphasizes that nuanced considerations about the plurality of possible ecological impacts of many grassland management practices are needed. The ecological impacts of several grassland management practices (e.g., mowing, grazing in meadows and in pastures, irrigation, spring cleaning, and burning), designated in the same manner in archival records, often differed significantly due to diverse site conditions and/or due to the details of carrying out these practices and in the timing of their application.[3895] Various sensitive details based on experience were only part of orally transmitted traditional ecological knowledge and not written down. For instance, although the frequent claim that soils were increasingly impoverished in non-fertilized meadows in the long run because hay was removed and no fertilizer was applied[3896] is usually correct as a general trend, the rate of this impoverishment of soil nutrient status varied greatly in different meadow agroecosystems. Namely, significant amounts of nutrients were made available to plants in many such meadows through the side impacts of land use and biological processes (Sections 6.2.2 and 6.3.1).

Patchworks of different sites with variable grassland species compositions in the Alps required significant time investment in walking, but the dispersed meadow parcels belonging to the same farm also had economic advantages. They mitigated the risk of a serious hay harvest failure in the same summer in many grassland parcels belonging to the same farm. This risk mitigation was a consequence of several factors, such as (1) decreased hazard and risk of contemporaneous damage in various meadows that belonged to the same farm, posed by some animals that came into conflict with grassland farming.

3893 Dahlströhm, Iuga, & Lennartsson, 2013, the entire article, quote from page 203.

3894 Dahlströhm, Iuga, & Lennartsson, 2013, 196–200, 202, 203 (quote).

3895 This conclusion was first presented at the conference of the European Society for Environmental History in Bristol in 2022 (Zwitter & Rasran, 2022). See also Peterken, 2013, 143–144, 147. Independently, Janišová et al., 2023 (especially p. 19) emphasized the same.

3896 E.g., Jacobsson, 2023.

Cockchafers, for instance, show a clear preference for certain habitat types—they are absent from higher montane, subalpine, and alpine meadows. In addition, they move very slowly through the landscape and inhabit only certain meadow habitats in the lower vegetation belts. (2) Some hailstorms destroyed the meadow yield in spatially limited belts in the landscape, thus affecting only some meadows in the case of a dispersed pattern of parcels. Another type of economic advantage of dispersed locations of meadows belonging to the same farm, located at different elevation zones and at dissimilar sites within the same elevation zone, was the significantly increased ability of farmers to mow several meadows that provided relatively good fodder quite soon after the hay plants matured, despite the high amount of time demanded by haymaking with a scythe (Sections 2, 3, 4.1.1.1, 4.1.3.1, 6.1.2, and 6.4.2).

The timing of haymaking was determined not only based on seed maturation—generative meadow reproduction, indeed crucial in species-rich meadows, received so much attention that even some barn sweepings full of seeds were sown in meadows where the turf had been devastated, instead of being used as highly nutritious fodder. Further factors influencing mowing timing included local legal norms, the duration of spring grazing in hay meadows, the presence or absence of fertilization or irrigation, the number of hay cuttings annually, total farm workload during the relevant time, weather (and Moon phase), the quality of hay from the local species composition and the related purpose of the use of the hay (fodder or animal bedding; if fodder, which was mostly the case—for which livestock type), flood hazard at highly flood-prone sites, and cultural history. As long as spring grazing was the first annual use of hay meadows, mowing dates alone, which are sometimes recorded in historical sources, say nothing about the state of plant development during haymaking. Namely, it was the combination of management (the interval between spring grazing and mowing; fertilization or irrigation or their absence; in some cases, historical records provide insight into all these management practices) that, in combination with environmental factors, determined the state of plant development or seed maturation. Regarding plant development in the mowing season, irrigation earlier in the vegetation period contributed not only through drought mitigation but also through supplying nutrients to plants and further aspects of increased soil fertility, and through the timing of the beginning of the vegetation period in spring where irrigation was carried out at that time (Sections 2, 3, 4.1.1.4, 6.1.2, 6.2, 6.3, 6.4, 6.5, and 10.1). Focusing on the cultural history impacts on mowing timing, one of the most obvious was the impact of the Christian interdiction of farm work (except for the unavoidable daily routine) on Sundays and other holy days on the timing of haymaking and storing of dry hay. There were, however thorough differences in time and space within the Alps concerning (1) the question whether and to what extent this interdiction was respected in the Christian cultural milieu, spanning from allowed to a taboo and also distancing from a taboo when the situation was critical; and (2) the number of holy days per summer concerned (Sections 4.1.1.4, 6.1.2.1.2, and 6.4.3.1). Exceptions were highly relevant during the cold and wet summers of the Little Ice Age.

The Little Ice Age brought many serious problems to grassland farming, occasionally challenging the core pillars of the economy of various farms. Some advantages could, however, also be identified, such as mitigation of the hazard of cockchafers coming into serious conflict with grassland farming, at least regionally. Less general and less relevant was the likely local increase in abundance of some snowbed species that are good fodder plants in some grasslands (Sections 4.1.1.2, 4.1.1.3, and 4.1.3.1).

The evaluation of fodder quality from species-rich seminatural grasslands based on traditional ecological knowledge followed more diversified criteria than usual in agronomy of the twentieth century, which paid less attention to the impact of local fodder on product taste and flavor and livestock health. Although grasses and legumes in hay and pastures matter greatly, they are not the only appreciated fodder plants in seminatural grasslands from a more complex perspective. A couple of other forbs, such as lady's mantles and alpine spicknel, produce highly nutritious hay. Some aromatic forbs in hay, such as thymes, result in highly fragrant dairy products that used to be highly esteemed in distant markets, and many plants from species-rich grasslands have favorable impacts on livestock health and promote appetite. Some sunny grasslands in the Alps, at high elevations in particular, provided more nutritious and healthier fodder than any lowland grasslands in Central Europe and contributed to higher product quality. This was long known and had a favorable influence on the market for dairy products, which were items of medium- or long-distance trade from the Late Midde Ages until at least the nineteenth century. However, haymaking at such sites has been highly labor-intensive (Sections 3.1, 4.1.1.2, 6.1.1.2, 6.3, 10.1.1, 10.1.2.2.2, 10.1.2.2.3, 10.1.2.2.4, and 10.2.1).

Nonetheless, at all elevations, there were also some meadows providing low-quality hay, and their share in the total meadow area differed at a local to regional scale due to dissimilar natural conditions and management. However, as long as long-distance overland transport of hay was limited, some low-quality hay types raised the resilience of the farm economy in the case of damp weather during haymaking because they were less prone to mold infestation, or nutrients were leached from such hay more slowly than from better fodder plants. Nonetheless, livestock's performance was very poor if they largely depended on such hay, such as fodder abounding in matgrass. Concerning the grasslands and wetlands providing low-quality hay, it needs to be emphasized that fodder quality changes greatly during the vegetation period in every grassland, and that there are differences among plant species in this regard. Many grassland and wetland monocots that become hard and unpalatable for livestock later in the year are soft, or even sweet and attractive, for livestock in early spring. At that time, they are appropriate even for milk cows, whose productivity strongly depends on fodder quality. This was particularly relevant for typical grazing in early spring in marshes, at sites characterized by middling or high plant nutrient availability. Grazing resources there were significantly more abundant than in mesic or dry meadows at the same time. The practice was so important that some cows whose lactation had stopped in winter due to too meagre feed started producing milk again based on early spring grazing in a marsh, exactly at the time when the milk crucially increased the amount of protein available to many farm families (Sections 3.1, 4.1.1.3, 6.1.2.5.1.1, 6.1.2.5.1.2, 6.2.2.1, 6.2.2.2, 6.3.1, 6.6.2.1, and 9).

Diverse livestock types within a farm (milk livestock, draft livestock, different animal species, etc.), historical livestock breeds, and meagre feeding standards in many cases allowed much more feasible economic use of hay from highly diverse seminatural hay meadows, including ones providing high-, medium-, and low-quality fodder, than more intensive and specialized livestock farming producing higher yields nowadays. What was not suitable for dairy cattle was acceptable for an ox or a horse on the days when they were not used for work, or for sheep or goats. A diet acceptable for an average nineteenth-century milk cow in the Alps is nowadays acceptable only for some cattle breeds to reach moderate performance. However, a significant percentage of the total fodder can nowadays originate from non-fertilized or moderately fertilized grasslands at many modern medium-intensive dairy farms if used in a well-considered, knowledge-intensive manner, and more such hay can be used at non-dairy farms (Sections 2, 5, 6.1.2.5.1, 6.3, 6.6.2.1, 8.2.2, 9, and 10.1).

In the long run, until the twentieth century, during winter barn feeding, livestock received not only first- and second-crop hay, straw of some cereal types, and some cereals for draft or breeding male animals, and, gradually, fodder legumes ("clovers"). Leaves from carefully selected tree species, from trees managed in a well-considered manner, as relatively nutritious fodder that contains minerals absent from or insufficient in grassland hay, played a prominent role in sheep and goat farming, and the addition of tree foliage to other hay contributed to the diversity of substances consumed by cattle. Fodder additions available in smaller quantities included barn sweepings, bran, in some regions of the Alps Iceland moss, and so on. Scalding or soaking barn sweepings, tree leaves, a limited fraction of the total hay, and some gathered organisms, Iceland moss in particular, increased the digestibility of fodder, which was all in all significantly lower than in intensive farming today, but more diverse and aromatic, and fully of local origin. During the warm part of the year, most gathering of specific, carefully selected plant species aimed to be used as fodder was intended for swine. Such gathering took place in significant quantities at appropriate sites at all imaginable elevations, from the colline to alpine vegetation belt, and specific knowledge was needed to turn some gathered organisms, such as Iceland moss, into tasty fodder (Sections 3.2.1, 6.3.1, 6.3.2, 6.5, and 8.2.2).

Gathering selected plant species in grasslands and wetlands in the Alps covered an extremely broad spectrum of aims, ranging from the aforementioned (swine) fodder all the way to aphrodisiacs, abortifacients, and building material. Plant gathering in the Alps generally focused on various plant organs: roots, rhizomes or bulbs, leaves, whole herbs, blossoms, fruits, such as berries, seeds, or rarely stalks. It motivated a significant long-term economic interest in some types of grasslands, even if they provided low-quality livestock fodder. In the history of grassland plant gathering, the Alps are a specific region in Central Europe because of their exceptional plant species richness and plant adaptations to harsher environmental conditions, at high elevations in particular. They allowed successful competition, thus the occurrence or even high abundance of several plant species absent or rare at lower elevations in most of Central Europe and beyond. Harsher environmental conditions also result in higher concentrations of essential oils. The same essential oils resulted in some healing plants being avoided by grazing livestock. Thus, trade in the organs of such plants, even rhizomes, went hand in hand with pasture clearing, such as in the case of yellow and dotted-flowered gentian in some microregions. The social history context contributed to the specific role of the Alps in the history of gathering in Europe. One such aspect is the location of the Alps being the closest mountain range to most of the rest of Central Europe. In most of the second half of the second millennium, the Alps have been surrounded by more densely populated regions than the Carpathians or the Dinaric Mountains; several regions surrounding the Alps were "among the most densely populated parts of Europe."[3897] Moreover, population density in the Alps was also very high for a mountain range in Europe in the long run. All this stimulated the local demand for gathered grassland plant organs and trade in these mountain plants or plant products to other regions of Europe and beyond (Sections 1.2, 2, 3.1, 6.1.2.5.2, 6.3.1, 6.6.1.1, 6.6.2.1, 6.6.2.2, and 8).

The economic and ecological relevance of grassland plant gathering in the Alps evolved over time, with a general increasing trend from the Late Middle Ages until the nineteenth or twentieth century, depending on the species and region, followed by a major twentieth-century decrease. It goes without saying that plant gathering in grasslands in the Alps had a longer historical tradition. Early fragmentary information on trade in plants from the Alps is already available from Classical Antiquity. This evidence is a consequence of the fact that gathering was aimed not only at covering the local demand, but also at regional or long-distance trade, and gradually even intercontinental trade.

[3897] On the history of population density in the Alps and Europe, see Viazzo, 1989; Livi Bacci, 2000; Ehmer, 2005; *Regions in Europe: 2022 Interactive Edition*, Population density. See also Aschwanden et al., 2022, 217.

Long-distance trade in large quantities of plants from the Alps first covered selected healing plants and those used in body care, relevant enough to result, for instance, in an inverse direction of the European Levantine trade than usual for aromatic merchandise, with the endemic spikenard valerian roots from the Eastern Alps being transported eastward. In this manner, the Islamic cultural and economic history of the Middle East had an impact on grassland agroecosystems at high elevations in many places in the Eastern Alps in the long run, usually raising (!) the abundance of spikenard valerian gathered there. Later, long-distance trade in healing plants from the Alps was joined by further trade aims, mainly the trade in forbs that served aesthetic purposes, most notably and copiously edelweiss in the nineteenth and early twentieth century. Edelweiss became rich in symbolic value, under the influence of wider cultural, tourism, and sports history developments that thoroughly changed the attitude toward this plant species (Sections 3.1 and 8).

History saw sustainable and unsustainable gathering, even including examples of sustainable but large-scale commercial uprooting, such as spikenard valerian, and yellow and dotted-flowered gentian in several regions (or by many individuals, in contrast to their colleagues who gathered unsustainably) and historical periods. Where this practice was carried out sustainably, it was often misunderstood by nature conservation in the nineteenth and twentieth century, and partly still nowadays. Namely, conventional nature conservation is rooted in the time of the highest agricultural density in the Alps, which resulted in particularly heavy gathering as part of the integrated farm economy, increasing the likelihood of excessive gathering in some places. Rooted in such a tradition and fears, conventional nature conservation has not sufficiently considered the specifics of individual species, populations, and habitats to be able to distinguish until recently between sustainable and excessive gathering in the case of some species abounding only in oligotrophic (and some mesotrophic) grasslands. Unnecessary limitations are sometimes counterproductive not only in the sense that they pose unneeded limits on farming and preservation of cultural heritage involving gathered plants, but in specific cases they even cause a decreased abundance of the protected species. However, in the context of twentieth- and twenty-first-century thorough reduction of the area of species-rich grassland habitats, plant nutrient intake from a polluted atmosphere, decline in connectivity between species-rich grassland habitats, and climate change, some gathering practices that were sustainable in history may prove unsustainable nowadays, and so these factors also have to be considered. Habitat preservation is usually more important than prohibition of gathering of above-ground plant organs to preserve protected species. In the case of a good condition of a population, adequate habitat connectivity, and sufficient population size, specific species, which are able to re-sprout from secondary roots after the main root or rhizome has been gathered, can even be uprooted in significant but limited quantities in monitored populations with no long-term decrease in the population size (Sections 3.1, 6.6.1.1, 6.6.2.1, 8, 10.1.2.2.4, and 10.2.3).

The aforementioned recent challenges in the preservation of oligotrophic and mesotrophic grasslands demonstrate that agrarian history faced not only long-term continuities, but also major discontinuities. Before addressing the contemporary history rupture in grassland history, a selection of relevant earlier discontinuities is dealt with here. They were results of either extensification of grassland use or intensification of farming. Before the second phase of the industrial sociometabolic regime (see the introduction to Section 1), farming intensification often relied on increased labor input to attain higher and/or more stable yields and/or a higher product quality[3898] (especially Sections 6.4 and 6.6).

An example of discontinuities caused by land-use extensification was temporary local or regional abandonment of grasslands due to a nexus of specific socioeconomic and environmental conditions, such as during the fourteenth- and fifteenth-century plague epidemics in parts of the Alps. In such cases, spontaneous reforestation, which can sometimes be proven by archival records or pollen analyses, was not the only serious challenge for restoration of productive agricultural grasslands. Fodder quality declined in a few years of absence of regular management. Earlier stages of ecological succession are less likely documented in old records or pollen cores than reforestation. However, restoration of an agriculturally more desired species composition after such a short-term abandonment phase was often already quite labor-intensive (Sections 3.2.2 and 6.6).

Focusing on discontinuities in grassland management caused by increased farming intensity, the eighteenth and nineteenth centuries in general saw much more significant innovations in farming in Europe than the preceding Late Middle Ages and Early Modern Period. This was a combined result of several effects. Population growth, faster advances in scientific knowledge, and intensified science outreach, in which mass media played a relevant role, were only three among them. Nonetheless, important innovations had already been characteristic of grassland farming earlier in some regions of the Alps, with a pronounced regional and chronological diversity in the spread of innovation or its absence; this was a general characteristic of innovation in European farming until the eighteenth or nineteenth century.[3899] Two early examples of such innovations in grassland history are dealt with in this volume. One of them is the Late Medieval spread of grassland irrigation that, as an innovation, included long main water channels that crossed significant elevation differences (in Wallis [Francoprovençal *Valês*, French *Valais*] they even passed along cliffs)

[3898] Krausmann, Weisz, & Eisenmenger, 2016, especially pp. 72–73, 75–77, 79.
[3899] Troßbach, 2008; Baumgartner, 2017.

and were often fed by glacial milk, which usually had a more relevant impact on soil fertility than irrigation water from nearby watercourses. Another example of a historical innovation relevant in grassland farming is the spread of roofed hayracks, which reached several regions of the Alps (in the southeast in particular) by the seventeenth century. The roofed hayrack is a construction mitigating the negative impact of rain during haymaking on hay quality, with a success rate incomparably higher than that of earlier strategies of coping with wet weather during haymaking. The hayrack is thus an expression of historical coping with a fully different challenge of grassland farming in much of the Alps, where the climate is moist, than in steppes. However, the territorial range of these early innovations was not comparable to the spread of farming innovation in the eighteenth and nineteenth centuries. The roofed hayrack, for instance, kept spreading to some further regions of the Alps later, in the eighteenth and nineteenth centuries. Its most developed form, the double hayrack, became frequent in southeastern Alps (whereas double hayracks have never become usual in the rest of the mountain range) only during the agricultural reforms of the second half of the eighteenth century and in the nineteenth century. This chronologically coincided with the nineteenth-century increase in quantities of longer hay that takes longer to dry—a consequence of the considerable increase in the area of moderately fertilized meadows. Longer hay required more frequent use of hayracks in hay farming than shorter hay from non-fertilized sites. Nonetheless, hayrack has never been used in all regions of the Alps (Sections 1.3, 4.1.1.3, 6.3, and 6.4).

Another example of a discontinuity in grassland history, which is dealt with in this volume, is the abandonment of spring grazing in hay meadows in the eighteenth or nineteenth century. This case emphasizes that historical abandonment of grassland management practices, generally seen as underdeveloped, could also affect customs that in some cases were economically or at least agro-ecologically relevant—a fact these conclusions will return to later. Spring grazing in hay meadows was very generally abandoned in the eighteenth and nineteenth centuries, only to again be recommended later for specific meadow agroecosystems (Sections 6.2 and 10.1.2.2.6).

Namely, the economic and ecological consequences of spring grazing in hay meadows varied greatly, and it was historically oriented toward several different economic and environmental goals. Historical dates when spring grazing in hay meadows ended were, however, adapted to the general impact of climate on plant development only in some places, whereas the fertilization regime had a stronger influence on the duration of spring grazing. Fertilized meadows were not grazed more than for a short time in early spring to allow making the highly nutritious second-crop hay (made only in fertilized or irrigated meadows) before the increased likelihood of rainy autumn weather. In non-fertilized meadows, spring grazing ended later than in fertilized ones. Where spring grazing in hay meadows lasted until late May, this reduced the hay quantity significantly, but at the same time it improved its quality, provided that livestock excrement did not appear in the hay. The climate and fertilization regime were not the only factors determining the end of historical spring grazing in hay meadows. The local legal tradition also mattered, and it was not entirely adapted to microenvironmental conditions throughout the Alps. The local legal tradition was inspired by various data sources, such as unequal traditional ecological knowledge and the religious symbolism of some feasts (such as St. George's Day or Ascension Day)—a type of orientation that through its symbolic value and/or respect for the same feasts in different environmental conditions helped influence the timing of several farm tasks in Central Europe. A certain extent of standardization of the historical timing of farm work in various microenvironments is particularly obvious in a region as environmentally diverse as the Alps. The economic and ecological impacts of spring grazing in hay meadows exceeded the relevance of fodder in livestock's diet. During grazing in hay meadows in spring and autumn, nutrients returned to the grassland through livestock's dung and urine, increasing the patchiness of grassland habitats. Livestock created gaps in meadows, destroying some species (including some plants undesired in livestock farming) but favoring some others because the gaps functioned as regeneration niches for their seedlings. Historical data, however, reveal a complex relation between spring grazing regimes and the occurrence and abundance of some plant species receiving special ecological attention, such as orchids and daffodils, and some plant species receiving special economic attention, such as sainfoin. This warns against oversimplifications regarding the impacts of spring grazing in hay meadows on plant species without empirical data, which can be historical. Trampling during spring grazing could be economically challenging, on steep slopes or in wet meadows in particular, but this was somewhat less of an issue until the nineteenth century, when the prevalent livestock were significantly lighter. Spring grazing affected wild fauna, including a reduction in the abundance of some animals that came into conflict with farming. Livestock also transported and deposited some germinable diaspores in meadows through endozoochory (dung) or epizoochory (i.e., external transport of plant diaspores, such as in hooves or in hair). Some of the plants introduced this way became established in the gaps in vegetation created by livestock (Sections 2, 5, 6.1.2.3, 6.2, 6.3, 6.4, and 7.1).

Indeed, movements of livestock that consumed hay from various species-rich meadows cut when seeds were ripe or that grazed in pastures with ripe diaspores were among the main vectors of nonnative plant species introduction to grasslands under the agrarian regime and the first phase of the industrial sociometabolic regime. The relevant middling- to long-distance livestock movements usually happened in the context of farming, trade, or major wars, such as the First World War—one of the conflicts in human history with the highest death toll, when the numbers of animals involved and the diversity of their uses reached a marked peak. That war was also the most ferocious slaughterhouse of horses in world history. Movements of livestock, fodder, and other supply goods under the war economy led to the occurrence of various plant species at sites otherwise not reached by them. Thus, certain plant species that were introduced to the new environment became established there, and their adaptation to the new ecosystems often lasted for many decades. Those species introduced during wars are not only botanical peculiarities, but they offer food for thought. Namely, such grassland species are living mementos of the great suffering by people and livestock, principally horses, and of major disruptions to the environment during the war. Later, a few species introduced in this manner in the second millennium became (potentially) invasive, which is a long-term ecological challenge triggered by historical wars. The case study of plant introductions through the Soča (Ital. *Isonzo*) Front supply during First World War focuses on a territory in the Julian Alps (Sln. *Julijske Alpe*, Ital. *Alpi Giulie*) that had already faced overgrazing by unprecedented horse numbers in 1915/16, followed by a disastrous fodder crisis in 1917, so that from May to August 1917 war-related starvation of horses was a tremendously greater killer of horses than the warfare. Within the total fodder, supplies of hay were particularly insufficient at that time, raising the grazing pressures in grassland agroecosystems. Overgrazed sites were full of gaps allowing colonization of plant species introduced during that war, such as yellow oxeye, most likely through hay or horses. However, in some cases, the same plant species was introduced by human activities to different parts of the Alps in a different manner—whereas yellow oxeye was introduced in part of the Julian Alps through First World War activities, it was planted for decoration in the Karawanks (Sln. *Karavanke*, Germ. *Karawanken*) and became naturalized in both regions. Indeed, cultural and intellectual objectives also led to the naturalization of some other plant species in grasslands in the Alps after they had been introduced to gardens in Europe for botanical or ornamental purposes, and a few such species also became invasive (Sections 6.2.2, 7.1, and 7.2).

The First World War, mentioned in the paragraph above, raised the globalization of the food trade, thus also strengthening farming interference in some marginal, unpredictable grassland environments in some landscapes on continents far from the battlefields (see also Section 7.2), most notably the drought-prone steppes of the Great Plains in the United States, which later faced dust storms (the Dust Bowl; however, the expansion of arable farming in the steppes was not the only reason for the Dust Bowl).[3900] Large-scale globalization of the market for farm products has older roots, in line with the general integration of European and global markets, and by the First World War it had already had profound impacts on grasslands in the Alps—especially

[3900] Isenberg, 2017, 134–135, 146–148; Leisering & Schwarzrock, 1997, 106; Schwarzrock, 1997, 108; Winiwarter & Knoll, 2007, 62.

since the second half of the nineteenth century, when technical advances made possible long-distance and cheap transport of agricultural goods not only on waterways, but also on railroads. In this respect, some interrelations can be identified between the history of seminatural grasslands in the Alps and the history of zonal steppes—the focus of global mainstream long-term grassland environmental history research (Sections 1.1 and 1.3). An example is the increase in cereal farming on steppes, from the second half of the nineteenth century until the mid-twentieth century,[3901] which went hand in hand with the gradual decrease in cereal farming in most of the Alps, made possible by long-distance and cheap cereal transport. Consequently, many former arable fields in the Alps were converted into moderately fertilized meadows (Sections 3.1, 6.3, and 10.1.2.2.3), and the same process played a role in the abandonment of remote, non-fertilized meadows (Sections 3.1 and 9). Another example of interrelations between the history of steppes and grasslands in the Alps in general is the major nineteenth- and early twentieth-century increase in sheep farming in the steppes of the Southern Hemisphere and the contemporaneous major decrease in sheep farming in Europe, including most regions of the Alps. One of the impacts of this process in seminatural grasslands in the Alps was the abandonment of the use of some former pasture complexes for sheep that were not suitable for cattle farming (Sections 3.2.2 and 6.3.1.1). Many others were converted into cattle pastures, which contributed to changes in species composition because the ecological impact of cattle grazing differs significantly from that of sheep grazing (Section 4.1.1.2). The gradual virtual abandonment of sheep farming in most of the Alps also made the short length of mountain hay from non-fertilized sites an unwanted characteristic (Section 6.3.1.1).

The real rupture in the twentieth-century history of grasslands in Central Europe followed later, as a consequence of interrelated socioeconomic transformations, resulting in large-scale abandonment or intensification of grassland management (the introduction to Section 1, Sections 1.2, 10.1.2, and 10.2.1). Whereas scarcity and poverty were "always pending in the agrarian [sociometabolic] regime," which was characterized by hard physical labor that made possible large-scale reliance on local resources in the countryside, "the dominant trend within mature industrial regim[e] is that of abundance (however unevenly distributed)."[3902] "Economic growth is a basic feature of the industrial regime . . . , in stark contrast to the agrarian regime, which was characterized by diminishing marginal returns and strong negative feedbacks from physical limitations, which prevented per capita economic growth for larger populations over longer periods."[3903] Nonetheless, mature industrial

3901 Aschwanden et al., 2022, 221; Isenberg, 2017, 146–147.
3902 Krausmann, Weisz, & Eisenmenger, 2016, 87 (quotes); Winiwarter & Bork, 2019, 130; Winiwarter, 2022, 23–24, 50–51.
3903 Krausmann, Weisz, & Eisenmenger, 2016, 87.

sociometabolic regime and economic growth as one of its principal features brought about new, complexly interrelated types of challenges.[3904] The nature of the challenges specific for the mature industrial sociometabolic regime and relevant in farming and food consumption in the Alps is often economic, environmental, societal, political, and/or psychosocial.[3905]

As in many domains of the real world heavily dominated by economic views, the interdisciplinary history of grasslands in the Alps is not only a story of long-term progress. "An increase in complexity can lead to crisis."[3906] Humanity in general still has the problem of acknowledging major environmental transformations "orchestrated by their economic activities" from which they benefit while other species are losing.[3907] The magnitude of such activities is so great that it leads to mass extinctions. Species-rich grasslands are one of the most threatened habitat types in Central Europe, inhabited by many specialist species that cannot adapt to intensively managed grasslands. The extent of recent mass extinction in species-rich grasslands (non-fertilized and moderately fertilized) is greater than required by well-considered, socially equitable, and highly ethical farming, trade, and consumption, supported by highly ethical policies, still making it possible to feed the world population in the coming decades and ensure an equitable position for farmers. Because seminatural grasslands are full of living beings, this is also a general ethical issue.[3908] The excessive spatial extent of highly intensive farming in Central Europe is threatening the wellbeing of future generations. A new middle way in farming in Europe is needed that includes enhancement of biodiversity-friendly structures and activities in both conventional and organic farming, including in areas without specific natural constraints on farming intensification, such as wide valley and basin bottoms in the Alps. The desired results are thoroughly interdisciplinarily and transdisciplinarily considered landscape compositions (landscape composition is the "number and proportion of different habitat types") and landscape configurations (landscape configuration is the distribution of habitat types in the landscape). In such complex "multifunctional agricultural landscapes," conventional farming (including conventional extensively and conventional intensively managed farming) and organic farming are practiced. Such landscapes include (in alphabetical order) (1) extensively managed farmland (as a must request including a significant percentage of seminatural meadows at the microregional level, not only at remote sites, to preserve "biodiversity and its multiple functions," involving the ecosystem services relevant in farming, recreation, and tourism) and extensively managed to non-managed off-field structures (such as hedges, stone walls, and exten-

3904 Compare Winiwarter, 2022; Hasenöhrl & Groß, 2022, 200.
3905 Compare Winiwarter, 2022.
3906 Compare Izdebski, 2022, 10.
3907 Paraphrased from Izdebski, 2022, 18 (the quote is also from there).
3908 E.g., Dietl & Lehmann, 2004, 11.

sively managed parcel margins), and (2) intensively managed farmland. Such landscapes result from the integration of high-nature-value farming "into a modern farm management" paradigm, "balancing socioeconomic and ecological goods."[3909] Thus, there is a need for careful exploration of the paths not traveled in grassland farming during the twentieth- and twenty-first-century conventional-farming intensification by innovatively maintaining a significant percentage of high-nature-value agroecosystems in all elevation belts of the landscapes in the Alps in a manner that reduces the injustices perpetrated against people, livestock, and wild living beings by modern society. This is a society causing challenging material and energy flows, without any hope of maintaining its high living standard in the long run unless made possible through well-considered, sustainable innovation and a changed attitude toward nature (Sections 1.2 and 10).

Since the Green Revolution reached grasslands in the Alps, there has been "more choice" in grassland farming than ever, as far as humanity is concerned. In general, there is also "more knowledge"—which, however, is highly unequally distributed, whereas large bodies of traditional ecological knowledge from before the industrialized farming intensification, relevant in high-nature-value grassland farming, such as those presented in this volume (Sections 4, 6, 8, and 10), have been lost or are threatened in many places. In addition to more choice and, in general, more knowledge, there are unprecedented challenges in the interactions between society and nature in grassland farming. All this results in "less room for chance," "more responsibility," and "greater worry"[3910] in the contact zone between (in alphabetical order) ecology, economics, and social equity.

In the framework of sustainable innovation, comparisons between the features of historical and recent farming and some historical practices of grassland management deserve more attention than usual in recent farming, such as (1) serious considerations about energy flows in farming in the agrarian sociometabolic regime and today; (2) the question of the relationship between the total environmental footprint of a farm and yields in the agrarian sociometabolic regime and today, not only in the microregion where the farm is located but also in distant environments where crops for industrially processed feed are grown, and in terms of synthesis of fertilizer, and other agrochemicals; (3) a partial return to the integrated approach to evaluating of fodder quality

3909 Jones & Poux, 2012, 457–458; Oppermann et al., 2012a, 472 (fourth quote); Kremen, 2015; Boetzl et al., 2021 (third quote from p. 7); Grass, Batáry, & Tscharntke, 2021 (first quote from p. 278); Tscharntke et al., 2021 (second and fifth quotes from p. 928); Leitgeb et al., 2023.

3910 Paraphrased from Massimo Livi Bacci's comments on recent human population numbers in Europe, contextualized in a long historical perspective (Livi Bacci, 2000, 189—quotes from there). Livi Bacci is a historical demographer and was president of the *International Union for the Scientific Study of Population* from 1989 to 1993.

in a significant percentage of grasslands, looking not only at the production of high-nitrogen biomass but also at the functional role of biodiversity in terms of livestock health and product quality; (4) the related observation of flowering grasslands and smelling the hay; (5) moderate fertilization combined with mowing timing allowing partial seed dispersal every second year at least; (6) absence of fertilization and mowing timing allowing seeds to ripen at sites with specific natural constraints on farming; (7) sainfoin sowing in some non-fertilized, relatively dry, base-rich or neutral meadow sites without plowing; (8) avoiding small sections of moist or wet meadows with nests of ground-breeding birds; and (9) mechanical, other grazing-related, and thermal suppression of plant species unwanted in animal farming. All these practices, however, require farmer-equitable and well-designed agri-environment payments (Sections 6.3.1, 6.3.2, and 10.1).

As regards the uses of hay from seminatural meadows nowadays, hay from moderately fertilized meadows is more widely useable than that from most non-fertilized sites, especially because moderately fertilized meadows also produce high-quality second-crop hay, and yields in moderately-fertilized meadows are higher and more stable than in non-fertilized ones. In non-fertilized meadows, where hay properties vary significantly due to the most dissimilar species compositions and strongest impacts of site conditions on hay quality and quantity, hay from some meadow agroecosystems from the patchwork landscapes in the Alps is useable for any livestock type, at least as a limited percentage of the total fodder; specific sites, usually at high elevations, produce particularly valuable fodder. In contrast, hay from several types of non-fertilized meadows is only useable for certain livestock types, and so even open-minded farmers cannot always use it feasibly in dairy farming nowadays. Traditional ecological knowledge helps recognize sites that produce viable hay in the long run even if not fertilized. Such hay can often be a suitable fodder addition to more calorific grassland fodder because such an addition is favorable for biodiversity, livestock's health, promotes appetite, and/or has favorable impacts on milk flavor. Well-considered selection of the hay available can at least allow viable use of such hay during specific phases of animal development (Sections 1.2, 2, 4.1.1.2, 5, 6.1.2.3, 6.1.2.5.1, 6.3, and 10.1).

Continuation with or restoration of the species-rich grassland management of the recent or more distant past has the advantage that "any features of value" in species-rich grasslands of the past "must have tolerated such management." In this manner, history-inspired management is likely to maintain or create favorable conditions for the target species historically occurring in the area.[3911] In this manner, historical knowledge can contribute to avoiding unnecessary nature-conservation limitations on farming, unless other components of environmental dynamics, such as climate change and general pollution, have

[3911] Adapted from Peterken, 2013, 390; see also Szabó, 2010.

rendered site conditions unsuitable. Thus, the impact of recent environmental change needs to be considered, and chance also plays a role in the competitive ability of plants and animals. Historical inspiration, however, does not mean that the grassland management has to be so "conservative that only one specific [historical] management treatment has to be applied." An open-minded approach welcomes various management regimes, leading to preservation or restoration of a significant percentage of species-rich grasslands, including some rare species or specific target species.[3912] This opens the ground for history-inspired innovation, which must be adapted to local conditions and avoids repeating the failures of the past (Sections 1.2 and 10.1).[3913]

There is, however, no one-size-fits-all definition of what the exact delimitation is between progress and decline[3914] in grassland management intensity in the Alps from the viewpoint of sustainability, taking into account human society and the environment. To make things more complicated, there are significant differences within the Alps (1) between the degree of environmental friendliness currently economically feasible (or perceived economically feasible) in different livestock farming types and tourism, and (2) in some aspects also between the interests of biodiversity and greenhouse gas emissions—both are equally relevant and none of them should receive disproportionately more attention. What is clear, however, is that the general prevalence of any kind of a lack of complex inter- and transdisciplinary, trans-sectoral actions, not fully considering the economic and environmental aspects, leads to a decline, and that the recent grassland farming transition has been excessively intensive in most of Central Europe as a consequence of actions or lack of actions by several highly diverse types of actors (Sections 1.2, 10.1, and 10.2).

Concerning recent controversies about the percentage of grasslands that could be species rich at the local, micro-, meso- and macroregional levels to feed the global population but at the same time to limit the farming-driven mass extinction in meadows, a historical parallel that reflects the knottiness of discussions on the scarcity of natural resources adds to the perception of the complexity of recent debates. This parallel focuses on the historical lament over a wood shortage under the agrarian sociometabolic regime in early modern Central Europe. At that time, wood was of vital importance, and biomass accounted for more than 95% of the total human domestic energy

[3912] Adapted from Poschlod, Baumann, & Karlík, 2009, 23 (quote); Peterken, 2013, 398–399; see, for instance such a sustainable adaptation of a management regime allowing the continuous occurrence of Alpine eryngo (*Eryngium alpinum*) on the slopes of Mount Porezen (pre-Alps south of the Julian Alps) in Bavcon, Ravnjak, & Praprotnik, 2019, 115.

[3913] Inspired by general considerations on relevance of history in ecology in Szabó, 2010, 382–383.

[3914] Compare Isenberg, 2017, 135–137; Winiwarter, 2022, 39–42.

consumption.[3915] Historical research has demonstrated that the early modern lament over a wood shortage was not always the result of a real or objectively feared wood shortage. It was also the consequence of specific interests and "the authority of interpretation of various forms of early modern expertise, which were linked, for instance, to questions of the status of forestry science, which became established around 1800" and which typically opposed traditional agroforestry practices, even if they were economically sound from the farmers' point of view and maintained high-nature-value, anthropogenically altered ecosystems.[3916] Similarly, the recent discussion on the share of the total meadow area that could viably be managed in an agroecological manner is not always a matter of objective standpoints, but to a significant extent also a consequence (1) of specific interests, even much more complex nowadays than in times of the early modern "wood shortage" question,[3917] and (2) of the history-rooted strong position of conventional agronomy, which has many merits but misses any tradition of complex ecosystem thinking that would also ethically take into account wild organisms as much as possible. Such an agronomy's stance would be strange as long as biodiversity was a side product of farming (Section 10.1.1),[3918] whereas fully integrating this aspect into modern agronomy (this is the topic of agroecological approaches, opposed by much of conventional agronomy—Sections 10.1.2.1, 10.1.2.2, and 10.2.1) is an ethical imperative during the recent mass extinction in most farmland[3919]—on a continent with millennia of tradition of land-sharing as the prevalent manner of preserving species-rich habitats.

Namely, in Europe, "limiting conservation to undisturbed wilderness areas would fail to cover the majority of endangered ecosystems and species."[3920] Only an approach also taking agroecology seriously has the potential to save a relevant percentage of most biodiverse habitats in the Alps—that is, the species-rich, seminatural meadows in forest clearings—and to preserve the ecosystem services of non-fertilized and moderately fertilized meadows in a significant percentage of this mountain range and beyond. However, just as conventional agronomy (in contrast to integrated agroecological approaches) lacks the tradition of taking wild organisms into account as much as possible, environmental sciences lack the tradition of thinking about the viable trade-offs between biodiversity on the one hand, and the farm economy and food production on the other, which are needed in the reality of the current en-

[3915] For typical energy flows in agrarian societies, see Krausmann, Weisz, & Eisenmenger, 2016, 67.
[3916] Grewe, 2004; Popplow, 2017, 200–203 (quote from pp. 202–203); see also Panjek, 2018.
[3917] Compare Kremen, 2015 and Winiwarter, 2022.
[3918] Compare Oppermann & Paracchini, 2012, 20–22; Grass, Batáry, and Tscharntke, 2021; Tscharntke et al., 2021.
[3919] Tscharntke et al., 2021.
[3920] Grass, Batáry, and Tscharntke, 2021 (quote from p. 256).

vironmental crisis, high population density, and low agricultural density in countries with a high living standard.[3921] There are currently major differences between Alpine countries in the degree of reconciliation between farming, agronomy, and nature conservation (Sections 10.1.2 and 10.2.1).

There is still a long way to a knowledge society in which an average consumer will be aware of the mass extinction occurring in grasslands in Central Europe and of the role of low-energy-input farming in maintaining a sufficient percentage of species-rich agroecosystems in the landscape at a small-scale level, including the desired ecosystem services.[3922] Concerning the Alps, instead of erroneously following the history-inspired notion that flowers from the alpine vegetation belt, such as edelweiss, are most threatened, the general public should learn that species of non-fertilized grasslands at lower elevations and those of moderately manured sites are most threatened nowadays[3923]—despite the airborne eutrophication of habitats from the polluted atmosphere, which is relevant at all elevations (Section 10.1.2.2.4). At lower and middle elevations, where biodiversity is most threatened, it is highest but least protected (Section 1.2).

There is still a long way to a knowledge society:

(1) In which there will be a legal obligation to inform consumers about the role (if any) of seminatural grasslands (floriferous: non-fertilized or moderately fertilized—the latter are typically purple, yellow, and white in May at mesic sites) in the diet of livestock that every kind of milk and meat, and every dairy and meat product on the market is based on (Section 10.2.1). Namely, species-rich grasslands are the most (and often the only) positive environmental impact of livestock farming;

(2) In which it will be prohibited to mislead consumers by directly or indirectly evoking the illusion that a product is based on species-rich grasslands through any part of product labels, stickers, and packaging, including bags (flowers on stickers, packages, or paper bags; labels such as "traditional," "selected quality," "organic," "from [a certain] national park," "from [a certain valley] in bloom," and so on), if the product does not support species-rich agroecosystems. Such unethical marketing, which in some cases is a consequence of the lack of awareness of the mass extinction going on in European grasslands, is far from innocent because it contributes to this mass extinction. Not only is this not prohibited—in some cases, such unethical marketing is even promoted by the authorities, such as through the agri-environment-climate payment based on the measure Hay Meat and Milk within the European

[3921] Compare Grass, Batáry, and Tscharntke, 2021; Tscharntke et al., 2021.
[3922] Compare Erhatič Širnik, 2011, especially p. 360; Oppermann, 2012, 231–232; Riedel, Walter, & Herzog, 2012, 432; Lomba et al., 2019, 38–39.
[3923] Bätzing, 2015, 278–279; Jaksetič (interview with Kaligarič), 2019, 22.

Common Agricultural Policy's Strategic Plan for Slovenia, 2023–2027, which uses the European label "Traditional Speciality Guaranteed" even for products that have nothing in common with traditional, species-rich grasslands. Only intensive, non-agroecological management is promoted in the handbook for hay production on the website for Slovenian hay farmers—in a country where the populations of the bird species that are grassland specialists declined by 40.8% in 2008–2018 (Sections 1.2 and 10.2.1). If such environmentally irresponsible marketing is formally promoted at this level, one can imagine that it has a broader impact on farmers' perception of what is ethically acceptable in the marketing of farm products because hard-working farmers have to swim in the ocean of globalized trade, which lacks any kind of solid environmental ethics. However, specific trans-sectoral and transdisciplinary initiatives, such as *ARGE Heumilch Österreich* (the agricultural cooperative for hay milk in Austria) and *Schweizerische Vereinigung integriert produzierender Bauern und Bäuerinnen* (The Swiss Association of Integrated Farming), can help create and maintain specific market niches facilitating intentional care for biodiversity in a limited but increasingly significant percentage of farmland at a large number of farms (Section 10.2.1).

Following the recommendations from points (1) and (2) will make it possible for informed consumers to decide whether to support or reduce grassland biodiversity through their diet. Only in this manner can equitable prices for products supporting non-fertilized and moderately fertilized grassland agroecosystems become viable. As long as buyers can easily find a cheaper product that unethically (tacitly) evokes the illusion that it is also based on species-rich grasslands (which is not the case), fair prices for products from high-nature value grassland farming are not possible (Section 10.2.1).

There is still a long way to a knowledge society in which the position of a high-nature-value farmer in the Alps will be more equitable and this profession more appreciated. There is still a long way to a knowledge society in which the perception of an average farmer in the Alps of what a modern farmer is will be a farmer that pays equal attention to in-depth knowledge gain in (in alphabetical order) the economic, environmental,[3924] and social aspects of farm production and farm life before making well-informed decisions about the future of the farm, taking into account at least the functional role of biodiversity and the rising importance of the manner in which the products are produced in the eyes of the buyers, and not only the link between product taste and quantity (Sections 10.1 and 10.2).

[3924] Jones & Poux, 2012, 457–458. In France, for instance, "delivering biodiversity is still seen as something rather effete, carrying little legitimacy within [most of] the agricultural community (Poux & Ramain, 2012, 219). Even in Switzerland, farmers' ecological knowledge is generally insufficient (Riedel, Walter, & Herzog, 2012, 432).

Such a knowledge society will perceive regions of the Alps that maintain a high percentage of species-rich, moderately fertilized, and non-fertilized grasslands as COLORFUL and BIODIVERSE, not just green destinations. Avoiding the homogenization of grassland agroecosystems (which is a consequence of management intensification) contributes to the attractiveness of landscapes for tourism, which is relevant because ecotourism is the most quickly growing tourism branch globally (Section 10.2.2). In this respect, tourism in the Alps should not passively rely on the established tradition of ecotourism reaching back to the nineteenth-century *belle epoque* ecotourism *avant la lettre*, or even longer (Section 8.1.3.1). At that time, biodiversity was a side product of farming. With rising awareness of the mass extinction occurring in meadows in Central Europe, the tourism relevance of blossoming meadows will likely increase, which is yet another argument calling for (1) preservation of such meadows in the Alps, where they are currently threatened in many places due to farming abandonment or intensification, and (2) closer financial links between tourism and high-nature value farming (Sections 1.2 and 10.2.2).

Embedded in the long-term historical perspective, an environmental historian can draw quite different conclusions about those landscapes of the Alps where the share of intensively managed grasslands is significantly higher than complex thinking would require. One option is to paraphrase the Austrian historical anthropologist Karl Kaser, who drew attention to the territories of the former Habsburg–Ottoman Military Frontier that were devastated in the wars of the 1990s after he had carried out in-depth research on the history of the Military Frontier from the sixteenth to nineteenth century. To translate Kaser's historical anthropological impressions of the devastated historical cultural heritage in the former Military Frontier to the meadows in the Alps, one can claim that—after a journey through a landscape where seminatural, non-fertilized, or moderately fertilized grasslands as natural and cultural heritage[3925] have recently been destroyed on a large scale and this is an ongoing process contributing to the recent mass extinction—an environmental historian that has dedicated a not-so-insignificant part of his scientific life to seminatural grasslands is left with only one thing: deep sadness.[3926]

Such negativism in environmental history conclusions is, however, counterproductive, as emphasized by Graeme Wynn's presidential address to the American Society for Environmental History in 2019.[3927] In the highly complex recent environmental crisis that goes hand in hand with a similarly complex socioeconomic transition in farming in the Alps and beyond, hope is not yet

[3925] It has often been emphasized that species-rich grasslands link natural, tangible, and intangible cultural heritage; for example, Luick, Jones, & Oppermann, 2012; Dahlströhm, Iuga, & Lennartsson, 2013; Babai & Molnár, 2014; Poschlod, 2015, 247–248.

[3926] Compare Kaser, 1997 (the paraphrased text is on p. 651).

[3927] Muchowski, 2022, 21–22.

lost for a more biodiverse future of a significant percentage of meadows and sustainable connectivity between species-rich grasslands within all vegetation belts in the Alps. This, however, requires much stronger engagement than experienced so far (Sections 10.1.2 and 10.2). A parallel between socioeconomic and intellectual history demonstrates that, since the nineteenth century, the developed world has become "capable of fully grasping, comprehending and—last but not least—critically reflecting on social change in its economic context."[3928] Actual success in delivering the downtrodden has, however, been limited. A selection of the fundamental factors that contributed to the limits of this success involves the complexity of "the protean interplay of political institutions, corporate power, [part of] technological innovation, . . . mass advertising," limits on objective information communication in the modern world posed by several types of information channels that lack a complex approach but take readers' time, and even intentional spreading of half-truths and lies. Despite the major differences in the complex recent social and environmental challenges, several types of solutions to the current environmental crisis have been facing obstacles posed by similar complexes of driving forces such as those presented, hindering a higher level of social equity.[3929] Nonetheless, it is encouraging that humanity is recently increasingly "grasping, comprehending," and critically reflecting on the environmental footprint of consumer society and the consequent environmental changes—mainly climate change, but gradually also biodiversity loss—in the context of socioeconomic transformations. Small and relevant but insufficient steps by various stakeholders, not only nature conservationists, in the direction of "how to deal with . . . mass interspecies death"[3930] in a socially equitable manner have already been taken,[3931] although with tremendous differences between Alpine countries (Sections 10.1.2, 10.2.1, and 10.2.2).

Thus, an alternative to the aforementioned pessimistic paraphrasing of Kaser is to claim that lifelong education is one of the fundamental drivers that can change human values and turn the human civilizing process in a sustainable direction: changes in value systems consist of interrelated rational, conscious

[3928] Izdebski, 2022, 18.

[3929] Compare Elias, 1994, 436–454; Gregory, 1999, 100–101 (quote from p. 100); Palmblad, 2024, 672–673; Beaufoy, Keenleyside, & Oppermann, 2012, 525; Jones & Poux, 2012, 456; Schader, Stolze, & Gattinger, 2012, 186; Kremen, 2015; Lomba et al., 2019, 36–37; Kupper, 2021, 184, 189; Hasenöhrl & Groß, 2022, 203–205; Winiwarter, 2022, especially pp. 20–22, 25–27, 38–51, 54, 58–63, 65–68; Schmitzberger et al., 2005, 282, 284.

[3930] The words of the playwright Miranda Rose Hall in *A Play for the Living in a Time of Extinction*. Hall, 2023, quote from pp. 14–15; gratitude is expressed to the theater director, playwright, writer, and translator Peter Petkovšek for recommending this work.

[3931] E.g., Elias, 1994, 358; Oppermann, Beaufoy, & Jones (eds.), 2012; Kupper, 2021, 172–173, 175–185, 190; Aschwanden et al., 2022, 219, 231, see also p. 221; Lomba et al., 2019, 37; Tscharntke et al., 2021.

components on the one hand, and unconscious ones on the other.[3932] In several aspects, the western world has not yet reached the optimal level of its state of civilization; our civilizing process is still in progress, and it is prone not only to a steady increase in the state of civilization, but also to decivilizing shocks, especially when the general level of incertitude increases.[3933] As a matter of the civilizing process, human societies have raised moral standards several times in history, requiring that individuals hold back momentary pleasure or inclination toward certain deeds in anticipation of the displeasure that would come if they gave in to the momentary pleasure or inclination to deeds considered immoral[3934] (see Section 10.1.1), such as buying or producing cheap food nowadays without paying attention to its entire environmental footprint, followed by the displeasure of the massive decline in grassland biodiversity. History emphasizes that periods of transition in the civilizing process, as likely the present one concerning the human attitude to the environmental footprint of food production, "offer special food for thought: previous [moral] standards [in this case, in the area of food production] became partly questionable, new advanced ones do not exist yet."[3935] The history of the human civilizing process shows that during such phases of transition in the civilizing process "people start questioning many behavioral patterns taken for granted by previous generations," and they start looking at human society in a more complex manner,[3936] such as the increasingly present questioning of the usual losses of biodiversity caused by an excessive percentage of today's usual intensive conventional farming in areas without special constraints on farming. In history, the civilizing process as a whole was unplanned in the long run, but it followed an inherent order. However, research results have long impacted the civilizing process, and increased knowledge about the functioning of the unplanned principles resulting in the civilizing process now increases the likelihood of planned impacts on this process[3937] to create a food system that will be more equitable and just toward the producers and the environment.[3938]

The education most relevant in this context must place adequate emphasis on interdisciplinary real-world challenges and be reflected in deeds, not only in words, in strengthened in-depth and thorough transdisciplinary and trans-sectoral actions, linking the natural sciences, the social sciences, the hu-

[3932] The general considerations concerning the civilizing process follow Elias, 1993 and Elias, 1994 (here pp. 377, 388–391); the application of this theory in environmental history that is not limited to human attitudes towards fauna is a novel contribution.

[3933] Elias, 1994, 451–454; Cvirn, 2001, 434.

[3934] Elias, 1994, 372.

[3935] Elias, 1994, 442.

[3936] Elias, 1994, 443.

[3937] Elias, 1994, 312–314, 316, 377, 395–397, 476–477; see also Elias, 1993, especially pp. 152–157, 171.

[3938] Compare Elias, 1993, 161–163.

manities, and various non-academic stakeholders[3939]—from farmers and farm advisors to consumers, trade actors, other business actors involved, political institutions, and nature conservationists. If humanity proves mature and responsible enough, consumers will feel uncomfortable buying more food than needed; they will pay attention to the ratio between food produced by intensive and extensive farming, and purchasing a higher proportion of intensively produced food than necessary to feed the population will also cause unease; not preserving enough of the species-rich habitats in the landscape will become unacceptable; and farmers will be paid equitably and more generally socially valued. Economic growth will, under such a scenario, likely be sacrificed for higher environmental and social ethical standards, resulting in more pride and personal happiness than possible in the current industrial sociometabolic regime—a goal potentially reachable based on much open-minded dialogue that requires "great stamina, a broad vision and a high frustration tolerance." Such a continuation of the civilizing process "can help us achieve social justice and the fair transformation that the Anthropocene requires from us" and "open our minds to move in the direction of true ecological justice, which might be our hope for survival and flourishing"[3940] (Sections 1.2 and 10).

[3939] Compare Favilli, 2016, 118; Winiwarter, 2022, especially pp. 14, 17, 19–22, 27, 32, 37–50, 54, 61–63, 65; Jones & Poux, 2012, especially pp. 451, 457–458.

[3940] Compare Polajnar Horvat, Smrekar, & Zorn, 2017, 30–31; Kupper, 2021, 168, 184, 190 (first quote from p. 190); Izdebski, 2022, 18 (second and third quotes); Winiwarter, 2022, especially pp. 39, 66–68. The link with the civilizing process is a novel contribution.

12 Sources and literature

12.1 Archival sources, objects, and oral history interviews

12.1.1 Archival sources

Agencija Republike Slovenije za okolje (ARSO; Slovenian Environment Agency, Ljubljana):

- Archives of phenological data: Koprivnik in Bohinj.

Arhiv knjižnice Oddeka za geografijo Filozofske fakultete Univerze v Ljubljani (AKOG FF UL; Archives of the Library of the Department of Geography, Faculty of Arts, University of Ljubljana):

- *Republika Slovenija – popis 1991* (Republic of Slovenia, 1991 Population Census): 2. 1. 03.
- *SR Slovenija – popis 1981: Popis prebivalstva, gospodinjstev in stanovanj* (Socialist Republic of Slovenia, 1981 Census of Population, Households, and Housing): 2–1–3. Part 3.

Arhiv Republike Slovenije (Archives of the Republic of Slovenia, Ljubljana):

SI AS 2, Deželni stanovi za Kranjsko (The Estates of the Province of Carniola):

- First registry, box 322.

SI AS 48, Deželna komisija za odkup in uravnavo zemljiških bremen za Kranjsko (Provincial Commission for Redemption and Regulation of *Profits à Prendre* in Carniola):

- Fascicle 56 (following my studies, this archival record group was moved from fascicles to boxes; the former fascicle 56 is now in boxes 84 and 85),
- Fascicle 163 (to identify the new box, order number 17–19 from the nineteenth-century organization of this fonds, which is also kept in the new boxes).

SI AS 137, Okrajno glavarstvo Radovljica (District Administrative Authority (Germ. *Bezirkshauptmannschaft*) of Radovljica, Germ. *Radmannsdorf*):

- Boxes 39, 121, 591, and 592.

SI AS 170, Soseska Koroška Bela (The Agrarian Community of Koroška Bela):

- Box 2.

SI AS 173, Imenjska knjiga za Kranjsko (*Gültbuch* for Carniola):

- *Gültbuch* No. 7, for Freeholders (*Freisasse*) (1662–1757),
- Rent-Roll (*Urbar*) of the Seigneury of Radovljica (Germ. *Radmannsdorf*), 1579.

SI AS 176, Franciscejski kataster za Kranjsko (Franciscan Land Cadaster for Carniola):

- L3, Cadastral Municipality of Planina pod Golico,
- L10, Cadastral Municipality of Jesenice,
- L16, Cadastral Municipality of Hrušica,
- L22, Cadastral Municipality of Podhom,
- L30, Cadastral Municipality of Nemški Rovt,
- L35, Cadastral Municipality of Blejska Dobrava,
- L56, Cadastral Municipality of Bohinjska Bistrica,
- L72, Cadastral Municipality of Gorjuše,
- L80, Cadastral Municipality of Hraše pri Radovljici,
- L90, Cadastral Municipality of Javornik,
- L102, Cadastral Municipality of Koroška Bela,
- L107, Cadastral Municipality of Češnjica v Bohinju,
- L139, Cadastral Municipality of Dovje,
- L157, Cadastral Municipality of Srednja Vas v Bohinju,
- L169, Cadastral Municipality of Nova Vas,
- L172, Cadastral Municipality of Nomenj,
- L179, Cadastral Municipality of Zgornje Gorje,
- L210, Cadastral Municipality of Poljšica,
- L227, Cadastral Municipality of Rečica,
- L242, Cadastral Municipality of Savica,
- L284, Cadastral Municipality of Studor,
- L287, Cadastral Municipality of Županje Njive,
- L310, Cadastral Municipality of Spodnje Gorje,
- L322, Cadastral Municipality of Višelnica,
- L335, Cadastral Municipality of Bohinjska Bela,
- L343, Cadastral Municipality of Selo pri Bledu.

SI AS 178, Franciscejski kataster za Koroško (Franciscan Land Cadaster for Carinthia):

- K312, Cadastral Municipality of Zgornje Jezersko,
- K470, Cadastral Municipality of Spodnje Jezersko.

SI AS 241, Ministrstvo za izvoz in uvod Ljudske republike Slovenije, 1945–1951 (Ministry of Export and Import of the People's Republic of Slovenia, 1945–1951):

- Box 1.

SI AS 721, Gospostvo Bled (The Seigneury of Bled, Germ. *Veldes*):

- Iustitialia, volumes 11, 12, 13, 14, and 21,
- Iustitialia, fascicles 25, 32, 33, and 34.

SI AS 736, Gospostvo Idrija (The Seigneury of Idrija, Germ. *Idria*):
- Box 1.

SI AS 1126, Državni sekretariat za blagovni promet Ljudske republike Slovenije, 1958–1965 (State Secretariat for Trade of the People's Republic of Slovenia, 1958–1965):
- Box 32.

SI AS 1131, Republiški sekretariat za trgovino Socialistične republike Slovenije, 1962–1965 (Secretariat for Trade of the Socialist Republic of Slovenia, 1962–1965):
- Box 15.

SI AS 2164, Zadružništvo na Slovenskem, 1922–1962 (Cooperatives in Slovenian Territory, 1922–1962):
- Boxes 2 and 3.

***Bundrov stan* Summer Cottage in the Laz Alp in the Julian Alps:**
- Notebook starting with "Planšarska staja v Lazu."

Geodetska uprava Republike Slovenije (GURS; Surveying and Mapping Authority of the Republic of Slovenia, Ljubljana):

Aerial Photos:
- Cadastral Municipality of Studor: years 1956, 1964, 1972, 1982, and 1992.

Kärntner Landesarchiv (Klagenfurt):

AT KLA 75, Franziszeischer Kataster:
- Cadastral Municipality no. 76203, Bad Vellach,
- Cadastral Municipality no. 76205, Ebriach,
- Cadastral Municipality no. 76211, Koprein Petzen,
- Cadastral Municipality no. 76213, Leppen,
- Cadastral Municipality no. 76214, Lobnig,
- Cadastral Municipality no. 76218, Remschenig.

AT KLA 112, Allgemeine Handschriftenreihe:
- Manuscripts 2435 and 2732.

AT KLA 129, Ansichtskartensammlung:
- E-Kahlkogel, 6.

AT KLA 207, Ständisches Archiv I:

Box 197.

Lechthaler Family Archive, Lasa (Germ. *Laas*) in the Venosta Valley (Germ. *Vinschgau*):
- Notebook Aufschrebung [!] für den Allitzbach im Jahre 1949.

Nadškofijski arhiv Ljubljana (Archdiocesan Archives in Ljubljana):

SI NŠAL 20, Gornji Grad Avstrija (Gornji Grad Austria):

- Boxes 17, 24, 26, 39, 40, 68, and 97.

SI NŠAL 100, Kapiteljski arhiv Ljubljana (Archives of the Ljubljana Cathedral Chapter):

- Fascicles 98 and 103.

Österreichisches Staatsarchiv, Kriegsarchiv (Vienna):

AT OeStA/KA FA (Feldakten) NFA (Neue Feldakten) HHK (Höhere Heereskommandos) AK/AGKdo (Armeekommandos und Armeegruppenkommandos (1914–1918)), 5. Armee (HGK (Heeresgruppenkommando) Boroević):

- Box 1108.

AT OeStA/KA FA NFA HHK HFK/HGK (Heeresfront- und Heeresgruppenkommandos) Kdo (Kommando der) Südwestfront II:

- Boxes 586, 587, 621.

Slovenski etnografski muzej, Dokumentacija (Slovenian Ethnographic Museum, Documentation, Ljubljana):

- Fieldwork Campaign 1, Šentjurij, Škocjan, Turjak (1948): Drawing 1/52b.
- Fieldwork Campaign 2, Šmarje–Sap, Polica (1949): Drawings 2/14 and 2/242.
- Fieldwork Campaign 4, Marezige (1950): Drawing 4/56.
- Fieldwork Campaign 5, Šentvid pri Stični (1950): Drawing 5/87.
- Fieldwork Campaign 7, Kobarid (1951): Drawings 7/120 and 7/161.
- Fieldwork Campaign 8, Trenta – Bovec (1952): Photo 8/138.
- Fieldwork campaign 10, Goriška Brda (year 1953): Photo 10/194.
- Fieldwork Campaign 11, Cerkno Region (1954): Drawings 11/31, 11/91, and 11/100, and Photo 11/53.
- Fieldwork Campaign 12, Brkini (1955): Drawing 12/7.
- Fieldwork Campaign 13, Kostanjevica (1956): Photo 13/266.
- Fieldwork Campaign 14, Žužemberk (1957): Drawing 14/18.
- Fieldwork Campaign 15, Vipava (1958): Drawing 15/13.
- Fieldwork Campaign 16, Črni Vrh, Vojsko (1959): Drawing 16/97 and Photo 16/135.
- Fieldwork Campaign 18, Šentrupert (1961): Photo 18/136.
- Fieldwork Campaign 20, Vitanje (1963): Photos 20/140 and 20/918.
- Fieldwork Campaign 21, Škocjan (1964): Notebook no. 3, and Photos 21/253, 21/413, and 21/436.
- Fieldwork Campaign 22, Drašiči (1965): Photo 22/264.
- Individual Fieldwork by Curators of the Slovenian Ethnographic Museum: Photos 1685/I and 1785/I.
- Jernej Šušteršič's Photo Collection: Photo 6141.

Steiermärkisches Landesarchiv (Graz):

AT StLA, Innerösterreichische Hofkammer, Sachabteilung:
- Box 69, folder 8.

AT StLA, Stadt Murau:
- Box 85, folder 2201.

Stiftsarchiv St. Paul im Lavanttal (Benediktinerstift St. Paul):

AT StiAStP, Grundherrschaft Eberndorf (Handschriften):
- Volumes 4, 179, 181, 183, 186, 187, 188, 189, 190, 191, 192, and 193.

Zavod Republike Slovenije za varstvo narave (ZRSVN; Institute of the Republic of Slovenia for Nature Conservation, Ljubljana):
- *Annex C.7 – Detailed Description of the Action C.7 [Project Life to Grasslands] Development and Economics Study of Dry Grasslands Preservation in the Haloze Project Sub-Area.* [2020].

12.1.2 Objects

Hodnik family barn in the Vogar montane meadow complex above Stara Fužina in Bohinj: old hayfork.

NMS (= National Museum of Slovenia, Ljubljana): Inv. nos. S 2591 and S 2592.

Zwitter Family Collection at Stara Fužina in Bohinj: painted souvenir plate from Obergurgl.

12.1.3 Oral history interviews

Anonymous Farmer, 2011. Farming by Residents of Kogelnik Farm in Planina pri Ljubnem, in the Mid-Twentieth Century. Planina pri Ljubnem (interviewed by Ž. Zwitter).

Anonymous Farmer, 2015. Grassland Management at Činžat in the Second Half of the Twentieth Century and Early Twenty-First Century. Činžat (interviewed by Ž. Zwitter).

Boltar, Angela, 2018. Traditional Grassland Management by Residents of Stara Fužina in the Mid-Twentieth Century. Stara Fužina (interviewed by Ž. Zwitter).

Ceklin, Tine, 2020. Traditional Farming by Residents of Stara Fužina Since the Mid-Twentieth Century. Stara Fužina (interviewed by Ž. Zwitter).

Cvetek, Jožef, 2021. Traditional Meadow Spring Cleaning. Stara Fužina (interviewed by Ž. Zwitter).

Frelih, Lojze, 2021. Hay Farming at the Zajamniki Alp. Zajamniki (interviewed by Ž. Zwitter).

Gartner, Gregor, 2017. Traditional Grassland Management by Residents of Studor in the Mid-Twentieth Century. Laz Alp in the Julian Alps (interviewed by Ž. Zwitter).

Gartner, Lucija, 2018 and 2022. Observations on Alpine Dock Consumption by Grazing Cows and Traditions on the Use of This Plant Species in Laz. Laz Alp in the Julian Alps (interviewed by Ž. Zwitter).

Gašperin, Janez, 2017. Traditional Grassland Management by Residents of Stara Fužina in the Mid-Twentieth Century. Stara Fužina (interviewed by Ž. Zwitter).

Grm, Antonija, 2017. Traditional Grassland Management by Residents of Srednja Vas v Bohinju in the Mid-Twentieth Century. Srednja Vas v Bohinju (interviewed by Ž. Zwitter).

Hodnik, Angela, 2019. Twentieth-Century History of Heartleaf Oxeye (*ruska roža*) in Bohinj. Srednja Vas v Bohinju (interviewed by Ž. Zwitter).

Hodnik, Lovro, 2018. Wetland Management on Vogar Hill in the Second Half of the Twentieth Century. Vogar Hill (interviewed by Ž. Zwitter).

Hodnik, Marjeta, 2018 and 2021. Traditional Grassland Management and Uses by Residents of Stara Fužina and Studor in Bohinj Since the Mid-Twentieth Century. Stara Fužina (interviewed by Ž. Zwitter).

Hodnik, Terezija, 2017. Grassland Management by Residents of Stara Fužina and Studor in the Second Half of the Twentieth Century. Stara Fužina (interviewed by Ž. Zwitter).

Hribar, Anton, 2017. Haymaking in Montane Meadows in Bohinj in the Second Half of the Twentieth Century. Bohinjska Češnjica (interviewed by Ž. Zwitter).

Hribar, Marjana, 2022. The Use of Hayracks. Bohinjska Češnjica (interviewed by Ž. Zwitter).

Janež, Ludvik, 2017. Traditional Grassland Management by Residents of Čadrg in the Mid-Twentieth Century. Čadrg (interviewed by Ž. Zwitter).

Kavčič, Anica, 2017. Mid-Twentieth-Century Farming and Gathering in the Grintojca and Laz Alps. Laz Alp in the Julian Alps (interviewed by Ž. Zwitter).

Kavčič, Janez, 2021. Traditional Grassland Management by Residents of Dolina near Tržič in the Mid-Twentieth Century. Dolina (interviewed by Ž. Zwitter).

Kranjc, Katarina, 2018 and 2020. Traditional Grassland Management by Residents of Stara Fužina in the Mid-Twentieth Century. Stara Fužina (interviewed by Ž. Zwitter).

Kunavec, Marjeta, 2017, 2018, 2019, 2021, and 2023. Traditional Grassland Management by Residents of Stara Fužina in the Mid-Twentieth Century. Stara Fužina (interviewed by Ž. Zwitter).

Lamot, Edvard, 2019. Spring Cleaning in a Former Meadow with Larch Trees on the Slope of the Pohorje Hills at Rakovec. Rakovec (interviewed by Ž. Zwitter).

Markeš, Martina, 2020. The Diversity of Attitudes toward Watercress Salad in Stara Fužina in the Second Half of the Twentieth Century. Stara Fužina (interviewed by Ž. Zwitter).

Mencinger, Ivanka, 2017, 2018, and 2020. Traditional Grassland Management by Residents of Stara Fužina in the Mid-Twentieth Century. Stara Fužina (interviewed by Ž. Zwitter).

Mlakar, Cilka, 2015, 2017, 2018, 2019, and 2020. Traditional Grassland Management in the Krstenica and Zajamniki Alps in the Mid-Twentieth Century. Krstenica and Zajamniki Alps (interviewed by Ž. Zwitter).

Mlakar, Franc, 2017 and 2018. Traditional Grassland Management by Residents of Bohinjska Češnjica in the Mid-Twentieth Century. Bohinjska Češnjica (interviewed by Ž. Zwitter).

Mlakar, Ivanka, 2020 and 2021. Traditional Grassland Management by Residents of Bohinjska Češnjica in the Mid-Twentieth Century. Zajamniki Alp (interviewed by Ž. Zwitter).

Odar, Minka, and Lovrenc Odar, 2017. Traditional Grassland Management by Residents of Studor in the Mid-Twentieth Century. Studor (interviewed by Ž. Zwitter).

Plesnik, Sophie, 2021. Wetland Farming and the Use of Wetland Resources at Lake Gösselsdorf (Sln. *Goslinsko jezero*) in the Mid-Twentieth Century. Sapotnik (Sln. *Zapotnik*) farm on the shore of Lake Gösselsdorf (interviewed by Ž. Zwitter).

Prepotnik, Ančka, and Marija Prepotnik, 2012. Traditional Farming by Residents of Haudej Farm in the Solčava Region in the Mid-Twentieth Century. Haudej farm (interviewed by Ž. Zwitter).

Rozman, Marija, 2020. Personal Experience with Autumn Crocus Gathering at Srednja Vas, Bohinj. Srednja Vas v Bohinju (interviewed by Ž. Zwitter).

Šetina, Marija, 2018. Traditional Grassland Management by Residents of the Isolated Farm at Holy Spirit Church in Bohinj in the Mid-Twentieth Century. Srednja Vas v Bohinju (interviewed by Ž. Zwitter).

Škantar, Janez, 2017. Grassland Management by Residents of Srednja Vas v Bohinju and Neighboring Villages in the Mid-Twentieth Century. Srednja Vas v Bohinju (interviewed by Ž. Zwitter).

Smukavec, Jože, 2017. Traditional Grassland Management by Residents of Zgornje Podjelje in the Mid-Twentieth Century. Zgornje Podjelje (interviewed by Ž. Zwitter).

Smukavec, Matej, 2021. Hay Farming at Javornica near Zgornje Podjelje. Javornica Alp (interviewed by Ž. Zwitter).

Smukavec, Vida, 2017. Traditional Grassland Management by Residents of Spodnje Podjelje in the Mid-Twentieth Century. Spodnje Podjelje (interviewed by Ž. Zwitter).

Sodja, Milena, 2017. Traditional Grassland Management by Residents of Srednja Vas v Bohinju in the Mid-Twentieth Century. Srednja Vas v Bohinju (interviewed by Ž. Zwitter).

Sodja, Minka, 2017. Traditional Grassland Management by Residents of Češnjica and Jereka in the Mid-Twentieth Century. Jereka (interviewed by Ž. Zwitter).

Sodja, Roman, 2022. Matgrass Farming at Ši above Srednja Vas. Stara Fužina (interviewed by Ž. Zwitter).

Soklič, Anton, 2017 and 2018. Grassland Management at the Isolated Gorjup Farm in Podjelje and Traditional Grassland Management by Residents of Podjelje in the Mid-Twentieth Century. Zgornje Podjelje (interviewed by Ž. Zwitter).

Stergar, Metka, 2018. Traditional Grassland Management by Residents of Stara Fužina in the Mid-Twentieth Century. Stara Fužina (interviewed by Ž. Zwitter).

Taler, Anton, and Matilda Taler, 2017. Traditional Grassland Management by Residents of Studor and Stara Fužina in the Mid-Twentieth Century. Studor (interviewed by Ž. Zwitter).

Urh, Julka, 2017 and 2022. Traditional Grassland Management by Residents of Stara Fužina in the Mid-Twentieth Century. Stara Fužina (interviewed by Ž. Zwitter).

12.2 Literature and published sources

15th International Wild Flower Festival: 28 May – 13 June 2021. https://www.bohinj.si/en/wild-flower-festival/ (September 25th, 2021).

18th International Wild Flower Festival. https://www.tnp.si/en/event/66/18th-international-wild-flower-festival.html (September 25th, 2024).

19th International Wild Flower Festival: 23. May – 8. June 2025. https://www.bohinj.si/en/international-wild-flower-festival/ (December 31st, 2024).

A., Z., 1967. "Svetski rat, prvi: Italijansko ratište" (First World War: Italian Battlefield). In Todorović, V., et al. (eds.), *Vojna enciklopedija 9: Slup–Teleskop*. Belgrade: Redakcija Vojne enciklopedije, 375–376, 381, 395–396, 404, and map supplements between pp. 374 and 375, 378 and 379, 396 and 397, 410 and 411.

Abbet, C., Mayor, R., Roguet, D., Spichinger, R., Hamburger, M., and O. Potterat, 2014. "Ethnobotanical Survey on Wild Alpine Food Plants in Lower and Central Valais (Switzerland)." *Journal of Ethnopharmacology* 151: 624–634 and supplementary materials.

Achilles, W., 1989. "Umwelt und Landwirtschaft in vorindustrieller Zeit." In Herrmann, B. (ed.), *Umwelt in der Geschichte: Beiträge zur Umweltgeschichte*. Göttingen: Vandenhoeck & Ruprecht, 77–88.

Acs, P., Wilhalm, T., and K. Oeggl, 2005. "Remains of Grasses Found with the Neolithic Iceman 'Ötzi.'" *Vegetation History and Archaeobotany* 14(3): 198–206.

Aeschimann, D., Lauber, K., Moser, D.M., and J.-P. Theurillat, 2004. *Flora alpina: Ein Atlas sämtlicher 4500 Gefäßpflanzen der Alpen*. Vols. 1 and 2. Bern, Stuttgart, and Vienna: Haupt.

Ágoston, G., 2009. "Warfare." In Ágoston, G., and B. Masters (eds.), *Encyclopedia of the Ottoman Empire*. New York: Facts On File, 591–595.

Ágoston, G., 2010. "Empires and Warfare in East-Central Europe, 1550–1750: The Ottoman–Habsburg Rivalry and Military Transformation." In Tallett, F., and D.J.B. Trim (eds.), *European Warfare 1350–1750*. Cambridge: Cambridge University Press, 110–134.

Agten, N., 2010. "Landwirtschaftliche Nutzungsmöglichkeit der Kulturlandschaft Turtmann - Unterems 'Kastleren.'" Kant: Dienststelle für Landwirtschaft Visp, CH. http://www.burger-turtmann.ch/starten/Nutzungsm%C3%B6glichkeit.pdf (September 5th, 2019).

Aichinger, E., 1933. *Vegetationskunde der Karawanken*. Jena: Gustav Fischer.

Akeret, Ö., and I. Geith-Chauviere, 2011. "Les macrorestes végétaux." In Grau Bitterli, M.H., and E. Fierz-Dayer (eds.), *Plateau de Bevaix, 6. Bevaix/Treytel-À Sugiez: Histoire d'un complexe mégalithique néolithique, témoins d'habitats du Campaniforme et du Bronze ancien*. Neuchâtel: Office et musée cantonal d'archéologie: 311–321.

Alič, V., 2011. "Spomenik posoškim pastirjem in sirarjem" (A Monument to Herders and Cheese-Makers in the Soča River Basin). *Dnevnik* (October 3rd, 2011): 7.

Alpine Convention: Biodiversity & Nature Conservation. https://www.alpconv.org/en/home/topics/biodiversity-nature-conservation/ (July 11th, 2024).

Ambrosoli, M., 1993. "Erbe selvatiche e leguminose nel lungo periodo: Il caso della *Medicago* prima del 1600." In Grieco, A.J., Redon, O., and L. Tongiorgi Tomasi (eds.), *Le Monde vegetal (XIIe-XVIIe siècles): Savoirs and usages sociaux*. Saint-Denis: Presses universitaires de Vincennes, 59–70.

Ambrožič, M., 2018. "Posledice terezijanskih reform in ukrepov v Cerkvi na Slovenskem" (Folgen theresianischer Reformen und Massnahmen auf die Kirche in slowenischen Gebieten). In Preinfalk, M., and B. Golec (eds.), *Marija Terezija: Med razsvetljenskimi reformami in zgodovinskim spominom*. Ljubljana: ZRC SAZU, Zgodovinski inštitut Milka Kosa, 217–237.

Amon, S., 1991. "Kmetijske in rokodelske novice" (Farmers' and Craftsmen's News). In Batis, J., et al. (eds.), *Enciklopedija Slovenije*. Vol. 5, *Kari–Krei*. Ljubljana: Mladinska knjiga, 117–118.

Andreae, T., 2022. “Die Schattenseiten von Hubatscheks Fotos.” https://tirol.orf.at/stories/3186072/ (December 14th, 2022).

Andrejka, R., 2013. “Ponebšek, Janko (1861–1935).” In *Slovenska biografija*. https://www.slovenska-biografija.si/oseba/sbi448296/ (December 12th, 2024).

Andrič, M., Martinčič, A., Štular, B., Petek, F., and T. Goslar, 2010. “Land-Use Changes in the Alps (Slovenia) in the Fifteenth, Nineteenth and Twentieth Centuries AD: A Comparative Study of the Pollen Record and Historical Data.” *The Holocene* 20(7): 1023–1037.

Andrič, M., Sabatier, P., Rapuc, W., Ogrinc, N., Dolenec, M., et al., 2020. “6600 Years of Human and Climate Impacts on Lake-Catchment and Vegetation in the Julian Alps (Lake Bohinj, Slovenia).” *Quaternary Science Reviews* 227: separate pagination, 1–18.

Andritsch, J., 1989. “Judenburg – der erste urkundlich genannte Markt in der Steiermark.” In Schöpfer, G. (ed.), *Menschen & Münzen & Märkte: Steirische Landesausstellung 1989 Judenburg*. Fohnsdorf: K. Podmenik, 27–37.

Angerer, T., 1987. *Čisava župa, pisana pogača in še kaj: Stare domače jedi iz južne Koroške* (Sour Soup, Colorful Cake and More: Old Traditional Dishes from Southern Carinthia). Klagenfurt: Mohorjeva založba.

Arcanum Maps, Cadastral Maps, Habsburg Empire – Cadastral Maps (XIX. [sic] Century). https://maps.arcanum.com/en/map/cadastral/?layers=3%2C4&bbox=1806636.9366629783%2C6135492.115519254%2C1834154.266845642%2C6145505.366224611 (November 23rd, 2022).

Arcanum Maps, Maps of Europe, Europe in the XVIII. [sic] Century. https://maps.arcanum.com/en/map/europe-18century-firstsurvey/?layers=163%2C165&bbox=5503.46603653254%2C5620109.546756446%2C2472279.2428557407%2C6532461.916368309 (November 20th, 2022).

Arcanum Maps, Maps of Europe, Europe in the XIX. [sic] Century (with the Third Military Survey). https://maps.arcanum.com/en/map/europe-19century-thirdsurvey/?layers=160%2C166&bbox=358336.78860090615%2C5755861.708990919%2C2119445.920291367%2C6396709.754133836 (November 23rd, 2022).

Archives Portal Europe: Eberndorf, Augustiner-Chorherrenstift (Urkunden). https://www.archivesportaleurope.net/advanced-search/search-in-archives/results-(archives)/?&repositoryCode=AT-KLA&term=&levelName=archdesc&t=fa&recordId=AT-KLA%20308 (December 31st, 2023).

Archivio di Stato di Trieste, Catasto franceschino, Mappe del Catasto franceschino, Distretto di Tolmino, Comune di Goriano di Circhina, Mappe colorate di primo rilevamento indicate come tipologia b (1822 – sec. XIX primo quarto). https://a4view.archiviodistatotrieste.it/patrimonio/a9afc623-0c94-47a7-99ea-7c8e28d3a4d8/212-b-mappe-colorate-di-primo-rilevamento-indicate-come-tipologia-b-1822-sec-xix-primo-quarto (June 15th, 2024).

ARSO (Agencija Republike Slovenije za okolje): Arhiv – opazovani in merjeni meteorološki podatki po Sloveniji (Slovenian Environment Agency, Archives—Observed and Measured Meteorological Data in Slovenia). https://meteo.arso.gov.si/met/sl/app/webmet/#webmet==8Sdwx2bhR2cv0WZ0V2bvEGcw9ydlJWblR3LwVnaz9SYtVmYh9iclFGbt9SaulGdugXbsx3cs9mdl5WahxXYyNGapZXZ8tHZv1WYp5mOnMHbvZXZulWYnwCchJXYtVGdlJnOn0UQQdSf; (December 19th, 2019).

Aschwanden, R., Buck, M., Kupper, P., and K.J. Schmidt, 2022. "Moving Mountains: The Protection of the Alps." In Wöbse, A.K., and P. Kupper (eds.), *Greening Europe: Environmental Protection in the Long Twentieth Century – A Handbook*. Munich and Vienna: De Gruyter Oldenbourg, 217–241.

Ash, H.B., 1993. "Introduction: Life and Works of Columella." In Ash, H.B. (ed.), *L.J.M. Columella, On Agriculture I–IV*. Vol. 1. Cambridge, MA, and London: Harvard University Press, XIII–XXV.

Atlas Slovenije (Atlas of Slovenia). 2005. Ljubljana, etc.: Mladinska knjiga and Geodetski zavod Slovenije.

Austad, I., and L. Hauge, 1989. "Restoration and Management of Historical Cultural Landscapes—An Important Aspect of Landscape Ecology." *Landschaft + Stadt* 21(4): 148–157.

Austrian Commission for UNESCO: Immaterielles Kulturerbe: Bräuche, Wissen, Handwerkstechniken: Wissen um die Standorte, das Ernten und das Verarbeiten des Punktierten Enzians: https://www.unesco.at/kultur/immaterielles-kulturerbe/oesterreichisches-verzeichnis/detail/article/wissen-um-die-standorte-das-ernten-und-das-verarbeiten-des-punktierten-enzians (June 3rd, 2021).

Axmanová, I., Kalusová, V., Danihelka, J., Dengler, J., Pergl, J., et al., 2021. "Neophyte Invasions in European Grasslands." *Journal of Vegetation Science* 32(2): separate pagination 1–17.

Babai, D., Molnár, Á., and Z. Molnár. 2014. *'Ahogy gondozza, úgy veszi hasznát': Hagyományos ökológiai tudás és gazdálkodás Gyimesben* ('As You Care for It, So You Benefit from It': Traditional Ecological Knowledge and Land Use in Ghimeş, Hung. *Gyimes*). Budapest: Vácrátót.

Babai, D., and Z. Molnár, 2013. "Multidimensionality and Scale in a Landscape Ethnoecological Partitioning of a Mountainous Landscape (Gyimes, Eastern Carpathians, Romania)." *Journal of Ethnobiology and Ethnomedicine* 9(11): separate pagination 1–21.

Babai, D., and Z. Molnár, 2014. "Small-Scale Traditional Management of Highly Species-Rich Grasslands in the Carpathians." *Agriculture, Ecosystems and Environment* 182: 123–130 and the electronic appendix.

Babai, D., and Z. Molnár, 2016. "Species-Rich Mountain Grasslands through the Eyes of the Farmer: Flora, Species Composition, and Extensive Grassland Management." *Martor* 21: 147–169.

Bailey, J.P., Child, L.E., and A.P. Conolly, 1996. "A Survey the Distribution of *Fallopia x bohemica* (Chrtek & Chrtková) J. Bailey (Polygonaceae) in the British Isles." *Watsonia* 21(2): 187–198.

Bajuk Senčar, T., 2021. "Evropeizacija in lokalizacija komemoracij ob stoletnici prve svetovne vojne" (Europeanization and Localization of First World War Centenary Commemorations). In Fikfak, J., and B. Jezernik (eds.), *Dediščina prva svetovne vojne: Reprezentacije in reinterpretacije*. Ljubljana: Znanstvena založba Filozofske fakultete Univerze v Ljubljani, 229–245.

Bakker, J.P., 1989. *Nature Management by Grazing and Cutting*. Dordrecht: Kluwer Academic Publishers.

Baldock, D., and J.E. Petersen, 2012. "Conclusions—What Does the Future Hold for HNV Farming?" In Oppermann, R., Beaufoy, G., and G. Jones (eds.), *High Nature Value Farming in Europe: 35 European Countries and Perspectives*. Ubstadt-Weiher, Heidelberg, and Basel: Regionalkultur, 537–539.

Balkovec, B., 2013. "The Slovenian—Slovenec (1873–1945): Political Newspaper for the Slovenian People: Brief Overview of the History to Mark the 140th Anniversary of the Beginning of its Publishing." In Tominšek Perovšek, M. (ed.), *Slovenec: Političen list za slovenski narod (1873–1945): Kratek pregled zgodovine ob 140-letnici njegovega izhajanja*. Ljubljana: Muzej novejše zgodovine Slovenije, 197.

Ballweber, L.R., 2024. "*Fasciola hepatica* in Ruminants." In *MSD Manual: Veterinary Manual*. https://www.msdvetmanual.com/digestive-system/fluke-infections-in-ruminants/fasciola-hepatica-in-ruminants (December 30th, 2024).

Balon, E.K., 1995. "Origin and Domestication of the Wild Carp, *Cyprinus carpio*: From Roman Gourmets to the Swimming Flowers." *Aquaculture* 129: 3–48.

Barão, M.J., and A. Soveral Dias, 2013. "The Ecology and Use of Edible Thistles in Évora, Alentejo, Southeastern Portugal." In Pardo-de-Santayana, M., Pieroni, A., and R.K. Puri (eds.), *Ethnobotany in the New Europe: People, Health and Wild Plant Resources*. New York and Oxford: Berghahn Books, 189–210.

Baravalle, R., 1929. "Zur Geschichte des Grazer Maßes." *Zeitschrift des Historischen Vereines für Steiermark* 25: 47–78.

Baravalle, R., 1931. "Zur Geschichte des Judenburger Maßes." *Zeitschrift des Historischen Vereines für Steiermark* 26: 190–199.

Bartosiewicz, L., 2006. "Animal Bones from the Medieval Settlement Otok (Gutenwerth) near Dobrava pri Škocjanu, Slovenia." *Arheološki vestnik* 57: 457–478.

Bartosiewicz, L., Biller, A.Z., Csippán, P., Daróczi-Szabó, L., Daróczi-Szabó, M., et al., 2018. "Animal Exploitation in Medieval Hungary." In Laszlovsky, J., Nagy, B., Szabó, P., and A. Vadas (eds.), *The Economy of Medieval Hungary*. Leiden and Boston: Brill, 113–165.

Baš, A., 1992. *Oblačilna kultura na Slovenskem v 17. in 18. stoletju* (La culture vestimentaire en Slovénie aux 17e et 18e siècles). Ljubljana: Državna založba Slovenije.

Baš, F., 1937. "Izgoni na Dravskem polju" (Die Bewässerung der Wiesenlandschaft am Draufelde). *Časopis za zgodovino in narodopisje* 32: 325–340.

Baš, F., 1984. "Kozolec na Slovenskem" (The Hayrack in Slovenian Territory). In Baš, A. (ed.), *Stavbe in gospodarstvo na slovenskem podeželju*. Ljubljana: Slovenska matica, 332–338.

Bätzing, W., 2007. "Ein Leben für die Bergbauern: Alpenlandschaften als Produkt bäuerlicher Arbeit: Die Fotos von Erika Hubatschek – mehr als die Dokumentation einer untergangenen Welt." In Merlin, F.W., Hellebart, S., and M. Machatschek (eds.), *Bergwelt im Wandel: Festschrift Erika Hubatschek zum 90. Geburtstag*. Klagenfurt: Kärntner Landesarchiv, 11–19.

Bätzing, W., 2015. *Die Alpen: Geschichte und Zunkunft einer europäischen Kulturlandschaft*. 4th revised and expanded ed. Munich: C.H. Beck.

Bauer, M., 2017. "Die Schätzungsoperate des Franziszeischen Katasters als agrarhistorische Quelle." In Boscani Leoni, S., and M. Stuber (eds.), *Wer das Gras wachsen hört: Wissensgeschichte(n) der pflanzlichen Ressourcen vom Mittelalter bis ins 20. Jahrhundert*. Innsbruck, Vienna, and Bolzano: Studienverlag, 216–250.

Baumgartner, S., 2017. "'Nützliche Gras-Arten und Kräuter': Die Zürcher *Ökonomische Kommission* und das Wissen vom Klee- und Wiesenbau." In Boscani Leoni, S., and M. Stuber (eds.), *Wer das Gras wachsen hört: Wissensgeschichte(n) der pflanzlichen Ressourcen vom Mittelalter bis ins 20. Jahrhundert*. Innsbruck, Vienna, and Bolzano: Studienverlag, 133–150.

Bavcon, J., Ravnjak, B., and N. Praprotnik, 2019. *Meadows—Steep and Colourful Grasslands*. Ljubljana: University of Ljubljana, Biotechnical Faculty, Department of Biology, University Botanic Gardens Ljubljana.

Beaufoy, G., Keenleyside, C., and R. Oppermann, 2012. "How should EU and National Policies Support HNV Farming?" In Oppermann, R., Beaufoy, G., and G. Jones (eds.), *High Nature Value Farming in Europe: 35 European Countries and Perspectives*. Ubstadt-Weiher, Heidelberg, and Basel: Regionalkultur, 525–535.

Beaufoy, G., Oppermann, R., and M.L. Paracchini, 2012. "European overview of HNV farmland types." In Oppermann, R., Beaufoy, G., and G. Jones (eds.), *High Nature Value Farming in Europe: 35 European Countries and Perspectives*. Ubstadt-Weiher, Heidelberg, and Basel: Regionalkultur, 27–31.

Beer, S., 1989. "Neubeginn und Wiederaufbau." In Schöpfer, G. (ed.), *Menschen & Münzen & Märkte: Steirische Landesausstellung 1989 Judenburg*. Fohnsdorf: Ing. K. Podmenik, 207–214.

Begon, M., Harper, J.L., and C.R. Townsend, 1991. *Okologie: Individuen, Populationen und Lebensgemeinschaften*. Basel: Birkenhäuser.

Beguš, I., 2015. *Avtonomija in ekonomija Nadiških dolin v Beneški republiki* (Autonomy and Economy of the Natisone River Catchment in the Republic of Venice). Koper: Univerza na Primorskem, Znanstveno-raziskovalno središče, Univerzitetna založba Annales.

Bekker, R.M., Verweij, G.L., Bakker, J.P., and L.F. Fresco, 2000. "Soil Seed Bank Dynamics in Hayfield Succession." *Journal of Ecology* 88(4): 594–607.

Bernhard, G. (ed.), 2006. *Documenta patriarchalia res gestas slovenicas illustrantia*. Vienna and Ljubljana: Slowenisches Wissenschaftsinstitut in Wien.

Bertoncelj, I., and T. Travnikar, 2024. "Farmer Participation in CAP Agri-Environment Measures for Biodiversity Conservation in Triglav National Park, Slovenia." *Italian Review of Agricultural Economics* 79(1): 75–83.

Beutler, C., 1973. "Un chapitre de la sensibilité collective: La littérature agricole en Europe continentale au XVIe siècle." *Annales: Économies, Sociétés, Civilisations* 28(5): 1280–1301.

Bevk, D., 2018. *Pestrost opraševalcev za zanesljivo pridelavo hrane* (Pollinator Diversity for Safe Food Production). Ljubljana: Nacionalni inštitut za biologijo.

Beyer-Thoma, H., 2007. "Schreber, Johann Christian Daniel Edler von." In *Neue Deutsche Biographie*. Vol. 23, 524–525. https://www.deutsche-biographie.de/sfz79145.html#ndbcontent (October 8th, 2021).

Bhatia, H., Manhas, R.K., Kumar, K., and R. Magotra, 2014. "Traditional Knowledge on Poisonous Plants of Udhampur District of Jammu and Kashmir, India." *Journal of Ethnopharmacology* 152: 207–216.

Bible: YouVersion, New American Standard Bible, NASB. 1995. https://www.bible.com/bible/compare/DEU.22.6-7 (December 14th, 2024).

"Bildteil." 1996. In Anderegg, K., and W. Bätzing (eds.), *Alpine Alltagskultur zwischen Beharrung und Wandel: Ausgewählte Arbeiten aus den Jahren 1956 bis 1991*. 2nd revised ed. Bern, Stuttgart, and Vienna: Paul Haupt, 385–468, 515–516.

Billamboz, A., 2014. "Dendroarchaeology and Cockchafers North of the Alps: Regional Patterns of a Middle Frequency Signal in Oak Tree-Ring Series." *Environmental Archaeology* 19(2): 114–123.

Bingener, A., 2010. "Stahl: 1. Technik." In Jaeger, F., et al. (eds.), *Enzyklopädie der Neuzeit*. Vol. 12, *Renaissance–Signatur*. Stuttgart and Weimar: J.B. Metzler, 809–815.

Birr, C., 2004. "Weistümer und 'Ländliche Rechtsquellen'." In Pauser, J., Scheutz, M., and T. Winkelbauer (eds.), *Quellenkunde der Habsburgermonarchie (16.–18. Jahrhundert): Ein exemplarisches Handbuch*. Vienna and Munich: R. Oldenbourg, 390–408.

Birrer, S., and R. Oppermann 2012. "Recreating HNV Farmland and Improving Nature Value—Farmers Taking Over New Challenges in Central Europe." In Oppermann, R., Beaufoy, G., and G. Jones (eds.), *High Nature Value Farming in Europe: 35 European Countries and Perspectives*. Ubstadt-Weiher, Heidelberg, and Basel: Regionalkultur, 484–490.

Bisse du Torrent Neuf/Savièse. https://www.les-bisses-du-valais.ch/en/Bisse/Bisse-du-Torrent-Neuf-Saviese/#situation (October 19th, 2020).

Bizjak, M., 2003. Ratio facta est: *Gospodarska struktura in poslovanje poznosrednjeveških gospostev na Slovenskem* (*Ratio facta est*: The Economic Structure and the Operation of Estates in Slovenian Territories in the Late Middle Ages). Ljubljana: ZRC.

Bizjak, M. 2014. "Medieval Account Books as a Source for Environmental History." In Štih, P., and Ž. Zwitter (eds.), *Man, Nature and Environment between the Northern Adriatic and the Eastern Alps in Premodern Times.* Ljubljana: Ljubljana University Press, Faculty of Arts, and Historical Association of Slovenia, 124–143.

Bizjak, M., 2018. "Integriranje dohodkov v kmečki ekonomiji v poznem srednjem veku" (Integrating Incomes in the Late Medieval Peasant Economy). In Panjek, A., and Ž. Lazarević (eds.), *Preživetje in podjetnost: Integrirana kmečka ekonomija na Slovenskem od srednjega veka do danes.* Koper: Univerza na Primorskem, 127–153.

Bizjak, M., Šilc, J., Seručnik, M., and N. Makuc, 2022. *Historična topografija Primorske (do leta 1500): Na podlagi gradiva Milka Kosa* (Historical Topography of the Littoral (until 1500): Based on Data Collected by Milko Kos). Ljubljana: ZRC SAZU, Zgodovinski inštitut Milka Kosa.

Blaj Hribar, N., Ojsteršek, M., Pančur, A., and M. Šorn, 2019. "V prvi svetovni vojni padli in umrli vojaki z ozemlja Slovenije" (Soldiers from the Territory of Slovenia that Fell and Died in the First World War). *Vojaška zgodovina* 15(2): 10–22.

Blanchard, I., 1986. "The Continental European Cattle Trades, 1400–1600." *Economic History Review* 2nd series 39(3): 427–460.

Blaznik, P. (ed.), 1963. *Urbaria aetatis mediae Sloveniam spectantia*. Vol. 4, *Urbaria episcopatus frisingensis*. Ljubljana: Academia scientiarum et artium slovenica.

Blaznik, P., 1970 (= Blaznik, 1970a). "Enote individualne posesti" (Einheiten des individuellen Besitzes). In Blaznik, P., et al. (eds.), *Gospodarska in družbena zgodovina Slovencev: Zgodovina agrarnih panog.* Vol. 1, *Agrarno gospodarstvo.* Ljubljana: Slovenska akademija znanosti in umetnosti and Državna založba Slovenije, 161–184.

Blaznik, P., 1970 (= Blaznik, 1970b). "Kolektivna kmečka posest" (Kollektiver Bauernbesitz). In Blaznik, P., et al. (eds.), *Gospodarska in družbena zgodovina Slovencev: Zgodovina agrarnih panog.* Vol. 1, *Agrarno gospodarstvo.* Ljubljana: Slovenska akademija znanosti in umetnosti and Državna založba Slovenije, 149–160.

Blaznik, P., 1970 (= Blaznik, 1970c). "Kolonizacija in populacija: Dodatna kolonizacija" (Kolonisation und Bevölkerung: Zusätzliche Siedlungen). In Blaznik, P., et al. (eds.), *Gospodarska in družbena zgodovina Slovencev: Zgodovina agrarnih panog.* Vol. 1, *Agrarno gospodarstvo.* Ljubljana: Slovenska akademija znanosti in umetnosti and Državna založba Slovenije, 88–90.

Blaznik, P., 1975. "Bitenj in franciscejski kataster" (Die Gemeinde Bitenj [i.e., Bitnje] und der Franziszeische Kataster). *Loški razgledi* 22(1): 83–93.

Blöschl, G., Kiss, A., Viglione, A., Barriendos, M., Böhm, O., et al., 2020. "Current European Flood-Rich Period Exceptional Compared with Past 500 Years." *Nature* 583: 560–566.

Boetzl, F.A., Krauss, J., Heinze, J., Hoffmann, H., Juffa, J., et al., 2021. "A Multitaxa Assessment of the Effectiveness of Agri-Environmental Schemes for Biodiversity Management." *Proceedings of the National Academy of Sciences of the United States of America* 118(10): e2016038118, separate pagination 1–9.

Bohner, A., 2008. "The Importance of Avalanches and Avalanche Tracks from an Ecological Point of View." In *Monitor/08 – Systems Behind a Safer Environment*. Irdning: Research and Education Centre Raumberg-Gumpenstein, Austria, 83–88.

Bohner, A., Habeler, H., Starlinger, F., and M. Suanjak, 2010. "Avalanches Keep Habitats Open and Species-Rich in the Montane and Subalpine Belt." *eco. mont* 2(1): 53–57.

Bonacci, O., 2004. "Poljes." In Gunn, J. (ed.), *Encyclopedia of Caves and Karst Science*. New York: Fitzroy Dearborn, 599–600.

Boršnik, M., 2013. "Tavčar, Ivan (1851–1923)." In *Slovenska biografija*. http://www.slovenska-biografija.si/oseba/sbi686990/#slovenski-biografski-leksikon (January 3rd, 2024).

Bortenschlager, S., 1999. "Die Umwelt des Mannes aus dem Eis und sein Einfluß darauf." In *Die Gletschermumie aus der Kupferzeit: Neue Forschungesergebnisse zum Mann aus dem Eis*. Bolzano and Vienna: Folio, 81–95.

Boscani Leoni, S., and M. Stuber (eds.), 2017. *Wer das Gras wachsen hört: Wissensgeschichte(n) der pflanzlichen Ressourcen vom Mittelalter bis ins 20. Jahrhundert*. Innsbruck, Vienna, and Bozen: Studienverlag.

Bosi, G., Labate, D., Rinaldi, R., Montecchi, M.C., Mazzanti, M., et al., 2019. "A Survey of the Late Roman Period (3rd–6th Century AD): Pollen, NPPs and Seeds/Fruits for Reconstructing Environmental and Cultural Changes after the Floods in Northern Italy." *Quaternary International* 499: 3–23.

Božič, L., 2019. "Kosec" (Corncrake). In Mihelič, T., et al. (eds.), *Atlas ptic Slovenije: popis gnezdilk 2002–2017*. Ljubljana: DOPPS, 138–139.

Bratož, R., 2000. "Socialni vidiki širjenja krščanstva (1.–3. stoletje)" (Social Aspects of the Spread of Christianity, 1st–3rd Century). *Zgodovina v šoli* 9(3–4): 16–25.

Bratož, R., 2017. "Zur Präsenz und Mobilität ethnischer Kleingruppen im Alpen-Adria-Raum während der Ostgotenherrschaft." In Dörfler, I., et al. (eds.), *Ad amussim: Festschrift zum 65. Geburtstag von Franz Glaser*. Klagenfurt am Wörtersee: Landesmuseum für Kärnten; Vienna: Österreichisches Archäologisches Institut an der Österreichischen Akademie der Wissenschaften, 215–248.

Breg Valjavec, M., Zorn, M., and D. Ribeiro, 2018. "Mapping War Geoheritage: Recognising Geomorphological Traces of War." *Open Geosciences* 10: 385–394.

Bregant, T., 2023. "Primer dobre prakse za ohranitev travnikov" (An Example of a Good Practice to Preserve Meadows). *Svet pod Triglavom* 31: 24–25.

Brekelmans, C., 1968. "Deuteronomium." In Haag, H. (ed.), *Bibel-Lexikon*. Einsiedeln, Zürich, and Cologne: Benzinger, 327–330.

Břicháček, P., and M. Beranová, 1993. "Beitrag zur Erkundung der landwirtschaftlichen Produktion in der späthallstattzeitlichen und latènezeitlichen Periode in Böhmen." *Archeologické rozhledy* 45: 251–267, 358–359.

Brinkschulte, E., 2010. "Schwangerschaftsabbruch." In Jaeger, F., et al. (eds.), *Enzyklopädie der Neuzeit*. Vol. 11, *Renaissane–Signatur*. Stuttgart and Weimar: J.B. Metzler, 962–964.

Brisebarre, A.-M. 1996. "Tradition and Modernity: French Shepherds' Use of Medicinal Bouquets." In McCorkle, C.M., Mathias, E., and T.W. Schillhorn van Veen (eds.), *Ethnoveterinary Research and Development*. London: Intermediate Technology Publications, 76–90.

Britannica Academic: Classical Economics. https://academic.eb.com/levels/collegiate/article/classical-economics/24233 (November 14th, 2023).

Britannica Academic: John Stuart Mill. https://academic.eb.com/levels/collegiate/article/John-Stuart-Mill/108730 (November 14th, 2023).

Britannica Money: glocalization. https://www.britannica.com/money/glocalization (July 12th, 2024).

Britovšek, M., 1958. "Uvajanje novih kultur na Kranjskem v drugi polovici 18. in v prvi polovici 19. stoletja" (Die Einführung der neueren Kulturpflanzen in Krain in der zweiten Hälfte des 18. und in der ersten Hälfte des 19. Jahrhunderts). *Zgodovinski časopis* 12: 111–149.

Broggi, M., 2016. "Foreword: Ecological Networks." In Plassmann, G., et al. (eds.), *Alpine Nature 2030: Creating [Ecological] Connectivity for Generations to Come*. Berlin: Federal Ministry for the Environment, Nature Conservation, Building and Nuclear Safety, 8–11.

Brönnimann, S., Pfister, C., and S. White, 2018. "Archives of Nature and Archives of Societies." In White, S., Pfister, C., and F. Mauelshagen (eds.), *The Palgrave Handbook of Climate History*. London: Palgrave Macmillan, 27–36.

Brönnimann, S., White, S., and V. Slonosky, 2018. "Climate from 1800 to 1970 in North America and Europe." In White, S., Pfister, C., and F. Mauelshagen (eds.), *The Palgrave Handbook of Climate History*. London: Palgrave Macmillan, 309–320.

Brooke, J.L., 2018. "The Holocene." In White, S., Pfister, C., and F. Mauelshagen (eds.), *The Palgrave Handbook of Climate History*. London: Palgrave Macmillan, 175–182.

Brown, V., 1976. "Columella, Lucius Junius Moderatus." In Cranz, F.E., and P.O. Kristeller (eds.), *Catalogus translationum et commentariorum: Mediaeval and Renaissance Latin Translations and Commentaries: Annotated Lists and Guides*. Vol. 3. Washington, D.C.: The Catholic University of America Press, 173–193.

Brückmüller, E., and H. Stekl, 1998. "Das Eisenbahnnetz der Habsburgermonarchie bis 1914." In Putzger, F.W., and E. Brückmüller (eds.), *Historischer Weltatlas zur allgemeinen und österreichischen Geschichte*. Vienna: Hölder-Pichler-Tempsky and ÖBV Pädagogischer Verlag, 67.

Brugger, O., and R. Wohlfarter, 1983. *Alpwirtschaft heute*. Graz and Stuttgart: Leopold Stocker.

Brus, R., 2008. *Sto grmovnih vrst na Slovenskem* (One Hundred Shrub Species in Slovenia). Ljubljana: Tehniška založba Slovenije.

Brus, R., 2012. *Drevesne vrste na Slovenskem* (Tree Species in Slovenia). 2nd ed. Ljubljana: Author.

Bucher, R., 2023. Contribution to the workshop "Introduction to Science for Policy: European Perspectives" at the *12th ESEH* [European Society for Environmental History] *Conference: Mountains and Plains: Past, Present and Future Environmental and Climatic Entanglements*. Bern, August 23rd, 2023.

Budkovič, T., 1999. *Bohinj 1914–1918: Med fronto in zaledjem* (Bohinj 1914–1918: Between the Front and its Hinterland). Klagenfurt, Ljubljana, and Vienna: Mohorjeva založba, 1999.

Budkovič, T., 2001. *Wochein/Bohinj 1914–1918: Das Aufmarschgebiet der Isonzofront*. Klagenfurt, Ljubljana, and Vienna: Hermagoras/Mohorjeva, 2001.

Budkovič, T., 2010. "Impact on the Environment of the Battles on the Isonzo (Soča) Front." In Štepec, M. (ed.), *1914[–]1918*. Ljubljana: National Museum of Contemporary History, 61–62.

Bundesamt für Naturschutz: Neobiota.de: Gebietsfremde und invasive Arten in Deutschland. https://neobiota.bfn.de/invasivitaetsbewertung/gefaesspflanzen.html (February 10th, 2022).

Bürgi, M., Bieling, C., von Hackwitz, K., Kizos, T., Lieskovský, J., et al., 2017. "Processes and Driving Forces in Changing Cultural Landscapes across Europe." *Landscape Ecology* 32: 2097–2112.

Bürgi, M., Hersperger, A.M., and N. Schneeberger, 2004. "Driving Forces of Landscape Change—Current and New Directions." *Landscape Ecology* 19: 857–868.

Butzer, K.W., 1996. "The Mountains of the Mediterranean World: An Environmental History. J. R. McNeill. New York: Cambridge University Press, 1994. vii and 423 pp., figs., tables, maps, refs., index. $69.95 cloth (ISBN 0-521-33248-6)." *Annals of the Association of American Geographers* 86(4): 780–782.

Butzer, K.W., 2003. "The Nature of Mediterranean Europe: An Ecological History. A. T. Grove and O. Rackham. New Haven, CT: Yale University Press, 2001. 384 pp., maps, photos, diags., biblio., and index. $75.00 cloth (ISBN 0-300-08443-9)." *Annals of the Association of American Geographers* 93(2): 494–498.

Caf, N., Sabatier, P., Cucinotta, J., Rapuc, W., Dolenec, M., et al., 2025. "Human Influence at the Holocene Treeline in the Julian Alps, Slovenia." *The Holocene* 35, forthcoming: separate pagination 1–14 (the manuscript from 2022 is cited).

Calic, M.-J., 2019. *Zgodovina Jugovzhodne Evrope* (Südosteuropa: Weltgeschichte einer Region). Ljubljana: Cankarjeva založba.

Camenisch, C., and C. Rohr, 2018. "When the Weather Turned Bad: The Research of Climate Impacts on Society and Economy During the Little Ice Age in Europe: An Overview." *Cuadernos de Investigación Geográfica* 44(1): 99–114.

Camuffo, D., 2018. "Evidence from the Archives of Societies: Early Instrumental Observations." In White, S., Pfister, C., and F. Mauelshagen (eds.), *The Palgrave Handbook of Climate History*. London: Palgrave Macmillan, 83–92.

Carp Reproduction, 2007. http://www.carp-fishing-tactics.com/science-carp-reproduction.html (August 19th, 2020).

Carrer, F., Walsh, K., and F. Mocci, 2020. "Ecology, Economy, and Upland Landscapes: Socio-Ecological Dynamics in the Alps during the Transition to Modernity." *Human Ecology* 48: 69–84.

Čepič, Z., 2005 (= Čepič, 2005a). "Posnemanje sovjetske družbene ureditve 1945–1948: Kmetijska politika in kmetijstvo" (Following the Example of the Soviet Social Order 1945–1948: Agricultural Policy and Agriculture). In Fischer, J., et al. (eds.), *Slovenska novejša zgodovina: Od programa Zedinjena Slovenija do mednarodnega priznanja Republike Slovenije 1848–1992*. Vol. 2. Ljubljana: Mladinska knjiga, 889–893.

Čepič, Z., 2005 (= Čepič, 2005b). "Posnemanje sovjetske družbene ureditve 1945–1948: Preskrba prebivalstva in obvezni odkupi" (Following the Example of the Soviet Social Order 1945–1948: Provisions and Compulsory Redemptions). In Fischer, J., et al. (eds.), *Slovenska novejša zgodovina: Od programa Zedinjena Slovenija do mednarodnega priznanja Republike Slovenije 1848–1992*. Vol. 2. Ljubljana: Mladinska knjiga, 893–900.

Čerček, E., 1948–1949. "Planine v južnih Kamniških Alpah" (The Alps in the Southern Kamnik Alps). *Geografski vestnik* 20–21: 37–86.

Cergol Paradiž, A., 2022. *"Bela kuga": Ilegalni abortusi in zmanjševanje rodnosti na Slovenskem v obdobju med obema vojnama* (The "White Plague": Illegal Abortions and Natality Decline in Slovenian Territory in the Interwar Period). Ljubljana: Univerza v Ljubljani.

Černe, A., 2001. "Development of Rail Network between 1841 and 1991." In Fridl, J., et al. (eds.), *National Atlas of Slovenia*. Ljubljana: Rokus, 161.

Cevc, T., 1992. *Bohinj in njegove planine: Srečanja s planšarsko kulturo* (Bohinj and Its Alps: Encounters with *Senn* Culture). Bled: Triglavski narodni park.

Cevc, T., and J. Čop, 1993. *Slovene Hay-Rack*. Žirovnica: Agens d.o.o.

Chauvel, B., Dessaint, F., Cardinal-Legrand, C., and F. Bretagnolle, 2006. "The Historical Spread of *Ambrosia artemisiifolia* L. in France from Herbarium Records." *Journal of Biogeography* 33(4): 665–673.

Cheyette, F.L., 2008. "The Disappearance of the Ancient Landscape and the Climatic Anomaly of the Early Middle Ages: A Question to be Pursued." *Early Medieval Europe* 16(2): 127–165.

Christanell, A., Vogl-Lukasser, B., Vogl. C.R., and M. Gütler 2010. "The Cultural Significance of Wild-Gathered Plant Species in Kartitsch (Eastern Tyrol, Austria) and the Influence of Socioeconomic Changes on Local Gathering Practices." In Pardo-de-Santayana, M., Pieroni, A., and R.K. Puri (eds.), *Ethnobotany in the New Europe: People, Health and Wild Plant Resources*. New York and Oxford: Berghahn Books, 51–75.

Cibic, T., 2011a. "Mrávlja" (Ant). In Baš, A., et al. (eds.), *Slovenski etnološki leksikon*. Ljubljana: Mladinska knjiga, 340.

Cibic, T., 2011b. "Pikapolónica" (Ladybug). In Baš, A., et al. (eds.), *Slovenski etnološki leksikon*. Ljubljana: Mladinska knjiga, 414–415.

Cigale, D., and B. Lampič, 2023. "Aspects of Tourism Sustainability on Organic Farms in Slovenia." *Journal of the Geographical Institute "Jovan Cvijić" SASA* 73(2): 251–267.

Cimperšek, M., 2014. "Eksploatacija pohorskih gozdov v preteklosti" (Exploitation of the Pohorje Forests in the Past). *Gozdarski vestnik* 72(9): 365–382.

Coates, P., 2004. "Emerging from the Wilderness (or, from Redwoods to Bananas): Recent Environmental History in the United States and the Rest of America." *Environmental History* 10(4): 407–438.

Collet, D., White, S., Bremer, S., Brenna, B., Glørstad, H., et al., 2025. "The Little Ice Age: The History and Future of a Traveling Concept." *Wiley Interdisciplinary Reviews* 16(5): separate pagination, 1–17.

Coulson, S.J., Bullock, J.M., Stevenson, M.J., and R.F. Pywell, 2001. "Colonization of Grassland by Sown Species: Dispersal versus Microsite Limitation in Responses to Management." *Journal of Applied Ecology* 38(1): 204–216.

Cova, U., 2000. "Wirtschaftliche Beziehungen zwischen Triest und Steiermark von der Zeit Kaiser Karls VI. bis 1809." *Zeitschrift des Historischen Vereines für Steiermark*, Sonderband 25, *Festschrift Gerhard Pferschy zum 70. Geburtstag*: 403–424.

Crane, E., 1999. *The World History of Beekeeping and Honey Hunting*. New York: Routledge.

Cremene, C., Groza, G., Rakosy, L., Schileyko, A.A., Baur, A., et al., 2005. "Alterations of Steppe-Like Grasslands in Eastern Europe: A Threat to Regional Biodiversity Hotspots." *Conservation Biology* 19(5): 1606–1618.

Csippán, P., and L. Ferenczi, 2020. "Fifty Shades of Grey? The Impact of the Hungarian Cattle Trade on Cattle Breeding in the Late Medieval and Early Modern Period." *Journal of Archaeological Science: Reports* 29: 102031, separate pagination 1–17.

Cultures for Health: How to Make Nettle Rennet for Cheesemaking. https://www.culturesforhealth.com/learn/cheese/make-nettle-rennet-cheesemaking/ (January 28th, 2022).

Cultures for Health: How to Make Thistle Rennet for Cheesemaking. https://www.culturesforhealth.com/learn/cheese/make-thistle-rennet-cheesemaking/ (January 28th, 2022).

Cunder, T., 2012. "Slovenia." In Oppermann, R., Beaufoy, G., and G. Jones (eds.), *High Nature Value Farming in Europe: 35 European Countries and Perspectives.* Ubstadt-Weiher, Heidelberg, and Basel: Regionalkultur, 374–381.

Cvetek, J., 1971. "Grbinasti travniki s posebnim ozirom na Bohinj" (Hummocky Meadows, Especially in Bohinj). *Geografski vestnik* 43: 67–78.

Cvijić, J., 1925. "Karst i čovek" (Karst and Man). *Glasnik geografskog društva* 11: 1–11.

Cvirn, J., 2001. "Civiliziranost kot habitus samonadzora" (The State of Being Civilized as a Habitus of Self Control). In Elias, N., *O procesu civiliziranja.* Vol. 2. Ljubljana: cf., 413–434.

Dahlström, A., Iuga, A.-M., and T. Lennartsson, 2013. "Managing Biodiversity Rich Hay Meadows in the EU: A Comparison of Swedish and Romanian Grasslands." *Environmental Conservation* 40: 194–205.

Daim, F., and F. Glaser, 1998. "Österreich zur Römerzeit: Österreich zur Zeit des Heiligen Severin (†482)." In Putzger, F.W., and E. Brückmüller (eds.), *Historischer Weltatlas zur allgemeinen und österreichischen Geschichte.* Vienna: Hölder-Pichler-Tempsky and ÖBV Pädagogischer Verlag, 19.

Dakskobler, I., 2023. "Vrstno bogata travišča s prevladujočim volkom na silikatnih tleh gorskega pasu" (Species-Rich Grasslands on Silicate Bedrock in Montane Belt, Dominated by Matgrass). *Svet pod Triglavom* 31: 8–9.

Dakskobler, I., Praprotnik, N., Seliškar, A., Strgar, Po., Strgar, Pe., et al., 2019. *Bohinj in Bloom: The Floral Richness of Bohinj.* Stara Fužina: Turizem Bohinj.

Das k.k. Hofgestüt zu Lippiza 1580 1880. 1880. Vienna: Kaiserlich-königliche Hof- und Staatsdruckerei.

Davidson, K.E., Fowler, M.S., Skov, M.W., Doerr, S.H., Beaumont, N., et al., 2017. "Livestock Grazing Alters Multiple Ecosystem Properties and Services in Salt Marshes: A Meta-Analysis." *Journal of Applied Ecology* 54: 1395–1405.

Davies, G.M., Domènech, R., Gray, A., and P.C. Johnson, 2016. "Vegetation Structure and Fire Weather Influence Variation in Burn Severity and Fuel Consumption during Peatland Wildfires." *Biogeosciences* 13(2): 389–398.

Davis, J.R., 2004. *The Rural Non-Farm Economy, Livelihoods and their Diversification: Issues and Options.* Chatham, UK: Natural Resources Institute.

de Fraine, J., 1968. "Gesetz." In Haag, H. (ed.), *Bibel-Lexikon*. Einsiedeln, Zürich, and Cologne: Benzinger, columns 576–579.

De Keyzer, M., 2016. "All We Are Is Dust in the Wind: The Social Causes of a 'Subculture of Coping' in the Late Medieval Coversand Belt." *Journal for the History of Environment and Society* 1: 1–35.

de Serres, O., 1603. *Le theatre d'agriculture et mesnage des champs.* 2nd revised ed. Paris: A. Saugrain.

Debeljak, N., Erbida, N., Makovec, G., Panjan, M., Pavlič, M., et al., 2020. *Life to Grasslands: Layman's Report.* Ljubljana: Institute of the Republic of Slovenia for Nature Conservation. https://www.lifetograsslands.si/wp-content/uploads/2016/05/LIFE-TO-GRASSLANDS-Laymans-report.pdf (December 30th, 2024).

Debuschewitz, E., and J. Sanders, 2022. "Environmental Impacts of Organic Agriculture and the Controversial Scientific Debates." *Organic Agriculture* 12(1): 1–15.

Deschmann, K., 1868. "Ueber einige in jüngster Zeit in Krain eingewanderte Pflanzen." *Laibacher Zeitung*, no. 78 (April 4th, 1868): 564–565; no. 79 (April 6th, 1868): 571.

Deutsche Bibliographie: Wolkenstein, Marx Sittich von. https://www.deutsche-biographie.de/pnd121369471.html (October 26th, 2021).

Deutsches Wörterbuch von Jacob Grimm und Wilhelm Grimm auf CD-ROM und im Internet: https://woerterbuchnetz.de/?sigle=DWB#0 (February 2nd, 2019).

Dictionary by Merriam-Webster. https://www.merriam-webster.com/ (January 31st, 2022).

Die Grenzen der Bistümer des Alten Reiches um 1500. Rekonstruktion der Germania Sacra anhand publizierter Karten und Quellenmaterialien (Datensammlungen der Germania Sacra. Materialien zur Kirche des Alten Reiches 4), Version 2, Göttingen 2022. https://rep.adw-goe.de/handle/11858/2223 (January 24th, 2024).

Die wichtigsten Grundsätze im Ameisenschutz. https://www.waldwissen.net/de/lebensraum-wald/tiere-im-wald/insekten-wirbellose/die-wichtigsten-grundsaetze-im-ameisenschutz (November 17th, 2023).

"Die Wirtschaft Mittel- und Westeuropas im Zeitalter des Frühkapitalismus (um 1550)." 1998. In Putzger, F.W., and E. Brückmüller (eds.), *Historischer Weltatlas zur allgemeinen und österreichischen Geschichte.* Vienna: Hölder-Pichler-Tempsky and ÖBV Pädagogischer Verlag, 42–43.

Dierschke, H., and G. Briemle, 2002. *Kulturgrasland: Wiesen, Weiden und verwandte Staudenfluren.* Stuttgart: Ulmer.

Dierssen, K., and B. Dierssen, 2001. *Moore: Ökosysteme Mitteleuropas aus geobotanischer Sicht.* Stuttgart: Ulmer.

Dietl, W., and J. Lehmann, 2004. *Ökologischer Wiesenbau: Nachhaltige Bewirtschaftung von Wiesen und Weiden.* Leopoldsdorf: Österreichischer Agrarverlag.

Dinklage, K., 1966 (= Dinklage, 1966a). "Das Mittelalter: Vom 12. bis zum Ende des 15. Jahrhunderts: Bäuerliche Wirtschafts- und Verfassungsverhältnisse im Spätmittelalter." In *Geschichte der Kärntner Landwirtschaft.* Klagenfurt: Johannes Heyn, 53–93.

Dinklage, K., 1966 (= Dinklage, 1966b). "Die Neuzeit: Vom 16. bis zum 18. Jahrhundert: Kärntens Landwirtschaft im Zeitalter des staatlichen Absolutismus." In *Geschichte der Kärntner Landwirtschaft.* Klagenfurt: Johannes Heyn, 94–148.

Dinklage, K., 1966 (= Dinklage, 1966c). "Die Neuzeit: Von der Mitte des 18. bis zur Mitte des 19. Jahrhunderts: Erste Landwirtschaftsförderung durch Gelehrte und Grundherren." In *Geschichte der Kärntner Landwirtschaft.* Klagenfurt: Johannes Heyn, 149–216.

Dobrovolný, P., Moberg, A., Brázdil, R., Pfister, C., Glaser, R., et al., 2010. "Monthly, Seasonal and Annual Temperature Reconstructions for Central Europe Derived from Documentary Evidence and Instrumental Records since AD 1500." *Climatic Change* 101: 69–107.

Dohrn-van Rossum, G., 2006. "Feiertage: 1. Christliche Feiertage." In Jaeger, F., et al. (eds.), *Enzyklopädie der Neuzeit.* Vol. 3, *Dynastie–Freundschaftslinien.* Stuttgart and Weimar: J.B. Metzler, 868–873.

Dolar, R., 2022. "Was ist ein historischer Umbruch." In Jerše, S., and K. Lahl (eds.), *Endpunkte. Und Neuanfänge: Geisteswissenschaftliche Annäherungen an die Dynamik von Zeitläufen.* Vienna and Cologne: Böhlau, 25–38.

Dolenc, E., 2000. "Učiteljski tovariš" (The Journal *Učiteljski tovariš*). In Batis, J., et al. (eds.), *Enciklopedija Slovenije.* Vol. 14, *U–We.* Ljubljana: Mladinska knjiga, 11.

Dolinar, F.M., 1980. "Podoba ljubljanske škofije v rimskih poročilih škofov Rinalda Scarlichija in Otona Friderika Buchheima" (Die Darstellung der Diözese Laibach nach den Romberichten der Bischöfe Rinaldo Scarlichi und Otto Friderick Graf von Buchheim). *Bogoslovni vestnik* 40(1): 26–45.

Doll, M., 2003. *Haustierhaltung und Schlachtsitten des Mittelalters und der Neuzeit: Eine Synthese aus archäozoologischen, bildlichen und schriftlichen Quellen Mitteleuropas.* Rahden/Westf.: Marie Leidorf GmbH.

Dolžan Eržen, T., 2024. "Katalog predmetov, ki so razstavljeni v Planšarskem muzeju v Stari Fužini" (Catalogue of Objects Exhibited in the Alpine Dairy

Museum in Stara Fužina). In Novak, A., *Planšarstvo v Bohinju: "Vse svoje življenje sem pustil v siru."* Kranj: Gorenjski muzej, 264–331.

Dornheim, S., 2016. "Wissenschaft versus Tradition? Die Physikalisch-oeconomische Bienengesellschaft der Oberlausitz und die Anfänge agrarischer Volksaufklärung in Sachsen." In Dauser, R., Fassl, P., and L. Schilling (eds.), *Wissenszirkulation auf dem Land vor der Industrialisierung.* Augsburg: Wißner, 85–97.

Dosedla, H.-C., 1971. "Streifgeräte zur Wiesenbearbeitung." In Wolfram, R. et al. (eds.), *Österreichischer Volkskundeatlas: Kommentar.* Vol. 3, Fascicle (*Lieferung*) 4. Wien: Hermann Böhlaus Nachfolger, separate pagination, 1–16.

Dražumerič, M., 1988. "Gospodarsko življenje v Beli krajini med 1. svetovno vojno" (The Economy of Bela Krajina during World War I). *Zgodovinski časopis* 42(1): 85–102.

Drobesch, W., 2009. "Bodenerfassung und Bodenbewertung als Teil einer Staatsmodernisierung: Theresianische Steuerrektifikation, Josephinischer Kataster und Franziszeischer Kataster." *Histoire des Alpes* 14: 165–183.

Drole, E., 2019. "Immovable Cultural Heritage of the Soča Front: Legal Protection and Conservation." In Košir, U., Črešnar, M., and D. Mlekuž (eds.), *Rediscovering the Great War: Archaeology and Enduring Legacies on the Soča and Eastern Fronts.* London and New York: Routledge, 199–213.

Društvo za opazovanje in proučevanje ptic Slovenije, 2017. "Kosec" (Corncrake). http://ptice.si/naravovarstvo-in-raziskave/karizmaticne-vrste/kosec/ (December 14th, 2019).

Duden Wörterbuch. https://www.duden.de/woerterbuch (April 28th, 2024).

Duh, M., 2023. *Vpliv izbruha vulkana Samalas na podnebje in preskrbo s hrano v Evropi in Aziji: Magistrsko delo* (The Impact of the Samalas Eruption on the Climate and Food Supply in Europe and Asia: Master's Thesis). Ljubljana: Univerza v Ljubljani, Filozofska fakulteta, Oddelek za zgodovino, Oddelek za geografijo.

Duncan, B.L., 2019. *Impacts of Cattle Grazing as a Tool to Control* Phragmites australis *in Wetlands on Nitrogen, Phosphorus, and Carbon: Master's Thesis.* Logan, UT, Utah State University.

Ebner, H., 1958. "Steiermarks Außenhandel 1766." *Blätter für Heimatkunde* 32: 83–90.

"Edelweissexport." 1893. *Mittheilungen des Deutschen und Oesterreichischen Alpenvereins* 19(3): 38.

Ehlert, K., Ehlert, T., Herold, P., Scharnhölz, R, and R. Oppermann, 2012. "Nature-Friendly Agricultural Machinery and Mechanical Operations." In Oppermann, R., Beaufoy, G., and G. Jones (eds.), *High Nature Value Farming in Europe: 35 European Countries and Perspectives.* Ubstadt-Weiher, Heidelberg, and Basel: Regionalkultur, 473–483.

Ehmer, J., 2005. “Bevölkerung.” In Jaeger, F., et al. (eds.), *Enzyklopädie der Neuzeit*. Vol. 2, *Beobachtung–Dürre*. Stuttgart and Weimar: J.B. Metzler, 94–119.

“Ein erfolgreiches Projekt ist abgeschlossen.” 2005. *Schütt Magazin* 6: 1–2. http://www.schuett.at/life/schuettmagazin.php (September 18th, 2019).

Eldem, E., 2009. “Istanbul (Byzantium, Constantinople).” In Ágoston, G., and B. Masters (eds.), *Encyclopedia of the Ottoman Empire*. New York: Facts On File, 286–290.

Elevationmap.net. https://elevationmap.net/ (May 1st, 2024).

Elias, N., 1993 and 1994. *Über den Prozeß der Zivilisation: Soziogenetische und psychogenetische Untersuchungen*. Vols. 1, *Wandlungen des Verhaltens in den weltlichen Oberschichten des Abendlandes*, and 2, *Wandlungen der Gesellschaft: Entwurf zu einer Theorie der Zivilisation*. Frankfurt am Main: Suhrkamp.

Ellenberg, H., 1952. *Landwirtschaftliche Pflanzensoziologie*. Vol. 2, *Wiesen und Weiden und ihre standörtliche Bewertung*. Stuttgart: Ulmer.

Ellenberg, H. & C. Leuschner, 2010. *Vegetation Mitteleuropas mit den Alpen: In ökologischer, dynamischer und historischer Sicht*. Stuttgart: Eugen Ulmer.

Emanuelsson, U., 2009. *The Rural Landscapes of Europe: How Man Has Shaped European Nature*. [city unknown]: Swedish Research Council Formas.

Embleton-Hamann, C., 2004. “Processes Responsible for the Development of a Pit and Mound Microrelief.” *Catena* 57: 175–188.

Encyclopaedia Britannica: *Georg August Schweinfurth: German Botanist*. https://www.britannica.com/biography/Georg-August-Schweinfurth (December 22nd, 2020).

Ender, M., 1998. “Vegetation von gemähten Bergwiesen (Bergmähdern) und deren Sukzession nach Auflassung der Mahd am Hoch-Tannberg (Vorarlberg).” *Vorarlberger Naturschau* 4: 169–246.

Engel, A., 2006 (= Engel, 2006a). “Farbstoffe: 3. Färbergewerbe.” In Jaeger, F., et al. (eds.), *Enzyklopädie der Neuzeit*. Vol. 3, *Dynastie–Freundschaftslinien*. Stuttgart and Weimar: J.B. Metzler, 825–827.

Engel, A., 2006 (= Engel, 2006b). “Farbstoffe: 4. Farbstoffhandel.” In Jaeger, F., et al. (eds.), *Enzyklopädie der Neuzeit*. Vol. 3, *Dynastie–Freundschaftslinien*. Stuttgart and Weimar: J.B. Metzler, 827–829.

Engels, B., 2016. “Tourism in the Alps—A Nature and Biodiversity Perspective.” In Plassmann, G., et al. (eds.), *Alpine Nature 2030: Creating [Ecological] Connectivity for Generations to Come*. Berlin: Federal Ministry for the Environment, Nature Conservation, Building and Nuclear Safety, 100.

Enoletnice – Index seminum 2018 (Annual Plants—Index seminum 2018). Laze: Društvo botanični vrt Raznolikost. http://www.raznolikost.si/katalog/2018/enoletnice_2018.pdf (October 15th, 2021).

Environmental Atlas of Slovenia. http://gis.arso.gov.si/atlasokolja/profile.aspx?id=Atlas_Okolja_AXL@ARSO&culture=en-US (February 8th, 2019).

Equadrat engineering GmbH: Sanierung Speikerhaus Oberwölz. https://www.equadrat.com/climadesign-und-fassade/sanierung-speikerhaus-oberwoelz (January 1st, 2021).

Ergebnisorientierter Naturschutzplan: Erfahrungen und Schlussfolgerungen: Fachbericht. 2019. Vienna. http://www.suske.at/files/fachbericht_de_final.pdf (August 10th, 2022).

Erhatič Širnik, R., 2011. *Varstvo ptičev v 19. in 20. stoletju na slovenskem ozemlju: Doktorska disertacija* (Protection of Birds in the Nineteenth and Twentieth Century in Slovenian Territory: Doctoral Dissertation). Ljubljana: Univerza v Ljubljani, Biotehniška fakulteta.

Eriksson, O., Cousins, S.A., and H.H. Bruun, 2002. "Land-Use History and Fragmentation of Traditionally Managed Grasslands in Scandinavia." *Journal of Vegetation Science* 13(5): 743–748.

Erjavec, M., and M. Peršič, 2005. "Living With the Lake, Living Without the Lake: An Introduction into the Research of the Way of Life by the Intermittent Karstic Lakes Petelinjsko Jezero and Palško Jezero." *Acta carsologica* 34(3): 784–814.

Erni, P., 1999. "Zur Methodik der Rekonstruktion einer historischen Kulturlandschaft: Das Beispiel Basadingen – ein Werkstattbericht zur kritischen Auswertung spätmittelalterlicher Urbarien." In Niederstätter, A. (ed.), *Aspekte der Landwirtschaft in der Bodenseeregion: Mittelalter und Frühe Neuzeit.* Dornbirn: Vorarlberger Verlagsanstalt, 8–21.

Erschbamer, B., Virtanen, R., and L. Nagy, 2003. "The Impact of Vertebrate Grazers on Vegetation in European High Mountains." In Nagy, L., et al. (eds.), *Alpine Biodiversity in Europe.* Berlin and Heidelberg: Springer, 377–396.

Escher, W., 1958. "Gestelle zum Austrocknen des Heus." In *Atlas der schweizerischen Volkskunde: Kommentar.* Part 1, Fascicle (*Lieferung*) 5. Basel: Schweizerische Gesellschaft für Volkskunde, 402–406.

Esri, Topographic [basemap]. 2018. www.arcgis.com (October 25th, 2018).

Esri, USGS, National Geographic Style Base [basemap]. 2025. https://tiles.arcgis.com/tiles/P3ePLMYs2RVChkJx/arcgis/rest/services/NatGeoStyleBase/MapServer (January 15th, 2025).

Essl, F., Rabitsch, W., and O. Breuss, 2002. *Neobiota in Österreich.* Vol. 43. Vienna: Umweltbundesamt.

Etnološka zbirka Nose (Nose Ethnological Collection). https://namuljavi.si/etnoloska-zbirka-nose/ (December 8th, 2023).

EU-DEM (*European Environment Agency,* European Digital Elevation Model [data file]). 2011. https://www.eea.europa.eu/en/datahub/datahubitem-view/d08852bc-7b5f-4835-a776-08362e2fbf4b (October 15th, 2018).

Europarc federation: Étude de cas: Concours des Prés Fleuris en France. https://www.europarc.org/case-studies/flowering-meadows-contest-france/ (October 12th, 2022).

European Commission, Environment, Corncrake. https://ec.europa.eu/environment/nature/conservation/wildbirds/threatened/c/crex_crex_en.htm (November 20th, 2019).

European Commission, Environment, Wild Birds: Threatened bird species in Annex I. https://ec.europa.eu/environment/nature/conservation/wildbirds/threatened/index_en.htm (November 20th, 2019).

Ewald, J., 2003. "The Calcareous Riddle: Why Are There So Many Calciphilous Species in the Central European Flora?" *Folia Geobotanica* 38(4): 357–366.

Fajgelj, V., 1953. "Planine v vzhodnem delu Savinjskih Alp" (Les alpages de la partie orientale des Alpes de Savinja). *Geografski vestnik* 25: 123–166.

Faroqhi, S., 2002. "Coping with the Central State, Coping with Local Power: Ottoman Regions and Notables from the Sixteenth to the Early Nineteenth Century." In Adanir, F., and S. Faroqhi (eds.), *The Ottomans and the Balkans: A Discussion of Historiography*. Leiden, Boston, and Cologne: Brill, 351–381.

Favilli, F., 2016. "Agriculture and Ecological Connectivity." In Plassmann, G., et al. (eds.), *Alpine Nature 2030: Creating [Ecological] Connectivity for Generations to Come*. Berlin: Federal Ministry for the Environment, Nature Conservation, Building and Nuclear Safety, 115–122.

Feng, Z., Bohleber, P., Ebser, S., Ringena, L., Schmidt, M., et al., 2019. "Dating Glacier Ice of the Last Millennium by Quantum Technology." *Proceedings of the National Academy of Sciences of the United States of America* 116(18): 8781–8786.

Ferle, M., 2004. "Bídermájer" (Biedermeier). In Baš, A., et al. (eds.), *Slovenski etnološki leksikon*. Ljubljana: Mladinska knjiga, 33.

Fikfak, J., and B. Jezernik, 2021. "Dediščina prva svetovne vojne: Reprezentacije in reinterpretacije" (Heritage of the First World War: Representations and Reinterpretations). In Fikfak, J., and B. Jezernik (eds.), *Dediščina prva svetovne vojne: Reprezentacije in reinterpretacije*. Ljubljana: Znanstvena založba Filozofske fakultete Univerze v Ljubljani, 9–40.

Filatova, S., Gyulai, F., and W. Kirleis, 2019. "Environmental Imposition or Ancient Farmers' Choice? A Study of the Presence of 'Inferior' Legumes in the Bronze Age Carpathian Basin (Hungary)." In dal Corso, M., et al. (eds.), *How's Life? Living Conditions in the 2nd and 1st Millennia BCE*. Leiden: Sidestone Press, 57–84.

Fiol, C., Prado, D., Mora, M., and J.I. Alava, 2016. "Nettle Cheese: Using Nettle Leaves (*Urtica dioica*) to Coagulate Milk in the Fresh Cheese Making Process." *International Journal of Gastronomy and Food Science* 4: 19–24.

Fischer, J., 2005. "Prva svetovna vojna: gospodarstvo v vojnih razmerah" (World War I: The War Economy). In Fischer, J., et al. (eds.), *Slovenska novejša zgodovina: Od programa Zedinjena Slovenija do mednarodnega priznanja Republike Slovenije 1848–1992*. Vol. 1. Ljubljana: Mladinska knjiga, 148–150.

Fischer, M., Rudmann-Maurer, K., Weyand, A., and J. Stöcklin, 2008. "Agricultural Land Use and Biodiversity in the Alps." *Mountain Research and Development* 28(2): 148–155.

Flückiger, E., and A. Radeff, 2000. "Globale Ökonomie im alten Staat Bern am Ende des Ancien Régime – Eine aussergewöhliche Quelle." *Berner Zeitschrift für Geschichte und Heimatkunde* 62(1): 5–40.

Food and Agriculture Organization of the United Nations, GIAHS [Globally Important Agricultural Heritage Systems] around the world, Traditional Hay Milk Farming in the Austrian Alpine Arc, Austria. https://www.fao.org/giahs/giahsaroundtheworld/designated-sites/europe-and-central-asia/hay-milk-farming-austria/en/ (July 21st, 2024).

Fornasin, A., 2017. "La demografia alpina in età preindustriale: Interpretazioni, problemi, prospettive." In Denzel, M.A., et al. (eds.), *Oeconomia Alpium I: Wirtschaftsgeschichte des Alpenraums in vorindustrieller Zeit: Forschungsaufriss, -konzepte und -perspektiven*. Berlin and Boston: De Gruyter and Oldenbourg, 57–71.

Frajman, B., 2003. "Žlezava nedotika (*Impatiens glandulifera Royle*) – invazivna vrsta v naši flori" (Himalayan Touch-Me-Not (*Impatiens glandulifera* Royle): An Invasive Species in the Flora of Slovenia). *Proteus* 65(6): 274–277.

Franz, C., and D. Fritz, 1978. "Cultivation aspects of *Gentiana lutea* L." *Acta Horticulturae* 73: 307–314.

Frequency of Easter Sundays. https://webspace.science.uu.nl/~gent0113/easter/easter_text2b.htm (August 10th, 2020).

Fresacher, W., 1965. "Eine Abrutschung von der Koralm im Jahre 1660." *Carinthia I* 155: 209–214.

Fresacher, W., 1974. *Die Freisassen in Kärnten*. Klagenfurt: Kärntner Landesarchiv.

Frey, D., 2003. *Patterns of Variation within the* Erigeron annuus *Complex in the United States and Europe: Doctoral Dissertation*. Zürich, ETH Zurich.

Friedrich, C., and W.-D. Müller-Jahncke, 2009. "Pharmazie." In Jaeger, F., et al. (eds.), *Enzyklopädie der Neuzeit*. Vol. 9, *Naturhaushalt–Physiokratie*. Stuttgart and Weimar: J.B. Metzler, 1080–1084.

Friess, T., 2001. "Die Wanzenfauna (Heteroptera) des Bergsturzgebietes Schütt/Dobratsch und seiner näheren Umgebung (Kärnten, Österreich): Faunistik, Zönotik und Naturschutz." *Carinthia II* 191: 357–388.

Fritsche, S., Gisler, M., Schwarz, G., Fäh, D., and P. Kästli, 2010. *COGEAR Module 1a: Historical Earthquakes in the Valais: Del. No.: 1a.2.1.* http://www.seismo.ethz.ch/export/sites/sedsite/research-and-teaching/.galleries/pdf_cogear/tdel_1a.2.1.pdf_2063069299.pdf (February 10th, 2017).

Frühneuhochdeutsches Wörterbuch. Vol. 1, *a–äpfelkern*. Berlin and New York: Walter de Gruyter, 1989.

Fuchs, R.-P., and W. Schulze, 2002. "Zeugenverhöre als historische Quellen – einige Vorüberlegungen." In Fuchs, R.-P., and W. Schulze (eds.), *Wahrheit, Wissen, Erinnerung: Zeugenverhörprotokolle als Quellen für soziale Wissensbestände in der Frühen Neuzeit*. Hamburg and London: Lit., 7–40.

Fueter, E.K., 1964. "Gesner, Konrad." https://www.deutsche-biographie.de/sfz20799.html#ndbcontent (December 31st, 2020).

Fynn, R.W., Morris, C.D., and T.J. Edwards, 2004. "Effect of Burning and Mowing on Grass and Forb Diversity in a Long-Term Grassland Experiment." *Applied Vegetation Science* 7(1): 1–10.

Gabrič, A., Gričnik, M., Mazej, T., Miletić, N., Pavković, T., et al., 2019. *Voda – strateška dobrina v Vitanju: [Projektno poročilo]* (Water—A Strategic Resource in Vitanje: Project Report). Ljubljana: Filozofska fakulteta Univerze v Ljubljani, Oddelek za geografijo.

Gabrovec, M., and P. Kumer, 2019. "Land-Use Changes in Slovenia from the Franciscean Cadastre Until Today." *Acta geographica Slovenica* 59: 63–81.

Galle Toplak, K., 2015. *Zdravilne rastine na Slovenskem* (Healing Plants in Slovenian Territory). Ljubljana: Mladinska knjiga.

Gams, I., 1960. "O višinski meji naseljenosti, ozimine, gozda in snega v slovenskih gorah" (On the Upper Limit of Rural Settlements, Autumn-Sown Cereals, the Forest Line, and the Snow Line in the Mountains of Slovenia). *Geografski vestnik* 32: 59–69.

[Gartner, M., and S. Uhan], 2020. *Gartner – Pr Odolneku Tourist Farm*: https://www.youtube.com/watch?v=tXNIO4KG66s (September 19th, 2021).

Gemeinde St. Kanzian am Klopeiner See: *Beiblatt zur Fischereiausübungsberechtigung Klopeiner See*. https://kanzian.at/wp-content/uploads/2017/05/Beiblatt-zur-Fischereiausuebungsberechtigung-Klopeiner-See-2017-A6_block.pdf (August 18th, 2020).

Gentner, G., 1932. "Einiges über den Speik." *Jahrbuch des Vereins zum Schutze der Alpenpflanzen* 4: 63–75.

Geografski atlas i statističko-geografski pregled svijeta (Geographical Atlas and World Geographical Statistics Overview). 1953. 3rd ed. Zagreb: Seljačka sloga.

Geografski terminološki slovar (Dictionary of Geographical Terminology). https://isjfr.zrc-sazu.si/sl/terminologisce/slovarji/geografski (February 9th, 2024).

Geršič, M., Gabrovec, M., and Ž. Zwitter, 2018. "Primerjava kulturne pokrajine Hraških listnekov in tamkajšnjega kmetovanja v prvi polovici 19. stoletja in danes" (Comparing the Cultural Landscape of the Hraše Leaf Litter Forests (Sln. *Hraški listneki*) and the Farming There in the First Half of the Nineteenth Century and Today). *Geografski vestnik* 90(1): 61–84.

Gestrin, F., 1962. "Économie et société en Slovénie au XVIe siècle." *Annales: Économies, sociétés, civilisations* 17(4): 663–690.

Gestrin, F., 1965. *Trgovina slovenskega zaledja s primorskimi mesti od 13. do konca 16. stoletja* (Le traffic commercial entre les contrées Slovènes de l'intérieur et les villes du littoral de l'Adriatique, du 13e au 16e siècle). Ljubljana: Academia scientiarum et artium Slovenica, Institutum historicum, Sectio historiae universalis et nationalis.

Gestrin, F., 1981. "Italijani v slovenskih deželah od 13. do 17. stoletja" (Gli Italiani nelle regioni slovene dal duecento al seicento). *Zgodovinski časopis* 35(3): 223–241.

Gingrich, S., Schmid, M., Gradwohl, M., and F. Krausmann, 2013. "How Material and Energy Flows Change Socio-Natural Arrangements: The Transformation of Agriculture in the Eisenwurzen Region, 1860–2000." In Singh, S.J., et al. (eds.), *Long-Term Socio-Ecological Research: Studies in Society-Nature Interactions Across Time and Space*. Dordrecht, etc.: Springer, 297–313.

Gladiolenwiese. Tourist information board in Gladiolenwiese at the northern Oberschütt (Sln. *Rogaje*), Carinthia, Austria.

Gladiolus illyricus MHNT.BOT.2009.17.12.jpg. https://commons.wikimedia.org/wiki/File:Gladiolus_illyricus_MHNT.BOT.2009.17.12.jpg (September 18th, 2019).

Glaser, F., 2013. "Landwirtschaftliche Basis spätantiker Höhensiedlungen in Noricum Mediterraneum." *Rudolfinum – Jahrbuch des Landesmuseums für Kärnten 2012*: 47–55.

Global Agenda for Sustainable Livestock: Progress Report 2021. 2022. [city unknown]: Global Agenda for Sustainable Livestock. http://www.livestockdialogue.org/fileadmin/templates/res_livestock/docs/2022/Report_GASL_2022_200dpi.pdf (September 25th, 2022).

Global Biodiversity Information Facility. https://www.gbif.org/occurrence/search?occurrence_status=present&q= (August 11th, 2022).

Godina Golija, M., 2011. "Hunger and Misery: The Influence of the First World War on the Diet of Slovenian Civilians." In Zweiniger-Bargielowska, I., Duffett, R., and A. Drouard (eds.), *Food and War in Twentieth Century Europe*. Farnham, UK, and Burlington: Ashgate, 85–97.

Golec, B., 2014. "Erze, Wasser, Feuer und Erdbeben als Mitgestalter des räumlichen, wirtschatlichen und sozialen Antlitzes der Städte und Märkte im Land Krain in der Vormodernen Epoche." In Štih, P., and Ž. Zwitter (eds.), *Man, Nature and Environment between the Northern Adriatic and the Eastern Alps in Premodern Times*. Ljubljana: University Press, Faculty of Arts and Historical Association of Slovenia, 156–180.

Golec, B., 2015. *Zemljiški gospod Janez Vajkard Valvasor in njegov urbar* (Johann Weikhard Valvasor als Grundherr und sein Urbar). Ljubljana: Arhiv Republike Slovenije and Javni zavod Bogenšperk.

Golob, N., 1985. *Dvanajstero mesecev: Povezave in ikonografija likovnih in literarnih ciklov od pozne antike do renesanse* (Zwölf Monate: Zusammenhänge und Ikonographie der bildenden und literarischen Zyklen von der Spätantike bis zur Renaissance). Ljubljana: Znanstveni inštitut Filozofske fakultete Univerze Edvarda Kardelja.

Golob, N., 1988. "Koledarček (MS 160) v ljubljanski Narodni in univerzitetni knjižnici" (Das Kalenderchen (MS 160) in der National- und Universitätsbibliothek zu Ljubljana). *Zbornik za umetnostno zgodovino* 24: 33–43, xiii–xx.

Gooders, J. (adapted by J. Gregori), 1998. *Ptiči Slovenije in Evrope* (Birds in Slovenia and in Europe). Ljubljana: Mladinska knjiga.

Google Maps. https://www.google.com/maps (April 29th, 2024).

Gorißen, S., 2010. "Sensenschmiede." In Jaeger, F., et al. (eds.), *Enzyklopädie der Neuzeit*. Vol. 11, *Renaissance–Signatur*. Stuttgart and Weimar: J.B. Metzler, 1095–1097.

Govedič, M., 2004. "Fishes from the Archaeological Site at Hočevarica." In Velušček, A. (ed.), *Hočevarica: An Eneolithic Pile Dwelling in the Ljubljansko Barje*. Ljubljana: Institute of Archaeology at ZRC SAZU and ZRC Publishing, 133–151.

Grabherr, G., Kusstatscher, K., and A. Mair, 1985. "Zur vegetationsökologischen Aufbereitung aktueller Naturschutzprobleme im Hochgebirge." *Verhandlungen der Zoologisch-Botanischen Gesellschaft in Österreich* 123: 269–291.

Gradivo za strokovni izpit za ribogojca: Osnutek gradiva (Study Material for a Professional Exam in Fish Breeding: A Draft). 2009. Spodnje Gameljne: Zavod za ribištvo Slovenije. https://www.zzrs.si/uploads/files/Strokovni_izpit_za_ribogojca_Gradivo.pdf (August 17th, 2020).

Grafenauer, B., 1970 (= Grafenauer, 1970a). "Gospodarski obrat kmetije" (Der Wirtschaftsbetrieb des Bauernhofes). In Blaznik, P., et al. (eds.), *Gospodarska in družbena zgodovina Slovencev: Zgodovina agrarnih panog*. Vol. 1, *Agrarno gospodarstvo*. Ljubljana: Slovenska akademija znanosti in umetnosti and Državna založba Slovenije, 617–651.

Grafenauer, B., 1970 (= Grafenauer, 1970b). "Nabiralništvo" (Die Sammelwirtschaft). In Blaznik, P., et al. (eds.), *Gospodarska in družbena zgodovina Slovencev: Zgodovina agrarnih panog*. Vol. 1, *Agrarno gospodarstvo*. Ljubljana: Slovenska akademija znanosti in umetnosti and Državna založba Slovenije, 465–467.

Grafenauer, B., 1970 (= Grafenauer, 1970c). "Poljedelski obdelovalni načini" (Die Formen des Ackerbaus). In Blaznik, P., et al. (eds.), *Gospodarska in družbena zgodovina Slovencev: Zgodovina agrarnih panog*. Vol. 1, *Agrarno gospodarstvo*. Ljubljana: Slovenska akademija znanosti in umetnosti and Državna založba Slovenije, 225–250.

Grafenauer, B., 1970 (= Grafenauer, 1970d). "Poljedelsko orodje" (Die Ackerbaugeräte). In Blaznik, P., et al. (eds.), *Gospodarska in družbena zgodovina*

Slovencev: Zgodovina agrarnih panog. Vol. 1, *Agrarno gospodarstvo*. Ljubljana: Slovenska akademija znanosti in umetnosti and Državna založba Slovenije, 201–218.

Granda, S., 1994. "Zgornja baška dolina v prvi polovici 19. stoletja" (The Upper Bača Valley in the First Half of the Nineteenth Century). *Kronika: Časopis za slovensko krajevno zgodovino* 42(1): 52–58.

Grass, F., 1950. *Pfarrei und Gemeinde im Spiegel der Weistümer Tirols*. Innsbruck: Tyrolia.

Grass, F., 1966. "Guarinonius, Hippolytus." In *Neue Deutsche Biographie*. Vol. 7, 247. https://www.deutsche-biographie.de/pnd118698788.html#ndbcontent (June 22nd, 2022).

Grass, I., Batáry, P., and T. Tscharntke, 2021. "Combining Land-Sparing and Land-Sharing in European Landscapes." In Bohan, D.A., and A.J. Vanbergen (eds.), *The Future of Agricultural Landscapes*. Part 2. London, etc.: Academic Press, 251–303.

Grass, N., 1948. *Beiträge zur Rechtsgeschichte der Alpwirtschaft*. Innsbruck: Universitätsverlag Wagner.

Grdina, I., and O. Luthar, 2010. "'Macht endlich Schluss!' Die Alltagserinnerung und das Bild des 'Anderen' in populären Interpretationen des Ersten Weltkrieges." In Rajšp, V. (ed.), *Isonzofront 1915–1917: Die Kultur des Erinnerns*. Vienna and Ljubljana: Slowenisches Wissenschaftliches Institut in Wien, 13–26.

Gregory, B.S., 1999. "*Is* Small Beautiful? Microhistory and the History of Everyday Life." *History and Theory* 38(1): 100–110.

Grele, R.J., 2007. "Oral History as Evidence." In Charlton, T.L., Myers, L.E., and R. Sharpless (eds.), *History of Oral History: Foundations and Methodology*. Lanham, MD: Rowman and Littlefield, 33–91.

Grewe, B.-S., 2004. "'Man sollte sehen und weinen!' Holznotalarm und Waldzerstörung vor der Industrialisierung." In Uekotter, F., and J. Hohensee (eds.), *Wird Kassandra heiser? Die Geschichte falscher Ökoalarme*. Stuttgart: Franz Steiner, 24–41.

Griesser, I., 1999. *Chronik von Niederthai: Besser gesagt: Was ich von Niederthai weiß*. Umhausen: Ötztaler Heimatverein, Gemeinde Umhausen.

Grime, J.P., Hodgson, J.G., and R. Hunt, 1988. *Comparative Plant Ecology: A Functional Approach to Common British Species*. London: Hyman.

Grodzins'kyj, A.M. (ed.), 1989. *Likars'ki roslyny: Encyklopedyčnyj dovidnyk* (Medicinal Plants: An Encyclopedic Guide). Kyiv: Holovna redaktsiya Ukrayins'koyi radjans'koji entsyklopediji.

Groß, R., 2015. "Damüls, Vorarlberg: 100 Years of Transformation from Meadows to Ski-Routes in an Alpine Environment." In Lytje, M., Nielsen, T.K., and M. Ottovay Jørgensen (eds.), *Challenging Ideas: Theory and Empirical*

Research in the Social Sciences and Humanities. Newcastle upon Tyne: Cambridge Scholars Publishing, 178–198.

Groß, R., 2019. *Die Beschleunigung der Berge: Eine Umweltgeschichte des Wintertourismus in Vorarlberg/Österreich (1920–2010)*. Cologne: Böhlau.

Gross, R., and V. Winiwarter, 2015. "Commodifying Snow, Taming the Waters: Socio-Ecological Niche Construction in an Alpine Village." *Water History* 7: 489–509.

Grotefend, H., and T. Ulrich, 1971. *Taschenbuch der Zeitrechnung des deutschen Mittelalters und der Neuzeit*. 11th ed. Hannover: Hahnsche Buchhandlung.

Grove, A.T., and O. Rackham, 2003. *The Nature of Mediterranean Europe: An Ecological History*. 2nd revised ed. New Haven, CT, and London: Yale University Press.

Grüne, N., 2016. "Wissenstransfer und politische Teilhabe: Agrarische Wissensbezüge als Partizipationsressource im 18. und 19. Jahrhundert." In Dauser, R., Fassl, P., and L. Schilling (eds.), *Wissenszirkulation auf dem Land vor der Industrialisierung*. Augsburg: Wißner, 23–35.

Günergun, F., and Ş. Etker, 2009. "Medicine." In Ágoston, G., and B. Masters (eds.), *Encyclopedia of the Ottoman Empire*. New York: Facts On File, 357–362.

GURS (Geodetska uprava Republike Slovenije), Državna meja (Surveying and Mapping Authority of the Republic of Slovenia, State Border) [data file]. 2018. https://ipi.eprostor.gov.si/jgp/data (October 15th, 2018).

Gustafsson, B., 1997. "Nature and Economy." In Teich, M., Porter, R., and B. Gustafsson (eds.), *Nature and Society in Historical Context*. Cambridge, New York, and Melbourne: Cambridge University Press, 347–363.

Guštin, D., 2005 (= Guštin, 2005a). "Prva svetovna vojna: Soška fronta" (World War I: The Isonzo (Sln. *Soča*) Front). In Fischer, J., et al. (eds.), *Slovenska novejša zgodovina: Od programa Zedinjena Slovenija do mednarodnega priznanja Republike Slovenije 1848–1992*. Vol. 1. Ljubljana: Mladinska knjiga, 131–137.

Guštin, D., 2005 (= Guštin, 2005b). "Soška fronta in njeno slovensko zaledje" (The Isonzo (Sln. *Soča*) Front and its Slovenian Hinterland). In Vodopivec, P., and K. Kleindienst (eds.), *Velika vojna in Slovenci*. Ljubljana: Slovenska matica, 62–74.

Habinc, M., 2011 (= Habinc, 2011a). "Klepánje" (Peening). In Baš, A., et al. (eds.), *Slovenski etnološki leksikon*. Ljubljana: Mladinska knjiga, 212.

Habinc, M., 2011 (= Habinc, 2011b). "Kosílnica" (Mower). In Baš, A., et al. (eds.), *Slovenski etnološki leksikon*. Ljubljana: Mladinska knjiga, 238.

Habjan, H., 2014. "Gospodarstvo in aprovizacija v prvem letu svetovne vojne" (The Economy and Purveyance in the First Year of the First World War). In Eržen Podlipnik, E., et al. (eds.), *Kuge, lakote in vojske – reši nas, o Gospod! Kranjska v prvem letu vélike vojne*. Ljubljana: Zgodovinski arhiv Ljubljana, 67–71.

Hable, E., and I. Präsent, 1985. "Seltene oder besonders interessante Pflanzen im Bezirk Murau: Unter Berücksichtigung ihres Vorkommens in den Schwarzenbergischen Besitzungen." *Schwarzenbergischer Almanach* 37: 383–426.

Hachtmann, R., 2007. *Tourismus-Geschichte*. Göttingen: Vandenhoeck & Ruprecht.

Hacquet, B., 1779. "Beobachtungen, und Heilungsmethoden einzelner Hornvie[sic!]krankheiten, welche durch Gifte aus den drey Naturreichen verursachet werden." *Sammlung Nützlicher Unterrichte* 4: 1–81.

Haidvogl, G., Pont, D., and Ž. Zwitter, 2019. "The History of Human Use and Interference: Alpine Rivers as Resource and Risk Factor." In Muhar, S., et al. (eds.), *Rivers of the Alps. Diversity in Nature and Culture*. Bern: Haupt, 36–44.

Hall, M.R., 2023. *A Play for the Living in a Time of Extinction*. London: Nick Hern Books.

Hanemann, B., 2014. *Die Eisenhortfunde der Pfalz aus dem 4. Jahrhundert nach Christus*. Vols. 1 and 2. Speyer: Generaldirektion Kulturelles Erbe Rheinland-Pfalz, Direktion Landesarchäologie, Aussenstelle Speyer.

Hanhart, R., and S. Sonderegger, 1998. *Appenzeller Bauernmalerei*. Herisau: Appenzeller Verlag.

Harmes, P., and P. Tout, 2015. *Slovenia—Alpine Flowers of Lake Bohinj and the Julian Alps: Naturetrek Tour Report*. [city unknown]: Naturetrek. https://d11gbzngntg4t4.cloudfront.net/reports/SVN01_report_150618_Slovenia_Flowers_of_Lake_Bohinj_the_Julian_Alps.pdf (October 15th, 2020).

Hartel, T., Craioveanu, C., and K.-O. Réti, 2016. "Tree Hay as Source of Economic Resilience in Traditional Social-Ecological Systems from Transylvania." *Martor* 21: 53–64.

Hartmann, A., Sudmanns, M., Augustin, H., Baraldi, A., and D. Tiede, 2023. "Estimating the Temporal Heterogeneity of Mowing Events on Grassland for Haymilk-Production Using Sentinel-2 and Greenness-Index." *Smart Agricultural Technology* 4: article no. 100157, separate pagination.

Hasenöhrl, U., and R. Groß, 2022. "Travelling (Western) Europe: Tourism, Regional Development, and Nature Protection." In Wöbse, A.K., and P. Kupper (eds.), *Greening Europe: Environmental Protection in the Long Twentieth Century—A Handbook*. Munich and Vienna: De Gruyter Oldenbourg, 185–215.

Hayek, A., 1931. *Prodromus Florae peninsulae Balcanicae*. Vol. 2, *Dicotyledoneae Sympetalae*. Dahlem bei Berlin: Repertorium, 1931.

Hayot Carbonero, C., Mueller-Harvey, I., Brown, T.A., and L. Smith, 2011. "Sainfoin (*Onobrychis viciifolia*): A Beneficial Forage Legume." *Plant Genetic Resources: Characterization and Utilization* 9(1): 70–85.

Hazler, V., 2021. "Dediščina vojaških pokopališč, kapelic, spomenikov in spominskih znamenj prve svetovne vojne na Slovenskem" (Heritage of First World

War Military Cemeteries, Chapel Shrines, Memorials, and Memorial Signs in Slovenian Territory). In Fikfak, J., and B. Jezernik (eds.), *Dediščina prva svetovne vojne: Reprezentacije in reinterpretacije*. Ljubljana: Znanstvena založba Filozofske fakultete Univerze v Ljubljani, 41–86.

Heinz, S., Mayer, F., and G. Kuhn, 2012. *Artenreiches Grünland Erkennen und Bewerten*. 2nd ed. Freising-Weihenstephan: Bayerische Landesanstalt für Landwirtschaft. https://www.lfl.bayern.de/mam/cms07/publikationen/daten/informationen/p_44089.pdf (February 22nd, 2022).

Hejcman, M., Hejcmanová, P., Pavlů, V., and J. Beneš, 2013. "Origin and History of Grasslands in Central Europe—A Review." *Grass and Forage Science* 68: 345–363.

Henning, J., 1987. *Südosteuropa zwischen Antike und Mittelalter*. Berlin: Akademie.

Henning, J., 1991. "Fortleben und Weiterentwicklung spätrömischer Agrargerätetraditionen in Nordgallien: Eine Mähsense der Merowingerzeit aus Kerkhove (Belgien)." *Acta Archaeologica Lovaniensia* 30: 49–59.

Henning, J., 1997. "Landwirtschaft der Franken." In Wieczorek, A., et al. (eds.), *Die Franken: Wegbereiter Europas*. Vol. 2. 2nd revised and expanded ed. Mannheim: Reiss-Museum, Museum für Vor- und Frühgeschichte; Berlin: Staatliche Museen, and Mainz: Philipp von Zabern, 774–785.

Henning, J., 2014. "Did the 'Agricultural Revolution' Go East with Carolingian Conquest? Some Reflections on Early Medieval Rural Economics of the Baiuvarii and Thuringi." In Fries-Knoblach, J., Steuer, H., and J. Hines (eds.), *The Baiuvarii and Thuringi: An Ethnographic Perspective*. Woodbridge: The Boydell Press, 331–360.

Herzon, I., Birge, T., Allen, B., Povellato, A., Vanni, F., et al., 2018. "Time to Look for Evidence: Results-Based Approach to Biodiversity Conservation on Farmland in Europe." *Land Use Policy* 71: 347–354.

Hessenberger, E., 2022. "Die traditionelle Berglandwirtschaft im Ötztal und rund um das heutige Museum in Lehn." In Hessenberger, E. (ed.), *Hoamatle. Heumahd. Heimat: Vom historischen Leben und Arbeiten im Ötztal*. Innsbruck: Studienverlag, 137–177.

Hiller, D., 2020. "Some Italian Wall Paintings of the Sunday Christ: Monitory and Redemptory Iconographies and the Effect of Position." *Perergon* 37(1): 79–112.

Hilpert, M., 2014. *Der Echte Speik: Flagship-Species der Region Nockberge: Seminararbeit Protected Areas*. Klagenfurt: Alpen-Adria Universität Klagenfurt-Graz-Wien, Institut für Geographie und Regionalforschung. https://issuu.com/nockberge/docs/koop14_taetigkeitsbericht_gesamt (December 28th, 2020).

Himmelreich, B., 2001. *Namesto žemlje črni kruh: Organizacija preskrbe z živili v Celju v času obeh svetovnih vojn* (Dark Bread Instead of Bread Rolls: Organization of Supplies of Celje (Germ. *Cilli*) with Foodstuffs during the First World War and Second World War). Celje: Zgodovinski arhiv.

Hinterwaldner, K., 2002. *Almwirtschaft und Almstreit in den Gerichten Ritten, Wangen und Villanders vom Mittelalter bis 1823*. Frankfurt am Main et al.: Peter Lang.

Hiss Reet: Die Schilfernte. https://www.hiss-reet.de/reetdach/wissenswertes-ueber-reet/schilfernte (May 29th, 2024).

Hlebanja, J., and N. Hlebanja, 2006. *Hlebanjev rod skozi stoletja v Srednjem Vrhu* (Centuries of the Hlebanja Family in Srednji Vrh). Ljubljana and Srednji Vrh: Hlebanja family.

Höbaus, E., Sommer, E., and D. Reinthaler, 2023. *Reinanke*. https://www.bmlrt.gv.at/land/lebensmittel/trad-lebensmittel/fisch/reinanke.html (January 14th, 2024).

Hodnik, U., 2016. *Planšarstvo v zgornjih bohinjskih planinah: Diplomsko delo* (Alpine Dairy Farming in Bohinj: Bachelor's Thesis). Ljubljana: Univerza v Ljubljani, Biotehniška fakulteta, Oddelek za zootehniko.

Hoffmann, R.C., and V. Winiwarter, 2010. Making Land and Water Meet: The Cycling of Nutrients between Fields and Ponds in Pre-Modern Europe. *Agricultural History* 84(3): 352–380.

Hofinger, W., and H. Bergmann. 2015. "Trientl, Adolf (1817–1897), Lehrer, Seelsorger und Naturwissenschaftler." In *Österreichisches Biographisches Lexikon*. https://www.biographien.ac.at/oebl/oebl_T/Trientl_Adolf_1817_1897.xml (October 29th, 2021).

Höfler, J., 1996. *Srednjeveške freske v Sloveniji* (Medieval Frescoes in Slovenia). Vol. 1, *Gorenjska*. Ljubljana: Družina.

Höfler, J., 2001. *Srednjeveške freske v Sloveniji* (Medieval Frescoes in Slovenia). Vol. 3, *Okolica Ljubljane z Notranjsko, Dolenjsko in Belo krajino* (Surroundings of Ljubljana, Notranjska, Dolenjska, and Bela Krajina). Ljubljana: Družina.

Hohensinner, S., Sonnlechner, C., Schmid, M., and V. Winiwarter, 2013. "Two Steps Back, One Step Forward: Reconstructing the Dynamic Danube Riverscape under Human Influence in Vienna." *Water History* 5: 121–143.

Horden, P., 2018. "Mediterranean Antiquity." In White, S., Pfister, C., and F. Mauelshagen (eds.), *The Palgrave Handbook of Climate History*. London: Palgrave Macmillan, 183–188.

Hörtenhuber, S., Lindenthal, T., Amon, B., Markut, B., Kirner, L., et al., 2010. "Greenhouse Gas Emissions from Selected Austrian Dairy Production Systems—Model Calculations Considering the Effects of Land Use Change." *Renewable Agriculture and Food Systems* 25(4): 316–329.

Horvat, J., 1995. "Gothic Wall Painting as a Source for Material Culture." In Lozar Štamcar, M. (ed.), *Gothic in Slovenia: The World of Objects*. Ljubljana: Narodni muzej, 136–155.

Hrvatin, M., and M. Zorn, 2020. "Climate and Hydrological Changes in Slovenia's Mountain Regions between 1961 and 2018." *Economic- and Ecohistory* 16: 201–218.

Hubatschek, E., 1992. *Bauernwerk in den Bergen: Arbeit und Leben der Bergbauern in Bilddokumenten aus einem halben Jahrhundert*. 5th ed. Innsbruck: Author.

Hubert, B., Meuret, M., and J. Bonnemaire, 2008. "Shepherds, Sheep and Forest Fires: A Reconception of Grazingland Management." In Hirsch Hadorn, G., et al. (eds.), *Handbook of Transdisciplinary Research*. [city unknown]: Springer, 103–126.

Humerca, R., 2019. *Paleobotanični pregled mokrišča Blate pri Rodinah: Seminarska naloga pri predmetu Arheologija okolja* (Palaeobotanical Overview of the Wetland Blate near Rodine: Term Paper). Jesenice: Univerza v Ljubljani, Filozofska fakulteta, Oddelek za arheologijo.

Humerca, R., 2020. *Topografski pregled Dežele: Diplomsko delo* (Topographic Survey of Dežela: Bachelor's Thesis). Jesenice: Univerza v Ljubljani, Filozofska fakulteta, Oddelek za arheologijo.

Ilešič, S., 1931. "Planine ob dolenji Zilji" (Die Almen des unteren Gailtales). *Geografski vestnik* 7: 120–148.

Ilić, M., Vertot, N., Vujasin, R., and A. Juvanc, *1994. Popis prebivalstva, gospodinjstev, stanovanj in kmečkih gospodarstev v Republiki Sloveniji v letu 1991* (Results of Surveys no. 617: Census of the Population, Households, Housings and Agricultural Holdings in the Republic of Slovenia in 1991). Ljubljana: Zavod Republike Slovenije za statistiko.

Info Flora 2004: Hypochaeris radicata L. https://www.infoflora.ch/en/flora/hypochaeris-radicata.html (April 21st, 2021).

*Institut für Seenforschung (KIS): Reinanke (*Coregonus *sp. "Kärnten")*. https://kis.ktn.gv.at/Flora-Fauna/Fische?ffid=59 (January 14th, 2024).

International Conference on Mountain Cheeses and XII Annual Meeting of FACE-network. https://mountaincheese.eu/#3.10 (August 16th, 2024).

Isenberg, A.C., 2017. "Seas of Grass: Grasslands in World Environmental History." In Isenberg, A.C. (ed.), *The Oxford Handbook of Environmental History*. Oxford and New York: Oxford University Press, 133–153.

Iuga, A., 2016. "Intangible Hay Heritage in Şurdeşti." *Martor* 21: 67–84.

Iuga, A., Iancu, B., and M. Stroe, 2016. "Introduction: A Place for Hay: Flexibility and Continuity in Hay-Meadow Management." *Martor* 21: 7–12.

Ivaşcu, C.M., Öllerer, K., and L. Rákosy, 2016. "The Traditional Perceptions of Hay and Hay-Meadow Management in a Historical Village from Maramureş County, Romania." *Martor* 21: 39–51.

Izdebski, A., 2022. "What Stories Should Historians Be Telling at the Dawn of the Anthropocene?" In Izdebski, A., Haldon, J., and P. Filipkowski (eds.), *Per-*

spectives on Public Policy in Societal-Environmental Crises: What the Future Needs from History. Cham: Springer: 9–19.

Izdebski, A., Guzowski, P., Poniat, R., Masci, L., Palli, J., et al., 2022. "Palaeoecological Data Indicates Land-Use Changes Across Europe Linked to Spatial Heterogeneity in Mortality During the Black Death Pandemic." *Nature Ecology & Evolution* 6: 297–306.

Jacobsson, O., 2023. "Meadows in Europe: Historical Perspectives on Sustainable Agricultural Land Use." In Nagy, R.-M., Fehér, A, and C. Izsán (eds.), *International Conference of the European Rural History Organisation: Rural History 2023*. Cluj-Napoca: Egyetemi Műhely Publisher, 248.

Jäger, G., 2006. "Hochweidewirtschaft, Klimaverschlechterung ('Kleine Eiszeit') und Gletschervorstöße in Tirol zwischen 1600 und 1850: Ein historisch-geographischer Beitrag zur Mensch-Umwelt-Forschung im mittleren Alpenraum." *Tiroler Heimat* 70: 5–84.

Jäger, G., 2007 (= Jäger, 2007a). "Bergbäuerliche Sozialgeschichte und Volkskultur: Das schwer verdiente Zubrot von Kleinhäuslerfrauen und Bergbäuerinnen in Tirol (vornmehmlich nach Reiseberichten und Beschreibungen aus dem 19. Jahrhundert)." In Merlin, F.W., Hellebart, S., and M. Machatschek (eds.), *Bergwelt im Wandel: Festschrift Erika Hubatschek zum 90. Geburtstag*. Klagenfurt: Kärntner Landesarchiv, 216–231.

Jäger, G., 2007 (= Jäger, 2007b). "Die Folgen der neuzeitlichen Klimaverschlechterung für die Hochweidestufe in Tirol zwischen 1550 und 1850." In Aistleitner, J. (ed.), *Alpine Kulturlandschaft im Wandel: Hugo Penz zum 65. Geburtstag*. Innsbruck: Geographie Innsbruck – Innsbrucker Geographische Gesellschaft, 91–112.

Jaksetič, J. (interview with the botanist M. Kaligarič), 2019. "Travniki našega otroštva" (Meadows of Our Childhood). *Delo, Sobotna priloga* (March 23rd, 2019), 20–22.

Janežič, F., 1971. *Majski hrošč v Sloveniji: Raziskovalno poročilo* (Maikäfer in Slowenien: Forschungsbericht). Ljubljana: Biotehniška fakulteta, Katedra za varstvo rastlin.

Janišová, M., Bojko, I., Ivaşcu, C.M., Iuga, A., Biro, A.-S., et al., 2023. "Grazing Hay Meadows: History, Distribution, and Ecological Context." *Applied Vegetation Science* 26(2): separate pagination, 1–24.

Janovsky, A. (interview with the wild food plant expert Dario Cortese), 2014. "Smrtno nevarne zamenjave rastlin" (Life-Threatening Plant Substitutions). *Slovenske novice* (April 3rd, 2014). https://old.slovenskenovice.si/lifestyle/vrt-dom/smrtno-nevarne-zamenjave-rastlin (December 22nd, 2024).

Jaritz, G., and V. Winiwarter, 1997. "On the Perceptions of Nature in a Renaissance Society." In Teich, M., Porter, R., and B. Gustafsson (eds.), *Nature and Society in Historical Context*. Cambridge, New York, and Melbourne: Cambridge University Press, 91–111.

Jenin, J., 2007. "Vom Leben am Steilhang: Ein Porträt von Erika Hubatschek." In Merlin, F.W., Hellebart, S., and M. Machatschek (eds.), *Bergwelt im Wandel: Festschrift Erika Hubatschek zum 90. Geburtstag*. Klagenfurt: Kärntner Landesarchiv, 25–31.

Jerše, S., 2017. "Slovenski kraji spomina: pojmi, teze in perspektive zgodovinskih raziskav" (Slovene Realms of Memory (*Les lieux de mémoire*): Concepts, Theses, and Perspectives of Historical Research). *Zgodovinski časopis* 71(2): 246–267.

Jesenšek, A., 2020. *Rastlinstvo, živalstvo in človek na območju perujskega dela inkovskega imperija v 16. stoletju: Diplomsko delo* (Flora, Fauna and Man in the Peruvian Territory of the Inca Empire in the Sixteenth Century: Bachelor's Thesis). Litija: Univerza v Ljubljani, Filozofska fakulteta, Oddelek za zgodovino.

Jevnikar, I., 2013. "Kugy, Julij (1858–1944)." In *Slovenska biografija*. Ljubljana: Slovenska akademija znanosti in umetnosti, Znanstvenoraziskovalni center SAZU. https://www.slovenska-biografija.si/oseba/sbi308558/ (September 2nd, 2024).

Jezernik, B., 2021. "Vojne rane se celijo počasi" (The Wounds of War Heal Slowly). In Fikfak, J., and B. Jezernik (eds.), *Dediščina prva svetovne vojne: Reprezentacije in reinterpretacije*. Ljubljana: Znanstvena založba Filozofske fakultete Univerze v Ljubljani, 115–174.

Jogan, N., 2006. "Japonski dresnik (*Fallopia japonica*) – rastlina leta 2006" (Japanese Knotweed (*Fallopia japonica*)—2006 Plant of the Year). *Proteus* 68(9–10): 437–440.

Jogan, N., Bačič, M., and S. Strgulc-Krajšek, 2012. "Tujerodne in invazivne rastline v Sloveniji" (Alien and Invasive Plants in Slovenia). In Jogan, N., Bačič, M., and S. Strgulc-Krajšek (eds.), *Neobiota Slovenije: Invazivne tujerodne vrste v Sloveniji ter vpliv na ohranjanje biotske raznovrstnosti in trajnostno rabo virov: Končno poročilo projekta*. Ljubljana: Univerza v Ljubljani, Biotehniška fakulteta, etc., 161–181.

Johansen, L., Westin, A., Wehn, S., Iuga, A., Ivascu, C.M., et al., 2019. "Traditional Semi-Natural Grassland Management with Heterogeneous Mowing Times Enhances Flower Resources for Pollinators in Agricultural Landscapes." *Global Ecology and Conservation* 19: 1–12.

Jones, A., 2012. *How to Harvest Water Reed for Thatching*. https://www.youtube.com/watch?v=hzRyVcQNZjU (May 26th, 2024).

Jones, G., and X. Poux, 2012. "Socio-Economic Perspectives on HNV Farming." In Oppermann, R., Beaufoy, G., and G. Jones (eds.), *High Nature Value Farming in Europe: 35 European Countries and Perspectives*. Ubstadt-Weiher, Heidelberg, and Basel: Regionalkultur, 451–458.

Jonstrup, A., Hedrén, M., and S. Andersson, 2016. "Host Environment and Local Genetic Adaptation Determine Phenotype in Parasitic *Rhinanthus angustifolius*." *Botanical Journal of the Linnean Society* 180(1): 89–103.

Jordan, B., 1945. "Planine v Karavankah" (La vie pastorale dans les Karavankes). *Geografski vestnik* 17: 49–104.

Juen senior, G., 2014. "Enzianschnaps: Ein immaterielles Kulturerbe." In Berndt, S. (ed.), *Ganz oben: Geschichten über Galtür und die Welt: Begleitbuch zur Ausstellung*. Galtür: Alpinarium, 56–57.

Julian Day and Civil Date Calculator. https://core2.gsfc.nasa.gov/time/julian.html (December 30th, 2024).

Jung, L.S., Winter, S., Eckstein, R.L., Kriechbaum, M., Karrer, G., et al., 2011. "Colchicum autumnale L." *Perspectives in Plant Ecology, Evolution and Systematics* 13(3): 227–244.

Jurca, J., 2003. "Balthasar Hacquet in veterinarstvo" (Balthasar Hacquet and Veterinary Medicine). *Hacquetia* 2(2): 75–84.

Just, T., 2004. "Grundherrschaftsprotokolle." In Pauser, J., Scheutz, M., and T. Winkelbauer (eds.), *Quellenkunde der Habsburgermonarchie (16.–18. Jahrhundert): Ein exemplarisches Handbuch*. Vienna and Munich: R. Oldenbourg, 443–447.

Juvanec, B., 2007. *Kozolec* (Hayrack). Ljubljana: i2 and Univerza v Ljubljani, Fakulteta za arhitekturo.

Juvanec, B., 2020. "Kozolec na Gorenjskem" (The Hayrack in Gorenjska). In Roškar, S. (ed.), *Od stoga do kozuca: Priročnik za rabo in vzdrževanje kozolcev na območju projekta Arhitektura gorenjskih vasi*. Jesenice: Razvojna agencija Zgornje Gorenjske, 10–23.

Kabusch, S., 2020. *Nutzung vs. Naturschutz am Beispiel der naturgeschützten Alpenpflanze Echter Speik (*Valeriana celtica *subsp.* norica*): Masterarbeit*. Vienna: Universität für Bodenkultur, Department für Nutzpflanzenwissenschaften.

Kačičnik Gabrič, A., 2008. "Kmečko življenje na visokogorskih kmetijah na primeru doline Tople v 19. stoletju" (Peasant Life on Highland Farms on the Case of the Topla Valley in the Nineteenth Century). *Kronika: Časopis za slovensko krajevno zgodovino* 56(2): 289–304.

Kačičnik Gabrič, A., 2009. "Jezersko – pozabljeni delček Koroške" (Jezersko: The Forgotten Part of Carinthia). *Kronika: Časopis za slovensko krajevno zgodovino* 57(1): 29–46.

Kačičnik Gabrič, A., 2014. *"To smemo že tako dolgo": Kmečke služnosti in njihova odprava* ("We Have Been Able to Do That for Such a Long Time": Peasant Easement and Its Abolishment). Ljubljana: Inštitut za novejšo zgodovino.

Kačičnik Gabrič, A., 2016. "Kmečko življenje v Zgornjesavski dolini v 18. in 19. stoletju" (Rural Life in the Upper Sava Valley in the Eighteenth and Nineteenth Centuries). *Kronika: Časopis za slovensko krajevno zgodovino* 64(3): 495–512.

KAGIS Maps (including *KAGIS Kartendienste*). https://kagis.ktn.gv.at/Karten/KAGIS-Karten-online (December 20th, 2022).

Kaligarič, M., Brecl, J., and S. Škornik, 2016. "High Potential of Sub-Mediterranean Dry Grasslands for Sheep Epizoochory." *Open Life Sciences* 11: 177–184 and supplemental data: https://www.degruyter.com/view/j/biol.2016.11.issue-1/biol-2016-0023/suppl/biol-2016-0023_suppl.pdf (June 20th, 2020).

Kaligarič, M., Čuš, J., Škornik, S., and D. Ivajnšič, 2019. "The Failure of Agri-Environment Measures to Promote and Conserve Grassland Biodiversity in Slovenia." *Land Use Policy* 80: 127–134.

Kambič, M., 2004 (Kambič, 2004a). "Die Entwicklung des Strafrechts im slowenischen Raum bis zum Jahre 1848." In Kozina, B., et al. (eds.), *Deren von Laibach Malefitz Freyhaittn: 1514*. Ljubljana: Zgodovinski arhiv Ljubljana; Graz: Institut für österreichische Rechsgeschichte und europäische Rechtsentwicklung der Karl-Franzens-Universität Graz, 164–182.

Kambič, M., 2004 (Kambič, 2004b). "Razvoj kazenskega prava na Slovenskem do leta 1848" (Die Entwicklung des Strafrechts im slowenischen Raum bis zum Jahre 1848). In Kozina, B., et al. (eds.), *Deren von Laibach Malefitz Freyhaittn: 1514*. Ljubljana: Zgodovinski arhiv Ljubljana; Graz: Institut für österreichische Rechsgeschichte und europäische Rechtsentwicklung der Karl-Franzens-Universität Graz, 145–163.

Kapfer, A., 2010. "Beitrag zur Geschichte des Grünlands Mitteleuropas: Darstellung im Kontext der landwirtschaftlichen Bodennutzungssysteme im Hinblick auf den Arten- und Biotopschutz." *Naturschutz und Landschaftsplanung* 42: 133–140.

Karlík, P., and P. Poschlod, 2009. "History or Abiotic Filter: Which is More Important in Determining the Species Composition of Calcareous Grasslands?" *Preslia* 81: 321–340.

Kärnten Atlas, Darstellung, Kartenhintergrund, Franziszeischer Kataster. https://gis.ktn.gv.at/atlas/(S(sa2jattmziyic0dshjyxaf4i))/init.aspx?karte=atlas_basis karten&ks=kaernten_atlas (November 16th, 2018).

Kärnten Atlas, Geländehöhe, Höhenschichten (10 m). https://gis.ktn.gv.at/atlas/(S(sa2jattmziyic0dshjyxaf4i))/init.aspx?karte=atlas_basiskarten&ks=kaern ten_atlas (November 16th, 2018).

Karrer, G., Unpublished data on the lifespan of *Bromus erectus*, *Arrhenatherum*, and *Dactylis*. Vienna: University of Natural Resources and Life Sciences.

Kaser, K., 1997. *Freier Bauer und Soldat: Die Militarisierung der agrarischen Gesellschaft an der kroatisch-slawonischen Militärgrenze (1535–1881)*. Vienna, Cologne, and Weimar: Böhlau.

Kaspret, A. (ed.), 1889 (= Kaspret (ed.), 1889a). "Unterrichtung herrn Hartmanns von Kreyg der K. M. Räten vnd Reformierern, auf das darin sie beschaid begeren." 1493–1501 [chronological orientation from p. 108]. In: "Ueber die Lage der oberkrainischen Bauernschaft beim Ausgange des XV. und im Anfange des XVI. Jahrh.," Beilage VI. *Mittheilungen des Musealvereines für Krain* 2: 120–132.

Kaspret, A. (ed.), 1889 (= Kaspret (ed.), 1889b). "Vermerckt die eingriff, so durch herrn Hartmann von Kreyg in die herrlichkait des ambts Radmannsdorf geschehen." 1493–1501 [chronological orientation from p. 108]. In: "Ueber die Lage der oberkrainischen Bauernschaft beim Ausgange des XV. und im Anfange des XVI. Jahrh.," Beilage IV. *Mittheilungen des Musealvereines für Krain* 2: 115–117.

Kaspret, A., 1929. [Carniola on] "Landgerichtskarte: Blatt 31, Laibach." In *Historischer Atlas der österreichischen Alpenländer*. Vienna: Akademie der Wissenschaften and Adolf Holzhausens Nachfolger.

Kauter, D., 2002. *"Sauergras" und "Wegbreit"? Die Entwicklung der Wiesen in Mitteleuropa zwischen 1500 und 1900*. Stuttgart: Günter Heimbach.

Kebe, J., 2011. *Cerkniško jezero in ljudje ob njem* (Lake Cerknica and the People Around It). Koper: Ognjišče.

Keenleyside, C., and R. Oppermann 2009. "A Farmer-Centred Outlook to Managing and to Maintaining High Nature Value Grasslands in Europe." In Veen, P., et al. (eds.), *Grasslands in Europe of High Nature Value*. Zeist: KNNV Publishing, 293–303.

Keller, E., 2004. *Bergmahdwirtschaft in Kärnten: Aktuelle Situation, Förderungsmöglichkeiten und Zukunftsperspektiven der Bergmähder: Endbericht*. [city unknown]: Arge NATURSCHUTZ.

Keller, E., 2007. "*A Fuadale Bergheu is bessa wia drei Fuadalan Landheu*: Über traditionelles und aktuelles Wissen rund um Almpflanzen, Almheu und Almmähder." In Merlin, F.W., Hellebart, S., and M. Machatschek (eds.), *Bergwelt im Wandel: Festschrift Erika Hubatschek zum 90. Geburtstag*. Klagenfurt: Kärntner Landesarchiv, 205–215.

Keller, T., 2009. "The Mountains Roar: The Alps during the Great War." *Environmental History* 14(2): 253–274.

Keller, T., 2018. "Mobilizing Nature for the First World War." In Tucker, R.P., et al. (eds.), *Environmental Histories of the First World War*. Cambridge: Cambridge University Press, 1–16.

Kerschbaumer, N., 2007. "Über alpine Brandwirtschaft in Kärnten." In Merlin, F.W., Hellebart, S., and M. Machatschek (eds.), *Bergwelt im Wandel: Festschrift Erika Hubatschek zum 90. Geburtstag*. Klagenfurt: Kärntner Landesarchiv, 141–148.

Kéry, M., Matthies, D., and H.H. Spillmann, 2000. "Reduced Fecundity and Offspring Performance in Small Populations of the Declining Grassland Plants *Primula veris* and *Gentiana lutea*." *Journal of Ecology* 88(1): 17–30.

Kirchinger, J., 2016. "Lokales, empirisches und theoretisches Wissen in der Gutswirtschaft des Augustiner-Chorherrenstifts Polling im Zeitalter der Aufklärung." In Dauser, R., Fassl, P., and L. Schilling (eds.), *Wissenszirkulation auf dem Land vor der Industrialisierung*. Augsburg: Wißner, 99–125.

Kiss, A., 2020. "Weather and Weather-Related Natural Hazards in Medieval Hungary IV: Documentary Evidence from 1400–1450." *Economic- and Ecohistory* 16: 9–54.

Klapp, E., and A. Stahlin, 1936. *Standorte, Pflanzengesellschaften und Leistung des Grünlandes. Am Beispiel thuringischer Wiesen.* Stuttgart: Eugen Ulmer.

Kleindinst, J., 1999. "Das Churrätische Reichsgutsurbar als agrargeschichtliche Quelle." In Niederstätter, A. (ed.), *Aspekte der Landwirtschaft in der Bodenseeregion: Mittelalter und Frühe Neuzeit.* Dornbirn: Vorarlberger Verlagsanstalt, 22–30.

Klemenčič, V., 1959. *Pokrajina med Snežnikom in Slavnikom: Gospodarska geografija* (La région entre le Snežnik et le Slavnik: Géographie économique). Ljubljana: Slovenska akademija znanosti in umetnosti.

Klinar, K., 2011. *Kako se pri vas reče? Hišna imena v naseljih Bohinjska Češnjica, Jereka, Podjelje, Srednja vas v Bohinju, Stara Fužina in Studor* (House Names in Bohinjska Češnjica, Jereka, Podjelje, Srednja Vas v Bohinju, Stara Fužina, and Studor). Bohinjska Bistrica: Občina Bohinj.

Klinar, K., 2020. "Kozolec kot del nesnovne kulturne dediščine" (The Hayrack as Part of Intangible Cultural Heritage). In Roškar, S. (ed.), *Od stoga do kozuca: Priročnik za rabo in vzdrževanje kozolcev na območju projekta Arhitektura gorenjskih vasi.* Jesenice: Razvojna agencija Zgornje Gorenjske, 42–59.

Klinar, K., and S. Roškar, 2020. "Tipološka karta kozolcev na območju LAS Gorenjska košarica" (Hayrack Type Map in the Area of the Local Action Group Gorenjska Košarica). In Roškar, S. (ed.), *Od stoga do kozuca: Priročnik za rabo in vzdrževanje kozolcev na območju projekta Arhitektura gorenjskih vasi.* Jesenice: Razvojna agencija Zgornje Gorenjske, 103.

Klinar, T., 2012. "Kmetijstvo v Rovtih nekoč in danes" (Farming in Rovti Past and Present). In Šubic, T. (ed.), *V rovtih nad Jesenicami: Narava in človek ustvarjata biotsko pestrost.* Ljubljana: Zavod Republike Slovenije za varstvo narave, 95–114.

Knific, T., 1999. "Arheološko najdišče Gradišče nad Bašljem" (The Archaeological Site Gradišče nad Bašljem). In Roblek, T. (ed.), *Preddvor v času in prostoru.* Preddvor: Občina, 55–67.

Knific, T., 2004. "Sekerovitá hrivna iz reke Ljubljanice (Slovenija)" (An Axe-Shaped Bar from the Ljubljanica River). In Fusek, G. (ed.), *Zborník na počesť Dariny Bialekovej.* Nitra: Archeologický ústav Slovenskej akadémie vied, 191–195.

Knittel, M., 2017. "Dominus creavit ex Terra Medicamenta." Heilpflanzenwissen in Johannes Gessners *Phytographia sacra.*" In: Boscani Leoni, S., and M. Stuber (eds.), *Wer das Gras wachsen hört: Wissensgeschichte(n) der pflanzlichen Ressourcen vom Mittelalter bis ins 20. Jahrhundert.* Innsbruck, Vienna, and Bozen: Studienverlag, 96–114.

Koblar, S., 2013. *Morfogeneza grbinastih travnikov: Zaključna seminarska naloga* (Morphogenesis of Hummocky Meadows: Bachelor's Thesis). Ljubljana: Univerza v Ljubljani, Filozofska fakulteta, Oddelek za geografijo, 2013.

Kočevar, V., 2014. "Uvedba gregorijanskega koledarja v notranjeavstrijskih deželah leta 1583" (Introduction of the Gregorian Calendar in the Inner Austrian Provinces in 1583). *Kronika: Časopis za slovensko krajevno zgodovino* 62(2), 173–192.

Kohli, L., 2011. "Stickstoffeintrag aus der Luft verändert Vielfalt." *BDM-Facts 3*: 1–4.

Kolar, A., [2019 or 2020]. *Certificiranje senenega mesa in mleka* (Certification of Hay Meat and Hay Milk). https://seneno.si/wp-content/uploads/2022/02/ZTPcertificiranje.pdf (July 21st, 2024).

Kolarek Novšek, P., 2024. "Presentation of the Hay Meat and Milk Quality Scheme." Presentation at the online workshop of the Working Group Mountain Agriculture and Mountain Forestry of the Alpine Convention: *Peatlands and Carbon Farming, Pastoralism, Forestry and Multifunctional Activities in the Mountains for the Sustainability of the Alpine Environment*. June 21st, 2024.

Komac, B., and M. Zorn, 2007. *Pobočni procesi in človek* (Mass Movements and People). Ljubljana: ZRC.

Komac, J., 2022. *'Srce se mi trga od žalosti': Vsakdanje življenje prebivalstva na Bovškem med véliko vojno* ("My Heart Breaks with Sorrow": Everyday Life of the Bovec Region Inhabitants During the Great War). Ljubljana: Inštitut za novejšo zgodovino.

Kompan, D., 1999. "Jezersko-solčavska ovca" (Jezersko–Solčava Sheep, Germ. *Seeländer Schaf*). In Kompan, D., Šalehar, A., and A. Holcman (eds.), *Ohranjene slovenske avtohtone domače živali: Opis in slikovni prikaz nekaterih vrst.* Domžale: Biotehniška fakulteta, Oddelek za zootehniko, 20–21.

Komposch, C., Frieß, T., and D. Kreiner, 2013. "Natural Hazards—Hazards for Nature? Avalanches as a Promotor of Biodiversity: A Case Study on the Invertebrate Fauna in the Gesäuse National Park (Styria, Austria)." In Bauch, K. (ed.), *5th Symposium for Research in Protected Areas: Conference Volume.* Matrei in Osttirol: Hohe Tauern National Park Council, 389–398.

Koncilija, Ž., 2009. "Prezrti heroji velike vojne: Konji, psi, golobi in ostale živali na frontah prve svetovne vojne" (Overlooked Heroes of the Great War: Horses, Dogs, Pigeons and Other Animals on the First World War Front). *Zgodovina za vse* 16(2): 79–95.

Konersmann, F., 2005. "Branntwein." In Jaeger, F., et al. (eds.), *Enzyklopädie der Neuzeit*. Vol. 2, *Beobachtung–Dürre*. Stuttgart and Weimar: J.B. Metzler, 404–406.

Koprivnik, J., 1923. *Pohorje*. Maribor: Sokolsko društvo.

Koren, R., 1982. "Kosilnice in škode na divjadi" (Mowers and Damage to Game). *Lovec* 65(6): 206–207.

Körner, C., 2003. *Alpine Plant Life: Functional Plant Ecology of High Mountain Ecosystems*. Berlin and Heidelberg: Springer.

Koropec, J., 1972. *Zemljiške gospoščine med Dravogradom in Mariborom do konca 16. stoletja* (Seigneuries between Dravograd and Maribor until the End of the Sixteenth Century). Maribor: Obzorja.

Kos, D., 2015 and 2016. *Zgodovina morale* (History of Morality). Vols. 1 and 2. Ljubljana: ZRC, ZRC SAZU.

Kos, M., 1960. "O nekaterih planinah v Bohinju in okoli Bleda" (Sur quelques alpages dans la région de Bohinj et aux environs de Bled). *Geografski vestnik* 32: 131–139.

Kosi, M., 1998. *Potujoči srednji vek: Cesta, popotnik in promet na Slovenskem med antiko in 16. stoletjem* (Roads, Travelers, and Traffic in Slovenian Territory from Antiquity to the Sixteenth Century). Ljubljana: ZRC.

Kosi, M., 2014. "Interactions Between the City and its Environment: The Case of Late Medieval Ljubljana." In Štih, P., and Ž. Zwitter (eds.), *Man, Nature and Environment between the Northern Adriatic and the Eastern Alps in Premodern Times*. Ljubljana: Ljubljana University Press, Faculty of Arts, and Historical Association of Slovenia, 102–123.

Košir, U., 2019. "Legacies of the Soča Front—From Rubbish to Heritage (1915–2017)." In Košir, U., Črešnar, M., and D. Mlekuž (eds.), *Rediscovering the Great War: Archaeology and Enduring Legacies on the Soča and Eastern Fronts*. London and New York: Routledge, 181–198.

Košir, U., Saunders, N.J., Črešnar, M., and G. Rutar, 2019. "Between Tourism and Oblivion: Rombon and Kolovrat—Conflict Landscapes on the Soča Front, 1915–2017." In Košir, U., Črešnar, M., and D. Mlekuž (eds.), *Rediscovering the Great War: Archaeology and Enduring Legacies on the Soča and Eastern Fronts*. London and New York: Routledge, 90–108.

"Košnja" (Haymaking). 1864. *Novice: Gospodarske, obrtniške in narodne* 22(25) (June 22nd): 199.

Kotnik, B., 2002/2003. *Zgodovina hiš južne Koroške* (Geschichte der Häuser in Südkärnten). Vol. 8, *Občina Sele* (Gemeinde Zell). Klagenfurt, Ljubljana, and Vienna: Mohorjeva/Hermagoras.

Kozorog, M., 2019. "On the Border: Perspectives on Memory Landscapes between Slovenia and Italy." In Košir, U., Črešnar, M., and D. Mlekuž (eds.), *Rediscovering the Great War: Archaeology and Enduring Legacies on the Soča and Eastern Fronts*. London and New York: Routledge, 63–76.

Kozorog Blatnik, T. (interview with the organic farmer and cheesemaker Toni Kukenberger), 2025. "Ljudi nič ne zanima, kaj se dogaja s korenčkom, samo da čim manj plačajo zanj" (People Do Not Care How Carrots Are Grown, They Just Want to Pay as Little as Possible for Them). *MMC*, RTV SLO. https://www.rtvslo.si/talenti-eu-regij/ljudi-nic-ne-zanima-kaj-se-dogaja-s-korenckom-samo-da-cim-manj-placajo-zanj/746372 (May 27th, 2025).

Kranemann, B., 1995. "Sonntag." In Angermann, N., et al. (eds.), *Lexikon des Mittelalters.* Vol. 7, *Planudes bis Stadt.* Munich: LexMa, 2050.

Krausmann, F., 2008. *Land Use and Socio-Economic Metabolism in Pre-Industrial Agricultural Systems: Four Nineteenth-Century Austrian Villages in Comparison.* Vienna: Institute of Social Ecology, 2008.

Krausmann, F., 2016. "From Energy Source to Sink: Transformations of Austrian Agriculture." In Haberl, H., et al. (eds.), *Social Ecology: Society-Nature Relations across Time and Space.* [Cham]: Springer, 433–445.

Krausmann, F., Schaffartzik, A., Mayer, A., Eisenmenger, N., Gingrich, S., et al., 2016. "Long-Term Trends in Global Material and Energy Use." In Haberl, H., et al. (eds.), *Social Ecology: Society-Nature Relations across Time and Space.* [Cham]: Springer, 199–211.

Krausmann, F., Weisz, H, and N. Eisenmenger, 2016. "Transitions in Sociometabolic Regimes throughout Human History." In Haberl, H., et al. (eds.), *Social Ecology: Society-Nature Relations across Time and Space.* [Cham]: Springer, 63–92.

Kravanja, B., 2021. "Dediščina soške fronte v luči heterotopij, utopij in drugih 'drugih prostorov'" (Isonzo (Sln. *Soča*) Front Heritage from the Perspective of Heterotopias, Utopias, and Other "Other Spaces"). In Fikfak, J., and B. Jezernik (eds.), *Dediščina prva svetovne vojne: Reprezentacije in reinterpretacije.* Ljubljana: Znanstvena založba Filozofske fakultete Univerze v Ljubljani, 197–228.

Kravanja, E., and T. Komac, 2023. "Tradicionalna priprava krme" (Traditional Fodder Preparation). *Svet pod Triglavom* 31: 18–19.

Kremen, C., 2015. "Reframing the Land-Sparing / Land-Sharing Debate for Biodiversity Conservation." *Annals of the New York Academy of Sciences* 1355: 52–76.

Krepfl, D., 2012. "Varstvo ključavnic ali narcis v rovtih" (Protecting Daffodils in Montane Meadows). In Šubic, T. (ed.), *V rovtih nad Jesenicami: Narava in človek ustvarjata biotsko pestrost.* Ljubljana: Zavod Republike Slovenije za varstvo narave, 7–18.

Kretschmer, I. and O. Nestroy, 1974. "Trocknungsgerüste für Futtergras." In Wolfram, R. et al. (eds.), *Österreichischer Volkskundeatlas: Kommentar.* Vol. 4, Fascicle (*Lieferung*) 5. Wien: Hermann Böhlaus Nachfolger, separate pagination, 1–15.

Kretzenbacher, L., 1997. "Neufunde spätmittelalterlicher Fresken vom 'Mahnbild'-Typus 'Feiertags-Christus' in Kärnten." *Österreichische Zeitschrift für Volkskunde* 100: 157–183.

Kuhar, A., and A. Simeonov, 2019. *Seneno meso in mleko: Pregled osnovne literature* (Hay Meat and Hay Milk: An Overview of Basic Literature). https://seneno.si/wp-content/uploads/2020/06/Pregled-literature-Kuhar.pdf (July 10th, 2024).

Kuhar, B., 1966. *Včeraj in danes v škocjanskih hribih: Spreminjanje vaške kulture* (The Recent Past and Present in the Hills around Škocjan: Changing Rural Culture). Ljubljana: Slovenski etnografski muzej.

Kulturportal der Flur- und Hausnamen, Maps: http://maps.flurnamen.at/ (October 5th, 2019).

Kumer, Z., 2005. "Robanov Joža." In Vršnik, J., *Preproste zgodbe s solčavskih planin*. Celje: Mohorjeva, 9–16.

Kupper, P., 2021. *Umweltgeschichte*. Göttingen: Vandenhoeck & Ruprecht.

Kupper, P., and A.-K. Wöbse, 2022. "Introduction: Writing a European History of Environmental Protection." In Wöbse, A.K., and P. Kupper (eds.), *Greening Europe: Environmental Protection in the Long Twentieth Century—A Handbook*. Munich and Vienna: De Gruyter Oldenbourg, 1–14.

Kuret, N., 1985 (= Kuret, 1985a). "Na pot" (Foreword). In Kuret, N., *Slovensko Štajersko pred marčno revolucijo 1848: Topografski podatki po odgovorih na vprašalnice nadvojvode Janeza (1811) in Georga Götha (1842)*. Vol. 1, Fascicle (*snopič*) 1. Ljubljana: Academia scientiarum et artium slovenica, 7–8.

Kuret, N., 1985 (= Kuret, 1985b). "Zbiranje podatkov z anketami" (Data Collection Using Questionnaires). In Kuret, N., *Slovensko Štajersko pred marčno revolucijo 1848: Topografski podatki po odgovorih na vprašalnice nadvojvode Janeza (1811) in Georga Götha (1842)*. Vol. 1, Fascicle (*snopič*) 1. Ljubljana: Academia scientiarum et artium slovenica, 9–50.

Kuret, N. (ed.), 1985, 1987, and 1989. *Slovensko Štajersko pred marčno revolucijo 1848: Topografski podatki po odgovorih na vprašalnice nadvojvode Janeza (1811) in Georga Götha (1842)* (Lower Styria before the March Revolution 1848: Topographic Data Collected through Questionnaires by Archduke John of Austria (1811) and Georg Göth (1842)). Vol. 1, Fascicles (*snopiči*) 1, 2, and 3. Ljubljana: Academia scientiarum et artium slovenica.

Kuret, N., 1989. *Praznično leto Slovencev* (Slovenian Festivities around the Year). Vol. 1. Ljubljana: Družina.

Kurinčič, M., 2000. "Zbirka Mirka Kurinčiča v Drežnici" (Mirko Kurinčič's Museum Collection in Drežnica). *Vojnozgodovinski zbornik* 4: 135–139.

Kus Veenvliet, J., 2012. "Prilagoditev protokola za presojo tveganja in analiza predpisov na področju tujerodnih vrst" (Adaptation of the Risk Assessment Protocol and Analysis of Alien Species Regulations). In Jogan, N., Bačič, M., and S. Strgulc-Krajšek, (eds.), *Neobiota Slovenije: Invazivne tujerodne vrste v Sloveniji ter vpliv na ohranjanje biotske raznovrstnosti in trajnostno rabo virov: Končno poročilo projekta*. Ljubljana: Univerza v Ljubljani, Biotehniška fakulteta, etc., 104–157.

Küster, H., and C. Keenleyside, 2009. "The Origin and Use of Agricultural Grasslands in Europe." In Veen, P., et al. (eds.), *Grasslands in Europe of High Nature Value*. Zeist: KNNV Publishing, 9–14.

Küzmič, F., and U. Šilc, 2017. "Alien Species in Different Habitat Types of Slovenia: Analysis of Vegetation Database." *Periodicum biologorum* 119(3): 199–208.

Kužnik, L., 2002. "'Branje' mravljinčjih jajc" (Gathering Ant Eggs). In Hazler, V. (ed.), *Vsakdanjiki in prazniki Skomarjanov: Zbornik člankov etnoloških raziskovalnih delavnic Skomarje 1997 in Skomarje 1998*. Zreče: LTO Rogla, 39–56.

Labouvie, E., 2006. "Empfängnisverhütung." In Jaeger, F., et al. (eds.), *Enzyklopädie der Neuzeit*. Vol. 3, *Dynastie–Freundschaftslinien*. Stuttgart and Weimar: J.B. Metzler, 261–263.

Laharnar, B., 2018. "Metal Finds, Glass Finds and Stone Moulds from the Iron Age Settlement at Most na Soči." In Dular, J., and S. Tecco Hvala (eds.), *The Iron Age Settlement at Most na Soči*. Ljubljana: ZRC SAZU, Inštitut za arheologijo, 195–247.

Lamb Care Australia: Toxic Poisoning, 2023. https://lambcareaustralia.org.au/toxic-poisoning/ (February 1st, 2024).

Landsteiner, E., 1993. "Bäuerliche Meteorologie: Zur Naturwahrnehmung bäuerlicher Weinproduzenten im niederösterreichisch-mährischen Grenzraum an der Wende vom 18. zum 19. Jahrhundert." *Historische Anthropologie* 1(1): 42–62.

Landsteiner, E., 2004. "Zehentdaten und Zehentregister." In Pauser, J., Scheutz, M., and T. Winkelbauer (eds.), *Quellenkunde der Habsburgermonarchie (16.–18. Jahrhundert): Ein exemplarisches Handbuch*. Vienna and Munich: R. Oldenbourg, 540–560.

Lane, F.C., 1973. *Venice: A Maritime Republic*. Baltimore and London: The Johns Hopkins University Press.

Lang, G., 1993. "Holozäne Veränderungen der Waldgrenze in den Schweizer Alpen – Methodische Ansätze und gegenwärtiger Kenntnisstand." In Brombacher, C., Jacomet, S., and J.N. Haas (eds.), *Festschrift Zoller: Beiträge zu Philosophie und Geschichte der Naturwissenschaften, Evolution und Systematik, Ökologie und Morphologie, Geobotanik, Pollenanalyse und Archäobotanik*. Berlin: J. Cramer, 317–327.

Lang, L., 2020. "Staatliche Ordnungsgesetzgebung als ein Element der neuzeitlichen Verwaltungskultur: Eine fallanalytische Annäherung an der Wende vom 18. zum 19. Jahrhundert." In Jerše, S., and F. Draženović (eds.), *Wege zum historischen Erkennen unserer Zeit*. Ljubljana: Znanstvena založba Filozofske fakultete Univerze v Ljubljani, 117–128.

Langenscheidt Dictionary. https://en.langenscheidt.com/german-english/ (February 3rd, 2021).

Langthaler, E., 2018. "Dissolution before Dissolution: The Crisis of the Wartime Food Regime in Austria-Hungary." In Tucker, R.P., et al. (eds.), *Environmental Histories of the First World War*. Cambridge: Cambridge University Press, 38–61.

Lauenroth, W.K., and P.B. Adler, 2008. "Demography of Perennial Grassland Plants: Survival, Life Expectancy and Life Span." *Journal of Ecology* 96(5): 1023–1032.

Lauk, C., and K.-H. Erb, 2016. "A Burning Issue: Anthropogenic Vegetation Fires." In Haberl, H., et al. (eds.), *Social Ecology: Society–Nature Relations across Time and Space.* [Cham]: Springer, 335–348.

Lavrenčič, A., 1998. "Ko žival postane človeška, se ne zmeni za korenine lastnega debla: Basni iz obdobja 1914–1918 / usode zamolčanih junakov vseh vojn" (Wenn das Tier zum Menschen wird, kümmert es sich nicht um die Wurzeln des eigenen Stammes: Tierfabeln aus der Zeit von 1914 bis 1918 / Die Schicksale der totgeschwiegenen Soldaten aller Kriege). *Zgodovina za vse* 5(1): 30–36.

Lavrič, J., 1960. "Ob popisu kmetijstva" (On the Occasion of the Agricultural Census). *Prikazi in študije* 6(6): 1–11.

Lazarević, Ž., 2005. "Velika gospodarska kriza" (The Great Economic Depression). In Fischer, J., et al. (eds.), *Slovenska novejša zgodovina: Od programa Zedinjena Slovenija do mednarodnega priznanja Republike Slovenije 1848–1992.* Vol. 1. Ljubljana: Mladinska knjiga, 473–479.

Lazarević, Ž., 2022. "Jugoslovansko ekonomsko zgodovinopisje med nacionalnim in mednarodnim prostorom po drugi svetovni vojni" (Yugoslav Economic Historiography between the National and International Context after World War II). *Prispevki za novejšo zgodovino* 62(3): 101–129.

Le tramway du Mont-Blanc, 110 ans d'histoire. 2021. https://actu.hautesavoie.fr/explorez-actu/le-tramway-du-mont-blanc-110-ans-dhistoire (August 15th, 2022).

Ledinek, Š., 1999. *Planina Krstenica: Diplomsko delo* (The Alp Krstenica: Bachelor's Thesis). Krstenica, Ljubljana, and Tacen: Univerza v Ljubljani, Filozofska fakulteta, Oddelek za etnologijo in kulturno antropologijo.

Ledinek Lozej, Š., 2021. "Labelling, Certification and Branding of Cheeses in the Southeastern Alps (Italy, Slovenia): Montasio, Bovec, Tolminc and Mohant Cheese." *Acta geographica Slovenica* 61(1): 153–168.

Ledinska imena v občini Jesenice (Flurnamen in der Gemeinde Jesenice). 2015. 1:18,000. Jesenice: Razvojna agencija Zgornje Gorenjske. https://www.flurnamen.at/wp-content/uploads/2018/07/Jesenice_12.pdf (July 15th, 2023).

Leibundgut, C., and I. Vonderstrass, 2016a. *Traditionelle Bewässerung – ein Kulturerbe Europas.* Vol. 1: *Grundlagen.* Langenthal: Merkur Druck AG.

Leibundgut, C., and I. Vonderstrass, 2016b. *Traditionelle Bewässerung – ein Kulturerbe Europas.* Vol. 2: *Regionale Dokumentation.* Langenthal: Merkur Druck AG.

Leisering, W., 1998. "Der Erste Weltkrieg 1914–1918." In Putzger, F.W., and E. Brückmüller (eds.), *Historischer Weltatlas zur allgemeinen und österreichi-*

schen Geschichte. Vienna: Hölder-Pichler-Tempsky and ÖBV Pädagogischer Verlag, 74.

Leisering, W., and G. Schwarzrock, 1997. "Weltwirtschaft 1914." In Leisering, W. (ed.), *Historischer Weltatlas*. 102nd ed. Berlin: Cornelsen, 106.

Leitgeb, F., Petrasek, R., Drapela, T., and T. Lindenthal, 2023. "Comparing the Ecological Sustainability Performance of Organic Farms and Conventional Production in Austria Using the SMART Farm Tool and Monte Carlo Simulation." *Organic Agriculture* 13: 173–191.

Lenartsson, T., Westin, A., Iuga, A., and C.M. Ivascu, 2023. "Bringing the Maps Alive: Historical Meadow Use in Sweden and Romania, a Comparison." Presentation at the *International Conference of the European Rural History Organisation: Rural History 2023*. Cluj-Napoca, September 12th, 2023.

Lexer, M., 1862. *Kärntisches Wörterbuch*. Leipzig: S. Hirzel.

Lexer, M., *Mittelhochdeutsches Handwörterbuch*: Saher. http://woerterbuchnetz.de/cgi-bin/WBNetz/wbgui_py?sigle=Lexer&mode=Vernetzung&lemid=LS00138#XLS00138 (July 10th, 2019).

LIDAR, 2018: see entry *Environmental Atlas of Slovenia*, Backgrounds, Lidar hillshade.

Life for Seeds. https://lifeforseeds.si/en/about-project/ (December 31st, 2024).

Liira, J., Issak, M., Jõgar, Ü., Mändoja, M., and M. Zobel, 2009. "Restoration Management of a Floodplain Meadow and Its Cost-Effectiveness—The Results of a 6-Year Experiment." *Annales Botanici Fennici* 46: 397–408.

"Lišaji kot živinska krma" (Lichens as Fodder). 1916. *Slovenski narod* 49(3) (January 5th, 1916): 4.

"Lišaji kot živinska poklaja" (Lichens as Fodder). 1916. *Slovenec: političen list za slovenski narod* 44(7) (January 11th, 1916): 5.

Livi Bacci, M., 2000. *The Population of Europe: A History*. Oxford: Blackwell Publishers.

Löffler, J., Anschlag, K., Baker, B., Finch, O.-D., Diekkrüger, B., et al., 2011. "Mountain Ecosystem Response to Global Change." *Erdkunde* 65(2): 189–213.

Lomba, A., Moreira, F., Klimek, S., Jongman, R.H.G., Sullivan, C., et al., 2019. "Back to the Future: Rethinking Socioecological Systems Underlying High Nature Value Farmlands." *Frontiers in Ecology and the Environment* 18(1): 36–42.

Lovrenčak, F., 1984. "The Timberline in the Yugoslav Alps." *Geographica Iugoslavica: Bulletin of the Union of the Geographical Societies of Yugoslavia* 5: 31–35.

Ložar, R., 1944. "Pridobivanje hrane in gospodarstvo (Obtaining Food and Economic Activity). In Ložar, R. (ed.), *Narodopisje Slovencev*. Vol. 1. Ljubljana: Klas, 98–191.

Łuczaj, Ł., Pieroni, A., Tardío, J., Pardo-de-Santayana, M., Sõukand, R., et al., 2012. "Wild Food Plant Use in 21st Century Europe: The Disappearance of Old Traditions and the Search for New Cuisines Involving Wild Edibles." *Acta Societatis Botanicorum Poloniae* 81(4): 359–370.

Łuczaj, Ł., Stawarczyk, K., Kosiek, T., Pietras, M., and A. Kujawa, 2015. "Wild Food Plants and Fungi Used by Ukrainians in the Western Part of the Maramureş Region in Romania." *Acta Societatis Botanicorum Poloniae* 84(3): 339–346.

Luick, R., Jones, G., and R. Oppermann, 2012. "Semi-Natural Vegetation: Pastures, Meadows and Related Vegetation Communities." In Oppermann, R., Beaufoy, G., and G. Jones (eds.), *High Nature Value Farming in Europe: 35 European Countries and Perspectives*. Ubstadt-Weiher, Heidelberg, and Basel: Regionalkultur, 32–57.

Maček, J., 1979. "Zgodovinski pogled na varstvo naravnega okolja" (Naturschutz im geschichtlichen Blickwinkel). *Zgodovinski časopis* 33(2): 299–309.

Machatschek, M., 1997 (= Machatschek, 1997a). "Einige Bewandtnisse der Wetterankündigungen." *Der Alm- und Bergbauer* 47(1–2): 15–18, 23–27.

Machatschek, M., 1997 (= Machatschek, 1997b). "'Graupenklauben' – Über die Sammelnutzung des Isländischen Mooses." *Der Alm- und Bergbauer* 47(4): 111–118.

Machatschek, M., 2002. *Laubgeschichten: Gebrauchswissen einer alten Baumwirtschaft, Speise- und Futterlaubkultur*. Vienna, Cologne, and Weimar: Böhlau.

Machatschek, M., 2006. "Überlegungen zum Futterangebot für Schweine auf Almweiden von der Vorzeit bis heute." In *Alpen: Archäologie, Almwirtschaftsgeschichte, Altwegeforschung, Dendrochronologie, Felsbildforschung, Geomorphologie, Geschichte, Gletscherforschung, Umweltforschung, Volkskunde, Zoologie: Festschrift 25 Jahre ANISA: Verein für alpine Forschung*. Haus im Ennstal: ANISA – Verein für alpine Forschung, 131–144.

Machatschek, M., 2007. "Historische Kulturtechniken im Alpenraum dargestellt an den Beispielen Dungmahder und Wasenhag." In Merlin, F.W., Hellebart, S., and M. Machatschek (eds.), *Bergwelt im Wandel: Festschrift Erika Hubatschek zum 90. Geburtstag*. Klagenfurt: Kärntner Landesarchiv, 149–155.

Machatschek, M., 2015. *Nahrhafte Landschaft*. Vol. 3, *Von Baumwässern, Fetthennen, Schaum- und Springkräutern, Ohrenpilzen, Kranawitt, süßen Eicheln und anderen wiederentdeckten Nutz- und Heilpflanzen*. Vienna, Cologne, and Weimar: Böhlau.

Machatschek, M., 2017. *Nahrhafte Landschaft*. Vol. 4, *Von Ferkelkräutern, Wiesenknopf, Ziegenmilch, Ruchgras, Rasch, Birnmehl, Kraterellen und anderen wildwachsenden Nutzpflanzen*. Vienna, Cologne, and Weimar: Böhlau.

Machatschek, M., and P. Kurz, 2006. *ALP Austria: Programm zur Sicherung und Entwicklung der alpinen Kulturlandschaft: Biodiversität*. [city unknown]:

Bundesministerium für Land- und Forstwirtschaft, Umwelt und Wasserwirtschaft, Land Kärnten, Land Oberösterreich, Land Salzburg, Land Steiermark, Land Tirol, and Land Vorarlberg.

Magda, D., de Sainte Marie, C., Plantureux, S., Agreil, C., Amiaud, B., et al., 2015. "Integrating Agricultural and Ecological Goals into the Management of Species-Rich Grasslands: Learning from the Flowering Meadows Competition in France." *Environmental Management* 56: 1053–1064.

Mahlerwein, G., 2016. "Agrarintensivierung und Wissenszirkulation: Vergleichende Beobachtungen an rheinhessischen, nordbadischen und schwäbischen Beispielen." In Dauser, R., Fassl, P., and L. Schilling (eds.), *Wissenszirkulation auf dem Land vor der Industrialisierung*. Augsburg: Wißner, 15–22.

Makarovič, G., 1995. *Slovenci in čas: Odnos do časa kot okvir in sestava vsakdanjega življenja* (The Slovenes and Time: The Attitude toward Time as a Framework and Element of Everyday Life). Ljubljana: Krtina.

Makarovič, G., 2007. "Kdaj so nastali kozolci?" (When Did Hayracks First Appear?). *Etnolog* 17: 209–249.

Makarovič, M., 1978. *Kmečko gospodarstvo na Slovenskem* (The Traditional Farm Economy in Slovenian Territory). Ljubljana: Mladinska knjiga.

Makarovič, M., 1982. *Strojna in Strojanci: Narodopisna podoba koroške hribovske vasi* (Strojna and Its Inhabitants: An Ethnological Study of a Highland Settlement in Carinthia). Ljubljana: Mladinska knjiga.

Makarovič, M., 2005. *Korte in Korčani* (Korte (Germ. *Trögern*) and Its Inhabitants). Korte/Trögern: Author.

Makarovič, M., 2016. [Grassland Management History Interviews with Farmers from Babno Polje, Jazbina, Josipdol, and Zgornja Orlica by the Ethnologist Marija Makarovič]. Typescript.

Makowiecki, D., Zabilska-Kunek, M., Seetah, K., Jarzebowski, M., and A. Pluskowski, 2019. "Farming, Hunting, and Fishing in Medieval Prussia: The Zooarchaeological Data." In Pluskowski, A. (ed.), *Environment, Colonization, and the Baltic Crusader States*. Turnhout: Brepols, 333–376.

Malbašić, J., 1998. *Arhivski inventar: Signatura fonda: AS 241: Ministrstvo za izvoz in uvoz Ljudske republike Slovenije: 1945–1951* (Archival Register: Fonds Signature: AS 241: Ministry of Export and Import of the People's Republic of Slovenia: 1945–1951). Ljubljana: [Arhiv Republike Slovenije].

Malbašić, J., 2005a. *Arhivski inventar: Signatura fonda: AS 1126: Državni sekretariat za blagovni promet Ljudske republike Slovenije: 1958–1965* (Archival Register: Fonds Signature: AS 1126: State Secretariat for Trade of the People's Republic of Slovenia: 1958–1965). Ljubljana: [Arhiv Republike Slovenije].

Malbašič, J., 2005b. *Arhivski inventar: Signatura fonda: AS 1131: Republiški sekretariat za trgovino Socialistične republike Slovenije: 1962–1965* (Archival Register: Fonds Signature: AS 1131: Secretariat for Trade of the Socialist Republic of Slovenia: 1962–1965). Ljubljana: [Arhiv Republike Slovenije].

Måren, I.E., Janovský, Z., Spindelböck, J.P., Daws, M.I., Kaland, P.E., et al., 2010. "Prescribed Burning of Northern Heathlands: *Calluna vulgaris* Germination Cues and Seed-Bank Dynamics." *Plant Ecology* 207(2): 245–256.

Marošek, F., 2007. "Kako smo v Vitanju prodajali 'seme divjega janeža'" (How "Wild Anise Seeds" Were Sold in Vitanje). *Vitanjčan: Glasilo Občine Vitanje*, no. 20 (December 20th, 2007): 14.

Martin, E.A., Dainese, M., Clough, Y., Báldi, A., Bommarco, R., et al. 2019. "The Interplay of Landscape Composition and Configuration: New Pathways to Manage Functional Biodiversity and Agroecosystem Services Across Europe." *Ecology Letters* 22: 1083–1094.

Martinčič, A., 2002. "Praprotnice in semenke" (Pteridophytes and Spermatophytes). In Gaberščik, A. (ed.), *Jezero, ki izginja: Monografija o Cerkniškem jezeru*. Ljubljana: Društvo ekologov Slovenije, 72–79.

Martinčič, A., 2007a. "1. družina: *Lycopodiaceae* – lisičjakovke" (Family 1: *Lycopodiaceae*). In Martinčič, A. (ed.), *Mala flora Slovenije: Ključ za določanje praprotnic in semenk*. 4th ed. Ljubljana: Tehniška založba Slovenije, 83–85.

Martinčič, A., 2007b. "48. družina: *Crassulaceae* – tolstičevke" (Family 48: *Crassulaceae*). In Martinčič, A. (ed.), *Mala flora Slovenije: Ključ za določanje praprotnic in semenk*. 4th ed. Ljubljana: Tehniška založba Slovenije, 234–237.

Martinčič, A., 2007c. "55. družina: *Fabaceae* (*Papilionaceae*) – metuljnice" (Family 55: *Fabaceae* (*Papilionaceae*)). In Martinčič, A. (ed.), *Mala flora Slovenije: Ključ za določanje praprotnic in semenk*. 4th ed. Ljubljana: Tehniška založba Slovenije, 286–328.

Martinčič, A., 2007d. "85. družina: *Apiaceae* – kobulnice" (Family 85: *Apiaceae*). In Martinčič, A. (ed.), *Mala flora Slovenije: Ključ za določanje praprotnic in semenk*. 4th ed. Ljubljana: Tehniška založba Slovenije, 379–412.

Martinčič, A., 2007e. "99. družina: *Malvaceae* – slezenovke" (Family 99: *Malvaceae*). In Martinčič, A. (ed.), *Mala flora Slovenije: Ključ za določanje praprotnic in semenk*. 4th ed. Ljubljana: Tehniška založba Slovenije, 470–474.

Martinčič, A., 2007f. "164. družina: *Cyperaceae* – ostričevke" (Family 164: *Cyperaceae*). In Martinčič, A. (ed.), *Mala flora Slovenije: Ključ za določanje praprotnic in semenk*. 4th ed. Ljubljana: Tehniška založba Slovenije, 793–821.

Martini, F., 1953. "Auerbach, Berthold." In *Neue Deutsche Biographie*. Vol. 1, 434–435. https://www.deutsche-biographie.de/pnd11865103X.html#ndbcontent (November 15th, 2019).

Masters, B., 2009 (= Masters, 2009a). "Aleppo (Alep; *Ar.*: Halab; *Turk.*: Haleb)." In Ágoston, G., and B. Masters (eds.), *Encyclopedia of the Ottoman Empire*. New York: Facts On File, 30–32.

Masters, B., 2009 (= Masters, 2009b). "Bathhouse." In Ágoston, G., and B. Masters (eds.), *Encyclopedia of the Ottoman Empire*. New York: Facts On File, 79–80.

Masters, B., 2009 (= Masters, 2009c). "Cairo (*Ar.*: al-Qahira; *Turk.*: Kahire)." In Ágoston, G., and B. Masters (eds.), *Encyclopedia of the Ottoman Empire*. New York: Facts On File, 112–114.

Masters, B., 2009 (= Masters, 2009d). "Coffee/Coffeehouses." In Ágoston, G., and B. Masters (eds.), *Encyclopedia of the Ottoman Empire*. New York: Facts On File, 138–139.

Masters, B., 2009 (= Masters, 2009e). "Damascus (*Ar.*: Dimashq, al-Sham; *Fr.* Damas; *Turk.*: Şam)." In Ágoston, G., and B. Masters (eds.), *Encyclopedia of the Ottoman Empire*. New York: Facts On File, 168–171.

Masters, B., 2009 (= Masters, 2009f). "Egypt (*anc.*: Aegyptus; *Ar.*: Misr; *Turk.*: Mısır)." In Ágoston, G., and B. Masters (eds.), *Encyclopedia of the Ottoman Empire*. New York: Facts On File, 204–206.

Masters, B., 2009 (= Masters, 2009g). "Syria (*Ar.*: Bilad al-Sham, Suriyya; *Heb.*: Aram; *Turk.*: Arabistan, Suriye)." In Ágoston, G., and B. Masters (eds.), *Encyclopedia of the Ottoman Empire*. New York: Facts On File, 550.

Mathieu, J., 1996. "Agrarintensivierung bei beschränktem Umweltpotential: Der Alpenraum vom 16. bis 19. Jahrhundert." *Zeitschrift für Agrargeschichte und Agrarsoziologie* 44: 137–161.

Mathieu, J., 2019. *The Alps: An Environmental History*. Cambridge, UK and Medford, OR: Polity Press.

Matičetov, M., 1948. "O etnografiji in folklori zapadnih Slovencev" (De l'ethnographie et du folklore des Slovènes occidentaux). *Slovenski etnograf* 1: 9–56.

Matičetov, M., 1976–1977. "Bedeníce" (Daffodils). *Traditiones: Acta Instituti ethnographiae Slovenorum* 5–6: 277–300.

Matouch, S., Traxler, A., and V. Grass, 2000. "Die Bergmähder des Kärntner Lesachtales – Biodiversität und Nutzungswandel." *Carinthia II* 190: 591–602.

Matthews, R., 1993. "Cetraria islandica." In *Fire Effects Information System*. U.S. Department of Agriculture, Forest Service, Rocky Mountain Research Station, Fire Sciences Laboratory. www.fs.fed.us/database/feis/lichens/cetisl/all.html (October 29th, 2021).

McClenachan, L., Rick, T., Thurstan, R.H., Trant, A., Alagona, P.S., et al., 2024. "Global Research Priorities for Historical Ecology to Inform Conservation." *Endangered Species Research* 54: 285–310.

McKittrick, M., 2015. "Industrial Agriculture." In McNeill, J.R., and E.S. Mauldin (eds.), *A Companion to Global Environmental History*. 2nd ed. Malden, MA, Oxford, UK, and Chichester, UK: Blackwell Publishing, 411–432.

McNeill, J.R., 2002. *The Mountains of the Mediterranean World: An Environmental History*. New York: Cambridge University Press.

McNeill, J.R., 2003. "Observations on the Nature and Culture of Environmental History." *History and Theory* 42: 5–43.

McNeill, J.R., 2010. "The State of the Field of Environmental History." *Annual Review of Environment and Resources* 35: 345–374.

Meilleur, B.A., 1982. "Du ramassage à la cueillette: L'example des Allues dans les Alpes du Nord." *Études rurales* 87–88: 165–174.

Meilleur, B.A., 1986. *Alluetain Ethnoecology and Traditional Economy: The Procurement and Production of Plant Resources in the Northern French Alps: A dissertation submitted in partial fulfillment of the requirements for the degree of Doctor of Philosophy*. [city unknown]: University of Washington.

Meiners, U., 2010. "Schneidegeräte: 1. Landwirtschaft und Ernte." In Jaeger, F., et al. (eds.), *Enzyklopädie der Neuzeit*. Vol. 11, *Renaissance–Signatur*. Stuttgart and Weimar: J.B. Metzler, 815–817.

Meixner, W., and G. Siegl, 2004. "Erwanderte Heimat: Hermann Wopfner und die Tiroler Bergbauern." In Bruckmüller, E., Langthaler, E., and J. Redl (eds.), *Agrargeschichte schreiben: Traditionen und Innovationen im internationalen Vergleich*. Innsbruck and Vienna: Studien Verlag, 228–239.

Meja za malo tvegano pitje alkohola (The Limit for Low-Risk Drinking). https://nalijem.si/o-alkoholu/meje (January 17th, 2024).

Mekinda-Majaron, T., 1995. *Klimatografija Slovenije: Temperatura zraka: Obdobje 1961–1990* (Climatography of Slovenia: Air Temperature: The Period 1961–1990). Ljubljana: Hidrometeorološki zavod Republike Slovenije.

Melik, A., 1927. "Morfologija in gospodarska izraba tal v Bohinju" (Morphologie et exploitation économique du sol de Bohinj). *Geografski vestnik* 3: 53–95.

Melik, A., 1931. *Kozolec na Slovenskem* (Le "kozolec" en pays slovènes). Ljubljana: Znanstveno društvo.

Melik, A., 1946. *Ljubljansko mostiščarsko jezero in dediščina po njem* (The Lake with Prehistoric Pile-Dwelling Settlements near Ljubljana and Its Legacy). Ljubljana: Slovenska akademija znanosti in umetnosti.

Melik, A., 1950. *Planine v Julijskih Alpah* (L'économie pastorale dans les Alpes Juliennes). Ljubljana: Slovenska akademija znanosti in umetnosti.

Melik, A., 1954. *Slovenski alpski svet* (Slovenian Alps). Ljubljana: Slovenska matica.

Melzer, H., 1965. "Neues und Kritisches zur Flora von Kärnten." *Carinthia II* 155: 172–190.

Menke, J.W., 1992. "Grazing and Fire Management for Native Perennial Grass Restoration in California Grasslands." *Fremontia* 20(2): 22–25.

Merchel, S., Mrak, I., Braucher, R., Benedetti, L., Repe, B., et al., 2014. "Surface Exposure Dating of the Veliki Vrh Rock Avalanche in Slovenia Associated with the 1348 Earthquake." *Quaternary Geochronology* 22: 33–42.

Mestnik, A., 2018. "Rastlinstvo v povesti Ivana Tavčarja *Cvetje v jeseni*: Pogled na rastlinstvo osrednjega dela Škofjeloškega hribovja skozi oči Ivana Tavčarja" (Plants in Ivan Tavčar's Novel *Autumn Blossoms*: A View of Plants in the Central Škofja Loka Hills through the Eyes of Ivan Tavčar). *Geografski obzornik* 65(1): 13–19.

Meuret, M., 1997. "Ho Do I Cope with that Bush? Optimizing Less Palatable Feeds at Pasture Using the MENU Model." *Options Méditerranéennes, Series A* 34: 53–57.

Meze, D., 1963. "Samotne kmetije na Solčavskem" (Les fermes isolées dans la région de Solčava). *Acta geographica* 8: 223–280.

Meze, D., 1965. "Samotne kmetije v Lučki pokrajini" (Les fermes isolées dans la région de Luče). *Acta geographica* 9: 181–244.

Meze, D., 1969. "Hribovske kmetije v vzhodnem delu Gornje Savinjske doline" (Les fermes de montagne de la partie orientale de le Valée supérieure de la Savinja). *Acta geographica* 11: 5–96.

Mihelič, S., 1970. "Čebelarstvo" (Die Bienenzucht). In Blaznik, P., et al. (eds.), *Gospodarska in družbena zgodovina Slovencev: Zgodovina agrarnih panog*. Vol. 1, *Agrarno gospodarstvo*. Ljubljana: Slovenska akademija znanosti in umetnosti and Državna založba Slovenije, 395–408.

Milotić, T., and M. Hoffmann, 2016a. "How Does Gut Passage Impact Endozoochorous Seed Dispersal Success? Evidence from a Gut Environment Simulation Experiment." *Basic and Applied Ecology* 17(2): 165–176.

Milotić, T., and M. Hoffmann, 2016b. "Reduced Germination Success of Temperate Grassland Seeds Sown in Dung: Consequences for Post-Dispersal Seed Fate." *Plant Biology* 18(6): 1038–1047.

Mirtl, M., Bahn, M., Battin, T., Borsdorf, A., Dirnböck, T., et al., 2015. *Research for the Future: LTER-Austria White Paper 2015: On the Status and Orientation of Process Oriented Ecosystem Research, Biodiversity and Conservation Research and Socio-Ecological Research in Austria*. Vienna: LTER-Austria.

Mlacović, D., 2023. "Natural Resources in Sources for Late-Medieval Adriatic." Presentation at the international doctoral workshop *Eastern Mediterranean Coastal and Island Environments: Natural Resources, Their Uses and Perceptions (15th–18th Centuries)*. Ljubljana, June 29th, 2023.

Mladenović, A., 2012. *Relief Map of Slovenia*. https://commons.wikimedia.org/wiki/File:Relief_map_of_Slovenia.png (October 5th, 2019).

Mlakar, P., 2018. "Narcise najlepše v prihodnjih desetih dneh" (Daffodils at Their Most Beautiful in Next Ten Days). *Dnevnik* (May 11th, 2018): 9.

Mlakar, P., 2022. "Travnike v vaseh pod Golico so že pobelile narcise" (Meadows Already White with Daffodils in Villages below Mt. Golica). *Dnevnik* (May 11th, 2022): 7.

Mlakar, V., 2015. *Rastlina je sveta od korenin do cveta: Tradicionalno znanje o rastlinskem svetu na Slovenskem* (A Plant is Awesome from Roots to Blossom: Slovenian Traditional Knowledge about Plants). Ljubljana: Author.

Mlaker-Šumenjak, M., 2011. *Čebela se predstavi* (A Bee Introduces Itself). 3rd ed. Brdo pri Lukovici: Čebelarska zveza Slovenije.

Mlekuž, D., Košir, U., and M. Črešnar, 2016. "Landscapes of Death and Suffering: Archaeology of Conflict Landscapes of the Upper Soča Valley, Slovenia." In: Stichelbaut, B., and D. Cowley (eds.), *Conflict Landscapes and Archaeology from Above*. London: Routledge, 127–145.

Mlinar, J., 2004. "Podoba srednjeveške poseljenosti območja današnje jeseniške občine" (Outline of Medieval Settlement of the Area of Today's Municipality of Jesenice). *Jeseniški zbornik* 9: 11–21.

Mlinar, J., 2018 (= Mlinar, 2018a). "Čas in okoliščine nastanka urbarja" (Time and Context of the Rent-Roll Creation). In *Urbarji belopeškega gospostva*. Ljubljana: ZRC, ZRC SAZU, 25–28.

Mlinar, J., 2018 (= Mlinar, 2018b). "Podložniške obveznosti in dajatve" (Dues and Other Duties by Subjects of the Seigneury). In *Urbarji belopeškega gospostva*. Ljubljana: ZRC, ZRC SAZU, 44–50.

Mlinar, J., 2018 (= Mlinar, 2018c). "Rokopisi" (Manuscripts). In *Urbarji belopeškega gospostva*. Ljubljana: ZRC, ZRC SAZU, 29–32.

Mlinar, J., 2018 (= Mlinar, 2018d). "Zgodovinski razvoj gospostva Bela Peč do 1636" (Historical Development of the Seigneury of Fusine (Friul. *Fusinis*, Germ. *Weißenfels*, Sln. *Bela Peč*) until 1636). In *Urbarji belopeškega gospostva*. Ljubljana: ZRC, ZRC SAZU, 19–24.

Mlinar, J. (ed.), 2018 (= Mlinar (ed.), 2018a). "Urbar gospostva Bela Peč 1498" (1498 Rent-Roll (*Urbar*) of the Seigneury of Fusine). In *Urbarji belopeškega gospostva*. Ljubljana: ZRC, ZRC SAZU, 61–85.

Mlinar, J. (ed.), 2018 (= Mlinar (ed.), 2018b). "Urbar gospostva Bela Peč 1636" (1636 Rent-Roll (*Urbar*) of the Seigneury of Fusine). In *Urbarji belopeškega gospostva*. Ljubljana: ZRC, ZRC SAZU, 87–171.

Mlinar, J., 2019. "Jesenice in okolica v luči belopeškega urbarja iz leta 1498" (Jesenice and Its Surrounding in the Light of the 1498 Rent Roll). *Jeseniški zbornik* 12: 32–49.

Mlinarič, J., 1979. "Območje Gornje Radgone do začetka XVII. stoletja" (The Territory around Gornja Radgona (Germ. *Oberradkersburg*) until the Early Seventeenth Century). *Arhivi: Glasilo arhivskega društva in arhivov Slovenije* 2(1–2): 8–12.

Moinuddin, A., Anchukaitis, K.J., Asrat, A., Borgaonkar, H.P., Braida, M., et al., 2013. "Continental-Scale Temperature Variability During the Past Two Millennia." *Nature Geoscience* 6, 339–346 (and "Corrigendum: Continental-Scale Temperature Variability During the Past Two Millennia." 2015. https://www.nature.com/articles/ngeo2566).

Molnár, Z., Gellény, K., Margóczi, K., and M. Biró, 2015. "Landscape Ethnoecological Knowledge Base and Management of Ecosystem Services in a Székely-Hungarian Pre-Capitalistic Village System (Transylvania, Romania)." *Journal of Ethnobiology and Ethnomedicine* 11: article no. 3, separate pagination 1–22.

Montana Field Guides: Common Lady's Mantle. https://fieldguide.mt.gov/speciesDetail.aspx?elcode=PDROS040D0 (October 27th, 2021).

Moon, D., 2013. *The Plough that Broke the Steppes: Agriculture and Environment on Russia's Grasslands, 1700–1914.* Oxford: Oxford University Press.

Moon, D., 2020. *The American Steppes: The Unexpected Russian Roots of Great Plains Agriculture, 1870s–1930s.* Cambridge: Cambridge University Press.

Moser, O., 1952. "Der Heurechen: Versuch einer volkskundlichen Bestandsaufnahme des Arbeitsgerätes in Kärnten." *Carinthia I* 142: 463–494.

Moser, O., 1974. "Grassense I—Grundtypen der Langstielsense." In Wolfram, R. et al. (eds.), *Österreichischer Volkskundeatlas: Kommentar.* Vol. 8, Fascicle (*Lieferung*) 5. Wien: Hermann Böhlaus Nachfolger, separate pagination, 1–46.

Moser, O., 1990. "Der 'Feiertagschristus' als Mahnbild und Quelle der Sachforschung: Zu zwei neuen Funden mittelalterlicher Fresken in Kärnten." *Österreichische Zeitschrift für Volkskunde* 93: 331–371 and Figures 1–6 between pp. 415 and 417.

Mrak, A., 2012. "Stara uspešnica v novi preobleki: 'Učilna zidana' je stara 135 let – na kakih 50 'podlage'" (A New Variation of an Old Favorite: The Poem "The Tit's Lament" Is 135 Years Old—And about Fifty More Have Passed since Its First Variant). *MMC*, RTV SLO. https://www.rtvslo.si/kultura/razglednice-preteklosti/stara-uspesnica-v-novi-preobleki/282526 (December 1st, 2022).

Muchowski, J., 2022. "The Anthropocene Contract. What Kind of Historian–Reader Agreement Does Environmental Historiography Need?" In Izdebski, A., Haldon, J., and P. Filipkowski (eds.), *Perspectives on Public Policy in Societal-Environmental Crises: What the Future Needs from History.* Cham: Springer, 21–31.

Mucina, L., Grabherr, G., and T. Ellmauer, 1993. *Die Pflanzengesellschaften Österreichs.* Vols. 1, *Anthropogene Vegetation*, and 2, *Natürliche waldfreie Vegetation.* Jena: Gustav Fischer.

Mühlburger, R., 2020. "Lebenswelt Villgraten: Neues Buch als Synthese von Literatur und Malerei." *Osttirol heute.* https://www.osttirol-heute.at/menschen/lebenswelt-villgraten-neues-buch-als-synthese-von-literatur-und-malerei/ (February 6th, 2022).

Myrdal, J., 2005. "The Perfect Scythe—And Other Implements." *Journal of Nordic Archaeological Science* 15: 5–17.

Nagy, L., Grabherr, G., Körner, C., and D.B.A. Thompson, 2003. "Alpine Biodiversity in Space and Time: A Synthesis." In Nagy, L., et al. (eds.), *Alpine Biodiversity in Europe*. Berlin and Heidelberg: Springer, 453–464.

Natural Earth Data, Rivers + Lake Centerlines (1:10 m) [data file]. 2025. https://www.naturalearthdata.com (January 15th, 2025).

Netting, R.M., 1974. "The System Nobody Knows: Village Irrigation in the Swiss Alps." In Downing, T.E., and M. Gibson (eds.), *Irrigation's Impact on Society*. Tucson: The University of Arizona Press, 67–75.

Netting, R.M., 1993. *Smallholders, Householders: Farm Families and the Ecology of Intensive, Sustainable Agriculture*. Stanford, CA: Stanford University Press.

New Entrant netWork: Business Models for Innovation, Entrepreneurship and Resilience in European Agriculture. https://www.newbie-academy.eu/ (September 19th, 2021).

Nicklas, L., 2018. "Winners and Losers of Climate Change in the Central and South Alps." Presentation at the conference AKH (Arbeitskreis Hochgebirge, Deutsche Gesellschaft für Geografie) Jahrestagung 2018 *Gebirgsräume im Wandel*. Innsbruck, February 3rd, 2018.

Nicolas-Obadia, G., 1974. *Atlas statistique agricole vaudois*. [Lausanne]: Office cantonal vaudois de l'urbanisme.

Nicolussi, K., 2013. "Zur Geschichte des Vernagtferners – Gletschervorstöße und Seeausbrüche im vergangenen Jahrtausend." *Publikationen Alpine Forschungsstelle Obergurgl* 3: 69–94.

Nicolussi, K., Kaufmann, M., Patzelt, G., van der Plicht, J., and A. Thurner, 2005. "Holocene Tree-Line Variability in the Kauner Valley, Central Eastern Alps, Indicated by Dendrochronological Analysis of Living Trees and Subfossil Logs." *Vegetation History and Archaeobotany* 14: 221–234.

Niederer, A., 1996 (= Niederer, 1996a). "Die alpine Alltagskultur zwischen Routine und der Adoption von Neuerungen." In Anderegg, K., and W. Bätzing (eds.), *Alpine Alltagskultur zwischen Beharrung und Wandel: Ausgewählte Arbeiten aus den Jahren 1956 bis 1991*. 2nd revised ed. Bern, Stuttgart, and Vienna: Paul Haupt, 116–138.

Niederer, A., 1996 (= Niederer, 1996b). "Frühere Formen kollektiver Arbeit in ländlichen Gemeinden." In Anderegg, K., and W. Bätzing (eds.), *Alpine Alltagskultur zwischen Beharrung und Wandel: Ausgewählte Arbeiten aus den Jahren 1956 bis 1991*. 2nd revised ed. Bern, Stuttgart, and Vienna: Paul Haupt, 337–353.

Niederer, A., 1996 (= Niederer, 1996c). "Gemeinwerk im Wallis: Bäuerliche Gemeinschaftsarbeit in Vergangenheit und Gegenwart." In Anderegg, K., and W. Bätzing (eds.), *Alpine Alltagskultur zwischen Beharrung und Wandel: Ausgewählte Arbeiten aus den Jahren 1956 bis 1991*. 2nd revised ed. Bern, Stuttgart, and Vienna: Paul Haupt, 19–88.

Niederer, A., 1996 (= Niederer, 1996d). "Traditionelle Wirtschafts- und Kulturformen in den Alpen: Teil I, Die materielle Kultur." In Anderegg, K., and W. Bätzing (eds.), *Alpine Alltagskultur zwischen Beharrung und Wandel: Ausgewählte Arbeiten aus den Jahren 1956 bis 1991*. 2nd revised ed. Bern, Stuttgart, and Vienna: Paul Haupt, 147–224.

Norz, C., 2014. *Sukzession in Lärchenwiesen in Tirol: Der Einfluss von Bewirtschaftung und Standortsfaktoren auf den Ablauf von Sukzession: Diplomarbeit*. Vienna: Universität für Bodenkultur, Institut für Waldökologie.

"Note on Transliteration and Spelling." 2009. In Ágoston, G., and B. Masters (eds.), *Encyclopedia of the Ottoman Empire*. New York: Facts On File, xxiii–xxiv.

Nothdurfter, J., 1979. *Die Eisenfunde von Sanzeno im Nonsberg*. Mainz am Rhein: Philipp von Zabern.

Notranjski regijski park: Mali škrobotec (Notranjska Regional Park: Yellow Rattle). https://www.notranjski-park.si/izobrazevalne-vsebine/rastlinski-svet/crnobinovke/mali-skrobotec (December 17th, 2023).

Notranjski regijski park: *Navadna turška detelja* (Notranjska Regional Park: Sainfoin). https://www.notranjski-park.si/izobrazevalne-vsebine/rastlinski-svet/metuljnice/navadna-turska-detelja (November 11th, 2022).

Novak, A., 1978. "Življenje in delo pastirjev v bohinjskih gorah" (Herders' Life and Work in the Mountains of Bohinj). *Snovanja* 12(4): 55–59.

Novak, A., 1987. "Planinsko sirarstvo" (Alpine Dairying). In Dežman, J. (ed.), *Bohinjski zbornik*. Radovljica: Skupščina občine Radovljica, 73–84.

Novak, A., 2024. *Planšarstvo v Bohinju: "Vse svoje življenje sem pustil v siru"* (Alpine Farming in Bohinj: "I Dedicated My Entire Life to Cheese"). Kranj: Gorenjski muzej.

Novak, J., Franz, C.M., and S. Novak, 1998. "Essential Oils of Rhizomes and Rootlets of *Valeriana celtica* L. ssp. *norica* Vierh. from Austria." *Journal of Essential Oil Research* 10(6): 637–640.

Novak, S., 1999/2000. "Der Speik – *Valeriana celtica* subsp. *norica*: Von der Kultur- zur Naturschutzpflanze." In Hülbusch, K.H. (ed.), *Gagel, Speik und Wegerich: Beiträge zur Landschafts- und Vegetationskunde*. Kassel: Arbeitsgemeinschaft Freiraum und Vegetation, 108–115.

Novak, V., 1942. "Ovčarstvo pod Stolom in v Planici" (Sheep Rearing below Mt. Stol and in the Planica Valley). *Etnolog* 15: 89–95.

Novak, V., 1955. "Odkup in ureditev služnostnih pašnih pravic v Bohinju (Slovenija)" (Die Ablösung und Regulierung der Weideservituten in Bohinj, Slowenien). *Zbornik Filozofske fakultete* 2: 257–315.

Novak, V., 1970. "Živinoreja" (Die Viehzucht). In Blaznik, P., et al. (eds.), *Gospodarska in družbena zgodovina Slovencev: Zgodovina agrarnih panog*. Vol. 1, *Agrarno gospodarstvo*. Ljubljana: Slovenska akademija znanosti in umetnosti and Državna založba Slovenije, 343–394.

Novak, V., 1975. "Aus der Geschichte des Alpwesens in der Brixner Grundherrschaft in Slowenien." In Griessmair, I. (ed.), *Alpes Orientales 7: Acta septimi Conventus de ethnographia Alpium Orientalium tractantis Brixiae anno MCMLXXII*. Munich: Rudolf Trofenik, 1–10.

Novobilský, A., Engström, A., Sollenberg, S., Gustafsson, K., Morrison, D.A, et. al., 2014. "Transmission Patterns of *Fasciola hepatica* to Ruminants in Sweden." *Veterinary Parasitology* 203(3–4): 276–286.

Nowosadtko, J., 2007. "Milzbrand, Tollwut, Wölfe, Spatzen und Maikäfer: Die gesellschaftliche Verteilung von Zuständigkeiten bei der Bekämpfung von Viehseuchen und schädlichen Tieren in der Frühen Neuzeit." In Engelken, K., Hünniger, D., and S. Windelen (eds.), *Beten, Impfen, Sammeln: Zur Viehseuchen- und Schädlingsbekämpfung in der Frühen Neuzeit*. Göttingen: Universitätsverlag, 79–98.

NŠAL 20: Gornji Grad Avstrija [arhivski popis] (NŠAL 20: Gornji Grad Austria [Archival Register]). Typescript. [Ljubljana: Nadškofijski arhiv].

NŠAL 100: Kapiteljski arhiv Ljubjana [arhivski popis] (NŠAL 100: Ljubljana Cathedral Chapter's Archive [Archival Register]). Typescript. [Ljubljana: Nadškofijski arhiv].

OBERLIN College and Conservatory: About OES: What is Sustainability. https://www.oberlin.edu/environmental-sustainability/about (September 2nd, 2021).

Oberschütt: Gladiolenwiese – Weinitzen. Tourist information board at the church in Oberschütt (Sln. *Rogaje*), Carinthia, Austria.

Odar, Š., 1993. *Okoljevarstvena problematika izravnavanja grbinastih travnikov v Bohinju: Diplomska naloga* (The Issues of Environmental Protection Related to the Leveling of Hummucky Meadows: Bachelor's Thesis). Ljubljana: Univerza v Ljubljani, Filozofska fakulteta, Oddelek za geografijo.

Ogrin, D., 2005. "Spreminjanje podnebja v holocenu" (Climatic Changes in the Holocene). *Geografski vestnik* 77(1): 57–66.

Oitzl, G., 2021. *Družbene in gospodarske posledice železarjenja na Slovenskem v poznem srednjem veku: Doktorska disertacija* (Social and Economic Effects of Ironmaking in Slovenian Territory in the Late Middle Ages: Doctoral Dissertation). Ljubljana: Univerza v Ljubljani, Filozofska fakulteta, Oddelek za zgodovino.

Oklahoma State University: Spontaneous Combustion a Possibility with Wet Hay. 2020. https://news.okstate.edu/articles/agriculture/2020/stotts_braums-fire.html (May 3rd, 2024).

Olivier de Serres: Le domaine du Pradel. http://www.olivier-de-serres.org/pradel.php (September 1st, 2020).

Oppermann, R., 2012. "Germany." In Oppermann, R., Beaufoy, G., and G. Jones (eds.), *High Nature Value Farming in Europe: 35 European Countries and Perspectives*. Ubstadt-Weiher, Heidelberg, and Basel: Regionalkultur, 220–233.

Oppermann, R., Beaufoy, G., and G. Jones (eds.), 2012. *High Nature Value Farming in Europe: 35 European Countries and Perspectives*. Ubstadt-Weiher, Heidelberg, and Basel: Regionalkultur.

Oppermann, R., Bosshard, A., Mestelan, P., de Sainte Marie, C., and J. Gelhausen, 2012 (= Oppermann et al., 2012a). "Awareness Raising among Farmers and in the Wider Public." In Oppermann, R., Beaufoy, G., and G. Jones (eds.), *High Nature Value Farming in Europe: 35 European Countries and Perspectives*. Ubstadt-Weiher, Heidelberg, and Basel: Regionalkultur, 466–472.

Oppermann, R., and J. Hoffmann, 2012. "Features of HNV Farmland Mosaic Landscapes." In Oppermann, R., Beaufoy, G., and G. Jones (eds.), *High Nature Value Farming in Europe: 35 European Countries and Perspectives*. Ubstadt-Weiher, Heidelberg, and Basel: Regionalkultur, 85–96.

Oppermann, R., Hoffmann, J., Weinhold, U., and J. Kleemann, 2012 (= Oppermann et al., 2012b). "Farmland as Habitat for Populations of Species of Conservation Interest." In Oppermann, R., Beaufoy, G., and G. Jones (eds.), *High Nature Value Farming in Europe: 35 European Countries and Perspectives*. Ubstadt-Weiher, Heidelberg, and Basel: Regionalkultur, 97–109.

Oppermann, R., and M.L. Paracchini, 2012. "HNV Farming—Central to European Cultural Landscapes and Biodiversity." In Oppermann, R., Beaufoy, G., and G. Jones (eds.), *High Nature Value Farming in Europe: 35 European Countries and Perspectives*. Ubstadt-Weiher, Heidelberg, and Basel: Regionalkultur, 17–23.

Orel, L., 2020. "Bohinjski obrazi – Simona Kejžar" (Faces from Bohinj: Simona Kejžar). *Bohinjske novice* 23(1): 7–8.

Orland, B., 2016. "Alpine Milk: Dairy Farming as a Pre-Modern Strategy of Land Use." In Johnson, S. (compiled), *Themes in Environmental History: Farming*. [Winwick, Cambridgeshire]: The White Horse Press, 22–56.

Orloff, S.B., Vargas, R.N., Cudney, D.W., Canevari, W.M., and J. Schmierer, 1989. "Dodder Control in Alfalfa." *California Agriculture* 43(4): 30–31.

Osnovna državna karta (Detailed Map of the State). 1967. Bled–40. 1:5,000. [Belgrade]: Zvezna geodetska uprava SFRJ.

Osnovna državna karta (Detailed Map of the State). 1968. Bohinjska Bistrica–13. 1:5,000. [Belgrade]: Zvezna geodetska uprava SFRJ.

Osnovna državna karta (Detailed Map of the State). 1972. Bled–7. 1:10,000. Ljubljana: Geodetska uprava SRS.

Padovan, D., Arrobbio, O., and A. Sciullo, 2022. "Social Metabolism." In Pellizzoni, L., Leonardi, E., and V. Asara (eds.), *Handbook of Critical Environmental Politics*. Cheltenham, UK, and Northampton, MA: Edward Elgar, 295–307.

Palmblad, J., 2024. "Environmental History's Usable Past: On Reviving the Scholarship of Synthesis." *Global Environment* 17: 666–675.

Palmer, R., 1985. "Medical Botany in Northern Italy in the Renaissance." *Journal of the Royal Society of Medicine* 78: 149–157.

Pančur, A., Blaj Hribar, N., and B. Györfi, 2017. *Seznam krajevnih imen na Kranjskem leta 1910* (*List of Place Names in Carniola in 1910*). [Ljubljana]: Inštitut za novejšo zgodovino. https://www.sistory.si/publication/36053 (April 28th, 2024).

Panjek, A., 2015 (= Panjek, 2015a). "Obmejni Benečani in Kranjci? Raba gozdov v Julijskih Alpah v zgodnjem novem veku" (Uses of Forest in the Julian Alps in the Early Modern Period and the Related Impact of the Border between the Venetian Republic and Carniola). In Kavrečič, P., et al. (eds.), *Vzhodno od Benetk, slovenski obmejni prostor: Gospodarstvo, družba, prebivalstvo in naravni viri v zgodnjem novem veku*. Koper: Univerzitetna založba Annales, 85–117.

Panjek, A., 2015 (= Panjek, 2015b). *Paesaggio culturale e ambiente del Carso: L'uso delle risorse naturali in età moderna*. Koper: Univerza na Primorskem.

Panjek, A., 2017. "The Integrated Peasant Economy as a Concept in Progress." In Panjek, A., Larsson, J., and L. Mocarelli (eds.), *Integrated Peasant Economy in a Comparative Perspective: Alps, Scandinavia and Beyond*. Koper: University of Primorska Press, 11–49.

Panjek, A., 2018. "Against the Desert in the Karst: A Paradigm-Shift from Ruined Landscape to Cultural Savannah." *Economic- and Ecohistory* 14: 52–71.

Papež Kristanc, A., 2012. *V sožitju z naravo: Priporočila za ohranjanje biotske pestrosti kulturne krajine Triglavskega narodnega parka* (Coexisting with Nature: Guidelines for the Preservation of the Biodiversity of the Cultural Landscape in Triglav National Park). [city unknown]: Triglavski narodni park.

Paracchini, M.L., and C. Capitani, 2012. "The Place of HNV Farmland in EU-Level Indicators for the Rural-Agrarian Landscape." In Oppermann, R., Beaufoy, G., and G. Jones (eds.), *High Nature Value Farming in Europe: 35 European Countries and Perspectives*. Ubstadt-Weiher, Heidelberg, and Basel: Regionalkultur, 517–523.

Paracchini, M.L., and R. Oppermann, 2012. "Public Goods and Ecosystem Services Delivered by HNV Farmland." In Oppermann, R., Beaufoy, G., and G. Jones (eds.), *High Nature Value Farming in Europe: 35 European Countries and Perspectives*. Ubstadt-Weiher, Heidelberg, and Basel: Regionalkultur, 446–450.

Pärtel, M., Mändla, R., and M. Zobel, 1999. "Landscape History of a Calcareous (Alvar) Grassland in Hanila, Western Estonia, during the Last Three Hundred Years." *Landscape Ecology* 14: 187–196.

Paschinger, H., 1939. "Landwirtschaftsgeographische Studien im Jaunfelde (Kärnten)." *Carinthia II* 49: 74–85.

Patz, H., 1946. "'Speik 500 Faß zu 50 Kilo greifbar': Als es in Steiermark noch keine Naturschutzverordnungen gab." *Murtaler Volkszeitung* 2(33) (August 17th, 1946): 3.

Patzelt, G., 1996. "Modelstudie Ötztal – Landschaftsgeschichte im Hochgebirgsraum." *Mitteilungen der Österreichischen Geographischen Gesellschaft* 138: 53–70.

Pauli, H., Gottfried, M., and G. Grabherr, 2003. "The Piz Linard (3411 m), the Grisons, Switzerland—Europe's Oldest Mountain Vegetation Study Site." In Nagy, L., et al. (eds.), *Alpine Biodiversity in Europe*. Berlin and Heidelberg: Springer, 443–448.

Pavković, T., 2021. *Govedo na Kranjskem, Koroškem in Štajerskem v 16. ter 17. stoletju: Magistrsko delo* (Cattle in Carniola, Carinthia, and Styria in the Sixteenth and Seventeenth Centuries: Master's Thesis). Ljubljana: Univerza v Ljubljani, Filozofska fakulteta, Oddelek za zgodovino.

Pavković, T., in progress. *Živinoreja v Notranji Avstriji v 16. ter 17. stoletju: Doktorska disertacija* (Animal Husbandry in Inner Austria in the Sixteenth and Seventeenth Centuries: Doctoral Dissertation). Ljubljana: Univerza v Ljubljani, Filozofska fakulteta, Oddelek za zgodovino.

Pavšek, M., 2002. *Snežni plazovi v Sloveniji: Geografske značilnosti in preventiva* (Snow Avalanches in Slovenia: Geographical Characteristics and Risk Mitigation). Ljubljana: ZRC, ZRC SAZU.

Pawera, L., Łuczaj, Ł., Pieroni, A., and Z. Polesny, 2017. "Traditional Plant Knowledge in the White Carpathians: Ethnobotany of Wild Food Plants and Crop Wild Relatives in the Czech Republic." *Human Ecology* 45: 655–671.

Pečnik, L., 2024. *Ženska v sodnih zapisnikih gospostva Bled v poznem 17. stoletju: Magistrsko delo* (Women in the Court Minutes of the Seigneury of Bled (Germ. *Veldes*) in the Late Seventeenth Century). Ljubljana: Univerza v Ljubljani, Filozofska fakulteta, Oddelek za zgodovino.

Pedani, M.P., 2009. "Venice." In Ágoston, G., and B. Masters (eds.), *Encyclopedia of the Ottoman Empire*. New York: Facts On File, 581–582.

Pe'er, G., Zinngrebe, Y., Hauck, J., Schindler, S., Dittrich, A., et al., 2017. "Adding Some Green to the Greening: Improving the EU's Ecological Focus Areas for Biodiversity and Farmers." *Conservation Letters: A Journal of the Society for Conservation Biology* 10(5): 517–530.

Peratoner, G., Figl, U., Florian, C., Klotz, C., and S. Gottardi, 2014. "Effectiveness of Grassland Management and Mechanical Methods for the Weed Control of *Colchicum autumnale* in Permanent Meadows." In Hopkins, A., et al. (eds.), *EGF at 50: The Future of European Grasslands: Proceedings of the 25th General Meeting of the European Grassland Federation Aberystwyth, Wales 7–11 September 2014*. Gogerddan: Aberystwyth University, 418–420.

Pergl, J., Petřik, P., Fleischhans, R., Adámek, M., and J. Brůna, 2020. "*Telekia speciosa* (Schreb.) Baum. in Human Made Environment: Spread and Persistence, Two Sides of the Same Coin." *BioInvasions Records* 9(1): 17–28.

Peršič, A., 2022. "Sabbatismo popolare friulano: Sopravvivenza di osservanze sabbatarie protocristiane aquileiesi?" In Cedilnik, A., and M. Lovenjak (eds.), *Na obzorju novega: Območje severnega Jadrana ter vzhodnoalpski in balkansko-podonavski prostor v obdobju pozne antike in zgodnjega srednjega veka.* Ljubljana: Znanstvena založba Filozofske fakultete Univerze v Ljubljani, Oddelek za zgodovino, 51–72.

Peskar, R., 1991. *Srednjeveške poslikave zunanjščin cerkva v osrednji Sloveniji: Diplomska naloga* (Medieval Paintings on the Exterior of Church Walls in Central Slovenia: Bachelor's Thesis). Ljubljana: Univerza v Ljubljani, Filozofska fakulteta, Oddelek za umetnostno zgodovino.

Petek, F., 2005. *Spremembe rabe tal v slovenskem alpskem svetu* (Land Use Changes in the Slovenian Alps). Ljubljana, ZRC.

Peterken, G., 2013. *Meadows.* London, etc.: British Wildlife Publishing.

Peternel, M.M., 2022. "Varstvo Alp!" (Protection of the Alps!). *Planinski vestnik* 122(2): 52–53.

Petkovšek, M., 2020 (= Petkovšek, 2020a). *Evropsko pomembni habitatni tipi* (Habitats of European Interest). Ljubljana: Agencija Republike Slovenije za okolje. https://kazalci.arso.gov.si/sl/content/evropsko-pomembni-habitatni-tipi-1 (December 30th, 2024).

Petkovšek, M., 2020 (= Petkovšek, 2020b). *Habitats of European Interest.* Ljubljana: Slovenian Environment Agency. https://kazalci.arso.gov.si/en/content/habi tats-european-interest-1 (December 30th, 2024).

Petkovšek, V., 1966. "Po sledovih prvih floristov v naših gorah" (Following the Footsteps of the First Experts on Flora in Slovenian Mountains). *Planinski vestnik* 66(7): 297–306.

Pfister, C., 1985 (= Pfister, 1985a). *Klimageschichte der Schweiz 1525–1860: Das Klima der Schweiz von 1525–1860 und seine Bedeutung in der Geschichte von Bevölkerung und Landwirtschaft.* Vol. 2, *Bevölkerung, Klima und Agrarmodernisierung.* 2nd ed. Bern, Stuttgart: Haupt.

Pfister, C., 1985 (= Pfister, Ch., 1985b). "Snow Cover, Snow Lines and Glaciers in Central Europe Since the 16th Century." In Tooley, M.J., and G.M. Sheail (eds.), *The Climatic Scene.* London, etc.: Allen & Unwin, 154–174.

Pfister, C., 1999. *Wetternachhersage: 500 Jahre Klimavariationen und Naturkatastrophen.* Bern, Stuttgrat, and Vienna: Haupt.

Pfister, C., 2011. *"The Monster Swallows You": Disaster Memory and Risk Culture in Western Europe, 1500–2000.* Munich: Rachel Carson Center for Environment and Society.

Pfister, C., 2018 (= Pfister, 2018a). "Evidence from the Archives of Societies: Documentary Evidence—Overview." In White, S., Pfister, C., and F. Mauelshagen (eds.), *The Palgrave Handbook of Climate History*. London: Palgrave Macmillan, 37–47.

Pfister, C., 2018 (= Pfister, Ch. 2018b). "Evidence from the Archives of Societies: Institutional Sources." In White, S., Pfister, C., and F. Mauelshagen (eds.), *The Palgrave Handbook of Climate History*. London: Palgrave Macmillan, 67–81.

Pfister, C., and D. Brändli, 1999. "Rodungen im Gebirge – Überschwemmungen im Vorland: Ein Deutungsmuster macht Karriere." In Sieferle, R.P., and H. Breuninger (eds.), *Natur-Bilder*. Frankfurt am Main and New York: Campus, 297–323.

Pfister, C., Brázdil, R., Luterbacher, J., Ogilvie, A.E.J., and S. White, 2018. "Early Modern Europe." In White, S., Pfister, C., and F. Mauelshagen (eds.), *The Palgrave Handbook of Climate History*. London: Palgrave Macmillan, 265–295.

Pfister, C., Camenisch, C., and P. Dobrovolný, 2018. "Analysis and Interpretation: Temperature and Precipitation Indices." In White, S., Pfister, C., and F. Mauelshagen (eds.), *The Palgrave Handbook of Climate History*. London: Palgrave Macmillan, 115–129.

Pfister, C., Holzhauser, H., and H.J. Zumbühl, 1994. "Neue Ergebnisse zur Vorstossdynamik der Grindelwaldgletscher vom 14. bis zum 16. Jahrhundert." *Mitteilungen der Naturforschenden Gesellschaft in Bern* 51: 55–79.

Pfister, C., and S. White, 2018. "Evidence from the Archives of Societies: Personal Documentary Sources." In White, S., Pfister, C., and F. Mauelshagen (eds.), *The Palgrave Handbook of Climate History*. London: Palgrave Macmillan, 49–65.

Pfister, U., 2007. "Industrialisierung." In Jaeger, F., et al. (eds.), *Enzyklopädie der Neuzeit*. Vol. 5, *Gymnasium–Japanhandel*. Stuttgart and Weimar: J.B. Metzler, 902–920.

Pfitzenmeyer, C.D.C., 1962. "*Arrhenatherum elatius* (L.) J. & C. Presl." *Journal of Ecology* 50: 235–245.

Pickl, O., 1982. "Brandwirtschaft und Umwelt seit der Besiedelung der Ostalpenländer." In Kellenbenz, H. (ed.), *Wirtschaftsentwicklung und Umweltbeeinflussung (14.–20. Jahrhundert)*. Wiesbaden: Franz Steiner, 27–55.

Piko, M., 1996. "Dobrla vas pripoveduje" (Eberndorf erzählt). In Makarovič, M. (ed.), *Dobrla vas in okolica: Iz preteklosti v sedanjost*. Klagenfurt, Ljubljana, and Vienna: Mohorjeva, 295–334.

Pils, G., 1994. *Die Wiesen Oberösterreichs: Eine Naturgeschichte des oberösterreichischen Grünlandes unter besonderer Berücksichtigung von Naturschutzaspekten*. Linz: Forschungsinstitut für Umweltinformatik.

Pinna, M., 1989 (= Pinna, 1989a). "Bilancio idrico." In Bozzini, M. et al. (eds.), *Atlante Tematico d'Italia*. Vol. 1. Milan: Touring Club Italiano, Consiglio Nazionale delle Ricerche, chapter 13, no page numbers.

Pinna, M., 1989 (= Pinna, 1989b). "Precipitazioni." In Bozzini, M. et al. (eds.), *Atlante Tematico d'Italia*. Vol. 1. Milan: Touring Club Italiano, Consiglio Nazionale delle Ricerche, chapter 12, no page numbers.

Pintar, L., 2012. "Novi nahajališči Isslerjevega dvorednika (*Diphasiastrum issleri* (Rouy) Holub)" (Two Unknown Localities of Issler's Alpine Clubmoss (*Diphasiastrum issleri* (Rouy) Holub)). *Proteus* 74: 426–428.

Pintar, M., 2003. "Prednosti in slabosti različnih načinov namakanja" (Advantages and Disadvantages of Different Irrigation Types). In Naglič, M., and A. Jagodic (eds.), *Kaj storiti za zmanjšanje posledic suše v kmetijstvu?* Ljubljana: Kmetijsko-gozdarska zbornica Slovenije, 23–32.

Pintar, M., 2005. "Namakanje in okolje" (Irrigation and the Environment). In Podgoršek, J. (ed.), *Vrtnarstvo: Kako do zdrave hrane in okolja? Slovenski vrtnarski posvet: Zbornik referatov.* Novo mesto: Kmetijsko gozdarski zavod Novo mesto, KZ Krka Novo mesto, and Kmetijska šola Grm Novo mesto, 331–338.

Pirc, S., and T. Budkovič, 1996. "Remains of World War I Geochemical Pollution in the Landscape." In Richardson, M. (ed.), *Environmental Xenobiotics*. Herfordshire: Taylor & Francis, 375–418.

Pivk, O., 2006. *Okrajno glavarstvo Radovljica 1819–1918: Arhivski inventar*, I[.] del (Radovljica District (*Bezirkshauptmannschaft*) 1819–1918: Archival Register, Part 1). Ljubljana: Arhiv Republike Slovenije.

Plantlife: Meadow Clary. https://www.plantlife.org.uk/uk/discover-wild-plants-nature/plant-fungi-species/meadow-clary (August 2nd, 2022).

Plants For A Future. https://pfaf.org/user/Default.aspx (January 13th, 2021).

Plassmann, G., 2016 (= Plassmann, 2016a). "Nature Protection in the Alps – Which Motivation?" In Plassmann, G., et al. (eds.), *Alpine Nature 2030: Creating [Ecological] Connectivity for Generations to Come*. Berlin: Federal Ministry for the Environment, Nature Conservation, Building and Nuclear Safety, 17–24.

Plassmann, G., 2016 (= Plassmann, 2016b). "The Conditions for Success of Nature Protection in the Alps." In Plassmann, G., et al. (eds.), *Alpine Nature 2030: Creating [Ecological] Connectivity for Generations to Come*. Berlin: Federal Ministry for the Environment, Nature Conservation, Building and Nuclear Safety, 32–36.

Plemel, V., 1862. "Beiträge zur Flora Krain's." *Jahresheft des Vereines des Krainischen Landes-Museums* 3: 120–164.

Pleteršnik, M., 2006. *Slovensko-nemški slovar* (Slovenian–German Dictionary). Part 2: P–Ž. Ljubljana: ZRC SAZU, ZRC.

Polajnar, J., 2008. *"Pfuj! To je gerdo!" K zgodovini morale na Slovenskem v dobi meščanstva* ("Fie! That's Ugly!" Toward a History of Morality in Slovenia in the Nineteenth Century). Celje: Zgodovinsko društvo Celje.

Polajnar Horvat, K., Smrekar, A., and M. Zorn, 2017. "Environmental Thought in Slovenia." In Petrić, H., and I. Žebec Šilj (eds.), *Environmentalism in Central and Southeastern Europe: Historical Perspectives*. Lanham etc.: Lexington Books, 21–33.

Polec, J., 1936. "Svobodniki na Kranjskem" (Die Freisassen in Krain). *Glasnik muzejskega društva za Slovenijo* 17(1–4): 5–142.

Poppendieck, H.-H., Bertram, H., Brandt, I., Kreft, K.-A., Kurz, H., et al., 2010. *Rote Liste und Florenliste der Gefäßpflanzen von Hamburg*. Hamburg: Behörde für Stadtentwicklung und Umwelt, Abteilung Naturschutz.

Popplow, M., 2016. "Wissenszirkulation auf dem Land vor der Industrialisierung: Schlusskommentar." In Dauser, R., Fassl, P., and L. Schilling (eds.), *Wissenszirkulation auf dem Land vor der Industrialisierung*. Augsburg: Wißner 255–260.

Popplow, M., 2017. "Wissensgeschichte der pflanzlichen Ressourcen in der *longue durée*." In Boscani Leoni, S., and M. Stuber (eds.), *Wer das Gras wachsen hört: Wissensgeschichte(n) der pflanzlichen Ressourcen vom Mittelalter bis ins 20. Jahrhundert*. Innsbruck, Vienna, and Bozen: Studienverlag, 199–210.

Poptcheva, K., Schwartze, P., Vogel, A., Kleinebecker, T., and N. Hölzel, 2009. "Changes in Wet Meadow Vegetation after 20 Years of Different Management in a Field Experiment (North-West Germany)." *Agriculture, Ecosystems and Environment* 134(1–2): 108–114.

Porenta, S., Štular, B., Toškan, B., Mileusnić, Z., and J. Dirjec, 2015. "The Late Medieval and Early Post-Medieval Site of Župnijski Dom in Šentvid pri Stični: Analysis of the Pottery and Animal Remains." *Arheološki vestnik* 66: 333–397.

Poschlod, P., 2015. *Geschichte der Kulturlandschaft: Entstehungsursachen und Steuerungsfaktoren der Entwicklung der Kulturlandschaft, Lebensraum- und Artenvielfalt in Mitteleuropa*. Stuttgart: Ulmer.

Poschlod, P., Baumann., A., and P. Karlík, 2009. "Origin and Development of Grasslands in Central Europe." In Veen, P., et al. (eds.), *Grasslands in Europe of High Nature Value*. Zeist: KNNV Publishing, 15–25.

Poschlod, P., Karlík, P., Baumann, A., and B. Wiedmann, 2008. "The History of Dry Calcareous Grasslands Near Kallmünz (Bavaria) Reconstructed by the Application of Palaeoecological, Historical and Recent-Ecological Methods." In Szabó, P., and R. Hédl (eds.), *Human Nature: Studies in Historical Ecology and Environmental History*. Brno: Institute of Botany of the ASCR, 130–143.

Potočnik Slavič, I., Cigale, D., Lampič, B., Perpar, A., and A. Udovč, 2016. *(Ne)raba razpoložljivih virov na kmetijah v Sloveniji* ((Non-)Utilisation of Available Resources on Farms in Slovenia). Ljubljana: Znanstvena založba Filozofske fakultete Univerze v Ljubljani.

Poux, X., and B. Ramain, 2012. "France." In Oppermann, R., Beaufoy, G., and G. Jones (eds.), *High Nature Value Farming in Europe: 35 European Countries and Perspectives*. Ubstadt-Weiher, Heidelberg, and Basel: Regionalkultur, 208–219.

Povž, M., and B. Sket, 1990. *Naše sladkovodne ribe* (Freshwater Fish Species in Slovenia). Ljubljana: Mladinska knjiga.

Poženel, A., 2007. "Izkušnje pri zatiranju poljskega majskega hrošča (*Melolontha melolontha* L.) na Idrijskem" (Experiences in Controlling Common Cockchafer (*Melolontha melolontha* L.) in the Idrija Region). In Maček, J. (ed.), *Zbornik predavanj in referatov 8. slovenskega posvetovanja o varstvu rastlin*. Ljubljana: Društvo za varstvo rastlin Slovenije, 72–77.

Praprotnik, A., 1849. "Lastovke: Povést" (Swallows: A Tale). *Vedež: Časopis za šolsko mladóst* 2(first half of the year, 16): 126–127.

Praprotnik, N., 1993 (= Praprotnik, 1993a). "Karavanške ključavnice" (Daffodils in the Karawanks). *Planinski vestnik* 93(5): 215–219.

Praprotnik, N., 1993 (= Praprotnik, 1993b). "Rastišča zapeljivih ključavnic" (Habitats of Seductive Daffodils). *Planinski vestnik* 93(6): 260–262.

Praprotnik, N., 2012. "Botanična pestrost v rovtih" (Plant Diversity in Montane and Subalpine Grasslands). In Šubic, T. (ed.), *V rovtih nad Jesenicami: Narava in človek ustvarjata biotsko pestrost*. Ljubljana: Zavod Republike Slovenije za varstvo narave, 29–48.

Praprotnik, N., 2015. "Botaniki, njihovo delo in herbarijske zbirke praprotnic in semenk v Prirodoslovnem muzeju Slovenije" (The Botanists, Their Work and Herbarium Collections of Vascular Plants in the Slovenian Museum of Natural History). *Scopolia: Revija Prirodoslovnega muzeja Slovenije* 83–84: 1–414.

"Pravilnik o uvrstitvi ogroženih rastlinskih in živalskih vrst v rdeči seznam" (Rules on the Inclusion of Endangered Plant and Animal Species on the Red List). 2002. *Uradni list RS* 12(82) (September 24th, 2002): 8893–8975, and http://www.pisrs.si/Pis.web/pregledPredpisa?id=ODRE1883# (November 18th, 2019).

Predan, D., 1956 (= Predan, 1956a). "Košnja" (Haymaking). In Kotnik, F., and V. Novak (eds.), *Kotnikov zbornik: Narodopisni zapiski z Gornjegrajskega in Kozjanskega*. Celje: Mestni muzej, 32.

Predan, D., 1956 (= Predan, 1956b). "Požarjenje" (Swidden Cultivation). In Kotnik, F., and V. Novak (eds.), *Kotnikov zbornik: Narodopisni zapiski z Gornjegrajskega in Kozjanskega*. Celje: Mestni muzej, 25–28.

Prešeren, J., 1933. "Imenoslovje okrog Begunjščice in Stola" (Place Names near Begunjščica and Stol). *Planinski vestnik* 33(4): 94–98.

Prezelj, K., 2012. *Vegetacija in vlažnostne razmere na grbinastih travnikih v Zgornji Radovni in Krmi: Diplomsko delo* (Vegetation and Humidity in the Hummocky Meadows in Zgornja Radovna and Krma: Bachelor's Thesis). Ljubljana: Univerza v Ljubljani, Biotehniška fakulteta, Oddelek za agronomijo.

Pribyl, H., 1998. "Der Sonntag als der Tag der wöchentlichen Arbeitsruhe." In Weiler, R. (ed.), *Der Tag des Herrn: Kulturgeschichte des Sonntags*. Vienna: Böhlau, 95–139.

Priesner, C., 2010. "Spiritus." In Jaeger, F., et al. (eds.), *Enzyklopädie der Neuzeit*. Vol. 12, *Silber–Subsidien*. Stuttgart and Weimar: J.B. Metzler, 371–373.

Prinčič, J., 1996. "Pretirano izkoriščanje slovenskih gozdov" (Over-Exploitation of Slovenian Forests). In Drnovšek, M., et al. (eds.), *Slovenska kronika XX. stoletja: 1941–1995*. Ljubljana: Nova revija, 166.

Prvi podatki popisa prebivalstva in stanovanj 31. III. 1971 (First Results of the Census of Population and Households with the Key Reference Date March 31st, 1971). 1971. Ljubljana: Zavod SR Slovenije za statistiko.

Pucher, E., 2014. "Sechs Jahrtausende alpine Viehwirtschaft." *Forschungesberichte der ANISA* 5: 73–100.

Rahmann, G., 2011. "Biodiversity and Organic Farming: What Do We Know?" *Landbauforschung: vTI [Johann Heinrich von Thünen Institute] Agriculture and Forestry Research* 61(3): 189–208.

Rake Matted Grass. https://www.fitturf.com/lawn-care-tips/rake-matted-grass/ (June 28th, 2019).

Ramšak, M., 2017. "Pohorske zooterapije v kontekstu ljudske medicine" (Pohorje Mountain Zootherapies in the Context of Folk Medicine). In Filej, B. (ed.), *Za človeka gre: Interdisciplinarnost, transnacionalnost in gradnja mostov: 5. mednarodna znanstvena konferenca: Zbornik prispevkov z recenzijo*. Maribor: Alma Mater Europea - ECM, 1061–1075.

Rasran, L., and H. Jeromin, 2010. "Problempflanzen im Fokus des Naturschutzmanagements von Dauergrünlandflächen." *Telma* 40: 119–136.

Rauchensteiner, M., 2010. "Geschichte der Erinnerung: Die Transformation des Ersten Weltkriegs." In Rajšp, V. (ed.), *Isonzofront 1915–1917: Die Kultur des Erinnerns*. Vienna and Ljubljana: Slowenisches Wissenschaftliches Institut in Wien, 55–70.

Ravazzoli, E., and P. Laner, 2018. "Language Groups, 2011." In Elmi, M. (ed.), *The Alps in 25 Maps*. Innsbruck and Bolzano: Permanent Secretariat of the Alpine Convention, 39.

Ravnik, L., 2008. "Običaji v Plavškem Rovtu" (Customs in Plavški Rovt). In Kokošinek, N., and Z. Torkar Tahir (eds.), *Kako so v vaseh pod Golico včasih živeli?* Jesenice: Ljudska univerza Jesenice, 37–39.

Realonline: 002351 Feiertagschristus/Christ with Craftsman's Tools. https://realonline.imareal.sbg.ac.at/en/detail/nr-002351 (February 8th, 2023).

Redman, M., 2012. "Rural Development Programmes and How They Can Support HNV Farming." In Oppermann, R., Beaufoy, G., and G. Jones (eds.), *High Nature Value Farming in Europe: 35 European Countries and Perspectives*. Ubstadt-Weiher, Heidelberg, and Basel: Regionalkultur, 491–501.

Regions in Europe: 2022 Interactive Edition, Population Density. https://ec.europa.eu/eurostat/cache/digpub/regions/#total-population (July 13th, 2024).

Reith, R., 2010 (= Reith, 2010a). "Schneidegeräte: 2. Gewerbe." In Jaeger, F., et al. (eds.), *Enzyklopädie der Neuzeit.* Vol. 11, *Renaissance–Signatur.* Stuttgart and Weimar: J.B. Metzler, 817–821.

Reith, R., 2010 (= Reith, 2010b). "Stahl: 2. Gewerbe." In *Enzyklopädie der Neuzeit.* Vol. 12, *Silber–Subsidien.* Stuttgart and Weimar: J.B. Metzler, 815–818.

Remec, M., 2015. *Podrgni, očedi, živali otrebi: Higiena in snaga v dobi meščanstva* (Rub, Tidy, Tend to Animals: Hygiene and Sanitation in the Bourgeois Period). Ljubljana: Inštitut za novejšo zgodovino.

Renes, H., Centeri, C., Eiter, S., Gaillard, B., Kruse, A., et al., 2020. "Water Meadows as European Agricultural Heritage." In Hein, C. (ed.), *Adaptive Strategies for Water Heritage: Past, Present, Future.* Cham: Springer Nature, 107–131.

Repaljščica – življenje na robu izumrtja (Whinchat—Life on the Brink of Extinction). 2015. https://www.ptice.si/2015/06/repaljscica-prvi-eu-simpozij/ (October 4th, 2021).

Repe, B., 2010. "Prepoznavanje osnovnih prsti slovenske klasifikacije" (Recognition of the Slovenian Soil Classification Types). *Dela* 34: 143–166.

Repič, J., 2021. "Spominska krajina in dediščina prve svetovne vojne v Julijskih Alpah" (First World War Memorial Landscape and Heritage in the Julian Alps). In Fikfak, J., and B. Jezernik (eds.), *Dediščina prva svetovne vojne: Reprezentacije in reinterpretacije.* Ljubljana: Znanstvena založba Filozofske fakultete Univerze v Ljubljani, 175–196.

Reshetnikova, N.M., 2016. "The Way of Emergence of Some Western European Plant Species in Kaluga Oblast—The Pathway of the German Army in 1941–1943." *Russian Journal of Biological Invasions* 7(1): 62–68.

Results-Based Nature Conservation Plan: Experiences and Conclusions from the Project 2013–2020: Expert Report. 2019 [publication year from the German report version; see *Ergebnisorientierter Naturschutzplan*]. Vienna: Suske consulting. https://www.rbpnetwork.eu/media/fachbericht_en_final_klein.pdf (August 10th, 2022).

Rhamm, K., 1908. *Urzeitliche Bauernhöfe in germanisch-slawischem Waldgebiet.* Part 1. Braunschweig: Friderich Vieweg und Sohn.

Riccabona, C., von Sauter, S., and A. Unterkircher, 2020. "Isidor Grießer." In *Lexikon Literatur in Tirol.* https://orawww.uibk.ac.at/apex/uprod/f?p=TLL:2:0::::P2 ID:209 (November 24th, 2021).

Riedel, S., Walter, T., and F. Herzog, 2012. "Switzerland." In Oppermann, R., Beaufoy, G., and G. Jones (eds.), *High Nature Value Farming in Europe: 35 European Countries and Perspectives.* Ubstadt-Weiher, Heidelberg, and Basel: Regionalkultur, 420–433.

Riedmann, J., 1985. "Mittelalter: Landwirtschaft." In Fontana, J. (ed.), *Geschichte des Landes Tirol*. Vol. 1. Bolzano: Athesia; Innsbruck and Vienna: Tyrolia, 492–497.

Rippmann Tauber, D., 2017. "Aneignung von 'Wildem' und Neuem durch Sprache – im Lichte von Agrarschriften und Kräuterbüchern." In: Boscani Leoni, S., and M. Stuber (eds.), *Wer das Gras wachsen hört: Wissensgeschichte(n) der pflanzlichen Ressourcen vom Mittelalter bis ins 20. Jahrhundert*. Innsbruck, Vienna, and Bozen: Studienverlag, 14–36.

Roehrs, H., Klooss, S., and W. Kirleis, 2013. "Evaluating Prehistoric Finds of *Arrhenatherum elatius* var. *bulbosum* in North-Western and Central Europe with an Emphasis on the First Neolithic Finds in Northern Germany." *Archaeological and Anthropological Sciences* 5(1): 1–15.

Rogelj, M., 2012. "Travniški habitatni tipi v rovtih nad Jesenicami" (Meadow Habitat Types in Montane and Subalpine Grasslands above Jesenice). In Šubic, T. (ed.), *V rovtih nad Jesenicami: Narava in človek ustvarjata biotsko pestrost*. Ljubljana: Zavod Republike Slovenije za varstvo narave, 49–64.

Rogelja Caf, N., and Š. Ledinek Lozej, 2021. "Hodim, pišem, raziskujem: Refleksija hoje in pisanja v etnološkem in antropološkem raziskovanju" (I Walk, I Write, I Research: A Reflection on Walking and Writing in Ethnological and Anthropological Research). *Etnolog: Glasnik Slovenskega etnografskega muzeja* 31: 67–86.

Rohr, C., 2007 (= Rohr, 2007a). *Extreme Naturereignisse im Ostalpenraum: Naturerfahrung im Spätmittelalter und am Beginn der Neuzeit*. Cologne, Weimar, and Vienna: Böhlau.

Rohr, C., 2007 (= Rohr, 2007b). "Zum Umgang mit Tierplagen im Alpenraum in der Frühen Neuzeit." In Engelken, K., Hünniger, D., and S. Windelen (eds.), *Beten, Impfen, Sammeln: Zur Viehseuchen- und Schädlingsbekämpfung in der Frühen Neuzeit*. Göttingen: Universitätsverlag, 99–133.

Rohr, C., 2012. "Stadterweiterung versus Naturgefahr: Die Stadt Wels während der Hochwasserkatastrophen von 1897 und 1899." *Jahrbuch des Oberösterreichischen Musealvereines – Gesellschaft für Landeskunde* 157: 551–574.

Rohr, C., 2013. "Floods of the Upper Danube River and Its Tributaries and Their Impact on Urban Economies (c. 1350–1600): The Examples of the Towns of Krems/Stein and Wels (Austria)." *Environment and History* 19: 133–148.

Rohr, C., 2017. "Klima und Umwelt als Rahmenbedingungen alpinen Wirtschaftens: Beispiele und Perspektiven." In Denzel, M.A., et al. (eds.), *Oeconomia Alpium I: Wirtschaftsgeschichte des Alpenraums in vorindustrieller Zeit: Forschungsaufriss, -konzepte und -perspektiven*. Berlin and Boston: De Gruyter and Oldenbourg, 73–101.

Rohr, C., Camenisch, C., and K. Pribyl, 2018. "European Middle Ages." In White, S., Pfister, C., and F. Mauelshagen (eds.), *The Palgrave Handbook of Climate History*. London: Palgrave Macmillan, 247–263.

Rösener, W., 1991. "Landesausbau und Kolonisation: I. Mitteleuropa." In Bautier, R.-H., et al. (eds.), *Lexikon des Mittelalters*. Vol. 5, *Hiera-Mittel bis Lukanien*. Munich and Zurich: Artemis, 1643–1646.

Rotherham, I.D., 2013. "Reinterpreting Wooded Landscapes, Shadow Woods and the Impacts of Grazing." In Rotherham, I.D. (ed.), *Trees, Forested Landscapes, and Grazing Animals: A European Perspective on Woodlands and Grazed Treescapes*. London and New York: Routledge, 72–86.

Rozman, F., 1970. "Socialna struktura naročnikov 'Novic' v letu 1845" (Social Structure of Subscribers to the Newspaper *Novice* in 1845). *Zgodovinski časopis* 24(1–2): 81–89.

Russell, P., 2010. "Mrs Beeton, the First Domestic Goddess." *The Financial Times* (December 3rd, 2010). https://web.archive.org/web/20151208154047/http://www.ft.com/cms/s/2/be9d91a6-fcd8-11df-ae2d-00144feab49a.html (December 28th, 2022).

Sadar, V., 1956. *Nauk o travinju: Skripta* (Grassland Agronomy: Textbook for Students). Ljubljana: Združenje agronomov – Odbor za izdajo skript.

SAGISonline, Basiskarten, Alle Themen. https://www.salzburg.gv.at/sagismobile/sagisonline/map/Basiskarten/Alle%20Themen (August 9th, 2022).

Šalehar, A., Kompan, D., Žan Lotrič, M., Kovač, M., Holcman, A., et al., 2014. *Razvoj pasem domačih živali v Sloveniji: Prvotne, izgubljene in pretopljene pasme* (Evolution of Livestock Breeds in Slovenia: Original, Lost, and Refined Breeds). 2nd revised ed. Rodica: Univerza v Ljubljani, Biotehniška fakulteta, Oddelek za zootehniko.

Samojlik, T., and D. Kuijper. 2013. "Grazed Wood-Pasture versus Browsed High Forests: Impact of Ungulates on Forest Landscapes from the Perspective of the Białowieża Primeval Forest." In Rotherham, I.D. (ed.), *Trees, Forested Landscapes, and Grazing Animals: A European Perspective on Woodlands and Grazed Treescapes*. London and New York: Routledge, 143–162.

Sandgruber, R., 1989. "Genußmittel und Kolonialwaren." In Schöpfer, G. (ed.), *Menschen & Münzen & Märkte: Steirische Landesausstellung 1989 Judenburg*. Fohnsdorf: K. Podmenik, 103–111.

Santana-Cordero, A.M., Szabó, P., Bürgi, M., and C.G. Armstrong, 2024. "The Practice of Historical Ecology: What, When, Where, How and What For." *Ambio: A Journal for Environment and Society* 53(5): 664–677.

Šarf, F., 1955. "Brana na Gorenjskem" (Die Egge in Oberkrain). *Slovenski etnograf* 8: 81–92.

Šarf, F., 1974. "Košnja na visokoležečih predplaninskih območjih" (Die Heumahd in den hochgelegenen Voralpenbereichen). *Traditiones: Acta instituti ethnographiae Slovenorum* 3: 133–148.

Šarf, F., 1991. "Košnja" (Haymaking). In Batis, J., et al. (eds.), *Enciklopedija Slovenije*. Vol. 5, *Kari–Krei*. Ljubljana: Mladinska knjiga, 332–333.

Šarf, F., 2011. "Kôšnja" (Haymaking). In Baš, A., et al. (eds.), *Slovenski etnološki leksikon*. Ljubljana: Mladinska knjiga, 239–240.

Šarf, F., Habinc, M., and I. Smerdel, 2011. "Víle" (Fork). In Baš, A., et al. (eds.), *Slovenski etnološki leksikon*. Ljubljana: Mladinska knjiga, 671–672.

Šašel, J., 1955. "Colatio kot rimska poštna postaja" (Colatio als römische Poststation). *Zbornik Filozofske fakultete* 2: 71–82.

Satellite World Map. https://satellites.pro/#46.858570,11.160285,16 (November 17th, 2018).

Saunders, N.J., Faulkner, N., Košir, U., Črešnar, M., and S. Thomas, 2013. "Conflict Landscapes of the Soča/Isonzo Front, 1915–2013: Archaeological-Anthropological Evaluation of the Soča Valley, Slovenia." *Arheo* 30: 47–66.

Sauvant, D., Baumont, R., and P. Faverdin, 1996. "Development of a Mechanistic Model of Intake and Chewing Activities of Sheep." *Journal of Animal Science* 74: 2785–2802.

Schader, C., Drapela, T., Markut, T., Meier, M.S., Lindenthal, T., et al., 2014. "Farm- and Product-Level Biodiversity Assessment of Conventional and Organic Dairy Production in Austria." *International Journal of Biodiversity Science, Ecosystem Services & Management* 10(1): 20–39.

Schader, C., Stolze, M., and A. Gattinger, 2012. "Environmental Performance of Organic Farming." In Boye, J.I., and Y. Arcand (eds.), *Green Technologies in Food Production and Processing*. New York: Springer, 183–210.

Scharr, K., 2001. *Leben an der Grenze der Dauersiedlung: Grund und Boden im "Ötztaler Gebirgsraum" (Ötztal – Schnals – Passeier) vom 13. bis zur Mitte des 19. Jahrhunderts*. Innsbruck: Universitätsverlag Wagner.

Scheidegger, T., 2008. *Mythos Edelweiss: Zur Kulturgeschichte eines alpinen Symbols*. Zürich: Botanischer Garten der Universität Zürich; Geneva: Conservatoire et Jardin botaniques Ville de Genève. https://www.isek.uzh.ch/dam/jcr:00000000-1d25-814c-ffff-ffffdedb53f5/MythosEdelweiss.pdf (June 15th, 2020).

Scheurer, A., 2010. "Craintes, fantasmes et croyances suscitées par la faune alpine: L'exemple du Valais romand (Suisse), XVIe–XXe siècles." *Histoire des Alpes* 15: 65–87.

Scheurer, T., 2016. "The Challenge of Engendering Ecological Connectivity—Topics and Impacts: Planning Dynamic Landscapes: Opportunities and Limitations of Spatial Planning in Creating Ecological Networks." In Plassmann, G., et al. (eds.), *Alpine Nature 2030: Creating [Ecological] Connectivity for Generations to Come*. Berlin: Federal Ministry for the Environment, Nature Conservation, Building and Nuclear Safety, 85–87.

Scheutz, M., 2004. "Gerichtsakten." In Pauser, J., Scheutz, M., and T. Winkelbauer (eds.), *Quellenkunde der Habsburgermonarchie (16.–18. Jahrhundert): Ein exemplarisches Handbuch*. Vienna and Munich: R. Oldenbourg, 561–571.

Schindler, C., 2010. "Columella (Lucius Iunius Moderatus Collumella): De re rustica." In Walde, C., and B. Egger (eds.), *Die Rezeption der antiken Literatur: Kulturhistorisches Werklexikon*. Stuttgart and Weimar: J.B. Metzler, 263–270.

Schirpke, U., Leitinger, G., Tasser, E., Schermer, M., Steinbacher, M., et al., 2013. "Multiple Ecosystem Services of a Changing Alpine Landscape: Past, Present and Future." *International Journal of Biodiversity Science, Ecosystem Services & Management* 9: 123–135.

Schlude, U., 2009. "Naturwissen und Schriftlichkeit: Warum eine Fürstin des 16. Jahrhunderts nicht auf den Mont Ventoux steigt und die Natur exakter begreift als die 'philologischen' Landwirte." In Ruppel, S., and A. Steinbrecher (eds.), *"Die Natur ist überall bey uns": Mensch und Natur in der Frühen Neuzeit*. Zürich: Chronos, 95–108.

Schmid, M., 2016. "Long-Term Risks of Colonization: The Bavarian 'Donau-Moos.'" In Haberl, H., et al. (eds.), *Social Ecology: Society-Nature Relations across Time and Space*. [Cham]: Springer, 391–410.

Schmidt-Wiegand, R., 1991. "Lex Baiuvariorum." In Bautier, R.-H., et al. (eds.), *Lexikon des Mittelalters*. Vol. 5, *Hiera-Mittel bis Lukanien*. Munich and Zurich: Artemis, 1928.

Schmitzberger, I., Wrbka, T., Steurer, B., Aschenbrenner, G., Peterseil, J., et al., 2005. "How Farming Styles Influence Biodiversity Maintenance in Austrian Agricultural Landscapes." *Agriculture, Ecosystems and Environment* 108: 274–290.

Schneiter, F., 1970. *Agrargeschichte der Brandwirtschaft*. Graz: Historische Landeskommission.

Schöggl-Ernst, E., 2004. "Historische Bodendokumentation: Urbare, Landtafeln und Grundbücher." In Pauser, J., Scheutz, M., and T. Winkelbauer (eds.), *Quellenkunde der Habsburgermonarchie (16.–18. Jahrhundert): Ein exemplarisches Handbuch*. Vienna and Munich: R. Oldenbourg, 516–529.

Schöller, R., 2010. "Schafe." In Jaeger, F., et al. (eds.), *Enzyklopädie der Neuzeit*. Vol. 11, *Renaissance–Signatur*. Stuttgart and Weimar: J.B. Metzler, 634–637.

Schreg, R., 2008. "Bevölkerungswachstum und Agrarisierung – Faktoren des früh- und hochmittelalterlichen Landesausbaus im Spiegel umweltarchäologischer Forschungen." In Herrmann, B. (ed.), *Beiträge zum Gottinger Umwelthistorischen Kolloquium 2007–2008*. Göttingen: Universitätsverlag, 117–146.

Schubert, R., Hilbig, W., and S. Klotz, 2001. *Bestimmungsbuch der Pflanzengesellschaften Deutschlands*. Heidelberg and Berlin: Spektrum Akademischer Verlag.

Schulte, L., Wetter, O., Wilhelm, B., Peña, J.C., Amann, B., et al., 2019. "Integration of Multi-Archive Datasets for the Development of a Four-Dimensional Paleoflood Model of Alpine Catchments." *Global and Planetary Change* 180: 66–88.

Schunko, C., and C.R. Vogl, 2010. "Organic Farmers Use of Wild Food Plants and Fungi in a Hilly Area in Styria (Austria)." *Journal of Ethnobiology and Ethnomedicine* 6, article no. 17: separate pagination 1–14.

Schwarzrock, G., 1997. "Wirtschaftliche Folgen des Ersten Weltkrieges." In Leisering, W. (ed.), *Historischer Weltatlas*. 102nd ed. Berlin: Cornelsen, 108.

"Schwendeaktion Gladiolenwiese." 2003. *Schütt Magazin* 2: 2. http://www.schuett.at/life/schuettmagazin.php (September 18th, 2019).

Schwienbacher, E., Marcante, S., and B. Erschbamer, 2010. "Alpine Species Seed Longevity in the Soil in Relation to Seed Size and Shape—A 5-Year Burial Experiment in the Central Alps." *Flora: Morphology, Distribution, Functional Ecology of Plants* 205(1): 19–25.

Sedmak, D., 1998 (= Sedmak 1998a). "Soška fronta" (The Isonzo (Sln. *Soča*) Front). In Batis, J., et al. (eds.), *Enciklopedija Slovenije*. Vol. 12, *Slovenska n–Sz*. Ljubljana: Mladinska knjiga, 159–167.

Sedmak, D., 1998 (= Sedmak 1998b). "Soška fronta: žrtve in vojna škoda" (The Isonzo (Sln. *Soča*) Front: Casualties and Destruction). In Batis, J., et al., *Enciklopedija Slovenije*. Vol. 12, *Slovenska n–Sz*. Ljubljana: Mladinska knjiga, 171–172.

Seed Gathering and Other Project Activities in Triglav National Park, 2024. https://lifeforseeds.si/en/seed-gathering-and-other-project-activities-in-triglav-national-park/ (December 31st, 2024).

Šekli, M., 2024. "The Etymological Origin of Slovenian Geographical Names with the Morpheme *-ov-/-ev*." *Voprosy onomastiki* 21(3): 178–192.

Seliškar, A., and T. Wraber, 1986. *Travniške rastline na Slovenskem: Sto pogostih vrst* (Grassland Plants in Slovenia: A Selection of a Hundred Common Species). Ljubljana: Prešernova družba.

Selišnik, I., and M. Verginella, 2021. "Kmetijske in rokodelske novice in razvoj bralne javnosti" (Farmers' and Craftsmen's News and the Development of the Reading Public). In Stanonik, M., and I. Slavec Gradišnik (eds.), *Janez Bleiweiss, Novice in modernizacija slovenske družbe*. Ljubljana: Slovenska akademija znanosti in umetnosti, 146–163.

Seneno, Gradiva (*From Hay*, Materials). https://seneno.si/gradiva/ (December 28th, 2024).

Šercelj, A., 1961. "Razvoj in propad gozda v Dolini Triglavskih jezer" (Entwicklung und Rückgang des Waldes im Tale der Triglaver-Seen). *Gozdarski vestnik* 19: 201–209.

Sever, A., 2005. *Razširjenost in gospodarski pomen jesenov (*Fraxinus *sp.) v območni enoti Kranj: Diplomsko delo* (Distribution and Economic Role of Ash Species (*Fraxinus* sp.) in Regional Unit Kranj: Bachelor's Thesis). Ljubljana: Univerza v Ljubljani, Biotehniška fakulteta, Oddelek za gozdarstvo in obnovljive gozdne vire.

Sevindik, M., Akgul, H., Pehlivan, M., and Z. Selamoglu, 2017. "Determination of Therapeutic Potential of *Mentha longifolia* ssp. *longifolia*." *Fresenius Environmental Bulletin* 26(7): 4757–4763.

Shrewdnia: Can Goats Eat Daffodils? 2024. https://shrewdnia.com/blogs/goats/can-goats-eat-daffodils (February 1st, 2024).

Siedler, M., 1992. *Marktgemeinde Eberndorf einst und heute.* Eberndorf: Marktgemeinde Eberndorf.

Šifrer, M., 1962. "Geografski učinki neurja med Peco in zgornjo Pako" (Geographic Impacts of the Storm between Mount Peca and the Upper Paka River). *Geografski zbornik* 7: 143–179.

Šilc, U., Vrbničanin, S., Božić, D., Čarni, A., and Z. Dajić Stevanović, 2012. "Alien Plant Species and Factors of Invasiveness of Anthropogenic Vegetation in the Northwestern Balkans—A Phytosociological Approach." *Central European Journal of Biology* 7(4): 720–730.

Sitzia, T., and G. Trentanovi, 2011. "Maggengo Meadow Patches Enclosed by Forests in the Italian Alps: Evidence of Landscape Legacy on Plant Diversity." *Biodiversity Conservation* 20: 945–961.

Škrubej, K., 2012. "Courtroom Oaths and Patrimonial Court in 18th-Century Carniola: Vestiges of 'ius proprium' and Local Autonomy: The Case of the Estate of Veldes/Bled." *Lex localis: Journal of Local Self-Government* 10(3): 203–228.

Slatnar, J., 2016. "Navadna turška detelja (Onobrychis vici[sic!]folia Scop.)" (Sainfoin (*Onobrychis viciifolia* Scop.)). *Naše travinje: Revija Društva za gospodarjenje na travinju Slovenije* 10: 9–11.

Slavin, P., 2018. "The 1310s Event." In White, S., Pfister, C., and F. Mauelshagen (eds.), *The Palgrave Handbook of Climate History.* London: Palgrave Macmillan, 495–515.

Šlebinger, J., 1937. "Slovenski časniki in časopisi: bibliografski pregled od 1797–1936" (Slovenian Newspapers 1797–1936: A Bibliographic Overview). In Borko, B. (ed.), *Razstava slovenskega novinarstva v Ljubljani 1937.* Ljubljana: Jugoslovensko novinarsko udruženje, ljubljanska sekcija, 1–175.

Slovar slovenskega knjižnega jezika (Standard Slovenian Dictionary). https://fran.si/iskanje?FilteredDictionaryIds=130&View=1&Query=%2A (August 20th, 2016).

Slovenia Green. https://www.slovenia-green.si/ (November 1st, 2023).

Slovenska historična topografija, Paleonimi (Slovenian Historical Topography, Palaeonyms). https://topografija.zrc-sazu.si/paleonimi.php (August 15th, 2018).

Slovenski etnografski muzej, Zbirke, Digitalne zbirke, Zbirke fotografij: Individualne terenske raziskave kustosov SEM / 630:LJU;F-0001685/I (Slovenian Ethnographic Museum, Collections, Digital Collections, Photo Collections:

Individual Fieldwork by Curators of the Slovenian Ethnographic Museum / 630:LJU;F-0001685/I). https://www.etno-muzej.si/sl/digitalne-zbirke/indivivudualne-terenske-raziskave/630ljuf-0001685i (January 3rd, 2024).

Slovenski etnografski muzej, Zbirke, Digitalne zbirke, Zbirke fotografij: Individualne terenske raziskave kustosov SEM / 630:LJU;F-0001785/I (Slovenian Ethnographic Museum, Collections, Digital Collections, Photo Collections: Individual Fieldwork by Curators of the Slovenian Ethnographic Museum / 630:LJU;F-0001785/I). https://www.etno-muzej.si/sl/digitalne-zbirke/indivivudualne-terenske-raziskave/630ljuf-0001785i (January 3rd, 2024).

Slovenski etnografski muzej, Zbirke, Digitalne zbirke, Zbirke fotografij: Jernej Šušteršič / 166 F0006141 (Slovenian Ethnographic Museum, Collections, Digital Collections, Photo Collections: Fieldwork Photos: Jernej Šušteršič / 166 F0006141). https://www.etno-muzej.si/sl/digitalne-zbirke/jernej-sustersic/f0006141 (August 10th, 2022).

Slovenski etnografski muzej, Zbirke, Digitalne zbirke, Zbirke fotografij, Košnja (Slovenian Ethnographic Museum, Collections, Digital Collections, Photo Collections, Haymaking). https://www.etno-muzej.si/sl/digitalne-zbirke/kljucne-besede/kosnja (December 15th, 2023).

Slovenski etnografski muzej, Zbirke, Digitalne zbirke, Zbirke fotografij: Terenske fotografije, Teren 1, 1. 8.–15. 9. 1948: Šentjurij – Škocjan – Turjak / 49 R0000001/052b (Slovenian Ethnographic Museum, Collections, Digital Collections, Photo Collections: Fieldwork Photos, Fieldwork Campaign 1, August 1st – September 15th, 1948: Šentjurij – Škocjan – Turjak / 49 R0000001/052b). https://www.etno-muzej.si/sl/digitalne-zbirke/sentjurij-skocjan-turjak/r0000001052b (December 15th, 2023).

Slovenski etnografski muzej, Zbirke, Digitalne zbirke, Zbirke fotografij: Terenske fotografije, Teren 2, 1. 8.–3. 9. 1949: Šmarje-Sap, Polica / 27 R0000002/14 (Slovenian Ethnographic Museum, Collections, Digital Collections, Photo Collections: Fieldwork Photos, Fieldwork Campaign 2, August 1st – September 3rd, 1949: Šmarje-Sap, Polica / 27 R0000002/14). https://www.etno-muzej.si/sl/digitalne-zbirke/smarje-sap-polica/r000000214 (December 15th, 2023).

Slovenski etnografski muzej, Zbirke, Digitalne zbirke, Zbirke fotografij: Terenske fotografije, Teren 2, 1. 8.–3. 9. 1949: Šmarje-Sap, Polica / 220 F0000002/242 (Slovenian Ethnographic Museum, Collections, Digital Collections, Photo Collections: Fieldwork Photos, Fieldwork Campaign 2, August 1st – September 3rd, 1949: Šmarje-Sap, Polica / 220 F0000002/242). https://www.etno-muzej.si/sl/digitalne-zbirke/smarje-sap-polica/f0000002242 (December 28th, 2023).

Slovenski etnografski muzej, Zbirke, Digitalne zbirke, Zbirke fotografij: Terenske fotografije, Teren 4, 1. 7.–30. 7. 1950: Marezige / 86 R0000004/56 (Slovenian Ethnographic Museum, Collections, Digital Collections, Photo Collections: Fieldwork Photos, Fieldwork Campaign 4, July 1st – July 30th, 1950:

Marezige / 86 R0000004/56). https://www.etno-muzej.si/sl/digitalne-zbirke/marezige/r000000456 (December 15th, 2023).

Slovenski etnografski muzej, Zbirke, Digitalne zbirke, Zbirke fotografij: Terenske fotografije, Teren 7, 3. 7.–3. 8. 1951: Kobarid / 101 R0000007/120 (Slovenian Ethnographic Museum, Collections, Digital Collections, Photo Collections: Fieldwork Photos, Fieldwork Campaign 7, July 3rd – August 3rd, 1951: Kobarid / 101 R0000007/120. https://www.etno-muzej.si/sl/digitalne-zbirke/kobarid/r0000007120 (August 10th, 2022).

Slovenski etnografski muzej, Zbirke, Digitalne zbirke, Zbirke fotografij: Terenske fotografije, Teren 7, 3. 7.–3. 8. 1951: Kobarid / 176 R0000007/161 (Slovenian Ethnographic Museum, Collections, Digital Collections, Photo Collections: Fieldwork Photos, Fieldwork Campaign 7, July 3rd – August 3rd, 1951: Kobarid / 176 R0000007/161. https://www.etno-muzej.si/sl/digitalne-zbirke/kobarid/r0000007161 (August 10th, 2022).

Slovenski etnografski muzej, Zbirke, Digitalne zbirke, Zbirke fotografij: Terenske fotografije, Teren 8, 1. 7.–31. 7. 1952: Trenta – Bovec / 140 F0000008/138 (Slovenian Ethnographic Museum, Collections, Digital Collections, Photo Collections: Fieldwork Photos, Fieldwork Campaign 8, July 1st – July 31st, 1952: Trenta – Bovec / 140 F0000008/138. https://www.etno-muzej.si/sl/digitalne-zbirke/trenta-bovec/f0000008138 (August 10th, 2022).

Slovenski etnografski muzej, Zbirke, Digitalne zbirke, Zbirke fotografij: Terenske fotografije, Teren 11, 30. 7.–31. 8. 1954: Cerkljansko / 29 R0000011/91 (Slovenian Ethnographic Museum, Collections, Digital Collections, Photo Collections: Fieldwork Photos, Fieldwork Campaign 11, July 30th – August 31st, 1954: Cerkno Region / 29 R0000011/91). https://www.etno-muzej.si/sl/digitalne-zbirke/cerkljansko/r000001191 (August 10th, 2022).

Slovenski etnografski muzej, Zbirke, Digitalne zbirke, Zbirke fotografij: Terenske fotografije, Teren 11, 30. 7.–31. 8. 1954: Cerkljansko / 34 F0000011/053 (Slovenian Ethnographic Museum, Collections, Digital Collections, Photo Collections: Fieldwork Photos, Fieldwork Campaign 11, July 30th – August 31st, 1954: Cerkno Region / 34 F0000011/053). https://www.etno-muzej.si/sl/digitalne-zbirke/cerkljansko/f0000011053 (August 10th, 2022).

Slovenski etnografski muzej, Zbirke, Digitalne zbirke, Zbirke fotografij: Terenske fotografije, Teren 11, 30. 7.–31. 8. 1954: Cerkljansko / 100 R0000011/31 (Slovenian Ethnographic Museum, Collections, Digital Collections, Photo Collections: Fieldwork Photos, Fieldwork Campaign 11, July 30th – August 31st, 1954: Cerkno Region / 100 R0000011/31). https://www.etno-muzej.si/sl/digitalne-zbirke/cerkljansko/r000001131 (August 10th, 2022).

Slovenski etnografski muzej, Zbirke, Digitalne zbirke, Zbirke fotografij: Terenske fotografije, Teren 11, 30. 7.–31. 8. 1954: Cerkljansko / 235 R0000011/100 (Slovenian Ethnographic Museum, Collections, Digital Collections, Photo Collections: Fieldwork Photos, Fieldwork Campaign 11, July 30th – August 31st, 1954: Cerkno Region / 235 R0000011/100). https://www.etno-muzej.si/sl/digitalne-zbirke/cerkljansko/r0000011100 (August 10th, 2022).

Slovenski etnografski muzej, Zbirke, Digitalne zbirke, Zbirke fotografij: Terenske fotografije, Teren 12, 1. 8.–31. 8. 1955: Brkini / 178 R0000012/7 (Slovenian Ethnographic Museum, Collections, Digital Collections, Photo Collections: Fieldwork Photos, Fieldwork Campaign 12, August 1st – August 31st, 1955: Brkini / 178 R0000012/7). https://www.etno-muzej.si/sl/digitalne-zbirke/brkini/r00000127 (December 15th, 2023).

Slovenski etnografski muzej, Zbirke, Digitalne zbirke, Zbirke fotografij: Terenske fotografije, Teren 13, 1. 8.–31. 8. 1956: Kostanjevica / 46 F0000013/266 (Slovenian Ethnographic Museum, Collections, Digital Collections, Photo Collections: Fieldwork Photos, Fieldwork Campaign 13, August 1st – August 31st, 1956: Kostanjevica / 46 F0000013/266). https://www.etno-muzej.si/sl/digitalne-zbirke/kostanjevica/f0000013266 (December 28th, 2023).

Slovenski etnografski muzej, Zbirke, Digitalne zbirke, Zbirke fotografij: Terenske fotografije, Teren 14, 1. 8.–28. 8. 1957: Žužemberk / 16 R0000014/18 (Slovenian Ethnographic Museum, Collections, Digital Collections, Photo Collections: Fieldwork Photos, Fieldwork Campaign 14, August 1st – August 28th, 1957: Žužemberk / 16 R0000014/18). https://www.etno-muzej.si/sl/digitalne-zbirke/zuzemberk/r000001418 (December 15th, 2023).

Slovenski etnografski muzej, Zbirke, Digitalne zbirke, Zbirke fotografij: Terenske fotografije, Teren 15, 1. 8.–28. 8. 1958: Vipava / 98 R0000015/13 (Slovenian Ethnographic Museum, Collections, Digital Collections, Photo Collections: Fieldwork Photos, Fieldwork Campaign 15, August 1st – August 28th, 1958: Vipava / 98 R0000015/13). https://www.etno-muzej.si/sl/digitalne-zbirke/vipava/r000001513 (December 15th, 2023).

Slovenski etnografski muzej, Zbirke, Digitalne zbirke, Zbirke fotografij: Terenske fotografije, Teren 16, 1. 8.–28. 8. 1959: Črni Vrh, Vojsko / 151 F0000016/135 (Slovenian Ethnographic Museum, Collections, Digital Collections, Photo Collections: Fieldwork Photos, Fieldwork Campaign 16, August 1st – August 28th, 1959: Črni Vrh, Vojsko / 151 F0000016/135). https://www.etno-muzej.si/sl/digitalne-zbirke/crni-vrh-vojsko/f0000016135 (December 15th, 2023).

Slovenski etnografski muzej, Zbirke, Digitalne zbirke, Zbirke fotografij: Terenske fotografije, Teren 18, 1. 8.–30. 8. 1961: Šentrupert / 3 F0000018/136 (Slovenian Ethnographic Museum, Collections, Digital Collections, Photo Collections: Fieldwork Photos, Fieldwork Campaign 18, August 1st – August 30th, 1961: Šentrupert / 3 F0000018/136). https://www.etno-muzej.si/sl/digitalne-zbirke/sentrupert/f0000018136 (December 15th, 2023).

Slovenski etnografski muzej, Zbirke, Digitalne zbirke, Zbirke fotografij: Terenske fotografije, Teren 20, 1. 7.–14. 7. 1963: Vitanje / 18 F0000020/918 (Slovenian Ethnographic Museum, Collections, Digital Collections, Photo Collections: Fieldwork Photos, Fieldwork Campaign 20, July 1st – July 14th, 1963: Vitanje / 18 F0000020/918). https://www.etno-muzej.si/sl/digitalne-zbirke/vitanje/f0000020918 (January 3rd, 2024).

Slovenski etnografski muzej, Zbirke, Digitalne zbirke, Zbirke fotografij: Terenske fotografije, Teren 20, 1. 7.–14. 7. 1963: Vitanje / 63 F0000020/140 (Slovenian

Ethnographic Museum, Collections, Digital Collections, Photo Collections: Fieldwork Photos, Fieldwork Campaign 20, July 1st – July 14th, 1963: Vitanje / 63 F0000020/140). https://www.etno-muzej.si/sl/digitalne-zbirke/vitanje/f0000020140 (January 3rd, 2024).

Slovenski etnografski muzej, Zbirke, Digitalne zbirke, Zbirke fotografij: Terenske fotografije, Teren 21, 1964: Škocjan / 66 F0000021/253 (Slovenian Ethnographic Museum, Collections, Digital Collections, Photo Collections: Fieldwork Photos, Fieldwork Campaign 21, 1964: Škocjan / 66 F0000021/253). https://www.etno-muzej.si/sl/digitalne-zbirke/drasici/f0000022264 (December 15th, 2023).

Slovenski etnografski muzej, Zbirke, Digitalne zbirke, Zbirke fotografij: Terenske fotografije, Teren 21, 1964: Škocjan / 138 F0000021/413 (Slovenian Ethnographic Museum, Collections, Digital Collections, Photo Collections: Fieldwork Photos, Fieldwork Campaign 21, 1964: Škocjan / 138 F0000021/413). https://www.etno-muzej.si/sl/digitalne-zbirke/skocjan/f0000021413 (December 15th, 2023).

Slovenski etnografski muzej, Zbirke, Digitalne zbirke, Zbirke fotografij: Terenske fotografije, Teren 22, 1965: Drašiči / 57 F0000022/264 (Slovenian Ethnographic Museum, Collections, Digital Collections, Photo Collections: Fieldwork Photos, Fieldwork Campaign 22, 1965: Drašiči / 57 F0000022/264). https://www.etno-muzej.si/sl/digitalne-zbirke/drasici/f0000022264 (December 15th, 2023).

Slovenski zgodovinski atlas (Atlas of the History of Slovenian Territory). 2011. Ljubljana: Nova revija.

Smerdel, I., 1994. *Oselniki* (Whetstone Holders). Ljubljana: Slovenski etnografski muzej.

Smernice dobre prakse v primarni pridelavi hrane in krme (Guidelines for Good Practice in Primary Production of Food and Feed). 2007. [Ljubljana]: Kmetijsko-gozdarska zbornica Slovenije.

Šmid Hribar, M., 2016. *Varovanje in trajnostni razvoj kulturne pokrajine na primeru Ljubljanskega barja* (Protection and Sustainable Development of a Cultural Landscape: The Case of the Ljubljana Marsh). Ljubljana: Geografski inštitut Antona Melika ZRC SAZU.

Šmid Hribar, M., and A. Lisec, 2011. "Protecting Trees Through an Inventory and Typology: Heritage Trees in the Karavanke Mountains, Slovenia." *Acta geographica Slovenica* 51(1): 169–180.

Smole, M., 1982. *Graščine na nekdanjem Kranjskem* (Seigneuries in the Historical Province of Carniola). Ljubljana: Državna založba Slovenije.

Smolej, F., 2008. "Pri Žlindru" (The Pri Žlindru Farm). In Kokošinek, N., and Z. Torkar Tahir (eds.), *Kako so v vaseh pod Golico včasih živeli?* Jesenice: Ljudska univerza Jesenice, 15–16.

Smrekar, A., Šmid Hribar, M., Tiran, J., and B. Erhartič, 2014. *Interpretacija okolja na primeru Ljubljanskega barja* (Environmental Interpretation: The Case of the Ljubljana Marsh). Ljubljana: ZRC SAZU, Založba ZRC.

Snoj, M., 2009. *Etimološki slovar slovenskih zemljepisnih imen* (Etymological Dictionary of Slovenian Geographical Names). Ljubljana: Modrijan and Založba ZRC.

Sokolov, P.D., 1993. *Rastitel'nye resursy SSSR: Tsvetkovye rastenija: Ih himičeskij sostav, ispol'zovanie: Semejstvo* Asteraceae (Compositae) (Plant Resources of the USSR: Flowering Plants: Their Chemical Composition and Use: Family *Asteraceae* (*Compositae*)). Saint Petersburg: Nauka.

Šorn, J., 1984. *Začetki industrije na Slovenskem* (The Beginnings of Industry in Slovenian Territory). Maribor: Obzorja.

Sosič, B., 2000. "Gradivo o ljudski medicini v zapisih terenskih ekip Slovenskega etnografskega muzeja" (Material on Folk Medicine Collected by the Field Teams of the Slovenian Ethnographic Museum). *Etnolog* 10: 213–254.

Sõukand, R., and A. Pieroni, 2016. "The Importance of a Border: Medical, Veterinary, and Wild Food Ethnobotany of the Hutsuls Living on the Romanian and Ukrainian Sides of Bukovina." *Journal of Ethnopharmacology* 185: 17–40.

Souvent, P., and S. Pirc, 2001. "Pollution Caused by Metallic Fragments Introduced into Soils because of World War I Activities." *Environmental Geology* 40(3): 317–323.

Special-Orts-Repertorium von Krain. 1884. Vienna: K.k. Statistische Central-Kommission.

Specifikacija: Izbrana kakovost: Mleko in mlečni izdelki (Specification: Selected Quality: Milk and Dairy Products). 2019. https://www.gov.si/assets/ministrstva/MKGP/PODROCJA/HRANA/SHEME-KAKOVOSTI/Specifikacije-izbrana-kakovost/IK_MLEKO_SPEC_notif_potrj_2_10_19.pdf (November 2nd, 2023).

Spehr, C., 2010. "Sonntag." In Jaeger, F., et al. (eds.), *Enzyklopädie der Neuzeit*. Vol. 12, *Silber–Subsidien*. Stuttgart and Weimar: J.B. Metzler, 209–211.

SPEICK Natural Cosmetics: An Exuberant Success Story. https://www.speick.de/en/company/history.html (December 28th, 2020).

Spode, H., 2005. "Branntwein." In Jaeger, F., et al. (eds.), *Enzyklopädie der Neuzeit*. Vol. 2, *Beobachtung–Dürre*. Stuttgart and Weimar: J.B. Metzler, 395–397.

"Sreča je izbrala Krkine nagrajence" (Luck Has Chosen the Krka Award Winners). 1969. *Dolenjski list* 20(48) (November 27th, 1969): 20.

Staab, F., 1991. "Kontinuität." In Bautier, R.-H., et al. (eds.), *Lexikon des Mittelalters*. Vol. 5, *Hiera-Mittel bis Lukanien*. Munich and Zurich: Artemis, 1417–1420.

Stadler-Planzer, H. Stadler, P., and R. Aschwanden, 2015. *Geschichte des Landes Uri*. Part 2b, *Von der helvetischen Umwälzung in die Gegenwart*. Schattdorf: Uranos.

Stadler-Planzer, H., Stadler, P., and B. Degler-Spengler, 2015. *Geschichte des Landes Uri*. Part 2a, *Frühe Neuzeit*. Schattdorf: Uranos.

Stampfli, A., 1995. "Species Composition and Standing Crop Variation in an Unfertilized Meadow and Its Relationship to Climatic Variability During Six Years." *Folia Geobotanica & Phytotaxonomica* 30(2): 117–130.

Steffen, W., Richardson, K., Rockström, J., Cornell, S.E., Fetzer, I., et al., 2015. "Planetary Boundaries: Guiding Human Development on a Changing Planet." *Science* 347(6223): article no. 1259855 and supplementary material.

Štepec, M., 2019. "The Heritage of the Soča Front and Collective Memory." In Košir, U., Črešnar, M., and D. Mlekuž (eds.), *Rediscovering the Great War: Archaeology and Enduring Legacies on the Soča and Eastern Fronts*. London and New York: Routledge, 214–233.

Stergar, R., 2015. "Hrana na bojiščih prve svetovne vojne: Izkušnje slovenskih vojakov" (Food on World War I Battlefields: Experiences of Slovenian Soldiers). *Prispevki za novejšo zgodovino* 55(2): 22–53.

Stergar, R., 2019. "L'expérience des soldats austro-hongrois sur le front austro-italien: Le problème du ravitaillement en vivres." In Fontana, G.L., and M. Mondini (eds.), *Soldati e quotidianità della guerra*. Ospedaletto: Pacini, 13–30.

Stickstoff – zuviel des Guten? Überlastung des Stickstoffkreislaufs zum Nutzen von Umwelt und Mensch wirksam reduzieren. No date. Dessau-Roßlau: Umweltbundesamt.

Štih, P., 1998. "Alpine Kolonisation und Migrationen im Mittelalter am Beispiel Sloweniens." *Histoire des Alpes* 3: 133–144.

Štih, P., 2010. "Wiped out by the Slavic Settlement? The Issue of Continuity Between Antiquity and the Early Middle Ages in the Slovene Area." In Curta, F. (ed.), *The Middle Ages between the Eastern Alps and the Northern Adriatic: Select Papers on Slovene Historiography and Medieval History*. Leiden and Boston: Brill, 87–99.

Štih, P., 2014 (= Štih, 2014a). "Die Integration der Karantanen und anderer Alpenslawen in das fränkisch-ottonische Reich: Einige Beobachtungen." *Carinthia I* 204(1): 43–59.

Štih, P., 2014 (= Štih, 2014b). "Mensch und Wald in den Ostalpen (bis zur Großen Kolonisation)." In Štih, P., and Ž. Zwitter (eds.), *Man, Nature and Environment between the Northern Adriatic and the Eastern Alps in Premodern Times*. Ljubljana: Ljubljana University Press, Faculty of Arts, and Historical Association of Slovenia, 36–51.

Štih, P., 2016. "Od 6. do konca 15. stoletja" (From the Sixth to the End of the Fifteenth Century). In Vidic, M. (ed.), *Slovenska zgodovina: Od prazgodovinskih kultur do začetka 21. stoletja*. Vol. 1. Ljubljana: Modrijan, 34–221.

Štih, P., 2018. "Als die Kirche slawisch zu sprechen begann: Zu den Hintergründen der Christianisierung in Karantanien und Pannonien." In Pohl, W., Diesenberger, M., and B. Zeller (eds.), *Neue Wege der Frühmittelalterforschung – Bilanz und Perspektiven*. Vienna: Österreichische Akademie der Wissenschaften, 339–356.

Štih, P., Simoniti, V., and P. Vodopivec, 2008. *A Slovene History: Society–Politics–Culture*. Ljubljana: Inštitut za novejšo zgodovino. http://www.sistory.si/11686/2250 (July 8th, 2018).

Strauss, S.Y., and A.A. Agrawal, 1999. "The Ecology and Evolution of Plant Tolerance to Herbivory." *Trends in Ecology & Evolution* 14(5): 179–185.

Strgar, V., 1981. "Genus *Reynoutria* v adventivni flori Slovenije" (The Genus *Reynoutria* Houtt. in the Adventive Flora of Slovenia). *Biološki vestnik* 29(2): 121–136.

Strgar, V., 1982. "Genus *Reynoutria* v adventivni flori Slovenije, II" (The Genus *Reynoutria* Houtt. in the Adventive Flora of Slovenia, II). *Biološki vestnik* 30(2): 151–154.

Štrumbl, Ž., 2017. *Arhivski popis: AS_2164: Zbirka zadružništvo na Slovenskem: 1922–1962* (Archival Register: AS_2164: Archival Collection Cooperatives in Slovenian Territory: 1922–1962). Ljubljana: [Arhiv Republike Slovenije].

Stuber, M., and L. Lienhard, 2007. "Nützliche Pflanzen: Systematische Verzeichnisse von Wild- und Kulturpflanzen im Umfeld der Oekonomischen Gesellschaft Bern 1762–1782." In Holenstein, A., Stuber, M., and G. Gerber-Visser (eds.), *Nützliche Wissenschaft und Ökonomie im Ancien Régime: Akteure, Themen, Kommunikationsformen*. Heidelberg: Palatina, 65–106.

Štucin, A., 2011. *Kraji in ljudje pod Poreznom: Sprehod skozi cerkljansko preteklost* (Settlements and People below Mt. Porezen: A Walk through the Past of the Region of Cerkno). Idrija: Mestni muzej Idrija – muzej za Idrijsko in Cerkljansko.

Studen, A., 2011. "Pričakovani in dejanski krepostni svet ljubezni, spolnosti in morale služkinj v meščanskih gospodinjstvih" (The Expected and Actual Virtuous World of Love, Sex, and Morality of Maidservants in the Bourgeois Households). *Prispevki za novejšo zgodovino* 51(2): 29–44.

Studen, A., 2024. "Scarcity Versus Abundance: A Case Study of the Hardship and Imminent Starvation in the District of Senožeče in 1864 and 1865." In Fuerst-Bjeliš, B., et al. (eds.), *Environmental Histories of the Dinaric Karst*. Cham: Springer Nature, 265–293.

Studen, A., and F. Kresal, 2005. "Prva svetovna vojna: Življenjske razmere" (The First World War: Living Conditions). In Fischer, J., et al. (eds.), *Slovenska novejša zgodovina: Od programa Zedinjena Slovenija do mednarodnega priznanja Republike Slovenije 1848–1992*. Vol. 1. Ljubljana: Mladinska knjiga, 150–153.

Štuhec, M., 1985. "Prebivalstvo Ljubljane v drugi polovici 17. stoletja na podlagi matičnih knjig" (Demography of Ljubljana in the Second Half of the Seventeenth Century Based on Church Registers). *Kronika: Časopis za slovensko krajevno zgodovino* 33(2–3): 124–136.

Štular, B., 2011. "The Use of Lidar-Derived Relief Models in Archaeological Topography: The Kobarid Region (Slovenia) Case Study." *Arheološki vestnik* 62: 393–432.

Sucholas, J., Molnár, Z., Łuczaj, Ł., and P. Poschlod, 2022. "Local Traditional Ecological Knowledge about Hay Management Practices in Wetlands of the Biebrza Valley, Poland." *Journal of Ethnobiology and Ethnomedicine* 18, article no. 9: separate pagination 1–42.

Suhi kraški travniki in pašniki: Priročnik za kmete (Dry Calcareous Meadows and Pastures: A Handbook for Farmers). 2023. [city unknown]: Zavod Republike Slovenije za varstvo narave; Republika Slovenija, Ministrstvo za kmetijstvo, gozdarstvo in prehrano. https://www.kgzs.si/uploads/eiv23/po/navodila/prirocnik_suhi_ktp_v1_2023_koncna.pdf (December 31st, 2024).

Sulzer, B., 2005. "Kühe sind nicht geschmacklos." In *Neues Handbuch Alp: Handfestes für Alpleute, Erstaunliches für Zaungäste*. Mollis: Zalpverlag, 62–67.

Šumrada, T., and E. Erjavec, 2023. "Will Farmers Cooperate to Conserve Biodiversity? The Use of Collective Bonus in the High Nature Value Farmland in Slovenia." *Acta geographica slovenica* 63(3): 69–83.

Šumrada, T., Erjavec, E., Šilc, U., and J. Žgajnar, 2024. "Socio-Economic Viability of the High Nature Value Farmland Under the CAP 2023–2027: The Case of a Sub-Mediterranean Region in Slovenia." *Agriculture* 14(10), article no. 1699: separate pagination, 1–32.

Šumrada, T., Japelj, A., Verbič, M., and E. Erjavec, 2022. "Farmers' Preferences for Result-Based Schemes for Grassland Conservation in Slovenia." *Journal for Nature Conservation* 66: article no. 126143, separate pagination 1–12.

Šumrada, T., Kmecl, P., and E. Erjavec, 2021. "Do the EU's Common Agricultural Policy Funds Negatively Affect the Diversity of Farmland Birds? Evidence from Slovenia." *Agriculture, Ecosystems and Environment* 306: article no. 107200, separate pagination 1–14.

Šumrada, T., and T. Trilar, 2019. "Jerebica" (Grey Partridge). In Mihelič, T., et al. (eds.), *Atlas ptic Slovenije: Popis gnezdilk 2002–2017*. Ljubljana: DOPPS, 66–67.

Šumrada, T., Vreš, B., Čelik, T., Šilc, U., Rac, I., et al., 2021. "Are Result-Based Schemes a Superior Approach to the Conservation of High Nature Value Grasslands? Evidence from Slovenia." *Land Use Policy* 111: article no. 105749, separate pagination 1–14.

Suske, W., Gantner, B., and S. Gattermaier, 2012. "Austria." In Oppermann, R., Beaufoy, G., and G. Jones (eds.), *High Nature Value Farming in Europe: 35*

European Countries and Perspectives. Ubstadt-Weiher, Heidelberg, and Basel: Regionalkultur, 118–127.

Svadlenak-Gomez, K., 2016. "The Global Framework for Nature Protection." In Plassmann, G., et al. (eds.), *Alpine Nature 2030: Creating [Ecological] Connectivity for Generations to Come*. Berlin: Federal Ministry for the Environment, Nature Conservation, Building and Nuclear Safety, 13–16.

Svadlenak-Gomez, K. and M. Badura, 2016. "Networking for Nature – The Challenges of Bringing the 'Right' People Together." In Plassmann, G., et al. (eds.), *Alpine Nature 2030: Creating [Ecological] Connectivity for Generations to Come*. Berlin: Federal Ministry for the Environment, Nature Conservation, Building and Nuclear Safety, 88–92.

Svoljšak, P., 1999. "Poizkus ocene vojaških in civilnih izgub (žrtev) med 1. svetovno vojno" (Versuch einer Schätzung der militärischen und zivilen Verluste (Opfer) während des Ersten Weltkriegs). In Granda, S., and B. Šatej (eds.), *Množične smrti na Slovenskem: Zbornik referatov*. Ljubljana: Zveza zgodovinskih društev Slovenije, 225–240.

Svoljšak, P., 2019. "Vračanje slovenskih beguncev v porušeno Posočje" (The Return of Slovenian Refugees in the Destroyed Soča (Ital. *Isonzo*) Region). In Dabo, M., and M. Radošević (eds.), *U sjeni Velikoga rata: Odraz ratnih zbivanja na život istarskoga civilnog stanovništva: Zbornik radova s međunarodnog znanstvenog skupa održanoga u Puli 13. – 15. listopada 2016*. Pula: Istarsko povijesno društvo and Povijesni i pomorski muzej Istre, 347–362.

Svoljšak, P., and G. Antoličič, 2018. *Leta strahote: Slovenci in prva svetovna vojna* (The Years of Horror: Slovenes and the First World War). Ljubljana: Cankarjeva založba.

Synthesis – Livestock and the Sustainable Development Goals, 2016. [city unknown]: Food and Agriculture Organization of the United Nations, Global Agenda for Sustainable Livestock. http://www.livestockdialogue.org/fileadmin/templates/res_livestock/docs/2016/Panama/FAO-AGAL_synthesis_Panama_Livestock_and_SDGs.pdf (September 25th, 2022).

Szabó, P., 2010. "Why History Matters in Ecology: An Interdisciplinary Perspective." *Environmental Conservation* 37: 380–387.

Szabó, P., 2013. "Rethinking Pannage: Historical Interactions between Oak and Swine." In Rotherham, I.D. (ed.), *Trees, Forested Landscapes, and Grazing Animals: A European Perspective on Woodlands and Grazed Treescapes*. London and New York: Routledge, 51–61.

Szabó, P., 2023. "'We Have Just Launched a Completely New and Unique Venture': Successes and Failures to Utilize Marginal Mountain Grasslands in Northern Moravia since the Middle Ages." In *12th ESEH Conference: Mountains and Plains: University of Bern: 22–26 August 2023*. Bern: [European Society for Environmental History], 96. https://www.eseh2023.unibe.ch/unibe/portal/fak_historisch/dga/micro_esh/content/e1228908/e1229995/e1405648/eseh2023_bern_programme_abstracts_3.2.pdf (July 5th, 2023).

Tacer, M., no date. "Brez malega blatnega polža ni metljavosti goveda in drobnice" (No Cattle and Sheep Infestation with Common Liver Fluke without the Snail *Lymnea stagnalis*). Tenetiše: Veterinarska praksa Tenetiše. https://www.vpt.si/priporocamo/ (August 3rd, 2021).

Tälle, M., Bergman, K.O., Paltto, H., Pihlgren, A., Svensson, R., et al., 2014. "Mowing for Biodiversity: Grass Trimmer and Knife Mower Perform Equally Well." *Biodiversity and Conservation* 23(12): 3073–3089.

Tappeiner, U., 2003. "Ein vielschichtiges Bild der Alpen." *GAIA: Ecological Perspectives for Science and Society* 12(4): 306–307.

Tappeiner, U., Borsdorf, A., and M. Bahn, 2013. "Long-Term Socio-Ecological Research in Mountain Regions: Perspectives from the Tyrolean Alps." In Singh, S.J., et al. (eds.), *Long-Term Socio-Ecological Research: Studies in Society-Nature Interactions across Time and Space*. Dordrecht et al.: Springer, 505–525.

Teibenbacher, P., 1989. "Steiermark und die europäische Expansion von 1500 bis 1800." In Schöpfer, G. (ed.), *Menschen & Münzen & Märkte: Steirische Landesausstellung 1989 Judenburg*. Fohnsdorf: K. Podmenik, 97–102.

Teibenbacher, P., Kramer, D., and W. Göderle, 2012. *An Inventory of Austrian Census Materials, 1857–1910: Final Report*. Rostock: Max Planck Institute for Demographic Research. https://censusmosaic.demog.berkeley.edu/sites/default/files/downloads/publications/mosaicWP/MOSAIC-WP-2012-007.pdf (July 20th, 2018).

Ter Borg, S.J., 1972. *Variability of* Rhinanthus serotinus *(Schönh.) Oborny in Relation to the Environment:* Doctoral Dissertation. Groningen: University of Groningen.

Testen, P., and T. Koren, 2015. "Učilnica na prostem – primer *Poti miru*" (Openair Classroom: The Example of the *Walk of Peace*). *Prispevki za novejšo zgodovino* 55(2): 183–198.

Teuscher, S., 2014. "Zur Mediengeschichte des 'mündlichen Rechts' im späteren Mittelalter." *Zeitschrift der Savigny-Stiftung für Rechtsgeschichte: Germanistische Abteilung* 131: 69–88.

The British Museum: Jean Poyer. https://www.britishmuseum.org/collection/term/BIOG42551 (September 18th, 2021).

The Challenge of Feeding the World Sustainably: Summary of the US–UK Scientific Forum on Sustainable Agriculture, 2021. Washington, DC: The National Academies Press.

The Future of Nature and Business: New Nature Economy Report II. 2020. Cologny: World Economic Forum. https://www3.weforum.org/docs/WEF_The_Future_Of_Nature_And_Business_2020.pdf (March 15th, 2022).

The Isonzo Front: Register of the Killed Soldiers. http://www.potmiru.si/eng/evidenca-padlih (April 30th, 2022).

The Morgan Library and Museum: Jean Poyer: Artist to the Court of Renaissance France Opens January 25. 2000. https://www.themorgan.org/press/2000/jean-poyer-exhibit (September 18, 2021).

The Walk of Peace, Bohinj: Military Supply Base for the Battlefield in the Krn Mountain Range. 2015. http://www.potmiru.si/pdf/Bohinj.pdf (May 1st, 2022).

The World Flora Online: Reynoutria x bohemica *Chrtek & Chrtková.* http://www.worldfloraonline.org/taxon/wfo-0000404407 (August 11th, 2022).

Theimer, M., 2000. *Literaturrecherche zur Winterhärte des Deutschen Weidelgrases: Deutsch- und englischsprachige Literatur: Diplomarbeit.* Freising: Fachhochschule Weihenstephan, Fachbereich Land- und Ernährungswirtschaft.

"They Used to Eat Corncrakes: Times Change, and Carnivores Such as Human Beings Go through Phases." 1997. *The Irish Times* (September 27th, 1997). https://www.irishtimes.com/opinion/they-used-to-eat-corncrakes-1.110351 (December 28th, 2022).

Thomas, K., 1987. *Man and the Natural World: Changing Attitudes in England 1500–1800.* Harmondsworth, etc.: Penguin Books, 1987.

Tiefenkarte des Klopeiner See. https://kis.ktn.gv.at/DE/repos/files/kis/import/Informationen/Tiefenkarten/Karten/klopeiner.pdf?exp=515918&fps=31589c123c7ad57cd9cad3fb1cf2acd5b7a1000c (August 18th, 2020).

Tinner, W., and J.-P. Theurillat, 2003. "Uppermost Limit, Extent, and Fluctuations of the Timberline and Treeline Ecocline in the Swiss Central Alps during the Past 11,500 Years." *Arctic, Antarctic and Alpine Research: An Interdisciplinary Journal* 35(2): 158–169.

tiris: Tiroler Rauminformationssystem, Hintergrund: Basiskarte Beschriftung, and Basisthemen: Kataster: https://maps.tirol.gv.at/synserver?user=guest&project=tmap_master (November 24th, 2021).

Titovšek, J., 1981. "Rdeče gozdne mravlje v Sloveniji" (Red Forest Ants in Slovenia). *Zbornik gozdarstva in lesarstva* 43: 77–107.

Tome, D., 2019. "Repaljščica" (Whinchat). In Mihelič, T., et al. (eds.), *Atlas ptic Slovenije: Popis gnezdilk 2002–2017.* Ljubljana: DOPPS, 426–427.

Torkar, G., and Ž. Zwitter, 2015. "Historical Impacts of Mercury Mining and Stocking of Non-Native Fish on Ichthyofauna in the Idrijca River Basin, Slovenia." *Aquatic Sciences* 77(3): 381–393.

Torkar Tahir, Z., 2012. "Življenje v Planini pod Golico nekdaj" (Life in Planina pod Golico in the Past). In Šubic, T. (ed.), *V rovtih nad Jesenicami: Narava in človek ustvarjata biotsko pestrost.* Ljubljana: Zavod Republike Slovenije za varstvo narave, 127–[143].

Toškan, B., 2013. "Domače govedo v romaniziranem vzhodnoalpskem prostoru: Arheozoološki pogled" (Domestic Cattle in the Romanized Southeast of the Alps: An Archaeozoological View). *Keria: Studia Latina et Graeca* 15(1): 35–71.

Toškan, B., 2016 (= Toškan, 2016a). "Govedoreja v jugovzhodnih Alpah med vzponom in padcem rimske države" (Cattle Husbandry in the Southeastern Alps Between the Rise and Fall of the Roman Empire). In Horvat, J. (ed.), *Okoljska arheologija in paleoekologija: Palinologija, arheobotanika, arheozoologija*. Ljubljana: ZRC SAZU, 239–244.

Toškan, B., 2016 (= Toškan, 2016b). "Mali sesalci kot orodje za prepoznavanje paleookoljskih sprememb" (Small Mammals as a Tool to Recognize Paleoenvironmental Changes). In Horvat, J. (ed.), *Okoljska arheologija in paleoekologija: Palinologija, arheobotanika, arheozoologija*. Ljubljana: ZRC SAZU, 220–225.

Toškan, B., 2016 (= Toškan, 2016c). "Tell Me What You Eat and I Will Tell You What You Are: Animal Remains from the Late-Medieval and Early Post-Medieval Grinfels Manor." In Djura Jelenko, S., et al. (eds.), *Grinfels Manor: Temporary Exhibition Catalogue*. Slovenj Gradec: Koroški pokrajinski muzej, 180–194, 201–202.

Toškan, B., 2017. "Živalski ostanki iz poznosrednjeveškega in zgodnjenovoveškega Slovenj Gradca" (Animal Remains from Late Medieval and Early Modern Slovenj Gradec [Germ. *Windischgraz*]). In Djura Jelenko, S., Brišnik, D., and M. Kajzer Cafnik (eds.), *Točka v času: Mednarodni arheološki simpozij: Arheološko najdišče Slovenj Gradec – glasbena šola 2009/2010*. Slovenj Gradec: Koroški pokrajinski muzej, 83–91.

Towards Sustainable Livestock. 2014. [city unknown]: Global Agenda for Sustainable Livestock. http://www.livestockdialogue.org/fileadmin/templates/res_livestock/docs/2014_Colombia/2014_Towards_Sustainable_Livestock-dec.pdf (September 25th, 2022).

Traditional Hay Milk Farming in the Austrian Alpine Arc. [2023]. [city unknown]: Food and Agriculture Organization of the United Nations. https://openknowledge.fao.org/server/api/core/bitstreams/3eaaf90d-48fe-4d9a-9571-ae18bd6cd14a/content (December 25th, 2024).

Transforming Food and Agriculture to Achieve the SDGs: 20 Interconnected Actions to Guide Decision-Makers. 2018. Rome: Food and Agriculture Organization of the United Nations. https://www.fao.org/3/I9900EN/i9900en.pdf (September 25th, 2022).

Tremel, F., 1947. "Der Handel der Stadt Judenburg im 16. Jahrhundert." *Zeitschrift des Historischen Vereines für Steriermark* 38: 95–164.

Trettler, H., 2024. "Hay Milk: Traditional Milk Production from Permanent Grassland in the Alpine Region." Presentation at the online workshop of the Working Group Mountain Agriculture and Mountain Forestry of the Alpine Convention: *Peatlands and Carbon Farming, Pastoralism, Forestry and Multifunctional Activities in the Mountains for the Sustainability of the Alpine Environment*. June 21st, 2024.

Triglavski narodni park: Kmetijstvo (Triglav National Park: Farming). https://www.tnp.si/sl/park/tradicionalne-dejavnosti/kmetijstvo/ (December 31st, 2024).

Triglavski narodni park: Razpis za kmetijstvo in podeželje na območju TNP (Triglav National Park: Call Concerning Agriculture and Rural Areas in TNP). https://www.tnp.si/sl/javni-zavod/medijsko-sredisce/novice/novica/145/razpis-za-kmetijstvo-in-podezelje-na-obmocju-tnp-2024.html (December 31st, 2024).

Trisorio, A., Borlizzi, A., and A. Povellato, 2012. "Italy." In Oppermann, R., Beaufoy, G., and G. Jones (eds.), *High Nature Value Farming in Europe: 35 European Countries and Perspectives*. Ubstadt-Weiher, Heidelberg, and Basel: Regionalkultur, 262–273.

Trombley, F.R., 2006. "Christianity: D. Expansion." In: Salazar, C.F. (ed.), *Brill's New Pauly*. https://referenceworks.brillonline.com/entries/brill-s-new-pauly/christianity-e233490?s.num=1&s.f.s2_parent=s.f.book.brill-s-new-pauly&s.q=christianity (February 11th, 2023).

Troßbach, W., 2006. "Farbstoffe: 2. Färbepflanzen." In Jaeger, F., et al. (eds.), *Enzyklopädie der Neuzeit*. Vol. 3, *Dynastie–Freundschaftslinien*. Stuttgart and Weimar: J.B. Metzler, 823–825.

Troßbach, W., 2008. "Landwirtschaft." In Jaeger, F., et al. (eds.), *Enzyklopädie der Neuzeit*. Vol. 7, *Konzert–Männlichkeit*. Stuttgart and Weimar: J.B. Metzler, 580–605.

Trtikova, M., Edwards, P.J., and S. Güsewell, 2010. "No Adaptation to Altitude in the Invasive Plant *Erigeron annuus* in the Swiss Alps." *Ecography* 33, 556–564.

Tscharntke, T., Grass, I., Wanger, T.C., Westphal, C., and P. Batáry, 2021. "Beyond Organic Farming—Harnessing Biodiversity-Friendly Landscapes." *Trends in Ecology & Evolution* 36(10): 919–930.

Tucker, R., 2015. "War and the Environment." In McNeill, J.R., and E.S. Mauldin (eds.), *A Companion to Global Environmental History*. 2nd ed. Malden MA, Oxford, UK, and Chichester, UK: Blackwell Publishing, 319–339.

Tuma, H., 1935. "Paberki iz Bohinja" (*Miscellanea* from Bohinj). *Planinski vestnik* 35(8): 250–251.

Uekötter, F., 2018. "Memories in Mud: The Environmental Legacy of the Great War." In Tucker, R.P., et al. (eds.), *Environmental Histories of the First World War*. Cambridge: Cambridge University Press, 278–295.

Umek, E., 1970. "Lov in lovstvo" (Jagd und Jagdwesen). In Blaznik, P., et al. (eds.), *Gospodarska in družbena zgodovina Slovencev: Zgodovina agrarnih panog*. Vol. 1, *Agrarno gospodarstvo*. Ljubljana: Slovenska akademija znanosti in umetnosti and Državna založba Slovenije, 469–494.

Umek, E., 1982. "Reformirani urbarji deželnoknežjih gospostev na Kranjskem" (Reformierte Urbare landesfürstlicher Herrschaften in Krain). *Zgodovinski časopis* 36(4): 311–320.

[United Nations] Food Systems Summit 2021: Dialogues: Synthesis of Independent Dialogues: Report 2. 2021. [city unknown]: United Nations. https://www.un.

org/sites/un2.un.org/files/unfss_independent_dialogue_synthesis_report_2_0.pdf (September 25th, 2022).

Universal Lexicon. 1731–1754. Halle: Johann Heinrich Zedler. https://www.zedler-lexikon.de/ (August 12th, 2020).

University of Natural Resources and Life Sciences, Vienna: Social Metabolism. https://boku.ac.at/en/wiso/isec/research/gesellschaftlicher-stoffwechsel (March 15th, 2024).

Uprava Republike Slovenije za varno hrano, veterinarstvo in varstvo rastlin: Kdo je krivec, da se jeseni že poleti sušijo? (Slovenian Administration for Food Safety, Veterinary Sector and Plant Protection: Who Is to Blame for Ash Trees Drying as Soon as in the Summer?). April 8th, 2020. https://www.gov.si/novice/2020-04-08-kdo-je-krivec-da-se-jeseni-ze-poleti-susijo/ (June 22nd, 2022).

Urbančič-Zemljič, M., and V. Škerlavaj, 1999. "Postaja japonski dresnik (*Reynoutria japonica* Houtt.) v Sloveniji problem?" (Is Japanese Knotweed (*Reynoutria japonica* Houtt.) Becoming a Problem in Slovenia?). In Maček, J. (ed.), *Zbornik predavanj in referatov 4. slovenskega posvetovanja o varstvu rastlin v Portorožu od 3. do 4. marca 1999*. Ljubljana: Društvo za varstvo rastlin Slovenije, 85–90.

"Uredba o zavarovanih prosto živečih rastlinskih vrstah" (Decree on Protected Wild Plant Species). 2004. *Uradni list RS* 14(46) (April 30th, 2004): 5933–5962.

"Uredba, s katero se spreminjajo dosedanji predpisi o prepovedanem lovskem času v Ljubljanski oblasti" (Decree Amending the Existing Regulations on the Closed Season for Hunting in the Ljubljana District (Sln. *oblast*)). 1929. *Samouprava: Uradni list Ljubljanske in Mariborske oblasti* 2(5): 130–131.

Ušeničnik, F., 1945. *Katoliška liturgika* (Roman Catholic Liturgics). 2nd ed. Ljubljana: Škofijski ordinariat.

Uzunoglu-Obenaus, S., 2008. "Archäobotanik – von Gräsern, Samen und Früchten." In Kern, A., et al. (eds.), *Salz-Reich: 7000 Jahre Hallstatt*. Vienna: Verlag des Naturhistorischen Museums, 98–99.

Valenčič, V., 1970 (= Valenčič, 1970a). "Kulturne rastline" (Kulturpflanzen). In Blaznik, P., et al. (eds.), *Gospodarska in družbena zgodovina Slovencev: Zgodovina agrarnih panog*. Vol. 1, *Agrarno gospodarstvo*. Ljubljana: Slovenska akademija znanosti in umetnosti and Državna založba Slovenije, 251–272.

Valenčič, V., 1970 (= Valenčič, 1970b). "Organizacije za napredek agrarne proizvodnje[,] specialno šolstvo in strokovna literatura" (Landwirtschaftliche Organisationen, Fachschulwesen und Fachliteratur). In Blaznik, P., et al. (eds.), *Gospodarska in družbena zgodovina Slovencev: Zgodovina agrarnih panog*. Vol. 1, *Agrarno gospodarstvo*. Ljubljana: Slovenska akademija znanosti in umetnosti and Državna založba Slovenije, 523–556.

Valentinitsch, H., 1989. “Die Verpachtung von Handelsmonopolen durch den Landesfürsten in Innerösterreich: Von der Mitte des 15. Jhs. bis zum Ende des 18. Jhs.” In Schöpfer, G. (ed.), *Menschen & Münzen & Märkte: Steirische Landesausstellung 1989 Judenburg*. Fohnsdorf: K. Podmenik, 85–95.

Valič, A., 1968. “Gradišče nad Pivko pri Naklem” (Une enceinte au-dessus de Pivka près de Naklo). *Arheološki vestnik* 19: 485–499.

Valvasor, J. W., 1689. *Die Ehre Deß Hertzogthums Crain*. Vol. 2. Ljubljana.

van der Knaap, W.O., van Leeuwen, J.F.N., Fankhauser, A., and B. Ammann, 2000. “Palynostratigraphy of the Last Centuries in Switzerland Based on 23 Lake and Mire Deposits: Chronostratigraphic Pollen Markers, Regional Patterns, and Local Histories.” *Review of Palaeobotany and Palynology* 108: 85–142.

van Eekeren, N., Iepema, G., and J. de Wit, 2014. “Does Early Spring Grazing Stimulate Spring Grass Production?” In Hopkins, A., et al. (eds.), *EGF at 50: The Future of European Grasslands: Proceedings of the 25th General Meeting of the European Grassland Federation Aberystwyth, Wales 7–11 September 2014*. Gogerddan: Aberystwyth University, 665–667.

Vera, F., 2013. “Can’t See the Trees for the Forest.” In Rotherham, I.D. (ed.), *Trees, Forested Landscapes, and Grazing Animals: A European Perspective on Woodlands and Grazed Treescapes*. London and New York: Routledge, 99–126.

Verbič, J., and J. Verbič, 2014–2020. *Pridelovanje kakovostnega sena in uporaba sena v prehrani goved* (The Production of Quality Hay and Use of Hay for Cattle). https://seneno.si/wp-content/uploads/2021/12/Priro%C4%8Dnik-seno-KIS-I.pdf (July 21st, 2024).

Verbič, M., 1960. [*Gospostvo Bled: Arhivski popis*] (The Seigneury of Bled: Archival Register). Typescript. Ljubljana: Arhiv Republike Slovenije.

Verginella, M., 2010. “Der Große Krieg zwischen Erinnern und Vergessen.” In Rajšp, V. (ed.), *Isonzofront 1915–1917: Die Kultur des Erinnerns*. Vienna and Ljubljana: Slowenisches Wissenschaftliches Institut in Wien, 87–97.

Véron, F., 2009. “Grasslands on the Hauts Plateaux of the Vercors, Alps, France.” In Veen, P., et al. (eds.), *Grasslands in Europe of High Nature Value*. Zeist: KNNV Publishing, 164–173.

Vertot, N., and M. Ilić, 2001. “Popis prebivalstva, gospodinjstev in stanovanj v Sloveniji v letu 2001 v luči popisov prebivalstva na Slovenskem v preteklosti in prihodnosti” (Population, Household and Housing Censuses in Slovenia in 2001 in the Light of Past and Future Population Censuses in Slovenia). Ljubljana: Statistični urad Republike Slovenije. https://www.stat.si/popis2002/si/clanki_o_popisu.html (October 15th, 2022).

Veterina Novo mesto: Fasciola hepatica – *veliki metljaj*. https://www.veterina-nm.si/fasciola-hepatica-veliki-metljaj/ (January 14th, 2024).

Viazzo, P.P., 1989. *Upland Communities: Environment, Population and Social Structure in the Alps Since the Sixteenth Century.* Cambridge, etc.: Cambridge University Press.

Vidrih, Z., 2005. *Pašnik: Najboljše za živali, zemljo in ljudi* (Pasture: The Best Grassland Use for Livestock, Soils and People. Slovenj Gradec: Kmetijska založba.

Vierhapper, F., 1925. "Zur Kenntnis der geographischen Verbreitung und Gliederung von Valeriana celtica." In Brockmann Jerosch, H. (ed.), *Festschrift Carl Schröter*. Zürich: Kommissionsverlag von Rascher & Co., 241–252.

Vilfan, S., 1954. "Prispevki k zgodovini mer na Slovenskem s posebnim ozirom na ljubljansko mero (16.–19. stoletje)" (Beiträge zur Geschichte der Masse in Slowenien unter besonderer Berücksichtigung des Masses von Ljubljana). *Zgodovinski časopis* 8: 27–86.

Vilfan, S., 1980. "Kmečko prebivalstvo po osebnem položaju: Osebni položaj v prvih dveh stoletjih novega veka (15. do srede 18. stol.)" (Die persönliche Lage der bäuerlichen Bevölkerung: Neuzeit bis zur Mitte des [18.] Jh.). In Blaznik, P., et al. (eds.), *Gospodarska in družbena zgodovina Slovencev: Zgodovina agrarnih panog*. Vol. 2, *Družbena razmerja in gibanja*. Ljubljana: Slovenska akademija znanosti in umetnosti and Državna založba Slovenije, 336–347.

Vilfan, S., 1986. "Temelji in razvoj denarnih sistemov v slovenskih deželah do 17. stoletja" (Die Grundlagen der Währungssysteme in den slowenischen Gebieten bis zum 17. Jahrhundert). *Zgodovinski časopis* 40(4): 397–412.

Vilfan, S., 1996. *Pravna zgodovina Slovencev: Od naselitve do zloma stare Jugoslavije* (Rechtsgeschichte der Slowenen: Bis zum Jahre 1941). Ljubljana: Slovenska matica.

Villanderer Alm. https://www.suedtirolerland.it/de/highlights/natur-und-landschaft/almen-und-hochplateaus/villanderer-alm/ (January 23rd, 2024).

Vilman, J., 2022. Discussion input after the presentation "Effect of Bohinjsko/From Bohinj Local Quality Brand on Local Production and Networking of Stakeholders" at the *International Conference on Mountain Cheese*. Bohinjska Bistrica, October 3rd, 2022.

Vintar Mally, K., Bobovnik, N., Lampič, B., and S. Kušar, 2021. "Attitudes of Farmers Towards Nature Conservation in Selected Areas of Dry Grasslands in Eastern Slovenia." *Dela* 56: 157–174.

Virtualna arhivska čitalnica: SI_ZAL_RAD/0014 Okrajni ljudski odbor Jesenice-Radovljica, 1945–1955 (Fond/zbirka) (Virtual Archive Reading Room: SI_ZAL_RAD/0014 District People's Committee Jesenice-Radovljica, 1945–1955). https://vac.sjas.gov.si/vac/search/details?id=1084226&refCode=SI_ZAL_RAD/0014&type=&subId=&subLink= (December 15th, 2023).

Vodopivec, P., 1991. "K razpravi o novi vsebinski zasnovi zgodovine v šoli" (Zur Diskussion über die neue inhaltliche Konzeption des Fachs Geschichte in der Schule). *Zgodovinski časopis* 45(1): 115–122.

Vodopivec, P., 2005. "Razvoj šolstva na Slovenskem" (Development of Education System in Slovenian Territory). In Fischer, J., et al. (eds.), *Slovenska novejša zgodovina: Od programa Zedinjena Slovenija do mednarodnega priznanja Republike Slovenije 1848–1992*. Vol. 1. Ljubljana: Mladinska knjiga, 59–62.

Vodopivec, P., 2008. "O kmetijskih družbah na Kranjskem, Štajerskem in Koroškem v predmarčni dobi" (On Agricultural Societies in Carniola, Styria, and Carinthia in the *Vormärz* Period). *Prispevki za novejšo zgodovino* 48(1): 7–20.

Vodopivec, P., 2011. "O Vačah pred velikim požarom leta 1834: Vače in okoliška naselja v gradivu franciscejskega katastra" (Vače before the Great Fire of 1834: Vače and the Surrounding Settlements in the Documents of the Franciscean Cadaster). *Kronika: Časopis za slovensko krajevno zgodovino* 59(3): 505–514.

Vogl-Lukasser, B., Vogl, C.R., Bizaj, M., Grasser, S., and C. Bertsch, 2006. *Lokales Erfahrungswissen über Pflanzenarten aus Wildsammlung mit Verwendung in der Fütterung und als Hausmittel in der Volksheilkunde bei landwirtschaftlichen Nutztieren in Osttirol: Endbericht*. Vienna: Universität für Bodenkultur Wien, Department für Nachhaltige Agrarsysteme, Institut für Ökologischen Landbau.

Vojvoda, M., 1965. *Razvoj kulturne pokrajine v Bohinju: Doktorska disertacija* (Development of the Cultural Landscape in Bohinj: Doctoral Dissertation). Ljubljana: [Univerza v Ljubljani, Filozofska fakulteta, Oddelek za geografijo].

Vojvoda, M. 1969. *Almgeographische Studien in den slowenischen Alpen*. Kallmünz and Regensburg: Michael Lassleben.

Volčjak, J. (ed.), 2017. *Vodnik po arhivskem gradivu 1. svetovne vojne* (Guide to World War I Archival Records). Ljubljana: Arhiv Republike Slovenije, Nadškofijski arhiv Ljubljana, and Zgodovinski arhiv Ljubljana; Celje: Zgodovinski arhiv Celje; Koper: Pokrajinski arhiv Koper and Škofijski arhiv Koper; Maribor: Nadškofijski arhiv Maribor and Pokrajinski arhiv Maribor; Nova Gorica: Pokrajinski arhiv v Novi Gorici; Ptuj: Zgodovinski arhiv na Ptuju.

von Glasenapp, M., and T.F. Thornton, 2011. "Traditional Ecological Knowledge of Swiss Alpine Farmers and Their Resilience to Socioecological Change." *Human Ecology* 39: 769–781.

von Hörmann, L., 1877. *Tiroler Volkstypen: Beiträge zur Geschichte der Sitten und Kleinindustrie in den Alpen*. Vienna: Carl Gerold's Sohn.

von Hörmann, L., 1996. *Tiroler Volksleben: Ein Beitrag zur deutschen Volks- und Sittenkunde. Faksimiledruck der 1909 erschienenen Ausgabe*. Bolzano: Athesia.

Vovk, J., 1978. "Ihtiološke raziskave Cerkniškega jezera" (Ichthyological Research on Lake Cerknica). *Acta carsologica* 8: 273–299.

Vreš, B., and N. Jogan, 2007. "*Rumex* L. (incl. *Acetosa* Mill., *Acetosella* (Meissn.) Fourr.) – kislica." In Martinčič A. (ed.), *Mala flora Slovenije*. Ljubljana: Tehniška založba Slovenije, 212–219.

Vršnik, J., 1962. "Novina" (Swidden). *Planinski vestnik* 62(1): 21–24.

Vršnik, J., 2005. *Preproste zgodbe s solčavskih planin* (Simple Stories from the Solčava Mountains). Celje: Mohorjeva.

Wagner, H., 1989. *Die natürliche Pflanzendecke Österreichs.* Vienna: Österreichische Akademie der Wissenschaften.

Wagner, K., 2005. "Rosegger, Peter." In *Neue Deutsche Biographie*. Vol. 22, 45–47. https://www.deutsche-biographie.de/pnd118602667.html#ndbcontent (December 27th, 2020).

Waibl, G., 2007. "Erika Hubatschek und ihr Fotoapparat." In Merlin, F.W., Hellebart, S., and M. Machatschek (eds.), *Bergwelt im Wandel: Festschrift Erika Hubatschek zum 90. Geburtstag.* Klagenfurt: Kärntner Landesarchiv, 37–44.

Walk of Peace – Map. https://www.thewalkofpeace.com/map/ (June 17th, 2022).

Weed of the Month: Canada Thistle. 2019. https://www.tcweed.org/weed-of-the-month-canada-thistle/ (June 5th, 2020).

Wehling, H.-W., 2007. *Grossbritannien: Geographie – Geschichte – Wirtschaft – Politik.* Darmstadt: Wissenschaftliche Buchgesellschaft.

Weinhold, K., 1869. *Die deutschen Monatnamen.* Halle: Buchhandlung des Waisenhauses.

Wendelberger, G., 1969. "'Das Speikkramperl': Vom Speikgraben und Speikhandel." *Jahrbuch des Vereins zum Schutze der Alpenpflanzen und -Tiere* 34: 110–114.

Westwood, J., 2009. *The Historical Atlas of World Railroads.* Amersham: Cartographica Press.

What Does the Bible Say about Ants? https://www.openbible.info/topics/ants (November 15th, 2022).

White, S., 2014. "The Real Little Ice Age." *Journal of Interdisciplinary History* 44: 327–352.

White, S., Brooke, J., and C. Pfister, 2018. "Climate, Weather, Agriculture, and Food." In White, S., Pfister, C., and F. Mauelshagen (eds.), *The Palgrave Handbook of Climate History.* London: Palgrave Macmillan, 331–353.

Wikimedia Commons, File:Illustration Hedysarum hedisaroides0.jpg. https://commons.wikimedia.org/wiki/File:Illustration_Hedysarum_hedisaroides0.jpg (December 30, 2024).

Wikimedia Commons: File:Illyrische Siegwurz (Gladiolus illyricus).jpg. https://commons.wikimedia.org/wiki/File:Illyrische_Siegwurz_(Gladiolus_illyricus).jpg (September 18th, 2019).

Wikimedia Commons: File:Monatsbilder Salzburger Handschrift 818.jpg. https://commons.wikimedia.org/wiki/File:Monatsbilder_Salzburger_Handschrift_818.jpg (October 11th, 2018).

Wilde, M., 2013. "Wild Vegetable Rennet." https://monicawilde.com/wild-vegetable-rennet/ (June 3rd, 2021).

Wildhaber, R., 1956. "Der 'Feiertagschristus' als ikonographischer Ausdruck der Sonntagsheiligung." *Zeitschrift für schweizerische Archäologie und Kunstgeschichte* 16: 1–34 and Tafeln 1–12.

Wilkes, A., Solymosi, K., and T. Tennigkeit, 2012. *Options for Support to Grassland Restoration in the Context of Climate Change Mitigation: Final Report [Ordered by FAO].* Freiburg: UNIQUE Forestry and Land Use GmbH. http://www.livestockdialogue.org/fileadmin/templates/res_livestock/docs/workshop/2012_07_May_Brasilia/background__paper__Brasilia_workshop.pdf (September 25th, 2022).

Willett, W., Rockström, J., Loken, B., Springmann, M., Lang, T., et al., 2019. "Food in the Anthropocene: The EAT-*Lancet* Commission on Healthy Diets from Sustainable Food Systems." *Lancet* 393: 447–492.

Willfort, R., 1971. *Zdravilne rastline in njih uporaba* (Health Through Medicinal Herbs). Maribor: Obzorja.

Willi, J., 2004. "Vorwort." In Dietl, W., and J. Lehmann, *Ökologischer Wiesenbau: Nachhaltige Bewirtschaftung von Wiesen und Weiden.* Leopoldsdorf: Österreichischer Agrarverlag, 6.

Winiwarter, V., 1999. "Böden in Agrargesellschaften: Wahrnehmung, Behandlung und Theorie von Cato bis Palladius." In Sieferle, R.P., and H. Breuninger (eds.), *Natur-Bilder: Wahrnehmungen von Natur und Umwelt in der Geschichte.* Frankfurt am Main and New York: Campus, 181–221.

Winiwarter, V., 2009. "*Haereticus scripsit hunc librum*: Die landwirtschaftliche Bibliothek des Benediktinerstifts Melk und ihre Leser." *Mitteilungen des Instituts für Österreichische Geschichtsforschung* 117: 225–244.

Winiwarter, V., 2016. "Method Précis: Working with Historical Material." In Haberl, H., et al. (eds.), *Social Ecology: Society-Nature Relations across Time and Space.* [Cham]: Springer, 411–415.

Winiwarter, V., 2022. *Der Weg zur klimagerechten Gesellschaft: Sieben Schritte in eine nachhaltige Zukunft.* Vienna: Picus.

Winiwarter, V., and H.-R. Bork, 2019. *Geschichte unserer Umwelt: 66 Reisen durch die Zeit.* 3rd expanded and updated ed. Darmstadt: Wissenschaftliche Buchgesellschaft.

Winiwarter, V., and M. Knoll, 2007. *Umweltgeschichte: Eine Einführung.* Cologne, Weimar, and Vienna: Böhlau.

Winiwarter, V., Schmid, M., Hohensinner, S., and G. Haidvogl, 2013. "The Environmental History of the Danube River Basin as an Issue of Long-Term Socio-Ecological Research." In Singh, S., et al. (eds.), *Long Term Socio-Ecological Research: Studies in Society–Nature Interactions Across Spatial and Temporal Scales.* Dordrecht, etc.: Springer, 103–122.

Winiwarter, V., and C. Sonnlechner, 2001. *Der soziale Metabolismus der vorindustriellen Landwirtschaft in Europa*. Stuttgart: Breuninger-Stiftung.

Winter, S., Jung, L.S., Eckstein, R.L., Otte, A., Donath, T.W., et al., 2014. "Control of the Toxic Plant *Colchicum autumnale* in Semi-Natural Grasslands: Effects of Cutting Treatments on Demography and Diversity." *Journal of Applied Ecology* 51: 524–533.

Winter, S., Penker, M., and M. Kriechbaum, 2011. "Integrating Farmers' Knowledge on Toxic Plants and Grassland Management: A Case Study on *Colchicum autumnale* in Austria." *Biodiversity Conservation* 20: 1763–1787.

Wirth, S.B., Glur, L., Gilli, A., and F.S. Anselmetti, 2013. "Holocene Flood Frequency Across the Central Alps—Solar Forcing and Evidence for Variations in North Atlantic Atmospheric Circulation." *Quaternary Science Reviews* 80: 112–128.

Wöbse, A.-K., 2022. "Counting Birds: Protecting European Avifauna and Habitats." In Wöbse, A.K., and P. Kupper (eds.), *Greening Europe: Environmental Protection in the Long Twentieth Century—A Handbook*. Munich and Vienna: De Gruyter Oldenbourg, 17–45.

Wohlmuth, J. (ed.), 2002. *Dekrete der ökumenischen Konzilien*. Vol. 3, *Konzilien der Neuzeit*. Paderborn: Ferdinand Schöningh.

Wopfner, H., 1997. *Bergbauernbuch: Von Arbeit und Leben des Tiroler Bergbauern in Vergangenheit und Gegenwart*. Vol. 3, *Wirtschaftliches Leben*. Innsbruck: Universitätsverlag Wagner.

Wraber, T., 1960. "Prispevki k poznavanju slovenske flore" (Beiträge zur Kenntnis der Flora Sloweniens). *Biološki vestnik* 7: 29–37.

Wraber, T., 1964. "Floristične novosti iz Julijski Alp" (Floristische Neufunde aus den Julischen Alpen). *Biološki vestnik* 12: 97–108.

Wraber, T., 1969. "Iz zgodovine o botaničnih raziskovanjih v Julijskih Alpah in Karavankah" (Aus der Geschichte der botanischen Forschungen in den Julischen Alpen und den Karawanken). *Jeklo in ljudje: Jeseniški zbornik* 2: 269–293.

Wraber, T., 1974. "Botanični sprehod skozi Dolino Triglavskih jezer" (A Botanical Walk through the Triglav Lakes Valley). *Proteus* 36: 405–410.

Wraber, T., 1977–1978. "Ceptec ni lepi čeveljc ali primer za napačno rabo ljudskega rastlinskega imena" ("Ceptec" Is Not Lady's-Slipper or an Example of an Erroneously Deciphered Folk Plant Name). *Proteus* 40(2): 70–72.

Wraber, T., 1993. "Sredozemsko rastlinstvo na Slovenskem" (Mediterranean Vegetation in Slovenia). *Časopis za kritiko znanosti* 158–159: 35–44.

Wraber, T., 2000. "Neofiti – rastline pritepenke: Severnoameriški rod rudbekij – že dolgo tudi v flori Slovenije" (Neophytes—Alien Plants: The North American Genus *Rudbeckia*—Also in the Flora of Slovenia for a Long Time). *Proteus* 63(2): 82–83.

Wraber, T., 2005. "O verjetni nesamoniklosti nekaterih semenk, primerov za *florulo castrensis*, v flori Slovenije" (On the Probable Alien Origin of Some Spermatophytes, Examples of the *florula castrensis* in the Flora of Slovenia). *Hladnikia* 18: 3–10.

Wraber, T., 2007 (= Wraber, 2007a). "114. družina: *Gentianaceae* – sviščevke" (Familiy 114: *Gentianaceae*). In Martinčič, A. (ed.), *Mala flora Slovenije: Ključ za določanje praprotnic in semenk*. 4th ed. Ljubljana: Tehniška založba Slovenije, 505–512.

Wraber, T., 2007 (= Wraber, 2007b). "136. družina: *Asteraceae* (*Compositae* subfam. *Asteroideae*) – nebinovke" (Family 136: *Asteraceae* (*Compositae* subfam. *Asteroideae*)). In Martinčič, A. (ed.), *Mala flora Slovenije: Ključ za določanje praprotnic in semenk*. 4th ed. Ljubljana: Tehniška založba Slovenije, 633–687

Wraber, T., 2007 (= Wraber, 2007c). "137. družina: *Cichoriaceae* – radičevke" (Family 137: *Cichoriaceae).* In Martinčič, A. (ed.), *Mala flora Slovenije: Ključ za določanje praprotnic in semenk*. 4th ed. Ljubljana: Tehniška založba Slovenije, 687–716.

Wraber, T., Gogala, M., Gregori, J., Adamič, F., 1989–1990. "Rastline in živali iz Slovenije v Valvasorjevi grafični zbirki" (Plants and Animals from Slovenia in Valvasor's Graphic Collection). *Proteus* 52(9–10): 343–356.

Wunderli, R., 2010. "Landwirtschaftlicher Strukturwandel und Pflanzendiversität im Urserntal (UR)." *Bauhinia* 22: 17–32.

Wutte, M., 1924. "Alte deutsche Berg- und Flußnamen in den Karawanken." In Deutsch-akademischer Geographenverein Graz (ed.), *Zur Geographie der deutschen Alpen: Professor Dr. Robert Sieger zum 60. Geburtstage gewidmet von Freunden und Schülern*. Vienna: L.W. Seidel & Sohn, 96–103.

Zalokar, J., 1854. *Umno kmetovanje in gospodarstvo* (Smart Farming and the Economy). Ljubljana: c. k. krajnska[sic!] kmetijska družba.

Zarabska-Bożejewicz, D., Studzińska-Sroka, E., and W. Fałtynowicz, 2015. "Transplantation of Lichen Thalli: A Case Study on *Cetraria islandica* for Conservation and Pharmaceutical Purposes." *Fungal Ecology* 16: 34–43.

Združenje kmečkih sirarjev Slovenije (Association of Farm Cheesemakers in Slovenia). https://www.kmeckisiri.si/ (November 2nd, 2023).

Zelnik, I., 2012 (= Zelnik, 2012a). "The Presence of Invasive Alien Plant Species in Different Habitats: Case Study from Slovenia." *Acta biologica slovenica* 55(2): 25–38.

Zelnik, I., 2012 (= Zelnik, 2012b). "Vpliv tujerodnih in invazivnih vrst rastlin na biodiverziteto" (Impact of Non-Native and Invasive Plant Species on Biodiversity). In Jogan, N., Bačič, M., and S. Strgulc-Krajšek, (eds.), *Neobiota Slovenije: Invazivne tujerodne vrste v Sloveniji ter vpliv na ohranjanje biotske raznovrstnosti in trajnostno rabo virov: Končno poročilo projekta*. Ljubljana: Univerza v Ljubljani, Biotehniška fakulteta, etc., 53–64.

Zenhäusern, G., 2008. "Witterung und Klima eines Walliser Alpentals nach Aufzeichnungen (1770–1812) des Weibels Johann Ignaz Inderschmitten von Binn." *Blätter aus der Walliser Geschichte* 40: 141–220.

Ziegenspeck, H., 1957. "Cordus, Valerius." In *Neue Deutsche Biographie*. Vol. 3, 359. https://www.deutsche-biographie.de/sfz69591.html#ndbcontent (December 31st, 2020).

Žontar, J., Cova, U., Dorsi, P., Spreitzhofer, K., and W. Wadl, 1988. "Die Verwaltung der Steiermark, Kärntens, Krains und des Küstenlandes 1747/47 bis 1848." In Žontar, J., et al. (eds.), *Handbücher und Karten zur Verwaltungsstruktur in den Ländern Kärnten, Krain, Küstenland und Steiermark bis zum Jahre 1918: Ein historisch-bibliographischer Führer*. Graz etc.: Steiermärkisches Landesarchiv, 31–49.

Žontar, M., 1987. "Gospodarski in družbeni razvoj Bohinja v drugi polovici 18. in v prvi polovici 19. stoletja" (Economic and Social Development in Bohinj in the Second Half of the Eighteenth Century and in the First Half of the Nineteenth Century). In Dežman, J. (ed.), *Bohinjski zbornik*. Radovljica: Skupščina občine Radovljica, 61–67.

Zorn, M., 2002. "Podori na Dobraču" (Rock Avalanches on Mt. Dobratsch). *Geografski vestnik* 74: 9–20.

Zorn, M., and M. Šmid Hribar, 2012. "A Landscape Altered by Man as a Protected Area: A Case Study of the Ljubljana Marsh (Ljubljansko barje)." *Economic- and Ecohistory* 8: 45–61.

Zorn, M., Smrekar, A., Skoberne, P., Šmuc, A., Brancelj, A., et al., 2015. *Dolina Triglavskih jezer* (The Triglav Lakes Valley). Ljubljana: ZRC.

Zrnec, C., 1998. "Fenološke karte" (Phenological Maps). In Fridl, J., et al. (eds.), *Geografski atlas Slovenije: Država v prostoru in času*. Ljubljana: Državna založba Slovenije, 112–113.

Zrnec, C., 2001. "Dandelion Blossoming." In Fridl, J., et al. (eds.), *National Altas of Slovenia*. Ljubljana: Rokus, 66.

Zukrigl, K., 1973. "Montane und subalpine Waldgesellschaften am Alpenostrand." *Mitteilungen der Forstlichen Bundesversuchsanstalt* 101: 1–387.

Zupan, A., 1934. "Kako popisuje Valvasor čebelarstvo v l. 1689" (Valvasor's 1689 Description of Bee Keeping). *Slovenski čebelar* 37(5–6): 95–97.

Zupančič, B., 2001. "Average Annual Precipitation between 1961 and 1990." In Fridl, J., et al. (eds.), *National Altas of Slovenia*. Ljubljana: Rokus, 59.

Župnija Poljane nad Škofjo Loko: Cerkev Žalostne Matere božje na Bukovem Vrhu (Parish of Poljane nad Škofjo Loko: Our Lady of Sorrows Church at Bukov Vrh). https://zupnija-poljane.rkc.si/cerkev-zalostne-matere-bozje-na-bukovem-vrhu/ (December 4th, 2023).

Žust, A., 2016. *Fenologija v Sloveniji: Priročnik za fenološka opazovanja* (Phenology in Slovenia: Manual for Phenological Observations). 1st reprint. Ljubljana: Ministrstvo za okolje in prostor, Agencija RS za okolje.

Zwitter, Ž., 2008. *Prispevek k poznavanju gospodarskih in socialnih razmer v Bohinju (1890–1910): Seminarska naloga* (A Contribution to the Knowledge of the Economic and Social Situation in Bohinj (1890–1910): Term Paper). Ljubljana: Univerza v Ljubljani, Filozofska fakulteta, Oddelek za zgodovino.

Zwitter, Ž., 2010. *Vpliv "male ledene dobe" na agrarno poselitev na ozemlju današnje Slovenije: Na primeru izbranih območij v Zgornji Savinjski dolini: Diplomsko delo* (The Influence of the "Little Ice Age" on the Agrarian Settlement on the Territory of Today's Slovenia: Selected Areas of the Upper Savinja Valley: Bachelor's Thesis). Ljubljana: Univerza v Ljubljani, Filozofska fakulteta, Oddelek za geografijo and Oddelek za zgodovino. https://repozitorij.uni-lj.si/Dokument.php?id=109061&lang=slv (December 27th, 2024).

Zwitter, Ž., 2013 (= Zwitter, 2013a). "Krčenje in agrarna raba gozdov na Gorenjskem v času višinske kolonizacije" (Forest Clearing and Agrarian Forest Uses in the Upper Carniola during Late-Medieval Highland Settlement). In Mrak, I., Potočnik Slavič, I., and B. Rogelj (eds.), *Gorenjska v obdobju glokalizacije*. Bled: Znanstvena založba Filozofske fakultete Univerze v Ljubljani, 85–98.

Zwitter, Ž., 2013 (= Zwitter, 2013b). "Vremenska in klimatska zgodovina v koledarjih in podložniških dnevnikih ljubljanskega škofa Tomaža Hrena (1597–1630)" (History of Weather and Climate in Calendars and Manorial Minutes of Thomas Chrön, Bishop of Ljubljana (1597–1630)). *Zgodovinski časopis* 67(3–4): 306–389.

Zwitter, Ž., 2014 (= Zwitter, 2014a). "Agrarna zgodovina podložnikov dveh gospostev med Podjuno in Menino v 16. in 17. stoletju" (Eine Agrargeschichte der Untertanen zweier Herrschaften zwischen Jauntal und Menina im 16. und 17. Jahrhundert). In Mihelič, D. (ed.), *Vizija raziskav slovenske gospodarske in družbene zgodovine*. Ljubljana: ZRC, 207–229.

Zwitter, Ž., 2014 (= Zwitter, 2014b). "Nekaj novih rezultatov o gospodarstvu in okolju severne in osrednje Goriške na začetku 18. stoletja" (Some New Results on the Economy and Environment in Northern and Central Province of Gorizia (Germ. *Görz*, Sln. *Goriška*) at the Beginning of the Eighteenth Century). In Čeč, D., Škoro Babić, A., and M. Košir (eds.), *Pogledi na tolminski punt in čas ob prelomu stoletja*. Nova Gorica: Zgodovinsko društvo za severno Primorsko; Ljubljana: Zveza zgodovinskih društev Slovenije, 38–50.

Zwitter, Ž., 2014 (= Zwitter, 2014c). "Od zgodnjega kapitalizma do konca prve svetovne vojne: razvoj cest in cestnega prometa v novem veku: Od začetka 16. stoletja do leta 1918" (From Early Capitalism to the End of the First World War: Developments of Roads and Road Traffic in Modern Times: From the Early Sixteenth Century to 1918). In Ficko, G., and S. Eržen (eds.), *Ceste na Slovenskem skozi čas*. Vol. 1. Ljubljana: Družba za raziskave v cestni in prometni stroki Slovenije, 141–237.

Zwitter, Ž., 2014 (= Zwitter, 2014d). "Okolje na Kranjskem v 17. stoletju po *Slavi vojvodine Kranjske*" (Die Umwelt in Krain im 17. Jahrhundert in der *Ehre Deß Hertzogthums Crain*). In Weiss, J. (ed.), *Studia Valvasoriana: Zbornik spremnih študij ob prvem integralnem prevodu* Die Ehre Deß Hertzogthums Crain *v slovenski jezik*. Ljubljana: Zavod Dežela Kranjska, 611–702.

Zwitter, Ž., 2014 (= Zwitter, 2014e). "Water and Forest in 17th-Century Jauntal/Podjuna (Carinthia): The Analysis of Patrimonial Court Records and a Description of Tenants' Holdings from the Seigneury of Eberndorf/Dobrla Vas." In Štih, P., and Ž. Zwitter (eds.), *Man, Nature and Environment Between the Northern Adriatic and the Eastern Alps in Premodern Times*. Ljubljana: Ljubljana University Press, Faculty of Arts, and Historical Association of Slovenia, 314–350.

Zwitter, Ž., 2015 (= Zwitter, 2015a). "Izbrani vidiki okoljske zgodovine kobilarne Lipica od ustanovitve do začetka tridesetih let 17. stoletja" (Selected Environmental History Aspects of the Lipica Stud Farm from Its Foundation to the Early 1630s). *Kronika: Časopis za slovensko krajevno zgodovino* 63(3): 457–474.

Zwitter, Ž., 2015 (= Zwitter, 2015b). "Material Responses to Natural Hazards in 16th and 17th Centuries: Cases from Present-Day Slovenia and Its Surroundings." *Dela* 43: 5–28.

Zwitter, Ž., 2015 (= Zwitter, 2015c). *Okoljska zgodovina srednjega in zgodnjega novega veka na stiku Alp, Panonske kotline, Dinarskega gorstva in Sredozemlja: Doktorska disertacija* (Environmental History of the Middle Ages and the Early Modern Period in the Contact Zone Between the Alps, Pannonian Basin, Dinaric Alps, and Mediterranean: Doctoral Dissertation). Ljubljana: Univerza v Ljubljani, Filozofska fakulteta, Oddelek za zgodovino.

Zwitter, Ž., 2015 (= Zwitter, 2015d). "Subsistence, Prosperity and Abandonment of Alpine Isolated Farms in the Dynamic 17th Century Environment: Case Study from the Upper Savinja Valley with Special Emphasis on Tenants' Inventories." *Economic- and Ecohistory* 11, special issue, *History and Sustainability*: 139–181.

Zwitter, Ž., 2016. "Urbarji iz 16. in 17. stoletja kot vir za okoljsko zgodovino" (Die Urbare aus dem 16. und 17. Jahrhundert als Quelle für die Umweltgeschichte). In Bizjak, M., and L. Žnidaršič Golec (eds.), *Urbarji na Slovenskem skozi stoletja*. Ljubljana: Arhiv Republike Slovenije and Zgodovinski inštitut Milka Kosa ZRC SAZU, 117–148.

Zwitter, Ž., 2018. "16th to Early 18th Century Proactive Coping with Natural Hazards in Slovenian Alps and Their Surroundings: With Lessons for the Future." In Jaćimović, M. (ed.), *Sustainable Development of Mountain Areas: Experiences, Challenges and Perspectives*. Podgorica: Crnogorska akademija nauka i umjetnosti, 163–183.

Zwitter, Ž., 2019. "Prispevek k poznavanju svobodnikov na Kranjskem od sredine 17. do sredine 18. stoletja" (Beitrag zur Geschichte der Freisassen in Krain seit der Mitte des 17. bis zur Mitte des 18. Jahrhunderts). In Golec, B. (ed.), *Družbena in identitetna mobilnost v slovenskem prostoru med poznim srednjim vekom in 20. stoletjem*. Ljubljana: ZRC, ZRC SAZU, 299–327.

Zwitter, Ž., 2020. "Eine Wissensgeschichte der Wiesen und Weiden im Südosten der Alpen im 16. und im frühen 17. Jahrhundert." *Mitteilungen des Instituts für Österreichische Geschichtsforschung* 128(1): 49–82.

Zwitter, Ž., 2021. *Okoljska zgodovina košnje, rib in ribolova na Cerkniškem jezeru, 1650–2000: Elaborat za Notranjski regijski park* (Environmental History of Mowing, Fish, and Fishing on Lake Cerknica, 1650–2000: Report for Notranjska Regional Park). Ljubljana: Filozofska fakulteta Univerze v Ljubljani, Oddelek za zgodovino.

Zwitter, Ž., 2024. "Historično prilagajanje ekstremnim okoljskim situacijam na Slovenskem s poudarkom na 16. in 17. stoletju: Z nauki za prihodnost" (Historical Adaptation to Extreme Environmental Situations in Slovenian Territory with a Focus on the Sixteenth and Seventeenth Centuries: With Lessons for the Future). In Lampič, B., and D. Ogrin (eds.), *Ljudje in okoljske spremembe skozi čas*. Ljubljana: Založba Univerze v Ljubljani, 35–79.

Zwitter, Ž., Fuerst-Bjeliš, B., Mrgić, J., Petrić, H., and M. Zorn, 2024. "The Dinaric Karst in Environmental History Perspective: Conclusions, Contextualisation, and Outlook." In Fuerst-Bjeliš, B., et al. (eds.), *Environmental Histories of the Dinaric Karst*. Cham: Springer Nature, 439–466.

Zwitter, Ž., and I. Mrak, 2017. "Postopno prilagajanje tradicionalne kulturne pokrajine pobočnim procesom: Primer iz Podolševe v Karavankah" (The Gradual Historical Adaptation of a Traditional Cultural Landscape to Slope Processes: A Case Study from Podolševa in the Karawanks). *Geografski obzornik* 64(3–4): 40–47.

Zwitter, Ž., and L. Rasran, 2022. "Monograph Species-Rich Grasslands in the Alps: Environmental History and Historical Ecology of Agroecosystems." Presentation at the *11th Biennial Conference of the European Society for Environmental History: Same Planet, Different Worlds: Environmental Histories Imagining Anew*. Bristol, July 4th–8th, 2022.